NEUROMETHODS

T0172270

Series Editor
Wolfgang Walz
University of Saskatchewan
Saskatoon, SK, Canada

For further volumes:
http://www.springer.com/series/7657

fMRI Techniques and Protocols

Second Edition

Edited by

Massimo Filippi

Neuroimaging Research Unit, INSPE, Division of Neuroscience, San Raffaele Scientific Institute and Vita-Salute San Raffaele University, Milan, Italy

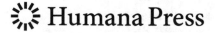 Humana Press

Editor
Massimo Filippi, MD, FEAN
Neuroimaging Research Unit
Institute of Experimental Neurology
Division of Neuroscience
San Raffaele Scientific Institute and
Vita-Salute San Raffaele University
Milan, Italy

Department of Neurology
Division of Neuroscience
San Raffaele Scientific Institute and
    Vita-Salute San Raffaele University
Milan, Italy

ISSN 0893-2336   ISSN 1940-6045   (electronic)
Neuromethods
ISBN 978-1-4939-8164-9   ISBN 978-1-4939-5611-1   (eBook)
DOI 10.1007/978-1-4939-5611-1

Printed on acid-free paper

This Humana Press imprint is published by Springer Nature
The registered company is Springer Science+Business Media LLC New York

Preface to the Series

Experimental life sciences have two basic foundations: concepts and tools. The *Neuromethods* series focuses on the tools and techniques unique to the investigation of the nervous system and excitable cells. It will not, however, shortchange the concept side of things as care has been taken to integrate these tools within the context of the concepts and questions under investigation. In this way, the series is unique in that it not only collects protocols but also includes theoretical background information and critiques which led to the methods and their development. Thus it gives the reader a better understanding of the origin of the techniques and their potential future development. The *Neuromethods* publishing program strikes a balance between recent and exciting developments like those concerning new animal models of disease, imaging, in vivo methods, and more established techniques, including, for example, immunocytochemistry and electrophysiological technologies. New trainees in neurosciences still need a sound footing in these older methods in order to apply a critical approach to their results.

Under the guidance of its founders, Alan Boulton and Glen Baker, the *Neuromethods* series has been a success since its first volume published through Humana Press in 1985. The series continues to flourish through many changes over the years. It is now published under the umbrella of Springer Protocols. While methods involving brain research have changed a lot since the series started, the publishing environment and technology have changed even more radically. Neuromethods has the distinct layout and style of the Springer Protocols program, designed specifically for readability and ease of reference in a laboratory setting.

The careful application of methods is potentially the most important step in the process of scientific inquiry. In the past, new methodologies led the way in developing new disciplines in the biological and medical sciences. For example, Physiology emerged out of Anatomy in the nineteenth century by harnessing new methods based on the newly discovered phenomenon of electricity. Nowadays, the relationships between disciplines and methods are more complex. Methods are now widely shared between disciplines and research areas. New developments in electronic publishing make it possible for scientists that encounter new methods to quickly find sources of information electronically. The design of individual volumes and chapters in this series takes this new access technology into account. Springer Protocols makes it possible to download single protocols separately. In addition, Springer makes its print-on-demand technology available globally. A print copy can therefore be acquired quickly and for a competitive price anywhere in the world.

Saskatoon, Canada   *Wolfgang Walz*

Preface

fMRI has gained a remarkable role as a tool for studying brain function due to its capability to provide an invaluable insight into the mechanisms through which the human brain works in healthy individuals and to explore the mechanisms associated with recovery of function or clinical deterioration in patients with different neurological and psychiatric conditions. The utility of this technique to monitor the effects of pharmacologic and rehabilitative treatments has also been recently demonstrated.

The second edition of this book aims at providing an up-to-date review of the main methodological aspects of fMRI, as well as a state-of-the-art summary of the achievements obtained by its application to the study of central nervous system functioning in the clinical arena. Future evolutions of fMRI techniques are also discussed. The contributors of this volume are all worldwide renowned scientists and physicians with a broad experience in the technical development and clinical use of fMRI. Although the field is ample, based on a series of very different disciplines and expanding at a dramatic pace every day, I believe that this book provides an adequate background against which to plan and design new studies to advance our knowledge on the physiology of the normal human brain and its change following tissue injury.

Part I of the volume is aimed at providing the basic knowledge for the understanding of the technical aspects of fMRI. It covers the basic principles of MRI and fMRI, the different options that can be used to set up an fMRI experiment, and the steps of fMRI analysis, from the preparation of data to the achievement of interpretable results. This part is therefore essential to introduce the readers to the "fMRI world" and make them able to interpret with enough criticism the results of their own experiments. A chapter is devoted to the advantages, caveats, and pitfalls of fMRI data acquired using high-field MR scanners. In addition, although still in its infancy, the assessment of brain connectomics with functional and structural imaging techniques is considered at length, given its potential for improving the understanding of normal and pathological brain function.

Part II provides an overview of the main results derived from the application of fMRI to the study of healthy individuals. Given its noninvasiveness, safety, and repeatability, fMRI is rapidly replacing, whenever possible, other functional techniques, such as positron emission tomography, to image the function of the normal brain. In addition, due to its spatial resolution, fMRI is commonly preferred to neurophysiological techniques to locate with precision the areas activated during the performance of experimental tasks. What has been achieved in the analysis of the main human functional systems with fMRI is illustrated, including, among many other aspects, behavior, language, memory, emotion, sensation, pain, vision, and hearing.

Part III is more clinically oriented and illustrates the main findings obtained by the application of fMRI to assess the role of brain plasticity in the major neurological and psychiatric conditions. The first chapter is devoted to fMRI studies of multiple sclerosis, since there is a growing body of evidence that brain functional reorganization has an important role, at least in same phases of the disease, in limiting the clinical consequences of MS-related irreversible tissue damage. Therefore, MS can be viewed as a "model" to understand how pathology can affect the patterns of brain recruitment. The results obtained in other white matter conditions, including isolated demyelinating myelitis and vasculitides, are then presented. The second chapter deals with stroke studies, which have shown consistently that reorganization of surviving neuronal networks is one of the key factors underlying recovery of function. The experimental caveats to be faced when studying patients with severe clinical impairment are also reviewed. The following chapters cover psychiatric and neurodegenerative diseases, a field where fMRI is providing important pieces of information not only for the understanding of the mechanisms underlying disease pathophysiology and genesis of symptomatology, but also for planning and monitoring novel treatment strategies. Then, two conditions, i.e., epilepsy and tumors, where fMRI is gaining an important role in the presurgical evaluation of patients, are discussed. The last contribution of this part describes the potential and some preliminary, but nevertheless promising, results on the use of fMRI in the monitoring of pharmacological treatments and motor rehabilitation.

Part IV is a glimpse into the future and presents novel approaches for the integration of fMRI data with measures of damage assessed using structural MR techniques and the application of fMRI to image spinal cord function. Finally, results derived from the application of graph analysis to assess network abnormalities in patients with several neurological and psychiatric disorders, including dementia, amyotrophic lateral sclerosis, multiple sclerosis, and schizophrenia, are presented.

The hope that has inspired this book is that it will be of help to clinicians and researchers in their daily life activity by providing a "user-friendly" summary of the field and the necessary background against which to plan and carry out future and successful studies. This is, indeed, an ever-growing and exciting field of research, where we have reached a lot in the past few years, but where there is still a long journey ahead of us.

Milan, Italy   *Massimo Filippi*

Contents

PART I   BOLD fMRI: BASIC PRINCIPLES

PART II   fMRI APPLICATION TO MEASURE BRAIN FUNCTION

Contributors

FEDERICA AGOSTA • *Neuroimaging Research Unit, Institute of Experimental Neurology, Division of Neuroscience, San Raffaele Scientific Institute and Vita-Salute San Raffaele University, Milan, Italy*

JOHN ASHBURNER • *Wellcome Trust Centre for Neuroimaging, Institute of Neurology, University College London, London, UK*

ERIK B. BEALL • *Imaging Institute, Cleveland Clinic, Cleveland, OH, USA*

CHRISTIAN F. BECKMANN • *Oxford University Centre for Functional Magnetic Resonance Imaging of the Brain (FMRIB), John Radcliffe Hospital, Oxford, UK*

JESSE J. BENGSON • *Center for Mind and Brain, University of California, Davis, CA, USA; Department of Psychology, Sonoma State University, Sonoma, CA, USA*

JEFFREY R. BINDER • *Language Imaging Laboratory, Department of Neurology and Biophysics, The Medical College of Wisconsin, Milwaukee, WI, USA*

MICHAEL BUCHFELDER • *Department of Neurosurgery, Universitätsklinikum Erlangen, Erlangen, Germany*

VINCE D. CALHOUN • *The Mind Research Network and Lovelace Biomedical and Environmental Research Institute, Albuquerque, NM, USA; Electrical and Computer, University of New Mexico, Albuquerque, NM, USA; Departments and Neurosciences, University of New Mexico, Albuquerque, NM, USA*

ELISA CANU • *Neuroimaging Research Unit, Institute of Experimental Neurology, Division of Neuroscience, San Raffaele Scientific Institute and Vita-Salute San Raffaele University, Milan, USA*

JESSICA M. CASSIDY • *Department of Neurology, University of California, Irvine, Orange, CA, USA*

ANTHONY CHEN • *Helen Wills Neuroscience Institute, University of California, Berkeley, CA, USA*

STEVEN C. CRAMER • *Department of Neurology, University of California, Irvine, Orange, CA, USA; Department of Anatomy & Neurobiology, University of California, Irvine, Orange, CA, USA*

MARK D'ESPOSITO • *Helen Wills Neuroscience Institute, University of California, Berkeley, CA, USA*

MELISSA A. DANIELS • *Brain Imaging Center, McLean Hospital, Belmont, MA, USA*

RALF DEICHMANN • *University Hospital, ZNN, Brain Imaging Center, Frankfurt, Germany*

NICHOLAS E. DIQUATTRO • *Center for Mind and Brain, University of California, Davis, CA, USA*

BRADFORD C. DICKERSON • *Gerontology Research Unit, Massachusetts General Hospital, Charlestown, MA, USA*

EMMA G. DUERDEN • *Groupe de recherche sur le système nerveux central, Université de Montréal, Montréal, QC, Canada*

GARY H. DUNCAN • *Groupe de recherche sur le système nerveux central, Université de Montréal, Montréal, QC, Canada*

SEAN P. FANNON • *Center for Mind and Brain, University of California, Davis, CA, USA; Department of Psychology, Folsom Lake College, Folsom, CA, USA*

STEFANIA FERRI • *Dipartimento di Neuroscienze, Universita degli Studi di Parma, Parma, Italy*

MASSIMO FILIPPI • *Neuroimaging Research Unit, Institute of Experimental Neurology, Division of Neuroscience, San Raffaele Scientific Institute and Vita-Salute San Raffaele University, Milan, Italy; Department of Neurology, Division of Neuroscience, San Raffaele Scientific Institute and Vita-Salute San Raffaele University, Milan, Italy*

ALEX FORNITO • *Brain and Mental Health Laboratory, Monash Institute of Cognitive and Clinical Neuroscience, School of Psychological Sciences and Monash Biomedical Imaging, Monash University, Melbourne, Australia*

KARL J. FRISTON • *Wellcome Centre for Neuroimaging, Institute of Neurology, University College London, London, UK*

OLIVER GANSLANDT • *Neurochirurgische Klinik, Klinikum Stuttgart, Stuttgart, Germany*

HUGH GARAVAN • *Department of Psychiatry, University of Vermont, Burlington, VT, USA*

JOY J. GENG • *Center for Mind and Brain, University of California, Davis, CA, USA; Department of Psychology, University of California, Davis, CA, USA*

MARIA LUISA GORNO-TEMPINI • *Department of Neurology, Memory and Aging Center, University of California, San Francisco, CA, USA*

LOUIS ANDRÉ VAN GRAAN • *Department of Clinical and Experimental Epilepsy, Institute of Neurology, University College London, London, UK*

WILLIAM GRISSOM • *Biomedical Engineering Department, Vanderbilt University, Nashville, TN, USA*

PETER GRUMMICH • *Department of Neurosurgery, Universitätsklinikum Erlangen, Erlangen, Germany*

ERIN L. HABECKER • *Brain Imaging Center, McLean Hospital, Belmont, MA, USA*

DEBORAH ANN HALL • *MRC Institute of Hearing Research, Nottingham, UK*

LUIS HERNANDEZ-GARCIA • *University of Michigan Functional MRI Laboratory and Biomedical Engineering Department, University of Michigan, Ann Arbor, MI, USA*

JOHN DARRELL VAN HORN • *USC Mark and Mary Stevens Neuroimaging and Informatics Institute, Keck School of Medicine of USC, University of Southern California, Los Angeles, CA, USA*

ALAYAR KANGARLU • *Columbia University and New York State Psychiatric Institute, New York, NY, USA*

ANDREW KAYSER • *Helen Wills Neuroscience Institute, University of California, Berkeley, CA, USA*

LOUIS LEMIEUX • *Department of Clinical and Experimental Epilepsy, Institute of Neurology, University College London, London, UK*

YUELU LIU • *Center for Mind and Brain, University of California, Davis, CA, USA*

MARK J. LOWE • *Imaging Institute, Cleveland Clinic, Cleveland, OH, USA*

GEORGE R. MANGUN • *Center for Mind and Brain, University of California, Davis, CA, USA; Department of Psychology, University of California, Davis, CA, USA; Department of Neurology, University of California, Davis, CA, USA*

PAUL M. MATTHEWS • *Department of Medicine and Centre for Neurotechnology, Hammersmith Hospital, Imperial College, London, UK; GlaxoSmithKline Clinical Imaging Centre, Hammersmith Hospital, Imperial College, London, UK*

ROBERTA MESSINA • *Neuroimaging Research Unit, Institute of Experimental Neurology, Division of Neuroscience, San Raffaele Scientific Institute and Vita-Salute San Raffaele University, Milan, Italy; Department of Neurology, Division of Neuroscience, San Raffaele Scientific Institute and Vita-Salute San Raffaele University, Milan, Italy*

EWALD MOSER • *High Field MR Centre, Medical University of Vienna, Austria; Division MR-Physics, Center for Medical Physics and Biomedical Engineering, Vienna, Austria*

KEVIN MURPHY • *School of Psychology, Cardiff University, Cardiff, Wales, UK*

THOMAS E. NICHOLS • *Oxford University Centre for Functional Magnetic Resonance Imaging of the Brain (FMRIB), John Radcliffe Hospital, Oxford, UK*

CHRISTOPHER NIMSKY • *Neurochirurgische Klinik, Klinikum Stuttgart, Stuttgart, Germany*

GUY A. ORBAN • *Dipartimento di Neuroscienze, Universita degli Studi di Parma, Parma, Italy*

ASPASIA ELENI PALTOGLOU • *MRC Institute of Hearing Research, Nottingham, UK*

SCOTT PELTIER • *University of Michigan Functional MRI Laboratory and Biomedical Engineering Department, University of Michigan, Ann Arbor, MI, USA*

MARTIN PEPER • *General and Biological Psychology Section, Faculty of Psychology, Philipps-Universität Marburg, Marburg, Germany*

ROBERT H. POWELL • *Department of Clinical and Experimental Epilepsy, Institute of Neurology, University College London, London, UK*

PERRY F. RENSHAW • *Brain Institute, University of Utah, Salt Lake City, UT, USA*

SIMON ROBINSON • *High Field MR Centre, Medical University of Vienna, Austria; Department of Biomedical Imaging and Image-guided Therapy, Medical University of Vienna, Austria*

MARIA A. ROCCA • *Neuroimaging Research Unit, Institute of Experimental Neurology, Division of Neuroscience, San Raffaele Scientific Institute and Vita-Salute San Raffaele University, Milan, Italy; Department of Neurology, Division of Neuroscience, San Raffaele Scientific Institute and Vita-Salute San Raffaele University, Milan, Italy*

STEPHEN M. SMITH • *Oxford University Centre for Functional Magnetic Resonance Imaging of the Brain (FMRIB), John Radcliffe Hospital, Oxford, UK*

PATRICK STROMAN • *Department of Diagnostic Radiology, c/o Center for Neuroscience Studies, Queen's University, Kingston, ON, Canada; Department of Physics, c/o Center for Neuroscience Studies, Queen's University, Kingston, ON, Canada*

JING SUI • *Brainnetome Center and National Laboratory of Pattern Recognition, Institute of Automation, Chinese Academy of Sciences, Beijing, China; The Mind Research Network and Lovelace Biomedical and Environmental Research Institute, Albuquerque, NM, USA*

RACHEL C. THORNTON • *Department of Clinical and Experimental Epilepsy, Institute of Neurology, University College London, London, UK*

ARTHUR W. TOGA • *USC Mark and Mary Stevens Neuroimaging and Informatics Institute, Keck School of Medicine of USC, University of Southern California, Los Angeles, CA, USA*

INDRE V. VISKONTAS • *Department of Neurology, Memory and Aging Center, University of California, San Francisco, CA, USA*

NICK S. WARD • *Wellcome Trust Centre for Neuroimaging, Institute of Neurology, University College London, London, UK*

MARK W. WOOLRICH • *Oxford University Centre for Functional Magnetic Resonance Imaging of the Brain (FMRIB), John Radcliffe Hospital, Oxford, UK*

Part I

BOLD fMRI: Basic Principles

Chapter 1

Principles of MRI and Functional MRI

Ralf Deichmann

Abstract

This chapter describes the basics of magnetic resonance imaging (MRI) and functional MRI (fMRI). It is aimed at beginners in the field and does not require any previous knowledge. Complex technical issues are made plausible by presenting plots and figures, rather than mathematical equations.

The part dealing with the basics of MRI covers spins, spin alignment in external magnetic fields, the magnetic resonance effect, field gradients, frequency encoding, phase encoding, slice selection, k-space, gradient echoes, and echo-planar imaging.

The part dealing with fMRI covers transverse relaxation times, the basics of the blood oxygen level-dependent (BOLD) contrast, and the hemodynamic response.

Key words Spin, Field gradients, Frequency encoding, Phase encoding, Slice selection, k-Space, Gradient echo, Echo-planar imaging, Transverse relaxation time, Blood oxygen level-dependent

1 Basic Physical Principles

1.1 Spins in an External Magnetic Field

The first question arising is "What do we actually see in MRI?"

In general, we see protons. A proton is the nucleus of the hydrogen atom. Hydrogen is the most common element in tissue, so if we are able to detect the presence of protons and display them with a certain spatial resolution, it is fair to say that we can "see" tissue.

The detection of protons is based on a physical property called the "spin." A correct description of spins is only possible with quantum mechanics and would be beyond the scope of this book, so may it suffice to say that a spin is quite similar to a compass needle. In particular, a compass needle carries a "magnetization" which enables it to align in an external magnetic field and produce a magnetic field itself (for example, a compass needle can be used to attract small iron particles). A spin possesses an elementary magnetization (albeit a tiny one), so it behaves in a similar way.

Let us consider a simple object containing protons, for example a container with water (Fig. 1). If there is no external magnetic field

Massimo Filippi (ed.), *fMRI Techniques and Protocols*, Neuromethods, vol. 119,
DOI 10.1007/978-1-4939-5611-1_1, © Springer Science+Business Media New York 2016

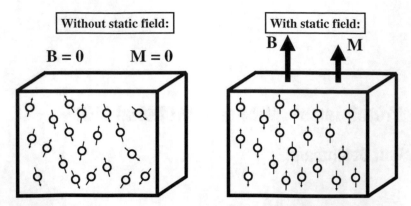

Fig. 1 (*Left*) Without an external magnetic field (labeled *B*) the spins point in different directions, the contributions of their respective magnetization vectors cancel out, so there is no net magnetization (labeled *M*). (*Right*) Inside an external magnetic field *B* the spins align in parallel or anti-parallel direction, with a slight majority of spins in parallel direction, giving rise to a net magnetization *M*

(usually labeled *B*), the spins will point in different directions, the contributions of their respective magnetization vectors will cancel out, so there is no net magnetization (usually labeled *M*). If, however, this object is placed into an external magnetic field *B* (e.g., into the bore of an MR scanner), the spins will align. According to the laws of quantum mechanics, this alignment is either parallel or anti-parallel to the external magnetic field, so once again one might assume that the single magnetization vectors will cancel each other. However, a slight majority of spins prefers the parallel direction. This results in a macroscopic net magnetization *M* which is parallel to *B* (Fig. 1, right). The idea is now: any measurement of *M* would correspond to the detection of the presence of protons in the object. If *M* is measured with a spatial resolution, we can display the result as an image. This is exactly what is done in MR imaging. The measurement of *M* is based on a physical effect which will be described in Subheading 1.2.

1.2 The Larmor Precession

Let us assume that we disturb the realigned spins in a way that (at least for a short time) the magnetization vector *M* is not parallel to *B* but tilted by a certain angle (how this can be achieved will be discussed in Subheading 1.3). In this case, an interesting process begins: the tilted magnetization vector rotates around the direction of the external magnetic field. This movement, which resembles closely the behavior of a spinning top, is called precession (Fig. 2). The frequency is called *Larmor frequency*. It is important to note that the Larmor frequency *f* is proportional to the external field *B*.

$$f = \gamma B. \tag{1}$$

In this formula, γ is the gyromagnetic ratio with a value of 42.58 MHz/T for hydrogen.

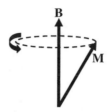

Fig. 2 Behavior of the tilted magnetization: Precession with the Larmor frequency

This effect is of major interest because during the precession, the spins send out an electromagnetic wave with the Larmor frequency. For field strengths that are common on clinical MR scanners (1.5 and 3 T), the Larmor frequencies are about 63 MHz and 127 MHz, respectively. As the reader may know, the tuning on standard frequency modulation (FM) radios ranges from 88 to 108 MHz, so it is fair to say that the signal sent out by the spins is similar to an ordinary FM radio broadcasting signal. For this reason, it is also called *radiofrequency (RF) signal*. It can be detected with an appliance very similar to an FM tuner. This is exactly the way how the magnetization M and thus the presence of protons are detected in MR scanners. In Subheading 1.3, it will be described how the magnetization can be tilted.

1.3 The Magnetic Resonance Effect

The next physical effect is more or less the opposite of the one that has just been discussed. This time, the object is exposed to an external RF signal which has *exactly* the Larmor frequency and is produced by a kind of "inbuilt FM broadcasting station" (Fig. 3, part 1). This signal will tilt the magnetization which subsequently starts to precede (Fig. 3, part 2). During precession, an RF signal is being sent out which can be detected with a kind of "FM tuner" (Fig. 3, part 3).

It should be noted that this concept only works if the incoming RF signal has *exactly* the spins' Larmor frequency. Otherwise, the magnetization will not be tilted and it is impossible to detect a signal. Thus, we are dealing with a resonance effect, and this explains why the imaging technique based on this effect is called *magnetic resonance imaging* (MRI). The nuclear magnetic resonance effect was first described independently by Bloch and Purcell in 1946 [1, 2].

We have now covered all the basic physical effects that are required to understand how MRI works. To summarize it, all MR experiments are based on the following principles:

- Put the object to be imaged into a strong external magnetic field B. The spins will align and create a net magnetization M which is parallel to B.

- Knowing B, calculate the Larmor frequency $f = \gamma B$ and send an external RF pulse which has exactly this frequency. This will tilt the magnetization. After a short time, the external RF has served its duty and can be switched off.

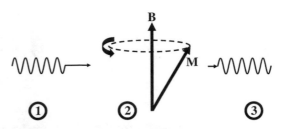

Fig. 3 An external radiofrequency (RF) pulse that has exactly the Larmor frequency tilts the magnetization (*1*). The tilted magnetization starts to precede (*2*) and sends out an RF pulse with the Larmor frequency itself (*3*)

- The magnetization vector now precedes. Switch on your FM tuner. If you detect a signal, you have detected the presence of protons.

Subheading 2 will deal with the question how spatial resolution can be achieved.

2   A One-Dimensional MR Experiment

Let us assume we have a closed box containing two glasses of water, one being half full and the other one full (Fig. 4a).

We now have to answer the following questions:

1. Are there any glasses inside the box or is it empty?
2. How many glasses are inside the box?
3. What is the exact location of the glasses?
4. What are the relative filling levels?

Of course: *we are not allowed to open the box*! Our investigation must rely purely on the physical principles we have discussed so far.

The whole procedure will be discussed step by step as follows:

Step 1. We put the box into the MR scanner, i.e., into a strong magnetic field *B*. In both glasses, spins will align, resulting in a net magnetization *M* in the first glass and, due to the larger number of protons, 2 *M* in the second glass (Fig. 4b).

Step 2. We calculate the Larmor frequency $f = \gamma B$ and send an external RF pulse which has exactly this frequency. This will cause a tilt of the magnetization vectors in both glasses which consequently start to precede, sending out an RF signal with the same frequency f (Fig. 4c). This signal can be detected with a kind of FM tuner, so we can answer at least the first question: the box is sending out a signal, so it cannot be empty. However, so far we are not able to comment on the number and locations of the glasses, because they send out signals with the same frequency and we can only detect the *sum signal* outside the box.

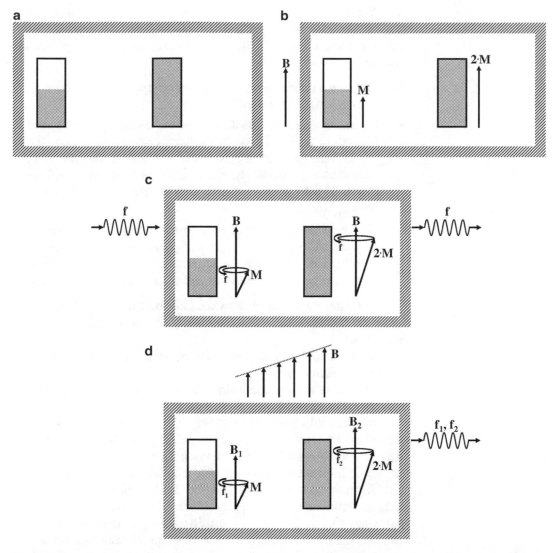

Fig. 4 (a) Two glasses of water inside a closed box. **(b)** Magnetization inside both glasses after placing the box inside an external magnetic field. **(c)** Reaction to an external radiofrequency (RF) pulse: the magnetization vectors in both glasses start to precede and send out RF signals with the Larmor frequency. **(d)** Effect of an external field gradient: the magnetic field strength at the position of both glasses is different, so radiofrequency signals with different Larmor frequencies are being sent

Step 3. This is the crucial step: we switch on a so-called *gradient field.* This simply means that we modify for a certain time the external magnetic field *B* in a way that it is no longer constant across the box but increases (slightly) from the left-hand to the right-hand side (Fig. 4d). In particular, the field strengths at the positions of the first and the second glass are now different, assuming the values B_1 and B_2, respectively. Since the Larmor frequency depends on the magnetic field strength, the magnetization vectors continue their precession with different frequencies f_1 and f_2. Once again, we measure the sum signal outside the box.

As the next step, a *frequency analysis* of this signal is performed. This means that the signal is decomposed into its spectrum of frequency components. In the present case, this analysis yields the following results:

- The signal contains two frequency components, so there must be two glasses inside the box.

- We can measure the absolute values of the frequencies f_1 and f_2. According to Eq. 1, we can calculate from these frequencies the field strengths B_1 and B_2 at the positions of the first and the second glass, respectively. Since we know how we modified the magnetic field B, we can deduce from this the exact positions of the glasses.

- The amplitude of the frequency component f_2 is twice the amplitude of the frequency component f_1. From this we can deduce that there must be twice the number of spins in the second glass, so we also obtain information about the relative filling levels.

In summary, we have answered all the questions above without opening the box, purely by using the concepts of magnetic resonance.

If you could follow this section, you have understood the basics of MR imaging.

The use of field gradients for spatial encoding was proposed by Lauterbur in 1973 [3].

In textbooks and publications, MR experiments are usually described by special plots, the so-called *pulse diagrams*. For the experiment described above, the respective pulse diagram is shown in Fig. 5, comprising an *RF axis* and a *gradient axis*. The entries on the RF axis correspond to the initial RF pulse which tilts the magnetization and the acquired signal. The entries on the gradient axis show that during signal acquisition the gradient Gx is switched on, i.e., during this time the external magnetic field is modified in a way that it increases linearly in a certain spatial direction (the x-direction of an arbitrarily chosen coordinate system). Because this gradient is switched on during the readout process, it is also referred to as *read gradient*.

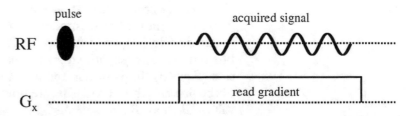

Fig. 5 Schematic plot of a one-dimensional MR experiment, showing a radiofrequency (RF) axis and a gradient (*Gx*) axis

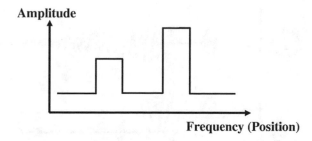

Fig. 6 Frequency spectrum resulting from the Fourier transform. The frequency corresponds to the position of the originating spin

3 The Fourier Transform

The frequency analysis of a signal is a mathematical process called *Fourier transform*. A Fourier transform yields the frequency spectrum of the signal, i.e., the amplitudes of the various frequency components. The frequency spectrum for the setup described above (two glasses in a box) is shown in Fig. 6.

As explained above, a spin's Larmor frequency depends on its local magnetic field strength and therefore on its position within the box, as long as the linear field gradient is switched on. It is therefore fair to say that the frequency spectrum shows a one-dimensional image of the scanned object. In the special case of Fig. 6, the two glasses, their respective positions, their relative filling levels, and even their diameters are clearly displayed.

4 The Gradient Echo

The *gradient echo* technique is of major importance in fMRI.

Let us consider the three experiments described in Fig. 7. Experiments will be discussed on a purely phenomenological basis first, in particular the experimental setup and the respective signal behaviors. The explanation will be given afterwards (Fig. 8) .

In the first experiment (Fig. 7, part 1), an initial RF pulse tilts the magnetization. As expected, precession starts and a signal can be acquired immediately after sending the RF pulse. This signal has a relatively long duration.

In the second experiment (Fig. 7, part 2), the signal is acquired while a gradient is switched on. In this case, the signal decays much faster. The reason for this will be given below. For the time being it is sufficient to note that obviously inhomogeneities of the static magnetic field, as created by a gradient, cause a more rapid signal decay.

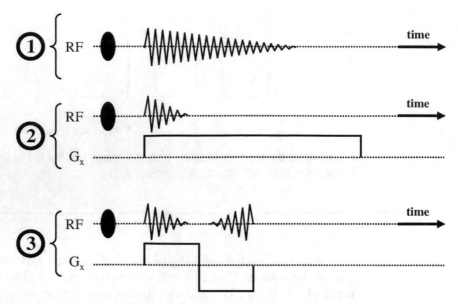

Fig. 7 Schematic sketch of three MR experiments: in the absence of any gradients, a long signal can be observed (*1*). In the presence of a gradient, the signal decays more rapidly (*2*). If the gradient is inverted, a gradient echo occurs (*3*)

The third experiment (Fig. 7, part 3) starts off like the second one, resulting in the same fast signal decay. However, after a while the gradient is *inverted* (a negative gradient *Gx* means that the magnetic field strength decreases in *x*-direction, rather than increasing). This leads to a striking phenomenon: the signal which seemed to have disappeared completely, suddenly comes back. This effect is called *gradient echo*.

It is relatively simple to explain these effects. For illustration, Fig. 8a shows the precession of four individual spins at different positions in the first experiment after the initial RF pulse (the respective magnetization vectors are seen "from top"). Although the spins are located at different positions, they are exposed to the same magnetic field because there are no gradients. Thus, they precede with the same Larmor frequency; their magnetization vectors are always parallel and add up to a relatively strong net magnetization.

In theory, this should go on forever. In practice, the signal decays due to transverse relaxation effects which will be discussed later.

Figure 8b shows the respective sketch for the second experiment. Due to the field gradient, the spins are exposed to different field strengths and precede with different Larmor frequencies. After a relatively short time, they are completely *dephased*, i.e., the magnetization vectors cancel each other and there is no net magnetization. The result is a fast signal decay, as depicted in Fig.7 (part 2) .

Fig. 8 (continued) dephase, reducing the duration of the signal. (**c**) Explanation of the third experiment: the gradient inversion leads to a change from anti-clockwise to clockwise rotation, so spins rephase. A strong signal, the so-called gradient echo can be observed when the magnetization vectors are parallel again

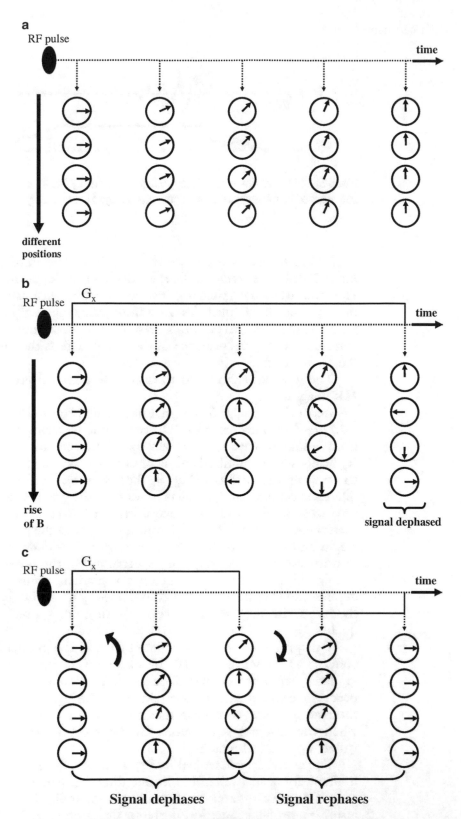

Fig. 8 (**a**) Explanation of the first experiment: spins at different positions still have the same Larmor frequencies as the magnetic field is homogeneous. As a consequence, their magnetization vectors remain parallel and sum up to a strong net magnetization over a relatively long time. (**b**) Explanation of the second experiment: spins at different positions have different Larmor frequencies due to the field gradient. Their magnetization vectors

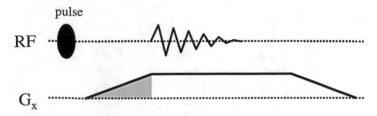

Fig. 9 Signal losses due to the finite gradient rise time: by the time the gradient has reached its full amplitude, the signal has decayed considerably

Figure 8c shows the respective sketch for the third experiment. Due to the gradient, the spins dephase, as described above. However, when the gradient is inverted the spins turn backwards, maintaining their frequencies. As a result, they rephase (i.e., they return to their original positions), resulting in a realignment which corresponds to a reappearance of the signal. This is the origin of the gradient echo.

The gradient echo helps to overcome a typical problem in MR imaging.

According to Fig. 5, a signal has to be acquired while the read gradient is switched on. However, due to technical limitations, gradients have a certain *rise time*, i.e., a certain time delay (typically several hundreds of μs or even several ms) is required to ramp up the gradient (Fig. 9, shaded area). This leads to a dilemma: on one hand, one has to wait for the gradient to reach its plateau level before signal acquisition can start, because only then there is a well-defined relationship between the position of a spin and its Larmor frequency, as required to deduce spatial information from the frequency spectrum. On the other hand, during the process of ramping up the gradient spins start to dephase, so we will have lost a considerable part of the signal by the time the acquisition starts, resulting in a poor image quality.

This problem can be overcome by using the gradient echo concept as shown in Fig. 10: by means of an initial negative gradient, spins are dephased deliberately. Rephasing and the occurrence of a gradient echo take place during the plateau of the read gradient, so we can measure a strong signal at a time when the gradient is constant. Gradient echo sequences are widely used in MR imaging.

The experiment described in Fig. 10 is one-dimensional, i.e., it allows for spatial resolution in one direction (the x-direction) only. If the gradient axes are chosen as shown in Fig. 11 (left), the result is a profile in anterior/posterior direction (Fig. 11, right). The full extent of the imaged object in y- and z-direction is projected onto the x-axis.

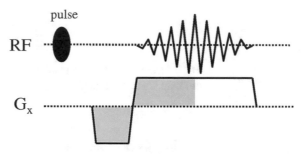

Fig. 10 Solution of the problem imposed by the finite gradient rise times: the initial negative gradient creates a gradient echo and thus a strong signal at a time when the read gradient has reached its full amplitude. The maximum of the echo occurs when the *shaded areas* are identical

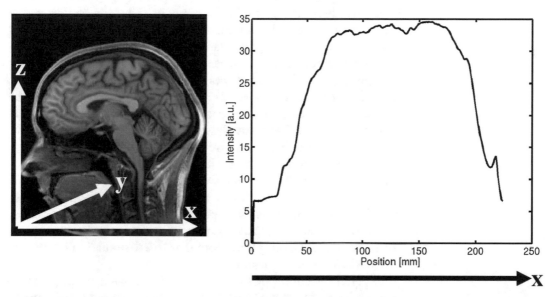

Fig. 11 If the gradient axes are chosen as shown on the *left-hand side*, the result of the one-dimensional experiment with a read gradient in *x*-direction is a profile in anterior/posterior direction

5   The *k*-Space

Before we move on to two-dimensional experiments (which allow us to obtain real images), the so-called *k-space* will be introduced. The *k*-space is a very useful concept when it comes to describing and understanding MR sequences.

Imagine a gradient is turned on for a certain time. During this time, a single data point is acquired (Fig. 12). The *k-value* of this data point is the area under the *preceding* gradient (shaded), i.e., the area under the gradient up to the time point of acquisition. This is simply a definition.

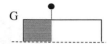

Fig. 12 Definition of a data point's *k*-value as the area under the gradient before the data point is sampled

Let us now analyze our one-dimensional MR experiment as depicted in Fig. 10. The signal acquisition consists basically of the acquisition of a series of discrete data points (Fig. 13) with different *k*-values.

The first data point is preceded by the negative dephasing gradient, so it has a negative *k*-value. The next data point "sees" a combined preparation: the negative dephasing gradient, followed by the first bit of the positive read gradient. Thus, its *k*-value (being the sum of individual areas under the preceding gradient) is still negative, but slightly higher than for the first data point. The subsequent data points have increasing *k*-values. The central data point is acquired when the shaded areas in Fig. 13 are identical. Thus, its *k*-value is zero. As described above, this is where the center of the gradient echo occurs, i.e., this data point will have the highest signal amplitude. The subsequent data points have increasing, positive *k*-values and the maximal *k*-value is attained for the last data point.

We learnt above that spatial resolution in one direction is achieved by acquiring a signal while a gradient in this direction is switched on. We further know that this gradient should be preceded by a negative dephasing gradient to obtain a gradient echo. The *k*-value concept allows us now to move on to an alternative formulation. However, please note that this formulation is basically identical to the previous one:

Spatial resolution in one direction is achieved by acquiring a series of data points with different *k*-values, ranging from a negative to a positive value. Maximum signal is attained for the data point for which the respective *k*-value is zero.

This concept was introduced because it makes it much easier to understand how two-dimensional imaging works. Basically, we need spatial resolution and thus gradients in two directions (the *x*- and the *y*-direction), so in general we have to attribute a *kx* and a *ky* value to each data point, corresponding to the areas under the respective preceding gradients (Fig. 14).

To visualize these *k*-values, we can create a coordinate system with the axes *kx* and *ky* and insert the data point at the respective position (Fig. 15). This is the so-called *k-space*. In the example of Fig. 14, there is a relatively large positive *kx* and a relatively small *ky*, so the position of this data point in *k*-space would be more or less as shown in Fig. 15.

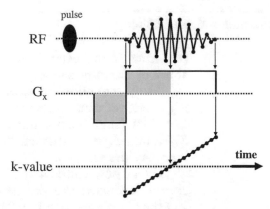

Fig. 13 Gradient echo experiment: the data points constituting the gradient echo have increasing k-values, ranging from a negative to a positive value. The k-value is zero for the central data point where the *shaded areas* are identical and the echo has maximum amplitude

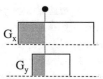

Fig. 14 Definition of a data point's *kx*- and *ky*-values as the areas under the respective gradients before the data point is sampled

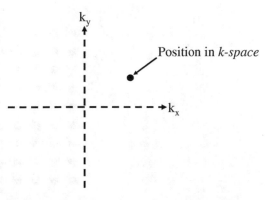

Fig. 15 Description of a data point's *kx*- and *ky*-values in *k*-space

6   Two-Dimensional Acquisition

It is now easy to extend the one-dimensional concept described above to two dimensions:

Spatial resolution in two directions (x, y) is achieved by acquiring a series of data points with different combinations of kx - and ky -values, filling the two-dimensional k-space as shown in Fig. 16. Maximum signal is attained for the central data point for which both k-values are zero.

Let us now consider the experiment depicted in Fig. 17. It closely resembles the experiment discussed in Fig. 13: the RF axis and the Gx axis are identical. This means that the data points have increasing kx-values, ranging from a negative to a positive value. However, there is in addition a Gy axis, showing a negative gradient which is switched on and off before the acquisition starts. Since this gradient is off during acquisition, all data points have the same (negative) ky-value, corresponding to the area under Gy. Thus, the ky-value is constant, whereas the kx-value increases. This means that the experiment acquires a single horizontal line in k-space.

To obtain spatial resolution in two dimensions, we have to acquire several horizontal lines in k-space. This means that we have to repeat the experiment from Fig. 17 several times with different amplitudes of the gradient Gy. In textbooks and publications, this is usually depicted as shown in Fig. 18: the gradient Gy appears as a "ladder" with an arrow, which means that the acquisition is repeated several times with different discrete Gy values, stepping from a minimum to a maximum value.

It should be noted that the gradient Gy is also referred to as *phase gradient* or *phase encoding gradient*.

The experiment described in Fig. 18 is two-dimensional, i.e., it allows for spatial resolution in two directions (the x- and

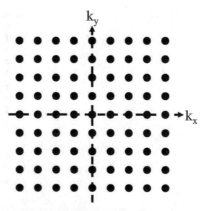

Fig. 16 The basis of two-dimensional imaging: several data points with different combinations of *kx*- and *ky*-values have to be sampled, filling the two-dimensional *k*-space

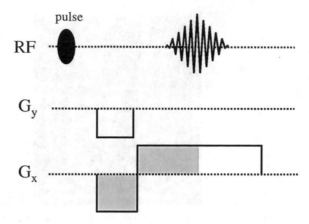

Fig. 17 A subset of a two-dimensional MR experiment. The echo covers a *single horizontal line* in *k*-space

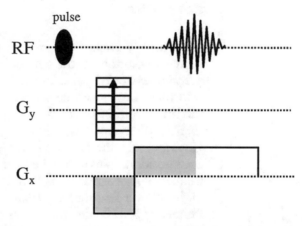

Fig. 18 Schematic description of a full two-dimensional MR experiment: the experiment shown in the previous figure is repeated several times with different values of the gradient *Gy*, so several *horizontal lines* in *k*-space are covered

y-direction). If the gradient axes are chosen as shown in Fig. 19 (left), the result is an image in the axial plane (Fig. 19, right). The full extent of the imaged object in *z*-direction is projected onto this plane, i.e., all axial slices are still added up.

7   Slice-Selective Excitation

So far, we have covered spatial resolution in two dimensions. To obtain spatial resolution in the third dimension, it would be useful to have a kind of "intelligent" excitation pulse which tilts the magnetization only inside a slice of interest. This would mean that only spins within this slice could contribute to the signal, and we could

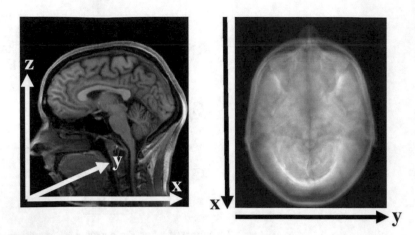

Fig. 19 If the gradient axes are chosen as shown on the *left-hand side*, the result of the two-dimensional experiment with a read gradient in *x*-direction and a phase gradient in *y*-direction is an image in the axial plane, showing an overlay of all axial slices

subsequently employ the experiment described above to obtain spatial resolution in the remaining two directions. These *slice-selective excitation pulses* exist. They are based on the following principles. Let us assume we want to excite an axial slice through the brain (Fig. 20). As a first step, we switch on a gradient in the spatial direction perpendicular to this slice (*z*-direction). Consequently, the Larmor frequency of the spins will depend on their position: spins inside the slice of interest have a certain Larmor frequency f_0, whereas spins in upper/lower parts of the brain have higher/lower Larmor frequencies. If we send an RF pulse with the frequency f_0 while the gradient is switched on, it will tilt the magnetization only inside the slice of interest (please remember that spin excitation is a resonance effect and affects only those spins whose Larmor frequency matches the frequency of the incoming RF pulse). In summary, we can say that an RF pulse which is transmitted while a field gradient is turned on causes a slice-selective excitation. After this special kind of excitation, we can proceed as shown in Fig. 18 to achieve spatial resolution in two dimensions within the slice of interest.

The complete imaging experiment with spatial resolution in three dimensions is shown in Fig. 21. Basically, it is similar to Fig. 18, but comprises a slice selective excitation, including the *slice gradient Gz.*

The latter requires some further explanations. As shown in Fig. 7 (part 2) and Fig. 8b, gradients cause dephasing of the spins and thus signal losses. The second half of the slice gradient (i.e., the part of the slice gradient that comes after sending the RF pulse) would have a similar effect. To avoid signal losses, a negative rephasing gradient has been added after the slice gradient which

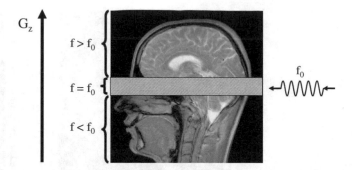

Fig. 20 The basis of slice selective excitation: due to the slice gradient *Gz* spins have spatially dependent Larmor frequencies. A radiofrequency pulse with frequency f_0 can only excite spins whose Larmor frequency corresponds to f_0. These spins are located in a plane perpendicular to the gradient direction

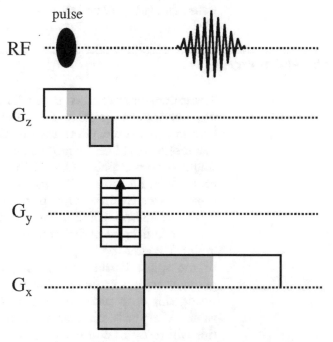

Fig. 21 Schematic description of a full three-dimensional MR experiment with slice selective excitation

compensates this effect, as described for the gradient echo. Full compensation is approximately achieved if the shaded areas on the *Gz* axis in Fig. 21 are identical.

The experiment described in Fig. 21 is three-dimensional, i.e., it allows for spatial resolution in all directions. If the gradient axes are chosen as shown in Fig. 22 (left), the result is an image in the axial plane (Fig. 22, right) with a finite slice thickness.

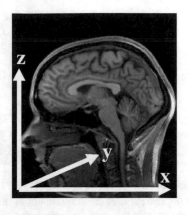 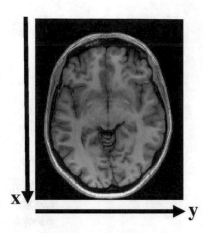

Fig. 22 If the gradient axes are chosen as shown on the *left-hand side*, the result of the three-dimensional experiment with a read gradient in *x*-direction, a phase gradient in *y*-direction, and a slice gradient in *z*-direction is an image in the axial plane with a finite slice thickness

8　Echo-Planar Imaging

The experiment described in Fig. 21 can be relatively time consuming, because it requires an RF pulse for each single echo, i.e., for each line in *k*-space. Given the duration of RF pulses (typically several ms) it would be far more efficient to acquire all echoes after a single excitation pulse. One of these *single shot sequences*, dubbed echo-planar imaging (EPI), was developed by Mansfield [4] and is nowadays widely used in functional imaging experiments. It is based on the acquisition of multiple gradient echoes as described in Fig. 23.

The initial part of this experiment is identical to the one shown in Fig. 13: after the RF pulse, a negative gradient causes dephasing of the spins. During the subsequent positive read gradient, a gradient echo is acquired. The *k*-values of the data points constituting this echo increase, ranging from a negative to a positive value. Afterwards, the read gradient is *inverted*. It is obvious that this will result in another gradient echo with *decreasing k*-values, which means that the second echo covers the same *k*-values as the first one, only *in reverse order*. After a further gradient inversion, a third echo can be acquired which covers the *k*-values in exactly the same way as the first one. Of course, this is only a one-dimensional experiment, because no gradients in *y*-direction are used.

A two-dimensional expansion is shown in Fig. 24. As before, a series of gradient echoes is acquired by means of an oscillating read gradient. However, a phase gradient with negative amplitude is switched before the first echo (shaded), so all data points constituting the first echo have the same (negative) *ky*-value, covering a horizontal line in *k*-space (Fig. 24, bottom). Between the acquisition of the first and the second echo, a very short phase gradient pulse is switched (a

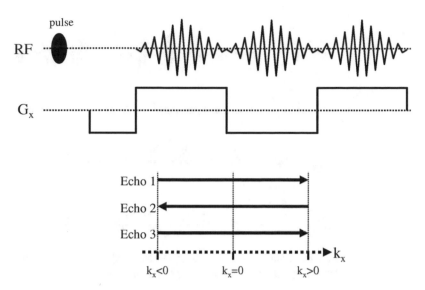

Fig. 23 Acquisition of multiple gradient echoes by successive read gradient inversion. All echoes cover the same *k*-values, but the order is reversed when comparing even and odd echoes

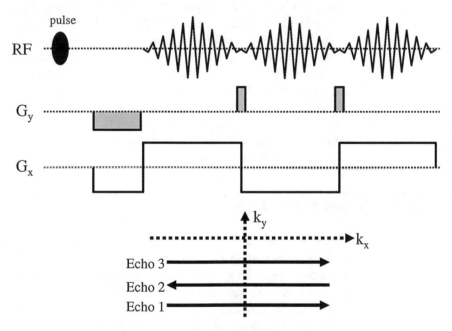

Fig. 24 The basis of echo planar imaging: due to intermediate "blips" (*shaded gradient pulses*) the echoes have increasing *ky*-values, thus covering different lines in *k*-space

so-called *blip*, shaded in Fig. 24). The *ky*-value of the second echo is determined by the sum of the areas under the initial phase gradient and this blip, so *ky* is increased and the second echo covers another horizontal, slightly "higher" line in *k*-space in reverse direction. This concept of intermediate blips is maintained throughout the remaining acquisition, resulting in a meander-like journey through *k*-space.

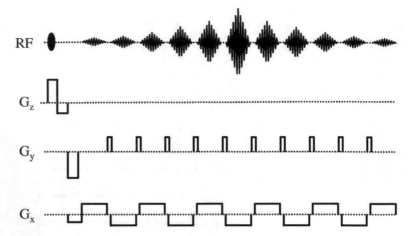

Fig. 25 A complete echo planar imaging experiment. The central echo which covers the center of *k*-space has the highest amplitude

Figure 25 shows the complete EPI experiment, including a slice gradient *Gz* (with a subsequent negative rephasing gradient) for slice-selective excitation. The echoes have different amplitudes due to their different positions in *k*-space, with the highest amplitude for the echo covering the center of *k*-space, as described before.

The main advantage of the EPI sequence is its speed due to the use of a single RF pulse only, with typical acquisition times of 50–100 ms per slice. Another advantage is its susceptibility to the blood oxygen level-dependent (BOLD) effect [5, 6] which is exploited in the majority of functional imaging studies and will be discussed in a later section.

9   The Transverse Relaxation Times T_2 and T_2^*

The first experiment in Fig. 7 (part 1) describes the acquisition of an MR signal after sending an excitation pulse, assuming the absence of any field gradients. According to Fig. 8a, spins at different locations will have the same Larmor frequency, so there are no dephasing effects, and in theory there should be no signal loss at all. However, the signal will still decay due to an effect called *transverse relaxation*. This is caused by the *spin-spin interaction*: in the classical view, the spins randomly exchange small amounts of energy, resulting in minor fluctuations of their Larmor frequencies and thus in gradual signal dephasing, even in otherwise perfectly homogeneous fields. The signal decays exponentially with a time constant called the *transverse relaxation time* T_2. In white matter and gray matter, T_2 has an approximate value of about 100 ms.

In reality, the signal would decay with a time constant even shorter than T_2. This is due to the following effect: tissue is not a homogeneous piece of matter, but consists of several, often microscopic components, for example arterioles and venules. In general, these components have slightly different magnetic properties, so they distort the magnetic field and create microscopic field gradients of varying amplitude and direction. As shown in Fig. 7 (part 2) and Fig. 8b, the presence of field gradients speeds up the signal decay due to spin dephasing. Thus, the signal decays with the *effective transverse relaxation time* T_2^* which is shorter than T_2. This time constant depends on the scanner field strength. In white and gray matter, T_2^* amounts to about 70 ms at 1.5 T and 45 ms at 3 T.

For several applications (including the most common fMRI techniques), it is useful to acquire so-called T_2^* *weighted images,* i.e., images where the local signal intensity depends on the local T_2^* value. The degree of T_2^* weighting can be influenced by modifying a certain acquisition parameter, the *echo time* (TE) .

10   The Echo Time

All MR experiments that were discussed so far are based on the same concept: an initial excitation pulse tilts the magnetization. After a certain time (during which one or more gradients are switched on and off) a signal is acquired. The time delay between excitation and acquisition is called the echo time (TE) (Fig. 26).

To achieve a certain degree of T_2^* weighting, TE must be chosen carefully. This is depicted in Fig. 27, showing a fast (dashed line) and a slow (bold line) T_2^* decay, and three different choices for TE.

The first choice (TE much shorter than T_2^*) would be problematic, because signal amplitudes are almost identical, so T_2^* contrasts would be poor.

The second choice (TE similar to T_2^*) would yield a much better T_2^* contrast.

The third choice (TE much longer than T_2^*) is again problematic: signal has decayed in both compartments, so the image would show noise only, but hardly any structures.

As an example, Fig. 28 shows T_2^* weighted brain images acquired with TE values of 10 ms, 50 ms, and 200 ms. At 10 ms, contrasts are relatively low. At 50 ms, a nice T_2^* contrast can be observed (arrow). This is due to an increased iron content which

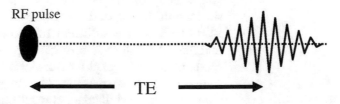

Fig. 26 Definition of the echo time (TE)

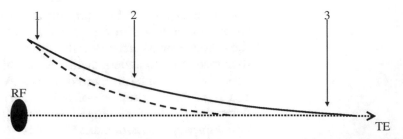

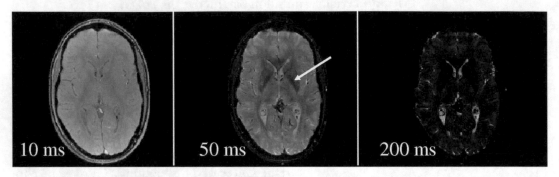

Fig. 27 Signal decay in compartments with a long T_2^* value (*solid line*) and a short T_2^* value (*dashed line*). If a short echo time (TE) is chosen (*1*), there is a high signal amplitude, but hardly any contrast between both compartments. For an intermediate TE (*2*), there is a good contrast and still a sufficient signal amplitude. For a long TE (*3*), the signal in both compartments has decayed

Fig. 28 T_2^* weighted brain images acquired with echo time (TE) values of 10, 50, and 200 ms. At 10 ms, contrasts are relatively low. At 50 ms, a nice T_2^* contrast can be observed (*arrow*). At 200 ms, signal has mostly decayed and only cerebrospinal fluid is visible

gives rise to magnetic field distortions and therefore reduces the T_2^* value. At a TE of 200 ms, signal has mostly decayed and only cerebrospinal fluid is visible due to its longer T_2^* value.

11   The Basis of the BOLD Contrast

The majority of fMRI techniques are based on the BOLD contrast which will be explained in this section.

The BOLD contrast is closely linked to two physical phenomena, called "diamagnetism" and "paramagnetism." A full discussion would be beyond the scope of this book. To understand the BOLD effect, it is sufficient to know the following facts: if a diamagnetic substance is brought into an external magnetic field, it tends to decrease slightly this field, whereas a paramagnetic substance tends to increase it. This means that the close vicinity of paramagnetic and diamagnetic substances causes a local distortion of the magnetic field near the interface.

Tissue is mainly *diamagnetic*. In contrast, blood contains a certain level of deoxyhemoglobin (i.e., hemoglobin that does not have oxygen attached) which is *paramagnetic*. Thus, the presence of blood in tissue means a close vicinity of substances with different magnetic properties, giving rise to microscopic field distortions. As explained above, the resulting field gradients cause spin dephasing and lower the T_2^* value. In summary one may say that due to the presence of deoxyhemoglobin, the signal intensity of tissue is slightly reduced in T_2^* weighted images.

After neuronal activation, blood is locally hyperoxygenated, corresponding to a wash-out of deoxyhemoglobin and an increased concentration of oxyhemoglobin. In contrast to deoxyhemoglobin, oxyhemoglobin is diamagnetic, so it has similar magnetic properties as tissue, leading to a more homogeneous magnetic field and an increased signal intensity in T_2^* weighted images.

Thus, the basic concept of the BOLD effect can be summarized as follows: the *hemodynamic response* to brain activation consists in a *decrease in deoxyhemoglobin* and an *increase in oxyhemoglobin*, resulting in an increased field homogeneity and thus a higher signal intensity in a series of T_2^* weighted images. Therefore, quantification of this signal enhancement allows for the detection of neuronal activation.

In reality, the physiology of the BOLD effect is more complex and depends on the following parameters: the cerebral blood flow (CBF), the cerebral blood volume (CBV), and the metabolic rate of oxygen consumption ($CMRO_2$). After a stimulus, the CBF goes up to deliver more oxygen to the site of neuronal activation, causing the BOLD effect as explained above. On the other hand, the $CMRO_2$ is increased, so more oxygen is consumed, which reduces the BOLD effect. The question arises if the first effect outpaces the second one, which is a prerequisite for blood hyperoxygenation and thus the detectability of the BOLD signal.

The exact physiology of the BOLD response is still controversial and several models have been proposed. For a more detailed overview, the reader is referred to the literature [7]. In the following section, one of the most common models will be explained.

Figure 29 is a simplified sketch, showing how the physiological parameters are affected by neuronal activation and how their interaction influences the signal intensity in T_2^* weighted images. Directly after the stimulus, $CMRO_2$ goes up, causing increased oxygen consumption and thus increased concentration of deoxyhemoglobin. As explained above, deoxyhemoglobin lowers the signal intensity in T_2^* weighted images, so there is an initial signal *reduction* with a duration of about 1 s, the so-called *initial dip*. It should be noted that this effect is small and not always present. About 1 s after the stimulus, the brain reacts by increasing the CBF, transporting oxygen to the site of activation. Fortunately, this effect outpaces the increase in $CMRO_2$, so blood becomes in fact hyperoxygenated. At the same time, the CBV is

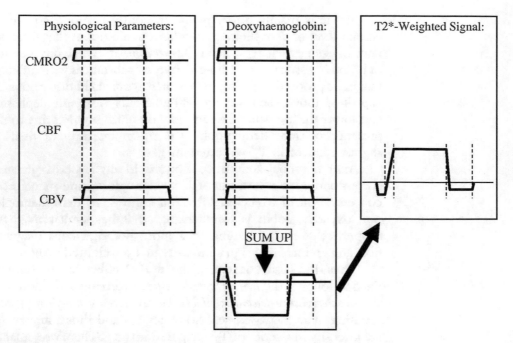

Fig. 29 Change of physiological parameters, the concentration of deoxyhemoglobin, and the T_2^* weighted signal amplitude in response to neuronal activation

increased. As this parameter determines the total amount of blood, an increased CBV causes an increased quantity of deoxyhemoglobin. However, this effect is again outpaced by the increase in CBF, and for a period of 4–6 s blood is hyperoxygenated, giving rise to a *positive BOLD response*. After this, $CMRO_2$ and CBF return to their baseline values. The relaxation of CBV is somewhat slower, so for a certain time there is an increased concentration of deoxyhemoglobin, resulting in a *post-stimulus signal undershoot* with a duration of about 30 s.

In summary, the right-hand side of Fig. 29 shows the complete signal behavior following neuronal activation, the so-called *hemodynamic response function* (HRF): after the initial dip, there is a strong positive BOLD response, followed by a small negative signal undershoot.

12   Choice of TE in fMRI Experiments

The question arises, which value of TE should be chosen to maximize the BOLD signal, and how to set up the parameters of the EPI sequence described above to achieve this value.

Figure 30 shows the theoretical dependence of the BOLD sensitivity on TE (which is given in units of T_2^*). The results are not surprising and correspond to the previous discussion (*see* Fig. 27): at short TE, the BOLD sensitivity is low because there is hardly any T_2^* weighting. At long TE, the BOLD sensitivity goes down because the signal decays due to transverse relaxation effects.

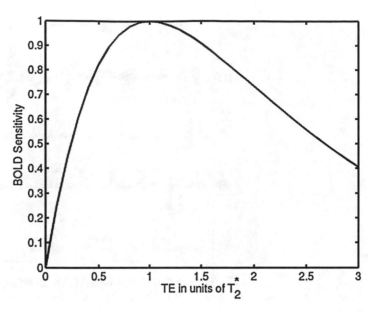

Fig. 30 Dependence of the blood oxygen level-dependent (BOLD) sensitivity on the chosen echo time (TE)

Maximum BOLD sensitivity is achieved when TE equals T_2^* (about 70 ms at 1.5 T and 45 ms at 3 T).

However, there is a fundamental problem with EPI sequences used in fMRI experiments: the magnetic field is usually distorted in brain areas that are close to air/tissue interfaces, in particular in the orbitofrontal cortex (due to the vicinity of the nasal sinus) and the temporal lobes (due to the vicinity of the ear canals). In these areas, local macroscopic field gradients lower the T_2^* value, resulting in signal losses in EPI sequences with relatively long TE values. This shows the basic dilemma with fMRI experiments based on the BOLD effect: on one hand, T_2^* weighting is required to be able to detect the hemodynamic response to neuronal activation. On the other hand, T_2^* weighting causes severe signal losses in certain brain areas. Therefore, it is advisable to keep TE as short as possible to avoid signal losses in these areas, but still as long as necessary to detect a BOLD signal. According to Fig. 30, a decent BOLD sensitivity can still be achieved if TE corresponds to about 2/3 of T_2^*. The general advice is therefore to use a TE of about 50 ms at 1.5 T, and 30 ms at 3 T.

The next question is how TE can be defined for the EPI sequence. As shown in Fig. 25, EPI implies the acquisition of a series of gradient echoes after a single excitation pulse, i.e., each echo has a different echo time. However, as explained above these echoes have different amplitudes, with maximum signal strength for the echo that covers the center of k-space (usually the central echo in the series if symmetric sampling is used). Therefore, the TE value of an EPI sequence is defined as the echo time of this central echo (Fig. 31). This shows that in EPI sequences TE is not simply

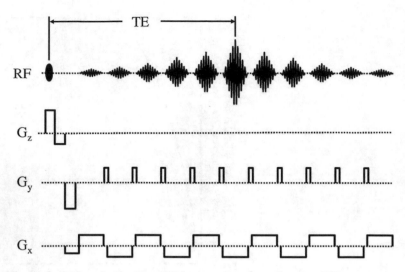

Fig. 31 Definition of echo time (TE) for an echo-planar imaging (EPI) sequence

a wasted waiting time (as one might have expected from Fig. 26), but can be used for the acquisition of the first half of the echo train, being another reason why EPI allows for a high temporal resolution in fMRI experiments.

References

1. Bloch F, Hansen WW, Packard M (1946) Nuclear induction. Phys Rev 69:127
2. Purcell EM, Torrey HC, Pound RV (1946) Resonance absorption by nuclear magnetic moments in a solid. Phys Rev 69:37–38
3. Lauterbur PC (1973) Image formation by induced local interactions: examples employing nuclear magnetic resonance. Nature 242:190–191
4. Mansfield P (1977) Multiplanar image formation using NMR spin echoes. J Phys C 10:L55–L58
5. Ogawa S, Lee TM, Nayak AS, Glynn P (1990) Oxygenation-sensitive contrast in magnetic resonance image of rodent brain at high magnetic fields. Magn Reson Med 14:68–78
6. Kwong KK, Belliveau JW, Chesler DA et al (1992) Dynamic magnetic resonance imaging of human brain activity during primary sensory stimulation. Proc Natl Acad Sci USA 89: 5675–5679
7. Buxton RB, Uludag K, Dubowitz DJ, Liu TT (2004) Modeling the hemodynamic response to brain activation. NeuroImage 23:S220–S233

Chapter 2

Introduction to Functional MRI Hardware

Luis Hernandez-Garcia, Scott Peltier, and William Grissom

Abstract

The chapter gives an overview of peripheral devices commonly used in fMRI experiments, and it addresses the principles, performance aspects, and specifications of fMRI hardware. The general guidelines for MR-compatible hardware are also discussed. The target audience is quite broad and mathematical descriptions are kept to a minimum and qualitative descriptions are favored whenever possible.

Key words Functional MRI, Hardware, Peripheral devices, MRI, Multimodal acquisition, Neuroimaging

1 Introduction

This chapter is concerned with both the MRI hardware components and the multitude of peripherals that are necessary for functional MRI. Our primary aim is to describe the different components of each subsystem and to identify the important features and parameters. We hope this chapter will be of some use to those who want to get a deeper understanding of the electronics involved, but we mostly want to convey how each of the parts is responsible for the quality (or lack thereof) of the images and the experiment in general. Thus we will try to give minimum requirements for the performance of each component and describe what happens when those requisites are not met.

There is a myriad of subsystems to explore and we cannot possibly do justice to all of them in this chapter, so we will limit ourselves to the main ones and to those that most commonly affect the performance of the system.

While this chapter primarily addresses the principles, performance aspects, and specifications of the functional MRI hardware, it is important that we keep in mind that the objective is to carry out experiments on human subjects performing cognitive tasks. Thus we will also keep in mind the ergonomics and safety characteristics of the equipment.

Massimo Filippi (ed.), *fMRI Techniques and Protocols*, Neuromethods, vol. 119,
DOI 10.1007/978-1-4939-5611-1_2, © Springer Science+Business Media New York 2016

Our target audience is fairly broad so we will try to keep mathematical descriptions to a minimum and give qualitative descriptions whenever possible. However, some of the descriptions and the requirements make a lot more sense in the context of the mathematical description of image acquisition and reconstruction. We also hope that this chapter can serve as a good introduction to each topic for those interested in the specific subjects. In a way, we approach this chapter as if we were trying to give advice to someone who is considering setting up a functional MRI facility and/or establishing a set of quality control protocols for such a facility. We will be careful to leave our descriptions as general as possible and to avoid endorsing specific vendors or commercial products.

In this second edition, we have done our best to update the contents to reflect the rapid progress in MR technology. In the last few years, we have seen considerable advances in the areas of parallel imaging, high-field magnets and gradient performance, and widespread adoption of these technologies. There are also exciting new developments in the construction of superconducting magnets that are reducing the need for liquid Helium. Complementary brain activity monitoring techniques are being used in conjunction with MRI, such as fNIRS, and new brain stimulation techniques are on the horizon, such as transcranial Direct Current Stimulation.

2 The MRI Scanner Environment

Let us begin by considering the surroundings of the MRI scanner. When deciding on the layout and location of an MRI scanner, there are several important questions one must ask. The first one is the scanner's purpose. Will it be used for clinical purposes or will it be dedicated to research on healthy subjects? In the case of research-dedicated scanners, one must think about a number of specific factors when designing the layout of the fMRI laboratory. This includes dressing rooms, and testing rooms where the subjects can be trained on the experimental paradigm prior to scanning. Whenever possible, it is important to make sure the control room is large enough to accommodate the needs of the researchers using it. It is not uncommon for fMRI experiments to require multiple pieces of custom stimulation/recording equipment in the control room and for several investigators to be present during the experiment, so that extra bench space and "elbow room" is very advantageous. When scanning clinical populations there are additional precautions and considerations, like the presence of MR compatible first aid equipment. We will not address that aspect in detail, as it can be an extensive discussion that will vary from case to case.

The first thing one notices about an MRI scanner is how big it is (the magnet alone can take up a space of roughly $4 \times 4 \times 6$ m) and

Fig. 1 These photos were taken during the installation of an MRI scanner at Resurgens Orthopaedics in Atlanta, GA on the 19th floor of their Crawford facility. Courtesy of Resurgens Orthopaedics, Atlanta, GA

one must find a site with sufficient space for the scanner. Access is also important, as the main magnet is typically delivered and installed in one piece. Furthermore, MRI magnets are filled with cryogens, which are delivered in large dewars. Thus there must be a wide path to the loading dock that avoids stairs and is clear of obstacles. Typically, MRI scanners are housed in basements or ground floors near the loading docks of hospitals, although exceptions exist, as the one shown in Fig. 1.

Next one must consider how the magnetic field will affect its surroundings. Before the magnet is installed, one must consider how far the magnetic field will extend. Most people who work with MRI are familiar with the enormous forces that the main magnetic field can exert on objects in its proximity (i.e. the magnet room). Modern MRI scanners are actively shielded and contain the magnetic field fairly well within the magnet room. Even with shielding, sometimes a subtle but significant magnetic field can extend beyond the walls of the magnet room. Thus it is important to keep in mind two questions: how well the magnetic field is contained, and what sort of equipment is in the rooms adjacent to the magnet room. The first question is usually answered in terms of the location of the "5 Gauss line."

The United States FDA regulates that the general public (anyone not working with an MRI scanner or being scanned) not be exposed to static magnetic fields over 5 G (5×10^{-4} T), and thus the 5 G boundary must be contained inside the magnet room. MRI scanner vendors will typically provide contour plots of the magnetic field superimposed on the blueprints of the room and provide consultation on the location of the scanner. One must realize, however, that smaller magnetic fields will extend beyond the walls of the magnet room, so it is important to note how quickly the field decays and what is located on those adjacent rooms. Subtle magnetic fields can affect electronic equipment in many ways. For example, a moving charge in the presence of a magnetic field will experience a force perpendicular to the magnetic field. This effect was most obvious in CRT monitors (now obsolete), whose images would be skewed by the magnetic field, and in the performance of computer hard drives and magnetic media. In fact, magnetic media, including the magnetic strips on credit cards, floppy disks, …etc. are typically erased when taken into the magnet room. Another example is that the life span of light bulbs near MRI scanners tends to be quite short because of the vibration of the filaments caused by switching the direction of the current in the presence of a large magnetic field. Hence, DC lights are often used in MRI scanner rooms to avoid this problem. Pace makers, neurostimulators, and implants must be kept outside the 5 G line, unless they have been tested and classified as MR compatible. A number of publications exist [1–3] and are updated regularly with classification of MR compatible devices.

Containment of the magnetic field can be achieved by two different kinds of shielding. Passive shields consist of building a symmetric box around the magnet out of thick iron walls (Fig. 2) that contain the magnetic field. Alternatively, active shields can be built as secondary electromagnets concentrically placed around the main magnet. The shield magnets are built such that the field

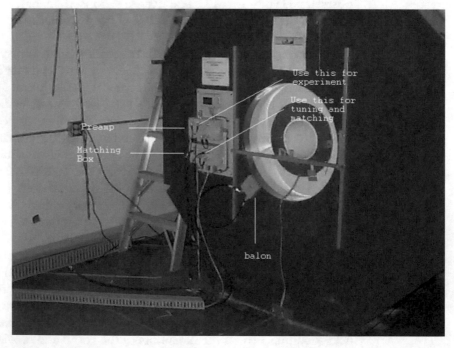

Fig. 2 A passively shielded 4.0 T magnet encased in a hexagonal iron cage to partially contain the main magnetic field

they produce points in the opposite direction of the main magnetic field. The strength of the shielding field is typically about half of the main field. For example, a 3 T actively shielded magnet can often be a 4.5 T magnet surrounded by a second, concentric and opposing 1.5 T magnet. Such a system would have a 3 T field inside of the bore, and because of the inverse square dependence of the magnetic field on the distance to the coil, the field outside of the magnets bore is dramatically reduced. Figure 3 shows a plot of the magnetic fields produced by the main magnet as a function of the distance to the center of the bore. The field produced by the shield and the net sum of both fields are superimposed on the same plot.

One must also consider how the environment will affect the MRI scanner. In this regard, the most important issue is the presence of electromagnetic noise sources. MRI scanners construct images from radio frequency (RF) electromagnetic signals. Thus, radio stations, cell phones, and other wireless communications will interfere with the MRI experiment and severely degrade image quality unless they are properly isolated. MRI rooms are usually encased in a copper shield box that blocks external RF radiation, and contains the MRI's RF radiation as well. The quality of the shield is critically important to the performance of the scanner and it must be tested carefully before the magnet is ramped up to field. Typically attenuations for RF shielding are 100 dB at the operating frequency range. A slightly defective soldered seam between copper sheets, or a nail going through the copper sheeting are

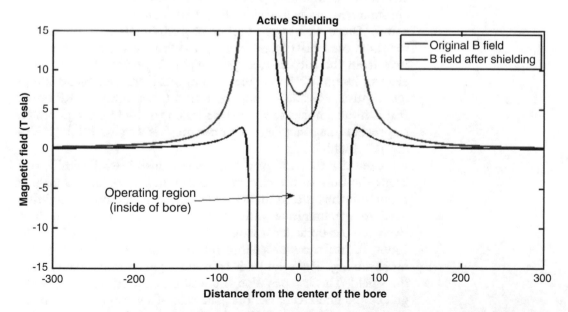

Fig. 3 A plot of the magnetic field strength along the radial direction. The thick vertical lines denote the location of the coil's windings. The thinner line denotes the operating region where the sample is placed. Ideally, this region should have a flat magnetic field, and additional fields (shims) are necessary

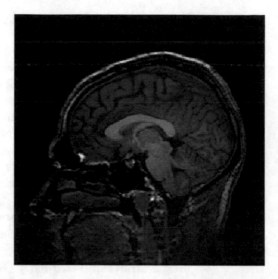

Fig. 4 The streaks in the image are caused by the presence of electromagnetic frequency noise at a single frequency. This is typically introduced by the presence of badly shielded electronic equipment in the room. The AC power supply running at 60 Hz to the device is in this case the culprit of the artifact

sufficient to let RF noise into the room that ruins images, so testing of the room shielding must be stressed. Figure 4 shows an example of the effects of an RF noise source on an MR image

Good shielding of the room is not enough. MR-related electronic equipment produces RF noise and can act as an antenna that passively carries RF noise from the outside into the room. There are also a number of peripherals that are needed for functional MRI in order to provide stimulation and record data from the subject (see Sect. 4). It is preferable to keep all electronics out of the magnet room, but if the electronic equipment must be inside, it must be tested thoroughly for RF noise. If it is indeed noisy, care must be taken to shield the equipment to prevent image artifacts (Copper mesh is very useful for building RF shields).

Consider the case of a button response box. Typically, one keeps the bulk of the electronic components out of the magnet room, but the buttons themselves must be in the scanner and they need to communicate with the response recording electronics. Even if fiber-optic technology is used to carry signals into the room, RF noise can enter through the same opening as the fiber optic cabling. The solution is to build a "penetration panel" into the shield. This is a panel on the wall that is outfitted with waveguides and filtered connectors. The role of a waveguide is to block any electromagnetic radiation that is not parallel with the direction of the waveguide, and at the same time, to guide the EM waves

Fig. 5 On the left is the penetration panel that connects the MRI scanner electronics to the magnet hardware. The image on the right is a second penetration panel used for all the additional stimulus/response equipment needed for fMRI

produced inside it. The filtered connectors typically remove high frequency radiation that may be carried by the cabling of the peripherals. Examples of penetration panels used for scanner cabling and for general, user-specific peripherals are shown in Fig. 5. Additional precautions against RF contamination is the used of twisted-pair and coaxial cabling to contain the fields produced inside the transmission lines.

MR scanner equipment must be kept in stable temperature and humidity conditions, as these can change the performance of the electronics and the field strength. All the amplifiers and computer equipment required for MR imaging can generate a significant amount of heat, so it is important that the environmental temperature-regulating equipment be powerful enough to handle it. Changes in the scanner performance can mask changes in brain activity so the scanner's performance be maintained as constant as possible.

To put things in perspective, the MRI electronics equipment produces approximately 50,000 BTU per hour. Typical requisites for the temperature and humidity in the room are variability of less than 3 °C per hour and 5 % per hour, respectively. Normal operating ranges are in the 15–32 °C temperature range and 30–70 % humidity range. Of particular interest are the gradient coils, since they can heat up significantly as a result of the large currents that run through them. As gradient coils heat up, their performance is degraded and thus they require its own cooling system to keep them stable. This system is usually a cooling loop involving a water chiller that can remove about 14,000 BTU per hour.[1]

It is also noteworthy that the scanner's performance can be affected by other unexpected environmental factors. Outside magnetic fields and vibrations can be an issue. For example, nearby construction can produce vibrations that will affect the MR signal's stability if the floor is not adequately mechanically damped. It is thus important to carry out vibration tests of the site before installation proceeds. Large moving objects, such as nearby trains can also generate magnetic fields that affect the scanner's stability [4]

One other issue that can cause a great deal of grief to investigators is the production of small electromagnetic spikes inside the magnet room. These occur when metal objects in the room vibrate (typically because of the gradients) and bang against each other or when arcing occurs across badly-soldered connections in home-made equipment. The RF receiver hardware is sensitive enough to pick up these spikes. Spikes in the k-space data translate into stripe patterns in the images (Fig. 6). It is thus very important to make sure that all metal equipment is well secured.

A useful option to consider for a functional MRI lab is a mock MRI scanner. This is advantageous since it allows subjects to get

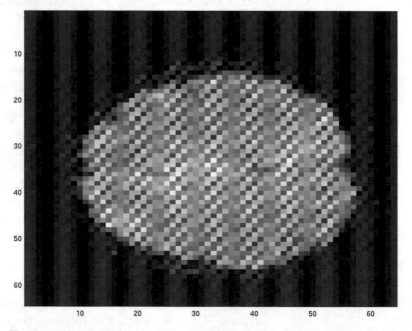

Fig. 6 A T2*-weighted image showing white pixel artifact. Spikes in *k*-space results in sinusoidal patterns in image spike

---

[1] These numbers are based on the specification of a 3 T scanner by General Electric (MR750).

used to the fMRI experience prior to their actual scans, in a safe environment. This can help alleviate subjects' apprehension and claustrophobia, and lead to reduced head motion and improved task performance.

For completeness, replication of the entire MR suite would be preferable, but usually this is limited by available space and resources. The most important factors to reproduce are the spatial dimensions, audio environment, and visual stimulus presentation of the MRI scanner.

It is important to have the subject familiarize themselves with the restrictive space in the MR scanner. This includes the inner diameters of both the bore of the main magnet and the head coil being used. These restrictions, combined with the distance the subject travels into the magnet from the home position of the patient bed, combine to give the overall physical experience. MR or CT patient beds and scanner housing may be recycled for this use, or the mock MR scanner can be built from scratch as the one shown in Figs. 7 and 8.

The audio environment of the scanner is also an important consideration. Having an audio recording of the actual scanner to play in the mock MR scanner will let the patient adjust to the jarring transition when the scanner starts playing sequences, and the to the volume of this noise throughout the scan. Inclusion of audio feedback can also demonstrate to the subjects that the scanner operator will be able to communicate with them between scans.

A duplication of the fMRI visual stimulus presentation can also serve to acclimatize the subjects. The standard forward- or rear-projection systems used to present visual stimuli are relatively easy to replicate in a non-MR environment, and allow the subject to get used to task presentation during the scan.

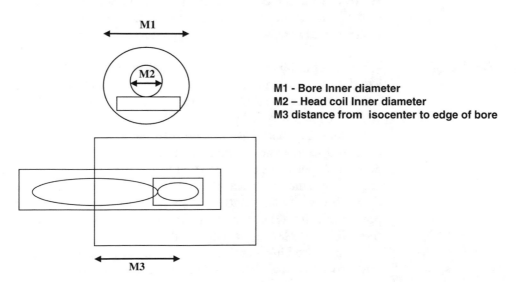

M1 - Bore Inner diameter
M2 – Head coil Inner diameter
M3 distance from isocenter to edge of bore

Fig. 7 Diagram of mock scanner showing critical dimensions

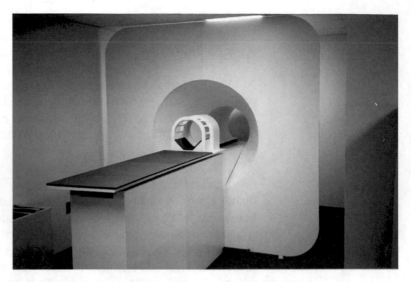

Fig. 8 Example of a mock MRI scanner (University of Michigan)

An MR mock scanner can help to increase patient comfort and task performance, especially in some target populations (e.g., individuals with autism, children), and can be used for screening (e.g., individuals with claustrophobia, large physical dimensions). It can also be useful for the training of fMRI lab personnel or the designing, troubleshooting, or rehearsing of complicated fMRI experiments.

3   The MRI Scanner

We can divide the components of an MRI scanner into four categories. The magnet, the magnetic field gradients, the RF Transmit/Receive hardware, and the data acquisition electronics.

3.1   The Magnet

We will begin by considering the magnet. Typical MRI magnets, like the one shown in Fig. 9, are large solenoid coils made of super-conducting metal (niobium alloys, typically). They are kept cooled at approximately 4 K by liquid Helium in order to achieve and maintain superconductivity.[2] The magnet is "ramped up" to field by introducing a current through a pair of leads that produces the desired magnetic field. Once the specified current has been built up, the circuit is closed such that the current "re-circulates" through the coil constantly and there is no need to supply more power to it. It is crucial to maintain the low temperature to prevent the coil from resisting the current flow.

A new, exciting development is the "dry magnet." The term refers to super-conducting magnets that are cooled without liquid

---

[2] For more information on superconductivity, see [5]

Fig. 9 A 3T magnet during installation at the University of Michigan's FMRI Laboratory

helium. This new generation of magnets uses cryo-coolers [6] that run constantly in order to maintain the temperature of the magnet in the super-conducting range. Cryo-coolers still use helium, although in its gaseous state and in much smaller quantities. A major advantage of this cooling technology is that it allows ramping the magnet up and down much more safely, quickly, and inexpensively, since no liquid helium fills are required in the process. At the time of this writing and to our knowledge, these are only currently available from one vendor of MRI systems (MR solutions) and only for preclinical systems. It is expected that this technology will be adopted for clinical MRI systems in the near future, but most MRI scanners in the world are still cooled by liquid Helium.

If the windings ever warm up and become resistive (i.e., they lose their super-conducting state), they dissipate the electric power as heat. This very undesirable event is termed "quenching." As the magnet's windings become more resistive, the very high current circulating through the coil produces heat, such that the liquid helium that is responsible for maintaining the superconducting temperature boils off. This rapid boiling of the helium quickly exacerbates the problem and the superconducting state is quickly lost. The very large currents can potentially produce enough heat to melt or severely damage the windings. The greatest danger however, is that the rapid rate of helium boiling can build up a great deal of pressure in the magnet and also flood the room with helium gas and suffocate whoever is there. While helium is not toxic, it displaces the oxygen in the room. Thus, it is crucial that the magnet be outfitted with emergency vents (manufacturers of MRI scanners typically include emergency quench ventilation systems). Additionally, magnet rooms are outfitted with oxygen sensors that sound an alarm when the oxygen level falls below safe levels. It must be stressed that all personnel be trained in emergency quench procedures in case the emergency systems fail.

Having considered what can happen when the magnet fails, let us now return to the more cheerful subject of what the magnet can do.

The key parameter in the magnet is its field strength (B_0), as it determines many properties of the images. Primarily, field strength determines the amount of spins that align with and against the field. The higher the field strength, the larger the population of aligned spins that can contribute to the MR signal. More specifically, the population of spins aligned with the magnetic field ($n+$) and against it ($n-$) is described by the Boltzmann equation

$$\left(\frac{n_+}{n_-}\right) = \exp\left(\frac{\gamma h B_0}{kT}\right) \tag{1}$$

where γ is the gyromagnetic constant for the material, h is Plank's constant, k is Boltzmann's constant, T is the temperature of the sample, and B_0 is the strength of the magnetic field. Hence, the higher the field, the more spins will contribute to the signal and thus yield a higher the signal to noise ratio (SNR).

The field strength also determines the resonance frequency of the spins, ω_0, in a linear fashion according to the Larmor equation.

$$\varpi_0 = \gamma B_0 \tag{2}$$

Where γ is again the gyromagnetic constant, which is specific for the nucleus in question. The resonant frequency will in turn determine the characteristics of the RF transmit and receiver subsystems (see Sect. 3.3). Field strength also affects both the longitudinal and transverse relaxation rates of the materials via the resonant frequency as predicted by the equations

$$\frac{1}{T_1} \alpha \left\{ \frac{\tau_c}{1 + \omega_0^2 \tau_c^2} + \frac{4\tau_c}{1 + 4\omega_0^2 \tau_c^2} \right\} \tag{3}$$

and

$$\frac{1}{T_2} \alpha \left\{ 3\tau_c + \frac{5\tau_c}{1 + \omega_0^2 \tau_c^2} + \frac{2\tau_c}{1 + 4\omega_0^2 \tau_c^2} \right\} \tag{4}$$

where τ_c is the correlation time (a measure of the tumbling rate and frequency of collisions between molecules) of the species. Note that T_1 is more heavily dependent on B_0 than T_2.

While T_1 and T_2 typically get longer, T_2^* gets shorter at higher fields. Recall that T_2^* is the rate of transverse signal loss accounting for both T_2 and macroscopic field inhomogeneity, that is

$$\frac{1}{T_2{}^*} = \frac{1}{T_2} + \frac{1}{T_2{}'} = \frac{1}{T_2} + \gamma \Delta B_0 \tag{5}$$

where T_2' is the relaxation due purely to the field inhomogeneity. This inhomogeneity in the magnetic field is usually produced by inhomogeneity in the magnetic susceptibility across the sample, e.g., the air in the ear canals has very different susceptibility than the water in brain tissue. The distortion of the magnetic field caused by magnetic susceptibility is described by

$$B_0{}' = B_0 (1 - \chi) \tag{6}$$

where χ is the magnetic susceptibility of the sample, B_0 is the original magnetic field and B_0' is the resulting magnetic field after considering the magnetic susceptibility. Thus the change in the field gets worse as the magnetic field increases.

The implications for functional imaging are that most imaging and spectroscopy applications benefit in terms of SNR, and that the bold oxygen level-dependent (BOLD) effect is more pronounced at higher fields. But, as is common in MR, there is a tradeoff. As the field increases, and T2* effects get shorter, susceptibility artifacts get much more pronounced. This is particularly significant, as the BOLD effect is observed by T2* weighted imaging, which is very sensitive to susceptibility artifacts.

Arterial Spin Labeling techniques [7] also benefit from higher field strength, as the longer T1 means longer duration of the label. Another major practical implication of working at a higher field is that the resonant frequency of protons is proportionally higher, and thus RF pulses deposit more power into the subject. The US FDA regulates the amount of RF power that can be used on a human subject cannot exceed 1.5 W/Kg.

The higher frequency of the pulses also means a shorter wavelength and the formation of standing waves in the sample during transmission. Hence, it is more challenging to achieve uniform excitation patterns across the imaging slice and parts of the imaging slice appear artificially brighter than others. Figure 10 (left) shows an example of this phenomenon (usually referred to as "dielectric effects") in brain images at 3 T [8–10]. The right panel of the figure shows the corrected image.

To put things in context, at the time of this writing, T2* weighted imaging techniques required for BOLD fMRI are fairly challenging at 7 T and not many groups are doing human work at these high fields, although there is an increasing trend toward 7 T. Presently there are only two research groups that have 9 T human imaging systems. At this time, 7 T magnets are predominantly used for small animal research systems. The "de facto" standard field strength for human fMRI systems in the last few years has become 3 T, although many sites still use 1.5 T scanners for fMRI.

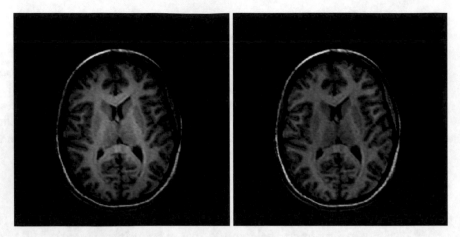

Fig. 10 Illustration of the dielectric effect on a high-resolution, T1 weighted, SPGR image. The center of the image appears brighter, because of the formation of standing waves in the RF pattern. As a result, the center of the field of view receives a higher flip angle than the periphery of the image. The image on the right has been corrected by removing the low frequency spatial oscillation with a 2D FIR filter

Besides the strength, one must consider the spatial homogeneity and temporal stability of the magnetic field. The magnetic field's homogeneity is crucial since the lack of it translates into severe image distortions. One challenge is that the shape of the magnetic field changes when an object (i.e., the subject) is introduced into the field. Consequently, MRI systems are usually outfitted with a set of small pieces of iron installed around the bore of the magnet, referred to as "passive shims." The location of the passive shims is carefully chosen to make the field more homogeneous. In order to adjust the field homogeneity for individual subjects, additional electromagnets whose field can be dynamically changed when the patient are used. These are referred to as active shims and the process of shaping the field is referred to as "shimming." There are many types of shim coils that are used to superimpose magnetic fields for shimming purposes. The shim coils are designed to produce spatial magnetic field gradients. These fields are typically shaped as linear, quadratic, and higher order functions of spatial position. While typical clinical scanning procedures require adjustments to the linear shims from patient to patient, it is our experience that T2* weighted (BOLD) fMRI benefits greatly from higher order shimming. Modern scanners are equipped with automatic shimming procedures [11] that can typically achieve homogeneities over a 1500 cm³ region of less than 20 Hz RMS, approximately.

In addition to being homogeneous, it is quite important that the magnetic field be as constant as possible over time. The field tends to drift over time due to a number of factors, among them temperature of the room and the equipment, as mentioned previously. These drifts are typically subtle and slow enough that they do not affect clinical / structural imaging. FMRI, however, is based on subtle signal changes over time and therefore, drifts act as significant confounds, especially in slow paradigms. Statistical and signal processing tools do exist to reduce these drifts effects, but it is much more desirable that they be reduced during acquisition. Unfortunately, there are many sources of drift in the MRI hardware, so it is important that the magnet undergo extensive stability testing before it becomes operational and that quality control tests including stability measurements be performed regularly. The scanner's stability can be measured on a phantom over a small region of interest. Figure 11 illustrates a typical stability test.

The physical configuration and shape of the magnet also plays an important role in many of these parameters. While "open" MRI systems exist and are used for large or claustrophobic subjects, their field strength is typically not sufficient for functional MRI applications and their use is limited to clinical applications that do not demand high-quality imaging. Among the closed bore systems, one can choose between short and long bore systems. Short bore systems are intended for head-only applications and can sometimes offer improved performance over smaller regions. Long

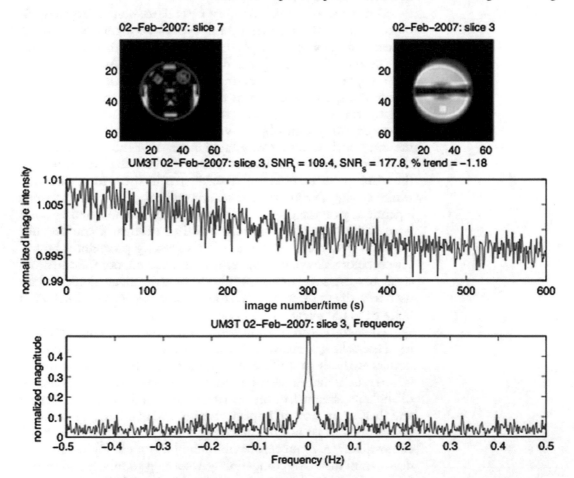

Fig. 11 A typical stability test showing the time course and its frequency content in a phantom (*above*)

bore systems, although more cumbersome, achieve greater field homogeneity over a larger area, which is beneficial for some applications, such as arterial spin labeling.

When considering magnets, it is crucial that we also consider safety issues. The most obvious issue is the very powerful force that magnetic fields of the magnitude required for MRI exert on ferromagnetic objects. These forces are inversely proportional to the square of the distance between the object and the dipole, and are directly proportional to the mass of the metal in question. recall that the magnetic field produced by a current is described by the Biot-Savart law:

$$B(r) = \frac{\mu_0}{4\pi} \int \frac{IdL \times r}{|r|^3} \tag{7}$$

where I is the current, L is a unit vector in the direction of the current, r is another unit vector pointing from the location of the wire to the location of interest, and $|r|$ is the distance from the location of interest to the current source.

One must obviously be very careful to keep ferromagnetic objects out of the magnet room. Typical accidents occur when someone forgets about small metallic objects in their pocket and they fly out of their pocket and strike someone. Accidents sometimes happen because there may be a very subtle force on the object at a specific location in the room leading an unsuspecting investigator to believe that the object is not ferromagnetic. However, moving the object a very small amount toward the magnet can translates into a very rapid increase of the magnetic field, since the magnetic field increases with the inverse of the square of the distance, as mentioned. A small step in the wrong direction while carrying a ferromagnetic object can be the difference between a gentle tug on the object and the object being launched into the bore of the magnet, to the horror of the investigator and the subject. It is thus paramount that strict screening procedures be followed before allowing people into the magnet room. Sometimes small bar magnets and airport security style metal detectors are used to verify the absence of ferromagnetic objects on the subject's body or to test allegedly MR compatible equipment.

These forces can also affect metal implants in the subject's bodies. Pacemakers, neurostimulators, and other implanted electronic devices are likely to malfunction putting the subject at great risk. It is thus crucial that subjects be thoroughly screened for the presence of implants, shrapnel, or other metal sources in their bodies. Having said that, many modern implants are built of titanium and nonreactive materials that are not ferromagnetic and are therefore "MR compatible." A number of publications exist cataloging medical devices and their MR compatibility according to model and manufacturer [1, 2] (and on the web: http://www.mrisafety.com/).

3.2   Magnetic Field Gradients

In order to produce images, an MRI scanner needs a spatially varying magnetic field (see image reconstruction chapter) under tight control by the user. This is accomplished by using an additional set of coils that add extra magnetic fields to the main field. A set of such coils is shown in Fig. 12. By supplying customized current waveforms to these coils, the user can change the distribution of the magnetic field's shape at will. In broad terms, by varying gradient's strength can be varied over time during the pulse sequence, one can obtain MR signals whose phase distribution is a function of the spatial distribution of the sample.

The ideal gradient set is capable of quickly changing the magnetic field as a linear function of spatial location along each of the Cartesian axes. Typical gradients in clinical and functional MRI are between 10 to 40 mT/m, but specialized gradient inserts exist that can produce larger gradients (in the range of 100 mT/m). Small bore animal systems can be outfitted with more powerful gradients (up to approximately 400 mT/m). The main challenges in MRI gradient design and construction usually consist of producing linear gradients in space and time, and the production of eddy currents. Motivated by the need to achieve finer spatial resolution and better axon fiber tracking through diffusion tensor images (DTI), Massachusetts General Hospital has developed a high-performance gradient insert that can achieve up to 300 mT/m in a human system, whereas standard clinical gradients rarely exceed 50 mT/m. This system is utilized at present primarily for experiments concerning the "Human Connectome Project" [12] (www.human-connectomeproject.org).

The spatial linearity of the gradients must be maintained over the volume of the sample to be imaged, or the images will appear warped (although these distortions can be corrected during reconstruction if the true shape of the gradient is known). The spatial

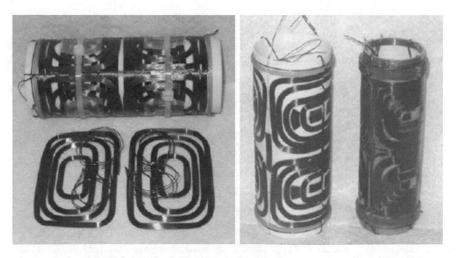

Fig. 12 Gradient coils from Doty Scientific (reproduced with permission of Doty Scientific)

linearity of the gradient fields is primarily a function of the shape of the gradient coils, and a great deal of effort goes into their design and construction (we will not go into those details here). Typical MRI scanner head gradients can maintain 95 % linearity over 30 cm.

The gradient coils must also be able to produce magnetic field gradients very quickly and accurately. The inductive nature of the coils causes their response to the input currents to be severely dampened. In order to correct this problem, typical gradient currents are "pre-compensated" in order to produce the desired waveform [13, 14]. An example of precompensation is illustrated in Fig. 13. Most maintenance or quality assurance protocols include gradient linearity and pre-compensation.

The rate at which a gradient is achieved is referred to as the slew rate. Slew rates of about 200 T/m/s can be generally achieved. However, there are FDA limitations (these are determined by the imaging sequence type and the duration of the stimulation. Typically the maximum allowed rate of change in magnetic field is approximately 20 T/s), since the sudden changes in the magnetic field can induce currents in the peripheral nervous system, causing muscle contractions and unpleasant or even painful sensations in the patient. This phenomenon is referred to as peripheral nerve stimulation (PNS) .

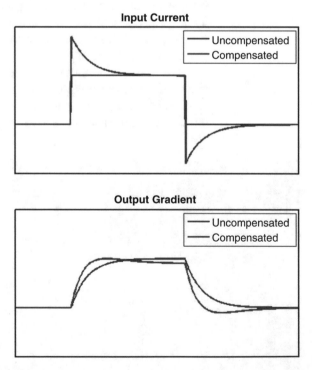

Fig. 13 Simplified illustration of gradient current compensation. The inductive effects of the gradient coils are partially corrected by modifying the input currents to the coil

Active shielding of gradients is necessary to contain the gradient fields and reduce the interactions between gradient coils and other conductors in the scanner. The principles are the same as the active shielding of the main magnetic field (see Sect. 3.1). In other words, a coil producing an opposite gradient field is built around the main gradients in order to cancel the fields outside the area of interest [15, 16].

Gradient vibration and noise are another issue to consider in gradient design. As the gradients are rapidly turned on and off, especially in echo planar sequences, they experience torques due to the presence of the main magnetic field. Thus the coils vibrate violently thus causing the familiar banging MRI sounds. The sound levels are quite loud and require that the subject's wear ear plugs. Active shielding can help reduce the vibrations and acoustic noise produced by the gradients [16].

3.3 RF Hardware

MRI scanners employ RF hardware to generate oscillating magnetic fields that cause the magnetization vector to tip into the transverse plane, and for signal reception. To create signal, an MRI scanner uses a powerful amplifier (generally 1–25 kW) to drive an excitation coil with a large pulse of electric current. In the reception stage, the receive coil is used to pick up the MR signal, which is then processed to create an image. The components of the RF chain that an fMRI user should pay attention to are the transmit and receive coils. Historically, the transmit and receive coils used for fMRI were the same. Today, given the widespread use of parallel imaging to improve image quality via reduced acquisition time, fMRI experiments typically use the scanner's body RF coil for transmit, and a set of multiple coils contained in a single housing that sit close to the head for receive. This approach is referred to as "parallel imaging" [17, 18] and, in FMRI, it's used to improve image quality via reduced acquisition time. Head coils can take many shapes and functional forms, however, for the purposes of SNR and image homogeneity comparisons, there are two main classes of coils: single-channel and multichannel or phased-array coils.

An RF coil is a resonant circuit, and can be modeled as a simple loop containing an inductor, a capacitor, and a resistor (Fig. 14). The inductor and capacitor represent actual circuit components lumped together with the inductance and capacitance of the sample. When a source of electrical current that oscillates at the circuit's resonant frequency is placed across its terminals, the impedances of the inductor and capacitor cancel, and the coil delivers the maximal amount of energy to the sample. In a reciprocal manner, current measured at the coil terminals due to energy radiated by the sample will be of maximum amplitude when that energy oscillates at the coil's resonant frequency. Because the frequency at which biological spins oscillate is determined by the main magnetic field strength via the Larmor relationship, an RF coil must be "tuned" to resonate at this frequency.

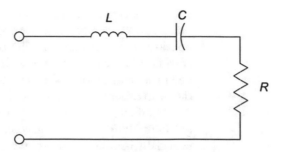

Fig. 14 Series Resistor-Inductor-Capacitor (RLC) circuit representing an RF coil and a biological sample

Single-channel surface coils are the simplest RF coil design. These are typically used to image small areas close to the surface of the skin, as their penetration depth drops sharply with distance from the coil. They consist of a simple loop that induces an oscillating magnetic field when a current is passed through it.

Although the phase information is routinely discarded in image reconstruction, MR signals are inherently vector quantities. RF coils are used to generate an oscillating magnetic field, and the signal emitted back by the sample is a rotating electromagnetic field. However, a simple surface coil can only produce and detect only one component of that vector: the component that is perpendicular to the plane of the coil. Quadrature coils typically consist of two perpendicular coils that transmit at 90° phase from each other. The resulting magnetic field is the vector sum of the two perpendicular fields. Quadrature reception can also be achieved with the same coils by adding phase to one of the channels in the receive chain.

The most popular quadrature-channel head coils generally take on a "birdcage" design (Fig. 15a). This classic design provides good SNR and image homogeneity characteristics, owing to the unique nature of the magnetic field that it creates [19]. Another type of single-channel coil commonly used in studies of the occipital cortex is a quadrature occipital coil (Fig. 15b), which provides high SNR in this localized region of the brain, and has a compact design compared to full head coils, allowing the experimenter greater freedom in stimulus hardware setup.

In contrast, multichannel or phased-array coils (Fig. 15c) are composed of a set of discrete and independent "surface" coils, arranged around the head. Generally, these coils can be brought into a tighter conformation around the head, which improves SNR. Taken alone, images obtained with individual surface coils will possess lower SNR and poor image homogeneity compared to a birdcage coil, however, when images from the coils are combined in a sum-of-squares reconstruction, excellent SNR can be achieved [20], though image homogeneity will still be worse than for a birdcage coil. Furthermore, most phased-array coils can be used only

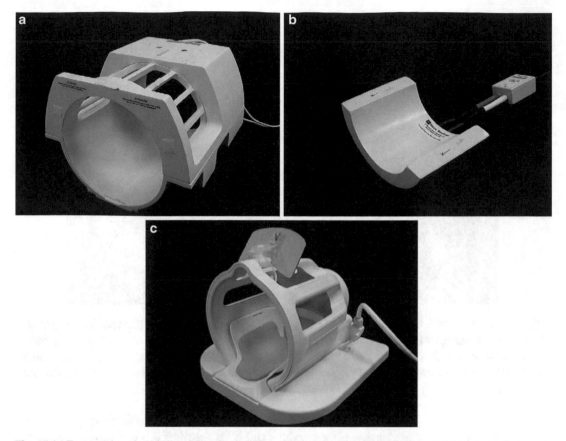

Fig. 15 (a) Transmit/receive birdcage head coil. (b)Receive-only quadrature occipital coil (c) 8-channel phased-array coil

for reception, so a separate coil (usually the scanner's body coil) must be employed for excitation, which can result in degraded image homogeneity, as well as increased SAR in areas outside the head. The central advantage to phased-array coils is that they allow the use of parallel imaging techniques, such as SENSE [17] and GRAPPA [21], when paired with RF signal chains capable of receiving multiple channels simultaneously, which are now commonly available from most MR vendors. Parallel imaging exploits the inhomogeneity of images obtained with surface coils to accelerate image acquisition, which results in a reduction of artifacts in fMRI images, as in Fig. 16. This comes at the cost of reduced SNR. Phased-array head coils used for fMRI commonly have 32 individual coil elements, and can be effectively used to reduce image acquisition time by a factor of 2–4. Scanners with up to 128 receive channels are available from MR vendors.

A birdcage coil generally provides a lot of flexibility in stimulus presentation setup, due to the large amount of room within the coil. One can use goggles, projector/mirror systems, and a range of other solutions in conjunction with a birdcage coil (as we discuss

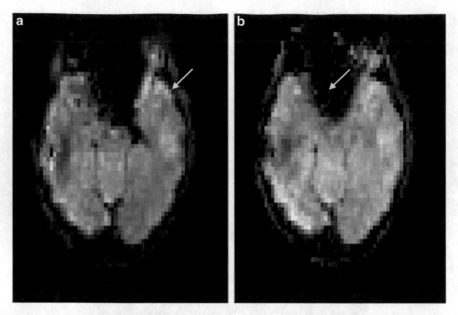

Fig. 16 Comparison of fMRI images obtained using conventional (**a**) and parallel imaging (**b**) in the inferior brain. In this example, parallel imaging with an 8-channel phased-array coil was used to reduce data readout time by a factor of 2. This reduced signal loss and image distortions due to susceptibility, particularly in the region indicated by the arrows. (Courtesy of Yoon Chung Kim, University of Michigan Functional MRI Laboratory)

later). In comparison, current phased-array coil designs limit presentation options, since the coils forming it are placed closer to the head, which may prevent the use of visors. However, phased-array coils are generally compatible with the most commonly used projector-based visual stimulus setups. When using visor-based stimulus systems, one must ensure that the electronics of the visor are properly shielded to prevent image artifacts of the type shown in Fig. 4. The "streaking" artifact shown in this image was caused by a visor with an electromagnetic "leak," that coupled to the receive coil and manifested as a false MR signal. The form of the artifact will depend on the pulse sequence used; in this example the streak is localized, while in another pulse sequence the artifact may be spread over the entire image.

In the near future, the design of multichannel receive coils is likely to be in part driven by simultaneous multislice imaging, a recent parallel imaging-based scan acceleration technique in which multiple slices in a volume are excited and read out simultaneously, leading to a scan acceleration factor equal to the number of simultaneously-excited slices [22, 23]. Whereas conventional parallel imaging is applied to accelerate the acquisition of each slice's data individually and thus requires a circumferential density of receive coils around the head in the plane of the slice, a simultaneous multislice acquisition requires a density of coils in the slice dimension (typically head-foot in fMRI) to successfully separate the simultaneously-excited slices' signals. Using appropriate coils and sequences, acceleration factors of 2–5 are common in simultaneous multislice exams [24].

Finally, it is now a common option for 3 T MRI scanners to be equipped with two transmit channels, which are used to independently adjust the signals transmitted into the two ports of the scanners' integrated quadrature body coil in order to produce more uniform RF fields, on a subject-to-subject basis. While this capability is primarily driven by the need for more uniform RF fields in body MRI, it may also provide some benefit in improving field uniformity in brain MRI [25]. Furthermore, multichannel transmit methods are currently being developed by several researchers to alleviate through-plane signal loss artifacts in the lower brain [26, 27].

3.4 Computing Resources

Functional MRI experiments generate large datasets. The raw data alone for a single subject, scanned for one hour, weighs in between 100 and 200 MB. Combine this with the space needed for image reconstructions and analysis, and one should budget a storage and scratch space of around 2 GB for every hour of scanning. In this section, we will provide some guidelines and suggestions on how to set up a computing environment to handle all this data. The two major factors influencing the design of your environment will be (1) money, and (2) the expertise available, in terms of computer systems and MRI data processing.

The data stream in a typical laboratory consists of four main stages. The first stage is the MRI scanner, which produces either raw, unreconstructed data, or reconstructed images that are ready for post-processing and analysis. If a lab has an MR physicist at its disposal, then the former case is often true, since an MR physicist may develop image reconstruction codes that provide improved image quality over vendor-provided software, and that can add in improved reconstruction techniques as they are developed. Assuming images have been reconstructed, the second stage consists of slice timing correction, realignment, coregistration, warping, and smoothing, which are all operations that prepare the dataset for statistical modeling and analysis. The third stage is statistical modeling and analysis. The fourth stage is data backup, though it is advisable to make backups of data at more than one point in the stream.

The majority of fMRI labs maintain one powerful workstation or computer cluster to do the bulk of their processing. Users can log into this computer from their own machines to initiate processing or view and download the preprocessed data for local analysis. The central advantage to this model is that the large and complicated software packages used to process fMRI need only be maintained on one machine, which greatly simplifies maintenance. A secondary advantage is that this model allows users greater flexibility in choosing the operating system of the computers they use; the workstation can be running a Unix derivative, which has benefits in terms of networking, stability, and the availability of fMRI software packages, while the users can be using Windows PC's or Macintoshes, which are generally more user-friendly systems. All computers that will be involved in processing or storing data should be connected with the fastest network possible, such as a gigabit network.

The central processing workstation should have as much RAM as possible. By today's computing standards, each CPU should have at least 2 GB RAM available to it, so that it may store the entire dataset and related files. The reason for this is to minimize the frequency with which the computer accesses its hard disk processing, which costs large amounts of time. The second main consideration is the number of CPU's the computer should have. This will largely depend on the number of simultaneous processing streams one expects to have running on the computer. The third consideration is storage. The main computer should be connected to a large data storage device, so that all current experiments can be instantly accessed, without requiring reloading of backed-up data. This device should be a set of hard disks configured in a redundant manner, such as a RAID array.

One of the central computing dilemmas that an fMRI lab will continually deal with is making data backups. There are two main questions to answer here: (1) at what points in the processing stream should backups be made, and (2) what form should backups take? Backups of the initial MR data are absolutely necessary, since an experimenter may be asked to reproduce their results at a later date, or bugs may be found in the post-processing stream (stage two), which will require re-processing the original data. After this stage in the stream, the choice of where to do backups will depend on the amount of backup space available and the speed with which the second and third stages may be executed, should the analyzed data be lost. The more points at which backups are made, the more quickly an experimenter could recover after a data loss or processing interruption. A laboratory also has many options in choosing backup forms, and it may be best to use a combination of them. Perhaps the simplest backup form is mirroring the data, i.e., storing the data in another set of hard disks whose sole purpose is to store backed-up and archived data. This solution is particularly simple in that backups can be fully automated, and instantly accessed. The other two main options are tape storage and optical media (DVD). While automated machines may be purchased to manage tape backups, hard disk capacity is rapidly outstripping tape capacity, and the mechanical nature of these machines makes them prone to frequent failure. On the other hand, DVD backups are more reliable and cheap, but they require human interaction to load DVD's and execute burning software. A backup schedule will also have to be worked out by the laboratory.

Because computer components (i.e., disks, CPU's, and video cards) frequently fail or need replacement, it is advisable to set aside part of the lab's initial capital for yearly computer maintenance. It is also advisable to purchase extended warranties for the computers, as they will be heavily used and if they fail, this can save the lab a significant amount of money in the long run. Furthermore, as technology advances, the lab should build into their operational budget the cost for new machines every few years.

4 Stimulus Presentation and Behavioral Data Collection Devices

There is a small but increasing number of manufacturers of stimulus presentation hardware and software. We will not review specific products or vendors but limit ourselves to describe the important characteristics of these devices.

Some components of the stimulation and response recording equipment must go inside and/or near the magnet. These must not be susceptible to magnetic forces for obvious safety reasons. Additionally, the presence of metals in headphones and head mounted displays can generate field distortions that cannot be compensated by shimming, even if they are not ferromagnetic. This results in severe image degradation and it is thus important that the devices be thoroughly tested on phantoms for image degradation.

As we alluded to before, electronic equipment in the scanner room must be adequately shielded to avoid introducing RF noise into the system. For example, LCD displays for visual stimulation inside the magnet are typically encased in a fine wire mesh that acts as a Faraday cage to contain RF leakage. Other audiovisual electronic equipment used in the MR environment, such as projectors and button response units are typically encased in brass or aluminum for the same reasons. Regardless of the manufacturers' best intentions, sometimes the shielding is not adequate or becomes damaged over time in subtle ways. Just as in the case of the room's RF shielding, an exposed wire or bad shielding connections can produce severe RF contamination of the images. Thus, it is paramount that the stimulation devices be checked upon purchase and periodically for RF leaks that may develop during delivery, installation, or daily use.

There are different technologies commercially available for MR compatible visual stimulation. The simplest approach is perhaps an LCD projector outfitted with narrow focus lenses that project the images into a back projection screen placed inside the bore of the magnet. The subject then can see the display through a set of mirrors that are mounted on the head coil assembly. The main advantages of this approach are simplicity and lower cost. The disadvantages are related to positioning issues and a reduced visual field for the subject.

Another approach to MR compatible visual stimulation is fiber optic display visors. This sort of display system is based on a optical signal converter that carries an SVGA quality image through an array of micro optic fibers to a head mounted display inside the scanner's head coil. This approach is very attractive in that there are no electronic components that need to be installed inside the scanner room and the display can be placed very accurately in front of the subject's eyes, maximizing the available visual field. The drawbacks are that in additional to being expensive, the fiber optics used in the array are very fine and brittle, so that regardless of the high quality of fabrication, there will always be a small number of broken fibers that result in dead pixels or small streaks in the image.

One of the more popular approaches is to display the images on a shielded LCD screen mounted in front of the subject's head. This screen can be either a large one that is mounted outside the scanner's RF coil, or a small one in a visor that the subject wears inside the coil. The advantages of this are the large visual field and ease of use of the system. The drawbacks are the high cost and the interactions between the display electronics and the magnet. Some of these devices become dimmer when placed inside the magnetic field. Additionally, if any RF leaks develop, they severely degrade the images, especially in the visor type systems since they sit inside of the RF coil.

Auditory stimulation is typically performed in the MR environment through two different kinds of headphones: pressure waveguide types, and shielded piezoelectrics. Both of these are highly effective devices. The pressure waveguide headphones keep the speakers outside of the magnet's bore and the sound is carried through rigid tubing into the headphones. The piezoelectric headphones are akin to standard speaker technology but use piezoelectrics to produce the vibrations. They require RF shielding of the cables and the electronics to prevent artifacts. Perhaps the biggest challenge for auditory stimulation is reduction of the MRI scanner's noise. There is very limited space inside the scanner's head coil for building an effective muffler into the headphones but fairly effective noise reduction (typically around 30 dB) can be achieved. The headphones' acoustic insulation is sometimes achieved by gel padding that attenuates the sound very effectively by forming a tight seal around the ear. Caution must be used as the gel in the padding produces an MR signal and is visible in the images so it must be taken into consideration during registration and normalization of structural images. The gel's resonant frequency is typically not the same as water and produces some off-resonance artifacts, but these tend to be mild.

MR compatible microphones for patient communication and verbal response recording are typically based on piezoelectric technology and require both electronic and acoustic shielding to reduce the scanner sound. To our knowledge of the present state of the art, the acoustic shielding of the microphone from the scanner sound is somewhat effective, but communication with the subject during the scan is still challenging. Some systems are equipped with active noise cancellation with limited success. Consequently, investigators often use pulse sequences with "quiet" (i.e., no gradient pulses) periods during the subject response time instead [28, 29].

Other response recording devices are primarily "button response units (BRU)" that are built into hand rests and strapped to the subject's hands. They typically carry only DC currents through twisted pair cables and RF noise is not an issue as the electronics to drive the system are kept outside the scanner room. There are a number of other response units, such as MR compatible joysticks and keyboards that are manufactured by small companies. While these are typically safe and effective, one should test all such equipment immediately upon purchase not only for functionality but for RF leakage, and ferromagnetic forces. Periodic RF testing of peripherals should be an integral of the fMRI facility's QA procedures.

5 Subject Monitoring

When running an fMRI experiment, it can be desirable to monitor and record subjects' status and peripheral signals during an fMRI experiment, to use as correlates of subject behavior or as nuisance signals in data correction. Some possibilities include monitoring cardiorespiratory rhythms, galvanic skin response, head motion, or eye-tracking. As stressed in the previous sections, all these considerations should fit with the comfort and safety of the subject.

In general, when considering recording peripheral signals on fMRI subjects, one should pay attention to: synchronization with the MR scanner, adequate sampling of the signal in question, and avoiding introducing signal noise in both the MR data and the recorded peripheral signals.

5.1 Scanner Synchronization

In order to match the recorded external signals with the fMRI data being recorded, synchronization with the start of the scan must be achieved. This can be done using a TTL pulse to/from the scanner from/to the external device or recording media. For instance, a logic pulse from the MR scanner to the computer recording physiological noise can be set to trigger the recording sequence. Commercial MR scanners from the main vendors (GE, Siemens, Philips) all have the capability to send or receive TTL sync pulses.

5.2 Physiological Monitoring

A limit to the effectiveness of functional MRI in detecting activation is the presence of physiological noise, which can equal or exceed the desired signal changes in an fMRI experiment [30]. These physiological fluctuations that are present during an fMRI scan can obscure the BOLD activity that the researcher is trying to detect. In addition, monitoring physiological rates can help as secondary reaction measures (such as monitoring the cardiac rate variability during a stress experiment).

5.3 Cardiac Monitoring

Monitoring the cardiac waveform can be achieved in several ways. The most common solutions are pulse oximeters or ECG patches. The primary cardiac harmonic frequency lies in the 0.5–2.0 Hz range, with both the first and secondary harmonics shown to affect the fMRI signal [31].

Pulse oximetry refers to indirectly monitoring the oxygen levels in the extremities to monitor the cardiac waveform. This is most often accomplished in fMRI labs by using an LED and photodiode that clips to the subject's finger, connected to a data acquisition board (see Fig. 17). Several MR scanner vendors offer this as part of the MR system (GE, Siemens), and stand-alone monitoring units from commercial vendors are also available (Invivo, Biopac). Normal setup with compliant subjects allows adequate sampling of cardiac rhythm, as seen in Fig. 18. Drawbacks include the fact that

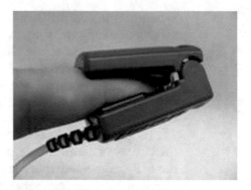

Fig. 17 Pulse oximeter for indirect measure of cardiac waveform

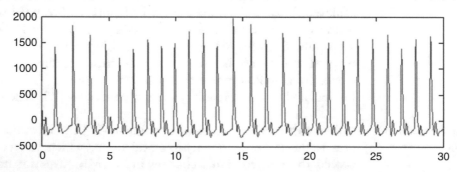

Fig. 18 Cardiac waveform acquired during an fMRI scan. (Data acquired on a 3.0 T GE scanner, using a pulse oximeter with a sampling rate of 40 Hz)

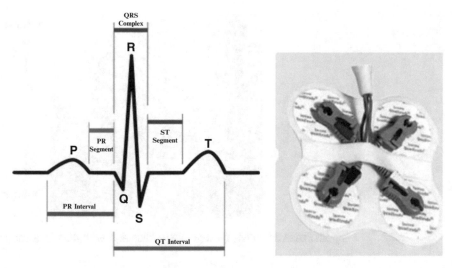

Fig. 19 Example of ECG patch (Courtesy of Invivo, www.invivocorp.com). With a schematic of a typical QRS waveform

subject motion may corrupt the signal, and motor tasks may be impeded with the oximeter placed on the finger (alternative placement on the ear or toe is possible).

ECG patches located over the heart allow high-fidelity monitoring of cardiac electrical activity. This allows identification of features beyond the simple cardiac peaks, such the QRS complex during the depolarization of the ventricles (Fig. 19). Disadvantages of ECG recording include increased setup complexity, and patient comfort.

5.4 Respiratory Monitoring

The respiratory rhythm has a normal frequency range of 0.1–0.5 Hz. Motion of the chest during respiration combined with the changes in oxygen saturation in the lungs lead to a modulation of the local magnetic field that can affect the phase of the MR signal at the position of the head during scanning (Fig. 20), which can lead to modulation of the recorded MR signal intensity. Mitigation of respiratory effects on the MR signal can include scanning during breath-holds, modified pulse sequences to sample and account for the modulation in magnetic field [32] and recording of the respiratory signal to use as nuisance covariates in post-processing analysis [33].

Monitoring respiratory rhythm can be accomplished by using a plethysmograph (pressure belt) around the waist of a subject, like the one shown in Fig. 21, or by using a nasal cannula to monitor expired CO_2 concentration. A sample respiratory waveform is shown in Fig. 22.

5.5 Galvanic Skin Response (GSR)

Galvanic skin response (GSR) is a measure of the electrical resistance of the skin, a physical property that has been shown to increase in response to subject arousal, mental effort, or stress. It is monitored

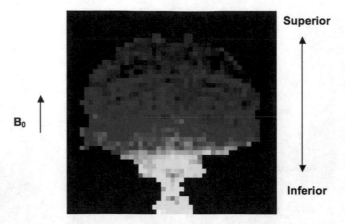

Fig. 20 Phase difference between inspiration and expiration for a coronal slice

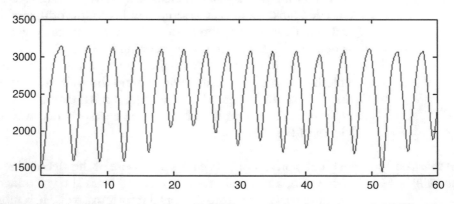

Fig. 21 Plethysmograph belt for measuring respiration

Fig. 22 Respiratory waveform acquired during an fMRI scan. (Data acquired on a 3.0 T GE scanner, using a pulse oximeter with a sampling rate of 40 Hz)

by measuring a voltage drop across the skin using paired electrodes. This can have the same motion sensitivity and motor task complication as the pulse oximeter if the electrodes are placed on the fingers. These problems are usually reduced by placing the electrodes between the second and third knuckles, instead of on the fingertip.

5.6 Head Motion Tracking

Severe head motion during an fMRI scan can severely corrupt the data. Besides minimizing patient motion using cushioning and restraints, a measure of head motion may also be collected to help correct data in post-processing. This may be done using modified pulse sequences (PACE), or by using external motion tracking [33–35]. Again, patient comfort and visual path should be taken into consideration.

5.7 Eye Tracking

Patient gaze and fixation time is important for several types of fMRI paradigms and pathological types. In addition, eye motion can be a source of variance in fMRI scans [36]. Thus, tracking eye position can be desirable. The common method is to monitor the position of infrared (IR) light that is reflected off the eye of the subject. This involves transmitting IR light to the subject's eye, and then recording it, using MR-compatible equipment. Several vendors provide hardware solutions including both long-range and short-range cameras. These systems typically have on the order of 0.1–1° spatial resolution and accuracy, with working ranges of 10–25° horizontally and vertically, and 60–120 Hz sampling rate.

In acquiring a physical eye-tracking system for an fMRI lab, consideration should be given to the optical path for the eye-tracking system, taking into account the MR bore, head coil, and visual stimulus presentation system; signal integrity of the MR data; ease of setup for the MR techs and scanners. This will also involve peripheral equipment located in the scanner room or control room: usually a camera, power supply, video monitor for real-time display of the subject's eye, and a PC.

6 Multimodal fMRI

Acquiring other neurophysiological measures with fMRI data can complement the excellent spatial resolution and depth penetration of fMRI with modalities that have superior temporal resolution and different sensitivities to the underlying neuronal activity. In the following sections, we will expand upon complementary modalities that allow investigations of response (e.g., EEG, fNIRS) and stimulation (e.g., TDCS, TMS).

6.1 FMRI-EEG

The simultaneous acquisiton of electroencephalography (EEG) data with fMRI data allows the higher temporal sampling rate of EEG (~5000 Hz) to be combined with the superior resolution (~ mm) and depth penetration of fMRI.

Several factors must be accounted for when setting up a simultaneous EEG/fMRI acquisition, chief among them safety and signal quality. In the following, we will try to touch on most of the considerations one will make when selecting and setting up EEG-fMRI hardware.

Several companies have MR-compatible EEG hardware commercially available (Brain Products, Neuroscan, EGI). A common EEG setup includes an electrode cap, connected to a signal amplifier and recording device. Any part of this setup that is inside the MR scanner room within the 5 G line must be nonferrous, and the amount of metal must be kept to a minimum, with care exercised around all metallic components (this includes electrode leads on the cap and skin, batteries in amplifiers, etc.). Fiber optics can be used for signal transmission after amplification, with recording devices located in the scanner control room.

Much planning is required to integrate the EEG with the MR. The head coil used for MR acquisition may affect the physical setup. For instance, an open birdcage coil may allow placement of the EEG cap cord and amplifier above the head of the subject, with no obstruction of the subject's field of view. With an alternative phased-array coil that is closed at one end, this setup may not be possible.

Also, for time synchronization, it may be necessary to use the TTL pulse from the scanner (mentioned in Sect. 5.1) to trigger the EEG recording device at each TR.

Finally, for fusion of the fMRI and EEG data, accurate electrode locations on the head should be recorded and transferred to the structural MR images to be used as references for localization. Commercial head position recording systems are available (Brainsight). These systems can record points on the subject's head using infrared positional markers, and coregister these coordinates with the structural MR scans of the subject. Subsequent coregistration of the fMRI and structural MR data allows direct overlay and source localization using both between the MR, fMRI, and EEG data.

The EEG equipment should not adversely affect the MR data, if proper materials and shielding are used. Images with and without the EEG equipment should be inspected for any introduction of AC line noise (~60 Hz) or localized variance in the structural and fMRI images.

The MR equipment will affect the EEG recording, due to the high-field environment, and the application of gradients during the MR acquisition (Fig. 23). However, the MR gradient artifact can be corrected for in post-processing, using either available software (EEGLAB, http://www.sccn.ucsd.edu/eeglab/) or adapting techniques such as PCA or ICA.

6.2 Functional Near-Infrared Spectroscopy (fNIRS)

Functional MRI allows investigation of the hemodynamic response to neural activity, thus allowing localization of brain regions involved in a cognitive task. However, fMRI is not a naturalistic setting.

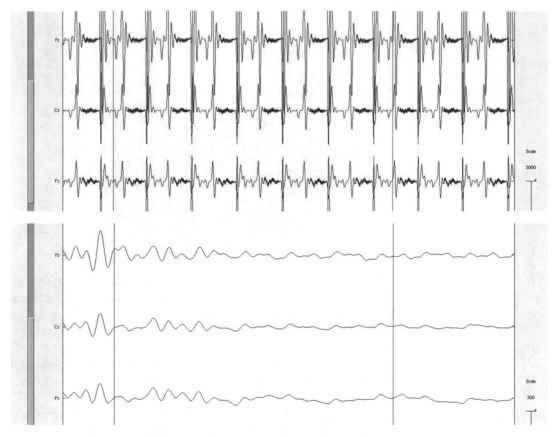

Fig. 23 EEG recorded during fMRI acquisition, before (*top*) and after (*bottom*) MR artifact correction

Functional near-infrared spectroscopy (fNIRS) offers a possible solution for portable neuroimaging in the field. It offers subsecond temporal sampling, with spatial resolution on the order of centimeters. For multimodal work with fMRI, it is attractive as it samples the same hemodynamic changes as BOLD fMRI imaging, namely the change in relative levels of oxygenated and deoxygenated blood, using the differential absorption of infrared laser light. An image of a commercially available FNIRS system can be seen in Fig. 24.

Generally, fNIRS operates by using infrared light source and detector pairs. While the light used is in the infrared range, and thus the majority of the setup may be thought of as fiber-optic and largely MR compatible, the internal construction of the probes may contain non-MR compatible materials (such as metal coatings in reflectors). Careful consideration must thus be used for a fNIRS-fMRI combined experiment. Several vendors offer MR-compatible probes (EGI, Imagent), or custom setups can be constructed in-house [37]. In terms of induced noise, attention must be paid to induced MR artifacts due to probe construction. However, due to

Fig. 24 Example MRI compatible fNIRS setup (Cortech Solutions, www.cortech-solutions.com/Products/NI/NI-OM), showing light guides, example subject setup configuration, and receiving equipment

the optical nature of the fNIRS signal, it is not affected by the MR environment, offering one advantage over EEG-fMRI acquisition.

6.3 Brain Stimulation and fMRI

A fundamental strategy used in cognitive neuroscience is to stimulate the brain in some way and then observe how it responds. Functional MRI provides such observations of the brain's responses. While most studies use sensory (audio, visual, tactile … etc.) cues and cognitive tasks as a form of stimulation, one can also stimulate human brains directly and noninvasively inside an MR scanner.

One such technique is transcranial magnetic stimulation (TMS), which has great potential not only as research tool but also as a therapeutic device [38–40]. The principle behind TMS is that a large current waveform is driven through a coil placed adjacent to the tissue of interest. The current in turn induces an electromagnetic field depending on the rate of change of the current, as predicted by classical electrodynamics. The induced electric field penetrates the tissue and induces eddy currents on conductors, such as nerve fibers. When a nerve fiber is aligned with the direction of the electric field, a large current is induced in the axon which causes its membrane to depolarize, effectively causing the transmission of an action potential [41–43]. After depolarization of the axonal membrane, the sodium–potassium pumps rebuild the membrane potential and the nerve fibers return to their original state within seconds.

The effects of these induced discharges are complex, depending upon the magnitude and timing of the TMS pulse affecting inhibitory and/or excitatory neuronal populations (for recent reviews, see [44–46]). If a single TMS pulse is applied in the hand region of the motor cortex, for example, motor neurons depolarize and the hand will twitch ('motor evoked potential', or MEP). Subthreshold stimulation, followed by supra-threshold stimulation, can inhibit or facilitate the MEP, which varies with the distance and location of the subthreshold stimulus [47].

By selectively applying a TMS pulse during the performance of a task, neural circuits are effectively jammed, and performance interruptions can be observed. In effect, one creates a controlled, completely reversible "lesion," enabling the study of brain function through perturbation of neuronal activity [44, 48]. Although these studies have rapidly become a popular investigative tool for cognitive neuroscientists, one important limitation stems from inadequate knowledge about the shape and magnitude of the induced current fields that introduce the perturbation. Hence, TMS and fMRI can be complementary for the study of brain function. TMS can interfere, or modulate the cognitive process under scrutiny by locally altering the responsiveness of the tissue, while fMRI can allow the investigators to precisely map out these effects. While TMS and fMRI experimental data can be coregistered and integrated after each experiment has been carried out separately [49], it is desirable to be able to observe the BOLD responses to TMS.

However, there are some clear challenges to carrying out joint TMS and fMRI experiments. Such experiments require TMS coils that contain no ferromagnetic parts and extra long cabling so that the amplifier/capacitor bank can be kept outside of the 5 G line. Specialized holders must be constructed to hold the TMS coil in the appropriate position during the duration of the scanning session (Fig. 25). Like all electronic equipment, the TMS hardware

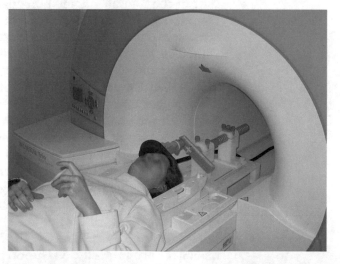

Fig. 25 MRI compatible TMS coil and holder apparatus (image courtesy of Dr. A. Thielscher of the Max Plank Institute for Biological Cybernetics http://www. kyb.mpg.de/)

must be shielded in a faraday cage to prevent RF contamination of the MRI signal [50]. Commercial TMS coils currently have these characteristics as optional features.

Another important aspect is the synchronization of TMS pulses and image acquisition. It is very important that the scanner's RF receiver chain be switched off while the stimulator is pulsing in order to protect it from the large signals that may severely damage it. At the same time the scanner's pulses may induce currents in the TMS hardware (although these are reportedly not harmful to the hardware). Isolation between the two is achieved by synchronizing the TMS pulses with the MRI scanner via TTL pulses. The imaging pulse sequence is typically designed with long gaps for the TMS pulses. These gaps include about 0.1 s to allow TMS induced eddy currents inside the bore to decay. There is also the challenge of the large torques that a large, sudden dipole exerts when in the presence of a large magnetic field. However, in the case of figure-eight coils, those torques cancel since the two "wings" of the coil are torqued in opposite directions. Furthermore, rapid bi-phasic pulses also cancel those torques and the subject does not perceive any such effects. The coil, however, experiences internal stresses from the magnetic forces [50–52].

Recently, transcranial direct current stimulation (TDCS) has been adopted by neuroscientists as a method to manipulate the excitability of neurons [53–56]. The technique is very simple and can be easily combined with functional MRI with minor safety concerns. In a nutshell, TDCS works by running a small direct current (1–2 mA) from an anode to a cathode placed across brain. While the mechanism is not entirely understood, the neurons beneath the anode experience enhanced excitability because of a shift in the transmembrane potential, and the opposite is true at the cathode [57]. At the time of writing, TDCS and its mechanism are a very exciting and active area of research. Indeed it is often combined with BOLD and ASL FMRI to study its effects [58, 59].

The main issues that must be addressed when conducting TDCS-FMRI experiments are the heating of the electrodes and wires by currents induced by the scanner's oscillating magnetic fields (RF and gradients). A solution to this problem is to increase the impedance of the electrodes and wires by placing resistors (>5 kOhm) in line. One must also be careful of the presence of the ancillary equipment, such as the power supply to the TDCS device, in the magnet room. In general it is good to keep that part outside the scanner room and feed the cabling through a penetration panel. In that case, the wires are likely to act as an antenna and bring in RF noise from outside the magnet room, which is detrimental for image quality. A solution to this problem is to also place a low-pass filter in line (between the power supply and the electrodes) to eliminate the RF noise in the system. One example of this approach can be seen in [58], where the investigators placed a

5.6 kOhm resistor on MR compatible rubber electrodes, and filters in line with the power supply that were rated to attenuate 60 dB between 20 and 200 MHz.

7 Conclusions

The hardware used in magnetic resonance imaging is quite extensive and we hope to have provided an adequate overview of the subsystems involved in the generation of MRI images. Functional MRI requires additional hardware for collection of behavioral data and stimulation of the subject while collecting the functional images. The greatest challenge is perhaps to coordinate all these devices while being mindful of the interactions between the devices and the MRI scanner. Failure to do so often results in severe artifacts in the desired measurements, or worse, the subject could be severely injured. In this chapter we have also explored the hardware requirements for multimodal imaging, such as EEG-fMRI or TMS-fMRI.

References

1. Shellock FG (2002) Reference manual for magnetic resonance safety, implants, and devices. Saunders, Oxford, UK

2. Shellock FG, Crues JV 3rd (2002) MR safety and the American College of Radiology white paper. AJR Am J Roentgenol 178:1349–1352

3. Train JJ (2003) Magnetic resonance compatible equipment. Anaesthesia 58:387, Author reply 387

4. Durand E, van de Moortele PF, Pachot-Clouard M, Le Bihan D (2001) Artifact due to B0 fluctuations in fMRI: correction using the k-space central line. Magn Reson Med 46:198–201

5. Tinkham M (2004) Introduction to superconductivity, 2nd edn, Dover Books on Physics. Dover Publications, Mineola, NY

6. Radebaugh R (2009) Cryocoolers: the state of the art and recent developments. J Phys Condens Matter 21:164219

7. Williams DS, Detre JA, Leigh JS, Koretsky AP (1992) Magnetic resonance imaging of perfusion using spin inversion of arterial water. Proc Natl Acad Sci U S A 89:212–216

8. Yang QX, Wang J, Zhang X et al (2002) Analysis of wave behavior in lossy dielectric samples at high field. Magn Reson Med 47:982–989

9. Collins CM, Liu W, Schreiber W, Yang QX, Smith MB (2005) Central brightening due to constructive interference with, without, and despite dielectric resonance. J Magn Reson Imaging 21:192–196

10. Tropp J (2004) Image brightening in samples of high dielectric constant. J Magn Reson 167:12–24

11. Schneider E, Glover G (1991) Rapid in vivo proton shimming. Magn Reson Med 18:335–347

12. Dylan Tisdall M, Witzel T, Tountcheva V, McNab JA, Adad JC, Kimmlingen R, Hoecht P, Eberlein E, Heberlein K, Schmitt F, Thein H, Wedeen Van J, Rosen BR, Wald LL (2012) Improving SNR in high b-value diffusion imaging using Gmax = 300 mT/m human gradients, Proc ISMRM 2012

13. Gach HM, Lowe IJ, Madio DP et al (1998) A programmable pre-emphasis system. Magn Reson Med 40:427–431

14. Wysong RE, Madio DP, Lowe IJ (1994) A novel eddy current compensation scheme for pulsed gradient systems. Magn Reson Med 31:572–575

15. Mansfield P, Chapman B (1986) Active magnetic screening of coils for static and time-dependent magnetic field generation in NMR imaging. J Phys E Sci Instrum 19:540–545

16. Edelstein WA, Kidane TK, Taracila V et al (2005) Active-passive gradient shielding for MRI acoustic noise reduction. Magn Reson Med 53:1013–1017

17. Pruessmann KP et al (1999) SENSE: sensitivity encoding for fast MRI. Magn Reson Med 42(5):952–962

18. Blaimer M, Breuer F, Mueller M et al (2004) SMASH, SENSE, PILS, GRAPPA: how to choose the optimal method. Top Magn Reson Imaging 15:223–236

19. Hoult DI, Chen CN, Sank VJ (1984) Quadrature detection in the laboratory frame. Magn Reson Med 1:339–353

20. Roemer PB, Edelstein WA, Hayes CE, Souza SP, Mueller OM (1990) The NMR phased array. Magn Reson Med 16:192–225

21. Griswold MA et al (2002) Generalized autocalibrating partially parallel acquisitions (GRAPPA). Magn Reson Med 47:1202–1210

22. Larkman D, Hajnal J, Herlihy A, Coutts G, Young I, Ehnholm G (2001) Use of multicoil arrays for separation of signal from multiple slices simultaneously excited. J Magn Reson Imaging 13(2):313–317

23. Setsompop K, Gagoski BA, Polimeni JR, Witzel T, Wedeen VJ, Wald LL (2012) Blipped-controlled aliasing in parallel imaging for simultaneous multislice echo planar imaging with reduced g-factor penalty. Magn Reson Med 67:1210–1224

24. Feinberg D, Moeller S, Smith S, Auerbach E, Ramanna S, Glasser M, Miller K, Ugurbil K, Yacoub E (2010) Multiplexed echo planar imaging for sub-second whole brain fmri and fast diffusion imaging. PLoS One 5(12), e15710

25. Zhang Z, Yip CY, Grissom W, Noll DC, Boada FE, Stenger VA (2007) Reduction of transmitter B1 inhomogeneity with transmit SENSE slice-select pulses. Magn Reson Med 57(5):842–847

26. Stenger VA, Boada FE, Noll DC (2000) Three-dimensional tailored RF pulses for the reduction of susceptibility artifacts in T2*-weighted functional MRI. Magn Reson Med 44(4):525–531

27. Yip CY, Fessler JA, Noll DC (2006) Advanced three-dimensional tailored RF pulse for signal recovery in T2*-weighted functional magnetic resonance imaging. Magn Reson Med 56(5):1050–1059

28. Jakob PM et al (1998) Functional burst imaging. Magn Reson Med 40:614–621

29. Edmister WB, Talavage TM, Ledden PJ, Weisskoff RM (1999) Improved auditory cortex imaging using clustered volume acquisitions. Hum Brain Mapp 7:89–97

30. Noll DC, Schneider W (1994) Theory, simulation, and compensation of physiological motion artifacts in functional MRI. Image processing, 1994. Proceedings ICIP-94. IEEE Int Conf 3:40–44

31. Hu X, Le TH, Parrish T, Erhard P (1995) Retrospective estimation and correction of physiological fluctuation in functional MRI. Magn Reson Med 34:201–212

32. Pfeuffer J, Van de Moortele PF, Ugurbil K, Hu X, Glover GH (2002) Correction of physiologically induced global off-resonance effects in dynamic echo-planar and spiral functional imaging. Magn Reson Med 47:344–353

33. Tremblay M, Tam F, Graham SJ (2005) Retrospective coregistration of functional magnetic resonance imaging data using external monitoring. Magn Reson Med 53:141–149

34. Zaitsev M, Dold C, Sakas G, Hennig J, Speck O (2006) Magnetic resonance imaging of freely moving objects: prospective real-time motion correction using an external optical motion tracking system. Neuroimage 31:1038–1050

35. Thesen S, Heid O, Mueller E, Schad LR (2000) Prospective acquisition correction for head motion with image-based tracking for real-time fMRI. Magn Reson Med 44:457–465

36. Chen W, Zhu XH (1997) Suppression of physiological eye movement artifacts in functional MRI using slab presaturation. Magn Reson Med 38:546–550

37. Harrivel AR et al (2009) Toward improved headgear for monitoring with functional near infrared spectroscopy. NeuroImage 47:S141

38. Barker AT (1991) An introduction to the basic principles of magnetic nerve stimulation. J Clin Neurophysiol 8:26–37

39. Barker AT (1999) The history and basic principles of magnetic nerve stimulation. Electroencephalogr Clin Neurophysiol Suppl 51:3–21

40. Jalinous R (1991) Technical and practical aspects of magnetic nerve stimulation. J Clin Neurophysiol 8:10–25

41. Ruohonen J, Ravazzani P, Tognola G, Grandori F (1997) Modeling peripheral nerve stimulation using magnetic fields. J Peripher Nerv Syst 2:17–29

42. Ilmoniemi RJ et al (1997) Neuronal responses to magnetic stimulation reveal cortical reactivity and connectivity. Neuroreport 8:3537–3540

43. Berne RM, Levy MN (1993) Physiology, Mosby year book. Mosby, St. Louis

44. George MS et al (2003) Transcranial magnetic stimulation. Neurosurg Clin N Am 14:283–301

45. Paus T (2005) Inferring causality in brain images: a perturbation approach. Philos Trans R Soc Lond B Biol Sci 360:1109–1114

46. Pascual-Leone A, Walsh V, Rothwell J (2000) Transcranial magnetic stimulation in cognitive neuroscience–virtual lesion, chronometry, and functional connectivity. Curr Opin Neurobiol 10:232–237

47. Rothwell JC (1999) Paired-pulse investigations of short-latency intracortical facilitation using TMS in humans. Electroencephalogr Clin Neurophysiol Suppl 51:113–119

48. Ilmoniemi RJ, Ruohonen J, Karhu J (1999) Transcranial magnetic stimulation–a new tool for functional imaging of the brain. Crit Rev Biomed Eng 27:241–284

49. Bastings EP et al (1998) Co-registration of cortical magnetic stimulation and functional magnetic resonance imaging. Neuroreport 9:1941–1946

50. Bohning DE et al (1998) Echoplanar BOLD fMRI of brain activation induced by concurrent transcranial magnetic stimulation. Invest Radiol 33:336–340

51. Bohning DE et al (1999) A combined TMS/fMRI study of intensity-dependent TMS over motor cortex. Biol Psychiatry 45:385–394

52. Bohning DE et al (2000) BOLD-fMRI response to single-pulse transcranial magnetic stimulation (TMS). J Magn Reson Imaging 11:569–574

53. Nitsche MA, Paulus W (2000) Excitability changes induced in the human motor cortex by weak transcranial direct current stimulation. J Physiol 527(Pt 3):633–639

54. Fregni F, Boggio PS, Nitsche M, Bermpohl F, Antal A, Feredoes E, Marcolin MA, Rigonatti SP, Silva MT, Paulus W, Pascual-Leone A (2005) Anodal transcranial direct current stimulation of prefrontal cortex enhances working memory. Exp Brain Res 166(1):23–30

55. Dieckhöfer A, Waberski TD, Nitsche M, Paulus W, Buchner H, Gobbelé R (2006) Transcranial direct current stimulation applied over the somatosensory cortex – differential effect on low and high frequency SEPs. Clin Neurophysiol 117(10):2221–2227

56. Wagner T, Valero-Cabre A, Pascual-Leone A (2007) Noninvasive human brain stimulation. Annu Rev Biomed Eng 9:527–565

57. Radman T, Ramos RL, Brumberg JC, Bikson M (2009) Role of cortical cell type and morphology in subthreshold and suprathreshold uniform electric field stimulation in vitro. Brain Stimul 2:215–228

58. Antal A et al (2011) Transcranial direct current stimulation over the primary motor cortex during fMRI. Neuroimage 55(2):590–596

59. Weber MJ et al (2014) Prefrontal transcranial direct current stimulation alters activation and connectivity in cortical and subcortical reward systems: A tDCS-fMRI study. Hum Brain Mapp 35(8):3673–3686

Chapter 3

Selection of Optimal Pulse Sequences for fMRI

Mark J. Lowe and Erik B. Beall

Abstract

In this chapter, we discuss technical considerations regarding pulse sequence selection and sequence parameter selection that can affect fMRI studies. The major focus is on optimizing MRI data acquisitions for blood oxygen level-dependent signal detection. Specific recommendations are made for generic 1.5, 3, and 7 T MRI scanners.

Key words MRI, fMRI, Pulse sequences, Blood oxygen level-dependent (BOLD), Echoplanar imaging, Spiral imaging, Multiband imaging, Motion

1   Introduction

NMR or MRI signals are generated by exposing nuclei placed in a static magnetic field to radiofrequency (RF) pulses in the presence of rapidly switching magnetic field gradients. These patterns of RF pulses and magnetic field gradients are referred to as *pulse sequences*. Pulse sequences dictate the contrast that will be present in MR images.

The issue of optimal pulse sequences, or more generally, optimal data acquisition strategies, for functional neuroimaging is a complex one. One cannot categorically say that a particular approach is superior to any other in all cases. In this chapter, we will examine the issues that affect the detection sensitivity of neuronal activation in MRI and discuss relevant data acquisition strategies that can be optimal in each situation. For those readers not interested in the technical details involved in fMRI pulse sequence optimization and wish to simply read a summary of recommended pulse sequence strategies for blood oxygen level-dependent (BOLD) fMRI, it is recommended that they skip to Sect. 5, which summarizes the issues and presents recommendations and caveats for each relevant sequence parameter.

Massimo Filippi (ed.), *fMRI Techniques and Protocols*, Neuromethods, vol. 119,
DOI 10.1007/978-1-4939-5611-1_3, © Springer Science+Business Media New York 2016

This chapter is organized in the following way:

1. Physics of functional contrast in MRI

   (a) Nuclear Magnetic Resonance Relaxometry

   (b) Exogenous Contrast

   (c) Endogenous Contrast

2. Ultrafast Spatial encoding

   (a) Echoplanar imaging

   (b) Spiral imaging

   (c) Parallel imaging

   (d) Partial Fourier imaging

   (e) Multiband echoplanar imaging

3. Artifacts

   (a) Nonphysiologic

   (b) Physiologic—head motion and cardiorespiratory noise

4. Optimization of sequence parameters

   (a) Relaxation parameters and functional contrast

   Field strength, intravascular, extravascular signal

   (b) Experimental design

   - Block design fMRI

   - Event-related fMRI

5. Summary

   Sequence recommendations

Many of these issues are covered in detail in other chapters. We introduce them here in the context of pulse sequence selection and optimization.

2 Physics of Functional Contrast in MRI

With a few notable exceptions, such as diffusion-weighted MRI, MRI contrast stems from taking advantage of the different NMR relaxation rates in different tissues and in the presence of pathology. Felix Bloch phenomenologically characterized the dynamic evolution of spin magnetization with two time constants, referred to as T_1 and T_2. An in-depth discussion of the Bloch equations is beyond the scope of this chapter, but for the purposes of understanding the interaction of pulse sequence parameters and functional contrast in MRI, it is useful to briefly describe the processes associated with these relaxation time constants.

T_1 relaxation is taken to be the time constant of the return of an excited ensemble of nuclei to the equilibrium state of the "lattice" or surroundings. So, before excitation, the ensemble will generally be in equilibrium with its surroundings. After excitation, T_1 governs the time for it to return to the state of equilibrium with the lattice. This is sometimes referred to as spin-lattice relaxation.

T_2 relaxation, which is technically an enhancement of T_1 relaxation (i.e., the upper limit of T_2 is T_1), is the time constant for an excited ensemble of nuclei to lose phase coherence through interactions with each other. This is sometimes referred to as spin-spin relaxation. T_2 relaxation in solids and tissue is typically much faster than T_1.

For functional imaging, another important parameter governing relaxation is $T_2{}^*$. $T_2{}^*$ is an enhancement of T_2 caused by magnetic field gradients inhomogeneities. $T_2{}^*$ is defined as:

$$\frac{1}{T_2^*} = \frac{1}{T_2} + \frac{1}{T_2'} \qquad (1)$$

where T_2' is the additional relaxation contribution from field inhomogeneities.

These relaxation processes are sensitive to the chemical environment of the nuclei. MRI utilizes this fact to produce images whose contrast is based on the different relaxation rates in different tissues.

2.1 Nuclear Magnetic Resonance Relaxometry

Exposing nuclei in a static magnetic field to RF radiation at the Larmor frequency, given by:

$$\nu_L = \gamma B \qquad (2)$$

will result in the absorption of energy by the nuclei. In Eq. (2), γ is the gyromagnetic ratio and is a property of the nucleus. Since B is the static field strength, we see from Eq. (2) that the Larmor frequency will rise linearly with field strength. For protons, $\gamma = 42.58$ MHz/T, so the Larmor frequency at 1.5 T is approximately 64 MHz.

When the RF radiation is stopped, the nuclei will gradually release the energy into the surrounding material until they return to the pre-excited state of equilibrium with their surroundings.

An MR pulse sequence is characterized mainly by two parameters that control the contrast of the acquired data. The first is called the repetition time, or TR, which dictates how frequently the nuclei in a particular location are excited. If they are excited much more rapidly than the T_1 relaxation rate of the tissue, the protons will not recover to equilibrium between excitations. After a few excitations, the nuclei in a given location will approach a steady-state. Figure 1 shows an example of the signal evolution in tissue with different T_1's as a function of TR.

The other important parameter that is used to control contrast is the echo time, or TE. This is the time after excitation that the observed

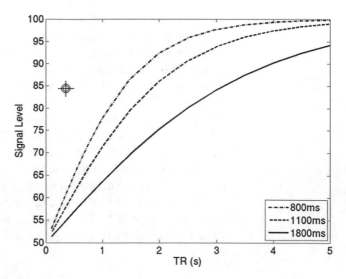

Fig 1 Steady-state MR signal as a function of repetition time for tissue with three different T_1 relaxation times

signal is spatially encoded. The amount of signal that can be spatially encoded is dictated by both TR and TE. Two of the most common methods for refocusing MR signal to allow spatial encoding are the spin echo (SE) and the gradient recalled echo (GRE) methods.

2.1.1   The Spin Echo

The time evolution of MR signal immediately after excitation is referred to as free induction decay (FID). During the FID, the processes governing the loss of signal coherence are a combination of T_1, T_2, and T_2'-related processes. In tissue, T_2' will have a large effect on the loss of signal. T_2' effects are what is referred to as reversible processes. The loss of phase coherence from these effects can be reversed by applying a refocusing RF pulse. Figure 2 illustrates the sequence timing, using a pulse sequence timing diagram. Application of a refocusing RF pulse at a time t after the initial excitation pulse, will result in a complete refocusing of the reversible dephasing effects at a time 2 t. This is referred to as a Spin Echo. The time $2t$ is usually called TE. The MR signal from a given pulse sequence can be derived from the Bloch equations. For a SE acquisition, the MR signal will be given by:

$$S_{SE} \propto \exp\left(\frac{-TE}{T_2}\right)\left\{1 - 2\exp\left(-\frac{TR - \dfrac{TE}{2}}{T_1}\right) + \exp\left(\frac{-TR}{T_1}\right)\right\} \quad (3)$$

As is clear from Eq. (3), the MR signal from a SE acquisition is moderated by T_1 and T_2. From this we can see that the T_2 will affect the encoded signal if the TE is comparable to, or longer than T_2. Figure 3 shows an example of the signal evolution for different TE's and T_2's.

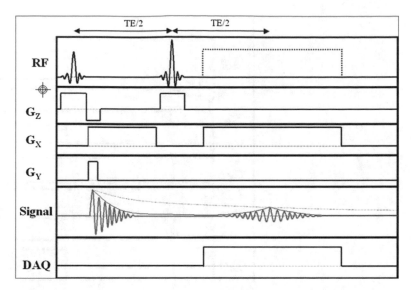

Fig 2 Pulse sequence diagram of a spin echo acquisition. The envelope of the signal indicates the FID, while the peak of the echo is modulated by T_2 according to Eq. (3)

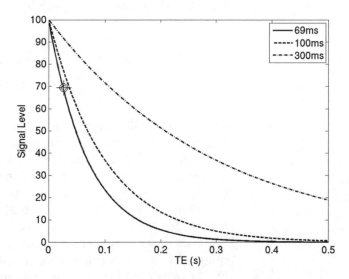

Fig 3 Steady-state MR signal as a function of echo time for tissue with three different T_2 relaxation times

2.1.2 The Gradient Recalled Echo

It is possible to perform the spatial encoding for MRI during the FID. An echo can be created by increasing the dephasing through application of a brief field gradient along a particular direction and then reversing it while acquiring the signal data. This is called a gradient recalled echo (GRE), or a field echo. The sequence

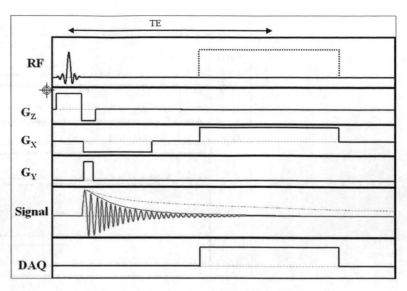

Fig 4 Pulse sequence diagram of a gradient recalled echo acquisition

diagram for this technique is shown in Fig. 4. The signal obtained from a GRE acquisition is given by:

$$S_{GRE} \propto \exp\left(\frac{-TE}{T_2^*}\right)\sin\alpha\,\frac{1-\exp\left(\dfrac{-TR}{T_1}\right)}{1-\cos\alpha\,\exp\left(\dfrac{-TR}{T_1}\right)} \tag{4}$$

where α is the flip angle, which is a sequence parameter that is a function of the amount of transmitted power[1].

It can be shown from Eqs. (3) and (4) that the TE that will maximize the difference in signal between two tissues with different T_2 or T_2^* is approximately the average of the T_2's or T_2^* of the two tissue types. Table 1 lists the T_1 and T_2 of gray matter and white matter in the human brain at 1.5 T and 3.0 T.

Historically, evidence of regionally specific functional contrast using MRI was first observed using an exogenous contrast agent [1]. However, this observation was very quickly followed by several groups, employing the phenomenon of BOLD contrast observed by Ogawa and colleagues [2], utilizing endogenous contrast to observe brain activation in several different brain regions [3–6].

2.2 Exogenous Functional Contrast

It is possible to generate dynamic MR images with functional contrast by utilizing the fact that regional blood flow increases proximal to activated neurons. This is typically done by using methods similar to those used to measure regional blood perfusion with

---

[1] The flip angle can also affect the contrast of the generated images, but for simplicity, we focus here on the more intuitive parameters TE and TR.

Table 1
Approximate relaxation times for gray and white matter at 1.5 T and 3 T

	1.5 T			3 T		
	T_1(ms)	T_2(ms)	T_2*(ms)	T_1(ms)	T_2(ms)	T_2*(ms)
White Matter	600	80	70	800	70	60
Gray Matter	900	100	60	1100	90	50

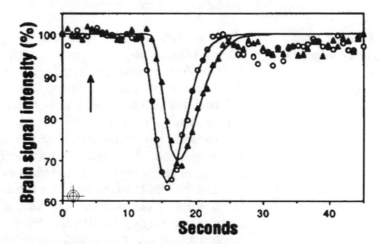

Fig 5 Changes in MR brain signal intensity during the first-pass transit of intravenously administered paramagnetic contrast agent. *Triangle* (Δ) symbols represent the time course of signal during photic stimulation and *circle* (o) symbols represent the time course during rest (darkness). Reproduced with permission from Ref. [1]

MRI. Gadolinium chelates will not cross the blood brain barrier. Thus, a bolus injection of such a paramagnetic material will cause a transient change in the T_2 relaxation near arterial blood vessels that are perfusing brain tissue. If this is done while rapidly acquiring T_2*-weighted MR images of the brain region that is active, one will observe a decrease in the measured signal intensity that is monotonically related to the volume of gadolinium passing through. One can infer directly the volume of blood perfusing this region from the signal decrease.

Functional contrast can thus be obtained in MRI by comparing the regional perfusion, measured with bolus contrast injection, while performing a task to that measured while at rest. Figure 5 is an example of the difference in the MRI signal evolution from the same brain region in visual cortex while undergoing photic stimulation and in darkness. One can see that the area under the curve for photic stimulation is larger than that for rest, indicating that the volume of blood perfusing the tissue was increased during stimulation.

In this manner, one can produce voxel level comparisons of the area under the bolus passage curve in the MR time courses and determine those whose measured volume change was statistically significant.

Due to the invasive nature of the necessary bolus injection, the (albeit low level) risk of adverse reaction to contrast agents, the lower signal-to-noise ratio (SNR) of perfusion measurement techniques, and to a lesser extent, the limited volume coverage of perfusion measuring techniques, exogenous contrast-enhanced fMRI is only rarely performed and usually for reasons specific to a particular experimental design.

2.3 Endogenous Functional Contrast

There are two principal mechanisms for generating contrast in MR images using endogenous features related to neuronal activation. Both of these mechanisms are related to the hemodynamic response to an increase in neuronal activation. One of these is a regional increase in blood flow and the other is a concomitant increase in the oxygenation content of the blood perfusing tissue near activated neurons.

Far and away the most commonly employed fMRI acquisitions utilize the fact that regional brain activation results in a local increase in blood oxygenation. This is called BOLD contrast. The contrast in BOLD stems from the fact that oxygenated hemoglobin is a weakly diamagnetic molecule, while deoxygenated hemoglobin is a strongly paramagnetic molecule. The relative increase in the concentration of oxygenated hemoglobin in the vessels perfusing activated tissue results in an increase in the T_2 and T_2^* relaxation times in the affected brain regions. Thus, methods utilizing BOLD contrast for fMRI employ acquisition techniques that are sensitive to changes in T_2 and T_2^*. Because of the flexibility of T_2 and T_2^* acquisition methods, this has become the contrast of choice for the vast majority of fMRI experiments. For this reason, the remainder of this chapter will focus on acquisition strategies to acquire BOLD-weighted MRI data and we will discuss methods to optimize these depending on experimental needs.

3 Ultrafast Spatial Encoding

It is possible to generate MR images that will demonstrate a change in signal in brain regions that transition from the inactive to active state. These transitions are typically very rapid and the advantage of MRI over other imaging techniques is the ability to acquire even whole brain images very rapidly. In this section, we introduce the concept of spatial encoding in MRI and discuss the most common ultrafast imaging pulse sequences used in fMRI. For simplicity, throughout this section, we refer to the net magnetization within a sample as "spin."

3.1   The Pulse Sequence

As stated above, the pulse sequence refers to the specific acquisition strategy in which spatial encoding and magnetization read-out is performed, providing the basic structure of the RF pulses and field gradients used. Because conventional MRI is based on Fourier spatial encoding, it has become convention within MRI to discuss pulse sequences in the context of k-space, another name for the Fourier conjugate of coordinate space. K-space is essentially the image in the spatial frequency domain, and most pulse sequences acquire image data in this domain. There are a variety of advantages to this; most importantly that a coordinate space image can be produced simply by performing a 2-dimensional Fourier transform (typically computed using the Fast Fourier Transform, or FFT) on sequentially acquired MRI data.

A pulse sequence for reading one arbitrary line of k-space with a gradient-recalled echo is shown in Fig. 6.

Starting with stage (1), waveforms are played out on the z-direction gradient, G_Z, and the RF transmit channel to excite a slice of proton spins. During stage (2), the readout gradient (G_x) prewind and phase-encode gradient selection (G_y) is performed while rephasing spins across the slice/slab with G_Z. During stage (3), the readout gradient is switched on while the emitted RF signal from the sample is recorded, denoted by the block of dotted lines. The diagrams shown are simplifications, where the timing and form of the gradients are changed according to various design considerations. The corresponding traversal of k-space for the pulse sequence in Fig. 6 is shown in Fig. 7.

Typical conventional (i.e., not single-shot) sequences repeat this process, for different lines of k_Y, or phase-encoding positions. This is shown in Fig. 6, step 2 with the G_Y gradient at multiple

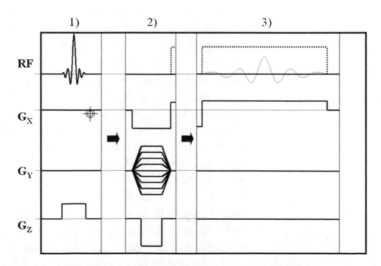

Fig 6 Pulse sequence diagram for reading one line of k-space, time increases from *left* to *right*. The proton echo from the sample is shown here in *light gray* during and under the readout window, which will be sampled by a receive coil

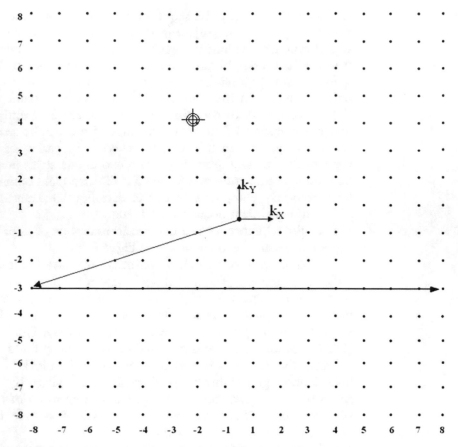

Fig 7 k-space diagram showing 16x16 matrix of data sampling points and trajectory of pulse sequence diagrammed in Fig. 6. k-space is first pre-wound in step 2 (G_X moves position in k_X from 0 to −8, G_Y moves position in k_Y from 0 down to line −3), then k_X is traversed in step 3 while sampling from −8 to +8

possible values containing the variation in the repeated lines of k-space sampling. It should be noted that the distance traveled in k-space is proportional to the time integral of the gradient strength in space, so it is possible to use a larger amplitude to shorten the time taken to traverse k-space[2].

The sequence described above pertains to a GRE, but without loss of generality, the same sequence applies to a SE sequence. A slice selective RF excitation pulse for both GRE and SE is typically a sinc function-shaped pulse, with amplitude set to rotate the slice magnetization 90° from the longitudinal magnetization direction into the plane transverse to the static field. A SE sequence is very similar, but with the addition of a refocusing pulse set at 180°, timed to play out midway between the centers of the RF excitation pulse in step 1 and the readout window in step 3 in Fig. 6. The differences between SE and GRE can be seen in Fig. 8. The timing of the inversion pulse after the excitation pulse dictates the TE, so a short TE can preclude the SE method. SE can be advantageous

[2] up to the limits of the gradient hardware and not without various drawbacks.

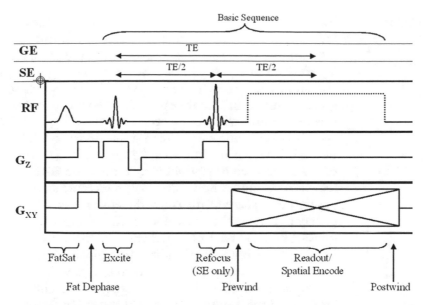

Fig 8 Generalized diagram for either gradient-echo (GE) or spin-echo (SE) ultrafast pulse sequences specific to ultrafast imaging. A fat saturation pulse (described later), followed by excitation, then a refocus pulse (if spin-echo only), prewind gradients to set position in k-space, then the readout and spatial encoding gradients. Finally there may be postwind spoiler/crusher gradients (RF may also be used at the end) to dephase residual signal

because it can cancel dephasing due to local field inhomogeneities, leading to an improved SNR, but single shot sequences get less of this benefit due to an effective spread of TEs which will be discussed later.

Strategies for what is referred to as "single-shot" imaging are critical to the high sampling rates necessary for dynamic imaging techniques such as fMRI. Echo-planar imaging (EPI) and spiral imaging are the most widely used of these. In addition, in-plane and multiband parallel imaging techniques, combined with the ubiquitous use of multichannel coil technology, will play an increasing role in fMRI.

Single-shot sequences differ from conventional pulse sequences in that the data for an entire slice are acquired in one readout window after one excitation. This has been made possible by fast gradient switching technologies, and single-shot sequences are available on all modern MRI scanners. Common to all fast imaging sequences are higher demands on the hardware, which increase vibration and heating of the scanner, leading to increasing inhomogeneity and field drift over time during long scans [7, 8]. Parallel imaging is a recent development, which reduces the readout time by acquiring data from multiple coils. These imaging strategies have various artifacts and tradeoffs, which will be discussed below following an introduction to the most common strategies.

3.2 Echo-Planar
Imaging

EPI follows the basic strategy of excitation of a slice or slab followed by readout of one line (in the read-out direction, or k_x) in k-space. The GRE sequence was shown in the pulse sequence timing diagram in Fig. 8 without the SE refocusing pulse, which was also shown in parts in Fig. 6. With the fast gradient switching speeds available in recent years, it has become possible to spatially encode an entire slice in one echo by performing multiple readouts and phase-encoding steps after a single excitation. The most common implementation, known as "Blipped EPI", involves excitation of a slice followed by readout of k_x line like the GRE sequence. The sequence continues however, after an increment, or "blip", of the position in k-space in the other dimension using a short duration gradient pulse (in the phase-encode direction, or k_y). Readout continues when the read-out gradient is reversed to read another k_x line in k-space immediately adjacent to the first line sampled but in the opposite direction. This is shown in Fig. 9.

These reversals and blips are repeated to adequately sample k-space and the resulting data can be treated in the same manner as multi-shot imaging, with the full readout of the slice or slab centered on the TE. The trajectory in k-space is shown in Fig. 10.

Three-dimensional acquisitions can be performed using an additional increment in the perpendicular dimension, or k_z, although most blipped-EPI sequences are two dimensional only due to the constraint of a shorter TE required. There exist many modifications to this basic structure, but all EPI strategies contain

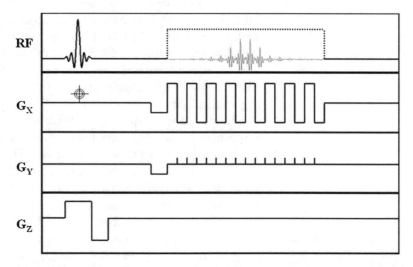

Fig 9 Blipped-EPI pulse sequence. Readout gradients are reversed following readout of each k_x line, along with a small increment of k-space in k_y direction, or a "blip" in G_y. 16 k_Y lines are read out, corresponding to the k-space diagram in Fig 7. Gradient-stimulated echo train is shown in *light gray*, which becomes stronger closer to center of k-space, and at center of each k_x-readout

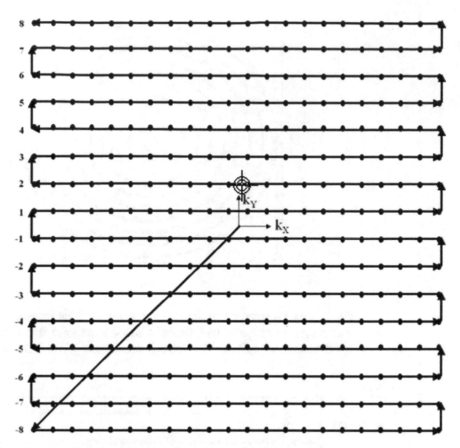

Fig 10 Cartesian trajectory in k-space for one-shot blipped-EPI sequence shown in Fig. 5. One-shot means full coverage of k-space is accomplished during echo of one excitation

a fast back-and-forth cycling of the gradients to produce a GRE train. The blipped-EPI strategy is commonly also referred to as Cartesian imaging, due to the rectangular trajectory of read-out in k-space. We will not discuss other non-Cartesian strategies that are no longer common such as constant-phase encode EPI or square-spiral EPI. The signal generation stage before the spatial encoding can include a refocusing pulse or not, depending on whether T_2 (SE-EPI) or T_2^* (GRE-EPI) weighting is desired.

3.3 Spiral Imaging

Another common strategy for single shot imaging is spiral imaging [9]. In this scheme, k-space is sampled in a spiral or circular manner, such as in Fig. 11, with less asymmetry between the rate of sampling in k_x- and k_y-space.

By applying sinusoidal gradients 90° out of phase to the read- and phase-encode gradients, k-space can be traversed in a circular manner by increasing the amplitude of the sinusoidal gradients. A typical sequence for spiral acquisitions is shown in Fig. 12.

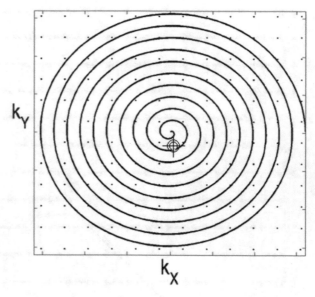

Fig 11 k-space sampling trajectory for a one-shot spiral imaging sequence shown over a rectangular grid. Central k-space is sampled first. Prior to FFT reconstruction, the data must be resampled from spiral grid to Cartesian grid

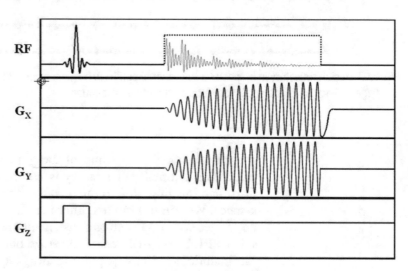

Fig 12 Pulse sequence timing diagram for spiral acquisition. Gradients during readout window are 90° phase-offset ramped sinusoids

This process continues until k-space is adequately sampled. There are many trajectory modifications to this scheme, but all have the basic property that the sampling of k-space is not Cartesian, but instead approximately radial-symmetric. Reconstruction of image space from k-space may be done using a Fourier transform after resampling the k-space data to a Cartesian grid, but there are implementations that reconstruct the image data using the discrete Fourier transform [10, 11].

Spiral imaging has advantages compared with blipped EPI, mostly related to the lower overall demand on the gradients. These include reduced gradient noise, improved SNR, lower induced eddy currents[3], and different geometric distortion artifacts [12]. Increased SNR is due to the earlier sampling of the center of k-space, but the correspondingly later acquisition of the outer regions of k-space mean the higher spatial frequencies have lower specificity than the readout direction of a blipped-EPI image. The typically sinusoidal gradient play-out means the gradients are switched at a lower rate of change, which reduces the induced eddy currents and gradient noise. Since most of the magnetization signal naturally lies near the center of k-space, which is sampled early, it is preferable to start the readout window at the TE. This modifies the timing from the Cartesian EPI sequence where the readout is centered on the TE, although newer spiral sequences (such as spiral-in/out) are available which also center the TE in the readout window [13]. A spiral-in/out sequence is shown in Fig. 13.

A Spiral-in/out sequence reduces the effects of the echo-shifting by centering the readout at the TE as in EPI. Readout begins prior to the TE, starting near the edges of k-space and spiraling in to the center, which is reached at the TE, before spiraling back out over new data points. After resampling the grid, every point in k-space now has two samplings symmetrically spaced about the TE, which are passed through FFT to give two images. These two images can be combined and the result is a reduced sensitivity to susceptibility signal loss and image data with an effective TE closer to that specified [14]. This is more demanding on the gradient hardware than spiral imaging and there can be drawbacks in image quality.

3.4   Parallel Imaging

Multiple receive coils have become a popular and widely available means to increase image SNR by providing multiple samples of a k-space trajectory. Because the coils cannot be located in the same place, they have varying spatial sensitivities to the tissue, which is maximal at the tissue nearest each coil element. This provides an alternative spatial encoding mechanism, where one sample of several parallel coil elements provides information about the magnetization density over several regions of tissue. Parallel imaging combines this spatial encoding with the gradient-mediated spatial encoding to skip some gradient-encoded lines in k-space and replace those gaps with information derived from the parallel coil elements [15]. The skipped lines in k-space reduce the field of view (FOV) seen by the coils by a reduction factor. The individual coils, if reconstructed with only the data acquired, would see the nearest portion of tissue inside that coil's FOV, but with aliased image overlap with other portions of tissue further away from the coil.

---

[3] Eddy currents are currents induced in gradient coils and other scanner components from the rapidly changing fields generated by the gradient coils.

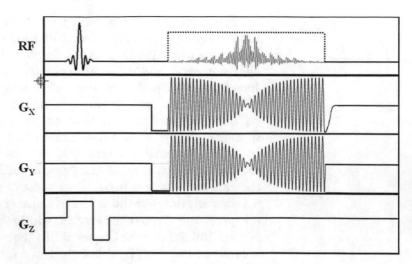

Fig 13 Spiral-in/out sequence acquires full k-space data prior to echo time and a second acquisition of k-space after echo time. The data from these echoes are combined in reconstruction

The methods used to un-alias the data can be separated into two strategies: image-space unfolding and k-space interpolation. While there are many methods, and more than a few hybrids, the most common implementations of each (sensitivity encoding, SENSE [16] and generalized autocalibrating partially parallel acquisitions, GRAPPA [17]) will be discussed, along with benefits/drawbacks.

3.4.1 SENSE

SENSE performs the reconstruction of parallel images in image space in an iterative manner using a seeded coil sensitivity matrix. Prior to the parallelized scan, the sensitivity of each coil in the full FOV is measured. These sensitivity maps are used as an initial guess for the "unfolding" matrix.

The under sampling of k-space shown in Fig. 14 leads to image aliasing when reconstructed. However, if the multiple coils are sensitive to spins from different aliased regions, then the portion of signal aliased or unaliased in each image can be differentiated using the sensitivity of the multiple coils to the different regions. The matrix inversion is performed iteratively after pre-processing the data to handle the problem of nonideal coil geometry.

3.4.2 GRAPPA

GRAPPA is a regenerative k-space method, using measured data to calculate missing phase-encoding lines. Outer regions of k-space have reduced sampling. The acceleration factor defines the number of lines skipped per line acquired. The central k-space lines, or autocalibration signal (ACS) lines, are fully sampled, which is shown in Fig. 15.

The ACS lines are used to interpolate the nonacquired lines of k-space by fitting the acquired lines to the ACS data. This is performed separately for each coil used, leading to weights specific to

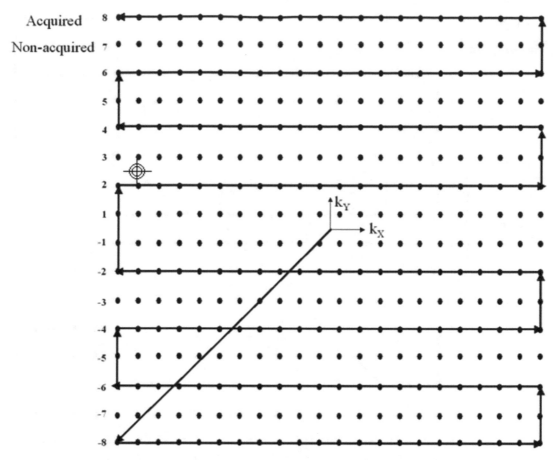

Fig 14 SENSE k-space traversal for acceleration factor 2. Odd lines of k-space are missed, even lines acquired. Acceleration factor equals acquired plus nonacquired number of lines, divided by acquired lines, in this case the full k-space matrix would have twice the number of lines as were actually acquired

each ACS line, for each coil. So for N coils, there will be N^2 weights resulting from the fitting procedure to use in interpolating the non-acquired lines. A particular coil's matrix is based on all coil signals, but masks out, or de-weights, signal from other regions outside the FOV of that coil, in k-space. The matrix weighting removes the aliasing seen in the original, undersampled images. The final, unaliased image data for each coil is combined by sum-of-square. The greatest advantage of GRAPPA over image-space methods is the determination of sensitivity from the k-space data itself, which is useful in images containing regions with poor homogeneity or low signal, both of which are the case with ultrafast imaging [18].

3.4.3 Tradeoffs

The primary benefit of parallel imaging is a reduction of the time spent spatially encoding (the readout window), but at a cost of SNR compared to the same sequence with a fully gradient-based spatially encoded image using the average signal from the parallel

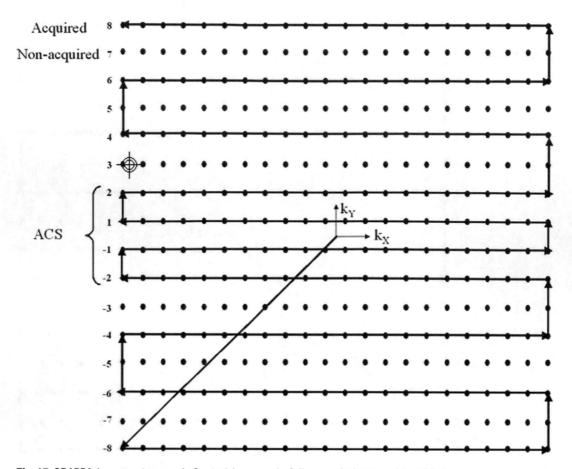

Fig 15 GRAPPA k-space traversal. Central k-space is fully sampled to provide ACS lines. Outer k-space is undersampled in phase-encoding direction by acceleration factor. Acceleration factor here is 2, so every other line is acquired. Same sampling is acquired for other coils

coils [19, 20]. The reduction in SNR is due to reduced coverage of k-space, or the square root of the acceleration factor. An additional cost for any parallel imaging is the coil geometry coverage, or g-factor. In SENSE imaging, the g-factor directly relates to the invertibility of the sensitivity matrix [21]. Since most ultrafast parallel blipped-EPI sequences are acquired in 2D only, the coil coverage should be optimized in the phase encoding direction [22]. Spiral-EPI with parallel imaging is more complicated than blipped-EPI and much more time consuming but recent advances have reduced the reconstruction time for parallel spiral-EPI [23–25].

3.4.4 Artifacts

Residual alias in the image is a common artifact seen with parallel imaging. The SENSE method requires the full image FOV to be greater than the object of interest in any accelerated directions; otherwise reconstruction will fail resulting in considerably aliased images. The only current solution with SENSE is to expand the FOV so there is no image wrapping [26]. This is not an issue with GRAPPA, because the spacing of k-space lines determines the FOV, but not the signal of a particular line. Therefore, k-space fitting under GRAPPA is not compromised by a smaller FOV, while image space fitting would be compromised by the aliasing image. With an ideal sensitivity map, it was believed SENSE could theoretically give better results than GRAPPA, however the accuracy of the maps are highly dependent on local field homogeneity and subject motion can invalidate them. There are now several sensitivity map methods, including ones based in part on GRAPPA autocalibrating methods that derive the maps from the data to get around these problems [27]. Finally, there is the issue of fitting the systems of equations in the presence of incomplete coil coverage. The SENSE method requires the solution of an inverse problem, but if there are regions of tissue that no coil has adequate sensitivity to, this is an ill-conditioned problem that cannot be exactly solved. All implementations of SENSE regularize or condition the data to work around this, but it means that the final reconstructed image may have local noise enhancements [28–30]. GRAPPA is also sensitive to this problem, but because fitting is done with k-space data, the noise enhancement is global rather than local [21].

3.4.5 Limitations

There is a practical limit to the number of coils and the acceleration factor, because adjacent coils will overlap spatially in their sensitivity and coverage of magnetized spins, reducing the ability to separate aliased signals. In typical applications of ultrafast imaging, the use of 2D EPI sequences leads to a limit on the acceleration factor of between 4 and 5 [19]. While not every scanner has multiple channels, it is becoming the standard for vendors to offer such capability. Many sites do not use parallel imaging for BOLD at 1.5 and 3 T due to higher than expected SNR loss at even the lowest acceleration factor [31], but at 7 T and higher field strengths parallel imaging is necessary in most BOLD acquisitions to reduce echo time and image warping. Future implementations of parallel imaging promise higher acceleration factors with less loss of SNR using hybrids of k-space and image-space methods with dynamically changing undersampling strategies, such as k–t SENSE [32] or k-t GRAPPA [33].

3.5 Partial Fourier Imaging

Ideal k-space data has complex conjugate symmetry, which can be exploited to reduce the acquisition time. Up to half of k-space can be interpolated from symmetry with the other half. This is referred to as partial Fourier imaging. The symmetry is only approximately true in real data due to scanner and tissue nonidealities, so algorithms to take advantage of this fact must use low-resolution approximations to account for nonzero phases in regions breaking this symmetry [34, 35]. With the use of partial Fourier acquisition, a higher spatial resolution can be acquired with less signal loss and blurring, with the result that the SNR does not drop along with the reduced acquisition time [36, 37].

3.6 Multiband Echoplanar Imaging

Multiband EPI refers to the simultaneous excitation of multiple 2D slices (multiband excitation), and the reconstruction of individual slices using parallel imaging reconstruction in the slice-encoding direction (slice-GRAPPA) [38, 39]. Multiband EPI is a recent addition to the arsenal of accelerated techniques to increase spatial and temporal resolution. Most importantly, multiband is not hindered by the SNR reduction associated with reduced sampling in traditional parallel imaging, because there is no reduced sampling of the signal with multiband. There is however a g-factor penalty from nonoptimal coil sensitivity profile. Similar to in-plane parallel imaging, the signal from N simultaneous slices can only be resolved if there are N coil measurements with varying sensitivity at each slice. This dependence is reduced considerably by using balanced blipped phase encoding in the slice direction, termed blipped-Controlled Aliasing in Parallel Imaging, or blipped-CAIPI, to improve signal separation in the slice direction [39]. Blipped-CAIPI consists of applying an alternating "balanced" blip on the slice-encoding gradient simultaneously with each phase-encoding blip on the phase-encoding gradient. By balanced blip, we mean that the sum of blips cancels out over the readout train. Using these blips, the signal from simultaneous slices can be shifted in the phase-encoding direction by a fraction of the FOV depending on where the slice is located and on the balancing of the slice blips, while sustaining minimal signal blurring from those slice-direction blips. The train of slice blips is designed to cancel out, either with alternating positive and negative polarity blips or a series of alternating fractions. This results in improved separation of the signals in k-space, leading directly to a reduced effective g-factor penalty and improved reconstruction. When using blipped-CAIPI multiband, the slice acceleration factor typically ranges from 2 to 8. Although higher acceleration factors have been reported [38], the reconstruction quality declines and artifacts increase as a nonlinear function of the acceleration factor. These methods have been used to acquire 2D EPI (BOLD, DTI, and ASL) data with a corresponding increase in the number of slices acquired per unit time. This can be used in either direction, e.g., either to obtain a lower TR or higher slice count in a given TR than previously possible.

Multiband EPI is becoming more practical and is widely used at both 3 and 7 T, but it is critically dependent upon the use of a coil with a sensitivity profile that varies across channels in the slice-encoding direction. Fortunately, most new coils being designed or sold for functional brain imaging are suitable for multiband imaging. Furthermore, multiband with blipped-CAIPI has been validated and shown to have minimal artifacts in optimized protocols [40, 41]. Nevertheless, there is one artifact most associated with multiband EPI: interslice leakage artifact. Interslice leakage refers to incomplete separation of slice signals over time, leading to spatial aliasing of BOLD effects from a given location across the other slices that had been excited at the same time as that given location. The level of artifact has been assessed [40–42] using simulation or reconstruction details not usually available to the average investigator. It may be possible to determine the level of artifact on time series image data alone, such as using seed voxel correlation [43] and computing the difference between correlation patterns to aliased slices and non-aliased slices, but there is at present no established method.

The flexibility afforded by these new techniques increases the likelihood that investigators will customize the protocol according to their specific needs. For example, some investigators are focusing efforts on whole brain BOLD acquisitions with short TR, while others focus on whole brain with thinner slices and are less concerned with TR. At present it is too early to recommend a specific protocol, but we will here describe two protocols with different goals that have been used for connectivity and fMRI studies. The first protocol was designed for a short TR and whole-brain coverage, the second protocol designed for high spatial resolution and whole-brain coverage. Both protocols use a 32 channel head coil.

3 T Short TR protocol: acceleration factor = 8, 2 mm isotropic voxel size, TR = 720 milliseconds. The short TR was intended both to directly sample physiologic noise artifacts from the cardiorespiratory cycles and to enhance statistical power using more BOLD samples in a given scan time [42]. Figure 16 shows an example functional connectivity study with this acquisition.

7 T high-resolution protocol: acceleration factor = 3, FOV/3 blipped-CAIPI FOV shifting, voxel size = $1.2 \times 1.2 \times 1.5$ mm, 81 1.5 mm thick slices, TR = 2.8 s. The high resolution protocol was intended to reduce slice thickness and increase resolution. Figure 17 shows an example functional connectivity study done using this protocol.

4 Artifacts

4.1 Non Physiologic

There are several potential artifacts from single-shot imaging techniques due to hardware realities, such as chemical shift (fat) artifact, eddy current artifacts induced by the fast gradient switching, imperfections in gradient ramping waveforms, and both blurring and signal loss due to nonuniform TEs combined with static field inhomogeneity.

4.1.1 Water-Fat Shift

Water-fat shift image artifact is a consequence of the off-resonance frequency of body fat that shifts the fat signal mostly in the phase-encode direction, misplacing it across the image. Fat suppression with an RF pulse at the resonance frequency of fat, often called chemical saturation, followed by a strong dephasing gradient is the standard countermeasure on EPI sequences. This RF pulse is done immediately before the initial excitation pulse and, due to the fact that the longitudinal signal of the fat is saturated, water protons in fat will experience no excitation. One immediate consequence of

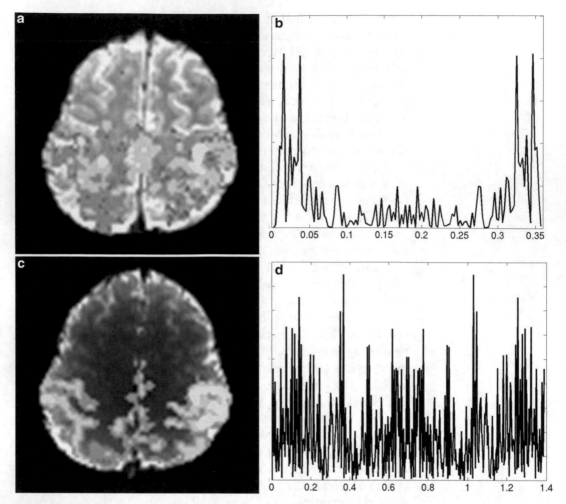

Fig 16 (a) Single-slice correlation map to seed located in left primary motor cortex, data consists of 132 volumes acquired with 2.8 s repetition time, (b) shows power spectrum of seed timeseries in Hz, (c) correlation map to same seed location in separate data consisting of 416 volumes acquired with 0.72 s repetition time and (d) shows power spectrum of seed timeseries of fast sampled data

this approach is an increase in the time taken by the sequence, as this off-resonance pulse must be performed once before every excitation pulse. Because the resonance frequency of fat protons is only 3.35 parts per million (ppm) in frequency away from water protons, the homogeneity of the static field must be very good to help ensure the suppression pulse acts only on fat protons and the excitation pulse acts only on water protons.

An alternative strategy to chemical saturation is the use of spatial-spectral RF excitation pulses [44]. These are patterned RF and gradient pulses played out over many milliseconds. Properly designed, the aggregate affect of the ensemble of pulses is to create discrete regions of excitation in space and frequency. It is possible to design these pulses such that the excited regions are separated by more than

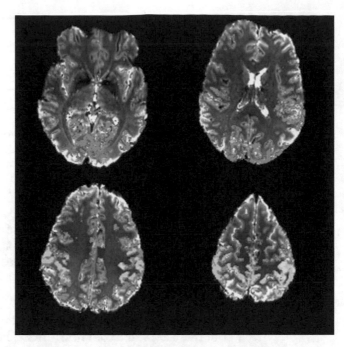

Fig 17 Correlation map to seed in left primary motor cortex in 7 T data acquired with multiband EPI acquisition with 1.2×1.2 mm $\times 1.5$ mm voxels

3.35 ppm in frequency, such that water protons in tissue in a given slice will be excited and water protons in fat will not. This strategy requires less field homogeneity than chemical saturation, but they tend to have a poor slice profile and can take up to twice as long as a good chemical saturation pulse to achieve the same result.

4.1.2 Gradient Nonidealities

Time-varying magnetic gradients induce eddy currents in nearby electrical conductors, such as the magnet cryostat. These eddy currents create magnetic fields that partially cancel the effect of the applied gradients. Fast gradient ramping is limited in hardware by the reactance of the gradient coils, creating effective upper limits on gradient switching that perturb the intended ramping waveform that the gradient coil is driven with. These waveform perturbations increase as gradient switching time decreases. The induced eddy currents and gradient ramping imperfections create phase errors in k-space magnetization read-out, which produces different artifacts depending on the k-space trajectory.

In blipped-EPI, artifact is magnified in the phase-encode direction. This creates what is known as a "phase ghost" or "N/2 ghost", an identical image at 2–5 % of the original image signal level but offset by 90° in the phase-encode direction. The ghost can have some overlap with the image of interest. In spiral EPI, the artifact is not as simple, but will result in an increase in noise level.

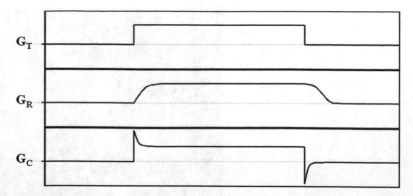

Fig 18 Gradient waveform calibration. G_T shows the theoretical, intended gradient, while G_R shows the real waveform due to eddy current damping the intended waveform. G_C shows a calibrated waveform to be played out on the gradient coils to produce the intended gradient despite the presence of eddy current

The corrective methods used vary by scanner manufacturer, but there are some commonalities. The first line of defense is in the screening of the gradient coils to reduce the field change and concomitant eddy currents. A second method employed is calibration of the gradient waveforms. To an extent, eddy currents can be predicted and compensated for by pre-emphasis of the gradient waveforms. This is shown in Fig. 18.

Generally, the effect of coil reactance is to dampen the intended gradient waveform by providing a resistance to it, so the waveform to be played out on each gradient coil is modified by a predetermined calibration. Changing the configuration of conductive objects in the scanner room can make this calibration obsolete, if they are near and large enough to be affected by the gradient fields. This could show up as a sudden increase in N/2 ghosting in blipped-EPI images, requiring a recalibration of the gradient waveform perturbation. Further anti-ghost calibrations to account for system timing offsets and residual eddy current effects may be performed, such as phase line correction. A calibration is typically taken during just prior to the readout window in the blipped-EPI sequence by sampling forwards and back across the middle of k-space. Eddy current and timing offsets result in a nonideal k-space trajectory that can be approximated as a simple shifting of each line of k-space forwards and back depending on direction of traversal. The calibration lines are used to resample every readout k_Y line to center the received echo [45, 46]. A failure of this online phase ghost correction algorithm would show up as a dramatic increase in phase ghost signal level, to a level comparable to the image of interest. An example of a failure of online phase ghost correction is shown in Fig. 19. In this case, signal changes seen as a result of phase ghost correction overwhelm the BOLD effect.

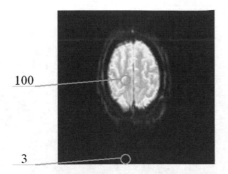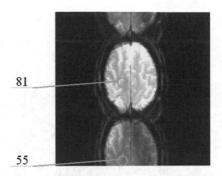

Fig 19 Phase ghost typical level on left. Phase ghost correction algorithm failure on right. All measured signal values normalized to first brain tissue measurement on *top left*

The effect of eddy current on spiral imaging is smaller since dB/dt is lower, but if uncompensated will result in warping, because it warps the k-space trajectory in both dimensions. Measuring the actual trajectory taken in k-space can be used in the resampling portion of reconstruction to correct this image warping similar to the blipped-EPI phase line correction [47].

4.1.3 Echo Shifting

The long readout time employed leaves single-shot images highly sensitive to static field homogeneity, leading to image distortion artifact in regions of inhomogeneity. The off-resonance frequency in these regions causes an accumulation of phase errors in those regions over the readout time. Phase errors specific to a region result in spatial encoding errors, which manifest as signal misplacement from that region. For blipped-EPI images, this is insignificant in the read-out, or k_x direction because it is sampled so quickly, but the phase-encode direction is sampled more slowly, resulting in spatial distortion, or blurring in the phase-encode direction in those regions. Spiral imaging samples the k_x and k_y dimensions at approximately the same rate, but the radial dimension is sampled more slowly, akin to the phase-encode direction in EPI. Spiral images are therefore blurred across both dimensions [48]. The geometric distortion can be "unwarped" from the images using the calculated pixel shifts from an acquired field map for both blipped-EPI [7, 49] and spiral imaging [35] (Fig. 20).

4.1.4 Signal Loss

Signal loss, or slice dropout, is caused by through-slice dephasing after the RF excitation. This signal loss cannot be recovered without modifying the pulse sequence. Strategies for overcoming this include: use of SE to refocus the dephasing effects, reducing the TE, reducing slice thickness and/or in-plane voxel size, and changing the scan plane. If hardware permits, the use of high order gradient shims and image-based shimming can help [50–52].

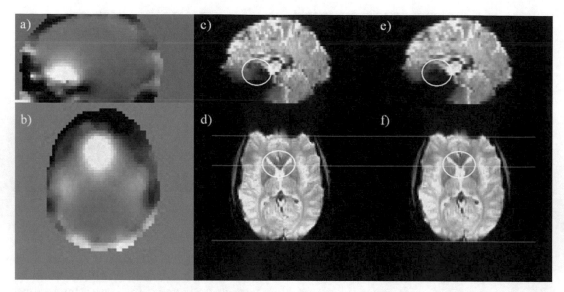

Fig 20 (a) Sagittal, (b) axial views of fieldmap. Other pictures show (c) and (d) blurred blipped-EPI signal and (e) and (f) unwarped images of the same using pixelshifts calculated from the fieldmap. Anterior regions shifted roughly 1–2 pixels, but signal loss near the frontal sinuses cannot be recovered, leaving a signal void

Another common method is z-shimming to unwrap some of the dephasing, but this has the disadvantage of reducing the SNR in unaffected regions. Z-shim methods rely on acquiring a fieldmap to estimate the gradient across a slice, and then applying an opposing (negative) slice gradient that is equal to the fieldmap measured gradient in the high susceptibility region [53, 54]. This reduces the dephasing in the high susceptibility region, but at the cost of increasing dephasing in other less-affected regions. To deal with this problem, typically two images are taken; one with the z-shim and one without and then these are added together to restore all signal loss, but at the cost of nearly doubled imaging time. This method is approximately equivalent to an older technique of tailoring RF pulses to apply a set dephase across the slice at time of excitation [55]. Alternatively, several images with a range of z-shim gradients linearly spaced between zero and the maximum measured gradient are taken, and these are averaged, however, this is more time consuming with little benefit over the more common method with two images.

4.1.5 Spiral Regridding   Apart from lower spatial specificity and increased acquisition time, problems typically associated with spiral imaging lie in the regridding of the spiral trajectory to a Cartesian coordinate system prior to Fourier transform. Regridding introduces subtle artifacts and reduces SNR, and is computationally intensive compared with the reconstruction methods used in blipped EPI, although computational improvements have been made [56, 57].

Until recently, the reconstruction had to be performed offline on a separate image-processing computer after the scan, which made it difficult to validate data online or use prospective motion correction. There are now online versions of spiral reconstruction, such that prospective motion implementations for spiral imaging have been used to monitor subject motion [58].

4.2   Physiologic Artifacts

There is another class of image artifacts present in functional neuro-imaging that have physiologic origins. Head motion [59–61] and physiologic noise from the heart and breathing cycles [62–64] are unavoidable non-neuronal sources of variance, changing underlying statistical distributions and introducing possible systematic effects in population studies. Accounting for these artifacts requires care in the acquisition of data and several stages of postprocessing of data (retrospective correction techniques will not be discussed here) after collection is complete. Preventive measures include head restraints or navigator echoes to reduce the effects of head motion, and routine scanner quality assurance measures to track the stability of the scanning hardware [65]. In addition, the collection of parallel measures of state during the image acquisition may be useful for artifact removal during postprocessing. These parallel measurements can include online motion detection parameters from navigator echo or prospective motion correction along with signals representing physiologic cardiac and respiratory cycles. Parallel measurements in general must be acquired as fast or faster than the slice acquisition sampling rate. It should be noted that there are post-acquisition methods for obtaining slice-sampling rate motion parameters [66] and physiologic cardiac and respiratory cycles [67].

4.2.1   Head Motion

Due to the fact that it is desired to maintain a high temporal resolution, the sampling rate in most fMRI acquisitions is fast compared to the T_1 of brain tissue. The consequence of this is that, after equilibrium is achieved after acquisition of a few volumes, the tissue is in a saturated state. This means that the magnetization is not completely recovered between excitations of a given slice. If a subject moves such that tissue from one slice moves into an adjacent slice, the tissue will, for the first excitation after the motion, be in a different state of saturation than the rest of the tissue in that slice. This leads to a signal change that is correlated with the motion, but will not be corrected by the traditional technique of retrospective realignment of the images. Prospective motion correction techniques are intended to deal with this problem in real-time.

Navigator-echoes can be used to obtain motion information during the acquisition of data [68, 69]. This technique uses the fact that the phase of MR data is sensitive to motion. Typically, low-power RF pulses are interspersed with the fMRI data acquisition and the phase information from the readout of the signal from these pulses is used to infer motion along a given direction. The drawbacks of the navigator echo approach is that, unless the power is very low, in which there is a limited ability to determine phase changes, the excitation pulses will affect the spin history of the fMRI data. Nevertheless, these approaches have been used with some success in fMRI.

An alternative approach that has been applied is to use real-time co-registration of a reference volume to the current volume to determine motion. The motion parameters are determined from the co-registration and are applied to the imaging system prior to acquiring the next volume [58, 70]. The drawback of this approach is that it is more computationally intensive that the navigator echo approach and it does not update the slice locations until after the motion occurs. Thus, registration-based prospective motion correction must be coupled with postprocessing motion correction. It should be noted that it is now possible to acquire motion parameters at the slice level using the image data itself [66], and this could be used in the future to further ameliorate head motion artifacts at the point of acquisition.

4.2.2 Physiologic Noise

Ongoing physiologic processes in living subjects present an additional potential artifact. Effects due to the cardiac and respiratory cycles have been identified as being significantly coupled to BOLD-weighted MR signal in voxels in the brain and spinal cord. The primary effect of the respiratory cycle on blipped-EPI fMRI data is an apparent shift in image position in the phase-encode direction. This is due to shifting of the resonant water frequency as the main field drifts due to chest expansion and contraction [71, 72]. The primary effect of the cardiac cycle is pulsatility artifact with each heartbeat, although the structure and timing of the artifact may vary across the brain due to the range of vessel sizes, stage in the vessel network, and location in the brain [73, 74]. The cardiac effects are more pronounced in certain regions such as the insula and brainstem, while respiratory effects are more global (Fig. 21).

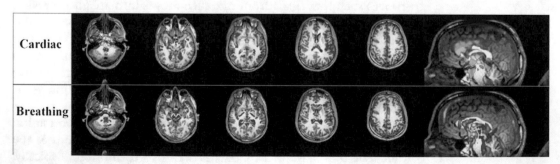

Fig 21 Averaged physiologic coupling maps in blipped-EPI with phase-encode direction in Anterior-Posterior axis, determined by temporal ICA. Cardiac coupling overlain on anatomy is shown on *top row*, respiratory coupling is shown on *bottom row*

Correcting for the effect of respiration can be accomplished using navigator echoes or off-resonance detection to follow the field shift during the scan, in the same way as described above for gross head motion. Gating acquisitions on either cardiac or respiratory rates is another method that has been used to reduce the variability in fMRI data [75]. This approach necessitates a correction for T_1 effects introduced by the variable TR.

Use of parallel measures can be used to effectively remove physiologic noise sufficient for most purposes in fMRI [76, 77]. Therefore, acquiring a pair of signals representing the cardiac cycle and respiratory cycle during the scan is desirable, although recent methods have been developed to estimate equivalent signals from the echo-planar data itself [67].

5   Optimization of Sequence Parameters

In order to generate MRI data with functional contrast, there are a number of experimental issues that need to be considered. Generally speaking, if the issue were simply rapidly generating images with BOLD contrast then the procedure would be to simply select sequence parameters such that the TR is a short as possible and the TE such that the expected changes in capillary or venous oxygenation result in a maximal signal change between rest and active neuronal state. However, the choice of optimal acquisition encompasses many other experimental issues and these should all be considered.

5.1 Relaxation Parameters and Functional Contrast

As described above, BOLD contrast in MRI is produced either through changes in T_2 or $T_2{}^*$. We are interested in optimizing MR signal differences between two states, rest and active. In this section, we discuss scanner and sequence issues that can affect the detection of these two states.

5.1.1 Field Strength and Relaxation Parameters

The effect of static field strength on T_2 relaxation in brain tissue, as evidenced from examination of Table 1, is generally that it is reduced. The effect of field strength on the BOLD signal is complex, and depends on the nature of the proton transport mechanism in the presence of the field defects introduced by the deoxygenated hemoglobin. Recent studies suggest that this mechanism is largely diffusive in nature at clinical field strengths, which would suggest a linear dependence of the BOLD signal on field strength. Experimental data bear out the linearity of the dependence of BOLD contrast on field strength [78]. Thus, BOLD contrast from extravascular protons can be taken to increase approximately linearly in the regime used by most commonly available MRI scanners (i.e., 0.3 T to 3.0 T).

The intravascular contribution to BOLD signal stems from the impact of the change in oxygenated hemoglobin concentration within the vessels and the consequent change in T_2 of the blood. The effect of this at the voxel level has been more difficult to describe than the extravascular effect due to the dependence on many factors such as blood volume, vessel size, and volume fraction.

5.2 Experimental Design

The goal in experimental design of fMRI studies is to take advantage of the fact that regional changes in blood flow and oxygenation result proximal to regions of increased neuronal activation. Historically, the basic methodology has been to acquire properly weighted MRI data of the brain regions of interest while a subject repeatedly performs tasks related to the brain function of interest. Initial methodology took advantage of the observation that continuously repeating a task during short intervals leads to an accumulation of signal from the overlapping of events in a time short compared to the hemodynamic response. This is a feature of the general linear model (GLM) of functional imaging [79].

Blocking activation in bursts of extended activity over many seconds, interleaved with long period of rest, leads to up to a much higher increase in hemodynamic response than short, isolated events. This fact makes it desirable, when possible, to use what is typically referred to as a *block design*.

In 1997, Josephs and colleagues proposed an alternative experimental design, intended to more specifically detect the MRI signal associated with neuronal events [80]. This experimental design takes advantage of the fact that, through synchronization of the time of stimuli and measurement of behavioral responses, functional imaging data can be analyzed for signal fluctuations correlated with brief, temporally separated neuronal events. This type of experimental design is referred to as *event-related* fMRI. Due to its suitability to address more complex neuroscience questions regarding brain activation and interactions, this has become a preferred experimental design among neuroscience researchers.

Since these two experimental approaches have different analysis strategies, the issues with regard to optimizing pulse sequences are different between them. In the sections below, we separately discuss these issues.

5.2.1 Block Design fMRI

As stated above, block design fMRI experiments are designed to create a large aggregate signal from activated neurons extended in time, interspersed with long periods of rest, or alternate task performance. Analysis of this type of data is typically performed with what is referred to as a reference function. The simplest method for analyzing these data is simply to calculate the cross correlation of the experimental reference function with the timeseries at each voxel [81]. Although more sophisticated methods have been developed that allow more complex analyses, accounting for nuisance effects and systematic effects of no-interest, for purposes of pulse sequence optimization, a correlation approach is sufficient to illustrate the issues.

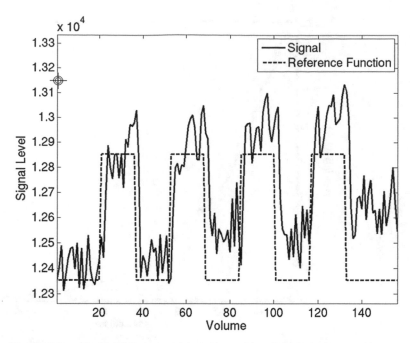

Fig 22 Example of fMRI timecourse from a voxel for a block-design experiment

Figure 22 shows an example of the signal time course from a voxel in response to a block design paradigm. The issues with regard to pulse sequence optimization are contrast-to-noise ratio (CNR), sampling rate, and number of samples. In principle, the TE and TR will control the CNR for a given pulse sequence (i.e., EPI, spiral, etc.). The sampling rate is the inverse of the TR. The detection efficiency of a pulse sequence will be determined by these and the number of samples. As an illustration of this, Fig. 23 shows the probability of getting a type II error at a false positive rate of 0.01 as a function of the number of samples.

5.2.2   Event-Related fMRI   Event-related fMRI relies on a different analysis strategy than block design fMRI. The principal difference as it relates to choice of pulse sequence is temporal resolution. A typical method for analyzing event-related fMRI is *deconvolution*. Deconvolution is an analysis method where rapidly repeated, although temporally separated, events can be extracted if the timing of the onset of the signal and either the duration or the hemodynamic response function is known. A detailed discussion of deconvolution techniques is beyond the scope of this chapter. The reader is referred to the chapter on statistical analysis in this volume for a more complete treatment.

Figure 24 shows a typical timing and signal response for a rapidly presented event-related fMRI experiment. Studies on the ability of deconvolution techniques to resolve neuronal timing shifts indicate that volume sampling rates (i.e., TR) of up to 3 s permit identification of neuronal timing shifts of order 100 ms [82].

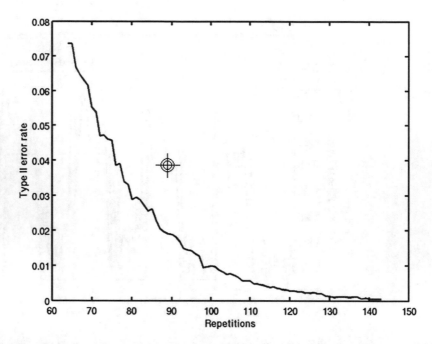

Fig 23 Probability of rejecting a true event (type II error) as a function of the number of samples at false positive rate of 0.01 for a two-cycle block design fMRI experiment. Result is from simulating image SNR = 50 with a 2 % BOLD signal change

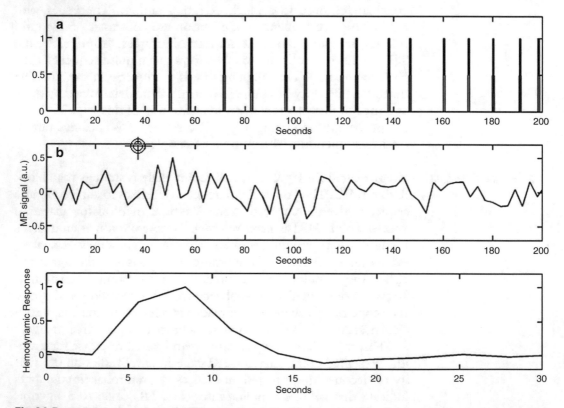

Fig 24 Example of event-related fMRI experiment. Rapid presented stimuli (**a**) results in single-voxel time-series shown in (**b**). hemodynamic response (**c**) is deconvolved from signal average over 100 voxels with temporal resolution 3 s. is shown in (**c**)

Therefore, if a goal of an experiment is to study relative timing of events, TR's of up to 3 s should be sufficient.

Another pulse sequence issue that should be of concern to researchers employing event-related experimental designs is SNR. As stated above, the CNR of event-related design is much lower than block design experiments. Therefore, it is recommended that researchers choose their acquisition strategy with this in mind. For instance, if a local cortical region is of interest, a surface coil array could be adopted to significantly increase SNR. Pulse sequence choices should be made carefully to avoid loss of SNR (shortest TE permissible for BOLD contrast, for example). In the next section of this chapter, issues with regard to SNR for the common pulse sequence parameters will be discussed in detail.

6   Summary Recommendations for Optimal BOLD fMRI

In this section, we will summarize the issues with regard to pulse sequence optimization in the performance of fMRI experiments. As stated above, currently, BOLD-weighted fMRI is the overwhelming method chosen for fMRI. For that reason, the recommendations presented in this summary will focus on BOLD-weighted acquisitions. In most cases, the technical discussions will be relevant for other types of image weighting.

In the context of the information presented previously in this chapter, we will present recommended acquisition strategies for a given field strength and we will then expand on the optimization issues with regard to each of the sequence parameters.

Table 2 lists basic pulse sequence recommendations for two field strengths and two experimental design strategies. The recommendations in Table 2 should be possible in almost any recently installed clinical MR scanner of the indicated field strength. Further recommendations made below may require special modifications to supplied clinical pulse sequences and it is strongly recommended that researchers seek the input of an MRI physicist experienced in fMRI.

6.1   Pulse Sequence

Various pulse sequences that have been proposed for fMRI acquisition were outlined in Sect. 2 of this chapter. Issues with regard to optimization of fMRI experiments are discussed here.

6.1.1   Echoplanar Imaging

This is now the most commonly available single shot imaging sequence available on MRI scanners. Data can be acquired in GRE mode and SE mode[4]. The issues with regard to optimization are

---

[4] In addition, a mixed mode EPI has been used in the literature known as ASE, or asymmetric spin echo. This is a SE EPI with the acquisition window shifted to be centered on a time point early on in the SE evolution. The result is an acquisition that has, in a sense, *adjustable* sensitivity to capillary and venous signal.

Table 2
Basic sequence parameters for BOLD fMRI acquisition on most clinical MRI scanners

	Field strength	
Sequence parameter	1.5 T	3.0 T
Sequence	GRE-EPI	GE-EPI
Scan plane	Axial	Axial
FOV	256 mm × 256 mm	256 × 256
Matrix	64 × 64	128 × 128
TE	50 ms	30 ms
TR	2000 ms	2800 ms
Flip angle	77°	80°
Receiver bandwidth (total)	125 kHz	250 kHz
Slice thickness	7 mm	4 mm
Slices (for whole-brain coverage)	18	32

that SE EPI employs, by necessity, a longer TE that will result in a smaller intravascular contribution, particularly at 3.0 T and higher, and refocuses dephasing effects from the static dephasing component of the extravascular signal that is specific to larger, distal vessels. The result is that SE EPI can be more spatially specific to the localization of neuronal activation. However, the SNR is much lower than GRE EPI. SE EPI is not recommended for field strengths below 3.0 T because the T_2 of blood is not short enough to be of benefit with regard to the intravascular BOLD signal and the SNR is too low to employ high enough spatial resolution for the spatial specificity to have a significant impact.

6.1.2 Spiral Imaging

Spiral imaging is becoming more common and is available as a product sequence on some clinical MRI scanners. As outlined above, the major advantage of spiral imaging with respect to EPI is that it is less demanding on the imaging gradients. Thus, there will be reduced image warping from eddy current effects. In addition, the nonuniform sampling of the Fourier domain image will lead to a different, possibly lower, sensitivity to motion effects and even some types of physiologic noise. Variants of the spiral technique have been proposed that are specifically designed to be more sensitive to the characteristics of the BOLD signal. This sequence is recommended for researchers employing systems with underpowered gradient systems and for situations where motion and/or physiologic noise or other types of image artifact, as discussed above, are a concern and alternate methods of addressing these are not available.

6.1.3 Parallel Imaging

MRI scanner manufacturers are increasingly moving to the use of head array RF coils *in lieu* of circularly polarized quadrature head RF coils for MRI. The cost, particularly at 1.5 T, is uniformity of SNR, and thus fMRI signal detection efficiency across the brain. This is less of an issue at high field strength, since dielectric effects in this frequency regime reduce the uniformity of the quadrature coil anyway. The advantage of head arrays is that parallel imaging methods can be used to accelerate spatial encoding of images. The result is dramatically increased image quality in regions where single shot imaging methods have historically been very poor in quality (e.g., orbitofrontal regions, mesial temporal lobe, brainstem). Some of these brain regions have important and interesting functions. Parallel imaging techniques with the head arrays available to most researchers will typically result in lower SNR throughout most of the brain, but these can still be effectively employed in situations where image artifact severely limits experimental options.

6.2 Scan Plane

Issues with regard to scan plane are largely esthetic. However, there are some technical issues that are worth discussing here.

Perhaps the most important issue is brain tissue coverage. The scan plane of choice can affect the coverage of brain tissue. Given a TE, receiver bandwidth, and TR, the number of slices available to be acquired in one TR is fixed on a given scanner.[5] The most efficient scan plane for acquiring most human brains is the sagittal plane. The brain in most adults is shortest in the right/left dimension, and so fewer slices will be required to cover the entire brain.

An axial acquisition plane is recommended in Table 2 due to the fact that it is a more intuitive scan plane to work in, both anatomically and from a physics perspective. Eddy current and higher order artifacts stemming from gradient and shim coil interactions, that are essentially related to coil geometry, are more easily understood in the axial plane. With that said, it is a simple extension to understand these effects in other scan planes. Axial imaging has the added advantage over sagittal and coronal imaging planes in that lateral, frontal, prefrontal, and posterior regions of the brain can be imaged entirely within a relatively few slices (i.e., the very top and very bottom of the brain are considered by many to be more "expendable" than these other regions). The axial plane is a very common imaging choice in fMRI and thus is listed in Table 2.

6.3 Field-of-View

FOV has an impact on fMRI signal optimization in three ways: (1) together with image matrix, it determines the in-plane voxel size and there are a number of issues with regard to this that will be

---

[5] There are, of course, other parameters that can affect this, such as gradient slew rate, partial fourier and/or field-of-view acquisition, etc.

outlined below, (2) image artifact reduction, particularly in the phase direction, with volume RF coils, and (3) brain coverage.

6.3.1 In-Plane Voxel Size

In-plane voxel size affects fMRI acquisition SNR (and thus fMRI signal detection efficiency) and image quality. At lower field strengths, the SNR issue will dominate and thus it is recommended to use a larger voxel size at the expense of spatial resolution in order to enhance signal detection efficiency. Further signal enhancement can be attained with minimal loss of spatial resolution at 1.5 T through special spatial filtering techniques [83, 84].

At field strengths of 3.0 T and higher, voxel size has an interaction with physiologic noise from cardiac and respiratory sources that can be detrimental to fMRI signal detection [64]. It is recommended that smaller voxels be employed at 3 T and higher to limit the impact on BOLD CNR from physiologic noise. If spatial resolution is not a concern, it is still recommended that data be acquired at a higher spatial resolution (i.e., smaller voxel size) and retrospective spatial filtering be employed to further increase the CNR [84].

6.3.2 Image Artifact Reduction

Due to the fact that the spatial encoding in the phase direction (i.e., the direction encoded using, for instance, the phase blipping described in Sect. 2.2 above) is not bandwidth limited, tissue outside of the FOV in the phase direction that experiences RF excitation, will be appear wrapped into the FOV with a signal intensity related to the leakage RF experienced by that tissue. For that reason, it is important for most acquisitions that the field of view in the phase direction is adequate to contain the entire brain volume and is oriented such that other tissue is not proximal to the FOV. An example would be a coronal plane acquisition with the phase encode direction in the inferior/superior direction. In this acquisition, even if the entire brain volume is within the FOV, tissue from the neck and trunk of the body that is within the sensitive volume of the transmit and receive RF coils will appear aliased into the top of the FOV. A more common problem is that the field of view is chosen too small and one side of the brain is wrapped into the other side of the brain (see Fig. 25). This is avoided most simply by adopting a large enough FOV in the phase direction. The recommendation in Table 2 is sufficient for most adults.

6.4 Image Matrix

As stated above, the principal impact of image matrix is on voxel size and these issues are outlined above. However, it will also affect the duration of the data acquistion for a single slice in single shot imaging. This duration, as discussed in Sect. 3.2 above, can have a detrimental effect on image quality in ultrafast imaging. Generally, the total readout time should be much less than T_2 or T_2^*. Typical methods of decreasing the scan duration while maintaining good

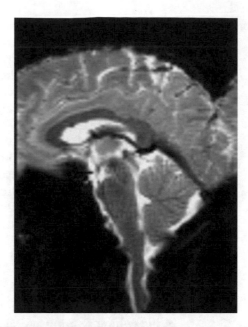

Fig 25 Sagittal EPI image with phase direction too small for the brain dimension in the anterior–posterior direction. The anterior portion is phase-wrapped into the tissue at the posterior part of the brain

spatial resolution are partial Fourier and partial field-of-view techniques, discussed briefly in Sect. 3.5. These are very commonly employed and can result in acceptable SNR tradeoffs that allow good image quality. For the 3.0 T acquisition recommended in Table 2, it is typical to use a partial Fourier, or partial echo, acquisition strategy. These are recommended such that reasonable SNR is maintained.

6.5   Echo Time (TE)

The TE is probably the most important consideration in optimizing a pulse sequence for BOLD contrast (or any T_2 or $T_2{}^*$ contrast for that matter). As discussed in Sect. 3.1, optimal BOLD contrast is obtained by selecting a TE that is the average of the T_2 or $T_2{}^*$ of the tissue in the active and inactive states. This will necessarily depend on the tissue characteristics and the field strength. The TEs recommended in Table 2 are typical echo times for a GRE-EPI acquisition that have a reasonable balance between tissue sensitivity and specificity. Adopting a longer TE can result in less sensitivity to intravascular signal, especially at 3.0 T and higher, while a shorter TE can result in higher SNR. Ranges of TE for $T_2{}^*$ BOLD imaging at 1.5 T include 40–65 ms, while ranges used experimentally at 3.0 T can range from 25 to 40 ms. One should be careful when adopting TEs outside of these ranges as BOLD contrast can be severely attenuated.

6.6 Repetition Time (TR)

With regard to BOLD contrast, TR has the fairly simple effect of increasing or lowering SNR based on the T_1 of the tissue. For a TR short with respect to the T_1 of the tissue of interest, the MR signal will be saturated. This will be discussed in some detail in the section regarding the flip angle. Here, we will simply point out that very short TRs can result a significant reduction in SNR, and can subsequently also result in a significant contribution from flow contrast, depending on the saturated state and whether a slice gap is included in the prescription.

Increased blood flow results in an apparent shortening of the T_1 relaxation time due to the effect of infusing blood on the net saturated state of the protons in a given voxel. Generally, the effect of increased flow on the T_1 of a given voxel can be expressed as:

$$\frac{1}{T_{1\text{eff}}} = \frac{1}{T_1} + \frac{f}{\lambda} \tag{5.1}$$

where $T_{1\text{eff}}$ is the observed T_1 in the presence of flow. λ is the tissue blood volume fraction and f is the rate of flow. Thus, an *increase* in blood flow will result in an apparent shortening of the observed T_1 in the affected brain region.

With regard to fMRI, the TR normally determines the sampling rate. This, combined with experimental design (stimulus presentation, duration, number of samples, etc.) determines the detection efficiency for BOLD signals (see also Sect. 4.2.1 above).

A note with regard to TR and fMRI is that sensitivity to out-of-plane motion, discussed above in Sect. 3.2.1, is a consequence of spin saturation in 2-dimensional single shot MRI. Longer TRs will lead to lower saturation, and thus lower sensitivity to out-of-plane motion[6].

6.7 Flip Angle

Technically, the flip angle relates to the amount of RF power applied at the excitation stage of a pulse sequence. For a given tissue type (i.e., T_1) and TE, MR signal is optimized at a flip angle referred to as the Ernst angle. The formula for the Ernst angle is given by:

$$\alpha_E = \cos^{-1}\left\{\exp\left(\frac{-\text{TR}}{T_1}\right)\right\} \tag{5.2}$$

Since the flip angle controls the amount of RF power transmitted to the tissue, reducing the flip angle in acquisitions with a short TR can reduce the amount of flow contribution observed in a BOLD-weighted acquisition, possibly increasing spatial specificity of detected neuronal activation.

---

[6]More accurately, retrospective motion correction techniques will be more effective.

6.8 Receiver Bandwidth

Receiver bandwidth (RBW), in a conventional MRI sequence, has an easily interpreted impact on images: since the receiver bandwidth is essentially the speed at which the MR signal is digitized, lower bandwidth results in higher SNR, but the resulting longer data acquisition can impact image quality. With single shot imaging techniques, the effect of receiver bandwidth is not as straightforward. At low RBW, the readout window length can be long enough such that SNR is reduced. At higher RBW, the reduced artifact from shorter readout time lessens or even completely eliminates the expected reduction in SNR. It is difficult to recommend an exact RBW for these types of acquisitions, since the optimal operating point will depend on scanner hardware characteristics such as slew rate that is highly variable between scanners. The parameters indicated in Table 2 should give reasonable results in most clinical MRI scanners. Increasing RBW can help to reduce susceptibility artifacts, such as image warping, in orbitofrontal or other regions in much the same way as discussed above under parallel imaging.

6.9 Slice Thickness

Choice of slice thickness has generally the same effect as spatial resolution mentioned above. Principal effects are brain coverage, SNR, and image quality. The recommended slice thickness in Table 2 should give nearly whole-brain coverage in most adults, with acceptable SNR and image artifact given field strength limitations.

As a further note, there is no mention of a slice gap in Table 2. A slice gap is not recommended in fMRI studies where the entire brain is desired. The RF excitation for a given slice will not be perfect, so if a small gap (<10% of slice thickness) between slices were permitted, the tissue in this gap would still be sampled, although less than if the slices were simply made thicker. Historically, slice gaps were included to improve SNR in 2D acquisition by reducing RF crosstalk between adjacent slices. The longer TR recommended in Table 2, along with an interleaved style pattern of slice excitation, should be sufficient to make this a negligible effect in most clinical scanners.

6.10 Number of Slices

The desired number of slices in an fMRI acquisition will affect temporal resolution and brain coverage. The recommended number of slices in Table 2 should permit whole brain coverage in most situations. Issues with regard to TR are discussed above.

7 Concluding Remarks

As stated at the beginning, and illustrated throughout this chapter, there are many experimental design issues in fMRI that will affect the exact pulse sequence prescription adopted for a particular study. The intent of this chapter is to give an overview of these issues and recommend a starting point for an sequence prescription that will be generally feasible on most modern MRI scanners, along with a sense of the impact of each of the sequence features. There are two caveats with regard to the content of this chapter: (1) it is strongly recommended that fMRI researchers work closely with an MR physicist experienced in fMRI when there are specific issues that may affect acquisition choices and (2) ideally, pulse sequence prescription and paradigm design should both be considered together when designing an fMRI experiment. Informing the data acquisition design based on the needs with regard to the experimental hypothesis or analysis methods is critical to a successful fMRI experiment.

References

1. Belliveau JW, Kennedy DN Jr, McKinstry RC et al (1991) Functional mapping of the human visual cortex by magnetic resonance imaging. Science 254:716–719

2. Ogawa S, Lee TM, Nayak AS, Glynn P (1990) Oxygenation-sensitive contrast in magnetic resonance image of rodent brain at high magnetic fields. Magn Reson Med 14:68–78

3. Bandettini PA, Wong EC, Hinks RS, Tikofsky RS, Hyde JS (1992) Time course EPI of human brain function during task activation. Magn Reson Med 25:390–397

4. Kwong KK, Belliveau JW, Chesler DA et al (1992) Dynamic magnetic resonance imaging of human brain activity during primary sensory stimulation. Proc Natl Acad Sci U S A 89:5675–5679

5. Ogawa S, Tank DW, Menon R et al (1992) Intrinsic signal changes accompanying sensory stimulation: functional brain mapping with magnetic resonance imaging. Proc Natl Acad Sci U S A 89:5951–5955

6. Frahm J, Bruhn H, Merboldt KD, Hanicke W (1992) Dynamic MR imaging of human brain oxygenation during rest and photic stimulation. J Magn Reson Imaging 2:501–505

7. Weisskoff RM, Davis TL (1992) Correcting gross distortion on echo planar images, Society of Magnetic Resonance in Medicine 11th annual meeting, Berlin

8. Foerster BU, Tomasi D, Caparelli EC (2005) Magnetic field shift due to mechanical vibration in functional magnetic resonance imaging. Magn Reson Med 54:1261–1267

9. Ahn CB, Kim JH, Cho ZH (1986) High-speed spiral-scan echo planar NMR imaging-I. IEEE Trans Med Imaging 5:2–7

10. Bruder H, Fischer H, Reinfelder HE, Schmitt F (1992) Image reconstruction for echo planar imaging with nonequidistant k-space sampling. Magn Reson Med 23:311–323

11. Pipe JG, Duerk JL (1995) Analytical resolution and noise characteristics of linearly reconstructed magnetic resonance data with arbitrary k-space sampling. Magn Reson Med 34:170–178

12. Bornert P, Schomberg H, Aldefeld B, Groen J (1999) Improvements in spiral MR imaging. Magma 9:29–41

13. Glover GH, Law CS (2001) Spiral-in/out BOLD fMRI for increased SNR and reduced susceptibility artifacts. Magn Reson Med 46:515–522

14. Preston AR, Thomason ME, Ochsner KN, Cooper JC, Glover GH (2004) Comparison of spiral-in/out and spiral-out BOLD fMRI at 1.5 and 3 T. Neuroimage 21:291–301

15. Sodickson DK, Manning WJ (1997) Simultaneous acquisition of spatial harmonics (SMASH): fast imaging with radiofrequency coil arrays. Magn Reson Med 38:591–603

16. Pruessmann KP, Weiger M, Scheidegger MB, Boesiger P (1999) SENSE: sensitivity encoding for fast MRI. Magn Reson Med 42:952–962

17. Griswold MA, Jakob PM, Heidemann RM et al (2002) Generalized autocalibrating partially parallel acquisitions (GRAPPA). Magn Reson Med 47:1202–1210

18. Heidemann RM, Griswold MA, Kiefer B et al (2003) Resolution enhancement in lung 1H imaging using parallel imaging methods. Magn Reson Med 49:391–394

19. Ohliger MA, Grant AK, Sodickson DK (2003) Ultimate intrinsic signal-to-noise ratio for parallel MRI: electromagnetic field considerations. Magn Reson Med 50:1018–1030

20. Wiesinger F, Boesiger P, Pruessmann KP (2004) Electrodynamics and ultimate SNR in parallel MR imaging. Magn Reson Med 52:376–390

21. Blaimer M, Breuer F, Mueller M, Heidemann RM, Griswold MA, Jakob PM (2004) SMASH,

SENSE, PILS, GRAPPA: how to choose the optimal method. Top Magn Reson Imaging 15:223–236

22. Ohliger MA, Sodickson DK (2006) An introduction to coil array design for parallel MRI. NMR Biomed 19:300–315

23. Pruessmann KP, Weiger M, Bornert P, Boesiger P (2001) Advances in sensitivity encoding with arbitrary k-space trajectories. Magn Reson Med 46:638–651

24. Weiger M, Pruessmann KP, Osterbauer R, Bornert P, Boesiger P, Jezzard P (2002) Sensitivity-encoded single-shot spiral imaging for reduced susceptibility artifacts in BOLD fMRI. Magn Reson Med 48:860–866

25. Heidemann RM, Griswold MA, Seiberlich N et al (2006) Direct parallel image reconstructions for spiral trajectories using GRAPPA. Magn Reson Med 56:317–326

26. Griswold MA, Kannengiesser S, Heidemann RM, Wang J, Jakob PM (2004) Field-of-view limitations in parallel imaging. Magn Reson Med 52:1118–1126

27. Griswold MA, Breuer F, Blaimer M et al (2006) Autocalibrated coil sensitivity estimation for parallel imaging. NMR Biomed 19:316–324

28. Sodickson DK (2000) Tailored SMASH image reconstructions for robust in vivo parallel MR imaging. Magn Reson Med 44:243–251

29. Sanchez-Gonzalez J, Tsao J, Dydak U, Desco M, Boesiger P, Paul PK (2006) Minimum-norm reconstruction for sensitivity-encoded magnetic resonance spectroscopic imaging. Magn Reson Med 55:287–295

30. Lin FH, Kwong KK, Belliveau JW, Wald LL (2004) Parallel imaging reconstruction using automatic regularization. Magn Reson Med 51:559–567

31. Block KT, Frahm J (2005) Spiral imaging: a critical appraisal. J Magn Reson Imaging 21:657–668

32. Tsao J, Boesiger P, Pruessmann KP (2003) k-t BLAST and k-t SENSE: dynamic MRI with high frame rate exploiting spatiotemporal correlations. Magn Reson Med 50:1031–1042

33. Huang F, Akao J, Vijayakumar S, Duensing GR, Limkeman M (2005) k-t GRAPPA: a k-space implementation for dynamic MRI with high reduction factor. Magn Reson Med 54:1172–1184

34. Cuppen JJ, Groen JP, Konijn J (1986) Magnetic resonance fast Fourier imaging. Med Phys 13:248–253

35. Noll DC, Nishimura DG, Macovski A (1991) Homodyne detection in magnetic resonance imaging. IEEE Trans Med Imaging 10:154–163

36. Jesmanowicz A, Bandettini PA, Hyde JS (1998) Single-shot half k-space high-resolution gradient-recalled EPI for fMRI at 3 Tesla. Magn Reson Med 40:754–762

37. Hyde JS, Biswal BB, Jesmanowicz A (2001) High-resolution fMRI using multislice partial k-space GR-EPI with cubic voxels. Magn Reson Med 46:114–125

38. Moeller S, Yacoub E, Olman CA et al (2010) Multiband multislice GE-EPI at 7 tesla, with 16-fold acceleration using partial parallel imaging with application to high spatial and temporal whole-brain fMRI. Magn Reson Med 63:1144–1153

39. Setsompop K, Gagoski BA, Polimeni JR, Witzel T, Wedeen VJ, Wald LL (2012) Blipped-controlled aliasing in parallel imaging for simultaneous multislice echo planar imaging with reduced g-factor penalty. Magn Reson Med 67:1210–1224

40. Cauley SF, Polimeni JR, Bhat H, Wald LL, (2014) Setsompop K Interslice leakage artifact reduction technique for simultaneous multislice acquisitions. Magn Reson Med 72:93–102

41. Xu J, Moeller S, Auerbach EJ, et al. (2013) Evaluation of slice accelerations using multiband echo planar imaging at 3 T. Neuroimage 83:991–1001

42. Ugurbil K, Xu J, Auerbach EJ, et al. (2013) Pushing spatial and temporal resolution for functional and diffusion MRI in the Human Connectome Project. Neuroimage 80:80–104

43. Jo HJ, Saad ZS, Simmons WK, Milbury LA, Cox RW (2010) Mapping sources of correlation in resting state FMRI, with artifact detection and removal. Neuroimage 52: 571–582

44. Meyer CH, Pauly JM, Macovski A, Nishimura DG (1990) Simultaneous spatial and spectral selective excitation. Magn Reson Med 15:287–304

45. Zhou XJ, Du YP, Bernstein MA, Reynolds HG, Maier JK, Polzin JA (1998) Concomitant magnetic-field-induced artifacts in axial echo planar imaging. Magn Reson Med 39:596–605

46. Reeder SB, Atalar E, Faranesh AZ, McVeigh ER (1999) Referenceless interleaved echo-planar imaging. Magn Reson Med 41:87–94

47. Duyn JH, Yang Y, Frank JA, van der Veen JW (1998) Simple correction method for k-space trajectory deviations in MRI. J Magn Reson 132:150–153

48. Yudilevich E, Stark H (1987) Spiral sampling in magnetic resonance imaging-the effect of inhomogeneities. IEEE Trans Med Imaging 6:337–345

49. Jezzard P, Balaban RS (1995) Correction for geometric distortion in echo planar images from B0 field variations. Magn Reson Med 34:65–73

50. Blamire AM, Rothman DL, Nixon T (1996) Dynamic shim updating: a new approach towards optimized whole brain shimming. Magn Reson Med 36:159–165

51. Wilson JL, Jenkinson M, de Araujo I, Kringelbach ML, Rolls ET, Jezzard P (2002) Fast, fully automated global and local magnetic field optimization for fMRI of the human brain. Neuroimage 17:967–976

52. Ward HA, Riederer SJ, Jack CR Jr (2002) Real-time autoshimming for echo planar timecourse imaging. Magn Reson Med 48:771–780

53. Yang QX, Williams GD, Demeure RJ, Mosher TJ, Smith MB (1998) Removal of local field gradient artifacts in T_2^*-weighted images at high fields by gradient-echo slice excitation profile imaging. Magn Reson Med 39:402–409

54. Constable RT, Spencer DD (1999) Composite image formation in z-shimmed functional MR imaging. Magn Reson Med 42:110–117

55. Chen N, Wyrwicz AM (1999) Removal of intravoxel dephasing artifact in gradient-echo images using a field-map based RF refocusing technique. Magn Reson Med 42:807–812

56. Oesterle C, Markl M, Strecker R, Kraemer FM, Hennig J (1999) Spiral reconstruction by regridding to a large rectilinear matrix: a practical solution for routine systems. J Magn Reson Imaging 10:84–92

57. Moriguchi H, Duerk JL (2001) Modified block uniform resampling (BURS) algorithm using truncated singular value decomposition: fast accurate gridding with noise and artifact reduction. Magn Reson Med 46:1189–1201

58. Nehrke K, Bornert P (2005) Prospective correction of affine motion for arbitrary MR sequences on a clinical scanner. Magn Reson Med 54:1130–1138

59. Hajnal JV, Myers R, Oatridge A, Schwieso JE, Young IR, Bydder GM (1994) Artifacts due to stimulus correlated motion in functional imaging of the brain. Magn Reson Med 31:283–291

60. Friston KJ, Williams S, Howard R, Frackowiak RS, Turner R (1996) Movement-related effects in fMRI time-series. Magn Reson Med 35:346–355

61. Bullmore ET, Brammer MJ, Rabe-Hesketh S et al (1999) Methods for diagnosis and treatment of stimulus-correlated motion in generic brain activation studies using fMRI. Hum Brain Mapp 7:38–48

62. Jezzard P, LeBihan D, Cuenod D, Pannier L, Prinster A, Turner R (1992) An investigation of the contribution of physiological noise in human functional MRI studies at 1.5 Tesla and 4 Tesla, Society of Magnetic Resonance in Medicine 12th annual meeting, New York

63. Lowe MJ, Mock BJ, Sorenson JA (1998) Functional connectivity in single and multislice echoplanar imaging using resting-state fluctuations. Neuroimage 7:119–132

64. Triantafyllou C, Hoge RD, Krueger G et al (2005) Comparison of physiological noise at 1.5 T, 3 T and 7 T and optimization of fMRI acquisition parameters. Neuroimage 26:243–250

65. Friedman L, Glover GH (2006) Report on a multicenter fMRI quality assurance protocol. J Magn Reson Imaging 23:827–839

66. Beall EB, Lowe MJ (2014) SimPACE: generating simulated motion corrupted BOLD data with synthetic-navigated acquisition for the development and evaluation of SLOMOCO: a new, highly effective slicewise motion correction. Neuroimage 101:21–34

67. Beall EB, Lowe MJ (2007) Isolating physiologic noise sources with independently determined spatial measures. Neuroimage 37:1286–1300

68. Fu ZW, Wang Y, Grimm RC et al (1995) Orbital navigator echoes for motion measurements in magnetic resonance imaging. Magn Reson Med 34:746–753

69. Lee CC, Jack CR Jr, Grimm RC et al (1996) Real-time adaptive motion correction in functional MRI. Magn Reson Med 36:436–444

70. Thesen S, Heid O, Mueller E, Schad LR (2000) Prospective acquisition correction for head motion with image-based tracking for real-time fMRI. Magn Reson Med 44:457–465

71. Zhao X, Bodurka J, Jesmanowicz A, Li SJ (2000) B(0)-fluctuation-induced temporal variation in EPI image series due to the disturbance of steady-state free precession. Magn Reson Med 44:758–765

72. Raj D, Anderson AW, Gore JC (2001) Respiratory effects in human functional magnetic resonance imaging due to bulk susceptibility changes. Phys Med Biol 46:3331–3340

73. Dagli MS, Ingeholm JE, Haxby JV (1999) Localization of cardiac-induced signal change in fMRI. Neuroimage 9:407–415

74. Bhattacharyya PK, Lowe MJ (2004) Cardiac-induced physiologic noise in tissue is a direct observation of cardiac-induced fluctuations. Magn Reson Imaging 22:9–13

75. Guimaraes AR, Melcher JR, Talavage TM et al (1998) Imaging subcortical auditory activity in humans. Hum Brain Mapp 6:33–41

76. Hu X, Le TH, Parrish T, Erhard P (1995) Retrospective estimation and correction of physiological fluctuation in functional MRI. Magn Reson Med 34:201–212

77. Glover GH, Li TQ, Ress D (2000) Image-based method for retrospective correction of physiological motion effects in fMRI: RETROICOR. Magn Reson Med 44:162–167

78. Stefanovic B, Pike GB (2004) Human whole-blood relaxometry at 1.5 T: assessment of diffusion and exchange models. Magn Reson Med 52:716–723

79. Friston KJ, Holmes AP, Worsley KJ, Poline J-B, Frith CD, Frackowiak R (1995) Statistical parametric mapping in functional imaging: a general linear approach. Hum Brain Mapp 2:189–210

80. Josephs O, Turner R, Friston KJ (1997) Event-related fMRI. Hum Brain Mapp 5:243–248

81. Bandettini PA, Jesmanowicz A, Wong EC, Hyde JS (1993) Processing strategies for time-course data sets in functional MRI of the human brain. Magn Reson Med 30:161–173

82. Miezin FM, Maccotta L, Ollinger JM, Petersen SE, Buckner RL (2000) Characterizing the hemodynamic response: effects of presentation rate, sampling procedure, and the possibility of ordering brain activity based on relative timing. Neuroimage 11:735–759

83. Lowe MJ, Sorenson JA (1997) Spatially filtering functional magnetic resonance imaging data. Magn Reson Med 37:723–729

84. Triantafyllou C, Hoge RD, Wald LL (2006) Effect of spatial smoothing on physiological noise in high-resolution fMRI. Neuroimage 32:551–557

Chapter 4

High-Field fMRI

Alayar Kangarlu

Abstract

Magnetic resonance imaging (MRI) allows detection of signal from constituent of biological tissues. Hydrogen (1H) is the most widely used element from which spectra and images are detected due to its abundance and high sensitivity manifested in its gyromagnetic ratio. The high contrast for soft tissue have afforded scientists invaluable information about brain structure and function. Among many parameters determining quality of MRI images, field strength is the most decisive one as it determines signal strength in fMRI images. Considering the low inherent sensitivity of fMRI, high magnetic field are the only way that activation contrast of neurofunctional studies could be increased. This is why there has been a relentless drive towards higher field strength in human imaging raising it up to 11.7 T to date. Technology of 7-T has become more widely available in scanners with fMRI capability. Development of many technologies such as multichannel RF coils, strong and fast gradients, simultaneous slice excitation, and brain-stimulation protocols have contributed to the expansion of fMRI as the method of choice for study of whole brain function. In this chapter, challenges of high-field fMRI in human studies are discussed among which signal to noise, susceptibility artifacts, multichannel RF coil designs are highlighted.

Key words High field, fMRI, Neuroimaging, Magnetic field, High resolution

1 Introduction

The high water content of biological tissues makes acquisition of anatomically accurate images of biological tissues possible [1, 2]. Imaging of structures are hinged on the contrast based on relaxation rates that are sensitive to the composition of tissues. The same mechanism is also used to visualize changes in signal as a function of physiology [3]. These facts have made fMRI a powerful tool for the study of neuroscience as it detects the changes in signal during the brain activities in response to specific stimulation designed to activate specific regions of the brain [4–10]. The mechanism based on which brain function is detected by fMRI depends on changes in the brain of magnetization caused by external stimulations of neurons. This process is successful only if responses to external cue stimulate enough neurons in the same

Massimo Filippi (ed.), *fMRI Techniques and Protocols*, Neuromethods, vol. 119,
DOI 10.1007/978-1-4939-5611-1_4, © Springer Science+Business Media New York 2016

region. Field strength of fMRI determines its sensitivity to the response of neuronal clusters. This makes differential response to the external stimulus a complex entity that can better visualize brain function. MR signal of brain activation is valid if it stems from a change caused by neuronal activity. Simultaneous detection of direct neuronal currents in the human brain has not yet been reported due to the sheer number of neurons involved and a lack of sensitivity to the neuron currents. However, neuronal activity produces a change in magnetic susceptibility of hemodynamics that modifies the magnetic field in the brain that can be detected as change in MRI signal [3]. Although, the mechanism that connects neuronal activity to hemodynamic response is still not well understood [11], the correlation of stimulus pattern with hemodynamic signals is well established in fMRI. This correlation has been observed for sensory, motor, and cognitive paradigms. These features have turned fMRI into a reliable tool for neuroscience and neurology research and with the improvement in MRI hardware and pulse sequences it is rapidly penetrating into psychiatry, neurosurgery, and psychology too.

Brain function can be noninvasively detected if the changes caused by its activity produce electromagnetic signals strong enough for high spatial localization and temporal resolution. Unlike invasive techniques that operate at the cellular or single neuronal level, whole brain access visualization with sensitivity for functional response is required to simultaneously detect all activated regions. The distinct advantage of fMRI is in its ability to acquire functional information from regions with vastly different anatomical geometry such as cortical regions and skull based brain tissues. Presently, only MRI can noninvasively access brain function and can be repeatedly applied on the same subject in multiple studies. The capability of fMRI to access different brain regions responding to a specific stimulation while simultaneously imaging the location of the functional regions makes it indispensible for the studies of brain normal function and dysfunction. But, we must keep in mind that fMRI signal is not a direct measure of neuronal activity. Computations for neuronal currents (nc) MRI has predicted a few part per billion (2–5 ppb) disturbance in MRI signal that is below noise floor. Such estimations demonstrate challenges involved in making of ncMRI a reality. Sensitivity of MRI to paramagnetic entities, however, brings hemodynamic and its coupling to neuronal activity to rescue [3, 11]. Diamagnetic nature of biological tissues makes blood with its rich iron content an ideal medium for the detection of physiological changes. The blood oxygen level dependent (BOLD) is an effect that measures changes in MR signal from deoxygenated hemoglobin (dHb) to oxygenated hemoglobin (O_2Hb) required by neuronal activation which modifies the magnetic field around the regions of oxygenation to the extent that changes in MR signal-to-noise ratio (SNR) can be measured in a comparative measurement. In fMRI studies, this change in signal ($\Delta R / R$) is taken as accurately representing neuronal activity. Higher magnetic susceptibility (χ) of dHb compared to that of O_2Hb is enough at fields above 1 Tesla (T) to raise $\Delta R / R$ to about 1 %/T which with modern instrumentation is detectable. Dependence of BOLD strength on the static magnetic field (**B0**) is a valuable feature of fMRI, which presently is benefiting from availability of 7.0 T whole body magnets for the study on human subjects.

The BOLD effect, however, depends on a number of physiological factors. The primary ones being cerebral blood flow (CBF), cerebral blood volume (CBV), and cerebral metabolic rate of oxygen (CMRO2). BOLD-based fMRI studies consider CBF, CBV, and CMRO2 mechanisms as being induced by changes in neuronal activities. To more effectively use fMRI in neuroscience research, the neurovascular coupling that relates neuronal activities to hemodynamics (BOLD) must be understood. As high magnetic field enhances BOLD signal it will play a vital role in determining the nature of MR signature of neuronal activities. The use of BOLD in study of diseases such as multiple sclerosis will become more widespread as evidence for involvement of gray matter in this disease becomes more available. Assessing the specifics of the cortical damage with fMRI depends on the ability to establish reliable correlation with specific physical and cognitive disability that needs high sensitivity and specificity to brain physiology. High-field fMRI could help with establishing an association of cortical activity with clinical relapses.

2   MR Signal

Strong magnets exert a torque on small magnets like protons. Such torque puts the spinning proton into a precession with a specific frequency called Larmor frequency. If an electromagnetic waves with the exact frequency as proton's Larmor frequency, usually in radio frequency (RF) range, is aimed at such proton, its energy will be absorbed to excite the proton from its ground state to an excited state [12, 13]. Magnets used in MRI scanners induce a resonance frequency in about 100 MHz range (10^8 Hz). At 3 T, for example, where proton Larmor frequency is 128 MHz, an RF wave with this exact frequency will be able to transfer its energy into the protons causing them to deflect from alignment with **B0**. Following this disturbance, proton magnetic dipole moment (μ) will return to its equilibrium position emitting an RF wave that will be picked up by the RF coil [14, 15]. The RF magnetic field (**B1**) that is induced into the coil circuit is mixed with signal from other events in the body that produce similar signals. The collective effect at a protons of μ magnetic moment at a frequency of ω is detected by the RF coil. The RF coil is trusted with the task of detecting the narrow frequency bandwidth that is created by the resonance condition.

The design of the RF coils, hence, is a rather critical matter and it is amply covered in the literature. While RF coils must operate at a narrow frequency range, they should be capable of operating in a very wide power range since kW RF power is required to excite the sample while a few milliseconds after excitation the coil should be able to detect a signal 1000's times weaker than the transmitted power. This remarkable resilience is constructed into a device that provides coverage to the entire head and accurately records the response of every cell that will get unraveled into images by computer programs of reconstruction routines.

Relaxation values govern the time course of the signal decay. So while the population of the protons (μ) aligned with **B0** is a function of **B0**, the signal will only persist within a time course comparable to spin–lattice (T_1) relaxation and spin–spin (T_2) relaxation [16]. Images can be produced where the tissue intensity represents relative T_1 values, the T_1-weighted (T_1W) contrast during which the T_2 relaxation must be kept at minimum. T_2 relaxation is a process of loss of coherence among aligned protons, while T_1 is the time during which excited dipoles return to their original orientation where they are unable to contribute to the signal. Considering a typical T_1 value of 1 s for brain tissues, a whole head image with 256×256 matrix takes about 5 min to acquire with no acceleration factors applied. The relatively long T_1 values set the acquisition time, as realignment of protons occur with that time scale. Acceleration of image acquisition is possible by simultaneous acquisition of multiple k-space lines for each excitation. In addition to T_1, spin–spin relaxation or T_2 decay, is also a mechanism that slows down MR image acquisition. Due to the inherent insensitivity of MRI, the population of magnetic moment required to produce detectable signal within a voxel is the difference (ΔN) between the parallel protons (N+) and anti-parallel protons (N-) which is relatively large. The sum of μ's ($\Delta N \mu$) within a voxel, i.e., magnetization vector or **M** determines the size of the signal. High magnetic field increases ΔN and through that the MR signal. Consequently, high fields can produce detectable signal from smaller voxels producing higher resolution images. As fMRI uses a fast imaging sequence, echo planar imaging (EPI), with T_2^* contrast it has high sensitivity to magnetic susceptibility. T_2^* depends on the sum of two mechanisms causing signal decay, i.e., spin–spin and local **B0** inhomogeneities that accelerate the loss of coherent precession of **M** over time [16, 17]. The contrast-to-noise ratio (CNR) of EPI-based BOLD signal used in fMRI depends on T_2^* changes caused by the difference in magnetic susceptibility of oxy- and deoxyhemoglobin. Since T_2^* is much shorter at high fields, high-field fMRI more accurately represents local magnetic field inhomogeneities caused by BOLD. As smaller voxels can be imaged with high-field fMRI, faster and stronger gradients are being designed and used to also increase temporal resolution to produce information more directly related to neuronal activity [8, 9].

3   B$_0$ Effects

Signal strength in MRI depends on **B0**, RF coil design, and relaxation values. **B0**, however, determines excess proton population that sets the limit of MR signal detectable by any magnet. In addition, the high magnetic susceptibility endows high field with dual advantages of high SNR and high susceptibility contrast. This fact is best at display with images acquired from 7 T scanners (Fig. 1). Among the parameters affecting image quality, **B0** is the single parameter whose effect on SNR and BOLD will expand the use of fMRI in assessment of physiological signatures of neurological disorders. It should be mentioned that dependence of MR signal on relaxations, susceptibility, CNR, and hardware creates both advantages and disadvantages at high field. Susceptibility artifacts in the regions with large cavities make the choice of premium field strength for fMRI studies a nontrivial matter. Susceptibility-based contrast can be used to image brain microstructure and to detect high brain iron as it has been suspected to play a role in many neurodegenerative diseases. However, the challenges

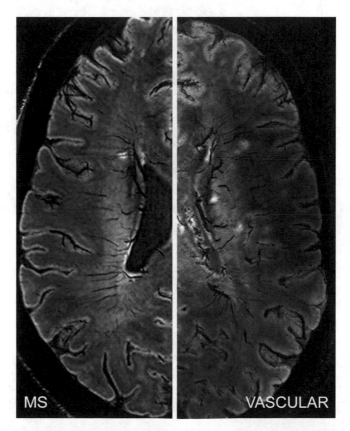

Fig. 1 A 7 T FLAIR* image of an MS patient vs. a vascular patient, in which a central vessel running through the MS lesion is clearly visible while absent through the vascular lesion. Courtesy of Prof. I.D.. Kilsdonk, VUMC, the Netherlands

involved in dealing the susceptibility artifacts of high-field fMRI has prevented its widespread use. Clever techniques such as high-order shimming, parallel transmit and receive (PTX), and short TE sequences could reduce the magnetic susceptibility artifacts at high field. These developments hold the promise to suppress the negative effects of high field and turn blights into blessings. Success in this front has far reaching implications on fMRI and its application in neurodegenerative diseases.

4   Relaxation Effects

In addition to high SNR, relaxation effects make 7.0 T an attractive field strength compared to 3.0 T and 1.5 T for fMRI studies (Fig. 2). For anatomical images, typical voxel size at 1.5 T is about 5 mm³, high SNR at 7.0 T allows image from biological tissues with 1-mm³ resolution in less than 10 min. The MR signal, however, is a function of the relaxation values, which will reduce the time that the magnetization vectors are available for sampling in transverse plane. High-field effects of susceptibility artifacts and BOLD effect make fMRI resolutions and its CNR a complex factor for researchers to optimize. While fMRI requires phase encoding an entire volume in one

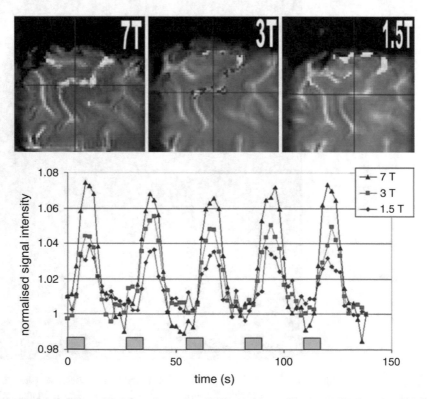

Fig. 2 fMRI study at 1.5, 3.0, and 7.0 T performed on Philips Achieva. Courtesy of University of Nottingham, UK

shot as in EPI, the artifacts of phase encoding gradients cannot be sufficiently refocused at high field. The high temporal resolution of single-shot EPI might have to be forsaken at high field because the spatial resolution and image quality suffer in single-shot EPI-at high-field fMRI. The geometric distortions caused by the phase encoding of single-shot EPI makes accurate coregistration of functional maps with anatomical images more complex than in lower fields. Furthermore, the point-spread functions (PSF) are broadened due to the long acquisition window during which higher signal decay occurs. One parameter that plays a critical role in determining many of these quantities is TE and its selection is critical in high field. But, the optimal TE is not always possible to find when large in-plane matrices are chosen to reach higher resolutions in fMRI studies. Variations of T_1 values with field strength are also important. T_1 values reported for gray matter (GM) at 7.0 T are about 2 s and for white matter values around 1.3 s [17–20]. Similar measurements at 3.0 T, have found T_1 values of about 1.5 s for GM and about 0.8 s for white matter [18–20]. T_1 values at 1.5 T reduce to about 1.2 s for GM and about 0.6 s for white matter [21]. Due to a need for fast encoding of the entire images in a time comparable to T_1 value, its absolute value does not have a large effect on fMRI images. However, T_2^* relaxation has a direct effect on the BOLD activation signal. This is especially true due to drastic change in T_2^* as a function of field strength. A multi-field measurement has reported [17] T_2^* for GM, WM, and putamen, respectively, to be 84.0, 66.2, and 55.5 ms at 1.5 T; 66.0, 53.2, and 31.5 ms at 3.0 T; and 33.2, 26.8., and 16.1 ms at 7.0 T. This shows that the difference between T_2^* of GM/WM has reduced from 18 ms at 1.5 T to 7 ms at 7.0 T. These results show a drop by a factor of about 2.5 in T_2^* of these tissues for a field increase from 1.5 to 7.0 T. Such changes have great implications on the outcomes of fast sequences at high field. For example, as T_2^* of some tissues of putamen drops to a value close to 10 ms, it will reduce the possibility of phase encoding the structure in a single-shot image for high-resolution images. This means that role of high field in producing high-resolution functional images must be further investigated.

Another challenge for high-field fMRI to exploit its advantages is the need for stronger gradients. Higher gradient amplitudes and faster switching rates can produce effects that can better be utilized at high fields. The new strong gradients of 80 mT/m that have become available on some 3 T and 7 T scanners will reduce the encoding time for a resolution of 1 mm to be 1 ms/line. This makes the total encoding time for 192 phase encoding steps to be 192 ms. The T_2^* of the human brain at 7 T is about 30 ms while T_2 of GM and WM are 93 ms and 76 ms, respectively. Thus, readouts of about 100 ms might be needed for SE EPI or partial k-space filling in GE EPI. Stronger gradients are going to be useful in reading a signal that lasts 30 ms. The fast decay induced by high magnetic susceptibility during readout causes strong blurring of images that must be suppressed before the advantages of EPI at high field could be exploited. More robust gradients are becoming available that are improving sensitivity and specificity of BOLD fMRI at high field. This will enable the high-field BOLD fMRI to more accurately localize and coregister with the high-resolution anatomical images at 7 T. BOLD sensitivity does increase at higher fields, which can only be fully utilized when their high-resolution maps are produced from the entire brain including tissues in the near proximity of the skullbase and brain regions near air/tissue boundaries.

5   Imaging the Brain Function

The best way to image brain function would be to directly detect neuronal activities. Absence of such a technique has provided an opportunity for indirect observation of brain function through BOLD, which can measure changes in hemodynamics as a result of controlled neuronal activation. The coupling of neuronal activity and vascular hemodynamics makes BOLD dependents on the details of communication between neurons, glia, and blood vessels. Furthermore, BOLD is an indication of the existence of a tight level of vascular reactivity between neuronal network and vascular system.

Accurate fMRI representation of the brain function depends on the understanding of mechanism of neurovascular coupling, i.e., how neuronal activity affects hemodynamic response and its MR signal. Such relationship will enable us to account for pharmacological or disease-induced modulations of neurovascular coupling that could use BOLD signal changes, or perfusion, for the assessment of drug efficacy and pathological modification of physiology. In addition, fMRI has become an important tool in studies in psychology, psychiatry research, and basic cognitive neuroscience research [11]. This is primarily due to the fact that fMRI has proven to be able to provide a veritable readout of mental contents.

Organization of the brain allows the functional studies to identify the neuronal basis of behavior, or at least the hemodynamic manifestation of that. The measure of neuronal activity is obtained by constructing activity maps from the functional units involved in various brain networks [4, 22, 23]. The functional units, commonly called cortical columns are made up of neuronal networks involved in the implementation of a specific function. They form an organized structure that interacts with other units of the system that repeatedly occur in the cortex. This organized structure and its columnar activation contribute to elucidate the function of specific cortical areas [24]. Imaging with a resolution allowing to detect the simultaneous activation of these units, i.e., the collective response of all columns involved in a specific external stimulation, would greatly enhance the credibility of fMRI studies. This is due to the fact that such resolution can establish the spatial localization of functional units. Resting-state fMRI, which can detect spatially dispersed but functionally connected regions that share information with each other, offers information on functional connectivity (FC) of the brain. Such FC maps are best utilized at high field, when they are capable of offering the temporal dependency of neuronal activation patterns of anatomically separate brain regions with high temporal resolution. High-field FC offers a measure of interaction between isolated clusters of columns involved in implementation of a function that will elucidate functional specialization of units and local networks at columnar level, as well as new insights in the overall organization of functional communication of brain networks. High-field fMRI is capable of whole brain imaging at both high temporal and spatial resolution, which together offer valuable information about the core aspects of the human brain, providing an overview of these novel imaging techniques and their implication to neuroscience. High-field fMRI offers the opportunity of the (1) use of spontaneous resting-state fMRI in determining functional connectivity, (2) to investigate the origin of these signals, (3) how functional connections are related to structural connections in the brain network and (4) how functional brain communication relate to cognitive performance. Analysis of functional connectivity patterns using graph theory, focusing on the overall organization of the functional brain network, is also a promising technique that takes advantage of these new functional connectivity tools in examining connectivity diseases, like multiple sclerosis, dementia, schizophrenia, and Alzheimer's disease. The potential to further empower FC fMRI with high-resolution maps based on functional units of the brain [5–8, 22] is another reason that makes high field an exciting technology for brain studies.

The primary advantage of high-field fMRI, however, remains the possibility of studying the brain physiology noninvasively with a high spatial and temporal resolution at the same time. FC fMRI reveals brain networks in resting state or based on experiments that measure brain activation due to the execution of a specific task. This is a unique capability that avails the entire brain for investigation at once. As such, it is critical that this capability is not compromised as field strength increases. Diagnoses based on functional mapping require high spatiotemporal resolution over the whole brain as field strength increases. The heterogeneity of the brain causes susceptibility-induced signal dropouts that worsen as the field strength increases. Unfortunately, this is the same mechanism that makes functional measurement of hemodynamics possible. So, we must develop reliable techniques to suppress signal dropouts while keeping susceptibility based CNR high. These conflicting needs may provide new incentives to move fMRI toward direct detection of neuronal correlates rather than the present mechanisms of BOLD or perfusion. This is where high field can change the paradigm rather than just improving resolution. In the meantime, efforts to boost both spatial and temporal resolution of functional brain studies are focusing on susceptibility suppression.

The potential of 7 T fMRI (Fig. 2) has been shown in its ability to take advantage of the magnetic susceptibility and the BOLD effect to obtain [25] high-resolution images and functional maps. Such advantage could increase linearly or even more than linearly with **B0** as more technology in hardware and software is developed for high field. For example, the high SNR of SE BOLD could eliminate contributions to the signal from large draining veins at fields of 3.0 T and higher. This way, BOLD from microvascular networks directly on or near the site of neuronal activity could be detected [26]. Such tool is capable of dealing with more fundamental question in quantification of the signal. An fMRI signal with resolution high enough to consistently quantify blood flow and energy consumption provides a valuable insight into the relationship between neuroenergetics and neuronal activity. Such relationship has not received attention in fMRI studies. Most of fMRI studies, instead, have concentrated on experimentally proving cognitive neuroscience theories. High field can provide more powerful tools and quantifiable measures for such endeavors.

Besides BOLD, arterial spin labeling (ASL) sequences have also provided reliable measurements of CBF. This feature makes perfusion-based fMRI a complement to BOLD. A version of ASL called continuous, or CASL, has shown particular potential to take advantages of high-field strengths to obtain high SNR and CNR. ASL is implemented by tagging (normally with inversion RF pulse) the blood flowing to the brain in the neck. After a delay time, the slice select RF pulse is followed by an acquisition sequence. The blood with its water magnetically labeled flows into the brain and has its transverse magnetization decaying at the rate of T_1. So, T_1 duration is important in detection of tissue perfusion. The signal in perfusion imaging is a function of regional blood flow and the longitudinal relaxation time T_1. The T_1-dependent part of perfusion signal makes perfusion SNR a function of magnetic field. At high field, perfusion will benefit from increase in T_1 as it provides more transverse magnetization in the image slice. At high field, perfusion can provide quantitative measures of absolute CBF, a more direct representation of neuronal activity than the BOLD signal.

5.1 Fast Imaging

Image acquisition in MRI is slower than in other techniques such as computed tomography and positron emission tomography. This is mostly due to the relaxation phenomenon. Fast imaging techniques are not widely popular for structural imaging due to the poor image qualities and technical limitations. Relaxations and dephasing requires refocusing of signals in the intervals of the order of TE and realignment of spins with the main magnetic field every TR seconds, where TR is called the repetition time. Refocusing can be achieved by gradient reversal or RF pulses. Depending on the acceleration rate and safety concerns, one or the other method can be used. For detection of physiological signals, however, the image acquisition rate should match the rate of physiological event. For brain functional imaging, this rate is of the order of a second. So, there is a need for imaging the entire brain within that timescale. For resolutions of the order of $5 \times 5 \times 5$ mm, the whole brain coverage requires 30–40 slices. For an image with 64 phase-encoding steps there is only 300 ms for refocusing and readout. These facts leave very few sequences for imaging at such rate. EPI is one such sequence. Its sequence details and the implications of its execution at high fields need close scrutiny in order to fully exploit its potentials in high-field functional imaging studies.

5.1.1 *Echo-Planar Imaging*

Fast imaging techniques achieve their speed by multiple refocusing of the spin ensemble during one TR. EPI as a GE-based technique is the fastest sequence and has a very low RF power content [27]. This aspect of EPI makes it suitable for high-field applications as RF absorption increases at high fields increasing the RF requirements. On the other hand, other aspects of EPI such as geometric distortion, blurring artifacts, and T_2^* signal loss are aggravated at higher fields [28, 29]. For instance, the geometric distortion that is caused by off-resonance effects will be further aggravated by long readout train of EPI. A phase offset that increases with TE will be created that will establish a linear phase gradient over k-space in the phase-encoding direction [30]. The image signal from these spins will get shifted as image is reconstructed. At high fields, this effect is proportionally stronger resulting in larger frequency shifts. However, the effect of long readouts can be drastically reduced by using parallel imaging. This will reduce geometrical distortions, but the T_2^* signal decay and blurring on the images will still remain. Other techniques have been introduced to deal with T_2^* relaxation causing distortion in images due to the decay in the signal along the k-space trajectory. Minimization of magnetic field inhomogeneities and susceptibility-induced effects requires the choice of TE close to T_2^*. As **B0** increases, T_2^* decreases and hardware and safety considerations often makes the minimum TE of single-shot EPI longer compared to T_2^*, which causes signal loss due to the phase dispersion caused by such choices of TE. Higher bandwidth could alleviate this problem but possibility of peripheral nerve stimulation will limit the use of much stronger gradients to achieve this. Other techniques have been proposed that will effectively restore T_2^* relaxation-induced signal loss and blurring. GE slice excitation profile imaging (GESEPI) is one such method that, combined with multichannel parallel receiver technology, such as sensitivity encoding (SENSE), will significantly enhance high-field EPI image qualities [31, 32].

Other EPI artifacts such as Nyquist ghost are independent of field strength and are inherent to the sequence k-space trajectory with various solutions applicable to their minimization at all field strengths [33]. Nyquist artifact is due to the time-reversal asymmetry of even and odd echoes and its ghosts overlap with the image causing a reduction in EPI SNR.

Next to its speed, the most important characteristic of EPI is the high magnetic susceptibility weighting it casts on images (Figs. 2 and 3). In fact, fMRI as the most important application of EPI takes advantage of EPI sensitivity to susceptibility change due to blood oxygenation. Unlike fMRI applications in which susceptibility enhances BOLD contrast, susceptibility weighting of EPI is not considered an advantage in applications such as diffusion-weighted imaging. As such, understanding of susceptibility is essential in enhancing its role where it helps fMRI and suppressing its undesirable aspects where it hurts data quality. A brief account of magnetic susceptibility of biological tissues is presented here to help appreciate the role of susceptibility in EPI-based BOLD signal changes.

5.2 Magnetic Susceptibility

Magnetic susceptibility, χ, is at the core of BOLD-based fMRI studies. When matter is exposed to strong magnetic field it will be magnetized [34]. In formation of χ, magnetic field ($\mathbf{H}$), magnetic induction ($\mathbf{B}$), and magnetization ($\mathbf{M}$) play roles. $\mathbf{H}$ is the entity that exists in vacuum and its penetration through space, i.e., free space of permeability $\mu_0 = 4\pi \times 10^{-7}$ H/m, is given by $\mathbf{B} = \mu_0 \mathbf{H}$. The magnetization, $\mathbf{M}$, represents the total magnetic moments per unit volume $\mathbf{M} = \sum \mu / v$. $\mathbf{M}$ is caused by $\mathbf{H}$ according to $\mathbf{M} = \chi \mathbf{H}$. $\mathbf{B}$ and $\mathbf{H}$ in SI unit system have units of Tesla and Ampere/meter, respectively. Inside a body placed in a magnetic field a magnetization $\mathbf{M}$ is generated that will produce a magnetic field of $\mathbf{B} = \mu_0 (\mathbf{M} + \mathbf{H})$. Replacing $\mathbf{M}$ in this expression will yield $\mathbf{B} = \mu \mathbf{H}$ where $\mu = \mu_0 (1 + \chi)$ will be the magnetic permeability of matter. As such, susceptibility of an object is a measure of enhancement of the magnetic field within its volume. This is important as it will determine how uniform a magnetic field ($\mathbf{B}0$) can be established inside the body in MRI. In μ, the characteristics of the free space and how its magnetic properties are modulated by matter through χ are hidden. $\mathbf{B}_0$ in turn, changes locally by χ causing the so-called susceptibility artifacts in MRI particularly in the areas of air/tissue interface [34, 35]. This effect causes a change in magnetic field as it is sensed inside a tissue and for heterogeneous tissues a contrast is generated between tissues, which are $\mathbf{B}0$ dependent. Difference in susceptibility, $\Delta\chi$, between adjacent tissues are small at low fields. If susceptibility-based inhomogeneity is smaller than inherent $\mathbf{B}0$ inhomogeneity it could be used for generating contrast for better visualization of tissues such as GM. High $\Delta\chi$ as exists at the air/tissue interfaces causes large variation in $\mathbf{B}0$ that is responsible for signal dropouts interfering with studies focused on these regions [29]. fMRI studies of regions near the ear canal, nasal cavity, and inferior frontal lobe suffer from this phenomenon.

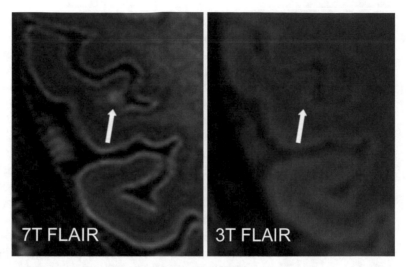

Fig. 3 An axial image showing cortical GM MS lesion acquired by 7 T FLAIR vs. 3 T FLAIR. (Courtesy of I.D. Klisdonk, VUMC, the Netherlands)

The most distinct role of susceptibility effect in MRI is in fMRI. It is based on the fact that **M** within a voxel is linearly proportional to **B**$_0$ determining the role of high field in susceptibility-based enhancements of CNR. Specifically, T_2^* values decrease allowing paramagnetic molecules such as dHb to generate more dephasing in collective proton precession at high magnetic fields. Figures 2 and 3 show how high activation induces high functional CNR on the images taken at 7.0 T compared to lower fields. The short T_2^* values due to paramagnetic properties of dHb causes the veins and structures with high-density vasculatures to have their dimensions exaggerated as shown in Fig. 1. This mechanism affects **B**$_0$ through χ making a larger variation in susceptibility, $\Delta\chi$, in brain tissues around activated neurons at high field subsequent to a perturbation. In brain activation studies, the stimulus causes change in volume and flow of oxygenated blood in the near proximity of activated brain regions. For the same activation, high field will use higher $\Delta\chi$ for better visualization of vasculature network which is coupled into the neuronal system in the brain.

Furthermore, high-field SNR allows the use of in vivo vascular imaging in establishing a relationship between brain tissue vascular density and functional imaging results. Independent information from vascular density could be attained from MR angiography to help better analysis of the fMRI data. In addition, such vascular density information could be used for the study of various topics from brain development and brain tumor staging to multiple sclerosis (MS). High-field fMRI in MS could better assess the effect of any changes in cortical activation during a particular task such as attention, memory, motor, etc. As high field enables better spatialized maps of the response to stimuli, fMRI could help assess the extent of neuropsychological problems. As fMRI becomes faster, more detailed brain activation in MS patients could be used to assess their normal motor function as is done clinically today. As the strength of response signal varies depending on the activated region of the brain and the accessibility to the detector, i.e., RF coil, high field could allow a wide range of regions and paradigms to be designed to compare performance of MS patients with healthy controls. Such quantitative assessment of functional performance of the brain will provide a valuable tool in enhancing the disease management. Structural MRI, however, has had great success in visualizing lesions of demyelination [36–38]. But, lack of specificity has prevented MRI from being established as a reliable one-shop stop for diagnosis of MS. A reliable fMRI technique with resolution to reveal accurate cerebral functional response to controlled stimulations will complement the existing structural MRI tools in better understanding of MS. Such potential is entirely due to inherent sensitivity of fMRI to hemodynamic changes induced by cognitive perturbation and will provide information independent of structural changes of the disease. In this regard, a unique aspect of high field, i.e., high susceptibility and SNR, offers a tool that, although is MR in nature, its attributes are not equally available at lower fields. The **B0**-dependent susceptibility contrast, furthermore, provides potential for depiction of microvascular structures that will further enrich the tool box of high-field magnets [39].

Magnetic susceptibility of blood is governed by the same effects as discussed above. It is high enough to generate the BOLD effect just due to the change in its oxygenation state. A dHb molecule contains four paramagnetic iron ions. During oxygenation, dHb combines with four oxygen molecules which results in an O_2Hb molecule, which normally has no net paramagnetic moment. O_2Hb is in fact slightly diamagnetic. This will cause the magnetic susceptibility of blood to change by about 10^{-6} if the blood is fully oxygenated. Taking the susceptibility of O_2Hb as zero, then change in blood magnetic susceptibility with oxygenation constitutes the basis of BOLD contrast in fMRI. A detailed account of the effect of **B0** on BOLD through the susceptibility mechanism will further elucidate the impact of high field in fMRI.

5.3   Blood Oxygen Level Dependent

A change in magnetic susceptibility of $\Delta\chi = 10^{-6}$ (SI system) in blood as a result of oxygenation is possible and forms the basis of fMRI. Through $\mu = \mu_0(1 + \chi)$, magnetic dipole strength of a voxel changes by $\Delta\chi$ and results in change in magnetization which constitute the basis of NMR signal. The maximum possible change in susceptibility due to blood oxygenation change is about one unit in SI. Assuming that $\Delta\chi = 10^{-6}$ is achieved during the activation, a corresponding 1.0×10^{-6} or 1 ppm change will result in magnetic field inhomogeneity. While at 1.5 T, 1 ppm inhomogeneity corresponds to about 63 Hz, at 7.0 T it could produce frequency shift of about 300 Hz. Such **B0** inhomogeneity will induce dephasing of spin coherence which will reduce the signal causing dark regions on T_2^*-weighted EPI images. Even spin-echo sequence will bear reminiscences of such susceptibility-induced signal loss near the veins. While for stationary tissues RF does refocus the resulted dephasing of spins, for moving water molecules in veins protons rephrasing is not complete, making BOLD effective as a T_2 as well as T_2^* effect.

fMRI signal is believed to largely originate from BOLD effect around small vessels, i.e., arterioles, capillaries, and venules [25]. The extravascular areas surrounding the small vessels represent loci of neuronal activity. But, there are contributions from large vessels to the BOLD signal as well. Such contribution must be quantified to ensure an accurate account of the role of small vessels vs. large vessels in fMRI. High magnetic fields provide a powerful tool in this regard. A known magnetic field at any position puts spins in a well-defined precession whose frequency provides knowledge of its location to produce a map of proton density. Spatial homogeneity and temporal stability of the field are important requirements for creating images faithful to the structures being studied. **B0** field homogeneity of high-field magnets is around 0.5 ppm that using high-order shimming could improve it to about 0.1 ppm over the head. Beyond this, as it was discussed, dHb produces high magnetic susceptibilities leading into comparable local inhomogeneities in the static field within the brain. At high field, regions in the brain, such as temporal lobes and basal ganglia, demonstrate high magnetic susceptibility providing a high contrast from the surrounding tissues [40]. Different scenarios for change in T_2^* are possible depending on the occupation of the voxel by capillaries, large vessels, and extravascular and intravascular BOLD [8]. In general, it can be stated that T_2^* signal differential between activation and rest period from these regions increases as a function of magnetic field. For example, if typical acquisition parameters for fMRI studies are receiver bandwidth of 2 kHz/pixel; TR 4000 ms; TE 40 ms; FOV 190×190 mm^2; 30–40 slices; slice thickness, 5 mm; then implications of these parameters at 7.0 T can be contrasted to 1.5 T through a simple frequency shift. A typical BOLD effect of 0.5 ppm or 150 Hz frequency shift at 7.0 T could result in as high as 7% change in signal. Considering that BOLD has typically produced SNR, $\Delta R/R$, of the order of 1% at 1.5 T, this fact indicates that a linear increase in $\Delta R/R$ with **B0** is possible.

BOLD contrast acts as a change in T_2^* rate, ΔR_2^*. What are the factors affecting ΔR_2^*? First, ΔR_2^* is directly influenced by the change in concentration of dHb. In fact, the volume susceptibility is directly proportional to volume of dHb and as such on ΔR_2^* [34]. Assuming that dHb is proportional to blood volume, the fraction of the blood volume *fdHb* occupied by dHb will have direct effect on the signal. Models have been proposed that assign dependence of magnetic susceptibility difference between blood O_2Hb and dHb, $\Delta\chi$, raised to a power between 1 and 2. For a venous oxygen saturation increasing from 60 to 95%, Davis et al. found a power of 1.5 fitting the simulated ΔR_2^* vs. $\Delta\chi$ curve best [41]. Such studies measure the oxygen consumption increase vs. blood flow increase as a result of functional stimulation of the brain, visual cortex in this case [41]. So, while there could be considerable differences between the increase in blood flow and oxygen consumption, there is a unanimous consent on the role of oxidative metabolism as a significant component of the metabolic response of the brain to externally induced neuronal activation.

A key role for oxidative metabolism during neuronal activation makes the role of high field more momentous in both settling such issues and enhancing the fMRI SNR. One needs to determine with certainty where the changes of the blood activation come from. They could come from the brain tissue or from the draining veins near the activated region. Many fMRI studies do not make any distinctions between these two contributions. This is partially due to the challenges involved in addressing the issue. As it was mentioned earlier, spin echo and diffusion weighting are used to differentiate contributions from different-sized vessels. Considering the small BOLD effect at low field, about 1% change in signal, an increase in fMRI signal is essential to enable suppression of BOLD signal through SE or diffusion in order to accurately investigate the source of activation. This is owing to the fact that SE EPI has more T_2 than T_2^* weighting reducing sensitivity to local susceptibility-based changes. The GE readout is responsible for the T_2^* contrast. The extent of T_2^* overlay on T_2 contrast of SE EPI is field dependent and drastic difference between T_2 and T_2^* at high field makes EPI readout in BOLD fMRI a good tool to investigate the exact location of the activated region.

Furthermore, changes in oxygenation induced by neuronal activation are complex. In the early stage of response, within the first 2–3 s, an increase in dHb is observed, which is called the "initial dip" [42]. At the end of this stage a decrease in dHb and an increase in O_2Hb are observed. High field can refine this hemodynamic behavior. The initial dip has not been so conspicuous at 1.5 T and as such not well documented. The strength of the initial dip has been reported to be more than five times stronger at 7.0 T compared to 1.5 T. Furthermore, the nature of the initial dip provides insight into the mechanism of oxygen utilization vs. cerebral blood flow. In this regard, the initial dip could be used as another tool at high field to study the correlation between hemodynamics and neuronal activities.

5.4 Physiological Noise

In the absence of physiological noise, fMRI at high field could produce functional maps of the brain with even higher resolution [43]. Scanners that already acquire submillimeter images at 7.0 T will approach microscopic resolution in the absence of the limiting noise. Unlike thermal noise, which is temperature dependent, flat in frequency, not encoded by gradients, and hence constant at room temperature, physiological noise is a function of biological activities with relatively strong MR implications. As acquisition time of individual slices is around 10 ms, physiological noise during that time is not as debilitating as it is in the time series. Intensity of this time series noise in fMRI has shown variations which correlate with the respiratory and cardiac cycles, indicating physiological modulation of BOLD by lung and cardiac function. In fact, these signals have variations independent of stimulation paradigm and thermal noise. Physiological noises and their correlation with various physiological functions are independent of field strength. Nevertheless, there are some indications that physiological noise might have some components in brain activation as well [43, 44]. Nevertheless, physiological noises have BOLD-like signal with low-frequency and TE-dependent variations [45]. It has also been shown that physiological noise could be dependent on the signal strength and its brain regional dependence. In this regard, it has been shown to have greater magnitude in cortical GM than in white matter [43]. The possibility of physiological noise dependence on the signal strength could not be related to magnetic field strength. However, conversion of brain metabolism into MR signal might produce a "resting-state" signal that will not correlate with external stimulations and consequently degrades the fMRI SNR. It has been proposed that relative strength of physiological noise could also be due to the choice of imaging resolution. This could be caused by a large voxel size which results in an increase in physiological noise which in turn degrades the activation signal [46]. These optimum voxels become smaller as field strength increases. However, if there is any vascular cause of physiological noise the inverse relation between the field strength and optimum voxel size will be limited.

6 High-Resolution fMRI

High spatial resolution (submillimeter voxels) is an expected outcome of imaging at high magnetic fields. The information content of fMRI data can best be extracted by using an accurate account of the effect of neural activity on fMRI signals. In order to make fMRI images directly depicting cortical information, it is crucial to image at the scale of functional units of cortical structure, i.e., cortical columns [47]. Details of structures of cortical columns are the most prominent features of the architecture of the cortex. The cortex is organized in layers parallel to its outer surface (horizontal layers). Layers are specialized in the cell types they contain. Both the cell types and their connections with other neurons are unique in each layer. Nevertheless, there are distinct units connecting neurons in the vertical direction (perpendicular to the outer surface). From the outer surface of the cortex inward, these neuronal units are piled up deep into the cortex and participate in producing response to the same external stimulations. The fact that these vertical structures penetrate though the entire cortical thickness gives them the attribute of cortical columns. The cortex is made up of about 20 billion neurons and contains progenitor cells and glial cells and their structure is organized in units of minicolumn each constituted of about 100 neurons. These minicolumns are tied to each other to form cortical columns [47]. fMRI at high field is capable of visualizing these columnar units.

To date, high-field fMRI of columnar organization has been concentrated mostly on the visual systems. Since neurons involved in the specific functions are incorporated in the same columns with average dimension of 0.5 mm along cortex surface, fMRI resolutions comparable to this dimension are essential for their observation. At lower fields, fMRI has shown to be able to detect site of the BOLD signal to within 5 mm at 1.5 T down to <1 mm at 7.0 T. But the point spread functions make the relationship between susceptibility-based BOLD and loci of neuronal activity a function of correlation between hemodynamics and neuronal response which is not known with certainty [48]. It has been reported that submillimeter in-plane resolution and the negative bold response (NBR) or "initial dip" can be used to locate the site of neural activation in the visual cortex (V2) of anesthetized cats at 7.0 T [49, 50]. Such findings at the columnar level will bestow fMRI a new capability in functional mapping of the brain. Also, it is clear that low-field fMRI cannot achieve similar results due to lower susceptibility and poor SNR and CNR in reduced voxel volumes. The spatial resolutions required for positive identification of sites of neuronal activities require resolutions in hundreds of microns range which are only possible at high magnetic field, i.e., >4.0 T. The neurophysiology of neuronal columns has to be reflected in BOLD response in a way to increase specificity and spatial resolution of fMRI. This places the spotlight on high magnetic fields. One major requirement of an imaging technique that is to elucidate the neurophysiology of the central nervous system using the BOLD dynamics is to reach high spatial and temporal resolution at the same time. High-field fMRI has shown to have such potential.

6.1 RF and Gradient Coil Technology

Gradients and RF coils are the two components of MRI scanners at the forefronts of signal generation and detection. As such, their less than ideal performance is the source of great many nuisances collectively referred to as artifacts [51]. To eliminate the high-field distortions of fMRI images, a variety of solutions are available [52]. Postprocessing techniques are proposed to correct for some of the distortions with known origins. Strong gradients also help reduce distortions as they increase the receiver bandwidth which in comparison, susceptibility induced changes will reduce. High-field works have also shown that multishot EPI has been able to reduce distortions causing an increase in acquisition time. For those measurements in which temporal resolution and spatial resolution do not have to be at maximum multiecho EPI is a viable approach. However, due to the low frequency nature of physiological noise, longer acquisition will increase the signal variations of physiological functions.

RF coil technology appropriate for high field has many design aspects in common with coils used in lower fields [15]. However, due to the nature of RF distribution at higher field, RF engineering needs many advances for adaptation to high field [53–55]. The popular bird-cage designs will be unable to take full advantage of high field. In particular, lumped element technology in which capacitors and inductors are used is a design based on circuit analysis using quasistatic field approximations [15]. But this analysis is only valid at low fields since the RF wavelength required for spin excitation decreases as the field strength increases. Specifically, RF wavelength in air at 1.5, 3.0, and 7.0 T are about 5, 2.5, and 1 m long. Taking into account the dielectric constant of biological tissues which are around 80, the wavelength inside the body reduces by a factor of inverse of square root of dielectric constant to around 50, 25, and 10 cm, respectively. Comparing the typical dimension of an RF coil, say 20 cm diameter, with these numbers makes it clear that quasistatic approximations are only valid for fields below 1.5 T where the RF coil dimension is much smaller than the wavelength of the RF field. At 7.0 T, the resonance frequency of 300 MHz makes the in-tissue wavelength to be about 10 cm. Since typical dimensions of the human head are comparable to this wavelength, the wave nature of the RF pulse becomes dominant within the head. Consequently, full wave Maxwell equation solutions are required to estimate the magnetic field ($B1$) and electric field (E_1) of the RF as it penetrates into the body during the spin excitations [54]. Such solutions are only possible through the use of sophisticated numerical computations, such as finite difference time domain (FDTD). This approach treats the RF coil interaction with the human body as a full wave electromagnetic modeling that not only provides an accurate map of distribution of $B1$ field over the subject but also offers a precise measure of specific absorption rate (SAR), which is an important indication of RF heating.

Inhomogeneous images acquired at high field (Fig. 1) demonstrate the effectiveness of the techniques developed for alleviation of inherent inhomogeneities of high-field images. These images point to a need for change in paradigm in the use of RF in high-field MRI. Use of computational tools for coil design is one pillar of the new paradigm. In addition, potential for excessive heat deposition predicted early in the history of MRI due to RF power constitutes a major safety issue that high field will have for a long time. Another issue that is highlighted at high field is dielectric effects that have revealed their presence in high-field images due to focusing of RF power at the central regions of the imaged body [56]. In studies at high field which are mostly done on the human heads, this effect shows strong inhomogeneous spread of RF reducing the power required in the peripheral regions for spin excitation. Dielectric effects or dielectric resonance problems at high field is an issue of concern for coil designers and must be taken into account in the use of high-field scanners and analysis of the data acquired by these systems.

Another pillar of the new paradigm is parallel imaging. Recent techniques for acceleration of image acquisition based on parallel imaging, SMASH-like methods and SENSE-like methods [57, 58], have shown promise in alleviating RF inhomogeneities at high field. Both methods use surface like coil element which have an RF profile stronger in the proximal regions than in the deep regions of the body. While the immediate use of multichannel coil technology (parallel imaging) is in the receiver mode to accelerate signal reception, parallel transmit will also play an important role by restoring RF distribution over the whole head [59]. Possibility of the use of multichannel receive and transmit technology will allow high-field fMRI to further accelerate and enhance image qualities with potential to achieve microscopic resolution BOLD and perfusion-based images with high temporal and spatial resolution.

Need for more powerful gradients is another necessity of high field that has been highlighted recently as the receiver bandwidth increases in high-field scanners. Although modern scanners are equipped with more robust gradients, the increase in gradient strength and slew rate continues. During the 1980s, clinical scanners were equipped with gradients of 20 mT/m strength and 50 T/m/s slew rate. Today, 40-mT/m gradients with 150 T/m/s slew rate are available in most clinical scanners. Such hardware has helped many fMRI studies at 3.0 T and has helped research in the development and use of more powerful gradients.

Gradients also are the source of many image artifacts. At high field, artifacts due to EPI are aggravated and research has achieved many successes in minimizing image artifacts. Advances have been also achieved in gradient coil design and gradient amplifiers. Technologies such as active shielding (AS) of gradient have been realized. Considerable reduction in eddy current and its artifacts are reported by the use of AS gradients. Improved technology in pre-emphasis also has contributed in making modern gradients capable of higher performance even compared to the recent generations. Manufacturers of specialty high-field gradients offer products with strengths of 50–100 mT/m with capability of 150–300 µs switch time. Such gradients can clock slew rate up to 200–500 T/m/s. High-field fMRI is the primary beneficiary of this technology, as strong gradients capable of faster switching rates can be used to recover signal losses due to inhomogeneities through suppression of T_2^* artifacts. There are, however, disadvantages such as $\mathrm{d}B/\mathrm{d}t$ which emerge as switching time reduces in gradients. Faster switching increases $\mathrm{d}B/\mathrm{d}t$, which induces stronger electric fields in conducting tissues of the body causing nerve stimulation in the subject. Both designers and users of MRI scanners are made aware of the potential hazards of high gradient-induced electric fields and their use is governed by software and hardware safety supervisors to prevent incidents, such as ventricular fibrillation. Fortunately, fMRI uses sequences such as EPI which is very similar to the conventional gradient recalled echo sequence and it acquires the entire image in a single shot. The artifacts due to susceptibility and fast switching have been addressed in solutions, such as interleaved EPI and discontinuity in k-space, which have been dealt with by flip angle adjustments. In all, solutions in cleaning up EPI and other fast imaging techniques are making strong gradients more useful for application in high-field fMRI studies.

Another approach for using strong gradients without their undesirable side effects is through asymmetric designs, where the gradient field is produced only over the intended body part. For fMRI of the brain, this is particularly useful as it allows the establishment of stronger and faster gradients while at the same time keeping the heart isolated from induced electric fields. As the field strength increases, head-only scanners are gaining more attention. While high-field advantage of SNR is independent of higher gradient strength or slew rate, the additional in-plane resolution and slice thickness that can be achieved by using powerful gradients will help achieve isotropic voxels and ultimately microscopic map of brain function.

7   Conclusion

Functional imaging has achieved much success due to the MRI inherent soft-tissue contrast and its capability of detecting paramagnetic-based brain activation signals. The proportionality of SNR with field strength is an opportunity that has the potential of achieving microscopic brain mapping. High-field fMRI uses the SNR currency to enhance sensitivity and specificity in probing neurophysiology. Many high-field advantages can be utilized through realizable improved ancillary technologies such as RF coils, new excitation/detection schemes, artifact reduction, gradient technology, and parallel imaging. Low-field fMRI has already produced data from brain function that allows much insight into cognitive neuroscience. High field, in turn, has shown potential of further unraveling brain mysteries by detecting activation caused by controlled external stimulations with resolution that is approaching dimensions of functioning units of the brain. Such is the true potential of high-field fMRI.

References

1. Lauterbur PC (1973) Image formation by induced local interactions: example employing nuclear magnetic resonance. Nature 242:190–191

2. Hoult DI, Lauterbur PC (1979) The sensitivity of the zeumatographic experiment involving human samples. J Magn Reson 34:425–433

3. Ogawa S, Tank DW, Menon R, Ellermann JM, Kim SG, Merkle H, Ugurbil K (1992) Intrinsic signal changes accompanying sensory stimulation: functional brain mapping with magnetic resonance imaging. Proc Natl Acad Sci U S A 89:5951–5955

4. Sadek JR, Hammeke TA (2002) Functional neuroimaging in neurology and psychiatry. CNS Spectr 7:286–290, 295–299

5. Yacoub E, Van De Moortele PF, Shmuel A, Uğurbil K (2005) Signal and noise characteristics of Hahn SE and GE BOLD fMRI at 7 T in humans. Neuroimage 24:738–750

6. Duong TQ, Yacoub E, Adriany G, Hu X, Ugurbil K, Vaughan JT, Merkle H, Kim SG (2002) High-resolution, spin-echo BOLD, and CBF fMRI at 4 and 7 T. Magn Reson Med 48:589–593

7. Pfeuffer J, Adriany G, Shmuel A, Yacoub E, Van De Moortele PF, Hu X, Ugurbil K (2002) Perfusion-based high-resolution functional imaging in the human brain at 7 Tesla. Magn Reson Med 47:903–911

8. Uğurbil K, Hu X, Chen W, Zhu XH, Kim SG, Georgopoulos A (1999) Functional mapping in the human brain using high magnetic fields. Philos Trans R Soc Lond B Biol Sci 354:1195–1213

9. Logothetis NK (2008) What we can do and what we cannot do with fMRI. Nature 12(453):869–878

10. Goense JB, Zappe AC, Logothetis NK (2007) High-resolution fMRI of macaque V1. Magn Reson Imaging 25:740–747

11. Shulman RD (2001) Functional imaging studies: linking mind and basic neuroscience. Am J Psychiatry 158:11–20

12. Bloch F (1946) Nuclear induction. Phys Rev 7:460–473

13. Pourcell EM, Torrey HC, Pound RV (1946) Resonance absorption by nuclear magnetic moments in a solid. Phys Rev 69:37–38

14. Hoult DI, Richards RE (1976) The signal-to-noise ratio of nuclear magnetic resonance experiment. J Magn Reson 24:71–85

15. Tropp J (1989) The theory of the bird-cage resonator. J Magn Reson 82:51–62

16. Bloembergen PEM, Pound RV (1948) Relaxation effects in nuclear magnetic resonance absorption. Phys Rev 73:679–746

17. Peters AM, Brookes MJ, Hoogenraad FG, Gowland PA, Francis ST, Morris PG, Bowtell R (2007) T2* measurements in human brain at 1.5, 3 and 7 T. Magn Reson Imaging 25:748–753

18. Wansapura JP, Holland SK, Dunn RS, Ball WS Jr (1999) NMR relaxation times in the human brain at 3.0 Tesla. J Magn Reson Imaging 9:531–538

19. Vymazal J, Righini A, Brooks RA, Canesi M, Mariani C, Leonardi M, Pezzoli G (1999) T1 and T2 in the brain of healthy subjects, patients with Parkinson disease, and patients with multiple system atrophy: relation to iron content. Radiology 211:489–495

20. Liu F, Garland M, Duan Y, Stark RI, Xu D, Dong Z, Bansal R, Peterson BS, Kangarlu A (2008) Study of the development of fetal baboon brain using magnetic resonance imaging at 3 Tesla. Neuroimage 40:148–159

21. Wright PJ, Mougin OE, Totman JJ, Peters AM, Brookes MJ, Coxon R, Morris PE, Clemence M, Francis ST, Bowtell RW, Gowland PA (2008) Water proton T (1) measurements in brain tissue at 7, 3, and 1.5T using IR-EPI, IR-TSE, and MPRAGE: results and optimization. MAGMA 21:121–130

22. Kim SG, Ugurbil K (2003) High-resolution functional magnetic resonance imaging of the animal brain. Methods 30:28–41

23. Meltzer HY, McGurk SR (1999) The effects of clozapine, risperidone, and olanzapine on cognitive function in schizophrenia. Schizophr Bull 25:233–255

24. Kim SG, Fukuda M (2008) Lessons from fMRI about mapping cortical columns. Neuroscientist 14:287–299

25. Yacoub E, Shmuel A, Logothetis N, Uğurbil K (2007) Robust detection of ocular dominance columns in humans using Hahn Spin Echo BOLD functional MRI at 7 Tesla. Neuroimage 37:1161–1177

26. Yacoub E, Shmuel A, Pfeuffer J, Van De Moortele PF, Adriany G, Andersen P, Vaughan JT, Merkle H, Ugurbil K, Hu X (2001) Imaging brain function in humans at 7 Tesla. Magn Reson Med 45:588–594

27. Mansfield P, Pykett IL, Morris PG (1978) Human whole body line-scan imaging by NMR. Br J Radiol 51:921–922

28. Goense JB, Logothetis NK (2008) Neurophysiology of the BOLD fMRI signal in awake monkeys. Curr Biol 18:631–640

29. Goense JB, Ku SP, Merkle H, Tolias AS, Logothetis NK (2008) fMRI of the temporal lobe of the awake monkey at 7 T. Neuroimage 39:1081–1093

30. Farzaneh F, Riederer SJ, Pelc NJ (1990) Analysis of T2 limitations and off-resonance effects on spatial resolution and artifacts in echo-planar imaging. Magn Reson Med 14:123–139

31. Yang QX, Smith MB, Briggs RW, Rycyna RE (1999) Microimaging at 14 Tesla using GESEPI for removal of magnetic susceptibility artifacts in T(2)(*)-weighted image contrast. J Magn Reson 141:1–6

32. Yang QX, Wang J, Smith MB, Meadowcroft M, Sun X, Eslinger PJ, Golay X (2004) Reduction of magnetic field inhomogeneity artifacts in echo planar imaging with SENSE and GESEPI at high field. Magn Reson Med 52:1418–1423

33. Chen NK, Wyrwicz AM (2004) Removal of EPI Nyquist ghost artifacts with two-dimensional phase correction. Magn Reson Med 51:1247–1253

34. Schenck JF (1996) The role of magnetic susceptibility in magnetic resonance imaging: MRI magnetic compatibility of the first and second kinds. Med Phys 23:815–850

35. Callaghan PT (1990) Susceptibility-limited resolution in nuclear magnetic resonance microscopy. J Magn Reson 87:304–318

36. Kangarlu A, Bourekas EC, Ray-Chaudhury A, Rammohan KW (2007) Cerebral cortical lesions in multiple sclerosis detected by MR imaging at 8 Tesla. AJNR Am J Neuroradiol 28:262–266

37. Filippi M, Rocca MA (2007) Conventional MRI in multiple sclerosis. J Neuroimaging 17(Suppl 1):3S–9S

38. Fazekas F, Soelberg-Sorensen P, Comi G, Filippi M (2007) MRI to monitor treatment efficacy in multiple sclerosis. J Neuroimaging 17(Suppl 1):50S–55S

39. Christoforidis GA, Bourekas EC, Baujan M, Abduljalil AM, Kangarlu A, Spigos DG, Chakeres DW, Robitaille PM (1999) High resolution MRI of the deep brain vascular anatomy at 8 Tesla: susceptibility-based enhancement of the venous structures. J Comput Assist Tomogr 23:857–866

40. Bourekas EC, Christoforidis GA, Abduljalil AM, Kangarlu A, Chakeres DW, Spigos DG, Robitaille PM (1999) High resolution MRI of the deep gray nuclei at 8 Tesla. J Comput Assist Tomogr 23:867–874

41. Davis TL, Kwong KK, Weisskopff RM, Rosen BR (1998) Calibrated functional MRI: mapping the dynamics of oxidative metabolism (hypercapniaycerebrovascular reactivity). Proc Natl Acad Sci U S A 95:1834–1839

42. Yacoub E, Shmuel A, Pfeuffer J, Van De Moortele PF, Adriany G, Ugurbil K, Hu X (2001) Investigation of the initial dip in fMRI at 7 Tesla. NMR Biomed 14:408–412

43. Krüger G, Glover GH (2001) Physiological noise in oxygenation-sensitive magnetic resonance imaging. Magn Reson Med 46:631–637

44. Wang SJ, Luo LM, Liang XY, Gui ZG, Chen CX (2005) Estimation and removal of physiological noise from undersampled multi-slice fMRI data in image space. IEEE EMBS 27:1371–1373

45. Hyde JS, Biswal BB, Jesmanowicz A (2001) High-resolution fMRI using multislice partial k-space GR-EPI with cubic voxels. Magn Reson Med 46:114–125

46. Glover GH, Krüger G (2002) Optimum voxel size in BOLD fMRI. Proc Int Soc Magn Reson Med 10:1395

47. Mountscale VB (1997) The columnar organization of the neocortex. Brain 120:701–722

48. Triantafylloua C, Hogea RD, Wald LL (2006) Effect of spatial smoothing on physiological noise in high-resolution fMRI. Neuroimage 32:551–557

49. Duong TQ, Kim DS, Ugurbil K, Kim SG (2001) Localized cerebral blood flow response at submillimeter columnar resolution. Proc Natl Acad Sci U S A 98:10904–10909

50. Kim DS, Duong TQ, Kim SG (2000) High-resolution mapping of isoorientation columns by fMRI. Nat Neurosci 3:164–169

51. Jezzard P, Clare S (1999) Sources of distortion in functional MRI data. Hum Brain Mapp 8:80–85

52. Speck O, Stadler J, Zaitsev M (2008) High resolution single-shot EPI at 7T. MAGMA Magn Reson Mater in Phys Biol Med 21:73–86

53. Baertlein BA, Ozbay O, Ibrahim T, Lee R, Yu Y, Kangarlu A, Robitaille PM (2000) Theoretical model for an MRI radio frequency resonator. IEEE Trans Biomed Eng 47:535–546

54. Ibrahim TS, Lee R, Baertlein BA, Kangarlu A, Robitaille PL (2000) Application of finite difference time domain method for the design of birdcage RF head coils using multi-port excitations. Magn Reson Imaging 18:733–742

55. Ibrahim TS, Kangarlu A, Chakeress DW (2005) Design and performance issues of RF coils utilized in ultra high field MRI: experimental and numerical evaluations. IEEE Trans Biomed Eng 52:1278–1284

56. Kangarlu A, Baertlein BA, Lee R, Ibrahim T, Yang L, Abduljalil AM, Robitaille PM (1999) Dielectric resonance phenomena in ultra high field MRI. J Comput Assist Tomogr 23:821–831

57. Pruessmann KP, Weiger M, Scheidegger MB, Boesiger P (1999) SENSE: sensitivity encoding for fast MRI. Magn Reson Med 42:952–962

58. Sodickson DK, Manning WJ (1997) Simultaneous acquisition of spatial harmonics (SMASH): fast imaging with radiofrequency coil arrays. Magn Reson Med 38:591–603

59. Katscher U, Börnert P, Leussler C, van den Brink JS (2003) Transmit SENSE. Magn Reson Med 49:144–150

Chapter 5

Experimental Design

Hugh Garavan and Kevin Murphy

Abstract

This chapter addresses issues particular to the optimal design of fMRI experiments. It describes procedures for isolating the psychological process of interest and gives an overview of block, event-related and participant-response-dependent designs. An additional focus is placed on data analysis with emphasis on optimizing and isolating the neuroimaging signal in activated brain regions. Finally, the chapter addresses a number of practical matters including optimal sample sizes and trial durations that confront all researchers when designing their experiments.

Key words Task design, Sample size, Scan durations, Analysis, Regression, Efficiency, Frequency

1 Overview

Noninvasive functional neuroimaging techniques enable researchers to study the neurobiological substrates of psychological processes. The large body of neuroimaging research has two fundamental purposes. The first is to identify the brain regions that underlie a particular psychological process while the second seeks to identify differential responses of these regions to various stimuli or task challenges. The latter focus yields insights into both how the brain accommodates varying task demands and how differences between individuals or between clinical and healthy comparison groups might be explained by differences in neurobiological functioning. To achieve these goals, it is essential that one be able to isolate the psychological process of interest and how best to do so, with particular regard to experimental design, is the focus of this chapter. Part II will describe issues particular to psychological experimental design, that is, experimental control over the cognitive or emotional process of interest. Part III focuses on data analysis with emphasis on optimizing and isolating the neuroimaging signal in the activated brain regions. Part III also addresses a number of practical matters that confront all researchers when designing their experiments.

Massimo Filippi (ed.), *fMRI Techniques and Protocols*, Neuromethods, vol. 119,
DOI 10.1007/978-1-4939-5611-1_5, © Springer Science+Business Media New York 2016

The distinction between isolating the psychological process of interest and isolating the signal associated with that process is made for pedagogical purposes. In practice, the two considerations are closely intertwined in that the experiment must be designed with a view to how the data are to be analyzed. In brief, a typical analysis decomposes the time-series data into their contributing sources of variance. These sources of variance, which are generally assumed to be linearly additive, can include signals of interest such as task-induced brain activity as well as nuisance signals such as those created by head-movement, scanner signal drift, physiological processes, or the intrusion of extraneous psychological processes. The most common method for analyzing these data is a linear decomposition of the various signal sources using for example, a multiple regression in which separate regressors and planned contrasts between regressors capture both the unwanted variance and the variance of interest. Clearly, the design of the experiment needs to take into consideration what regressors and what contrasts of interest will be included in the analyses in order to ensure that the final brain activation map can be attributed to the psychological process of interest.

2   Task Design

As fMRI data are inherently noisy, it is important to induce as strong a signal as possible. This serves to maximize the contrast between the active task state and a comparison state (e.g., between a cognitive task and a visuomotor control task). In addition to maximizing contrast within an individual, it is also important to maximize contrast between individuals (e.g., between a clinical group and healthy controls) or between two times of testing (e.g., a pre-post comparison of treatment effects). To maximize the contrast between groups, it is advisable to isolate the psychological process that best discriminates the two groups. In this regard, neuroimaging researchers would be well-served by grounding their experimental methods in the relevant psychological literature that identifies the key functions that distinguish the clinical and control groups and that provides a wealth of research methods detailing how to isolate those functions experimentally.

Experimental designs in fMRI can be categorized into block, event-related, and a third, broader category, labeled participant-response dependent, in which a continuous measure obtained from the participant provides a regressor for probing brain activity. The block design averages brain activation over a sustained period of time (20–30 s would be typical durations) and contrasts this with similar periods of either a resting state or a comparison task which is typically chosen to contain all task demands except the psychological process of interest. Brain regions that differ between these conditions may then be attributed to the psychological process. In an effort to exclude signals associated with confounding physiological processes (described in detail below), aperiodic block durations, in which the alternating ON and OFF periods vary in durations, may be advisable.

This standard block design can be supplemented by including gradations of task challenge. This type of parametric manipulation can be quite advantageous: whereas the standard two-condition comparison (e.g., task A vs. task B or task A vs. rest) is open to the criticism of pure insertion (i.e., whether it is possible to selectively include and exclude a psychological function without affecting other task-related processes), the parametric manipulation assumes that the process is always present but to varying degrees in accordance with the demands placed on that process. Examples would include presenting various intensity levels of a sensory stimulus [1] or manipulating the number of memoranda in a working memory task [2]. Block designs can also be enhanced by a sort of psychological triangulation in which the conjunction of distinct block design contrasts allows one to isolate a psychological process that can be separated from irrelevant surface features of the tasks [3]. For example, if one wishes to isolate the neuroanatomy of the mental rehearsal component of verbal working memory, one might design an experiment using quite distinct classes of stimuli with each class accompanied by its own control comparison. One task might require participants to store a list of common nouns over a rehearsal period and recall the words after that rehearsal period. A reasonable control condition for this task might be one in which the word list remains on-screen for the duration of the rehearsal period and participants read, rather than recall, the words after the rehearsal period. The second task might present a list of nonsense syllables through earphones. At the end of the rehearsal period a single nonsense syllable is presented and participants report, using a button box, if the single item was one of the rehearsed items. A control condition for this second task might simply prompt the participant to make a predetermined button press response at the end of a delay period that was of similar duration to the rehearsal period. The conjunction between the two activation maps, in which activation for each task is first subtracted from its control condition, may be argued to represent core regions responsible for verbal working memory for which the influence of extraneous task features (e.g., linguistic stimulus properties, response modalities, recall vs. recognition) are minimized. This strength of the conjunction approach, however, may often need to be balanced against the time costs involved in testing all the required conditions.

In circumstances in which a psychological process can be isolated temporally then event-related designs are particularly useful. Here, brain activation time-locked to the events of interest can be selectively averaged enabling the researcher to embed trials of interest amidst other control trials and to categorize the trials after the participant has completed the experiment. Error trials can be excluded (or averaged separately) and events can be coded by whether a participant detected a target or not, responded relatively fast or not, produced a subsequent behavior or not and so on [4]. This affords the researcher increased flexibility in probing the data-set and has obvious advantages over the block design in circumstances in which the psychological process cannot be presented in blocks as in, for example, an oddball paradigm in which the nature of the phenomenon mandates that events are infrequent and unpredictable. The block and event-related designs can also be combined such that events of interest during an active task period can be isolated while the task period itself can be simultaneously contrasted against a control period [5, 6]. This type of mixed design provides additional information in that one can determine the inter-relationships between tonic activity levels (e.g., sustained attention or an induced emotional state) and the processing of a discrete trial (e.g., detection of a fearful face).

A final category of experimental design is what we have labeled participant-response dependent. Here, the participant provides a continuous measure that can, for example, be used to generate a regressor to correlate against brain activity measures. Despite a loss of experimental control over the participant's behavior, this category of design affords much flexibility when the phenomenon of interest is either not strictly task-dependent or is difficult to experimentally induce. Examples include resting state acquisitions (in which correlated patterns of brain activity can be detected while the participant simply rests) [7], biofeedback (in which, for example, a participant learns to control their level of brain activity) [8], passive viewing of a movie clip (in which there may be multiple sources of stimulation with each varying with a different time-course) [9] or in which performance varies in an unpredictable manner [10]. Performance modulations for which one could assess brain activation changes can be quite wide-ranging including response times or response time variability on a continuous performance task [11], frequently sampled self-report measures of mood [12] and physiological measures such as heart rate or pupil-diameter [13]. In these examples the discrete measurements can be interpolated to provide a continuous time-series that can be correlated with the brain activation time-series data.

A related approach, albeit one that is manipulated to a degree by the experimenter, is one in which the researcher derives a computational model of a subject's performance. Here, a formal model of the processes hypothesized to underlie task performance is developed by the experimenter. For example, on a forced-choice reward task in which the subject attempts to maximize their winnings, one might hypothesize that subjects develop expectations of reward based on previous trial outcomes, experience prediction errors if those expectations are violated, learn at different rates and so on. The parameterization in a formal model of these psychological processes allows novel regressors to be created that track those processes over time. Importantly, identifying brain regions where activity covaries with those regressors serves to validate the theoretical computational model and its underlying mechanisms and, in this regard, is a valuable advance beyond the more commonplace localization of functions [14].

An important consideration permeating all experimental designs is the choice of baseline against which activation is contrasted. These baselines can be explicit as in the block design in which specific blocks are chosen for comparison or implicit as in the event-related and participant-response dependent designs in which the baseline is all task-related activity that is not accommodated by regressors in the data analysis. The choice of baseline determines the interpretation of what processes are captured in an activation map and requires very careful consideration by the experimenter.

2.1 Choosing an Experimental Design

The choice of which design to employ will be dictated by the particulars of the psychological process to be investigated and how easy it is to isolate. Block designs can be employed if the psychological function is easy to isolate or if it is of particular interest to compare two tasks. A simple example would be a contrast between unilateral and bilateral finger movements. Here, blocks of finger movement in just one hand could be alternated with blocks of finger movements in two hands. Rest periods might also be included in order to provide a low-level baseline against which any task-related activity could be assessed. The inclusion of a resting state baseline is generally advantageous as contrasts between two task-active periods can often be ambiguous in that greater activation in condition A relative to condition B could result from either more positive activation in A or a greater deactivation in B. A resting state baseline allows one to resolve this ambiguity by showing if activation increases or decreases in any one condition relative to the resting baseline. However, care must be taken when comparing task to rest as in the last few years a large body of literature has demonstrated that rest itself is not the absence of neural activity. For example, it has been shown that the default mode network is more active during "rest" than during a task (see further description of resting-state phenomena and analyses in the chapter by Fornito—Chap. 10).

If the psychological function is not easily isolated then a conjunction analysis may be useful. As can be seen in the verbal working memory example given above, the conjunction design enables the researcher to identify the core functional neuroanatomy that is common across different operationalizations of a psychological process. In addition, it can also reveal task-specific activations enabling, for example, one to determine how verbal working memory rehearsal for linguistic information differs to that of nonlinguistic information. An alternative approach may parametrically manipulate verbal working memory demands by asking participants to rehearse items of varying set sizes. The presumption here is that more items will engage verbal working memory rehearsal to a greater extent resulting in changes in activation corresponding to the increased memory loads.

Although block designs suffer from an inability to isolate cognitive events that are temporally proximal by virtue of averaging over a prolonged duration, and may provide activation measures contaminated by extraneous tonic processes or isolated events (e.g., errors), they nonetheless have some advantages. For example, if the psychological process of interest by its very nature exists over a prolonged duration (e.g., sustained attention) or does not exist as a temporally discrete event (e.g., an emotional reaction), then it may be assayed best by a block design.

Conversely, the event-related design is particularly useful if one's goal is to isolate distinct cognitive events. In between-group comparisons (or time 1 vs. time 2 comparisons) the event-related design has the added advantage of being able to equate performance levels by comparing the groups on correct trials only. That is, one can compare correct performance trials of one group against the correct performance trials of the second group even if the absolute numbers of correct trials differ between the groups. In this regard, contamination from activity specific to error-related processes will not confound the between-group comparison [15]. In a similar manner, selective averaging of trials may make it possible to eliminate other group differences (e.g., response speed) assuming that there are sufficient numbers of trials for this type of a matched-trial analysis. This is a particularly welcome feature as activation differences between groups that one may wish to attribute to a psychological difference can be confounded by secondary behavioral or performance differences [16]. Indeed, the relationship between performance and activation levels is not straightforward. Often, researchers wish to ensure that the task produces performance differences between groups (or within a group following some experimental manipulation) in order to justify that choice of task or the focus on the psychological process engaged by the task; why study the neurobiology of attention between healthy controls and children with attention deficit hyperactivity disorder (ADHD) if the latter are not shown to be worse on the attention task? However, this can be a double-edged sword in that performance differences and knock-on effects such as differences in frustration or anxiety levels can confound interpretation of activation levels. One proposed solution is to administer a task that is within the level of competence of all participants and which, therefore, may not produce group differences in performance. Such a task can be considered a probe of the neurocognitive functioning of the groups and substantial empirical evidence shows that brain activation differences are often observed in the absence of performance differences. The typical interpretation of activation differences when there are no performance differences is that reduced activation reflects better neural efficiency and less "effort." This interpretation is supported by studies showing greater levels of activity as task difficulty increases or those that show reduced activation following practice of a psychological process [17].

Finally, the participant-dependent response design may be a sensible choice when one can obtain a continuous measurement from the participant but cannot exercise full experimental control over behavior. For example, although emotional states are difficult to induce (and extinguish) experimentally, a physiological, self-report, or task-induced measure can provide a time-course of that emotional state which can be used to detect correlated brain regions.

3   Optimizing Experimental Task Designs

The key issue when optimizing experimental task design in fMRI is statistical power. There are many important basic variables to be chosen which, if selected wisely, will lead to high power and thus robust and reliable results. Too often these variables are chosen arbitrarily leading to poor experimental designs that fail to yield the expected outcomes.

When designing a task one needs to consider practical issues such as the number of participants or events required to give reliable results along with more analytic issues such as the estimation efficiency of the task design. These pragmatics are often dictated by feasibility constraints such as the availability of participants (e.g., how much access to the clinical population under study does one have?), the cost of scan time or the amount of available time in which the participant will remain comfortable and compliant. Despite the ubiquity of these concerns, surprisingly few studies have addressed them and, instead, more emphasis has been placed on the analytic issues of presentation rate, duty cycles, sampling procedures, detectability of activation and efficiency of response estimation. These analytic issues relate the task that will be performed to the analysis methods that will be employed and provide guidance on the design details of an experiment. It is important, however, that analytic considerations are not allowed to dictate the design of a task such that it is no longer appropriate for measuring/engaging the psychological process under study.

3.1 Practical Issues   There has been increased concern in recent years about the power of many (most?) fMRI studies and a number of cogent critiques suggest that low sample sizes have generated a disconcerting number of false positive results [18, 19]. However, only a handful of studies have addressed how many participants are required to yield stable activation maps. The first paper addressing this issue showed that conjunction analysis with a fixed-effect model is sufficient to make inferences about population characteristics thus reducing the number of participants required to infer differences between populations [20]. Although quite useful, this conclusion does not give a clear indication of the number of participants required. By estimating the mean differences and variability between two block conditions, Desmond and Glover were able to perform simulation experiments generating power curves from which they could calculate the required number of participants [21]. They found that for a liberal threshold of p=0.05, 12 participants were required to yield 80% power in a single voxel for typical block design activation levels. However, in fMRI the multiple comparisons problem and the associated potential for high levels of false positives requires us to go to stricter thresholds where they demonstrated that twice the number of participants would be needed to maintain the same level of statistical power. This recommended number of participants is higher than the majority of fMRI studies but is similar to independent assessments based on empirical data from a visual/audio/motor task [22] and from an event-related cognitive task [23].

The Murphy and Garavan study [23] found that statistical power is surprisingly low at typical sample sizes ($n < 20$) but that voxels that were significantly active from these smaller sample sizes tended to be true positives. Although voxelwise overlap may be poor in tests of reproducibility, the locations of activated areas provide some optimism for studies with typical sample sizes. It was found that the similarity between centers-of-mass for activated regions does not increase after more than 20 participants are included in the statistics. The conclusion can be drawn from this paper that a study with fewer numbers of participants than Desmond and Glover propose is not necessarily inaccurate but it is incomplete: activated areas are likely to be true positives but there will be a sizable number of false negatives. Arriving at a similar conclusion, Thirion and colleagues argue that the reliability of group analyses is strongly affected by inter-subject variability and recommended that 20 subjects or more should be included in fMRI studies [22]. Needless to say, the required number of participants is influenced by the effect size which, in turn, is affected by the sensitivity of the experiment (e.g., the strength of the experimental manipulation, the quality of the data acquisition and the accuracy of the data analyses) and the contrast-to-noise of the signal that scales with field strength of the MRI scanner. These considerations may be even more important if one's intention is to detect what is likely to be an even smaller effect size of a between-group comparison. However, with the push to higher field strengths (e.g., 7 T), smaller effect sizes should be detectable with similar numbers of participants.

Little research has addressed the optimal number of scans/events needed for a successful fMRI study. A simple reason for this is that there is no standard metric for determining the required number of scans/events and no gold standard for determining when the optimal number of events has been reached. One metric that has been utilized is the spatial extent of activation under the assumption that as more scans/events are included in the analysis, the spatial extent of activation will increase until all activated cortex is deemed above significance. When this occurs the spatial extent should asymptote providing an estimate of the required number of scans/events. Using this approach in a block design experiment, Saad and colleagues demonstrated that the spatial extent of activation increased monotonically with the number of scans included in the analysis and failed to asymptote after twenty-two 200 s long scans [24]. Similarly, Huettel and McCarthy found that the spatial extent of activation failed to asymptote even after 150 events in an event-related design [25]. However, this failure to asymptote may be a consequence of the analysis method employed [26]. The correlation method does not asymptote because the goodness-of-fit to the regressor continues to rise with increasing degrees-of-freedom (df), which implies that the correlation measure will never plateau by adding more time points. The Huettel and McCarthy result [25] was replicated by Murphy and Garavan [26] but they also demonstrated that when using a standard general linear modeling (GLM) analysis rather than a correlation, the spatial extent of activation asymptotes after roughly 25 events in a properly jittered event-related design. This is certainly a more attainable number of events in the available scan time of standard fMRI studies. It can be assumed that at least 25 of each type of event are needed if there is more than one psychological process under study. Also, these results have been derived from primary sensorimotor processes in the brain so it is unclear whether they will still hold for more subtle cognitive activations. Again, differences in activation have not been addressed either: it is quite possible that many more events would be required to distinguish two processes with slightly varying activation levels since these differences could be dominated by noise.

A related concern is the optimal duration of a scan. How long a scan should last is obviously dependent on how densely the required number of events can be distributed. For example, a GO/NOGO task must sparsely distribute NOGO events due to the need to build up a prepotency to respond while a simple motor response task can present the events more frequently. Other issues that limit how long a scan can last include participant comfort and ability to stay engaged in the task along with technical concerns such as throughput of data and image reconstruction times. For these reasons and more, it is common to split a scanning session into separate scans lasting 5–10 min each after which they can be concatenated into one single time-series and treated as a single scan in the analyses. However, breaks in scanning reduce the efficiency of any temporal filtering that is used and can also introduce unwanted session effects. If the goal is to detect activation then a block design is the most efficient approach (see below). In this case, the length of the scan is dependent on the amount of noise in the time series (which can be measured by calculating the temporal signal-to-noise ratio (TSNR) defined as the mean of the time series divided by its standard deviation), the size of the effect to be measured (*eff*) and the significance level (*P*) at which the activation is to be detected [27]. These authors derive an equation that determines how long one needs to scan to detect activation with a block design for volumes with high spatial resolution and suggest how this can be extended to an event-related design:

$$N_G = 8 \left[1.5 \left(1 + e^{\log_{10} P/2} \right) \left(\frac{\text{erfc}^{-1}(P)}{(\text{TSNR})(\text{eff})} \right) \right]^2 \tag{1}$$

where N_G is the number of time points required for activation detection. Estimates of the size of the effect can be obtained from previous studies and TSNR measurements can be made using a short resting scan. Since these variables differ widely across types of task, brain regions, and scanners, it is impractical to suggest an optimal scan duration here.

3.2 Analytic Issues

The purpose of an experimental design is to alter neural activity, and hence the blood oxygen level dependent (BOLD) signal, as effectively as possible and in a predicted way thereby enabling the researcher to detect the resulting brain changes. Using this prediction, one looks for corresponding patterns in the fMRI time series to determine which voxels were engaged in the task. A simple reference time series can be produced by convolving the stimulus timing function (which is equal to 0 when no stimulus is applied and 1 when a stimulus is presented) with a hemodynamic response function (HRF) that accurately represents the shape of the BOLD response after a single event. The gamma variate function, $y(t) = t\, r\, e^{-t/b}$, has been shown to effectively model the hemodynamic response to brief stimuli [28], with parameters $r = 8.6$ and $c = 0.51$, and is a popular choice for modelling the hemodynamic shape. The difference between two gamma-variates is also used in order to model the post-stimulus undershoot. It is important that the chosen HRF model accurately reflects the true shape of the response. If, for some reason, the hemodynamic shape of a participant is atypical (e.g., following treatment with a substance that directly affects the vasculature or a patient group with vascular damage), then the results of the analysis could be confounded by this difference in shape. (It should be noted that there are more advanced approaches to reference time-series formation, such as ones that use basis functions rather than a predetermined HRF shape and these are addressed in a later chapter).

The simplest type of analysis is a linear least squares regression of the equation:

$$y(t) = \beta \cdot x(t) + \alpha + \varepsilon(t) \tag{2}$$

where $y(t)$ is the voxel time-series data, $x(t)$ is the reference function (i.e., the expected BOLD response to the stimulus) with β its scaling factor, α is a constant, and $\varepsilon(t)$ is a random Gaussian white-noise term. Both β and α are unknown parameters that are fit by the linear regression method. This equation can be extended to include extra regressors to remove unwanted trends in the data, such as baseline drift and physiological nuisance regressors, whilst simultaneously computing the scaling factor. This scaling factor, β, can then be used as an activation measure for each voxel.

Multiple reference waveforms can easily be included in this type of analysis, denoted by the term *multiple linear regression*. In an experiment with two active conditions [1, 2] the equation:

$$y(t) = \beta_1 \cdot x_1(t) + \beta_2 \cdot x_2(t) + \alpha + \delta \cdot t + \varepsilon(t) \tag{3}$$

is fitted to the data. For this model, four parameters are estimated, the two scaling factors $\beta 1$ and $\beta 2$ and the baseline α and also a baseline drift rate δ which accommodates for linear changes in the baseline over time. It is possible to investigate whether $\beta 1$ or $\beta 2$ are nonzero and whether $\beta 1$ is different from $\beta 2$ with statistical significance calculated using F-tests. This method allows one to identify active areas in the brain and calculate if an area is more active in one condition than another, thereby satisfying the two primary purposes of fMRI. This equation can be further generalized to $Y = X \beta + \varepsilon$, where Y is a column vector of the voxel's time-series data, X is the design matrix, β is a column vector of scaling factors and ε is a column vector of Gaussian white noise terms [29, 30]. This equation is called the *General Linear Model* (GLM) and is the basis for most fMRI analytic techniques. The columns of the design matrix X model the effects of interest and also confounding variables and are, in essence, the reference waveforms mentioned above.

When designing an experimental task, one is essentially specifying these reference waveforms/regressors. To maximize statistical power, these regressors must be chosen wisely. For example, F-tests are used to determine if there are significant differences between the regressors. To increase statistical power, one can increase the df by lengthening the task (for long TRs, each additional timepoint adds a new df). It might seem like a good idea to use extremely short TRs to increase the number of time points and hence the statistical power. However, to gain an extra df for each additional timepoint, each timepoint must be statistically independent from every other. Unfortunately, due to autocorrelations introduced into the fMRI data by physiological noise and scanner drifts, this is not the case. This example demonstrates that knowledge of the underlying mechanisms of fMRI along with the analysis methods is required when choosing even the simplest variables (such as the number of time points and the TR) for experimental design.

Efficiency of a task design is a measure of how accurately the GLM can estimate the β weights for each of the regressors, that is, how small the predicted variance of the β estimates will be. For example, assume a block design task that induces exactly a 2 % signal change in hundreds of voxels, all with different noise properties but with the same noise variance. If a GLM analysis is performed, the β estimate for every voxel will be approximately 2 % for all voxels with very little deviation. Since the variance of the estimates does not differ widely with noise distributions, this would be considered an efficient task. However, matters become more complicated when there is more than one regressor. Assume that there are two block conditions, A and B, where A and B are identical with the exception that B is delayed with respect to A by one TR. These two regressors are highly correlated so if a voxel responds only to condition A, it will be extremely difficult for the GLM to distinguish this from a voxel that responds only to B. For this reason, the β estimates for each of the conditions will vary substantially and this would be considered an inefficient design. If the conditions are designed so that they have zero correlation (this is achieved by delaying B by half a block length relative to A), it would be very easy to distinguish voxels that respond to each of the conditions individually or both of the conditions together. Therefore, the variance of the β estimates would be quite small and so the design is efficient. These simple examples show that efficient task designs come from regressors that are not correlated with each other. This can be slightly complicated by the contrasts of interest. For example, say we have a jittered event-related design where conditions A and B are randomly presented. If we want to find voxels that respond only to A (i.e., a contrast matrix of C = [1 0]), only to B (i.e., C = [0 1]) or differ in their response from A to B (i.e., C = [1 –1]), this design is very efficient. However, if we want to determine voxels that respond equally to both A and B (i.e., C = [1 1]), then the design is very inefficient because such a voxel will always have an elevated activation level and therefore will be indistinguishable from a voxel that does not respond to either task. The simple idea that regressors must be minimally correlated becomes more complicated when multiple conditions, nuisance regressors, and contrasts are placed into a GLM analysis. The efficiency of a task is related to the covariance of the design matrix X (i.e., all regressors expressed as columns of a matrix) and is given the formula:

$$e = \text{trace}\left(C' \times (X'X)^{-1} \times C \right)^{-1} \qquad (4)$$

where C is the matrix of contrast weights and $'$ denotes the transpose of a matrix. Efficiency calculations should be carried out on all experimental designs before scanning to check that the regressors are sufficiently independent. A paper by Smith and colleagues argues against this efficiency calculation since it relates to computational precision rather than image noise [31]. This paper formulates the standard efficiency equations in terms of the required BOLD effect which takes into account the strength and smoothness of the time-series noise.

The question "how do we design a good fMRI task?" is really asking "what experimental timing will produce the most efficient design?". There are two variables under our control, the stimulus duration (SD: defined as the length of time the stimulus is displayed) and the interstimulus interval (ISI: defined as the length of time between the offset of one event and the onset of another). Another common term is stimulus onset asynchrony (SOA) defined as SOA = SD + ISI. (Sometimes, ISI is used to mean SOA so care must be taken to understand the true meaning when reading the literature.) To maximize efficiency (i.e., minimize correlations between regressors by ensuring a clear temporal separation between the event types) one can use either a fixed ISI but vary the order of events from different conditions or one can fix the order of the conditions and vary the ISI. For example, if an event from either condition A or condition B is to be presented every TR, it is very inefficient to present the events in an alternating fashion A,B,A, … However, efficiency is increased if the order is randomized. On the other hand, if B must follow A (e.g., A is a picture of an object and the participant must respond to B, a word, deciding whether it matches the object or not), then randomizing the order is not possible. Therefore, we must vary the ISI between successive As and Bs to increase the efficiency of the design.

The issue of experimental timing is very important in fMRI tasks due to the relatively poor temporal resolution of the technique. Bandettini and Cox have shown that with a 2 s SD the optimal ISI is 12 to 14 s when the ISI is kept constant [32]. At this optimal ISI, the experimentally determined functional contrast (i.e., the ability to detect activation) of an event-related task is only 35 % lower than that of a block-design (which, as explained below, is the most efficient design). Simulations assuming a linear system showed that this should be 65 % lower suggesting the HRF is a nonlinear system. Most techniques in event-related fMRI analysis assume that the hemodynamic shape of the BOLD signal is linearly additive. It has also been shown that when the ISI is allowed to vary, the hemodynamic response shows a 17–25 % reduction in amplitude when trial onsets are spaced (on average) 5 s apart compared to those spaced 20 s apart [33]. However, power analysis indicated that the increased number of trials at fast rates outweighs this decrease in amplitude if statistically reliable response detection is the goal. So, despite the HRF being nonlinear at fast presentation rates, the mismatch with the regressor is compensated by the increase in trial numbers. Dale also demonstrated that if the ISI varies, the statistical efficiency improves monotonically with decreasing mean ISI and that the efficiency can be up to ten times greater than that of a fixed ISI design [34]. These lessons on stimulus timing suggest that even though the HRF is nonlinear at short ISIs, closely packed, randomly presented events produce highly efficient designs.

There are two fundamentally different goals when analyzing event-related fMRI tasks: detection of signal change (which has been the focus thus far) and estimation of the HRF. Detection of the signal change involves determining one variable: the amplitude of the hemodynamic response. More information can be gleaned by estimating the HRF (e.g., time to onset, rise time, fall time, area under the curve) which can be used to determine subtle differences between groups or conditions that may not show up in an amplitude measure. However, this information comes at a cost: the experimental task can be optimized for either detection or estimation but not both. Birn and colleagues showed that the estimation of the HRF is optimized when stimuli are frequently alternated between task and control states, whereas detection of activated areas is optimized by block designs [35]. Liu and colleagues have developed a method that can simultaneously achieve the estimation efficiency of randomized designs and the detection power of block designs at a cost of increasing the length of the experiment by less than a factor of two [36]. There are many programs that allow one to randomly (or not so randomly) generate thousands of task designs in order to choose the most efficient for the task at hand, be it detection or estimation. Genetic algorithms (optimization algorithms that code different designs like chromosomes and allow them to "crossover" and "point mutate" as they "replicate") that can produce designs that outperform random designs on estimation efficiency, detection efficiency, and design counterbalancing have also been developed [37]. Further work has also shown that using advanced mathematical techniques, block designs, rapid event-related designs, m-sequence designs (reference time series with an autocorrelation of zero) and mixed designs can nearly achieve their theoretically predicted efficiency and can be used in practice to obtain advantageous trade-offs between efficiency and detection power [38]. It is important when using programs to design experiments to realize that they may converge on a structure that may be problematic for the psychological process under investigation (e.g., the most efficient task for detecting activation is a block design, however, as noted above, if we want to design an oddball study the oddball events of interest should not occur in a block).

When designing a task, one must also consider the frequencies at which the events of interest are presented. Analysis packages often perform high pass filtering to remove low-frequency drifts from the data. If all frequencies below the limit of, say, 0.01 Hz are removed then the activation to a task with a block lasting greater than 100 s will also be removed. Similarly, this would be true for event-related tasks if the events were presented at the same low frequency. Other frequencies exist in the data that one must consider. It is possible to remove the influence of physiological noise from fMRI data using techniques such as RETROICOR [39]. These physiological noise sources are known to produce fluctuations in the data at the cardiac frequency ~1.1 Hz, at the respiration frequency ~0.3 Hz and also at the respiration volume variation frequency ~0.03 Hz [40]. If these techniques are to be used and the task predominantly displays power at one of these frequencies (e.g., blocks lasting 33 s have a frequency of 0.03 Hz), then the correction techniques may remove the activations of interest and not just the fluctuations due to unwanted physiological processes. Conversely, if these corrections are not used, then the GLM may denote these physiological fluctuations as activations (if the phase of the fluctuations matches the phase of the task). Indeed, it has been demonstrated that task-related breathing fluctuations cause changes in fMRI signals across the whole brain that are time-locked to the task but are unrelated to neural activity [41]. One must also bear in mind that when using a long TR, all frequencies will be aliased into a narrow frequency band (e.g., with a TR of 2 s all frequencies above 0.25 Hz will be aliased into the range of 0–0.25 Hz). This means that although the frequencies may seem far apart, the task and the physiological noise may alias to the same frequency (e.g., for a TR of 2 s, the respiratory frequency 0.3 Hz will be aliased to 0.2 Hz as will a task frequency of 0.7 Hz, that is, one event every 1.4 s). To avoid this problem it is best not to have the events regularly spaced so they reside at one frequency but to have random ISIs thus spreading the power to different frequencies. The most efficient tasks are ones whose power is spread widely across the whole available frequency spectrum.

4   Conclusions

Designing fMRI tasks can be difficult with logistical constraints (e.g., how many participants and how much time per participant can one afford) obliging the experimenter to optimize the study design. The emphasis here has been on the experimental and analytic means of isolating a psychological process and its associated fMRI signal. Both considerations are central: optimal efficiency is of little comfort if one measures the wrong thing but there is little to be gained from an inaccurate measurement of a robust psychological phenomenon. General recommendations include the importance of grounding one's experiment in the appropriate theoretical framework and using appropriate experimental methods, generating designs that are tested for their efficiency prior to data collection, ensuring that a sufficiently large sample is tested and being clear on whether one's goal is the detection of a response or the estimation of that response.

References

1. Helmchen C, Mohr C, Erdmann C, Binkofski F, Buchel C (2006) Neural activity related to self- versus externally generated painful stimuli reveals distinct differences in the lateral pain system in a parametric fMRI study. Hum Brain Mapp 27:755–765

2. Braver TS, Cohen JD, Nystrom LE, Jonides J, Smith EE, Noll DC (1997) A parametric study of prefrontal cortex involvement in human working memory. Neuroimage 5:49–62

3. Price CJ, Friston KJ (1997) Cognitive conjunction: a new approach to brain activation experiments. Neuroimage 5:261–270

4. Garavan H, Ross TJ, Murphy K, Roche RA, Stein EA (2002) Dissociable executive functions in the dynamic control of behavior: inhibition, error detection, and correction. Neuroimage 17:1820–1829

5. Donaldson DI, Petersen SE, Ollinger JM, Buckner RL (2001) Dissociating state and item components of recognition memory using fMRI. Neuroimage 13:129–142

6. Simões-Franklin C, Hester R, Shpaner M, Foxe JJ, Garavan H (2010) Executive function and error detection: the effect of motivation on cingulate and ventral striatum activity. Hum Brain Mapp 31:458–469

7. Margulies DS, Kelly AM, Uddin LQ, Biswal BB, Castellanos FX, Milham MP (2007) Mapping the functional connectivity of anterior cingulate cortex. Neuroimage 37:579–588

8. Weiskopf N, Veit R, Erb M et al (2003) Physiological self-regulation of regional brain activity using real-time functional magnetic resonance imaging (fMRI): methodology and exemplary data. Neuroimage 19:577–586

9. Hasson U, Nir Y, Levy I, Fuhrmann G, Malach R (2004) Intersubject synchronization of cortical activity during natural vision. Science 303:1634–1640

10. Slotnick SD, Yantis S (2005) Common neural substrates for the control and effects of visual attention and perceptual bistability. Brain Res Cogn Brain Res 24:97–108

11. Hahn B, Ross TJ, Stein EA (2007) Cingulate activation increases dynamically with response speed under stimulus unpredictability. Cereb Cortex 17:1664–1671

12. Risinger RC, Salmeron BJ, Ross TJ et al (2005) Neural correlates of high and craving during cocaine self-administration using BOLD fMRI. Neuroimage 26:1097–1108

13. Kampe KK, Frith CD, Frith U (2003) "Hey John": signals conveying communicative intention toward the self activate brain regions associated with "mentalizing," regardless of modality. J Neurosci 23:5258–5263

14. O'Doherty JP, Dayan P, Friston K, Critchley H, Dolan RJ (2003) Temporal difference models and reward-related learning in the human brain. Neuron 38(2):329–337, PMID: 12718865

15. Murphy K, Garavan H (2004) Artifactual fMRI group and condition differences driven by performance confounds. Neuroimage 21:219–228

16. Poldrack RA (2000) Imaging brain plasticity: conceptual and methodological issues--a theoretical review. Neuroimage 12:1–13

17. Kelly AM, Garavan H (2005) Human functional neuroimaging of brain changes associated with practice. Cereb Cortex 15:1089–1102

18. Ioannidis JP (2005) Why most published research findings are false. PLoS Med 2:e124

19. Button KS, Ioannidis JP, Mokrysz C, Nosek BA, Flint J, Robinson ES, Munafò MR (2013) MR. Power failure: why small sample size undermines the reliability of neuroscience. Nat Rev Neurosci 5:365–76. doi:10.1038/nrn3475, Epub 2013 Apr 10. Erratum in: Nat Rev Neurosci. 6, 451

20. Friston KJ, Holmes AP, Worsley KJ (1999) How many subjects constitute a study? Neuroimage 10:1–5

21. Desmond JE, Glover GH (2002) Estimating sample size in functional MRI (fMRI) neuroimaging studies: statistical power analyses. J Neurosci Methods 118:115–128

22. Thirion B, Pinel P, Meriaux S, Roche A, Dehaene S, Poline JB (2007) Analysis of a large fMRI cohort: statistical and methodological issues for group analyses. Neuroimage 35:105–120

23. Murphy K, Garavan H (2004) An empirical investigation into the number of subjects required for an event-related fMRI study. Neuroimage 22:879–885

24. Saad ZS, Ropella KM, DeYoe EA, Bandettini PA (2003) The spatial extent of the BOLD response. Neuroimage 19:132–144

25. Huettel SA, McCarthy G (2001) The effects of single-trial averaging upon the spatial extent of fMRI activation. Neuroreport 12:2411–2416

26. Murphy K, Garavan H (2005) Deriving the optimal number of events for an event-related fMRI study based on the spatial extent of activation. Neuroimage 27:771–777

27. Murphy K, Bodurka J, Bandettini PA (2007) How long to scan? The relationship between fMRI temporal signal to noise ratio and necessary scan duration. Neuroimage 34:565–574

28. Cohen MS (1997) Parametric analysis of fMRI data using linear systems methods. Neuroimage 6:93–103

29. Friston KJ, Holmes AP, Poline JB et al (1995) Analysis of fMRI time-series revisited. Neuroimage 2:45–53

30. Worsley KJ, Friston KJ (1995) Analysis of fMRI time-series revisited–again. Neuroimage 2:173–181

31. Smith S, Jenkinson M, Beckmann C, Miller K, Woolrich M (2007) Meaningful design and contrast estimability in FMRI. Neuroimage 34:127–136

32. Bandettini PA, Cox RW (2000) Event-related fMRI contrast when using constant interstimulus interval: theory and experiment. Magn Reson Med 43:540–548

33. Miezin FM, Maccotta L, Ollinger JM, Petersen SE, Buckner RL (2000) Characterizing the hemodynamic response: effects of presentation rate, sampling procedure, and the possibility of

34. Dale AM (1999) Optimal experimental design for event-related fMRI. Hum Brain Mapp 8:109–114

35. Birn RM, Cox RW, Bandettini PA (2002) Detection versus estimation in event-related fMRI: choosing the optimal stimulus timing. Neuroimage 15:252–264

36. Liu TT, Frank LR, Wong EC, Buxton RB (2001) Detection power, estimation efficiency, and predictability in event-related fMRI. Neuroimage 13:759–773

37. Wager TD, Nichols TE (2003) Optimization of experimental design in fMRI: a general framework using a genetic algorithm. Neuroimage 18:293–309

38. Liu TT (2004) Efficiency, power, and entropy in event-related fMRI with multiple trial types. Part II: design of experiments. Neuroimage 21:401–413

39. Glover GH, Li TQ, Ress D (2000) Image-based method for retrospective correction of physiological motion effects in fMRI: RETROICOR. Magn Reson Med 44:162–167

40. Birn RM, Diamond JB, Smith MA, Bandettini PA (2006) Separating respiratory-variation-related fluctuations from neuronal-activity-related fluctuations in fMRI. Neuroimage 31:1536–1548

41. Birn RM, Murphy K, Handwerker DA, Bandettini PA (2009) fMRI in the presence of task-correlated breathing variations. NeuroImage 47(3):1092–1104

ordering brain activity based on relative timing. Neuroimage 11:735–759

Chapter 6

Preparing fMRI Data for Statistical Analysis

John Ashburner

Abstract

This chapter describes the procedures applied to fMRI data prior to their statistical analysis. This usually begins with converting the data from original MR format to a form that can be used by the analysis software. The data are then motion corrected. If an anatomical scan is collected for the subject, then it would be coregistered with the fMRI, and may serve to estimate the warps needed to spatially normalize the fMRI to some standard space. The final processing step is usually to smooth the data.

Key words Generative model, fMRI, Registration, Artifact correction, Spatial normalization, Smoothing

1   Introduction

This chapter provides a brief overview of the image processing steps currently used for transforming fMRI data into a form suitable for analysis using some form of statistical parametric mapping. Processing strategies for fMRI data are not fixed, and the particular procedures used depend on the data and the aims of the analysis. Pragmatic motivations, such as software availability and ease of use, also play a major role in determining how fMRI data are processed. This chapter focuses on the main steps that are usually applied to the data, which mostly involve various forms of image registration. The first step is usually to convert from DICOM format, to a file format that is more manageable. This is followed by motion correcting the data, which may include a distortion correction procedure. Often, there is also an anatomical scan collected for each subject, and this would be brought into alignment with the fMRI by a coregistration step. The anatomical image is sometimes useful in order to "spatially normalize" the fMRI data. The warps, needed to deform the fMRI to some standard space, can be estimated using the anatomical image. Once these warps have been estimated, they can be applied to the motion corrected fMRI data, to spatially normalize them. The final step, before statistical analysis, is usually to smooth the data.

Massimo Filippi (ed.), *fMRI Techniques and Protocols*, Neuromethods, vol. 119,
DOI 10.1007/978-1-4939-5611-1_6, © Springer Science+Business Media New York 2016

There are many variations on this sequence of operations. For example, a two-level approach for multi-subject analysis (random effects model) may be performed by generating parameter-estimate images from fMRI data that have not been spatially normalized. These parameter images could then be warped to the standard space and the statistical analysis performed on them. The smoothing step would be omitted if the statistical analysis included a model of spatial smoothness. This chapter says nothing about "slice-time correction", and assumes that the model used in the subsequent statistical analysis accounts for the fact that slices of fMRI data are not acquired simultaneously. Surface based approaches, in which the fMRI data is projected on to a representation of the cortical surface, are not covered.

2   File Format Conversion

Most MRI scanners produce image data in a format that conforms to the *DICOM Standard*. This stands for "*Digital Imaging and Communications in Medicine*", and it is the standard used in virtually all hospitals worldwide. To keep up with technological advances, the DICOM Standard is re-published approximately every year or two. The standard is also extensible, and scanner manufacturers customize file formats to suit their own particular needs. Several hundred pages of documentation describing the basic file format are available from http://dicom.nema.org/.

Most fMRI analysis is currently performed within an academic setting, where the complexity of DICOM is un-necessary. Neuroimaging analysis tools are written by scientists and engineers who wish to avoid working with complex and difficult formats. As a result, several different file formats for fMRI data arose, many of which were variants of the ANALYZE™ 7.5 format, which consists of an ".img" file containing the image data itself, plus a ".hdr" file containing various pieces of descriptive information. For a number of years, various fMRI analysis software developers used the ANALYZE format in slightly different ways, or had their own file formats. This made inter-operability among packages very difficult, which precluded the use of tools developed at one site with tools developed at another. One classic example of such problems was the different ways in which the *voxels*[1] of an image are ordered, which often caused uncertainty about the laterality of the brain.

The NIfTI-1 data format was developed in order to facilitate inter-operability among fMRI data analysis packages[2], and it was recently extended to allow much larger images to be stored using

---

[1] A voxel is a three dimensional pixel, and can be thought of as a "volume element", as opposed to a "picture element".

[2] See http://nifti.nimh.nih.gov/.

the NIfTI-2 format. Standards have been agreed on how the data format should be used, with the aim of making it easier to mix different software packages. Providing that only NIfTI compliant software is used, there should no longer be any confusion about the orientation of the brains within the images.

Images are usually treated as an array of voxels. For example, an anatomical image is generally treated as a 3D array, and most packages require this volume to be stored in a single file. DICOM usually stores each slice separately (as a series of 2D arrays), but most file format conversion routines will stack these slices together into a 3D volume. A run of fMRI data is usually considered as a 4D array, although for many procedures, it is often convenient to treat it as a time-series of 3D arrays. Some packages assume that the entire run is saved in a single file (e.g., FSL), whereas others treat the data as a series of files containing 3D volumes (e.g., SPM).

The NIfTI format allows storage on disk to be in either a left- or right-handed coordinate system. However, the format includes an implicit spatial transformation into a right-handed coordinate system. This transform maps from data coordinates (e.g., column i, row j, slice k), into some real world (x,y,z) positions in space. These positions could relate to Talairach & Tournoux (T&T) space [1], Montreal Neurological Institute (MNI) space [2, 3], or patient-based scanner coordinates. For T&T and MNI coordinates, *x* increases from left to right, *y* increases from posterior to anterior, and *z* increases in the inferior to superior direction. MRI data are usually exported from the scanner as DICOM format, which encodes the positions and orientations of the slices. When data are converted from DICOM to NIfTI format, the relevant position and orientation information can be determined from the *"Pixel Spacing"*, *"Image Orientation"* and *"Image Position"* fields of the DICOM files.

Terms such as "neurological" and "radiological convention" relate only to visualization of axial images. They are unrelated to how the data are stored on disk, or even how the real-world coordinates are represented. It is more appropriate to consider whether the real-world coordinates system is left- or right-handed. T&T use a right-handed system, whereas the storage convention of ANALYZE files is usually considered as left-handed (*x* increases from right to left). These coordinate systems are mirror images of each other, so transforming between left- and right-handed systems involves flipping, and cannot be done by rotations alone.

3   Corrections to fMRI Data

Most processing of fMRI data involves some form of spatial registration. The head of a single individual is generally considered to be fairly rigid, so the initial aim is to bring all the image volumes of each individual subject into alignment (intra-subject registration), prior to registering data from all the subjects together (inter-subject registration). Various artifact corrections may be incorporated within the intra-subject registration procedures.

3.1 Artifact Correction

There are a number of image artifacts that result from the very fast acquisition times required for fMRI. Many of these have detrimental effects if not properly modelled. Some groups have developed in-house software to improve on the algorithms supplied by scanner manufacturers for reconstructing images from the original K-space data. The objectives of these custom reconstruction algorithms include reducing Nyquist ghosting artifacts and ensuring that the model uses a better trajectory through K-space. Sites that perform their own image reconstruction require the original complex K-space data, for which there is no clearly defined DICOM standard.

The introduction of a subject into the scanner causes distortions of the magnetic field [4]. The field may be uniform when there is no subject present; but with a subject in the scanner, the field is influenced (through Maxwell's equations) by the varying magnetic susceptibilities of different tissues. These effects are especially prominent at the interface between tissue and air, resulting in (for example) drop-out and distortion in the frontal lobe in regions close to the nasal sinuses. For echo-planar images (EPI), the main effects of magnetic field inhomogeneity are spatial distortions in the phase-encoding direction of the images, and dropouts (signal loss) that arise because of through-plane de-phasing. Some of the distortions can be reduced by active shimming (changing the field of the scanner via the shim coils), or by passive shimming (introducing diamagnetic material into the orifices of the subject), but these measures only reduce the effects of distortions and dropouts, and cannot completely counteract them.

The models used for intra-subject registration of the head often assume rigid-body movement. Obtaining accurate alignment of a relatively distortion-free anatomical image with highly distorted fMRI data is not possible, unless the geometric distortions are corrected. Therefore, one of the first steps is often a correction for these distortions. Retrieving signal that is lost in dropouts is not possible, but there are a number of possible *post-hoc* approaches for correcting geometric distortions in the images.

- It is possible to compute field maps from additional scans that are normally collected just prior to the fMRI runs [4, 5] (see Fig. 1). This involves processing complex data (i.e., real and imaginary, or phase and magnitude) from these measurements in order to compute an unwrapped version of the phase. Phase measurements are in the range of $-\pi$ to π radians, or are from 0 to 2π radians. Phase unwrapping [6, 7] involves trying to add or subtract multiples of 2π to the values, such that the result is as spatially smooth as possible. With appropriate rescaling, this unwrapped phase map is converted into a voxel-displacement map for correcting the fMRI.

Unwarped EPI   Warped EPI   Fieldmap in Hz

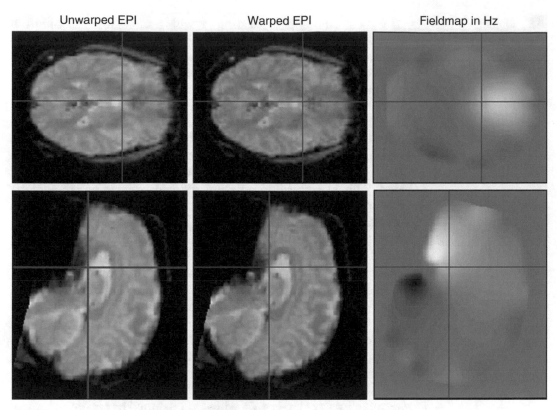

Fig. 1 Distortions in EPI can be corrected by field maps

- If the air, bone and other tissue can be segmented from the anatomical images, then it becomes possible to simulate field maps [8] by solving Maxwell's equations. Separating air from bone is quite difficult from MRI scans, as both generally appear dark. Air appears dark because of its low proton density, whereas hard tissue such as bone has a very short T2 relaxation time, so most of the signal has decayed before it is detected. Additional prior information generated from computed tomography (CT) scans is generally needed in order to attempt such segmentation [9, 10]. For this reason, the approach has not been widely adopted.

- Image registration procedures can also be used to estimate the warps that align a distorted fMRI scan with a (relatively) distortion-free anatomical image [11, 12]. Contrast differences between the images mean that some form of information-theoretic objective function is usually required, and the effects of signal dropout in the fMRI should also be taken into consideration [13].

The field map approach is generally the most accurate way to correct geometric distortions, although a correction that combines all of the above strategies into a single model is likely to be the most accurate.

3.2 Motion Correction

The most common application of within-modality registration in functional imaging is to reduce motion artifacts by realigning the volumes in the image time-series. The objective of realignment is to determine the rigid-body transformations that best align the series of functional image volumes to the same space. Blood oxygenation-level-dependent (BOLD) signal changes elicited by the haemodynamic response tend to be small compared to apparent signal differences that can result from subject movement [14]. Subject head movement in the scanner cannot be completely eliminated, so retrospective motion correction is usually performed as a processing step. This is especially important for experiments where subjects may move in a way that is correlated with the different experimental conditions. Even tiny systematic differences can result in a significant signal accumulating over numerous scans. Without suitable corrections, artifacts arising from subject movement, which correlate with the experiment, may appear as activations. A second reason why motion correction is important is that it increases sensitivity. The *t*-test is based on the signal change relative to the residual variance. The residual variance is computed from the sum of squared differences between the data and the linear model to which it is fitted; movement artifacts add to this residual variance, and so reduce the sensitivity of the test to true activations.

At its simplest, image registration involves estimating a mapping between a pair of images. One image is assumed to remain stationary (the reference image), whereas the other (the moved image) is spatially transformed to match it. In order to transform the moved image to match the reference, it is necessary to determine a mapping from the location of each voxel in the reference to a corresponding location in the moved image. The moved image is then re-sampled at the new locations. The mapping can be thought of as a function of a set of estimated transformation parameters. The shape of a human brain changes very little with head movement, so rigid-body transformations can be used to model different head positions and orientations of the same subject. Matching of two images is performed by finding the spatial transformation (mapping) that optimizes some mutual function of the images. For the case of a rigid-body transformation in three dimensions, the mapping is defined by six parameters: three translations and three rotations.

There are two steps involved in registering a pair of images together. There is the *registration* itself, whereby the set of parameters describing the mapping is estimated. Then there is the *transformation*, where one of the images is transformed according to the estimated parameters. For rigid registration, this step is often referred to as "reslicing".

Registration involves estimating the parameters of a spatial transformation that "best" match the images. The quality of the match is based on an *objective function*, which is maximized or minimized using some *optimization algorithm*. It is not computationally feasible to search over all possible parameter settings to find the global optimum of the objective function. Instead, registration is usually performed as a local optimization, which involves assigning an initial guess to the parameters, and then trying to improve the estimates in an iterative way. This involves iteratively transforming the moved image many times, using different parameter values, until the objective function can no longer be improved upon. Whether or not the final estimate is globally optimal will depend on how far the optimal solution is from the starting estimate. For this reason, packages such as SPM sometimes require images to be manually repositioned prior to performing any image registration.

Alignment of fMRI data is usually achieved by minimizing the mean squared difference between each of the images and a reference image, where the reference image could be one of the images in the series. For slightly better results, this procedure could be repeated, but instead of matching to one of the images from the series, the images would be registered to their mean (after a first-pass alignment). In general, the best objective function to use for image registration depends on what assumptions can be made about the data. In the case of the mean squared difference objective function, the assumption is that the image noise is approximately Gaussian, and does not vary over the image.

Even after rigid realignment, there may still be some motion-related artifacts remaining in the data. There are many sources of such residual artifacts, and the most obvious ones are:

- Interpolation error from the re-sampling algorithm [15] used to transform the images can be a source of motion-related artifacts. For this reason, and also for speed and efficiency, some registration algorithms use a Fourier interpolation method [16, 17].

- When MR images are reconstructed, the final images are usually the modulus of the initially complex data. This results in voxels that should be negative being rendered positive. This has implications when the images are re-sampled, because it leads to errors at the edge of the brain that cannot be corrected, irrespective of how accurate the interpolation method is. Possible ways to circumvent this problem are to work with complex data, or apply a low-pass filter to the complex data before taking the modulus.

- The sensitivity (slice-selection) profile of each slice also plays a role in introducing artifacts [18]. Gaps between slices are difficult to deal with as it is not possible to recover information that was not actually acquired.

- fMRI images are spatially distorted, and the amount of distortion depends partly upon the position of the subject's head within the magnetic field. Interactions between image distortion and the orientation of a subject's head in the scanner can also cause other problems because purely rigid alignment does not take this into account. Relatively large subject movements result in the brain images changing shape, and these shape changes cannot be corrected by a rigid-body transformation alone. The interaction between image distortion and head orientation illustrates a limitation of conceptualizing processing as the application of a series of tools to the data. These issues are better resolved by a generative model that combines both a model for image distortions, and a model of subject motion [19].

- Each volume of a series of fMRI data is currently acquired a plane at a time over a period of about a second. Subject movement between acquiring the first and last plane of any volume is another reason why the image volumes may not strictly obey the rules of rigid-body motion [20]. A better model would allow each slice to move separately—but it may lead to technical problems if there was too much movement. For example, the model would allow some points in the brain to be scanned more than once during the acquisition of a volume, and some points not to be scanned at all. Filling in the appropriate values in the corrected images is difficult if there is no actual data to sample.

- After a slice is magnetized, the excited tissue takes time to recover to its original state, and the amount of recovery that has taken place will influence the intensity of the tissue in the image. This effect can be seen in the first few scans of an fMRI time series, and is the reason why a few "dummy scans" are collected at the start of an fMRI run in order for the intensities to stabilize. Out of plane movement will result in a slightly different part of the brain being excited during each repeat. This means that the spin-excitation will vary in a way that is related to head motion, and so leads to more movement related artifacts [21].

- Nyquist ghost artifacts in MR images do not obey the same rigid-body rules as the head, so a rigid rotation to align the head will not mean that the ghosts are aligned. The same also applies to other image artifacts, such as those arising due to chemical shifts.

- The accuracy of the estimated registration parameters is normally in the region of tens of μm. This is dependent upon many factors, including the effects just mentioned. Even the signal changes elicited by the experiment can have a slight effect (a few μm) on the estimated parameters [22], so this in turn may have consequences in terms of how significant differences are interpreted

These problems cannot be corrected by simple rigid-body realignment, and so may be sources of stimulus correlated motion artifacts. Systematic movement artifacts resulting in a signal change of only one or two percent can lead to highly significant false positives[3] over an experiment with hundreds of scans. This is especially

---

[3] These would actually be Type III errors, rather than Type I, because the null hypothesis would be correctly rejected (there is a statistically significant effect in the data) but for the wrong reason (the effect is due to motion, rather than BOLD signal).

important for experiments where some conditions may cause slight head movements (such as motor tasks, or speech), because these movements are likely to be highly correlated with the experimental design. In cases like this, it is difficult to separate true activations from stimulus correlated motion artifacts. All that can be concluded is whether or not there is a difference among the data. The specific causes of any difference remain unknown, but it is generally hoped that they relate to BOLD signal changes. Providing there are enough images in the series and the movements are small, some of these artifacts can be removed during the subsequent statistical analysis by regressing out any signal that is correlated with functions of the estimated movement parameters [21]. However, when the estimates of the movement are related to the experimental design, it is likely that much of the interesting BOLD signal will also be regressed out of the data. For retrospective motion correction, these issues remain unresolved, and will remain unresolved until interactions among processing steps (including the statistical analysis) are properly modeled.

Prospective motion correction approaches are now becoming more practical, and their application in fMRI studies is increasingly widespread. These methods involve tracking the motion of the subject in the scanner, and adjusting the data acquisition accordingly. A recent review of this subject area can be found in [23].

4   Inter-Modality Registration

For studies of a single subject, sites of activation can be localized more clearly by superimposing them on a high-resolution anatomical (structural) image of the subject (typically a T1-weighted MRI). This requires registration of the functional images with the anatomical image. A further use for this registration is that a more precise spatial normalization can be achieved by estimating the requisite warps from a more detailed anatomical image. If the functional and anatomical images are in register, then a warp estimated from the anatomical image can also be applied to the functional images. In practice, this requires the usual geometric distortions found in fMRI to have been corrected.

As in the case of movement correction, this registration is normally performed by optimizing a set of parameters describing a rigid-body transformation, but the matching criterion needs to be more complex because the anatomical and functional images normally have very different patterns of intensity. A simple mean-squared difference model will no longer be effective, so alternative objective functions (similarity measures) are needed. Inter-modal registration approaches initially involved the use of landmarks, which were manually defined on the images. The images were registered by bringing the landmarks into alignment. An early automated approach based on a similarity measure between images was the AIR (automated image registration) algorithm [24]. The objective function was obtained by dividing the intensities of one image into a number of bins. For the voxels associated with each bin, the idea was to minimize the variance of the corresponding voxel intensities of the other image. The algorithm was originally intended for registering positron emission tomography (PET) and anatomical MRI, and worked well—providing the MRI had non-brain tissue removed.

Many of the more recent similarity measures used for inter-modal (as well as intra-modal [25]) registration are based on *information theory*. These measures are based on joint probability distributions of intensities in the images, usually discretely represented in the form of 2D joint histograms, which are normalized to sum to one. The most commonly used measure of image alignment is *mutual information* (MI) [26, 27] (also known as *Shannon information*). MI is a measure of dependence of one image on the other, and can be considered as a distance (*Kullback-Leibler divergence*) between the joint distribution and the equivalent distribution assuming complete independence. Another perspective is that MI is a measure of the reduction of uncertainty about one image given the other. Registration algorithms work under the assumption that the MI between the images is maximized when they are in register (Fig. 2). A number of other information theoretic measures have since been devised [28, 29], and a more complete review of information theoretic image registration approaches is given by [30]. A number of inter-modality registration algorithms have been thoroughly evaluated on the same data [31], and alignment accuracy is generally found not to be as high as that obtained by within modality registration.

Artifacts in MRI can lead to problems when attempting to align images using information theoretic approaches. In particular, intensity nonuniformity artifacts (also known as "bias" or "inhomogeneity") can severely degrade the accuracy of image registration procedures [32]. There are a number of intensity inhomogeneity correction methods, which are typically based on information theoretic measures [33, 34]. It would seem natural to see image registration and inhomogeneity correction combined into a common information theoretic framework.

5   Spatial Normalization

Currently, the main application for deformable image registration within imaging neuroscience is the procedure known as *spatial normalization*. This involves warping the brain images from different subjects in a study into roughly the same standard space to allow signal averaging across subjects. In functional imaging studies, spatial normalization is useful for determining what happens generically

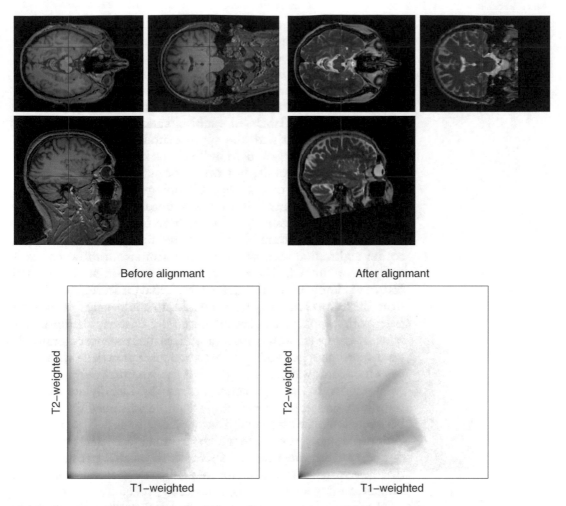

Fig. 2 The *top row* shows orthogonal sections of two MR images of different contrasts. Below this are joint intensity histograms of the image pair, both before and after image registration (note that the pictures show $\log(1 + N)$, where N is the count in each histogram bin)

over individuals. A further advantage is that activation sites can be reported according to their Euclidean coordinates within a standard coordinate system [35]. The most commonly adopted coordinate system within the brain imaging community is that described by Talairach and Tournoux [1], although new standards are emerging, which are based on digital atlases [2, 3, 36].

This section provides a brief overview of the ideas underlying deformable image registration. This is a large area of research, but a more comprehensive review may be found in [37, 38]. The previous sections described rigid-body approaches for registering brain images of the same subject, where it was assumed that there are no differences among the shapes of the brains. This is often a reasonable assumption to make for intra-subject registration, but is not appropriate for aligning brain images of different subjects. In addition to estimating an unknown pose and position, inter-subject registration approaches also need to model the different sizes and shapes of the subjects' heads and brains.

Methods of spatially normalizing images can be divided broadly into *label based* and *intensity based*. Label based techniques involve identifying features (labels) in a subject's image, and then bringing these into alignment with the appropriate location in some atlas. The original strategy proposed by Talairach and Tournoux involved matching discrete points, but other forms of label could also be used, such as lines or surfaces. Homologous features are often identified manually, but this process is time consuming and subjective. Another disadvantage of using points as landmarks is that there are very few readily identifiable discrete points in the brain, so the registration accuracy in regions away from those points is likely to be limited. The required transformation at the defined features is known, but the deforming behavior in regions distant from the features can only be estimated, so it is usually forced to be as smooth as possible. There are a number of interpolation methods that ensure smooth spatial transforms, but the most commonly used approaches generally involve some form of radially symmetric basis functions, which are centered at the landmarks.

Intensity based approaches operate by identifying a spatial transformation that optimizes some voxel-similarity measure between template data and the subject's image. The template defines the standard space to which all the subjects' data are warped during spatial normalization. Typically, the spatial transformation that best matches the template to a subject's anatomical image is estimated using an iterative optimization procedure.

Image registration uses a mathematical model to explain the data. Such a model will contain a number of unknown parameters that describe how an image is deformed, or warped. The objective is usually to determine the best possible values for these parameters by optimizing some objective function; in other words, to find the single most probable deformation, given the data. In such cases, the objective function can be considered as a measure of this probability. A key element of probability theory is Bayes' theorem:

$$p(\theta \mid \mathbf{D}) = p(\mathbf{D} \mid \theta) p(\theta) / p(\mathbf{D})$$

This *posterior probability* of the parameters, given the image data $(p(\theta|\mathbf{D}))$ is proportional to the probability of the image data given the parameters $(p(\mathbf{D}|\theta)$—the *likelihood*), times the *prior probability* of the parameters $(p(\theta))$. The probability of the data $(p(\mathbf{D}))$ is treated as a constant because the data are fixed and known. The objective is usually to find the most probable parameter values, and not the actual probability density, so this factor can be ignored. The most probable set of values for the parameters is known as the *maximum a posteriori* (MAP) estimate. In practice, the objective function is normally the logarithm of the posterior probability (in which case it is maximized) or the negative logarithm (which is minimized). The objective function can therefore be considered as the sum of two terms: a likelihood term, and a prior term:

$$-\log p(\theta,\mathbf{D}) = -\log p(\mathbf{D}\,|\,\theta) - \log p(\theta)$$

The likelihood term is a measure of the probability of observing an image given some set of model parameters. A simple example would be where an image is modeled as a warped version of a template image, but with Gaussian random noise added. Such a model reduces to minimizing the sum of squared differences between the image and warped template. It is possible that parts of the image correspond to a region that falls outside the field of view of the template, so it is usual to simply use the mean-squared difference in the overlapping region.

The prior term reflects the prior probability of a deformation occurring—effectively biasing the deformations to be realistic. If one considers a model whereby each voxel can move independently in three dimensions, then there would be three times as many parameters to estimate as there are observations. This would simply not be achievable without *regularizing* the parameter estimation by modeling a prior probability [39].

Registration is usually considered as an optimization procedure, which involves searching for the model parameters that maximize or minimize the objective function. If the registration is based on matching landmarks together, then it is often possible to register the images in a single step, because the problem can be solved by a single matrix inversion. In contrast, if image registration is based on matching intensities, then some form of iterative scheme is required. These procedures are usually very susceptible to poor starting estimates, so a number of hybrid approaches have emerged [40, 41] that combine intensity based methods with feature matching (typically sulci). Registration methods usually attempt to find the single most probable realization of all possible transformations. Robust methods that almost always find the global optimum would take an extremely long time to run with a model that uses millions of parameters, so these methods are simply not feasible for problems of this scale. However, if sulci and gyri can be labeled easily from the brain images, then these features can be used to bias the registration, therefore increasing the likelihood of obtaining a more globally optimal solution.

In practice, the parameters describing the spatial transformations, which map between a subject's images and a standard coordinate system, are usually estimated by matching a template with an anatomical (typically T_1-weighted) image. Providing the fMRI data are in accurate alignment with the anatomical image, then the spatial transformation that would warp them to the standard coordinate system is also known. Spatially normalized versions of the fMRI data can simply be created using the same set of estimated parameters.

5.1 Matching Criteria

The matching criterion is often based upon minimizing the mean-squared differences or maximizing the correlation between the image and template. For this criterion to be successful, it requires the individual's image to have the visual appearance of a warped version of the template. In other words, there must be correspondence in the gray levels of the different tissue types between the images. The mean-squared difference objective function makes a number of assumptions. If the image data do not meet these assumptions, then the objective function may not accurately reflect the goodness of fit, and the estimated deformations will be suboptimal. Under some circumstances, it may be better to weight different regions to a greater or lesser extent. For example, when spatially normalizing a brain image containing a lesion, the mean squared difference around the lesion should contribute little or nothing to the objective function [42]. This is currently achieved by assigned lower weights for the matching criterion in these regions, so that they have much less influence on the final solution.

In addition to modeling geometric deformations of the template, there may also be extra parameters within the model that describe intensity variability. A very simple example would be the inclusion of an additional intensity scaling parameter, but the models can be much more complicated. There are many possible objective functions, each making a different assumption about the data and requiring different parameterizations of the template intensity distribution. For example, matching can be based on feature vectors derived from the images [43], or can rely on some information theoretic model [44]. There is no single universally best criterion to use for all data.

Often, it is necessary to process the anatomical images prior to any attempt to align them with a template. This may involve stripping off non-brain tissue from the image, which can improve the accuracy with which the brains themselves are registered. Because the interesting signal predominantly arises in gray matter, another strategy for increasing spatial normalization accuracy is to simply spatially normalize the data by aligning gray matter with a gray matter template image. A slightly better approach would involve simultaneously matching gray with gray and white with white. There are a number of readily available tissue segmentation algorithms that can be used for identifying gray matter in brain MRI.

Another strategy that can be of great benefit for increasing registration accuracy is the correction of intensity nonuniformity artifact [33, 34], which would otherwise prevent accurate alignment. Such correction algorithms may be standalone [45], or they may be incorporated within tissue segmentation procedures [46, 47]. Some deformable registration procedures explicitly incorporate bias correction [48], but recent developments combine bias correction, warping and tissue segmentation into the same model [49–53]. Within such unified generative models, the registration and bias correction inform the tissue segmentation, and the tissue segmentation informs the registration and bias correction.

Currently, most spatial normalization algorithms use only a single image from each subject, which is typically T_1-weighted. Such images only really delineate different tissue types. Further information that may help the registration could be obtained from other data such as diffusion weighted images [54]. These provide anatomical information more directly related to connectivity and implicitly function, possibly leading to improved registration of functionally specialized areas. Matching diffusion images of a pair of subjects together is likely to give different deformation estimates than would be obtained through matching T_1 weighted images of the same subjects. The only way to achieve an internally consistent match is through performing the registrations simultaneously, within the same model. Similarly, the patterns of bold signal across subjects could, in principle, be used to drive the registration—although a naïve implementation of such an approach may cause problems for interpreting group results. We are now beginning to see an interest in driving deformable brain registration using resting-state data (e.g., [55]).

The choice of template data used for spatial normalization is important. It is sometimes tempting to base a template on the brain of a single individual, but such a procedure would produce different results depending upon the choice of whose brain was used. One could consider an optimal template being some form of average [56–58]. On average, registering such a template with a brain image generally requires smaller (and therefore less error prone) deformations. Such averages generally lack some of the detail present in the individual subjects. Structures that are more difficult to match are generally slightly blurred in the average, whereas the structures that can be more reliably matched are sharper. Such an average generated from a large population of subjects would be ideal for use as a general purpose template. Another reason for using a template that better represents the study population is that it does not bias the results more towards some brain regions than others. During spatial normalization of a brain image, some regions need to be expanded and other regions need to contract in order to match. If some brain structure is especially small in the template, then this region will be contracted in the brains in the study, leading to a systematic reduction in the amount of BOLD signal detected from it.

5.2 Deformation Models

At its simplest, deformable image registration involves estimating a smooth, continuous mapping between the points in one image and those in another. This mapping allows one image to be re-sampled so that it is warped (deformed) to match another (see Fig. 3). There are many ways of modeling such mappings, but these fit into two broad categories [59].

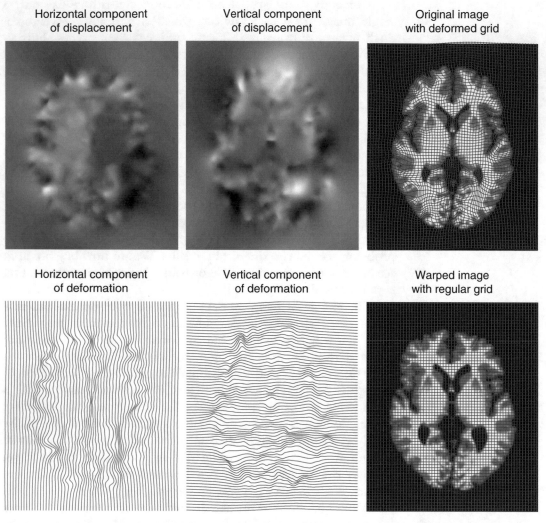

Fig. 3 2D displacements generated from two scalar fields. The first two panels pf the *top row* show displacements represented as *images*. Below these are different representations of these components. The deformation field resulting from combining the components is overlaid on the top-right image, in order to deform the image as shown at the *bottom-right*

- The *small-deformation* framework does not necessarily preserve topology[4]—although if the deformations are relatively small, then there may still be a one-to-one mapping between the images. This framework usually models deformations by a smooth displacement field.

- The *diffeomorphic* framework generates deformations that have a number of useful properties, such as enforcing the preservation of topology [60]. Within this framework, deformations are parameterized in terms of smooth velocity fields.

Images can be treated as continuous functions of space. Reading the value at some arbitrary point involves interpolating between the original voxels. For many interpolation methods, the functions are parameterized by linear combinations of basis functions, such as B-spline bases, centered at each original voxel. Similarly, the deformations themselves can also be parameterized by linear combinations of smooth, continuous basis functions.

A potentially enormous number of parameters are required to describe the deforming transformations that align two images (i.e., the problem can be very high-dimensional). However, much of the spatial variability can be captured using just a few parameters. Sometimes only an affine transformation is used to approximately register images of different subjects. This accounts for differences in position, orientation and overall brain dimensions, and often provides a good starting point for higher-dimensional registration models.

Some of the more primitive deformable registration algorithms encode displacements via a linear combination of a relatively small numbers of basis functions. One such approach is part of the AIR package [61, 62], which uses polynomials (see Fig. 4) to model shape variability. Other models parameterize a displacement field, which is added to an identity transform. Families of basis functions for such models include Fourier bases [63], sine and cosine transform basis functions, which were used by early versions of the Statistical Parametric Mapping (SPM) software [64] (see Fig. 4). These models involve in the order of about 1000 parameters, and only permit the global head or brain shape to be modeled.

Radial basis functions are another family of parameterizations, which are often used in conjunction with an affine transformation. Each radial basis function is centered at some point and the amplitude is then a function of the distance from that point. Thin-plate splines are one of the most widely used radial basis functions for image warping and are especially suited to manual landmark matching [65, 66]. The landmarks may be known, but interpolation is

---

[4] The word "topology" is used in the same sense as in "Topological Properties of Smooth Anatomical Maps" [15]. If spatial transformations are not one-to-one and continuous, then the topological properties of different structures can change.

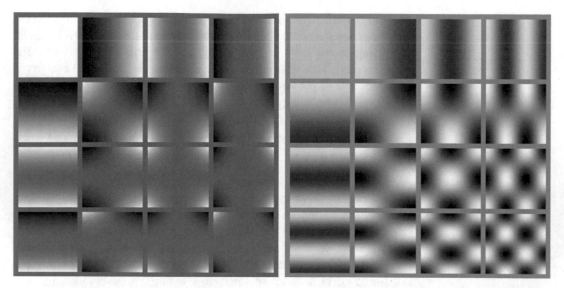

Fig. 4 Polynomial basis functions (*left*) and Cosine transform basis functions (*right*). Horizontal and vertical (and through-plane for 3D) displacement fields may be modeled by linear combinations of such functions (see Fig. 3)

needed in order to define the mapping between these known points. By modeling it with thin-plate splines, the mapping function has the smallest bending energy. Other choices of basis function reduce other energy measures, and these functions relate to the convolution filters that are sometimes used for image matching [67, 68].

B-spline bases are also used for parameterizing displacements [11, 69, 70] (see Fig. 5). They are related to the radial basis functions in that they are centered at discrete points, but the amplitude is the product of functions of distance in the three orthogonal directions (i.e., they are *separable*). The separability and local support of these functions confers certain advantages in terms of being able to rapidly generate displacement fields through a convolution-like procedure. Very detailed displacement fields can be generated by modeling an individual displacement at each voxel. This may not appear to be a basis function approach, but the assumptions within such models are often that the fields are tri-linearly interpolated. This is the same as a first degree B-spline basis function model.

Regularization is generally based on some measure of deformation smoothness. Smoother deformations are deemed to be more probable—a priori—than deformations containing a great deal of detailed information. The regularization term (prior term) of the objective function is often thought of as an "*energy density*". Commonly used forms for this are the *membrane energy, bending energy* or *linear-elastic energy*. The form of the prior used by the registration will influence the estimated deformations (see Fig. 6).

Small deformation approaches do not guarantee that the estimated warps will be one-to-one and invertible. It is easy to

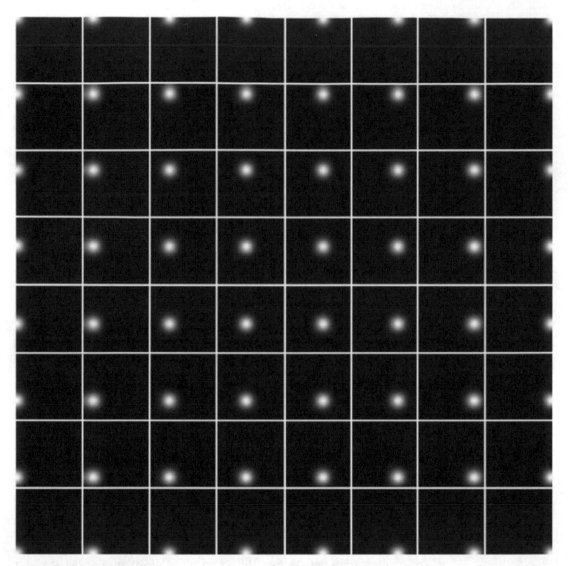

Fig. 5 B-spline basis functions allow more detailed warps to be estimated than polynomial or cosine transform basis functions

introduce folding within this simple setting, where a point in one image may appear to align with two or more points in the other. For more biologically plausible results, it is useful to constrain the warps to be one-to-one by working within a *diffeomorphic*[5] setting. The key concept of this framework is that the deformations are generated by the composition of a series of much smaller deformations (i.e., warped warps). For deformations, the composition operation is

---

[5] A diffeomorphism is a globally one-to-one (bijective) smooth and continuous mapping with derivatives that are invertible (i.e., non-zero Jacobian determinant).

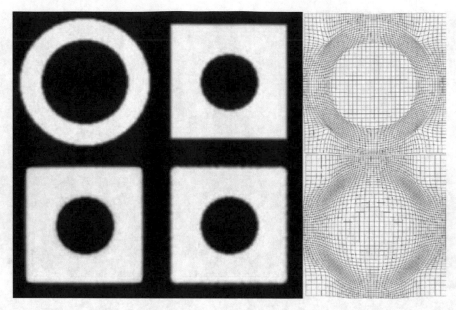

Fig. 6 This figure illustrates the effect of different types of regularization. The *top row on the left* shows simulated 2D images of a circle and a square. Below these is the circle after it has been warped to match the square, using both membrane and bending energy priors. These warped images are almost visually indistinguishable, but the resulting deformation fields using these different priors are quite different. These are shown on the *right*, with the deformation generated with the membrane energy prior shown above the deformation that used the bending energy prior

achieved by re-sampling one deformation field by another. Providing the constituent deformations are small enough, then they are likely to be one-to-one. A composition of a pair of one-to-one mappings will produce a new mapping that is also one-to-one. Multiple nesting can be achieved, so that large one-to-one deformations can be obtained from the composition of many very small ones. From a mathematical perspective, diffeomorphic deformations are the result of integrating differential equations over a unit of time, in which the deformations are a function of smooth continuous velocity fields. The composition of a series of small deformations can be viewed as an *Euler integration* of these differential equations.

Early diffeomorphic registration approaches were based on the *greedy "viscous fluid"* registration method of Christensen and Miller [71, 72], which models one image as it "flows" to match the shape of the other. These methods can account for large deformations while ensuring that the topology of the warped image is preserved. Their disadvantage is that they are not formulated to find the smoothest deformation. More recent algorithms for large deformation registration do aim to find the smoothest solution. For example, the *LDDMM* (Large Deformation Diffeomorphic Metric Mapping) algorithm [73] does not fix the deformation parameters once they have been estimated. It continues to update them such that the objective function is properly optimized. Such approaches essentially parameterize the model by velocities, and compute the deformation as the medium warps over unit time. A number of other variations on the diffeomorphic framework have emerged in recent years. These include the very popular *ANTS/Syn* software [74], a number of algorithms that compute deformations by applying a "scaling and squaring" procedure to a single velocity field [75–78], as well as some approaches that try to replicate the deformations of LDDMM using a "geodesic shooting" approach [79–81].

Optimization problems for complex nonlinear models, such as those used for image registration, can easily get caught in local optima; so there is no guarantee that the estimate determined by the algorithm is globally optimal. If the starting estimates are sufficiently close to the global optimum, then a local optimization algorithm is more likely to find the globally optimal solution. Therefore, the choice of starting parameters can influence the accuracy of the final registration result. One method of increasing the likelihood of achieving a good solution is to gradually reduce the amount of regularization. Registration is first performed using heavy regularization. Once this solution is found, then the procedure is repeated using less regularization, and so on. This has the effect of making the registration estimate the more global deformations before estimating more detailed ones. The images could also be smoother for the earlier iterations in order to reduce the amount of confounding information and the number of local optima. A review of such approaches can be found in [82].

The accuracy of a number of deformable registration methods has been assessed using T1-weighted brain MRI, using manually labelled brain structures as a ground truth with which to compare [83]. The general trend was that those registration methods with most flexibility (i.e., lots of parameters) tended to outperform those using relatively few parameters. Figure 7 shows the average of 550 T1-weighted MRI scans, which have been spatially normalized using a relatively flexible registration approach.

6 Smoothing

Usually, the final step of the processing pipeline is to smooth the images, which involves convolving the data with a three dimensional Gaussian kernel (see Fig. 8). The amount of smoothing is defined by the *full width at half maximum* (*FWHM*) of the smoothing kernel. A broader FWHM produces smoother results, and the choice of FWHM is determined by many factors.

More smoothing is usually used prior to statistical analyses of group studies, than would be used for studies of single individuals. Inter-subject registration is generally less accurate than the rigid-body

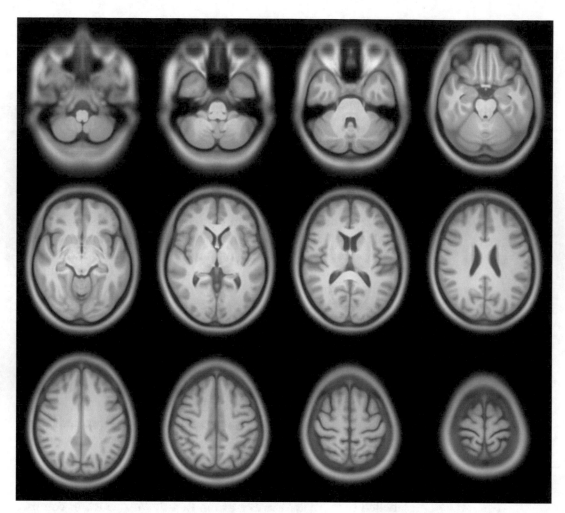

Fig. 7 An average of 550 spatially normalized T$_1$-weighted images

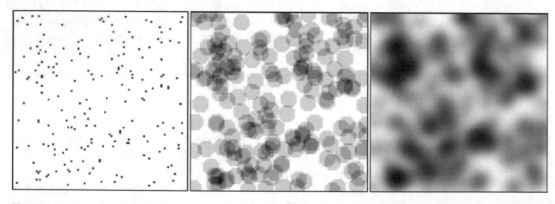

Fig. 8 This figure shows the effect of convolving the image on the *left* with different kernels. In the *center*, the image has been convolved with a circular kernel. The result is an image in which each pixel is the average of the values from the original image within the radius of the kernel. At the *right* is a result from convolving with a Gaussian kernel. Pixels in this image are weighted averages, where the weights depend on the distance from the center of the kernel

registration that is used within subject. If homologous functional regions in different subjects are not well aligned, then the activations will appear in different places in different subjects. Smoothing is used in order to increase the amount of overlap across subjects, and so increase the significance of the results. The optimal amount of smoothing therefore depends upon the alignment accuracy.

Less smoothing would typically be used for single-subject analyses. There are various reasons for smoothing such data, but the main one relates to the spatial frequencies of the interesting signal, compared to the noise. If the noise contains proportionally more high frequency than the signal, then it makes sense to remove some of the high frequencies by smoothing. Another reason for smoothing is that it reduces the effective number of independent statistical tests that are performed. This can lead to greater sensitivity in results that are corrected for multiple comparisons. The disadvantage would be that localization of activations is less precise.

Some of the more recent approaches for the analysis of fMRI data involve spatiotemporal models of activation. Such analyses do not require the data to be smoothed, as the models themselves deal with this issue.

7   Summary and Conclusions

The current fMRI data analysis paradigm involves applying a pipeline of tools to the data. One procedure is applied to the images, to produce some output. Then another procedure is applied to the output of that, and so on. The end result is a version of the data that has been massaged into a form suitable for applying simple statistical tests to. A number of centers have developed pipeline environments [84–86] to facilitate such processing streams.

An alternative and more principled approach is to consider a full model of how the data could have arisen. Such a *generative model* would involve components for modeling the physics of the scanner, the motion of the subjects, the brain-shape variability among the population, and models for how the experiment elicits changes in the data. It would then be used to model the raw data, in order to make the kinds of inferences in which neuroscientists are interested. A full model for everything is a long way off, but scientists are making progress in terms of simulating fMRI data in individual subjects [87]. Once a simulation model can be made, then it is simply a matter of inverting it, to make the necessary inferences. Such models usually include a number of parameters that influence the data, but are of no interest in themselves. Accurate model inversion would require the effects of these uninteresting variables to be "integrated out", which is not a straightforward procedure and is an area that occupies much of current methodological research. Choosing an optimal model for a data-set would essentially be a form of *model selection*, and could be done empirically by determining which has the greatest supporting evidence.

Many of the issues that concern users are related to the un-modeled interactions that arise through the sequential application of tools to the data. For example, when motion correction is applied to an fMRI time series, the interesting signal elicited by the experiment will influence how the algorithm estimates subject motion. Similarly, interactions between image artifacts and subject movement can also lead to problems, which can only be reduced by including artifact correction within the motion correction. There are many other examples of where combining procedures could produce similar benefits.

References

1. Talairach J, Tournoux P (1988) Coplanar stereotaxic atlas of the human brain. Thieme Medical, New York

2. Evans AC, Collins DL, Milner B (1992) An MRI-based stereotactic atlas from 250 young normal subjects. Soc Neurosci Abstr 18:408

3. Evans AC, Collins DL, Mills SR, Brown ED, Kelly RL, Peters TM. 3D statistical neuroanatomical models from 305 MRI volumes. In: Proc IEEE Nuclear Science Symposium and Medical Imaging Conference. 1993. pp 1813–1817

4. Jezzard P, Clare S (1999) Sources of distortion in functional MRI data. Hum Brain Mapp 8(2):80–85

5. Hutton C, Bork A, Josephs O, Deichmann R, Ashburner J, Turner R (2002) Image distortion correction in fMRI: a quantitative evaluation. Neuroimage 16(1):217–240

6. Cusack R, Papadakis N (2002) New robust 3-D phase unwrapping algorithms: application to magnetic field mapping and undistorting echo-planar images. Neuroimage 16(3):754–764

7. Jenkinson M (2003) Fast, automated, N-dimensional phase-unwrapping algorithm. Magn Reson Med 49(1):193–197

8. Jenkinson M, Wilson J, Jezzard P (2004) A perturbation method for magnetic field calculations of non-conductive objects. Magn Reson Med 52(3):471–477

9. Poynton C, Jenkinson M, Whalen S, Golby AJ, Wells W III (2008) Fieldmap-free retrospective registration and distortion correction for EPI-based functional imaging. In: Proc Medical Image Computing and Computer-Assisted Intervention (MICCAI). Springer, Berlin-Heidelberg, pp 271–279

10. Poynton C, Jenkinson M, Wells W III (2009) Atlas-based improved prediction of magnetic field inhomogeneity for distortion correction of EPI data. In: Proc Medical Image Computing and Computer-Assisted Intervention (MICCAI). Springer, Berlin, pp 951–959

11. Studholme C, Constable RT, Duncan JS (2000) Accurate alignment of functional EPI data to anatomical MRI using a physics-based distortion model. IEEE Trans Med Imaging 19(11):1115–1127

12. Kybic J, Thévenaz P, Nirkko A, Unser M (2000) Unwarping of unidirectionally distorted EPI images. IEEE Trans Med Imaging 19(2):80–93

13. Li Y, Xu N, Fitzpatrick JM, Morgan VL, Pickens DR, Dawant BM (2007) Accounting for signal loss due to dephasing in the correction of distortions in gradient-echo EPI via nonrigid registration. IEEE Trans Med Imaging 26(12):1698–1707

14. Hajnal JV, Mayers R, Oatridge A, Schwieso JE, Young JR, Bydder GM (1994) Artifacts due to stimulus correlated motion in functional imaging of the brain. Magn Reson Med 31:289–291

15. Thévenaz P, Blu T, Unser M (2000) Interpolation revisited. IEEE Trans Med Imaging 19(7):739–758

16. Eddy WF, Fitzgerald M, Noll DC (1996) Improved image registration by using Fourier interpolation. Magn Reson Med 36:923–931

17. Cox RW, Jesmanowicz A (1999) Real-time 3D image registration for functional MRI. Magn Reson Med 42:1014–1018

18. Noll DC, Boada FE, Eddy WF (1997) A spectral approach to analyzing slice selection in planar imaging: optimization for through-plane interpolation. Magn Reson Med 38:151–160

19. Andersson JLR, Hutton C, Ashburner J, Turner R, Friston KJ (2001) Modeling geometric deformations in EPI time series. NeuroImage 13:903–919

20. Bannister PR, Brady JM, Jenkinson M (2007) Integrating temporal information with a non-rigid method of motion correction for functional magnetic resonance images. Image Vis Comput 25(3):311–320

21. Friston KJ, Williams S, Howard R, Frackowiak RSJ, Turner R (1996) Movement-related effects in fMRI time-series. Magn Reson Med 35:346–355

22. Freire L, Mangin JF (2001) Motion correction algorithms of the brain mapping community create spurious functional activations. In: Insana MF, Leahy RM (eds) Proc Information Processing in Medical Imaging (IPMI), vol 2082, Lecture Notes in Computer Science. Springer, Berlin, pp 246–258

23. Maclaren J, Herbst M, Speck O, Zaitsev M (2013) Prospective motion correction in brain imaging: a review. Magn Reson Med 69(3):621–636

24. Woods RP, Mazziotta JC, Cherry SR (1993) MRI-PET registration with automated algorithm. J Comput Assist Tomogr 17:536–546

25. Holden M, Hill DLG, Denton ERE, Jarosz JM, Cox TCS, Rohlfing T, Goodey J, Hawkes DJ (2000) Voxel similarity measures for 3D serial MR brain image registration. IEEE Trans Med Imaging 19(2):94–102

26. Collignon A, Maes F, Delaere D, Vandermeulen D, Suetens P, Marchal G (1995) Automated multi-modality image registration based on information theory. In: Bizais Y, Barillot C, Di Paola R (eds) Proc Information Processing in Medical Imaging (IPMI). Kluwer Academic, Dordrecht, pp 263–274

27. Wells WM III, Viola P, Atsumi H, Nakajima S, Kikinis R (1996) Multi-modal volume registration by maximisation of mutual information. Med Image Anal 1(1):35–51

28. Maes F, Collignon A, Vandermeulen D, Marchal G, Seutens P (1997) Multimodality image registration by maximisation of mutual information. IEEE Trans Image Process 16:187–197

29. Studholme C, Hill DLG, Hawkes DJ (1999) An overlap invariant entropy measure of 3D medical image alignment. Pattern Recogn 32:71–86

30. Pluim JPW, Maintz JBA, Viergever MA (2003) Mutual-information-based registration of medical images: a survey. IEEE Trans Med Imaging 22(8):986–1004

31. West J, Fitzpatrick JM, Wang MY, Dawant BM, Maurer CR, Kessler RM, Maciunas RJ, Barillot C, Lemoine D, Collignon A, Maes F, Suetens P, Vandermeulen D, van den Elsen PA, Napel S, Sumanaweera TS, Harkness B, Hemler PF, Hill DLG, Hawkes DJ, Studholme C, Maintz JBA, Viergever MA, Malandain G, Pennec X, Noz ME, Maguire GQ, Pollack M, Pelizzari CA, Robb RA, Hanson D, Woods RP. Comparison and evaluation of retrospective intermodality brain image registration techniques. J Comput Assist Tomo 1997;21:554–566.

32. Gonzalez-Castillo J, Duthie KN, Saad ZS, Chu C, Bandettini PA, Luh WM (2013) Effects of image contrast on functional MRI image registration. NeuroImage 67:163–174

33. Hou Z (2006) A review on MR image intensity inhomogeneity correction. Int J Biomed Imag 2006, Article ID 49515, 11 pages. doi:10.1155/IJBI/2006/49515

34. Vovk U, Pernus F, Likar B (2007) A review of methods for correction of intensity in homogeneity in MRI. IEEE Trans Med Imaging 26(3):405–421

35. Fox PT (1995) Spatial normalization: origins, objectives, applications, and alternatives. Hum Brain Mapp 3:161–164

36. Mazziotta JC, Toga AW, Evans A, Fox P, Lancaster J (1995) A probablistic atlas of the human brain: theory and rationale for its development. NeuroImage 2:89–101

37. Sotiras A, Davatzikos C, Paragios N (2013) Deformable medical image registration: a survey. IEEE Trans Med Imaging 32(7):1153–1190

38. Oliveira FP, Tavares JMR (2014) Medical image registration: a review. Comput Methods Biomech Biomed Engin 17(2):73–93

39. Rohlfing T (2012) Image similarity and tissue overlaps as surrogates for image registration accuracy: widely used but unreliable. IEEE Trans Med Imaging 31(2):153–163

40. Auzias G, Colliot O, Glaunes JA, Perrot M, Mangin JF, Trouvé A, Baillet S (2011) Diffeomorphic brain registration under exhaustive sulcal constraints. IEEE Trans Med Imaging 30(6):1214–1227

41. Du J, Younes L, Qiu A (2011) Whole brain diffeomorphic metric mapping via integration of sulcal and gyral curves, cortical surfaces, and images. NeuroImage 56(1):162–173

42. Brett M, Leff AP, Rorden C, Ashburner J (2001) Spatial normalization of brain images with focal lesions using cost function masking. NeuroImage 14(2):486–500

43. Shen D, Davatzikos C (2002) HAMMER: hierarchical attribute matching mechanism for elastic registration. IEEE Trans Image Process 21(11):1421–1439

44. D'Agostino E, Maes F, Vandermeulen D, Suetens P (2004) Non-rigid atlas-to-image registration by minimization of class-conditional image entropy. In: Barillot C, Haynor DR, Hellier P (eds) Proc Medical Image Computing and Computer-Assisted Intervention (MICCAI), vol 3216, Lecture notes in computer science. Springer, Berlin-Heidelberg, pp 745–753

45. Sled JG, Zijdenbos AP, Evans AC (1998) A non-parametric method for automatic correction of intensity non-uniformity in MRI data. IEEE Trans Med Imaging 17(1):87–97

46. Wells WM III, Grimson WEL, Kikinis R, Jolesz FA (1996) Adaptive segmentation of MRI data. IEEE Trans Med Imaging 15(4):429–442

47. van Leemput K, Maes F, Vandermeulen D, Suetens P (1999) Automated model-based bias field correction of MR images of the brain. IEEE Trans Med Imaging 18(10):885–896

48. Studholme C, Cardenas V, Song E, Ezekiel F, Maudsley A, Weiner M (2004) Accurate template-based correction of brain MRI intensity distortion with application to dementia and aging. IEEE Trans Med Imaging 23(1):99–110

49. Fischl B, Salat DH, Busa E, Albert M, Dieterich M, Haselgrove C, van der Kouwe A, Killiany R, Kennedy D, Klaveness S, Montillo A, Makris N, Rosen B, Dale AM (2002) Whole brain segmentation: automated labeling of neuroanatomical structures in the human brain. Neuron 33:341–355

50. Fischl B, Salat DH, van der Kouwe AJW, Makris N, Ségonne F, Quinn BT, Dale AM (2004) Sequence-independent segmentation of magnetic resonance images. NeuroImage 23:S69–S84

51. Ashburner J, Friston KJ (2005) Unified segmentation. NeuroImage 26:839–851

52. Pohl KM, Fisher J, Levitt JJ, Shenton ME, Kikinis R, Grimson WEL, Wells WM II (2005) A unifying approach to registration, segmentation, and intensity correction. In: Medical Image Computing and Computer-Assisted Intervention (MICCAI). Springer, Berlin, pp 310–318

53. D'Agostino E, Maes F, Vandermeulen D, Suetens P (2006) A unified framework for atlas based brain image segmentation and registration. In: Pluim JPW, Likar B, Gerritsen FA (eds) Proc Third International Workshop on Biomedical Image Registration (WBIR), vol 4057, Lecture notes in computer science. Springer, Berlin-Heidelberg, pp 136–143

54. Zhang H, Yushkevich PA, Gee JC (2004) Registration of diffusion tensor images. In: Proc IEEE Computer Society conference on Computer Vision and Pattern Recognition (CVPR).pp 842–847.

55. Khullar S, Michael AM, Cahill ND, Kiehl KA, Pearlson G, Baum SA, Calhoun VD (2011) ICA-fNORM: Spatial normalization of fMRI data using intrinsic group-ICA networks. Front Sys Neurosci 5(93):1–18.

56. Joshi S, Davis B, Jomier M, Gerig G (2004) Unbiased diffeomorphic atlas construction for computational anatomy. NeuroImage 23:S151–S160

57. Davis B, Lorenzen P, Joshi S (2004) Large deformation minimum mean squared error template estimation for computational anatomy. In: Proc IEEE International Symposium on Biomedical Imaging (ISBI). pp 173–176

58. Lorenzen P, Davis B, Gerig G, Bullitt E, Joshi S (2004) Multi-class posterior atlas formation via unbiased Kullback-Leibler template estimation. In: Barillot C, Haynor DR, Hellier P (eds) Proc Medical Image Computing and Computer-Assisted Intervention (MICCAI), vol 3216, Lecture notes in computer science. Springer, Berlin, pp 95–102

59. Miller MI (2004) Computational anatomy: shape, growth, and atrophy comparison via diffeomorphisms. NeuroImage 23:S19–S33

60. Christensen GE, Rabbitt RD, Miller MI, Joshi SC, Grenander U, Coogan TA, Van Essen DC (1995) Topological properties of smooth anatomic maps. In: Bizais Y, Barillot C, Di Paola R (eds) Proc Information Processing in Medical Imaging (IPMI). Kluwer Academic, Dordrecht, pp 101–112

61. Woods RP, Grafton ST, Holmes CJ, Cherry SR, Mazziotta JC (1998) Automated image registration: I. General methods and intrasubject, intramodality validation. J Comput Assist Tomogr 22(1):139–152

62. Woods RP, Grafton ST, Watson JDG, Sicotte NL, Mazziotta JC (1998) Automated image registration: II. Intersubject validation of linear and nonlinear models. J Comput Assist Tomogr 22(1):153–165

63. Christensen GE (1999) Consistent linear elastic transformations for image matching. In: Kuba A, Sámal M, Todd-Pokropek A (eds) Proc Information Processing in Medical Imaging (IPMI), vol 1613, Lecture notes in computer science. Springer, Berlin, pp 224–237

64. Ashburner J, Friston KJ (1999) Nonlinear spatial normalization using basis functions. Hum Brain Mapp 7:254–266

65. Bookstein FL (1989) Principal warps: thin-plate splines and the decomposition of deformations. IEEE Trans Pattern Anal Mach Intell 11(6):567–585

66. Bookstein FL (1997) Quadratic variation of deformations. In: Duncan J, Gindi G (eds) Proc Information Processing in Medical Imaging (IPMI), vol 1230, Lecture notes in computer science. Springer, Berlin-Heidelberg, pp 15–28

67. Bro-Nielsen M, Gramkow G (1996) Fast fluid registration of medical images. In: Höhne KH, Kikinis R (eds) Proc Visualization in Biomedical Computing (VBC), vol 1131, Lecture notes in computer science. Springer, Berlin, pp 267–276

68. Thirion JP (1995) Fast non-rigid matching of 3D medical images. Technical report no 2547. Institut National de Recherche en Informatique et en Automatique

69. Rueckert D, Sonoda LI, Hayes C, Hill DLG, Leachand MO, Hawkes DJ (1999) Nonrigid registration using free-form deformations: application to breast MR images. IEEE Trans Image Process 18(8):712–721

70. Thévenaz P, Unser M (2000) Optimization of mutual information for multiresolution image registration. IEEE Trans Image Process 9(12):2083–2099

71. Christensen GE (1994) Deformable shape models for anatomy. Doctoral Thesis. Washington University, Sever Institute of Technology

72. Christensen GE, Rabbitt RD, Miller MI (1996) Deformable templates using large deformation kinematics. IEEE Trans Image Process 5:1435–1447

73. Beg MF, Miller MI, Trouvé A, Younes L (2005) Computing large deformation metric mappings via geodesic flows of diffeomorphisms. Int J Comput Vis 61(2):139–157

74. Avants B, Gee JC (2004) Geodesic estimation for large deformation anatomical shape averaging and interpolation. NeuroImage 23:S139–S150

75. Ashburner J (2007) A fast diffeomorphic image registration algorithm. Neuroimage 38(1):95–113

76. Hernandez M, Bossa MN, Olmos S (2007) Registration of anatomical images using geodesic paths of diffeomorphisms parameterized with stationary vector fields. In: Proc IEEE 11th International Conference on Computer Vision (ICCV). IEEE. pp. 1–8.

77. Vercauteren T, Pennec X, Perchant A, Ayache N (2008) Symmetric log-domain diffeomorphic registration: a demons-based approach. In: Proc Medical Image Computing and Computer-Assisted Intervention (MICCAI). Springer, Berlin, pp 754–761

78. Modat M, Daga P, Cardoso MJ, Ourselin S, Ridgway GR, Ashburner J (2012) Parametric non-rigid registration using a stationary velocity field. In: Proc IEEE Workshop on Mathematical Methods in Biomedical Image Analysis (MMBIA). IEEE. pp 145–150

79. Miller MI, Trouvé A, Younes L (2006) Geodesic shooting for computational anatomy. J Math Imag Vis 24(2):209–228

80. Ashburner J, Friston KJ (2011) Diffeomorphic registration using geodesic shooting and Gauss–Newton optimisation. NeuroImage 55(3):954–967

81. Vialard FX, Risser L, Rueckert D, Cotter CJ (2012) Diffeomorphic 3D image registration via geodesic shooting using an efficient adjoint calculation. Int J Comput Vis 97(2):229–241

82. Lester H, Arridge SR (1999) A survey of hierarchical non-linear medical image registration. Pattern Recogn 32:129–149

83. Klein A, Andersson J, Ardekani BA, Ashburner J, Avants B, Chiang MC, Christensen GE, Collins DL, Gee J, Hellier P, Song JH, Jenkinson M, Lepage C, Rueckert D, Thompson P, Vercauteren T, Woods RP, Mann JJ, Parsey RV (2009) Evaluation of 14 nonlinear deformation algorithms applied to human brain MRI registration. Neuroimage 46(3):786–802

84. Fissell K, Tseytlin E, Cunningham D, Carter CS, Schneider W, Cohen JD (2003) Fiswidgets: a graphical computing environment for neuroimaging analysis. Neuroinformatics 1(1):111–125

85. Rex DE, Maa JQ, Toga AW (2003) The LONI pipeline processing environment. NeuroImage 19(3):1033–1048

86. Zijdenbos AP, Forghani R, Evans AC (2002) Automatic 'pipeline' analysis of 3-D MRI data for clinical trials: application to multiple sclerosis. IEEE Trans Med Imaging 21(10):1280–1291

87. Drobnjak I, Gavaghan D, Suli E, Pitt-Francis J, Jenkinson M (2006) Development of a fMRI simulator for modelling realistic rigid-body motion artifacts. Magn Reson Med 56(2):364–380

Chapter 7

Statistical Analysis of fMRI Data

Mark W. Woolrich, Christian F. Beckmann, Thomas E. Nichols, and Stephen M. Smith

Abstract

fMRI is a powerful tool used in the study of brain function. It can noninvasively detect signal changes in areas of the brain where neuronal activity is varying. This chapter is a comprehensive description of the various steps in the statistical analysis of fMRI data. This will cover topics such as the general linear model (including orthogonality, hemodynamic variability, noise modeling, and the use of contrasts), multisubject statistics, and statistical thresholding (including random field theory and permutation methods).

Key words fMRI, Analysis, Statistics, General linear model, Multisubject statistics, Statistical thresholding

---

1 Introduction

fMRI is a powerful tool used in the study of brain function. It can noninvasively detect signal changes in areas of the brain where neuronal activity is varying. fMRI can therefore give high-quality visualization of the location of activity in the brain resulting from sensory stimulation or cognitive function. It allows, for example, the study of how the healthy brain functions, how different diseases affect the brain, or how drugs can modulate activity or post-damage recovery.

After an fMRI experiment has been designed and carried out, the resulting data must be passed through various analysis steps before the experimenter can get answers to questions about experimentally related activations at the individual or multisubject level. This chapter focuses on the statistical aspects of such analysis.

We need a statistical approach for two reasons. First, fMRI data is very noisy. The noise is often of the same order of magnitude as the fMRI signal changes we are trying to detect, and as such we can only approximately estimate the signal changes. Statistics are therefore needed to ask if the estimated signal changes are significant, given the quality of the approximation. Second, many fMRI studies are carried out with the intention of answering some

Massimo Filippi (ed.), *fMRI Techniques and Protocols*, Neuromethods, vol. 119,
DOI 10.1007/978-1-4939-5611-1_7, © Springer Science+Business Media New York 2016

question about a population of individuals. For example, we might want to know what the difference in neural activation is between a patient and control group. Such population differences in neural activation can only ever be approximated, because not only is the fMRI data noisy but also we only ever have a sample of subjects from the populations. This issue is somewhat exacerbated in fMRI studies as the number of subjects sampled is typically quite small! Statistics are needed to see if the approximated population differences are significant given the quality of those approximations.

Figure 1 shows an illustration of the main analysis steps carried out in a typical fMRI study. There are three main components in the process. First, the individual subjects' fMRI data must be processed. Second, the information gleaned from this about the effect sizes (the size of the fMRI signal change in response to the experimental task) for each subject is then combined in a group analysis. Finally, the group effect sizes are statistically thresholded to ask questions such as "Where is there significant activity in response to the experimental task for the population?" or "Where are there significant differences between populations (e.g., controls versus patients)?." This final thresholding is carried out on statistic images as it takes into account the spatial characteristics of the data. Note that one could also perform thresholding on the effect size statistics from a single-subject's analysis, allowing one to ask questions such as "Where is there significant activity in response to the experimental task for this subject?" The various steps in the analysis will be described in detail throughout the chapter.

2   Statistical Analysis of a Single fMRI Dataset

Later in the chapter, we will see how we go about asking statistical questions about populations of subjects. However, before this can be done, the individual subject fMRI data must be analyzed. The most common way of statistically analyzing fMRI data is by using a general linear model (GLM). As we shall see, this is a powerful framework that allows a wide range of different statistical questions to be asked about the data.

2.1   fMRI Data

In a typical fMRI imaging session, a low-resolution functional volume is acquired every few seconds (MR volumes are often also referred to as "images" or "scans"). Over the course of the experiment, 100 volumes or more are typically recorded. In the simplest possible experiment, some images will be taken while stimulation[1] is applied, and some will be taken with the subject at rest. Because the images are taken using an MR sequence which is sensitive to changes

---

[1] For the remainder of this chapter, reference to "stimulation" should be taken to include also the carrying out of physical or cognitive activity.

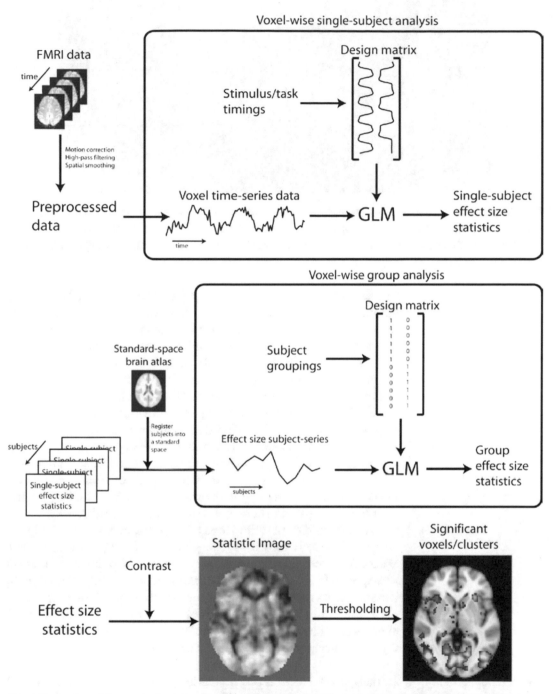

Fig. 1 Illustration of the analysis steps carried out in a typical fMRI group study. There are three main components in the process. First, the individual subjects' fMRI data must be processed (*top*). The information gleaned from this about the effect sizes (the size of the fMRI signal change in response to the experimental task) for each subject are then combined in a group analysis (*middle*). The group effect sizes statistic images are then statistically thresholded to find significant brain areas (*bottom*)

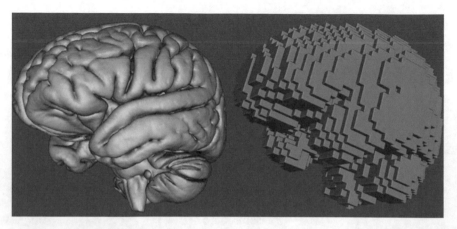

Fig. 2 What are voxels? Shown here are surface renderings of 3D brain images. On the *left* is a high-resolution image, with small (0.5 × 0.5 × 0.5 mm) voxels; the voxels are too small to see. On the *right* is a low-resolution image of the same brain, with large (5 × 5 × 5 mm) voxels, clearly showing the voxels making up the image

in local blood oxygenation level, parts of the images taken during stimulation should show increased intensity, compared with those taken while at rest. The parts of these images that show increased intensity should correspond to the brain areas that are activated by the stimulation. The goal of fMRI analysis is to detect, in a robust, sensitive, and valid way, those parts of the brain that show changes in intensity at the points in time that the stimulation was applied.

A single volume is made up of individual cuboid elements called voxels (Fig. 2). An fMRI dataset from a single session can be thought of either as t volumes, one taken every few seconds, or as v voxels, each with an associated time series of t time points. It is important to be able to conceptualize both of these representations, as some analysis step make more sense when thinking of the data in one way and others make more sense the other way.

An example time series from a single voxel is shown in Fig. 3. Image intensity is shown on the y-axis and time (in scans) on the x-axis. As described above, for some of the time points, stimulation was applied (the higher intensity periods), and at some time points, the subject was at rest. As well as the effect of the stimulation being clear, the high-frequency noise is also apparent. The aim of fMRI analysis is to identify in which voxels' time series the signal of interest is significantly greater than the noise level.

2.2 Preparing fMRI Data for Statistical Analysis

Initially, a 4D dataset is pre-processed. This pre-processing is aimed at not only removing artifacts and reducing noise but also conditioning the data so that it is more amenable to the statistical analysis that is to follow.

The most basic required steps that will be typically carried out are as follows. Once data has been acquired by the MR scanner, the pre-processing starts by *reconstructing* the raw "k-space" data into

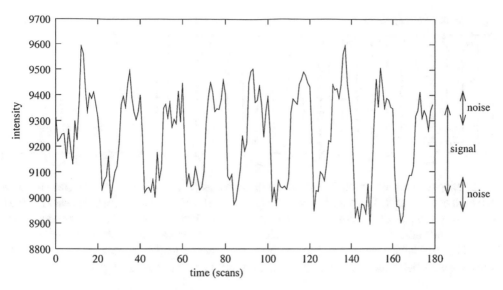

Fig. 3 An example time series at a strongly activated voxel from a visual stimulation experiment. Here the signal is significantly larger than the noise level. Periods of stimulation are alternated with periods of rest—a complete stimulation—rest cycle lasts 20 scans

images that actually look like brains. The data is then *motion corrected*, where each volume is transformed (using rotation and translation) so that the image of the brain within each volume is aligned with that in every other volume. *Spatial smoothing* is then carried out, principally to reduce noise, hopefully without significantly affecting the activation signal. Finally, each voxel's time series is *temporally high-pass filtered* with a filter designed to remove the large amount of low-frequency temporal noise found in FMRI data, without removing the signal of interest.

Chapter 6 has already covered fMRI pre-processing in much more detail, including other optional steps that have not been mentioned here.

2.3 Predicting the Response

The first step in the statistical analysis is to come up with a good prediction, or model, of what we think the measured fMRI signal response will look like in voxels that are active. In the simplest type of fMRI experiment, we alternate periods of stimulation with periods of rest, in what we will refer to as a square-wave block design, as shown in Fig. 4 (left). We expect that a voxel which is active in response to the stimulus will contain an fMRI signal that generally fluctuates up and down with a time course that is similar to the stimulus time course (Fig. 4 left), whereas an inactive voxel will not.

2.4 Hemodynamic Response Function

However, can we come up with a better prediction of the fMRI signal than Fig. 4 (left)? In particular, we know, from experiment, that the response to a very short stimulus looks like the curve shown in Fig. 5. We refer to this response to an impulse of stimulus

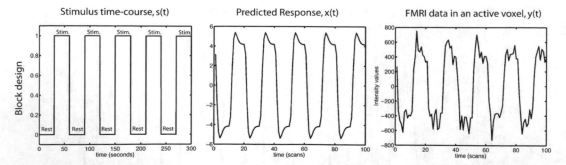

Fig. 4 Predicting the response using the known stimulus timings. This example is a square-wave block design where blocks of stimulation are alternated with blocks of rest. The square-waveform (*left*) describes the input stimulus timing; the predicted response (*middle*) results from convolving the stimulus time course with the hemodynamic response function and then sampling it at the temporal resolution of the experiment. This experiment has a repetition time (*TR*) of 2 s. This process produces a model, or predicted response, that looks much more like the data measured in voxels that are responding to the stimulus (*right*)

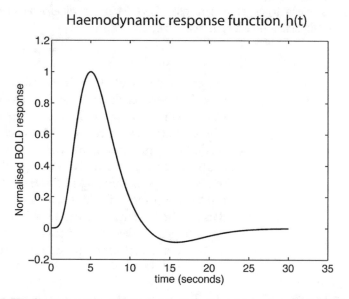

Fig. 5 The hemodynamic response function. A brief impulse of stimulation at $t = 0$ s causes a blood oxygen level dependent (*BOLD*) signal that is delayed and blurred. Here it is modeled as a double-gamma function

as the hemodynamic response function (HRF), and it is basically a delayed and blurred version of the short stimulus burst. This is because the variations that we can detect in blood oxygen level dependent (BOLD) fMRI signal are due to processes taking place in the vasculature: things such as the amount of blood oxygenation, blood flow, and blood volume change when neural activation increases or decreases. Unsurprisingly, these vascular changes occur on a slower timescale than the neural activity. Note that there are also more subtle characteristics of the HRF. For example, as can be seen in Fig. 5, there can be a post-stimulus undershoot as the HRF temporarily drops below baseline before rising back to zero. The HRF in Fig. 5 is a commonly used HRF, and is a double-gamma function of the form:

$$h(t) = \frac{G(\mu_1, \sigma_1^2) - G(\mu_2, \sigma_2^2)}{\rho} \tag{1}$$

where $h(t)$ is the HRF as a function of time, $G(\mu, \sigma^2)$ is a Gamma distribution parameterized by its mean, μ, and variance, σ^2 (note that we can convert these to the traditional Gamma distribution parameters using $\alpha = \mu^2/\sigma^2$ and $\beta = \sigma^2/\mu$), and ρ is the ratio of the height of the positive Gamma to the negative Gamma.[2]

The most straightforward way of incorporating the HRF into our predicted response is to apply its delaying and blurring effect to the raw stimulus time course that we have in our fMRI experiment. This is achieved by the mathematical operation of convolution:

$$x(t) = \int_0^\infty h(\tau) s(t - \tau) \, d\tau \tag{2}$$

This essentially assumes that the effects of the different impulses that make up the stimulus time course add together in an additive, linear fashion [1–4]. Figure 4 (middle) shows the result of convolving the HRF in Fig. 5 with the square-wave stimulus in Fig. 4 (left) to form our new improved predicted response. Strictly speaking, making the assumption of linearity is incorrect. We will see later how we can address this issue, and also discuss how to tell if that is necessary.

For now, using the HRF and convolving it with the stimulus time course provides us with a way in which we can come up with a reasonable prediction of the response for any general stimulus type. For example, Fig. 6 shows the predicted response for a sparse single-event design and a dense randomized single-event design. Armed with our predicted response, we can then look to find those voxels that have fMRI time courses that match the predicted response well and, if they pass a statistical test, label them as being active voxels.

3   General Linear Modeling

We have so far discussed how to come up with a prediction of the fMRI signal in response to an experimental stimulus. However, what do we do when we have more than one stimulus switching on or off throughout the experiment? The answer is to use a GLM. This assumes that each of the stimuli have their own predicted response, and that these predicted responses then add together linearly in

---

[2] The particular HRF in Fig. 5 has parameter values $\mu_1 = 6$ s, $\sigma_1 = 2.45$ s, $\mu_2 = 16$ s, $\sigma_2 = 4$ s, and $\rho = 6$.

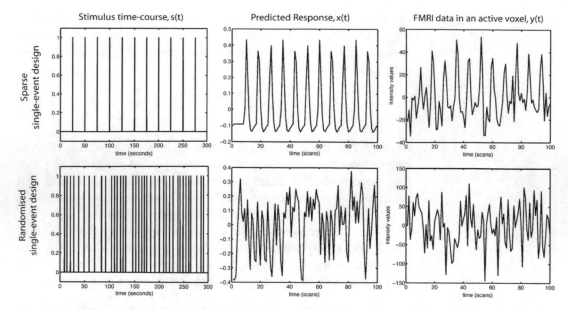

Fig. 6 Using the hemodynamic response function (*HRF*) and convolving it with the stimulus time course provides a way in which we can predict the response for any general stimulus type. Here we can see the predicted response for a sparse single-event design, where short bursts of 0.1 s stimulation are 25 s apart (*top*), and for a randomized single-event design, where short bursts of 0.1 s stimulation are separated with inter-stimulus intervals (*ISIs*) sampled from a Poisson distribution with mean of 7 s (*bottom*). Randomized single-event designs are an excellent way of working with stimuli that by their nature need to be single events, as they generally have better sensitivity than sparse single-event designs [5]

some combination unique to each voxel, to explain the data measured in that voxel. For example, consider that we have two experimental stimuli: one auditory and one visual. Both are square-wave block designs, but they switch on and off at different times. The overall predicted response is given by a linear combination of the predicted responses:

$$y(t) = \beta_1 x_1(t) + \beta_2 x_2(t) + c + e(t) \tag{3}$$

where $y(t)$ is the data in one voxel, and is a 1D vector (time course) of intensity values with one value for each time point. $x_1(t)$ and $x_2(t)$ are the predicted responses for our auditory and visual experimental stimuli, respectively, and both are also 1D time courses with one value for each time point. c is a constant and would correspond to the mean intensity value in the data. The linear combination of the predicted responses needed to explain the data in a particular voxel is described by the parameters β_1 and β_2. $e(t)$ models the noise that is present in fMRI data.

Model fitting involves adjusting the mean level, c, and the parameters β_1 and β_2, to best fit the data. For example, if a particular voxel responds strongly to model x_1, the model-fitting will find a large value for β_1; if the data instead looks more like the second model time course, x_2, then the model-fitting will give β_2 a large value. The GLM is used to analyze each voxel's time series independently. This is often referred to as a mass univariate analysis, and outputs statistics independently at each voxel.

There are a number of different of names that get used to describe the different components of the GLM. The predicted responses within a GLM are often referred to as *explanatory variables* (*EVs*), as they explain different processes in the data. They can also be referred to as *regressors*, and the βs as *regression parameters*, as we are performing what is also known as a multiple regression. The regression parameters, β, are also sometimes referred to as *effect sizes*, as they describe the size of the response to the corresponding underlying experimental stimuli.[3]

3.1   Design Matrix

The GLM is often formulated in matrix notation. All of the parameters are grouped together into a $P \times 1$ vector β (where P is the number of EVs) and all of the EVs are grouped together into an $N \times P$ matrix X, often referred to as the design matrix, where N is the number of time points in the experiment. This gives us the GLM in the following form:

$$Y = X\beta + e \tag{4}$$

where Y is the $N \times 1$ vector of intensity values in the data, and e is the $N \times 1$ noise vector. You may wonder what has happened to the mean parameter, c, in this new equation. There are two common ways in which this is handled. The first is to remove the mean, or de-mean, the data Y, and to also separately de-mean all of the EVs in the design matrix. This is appropriate as there is no information in the mean signal intensity of the fMRI data that can aid us in our statistical analysis. The second is to leave the mean as part of the GLM by creating an EV that has the value of 1 at every time point. The β or regression parameter for this EV will be determined when we fit the GLM to the data and will relate to the estimate of the mean parameter.

Figure 7 shows a design matrix for our example experiment with two stimuli (auditory and visual). Each column in the design matrix is a different part of the model. The left column (x_1 or EV1) models the auditory stimulation, and the right column (x_2 or EV2) models the visual stimulation.

3.2   Fitting the GLM to the Data

The GLM is fit to the data at each voxel separately. This is achieved by adjusting the estimates of the regression parameters to find the best fit of the model to the data. Typically, it is assumed that the fMRI noise, e, is well modeled by a Gaussian distribution with a

---

[3] Note that this common usage is slightly different from the definition sometimes used in the statistics literature, where effect size means β divided by the noise level.

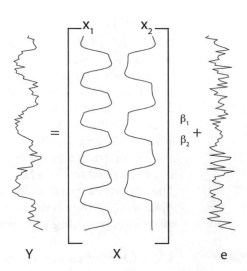

Fig. 7 Example of the general linear model (*GLM*) for an experiment containing auditory and visual stimuli which have different stimulus timings. The design matrix, *X*, contains two predicted responses (also known as regressors or explanatory variables): x_1 for the auditory stimulus and x_2 for the visual stimulus—*see* Eq. (4). In this, visualization time is running downwards. It can be seen that the particular voxel data shown, *Y*, is from a voxel that is strongly activating in response to the visual stimulus modeled by x_2, but not to the auditory stimulus modeled by x_1

standard deviation, σ, unique to each voxel. When this assumption is made, the best fit of the model to the data is equivalent to minimizing the sum (over all time points) of the squared difference between the data, Υ, and the signal model, $X\beta$. That is we choose the β values that minimize:

$$\sum_t \left(\Upsilon_t - X_t \beta \right)^2 \tag{5}$$

Mathematically, it can be shown that this is minimized when we estimate the βs as:

$$\hat{\beta} = \left(X^T X \right)^{-1} X^T \Upsilon \tag{6}$$

where $\hat{\beta}$ is our regression parameter estimate. In the example visual/auditory experiment in Fig. 7, it can be seen that the particular voxel data shown, Υ, is from a voxel that is strongly activating in response to the visual stimulus modeled by x_2, but not to the auditory stimulus modeled by x_1. This would result in a large value for β_2 and a low value for β_1 when the GLM is fit to this data, suggesting that there is visual activation but no auditory activation in this voxel.

3.3 Temporal Autocorrelation

Until now we have considered a rather simple approach to dealing with the noise that is present in fMRI, by assuming that it is well modeled as coming from a Gaussian distribution. Unfortunately, in practice, this is not the whole story. This is because fMRI noise, that is, the signal we record in the absence of any stimulation, is temporally autocorrelated. In particular, in the gray matter, this corresponds to the fMRI noise being temporally smooth. This is because the nature of the many artifacts that make up fMRI noise, for example, thermal noise, cardiac and respiratory rhythms, autoregulatory oscillations, and networks of spontaneous neural activity, tend to occur more at low frequency than at high frequency. We can see this imbalance between low and high frequency if we look at a plot of the power spectrum (the absolute value of a Fourier transform) of fMRI noise, an example of which is shown in Fig. 8a.

The presence of temporal autocorrelation is a concern because it affects the choice of the optimal estimation (model fitting) method, and perhaps more important, the accuracy of the

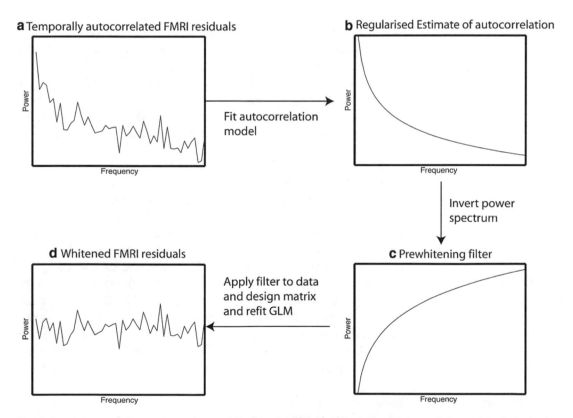

Fig. 8 A summary of the process of pre-whitening. (**a**) Plot of the power spectrum of the residuals from an initial general linear model (*GLM*) fit to the data from one voxel. (**b**) Estimated power spectrum (representing the temporal autocorrelation estimate) obtained from fitting an autocorrelation model. (**c**) This spectrum is inverted to create the frequency characteristics of a temporal filter designed to "undo" the autocorrelation. (**d**) The pre-whitening temporal filter is applied to both the data and the explanatory variables (*EVs*) in the design matrix, and then this pre-whitened GLM is refit to the pre-whitened data. The residuals that result from this refit of the GLM should now be approximately white, that is have a flat power spectrum

subsequent statistical tests. For example, if we ignore the fact that there is an increased amount of noise at low frequency, then we can underestimate the variability in the data and produce under-conservative statistical tests.

A number of strategies have been proposed for dealing with the problem of temporally autocorrelated noise. Traditionally, the most commonly used approach is pre-whitening [6–8]. The process of pre-whitening is summarized in Fig. 8. The first step is to estimate the temporal autocorrelation on the residuals, $r = \Upsilon - X\hat{\beta}$, from an initial GLM fit. The temporal autocorrelation estimate is then used to construct a pre-whitening temporal filter designed to "undo" the autocorrelation. In other words, the filter is designed to re-dress the imbalance between high and low frequency in the fMRI noise power spectrum, so that the new power spectrum has equal power at all frequencies. By analogy, since white light is a result of there being equal amounts of light at all frequencies/colors, we refer to noise with equal power at all frequencies as white noise—and there-fore to the whole approach as pre-whitening. The pre-whitening temporal filter is applied to both the data and the EVs in the design matrix, and then this pre-whitened GLM is refitted.

A crucial step in pre-whitening is the estimation of the tempo-ral autocorrelation. A wide range of approaches have been pro-posed, including the use of auto-regressive (AR) models [6, 9], AR plus white noise models [10], spectral smoothing [8, 11], and spa-tial regularization of autocorrelation estimates [8, 12, 13].

The pre-whitening approach described above is one that was designed to work in a classical statistical ("frequentist") frame-work. It is possible that inferring the autocorrelation on the residu-als from an initial GLM fit can introduce inaccuracies, since the residuals, r, only serve as an approximation to the true error, e. More recently, alternative Bayesian strategies have been developed to deal with this problem [14, 15]. These have the advantage of inferring the autocorrelation characteristics at the same time as the GLM regression parameters, and to take into account the uncer-tainty in the temporal autocorrelation estimation. However, these issues aside, they are essentially performing the same pre-whitening approach we have already described. Although computationally more demanding these techniques are being increasingly used.

3.4 Inferring Neural Activity

When we fit a particular GLM to a particular voxel's data, we get regression parameter estimates that indicate how much of each EV is needed to explain what we see in the data. If the parameter estimate of β for any particular EV is nonzero, then it might seem reasonable to assume that the voxel in question is neuronally responding to the stimulus that the EV represents. However, we only have estimates/approximations of the true β obtained from a limited amount of noisy fMRI data. So, given the amount of noise and the estimate of β obtained, how much can we trust that any particular β is nonzero?

It is only those voxels where we can satisfy ourselves in a statistical manner that this is the case, that we label as being active. The first step towards this is to convert the parameter estimates of β into a useful statistic. Most commonly, we use a T-statistic, given by:

$$t = \frac{\hat{\beta}}{\text{std}\left(\hat{\beta}\right)} \qquad (7)$$

where the denominator is the standard deviation (uncertainty) of our parameter estimate. If the parameter estimate is low relative to its estimated uncertainty, the T-statistic, t, will be low, implying that β is unlikely to be significantly nonzero (and vice versa). We will see later what the standard deviation of our parameter estimate depends upon.

The question remains as to how we determine that a T-statistic is significantly nonzero. This is achieved by comparing the calculated T-statistic to the distribution of T-statistics we would expect to get if the true β value was zero. This is a *null hypothesis test* (the null hypothesis is that β is zero).

As mentioned earlier, we typically assume that the noise in fMRI is Gaussian distributed. This means that our expected distribution of T-statistics under the null hypothesis is T-distributed. This is a standard statistical distribution for which the probability, or P-value, of getting a T-value greater than the one we have calculated if the null hypothesis were true, can easily be calculated. As illustrated in Fig. 9, a low probability (low P-value) of the null

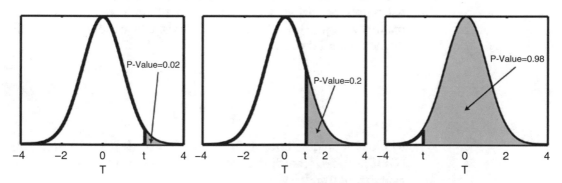

Fig. 9 Performing a T-test. The T-statistic we calculate for each regression parameter estimate is compared with the distribution of T-statistics we would expect if the true regression parameter were zero. This "null distribution" is a standard T-distribution. Here we are using a T-distribution with 100 DOF (this would correspond to about 100 time points in the data). We calculate a probability, or P-value, as the proportion of the area under the curve in the positive tail of the distribution defined by our T-statistic, t. A low probability, or P-value, (*left plot*) of the null hypothesis being true means that we can more confidently reject the null hypothesis and label the voxel as having a nonzero β, and therefore as being neuronally activated by the stimulus that the β in question represents. T-statistics that have larger P-values (*middle plot*) or are deep into the negative tail of the distribution (*right plot*) have high P-values. In the latter case, this might seem a bit counterintuitive as we have a T-statistic that is in the extremities of the null distribution. However, this is because the test is directional and so we calculate the P-value by looking in the positive tail of the distribution only

hypothesis being true means that we can more confidently reject the null hypothesis and label the voxel as having a nonzero β. We will see in Sect. 7 how we choose a threshold for the P-values.

3.5 Contrasts

So far, we have addressed how we might go about producing T-statistics and P-values, which describe how strongly each voxel is related to each EV in our design matrix. Contrasts are a framework whereby we can ask not just questions about each EVs parameter estimate (PE, or β) in isolation, but also a wider range of questions that compare the different parameter estimates with each other.

In general, a contrast is defined by a $P \times 1$ vector, c, (recall that P is the number of EVs in the design matrix). This is multiplied by the $P \times 1$ vector of parameter estimates, $\widehat{\beta}$, to give what is known as a contrast of parameter estimates (COPEs), $c^T \widehat{\beta}$. The COPE is therefore just a linear combination of the parameters estimates; it is equal to the sum of each PE multiplied by the relevant number in the contrast vector.

For example, it may be desirable to compare two different PEs to test directly whether one EV is more "relevant" to the data than another EV. In our combined auditory and visual experiment, this would be asking "Where does the brain respond more strongly to the auditory stimulus compared with the visual stimulus?". In this example, we have two EVs, one for the auditory stimulus and one for the visual (recall Fig. 7). So our contrast to answer this question would be $c^T = [1 - 1]$ (c is a column vector, hence the transpose here), resulting in a COPE, $c^T \beta = 1\beta_1 - 1\beta_2$.

The COPE is then simply treated as if it were itself an individual regression parameter estimate. In other words, in the same manner as Eq. (7), we calculate a T-statistic by dividing the COPE by its standard deviation:

$$t = \frac{c^T \widehat{\beta}}{\sqrt{\mathrm{var}\left(c^T \widehat{\beta}\right)}} \qquad (8)$$

where the T-distribution in question has degrees of freedom (DOF) of $N - P$ (recall that N is the number of time points in the experiment), and:

$$\mathrm{var}\left(c^T \widehat{\beta}\right) = \widehat{\sigma}^2 c^T \left(X^T X\right)^{-1} c \qquad (9)$$

where σ^2 is the estimate of the variance of the fMRI noise:

$$\widehat{\sigma}^2 = \frac{r^T r}{(N - P)} \qquad (10)$$

where r is the residual, that is, an estimate of the error, e, and is what is left over after the model is fit to the data, and is given by $r = \Upsilon - X\widehat{\beta}$.

As with the T-test on a single PE, we can calculate a probability, or P-value, of getting the calculated T-statistic under the null hypothesis in the same manner as in Fig. 9. The null hypothesis is that the COPE $= 0$, so if the calculated P-value is low then this suggests that the it is unlikely that the COPE $= 0$. If we apply a $c^T = [1 - 1]$ contrast in the auditory/visual experiment, then this is equivalent to saying "the voxel is responding more strongly to the auditory stimulus compared with the visual stimulus." We can infer that it is the auditory that is stronger than the visual, and not vice versa, because these T-tests on these contrasts are directional. Technically, this is because we are doing the null-hypothesis test in Fig. 9 on one tail (in particular, the right-hand tail) of the distribution only. As shown in Table 1, if we want to ask "Where does the brain respond more strongly to the visual stimulus compared with the auditory stimulus?," then we would use $c^T = [-1 \ 1]$. All that remains to complete the null hypothesis test is a choice of threshold, such that if the calculated P-value drops below that threshold, we reject the null hypothesis and say that the contrast is significant. We shall see later in Sect. 7 how we go about doing this while also taking into account the spatial nature of the data.

Even in the relatively simple auditory/visual experiment, there are a number of different questions that can be answered. *See* Table 1 for some other examples.

Even in this relatively simple experiment, there are a number of different questions that can be answered. In particular, it is important to remember that the directionality of the T-test is important. Note that:

EV1 is the auditory predicted response and EV2 is the visual predicted response.

COPE stands for contrast of parameter estimate and *EV* stands for explanatory variable

3.5.1 F-Tests

The last example contrast in Table 1 was a $[1 \ 1]$ contrast. It is a common misconception that such a contrast asks "Where is there significant activation due to either the visual or the auditory stimulation?" In fact, this contrast calculates $\text{COPE} = \hat{\beta}_1 + \hat{\beta}_2$, which is proportional to the average value of the two regression parameters, $\left(\hat{\beta}_1 + \hat{\beta}_2\right) / 2$, and hence is actually asking "Where is there significant activation averaged across both conditions?"

So how do we go about asking the question "I want to find where there is significant activity due to either the visual or the auditory stimulation"? The answer is to use F-tests. An F-test is defined by specifying a set of contrasts that we want to test simultaneously. This then tests the null hypothesis that all of the COPEs that are in the F-test are equal to zero. Therefore, we can find significance (reject the null hypothesis) if any of the COPEs is nonzero. Another perspective is that the F-test will find significance if

Table 1
Examples of contrasts that might be used in the two stimulus auditory/visual experiment

Contrast, cT	COPE, cTβ	Meaning
[1 0]	β_1	Where is there significant auditory activation?
[0 1]	β_2	Where is there significant visual activation?
[−1 0]	$-\beta_1$	Where is there significant negative auditory activation?
[0 −1]	$-\beta_2$	Where is there significant negative visual activation?
[1 −1]	$\beta_1 - \beta_2$	Where is there auditory activation significantly greater than visual activation?
[−1 1]	$\beta_2 - \beta_1$	Where is there visual activation significantly greater than auditory activation?
[1 1]	$\beta_1 + \beta_2$	Where is there significant activation averaged across both conditions?

there is any linear combination of the COPEs that explains a significant amount of variance in the data.

So to ask "Where is there significant activity due to either the visual or the auditory stimulation?," we simply need to include in an F-test the contrast that asks where there is significant activity due to the auditory stimulation, [1 0], along with the contrast that asks where there is significant activity to the visual stimulation, [0 1]. Formally, this is done with an F-test contrast matrix, c, that contains both of these contrasts:

$$c^T = \begin{pmatrix} 1 & 0 \\ 0 & 1 \end{pmatrix}.$$

Note that in the T-tests, our contrasts were $P \times 1$ vectors. Now, F-tests are generally described as $P \times K$ contrast matrices, c, where K is the number of contrasts in the F-test, and recall that P is the number of regression parameters in the GLM. Using this contrast matrix, we can then calculate an F-statistic:

$$f = \frac{\hat{\beta}^T c \, \mathrm{var}\left(c^T \hat{\beta}\right) c^T \hat{\beta}}{K}. \tag{11}$$

Analagous to how the T-statistics were T-distributed under the null hypothesis that the COPE is zero, this F-statistic is F-distributed (with DOF K and $N-P$) under the null hypothesis that all of the contrasts in the F-test are zero $\left(c^T \hat{\beta} = 0\right)$. As such, we can proceed with a null-hypothesis test in exactly the same manner as we did with the T-test.

An important characteristic of *F*-tests is that they are blind to the directionality of the contrasts that make up the test. In other words, in our auditory/visual experiment example, the following *F*-tests are all equivalent:

$$\begin{pmatrix} 1 & 0 \\ 0 & 1 \end{pmatrix}, \begin{pmatrix} -1 & 0 \\ 0 & -1 \end{pmatrix}, \begin{pmatrix} -1 & 0 \\ 0 & 1 \end{pmatrix}, \begin{pmatrix} 1 & 0 \\ 0 & -1 \end{pmatrix}.$$

This is because *F*-tests are testing if there is a linear combination of the COPEs that explain a significant amount of variance in the data, and when we consider this variance, we are ignoring the sign of the COPEs.

It is instructive to consider how *F*-tests relate to *T*-tests by considering an *F*-test that consists of just one contrast. As illustrated in Fig. 10, such an *F*-test is equivalent to a two-tailed *T*-test on the contrast in question, with the relationship $f = t^2$. Hence, *F*-contrasts containing single contrasts can be used to mimic two-tailed *T*-tests. The *F*-test's blindness to the directionality of the contrasts is readily apparent when we consider the equivalent two-tailed *T*-test. It is because we can get significance with either a significantly positive or negative COPE in either tail. As a result of calculating the *P*-value under both tails, a two-tailed *T*-test (or equivalently the *F*-test) is more conservative (with respect to positive activation) than the one-tailed *T*-test on the same contrast.

3.6 Interaction Example

It is possible that the response to two different stimuli, when applied simultaneously, is greater than that predicted by adding up the responses to the stimuli when applied separately. If this is the case, then such "nonlinear interactions" may need to be allowed for in the model. The simplest way of doing this is to set up the

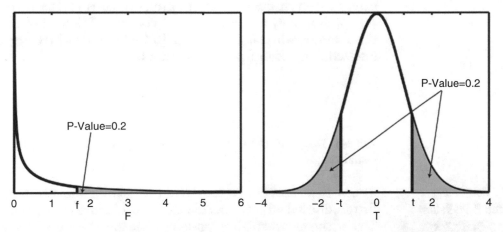

Fig. 10 Here we are considering an *F* -test that consists of just one contrast. In this case, the *F*-test (*left*) is equivalent to a two-tailed *T*-test (*right*) on the same contrast, with the relationship $f = t^2$

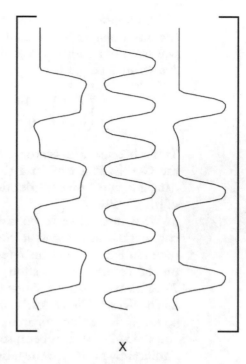

X

Fig. 11 Example of modeling a nonlinear interaction between stimuli. The first two explanatory variables (*EVs*) model the separate stimuli and the third models the interaction, that is, accounts for the "extra" response when both stimuli are applied together

two originals EVs, and then add an interaction EV, which will only be "up" when both of the original EVs are "up" and "down" otherwise. In Fig. 11, EV1 could represent the application of a drug and EV2 could represent visual stimulation. EV3 will model the extent to which the response to drug + visual is greater than the sum of drug-only and visual-only. A contrast of $[0\ 0\ 1]$ will show this measure, whereas a contrast of $[0\ 0-1]$ shows where negative interaction is occurring. An F-contrast of

$$\begin{pmatrix} 1 & 0 & 0 \\ 0 & 1 & 0 \end{pmatrix}$$

will ask where is there significant activity to either drug-only or visual-only.

3.7 Converting T- and F-Statistics into Z-Statistics

T- and F-statistics can be converted to Z-statistics, that is, statistics that are distributed as a standardized Normal (Gaussian) distribution. This is simply achieved by ensuring that the P-value is the same regardless of which statistic is used, so to convert from a T- to Z-statistic, we calculate the P-value for the given T-statistic and then determine the Z-statistic as being the one that gives the same P-value. One reason for doing this is so that we can compare statistics using a common currency. Another reason is so that we can perform generic thresholding techniques (such as described in Sect. 7), using Z-statistic maps, regardless of whether we have done T- or F-tests.

3.8 Percent Signal Changes

It is useful to be able to convert regression parameter estimates into percent BOLD changes. This is because percent BOLD change can be a common currency across different experiments (though BOLD is not a quantitative measure and depends on many experimental factors, and so should not be treated as comparable without a good deal of care). It does not depend on things like the arbitrary scaling of intensity values output from the scanner or arbitrary scaling of the EVs. As illustrated in Fig. 12, this is simple to do and just requires that we have an estimate of the baseline signal intensity, C, from the voxel in question, and that we know the peak-to-peak height, H, of the relevant EV. The percent BOLD change is then calculated as:

$$\%\text{change} = 100\frac{H\widehat{\beta}}{C}. \tag{12}$$

Since the signal fluctuation $\widehat{\beta} \times H$ is always small with respect to the baseline C, it is common to approximate C as more simply the mean of the time series, C'.

It is possible to do this for contrasts as well. However, it is not immediately obvious what the peak-to-peak height is in the context of a general contrast. This can be dealt with by determining

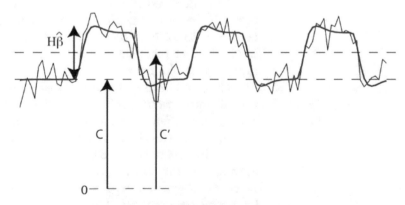

Fig. 12 Illustration of the calculation of percent blood oxygen level dependent (*BOLD*) signal change from a general linear model (*GLM*) fit. % change $= 100H\widehat{\beta} / \text{C}$ where $\widehat{\beta}$ is the regression parameter estimate (effect size) and H is the peak-to-peak height for the relevant explanatory variable (*EV*). C is the baseline intensity of the fMRI data, but is common to approximate this with the mean of the time series, C'

the effective regressor for a contrast. The effective regressor is the regressor in a new version of the design matrix whose regression parameter estimate (and its variance) is equal to the original COPE. This is given by [16]:

$$X_{\text{eff}} = XQc\left(c^{\mathrm{T}}Qc\right)^{-1}, \text{where } Q = \left(X^{\mathrm{T}}X\right)^{-1} \qquad (13)$$

The peak-to-peak height of this effective regressor can then be used in the percent signal change calculation above.

3.9 Issues with Orthogonality and Estimating Contrasts

We described in Sect. 3.2 how we obtain regression parameter estimates by finding the best fit of the GLM to the data (by using Eq. (6). The regression parameter estimates describe how much we need of each EV to explain what we see in the data. However, consider what would happen in a poorly designed experiment, where two different stimuli are switched on and off at very similar times. The resulting predicted responses (EVs) are highly correlated. Note that we use the terms correlated and non-orthogonal (similarly uncorrelated and orthogonal) interchangeably. The top of Fig.13 shows a design matrix containing our two very similar EVs. The model fitting will determine the regression parameter estimates $\hat{\beta}_1$ and $\hat{\beta}_2$ and describe how much we need of each EV to explain what we see in the data. However, because EV1 and EV2 are so similar, we can equally well use either EV1 or EV2 to explain what we see in the data. The result is that we cannot estimate either $\hat{\beta}_1$ or $\hat{\beta}_2$, separately from each other, very well. Mathematically, we say that the design matrix is not of "full rank," or that it is "rank deficient."

To understand this, consider solving two simultaneous equations. If we have two unknowns to solve for, then we need two equations to solve for them. However, if the two equations are the same, then we really have only one equation and we cannot solve for the two unknowns.

The good news is that our statistical tests (T- and F-tests) take this all into account. When two EVs are highly correlated, the appropriate variances of the regression parameter estimates (*see* Eq. (9)) are automatically increased—acknowledging the fact that we cannot determine which EV should explain what in the data. So, even though the statistics accounts for orthogonality, it is clear that when we design our experiments, we want to avoid this being an issue whenever possible. This can be achieved by using approaches that assess the efficiency of experimental designs such as [16–18].

In Fig. 14, we can see three different design matrices. The first contains two EVs that are highly correlated, the second shows two EVs that are partly correlated, and the third shows two EVs that are completely uncorrelated. As discussed, the first design matrix is "rank deficient." The third design matrix is the ideal scenario in

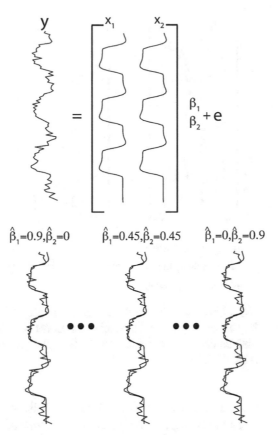

Fig. 13 Rank-deficient design matrices. At the top, we can see a general linear model (*GLM*) with a design matrix containing two identical explanatory variables (*EVs*). Since EV1 and EV2 are so similar, we can equally well use either EV1 or EV2 to explain what we see in the data. At the *bottom* of the figure, we can see examples of identically good model fits with any linear combination of the EVs as long as $\widehat{\beta}_1 + \widehat{\beta}_2 = 0.9$. The result is that we cannot estimate either $\widehat{\beta}_1$ or $\widehat{\beta}_2$ with any certainty (corresponding to [1 0] or [0 1] contrasts), but we can estimate $\widehat{\beta}_1 + \widehat{\beta}_2$ (corresponding to a [1 1] contrast)

that the EVs are uncorrelated and there is no ambiguity in how to determine which EV explains what in the data. We refer to this as a "well-conditioned" design matrix. But what happens in the case of the second design matrix where the EVs are partially correlated? It is useful to think of the EVs as having uncorrelated (orthogonal) and correlated (non-orthogonal) components. The correlated (or non-orthogonal) components are of no use, as they are, by definition, identical and cannot be used to disambiguate which EV explains what in the data. Hence, the model fitting can only be driven by the uncorrelated, or orthogonal, components of the EVs. As long as there is a substantial orthogonal component, then there is sufficient information to get an efficient estimate of the regression parameters, and we can successfully infer on such GLMs.

The idea that the model fitting can only be driven by the uncorrelated, or orthogonal, components of the EVs is a crucial

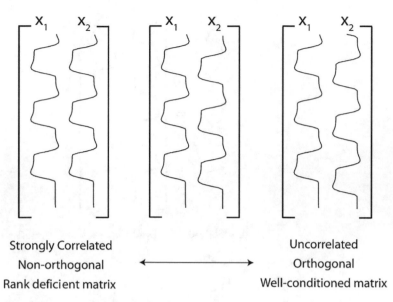

Fig. 14 Examples of different design matrices. Design matrix with two explanatory variables (*EVs*) that are highly correlated (*left*). Design matrix with two EVs that are partially correlated (*middle*). Design matrix with two EVs that are uncorrelated (*right*)

one, and is well illustrated by the following example. What happens to the regression parameter estimates when we have two partially correlated EVs, and orthogonalize one with respect to the other? Figure 15 illustrates such a case. On the right, EV2 has been "orthogonalized with respect to" EV1, which just means that the part of EV2 which is correlated with EV1 has been subtracted from it. The counterintuitive result is that, even though it is EV2 that has been changed and EV1 has remained the same, it is β_1 that has changed and β_2 that remains the same. To understand this, remember that the model fitting can only be driven by the orthogonal components of the EVs. Although EV2 has changed, it has changed to be equal to the original orthogonal component and hence its orthogonal component is unchanged; subsequently, its regression parameter estimate is still the same. In contrast, although EV1 has not changed, because EV2 has been orthogonalized with respect to EV1, the orthogonal component of EV1 has changed; subsequently, its regression parameter estimate is different.

Everything we have considered up to now has been within the context of considering problems where we have two (or more) EVs that are correlated with one another. In fact, the problem is more general than this. We can see this by extending the analogy of solving simultaneous equations. In general, one encounters the same problems whenever it is possible to find a linear combination of the equations that is equal to another of the equations. For example, consider that we have three equations to solve for three unknowns,

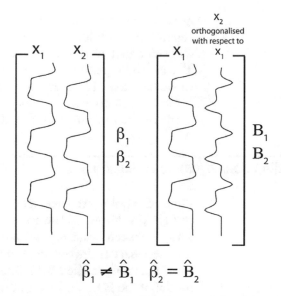

$$\hat{\beta}_1 \neq \hat{B}_1 \quad \hat{\beta}_2 = \hat{B}_2$$

Fig. 15 Effects of orthogonalization. We, first, fit the design matrix (containing two partially correlated explanatory variables, *EVs*) on the left to the data from a voxel, and obtain regression parameter estimates $\hat{\beta}_1$ and $\hat{\beta}_2$. We then construct a new design matrix, shown on the right, where EV1 is unchanged and EV2 is the old EV2 orthogonalized with respect to EV1. We then fit this new design matrix to the same data and obtain new regression parameter estimates $\hat{\beta}_1$ and $\hat{\beta}_2$. The counterintuitive result is that even though it is EV2 that has been changed and EV1 that has remained the same, it is $\hat{\beta}_1$ that has changed and $\hat{\beta}_2$ that remains the same. The underlying reason for this is that the model fitting can only ever be driven by the orthogonal components of the EVs

but it turns out that if we take two times Eq. (1) and subtract Eq. (2), then we get exactly Eq. (3). In that case, we actually only really have two equations to solve for our three unknowns, and so we are in trouble. In the GLM, the EVs are analogous to the equations, and the regression parameters are the unknowns. And so we have a problem if any EV is the same (or close to being the same) as a weighted sum of the other EVs in the design matrix. Again, we describe such a design matrix as being (or close to being) "rank deficient."

Even if we do have a design matrix that is close to being rank deficient, and therefore, there are some regression parameters that cannot be very well estimated, there may well be other regression parameters that can be estimated. There may even be contrasts that actually include the hard-to-estimate regression parameters, but that can still be estimated [16]. At first glance, this may seem a little counterintuitive. However, things should become clear if we consider a simple example of this. This occurs when we have the situation shown in Fig. 13 where we had a design matrix with two very similar EVs. As already discussed, we cannot estimate at all well the individual parameters β_1 and β_2 with [1 0] and [0 1] contrasts. However, we can estimate a [1 1] contrast since this does not require us to separate out which EV explains what in the data. Again the simultaneous the first equation analogy is useful. Consider that we have two equations, $2x + 2y = 2$ and $4x + 4y = 4$. The second equation is simply two times of the first equation; and so we have a problem, and we cannot solve for x and y individually. However, we can solve for $x + y$; it is equal to one. Solving for $x + y$ is analogous to estimating the [1 1] contrast in our GLM.

4   Modeling Hemodynamic Variability

Up to this point, we have been working under the assumption that the HRF is known. However, it is well established that the HRF varies between brain regions and subjects [19] and so in practice, it is necessary to incorporate into our modeling of the fMRI data some flexibility in the HRF. One option is to have a parameterized model of the HRF and then estimate the HRF shape parameters at the same time as we estimate the GLM regression parameters that represent the size of the response. For example, we could use the double gamma HRF illustrated in Fig. 5, but instead of fixing the five parameters that describe the shape, we now look to estimate those parameters from the fMRI data. The problem with this approach is that it is not straightforward to estimate these HRF shape parameters within the GLM framework, as they generally require nonlinear estimation approaches. A number of these approaches have been proposed, predominantly using Bayesian techniques [20–24]. However, these approaches are computationally demanding and are not yet in common use.

4.1   HRF Basis Sets

A popular alternative is to use the approach of basis functions. These allow HRF modeling flexibility but within the computationally undemanding GLM framework [19]. Figure 16a shows just one example of an HRF basis set that contains three basis functions. The choice of basis set is clearly important and we will come to that later. Whatever basis set is used, the principle is the same: different linear combinations of the basis sets can be used to give different HRF shapes. This is illustrated in Fig. 16b.

But how do we use these HRF basis functions in combination with our known stimulus timings to create predicted responses that can be used in the GLM? The answer is to separately convolve each of the HRF basis functions with the stimulus function to create an EV for each of the basis functions, as shown in Fig. 17. When the resulting design matrix is fit to the fMRI data, the required linear combination of these EVs is determined. If desired, the same linear combination can then be applied to the HRF basis functions to show the implied HRF shape.

The question remains as to how we set up a statistical test to ask, for example, "Where is there significant activation due to

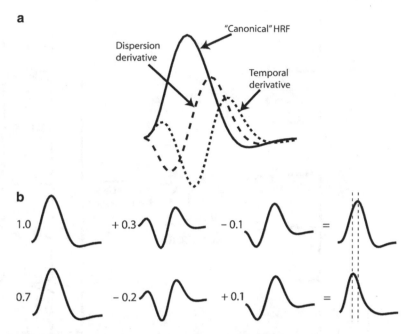

Fig. 16 (**a**) Example of a hemodynamic response function (*HRF*) basis set containing three basis functions. (**b**) Different linear combinations of the basis functions in the basis set can be used to obtain different of HRF shapes

condition A?" when we model the response to condition A using HRF basis functions. One answer is to use an *F*-test. Recall that one way to think about an *F*-test is that it will find significance if there is any linear combination of the contrasts (in the *F*-test) that explain a significant amount of variance in the data. So if we simply create an *F*-test made from the contrasts that pick out each of the regression parameter estimates for each of our basis function EVs for condition A, then we will find where there are any linear combinations of the basis set EVs that can be used to give HRF shapes that explain significant amounts of variation in the data. In other words, if we have an experiment with just condition A, and we model it using three basis functions (Fig. 17), then we can ask "Where is there significant activation to condition A?" with the *F*-test contrast matrix:

$$c^T = \begin{pmatrix} 1 & 0 & 0 \\ 0 & 1 & 0 \\ 0 & 0 & 1 \end{pmatrix}.$$

An important point to remember is that, as we are using an *F*-test, we lose directionality of the test. So we cannot tell if we are finding significance with a positive or negative response. However, it is possible to recover this post hoc if we are using a basis set that contains a canonical HRF, by looking at the sign of the regression parameter estimate for the corresponding canonical HRF EV. We

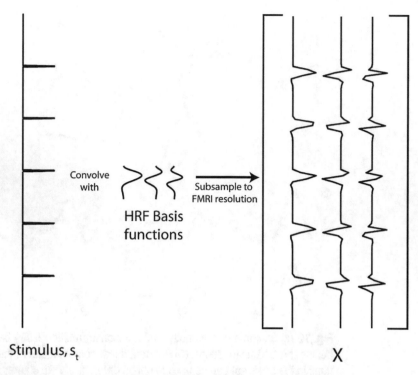

Stimulus, s_t X

Fig. 17 Setting up design matrix explanatory variables (*EVs*) using a hemodynamic response function (*HRF*) basis set. Each HRF basis function is separately convolved with the stimulus function to create an EV for each of the basis functions

can also make comparisons between two different conditions by pairing up the corresponding HRF EVs for the two conditions and setting up an *F*-test that looks for linear combinations of differences between corresponding HRF EV regression parameters. In other words, with three basis functions in our basis set, EVs 1–3 model the HRF EVs for condition A, and EVs 4–6 model the HRF EVs for condition B, then we can ask "Where is there significantly different activation between condition A and condition B" with the *F*-test contrast matrix:

$$c^T = \begin{pmatrix} 1 & 0 & 0 & -1 & 0 & 0 \\ 0 & 1 & 0 & 0 & -1 & 0 \\ 0 & 0 & 1 & 0 & 0 & -1 \end{pmatrix}.$$

More precisely, this is looking for where there is a significant amount of variance being explained by any linear combination of differences between condition A and condition B for corresponding basis function EVs. This means that we can find a significant difference due to either a "shape" or "size" change. Note that for this approach to be sensible, we need to use the same basis set for both conditions.

4.1.1 Choosing a Basis Set

Figure 16a showed an example basis set that had been derived from a parameterized HRF that was made up of a series of half-cosine functions [15]. This was obtained by sampling thousands of example HRFs from within a range of plausible parameter values for the HRF model (Fig. 18), and then a principal component analysis was carried out on these samples to determine the principal modes/components of variation in the HRF shape. The three highest principle components were then used as the basis set. Note that this can equally well be done on any form of parameterized HRF, for example, a double gamma HRF or a biophysical model such as the balloon model [25].

When this approach is taken with any plausible HRF model, it is typical for the first basis function to turn out to be the mean HRF shape, or a "canonical" HRF, for the second to approximate the temporal derivative (i.e., linear combinations of the first and second basis functions result in versions of the canonical HRF shifted in time), and for the third to approximate a dispersion derivative (i.e., linear combinations of the first and third basis functions result in versions of the HRF with different widths of the main positive response). Although other basis sets that have been proposed (e.g., sets of Gamma functions and finite impulse response functions), this kind of basis set is highly recommended since it parsimoniously captures shape variations. It also has the advantage that the first (canonical) basis function will tend to dominate the fit to the data and can then be used to determine the positivity or negativity of the HRF. Note that it is quite common for people to neglect the dispersion derivative and use just a temporal derivative since temporal shifts represent the most important variation in the HRF shape, particularly when working with boxcar stimuli.

Thus far, we have considered basis sets made up of two or three basis functions. But why not use more? The reason for this,

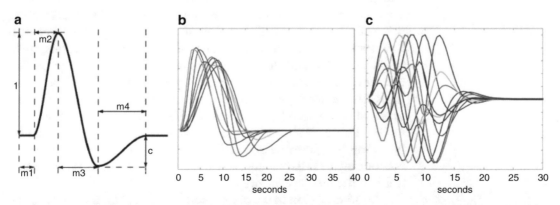

Fig. 18 (a) Parameterized hemodynamic response function (*HRF*) model. (b) Example HRFs sampled from this parameterized model of the HRF for plausible parameter values. (c) Samples of the HRF obtained from random linear combinations of the basis set shown in Fig. 16a

and in general the reason why basis sets with large numbers of basis functions (e.g., finite impulse response basis sets) are suboptimal, is that the GLM becomes unrealistically flexible. This problem is evident even with just three basis functions. Figure 18c shows possible HRFs that result from random linear combinations of the basis set in Fig. 16a. Clearly, many of these HRFs are nonsensical. The problem is that random fluctuations in the fMRI noise can, by chance, look like these nonsensical HRFs and so we "over-fit" the model to the noise. The statistical inference (e.g., via F-tests across the basis functions) is still valid, but we lose sensitivity; it becomes harder to detect genuine activations. On the contrary, if we use too few basis functions then we can fail to estimate the true HRF and our model is then a poor match to the data and again we suffer a reduction in sensitivity. So there is a trade-off between providing enough basis functions to provide enough HRF variability, while not having so many that over-fitting becomes a problem. A general rule of thumb is that three basis functions are good for single-event designs and two basis functions are good for boxcar designs.

An increasingly used approach to overcome this problem is to infer on models that incorporate HRF variability using a Bayesian framework. One advantage of a Bayesian approach is that prior information can be included. Priors can be used that prohibit nonsensical HRFs. Subsequently, more flexibility can be allowed while protecting against over-fitting [15, 22–24].

4.1.2 Basis Functions and Group Inference

In Sect. 6, we will discuss how we model multisession/subject fMRI data. However, it is worth mentioning how basis functions are best used when we are ultimately doing a group analysis. In particular, we consider this in the context of the simple case of inferring a population group mean. One option might be to pass up the regression parameter estimates for all basis functions into the higher-level group analysis, obtain the group average for each basis function separately, and then perform an F-test across them at the group level (in the same manner as we would do in a single-session analysis). However, it is not clear what benefit there would be of doing this. When our basis set contains a "canonical" HRF, the other basis functions, such as the temporal and dispersion derivatives, tend to average out to zero at the group level due to the different subject HRF shape variations. Subsequently, an often-recommended approach is to only pass up to the group level the canonical HRF regression parameter estimates. This makes for a simple group analysis, and the benefits of including the basis function at the first level are still felt in terms of accounting for HRF variability that would otherwise cause increased noise in the first-level analysis.

Another option that can be taken is to calculate a size summary statistic from the single-session analyses (e.g., the root mean square of the basis function regression parameter estimates), and pass that up to the group level. However, it is important to note that this would then require different inference methods at the group level (e.g., *see* Sect. 7.4.2) than is generally used, as the population distribution of such a summary statistic is likely to be non-Gaussian.

4.2 Nonlinearities

So far we have assumed linearity of the HRF. That is, we have assumed that the response to a stimulus is well modeled by (linear) convolution of the stimulus with the HRF. Typically, this is the approach that people take in the majority of fMRI analyses. However, it has been shown that this assumption is poor in certain situations. For example, it can be shown that the response to a prolonged stimulus is not as large as the one we would predict from extrapolating results from applying a short stimulus [26, 27], and nonlinearities are predominant when there are short separations (less than ~3 s) between stimuli [28]. Normally, these situations are intentionally avoided by designing experiments appropriately. For example, we avoid experiments where single events are occurring less than approximately 3 s apart, or experiments that require comparisons between a mix of short (e.g., single-event) and prolonged (e.g., boxcar) stimuli. However, if these situations are unavoidable, then it becomes necessary to model the nonlinearities.

Such nonlinearities are predicted by nonlinear biophysical models, for example, the balloon model [25]. Hence, one solution is to model fMRI data using these nonlinear biophysical models [22]. Another approach that can be used in the GLM setting is to extend the idea of convolution to include second-order nonlinear terms using Volterra kernels [28].

4.2.1 Volterra Kernels

Volterra kernels are a generalization of convolution to include higher order nonlinear terms. In fMRI, we need only to add the second-order terms to the first-order convolution terms to get the most important nonlinear behavior. A second-order Volterra kernel model is given by:

$$x(t) = \int_0^\infty h_1(\tau) s(t-\tau) \, d\tau +$$

$$\int_0^\infty \int_0^\infty h_2(\tau_1, \tau_2) s(t-\tau_1) s(t-\tau_2) \, d\tau_1 d\tau_2 \tag{15}$$

where $s(t)$ is the stimulus, the first term contains the traditional linear HRF first-order kernel, $h_1(\tau)$, and the second term includes the second order kernel, $h_2(\tau_1, \tau_2)$.

Many of the issues surrounding the use of second-order Volterra kernel basis functions are the same as they are for linear basis functions. For example, Volterra kernels can be determined empirically [25, 28], or derived from nonlinear biophysical models [22]. Either way, as with linear basis functions, there is variability in the response between different subjects and brain regions, and this variability can be parsimoniously captured within the GLM by using basis functions. We can obtain parsimonious basis sets for first- and second-order kernels by using principle component analysis on samples of the response from empirical data or from parameterized models. Furthermore, we can infer on the Volterra kernels in a Bayesian framework with priors that prohibit nonsensical responses, so that more flexibility can be allowed while protecting against over-fitting [22].

5   De-Noising fMRI Data

fMRI data are inherently noisy and contain a variety of fluctuations induced by processes beyond the control of the experimenter. Examples of such effects include artifacts related to the MR physics (such as slice-dependent signal dropout due to imperfect switching of the slice-select gradients, EPI "ghosting," and thermal noise), subjects' head motion effects, fluctuations induced by the cardiac and respiratory cycles, and spontaneous low-frequency fluctuations of the baseline signal.

Under the assumptions of the GLM, any fluctuation in the measured BOLD signal that is not modeled by the EVs in the design matrix is deemed to be noise. In Sect. 3.3, we discussed how such artifacts are more likely to occur at low frequencies and how we can use pre-whitening to deal with this. However, this assumed that all these fluctuations are stochastic, and that within the framework of the GLM, this stochastic noise is well described by a Gaussian distribution.

In practice, however, some of the underlying random noise fluctuations will have very distinct spatial and/or temporal structure. As an example, Fig. 19 shows a variety of such structured noise components identified from a single fMRI dataset using an independent component analysis (ICA) decomposition [29]. This suggests that such effects are structured rather than random stochastic noise and the challenge is to account for their existence in order to obtain optimal estimates of the GLM model parameters.

5.1   Structured Noise and the GLM

The main problem with such structured noise artifacts is that they can severely impact our GLM-based analysis. The part of the artifact that is orthogonal (uncorrelated) with all of the EVs will simply not be modeled by the design matrix regressors, and therefore, the presence of the structured noise effect will be reflected by an increase in the residual GLM noise variance. This, in turn, will decrease any T- or F-statistics value, making it harder for us to detect true activations. The non-orthogonal (correlated) part of such an artifact, however, will result in wrong parameter estimates for those EVs that correlate with the artifact. If the correlation is

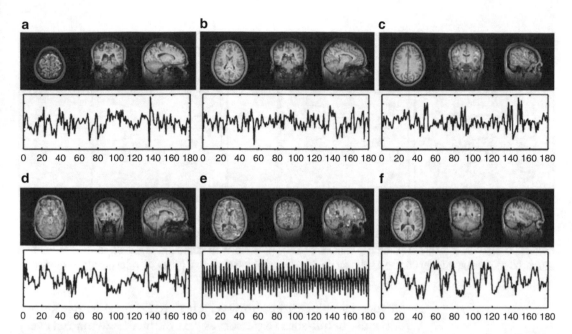

Fig. 19 Examples of "structured noise" identified in a single fMRI dataset using independent component analysis [29]: (**a**) residual head motion, (**b**) signal fluctuations in the ventricles, (**c**) spontaneous fluctuations in the bilateral sensory motor cortex, (**d**) fluctuations close to the sinuses (likely due to interactions between B_0 field inhomogeneities and head motion), (**e**) high-frequency image ghosting, and (**f**) more spontaneous low-frequency fluctuations

positive, we will overestimate the parameter for the EV, making it more likely that we wrongly detect activations where there are none. If, on the other hand, the correlation is negative, then estimated effect sizes will be underestimated, making it harder to detect true activations.

In short, the impact on the GLM estimates and the statistics values can be profound. Therefore, if we can characterize the spatial and/or temporal structure of such artifacts, it is desirable to incorporate this knowledge into the data analysis and to explicitly account for the presence of these effects in the data.

5.2 Nuisance Regressors in the GLM

One possible way of correcting for the negative impact on GLM statistics is to introduce additional "nuisance" or "confound" regressors in the GLM design matrix. Remember from Sect. 3.9 that in the case of multiple regressors, the parameter estimates for each of the EVs can only be driven by the uncorrelated (orthogonal) component of an EV. If we can find a suitable characterization of the temporal structure of an artifact, we can add this as a new regressor to the design matrix in order to use this to "explain" some of the measured variation in the data. The parameter estimate for EVs of interest will then only reflect the amount of variation that the EV can explain over and above what can already be explained by

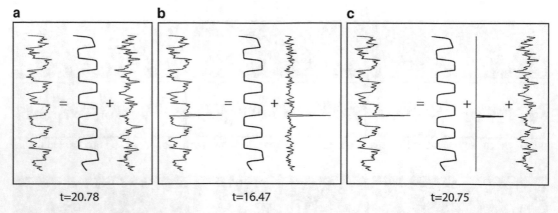

Fig. 20 Example of the utility of nuisance regressors in the general linear model (*GLM*): (**a**) data without artifact regressed against a single explanatory variables (*EVs*), (**b**) data with confound analyzed in a GLM without nuisance regressor, and (**c**) confounded data analyzed using both the EV of interest and a nuisance regressor

nuisance variables. Note, however, that we do need to pay a price for the use of nuisance regressors as part of the design matrix: every new regressor does reduce the number of DOF for our final statistical comparison. As such, it is desirable to keep the number of nuisance regressors to a minimum while trying to maximize the amount of structured noise variance captured by these regressors.

Figure 20 illustrates the use of a nuisance regressor. In this example, an fMRI time series exhibits an intensity jump due to the presence of a scanner-induced image artifact such that during one of the TRs, the measured image intensity for this time point is about 10 % above the mean intensity level. In Fig. 20a, the voxel's time series without the artifact is analyzed using a simple GLM with one single EV. The analysis represents the data as a linear combination of the EV of interest and residual noise. The level of activation in this voxel is high, resulting in a very significant *T*-statistic. In Fig. 20b, the same GLM design matrix is now fitted to the voxel's time series with the artifact present. The timing of the artifact is almost entirely uncorrelated with the primary EV and the presence of the artifact will therefore result in an inflated residual variance, causing a significant drop in the *T*-statistics value of more than 20 %. If information about the temporal characteristics of the artifact is available, then we can model the intensity variation at the specific time of the artifact by introducing a nuisance variable into the GLM design.

5.2.1 Deriving Nuisance Regressors

There are various ways of deriving useful nuisance variables. In general, these additional EVs should reflect the temporal characteristics of structured noise that is thought to exist in the data.

Motion of the subject in the scanner is a typical problem in fMRI, and there are often intensity fluctuations related to head motion still present in the data even after alignment-based motion correction. A common set of nuisance regressors (used to help model out such residual effects of motion) is the set of six time series obtained as the parameters of the head motion correction procedure. In this case, the intensity variations in the data are expected to correlate with the size of the three translations and three rotations. When included, these regressors can jointly "explain" any signal variation in the data correlated with head motion.

Other sources of structured noise effects are the subjects' cardiac and respiratory cycles. A popular approach is to use retrospective image correction (RETROICOR [30]) in order to correct for these effects. Using additional measurements of the heart and respiration cycles, one can derive nuisance regressors that permit one to remove all signal variation in the fMRI data that temporally correlates with the relative phase of these physiological cycles. The regressors are based on these additional measurements as low-order Fourier terms, and can significantly reduce the amount of structured noise induced by physiological fluctuations.

5.2.2   ICA-Based De-noising

The ability to correct for additive structured noise depends on the ability to characterize these noise components in terms of their temporal evolution. In the previous two examples, this was obtained by accurately estimating rigid-body motion or by using secondary measurements of physiological processes. For other types of noise, it is often not easy to predict such nuisance regressors based on the understanding of the biophysics and of the imaging process. One possibility is to use a model-free data analysis approach, such as ICA, in order to identify structured noise effects in the data prior to the GLM analysis. ICA and related techniques decompose the fMRI data into modes of variation that define the spatial and temporal extent of underlying fluctuations [29]. The estimated time courses of a component can then be used as nuisance regressors as part of a GLM analysis. An alternative is to explicitly regress out such effects prior to a GLM-based analysis, effectively running the model-based analysis on the residuals of a prior linear regression model designed to de-noise the data. Currently, as no well-established techniques exist for automatically identifying such noise components, such an approach relies on the experimenter to visually inspect all components. Further research is required to integrate such a model-free identification (e.g., using ICA) into the standard GLM in an unbiased objective way.

5.3   Example

Figure 21 demonstrates the impact of structured noise on the GLM and highlights the utility of such a de-noising approach. Subjects were requested to perform simple finger tapping using either the left or right hand. All subjects were right-handed and one of the contrasts of interest involved the left-right comparison "Where is the activity larger when using the left hand when compared with the right hand?" Prior to the noise removal, the

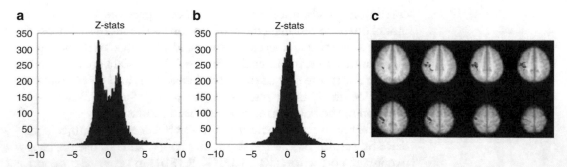

Fig. 21 Example of the effect of fMRI de-noising in a simple finger tapping experiment: (**a**) histogram of the Z-statistic image for the differential left- versus right-hand finger tapping contrast. Because of the presence of structured noise, the histogram is far from being Gaussian distributed; (**b**) after regressing out a variety of structured noise effects, the histogram of Z-statistic values becomes unimodal and much closer to a Gaussian distribution; and (**c**) map of significant voxels after regressing nuisance effects out of the data

histogram of the Z-statistic values for this contrast is highly non-Gaussian. The contrast map itself did not reveal any significant differences when using standard thresholding. After de-noising, the Z-statistic image of this contrast identifies significant differences in the BOLD, particularly in right motor cortical areas.

6   Multisubject Statistics

We have so far only focused on ways of modeling and fitting the (time series) signal and residual noise at the individual single-session level, in order to derive effect size estimates from a single fMRI dataset. The majority of fMRI studies, however, are used to address questions about activation effects in populations of subjects. This generally involves a multisubject and/or multisession approach where data are analyzed in such a way as to allow for hypothesis tests at the group level [12, 31], for example, in order to assess whether the observed effects are common and stable across or between groups of interest.

Figure 22 illustrates an example scenario where the question of interest involves estimating the difference in activation between two groups of subjects. This question is addressed by having different GLMs at the session, subject, and group level in a hierarchical fashion. At the lowest level of the analysis, the single-session time series data is modeled in the way described previously. At the subject and group level, there are GLMs that, for example, model the subject (cross-session) mean and group (cross-subject) mean effect sizes, respectively. At the top-level of the hierarchy, a set of statistic images is created that can be used for final statistical inference.

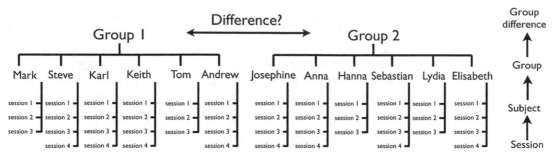

Fig. 22 Hierarchical general linear model (*GLM*) for the analysis of group fMRI data. Within a summary statistics approach, the GLMs are estimated one level at a time and summary statistics are passed up to the next level of the hierarchy

6.1 Brain Atlases

Registration (aligning different brain images) is typically used when combining fMRI data from different sessions or subjects in a multisubject analysis. This allows us to assume that the data we are comparing across sessions or subjects come from approximately corresponding areas of the brain. In doing this, it is typical to transform the data into a common "standard brain space," for example, the co-ordinate system specified by Talairach and Tournoux [32]. These standard spaces can be either what are known as templates or atlases.

A *template* is typically an average of many brains, all registered into a common co-ordinate system. An example is the MNI 305 average [33]. An *atlas* is also based in a common co-ordinate system, but contains richer information about the brain at each voxel, for example, information about tissue type, local brain structure, or functional area. Atlases can inform interpretation of fMRI experiments in a variety of ways, helping the experimenter gain the maximum value from the data.

6.2 Fixed- Versus Mixed-Effects Models

An important question is that of how to model and estimate effects at the intermediate and higher level of the hierarchy. If we were only concerned about the particular set of subjects in our study, then we would use a fixed-effects model. More typically, however, we would want to generate results that extend beyond the particular population of subjects scanned as part of the study, into the wider population. In this case, we also need to account for the fact that the individual subjects themselves are sampled from the wider population and thus are random quantities with associated variances. It is exactly this step that marks the transition from a simple fixed-effects model to a mixed-effects model and it is imperative to formulate a model at the group level that allows for the explicit modeling and estimation of these additional variance terms.

As an example, consider the simplest case of estimating the effect size of a group of M subjects, where for each subject k, the pre-processed fMRI data is Y_k, the design matrix is X_k and the parameter estimates are βk (for $k=1,\dots, M$). The individual first-level GLMs relate first-level regression parameters to the M individual datasets: $Yk = Xk\beta k + \varepsilon k$, where εk specifies the single-subject residuals. If we are only concerned about the exact population of subjects scanned under our fMRI paradigm, then the estimate of the group mean effect size is simply the average over all the lower-level parameter estimates: $\beta_g = (1/M)\sum_{k=1}^{M}\beta_k$, that is, our second-level GLM is simply

$$\beta_k = X_g \beta_g \tag{16}$$

where $X_g = [1/M, \dots, 1/M]^T$, and we have concatenated all of the first-level regression parameters into one vector:

$$\beta_k = \begin{bmatrix} \beta_1 \\ \beta_2 \\ \vdots \\ \beta_2 \end{bmatrix}.$$

In this case, we simply need to average the first-level regression parameters and the only variance to consider is the average first-level variance.

If, however, we want to generalize our findings to the wider population then the second-level analysis needs to account for the sampling of the subjects, and we need to proceed by modeling the group effect size of interest as

$$\beta_k = X_g \beta_g + \varepsilon_g \tag{17}$$

where ε_g accounts for the variation of the different subjects' means from the overall group mean. Both the within- and the between-subject variations contribute to the total mixed-effects variance against which the mean effect size is tested during the statistical inference procedure.

The difference between the two approaches is illustrated in Fig. 23. In the fixed-effects model (a), only the first-level variances need to be considered, whereas in the case of the mixed-effects analysis (b), both the first-level fixed-effects variances and the higher-level random-effects variance contribute to the total mixed-effects variance used for inference. The between-subject variance, σ_g^2, then accounts for the random sampling of the particular subjects from the wider population.

In the following sections, we assume that a mixed-effects analysis is being performed, as this is typically what is required. Fixed-effect analyses are performed in a similar manner but without the complication of needing to estimate the random-effect variances.

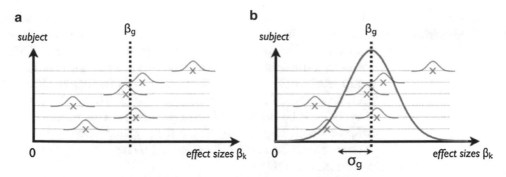

Fig. 23 Illustration of a simple group analysis using the fixed-effect and the mixed-effects models: (**a**) in the fixed-effects analysis, the only variance contribution to consider is the lower-level within-subject variance; (**b**) in the mixed-effects analysis, the between-subject random-effects variance, σ^2, accounts for the random sampling of the subjects themselves and contributes to the overall mixed-effects variance

6.3 Summary Statistics Approach

We now come to the question of how we infer on the multilevel hierarchy of a group study (an example of which was shown in Fig. 22), in order to ask questions such as "Where is there significant activity in response to the experimental task for the population?" or "Where is there significant differences between populations (e.g., controls versus patients)?" Recall that each level in the hierarchy is represented by its own GLM. Hence, one approach is to formulate a single complete GLM that combines together the first-level and higher-level GLMs. An example of such an approach is presented in Friston et al. [34] where the group analysis is carried out "all-in-one" using the within-session fMRI time series data as input. However, in fMRI, where the human and computational costs involved in data analysis are relatively high, it is desirable to be able to make group-level inferences using the results of separate first-level analyses. This approach is commonly referred to as the "*summary statistics*" approach to fMRI analysis [31]. Within such an approach, group parameters of interest can easily be refined as more data become available.

In Holmes and Friston [31], the regression parameter estimates were used as summary statistics. The regression parameter estimates from the lower level are used as the "data" at the next level up. For example, the estimates of βk are used in place of βk in Eq. (17). This approach was shown to be equivalent to inferring all-in-one under certain conditions [31]. For example, it requires balanced designs, that is, all lower-level design matrices need to be identical, preventing the use of behavioral scores or subject-specific confound regressors.

However, top-level inference using the summary statistics approach can be made equivalent to the all-in-one approach without such restrictions [35, 36], if we pass up the correct summary statistics. In particular, it is important to pass up information about not only the effect sizes from the lower levels, but also their variances. We shall explore the benefits of doing this in Sect. 6.4.

**6.4 Estimation
of the Mixed-Effects
Model**

When we use a summary statistic approach, we infer on each level of the hierarchy one at a time. The first-level inference is as we described earlier in the chapter. At higher levels in the hierarchy, however, estimating the regression parameters and variances of group-level GLMs offers a different set of challenges. At the first level, there typically exists a large number of observations (typically more than 100), so that relevant parameters and variances can be estimated with high DOF. In contrast, group-level variance component estimation is typically troubled by having very few observations (i.e., low DOF).

A key issue in estimating mixed-effects models within the "summary statistics" approach is whether the variance information from the lower levels (e.g., first level) is used at the higher levels (e.g., group level). Approaches that use the variance information from the lower levels have a number of substantial advantages. First, such approaches do not require balanced designs and so permit the analysis of fMRI data where the first-level design matrices have different structure from each other (e.g., contain behavioral scores as regressors) or where the data contains different numbers of observations (e.g., different numbers of sessions for each subject). Second, they provide more accurate variance estimation (and therefore more accurate inference) by ensuring that at every level, only positive estimates of the random-effects variances contribute to the overall mixed-effects variance. Finally, such approaches increase the ability to detect real activation, by weighting the different contributions from the lower levels by using the lower-level variance information. For example, effect sizes from subjects with high first-level variance get down-weighted compared with those with low first-level variance, when inferring at the group level.

Estimating mixed-effects models when the lower-level variance information is ignored can be carried out easily using ordinary least squares [31]. Approaches that use the lower-level variance information and provide the advantages described above are a little more involved. For example, Worsley et al. [12] used an expectation maximization approach. Woolrich et al. [36] used a fully Bayesian framework using appropriate noninformative priors. This approach had the added advantage that one can model different variance components for different groups. For example, one can contrast effect sizes in a population of patients relative to a population of controls under the assumption that these two groups have different within-group variance. This is important as, empirically, patient populations exhibit larger within-group variability than a carefully selected population of controls.

**6.5 Handling Outlier
Subjects**

The approaches described so far assume that the population distributions of the effect sizes are well modeled using a Gaussian distribution. However, in practice group studies can include "outlier" subjects whose effect sizes are completely at odds with the general population for reasons that are not of experimental interest. For example, it could be due to excessive subject motion or misunderstanding by the subject of the experiment instructions. The estimate of the population variance can be inflated by outlier subjects, and the population mean estimates can be under-or over-estimated. This is analogous to the presence of structured noise in the context of single-session analysis, as discussed in Sect. 5.

A number of approaches have been proposed to deal with this problem. One option is to visually inspect both the data and the results of a group analysis to deduce outliers. These outlier subjects can then be removed and the group study re-analyzed without them. Although useful exploratory approaches have been proposed that aid in this process [37–39], human intervention is often still required. It is preferable to use approaches that are automatic, and soft-assign outlier behavior in a spatially localized manner [40, 41]. The approach of Woolrich [41] has the added benefit that it uses the lower-level variance information.

Another possibility to dealing with outliers is to use nonparametric statistics (*see* Sect. 7.4.2). For example, Meriaux et al. [42] and Roche et al. [43] use permutation tests that take advantage of the lower-level variance information. These approaches protect the validity of the statistics but can be less sensitive compared with techniques that explicitly model the outliers [41]. However, they are also potentially able to handle other deviations from Gaussian population distributions (e.g., populations with two sub-populations), and there is evidence that nonparametric statistics in general handle the multiple comparison problem better than random field theory (RFT) [44] (this is discussed further in Sect. 7.4.2).

6.6 Creating Higher-Level GLMs

When creating higher-level GLM design matrices, it is important to ensure that the design matrix at least models the cross-subject mean effect. This is different from a first-level analysis where the overall time series mean is normally not of interest and might actually be removed prior to the first-level GLM. In the case of a higher-level analysis, however, the mean lower-level effect often is exactly what is of interest and therefore needs to be explicitly modeled as part of the design matrix, either as a single EV or as a linear combination of EVs.

Figure 24 gives a selection of typical fMRI higher-level designs. In the simplest case (a), the higher-level design only models a single group mean effect and a simple [1] contrast then tests if the mean effect is greater than 0.

In some cases, additional subject-specific behavioral scores need to be included as additional EVs (b). This can be either to remove some higher-level variation of no interest by including these EVs as nuisance regressors (e.g., by regressing out subjects' age or gender or reaction time), or because these regressors are part of a testable hypothesis and need to be included in a contrast

$$
\mathbf{a} \quad \mathbf{b} \qquad\qquad \mathbf{c} \qquad\qquad \mathbf{d}
$$

$$
\begin{bmatrix} 1 \\ 1 \\ 1 \\ 1 \\ 1 \\ 1 \\ 1 \\ 1 \end{bmatrix}
\begin{bmatrix} 1 & -2 \\ 1 & -7 \\ 1 & 4 \\ 1 & 0 \\ 1 & 3 \\ 1 & 4 \\ 1 & -5 \\ 1 & 3 \end{bmatrix}
\begin{bmatrix} 1 & 0 \\ 1 & 0 \\ 1 & 0 \\ 1 & 0 \\ 0 & 1 \\ 0 & 1 \\ 0 & 1 \\ 0 & 1 \end{bmatrix}
\begin{bmatrix} 1 & 1 & 0 & 0 & 0 \\ -1 & 1 & 0 & 0 & 0 \\ 1 & 0 & 1 & 0 & 0 \\ -1 & 0 & 1 & 0 & 0 \\ 1 & 0 & 0 & 1 & 0 \\ -1 & 0 & 0 & 1 & 0 \\ 1 & 0 & 0 & 0 & 1 \\ -1 & 0 & 0 & 0 & 1 \end{bmatrix}
$$

Fig. 24 Typical higher-level general linear model (*GLM*) design matrices: (**a**) group mean effect size over eight subjects, (**b**) group mean and confounds over eight subjects, (**c**) unpaired group difference over two groups of four subjects, (**d**) paired group difference test over two conditions for four subjects. Note that for the sake of space, the number of subjects assumed here is lower than what would be typically expected in a group study

of interest (e.g., a researcher might be interested in effects which correlate significantly with duration of treatment in a clinical population). In both cases, the additional EVs would need to be orthogonalized relative to the EV that is modeling the group mean. This is so that the regression parameter for the group mean EV can indeed be interpreted as the overall group mean effect.

The simplest multiple-group design involves just two groups where the question of interest involves assessing the between-group difference (c). In this case, each group's mean effect is modeled using a separate EV, a $[1-1]$ contrast can then be used to assess $A > B$ differences; the negative $[-1\ 1]$ contrast tests for $B > A$ differences.

Another typical design involves testing for differences between a set of data generated under different conditions A and B in the same population (**d**), for example, where subjects get scanned before and after a period of learning. This is often referred to as a *paired T-test*. Every subject has two observations and we need to account for the within-subject covariance by means of subject-specific confound regressors. In this case, the first EV models the $A - B$ differences for the M subjects, while the M additional EVs account for the subject-specific mean effects. For example, the second EV in Fig. 24d models the mean effect for the first subject. A $[1\ 0\ 0\ 0\ 0]$ contrast can then be used to assess the $A - B$ paired difference.

7   Inference ("Thresholding")

As we saw in Sect. 3.5, a result of fitting a GLM is a T-statistic image for each contrast, where the intensity at each voxel assesses the evidence for a nonzero effect. Ideally, the statistic image would be zero where there is no effect and very large where there is an effect. Of course, because of the noise, this is not the case, and we must make inference—a statistically calibrated decision—on which voxels exhibit a signal and which voxels are just consistent with noise. Here we are talking about inferring on T-statistic images; however, the issues involved with F-statistic images generated from F-tests, or Z-statistic images (*see* Sect. 3.7), are similar.

The simplest inference method is voxel-wise thresholding. If the value of a T-statistic image is t at a given voxel, then we reject the null hypothesis of no experimental effect if $t > u$ (where u is a significance threshold). However, one should first ask: Why threshold?

7.1 Inference with the Mass Univariate Model

The natural questions that any user of FMRI has of their data are: "What is the location of my signal?," "What is the extent of my signal about that location?," "What is the magnitude of my signal?." For each of these, our statistical model should provide an estimate, a measure of uncertainty of the estimate (i.e., a standard error or a confidence interval), and a significance measure, like a P-value (i.e., could our result be explained by chance alone). For example, if a visual effect produced a cluster (a contiguous groups of suprathreshold voxels, more on this in Sect. 7.2) with a peak at a certain location, how certain can I be that the true center of activation is near that location? Or, if a cluster had a volume of 500 voxels, what is my confidence that the true signal extent is 500 voxels?

Surprisingly, such basic questions cannot be answered with standard fMRI methods. In fMRI, we generally use a mass univariate model, where a GLM is fit independently at each voxel. No information is shared over space, and, specifically, no explicit spatial model is used to express the extended signals that we expect. While more advanced methods that address these issues exist [45], the only inferential questions that a mass univariate model can answer are (a) "What is the signal magnitude at each voxel (with standard errors and P-values)?" and (b) "What is the signal extent for a given cluster-defining threshold (P-values only)?." However, standard errors and P-values on locations (e.g., confidence intervals on local maxima or center of mass of a cluster) are not available.

The remainder of Sect. 7.2 focuses on these two types of inferences: voxel-wise and cluster-wise.

7.2 Voxel-Wise Versus Cluster-Wise Inference

The result of applying a contrast to the GLM fit at each voxel is a statistic image. This is anywhere from $I = 20{,}000\text{--}100{,}000$ brain voxels in a statistic image. The value in the image at each voxel is a T-, F-, or Z-statistic that measures the evidence for an effect defined by the contrast. The process of applying threshold u, and retaining all voxels with statistic value greater than u is known as voxel-wise inference. Precisely, we are performing I statistical tests of significance, rejecting the null hypothesis at voxel i if $t_i \geq u$, where t_i is the statistic value at voxel i.

Alternatively, we can apply a cluster-forming threshold, u_{clus}, create a binary image of voxels $t_i \geq u_{clus}$, and identify clusters, that is, contiguous groups of supra-threshold voxels, with cluster having size S. The process of retaining all clusters with size greater than k is known as cluster-wise inference. Precisely, we are performing L statistical tests, one for each of the L clusters in the image, rejecting the cluster null hypothesis if $S \geq k$. The cluster null hypothesis is that all of the voxels in cluster have no signal, and the inference is cluster-by-cluster. Hence, an unusually large cluster extent only tells us that there exists one or more signal voxels within the cluster, but not which voxels within the cluster have signal.

Figure 25 illustrates the difference between voxel- and cluster-wise inference. While voxel-wise inference rejects the null hypothesis for individual voxels, cluster inference jointly rejects the null hypothesis for a set of voxels within a cluster as significant. As such, we say that cluster-wise inference has less spatial specificity than voxel-wise inference. On the contrary, since there are many fewer clusters than voxels, the multiple testing problem (*see* Sect. 7.3) is more severe with voxel-wise inference.

What cluster-forming threshold should be used? Using a relatively low threshold will allow clusters with small statistic values to be formed, but the clusters may not be significant, as a low threshold will also result in large clusters by chance alone. A high threshold ensures that clusters due to chance noise are small, but may then miss true signals that have relatively small magnitude. Below we will introduce two methods for finding P-values for cluster size, one of which requires relatively high thresholds: Random Field Theory requires relatively high u_{clus} values (uncorrected P-value of 0.001 or smaller) to give accurate inferences [46], while permutation is valid with any chosen u_{clus} threshold. Regardless of which threshold is chosen, the threshold should be picked before examining the data. Trying many thresholds introduces a multiplicity that is not easily accounted for, and will reduce the confidence of any significant findings. Later, in Sect. 7.5, we will consider the approach of threshold-free cluster enhancement, which provides cluster-like inference without the dependence on u_{clus}.

Should we be using voxel-wise or cluster-wise inference, or both? If one examines both cluster-wise and voxel-wise results yet another multiple testing problem is introduced, and so one method does need to be chosen a priori. Friston et al. [47] shows that when the anticipated signals are broad or spatially extended, cluster size inference is best, and when focal, intense signals are anticipated, voxel-wise inference is best. Both methods are widely used, with cluster-wise inference being slightly more common, probably due to the large extent of effects typical after spatial smoothing.

7.3 Correction for Multiple Tests

Statistical hypothesis testing provides a means to test a default, or null, hypothesis with a pre-specified false positive rate. At a single voxel, a test statistic can be compared to an $\alpha = 0.05$ significance threshold, denoted $u\alpha$, where α is the allowed risk of false positives. For example, a Z-statistic at one voxel will, with many repetitions of a null experiment, exceed $u\alpha = 2$ about 2% of the time.

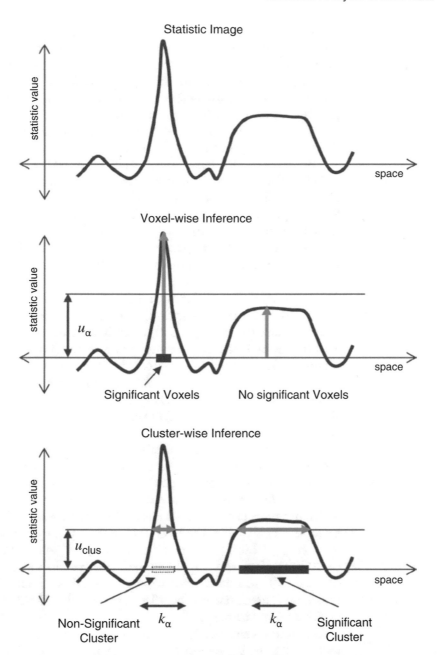

Fig. 25 Voxel-wise versus cluster-wise inference. Inferences on fMRI statistic images are made either through voxel-wise or through cluster-wise methods. The *top panel* illustrates the values in a statistic image through one line of space, where large values indicate evidence for an experimental effect. The *middle panel* illustrates voxel-wise inference, where a significance threshold $u\alpha$ is applied to the image, and voxels above that threshold are labeled as significant. The advantage of voxel-wise inference is that individual voxels are marked as significant, but the disadvantage is no spatial information is considered, and unusually expansive effects may be missed. The *bottom panel* illustrates cluster-wise inference, where a cluster-forming threshold u_{clus} is applied to the image, and contiguous voxels are formed into clusters. Clusters that exceed a significance threshold $k\alpha$ in size are marked as significant. The advantage of cluster-wise inference is that low, spatially extended signals can be detected. The disadvantage is that clusters as a whole are marked as significant, and individual voxels within a cluster cannot be marked individually as significant

For a particular observed statistic value, say $t = 3.3$, we measure the evidence against the null hypothesis with a P-value, here $p = 0.0005$, which is the chance of obtaining, over repeated null experiments, a result greater or equal to $t = 3.3$. (Take care not to confuse P-values with posterior probabilities: Bayesian methods give the posterior probability that the null is true, conditional on the data; classical P-values, in contrast, are the probability of the data, conditional on the null hypothesis).

But if $I = 10,000$ voxels (or $K = 100$ clusters) are tested at level $\alpha = 0.05$, the false positive risk is only controlled at each voxel. Then over all voxels, a total of $I \times \alpha = 500$ false positive voxels (or five false positive clusters) would be expected. The problem is that standard hypothesis testing only accounts for the risk of false positives for one test. If multiple tests are to be considered, some measure of the risk of false positives over multiple tests is required.

7.3.1 Measures of Multiple False Positives: FWE and FDR

To carefully define different types of false positive measures, Table 2 shows a cross classification of each of the I voxels in an image. Voxels can be truly null, or truly have nonzero signal, and additionally each voxel can be detected by some (imperfect) thresholding method, or fail to be detected.

To fill in the values for true null and true signal voxels, we need to somehow know the underlying truth. Voxels that are marked as significant are "detected" and voxels that are not marked as significant are "not detected"

The standard measure of false positives in multiple testing is the *family-wise error (FWE) rate*, that is, the chance of one or more false positive voxels (or clusters) anywhere in the image [FWE $= P (I_{N+} > 0)$]. *Bonferroni* is the most widely known FWE method, which produces critical thresholds u^{FWE} that controls the FWE. This method simply divides the FWE rate (e.g., 0.05) by the number of tests (e.g., 10,000), to give a voxel-wise P-value threshold α (0.000005) that, when applied voxel-wise, results in the originally desired FWE control. Regardless of the method, if an $\alpha^{FWE} = 0.05$ threshold is used, one can be 95 % confident that there are no false positives in the image at all.

A more recent measure of false positives is the false discovery rate (FDR), the expected false discovery proportion (FDP), where FDP is the proportion of false positives among all reported positives. Precisely, if I_+ voxels are detected, FDP is the proportion of these that are false "discoveries," FDP $= I_{N+}/I_+$, where FDP is defined to be 0 if no voxels are detected. FDP is a random quantity that cannot be known for any particular real dataset, so the FDR is defined as the expected value of FDP over many datasets, FDR $= E(\text{FDP})$. Figure 26 shows the difference between FDR and FWE inference. Imagine the ten images shown in the figure as the next ten experiments you will analyze; of course, you only consider a single dataset at a time, but this illustrates how the methods are calibrated over many (idealized) repetitions of an experiment.

Table 2

Cross-tabulation of the number voxels in difference inference categories

	Voxels not detected	Voxels detected	
True null voxels	I_{N-}	I_{N+}	I_N
True signal voxels	I_{S-}	I_{S+}	I_S
	I_-	I_+	I

Signal+Noise

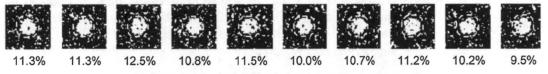

10 realizations of signal corrupted by noise

Control of Per Comparison Rate at 10% - Uncorrected

11.3%   11.3%   12.5%   10.8%   11.5%   10.0%   10.7%   11.2%   10.2%   9.5%

Percentage of Null Pixels that are False Positives

Control of Familywise Error Rate at 10%

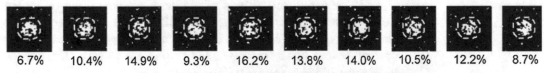

FWE

Occurrence of Familywise Error

Control of False Discovery Rate at 10%

6.7%   10.4%   14.9%   9.3%   16.2%   13.8%   14.0%   10.5%   12.2%   8.7%

Percentage of Activated Pixels that are False Positives

Fig. 26 Comparison of uncorrected versus family-wise error (*FWE*)-corrected versus false discovery rate (*FDR*)-corrected inferences. The *top rows* shows ten realizations of a central, circular signal added to smooth noise, and the next three rows show different possible voxel-wise thresholding methods applied to each realization. The *dashed circles* indicate the extent of the signal. The *second row* shows the result of an $\alpha = 10\%$ uncorrected threshold; most of the signal is correctly detected, but much of the background is also incorrectly detected. On average, 10% of the background consists of false positives, but notice that the exact proportion of false positives varies from realization to realization. The *third row* shows the result of using an $\alpha^{FWE} = 10\%$ threshold (e.g., a threshold from Bonferroni or random field theory). Much less of the signal is detected, but there are many fewer false positives, only 1-in-10 of the datasets considered had any false positives, this is a FWE. Of course in practice, we never know if the dataset in our hands is the 1-in-10 (or 20) that contains a family-wise error. The *bottom row* shows the result of using an $\alpha^{FDR} = 10\%$ threshold. While we are guaranteed that the percentage of detected voxels that are false positive does not exceed 10% on average, the actual percentage can vary considerably

FDR is a more lenient measure of false positives allowing some false positives—as a fraction of the number of detections—while FWE regards any false positives as an error. One special case is notable when the definitions of FDR and FWE coincide. If there are no signal voxels at all ($I_S = 0$), then FDP is 1 whenever there is an FWE, and the two methods give the same control of false positives. (In technical terms, it is said that FDR has weak control of FWE). This hints at the adaptive nature of FDR: when there is no signal, it behaves like FWE; as there are more and more signal voxels, it admits more and more false positives, yielding increased power while still controlling false positives in proportion.

The voxel-wise P-values referred to in previous sections are more accurately referred to as "uncorrected P-values," as they do not account for the multiple testing problem. "Corrected P-values" refer to the FWE rate (or FDR if this is used instead).

7.4 Corrected Inference Methods

So far we have only defined measure of false positives, but we have not described how we obtain thresholds that control these false positive measures. In addition to Bonferroni, there are two further methods that are commonly used in fMRI for controlling FWE; these are RFT and permutation, whereas there is generally just a single method for FDR.

7.4.1 Controlling FWE with RFT

The Bonferroni method for controlling FWE uses a significance threshold corresponding to $\alpha = \alpha^{FWE}/I$, the nominal FWE test level divided by the number of tests. Bonferroni becomes quite conservative when the data is smooth, and has no way to adapt to the data in anyway. For example, imagine an extreme case where FMRI data is smoothed with a 1 m wide Gaussian smoothing kernel; such data will produce a statistic image with essentially a single constant value, meaning there is no multiple testing problem anymore; however, the Bonferroni threshold will still prescribe dividing α^{FWE} by, say, 10,000.

RFT uses the smoothness of the data to adjust the significance threshold while still controlling FWE. The mathematics involved are elegant yet quite involved, and in what follows, we only give the most cursory review. For a more detailed review, see [48], or for a more technical overview, see [49]. The original Gaussian RFT paper for PET imaging remains a useful introduction [50], though also see [51] for more up-to-date results including T- and F-statistic RFT results.

To use RFT results, we must know the smoothness of the data, precisely the smoothness of the standardized noise images (e/σ in the notation from Sect. 3). Smoothness is parameterized by the full width at half maximum (FWHM) of the Gaussian kernel required to simulate images with the same apparent spatial smoothness as our data (Fig. 27). For example, if we say that our data has 6-mm FWHM smoothness, it means that if we were to simulate our data, we would generate noise data with no correlation and then convolve it with a Gaussian kernel with FHWM of 6 mm. The exact

Full Width at Half Maximum

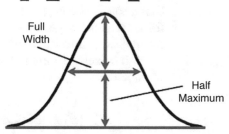

Full
Width

Half
Maximum

Fig. 27 Full width at half maximum (*FWHM*) is a generic way to describe the spread of a distribution, and is the way that smoothness is measured for random field theory

form of the spatial dependence of our data does not have to follow a Gaussian kernel, but for convenience, we describe the strength of the dependence in terms of the size a Gaussian kernel.

It may seem that if we take our fMRI data fresh from the scanner and convolve it with a 6-mm Gaussian kernel, our FWHM for RFT would be 6 mm. This is incorrect, however, as the noise smoothness includes both intrinsic sources of smoothness (imperfect MRI resolution, physiological artifacts, etc.) and smoothness induced by the applied smoothing. As a result, the smoothness for RFT is not a user-specified parameter, but rather estimated from the residuals of the GLM $\left(\Upsilon - X\hat{\beta}\right)$.

7.4.2 RESELs

The definition of FWHM smoothness creates a notion of a smoothness-equivalent volume, a resolution element or RESEL. If the smoothness of the data is FWHMx, FWHMy, FWHMz in each of the principal directions, then a volume of space with dimensions FWHMx× FWHMy× FWHMz is one RESEL. In very approximate terms, the RESEL count captures the amount of independent information in the image; fewer RESELs = smoother data = less information = less severe multiple testing problem; more RESELS = rougher data = more severe multiple testing problem. The total RESEL count for a search volume is:

$$\text{RESELcount} = I \,/ \left(\text{FWHM}_x \times \text{FWHM}_y \times \text{FWHM}_z\right) \quad (18)$$

where I is the total number of voxels in the brain and FHWM is expressed in units of voxels. The RESEL count is important because it is the summary measure that determines the RFT threshold, as illustrated next.

7.4.3 RFT-Corrected P-Values

RFT can be used with any type of statistic image a GLM can create, including T-, F-, and Z-statistic images [51]. The formulas provide FWE-corrected P-values for voxel-wise thresholds and cluster sizes and include corrections to account for edge effects (i.e., when blobs touch the edge of the search volume). The simplest result, for Z-statistic images with no edge corrections, can be used to gain some insight into the method.

For voxel-wise inference, a voxel with value z has a corrected P-value of:

$$P_{\text{vox}}^{\text{FWE}}(z) = \text{RESELcount}(2\pi)^{-2}(z^2 - 1)\exp\left(-\frac{z^2}{2}\right)$$

This shows that as z grows, the corrected P-value shrinks (the exponential term dominates), which of course makes sense, as larger statistic values should produce smaller P-values. As the RESEL count grows, the corrected P-value grows. The RESEL count can increase because the search volume increases, which again is sensible, as a greater search volume demands a greater correction for multiple testing, and hence a less significant P-value. The RESEL count can also increase if the smoothness decreases, as per [18], which increases the amount of information in the image, again demanding a greater correction for multiple testing. For cluster-wise inference, the equation for the FWE-corrected P-value is more involved, but it also accounts for the image search space and smoothness.

7.4.4 Small-Volume Correction

The first RFT results published (and the equation shown above) assumed that the search region was large relative to the smoothness of the image. This assumption was needed to avoid dealing with the case when a cluster touches the boundary of the search region. To see why this could be a problem, imagine two statistic images with the same smoothness, one the size and shape of the brain, the other with the same total volume but the shape of a long, narrow sausage. In the latter case, it is more likely that clusters will touch the edge of the image, and, relative to other clusters, have smaller volume. The results in [51] can be used to produce P-values that are accurate even with small search regions. When these results are used, they are sometimes referred to as "small volume correction."

7.4.5 RFT Assumptions

The use of RFT results is based on several assumptions and approximations. The essential assumptions are:

1. *Gaussian data*. For any collection of voxels, the distribution of the data is multivariate Gaussian.

2. *Sufficient smoothness*. The data must be sufficiently smooth to approximate continuous random fields (upon which the theory is based).

3. *Known smoothness*. The results assume that the FWHM smoothness parameters are exact and contain at most negligible error.

4. *Constant smoothness for cluster-wise inference only.* The standard cluster-wise results assume that the smoothness is the same everywhere in the image. If the data are "nonstationary," regions of the brain that are smoother than others will generate large clusters just by chance. Updated methods are available [52] which account for varying nonstationarity, but they have reduced sensitivity unless the nonstationarity is severe; hence, this is principally suitable for voxel-based morphometry data. Generally, FMRI data does not exhibit severe nonstationarity.

In the light of these extensive assumptions, there can be good reason to seek alternative methods that do not require as many assumptions.

7.4.6 Controlling FWE with Permutation

Nonparametric methods are generally used when the standard parametric assumptions are known to be false or cannot be verified. In the case of RFT, the assumptions are nearly impossible to verify, but more importantly, it has been found that voxel-wise RFT results are quite conservative for small group studies (e.g., when the number of subjects is less than about 40; [48, 53]). Hence, there has been interest in using alternative methods.

Instead of making assumptions about the distribution of the data, permutation testing uses the distribution of the data itself to find P-values and thresholds. Figure 28 illustrates the reasoning of the permutation test in the two-group setting.

While nonparametric tests are sometimes referred to as assumption-free, in fact they also have assumptions, just much weaker ones than standard parametric methods. The essential assumption for the permutation test is exchangeability under the null hypothesis. Exchangeability means that the data can be permuted (relative to the model) without altering its joint distribution. fMRI data presents a challenge for permutation testing. At the first level, temporal autocorrelation renders the data nonexchangeable and permutation methods cannot be directly applied (the data must be de-correlated, then permuted, and then re-correlated (*see* [6, 54] for more details). However, in second-level analyses, exchangeability is generally not a problem. In the example in the figure, we assume that, under the null hypothesis, all six subjects are exchangeable—this is very reasonable because if there is no group effect (this is the null hypothesis), then there is nothing special about the first three subjects versus the last three. For this example, there are 20 possible ways of permuting the groups (including the correct labeling). For an arbitrary dataset with group sizes n_1 and n_2, the number of possible permutations is $(n_1 + n_2)!/(n_1!n_2!)$.

By permuting the data many times, and for each permutation, assuming that the resulting test statistic (in this case, the group-difference T-statistic) is an sample of what we would see if there were no real effect present, we build up a histogram of test

Illustration of Permutation Inference: Two-Sample T

	Group A			Group B		
a Data. Single voxel %BOLD change for 6 subjects, 3 from each group	2.60	1.99	1.95	0.10	−0.43	1.20

b Possible permutations. Twenty possible ways of shuffling A and B group labels.   Under the null hypothesis group labels are meaningless, and all of these labellings are equivalent.

AAABBB	ABABAB	BAAABB	BABBAA
AABABB	ABABBA	BAABAB	BBAAAB
AABBAB	ABBAAB	BAABBA	BBAABA
AABBBA	ABBABA	BABAAB	BBABAA
ABAABB	ABBBAA	BABABA	BBBAAA

c Permutation distribution. Twenty T statistic values from each possible permutation.   Correctly labelled data gives T = 3.61.   Only that permutation is as large or larger, so P-value is 1/20 = 0.05

AAABBB	3.61	ABABAB	0.26	BAAABB	0.21	BABBAA	−0.61
AABABB	0.64	ABABBA	1.63	BAABAB	−0.12	BBAAAB	−1.68
AABBAB	0.28	ABBAAB	−0.99	BAABBA	0.99	BBAABA	−0.28
AABBBA	1.68	ABBABA	0.12	BABAAB	−1.63	BBABAA	−0.64
ABAABB	0.61	ABBBAA	−0.21	BABABA	−0.26	BBBAAA	−3.61

d Permutation distribution in histogram form.

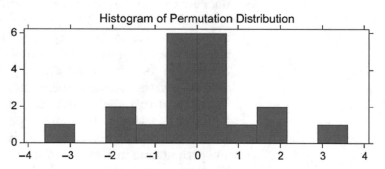

Fig. 28 Permutation test applied to data from a single voxel for a hypothetical two-group fMRI second-level analysis

statistic values that will serve as the null distribution of that test statistic. We can then look to see how much "area under the tail" lies to the right of the actual test statistic value that we originally observed (under the correct labeling of the data), and hence estimate our *P*-value; this is the same principle for relating the null distribution of the test statistic to the *P*-value as we saw in Fig. 9, but in this case, the null distribution has been generated via a completely different methodology.

The permutation test for the two group case can be generalized to three or more groups. In that case, we are testing the null hypothesis that all groups are the same, and use an F-test to measure the evidence of any difference. Under the null hypothesis, all subjects can be freely permuted. Likewise for a simple correlation model, the null hypothesis of no association justifies the free permutation of all of the subjects. The permutation test for the one group case, however, is problematic without further assumptions: If all we have is a single group, what is there to permute? Shuffling the order of subjects will not change the value of a one-sample T-test.

In a second-level single group mean, the COPE images are always created as relative differences between baseline and active data. Even if an event-related design is used, and a contrast selects a single predictor, the effective predictor is a subtraction of event and baseline data. This is the case due to the relative, nonquantitative nature of the BOLD signal. As a result, we use here a slightly different assumption to generate "permutations." Under the null hypothesis, we assume that each individual's second-level COPE data are mean zero and have a symmetric distribution. Assuming mean zero data is reasonable, as, under the null, we expect no activation, positive or negative. Assuming a symmetric distribution is a weakened form of normality, and is exactly satisfied for any balanced effect (i.e., a COPE constructed as difference of two averages, where an equal number of scans contributed to each average).

The one-sample, group-level fMRI permutation thus works as follows. Assuming mean zero, symmetrically distributed COPE data, we randomly multiply each subject's data by 1 or -1, or, equivalently, randomly flip the signs of each subject's COPE image. Since the data are symmetrically distributed about zero, multiplication by -1 does not alter the distribution, and we generate a realization that is equivalent to the original data. If there are n subjects in the analysis, there are 2^n possible ways to flip the signs of the group-level data.

The permutation methods described so far will create uncorrected P-values at each voxel. Control of the FWE rate is easily obtained with permutation testing via the following observation: In complete-null data, an FWE occurs whenever one or more voxels exceed the threshold, which occurs exactly when the voxel with the largest statistic exceeds the threshold. Hence, inference based on the permutation distribution of the largest (maximum) statistic provides valid FWE inferences. Specifically, at each permutation, the maximum statistic value (across all voxels in the brain) is noted, creating a null distribution of the maximum-across-space test statistic. The 95th percentile of that distribution gives an FWE-corrected threshold ("$p < 0.05$, corrected"), and any particular statistic value can be compared to the maximum permutation distribution to obtain an FWE-corrected P-value. Similarly, the maximal cluster size distribution can be created to provide FWE cluster-wise inferences.

It is important to note that, while permutation may appear to be a completely different approach than those we described earlier, in fact all pre-processing and modeling are generally the same, and it is only the *P*-value computation that differs. This is because the standard statistical models used generally have good sensitivity and robustness properties and should be used unaltered. For example, above we only discussed one- and two-sample *T*-tests, and did not mention other traditional nonparametric test statistics based on ranks, like the Wilcoxon Mann-Whitney test, as they often have much less power. There is an exception, however with small group data, with 20 or fewer subjects. With such small group data, there can be substantial sensitivity gains by using a nonstandard statistic, namely the smoothed variance *T*-test. As the fMRI data is generally smoothed before statistical modeling, we expect the variance image to be smooth as well. However, when the number of subjects is very small, the DOF available to estimate the variance is very low, and this can result in a noisy sample variance image. Smoothing regularizes the estimated variance image, effectively increasing the DOF and increasing sensitivity. While the null distribution of the smoothed estimated variance *T*-statistic image is not known, and so parametric methods cannot be used, nonparametric permutation methods can easily be used to generate FWE inferences based on the smoothed variance results.

Finally, if the number of possible permutations is very large, it can be impossible to compute them all. For example, for a 20 subject one-group analysis, there are over one million possible sign-flips of the data. In fact, it is sufficient to run a random subset of all possible permutations. If only k of a large number of possible permutations is used, the margin of error on the *P*–values is approximately $\pm 2p\, p(1-p)\,/\,k$, where p is the true *P*-value. For a nominal p of 0.05, this suggests that 1000 permutations is nearly sufficient (ME $= \pm$ 0.014, or 28% of 0.05), while 10,000 is probably more than enough (ME $= \pm$ 0.0044, or 8.7% of 0.05).

7.4.7   Controlling FDR

The method for finding a threshold that controls FDR is surprisingly simple. It is based only on the uncorrected voxel-wise *P*-values in the statistic image. Let Pi be the *P*-value at voxel i, and $P_{(i)}$ be the ordered *P*-values, $P_{(1)} \leq P_{(2)} \leq \cdots \leq P_{(I)}$. Then the largest index i that satisfies

$$P_{(i)} \leq \frac{i}{I}\alpha^{\text{FDR}} \qquad (19)$$

defines the FDR threshold as $P_{(i)}$ [55]. This method works even when there is positive dependence between voxels [56, 57].

7.4.8   Controlling False Positives and True Negatives: Mixture Modeling

So far we have considered techniques that control the rate of false positives. This depends on knowing the null distribution (or non-activation distribution) for relevant statistics under the null hypothesis. In contrast, mixture modeling provides us with a way of estimating the "activating" and "nonactivating" distributions from the data itself. For example, nonactivating voxel statistics may be modeled as coming from a (zero, or close-to-zero, mean) Gaussian distribution, activating voxels statistics as coming from a Gamma distribution, and de-activating voxels statistics as coming from a negative Gamma distribution [58–60]. The means and variances of these distributions are estimated from the whole statistic image. An example is shown in Fig. 29. Note that the mixture modeling approach is similar to the permutation approach (discussed in Sect. 7.4.2) in that both approaches extract information about the null distribution (or nonactivation distribution) from the data itself. However, permutation methods extract information about only the null distribution without making strong distributional assumptions, whereas mixture modeling extracts information about both the nonactivating and activating distributions by making strong distributional assumptions.

Mixture modeling can provide a number of advantages over null hypothesis testing. First, there is a well-known problem in null hypothesis testing of FMRI in that if enough observations (e.g., time points) are made, then every voxel in the brain will reject the null hypothesis [34]. This is because in practice no voxels will show completely zero response to the stimulus, if only due to modeling inadequacies such as unmodeled stimulus-correlated motion or the point spread function of the scanner. By doing mixture modeling,

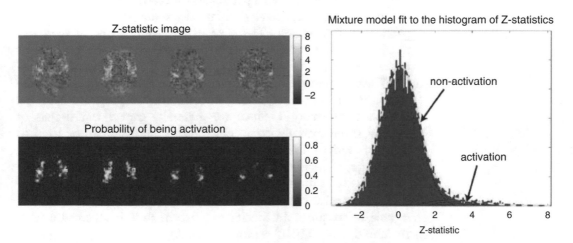

Fig. 29 Mixture modeling of a *Z*-statistic image. *Top-left*: Four example slices of a *Z*-statistic image obtained from fitting a general linear model (*GLM*) to the fMRI data at each voxel from an individual subject. The experiment was a pain stimulus applied using a sparse single-event design. *Right*: Mixture model fit to the histogram of *Z*-statistics. Nonactivating voxel are modeled as coming from a close-to-zero mean Gaussian distribution, activating voxels as coming from a Gamma distribution, and de-activating voxels as coming from a negative Gamma distribution. However, note that there were found to be no de-activating voxels in this case. *Bottom-left*: Image showing the probability that a voxel is activating—this information that can be extracted from the mixture model fit and can be used in thresholding to approximately control the true positive rate (*TPR*) as an alternative to null hypothesis testing

we can overcome this by instead of asking the question "Is the activation zero or not?," we ask the question "Is the activation bigger than the overall background level of 'activation'?."

Mixture modeling also provides inference flexibility. Because we have both the "activating" and "nonactivating" distributions, we can calculate the probability of a voxel being "activating" and the probability of a voxel being "nonactivating." This provides us with far more inference flexibility compared with null hypothesis testing. We can still look to control the FPR by thresholding using the probability of a voxel being "nonactivating." But now we could also look to approximately control the true positive rate (TPR) by thresholding using the probability of a voxel being "activating." Controlling the TPR may be of real importance when using fMRI for pre-surgery planning [61].

The wider-spread use of mixture modeling is currently somewhat hampered by the violation of the strong distributional assumptions that need to be made. In the future, this may be alleviated by the improvement of techniques such as ICA de-noising (see Sect. 5.2.2) rendering the distributional assumptions valid.

7.5 Enhancing Statistic Images

Aside from thresholding the final statistic images (either voxel-wise or cluster-wise), it is not advisable to make image-processing adjustments to statistic images. For example, smoothing a T-statistic image would be disastrous: While a T-statistic has approximately unit variance and follows a particular null distribution, a smoothed T-statistic image will have dramatically reduced variance with no particular distribution. Two exceptions to this are wavelet de-noising methods and a recently proposed threshold-free cluster enhancement (TFCE) method.

Wavelet methods transform the data in a scale-dependent fashion, so that all of the large-scale information is segregated from the fine-scale information. Since we generally expect the signals of interest to be spatially extended, wavelet methods can be used to "shrink" variation associated with the finest scales, "de-noising" the image, while preserving the large-scale structure. For an overview of wavelet methods applied to fMRI, see [62].

Cluster-wise inference also tries to capture spatially extended signals, but requires the specification of an arbitrary cluster-forming threshold u_{clus}. TFCE removes this dependence by, in essence, using all possible u_{clus} values and then merging all the results into a single image. Specifically, at each voxel, let $ei(h)$ be the extent of the cluster that voxel i belongs to with cluster-forming threshold h (or 0 if $t_i < h$). Then, the TFCE image is defined by $\sum_{h>0} e_i(h)^E h^H$, where the sum is computed for a discrete set of h values, from 0 to the maximum statistic value, and E and H are tuning parameters. In [63], $E = 0.5$ and $H = 2$ were found to give generally good performance for a range of classes of signals. TFCE seems to succeed in matching or exceeding the sensitivity of optimized cluster-based thresholding without the arbitrariness and instability of the smoothing and initial thresholding. There is no known distribution for the TFCE image, and so permutation testing is used to convert the TFCE image into *P*-values.

References

1. Friston K, Worsley K, Frackowiak R, Mazziotta J, Evans A (1994) Assessing the significance of focal activations using their spatial extent. Hum Brain Mapp 1:214–220

2. Hykin J, Bowtell R, Glover P, Coxon R, Blumhardt L, Mansfield P (1995) Investigation of the linearity of functional activation signal changes in the brain using echo planar imaging (EPI) at 3.0 T. In: Proc of the SMR and ESMRB Joint Meeting. p 795

3. Cohen M (1997) Parametric analysis of fMRI data using linear systems methods. NeuroImage 6:93–103

4. Dale A, Buckner R (1997) Selective averaging of rapidly presented individual trials using fMRI. Hum Brain Mapp 5:329–340

5. Burock MA, Buckner RL, Woldorff MG, Rosen BR, Dale AM (1998) Randomized event-related experimental designs allow for extremely rapid presentation rates using functional MRI. NeuroReport 9:3735–3739

6. Bullmore E, Brammer M, Williams S et al (1996) Statistical methods of estimation and inference for functional MR image analysis. Magn Reson Med 35:261–277

7. Friston K, Josephs O, Zarahn E, Holmes A, Rouquette S, Poline J-B (2000) To smooth or not to smooth? NeuroImage 12:196–208

8. Woolrich M, Ripley B, Brady J, Smith S (2001) Temporal autocorrelation in univariate linear modelling of FMRI data. NeuroImage 14:1370–1386

9. Locascio J, Jennings P, Moore C, Corkin S (1997) Time series analysis in the time domain and resampling methods for studies of functional magnetic resonance brain imaging. Hum Brain Mapp 5:168–193

10. Purdon P, Weisskoff R (1998) Effect of temporal autocorrelation due to physiological noise and stimulus paradigm on voxel-level false-positive rates in fMRI. Hum Brain Mapp 6:239–249

11. Marchini J, Ripley B (2000) A new statistical approach to detecting significant activation in functional MRI. NeuroImage 12:366–380

12. Worsley K, Liao C, Aston J et al (2002) A general statistical analysis for fMRI data. NeuroImage 15:1–15

13. Gautama T, Van Hulle MM (2004) Optimal spatial regularisation of autocorrelation estimates in fMRI analysis. Neuroimage 23:1203–1216

14. Penny W, Kiebel S, Friston K (2003) Variational Bayesian inference for fMRI time series. NeuroImage 19:1477–1491

15. Woolrich M, Behrens T, Smith S (2004) Constrained linear basis sets for HRF modelling using Variational Bayes. NeuroImage 21:1748–1761

16. Smith S, Jenkinson M, Beckmann C, Miller K, Woolrich M (2007) Meaningful design and contrast estimability in fMRI. NeuroImage 34:127–136

17. Dale A, Greve D, Burock M (1999) Optimal stimulus sequences for event-related fMRI. NeuroImage 9:S33

18. Wager T, Nichols T (2003) Optimization of experimental design in fMRI: a general framework using a genetic algorithm. Neuroimage 18:293–309

19. Josephs O, Turner R, Friston K (1997) Event-related fMRI. Hum Brain Mapp 5:1–7

20. Lange N, Zeger S (1997) Non-linear Fourier time series analysis for human brain mapping by functional magnetic resonance imaging. Appl Stat 46:1–29

21. Genovese C (2000) A Bayesian time-course model for functional magnetic resonance imaging data (with discussion). J Am Stat Assoc 95:691–703

22. Friston KJ (2002) Bayesian estimation of dynamical systems: an application to fMRI. NeuroImage 16:513–530

23. Marrelec G, Benali H, Ciuciu P, Pélégrini-Issac M, Poline J-B (2003) Robust Bayesian estimation of the hemodynamic response function in event-related BOLD MRI using basic physiological information. Hum Brain Mapp 19:1–17

24. Woolrich M, Jenkinson M, Brady J, Smith S (2004) Fully Bayesian spatio-temporal modelling of FMRI data. IEEE Trans Med Imaging 23:213–231

25. Buxton R, Uludag K, Dubowitz D, Liu T (2004) Modeling the hemodynamic response to brain activation. NeuroImage 23(S1):220–233

26. Boynton G, Engel S, Glover G, Heeger D (1996) Linear systems analysis of functional magnetic resonance imaging in human V1. J Neurosci 16:4207–4221

27. Glover G (1999) Deconvolution of impulse response in event-related BOLD fMRI. NeuroImage 9:416–429

28. Friston K, Josephs O, Rees G, Turner R (1998) Nonlinear event-related responses in fMRI. Magn Reson Med 39:41–52

29. Beckmann C, Smith S (2004) Probabilistic independent component analysis for functional magnetic resonance imaging. IEEE Trans Med Imaging 23:137–152

30. Glover G, Li T, Ress D (2000) Image-based method for retrospective correction of physiological motion effects in fMRI: Retroicor. Magn Reson Med 44:162–167

31. Holmes A, Friston K (1998) Generalisability, random effects & population inference. Fourth Int Conf on Functional Mapping of the Human Brain. NeuroImage 7:S754

32. Talairach J, Tournoux P (1988) Co-planar stereotaxic atlas of the human brain. Thieme, New York

33. Collins D, Neelin P, Peters T, Evans A (1994) Automatic 3D intersubject registration of MR volumetric data in standardized Talairach space. J Comput Assist Tomo 18:192–205

34. Friston KJ, Penny W, Phillips C, Kiebel S, Hinton G, Ashburner J (2002) Classical and Bayesian inference in neuroimaging: theory. NeuroImage 16:465–483

35. Beckmann C, Jenkinson M, Smith S (2003) General multi-level linear modelling for group analysis in FMRI. NeuroImage 20:1052–1063

36. Woolrich M, Behrens T, Beckmann C, Jenkinson M, Smith S (2004) Multi-level linear modelling for FMRI group analysis using Bayesian inference. NeuroImage 21:1732–1747

37. Kherif F, Poline J-B, Meriaux S, Benali H, Flandin G, Brett M (2003) Group analysis in functional neuroimaging: selecting subjects using similarity measures. Neuroimage 20:2197–2208

38. Luo W-L, Nichols TE (2003) Diagnosis and exploration of massively univariate neuroimaging models. Neuroimage 19:1014–1032

39. Seghier M, Friston K, Price C (2007) Detecting subject-specific activations using fuzzy clustering. Neuroimage 36:594–605

40. Wager T, Keller M, Lacey S, Jonides J (2005) Increased sensitivity in neuroimaging analyses using robust regression. NeuroImage 26:99–113

41. Woolrich M (2008) Robust group analysis using outlier inference. NeuroImage 41:286–301

42. Meriaux S, Roche A, Dehaene-Lambertz G, Thirion B, Poline J (2006) Combined permutation test and mixed-effect model for group average analysis in fMRI. Hum Brain Mapp 27:402–410

43. Roche A, Meriaux S, Keller M, Thirion B (2007) Mixed-effect statistics for group analysis in fMRI: a nonpara-metric maximum likelihood approach. Neuroimage 38:501–510

44. Thirion B, Pinel P, Mriaux S, Roche A, Dehaene S, Poline J (2007) Analysis of a large fMRI cohort: statistical and methodological issues for group analyses. Neuroimage 35:105–120

45. Hartvig NV, Jensen JL (2000) Spatial mixture modeling of fMRI data. Hum Brain Mapp 11:233–248

46. Hayasaka S, Nichols TE (2003) Validating cluster size inference: random field and permutation methods. NeuroImage 20:2343–2356

47. Friston KJ, Holmes A, Poline J-B, Price CJ, Frith CD (1996) Detecting activations in PET and fMRI: levels of inference and power. NeuroImage 4:223–235

48. Nichols TE, Hayasaka S (2003) Controlling the familywise error rate in functional neuroimaging: a comparative review. Stat Methods Med Res 12:419–446

49. Cao J, Worsley KJ (2001) Applications of random fields in human brain mapping. In: Moore M, (ed) Spatial statistics: methodological aspects and applications, vol 159, Springer lecture notes in statistics. Springer. pp 169–182

50. Worsley KJ, Evans AC, Marrett S, Neelin P (1992) Three-dimensional statistical analysis for cbf activation studies in human brain. J Cerebr Blood F Met 12:900–918

51. Worsley KJ, Marrett S, Neelin P, Vandal AC, Friston KJ, Evans AC (1996) A unified statistical approach for determining significant signals in images of cerebral activation. Hum Brain Mapp 4:58–73

52. Hayasaka S, Luan Phan K, Liberzon I, Worsley KJ, Nichols TE (2004) Nonstationary cluster-size inference with random field and permutation methods. NeuroImage 22:676–687

53. Nichols T, Holmes A (2001) Nonparametric permutation tests for functional neuroimaging: a primer with examples. Hum Brain Mapp 15:1–25

54. Bullmore E, Long C, Suckling J et al (2001) Colored noise and computational inference in neurophysiological (fMRI) time series analysis: resampling methods in time and wavelet domains. Hum Brain Mapp 12:61–78

55. Benjamini Y, Hochberg Y (1995) Controlling the false discovery rate: a practical and powerful approach to multiple testing. J R Stat Soc Ser B Methodol 57:289–300

56. Genovese C, Lazar N, Nichols T (2002) Thresholding of statistical maps in functional neuroimaging using the false discovery rate. NeuroImage 15:870–878

57. Benjamini Y, Yekutieli D (2001) The control of the false discovery rate in multiple testing under dependency. Ann Stat 29:1165–1188

58. Everitt B, Bullmore E (1999) Mixture model mapping of brain activation in functional magnetic resonance images. Hum Brain Mapp 7:1–14

59. Hartvig N (2000) A stochastic geometry model for fMRI data. Technical Report 410. Department of Theoretical Statistics, University of Aarhus

60. Woolrich M, Behrens T (2006) Variational Bayes inference of spatial mixture models for segmentation. IEEE Trans Med Imaging 25:1380–1391

61. Bartsch A, Homola G, Biller A, Solymosi L, Bendszus M (2006) Diagnostic functional MRI: illustrated clinical applications and decision-making. J Magn Reson Imaging 23:921–932

62. Van De Ville D, Blu T, Unser M (2006) Surfing the brain – an overview of wavelet-based techniques for fMRI data analysis. IEEE Eng Med Biol 25:65–78

63. Smith SM, Nichols TE (2008) Threshold-free cluster enhancement: addressing problems of smoothing, threshold dependence and localisation in cluster inference. NeuroImage. doi:10.1016/j.neuroimage.2008.03.061, In press; Epub ahead of print April 11, 2008

Chapter 8

Dynamic Causal Modeling of Brain Responses

Karl J. Friston

Abstract

This chapter is about modeling-distributed brain responses and, in particular, the functional integration among neuronal systems. Inferences about the functional organization of the brain rest on models of how measurements of evoked responses are caused. These models can be quite diverse, ranging from conceptual models of functional anatomy to mathematical models of neuronal and hemodynamics. The aim of this chapter is to introduce dynamic causal models. These models can be regarded as generalizations of the simple models employed in conventional analyses of regionally specific brain responses. In what follows, we will start with anatomical models of functional brain architectures, which motivate some of the basic principles of neuroimaging. We then review briefly statistical models (e.g., the general linear model) used for making classical and Bayesian inferences about *where* neuronal responses are expressed. By incorporating biophysical constraints, these basic models can be finessed and, in a dynamic setting, rendered causal. This allows us to infer *how* interactions among brain regions are mediated. This chapter focuses on causal models for distributed responses measured with fMRI and electroencephalography. The latter is based on neural-mass models and affords mechanistic inferences about how evoked responses are caused, at the level of neuronal subpopulations and the coupling among them.

Key words Functional connectivity, Effective connectivity, Dynamic causal modeling, Causal, Dynamic, Nonlinear

1 Introduction

Neuroscience depends on conceptual, anatomical, statistical, and causal models that link ideas about how the brain works to observed neuronal responses. Here, we highlight the relationships among the sorts of models that are employed in imaging, with a special focus on dynamic causal models of functional brain architectures. We will show how simple statistical models used to identify where evoked brain responses are expressed (cf, neo-phrenology) can be elaborated to provide models of how neuronal responses are caused (e.g., dynamic causal modeling—DCM). We will review a series of models that range from conceptual models, motivating experimental design, to detailed biophysical models of coupled neuronal ensembles that enable questions to be asked, at a physiological and computational level.

Massimo Filippi (ed.), *fMRI Techniques and Protocols*, Neuromethods, vol. 119,
DOI 10.1007/978-1-4939-5611-1_8, © Springer Science+Business Media New York 2016

Anatomical models of functional brain architectures motivate the fundaments of neuroimaging. In the first section, we review the distinction between functional *specialization* and *integration* and how these principles serve as the basis for most models of neuroimaging data. The next section turns to simple statistical models (e.g., the general linear model—GLM) used for making classical and Bayesian inferences about functional specialization, in terms of where neuronal responses are expressed. By incorporating biological constraints, simple observation models can be made more realistic and, in a dynamic framework, causal. This section concludes by considering the biophysical modeling of hemodynamic responses. All the models considered in this section pertain to regional responses. In the final section, we focus on models of distributed responses, where the interactions among cortical areas or neuronal subpopulations are modeled explicitly. This section covers the distinction between *functional* and *effective connectivity* and reviews DCM of functional integration, using fMRI and electroencephalogram (EEG). We conclude with an example from event-related potential (ERP) research and show how the mismatch negativity (MMN) can be explained by changes in coupling among neuronal sources that may underlie perceptual learning.

2   Anatomical Models

2.1   Functional Specialization and Integration

From a historical perspective, the distinction between functional specialization and functional integration relates to the dialectic between *localizationism* and *connectionism* that dominated thinking about brain function in the nineteenth century. Since the formulation of phrenology by Gall, who postulated fixed one-to-one relations between particular parts of the brain and specific mental attributes, the identification of a particular brain region with a specific function has become a central theme in neuroscience. Somewhat ironically, the notion that distinct brain functions could be localized in the brain was strengthened by early scientific attempts to refute the phrenologists' claims. In 1808, a scientific committee of the Athénée at Paris, chaired by Cuvier, declared that phrenology was an unscientific and invalid theory [1]. This conclusion, which was not based on experimental results, may have been enforced by Napoleon Bonaparte (who, allegedly, was not amused after Gall's phrenological examination of his own skull did not give the flattering results expected). During the following decades, lesion and electrical stimulation paradigms were developed to test whether functions could indeed be localized in animal models. Initial lesion experiments by Flourens on pigeons were incompatible with phrenologist predictions, but later experiments, including stimulation experiments in dogs and monkeys by Fritsch, Hitzig, and Ferrier, supported the idea that there was a relation between distinct brain regions and certain cognitive or motor functions. Additionally, clinicians like Broca and Wernicke showed that patients with focal brain lesions in particular locations showed specific impairments. However, it was realized early on that, in spite of these experimental findings, it was generally difficult to attribute a specific function to a cortical area, given the dependence of cerebral activity on the anatomical connections between distant brain regions; for example, a meeting that took place on 4 August 1881 addressed the difficulties of attributing function to a cortical area, given the dependence of cerebral activity on underlying connections [2]. This meeting was entitled "localisation of function in the cortex cerebri." Goltz [3], although accepting the results of electrical stimulation in dog and monkey cortex, considered that the excitation method was inconclusive, in that movements elicited might have originated in related pathways, or current could have spread to distant centers. In short, the excitation method could not be used to infer functional localization because localizationism discounted interactions or functional integration among different brain areas. It was proposed that lesion studies could supplement excitation experiments. Ironically, it was the observations on patients with brain lesions some years later (*see* Ref. [4]) that led to the concept of *disconnection syndromes* and the refutation of localizationism as a complete or sufficient explanation of cortical organization. Functional localization implies that a function can be localized in a cortical area, whereas specialization suggests that a cortical area is specialized for some aspects of perceptual or motor processing, and that this specialization is anatomically *segregated* within the cortex. The cortical infrastructure supporting a single function may then involve many specialized areas whose union is mediated by the functional integration among them. In this view, functional specialization is only meaningful in the context of functional integration and vice versa.

2.2 Functional Specialization and Segregation

The functional role of any component (e.g., cortical area, sub-area, or neuronal population) of the brain is defined largely by its connections. Certain patterns of cortical projections are so common that they could amount to rules of cortical connectivity. "These rules revolve around one, apparently, over-riding strategy that the cerebral cortex uses—that of functional segregation" [5]. Functional segregation demands that cells with common functional properties be grouped together. This architectural constraint necessitates both convergence and divergence of cortical connections. Extrinsic connections among cortical regions are not continuous but occur in patches or clusters. This patchiness has, in some instances, a clear relationship to functional segregation. For example, when recordings are made in V2, directionally selective (but not wavelength or color selective) cells are found exclusively in its thick stripes. Retrograde (i.e., backward) labeling of cells in V5 is limited to these thick stripes; all the available physiological evidence suggests that V5 is a functionally homogeneous area that is specialized for visual motion. Evidence of this nature supports the notion that patchy connectivity is the anatomical infrastructure that mediates functional segregation and specialization. If it is the case that neurons in a given cortical area share a common responsiveness, by virtue of their extrinsic connectivity, to some sensorimotor or cognitive attribute, then this functional segregation is also an anatomical one.

In summary, functional specialization suggests that challenging a subject with the appropriate sensorimotor attribute or cognitive process should lead to activity changes in, and only in, the specialized areas. This is the anatomical and physiological model upon which the search for regionally specific effects is based. We will deal briefly with models of regionally specific responses and return to models of functional integration.

3 Statistical Models

3.1 Statistical Parametric Mapping

Functional mapping studies are usually analyzed with some form of statistical parametric mapping (SPM). SPM entails the construction of continuous statistical maps (e.g., t-maps) to test hypotheses about regionally specific effects [6]. SPM uses the GLM and random field theory (RFT) to analyze and make classical inferences about brain responses. Parameters of the GLM are estimated in exactly the same way as in conventional analysis of discrete data. RFT is used to resolve the multiple-comparisons problem induced by making inferences over a volume of the brain. RFT provides a method for adjusting p-values for the search volume of an SPM to control false positive rates. It plays the same role for continuous data (i.e., images or time series) as the Bonferroni correction for a family of discontinuous or discrete statistical tests.

There is a Bayesian alternative to classical inference with SPMs. This rests on conditional inferences about an effect, given the data, as opposed to classical inferences about the data, given the effect is zero. Bayesian inferences about effects that are continuous in space use posterior probability maps (PPMs). Although less established than SPMs, PPMs are potentially useful, not least because they do not have to contend with the multiple-comparisons problem induced by classical inference (see Ref. [7]). In contradistinction to SPM, this means that inferences about a given regional response do not depend on inferences about responses elsewhere. Bayesian inference is particularly relevant to dynamic casual modeling because the Bayesian formulation is an essential part of model specification and inversion. Before looking at the models underlying Bayesian inference, we briefly review estimation and classical inference in the context of the GLM and show how this can be generalized to give a Bayesian approach.

3.2 General Linear Model

The GLM is a simple equation

$$y = X\beta + \varepsilon \tag{1}$$

that expresses an observed response y in terms of a linear combination of explanatory variables in the design matrix X, plus a well-behaved error term. The GLM is variously known as analysis of variance or multiple-regression and subsumes simpler variants, like the t-test for a difference in means, to more elaborate linear convolution models (see below). Each column of the design matrix models a cause of the data. These are referred to as explanatory variables, covariates, or regressors. Sometimes the design matrix contains covariates or indicator variables that take values of zero or one, to indicate the presence of a particular level of an experimental factor (cf, analysis of variance). The relative contribution of each of these columns to the response is controlled by the parameters, β. Inferences about the parameter estimates are made using t or F-statistics, as in conventional statistics. Having computed the statistic, RFT is used to assign adjusted p-values to topological features of the SPM, such as the height of peaks or the spatial extent of blobs. This p-value is a function of the search volume and smoothness. The intuition behind RFT is that it controls the false positive rate of peaks corresponding to regional effects. A Bonferroni correction would control the false positive rate of voxels, which is inexact and unnecessarily severe. The p-value is the probability of getting a peak in the SPM, or higher, by chance over the search volume. If sufficiently small (usually less than 0.05), the regional effect is declared significant.

3.3 Classical and Bayesian Inference

Inference in neuroimaging is restricted largely to classical inferences based upon SPMs. The statistics that comprise these SPMs are essentially functions of the data. The probability distribution of the chosen statistic, under the null hypothesis (i.e., the null distribution), is used to compute a p-value. This p-value is the probability of obtaining the statistic, or the data, given that the null hypothesis is true. If sufficiently small, the null hypothesis is rejected and an inference is made. The alternative approach is to use Bayesian or conditional inference based upon the posterior distribution of the activation given the data [8]. This necessitates the specification of priors (i.e., the probability distribution of the activation or model parameter). Bayesian inference requires the conditional or posterior distribution and therefore rests upon a posterior density analysis. A useful way to summarize this posterior density is to compute the probability that the activation exceeds some threshold. This represents a Bayesian inference about the effect, in relation to the specified threshold. By computing posterior probability for each voxel, we can construct PPMs that are a useful complement to classical SPMs.

The motivation for using conditional or Bayesian inference is that it has high face validity. This is because the inference is about an effect, or activation, being greater than some specified size that has some meaning in relation to underlying neurophysiology. This contrasts with classical inference, in which the inference is about the effect being significantly different than zero. The problem for classical inference is that trivial departures from the null hypothesis can be declared significant, with sufficient data or sensitivity. From the point of view of neuroimaging, posterior inference is especially useful because it eschews the multiple-comparisons problem. In classical inference, one tries to ensure that the probability of rejecting the null hypothesis incorrectly is maintained at a small rate, despite making inferences over large volumes of the brain. This induces a multiple-comparisons problem that, for spatially continuous data, requires an adjustment or correction to the p-value using RFT as mentioned earlier. This correction means that classical inference becomes less sensitive or powerful with large search volumes. In contradistinction, posterior inference does not have to contend with the multiple-comparisons problem because there are no false-positives. The probability that activation has occurred, given the data, at any particular voxel is the same, irrespective of whether one has analyzed that voxel or the entire brain. For this reason, posterior inference using PPMs represents a relatively more powerful approach than classical inference in neuroimaging.

3.3.1 Hierarchical Models and Empirical Bayes

PPMs require the posterior distribution or conditional distribution of the activation (a contrast of conditional parameter estimates), given the data. This posterior density can be computed, under Gaussian assumptions, using Bayes rule. Bayes rule requires the specification of a likelihood function and the prior density of the model parameters. The models used to form PPMs and the likelihood functions are exactly the same as in classical SPM analyses, namely, the GLM. The only extra information that is required is the prior probability distribution of the parameters. Although it would be possible to specify those using independent data or some plausible physiological constraints, there is an alternative to this fully Bayesian approach. The alternative is *empirical Bayes* in which the prior distributions are estimated from the data. Empirical Bayes requires a *hierarchical observation model* where the parameters and hyper-parameters at any particular level can be treated as priors on the level below. There are numerous examples of hierarchical observation models in neuroimaging. For example, the distinction between fixed- and mixed-effects analyses of multisubject studies relies upon a two-level hierarchical model. However, in neuroimaging, there is a natural hierarchical observation model that is common to all brain mapping experiments. This is the hierarchy induced by looking for the same effects at every voxel within the brain (or gray matter). The first level of the hierarchy corresponds to the experimental effects at any particular voxel and the second level comprises the effects over voxels. Put simply, the variation in a contrast, over voxels, can be used as the prior variance of that contrast at any particular voxel. Hierarchical linear models have the following form:

$$
\begin{aligned}
y &= X^{(1)}\beta^{(1)}\varepsilon^{(1)} \\
\beta^{(1)} &= X^{(2)}\beta^{(2)}\varepsilon^{(2)} \\
\beta^{(2)} &= \ldots
\end{aligned}
\tag{2}
$$

This is exactly the same as Eq. (1) but now the parameters of the first level are generated by a supra-ordinate linear model and so on to any hierarchical depth required. These hierarchical observation models are an important extension of the GLM and are usually estimated using expectation maximization (EM) [9]. In the present context, the response variables comprise the responses at all voxels and $\beta^{(1)}$s are the treatment effects we want to make an inference about. Because we have invoked a second level, the first-level parameters embody random effects and are generated by a second-level linear model. At the second level, $\beta^{(2)}$ is the average effect over voxels and $\varepsilon^{(2)}$ is its voxel-to-voxel variation. By estimating the variance of $\varepsilon^{(2)}$, one is implicitly estimating an empirical prior on the first-level parameters at each voxel. This prior can then be used to estimate the posterior probability of $\beta^{(1)}$ being greater than some threshold at each voxel. An example of the ensuing PPM is provided in Fig. 1 along with the classical SPM.

In summary, we have seen how the GLM can be used to test hypotheses about brain responses and how, in a hierarchical form, it enables empirical Bayesian or conditional inference. Then, we deal with the dynamic systems and how they can be formulated as GLMs. These dynamic models take us closer to how brain responses are actually caused by experimental manipulations and represent the next step towards dynamic causal models of brain responses.

3.4 Dynamic Models

3.4.1 Convolution Models and Temporal Basis Functions

In Friston et al. [10], the form of the impulsed hemodynamic response function (HRF) was estimated using a least squares deconvolution and a linear time invariant model, where evoked neuronal responses are *convolved* or smoothed with an HRF to give the measured hemodynamic response (*see also* Ref. [11]). This simple linear convolution model is the cornerstone for making statistical inferences about activations in fMRI with the GLM. An impulse response function is the response to a single impulse, measured at a series of times after the input. It characterizes the input–output behavior of the system (i.e., voxel) and places important constraints on the sorts of inputs that will excite a response.

Knowing the form of the HRF is important for several reasons, not least because it furnishes better statistical models of the data.

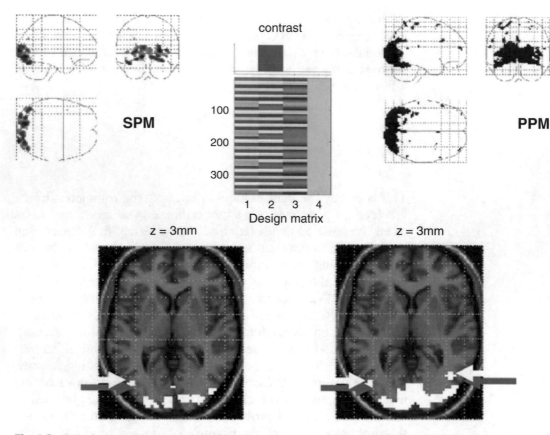

Fig. 1 Statistical parametric mapping (*SPM*) and posterior probability map (*PPM*) for an fMRI study of attention to visual motion. The display format (*lower panel*) uses an axial slice through extra-striate regions but the thresholds are the same as employed the in maximum-intensity projections (*upper panels*). *Upper right*: The activation threshold for the PPM was 0.7 au, meaning that all voxels shown had a 90 % chance of an activation of 0.7 % or more. *Upper left*: The corresponding SPM using an adjusted threshold at *p* = 0.05. Note the bilateral foci of motion-related responses in the PPM that are not seen in the SPM (*gray arrows*). As can be imputed from the design matrix (*upper-middle panel*), the statistical model of evoked responses comprised boxcar regressors convolved with a canonical hemodynamic response function. The *middle column* corresponds to the presentation of moving dots and was the stimulus attribute tested by the contrast

The HRF may vary from voxel to voxel and this has to be accommodated in the GLM. To allow for different HRFs in different brain regions, temporal basis functions were introduced [12] to model evoked responses in fMRI and applied to event-related responses in Josephs et al. [13] (*see also* Ref. [14]). The basic idea behind temporal basis functions is that the hemodynamic response, induced by any given trial type, can be expressed as the linear combination of (basis) functions of peri-stimulus time. The convolution model for fMRI responses takes a stimulus function encoding the neuronal responses and convolves it with an HRF to give a regressor that enters the design matrix. When using basis functions, the stimulus function is convolved with each basis function to give a series of regressors. Mathematically, we can express this model as

$$y(t) = X\beta + \varepsilon \qquad\qquad y(t) = (t) \otimes h(t)$$

$$\Leftrightarrow \tag{3}$$

$$X_i = T_i(t) \otimes u(t) \qquad h(t) = \beta_1 T_1(t) + \beta_2 T_2(t) + \dots$$

where $\otimes$ means convolution. This equivalence shows how any convolution model (right) can be converted into a GLM (left), using temporal basis functions. The parameter estimates are the coefficients or weights that determine the mixture of basis functions of time $T_i(t)$ that models $h(t)$, the HRF for the trial type and voxel in question. We find the most useful basis set to be a canonical HRF and its derivatives with respect to the key parameters that determine its form (see below). Temporal basis functions are important because they provide a graceful transition between conventional multilinear regression models with one stimulus function per condition and finite impulse response (FIR) models with a parameter for each time point, following the onset of a condition or trial type. Figure 2 illustrates this

Temporal basis functions

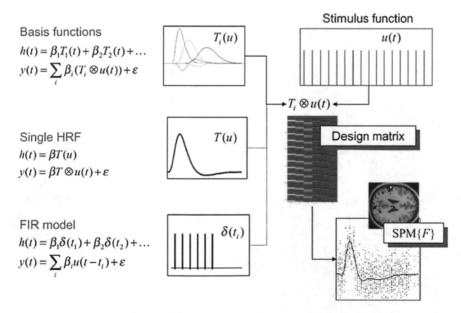

Fig. 2 Temporal basis functions offer useful constraints on the form of the estimated response that retain the flexibility of finite impulse response (*FIR*) models and the efficiency of single regressor models. The specification of these constrained FIR models involves setting up stimulus functions $u(t)$ that model expected neuronal changes, for example, boxcar-functions of epoch-related responses or spike-(δ)-functions at the onset of specific events or trials. These stimulus functions are then convolved with a set of basis functions $Ti(t)$ of peristimulus time that, in some linear combination, model the HRF. The ensuing regressors are assembled into the design matrix. The basis functions can be as simple as a single canonical HRF (*middle*), through to a series of top-hat-functions $\delta i(t)$ (*bottom*). The latter case corresponds to an FIR model and the coefficients constitute estimates of the impulse response function at a finite number of discrete sampling times. Selective averaging in event-related fMRI [39] is mathematically equivalent to this limiting case

graphically. In short, temporal basis functions offer useful constraints on the form of the estimated response that retain the flexibility of FIR models and the efficiency of single regressor models.

3.5 Biophysical Models

3.5.1 Input-State-Output Systems

By adopting a convolution model for brain responses in fMRI, we are implicitly positing a dynamic system that converts neuronal responses into observed hemodynamic responses. Our understanding of the biophysical and physiological mechanisms that underpin the HRF has grown considerably in the last few years (e.g., [15–17]). Figure 3 shows some simulations based on the hemodynamic model described in Friston et al. [18]. Here, neuronal activity induces some autoregulated vasoactive signal that causes transient increases in regional cerebral blood flow (rCBF). The resulting flow increases dilate a venous balloon, increasing its volume, and diluting venous blood to decrease deoxyhemoglobin content. The blood oxygen level dependent (BOLD) signal is roughly proportional to the concentration of deoxyhemoglobin and follows the rCBF response with about a second delay. The model is framed in terms of differential equations, examples of which are provided in left panel.

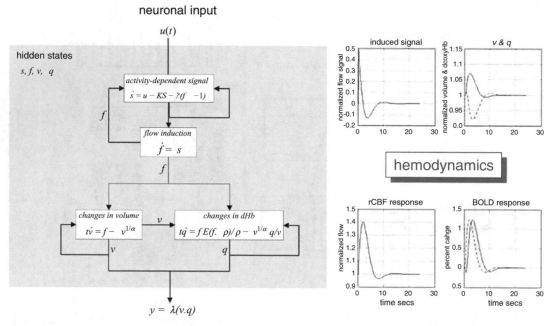

Fig. 3 *Right*: Hemodynamics elicited by an impulse of neuronal activity as predicted by a dynamical biophysical model (*left*). A burst of neuronal activity causes an increase in flow-inducing signal that decays with first-order kinetics and is downregulated by local flow. This signal increases regional cerebral blood flow (*rCBF*), which dilates the venous capillaries, increasing volume *v*. Concurrently, venous blood is expelled from the venous pool decreasing deoxyhemoglobin content *q*. The resulting fall in deoxyhemoglobin concentration leads to a transient increases in blood oxygen level dependent (*BOLD*) signal and a subsequent undershoot. *Left*: Hemodynamic model; on which these simulations were based

Note that we have introduced variables like volume and deoxyhemoglobin concentrations that are not actually observed. These are referred to as the *hidden states* of input-state-output models. The state and output equations of any analytic dynamical system are

$$\dot{x}(t) = f(x,,u,,\theta)$$
$$y(t) = g(x,,u,,\theta) + \varepsilon \qquad (4)$$

The first line is an ordinary differential equation and expresses the rate of change of the states as a parameterized function of the states and inputs. Typically, the inputs $u(t)$ correspond to designed experimental effects (e.g., the stimulus function in fMRI). There is a fundamental and causal relationship [19] between the outputs and the history of the inputs in Eq. (4). This relationship conforms to a Volterra series, which expresses the output as a generalized convolution of the input, critically without reference to the hidden states $x(t)$. This series is simply a functional Taylor expansion of the outputs with respect to the inputs [20]. The reason for it is a *functional* expansion that the inputs are a function of time.[1]

$$y(t) = \sum_i \int_0^t \dots \int_0^t \kappa_i(\sigma_1,,\dots,,\sigma_i) u(t-\sigma_1),\dots,u(t-\sigma_i)$$

$$d\sigma_1,\dots,d\sigma_i \kappa_i(\sigma_1,,\dots,,\sigma_i) = \frac{\partial^i y(t)}{\partial u(t-\sigma_1),\dots,\partial u(t-\sigma_i)} \qquad (5)$$

where $k_i(\sigma_1,\dots,\sigma_i)$ is the *i*th-order kernel. In Eq. (5), the integrals are restricted to the past. This renders the system causal. The key thing here is that Eq. (5) is simply a convolution and can be expressed as a GLM as in Eq. (3). This means that we can take a neurophysiologically realistic model of hemodynamic responses and use it as an observation model to estimate parameters using observed data. Here the model is parameterized in terms of kernels that have a direct analytic relation to the original parameters θ of the biophysical system. The first-order kernel is simply the conventional HRF. High-order kernels correspond to high-order HRFs and can be estimated using basis functions as described above. In fact, by choosing basis functions according to

$$A(\sigma)_i = \frac{\partial \kappa(\sigma)_1}{\partial \theta_i}, \qquad (6)$$

one can estimate the biophysical parameters because $\beta_i = \theta_i$ to a first-order approximation. The critical step we have taken here is to start with a dynamic causal model of how responses are generated

---

[1] For simplicity, here and in Eq. (7), we deal with only one experimental input.

and construct a general linear observation model that allows us to estimate and infer things about the parameters of that model. This is in contrast to the conventional use of the GLM with design matrices that are not informed by a forward model of how data are caused. This approach to modeling brain responses has a much more direct connection with underlying physiology and rests upon an understanding of the underlying system.

3.5.2 Nonlinear System Identification

Once a suitable causal model has been established (e.g., Fig. 3), we can estimate second-order kernels. These kernels represent a non-linear characterization of the HRF that can model interactions among stimuli in causing responses. One important manifestation of the nonlinear effects, captured by the second-order kernels, is a modulation of stimulus-specific responses by preceding stimuli that are proximate in time. This means that responses at high-stimulus presentation rates saturate and, in some instances, show an inverted U behavior. This behavior appears to be specific to BOLD effects (as distinct from evoked changes in CBF) and may represent a *hemodynamic refractoriness*. This effect has important implications for event-related fMRI, where one may want to present trials in quick succession.

In summary, we started with models of regionally specific responses, framed in terms of the GLM, in which responses were modeled as linear mixtures of designed changes in explanatory variables. Hierarchical extensions to linear observation models enable random-effects analyses and, in particular, empirical Bayes. The mechanistic utility of these models is realized though the use of forward models that embody causal dynamics. Simple variants of these are the linear convolution models used to construct explanatory variables in conventional analyses of fMRI data. These are a special case of generalized convolution models that are mathematically equivalent to input-state-output systems comprising hidden states. Estimation and inference with these dynamic models tells us something about *how* the response was caused, but only at the level of a single voxel. Section 4 retains the same perspective on models, but in the context of distributed responses and functional integration.

4   Models of Functional Integration

4.1   Functional and Effective Connectivity

Imaging neuroscience has established functional specialization as a principle of brain organization in man. The integration of specialized areas has proven more difficult to assess. Functional integration is usually inferred on the basis of correlations among measurements of neuronal activity. Functional connectivity is defined as statistical dependencies or correlations *among remote neurophysiological events*. However, correlations can arise in a variety of ways. For example, in multiunit electrode recordings, they can result from stimulus-locked transients evoked by a common input or reflect stimulus-induced oscillations mediated by synaptic connections [21]. Integration within a distributed system is usually better understood in terms of effective connectivity. Effective connectivity refers explicitly to *the influence that one neural system exerts over another*, either at a synaptic (i.e., synaptic efficacy) or at a population level. It has been proposed that "the (electrophysiological) notion of effective connectivity should be understood as the experiment- and time-dependent, simplest possible circuit diagram that would replicate the observed timing relationships between the recorded neurons" [22]. This speaks about two important points: (a) effective connectivity is dynamic, that is, activity-dependent and (b) it depends upon a model of the interactions. The estimation procedures employed in functional neuroimaging can be divided into linear nondynamic models (e.g., [23]) or nonlinear dynamic models.

There is a necessary link between functional integration and multivariate analyses because the latter are necessary to model interactions among brain regions. Multivariate approaches can be divided into those that are inferential in nature and those that are data-led or exploratory. We will first consider multivariate approaches that look at functional connectivity or covariance patterns (and are generally exploratory) and then turn to models of effective connectivity (that allow for inference about their parameters).

4.1.1 Eigenimage Analysis and Related Approaches

In Friston et al. [24], we introduced voxel-based principal component analysis (PCA) of neuroimaging time series to characterize distributed brain systems implicated in sensorimotor, perceptual, or cognitive processes. These distributed systems are identified with principal components or *eigenimages* that correspond to spatial modes of coherent brain activity. This approach represents one of the simplest multivariate characterizations of functional neuroimaging time series and falls into the class of exploratory analyses. Principal component or eigenimage analysis generally uses singular value decomposition to identify a set of orthogonal spatial modes that capture the greatest amount of variance expressed over time. As such, the ensuing modes embody the most prominent aspects of the variance–covariance structure of a given time series. Noting that covariance among brain regions is equivalent to functional connectivity renders eigenimage analysis particularly interesting because it was among the first ways of addressing functional integration (i.e., connectivity) with neuroimaging data. Subsequently, eigenimage analysis has been elaborated in a number of ways. Notable among these is canonical variate analysis (CVA) and multidimensional scaling [25, 26]. CVA was introduced in the context of multiple analysis of covariance and uses the generalized eigenvector solution to maximize the variance that can be explained by some explanatory variables relative to error. CVA can be thought of as an extension of eigenimage analysis that refers explicitly to some explanatory variables and allows for statistical inference.

In fMRI, eigenimage analysis (e.g., [27]) is generally used as an exploratory device to characterize coherent brain activity. These variance components may, or may not, be related to experimental design. For example, endogenous coherent dynamics have been observed in the motor system at very low frequencies [28]. Despite its exploratory power, eigenimage analysis is limited for two reasons. First, it offers only a linear decomposition of any set of neurophysiological measurements and second, the particular set of eigenimages or spatial modes obtained is determined by constraints that are biologically implausible. These aspects of PCA confer inherent limitations on the interpretability and usefulness of eigenimage analysis of biological time series and have motivated the exploration of nonlinear PCA and neural network approaches.

Two other important approaches deserve to be mentioned here. The first is independent component analysis (ICA). ICA uses entropy maximization to find, using iterative schemes, spatial modes or their dynamics that are approximately *independent*. This is a stronger requirement than *orthogonality* in PCA and involves removing high-order correlations among the modes (or dynamics). It was initially introduced as *spatial* ICA [29], in which the independence constraint was applied to the modes (with no constraints on their temporal expression). More recent approaches use, by analogy with magneto- and electrophysiological time series analysis, *temporal* ICA where the dynamics are enforced to be independent. This requires an initial dimension reduction (usually using conventional eigenimage analysis). Finally, there has been an interest in cluster analysis [30]. Conceptually, this can be related to eigenimage analysis through multidimensional scaling and principal co-ordinate analysis.

All these approaches are interesting, but they are not used very much. This is largely because they tell you nothing about how the brain works or allow one to ask specific questions. Simply demonstrating statistical dependencies among regional brain responses or endogenous activity (i.e., demonstrating functional connectivity) does not address how these responses were caused. To address this, one needs explicit models of integration or more precisely, effective connectivity.

4.2 Dynamic Causal Modeling with Bilinear Models

This section is about modeling interactions among neuronal populations, at a cortical level, using neuroimaging time series and dynamic causal models that are informed by the biophysics of the system studied. The aim of DCM [31] is to estimate, and make inferences about, the coupling among brain areas and how that coupling is influenced by experimental changes (e.g., time or cognitive set). The basic idea is to construct a reasonably realistic neuronal model of interacting cortical regions or nodes. This model is then supplemented with a forward model of how neuronal or synaptic activity translates into a measured response (see previous section). This enables the parameters of the neuronal model (i.e., effective connectivity) to be estimated from observed data.

Intuitively, this approach regards an experiment as a designed perturbation of neuronal dynamics that are promulgated and distributed throughout a system of coupled anatomical nodes to change region-specific neuronal activity. These changes engender, through a measurement-specific forward model, responses that are used to identify the architecture and time constants of the system at a neuronal level. This represents a departure from conventional approaches (e.g., structural equation modeling and auto-regression models; [32, 33]), in which one assumes that the observed responses are driven by endogenous or intrinsic noise (i.e., innovations). In contrast, dynamic causal models assume that the responses are driven by designed changes in inputs. An important conceptual aspect of dynamic causal models pertains to how the experimental inputs enter the model and cause neuronal responses. Experimental variables can elicit responses in one of two ways. First, they can elicit responses through direct influences on specific anatomical nodes. This would be appropriate, for example, in modeling sensory-evoked responses in early visual cortices. The second class of input exerts its effect vicariously, through a modulation of the coupling among nodes. This sort of experimental variable would normally be more enduring, for example, attention to a particular attribute or the maintenance of some perceptual set. These distinctions are seen most clearly in relation to particular forms of causal models used for estimation, for example, the bilinear approximation

$$\dot{x} = f(x,u) = Ax + uBx + Cu$$
$$y = g(x) + \varepsilon \tag{7}$$
$$A = \frac{\partial f(0,0)}{\partial x} \quad B = \frac{\partial^2 f(0,0)}{\partial x \partial u} \quad C = \frac{\partial f(0,0)}{\partial u}$$

where $\dot{x} = \partial x / \partial t$. This is an approximation to any model of how changes in neuronal activity in one region x_i are caused by activity in the other regions. Here the output function $g(x)$ embodies a hemodynamic convolution, linking neuronal activity to BOLD, for each region (e.g., that in Fig. 3). The matrix A represents the coupling among the regions in the absence of input $u(t)$. This can be thought of as the endogenous coupling in the absence of experimental perturbations. The matrix B is effectively the change in coupling induced by the input. It encodes the input-sensitive changes in A or, equivalently, the modulation of coupling by experimental manipulations. Because B is a second-order derivative, it is referred to as *bilinear*. Finally, the matrix C embodies the exogenous influences of inputs on neuronal activity. The parameters $\theta = A, B$, and C are the connectivity or coupling matrices that we wish to identify and define the functional architecture and interactions among brain regions at a neuronal level. They play the same role as rate constant in kinetic models and therefore have units of Hertz or per second.

Because Eq. (7) has exactly the same form as Eq. (4), we can express it as a GLM and estimate the parameters using EM in the usual way (*see* Ref. [31]). Generally, estimation in the context of highly parameterized models like DCMs requires constraints in the form of priors. These priors enable conditional inference about the connectivity estimates. The sorts of questions that can be addressed with DCMs are now illustrated by looking at how attentional modulation is mediated in sensory processing hierarchies in the brain.

4.2.1 DCM and Attentional Modulation

It has been established that the superior parietal cortex (SPC) exerts a modulatory role on V5 responses using Volterra-based regression models [34] and that the inferior frontal gyrus (IFG) exerts a similar influence on SPC using structural equation modeling [32]. The example here shows that DCM leads to the same conclusions but starting from a completely different construct. The experimental paradigm and data acquisition are described in the legend to Fig. 4. This figure also shows the location of the regions that entered the DCM. These regions were based on maxima from conventional SPMs testing for the effects of photic stimulation, motion, and

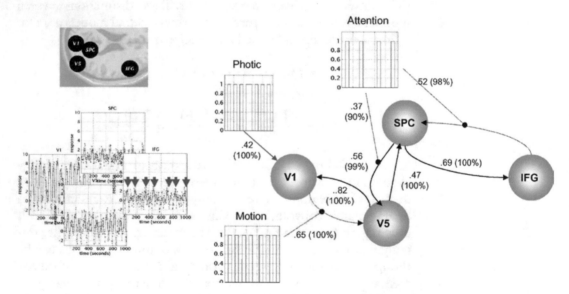

Fig. 4 Results of a dynamic causal modeling (*DCM*) analysis of attention to visual motion with fMRI. *Right panel*: Functional architecture based upon the conditional estimates shown alongside their connections, with the percent confidence that they exceeded threshold in brackets. The most interesting aspects of this architecture involve the role of motion and attention in exerting bilinear effects. Critically, the influence of motion is to enable connections from V1 to the motion-sensitive area V5. The influence of attention is to enable backward connections from the inferior frontal gyrus (*IFG*) to the superior parietal cortex (*SPC*). Furthermore, attention increases the influence of SPC on V5. *Dotted arrows* connecting regions represent significant bilinear effects in the absence of a significant intrinsic coupling. *Left panel*: Fitted responses based upon the conditional estimates and the adjusted data are shown for each region in the DCM. The insert (*upper left*) shows the location of the regions

attention. Regional time courses were taken as the first eigenvariate of 8-mm-spherical volumes of interest, centered on the maxima shown in the figure. The exogenous inputs, in this example, comprise one sensory perturbation and two contextual inputs. The sensory input was simply the presence of photic stimulation and the first contextual one was presence of motion in the visual field. The second contextual input, encoding attentional set, was one during attention to speed changes and zero otherwise. The outputs corresponded to the four regional eigenvariates in Fig. 4 (left panel). The intrinsic connections were constrained to conform to a hierarchical pattern in which each area was reciprocally connected to its supra-ordinate area. Photic stimulation entered at, and only at, V1. The effect of motion in the visual field was modeled as a bilinear modulation of the V1 to V5 connectivity and attention was allowed to modulate the backward connections from IFG and SPC.

Subjects were studied with fMRI under identical stimulus conditions (visual motion subtended by radially moving dots) while manipulating the attentional component of the task (detection of velocity changes). The data were acquired from a normal subject at 2 T. Each subject had four consecutive 100-scan sessions comprising a series of ten-scan blocks under five different conditions D F A F N F A F N S. The first condition (D) was a dummy condition to allow for magnetic saturation effects. F (fixation) corresponds to a low-level baseline where subjects viewed a fixation point at the center of a screen. In condition A (attention), subjects viewed 250 dots moving radially from the center at 4.7°/s and were asked to detect changes in radial velocity. In condition N (no attention), the subjects were asked simply to view the moving dots. In condition S (stationary), subjects viewed stationary dots. The order of A and N was swapped for the last two sessions. In all conditions, subjects fixated the center of the screen. During scanning, there were no speed changes. No overt response was required in any condition.

The results of the DCM are shown in Fig. 4 (right panel). Of primary interest here is the modulatory effect of attention that is expressed in terms of the bilinear coupling parameters for this input. As expected, we can be highly confident that attention modulates the backward connections from IFG to SPC and from SPC to V5. Indeed, the influences of IFG on SPC are negligible in the absence of attention (dotted connection). It is important to note that the only way that attentional manipulation can affect brain responses was through this bilinear effect. Attention-related responses are seen throughout the system (attention epochs are marked with arrows in the plot of IFG responses in the left panel). This attentional modulation is accounted for, sufficiently, by changing just two connections. This change is, presumably, instantiated by instructional set at the beginning of each epoch.

The second thing, this analysis illustrates, is how functional segregation is modeled in DCM. Here one can regard V1 as "segregating" motion from other visual information and distributing it to the motion-sensitive area V5. This segregation is modeled as a bilinear "enabling" of V1–V5 connections when, and only when, motion is present. Note that in the absence of motion, the intrinsic V1–V5 connection was trivially small (in fact the estimate was –0.04 Hz). The key advantage of entering motion through a bilinear effect, as opposed to a direct effect on V5, is that we can finesse the inference that V5 shows motion-selective responses with the assertion that these responses are mediated by afferents from V1. The two bilinear effects above represent two important aspects of functional integration that DCM is able to characterize.

4.2.2 Structural Equation Modeling as a Special Case of DCM

The central idea, behind DCM, is to treat the brain as a deterministic nonlinear dynamic system that is subject to inputs and produces outputs. Effective connectivity is parameterized in terms of coupling among unobserved brain states (e.g., neuronal activity in different regions). The objective is to estimate these parameters by perturbing the system and measuring the response. This is in contradistinction to established methods for estimating effective connectivity from neurophysiological time series, which include SEM and models based on multivariate auto-regressive processes. In these models, there is no designed perturbation and the inputs are treated as unknown and stochastic. Furthermore, the inputs are often assumed to express themselves instantaneously such that, at the point of observation, the change in states is zero. From Eq. (7), in the absence of bilinear effects, we have

$$\dot{x} = 0 = Ax + Cu$$
$$x = -A^{-1}Cu \tag{8}$$

This is the regression equation used in SEM where $A = D - I$ and D contains the off-diagonal connections among regions. The key point here is that A is estimated by assuming that $u(t)$ is some random innovation with known covariance. This is not really tenable for designed experiments when $u(t)$ represent carefully structured experimental inputs. Although SEM and related auto-regressive techniques are useful for establishing dependence among regional responses, they are not surrogates for informed causal models based on the underlying dynamics of these responses.

In this section, we have covered multivariate techniques ranging from eigenimage analysis that does not have an explicit forward or causal model to DCM that does. The bilinear approximation to any DCM has been illustrated though its use with fMRI to study attentional modulation. The parameters of the bilinear approximation include first-order effective connectivity A and its experimentally induced changes B. Although the bilinear approximation is useful, it is possible to model coupling among neuronal subpopulations explicitly. We conclude with a DCM that embraces a number of neurobiological facts and takes us much closer to a mechanistic understanding of how brain responses are generated. This example uses responses measured with EEG.

4.3 Dynamic Causal Modeling with Neural Mass Models

ERPs have been used for decades as electrophysiological correlates of perceptual and cognitive operations. However, the exact neurobiological mechanisms underlying their generation are largely unknown. In this section, we use neuronally plausible models to understand event-related responses. Our example shows that changes in connectivity are sufficient to explain certain ERP components. Specifically, we will look at the MMN, a component associated with rare or unexpected events. If the unexpected nature of rare stimuli depends on learning which stimuli are frequent, then the MMN must be due to plastic changes in connectivity that mediate perceptual learning. We conclude by showing that advances in the modeling of evoked responses now afford measures of connectivity among cortical sources that can be used to quantify the effects of perceptual learning.

4.3.1 Neural Mass Models

The minimal model we have developed [35] uses the connectivity rules described by Felleman and Van Essen [36] to assemble a network of coupled sources. These rules are based on a partitioning of the cortical sheet into supra-, infra-granular, and granular layer (layer 4). Bottom-up or forward connections originate in agranular layers and terminate in layer 4. Top-down or backward connections target agranular layers. Lateral connections originate in agranular layers and target all layers. These long-range or extrinsic cortico-cortical connections are excitatory and arise from pyramidal cells.

Each region or source is modeled using a neural mass model described by David and Friston [35], based on the model of Jansen and Rit [37]. This model emulates the activity of a cortical area using three neuronal subpopulations, assigned to granular and agranular layers. A population of excitatory pyramidal (output) cells receives inputs from inhibitory and excitatory populations of inter-neurons, via intrinsic connections (intrinsic connections are confined to the cortical sheet). Within this model, excitatory inter-neurons can be regarded as spiny stellate cells found predominantly in layer 4 and in receipt of forward connections. Excitatory pyramidal cells and inhibitory inter-neurons are considered to occupy agranular layers and receive backward and lateral inputs (Fig. 5).

To model event-related responses, the network receives inputs via input connections. These connections are exactly the same as forward connections and deliver inputs to the spiny stellate cells in layer 4. In the present context, inputs $u(t)$ model subcortical auditory inputs. The vector C controls the influence of the input on each source. The lower, upper, and leading diagonal matrices A^F, A^B, A^L encode forward, backward, and lateral connections, respectively. The DCM here is specified in terms of the state equations shown in Fig. 5 and a linear output equation

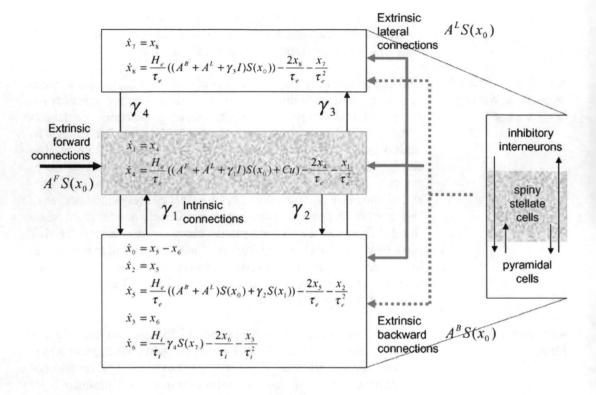

Neuronal model

Fig. 5 Schematic of the dynamic causal modeling (*DCM*) used to model electrical responses. This schematic shows the state equations describing the dynamics of sources or regions. Each source is modeled with three subpopulations (pyramidal, spiny stellate, and inhibitory inter-neurons) as described in the main text. These have been assigned to granular and agranular cortical layers that receive forward and backward connections, respectively

$$\dot{x} = f(x,u)$$
$$y = Lx_0 + \varepsilon \tag{9}$$

where $x0$ represents the trans-membrane potential of pyramidal cells and L is a lead field matrix coupling electrical sources to the EEG channels. This should be compared with the DCM above for hemodynamics; here, the equations governing the evolution of neuronal states are much more complicated and realistic, as opposed to the bilinear approximation in Eq. (7). Conversely, the output equation is a simple linearity, as opposed to the nonlinear observer used for fMRI. As an example, the state equation for the inhibitory subpopulation is [2]

---

[2] Propagation delays on the extrinsic connections have been omitted for clarity here and in Fig. 5.

$$\dot{x}_7 = x_8$$

$$\dot{x}_8 = \frac{H_e}{\tau_e}\left[\left(A^B + A^L + \gamma_3 I\right)S(x_0)\right] - \frac{2x_8}{\tau_e} - \frac{x_7}{\tau_e^2} \qquad (10)$$

Within each subpopulation, the evolution of neuronal states rests on two operators. The first transforms the average density of pre-synaptic inputs into the average postsynaptic membrane potential. This is modeled by a linear transformation with excitatory and inhibitory kernels parameterized by $H_{\varepsilon,i}$ and $\tau_{e,j}$. $H_{e,j}$ controls the maximum post-synaptic potential and $\tau_{e,j}$ represents a lumped rate constant. The second operator S transforms the average potential of each subpopulation into an average firing rate. This is assumed to be instantaneous and is a sigmoid function. Interactions, among the subpopulations, depend on constants, $\gamma_{1,2,3,4}$, which control the strength of intrinsic connections and reflect the total number of synapses expressed by each subpopulation. In Eq. (10), the top line expresses the rate of change of voltage as a function of current. The second line specifies how current changes as a function of voltage, current, and pre-synaptic input from extrinsic and intrinsic sources. Having specified the DCM in terms of these equations, one can estimate the coupling parameters from empirical data using *EM* as described above. See Ref. [38] for more details.

4.3.2 Perceptual Learning and the MMN

The example shown in Fig. 6 is an attempt to model the MMN in terms of changes in backward and lateral connections among cortical sources. In this example, two (averaged) channels of EEG data were modeled with three cortical sources. Using this generative or forward model, we estimated differences in the strength of these connections for rare and frequent stimuli. As expected, we could account for detailed differences in the ERPs (the MMN) by changes in connectivity (see figure legend for details). Interestingly, these differences were expressed selectively in the lateral connections. If this model is a sufficient approximation to the real sources, these changes are a noninvasive measure of plasticity, mediating perceptual learning, in the human brain.

5 Conclusion

In this chapter, we have reviewed some key models that underpin image analysis and have touched briefly on ways of assessing specialization and integration in the brain. These models can be regarded as a succession of modeling endeavors, that drawing more and more on our understanding of how brain-imaging signals are generated, both in terms of biophysics and the underlying neuronal interactions. We have seen how hierarchical linear observation models encode the treatment effects elicited by experimental

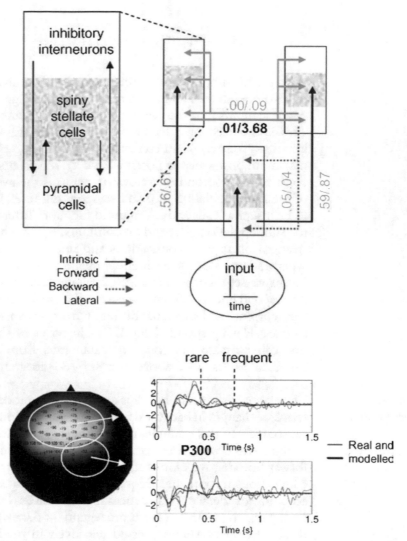

Fig. 6 Summary of a dynamic causal modeling (*DCM*) analysis of event-related potentials (*ERPs*) elicited during an auditory oddball paradigm, employing rare and frequent pure tones. *Upper panel*: Schematic showing the architecture of the neuronal model used to explain the empirical data. Sources were coupled with extrinsic cortico-cortical connections following the rules of Felleman and van Essen. The free parameters of this model included intrinsic and extrinsic connection strengths that were adjusted to best explain the data. In this example, the lead field was also estimated, with no spatial constraints. The parameters were estimated for ERPs recorded during the presentation of rare and frequent tones and are reported beside their corresponding connection (frequent/rare). The most notable finding was that the mismatch response could be explained by a selective increase in lateral connection strength from 0.1 to 3.68 Hz (highlighted in *bold*). *Lower panel*: The channel positions (*left*) and ERPs (*right*) averaged over two subsets of channels (*circled on the left*). Note the correspondence between the measured ERPs and those generated by the model. Auditory stimuli, 1,000 or 2,000 Hz tones with 5 ms rise and fall times and 80 ms duration, were presented binaurally. The tones were presented for 15 min, every 2 s in a pseudo-random sequence with 2000-Hz tones occurring 20 % of the time and 1,000-Hz tones occurring 80 % of the time. The subject was instructed to keep a mental record of the number of 2000-Hz tones (nonfrequent target tones). Data were acquired using 128 EEG electrodes with 1,000 Hz sample frequency. Before averaging, data were referenced to mean earlobe activity and band-pass filtered between 1 and 30 Hz. Trials showing ocular artifacts and bad channels were removed from further analysis

design. GLMs based on convolution models imply an underlying dynamic input-state-output system. The form of these systems can be used to constrain convolution models and explore some of their simpler nonlinear properties. By creating observation models based on explicit forward models of neuronal interactions, one can model and assess interactions among distributed cortical areas and make inferences about coupling at the neuronal level. The next years will probably see an increasing realism in the dynamic causal models introduced above (*see* Ref. [39]). These endeavors are likely to encompass fMRI signals enabling the conjoint modeling, or fusion, of different modalities and the marriage of computational neuroscience with the modeling of brain responses.

References

1. Staum M (1995) Physiognomy and phrenology at the Paris Athénée. J Hist Ideas 6:443–462

2. Phillips CG, Zeki S, Barlow HB (1984) Localisation of function in the cerebral cortex: past present and future. Brain 107:327–361

3. Goltz F (1881) Transactions of the 7th international medical congress (W. MacCormac, Ed.), Vol. I, JW Kolkmann: London, 1881:218–228

4. Absher JR, Benson DF (1993) Disconnection syndromes: an overview of Geschwind's contributions. Neurology 43:862–867

5. Zeki S (1990) The motion pathways of the visual cortex. Vision: coding and efficiency (C. Blakemore, Ed.). Cambridge University Press, pp 321–345

6. Friston KJ, Frith CD, Liddle PF, Frackowiak RSJ (1991) Comparing functional (PET) images: the assessment of significant change. J Cereb Blood Flow Metab 11:690–699

7. Berry DA, Hochberg Y (1999) Bayesian perspectives on multiple comparisons. J Stat Plan Infer 82:215–227

8. Holmes A, Ford I (1993) A Bayesian approach to significance testing for statistic images from PET. In: Uemura K, Lassen NA, Jones T, Kanno I (eds) Quantification of brain function, tracer kinetics and image analysis in brain PET. Excerpta Medica, Int. Cong. Series No. 1993. 1030:521–534

9. Dempster AP, Laird NM, Rubin (1977) Maximum likelihood from incomplete data via the EM algorithm. J Roy Stat Soc B 39:1–38

10. Friston KJ, Jezzard P, Turner R (1994) Analysis of functional MRI time series. Human Brain Map 1:153–171

11. Boynton GM, Engel SA, Glover GH, Heeger DJ (1996) Linear systems analysis of functional magnetic resonance imaging in human V1. J Neurosci 16:4207–4221

12. Friston KJ, Frith CD, Turner R, Frackowiak RSJ (1995) Characterising evoked hemodynamics with fMRI. NeuroImage 2:157–165

13. Josephs O, Turner R, Friston KJ (1997) Event-related fMRI Hum. Brain Mapp 5:243–248

14. Lange N, Zeger SL (1997) Non-linear Fourier time series analysis for human brain mapping by functional magnetic resonance imaging (with discussion). J Roy Stat Soc Ser C 46:1–29

15. Buxton RB, Frank LR (1997) A model for the coupling between cerebral blood flow and oxygen metabolism during neural stimulation. J Cereb Blood Flow Metab 17:64–72

16. Mandeville JB, Marota JJ, Ayata C, Zararchuk G, Moskowitz MA, Rosen B, Weisskoff RM (1999) Evidence of a cerebrovascular postarteriole Windkessel with delayed compliance. J Cereb Blood Flow Metab 19:679–689

17. Hoge RD, Atkinson J, Gill B, Crelier GR, Marrett S, Pike GB (1999) Linear coupling between cerebral blood flow and oxygen consumption in activated human cortex. Proc Natl Acad Sci 96:9403–9408

18. Friston KJ, Mechelli A, Turner R, Price CJ (2000) Nonlinear responses in fMRI: the Balloon model, Volterra kernels, and other hemodynamics. NeuroImage 12:466–477

19. Fliess M, Lamnabhi M, Lamnabhi-Lagarrigue F (1983) An algebraic approach to nonlinear functional expansions. IEEE Trans Circuits Syst 30:554–570

20. Bendat JS (1990) Nonlinear system analysis and identification from random data. John Wiley, New York

21. Gerstein GL, Perkel DH (1969) Simultaneously recorded trains of action potentials: analysis

and functional interpretation. Science 164:828–830

22. Aertsen A, Preissl H (1991) Dynamics of activity and connectivity in physiological neuronal Networks. In: Schuster HG (ed) Non linear dynamics and neuronal networks. VCH, New York, pp 281–302

23. McIntosh AR, Gonzalez-Lima F (1994) Structural equation modelling and its application to network analysis in functional brain imaging. Hum Brain Mapp 2:2–22

24. Friston KJ, Frith CD, Liddle PF, Frackowiak RSJ (1993) Functional connectivity: the principal component analysis of large data sets. J Cereb Blood Flow Metab 13:5–14

25. Friston KJ, Poline J-B, Holmes AP, Frith CD, Frackowiak RSJ (1996) A multivariate analysis of PET activation studies. Hum Brain Mapp 4:140–151

26. Friston KJ, Frith CD, Fletcher P, Liddle PF, Frackowiak RSJ (1996) Functional topography: multidimensional scaling and functional connectivity in the brain. Cereb Cortex 6:156–164

27. Sychra JJ, Bandettini PA, Bhattacharya N, Lin Q (1994) Synthetic images by subspace transforms. I. Principal component images and related filters. Med Phys 21:193–201

28. Biswal B, Yetkin FZ, Haughton VM, Hyde JS (1995) Functional connectivity in the motor cortex of resting human brain using echoplanar MRI. Magn Reson Med 34:537–541

29. McKeown M, Jung T-P, Makeig S, Brown G, Kinderman S, Lee T-W, Sejnowski T (1998) Spatially independent activity patterns in functional MRI data during the Stroop colour naming task. Proc Natl Acad Sci 95:803–810

30. Baumgartner R, Scarth G, Teichtmeister C, Somorjai R, Moser E (1997) Fuzzy clustering of gradient-echo functional MRI in the human visual cortex. Part 1: reproducibility. J Magn Reson Imaging 7:1094–1101

31. Friston KJ, Harrison L, Penny W (2003) Dynamic causal modelling. NeuroImage 19:1273–1302

32. Büchel C, Friston KJ (1997) Modulation of connectivity in visual pathways by attention: cortical interactions evaluated with structural equation modelling and fMRI. Cereb Cortex 7:768–778

33. Harrison LM, Penny W, Friston KJ (2003) Multivariate autoregressive modelling of fMRI time series. NeuroImage 19:1477–1491

34. Friston KJ, Büchel C (2000) Attentional modulation of effective connectivity from V2 to V5/MT in humans. Proc Natl Acad Sci U S A 97:7591–7596

35. David O, Friston KJ (2003) A neural mass model for MEG/EEG: coupling and neuronal dynamics. NeuroImage 20:1743–1755

36. Felleman DJ, Van Essen DC (1992) Distributed hierarchical processing in the primate cerebral cortex. Cereb Cortex 1:1–47

37. Jansen BH, Rit VG (1995) Electroencephalogram and visual evoked potential generation in a mathematical model of coupled cortical columns. Biol Cybern 73:357–366

38. David O, Kiebel SJ, Harrison LM, Mattout J, Kilner JM, Friston KJ (2006) Dynamic causal modelling of evoked responses in EEG and MEG. NeuroImage 30:1255–1272

39. Horwitz B, Friston KJ, Taylor JG (2001) Neural modelling and functional brain imaging: an overview. Neural Netw 13:829–846

Chapter 9

Brain Atlases: Their Development and Role in Functional Inference

John Darrell Van Horn and Arthur W. Toga

Abstract

Imparting functional meaning to neuroanatomical location has been among the greatest challenges to neuroscientists. The characterization of the brain architecture responsible in human cognition received a boost in momentum with the emergence of in vivo functional and structural neuroimaging technology over the past 30 years. Yet, individual variability in cortical gyrification as well as the patterns of blood flow-related activity measured using fMRI and positron emission tomography complicated direct comparisons across subjects without spatially accounting for overall brain size and shape. This realization resulted in considerable effort now involving the collective efforts of neuroscientists, computer scientists, and mathematicians to develop common brain atlas spaces against which the regions of activity may be accurately referenced. We examine recent developments in brain imaging and computational anatomy that have greatly expanded our ability to analyze brain structure and function. The enormous diversity of brain maps and imaging methods has spurred the development of population-based digital brain atlases. Atlases store information on how the brain varies across age and gender, across time, in health and disease, and in large human populations. We describe how brain atlases, and the computational tools that align new datasets with them, facilitate comparison of brain data across experiments, laboratories, and from different imaging devices. The major philosophies are presented that underlie the construction of probabilistic atlases, which store information on anatomic and functional variability in a population. Algorithms which create composite brain maps and atlases based on multiple subjects are examined. We show that group patterns of cortical organization, asymmetry, and disease-specific trends can be resolved that may not be apparent in individual brain maps. Finally, we describe the development of four-dimensional maps that store information on the dynamics of brain change in development and disease.

Key words Brain atlases, Neuroanatomy, Diffeomorphism, Warping, Functional activity, Inference

1 Introduction

Over a century ago, in a horrific accident, damage to the frontal lobe of Phineas Gage produced profound changes in his personality and cognitive function, the beginning of what may be considered as the modern era of the localization of brain function [1]. Nearly a decade later, the examination by Paul Broca of aphasic patients having damage to the left inferior frontal areas (pars triangularis, pars

Massimo Filippi (ed.), *fMRI Techniques and Protocols*, Neuromethods, vol. 119,
DOI 10.1007/978-1-4939-5611-1_9, © Springer Science+Business Media New York 2016

opercularis) solidified the notion that function could be linked to specific areas of cortical tissue [2]. At the turn of the twentieth century, Dr. Alois Alzheimer, a German psychiatrist, identified the first case of what became known as Alzheimer's Disease (AD) [3], now known to severely affect the thickness of the cortical mantle as well as the hippocampal area and its surrounding tissues. Since those early reports, a principle goal in neuroscience has been to classify the specific brain regions possessing unique functional components of complex thought and how these functions might be altered in response to injury or as a result of disease.

Over the past three decades, with the emergence of neuroimaging as the primary tool for the examination of the brain in vivo during cognitive task performance, the mapping of mental function has given rise to an explosion of functional experimentation and an ever-widening interest in understanding brain processes from fields beyond traditional neuroscience (e.g., economics, criminology, social science). This intense effort and expanse of data has emphasized the realization that there is considerable individual variation in brain size and shape that must be accounted for in the processing of brain imaging data and the assignment of functional significance. Evaluation and comparison of brain imaging data with respect to and against well-defined anatomical references is now a critical element in the localization of essential cognitive functions in nearly all functional imaging investigations.

Brain atlases can now comprise imaging data describing multiple aspects of brain structure or function at different scales from different subjects, yielding a truly integrative and comprehensive description of this organ in health and disease [4, 5]. However, such complexity and variability of brain structure, especially in the gyral patterns of the human cortex, present challenges in creating standardized brain atlases that reflect the anatomy of a population [6]. This chapter discusses the concepts behind population-based, age-, and disease-specific brain atlas construction that can be used to reflect the specific anatomy and physiology of a particular clinical subpopulation. Based on well-characterized subject groups, age-specific atlases can potentially contain thousands of structure models, composite maps, average templates, and visualizations of structural variability, asymmetry, and group-specific differences. They correlate the structural, metabolic, molecular, and histologic hallmarks of the disease [7, 8]. Rather than simply fusing information from multiple subjects and sources, new mathematical strategies are being introduced to resolve group-specific features not apparent in individual scans [9]. High-dimensional elastic mappings, based on covariant partial differential equations, are developed to encode patterns of cortical variation [10–12]. In the resulting brain atlas, age-stratified features and regional asymmetries emerge that are not apparent in individual anatomies. The consequential probabilistic atlas spaces can be used to identify patterns of altered structure and function, and can guide algorithms for knowledge-based image analysis, automated image labeling, tissue classification, data mining, and functional image analysis. These integrative techniques have provided significant motivation for human brain mapping initiatives and have important applications in health and disease.

2 Methods for Brain Atlas Construction

Creating atlases relies on the accumulation and compilation of many image sets along with appropriate registration and warping strategies, indexing schemes, and nomenclature systems. The processing of multimodal brain images in the context of an atlas enables a more meaningful interpretation (Fig. 1). The complexity and variability of human brain (as well as other species) across subjects is so great that reliance on atlases is essential to effectively manipulate, analyze, and interpret brain data. Central to these tasks is the construction of averages, templates, and models to describe how the brain and its component parts are organized. Design of appropriate reference systems for human brain data presents considerable challenges, since these systems must capture how brain structure and function vary in large populations, across age and gender, in different disease states, across imaging modalities, and even across species.

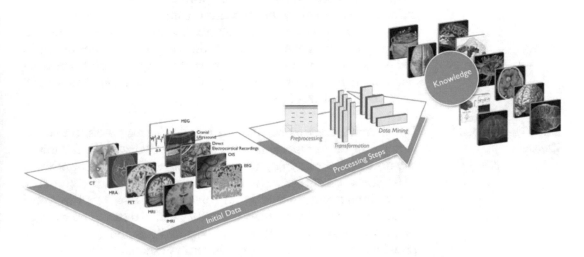

Fig. 1 A variety of neuroimaging methods permit the acquisition of brain data over time and space having a range of resolution granularity. Moreover, variation across individuals and how this changes over the lifespan must be accounted for in statistical examination of the data. Mapping these data to known spatial coordinate systems enables highly accurate inference concerning the brain's structural change over time, between populations, or in terms of localizing functional change. Atlases denoting this variation after spatial warping, the characterization of shape, three-dimensional (3D) distortion, etc. will be essential in describing structural and functional alteration associated with normal aging as well as in disease

2.1 Basic Image Registration

Image registration is an elemental step in many of the analytic strategies involving brain imaging today [13]. Initially developed as an image processing technique to spatially align one image to match another, image registration now has a vast range of applications, such as automated image labeling and for pathology detection in individuals or groups [14]. Registration algorithms can encode patterns of anatomic variability in large human populations, and can use this information to create disease-specific, population-based brain atlases [15]. They may also blend data from multiple imaging devices to correlate different measures of brain structure and function. Finally, modern registration algorithms and workflows can serve as a basic measure for patterns of structural change during brain development, tumor growth, or degenerative disease processes [16].

2.2 Geodesic Averaging of Brain Shape

The objective in geodesic approaches has been to encourage variational methods for anatomical averaging that operate within the space of the underlying image registration problem [17]. This approach is effective when using a large deformation viscous framework, where linear averaging might not be appropriate. The theory behind it is similar to registration-based techniques but with single image force replaced by the average forces from multiple sources. These group forces drive an average transport ordinary differential equation allowing one to estimate the geodesic that moves an image toward the mean shape configuration. This model provides large deformation atlases that are optimal with respect to the shape manifold as defined by the data and the image registration assumptions. These procedures generate refined average representations of highly variable anatomy from distinct populations. For example, the population statistics have been used to show a significant doubling of the relative prefrontal lobe size in humans, as compared to nonhuman primates [18].

2.3 Density-Based Atlases

Initial approaches to population-based atlasing concentrated on generating mean representations of anatomy through the "intensity pooling" of multiple MRI scans. This involves large number of MRI scans which are each linearly transformed into stereotaxic space, intensity-normalized, and averaged on a voxel-by-voxel basis, producing an average intensity MRI dataset. The average brains that result have large areas, especially at the cortex, where individual structures are blurred because of spatial variability in the population. While this blurring limits their usefulness as a quantitative tool, the templates can be used as targets for the automated registration and mapping of MR and co-registered functional data into stereotaxic space [19].

2.4 Label-Based Atlases

In label-based approaches, large ensembles of brain data are labeled or "segmented" by a human operator or algorithmically into subvolumes after mapping individual datasets into stereotaxic space. A probability map is then constructed for each segmented structure, by determining the proportion of subjects assigned a given anatomic label at each voxel position in stereotactic space. The prior information which these probability maps provide on the location of various tissue classes in stereotactic space has been useful in designing automated tissue classifiers and approaches to correct radio-frequency and intensity inhomogeneities in MR scans. Statistical data on anatomic labels and tissue types normally found at given positions in stereotactic space provide a vital independent source of information to guide and inform mathematical algorithms, which analyze neuroanatomical data in stereotactic space.

2.5 Encoding Brain Variation

Measuring and accounting for the considerable variability in brain shape across human populations necessitates realistically complex mathematical strategies to encode comprehensive information on structural variability [20]. Particularly relevant is a three-dimensional (3D) statistical information on group-specific patterns of variation and how these patterns are altered in disease. This information can be represented such that it can be exploited by expert diagnostic systems, whose goal is to detect subtle or diffuse structural alterations in disease [21]. Strategies for detecting structural anomalies can leverage information in anatomical databases by invoking encoded knowledge on the variations in geometry and location of neuroanatomic regions and critical functional interfaces, especially at the cortex.

2.6 Shape and Pattern Theory

Of particular relevance in dealing with brain substructures are methods used to define a mean shape in such a way that departures from this mean shape can be treated as a linear process [22]. Linearization of the pathology detection problem, by constructing various shape manifolds and their associated tangent spaces, allows the use of conventional statistical procedures and linear decomposition of departures from the mean to characterize shape change. These approaches have been applied to detect structural anomalies in schizophrenia by identification of statistical differences in mean shape of brain structures [23–25] (Fig. 2).

2.7 Deformation Atlases

When applied to two different 3D brain scans, a nonlinear registration or warping algorithm calculates a deformation map that matches up brain structures in one scan with their counterparts in the other. The deformation map indicates 3D patterns of anatomic differences between the two subjects or populations [26]. In probabilistic atlases based on deformation maps, statistical properties of these deformation maps are encoded locally to determine the magnitude and directional biases of anatomic variation [27]. Encoding of local variation can then be used to assess the severity of structural variants outside of the normal range, which may be a sign of

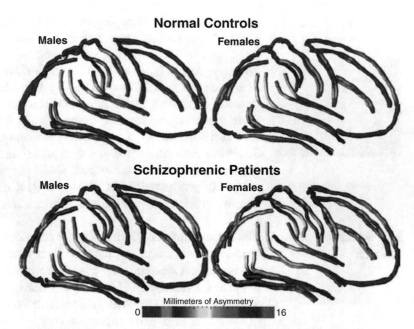

Fig. 2 Variability in brain cortical architectural asymmetry leads directly to the consideration of probabilistic atlases and the use of cortically derived landmarks, for example, sulcal lines. In this figure, sulcal lines were manually determined for male and female normal subjects and those diagnosed with schizophrenia. Asymmetry maps were created in each group as defined by sex and diagnosis (*NC* normal controls, *SZ* schizophrenic patients). Sulcal mesh averages for each hemisphere were subtracted from a reflected version of the same structure in the other hemisphere to create displacement vectors. Thus, the mappings measure the degree of lateralization in terms of millimeters of displacement for the line under diffeomorphic and atlas space constraints. These maps, therefore, represent the magnitude of average asymmetry in sulcal anatomy between the two hemispheres between males and females and schizophrenic and normal subjects (Figure adapted from Narr et al. [72])

disease. A major goal in designing this type of pathology detection system is to recognize that both the magnitude and local directional biases of structural variability in the brain may be different at every single anatomic point. Such atlases are not only cortically based but can also be done, for instance, on substructures and cerebellum [28].

2.8 Disease-Specific Atlases

Disease-specific atlases are designed to reflect the unique anatomy and physiology of a particular clinical subpopulation. Based on well-characterized patient groups, these atlases contain thousands of structure models, as well as composite maps, average templates, and visualizations of structural variability, asymmetry, and group-specific differences. They act as a quantitative framework that correlates the structural, metabolic, molecular, and histologic hallmarks of the disease. Because they retain information on group anatomical variability, disease-specific atlases are a type of probabilistic atlas specialized to represent a particular clinical group. The resulting atlases can identify patterns of altered structure or function, and can guide algorithms for knowledge-based image analysis, automated image labeling, tissue classification, and functional image analysis.

2.9 Genetic Atlases

Inclusion of genetic data in an atlas makes it possible to go beyond simply describing the effects of a disease on the brain to investigating its fundamental causes. This not only allows the direct mapping of genetic influences on brain structure, but also allows us to quantify heritability for different features of the brain. Familial, twin, and genetic linkage studies have recently begun to expand the atlas concept to tie together genetic and imaging studies of disease [7, 29]. Atlases that contain genetic brain maps, and a means to analyze them, can help screen relatives for inherited disease. They also offer a framework to mine large imaging databases for risk genes and quantitative trait loci, as well as genetic and environmental triggers of disease.

2.10 Age and Developmental Stratification

The brain changes remarkably in its size and complexity over the lifespan. There is considerable need to account for the age of particular populations in the context of brain maturation and the development of age-stratified normal brain atlas spaces [30]. People who are mildly cognitively impaired, for instance, are at a fivefold increased risk of imminent conversion to dementia, and present specific structural brain changes that are predictive of imminent disease onset [31, 32]. Language impairment in AD patients is also correlated with cortical atrophy in the left temporal and parietal lobes, bilateral frontal lobes, and the right temporal pole [33]. However, characterizing such change presents particular computational challenges. The fitting of brain anatomy to a single template of undetermined age specification may lead to errors in inference about brain morphometry of function relative to an inappropriate underlying template. Alternative approaches can also be fruitful and metrics, such as shape [34], cortical thickness mapping, tensor-based morphometry (TBM), may be better suited for shedding light on the neuroscience of aging and brain degeneration in AD and mild cognitive impairment (MCI) [35].

3 Openly Available Atlases of the Brain

An increasing number and variety of brain atlases for humans, as well as other species, are being made openly available online for the neuroscience community to use as authoritative references, for inclusion in data processing workflows, or for the display of results. These include probabilistic anatomical atlases [15, 36, 37], white matter fiber atlases [38], and cortical surface atlases [39]. A brief listing of several from human, nonhuman primate, and the mouse are provided in Table 1.

Table 1
Human, nonhuman primate, and mouse probabilistic anatomical atlases

Name	URL(s)	Comment
Adult C57BL/6 J Mouse Brain	https://www.bnl.gov/medical/RCIBI/mouse/Atlas_creation.asp	Anatomical atlas of the C57BL/6 J mouse brain
Atlases of the Brain	https://msu.edu/~brains/brains/human/index.html	Anatomical sections and MR image data of human brain
The Allen Brain Atlas	http://www.alleninstitute.org/ http://www.brain-map.org/	Atlas of anatomy and gene expression in the mouse
Brainmaps.org	http://brainmaps.org/	Digitally scanned images of serial sections of both primate and nonprimate brains
Comparative Mammalian Brain Collections	http://www.brainmuseum.org/	Digital photos of whole brain and serial sections from a range of primate brains
Digital Anatomist	http://da.biostr.washington.edu/da.html	Interactive brain atlas surface models and neuroimaging data
The Human Brain Atlas	http://www.thehumanbrain.info/	Interactive online atlas of anatomical preparations and accompanying labeled sections
ICBM/LONI Probabilistic Atlas Series	http://www.loni.usc.edu/atlases/	Human brain atlases based upon probabilistic metrics of regional location
The Mouse Brain Library	http://www.mbl.org/	High-resolution images and database of brains from many genetically characterized strains of mice
The Mouse Connectome Atlas	http://www.mouseconnectome.org/	Database of neuroanatomical connectivity in the mouse brain
Mouse Lemur Brain	http://atlasserv.caltech.edu/Lemur/Start_lemur.html	Downloadable MR image volumes of the mouse lemur (*Microcebus murinus*) brain
Surface Data Management System (SuMs) Atlases	http://sumsdb.wustl.edu/sums/humanpalsmore.do;jsessionid=2gfybndwl1	Standardized brain surface models with links to associated functional data
White Matter Atlas	http://www.dtiatlas.org/	Human brain white matter maps obtained from diffusion tensor imaging (DTI)
The Whole Brain Atlas	http://www.med.harvard.edu/AANLIB/home.html	Human brain atlas including images from postmortem serial sections and MRI. Includes aging brain images

4   Applications of Atlases for Regional Parcellation and Functional Inference

Without reference to known geometries or atlas spaces, precise functional localization is not formally possible at the population level. For instance, in positron emission tomography (PET)/fMRI studies of human cognition analyzed using the statistical parametric mapping (SPM) software package relying on the Montreal Neurological Institute (MNI) atlas as the basis for within group and between group statistical comparisons, typically, each subject's high-resolution anatomical image is warped to the MNI multisubject T1 whole brain template using nonlinear and affine methods [40]. This transformation is then applied to the collection of linearly aligned fMRI time series or task-condition-specific PET images [41]. Employing regional labeling based upon the chosen atlas space, the process of localization analysis is enhanced by reference to known anatomical delineations. The process is typically decomposed into a series of steps: data warping, feature extraction, identification of loci, fitting of labels, and region activity value extraction ([42], for discussion).

Obtaining reliable spatial registration with the chosen atlas space is essential to accurate localization of functional activity. The alignment accuracy and impact on functional maps of four spatial normalization procedures have been compared using a set of high-resolution brain MRIs and functional PET volumes [43], suggesting that the functional variability is much larger than that comprised anatomically and that precise alignment of anatomical features has low influence on the resulting inter-subject functional maps. At larger spatial resolutions, however, differences in localization of activated areas appear to be a consequence of the particular spatial normalization procedure employed. Despite these concerns, however, for typical sample sizes and numbers of observations per subject, reliable functional localization is achieved when performed for each individual using data in atlas space [44].

The use of atlases provides more than just a space in which to morph images for computing population averages and inferential statistics on function, but is also useful for the precise labeling of cortical regions. Figure 3 shows an example of a LONI Pipeline (http://pipeline.loni.usc.edu) workflow for the process of brain extraction using FSL's BET followed by warping to the ICBM452 atlas using FSL FLIRT, and brain surface parcellation using the Brain Parser algorithm [45]. Unlike alternative methods for detecting the major cortical sulci, which use a set of predefined rules based on properties of the cortical surface such as the mean curvature, this approach learns a discriminative model using the probabilistic boosting tree (PBT) algorithm, a supervised learning approach which selects and combines hundreds of features at different scales, such as curvatures, gradients, and shape index [46].

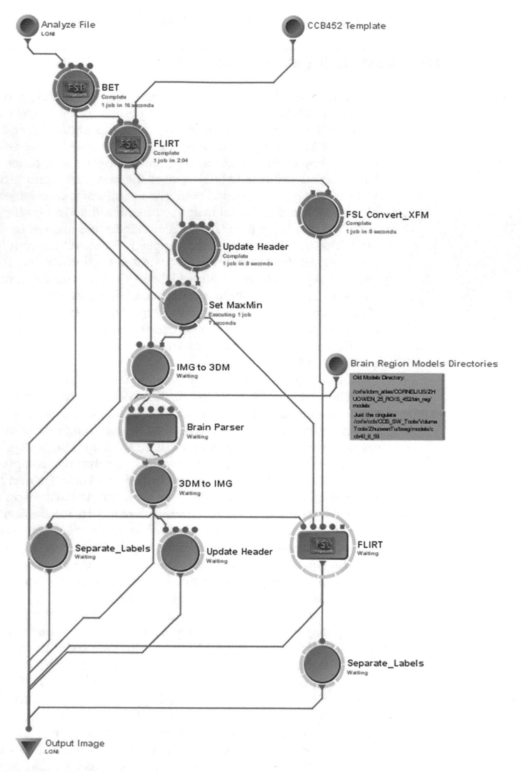

Fig. 3 (a) A LONI Pipeline workflow diagram that reads in an MR structural image volume, performs skull stripping, fits the data to the ICBM452 standardized atlas, performs Bayesian boost-tree region classification, and **(b)** returns the regional delineation results to the original MR image space. Once obtained, the region labeling can be used to mask functional and/or extract functionally related signal from blood oxygen level dependent data. Such automated labeling of brain regions would be made considerably more challenging without the use of standard atlas spaces

Example output from this method can then be used as regions of interest (ROIs) from which blood oxygen level dependent (BOLD) values may then be obtained.

Historically, standardized atlases have demonstrated their greatest utility in the fitting of experimental data from functional imaging studies with PET [40] and fMRI [47]. Population-based inference [48], the identification of individual differences [49], meta-analytic comparisons [50], and other applications performed across subjects depend upon accurate atlas-based normalization. There is no doubt that this is the most scientifically beneficial justification of atlas construction and demonstrates their ability to form the spatial basis for comparing subjects with respect to cognitive operations or in comparisons between patient samples.

Continued improvement and enhancement of extant atlases, such as the Talairach atlas [51], the several iterations of the MNI atlas, cytoarchitectonic atlases [52], as well as those of the ICBM probabilistic atlas [53], provide greater accuracy with respect to functional data and, hence, localization power. Improvement and enhancement of MNI or Talairach landmarks, for example, may enable more rapid calculation of spatial transformations, thereby providing flexibility for specific applications [54].

5   Brain Atlas Revision and Evolution

Spatial mapping of any form is an ongoing process of determining accurate coordinate locations for content appropriate to that mapping's purpose. As previous content in a map changes or as new information is obtained, these maps need to be updated, corrected, and/or modified to reflect these changes in knowledge and the importance of what new knowledge is being conveyed. For instance, the information contained in US aeronautical charts is republished approximately every 3 months partly to reflect changes in the Earth's magnetic field isogonic declination lines that point toward the magnetic North Pole. These field lines vary across North America, from approximately −19° in the Western US through to nearly +20° in the East, and must be accounted for to locate the direction of the true North Pole when charting a navigational course. However, the Earth's magnetic field has been drifting slowly westward since measurements began around 1850, shifting roughly 0.1° or 40 km/year. Failure to periodically update published maps that incorporate this shift in magnetic declination, as well as other information concerning changes in the Earth's topographical features, the construction of tall buildings in urban areas, alterations in air traffic routing, errors in earlier revisions, etc., could result in pilots or computers making substantial navigational errors due to improper compass and course directional settings in the absence of global positioning system (GPS) satellite information.

Errors often appear in maps resulting from the information that was used to create them. Landmarks of note may be mislocated, their spatial extents distorted, and place names misspelled or mistranslated. This would also be true for brain mapping atlases where previous inaccuracies must be addressed, additional data included, or data from other modalities considered. Cytoarchitectonic maps from the classical period of describing brain anatomy have been noted as failing to incorporate sulcal pattern, variation in cell orientation, and being presented as idealized versions of brain structure [15].

However, even modern approaches, using multimodal methods, large databases, and sophisticated computer methods, are not immune from introducing errors. Data misaligned with respect to a standard atlas space may result in gross inaccuracies and considerable problems when trying to make inferences between diagnostic groups. For instance, the statistics of resulting voxel-based morphometric comparisons may be uninformative about group differences wherever the spatial normalization algorithm has failed to register on any robustly appearing image gradient [55]. This has severe consequences for random-effects-based analyses of morphometric changes due to disease or clinical outcome.

Electronic versions of the atlas of Talairach and Tournoux [56], including the Talairach Daemon (http://www.talairach.org/) and the official published versions by Thieme, have been found to contain a discrepant region of the precentral gyrus on axial slice +35 mm that extends far forward into the frontal lobe. This region has been found to be anatomically incorrect and internally inconsistent within the digital atlas software applications that employ multiplanar cross-referencing tools [57]. This may be a case of simple mislabeling but other forms of atlas warping are known to result in distortions which must be predicated in context with the accurate interpretation of location. As new data are included and novel techniques are developed to inform atlases that are open to scrutiny by researchers, with ongoing updates and corrections, will they become most widely valuable.

Workers in our lab have pursued the construction of improved digital brain atlases composed of data from manually delineated high-resolution MRI [58]. A total of 56 structures were labeled on the MR volumes of 40 normal healthy volunteers. The labeling was performed according to a set of protocols developed specifically for this project. In brief, pairs of raters were assigned to each structure and trained on the protocol for delineating that structure. Each rater pair was tested for concordance on 6 of the 40 brains; once they had achieved reliability standards, they divided the task of delineating the remaining 34 brains. The data were then spatially normalized to well-known atlas-based templates using each of three popular algorithms: AIR's nonlinear warp [59] paired with the ICBM452 Warp 5 atlas [60], FSL's FLIRT [61] was paired with its own template, a skull-stripped version of the ICBM152 T1 average; and SPM's unified segmentation method [62] was paired with its canonical brain, the whole head ICBM152 T1-weighted average. In the end, these approaches produced three variants of a resultant atlas, where each was constructed from 40 representative samples of a data processing stream that one might use for analysis. For each normalization algorithm, the individual structure delineations were then resampled according to the derived transformations and computed averages were obtained at each voxel location to estimate the probability of that voxel belonging to each of the 56 structures. Each version of the atlas contains, for every voxel, probability densities for each region, thereby providing a resource for automated probabilistic labeling of external data types registered into standard spaces. Additionally, computed average intensity images and tissue density maps based on the three methods and target spaces were also obtained. These atlases are publicly available on the LONI Web site (http://loni.usc.edu) and, we believe, will serve as critical resources for diverse applications including meta-analysis of functional and structural imaging data and other bioinformatics applications where display of arbitrary labels in probabilistically defined anatomic space will facilitate both knowledge-based development and visualization of findings from multiple disciplines. However, in time these, too, will be replaced by still more accurate atlases of larger sample size, improved spatial resolution, with finer anatomical detail.

6   Conclusions

The evolution of brain atlases has seen tremendous advances; they can now accommodate observations from multiple modalities and from populations of subjects collected at different laboratories. The probabilistic systems described here show promise for identifying patterns of structural, functional, and molecular variation in large image databases for pathology detection in individuals and groups and for determining the effects of age, gender, handedness, and other demographic or genetic factors on brain structures in space and time. Integrating these observations to enable statistical comparison has already provided a deeper understanding of the relationship between brain structure and function. Importantly, the utility of an atlas depends on appropriate coordinate systems, registration, and deformation methods to allow the statistical combination of multiple observations in an agreed, but expandable, digital reference framework. In this review, we highlighted two sources of data that will have an increasingly important role in integrative brain atlases: molecular architectonics and diffusion tensor imaging (DTI). Once stored in a population-based atlas, information from these techniques can help to interpret more conventional functional and structural brain maps by integrating them with data on molecular content, physiology, and fiber connections – a development that can help to formulate and test new types of neuroscientific models. A goal of systems neuroscience is to establish brain systems that underlie cognitive processes and the factors that influence them. DTI data on fiber connectivity, stored in an atlas coordinate system, can offer a rigorous computational basis to test how identifiable anatomical systems (e.g., visual, limbic, or corticothalamic pathways) interact. This atlas information can be invoked as ROI that are incorporated into the statistical design of functional brain mapping studies (e.g., with fMRI or electroencephalography), even when underlying fiber connections are not evident in the data being collected for a particular study. Molecular architectonic mapping also provides a complementary perspective in which known neurotransmitter and receptor pathways—the physiology and molecular features of which are now well understood—can be associated with functional subdivisions of the cortex, identified with tomographic imaging. For example, an fMRI study of inhibitory cognitive processes in drug abusers might be informed by other modalities of data on limbic–prefrontal connectivity (from DTI), or on cortical monoamine receptor distributions (from architectonic mapping). In each of these contexts, the coordinate system of the atlas, and the transformations that equate different modality data in the same reference frame, provide the means to build and test system-level models of cognition or disease, incorporating data from traditionally separate domains of neuroscience.

As brain atlases soon begin to incorporate data from thousands of subjects, new questions in basic and clinical neuroscience can be addressed that were previously out of reach. For example, quantitative genetic studies are underway to link functional, structural, and connectivity information with variations in candidate genetic polymorphisms that could influence them. As polygenic disorders involve the interaction of multiple genetic variations, each with a small effect on the overall phenotype, digital atlases provide the ideal setting to mine large numbers of images computationally with hybrid techniques from computational anatomy and quantitative genetics (such as linkage and association studies in which a statistic is computed at each voxel location in the brain).

Should atlases be constructed specific to different age groups or different age-related diseases? Several other authors [30, 63] have come to the same conclusion and many population-based atlases have emerged in response to this need [64]. But the same logic can be carried to the next level by creating many different population-based atlases, each specific to the group demographics, disease, age, or other characteristics of the subjects being studied. These provide, not only population statistics within the map, but arguably better represent the morphological signature of that particular cohort. What must be included in all analyses are confidence statistics on where the activity takes place. Whether this entails a statistic on probability, percentile, or other metric may depend on the experimental design and other factors. Adoption of a single normal atlas, even a probabilistic version, for all subject studies provides for the nominal capability for easier comparisons but in doing so fails to adequately measure the nuances within or between each group (Fig. 3). It, therefore, seems that it might be prudent to avoid dependency on a single modality, single group representation for every study. Any given imaging experiment will be better served by mapping to a population-based atlas that closely resembles the cohort under study. We suggest that population atlases for groups, such as AD [35, 65], schizophrenia [66, 67], pediatric populations [68, 69], autism [70], even decades of life [71], should be utilized, as appropriate, for that specific subject group.

The next generation of population-based atlases [73] will provide the necessary statistical power to identify demographic, genetic, and environmental factors that influence therapeutic response. These will be essential in the study of the normal and abnormal human brain. Most important of all, brain atlases are now being enriched with data from genetics, protein expression, as well as reflecting associations with phenotypic behaviors. These efforts can be expected to yield entirely new avenues of research into the functional organization of the brain and how this is altered in disease will be of interest not only just to specialists in neuroimaging, but also to all basic and clinical neuroscientists.

References

1. Haas LF (2001) Phineas Gage and the science of brain localisation. J Neurol Neurosurg Psychiatry 71:761

2. Cowie SE (2000) A place in history: Paul Broca and cerebral localization. J Invest Surg 13:297–298

3. Goedert M, Ghetti B (2007) Alois Alzheimer: his life and times. Brain Pathol 17:57–62

4. Roland PE, Zilles K (1994) Brain atlases – a new research tool. Trends Neurosci 17:458–467

5. Toga AW, Thompson PM (2001) Maps of the brain. Anat Rec 265:37–53

6. Toga AW, Thompson PM (2002) New approaches in brain morphometry. Am J Geriatr Psychiatry 10:13–23

7. Thompson P, Cannon TD, Toga AW (2002) Mapping genetic influences on human brain structure. Ann Med 34:523–536

8. Narr KL, Thompson PM, Sharma T, Moussai J, Cannestra AF, Toga AW (2000) Mapping morphology of the corpus callosum in schizophrenia. Cereb Cortex 10:40–49

9. Davatzikos C (1996) Spatial normalization of 3D brain images using deformable models. J Comput Assist Tomogr 20:656–665

10. Davatzikos C (1997) Spatial transformation and registration of brain images using elastically deformable models. Comput Vis Image Underst 66:207–222

11. Thompson PM, Woods RP, Mega MS, Toga AW (2000) Mathematical/computational challenges in creating deformable and probabilistic atlases of the human brain. Hum Brain Mapp 9:81–92

12. Weaver JB, Healy DM Jr, Periaswamy S, Kostelec PJ (1998) Elastic image registration using correlations. J Digit Imaging 11:59–65

13. Barillot C, Lemoine D, Le Briquer L, Lachmann F, Gibaud B (1993) Data fusion in medical imaging: merging multimodal and multipatient images, identification of structures and 3D display aspects. Eur J Radiol 17:22–27

14. Woods RP, Grafton ST, Holmes CJ, Cherry SR, Mazziotta JC (1998) Automated image

registration. I. General methods and intrasubject, intramodality validation. J Comput Assist Tomogr 22:139–152

15. Toga AW, Thompson PM, Mori S, Amunts K, Zilles K (2006) Towards multimodal atlases of the human brain. Nat Rev Neurosci 7:952–966

16. Woods RP (2003) Characterizing volume and surface deformations in an atlas framework: theory, applications, and implementation. Neuroimage 18:769–788

17. Avants B, Gee JC (2004) Geodesic estimation for large deformation anatomical shape averaging and interpolation. Neuroimage 23(Suppl 1):S139–S150

18. Avants BB, Schoenemann PT, Gee JC (2006) Lagrangian frame diffeomorphic image registration: morphometric comparison of human and chimpanzee cortex. Med Image Anal 10:397–412

19. Evans AC, Collins DL, Milner B (1992) An MRI-based stereotactic atlas from 250 young normal subjects. J Neurosci Abstr 18:408

20. Durrleman S, Pennec X, Trouve A, Ayache N (2007) Measuring brain variability via sulcal lines registration: a diffeomorphic approach. Med Image Comput Comput Assist Interv 10(Pt 1):675–682

21. Alayon S, Robertson R, Warfield SK, Ruiz-Alzola J (2007) A fuzzy system for helping medical diagnosis of malformations of cortical development. J Biomed Inform 40:221–235

22. Rohlfing T, Maurer CR Jr (2007) Shape-based averaging. IEEE Trans Image Process 16:153–161

23. Narr KL, Bilder RM, Luders E et al (2007) Asymmetries of cortical shape: effects of handedness, sex and schizophrenia. Neuroimage 34:939–948

24. Thompson PM, Giedd JN, Woods RP, MacDonald D, Evans AC, Toga AW (2000) Growth patterns in the developing brain detected by using continuum mechanical tensor maps. Nature 404:190–193

25. Corouge I, Dojat M, Barillot C (2004) Statistical shape modeling of low level visual area borders. Med Image Anal 8:353–360

26. Cardenas VA, Boxer AL, Chao LL et al (2007) Deformation-based morphometry reveals brain atrophy in frontotemporal dementia. Arch Neurol 64:873–877

27. Leow AD, Klunder AD, Jack CR Jr et al (2006) Longitudinal stability of MRI for mapping brain change using tensor-based morphometry. Neuroimage 31:627–640

28. Diedrichsen J (2006) A spatially unbiased atlas template of the human cerebellum. Neuroimage 33:127–138

29. Toga AW, Thompson PM (2005) Genetics of brain structure and intelligence. Annu Rev Neurosci 28:1–23

30. Toga AW, Thompson PM, Sowell ER (2006) Mapping brain maturation. Trends Neurosci 29:148–159

31. Apostolova LG, Thompson PM (2007) Brain mapping as a tool to study neurodegeneration. Neurotherapeutics 4(3):387–400

32. Apostolova LG, Akopyan GG, Partiali N et al (2007) Structural correlates of apathy in Alzheimer's disease. Dement Geriatr Cogn Disord 24:91–97

33. Apostolova LG, Lu P, Rogers S et al (2008) 3D mapping of language networks in clinical and pre-clinical Alzheimer's disease. Brain Lang 104:33–41

34. Scher AI, Xu Y, Korf ES et al (2007) Hippocampal shape analysis in Alzheimer's disease: a population-based study. Neuroimage 36:8–18

35. Thompson PM, Hayashi KM, Dutton RA et al (2007) Tracking Alzheimer's disease. Ann N Y Acad Sci 1097:183–214

36. Mazziotta JC, Toga AW, Evans AC, Fox PT, Lancaster JL (1995) Digital brain atlases. Trends Neurosci 18:210–211

37. Toga AW, Thompson PM, Mega MS, Narr KL, Blanton RE (2001) Probabilistic approaches for atlasing normal and disease-specific brain variability. Anat Embryol (Berl) 204:267–282

38. Wakana S, Jiang H, Nagae-Poetscher LM, van Zijl PC, Mori S (2004) Fiber tract-based atlas of human white matter anatomy. Radiology 230:77–87

39. Van Essen DC (2005) A Population-Average, Landmark- and Surface-based (PALS) atlas of human cerebral cortex. Neuroimage 15:635–662

40. Fox PT, Perlmutter JS, Raichle ME (1984) Stereotactic method for determining anatomical localization in physiological brain images. J Cereb Blood Flow Metab 4:634

41. Evans AC, Marrett S, Neelin P et al (1992) Anatomical mapping of functional activation in stereotactic coordinate space. Neuroimage 1:43–53

42. Nowinski WL, Thirunavuukarasuu A (2001) Atlas-assisted localization analysis of functional images. Med Image Anal 5:207–220

43. Crivello F, Schormann T, Tzourio-Mazoyer N, Roland PE, Zilles K, Mazoyer BM (2002) Comparison of spatial normalization procedures and their impact on functional maps. Hum Brain Mapp 16:228–250

44. Swallow KM, Braver TS, Snyder AZ, Speer NK, Zacks JM (2003) Reliability of functional

localization using fMRI. Neuroimage 20:1561–1577

45. Tu Z, Zheng S, Yuille AL et al (2007) Automated extraction of the cortical sulci based on a supervised learning approach. IEEE Trans Med Imaging 26:541–552

46. Luders E, Thompson PM, Narr KL, Toga AW, Jancke L, Gaser C (2006) A curvature-based approach to estimate local gyrification on the cortical surface. Neuroimage 29:1224–1230

47. Ashburner J, Friston KJ (1999) Nonlinear spatial normalization using basis functions. Hum Brain Mapp 7:254–266

48. Friston KJ, Stephan KE, Lund TE, Morcom A, Kiebel S (2005) Mixed-effects and fMRI studies. Neuroimage 24:244–252

49. Miller MB, Van Horn JD, Wolford GL et al (2002) Extensive individual differences in brain activations associated with episodic retrieval are reliable over time. J Cogn Neurosci 14:1200–1214

50. Fox PT, Parsons LM, Lancaster JL (1998) Beyond the single study: function/location metanalysis in cognitive neuroimaging. Curr Opin Neurobiol 8:178–187

51. Nowinski WL (2005) The cerefy brain atlases: continuous enhancement of the electronic talairach-tournoux brain atlas. Neuroinformatics 3:293–300

52. Amunts K, Schleicher A, Zilles K (2007) Cytoarchitecture of the cerebral cortex – more than localization. Neuroimage 37:1061–1065, discussion 6–8

53. Mazziotta J, Toga AW, Evans A et al (2001) A probabilistic atlas and reference system for the human brain: International Consortium for Brain Mapping (ICBM). Philos Trans R Soc Lond B Biol Sci 356:1293–1322

54. Nowinski WL (2001) Modified Talairach landmarks. Acta Neurochir (Wien) 143:1045–1057

55. Bookstein FL (2001) Voxel-based morphometry" should not be used with imperfectly registered images. Neuroimage 14:1454–1462

56. Talairach J, Tournoux P (1988) Co-planar stereotactic atlas of the human brain. Tieme, New York

57. Maldjian JA, Laurienti PJ, Burdette JH (2004) Precentral gyrus discrepancy in electronic versions of the Talairach atlas. Neuroimage 21:450–455

58. Shattuck DW, Mirza M, Adisetiyo V et al (2008) Construction of a 3D probabilistic atlas of human cortical structures. Neuroimage 39:1064–1080

59. Woods RP, Grafton ST, Watson JD, Sicotte NL, Mazziotta JC (1998) Automated image registration. II. Intersubject validation of linear and nonlinear models. J Comput Assist Tomogr 22:153–165

60. Rex DE, Ma JQ, The TAW, LONI (2003) Pipeline processing environment. Neuroimage 19:1033–1048

61. Smith SM, Jenkinson M, Woolrich MW et al (2004) Advances in functional and structural MR image analysis and implementation as FSL. Neuroimage 23(Suppl 1):S208–S219

62. Ashburner J, Friston KJ (2005) Unified segmentation. Neuroimage 26:839–851

63. Van Essen DC (2002) Windows on the brain: the emerging role of atlases and databases in neuroscience. Curr Opin Neurobiol 12:574–579

64. Mazziotta J, Toga A, Evans A et al (2001) A four-dimensional probabilistic atlas of the human brain. J Am Med Inform Assoc 8:401–430

65. Mega MS, Dinov ID, Mazziotta JC et al (2005) Automated brain tissue assessment in the elderly and demented population: construction and validation of a sub-volume probabilistic brain atlas. Neuroimage 26:1009–1018

66. Yoon U, Lee JM, Koo BB et al (2005) Quantitative analysis of group-specific brain tissue probability map for schizophrenic patients. Neuroimage 26:502–512

67. Cannon TD, Thompson PM, van Erp TG et al (2006) Mapping heritability and molecular genetic associations with cortical features using probabilistic brain atlases: methods and applications to schizophrenia. Neuroinformatics 4:5–19

68. Wilke M, Schmithorst VJ, Holland SK (2002) Assessment of spatial normalization of whole-brain magnetic resonance images in children. Hum Brain Mapp 17:48–60

69. Jelacic S, de Regt D, Weinberger E (2006) Interactive digital MR atlas of the pediatric brain. Radiographics 26:497–501

70. Joshi S, Davis B, Jomier M, Gerig G (2004) Unbiased diffeomorphic atlas construction for computational anatomy. Neuroimage 23(Suppl 1):S151–S160

71. Mazziotta J, Toga A, Evans A et al (2001) A four-dimensional probabilistic atlas of the human brain. J Am Med Inform Assoc 8:401–430

72. Narr K, Thompson P, Sharma T et al (2001) Three-dimensional mapping of gyral shape and cortical surface asymmetries in schizophrenia: gender effects. Am J Psychiatry 158:244–255

73. Amunts K, Hawrylycz MJ, Van Essen DC et al (2014) Interoperable atlases of the human brain. Neuroimage 99:525–532

Chapter 10

Graph Theoretic Analysis of Human Brain Networks

Alex Fornito

Abstract

The human brain is a highly interconnected network. It is thus suitable for investigation with graph theory, a branch of mathematics concerned with understanding systems of interacting elements. Graph theory has become a popular tool for analyzing human MRI data. In this work, brain networks are modeled as graphs of nodes connected by edges. The nodes represent distinct brain regions and the edges represent some measure of structural or functional interaction between regions. This representation enables the computation of a broad range of metrics that quantify diverse aspects of network organization, thus offering a powerful framework for understanding brain structure and function in both health and disease. This chapter overviews the principles and methods involved in building and analyzing graph theoretic models of the brain using MRI. It explains basic concepts, provides examples of how graph theory has shed new light on brain organization, and considers some limitations of current applications.

Key words Connectome, Connectivity, Graph analysis, Network, Complexity, MRI, DTI, fMRI

---

1 Introduction

The human brain is a complex, interconnected network. At microscopic resolutions, axons and dendrites sprout from neuronal soma to enable communication with several thousand other neurons [1]. At macroscopic resolutions, the axons of populations of adjacent neurons coalesce to form fiber bundles that project through the white matter volume of the brain to connect distal areas. This interconnectivity allows the integration of segregated and functionally specialized neuronal systems distributed throughout the brain. The network organization of the brain thus fundamentally shapes its function, and generating comprehensive maps of brain connectivity—so-called connectomes [2]—has become a major goal of neuroscience [3–5]. The burgeoning field of neural connectomics is thus generating rich data sets describing brain network organization across multiple species and multiple scales of resolution [6–12] using various microscopic, genetic, tract-tracing, informatic, and neuroimaging methods.

Massimo Filippi (ed.), *fMRI Techniques and Protocols*, Neuromethods, vol. 119,
DOI 10.1007/978-1-4939-5611-1_10, © Springer Science+Business Media New York 2016

Graph theory provides an ideal framework for characterizing, comparing, and integrating results across these diverse data. Graph theory is a branch of mathematics concerned with studying systems of interacting elements. The central assumption of the approach is that any such system can be represented as a graph of nodes (also called vertices) connected by edges (also called links, arcs, or connections). Equivalently, the network can be represented as a matrix, in which each i-th row and j-th column represents a distinct node and each ij-th element encodes the type and strength of connectivity between each node pair (Fig. 1). The simplicity of this approach is matched by its versatility—nodes in brain graphs could represent individual neurons, neuronal populations, or macroscopic regions, while edges could represent axonal, dendritic, or synaptic contacts, or large-scale fiber bundles. More broadly, graph theory has also been used to model other networks found in nature, including social

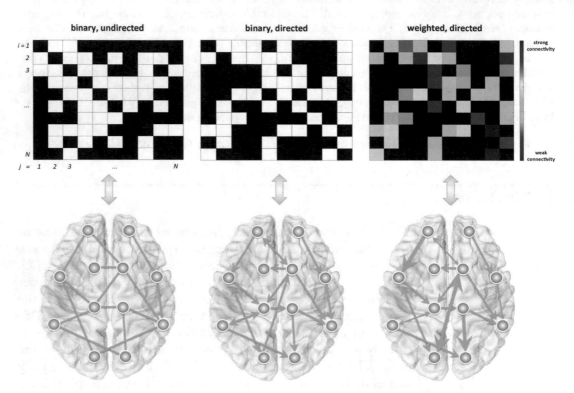

Fig. 1 Matrix and graph-based representations of brain networks. In matrix form (*top row*), each row and column represents a different region and each element represents the connectivity between region pairs. Matrices can be either binary, representing only the presence or absence of a connection (*left* and *middle*), or weighted to represent variations in the strength of inter-regional connectivity (right). The matrices can also be symmetric, representing an undirected network (left; note how the *top-right* and *bottom left* triangles of this matrix are mirror images of each other), or asymmetric, to represent a directed network (*middle* and *right*). In graph form (*bottom row*), brain regions are represented as nodes or circles and connectivity as edges. Arrowheads can be used to represent the directionality of connectivity in directed networks (*middle, right*). Edge thickness can be used to represent variations in edge weight (*right*). These graphs are used for illustrative purposes and do not directly map onto the matrices shown in the *top row*. Weighted, undirected networks are not shown

networks (e.g., nodes are people, edges are social or professional ties), economic networks (e.g., nodes are companies or countries and edges represent financial or trade transactions), technological networks such the world wide web (e.g., nodes are websites and edges are hyperlinks), transportation networks such as the global air transportation network (e.g., nodes are airports and edges are connecting flights), ecological networks (e.g., nodes are species and edges are predator-prey interactions) and biological networks other than the brain (e.g., nodes can be proteins or genes and edges can encode molecular interactions or coexpression patterns) (see [13, 14] for reviews). Graph theory thus provides a standardized approach for representing and analyzing diverse types of network data.

In recent years, processing pipelines have been established to allow the application of graph theoretic methods to human neuroimaging data (Fig. 2). This is an important advance, since in vivo

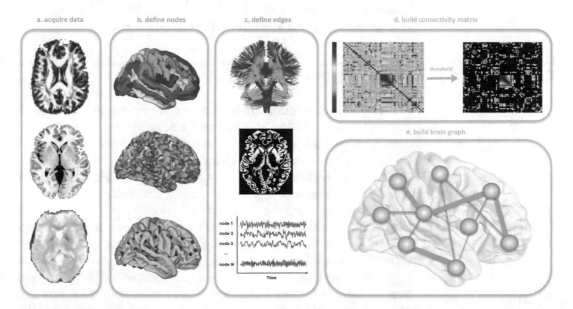

Fig. 2 Basic processing pipeline for graph theoretic analysis of MRI data. (**a**) Imaging data are first acquired. Structural connectivity is often assessed using either diffusion MRI (*top*) or T1-weighted MRI (*middle*). Functional and effective connectivity are typically investigated using fMRI (*bottom*). (**b**) Once the imaging data have been acquired, the brain must be parcellated into distinct regions, which act as network nodes. Shown here are examples of an anatomical parcellation (*top*), a random parcellation (*middle*), and a functional parcellation (*bottom*). (**c**) The next step is to define some measure of connectivity between nodes, which will represent the edges of our brain graph. With diffusion MRI, inter-regional connectivity is measured using tractography (*top*). With T1-weighted MRI, structural connectivity is indirectly measured using inter-subject covariations in gray matter morphometry (*middle*). With fMRI, functional connectivity is measured as a statistical dependence between regional time series (*bottom*). (**d**) The connectivity between all pairs of brain regions can be represented as a connectivity matrix. MRI analyses typically yield weighted and symmetric matrices (*left*). These matrices can be thresholded to emphasize the strongest links in the network (*right*). (**e**) The connectivity matrix can be used to generate a graph-based representation of the network (i.e., a brain graph), in which regions are represented as nodes and connections as edges. See also Figs. 1 and 3. Parts of this figure have been reproduced from [109] with permission

imaging techniques such as magnetic resonance imaging (MRI) currently represent the most cost-effective and tractable method for generating connectomic maps in humans [2] (see also [15, 16] for developments in the analysis of post-mortem tissue). Accordingly, there now exists a large literature using graph theory for the analysis of MRI data. Diffusion MRI is commonly used to assess structural connectivity of the brain, while functional MRI is used to characterize functional interactions between brain regions [17, 18]. Either technique can be used to generate a graph-based representation of brain network connectivity.

In this chapter, we consider some of the basic principles and methods of graph theoretic analysis of human MRI data. We first offer a brief history and rationale supporting the use of graph theory in neuroscience, and in the analysis of human neuroimaging data in particular. We then discuss how such analyses are performed with respect to the two major steps involved: building a brain graph and analyzing a brain graph. We close by considering some emerging trends and areas for improvement in the field.

2   A Brief History of Graph Theory and MRI

Leonhard Euler is often credited as the founding father of graph theory [19]. In 1736, he published a solution to a major unresolved mathematical problem of the time: whether it was possible to find a route that traversed each of the seven bridges of the Prussian city of Konigsberg (now Kaliningrad in Russia) without crossing any single bridge more than once. Euler simplified the problem by depicting the geography of the city as a graph, in which the nodes were distinct landmasses and the edges were the bridges that connected these masses. Using this abstraction, he was able to show that no such path was possible.

Following Euler's success, the application of graph theory was largely restricted to mathematical studies of topology and certain areas of theoretical chemistry. It was not until the middle of the twentieth century that the broader applicability of graph theory was realized first in the social sciences [20–23] and then in other fields following the seminal work of Paul Erdős and Alfred Rényi [24]. These two authors explored the mathematical properties of probabilistic random graphs in which nodes were connected with uniform probability. Such random graphs were used to model a wide variety of real-world biological, technological, and social systems until two analyses, both published in 1998, demonstrated that many such systems display a more complex pattern of connectivity than implied by the Erdős-Rényi model.

One analysis, by Barabási and Albert [25], showed that networks as diverse as the world wide web, film actor collaborations, and the western US electricity power grid show a heterogeneous distribution of connectivity across nodes. This distribution was characterized by a large number of vertices with a small number of connections and a small subset of vertices that were very highly interconnected with the rest of the network. These highly connected nodes represented putative network hubs. The hub dominance of these networks departs from the expectations of the Erdős-Rényi random graph model, in which each node has an equal probability of being connected.

In parallel, Watts and Strogatz [26] published an analysis that also suggested the Erdős-Rényi model was an insufficient model for real-world systems. They contrasted the organization of the Erdős-Rényi model with a regular, lattice-like graph. The nodes of this regular graph were arranged around a ring and were linked only to their k nearest neighbors. Two properties of these networks were examined: the clustering coefficient and the characteristic path length. The clustering coefficient quantifies the probability that two nodes connected to a third are also connected with each other (Fig. 4c). This metric captures a phenomenon that is well known in social networks, where two people are more likely to be friends if they share a third friend in common. The characteristic path length of a network is the average number of connections required to travel from one node to any other node in the network (Fig. 4b). A shorter average path length implies that the network is integrated and that information is able to spread more rapidly throughout the network. Watts and Strogatz found that the short-range connectivity of the regular graph led to high clustering but high path length, since many short-range links were required to travel from one end of the network to the other. In contrast, the Erdős-Rényi random graph had lower path length but also had lower clustering, since all nodes had an equal probability of being connected. Compared to the extreme cases presented by the random and regular graphs, Watts and Strogatz found that many real world networks, including the neuronal network of *C elegans*, showed high clustering, much like a regular graph, coupled with a short average path length comparable to a random network. Indeed, they found that randomly rewiring just a small fraction of edges in a regular graph created network "short-cuts" that produced a dramatic reduction in the characteristic path length of the network with negligible impact on clustering, leading to a regime characterised by both high clustering and short average path length. They termed this organization "small-world," in reference to the six degrees of separation phenomenon thought to characterize social networks.

It is easy to see how the findings of Barabási and Albert [25] and Watts and Strogatz [26] can provide important insights into brain network organization. For example, a heterogeneous distribution of connectivity across nodes within the brain would point to highly connected hub regions that play a particularly prominent role in information-processing and integrating diverse network elements. Such a role has been proposed, for example, for association cortices [27]. Similarly, the small-world properties of high clustering and short path length provide an ideal foundation for functional specialization (tightly clustered connectivity) and functional integration (low average path length)—two fundamental principles of brain function [28, 29]. Accordingly, graph theory was employed in some of the earliest analyses of connectomes inferred from the synthesis of published tract-tracing studies [9, 30]. This work paved the way for a substantial body of subsequent work in these and other connectivity datasets [31–34] (reviewed in Ref. [35]).

Graph theory was first applied to human electroencephalography (EEG) and magnetoencephalography (MEG) data in 2004 [36], followed by several fMRI analyses published shortly thereafter [37–40]. Each of these studies focused on functional connectivity networks. Connectivity was measured either between individual voxels [37], electrodes/sensors [36], or large-scale anatomical regions [38, 39]. This work presented evidence for a heterogeneous distribution of connectivity across nodes, pointing to the presence of highly connected hub regions, as well as the high clustering and short average path length consistent with small-world organization. Subsequent graph theoretic analysis of human structural connectivity data measured with diffusion MRI yielded the same conclusions [12, 41–43].

These developments paralleled an explosion in the graph theoretic characterization and modeling of diverse types of complex systems, and the emergence of a formal science of complex networks [13, 14, 44]. As this field has matured, it has generated a large repertoire of diverse measures and theoretical concepts for understanding different aspects of network organization, many of which have great appeal for neuroscience. Graph theory thus not only offers a standardized, flexible, and scalable method for representing brain networks, it also provides a rich range of metrics for making sense of brain network data.

In the following sections, we consider some of the basic graph theoretic measures applied to human neuroimaging data and discuss what they have taught us about brain network organization. We focus principally on concepts and measures that have been most commonly applied in neuroimaging contexts. We discuss more advanced topics in the final section. As a first step, we consider issues associated with building a brain graph, as these represent an important foundation for any subsequent analysis.

3   Building a Brain Graph

Nodes and edges are the fundamental building block of any network graph. The basic unit of analysis in an MRI experiment is a voxel, which is typically 1–3 mm³ in volume. Voxels thus represent an aggregation of large populations of neural elements; on average an estimated 20,000–30,000 neurons and billions of synapses [45]. This coarse resolution creates ambiguities when attempting to define appropriate nodes and edges for network analysis. This problem is critical as invalid node definitions can alter, distort, or bias results, and accurate mapping of connectivity (edges) is essential for any valid analysis of network organization [46–50].

4   Defining Nodes

The central problem for node definition in MRI concerns how voxels should be aggregated to define a valid parcellation of the brain. Individual neurons and neuronal columns have both been proposed as fundamental units for brain network organization [7, 51], but cannot be resolved with typical human MRI acquisitions. Similarly, cytoarchitectonic regions, such as those delineated in Brodmann's classic map, cannot be resolved with MRI and the boundaries of these regions are often poorly correlated with macroscopic landmarks (e.g., sulci and gyri) [52, 53]. Due to these limitations, several different heuristic approaches have been used for node definition with MRI.

To this end, three criteria for an ideal node for a brain graph have been proposed [18]: (1) spatial embedding; (2) intrinsic homogeneity; and (3) extrinsic heterogeneity. The first criterion simply means that spatial relationships between nodes should be taken into account. The brain is embedded within the three-dimensional volume of the skull and this embedding places important constraints on network wiring [54, 55]. Fortunately, spatial relations between nodes are easily accounted for in MRI analysis with standard stereotactic mapping techniques, since the location of each region can be easily represented with reference to the Montreal Neurological Institute (MNI) or Talairach and Tourneoux coordinate systems.

The second and third criteria simply mean that each node should represent a structurally or functionally homogeneous entity (intrinsic homogeneity) and should be distinguishable from other nodes (extrinsic heterogeneity). For example, a cytoarchitectonic region is intrinsically homogeneous, to the extent that it defines a population of neurons with shared histological properties. Different regions are extrinsically heterogeneous, to the extent that they serve different functional roles in the network (e.g., areas of visual cortex processes visual information, regions of parietal cortex processes spatial information, and so on).

In practice, the extrinsic heterogeneity criterion is difficult to fulfill. The unique functional role of an individual node in a brain graph is determined by its connectivity with other areas (i.e., its "connectional fingerprint") [56], as well as its own intrinsic circuitry, cellular composition, gene expression patterns, physiological properties, and so on. Comprehensively understanding how the intrinsic properties of each node vary, and how this variation defines the functional role of that node, is an unresolved question in neuroscience. Consequently, nearly all graph theoretic studies of brain networks treat nodes as uniform network elements, and the primary distinctions between nodes are based on variations in their connectivity profiles with other areas. (Note however, that some computational models do explicitly account for variations in the functional roles of different networks nodes [57].)

Given that the spatial embedding criterion is easily accommodated with standard imaging techniques and the extrinsic heterogeneity criterion is difficult to meet in practice, the primary emphasis in developing methods for defining nodes for graph theoretic analysis of MRI data has been on the intrinsic homogeneity criterion. Five broad approaches have been used, which we refer to here as voxel-based, anatomical, random, data-driven, and quantitative.

4.1 Voxel-Based Parcellation

A simple solution to the node identification problem in MRI is to use the best resolution possible and thus treat each individual voxel as a distinct network node. This approach has been used in fMRI analyses (e.g. [49, 58, 59],) but suffers from two major drawbacks. First, there is no guarantee that voxel borders represent the appropriate boundaries for delineating homogeneous neuronal populations in the brain [60]. In other words, functionally homogeneous populations could extend over spatial scales that are either smaller (e.g., cortical columns) or larger (cytoarchitectonic divisions) than the volume occupied by a single voxel. The second limitation of a voxel-based approach is that it is computationally intensive. Voxel-based networks typically comprise 10^4–10^5 nodes and millions of connections. Such large networks can pose problems for computational tractability.

4.2 Anatomical atlases

Anatomical atlases offer an alternative method for brain parcellation that minimizes computational burden. For example, one popular atlas, the Automated Anatomical Labelling (AAL) atlas [61], parcellates the brain into 116 regions largely defined according to the sulcal and gyral landmarks of an individual brain. The generalizability of this template is thus questionable. Alternative atlases, based on probabilistic maps of sulcal and gyral regions, such as the Harvard-Oxford atlas (http://fsl.fmrib.ox.ac.uk/fsl/fslwiki/

Atlases) and the Desikan-Killaney atlas [62](see Fig. 2b, top), overcome this limitation. However, as previously stated, sulcal and gyral landmarks often correspond poorly with the borders of actual functional subdivisions of the brain. This poor correspondence limits the validity of these anatomical atlases. Moreover, the size of the regions in anatomical parcellations can vary considerably, and these variations can affect connectivity estimates and thus bias subsequent analyses [48, 63].

4.3 Random Parcellation

One way to ensure that a parcellation comprises regions with homogeneous volume is to divide the brain into random parcels of similar size (e.g., Fig. 2b, middle). These parcellations can be performed at varying resolutions, typically yielding networks between 10^2 and 10^4 nodes [12, 48, 50]. This approach ensures homogeneity of regional volume, but there is no guarantee that the random parcels accurately capture true functional subdivisions of the brain. Replication of results across several iterations of a random template may also be necessary to ensure that any findings are not due to the specific characteristics of any single instance of a random parcellation.

4.4 Functional Parcellation

A more hypothesis-driven method uses regions-of-interest defined according to some functional property of interest (e.g., Fig. 2b, bottom). For example, one study used a meta-analysis of task-based fMRI activation studies across a range of cognitive processes to identify 160 stereotactic coordinates of task-related activation peaks [64]. Spherical regions-of-interest centered on these coordinates were then created and used as nodes in a graph theoretic analysis of developmental effects on human brain functional connectivity (see also [65]). Other studies have used similar approaches after defining spherical regions-of-interest centered on peak coordinates derived either from resting-state functional connectivity analyses [66] or task-related activation mapping [67, 68] (see also [69]). This approach is well suited for testing hypotheses about specific systems of interest with functional MRI. However, this method is harder to use with diffusion imaging, where larger areas of gray and white matter may need to be sampled to adequately measure the tracts projecting into and out of a region.

4.5 Data-Driven Parcellation

In contrast to hypothesis-driven methods for defining regions-of-interest, data-driven approaches attempt to define functionally homogeneous collections of voxels using specific characteristics of the imaging data. A classic example is spatial independent component analysis (ICA), which decomposes fMRI data into a set of components (networks) whose voxels are correlated with each other, and maximally spatially independent of other components [70, 71]. Lower-order decompositions typically recover canonical neural networks such as the default mode network (DMN), frontoparietal network, and so on [72]. Graph analysis between these networks can be performed (called "functional network connectivity" analysis; e.g. [73],). However, as these networks commonly involve multiple, spatially distributed brain regions, they do not conform to a traditional conception of a brain network node as consisting of an anatomically contiguous region. Higher order decompositions can separate the regions comprising these distributed networks into separate components [74], although it is often difficult to know, a priori, the dimensionality of the decomposition required to obtain such a solution.

Other data-driven approaches examine the connectivity of each voxel or small region to all other areas, and cluster voxels or regions with similar connectivity profiles into single parcellation units [59, 75–77]. When combined with spatial constraints [78], these methods can yield parcellations of the brain that guarantee spatial contiguity of all voxels within a region, while also ensuring that such voxels fulfill the criterion of intrinsic homogeneity (to the extent that they share similar inter-regional connectivity profiles). Whole-brain parcellations using such methods applied to task-free, resting-state fMRI are robust across samples [59, 77]. Similar approaches have been applied to diffusion MRI data to parcellate-specific regions based on structural connectivity profiles [79–81]. Diffusion MRI, being a measure of brain anatomy rather than function, should provide a more stable parcellation than those based on fMRI, but whole-brain parcellation of diffusion MRI has not yet been extensively validated.

4.6 Quantitative Parcellations

A related data-driven approach involves using quantitative, biological criteria to define nodes. Recently, it has been suggested that the ratio of T1- to T2-weighted imaging contrast can be used to generate myelin maps of the brain, and that gradient detection algorithms can be used to identify boundaries where there are sharp transitions in myeloarchitecture [82]. In some areas, these boundaries correspond well with known functional boundaries [82]. However, it is as yet unclear whether this approach is scalable to whole-brain parcellations. Alternative quantitative parcellations involve the projection of data from postmortem analyses into stereotactic space, yielding probabilistic cytoarchitectonic atlases [83] and maps of regional variations in chemoarchitecture [84]. However, such data are presently only available for limited regions of cortex.

4.7 Summary

There is a variety of methods for delineating brain network nodes in MRI data. Each approach has distinct strengths and weaknesses and there is no gold standard. Ultimately, investigators must choose an approach best suited to the specific hypothesis being tested, and ensure that analyses are interpreted with respect to the limitations of the specific method employed.

5   Defining edges

The edges of a brain graph represent the connectivity between pairs of brain regions. There are three broad classes of brain connectivity: structural, functional, and effective. Both the specific class of connectivity studied and the method used to measure it have a major impact on the structure of the resulting brain graph and the types of analyses that can be performed. This section discusses issues associated with measuring each of these types of connectivity.

5.1   Structural Connectivity

Structural connectivity refers to the anatomical connections linking distinct neural elements. At resolutions accessible with MRI, structural connectivity refers to axonal fiber bundles intersecting macroscopic brain regions. Structural connectivity has been measured with MRI using two approaches. A relatively indirect method involves analyzing correlations in the gray matter volume (or density or cortical thickness) of different regions across subjects (e.g., Fig. 2c, middle). Volumes of different brain regions vary across individuals. If these inter-individual variations of volume are correlated between two regions, the regions are said to be "connected." It is generally assumed that these correlations reflect anatomical connectivity or mutually trophic influences [85, 86].

A more direct approach uses diffusion MRI. Specifically, tractographic analyses attempt to reconstruct the trajectories of major fiber bundles based on the preferred direction of water diffusion in each voxel. Axons present barriers to water diffusion. The direction of preferred water diffusion in the brain is thus constrained by the trajectory of its axonal fibers. These trajectories are reconstructed using streamlines that propagate through the white matter on a voxel-by-voxel basis according to specific algorithmic rules (e.g., Fig. 2c, top). A wide range of tractography algorithms exists and each is associated with distinct strengths and weaknesses [87] (see also [18]). The accuracy of the algorithm critically determines the validity of the resulting structural brain graph.

Once putative fiber tracts have been reconstructed, connectivity between regions is commonly measured using one of two approaches. One method estimates the strength of connectivity as the number of reconstructed trajectories intersecting each pair of brain regions. The raw streamline count is also often normalized by the size of the connected regions, since larger regions will generally intersect a larger number of streamlines [12]. Measuring structural connectivity based on streamline counts assumes that there is a correlation between the number of axons comprising a fiber bundle and the number of streamlines required to reconstruct that bundle. However, a streamline is not tantamount to an axon; rather it is an abstract structure that is used to track a trajectory of water diffusion through the brain. It therefore provides an indirect measure of axonal structure. In reality, reconstructed bundles may vary in the number of streamlines they posses because of differences in the axonal density of the actual fiber pathway, variations in the orientation and organization of the fibers, or variations in the signal-to-noise characteristics of the image [88].

An alternative approach is to compute some measure of fiber integrity averaged over the extent of the reconstructed fiber tract. One commonly used index is fractional anisotropy (FA), which measures the degree to which water diffusion is constrained within each voxel. Disorganized or damaged axons offer reduced barriers to water diffusion. Lower FA values are thus often used as a marker of impaired white matter integrity, and the average FA of voxels within a tract can be used to index of the integrity of that tract. Alternative diffusivity measures, such as mean, radial, and axial measures of diffusivity, can also be used for this purpose. This approach attempts to derive a more biologically meaningful index of connectivity than streamline count, although diffusion-based measures of white matter integrity can also be affected by differences in white matter organization and image signal-to-noise, making their interpretation ambiguous [88]. An alternative approach involves combining tractography results with magnetization transfer images, which provide a more direct index of the myelin content of brain voxels [89]. Another promising line of work is developing novel diffusion imaging sequences for measuring axonal diameter [90].

An important limitation of diffusion tractography is that it cannot resolve the source and target of a fiber pathway. The resulting connectivity measures are therefore *undirected* – they tell us whether a connection between two regions exists, but they do not tell us whether region i connects to j or vice-versa. As discussed below, this simplification limits the types of graph theoretic analyses that can be performed and precludes a consideration of the directionality of information flow in the brain.

5.2 Functional Connectivity

Functional connectivity refers to a statistical dependence between neurophysiological recordings measured in distinct brain regions [91] (e.g., Fig. 2c, bottom). It thus quantifies functional interactions within brain networks. Most commonly, functional connectivity is assessed via a simple Pearson correlation between time courses extracted from two or more regions, although alternative measures such as partial correlation, mutual information, coherence, and so on can be used. Studies of simulated fMRI data have shown that these measures, which result in undirected measures of connectivity, vary in their ability to accurately reconstruct the true edges of a graph, although simple measures such as the Pearson correlation and partial correlation perform reasonably well [47].

Nonetheless, these measures do have limitations. For example, correlations do not distinguish between direct and indirect connections, meaning that two regions lacking a structural connection may still show highly correlated activity because they are indirectly connected via a third area. Indeed, it is well known that correlation-based measures of functional connectivity are sensitive to polysynaptic connections [92]; as a result, functional connectivity brain graphs tend to be more densely connected than structural connectivity graphs [93]. This sensitivity to indirect connections can introduce non-trivial structure into the network [94]. Partial correlations partly correct this problem, but can also be associated with bias in large-scale brain network analyses [94].

Functional connectivity is often assessed during task performance or task-free "resting" states [95, 96]. During resting-state designs, functional connectivity has traditionally been measured using temporal correlation taken across the duration of the fMRI acquisition protocol. In effect, this approach summarizes brain activity occurring over that period with a single scalar value. It therefore assumes stationarity of brain dynamics. Recent work has shown that brain dynamics show significant non-stationarities [97, 98] (see [99] for a review). These non-stationarities can be assessed using sliding window analyses [97] or multivariate decompositions (e.g., ICA) in the temporal domain [98]. The result is a time series of networks, which can then be analyzed to understand how brain network organization evolves over time. Recent multiband fMRI acquisition sequences that enable more rapid sampling of brain activity increase the power of such analyses [100].

A parallel line of work investigates functional connectivity during task performance. In such analyses, we are often interested in isolating brain functional networks that are modulated by changing task conditions. Two approaches that are scalable to whole-brain networks have been developed. One method, termed beta series correlation, attempts to model regional activation changes to each and every task event. Events are then sorted by condition and concatenated to generate condition-specific pseudo-time series (termed beta series, because event-related activity is modeled using a beta coefficient estimated with a general linear model) reflecting trial-to-trial variations of evoked activity. These pseudo-time series are then correlated between regions to quantify task-related functional connectivity as covariations in trial-to-trial fluctuations of brain activity evoked by each task condition [101, 102]. An alternative method is an adaptation of the psychophysiological interaction (PPI) framework first proposed by Friston et al. [103]. PPI analysis multiples a task regressor of interest by the time course in a region-of-interest to generate a psychophysiological interaction term representing task-related modulations of that region's activity. These terms can be correlated between regions to estimate task-related functional connectivity. Partialling out the raw regional time courses and task regressors allows task-related functional connectivity to be isolated from task-unrelated (intrinsic) dynamics [69]. Several variants of this method that are scalable to whole-brain networks have been used [67–69, 104].

5.3 Effective Connectivity

Effective connectivity is the influence of one neural system over another [91]. It thus quantifies causal interactions amongst brain regions and the resulting connectivity estimates are directed (Fig. 1). Critically, the causal interactions that define effective connectivity must be specified at the neuronal level. Because neuronal dynamics are not directly observable with fMRI, estimating effective connectivity with this imaging technique requires a model that maps the observed hemodynamic signal changes to the underlying neuronal dynamics from which they were generated [105].

The most popular framework for effective connectivity analysis of fMRI data is dynamical causal modeling (DCM) [106]. DCM uses a model of neurovascular coupling to specify the mapping between neuronal activity and hemodynamics. Different graph models of causal interactions (i.e., directed graph configurations) between neuronal systems are specified and compared to determine which model best accounts for the observed fMRI data. As the number of possible graph models rapidly increases with network size, DCM has traditionally only been applied to relatively small subnetworks comprising a few regions. Work is under way to scale these methods to larger systems [107]. DCM is applicable to both task and resting-state fMRI data [108].

5.4 Summary

MRI-based measures of structural and functional brain connectivity are indirect. As a result, connectivity analyses of MRI data must be interpreted with regards to the limitations of the measurement technique. In the context of graph theoretic analyses, a major limitation is that most structural and functional connectivity measures are undirected, limiting our capacity to resolve directions of information flow in the brain and to uncover hierarchical organizational features of brain networks (e.g., top-down vs. bottom-up connections). Nonetheless, many of the organizational properties of the human brain identified with MRI have been replicated in analyses of other species, where connectivity data have been acquired using more precise and invasive techniques. We consider some of these properties in Sect. 9.

6 From Connectivity Matrix to Brain Graph

Once structural, functional, or effective connectivity between every pair of brain regions in a given parcellation has been measured, the data can be succinctly represented as a connectivity matrix. In graph theory, this matrix is often called an adjacency matrix, denoted A (Fig. 1). These matrices can be binary or weighted, symmetric, or asymmetric. In a binary matrix, $A_{ij} = 1$ if nodes i and j are connected and $A_{ij} = 0$ otherwise. All connections are treated equally and no distinction is made between connections with different strength or weighting. Binary matrices thus represent the presence or absence of a neural connection. By contrast, the elements of a weighted matrix can span a range of values that is determined by the method used to measure connectivity. For example, if functional connectivity is estimated using a Pearson correlation, the values will be bounded in the range $[-1, 1]$. If structural connectivity is estimated using streamline counts, the values will positive integers.

If a connectivity matrix is symmetric, the values in the upper triangle of the matrix are the same as the values in the lower triangle (Fig. 1). In other words, $A_{ij} = A_{ji}$. No distinction is made concerning the source and target of a connection. Such matrices are typical of diffusion MRI or correlation-based analyses. Asymmetric matrices explicitly encode asymmetries (and thus, directionality) in connectivity (i.e., $A_{ij} \neq A_{ji}$). Asymmetric matrices are used to represent effective connectivity (Fig. 1). Both symmetric and asymmetric matrices can be either weighted or unweighted.

MRI-based estimates of connectivity are inherently noisy and it is often useful to threshold the connectivity matrix to distinguish real or probable connections from spurious or improbable connections (Fig. 2d). Thresholding should be done with care, as networks should only be compared if they have the same number of nodes and edges. Any thresholding procedure should thus try to respect this condition. One common approach is to apply an adaptive threshold to different networks to achieve a pre-specified connection density, which represents the number of edges present in a network relative to the total possible number of edges. For example, we could apply a threshold that ensures only the top 10% of connections across a sample of individuals are retained. However, systematic differences in connectivity strength between individuals or groups can bias this approach. If one group has lower mean connectivity than another group, a lower threshold will be required to achieve the desired connection density. This lowered threshold may result in the retention of a larger number of low-weight and potentially spurious connections, altering network structure [109]. A variety of alternative thresholding methods are available, but no method is completely free of bias [110]. It is therefore prudent to repeat analyses across a range of thresholds and using alternative strategies to ensure that any results obtained are robust to this methodological parameter.

Once the final structure of the connectivity matrix has been determined, the network can be represented as a graph (Figs. 1 and 2d).

In graph form, the rows and columns of the matrix—that is, the brain regions—are depicted as nodes (often circles), and the matrix elements A_{ij} determine which pairs of nodes are linked by edges. Variations in connectivity strength encoded in weighted matrices are commonly represented as variations in edge thickness (Fig. 1). The directionality of connectivity that is encoded in asymmetric matrices is depicted by arrowheads attached to the edges (Fig. 1).

Brain graphs can be projected in different ways, in order to highlight specific relations between nodes. Common displays in neuroscience include anatomical projections, in which nodes are positioned according to their stereotactic coordinates; topological projections, in which node positions are determined based on some topological relation between nodes; and circular projections (also called connectograms [111]), which allow a simplified view of the connectivity of the entire network. Examples of each of these projections are presented in Fig. 3.

7   Analyzing Brain Graphs

Once the network has been mapped, it can be analysed with respect to either its connectivity or topology. Connectivity analysis concentrates on variations in the type and strength of connectivity between brain regions. Topological analysis is concerned with understanding how connections are arranged with respect to each other, and provides insight into key organizational principles of the connectome.

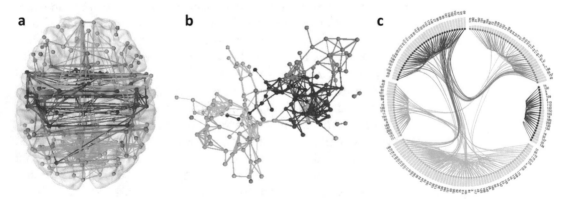

Fig. 3 Different graph projections of a brain network. (**a**) A projection in anatomical space, where nodes are positioned according to their stereotactic coordinates. Nodes and edges are colored according to the module to which they have been assigned. (**b**) A topological projection, in which nodes are located more closely in space if they have short path length between them. Nodes near the center of the graph have a short average path length to other nodes, and thus represent central elements of the network. Nodes colors are the same as in (**a**). (**c**) A ring projection, which is also called a connectogram. Nodes are clustered (and *colored*) by the modules to which they belong. Within each grouping, the nodes have been ordered according to their averaged connectivity strength with other nodes (*yellow bars*). The nodes have been labeled using arbitrary numbers

8   Connectivity Analysis

Connectivity can be studied at the level of specific neural systems, called candidate systems analysis, or across the entire brain, called connectome-wide analysis [112]. Candidate system analyses do not require a comprehensive map of connectivity between all pairs of brain regions and typically focus on one or a few networks of interest. Such analyses are exemplified by seed-based connectivity approaches, in which the structural or functional connectivity of specific seed regions to the rest of the brain is assessed [80, 113], and studies using ICA to investigate brain network connectivity.

Connectome-wide analyses interrogate effects at each and every element of the connectivity matrix. These analyses pose a major multiple comparisons problem. In an undirected network with N nodes, there are $\dfrac{N(N-1)}{2}$ possible connections. Thus, an analysis of an undirected network with 10^3 nodes will require correction over $499, 500$ comparisons. A simple Bonferroni correction in such circumstances will be too conservative. Fortunately, more powerful correction procedures are available [114–116]. In one such approach, called the network-based statistic (NBS) [116], a test statistic of interest (e.g., t-test, correlation, etc.) is computed at each and every edge, resulting in a statistic matrix with the same dimensions as the connectivity matrix. A primary threshold is applied to this matrix to define a pseudo-network of statistic values. The size of the connected components of this pseudo-network are computed and evaluated with respect to an empirical null distribution, which is generated by repeating the analysis many times after appropriate permutation of the data. In this context, a connected component refers to a collection of nodes that can be linked by a set of supra-threshold edges. The probability of observing components as large as those seen in the data by chance is computed with respect to the empirical null distribution of maximum component sizes, ensuring that the resulting probability values are corrected for multiple comparisons [117]. In this manner, the NBS identifies sets of nodes and edges that show a common effect of interest. Simulation studies have shown that, compared to traditional correction methods such as the False Discovery Rate [118], the NBS can offer great gains in sensitivity when effects are distributed across multiple edges [116, 119].

An alternative to analyzing connectivity at each and every edge is to examine the average connectivity of each node to all other brain regions. This measure has been variously referred to as connectivity strength, connectivity density, or global connectivity [120–122]. Since the analysis is conducted across nodes rather than edges, the multiple comparison correction required is much smaller (i.e., on the order of N, rather than $N(N-1)$). The analysis is also readily scalable to voxel-wise networks and can be useful for mapping areas in which connectivity varies with some variable of interest (e.g., diagnosis or cognitive performance). However, as the connectivity of each node is averaged across all other brain regions, more subtle effects specific to particular pair-wise links or circuits may be missed.

9   Topological Analysis

Topological analysis of brain networks draws most heavily on graph theory. The goal of topological analysis is to characterize how connections and nodes relate to each other, thereby shedding light on principles of brain network organization. Studies of brain network topology have revealed several non-trivial topological properties. In this section, we overview some of the most commonly applied metrics. Formal definitions are presented for binary, undirected networks. Generalizations for weighted and undirected networks are also available (see [123]).

9.1   Hub Dominance

Barabási and Albert's [25] discovery that many real-world networks have a heterogeneous distribution of connectivity across nodes suggests that such networks possess highly connected hubs that exert a disproportionate influence over the network. In network parlance, the total number of connections attached to a node is called its connectivity degree, denoted k, and the distribution of degree values across nodes is the degree distribution of a network. In the real-world networks they studied, Barabási and Albert found that the probability of finding a node with increasing k decayed as a power-law of the form $P(k) \sim k^{-\alpha}$, where α is the scaling exponent that determines the rate of decay. This decay is generally much slower than the decay observed in homogeneous random networks such as those studied by Erdős and Rényi, where the degree distribution approximates a Gaussian (more accurately, it conforms to a binomial distribution); i.e., most nodes have a degree value close to the mean, and the probability of finding large deviations from this mean is very low. This clustering around a mean value endows these networks with a single, characteristic scale. In contrast, power-law degree distributions are heavily skewed with an extended tail, pointing to a higher probability of finding nodes with very high connectivity values despite most nodes having low degree. There is no meaningful average degree or characteristic scale in these networks, so they are sometimes called scale-free.

Early work suggested that brain networks, at least when constructed at high resolution, conform to a power-law degree distribution [37, 49, 58]. Other studies of both structural and functional connectivity networks have more commonly reported evidence that the degree distribution of the human brain follows an exponentially truncated power law [12, 40, 48]. Truncated power-law distributions show power-law scaling over a limited regime, pointing to the existence of highly connected hub nodes (e.g., Fig. 4a). However, they show a rapid decay in the probability of finding nodes with very high degree beyond a certain cut-off. For this reason, networks with truncated power-law distributions are sometimes referred to as broad-scale [124]. This organization seems plausible for the brain, as metabolic and spatial constraints place an upper limit on the total number of connections that any single region can possess.

The presence of highly connected hubs in the brain suggests that certain brain regions play a critical role in integrative network function. Studies of both structural and functional connectivity converge to suggest that brain network hubs are predominantly located in multimodal association cortex, although hubs in the striatum and thalamus have also been noted [125–127]. These findings are consistent with the proposed role of association cortices and subcortical nuclei in integrating information from diverse modalities. These brain network hubs are more highly interconnected with each other than expected by chance, forming a so-called rich-club of connectivity: a densely connected core of high-degree nodes that acts as a central information-processing backbone that absorbs a large bulk of neuronal traffic [128, 129]. The rich-club organization of the brain facilitates the rapid transfer and integration of information between otherwise segregated neural systems [130].

9.2 Robustness to Damage

The degree distribution of a network has important implications for its robustness to damage. Network robustness can be assessed by removing a node (or edge) and its incident connections according to different rules. The properties of the remaining network can be analyzed after removal of each node to uncover which elements are most critical for network integrity. One commonly studied property in this context is the size of the largest connected component, S. In an intact network, $S = N$, where N is the number of nodes in the network. The value S can be computed after each node is removed to determine the number of nodes that is required to fragment the network (determined by the point at which $S < N$).

By removing nodes in different orders, we can simulate different types of network disruption. Stochastic failures can be simulated by random removal of nodes, whereas attacks can be simulated by targeted deletion of nodes based on their connectivity degree or some other property of interest. Compared to single-scale networks, scale-free systems are more resilient to random failure but highly vulnerable to targeted attack [131]. This vulnerability to attack arises because the concentration of connectivity on the hub nodes of a scale-free network means that only a few hubs need to be removed

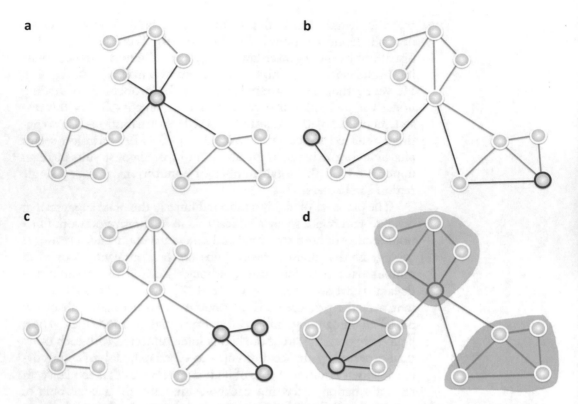

Fig. 4 Example of key topological properties of brain networks. (**a**) illustration of a single network hub (*red*) with high degree compared to other nodes. (**b**) example of the shortest path between two nodes at opposite ends of the network (*red*). In this case, the shortest path traverses five edges, so the path length between these two nodes is five. The characteristic path length of a network is the average path length between every pair of nodes. (**c**) Example of a connected triangle of nodes (*red*). The clustering coefficient of a node is computed as the number of such triangles attached to that node, relative to the total possible number of triangles. (**d**) Illustration of a modular decomposition of the network. Three modules have been identified, as represented by the different background colors. Nodes within a module are strongly connected with each other and sparsely connected with nodes in other modules. Such a decomposition allows analysis of node roles and hub category. *Red* highlights a provincial hub which is highly connected within its own module. *Orange* highlights a connector hub, which has connections distributed across all modules

to fragment the network. However, since hubs are relatively rare, the probability that they will be affected by random failure is low; hence the greater resistance to random node deletion relative to single-scale systems. Compared to scale-free systems, networks with truncated power-law degree distributions, such as those thought to characterize human brain networks, show comparable resilience to failure and better robustness in the face of targeted attack [40]. This enhanced robustness occurs because the concentration of connectivity on hub nodes is less extreme than in scale-free systems.

Nonetheless, damage to hub nodes, or the links between them, exerts a more severe impact on network function than damage to peripheral brain regions or connections [125]. The clinical implications of this conclusion should be obvious: we should expect that brain disorders affecting hub regions will present with more severe symptoms and/or a higher degree of impairment [132]. The high connectivity of brain network hubs may also render them more susceptible to disease processes originating elsewhere in the brain [55, 132]. Consistent with this view, pathology of hub regions is over-represented in a wide variety of brain disorders [133].

9.3   Small-Worldness

The small-world class of networks discovered by Watts and Strogatz [26] provides an appealing model for the brain. Clustered connectivity offers a substrate for functional specialization, whereas short average path length facilities functional integration. Clustering is formally quantified using the clustering coefficient, which computes the probability that two nodes linked to an index node are also connected with each other. In other words, it counts the number of closed triangles attached to a node (Fig. 4c). The path length between two nodes is simply the number of edges on the shortest path intersecting those nodes (Fig. 4b). The characteristic path length of a network is the average path length computed across all node pairs.

Empirically, the small-worldness of a network can be quantified by comparison to an ensemble of random graphs matched for the number of nodes, edges, and degree distribution. Various algorithms are available for constructing such graphs by rewiring the connections of an observed network [122, 134]. Such surrogate networks provide a useful baseline for evaluating the degree to which a particular topological property is expressed in the brain.

Formally, a network is considered small-world if the scalar quantity $\sigma > 1$, where $\sigma = \gamma / \lambda$ [135]. The quantity γ is computed as a ratio of the observed clustering coefficient to the average clustering of an ensemble of matched randomized networks (i.e., $\gamma = C_{obs} / C_{rand}$). The quantity λ is the ratio of the observed network path length to average path length computed in the same ensemble of randomized graphs (i.e., $\lambda = L_{obs} / L_{rand}$). Small-world networks will have greater clustering and similar path length to randomized surrogates. As such, $\gamma > 1$ and $\lambda \sim 1$, yielding $\sigma > 1$. Since σ is a ratio, variations in this value may be driven by changes in either γ or λ. It is therefore often more useful to understand variations in those parameters before considering σ.

9.4   Cost and Efficiency

The characteristic path length of a network is closely related to its topological efficiency. Communication in a network will be more efficient when fewer connections are required to transfer information between any two nodes; i.e., when the characteristic path length is low. We can thus define a topological measure of the global efficiency of a network as being inversely related to the characteristic path length of the network [136]:

$$E_{glob} = \frac{1}{N(N-1)} \sum_{i,j} \frac{1}{L_{ij}},$$

where L_{ij} is the minimum path length between nodes i and j. Similarly, a local measure of local communication efficiency can be computed as the efficiency of the subgraph defined by an index node's neighbors (i.e., the nodes to which it directly connects), after removal of that node [137]. It should be evident from the above definition of E_{glob} that the efficiency of a network can be improved simply adding more direct connections between nodes, since each direct link between nodes reduces network path length. However, in many real-world networks, there is often a cost associated with forming and maintaining each connection. This is certainly true for the brain, where axons, dendrites and synapses consume precious and limited metabolic resources. Accordingly, minimization of wiring costs is known to be an important pressure on brain organization, although wiring costs in the brain are not absolutely minimized [138]. If this were the case, brain connectivity would have a lattice-like arrangement, in which links were only formed between spatially adjacent nodes via short-range connections [55]. Instead the brain forms certain, long-range and high-cost connections that promote integration and communication efficiency, and give rise to its small-world organization [128, 137, 138]. Brain networks thus appear to be configured, at least in part, to satisfy competitive pressures to minimize cost and support efficient, integrated and complex function [54, 55, 139]. One fMRI study of healthy twins found evidence that this trade-off between cost and efficiency is strongly heritable [140], suggesting that it represents an important selection pressure on brain network evolution.

9.5 Modularity

The clustered connectivity of many real-world networks often means that they can be (nearly) decomposed into subsets of nodes, termed modules, which show higher connectivity with each other than with other network elements (Fig. 4d). There exists a wide variety of algorithms for decomposing networks into modules (reviewed in [141]). These algorithms commonly attempt to find a decomposition of the network that maximizes some quality function. The most commonly used function of this type is the Newman-Girvan Q-statistic [142], defined as

$$Q = \frac{1}{2m} \sum_{i,j \in N} \left(A_{ij} - \frac{k_i k_j}{2m} \right) \delta_{s_i, s_j},$$

where m is the total number of edges in the network, $A_{ij} = 1$ if nodes i and j are connected and zero otherwise, k is the node degree and $\delta_{s_i,s_j} = 1$ if nodes i and j belong to the same module and 0 otherwise. The term $\dfrac{k_i k_j}{2m}$ represents the expected connectivity between nodes i and j if the positions of the edges in the graph were completely randomized. Thus, the modularity of a network is defined as the mean difference between the actual and chance-expected connectivity between node pairs belonging to the same module.

The goal in modularity analysis is to find a decomposition that maximizes the Q-statistic. Optimization via exhaustive search is intractable for anything but the smallest networks, so various heuristic algorithms have been proposed (reviewed in [141]; see [143] for a comparative evaluation). Because such heuristics are used, there may be many solutions that yield similar Q-values (i.e., the solutions may be degenerate [144]). Consequently, multiple runs of the algorithm should be performed to derive a consensus partition ([145]; see also [67, 69, 122]). Furthermore, since even random networks can show some modular structure [146], it is often useful to compare the Q-statistic of the observed network to an ensemble of random networks matched for the number of nodes, edges, and degree distribution. Such an analysis allows statistical inference on whether modularity is a significant characteristic of the empirical network.

Modularity analysis offers a powerful tool for characterizing topologically separable systems in the brain. Most such analyses of MRI data typically yield 4-6 modules, in which spatially adjacent nodes are often grouped together [125, 147, 148]. This spatial clustering may occur because brain connectivity decays rapidly with physical separation between nodes [149, 150]. This penalty on long-distance connectivity is compounded by imaging techniques such as diffusion MRI, which have a limited ability to reconstruct long-range connections due to difficulty tracking trajectories through voxels with crossing fibers [42]. Careful processing of functional MRI data has been shown to uncover larger numbers of spatially distributed networks that mirror those identified with other decomposition techniques, such as ICA [59].

A common assumption is that the high interconnectivity of nodes belonging to the same module implies some commonality of function. Evidence in support of this hypothesis was provided by meta-analytic work demonstrating that inter-regional coactivation during specific types of tasks was preferentially expressed within distinct modules [151]. This result links topological modules of the brain to the psychological modules long thought to support cognition [152]. Topological modularity also enables robustness to damage while also promoting functional diversity and adaptability [153, 154]: damage sustained in a modular system will often be limited to the affected module, and the diverse function of different modules facilitate adaptation to different environmental challenges.

A particular strength of modularity analysis is that it allows one to characterize the role played by each node in the network with respect to its degree of intra-modular and inter-modular connectivity [155]. Intra-modular connectivity is commonly computed as a z-score of each node's intra-module connectivity relative to other nodes in the same module [155]. Nodes with high intra-modular connectivity are called "provincial hubs," and are thought to play an important role in functional specialization, acting as central components of the module to which they belong (Fig. 4d). Inter-module connectivity is commonly measured using the participation coefficient, which indexes how a node's connectivity is distributed across different modules [155]. Nodes with a relatively even distribution of connectivity across different modules are called "connector hubs" and play an important role in functional integration because they support communication between different modules (Fig. 4d). Node role analysis can be used to identify different types of brain hubs [127] and to characterize the topological role of brain nodes under different contexts [67, 69].

Most methods for modularity decomposition yield a hard segmentation of the brain such that nodes can belong to only one module. In reality, it is likely brain nodes can belong to more than one functional system. Indeed, this is a defining feature of association cortex. Algorithms for fuzzy or overlapping modular decompositions are available, although evaluations with respect to benchmark networks have found their performance to be lacking in many circumstances [156]. The modular organization of the brain can also show a hierarchical structure, comprising modules within modules across several scales of resolution [157], although such multiscale organization has seldom been investigated in brain networks (for an exception, see [158]).

9.6   Summary

This section has presented a brief overview of some of the basic graph theoretic concepts applied to neuroimaging data, and the insights they have provided into brain network organization. Many other metrics are available, enabling a diverse range of analyses (see [159] for an introduction). It is important to bear in mind that many such methods were developed in the physical or social sciences with networks other than the brain in mind. As such, they may not be directly portable to neuroscientific contexts. Indeed, the first wave of graph theoretic studies of neuroimaging data have concentrated on measuring canonical topological properties such as those discussed here, but alternative measures may provide more appropriate models of brain function. We consider some of these issues in the next section.

10   Issues for Consideration and Developing Trends

In this chapter, we have considered the basic methodologies behind building and analyzing a graph theoretic model of the human connectome with MRI data. We have also highlighted the insights into brain organization that have been gained by this perspective and have drawn attention to the limitations of current methodologies where appropriate. Some additional considerations should be taken into account.

As already stated, many graph theoretic measures were developed for the analysis of networks other than the brain. Consequently, not all typical graph theoretic metrics may be appropriate for the characterization of brain networks. For example, topological measures based on shortest paths, such as the characteristic path length and global efficiency, assume that information in the brain travels along the shortest path between regions. In order to find the shortest path in a network, one must have global knowledge of network topology to find the optimal route. It is unlikely that any individual neural element possesses such knowledge. This limitation was elegantly shown in a recent study examining the relationship between structural and functional connectivity as measured using diffusion MRI and functional MRI, respectively [160]. Specifically, it was shown that characteristics of the shortest *structural* path between nodes strongly predicted the *functional* connectivity between those regions. These characteristics indexed how easy it is to find and/or remain on the shortest path. We should not expect such properties to relate to functional connectivity if neuronal signaling propagates exclusively along shortest paths (i.e., in such a scenario, the ease with which such a path can be found should be irrelevant). As such, alternative measures of brain network topology that assume dynamics which more closely approximate information transmission in the brain, such as those based on locally guided diffusion processes, may offer useful alternative measures of network communication processes [129, 160, 161]. Moreover, methods for characterizing the temporal evolution of topological properties on non-stationary networks are being developed for analyses of dynamic functional connectivity [162].

Another consideration in graph theoretic analysis of MRI data specifically concerns functional connectivity networks. In such networks, edges are determined by some measure of statistical dependence, such as the correlation coefficient. In this sense, the edges are somewhat abstract quantities. A structural link measured with diffusion MRI unambiguously indexes a physical connection between nodes (notwithstanding the limitations of the measurement technique; [88]). A functional link on the other hand, is a statistical measure of covariation in some physiological process. Due to the statistical nature of functional connectivity measures, certain topological properties computed on brain functional networks may be difficult to interpret. For example, topological measures that consider indirect paths between nodes (e.g., those based on path length) are unlikely to represent viable measures of information exchanged between two regions, since the measured functional connectivity of those regions provides a direct estimate of their functional interaction [123]. Popular statistical measures such as the correlation coefficient are also often signed (that is, edge weights can be either positive or negative). Signed edges imply a qualitatively distinct type of interaction between brain regions [18]. This information is often ignored in imaging analyses, where weights are either converted to absolute values or thresholded to focus only on positive weights. The adaptation of graph theoretic measures to deal with signed weights will assist in overcoming this problem [123].

Finally, graph theory is emerging as a useful tool for integrating theory and experiment. Neural dynamics can be simulated on empirically derived network structures (e.g., a structural connectivity network measured with diffusion MRI) to generate large-scale models of network functional connectivity [163]. These models allow analysis of how simulated lesions impact network dynamics [164, 165]. Alternatively, models of disease processes can be simulated on the network structure to determine whether they accurately predict empirical patterns of disorder-related neuropathology [166, 167]. Similarly, developmental processes can be investigated using network growth models [168]. In these models, networks are grown by adding nodes and edges according to specific rules. The properties of the resulting network are compared to empirical data; a good match between model and data implies that the growth rules represent an important factor in brain network development. Growth models have been used to uncover key organizational imperatives for brain networks, largely involving trade-offs between wiring costs and complex topological properties [139, 150, 168, 169], and for modeling developmental abnormalities in brain disorders [168] (see also [170, 171]).

Addressing the issues raised here and capitalizing on these emerging trends will ensure more accurate modeling of brain imaging data, while also establishing graph theory as a flexible and powerful framework for the integration of theory and experiment in neuroscience. This integration will be necessary to move beyond the simple description of empirical findings to formulate and test competing hypotheses about the underlying mechanisms that generated the observed data.

References

1. Cherniak C (1990) The bounded brain: toward quantitative neuroanatomy. J Cogn Neurosci 2:58–68
2. Sporns O, Tononi G, Kötter R (2005) The human connectome: a structural description of the human brain. PLoS Comput Biol 1:e42
3. Van Essen DC et al (2012) The Human Connectome Project: a data acquisition perspective. Neuroimage 62:2222–2231
4. Bohland JW et al (2009) A proposal for a coordinated effort for the determination of brainwide neuroanatomical connectivity in model organisms at a mesoscopic scale. PLoS Comput Biol 5:e1000334
5. Kandel ER, Markram H, Matthews PM, Yuste R, Koch C (2013) Neuroscience thinks big (and collaboratively). Nat Rev Neurosci 14:659–664
6. White JG, Southgate E, Thomson JN, Brenner S (1986) The structure of the nervous system of the nematode Caenorhabditis elegans. Philos Trans R Soc Lond B Biol Sci 314:1–340
7. Lichtman JW, Pfister H, Shavit N (2014) The big data challenges of connectomics. Nat Neurosci 17:1448–1454
8. Chiang A-S et al (2011) Three-dimensional reconstruction of brain-wide wiring networks in Drosophila at single-cell resolution. Curr Biol 21:1–11
9. Scannell JW, Young MP (1993) The connectional organization of neural systems in the cat cerebral cortex. Curr Biol 3:191–200
10. Shanahan M, Bingman VP, Shimizu T, Gunturkun O (2013) Large-scale network organization in the avian forebrain: a connectivity matrix and theoretical analysis. Front Comput Neurosci 7:1–17
11. Stephan KE (2013) The history of CoCoMac. NeuroImage 80:46–52
12. Hagmann P et al (2007) Mapping human whole-brain structural networks with diffusion MRI. PLoS One 2:e597
13. Newman MJE (2003) The structure and function of complex networks. SIAM Rev 45:167–256
14. Boccaletti S, Latora V, Moreno Y, Chavez M, Hwang DU (2006) Complex networks: structure and dynamics. Phys Rep 424:175–308
15. Axer M et al (2011) A novel approach to the human connectome: ultra-high resolution mapping of fiber tracts in the brain. Neuroimage 54:1091–1101
16. Chung K, Deisseroth K (2013) CLARITY for mapping the nervous system. Nat Meth 10:508–513
17. Bullmore E, Sporns O (2009) Complex brain networks: graph theoretical analysis of structural and functional systems. Nat Rev Neurosci 10:186–198
18. Fornito A, Zalesky A, Breakspear M (2013) Graph analysis of the human connectome: promise, progress, and pitfalls. Neuroimage 80:426–444
19. Euler L (1736) Solutio problematis ad geometriam situs pertinentis. Commentarii Academiae Scientiarum Imperialis Petropolitanae 8:128–140
20. Luce RD, Perry AD (1949) A method of matrix analysis of group structure. Psychometrika 14:95–116
21. Katz L (1947) On the matric analysis of sociometric data. Sociometry 10:233–241
22. Forsyth E, Katz L (1946) A matrix approach to the analysis of sociometric data: preliminary report. Sociometry 9:340–347
23. Harary F, Norman RZ (1953) Graph theory as a mathematical model in social science. University of Michigan Press
24. Erdos P, Renyi A (1959) On random graphs. Publ Math Debrecen 6:290–297
25. Barabasi A, Albert R (1999) Emergence of scaling in random networks. Science 286:509–512
26. Watts DJ, Strogatz SH (1998) Collective dynamics of 'small-world' networks. Nature 393:440–442
27. Mesulam MM (1998) From sensation to cognition. Brain 121:1013–1052
28. Tononi G, Sporns O, Edelman GM (1994) A measure for brain complexity: relating functional segregation and integration in the nervous system. Proc Natl Acad Sci U S A 91:5033–5037
29. Friston KJ (2011) Functional and effective connectivity: a review. Brain Connect 1:13–36
30. Felleman DJ, Van Essen DC (1991) Distributed hierarchical processing in the primate cerebral cortex. Cereb Cortex 1:1–47
31. Scannell JW, Blakemore C, Young MP (1995) Analysis of connectivity in the cat cerebral cortex. J Neurosci 15:1463–1483
32. Hilgetag CC, Burns GA, O'Neill MA, Scannell JW, Young MP (2000) Anatomical connectivity defines the organization of clusters of cortical areas in the macaque monkey and the cat. Philos Trans R Soc Lond B Biol Sci 355:91–110
33. Sporns O, Tononi G, Edelman GM (2000) Theoretical neuroanatomy: relating anatomical and functional connectivity in graphs and cortical connection matrices. Cereb Cortex 10:127–141
34. Scannell JW, Burns GA, Hilgetag CC, O'Neil MA, Young MP (1999) The connectional

organization of the cortico-thalamic system of the cat. Cereb Cortex 9:277–299

35. Sporns O, Chialvo DR, Kaiser M, Hilgetag CC (2004) Organization, development and function of complex brain networks. Trends Cogn Sci 8:418–425

36. Stam CJ (2004) Functional connectivity patterns of human magnetoencephalographic recordings: a 'small-world' network? Neurosci Lett 355:25–28

37. Eguiluz VM, Chialvo DR, Cecchi GA, Baliki M, Apkarian AV (2005) Scale-free brain functional networks. Phys Rev Lett 94:018102

38. Salvador R, Suckling J, Schwarzbauer C, Bullmore E (2005) Undirected graphs of frequency-dependent functional connectivity in whole brain networks. Philos Trans R Soc Lond B Biol Sci 360:937–946

39. Salvador R et al (2005) Neurophysiological architecture of functional magnetic resonance images of human brain. Cereb Cortex 15:1332–1342

40. Achard S, Salvador R, Whitcher B, Suckling J, Bullmore E (2006) A resilient, low-frequency, small-world human brain functional network with highly connected association cortical hubs. J Neurosci 26:63–72

41. Iturria-Medina Y et al (2007) Characterizing brain anatomical connections using diffusion weighted MRI and graph theory. NeuroImage 36:645–660

42. Zalesky A, Fornito A (2009) A DTI-derived measure of cortico-cortical connectivity. IEEE Trans Med Imaging 28:1023–1036

43. Skudlarski P et al (2008) Measuring brain connectivity: diffusion tensor imaging validates resting state temporal correlations. Neuroimage 43:554–561

44. Albert R, Barabasi AL (2002) Statistical mechanics of complex networks. Rev Mod Phys 74:47–97

45. Logothetis NK (2008) What we can do and what we cannot do with fMRI. Nature 453:869–878

46. Butts CT (2009) Revisiting the foundations of network analysis. Science 325:414–416

47. Smith SM et al (2011) Network modelling methods for FMRI. Neuroimage 54:875–891

48. Fornito A, Zalesky A, Bullmore ET (2010) Network scaling effects in graph analytic studies of human resting-state FMRI data. Front Syst Neurosci 4:22

49. Hayasaka S, Laurienti PJ (2010) Comparison of characteristics between region-and voxel-based network analyses in resting-state fMRI data. Neuroimage 50:499–508

50. Zalesky A et al (2010) Whole-brain anatomical networks: does the choice of nodes matter? Neuroimage 50:970–983

51. Mountcastle VB (1997) The columnar organization of the neocortex. Brain 120(Pt 4):701–722

52. Rademacher J, Caviness VS Jr, Steinmetz H, Galaburda AM (1993) Topographical variation of the human primary cortices: implications for neuroimaging, brain mapping, and neurobiology. Cereb Cortex 3:313–329

53. Welker W (1990) 8b: Comparative structure and evolution of cerebral cortex. In: Jones EG, Peters A (eds) Cerebral cortex. Plenum, New York, pp 3–136

54. Bassett DS et al (2010) Efficient physical embedding of topologically complex information processing networks in brains and computer circuits. PLoS Comput Biol 6:e1000748

55. Bullmore E, Sporns O (2012) The economy of brain network organization. Nat Rev Neurosci 13:336–349

56. Passingham RE, Stephan KE, Kotter R (2002) The anatomical basis of functional localization in the cortex. Nat Rev Neurosci 3:606–616

57. Eliasmith C et al (2012) A large-scale model of the functioning brain. Science 338:1202–1205

58. van den Heuvel MP, Stam CJ, Boersma M, Pol HEH (2008) Small-world and scale-free organization of voxel-based resting-state functional connectivity in the human brain. Neuroimage 43:528–539

59. Power JD et al (2011) Functional network organization of the human brain. Neuron 72:665–678

60. Wig GS, Schlaggar BL, Petersen SE (2011) Concepts and principles in the analysis of brain networks. Ann N Y Acad Sci 1224:126–146

61. Tzourio-Mazoyer N et al (2002) Automated anatomical labeling of activations in SPM using a macroscopic anatomical parcellation of the MNI MRI single-subject brain. NeuroImage 15:273–289

62. Desikan RS et al (2006) An automated labeling system for subdividing the human cerebral cortex on MRI scans into gyral based regions of interest. NeuroImage 31:968–980

63. Salvador R et al (2008) A simple view of the brain through a frequency-specific functional connectivity measure. NeuroImage 39:279–289

64. Dosenbach NU et al (2010) Prediction of individual brain maturity using fMRI. Science 329:1358–1361

65. Fair DA et al (2007) Development of distinct control networks through segregation and integration. Proc Natl Acad Sci U S A 104:13507–13512

66. Andrews-Hanna JR, Reidler JS, Sepulcre J, Poulin R, Buckner RL (2010) Functional-anatomic fractionation of the brain's default network. Neuron 65:550–562

67. Dwyer DB et al (2014) Large-scale brain network dynamics supporting adolescent cognitive control. J Neurosci 34:14096–14107

68. Cocchi L et al (2014) Complexity in relational processing predicts changes in functional brain network dynamics. Cereb Cortex 24:2283–2296

69. Fornito A, Harrison BJ, Zalesky A, Simons JS (2012) Competitive and cooperative dynamics of large-scale brain functional networks supporting recollection. Proc Natl Acad Sci U S A 109:12788–12793

70. Beckmann CF, DeLuca M, Devlin JT, Smith SM (2005) Investigations into resting-state connectivity using independent component analysis. Philos Trans R Soc Lond B Biol Sci 360:1001–1013

71. Calhoun VD, Adali T, Pekar JJ (2004) A method for comparing group fMRI data using independent component analysis: application to visual, motor and visuomotor tasks. Magn Reson Imaging 22:1181–1191

72. Smith SM et al (2009) Correspondence of the brain's functional architecture during activation and rest. Proc Natl Acad Sci U S A 106:13040–13045

73. Yu Q et al (2011) Altered topological properties of functional network connectivity in schizophrenia during resting state: a small-world brain network study. PLoS One 6:e25423

74. Kiviniemi V et al (2009) Functional segmentation of the brain cortex using high model order group PICA. Hum Brain Mapp 30:3865–3886

75. Nelson SM et al (2010) A parcellation scheme for human left lateral parietal cortex. Neuron 67:156–170

76. Cohen AL et al (2008) Defining functional areas in individual human brains using resting functional connectivity MRI. NeuroImage 41:45–57

77. Yeo BT et al (2011) The organization of the human cerebral cortex estimated by intrinsic functional connectivity. J Neurophysiol 106:1125–1165

78. Craddock RC, James GA, Holtzheimer PE, Hu XP, Mayberg HS (2012) A whole brain fMRI atlas generated via spatially constrained spectral clustering. Hum Brain Mapp 33:1914–1928

79. Johansen-Berg H et al (2004) Changes in connectivity profiles define functionally distinct regions in human medial frontal cortex. Proc Natl Acad Sci U S A 101:13335–13340

80. Behrens TE et al (2003) Non-invasive mapping of connections between human thalamus and cortex using diffusion imaging. Nat Neurosci 6:750–757

81. Anwander A, Tittgemeyer M, von Cramon DY, Friederici AD, Knosche TR (2007) Connectivity-based parcellation of Broca's area. Cereb Cortex 17:816–825

82. Glasser MF, Van Essen DC (2011) Mapping human cortical areas in vivo based on myelin content as revealed by T1- and T2-weighted MRI. J Neurosci 31:11597–11616

83. Eickhoff SB et al (2005) A new SPM toolbox for combining probabilistic cytoarchitectonic maps and functional imaging data. NeuroImage 25:1325–1335

84. Zilles K et al (2002) Architectonics of the human cerebral cortex and transmitter receptor fingerprints: reconciling functional neuroanatomy and neurochemistry. Eur Neuropsychopharmacol 12:587–599

85. Alexander-Bloch A, Giedd JN, Bullmore E (2013) Imaging structural co-variance between human brain regions. Nat Rev Neurosci 14:322–336

86. Lerch JP et al (2006) Mapping anatomical correlations across cerebral cortex (MACACC) using cortical thickness from MRI. NeuroImage 31:993–1003

87. Bastiani M, Shah NJ, Goebel R, Roebroeck A (2012) Human cortical connectome reconstruction from diffusion weighted MRI: the effect of tractography algorithm. Neuroimage 62:1732–1749

88. Jones DK, Knösche TR, Turner R (2013) White matter integrity, fiber count, and other fallacies: the do's and dont's of diffusion MRI. Neuroimage 73:239–254

89. van den Heuvel MP, Mandl RCW, Stam CJ, Kahn RS, Hulshoff Pol HE (2010) Aberrant frontal and temporal complex network structure in schizophrenia: a graph theoretical analysis. J Neurosci 30:15915–15926

90. Alexander DC et al (2010) Orientationally invariant indices of axon diameter and density from diffusion MRI. Neuroimage 52:1374–1389

91. Friston KJ (1994) Functional and effective connectivity in neuroimaging: a synthesis. Hum Brain Mapping 2:56–78

92. Vincent JL et al (2007) Intrinsic functional architecture in the anaesthetized monkey brain. Nature 447:83–86

93. Honey CJ et al (2009) Predicting human resting-state functional connectivity from structural connectivity. Proc Natl Acad Sci U S A 106:2035–2040

94. Zalesky A, Fornito A, Bullmore E (2012) On the use of correlation as a measure of network connectivity. Neuroimage 60:2096–2106

95. Fox MD, Raichle ME (2007) Spontaneous fluctuations in brain activity observed with functional magnetic resonance imaging. Nat Rev Neurosci 8:700–711

96. Fornito A, Bullmore ET (2010) What can spontaneous fluctuations of the blood oxygenation-level-dependent signal tell us about psychiatric disorders? Curr Opin Psychiatry 23:239–249

97. Zalesky A, Fornito A, Cocchi L, Gollo LL, Breakspear M (2014) Time-resolved resting-state brain networks. Proc Natl Acad Sci U S A 111:10341–10346

98. Smith SM et al (2012) Temporally-independent functional modes of spontaneous brain activity. Proc Natl Acad Sci U S A 109:3131–3136

99. Hutchison RM et al (2013) Dynamic functional connectivity: promise, issues, and interpretations. NeuroImage 80:360–378

100. Feinberg DA et al (2010) Multiplexed echo planar imaging for sub-second whole brain FMRI and fast diffusion imaging. PLoS One 5:e15710

101. Rissman J, Gazzaley A, D'Esposito M (2004) Measuring functional connectivity during distinct stages of a cognitive task. NeuroImage 23:752–763

102. Fornito A, Yoon J, Zalesky A, Bullmore ET, Carter CS (2011) General and specific functional connectivity disturbances in first-episode schizophrenia during cognitive control performance. Biol Psychiatry 70:64–72

103. Friston KJ et al (1997) Psychophysiological and modulatory interactions in neuroimaging. NeuroImage 6:218–229

104. Cole MW et al (2013) Multi-task connectivity reveals flexible hubs for adaptive task control. Nat Neurosci 16:1348–1355

105. Friston K, Moran R, Seth AK (2013) Analysing connectivity with Granger causality and dynamic causal modelling. Curr Opin Neurobiol 23:172–178

106. Friston KJ, Harrison L, Penny W (2003) Dynamic causal modelling. NeuroImage 19:1273–1302

107. Seghier ML, Friston KJ (2013) Network discovery with large DCMs. Neuroimage 68:181–191

108. Friston KJ, Kahan J, Biswal B, Razi A (2014) A DCM for resting state fMRI. Neuroimage 94:396–407

109. Fornito A, Zalesky A, Pantelis C, Bullmore ET (2012) Schizophrenia, neuroimaging and connectomics. Neuroimage 62:2296–2314

110. van Wijk BCM, Stam CJ, Daffertshofer A (2010) Comparing brain networks of different size and connectivity density using graph theory. Plos One 5:e13701

111. Irimia A, Chambers MC, Torgerson CM, Van Horn JD (2012) Circular representation of human cortical networks for subject and population-level connectomic visualization. Neuroimage 60:1340–1351

112. Fornito A, Bullmore ET (2015) Connectomics: a new paradigm for understanding brain disease. Eur Neuropsychopharmacol. 25:733–748

113. Fornito A et al (2013) Functional dysconnectivity of corticostriatal circuitry as a risk phenotype for psychosis. JAMA Psychiatry 70:1143–1151

114. Meskaldji DE et al (2011) Adaptive strategy for the statistical analysis of connectomes. Plos One 6:e23009

115. Ginestet CE, Simmons A (2011) Statistical parametric network analysis of functional connectivity dynamics during a working memory task. Neuroimage 55:688–704

116. Zalesky A, Fornito A, Bullmore ET (2010) Network-based statistic: Identifying differences in brain networks. Neuroimage 53:1197–1207

117. Nichols TE, Holmes AP (2002) Nonparametric permutation tests for functional neuroimaging: a primer with examples. Hum Brain Mapp 15:1–25

118. Benjamini Y, Hochberg Y (1995) Controlling the false discovery rate: a practical and powerful approach to multiple testing. J R Stat Soc Ser B 57:289–300

119. Zalesky A, Cocchi L, Fornito A, Murray MM, Bullmore E (2012) Connectivity differences in brain networks. Neuroimage 60:1055–1062

120. Tomasi D, Volkow VD (2010) Functional connectivity density mapping. Proc Natl Acad Sci U S A, 107:9885–9890

121. Cole MW, Anticevic A, Repovs G, Barch D (2011) Variable global dysconnectivity and individual differences in schizophrenia. Biol Psychiatry 70:43–50

122. Rubinov M, Sporns O (2011) Weight-conserving characterization of complex functional brain networks. Neuroimage 56:2068–2079

123. Rubinov M, Sporns O (2010) Complex network measures of brain connectivity: uses and interpretations. Neuroimage 52:1059–1069

124. Amaral LA, Scala A, Barthelemy M, Stanley HE (2000) Classes of small-world networks. Proc Natl Acad Sci U S A 97:11149–11152

125. van den Heuvel MP, Sporns O (2011) Rich-club organization of the human connectome. J Neurosci 31:15775–15786

126. Buckner RL et al (2009) Cortical hubs revealed by intrinsic functional connectivity: mapping, assessment of stability, and relation to Alzheimer's Disease. J Neurosci 29:1860–1873

127. Power JD, Schlaggar BL, Lessov-Schlaggar CN, Petersen SE (2013) Evidence for hubs in human functional brain networks. Neuron 79:798–813

128. van den Heuvel MP, Kahn RS, Goni J, Sporns O (2012) High-cost, high-capacity backbone for global brain communication. Proc Natl Acad Sci U S A 109:11372–11377

129. Mišić B, Sporns O, McIntosh AR (2014) Communication efficiency and congestion of signal traffic in large-scale brain networks. PLoS Comput Biol 10:e1003427

130. van den Heuvel MP, Sporns O (2013) An anatomical substrate for integration among functional networks in human cortex. J Neurosci 33:14489–14500

131. Albert R, Jeong H, Barabasi AL (2000) Error and attack tolerance of complex networks. Nature 406:378–382

132. Fornito A, Breakspear M, Zalesky A (2015) The connectomics of brain disorders. Nat Rev Neurosci 16:159–172

133. Crossley NA et al (2014) The hubs of the human connectome are generally implicated in the anatomy of brain disorders. Brain 137:2382–2395

134. Maslov S, Sneppen K (2002) Specificity and stability in topology of protein networks. Science 296:910–913

135. Humphries MD, Gurney K, Prescott TJ (2006) The brainstem reticular formation is a small-world, not scale-free, network. Proc Biol Sci 273:503–511

136. Latora V, Marchiori M (2001) Efficient behavior of small-world networks. Phys Rev Lett 87:198701

137. Latora V, Marchiori M (2003) Economic small-world behavior in weighted networks. Eur Phys J B 32:249–263

138. Kaiser M, Hilgetag CC (2006) Nonoptimal component placement, but short processing paths, due to long-distance projections in neural systems. PLoS Comput Biol 2:e95

139. Chen Y, Wang S, Hilgetag CC, Zhou C (2013) Trade-off between multiple constraints enables simultaneous formation of modules and hubs in neural systems. PLoS Comput Biol 9:e1002937

140. Fornito A et al (2011) Genetic influences on cost-efficient organization of human cortical functional networks. J Neurosci 31:3261–3270

141. Fortunato S (2010) Community detection in graphs. Phys Rep 486:75–174

142. Newman M, Girvan M (2004) Finding and evaluating community structure in networks. Phys Rev 69:026113

143. Lancichinetti A, Fortunato S (2009) Community detection algorithms: a comparative analysis. Phys Rev E 80:056117

144. Good BH, de Montjoye YA, Clauset A (2010) Performance of modularity maximization in practical contexts. Phys Rev E 81:046106

145. Lancichinetti A, Fortunato S (2012) Consensus clustering in complex networks. Sci Rep 2:336

146. Guimerà R, Sales-Pardo M, Amaral L (2004) Modularity from fluctuations in random graphs and complex networks. Phys Rev E 70:025101

147. Hagmann P et al (2008) Mapping the structural core of human cerebral cortex. PLoS Biol 6:e159

148. Meunier D, Achard S, Morcom A, Bullmore E (2009) Age-related changes in modular organization of human brain functional networks. NeuroImage 44:715–723

149. Buzsaki G, Geisler C, Henze DA, Wang XJ (2004) Interneuron diversity series: circuit complexity and axon wiring economy of cortical interneurons. Trends Neurosci 27:186–193

150. Ercsey-Ravasz M et al (2013) A predictive network model of cerebral cortical connectivity based on a distance rule. Neuron 80:184–197

151. Crossley NA, Mechelli A, Vertes PE (2013) Cognitive relevance of the community structure of the human brain functional coactivation network. Proc Natl Acad Sci U S A 110:11583–11588

152. Fodor JA (1983) Modularity of mind: an essay on faculty psychology. MIT Press

153. Simon HA (1962) The architecture of complexity. Proc Am Philos Soc 106:467–482

154. Kitano H (2004) Biological robustness. Nat Rev Genet 5:826–837

155. Guimera R, Nunes Amaral LA (2005) Functional cartography of complex metabolic networks. Nature 433:895–900

156. Xie J, Kelley S, Szymanski BK (2013) Overlapping community detection in networks. ACM Comput Surv 45:1–35

157. Meunier D, Lambiotte R, Bullmore ET (2011) Modular and hierarchically modular organization of brain networks. Front Neurosci 4:200

158. Meunier D, Lambiotte R, Fornito A, Ersche KD, Bullmore ET (2009) Hierarchical modularity in human brain functional networks. Front Neuroinform 3:37

159. Newman MEJ (2010) Networks. a Introduction. Oxford University Press

160. Goñi J et al (2014) Resting-brain functional connectivity predicted by analytic measures of network communication. Proc Natl Acad Sci U S A 111:833–838

161. Betzel RF et al (2014) Multi-scale community organization of the human struc-

tural connectome and its relationship with resting-state functional connectivity. Net Sci 1:353–373

162. Bassett DS et al (2011) Dynamic reconfiguration of human brain networks during learning. Proc Natl Acad Sci U S A 108:7641–7646

163. Deco G, Jirsa VK, McIntosh AR (2011) Emerging concepts for the dynamical organization of resting-state activity in the brain. Nat Publ Group 12:43–56

164. Alstott J, Breakspear M, Hagmann P, Cammoun L, Sporns O (2009) Modeling the impact of lesions in the human brain. PLoS Comput Biol 5:e1000408

165. Honey CJ, Sporns O (2008) Dynamical consequences of lesions in cortical networks. Hum Brain Mapp 29:802–809

166. Raj A, Kuceyeski A, Weiner M (2012) A network diffusion model of disease progression in dementia. Neuron 73:1204–1215

167. de Haan W, Mott K, van Straaten ECW, Scheltens P, Stam CJ (2012) Activity dependent degeneration explains hub vulnerability in Alzheimer's disease. PLoS Comput Biol 8:e1002582

168. Vertes PE et al (2012) Simple models of human brain functional networks. Proc Natl Acad Sci U S A 109:5868–5873

169. Song HF, Kennedy H, Wang X-J (2014) Spatial embedding of structural similarity in the cerebral cortex. Proc Natl Acad Sci U S A 111:16580–16585

170. Goni J et al (2013) Exploring the morphospace of communication efficiency in complex networks. PLoS One 8, e58070

171. Avena-Koenigsberger A et al (2014) Using Pareto optimality to explore the topology and dynamics of the human connectome. Philos Trans R Soc Lond B Biol Sci 369:20130530

Part II

fMRI Application to Measure Brain Function

Chapter 11

Functional MRI: Applications in Cognitive Neuroscience

Mark D'Esposito, Andrew Kayser, and Anthony Chen

Abstract

Neuroimaging, in many respects, revolutionized the study of cognitive neuroscience, the discipline that attempts to determine the neural mechanisms underlying cognitive processes. Early studies of brain–behavior relationships relied on a precise neurological exam as the basis for hypothesizing the site of brain damage that was responsible for a given behavioral syndrome. The advent of structural brain imaging, first with computerized tomography and later with magnetic resonance imaging, paved the way for more precise anatomical localization of the cognitive deficits that manifest after brain injury. Functional neuroimaging, broadly defined as techniques that provide measures of brain activity, further increased our ability to study the neural basis of behavior. Functional MRI (fMRI), in particular, is an extremely powerful technique that affords excellent spatial and temporal resolution. This chapter focuses on the principles underlying fMRI as a cognitive neuroscience tool for exploring brain–behavior relationships.

Key words Functional MRI, Cognitive neuroscience, Experimental design, Statistics

1 Introduction

Cognitive neuroscience is a discipline that attempts to determine the neural mechanisms underlying cognitive processes. Specifically, cognitive neuroscientists test hypotheses about brain–behavior relationships that can be organized along two conceptual domains: *functional specialization*—the idea that functional modules exist within the brain, that is, areas of the cerebral cortex that are specialized for a specific cognitive process, and *functional integration*—the idea that a cognitive process can be an emergent property of interactions among a network of brain regions, which suggests that a brain region can play a different role across many functions.

Early investigations of brain–behavior relationships consisted of careful observation of individuals with neurological injury resulting in focal brain damage. The idea of functional specialization evolved from hypotheses that damage to a particular brain region was responsible for a given behavioral syndrome that was characterized by a precise neurological examination. For instance,

Massimo Filippi (ed.), *fMRI Techniques and Protocols*, Neuromethods, vol. 119,
DOI 10.1007/978-1-4939-5611-1_11, © Springer Science+Business Media New York 2016

the association of aphasia with right-sided limb weakness implicated the left hemisphere as the site of language abilities. Moreover, upon the death of a patient with a neurological disorder, clinico-pathological correlations provided confirmatory information about the site of damage causing a specific neurobehavioral syndrome such as aphasia. For example, in 1861 Paul Broca's observations of nonfluent aphasia in the setting of a damaged left inferior frontal gyrus (IFG) cemented the belief that this brain region was critical for speech output [1]. The introduction of structural brain imaging more than 100 years after Broca's observations, first with computerized tomography (CT) and later with magnetic resonance imaging (MRI), paved the way for more precise anatomical localization in the living patient of the cognitive deficits that develop after brain injury. The superb spatial resolution of structural neuroimaging has reduced the reliance on the infrequently obtained autopsy for making brain–behavior correlations.

Functional neuroimaging, broadly defined as techniques that measure brain activity, expanded our ability to study the neural basis of cognitive processes. One such method, fMRI is as an extremely powerful technique that affords excellent spatial and temporal resolution. Measuring regional brain activity in healthy subjects while they perform cognitive tasks links localized brain activity with specific behaviors. For example, functional neuroimaging studies have demonstrated that the left IFG is consistently activated during the performance of speech production tasks in healthy individuals [2]. Such findings from functional neuroimaging are complementary to findings derived from observations of patients with focal brain damage. This chapter focuses on the principles underlying fMRI as a cognitive neuroscience tool for exploring brain–behavior relationships.

2   Inference in Functional Neuroimaging Studies of Cognitive Processes

Insight regarding the link between brain and behavior can be gained through a variety of approaches. It is unlikely that any single neuroscience method is sufficient to fully investigate any particular question regarding the mechanisms underlying cognitive function. From a methodological point of view, each method will offer different temporal and spatial resolution. From a conceptual point of view, each method will provide data that will support different types of inferences that can be drawn from it. Thus, data obtained addressing a single question but derived from multiple methods can provide more comprehensive and inferentially sound conclusions.

Functional neuroimaging studies support inferences about the association of a particular brain system with a cognitive process. However, it is difficult to prove in such a study that the observed activity is necessary for an isolated cognitive process because perfect control over a subject's cognitive processes during a functional neuroimaging experiment is never possible. Even if the task performed by a subject is well designed, it is difficult to demonstrate conclusively that he or she is differentially engaging a single, identified cognitive process. The subject may engage in unwanted cognitive processes that either have no overt, measurable effects or are perfectly confounded with the process of interest. Consequently, the neural activity measured by the functional neuroimaging technique may result from some confounding neural computation that is itself not necessary for executing the cognitive process seemingly under study. In other words, functional neuroimaging is an observational, correlative method [3]. It is important to note that the inferences that can be drawn from functional neuroimaging studies such as fMRI apply to all methods of physiological measurement (e.g., electroencephalography, EEG, or magnetoencephalography, MEG).

The inference of necessity cannot be made without showing that a focal brain lesion disrupts the cognitive process in question. However, unlike precise surgical or neurotoxic lesions in animal models, lesions in patients are often extensive, damaging local neurons and "fibers of passage." For example, damage to prominent white matter tracts can cause cognitive deficits similar to those produced by cortical lesions, such as the amnesia resulting from lesions of the fornix, the main white matter pathway projecting from the hippocampus [4]. In addition, connections from region "A" may support the continued metabolic function of region "B," but region A may not be computationally involved in certain processes undertaken by region B. Thus, damage to region A could impair the function of region B via two possible mechanisms: (1) diaschisis [5, 6] and (2) retrograde trans-synaptic degeneration. Consequently, studies of patients with focal lesions cannot conclusively demonstrate that the neurons within a specific region are themselves critical to the computational support of an impaired cognitive process.

Empirical studies using lesion and electrophysiological methods demonstrate these issues regarding the types of inferences that can be logically drawn from them. For example, in monkeys, single-unit recording reveals neurons in the lateral prefrontal cortex (PFC) that increase their firing during the delay between the presentation of information to be remembered and a few seconds later when that information must be recalled [7, 8]. These studies are taken as evidence that persistent neural activity in the PFC is involved in temporary storage of information, a cognitive process known as working memory. The necessity of PFC for working memory was demonstrated in other monkey studies showing that PFC lesions impair performance on working memory tasks, but not on tasks that do not require temporarily holding information in memory [9]. Persistent neural activity during working memory tasks is also found in the hippocampus [10, 11]. Hippocampal lesions, however, do not impair performance on most working memory tasks [12], which suggests that the hippocampus is *involved* in maintaining information over short periods of time, but is not *necessary* for this cognitive operation. Observations in humans support this notion. For example, the well-studied patient H.M., with complete bilateral hippocampal damage and the severe inability to learn new information, could nevertheless perform normally on working memory tasks such as digit span [13]. The hippocampus is implicated in long-term memory especially when relations between multiple items and multiple features of a complex, novel item must be retained. Thus, the hippocampus may only be engaged during working memory tasks that require someone to subsequently remember novel information [14].

When the results from lesion and functional neuroimaging studies are combined, a stronger level of inference emerges [15]. As in the examples of Broca's aphasia or working memory, a lesion of a specific brain region causes impairment of a given cognitive process and when engaged by an intact individual, that cognitive process evokes neural activity in the same brain region. Given these findings, the inference that this brain region is computationally necessary for the cognitive process is stronger than the data derived from each study performed in isolation. Thus, lesion and functional neuroimaging studies are complementary, each providing inferential support that the other lacks.

Other types of inferential failure can occur in the interpretation of functional neuroimaging studies when other common assumptions do not hold true. First, it is assumed that if a cognitive process activates a particular brain region (evoked by a particular task), the neural activity in that brain region must depend on engaging that particular cognitive process. For example, a brain region showing greater activation during the presentation of faces than to other types of stimuli, such as photographs of cars or buildings, is considered to engage face perception processes. However, this region may also support other higher-level cognitive processes such as memory processes, in addition to lower level perceptual processes [16]. See ref. [17] for a further discussion of this issue.

The opposite type of inference is made when it is assumed that if a particular brain region is activated during the performance of a cognitive task, the subject must have engaged the cognitive process supported by that region during the task (referred to as a "reverse inference"). For example, when activation of the frontal lobes was observed during a mental rotation task, it was proposed that subjects engaged working memory processes to recall the identity of the rotated target [18]. (They derived this assumption from other imaging studies showing activation of the frontal lobes during working memory tasks.) However, in this example, because some other cognitive process supported by the frontal lobes could have activated this region [19], one cannot be sure that working memory was engaged leading to the activation of the frontal lobes. Unfortunately, this potentially faulty logic is a fairly common practice in fMRI studies. See ref. [20] for a further discussion of this issue.

In summary, interpretation of the results of functional neuroimaging studies attempting to link brain and behavior rests on numerous assumptions. Familiarity with the types of inferences that can and cannot be drawn from these studies is helpful for assessing the validity of the findings reported by such studies.

3   Functional MRI as a Cognitive Neuroscience Tool

Functional MRI has become the predominant functional neuroimaging method for studying the neural basis of cognitive processes in humans. Compared to its predecessor, positron emission tomography (PET) scanning, fMRI offers many advantages. For example, MRI scanners are much more widely available, and imaging costs are less expensive since MRI does not require a cyclotron to produce radioisotopes. MRI is also a noninvasive procedure since there is no requirement for injection of a radioisotope into the bloodstream. Also, given the half-life of available radioisotopes, PET scanning is unable to provide comparable temporal resolution to that of fMRI which can provide images of behavioral events occurring on the order of seconds rather than the summation of many behavioral events over tens of seconds.

In selected circumstances, however, PET scanning can provide an advantage over fMRI for studying certain questions concerning the neural basis of cognition. For example, a particular advantage of PET scanning in the study of cognition that can nicely complement fMRI studies is its ability to assess neurochemical (neurotransmitter and neuromodulator) systems. Radioactively labeled ligands may be used to directly measure density and distribution of particular receptors and even receptor subtypes, distribution of presynaptic terminals or enzymes involved in the production or breakdown of particular neurochemicals [21]. For example, one study measured dopamine synthesis capacity in the striatum with PET and used fMRI to measure brain activity during a working memory task. It was found that activity in frontal cortex during the working memory task was related to caudate dopamine levels as well as task accuracy. Thus, combining PET and fMRI data in this unique way allowed the investigators to test a question regarding the neurochemical basis of cognition [22].

The MRI scanner, compared to a behavioral testing room, is less than ideal for performing most cognitive neuroscience experiments. Experiments are performed in the awkward position of lying on one's back, often requiring subjects to visualize the presentation of stimuli through a mirror, in an acoustically noisy environment. Moreover, most individuals develop some degree of claustrophobia due to the small bore of the MRI scanner and find it difficult to remain completely motionless for a long duration of time that is required for most experiments (e.g., usually 60–90 min). These constraints of the MRI scanner make it especially difficult to scan children or certain patient populations (e.g., Parkinson's disease patients), which has resulted in many fewer fMRI studies involving children than adults and neurological patients in general. However, mock scanners have been built in many imaging centers, with motion devices that acclimate children to the scanner environment before they participate in an fMRI study. This approach has led to an increasing number of fMRI studies of children being reported in the literature that are providing tremendous insight regarding the mechanisms underlying the developing brain (*for review, see* [23]).

All sensory systems have been investigated with fMRI including the visual, auditory, somatosensory, olfactory, and gustatory systems. Each system requires different technologies for successful presentation of relevant stimuli within an MRI environment. At the time of this writing, there are now many off-the-shelf commercial products that exist that are MRI-compatible. Acquiring ancillary electrophysiological data such as electromyographic recordings to measure muscle contraction or electrodermal responses to measure autonomic activity enhances many cognitive neuroscience experiments. Devices have been developed that are MR compatible for these types of measurements as well as other physiological measures such as heart rate, electrocardiography, oxygen saturation, and respiratory rate. The recording of eye movements is commonplace in MRI scanners predominantly with the use of infrared video cameras equipped with long range optics. Video images of the pupil–corneal reflection can be sampled at 500–1000 Hz allowing for the accurate ($<0.5°$) localization of gaze within 50 horizontal and 40 vertical degrees of visual angle.

EEG recordings have also been successfully performed during MRI scanning [24, 25]. Both measures of event-related potentials (ERPs) and spectral EEG power in specific frequency bands and have been successfully recorded and related to variations in underlying BOLD activity and behavior [26–29]. However, the recording of low amplitude EEG events, such as ERPs and transient changes in spectral EEG power, can be more difficult in a magnetic field due to large artifacts induced by gradient switching and head movement and voltage changes from cardiac pulsation. The optimization of data acquisition methods and post-processing algorithms to remove artifacts have allowed for reliable measurements of ERPs and transient EEG events during fMRI scanning [30–33]. In summary, most challenges facing cognitive experiments and the study of spontaneous activity within the MRI environment have been overcome, creating an environment that is comparable to standard psychophysical testing labs outside of a scanner. Recent work has focused on minimizing exacerbated EEG artifacts present during high-field MRI scanning [34]. Although individual laboratories have achieved most of these advancements, MRI scanners originally designed for clinical use by manufacturers are now being designed with consideration of many of these research-related issues.

Another promising technique is the delivery of transcranial magnetic stimulation (TMS) during MRI scanning [35, 36]. TMS induces depolarization of neurons under the coil and, when combined with functional MRI, can be used to reveal patterns of remote connectivity, such as between the frontal eye field (FEF) and early visual cortex [36], the lateral prefrontal cortex and face- and house-selective regions in temporal cortex [35], and within and between large-scale brain networks [37]. There are many challenges in combining TMS and MRI such as the need for a large MRI head coil to accommodate the presence of the TMS coil, the difficulty of precise localization [38], and the increased subject discomfort. However, perhaps the largest challenge of delivering TMS in a manner that does not lead to artifacts in the MRI signal has been largely overcome by new commercially available TMS coils

3.1 Temporal Resolution

Two types of temporal resolution need to be considered for cognitive neuroscience experiments. First, what is the briefest neural event that can be detected as an fMRI signal? Second, how close together can two neural events occur and be resolved as separable fMRI signals?

The time scale on which neural changes occur is quite rapid. For example, neural activity in the lateral intraparietal area of monkeys increases within 100 ms of the visual presentation of a saccade target [39]. In contrast, the fMRI signal gradually increases to its peak magnitude within 4–6 s after an experimentally induced brief (<1 s) change in neural activity, and then decays back to baseline after several more seconds [40–42]. This slow time course of fMRI signal change in response to such a brief increase in neural activity is informally referred to as the blood oxygen level-dependent (BOLD) fMRI hemodynamic response or simply, the hemodynamic response (Fig. 1). Thus, neural dynamics and neurally evoked hemodynamics, as measured with fMRI, are on quite different time scales.

The sluggishness of the hemodynamic response limits the temporal resolution of the fMRI signal to hundreds of milliseconds to seconds as opposed to the millisecond temporal resolution of electrophysiological recordings of neural activity, such as from single-unit recording in monkeys and EEG or MEG in humans. However, it has been clearly demonstrated that brief changes in neural activity can be detected with reasonable statistical power using fMRI. For example, appreciable fMRI signal can be observed in sensorimotor cortex in association with single finger movements [43] and in visual cortex during very briefly presented (34 ms) visual stimuli [44]. In contrast, the temporal resolution of fMRI

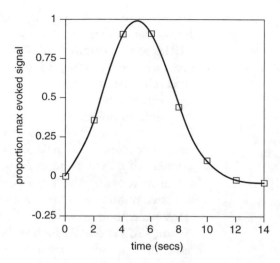

Fig. 1 A typical hemodynamic response (i.e., fMRI signal change in response to a brief increase of neural activity) from the primary sensorimotor cortex. The fMRI signal peaked approximately 5 s after the onset of the motor response (at time zero)

limits the detection of sequential changes in neural activity that occur rapidly with respect to the hemodynamic response. That is, the ability to resolve the changes in the fMRI signal associated with two neural events often requires the separation of those events by a relatively long period of time compared with the width of the hemodynamic response. This is because two neural events closely spaced in time will produce a hemodynamic response that reflects the accumulation from both neural events, making it difficult to estimate the contribution of each individual neural event. In general, evoked fMRI responses to discrete neural events separated by at least 4 s appear to be within the range of resolution [45]. However, provided that the stimuli are presented randomly, significant differential functional responses between two events (e.g., flashing visual stimuli) spaced as closely as 500 ms apart can be detected [46–48]. The effect of fixed and randomized intertrial intervals on the BOLD signal is illustrated in Fig. 2.

In some tasks, the order of individual trial events cannot be randomized. For example, in certain types of working memory tasks, the presentation of the information to be remembered during the delay period, and the period when the subject must recall the information, are individual trial events whose order cannot be randomized. In these types of tasks, short time scales (<4 s) cannot be temporally resolved. These temporal resolution issues in fMRI have been extensively considered regarding their impact on experimental design [49, 50].

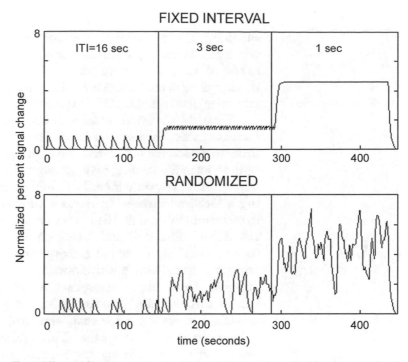

Fig. 2 Effect of fixed vs. randomized intertrial intervals on the blood oxygen level-dependent (BOLD) fMRI signal [46]

3.2 Spatial Resolution

As approaches are sought that maximize both BOLD signal strength and in-plane resolution, fMRI studies in humans have recently been extended to higher magnetic field strengths (7.0 T and 9.4 T) [51–53]. Such studies have the power to potentially evaluate much finer cortical details, such as the representation of individual fingertips within primary somatosensory cortex [51]. However, as the field strength increases, factors that are less consequential at 3.0 T—including magnetic field inhomogeneities [52] and the contribution of macrovascular structures to the typical gradient-echo signal [53]—become significantly more problematic, requiring further innovations in pulse sequence development. Single-echo gradient-echo sequences using echo times (TE) that exceed the repetition time (TR), for example, take advantage of reduced distortion relative to single-shot gradient-echo sequences, while also avoiding the prolonged acquisition times of typical single-echo sequences. During functional MRI of a simple finger tapping task at 9.4 T using such a sequence, researchers were able to obtain 0.4 × 0.4 mm in-plane resolution within presumptive primary motor cortex [53]. Similarly, spin-echo sequences, which have a reduced signal-to-noise ratio relative to gradient-echo sequences but greater spatial specificity, become feasible for use at 9.4 T. Within a finger-tapping paradigm, a study taking this approach reduced the influence of macrovascular contributions to the BOLD signal relative to a gradient-echo sequence, while obtaining 1 mm isotropic resolution. As such techniques are validated and extended, they may someday allow for imaging of thousands of neurons per voxel, as opposed to the hundreds of thousands of neurons per voxel currently more typical for a human cognitive neuroscience fMRI experiment.

Virtually all fMRI studies model the large BOLD signal increase, which is due to a local low-deoxyhemoglobin state, in order to detect changes correlating with a behavioral task. However, optical imaging studies have demonstrated that preceding this large positive response there is an initial negative response reflecting a localized increase in oxygen consumption causing a high-deoxyhemoglobin state [54]. This early hemodynamic response is called the "initial dip" and is thought to be more tightly coupled to the actual site of neural activity evoking the BOLD signal as compared to the later positive portion of the BOLD response. For example, Kim et al., scanning cats in a high field scanner, demonstrated that the early negative BOLD response (e.g., initial dip) produced activation maps that were consistent with orientation columns within visual cortex. This finding is quite remarkable given that the average spacing between two adjacent orientation columns in cortex is approximately 1 mm. In contrast, the activation maps produced by the delayed positive BOLD response appeared more diffuse and cortical columnar organization could not be identified [55]. Thus, empirical evidence suggests that deriving activation maps by correlating behavioral responses with the initial dip may markedly improve spatial resolution.

Another unique method for improving spatial resolution has been called functional magnetic resonance-adaptation (fMR-A), which could provide a means for identifying and assessing the functional attributes of sharply defined neuronal populations within a given region of the brain [56]. Even if the spatial resolution of fMRI evolves to the point of being able to resolve a population of a few hundred neurons within a voxel, it is still likely that this small population will contain neurons with very different functional properties that will be averaged together. The adaptation method is based on several basic principles. First, repeated presentation of the same type of stimuli (i.e., a picture of the one object) causes neurons to adapt to those stimuli (i.e., neuronal firing is reduced). Second, if these neurons are then exposed to a different type of stimulus (i.e., a picture of another object) or a change in some property of the stimulus (i.e., the same object in a different orientation), then recovery from adaptation can be assessed (i.e., whether or not the BOLD signal returns to its original state). If the signal remains adapted it implies that the neurons are invariant to the attribute that was changed or if the signal recovers from the adapted state it would imply that the neurons are sensitive to that attribute. For example, Grill-Spector et al. demonstrated that an area of lateral occipital cortex thought to be important for object recognition was less sensitive to changes in object size and position as compared to changes in illumination and viewpoint [57]. Thus, with this method it is possible to investigate the functional properties of neuronal populations with a level of spatial resolution that is beyond that obtained from conventional fMRI data analysis methods.

Considering all the neuroscientific methods available today for studying human brain–behavior relationships, fMRI provides an excellent balance of temporal and spatial resolution. Improvements on both fronts will clearly add to type of basic and clinical neuroscientific questions that can be addressed with this method.

4   Issues in Functional MRI Experimental Design

Numerous options exist for designing experiments using fMRI. The prototypical fMRI experimental design consists of two behavioral tasks presented in blocks of trials alternating over the course of a scanning session, and the fMRI signal between the two tasks is compared. This is known as a blocked design. For example, a given block might present a series of faces to be viewed passively, which evokes a particular cognitive process, such as face perception. The "experimental" block alternates with a "control" block that is designed to evoke all of the cognitive processes present in the experimental block except for the cognitive process of interest. In this experiment the control block may comprise a series of objects. In this way, the stimuli used in experimental and control tasks have similar visual attributes, but differ in the attribute of interest (i.e., faces). The inferential framework of "cognitive subtraction" [58] attributes differences in neural activity between the two tasks to the specific cognitive process (i.e., face perception). Cognitive subtraction was originally conceived by Donders in the late 1800s for studying the chronometric substrates of cognitive processes [59] and was a major innovation in imaging [58, 60].

The assumptions required for cognitive subtraction may not always hold and could produce erroneous interpretation of functional neuroimaging data [45]. Cognitive subtraction relies on two assumptions: "pure insertion" and linearity. Pure insertion implies that a cognitive process can be added to a preexisting set of cognitive processes without affecting them. This assumption is difficult to prove because one needs an independent measure of the preexisting processes in the absence and presence of the new process [59]. If pure insertion fails as an assumption, a difference in the neuroimaging signal between the two tasks might be observed, not because a specific cognitive process was engaged in one task and not the other, but because the added cognitive process and the preexisting cognitive processes interact.

An example of this point is illustrated in working memory studies using delayed-response tasks [61]. These tasks [62] typically present information that the subject must remember (engaging an *encoding* process), followed by a delay period during which the subject must hold the information in memory over a short period of time (engaging a *memory* process), followed by a probe that requires the subject to make a decision based on the stored information (engaging a *retrieval* process). The brain regions engaged by evoking the *memory* process theoretically are revealed by subtracting the BOLD signal measured by fMRI during a block of trials that the subject performs that do not have a delay period (only engaging the *encoding* and *retrieval* processes) from a block of trials with a delay period (engaging the *encoding*, *memory*, and *retrieval* processes). In this example, if the addition or "insertion" of a delay period between the *encoding* and *retrieval* processes affects these other behavioral processes in the task, the result is failure to meet the assumptions of cognitive subtraction. That is, these "nonmemory" processes may differ in delay trials and no-delay trials, resulting in a failure to cancel each other out in the two types of trials that are being compared.

Empirical evidence of such failure exists [63]. For example, Figure 3 demonstrates BOLD signal derived from the PFC from a subject performing a delayed-response task similar to the tasks

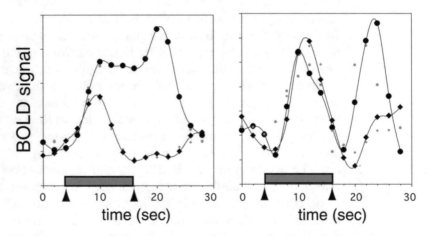

Fig. 3 Data derived from the performance of a normal subject on a spatial delayed-response task [64]. This task comprised both delay trials (*circles*) as well as trials without a delay period (no-delay trials; *diamonds*). (**a**) Trial averaged fMRI signal from prefrontal cortex that displayed delay-correlated activity. The gray bar along the x-axis denotes the 12 s delay period during delay trials. The delay trials display a level of fMRI signal greater than baseline throughout the period of time corresponding to the retention delay (taking into account the delay and dispersion of the fMRI signal). The peaks seen in the signal correspond to the encoding and retrieval periods. (**b**) Trial averaged fMRI signal from a region in prefrontal cortex that did not display the characteristics of delay-correlated activity. This region displays a significant functional change associated with the no-delay trials, and a significant functional change associated with the encoding and retrieval periods of the delay trials, but not one associated with the retention delay of delay trials

described above. The left side of the figure illustrates BOLD signal consistent with delay period activity whereas the right side of the figure illustrates BOLD signal from another region of PFC that did not display sustained activity during the delay yet showed greater activity in the delay trials as compared to the trials without a delay. In any blocked functional neuroimaging study that compares delay vs. no-delay trials with subtraction, such a region would be detected and likely assumed to be a "memory" region. Thus, this result provides empirical grounds for adopting a healthy doubt regarding the inferences drawn from imaging studies that rely exclusively on cognitive subtraction.

In functional neuroimaging, the transform between the neural signal and the hemodynamic response (measured by fMRI) must also be linear for the cognitive subtractive method to yield valid results. In other words, it is assumed that the BOLD signal being measured is approximately proportional to the local neural activity that evokes it. Surprisingly, although thousands of empirical studies using fMRI to study brain–behavior relationships have been published, only a handful exist that have explored the neurophysiological basis of the BOLD signal (for reviews *see* refs. [64, 65]). In several studies it has been demonstrated that linearity does not strictly hold for the BOLD fMRI system but a linear transform model is reasonably consistent with the data. For example, Boynton et al. tested whether the BOLD signal in response to long duration stimuli can be predicted by summing the responses to shorter duration stimuli [42]. Using pulses of flickering checkerboard patterns and measuring within human primary visual cortex, these investigators found that the BOLD signal response to various durations of stimulus presentation (6, 12, or 24 s) could be predicted from the responses they obtained from shorter stimulus presentations. For example, the BOLD signal response to a 6 s pulse could be predicted from the summation of the BOLD signal response to the 3 s pulse with a copy of the same response delayed by 3 s. However, temporal summation did not always hold, and there are clearly nonlinear effects in the transform of neural activity to a hemodynamic response that must be considered [66–69]. If these nonlinearities lead to saturation of the BOLD effect at a certain stimulus intensity, erroneous interpretation of particular results of fMRI experiments may occur.

Another class of experimental designs, called event-related fMRI, attempts to detect changes associated with individual trials, as opposed to the larger unit of time comprising a block of trials [70, 71]. Each individual trial may be composed of one behavioral "event," such as the presentation of a single stimulus (e.g., a face or object to be perceived) or several behavioral events such as in the delayed-response task described above (e.g., an item to be remembered, a delay period, and a motor response in a delayed-response task). For example, with an event-related design, activity within the PFC has consistently been shown to correlate with the delay period, supporting the role of the PFC in temporarily maintaining information [63]. This finding is consistent with single-neuron recording studies in the PFC of monkeys [7]. An event-related design offers numerous advantages. For example, it allows for stimulus or trial randomization avoiding the behavioral confounds of blocked trials. It also permits the separate analysis of functional responses that are identified only in retrospect (i.e., trials on which the subject made a correct or incorrect response). Of course, an experiment does not have to be limited to either a block or event-related designs—a mixed-type (both event-related and blocked) design where particular trial types are randomized within a block is perfectly feasible. In this type of design, both item-related processes (e.g., transient responses to stimuli) as well as state-related processes (processes sustained throughout a block of trials or a task) are perfectly feasible [72, 73].

Overall, much flexibility exists in the type of experimental design that can be utilized in fMRI experiments and continued innovation in this area will greatly expand the types of neuroscientific questions that can be addressed.

5   Issues in Interpretation of fMRI Data

5.1   Statistics

Many statistical techniques are used for analyzing fMRI data, but no single method has emerged as the ideal or "gold standard." The analysis of any fMRI experiment designed to contradict the null hypothesis (i.e., there is no difference between experimental conditions) requires inferential statistics. If the difference between two experimental conditions is too large to be reasonably due to chance, then the null hypothesis is rejected in favor of the alternative hypothesis, which typically is the experimenter's hypothesis (e.g., the fusiform gyrus is activated to a greater extent by viewing faces than objects). Unfortunately, since errors can occur in any statistical test, experimenters will never know when an error is committed and can only try to minimize them [74]. Knowledge of several basic statistical issues provides a solid foundation for the correct interpretation of the data derived from fMRI studies.

Two types of statistical errors can occur. A type I error is committed when the null hypothesis is falsely rejected, that is, a difference between experimental conditions is found but a difference does not truly exist. This type of error is also called a false-positive error. In an fMRI study, a false-positive error would be finding a brain region activated during a cognitive task, when actually it is not. A type II error is committed when the null hypothesis is accepted when it is false, that is, no difference between experimental conditions exists when a difference does exist. This type of error is also called a false-negative error. A false-negative error in an fMRI study would be failing to find a brain region activated during the performance of a cognitive task when actually it is. The concept of type II error is closely related to the idea of statistical power. If the false-negative rate for a given study design is 20%, for instance, then the "power" of that design to detect an activation is 100–20% or 80%.

In cognitive neuroscience studies, much emphasis has been placed on avoiding type I errors. The negative effects of incorrectly identifying a brain region as task-active include the expenditures of time, money, and effort spent in replicating and/or expanding upon a false positive result. Type II error, on the other hand, is seen as less damning; failure to detect brain activity in a research study has fewer implications for future research, provided that one is careful to interpret so-called null results correctly. For example, cognitive neuroscience studies (due to factors such as the expense and the difficulty of finding research participants, for example) tend to employ a small number of subjects—15 would typical—and therefore frequently lack power to detect significant brain activations. One must consequently be careful to avoid interpreting a lack of activation in one part of the brain as true inactivity during the task.

In a clinical research study, on the contrary, the emphasis may be different, especially when fMRI studies are being used diagnostically in individual patients. A type II error—failing to detect active brain regions related to movement or language in the vicinity of a brain tumor, for example—may lead to a larger surgical resection that leaves the patient with avoidable residual deficits. On the contrary, a type I error—for example, identifying motor activity adjacent to a tumor when in fact none exists—may erroneously lead to a more cautious surgical resection, or to use of a different treatment modality. Which error is deemed more tolerable may depend on the clinical situation.

In fMRI experiments, like all experiments, a tolerable probability for type I error, typically less than 5%, is chosen for adequate control of specificity, that is, control of false-positive rates. Two features of fMRI data can cause unacceptable false-positive rates, even with traditional parametric statistical tests. First, there is the problem of multiple comparisons. For the typical resolution of images acquired during fMRI scans, the full extent of the human brain could comprise as many as 15,000 voxels. Thus, with any given statistical comparison of two experimental conditions, there are actually 15,000 statistical comparisons being performed. With such a large number of statistical tests, the probability of finding a false-positive activation, that is, committing a type I error, somewhere in the brain increases. Several methods exist to deal with this problem. One method, a Bonferroni correction, assumes that each statistical test is independent and calculates the probability of type I error by dividing the chosen probability ($p = 0.05$) by the number of statistical tests performed. Another method is based on Gaussian field theory [75], and calculates the probability of type I error when imaging data are spatially smoothed. Many other methods for determining thresholds of statistical maps are proposed and utilized [76, 77] but unfortunately, no single method has been universally accepted. Nevertheless, all fMRI studies must apply some type of correction for multiple comparisons to control the false-positive rate.

The second feature that might increase the false-positive rate is the "noise" in fMRI data. Data from BOLD fMRI are temporally autocorrelated, with more noise at some frequencies than at others. The shape of this noise distribution is characterized by a 1/frequency function with increasing noise at lower frequencies [78]. Traditional parametric and nonparametric statistical tests assume that the noise is not temporally autocorrelated, that is, each observation is independent. Therefore, any statistical test used in fMRI studies must account for the noise structure of fMRI data. If not, the false-positive rates will inflate [78, 79].

Type II error is rarely considered in functional neuroimaging studies. When a brain map from an fMRI experiment is presented, several areas of activation are typically attributed to some experimental manipulation. The focus of most fMRI studies is on brain activation whereas it is often implicitly assumed that all of the other areas (typically most of the brain) were not activated during the experiment. Power as a statistical concept refers to the probability of correctly rejecting the null hypothesis [74]. As the power of an fMRI study to detect changes in brain activity increases, the false-negative rate decreases. Unfortunately, power calculations for particular fMRI experiments are rarely performed, although methods exist to address this issue [80–82]. Reports that specific brain areas were not active during an experimental manipulation should provide an estimate of the power required for detection of a change in the region. All experiments should be designed to maximize power. Relatively simple strategies can increase power in an fMRI experiment in certain circumstances, such as increasing the amount of imaging data collected or increasing the number of subjects studied. It is also important to note that task designs can affect sensitivity [83]. For example, since BOLD fMRI data are temporally autocorrelated, experiments with fundamental frequencies in the lower range (e.g., a boxcar design with 60 s epochs) will have reduced sensitivity, due to the presence of greater noise at these lower frequencies. Finally, in a study that simultaneously measured neural signal via intracortical recording and BOLD signal in a monkey, it was observed that the SNR of the neural signal was on average at least one order of magnitude higher than that of the BOLD signal. The investigators of this study concluded that "the statistical and thresholding methods applied to the hemodynamic responses probably underestimate a great deal of actual neural activity related to a stimulus or task" [84]. Thus, the magnitude of type II error in BOLD fMRI may currently be underestimated and warrants further consideration in the interpretation of almost any cognitive neuroscience experiment.

5.2 Altered Hemodynamic Response

When comparing changes in fMRI BOLD signal levels within the brain of an individual subject across different cognitive tasks and making conclusions regarding changes in neural activity and the pattern of activity, numerous assumptions are made regarding the steps comprising neurovascular coupling (stimulus → neural activity → hemodynamic response → BOLD signal) and the regional variability of the metabolic and vascular parameters influencing the BOLD signal. It should be obvious that fMRI studies of cognition of individuals with local vascular compromise or diffuse vascular disease (e.g., patients with strokes or normal elderly) are potentially problematic. For example, many fMRI studies have sought to identify age-related changes in the neural substrates of cognitive processes. Those studies that directly compare changes in fMRI BOLD signal intensity across age groups rely upon the assumption of age-equivalent coupling of neural activity to BOLD signal. However, there is empirical evidence that suggests that this general assumption may not hold true. Extensive research on the aging neurovascular system has revealed that it undergoes significant changes in multiple domains in a continuum throughout the human lifespan, probably as early as the fourth decade (for review *see* ref. [85]). These changes affect the vascular ultrastructure [86], the resting cerebral blood flow [87, 88], the vascular responsiveness of the vessels [89], and the cerebral metabolic rate of oxygen consumption [90, 91]. Aging is also frequently associated with comorbidities such as diabetes, hypertension, and hyperlipidemia, all of which may affect the fMRI BOLD signal by affecting cerebral blood flow and neurovascular coupling [92]. Any one of these age-related differences in the vascular system could conceivably produce age-related differences in BOLD fMRI signal responsiveness, greatly affecting the interpretation of results from such studies.

Our laboratory compared the hemodynamic response function (HRF) characteristics in the sensorimotor cortex of young and older subjects in response to a simple motor reaction-time task [70]. The provisional assumption was made that there was identical neural activity between the two populations based on physiological findings of equivalent movement-related electrical potentials in subjects under similar conditions [93]. Thus, we presumed that any changes that we observed in BOLD fMRI signal between young and older individuals in motor cortex would be due to vascular, and not neural activity changes in normal aging. Several important similarities and differences were observed between age groups. Although there was no significant difference in the shape of the hemodynamic response curve or peak amplitude of the signal, we found a significantly decreased SNR in the fMRI BOLD signal in older individuals as compared to young individuals. This was attributed to a greater level of noise in the older individuals. We also observed a decrease in the spatial extent of the BOLD signal in older individuals compared to younger individuals in sensorimotor cortex (i.e., the median number of suprathreshold voxels). Similar results have been replicated by other laboratories (**e.g.**, [94, 95]). These findings suggest that there is some property of the coupling between neural activity and fMRI BOLD signal that changes with age.

The notion that vascular differences among individuals may affect BOLD signal is especially a concern when considering studies of patient populations with known vascular changes such as stroke. For example, in a fMRI study of patients with an isolated subcortical lacunar stroke compared to a group of age-matched controls, a decrease in the rate of rise and the maximal fMRI BOLD HRF to a finger- or hand-tapping task in both the sensorimotor cortex of the hemisphere affected by the stroke and the unaffected hemisphere was found [96]. These investigators proposed that given the widespread changes of these fMRI BOLD signal differences, the change was unlikely a direct consequence of the subcortical lacunar stroke, but rather a manifestation of preexisting diffuse vascular pathology.

In summary, comparing BOLD signal in two different groups of individuals that may differ in their vascular system should be done with caution [97]. For example, in one scenario, a comparison of activation of young and elderly individuals during a cognitive task may show less activation by elderly (as compared to young subjects) in some brain regions, but greater activation in other regions. In this scenario, it is unlikely that regional variations in the hemodynamic coupling of neural activity to fMRI signal would account for such age-related differences in patterns of activation. In another scenario, a comparison of young and elderly subjects may show less activation by elderly (as compared to young subjects) in some brain regions, but no evidence of greater activation in any other region. In this case, it is possible that the observed age-related differences are not due to differences in intensity of neural activity, but rather to other nonneuronal contributions to the imaging signal, i.e., neurovascular coupling.

In summary, fMRI BOLD contrast methods yield signal changes that result from a complex mix of vascular effects and provide only relative, rather than absolute, measures. One approach to accounting for the influence of purely vascular effects is to directly measure regional and individual variability in vascular reactivity via a breathholding task, which increases carbon dioxide concentration in the blood and leads to vascular dilatation [98]. The task-related BOLD signal in each subject can then be corrected for particular region- and subject-specific vascular effects. One alternative functional neuroimaging approach, based on more direct measurements of cerebral blood flow to active brain areas, is known as arterial spin labeling (ASL). In the various ASL techniques, the MRI scanner selectively magnetizes blood flow with a particular range of locations and/or velocities, then waits for the appearance of the magnetic "tag" in downstream vessels. It thus becomes possible to

obtain absolute measures of cerebral perfusion [99], thereby opening up the possibility of more quantitatively distinguishing between
the differential influence of a disease on blood flow, and its effect on
brain activity [100]. Additionally, relative to BOLD contrast these
absolute measurements appear to be more stable over long experiments because of better signal-to-noise at very low frequencies
[101], to show less between-subject and between-session variability
[102], and to produce decreased susceptibility artifact in areas such
as medial temporal lobe [103]. A significant limitation is temporal
resolution: one must both wait for the generation of sufficient magnetic label, and also acquire two scans, a reference scan and a postlabeling scan, to produce a single data point. However, a recently a
new MRI acquisition method has been developed that allows for
more slices for measuring perfusion in a larger region of the brain
than currently possible with previous methods [104]. Another
potential disadvantage somewhat related to the temporal issues is
the lower SNR of ASL relative to BOLD, but this decline may be
compensated in group studies by the observation that ASL methods appear to be less variable across subjects [99].

6 Types of Hypotheses Tested Using fMRI

Functional neuroimaging experiments test hypotheses regarding
the anatomical specificity for cognitive processes (functional specialization) or direct or indirect interactions among brain regions
(functional integration). The experimental design and statistical
analyses chosen will determine the types of questions that can be
addressed. Ultimately, the most powerful approach for the testing
of theories on brain–behavior relationships is the analysis of converging data from multiple methods.

6.1 Functional Specialization

The major focus of fMRI studies of cognition is testing theories on
functional specialization. The concept of functional specialization
is based on the premise that functional modules exist within the
brain, that is, areas of the cerebral cortex are specialized for a specific cognitive process. For example, facial recognition is a critical
primary function likely served by a functional module.
Prosopagnosia is the selective inability to recognize faces. Patients
with prosopagnosia, however, can recognize familiar faces, such as
those of relatives, by other means, such as the voice, dress, or
shape. Other types of visual recognition, such as identifying common objects, are normal. Prosopagnosia arises from lesions of the
inferomedial temporo-occipital lobe, which are usually due to a
stroke within the posterior cerebral artery circulation. No lesion
studies have precisely localized the area crucial for facial perception. However, they provide strong evidence that a brain area is
specialized for processing faces. Functional imaging studies have

provided anatomical specificity for such a module. For example, Kanwisher et al. [105] used fMRI to test a group of healthy individuals and found that the fusiform gyrus was significantly more active when the subjects viewed faces than when they viewed assorted common objects. The specificity of a "fusiform face area" was further demonstrated by the finding that this area also responded significantly more strongly to passive viewing of faces than to scrambled two-tone faces, front-view photographs of houses, and photographs of human hands. These elegant experiments allowed the investigators to reject alternative functions of the face area, such as visual attention, subordinate-level classification, or general processing of any animate or human forms, demonstrating that this region selectively perceives faces.

Of course, the existence of brain areas specialized for certain functions does not exclude the strong possibility that those areas show either finer, voxel-level structure or are part of larger networks. Recent neuroimaging work has focused on pattern classification methods—that is, on techniques to explore whether a distributed spatial pattern of brain activity, both within a single region and across larger brain areas, corresponds to object (or more abstract) representations. This area of research draws on results from physics, computer science, and statistics, among other disciplines, to search for more broadly distributed structure in neuroimaging data. As such, the techniques themselves differ. For example, to distinguish between voxel activity patterns across experimental conditions, various reports have used correlations between the set of activations in visual responses to faces and other objects [106]; neural network classifiers to identify particular patterns correlated with particular memories [107]; and variants of a matrix algebra transformation known as singular value decomposition to look for distributed spatial correlates of memory storage and search [108]. By establishing sophisticated models of the relationships between brain activity and visual stimuli in visual cortex, representations of natural images may even be successfully decoded [109]. A large number of other techniques—too large to be reviewed here—are also continually being developed [110, 111]. As such research demonstrates that task-relevant brain activity can be detected even in the absence of classic univariate activity changes. However, it will remain important to control for potential confounds in brain activity data, with validation via comparison with behavioral responses, in order to ensure that these patterns are not epiphenomenal or a result of confounds such as reaction time [112]. At a higher tier of analysis, information decoding techniques are being used to examine mechanisms by which higher order cognition can modulate information representations. A step beyond simply detecting the existence of a particular representational code, one can now ask, for example, to what extent goal-direction (attention) might change the tuning of neural network codes to better represent information related to a goal [113, 114].

6.2 Functional Integration

Functional neuroimaging experiments can also test hypotheses about interactions between brain regions by focusing on covariances of activation levels between regions [115, 116]. These covariances reflect "functional connectivity," a concept that was originally developed in reference to temporal interactions among individual neurons [117].

In addition to providing information about the specialization of various brain regions, functional neuroimaging can also address the interactions between brain regions that underlie cognitive processing. Understanding the various techniques that permit these types of analysis comprises a very active area of current research [118]. However, most, if not all, of the techniques used to test for regional interactions are ultimately based on the covariance of activation levels in different brain regions across time—in other words, on the way in which activity levels in different areas of the brain rise or fall in relation to each other. Such statistical techniques are commonly known as "multivariate," both because they rely on interactions between two or more brain areas, and to distinguish them from the "univariate" methods applied in most tests of functional specialization.

The universe of multivariate techniques is further subdivided into two types, determined by whether the method in question is designed to assess connectivity in a model-free ("functional connectivity") or model-based ("effective connectivity") fashion. The former refers simply to methods that measure the temporal covariance in activity between brain areas without a priori notions about which brain areas are relevant or how they should interact. Examples of model-free techniques would include correlation and its frequency-based analogue, coherence, which can be applied irrespective of hypotheses about the neural events that produced them. On the contrary, model-based, or effective connectivity, approaches begin with hypotheses about the interactions between different brain regions, and attempt to support/refute them by evaluating the presence/absence of specific activity covariance patterns. Examples of these techniques would include structural equation modeling and dynamic causal modeling, both of which start by postulating the existence of influences (potentially complex, potentially time-varying) between specific brain regions. Both types of statistical techniques have value, of course; their use is determined by the problem at hand. Model-free approaches are more general, and more easily deployed in exploratory analyses. However, they are not as powerful as model-based methods that address specific hypotheses about how regions interact—but which fail if the model is misspecified. Model-free methods, for example, may be more useful when attempting to determine which networks of brain areas might be involved in a task, whereas model-based methods may be most appropriate when the nodes of the network are known, and specific notions about how they interact need to be tested.

In our own laboratory, we have developed and used functional connectivity techniques to understand how brain interactions change under different task conditions, and over time [119, 120]. For example, we have shown that functional connectivity changes as subjects learn a complex finger tapping task [121]. In the early phases of learning, the data show that subjects not only activate wide areas of primary sensorimotor cortex, premotor cortex, and the supplementary motor area, but also that the coherence between these areas is increased relative to later stages. Such changes were not observed when subjects performed an already learned motor skill; and more importantly, they were not found in the univariate responses, whose means were unchanged despite the changes in the subjects' facility at the task. Similarly, in a working memory task for faces [122], we have found an interesting dissociation between their univariate and multivariate analyses in the networks that support so-called delay period activity (see below). In our protocol, subjects encoded a cue face, maintained the image across a delay of several seconds, and then decided whether a subsequently presented probe face matched the initial one. Interestingly, we found that despite a general decrease in the univariate activity from the cue to the delay period, there was a robust increase in the correlation between activity in the right fusiform face area (a brain region known to be sensitive to face stimuli) and a diffuse set of brain regions including the frontal and parietal cortices as well as the basal ganglia.

In such known networks, effective connectivity techniques can be employed to more specifically evaluate the influence of the nodes of the network on each other. McIntosh et al., for example, were able to exploit their own functional neuroimaging research on working memory networks to formulate a hypothesis about the interactions of the PFC, cingulate cortex, and other brain regions during task performance [116]. Using structural equation modeling, the authors found shifting prefrontal and limbic interactions in a working memory task for faces as the retention delay increased (Fig. 4). The different interactions between brain regions at short and long delays were interpreted as a functional change. For example, strong corticolimbic interactions were found at short delays, but at longer delays, when the image of the face was more difficult to maintain, strong fronto-cingulate-occipital interactions were found. The investigators postulated that the former finding was due to maintaining an iconic facial representation, and the latter due to an expanded encoding strategy, resulting in more resilient memory. As in our own previous studies, information that was not seen in the univariate analysis was captured by an approach sensitive to regional interactions. In addition to structural equation modeling, other approaches have been applied to fMRI datasets to capture information regarding the relative timing of activation across brain regions such as Granger causality, information analysis, and coherence (*see* [119, 120, 123]).

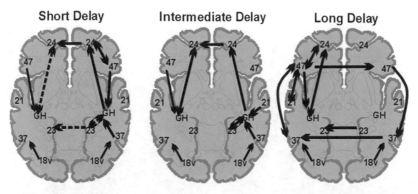

Short Delay   **Intermediate Delay**   **Long Delay**

Fig. 4 Network analysis of fMRI data using structural equation modeling during performance of a working memory task across three different delay periods [111]. Areas of correlated increases in activation (*solid lines*) and areas of correlated decreases in activation (*dotted lines*) are shown. Note the different pattern of interactions among brain regions at short and long delays

Mathematical tools based on graph theory have recently emerged as a method to quantify large-scale network properties of the brain as well as to identify the role of individual brain regions within these large-scale networks. These tools, developed for analyzing a wide variety of networks (e.g., social networks, the internet, protein associations), allow one to make quantitative measurements of brain network structure. Typically, these methods are used to analyze the spontaneous coherent fluctuations in BOLD signal measured by fMRI at rest, which consistently identifies stable intrinsic functional networks, that, in a short fMRI recording session, recapitulate a number of sub-networks normally engaged by a variety of different tasks (see Fig. 5).

6.3   Cognitive Theory   Experiments using fMRI can also test theories of the underlying mechanisms of cognition. For example, an fMRI study [124] attempted to answer the question, "To what extent does perception depend on attention?" One hypothesis is that unattended stimuli in the environment receive very little processing [125], but another hypothesis is that the processing load in a relevant task determines the extent to which irrelevant stimuli are processed [126]. These alternative hypotheses were tested by asking normal individuals to perform linguistic tasks of low or high load while ignoring irrelevant visual motion in the periphery of a display. Visual motion was used as the distracting stimulus, because it activates a distinct region of the brain (cortical area MT or V5, another functional module in the visual system). Activation of area MT would indicate that irrelevant visual motion was processed. Although task and irrelevant stimuli were unrelated, fMRI of motion-related activity in MT showed a reduction in motion processing during the high-processing load condition in the linguistic task. These findings supported the hypothesis that perception of irrelevant environmental

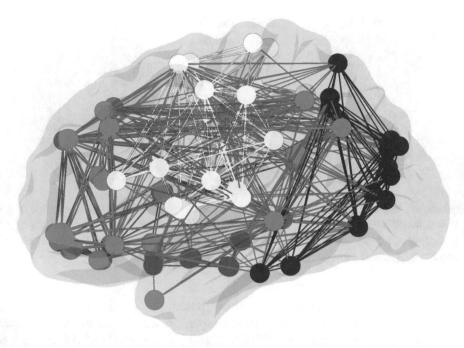

Fig. 5 A brain graph derived from resting state fMRI data collected from healthy young subjects illustrating identified modules, represented as different *shades of color*. There are four distinct modules identified in this graph

information depends on the information processing load that is currently relevant and being attended to. Thus, by the finding that perception depends on attention, this fMRI experiment provides insight regarding underlying cognitive mechanism.

7   Integration of Multiple Methods

The most powerful approach toward understanding brain–behavior relationships comes from analyzing converging data from multiple methods. There are several ways in which different methods can provide complementary data. For example, one method can provide superior spatial resolution (e.g., fMRI) whereas the other can provide superior temporal resolution (e.g., ERP). Also, the data from one method may allow for different conclusions to be drawn from it such as whether a particular brain region is necessary to implement a cognitive process (i.e., lesion methods) or whether it is only involved during its implementation (i.e., physiological methods). The following sections describe examples of such approaches.

7.1   Combined fMRI/ Lesion Studies

The combined use of functional neuroimaging and lesions studies can be illustrated with studies of the neural basis of semantic memory, the cognitive system that represents our knowledge of the world. Early studies of patients with focal lesions supported the notion that the temporal lobes mediate the retrieval of semantic knowledge [127]. For example, patients with temporal lobe lesions may show a disproportionate impairment in the knowledge of living things (e.g., animals) compared with nonliving things. Other patients have a disproportionate deficit in the knowledge of nonliving things [128]. These observations led to the notion that the semantic memory system is subdivided into different sensorimotor modalities, that is, living things, compared with nonliving things, are represented by their visual and other sensory attributes (e.g., a banana is yellow), while nonliving things are represented by their function (e.g., a hammer is a tool but comes in many different visual forms). The small number of patients with these deficits, and often large lesions, limits precise anatomical-behavioral relationships. However, functional neuroimaging studies in normal subjects can provide spatial resolution that the lesion method lacks [129].

These original observations regarding the neural basis of semantic memory conflicted with functional neuroimaging studies consistently showing activation of the left IFG during the retrieval of semantic knowledge. For example, an early cognitive activation PET study revealed IFG activation during a verb generation task compared with a simple word repetition task [60]. A subsequent fMRI study [130] offered a fundamentally different interpretation of the apparent conflict between lesion and functional neuroimaging studies of semantic knowledge: left IFG activity is associated with the need to select some relevant feature of semantic knowledge from competing alternatives, not retrieval of semantic knowledge per se. This interpretation was supported by an fMRI experiment in normal individuals in which selection, but not retrieval, demands were varied across three semantic tasks. In a verb generation task, in a high selection condition, subjects generated verbs to nouns with many appropriate associated responses without any clearly dominant response (e.g., "wheel"), but in a low selection condition nouns with few associated responses or with a clear dominant response (e.g., "scissors") were used. In this way, all tasks required semantic retrieval, and differed only in the amount of selection required. The fMRI signal within the left IFG increased as the selection demands increased (Fig. 6). When the degree of semantic processing varied independently of selection demands, there was no difference in left IFG activity, suggesting that selection, not retrieval, of semantic knowledge drives activity in the left IFG.

To determine if left IFG activity was correlated with but not necessary for selecting information from semantic memory, the same task used during the fMRI study was used to examine the ability of patients with focal frontal lesions to generate verbs [131]. Supporting the earlier claim regarding left IFG function derived from an fMRI study [130], the overlap of the lesions in patients with deficits on this task corresponded to the site of maximum

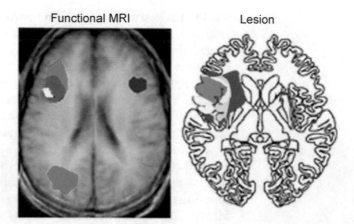

Functional MRI   Lesion

Fig. 6 Regions of overlap of fMRI activity in healthy human subjects (*left side of figure*) during the performance of three semantic memory tasks, with the convergence of activity within the left inferior frontal gyrus (*white region*) [125]. Regions of overlap of lesion location in patients with selection-related deficits on a verb generation task (*right side of figure*) with maximal overlap within the left inferior frontal gyrus [126]

fMRI activation in healthy young subjects during the verb generation task (Fig. 6). In this example, the approach of using converging evidence from lesion and fMRI studies differs in a subtle but important way from the study described earlier that isolated the face processing module. Patients with left IFG lesions do not present with an identifiable neurobehavioral syndrome reflecting the nature of the processing in this region. Guided by the fMRI results from healthy young subjects, the investigators studied patients with left IFG lesions to test a hypothesis regarding the necessity of this region in a specific cognitive process. Coupled with the well-established finding that lesions of the left temporal lobe impair semantic knowledge, these studies further our understanding of the neural network mediating semantic memory.

7.2 Combined fMRI/ Transcranial Magnetic Stimulation Studies

Transcranial magnetic stimulation (TMS) is a noninvasive method that can induce a reversible "virtual" lesion of the cerebral cortex in a normal human subject [132]. Using both fMRI and TMS provides another means of combining brain activation data with data derived from the lesion method. There are several advantages for using TMS as a lesion method. First, brain injury likely results in brain reorganization after the injury and studies of patients with lesions assume that the nonlesioned brain areas have not been affected, whereas TMS is performed on the normal brain. Another advantage for using TMS is that it has excellent spatial resolution and can target specific locations in the brain whereas lesions in patients with brain injury are markedly variable in location and size across individuals. Such an approach can be illustrated in an investigation of the role of the medial frontal cortex in task switching [133]. In this study, subjects first performed an fMRI study that identified the regions that were active when they stayed on the current task vs. when they switched to a new task. It was found that medial frontal cortex is activated when switching between tasks. In order to determine if the medial frontal cortex was necessary for the processes involved in task switching, the same paradigm was utilized during inactivation of the medial frontal cortex with TMS. Guided by the locations of activation observed in the fMRI study, and using an MRI guided frameless stereotaxic procedure, it was found that applying a TMS pulse over the medial frontal cortex disrupted performance only during trials during which the subject was required to switch between tasks. TMS over adjacent brain regions did not show this effect. Also, the excellent temporal resolution of TMS allowed the investigators to stimulate during precise periods of the task, determining that the observed effect was during the time when the subjects were presented a cue indicating they must switch tasks prior to the actual performance of the new task. Thus, combining the results from both fMRI and TMS, it was concluded that medial PFC was essential for allowing individuals to intentionally switch to a new task.

It is possible to perform TMS studies not only as an adjunct to, but also concurrently with, fMRI. The advantage of this approach is clear: applying TMS at various times *during* (rather than after) fMRI scans permits it to be causally linked with functional changes in the brain, even independently of behavior. In an early study employing this technique, Ruff, Driver and colleagues [36, 134] examined the influence on early visual cortex of a parietal region (the anterior intraparietal sulcus, or aIPS) implicated in the generation of both covert spatial attention and eye movements. They chose a range of TMS)stimulus intensities, all of which were thought to be in an effectively stimulatory rather than inhibitory range, and applied them to the aIPS while subjects fixated the center of a viewing screen. On some trials, a randomly moving visual stimulus was present; subjects had no other task than to maintain fixation. Using this approach, the authors were able to demonstrate a parametric, so-called top-down effect from aIPS following TMS—an increase in the BOLD response in early visual cortex with increasing TMS intensity— that could be found only when visual stimuli were absent, and that did not vary with retinotopic eccentricity. In distinction, their previous work (extended here) had shown that TMS of the frontal eye field (FEF) led to a decrease in BOLD response in the central visual field but to an increase in BOLD response in the peripheral visual field, irrespective of the presence or absence of a visual stimulus. The authors were consequently able to conclude that the aIPS and the FEF have distinct top-down effects on visual cortex, a finding that would not have been possible without concurrent TMS.

7.3 Combined fMRI/ Event-Related Potential Studies

The strength of combining these two methods is coupling the superb spatial resolution of fMRI with the superb temporal resolution of ERP recording. An example of such a study was reported by Dehaene et al. who asked the question "Does the human capacity for mathematical intuition depend on linguistic competence or on visuospatial representations?" [135]. In this study, subjects performed two addition tasks—one in which they were instructed to select the correct sum from two numerically close numbers (exact condition) and one in which they were instructed to estimate the result and select the closest number (approximate condition). During fMRI scanning greater bilateral parietal lobe activation was observed in the approximation condition as compared to the exact condition. Since this activation was outside the perisylvian language zone, it was taken as support that visuospatial processes were engaged during the cognitive operations involved in approximate calculation. Greater left lateralized frontal lobe activation was observed to be greater in the exact condition as compared to the approximate condition, which was taken as evidence for language dependent coding of exact addition facts. In order to consider an alternative explanation of the fMRI findings, the investigators also performed an ERP study. The alternative explanation was that in both the exact and approximate tasks, subjects would compute the exact result using the same representation for numbers but later processing, when they had to make a decision as to the correct choice, was what led to the differences in brain activation. Since fMRI does not offer adequate temporal resolution to resolve these two behavioral events on such a brief time scale, ERP was the appropriate method to test this hypothesis. In the ERP study it was demonstrated that the evoked neural response during exact and approximate trials already differed significantly during the first 400 ms of a trial before subjects had to make a decision.

7.4 Combined fMRI/ Pharmacological Studies

Combining pharmacological challenges during the performance of cognitive tasks during fMRI scanning may yield significantly different information than either method alone. In isolation, fMRI cognitive task paradigms provide little information with respect to the underlying pharmacologic systems involved in cognition. On the contrary, drug administration without a brain measure cannot determine underlying neural mechanisms of the effects of neuromodulatory systems on cognition. Combining the two approaches allows the potential of probing the pharmacologic bases of behavior. One may measure the interactive effects of drug (compared to placebo, or a range of doses) with cognitive task-related modulation of brain activity. It is fair to infer that drug×task interactions reflect modulation of the underlying anatomical and chemical brain systems, and do not simply reflect nonspecific vascular effects. For example, dopaminergic agonists have been shown to have task-specific effects [136–138], and different component processes of working memory are differentially affected by a dopaminergic drug, with effects that may differ between individuals depending on their baseline state [139]. This latter study demonstrated that a dopamine agonist improved the flexible updating (switching) of relevant information in working memory. However, the effect only occurred in individuals with low working memory capacity, but not in individuals with higher working memory capacity. This behavioral effect was accompanied by dissociable effects of the dopaminergic agonist on fronto-striatal activity. The dopamine agonist modulated the striatum during switching but not during distraction from relevant information in working memory, while the lateral frontal cortex was modulated by the drug during distraction but not during switching.

8   Application of a Cognitive Neuroscience Approach Toward Clinical Studies

8.1   Use of Biomarkers Derived from Cognitive Neuroscience Studies

Cognitive neuroscience studies using fMRI may provide an important foundation for clinical studies. A biomarker is an indicator that reflects a process, event, or condition in a biological system. Biomarkers may be useful for providing a measure of exposure, effect, or susceptibility. Reliable biomarkers of a neural system could reliably quantify how such a neural system is affected by almost any input. The input may be the effects of a drug, the effects of cognitive therapy, or the effects of a disease process. For a measurement to be useful as a biomarker in clinical studies, it needs to have well-defined significance based on preclinical studies. That is, a change in an fMRI measurement would ideally reflect a change in a well-understood process, thus providing a clear a priori hypothesis and interpretation of the findings. Once the processes are established, fMRI biomarkers may then be useful for addressing a number of clinical questions. For any neurophysiologic measurement to be a *surrogate* marker, a stable, reliable relationship between the fMRI measurement and a defined clinical outcome needs to be defined. Only in that scenario would an fMRI measurement provide a suitable surrogate for other clinical outcomes. Cognitive neuroscience studies provide the foundation for fMRI biomarkers, but the studies necessary for defining fMRI surrogate markers are rarely done.

Questions regarding the mechanisms of brain function disrupted by pathologic states, processes affected by treatment interventions, or the nature of post-injury reorganization of function are examples of clinical questions that can be tested with fMRI. For example, attentional modulation of information processing-related activity in visual cortex is a well-established phenomenon in cognitive neuroscience studies, with effects measurable using fMRI. For example, it has been shown that activity in category-selective regions of inferior temporal cortex is modulated based on the target of attention, relatively up-modulated if the target is relevant to the region and down-modulated if not relevant [140, 141]. This finding provides a biomarker of attentional control over visual processing, and as noted below, could serve as a useful biomarker for clinical interventions such as cognitive training in individuals with attentional deficits.

8.2 Functional MRI for Measuring the Effect of Clinical Interventions

Functional MRI may be useful not only in defining "static" brain–behavioral relationships, but also may be applied to defining the neural mechanisms that underlie learning, experience, or injury. Two general categories of questions may be investigated. First, fMRI can be used to examine factors that influence response to the perturbations of training (learning), experience or injury. Second, fMRI can be used to examine changes that underlie or are the result of these various perturbations.

Investigation of baseline factors that may influence response to training has particular clinical relevance. A better understanding of pre-training neural characteristics that influence response to rehabilitation training could have major clinical value in guiding treatment decisions. fMRI could provide a number of possible measurements that could mark an important neural process. For example, certain parameters of brain network organization may be particularly important in supporting the potential for learning and plasticity. For example, parameters of the functional organization of whole brain networks have been shown to predict response to training of attention regulation after injury [142]. In another example, a simple measurement of the quantity of activation in prefrontal cortex has been shown to predict response to training to use a verbal memory strategy [143]. Such approaches may help elucidate either personal factors or strategic approaches that underlie variations in learning or response to interventions.

Investigation of *changes* over time is particularly relevant for understanding neural mechanisms of post-injury rehabilitation. In order to assess changes with intervention, longitudinal or repeated measurements are required. Because fMRI involves no exposure-limiting factors such as radiation, it is suitable for repeated measurements. However, multi-session studies are also significantly more complicated to design, analyze, and interpret due to a number of issues discussed below.

There are at least two distinct approaches relevant to assessing changes within an individual. First, fMRI may be used for determining the after-effects of a learning intervention. Functional MRI measures pre- and post-intervention may be used to address this question. For example, after two pieces of information have been strongly associated over repetitive exposures, one may find reduced activation in response to presentation of that information, but increased functional connectivity between regions of the brain that process the two types of information [115]. Second, fMRI may be used for determining the processes that occur during an intervention, such as cognitive training. To do this one would need to acquire fMRI data *during* the process of training. An alternative approach is to use a cross-sectional approach to examine differences across individuals rather than within individuals [144]. For example, brain activation differences between experts in a particular skill (e.g., long-term meditation practitioners, pianists) and novices may be used to infer the neural effects of training to achieve expertise. However, other confounding effects of differences between cohorts are difficult to exclude (e.g., self-selection in persevering to achieve expertise), and a stronger inference for causation requires longitudinal, prospective studies.

The use of fMRI to define changes over time requires consideration of certain additional methodological issues. Test–retest reliability needs to be considered. Estimates of reliability depend on what is being measured. For example, in statistical parametric mapping, the question may be whether particular brain regions are stably labeled as "active" or not in serial sessions. A handful of studies have addressed this question. For example, one group showed that with a classification learning task, scans 1 year apart resulted in highly concordant results with defined regions of interest [145]. Another group showed that maps obtained from a working memory task were similar across time [146], but with a motor task, there appeared to be significant variation over time in volume and spatial location of activation [147].

In longitudinal studies, sources of variability may be both physiologic and nonphysiologic (e.g., MRI hardware). In some cases, the magnitudes of activation in specific brain regions of interest are themselves an outcome of interest. In these instances the stability of BOLD signal measurements becomes an even more salient issue. It may be worthwhile to utilize within-session indices that effectively normalize parameters of interest. For example, rather than comparing estimates of the magnitudes of activation, it may be worthwhile utilizing an index of activity for one condition compared to a second controlled condition with each session. An additional statistical approach that could account for potential variability in SNR is to combine data sets across sessions and then "whiten" the noise, effectively normalizing noise contribution across sessions. Another promising future direction is the use of quantitative techniques such as arterial spin labeling (ASL), mentioned earlier in this chapter, to help reduce nonphysiologic sources of variability. This type of quantitative index may be particularly valuable in studies that attempt to examine brain functioning longitudinally.

Other factors that concurrently change over time can produce confounds to the interpretation of longitudinal studies. For example, performance may change, resulting in changes in reaction time or accuracy. All of these may alter measured responses making determination of the neural bases of the process of interest, such as a treatment intervention, more difficult. These and a number of other theoretical issues are discussed by Poldrack in consideration of learning-related (though not post-injury) changes [144].

Other analytic approaches may be taken that are less sensitive to nonphysiologic instabilities. For example, one could test for changes in the spatial *pattern* of activation, which is not necessarily affected by signal magnitude changes. For example, one could test whether the patterns of activity are identical to within a scaling factor [108]. Furthermore, one could examine the more fundamental measurement of the information coded within brain activity patterns. These measurements may provide more informative indices of particular neural process, while also being more robust for longitudinal studies.

9   Conclusions

Functional MRI is an extremely valuable tool for studying brain–behavior relationships, as it is widely available, noninvasive, and has superb temporal and spatial resolution. New approaches in fMRI experimental design and data analysis continue to appear at an almost exponential rate, leading to numerous options for testing hypotheses on brain–behavior relationships. Combined with information from complementary methods, such as the study of patients with focal lesions, healthy individuals with TMS, pharmacological interventions, or ERP, data from fMRI studies provide new insights regarding the organization of the cerebral cortex as well as the neural mechanisms underlying cognition. Moreover, cognitive neuroscience approaches that have been developed for fMRI provide an excellent foundation for its use as a clinical tool.

References

1. Broca P (1861) Remarques sur le siege de la faculte du langage articule suivies d'une observation d'amphemie (perte de al parole). Bull Mem Soc Anat Paris 36:330–357

2. Buckner RL, Raichle ME, Petersen SE (1995) Dissociation of human prefrontal cortical areas across different speech production tasks and gender groups. J Neurophysiol 74(5):2163–2173

3. Sarter M, Bernston G, Cacioppo J (1996) Brain imaging and cognitive neuroscience: toward strong inference in attributing function to structure. Am Psychol 51:13–21

4. Gaffan D, Gaffan EA (1991) Amnesia in man following transection of the fornix: a review. Brain 114:2611–2618

5. Feeney DM, Baron JC (1986) Diaschisis. Stroke 17(5):817–830

6. Carrera E, Tononi G (2014) Diaschisis: past, present, future. Brain 137:2408–2422

7. Fuster JM, Alexander GE (1971) Neuron activity related to short-term memory. Science 173:652–654

8. Funahashi S, Bruce CJ, Goldman-Rakic PS (1989) Mnemonic coding of visual space in the monkey's dorsolateral prefrontal cortex. J Neurophysiol 61:331–349

9. Funahashi S, Bruce CJ, Goldman-Rakic PS (1993) Dorsolateral prefrontal lesions and oculomotor delayed-response performance: evidence for mnemonic "scotomas". J Neurosci 13:1479–1497

10. Watanabe T, Niki H (1985) Hippocampal unit activity and delayed response in the monkey. Brain Res 325(1–2):241–254

11. Cahusac PM, Miyashita Y, Rolls ET (1989) Responses of hippocampal formation neurons in the monkey related to delayed spatial response and object-place memory tasks. Behav Brain Res 33(3):229–240

12. Alvarez P, Zola-Morgan S, Squire LR (1994) The animal model of human amnesia: long-term memory impaired and short-term memory intact. Proc Natl Acad Sci U S A 91(12):5637–5641

13. Corkin S (1984) Lasting consequences of bilateral medial temporal lobectomy: clinical course and experimental findings in H.M. Semin Neurol 4:249–259

14. Ranganath C, D'Esposito M (2001) Medial temporal lobe activity associated with active maintenance of novel information. Neuron 31(5):865–873

15. D'Esposito M (2010) Why methods matter in the study of the biological basis of the mind: a behavioral neurologist's perspective. In: Reuter-Lorenz PA, Baynes K, Mangun GR, Phelps EA (eds) The cognitive neurosocience of mind: a tribute to Michael Gazzaniga. MIT Press, Cambridge, MA

16. Druzgal TJ, D'Esposito M (2001) Activity in fusiform face area modulated as a function of working memory load. Brain Res Cogn Brain Res 10(3):355–364

17. Henson R (2006) Forward inference using functional neuroimaging: dissociations versus associations. Trends Cogn Sci 10(2):64–69

18. Cohen MS, Kosslyn SM, Breiter HC et al (1996) Changes in cortical activity during mental rotation: a mapping study using functional MRI. Brain 119:89–100

19. D'Esposito M, Ballard D, Aguirre GK, Zarahn E (1998) Human prefrontal cortex is not specific for working memory: a functional MRI study. Neuroimage 8(3):274–282

20. Poldrack RA (2006) Can cognitive processes be inferred from neuroimaging data? Trends Cogn Sci 10(2):59–63

21. Grasby PM (2002) Imaging the neurochemical brain in health and disease. Clin Med 2(1):67–73

22. Landau SM, Lal R, O'Neil JP, Baker S, Jagust WJ (2009) Striatal dopamine and working memory. Cereb Cortex 19(2):445–454

23. Blakemore SJ (2012) Imaging brain development: the adolescent brain. Neuroimage 61(2):397–406

24. Ritter P, Villringer A (2006) Simultaneous EEG-fMRI. Neurosci Biobehav Rev 30(6):823–838

25. Jorge J, van der Zwaag W, Figueiredo P (2013) EEG-fMRI integration for the study of human brain function. Neuroimage 102:24–34

26. Sadaghiani S, Scheeringa R, Lehongre K, Morillon B, Giraud A-L, Kleinschmidt A (2010) Intrinsic connectivity networks, alpha oscillations, and tonic alertness: a simultane-ous electroencephalography/functional magnetic resonance imaging study. J Neurosci 30(30):10243–10250

27. Sadeh B, Podlipsky I, Zhdanov A, Yovel G (2010) Event-related potential and functional MRI measures of face-selectivity are highly correlated: a simultaneous ERP-fMRI investigation. Hum Brain Mapp 31(10):1490–1501

28. Becker R, Reinacher M, Freyer F, Villringer A, Ritter P (2011) How ongoing neuronal oscillations account for evoked fMRI variability. J Neurosci 31(30):11016–11027

29. Bergmann TO, Mölle M, Diedrichs J, Born J, Siebner HR (2012) Sleep spindle-related reactivation of category-specific cortical regions after learning face-scene associations. Neuroimage 59(3):2733–2742

30. Mantini D, Perrucci MG, Cugini S, Ferretti A, Romani GL, Del Gratta C (2007) Complete artifact removal for EEG recorded during continuous fMRI using independent component analysis. Neuroimage 34:598–607

31. Debener S, Strobel A, Sorger B et al (2007) Improved quality of auditory event-related potentials recorded simultaneously with 3-T fMRI: Removal of the ballistocardiogram artefact. Neuroimage 34:587–597

32. Moosmann M, Schönfelder VH, Specht K, Scheeringa R, Nordby H, Hugdahl K (2009) Realignment parameter-informed artefact correction for simultaneous EEG-fMRI recordings. Neuroimage 45(4):1144–1150

33. Mullinger KJ, Yan WX, Bowtell R (2011) Reducing the gradient artefact in simultaneous EEG-fMRI by adjusting the subject's axial position. Neuroimage 54:1942–1950

34. Neuner I, Arrubla J, Felder J, Shah NJ (2014) Simultaneous EEG-fMRI acquisition at low, high and ultra-high magnetic fields up to 9.4T: perspectives and challenges. Neuroimage 102(P1):71–79

35. Feredoes E, Heinen K, Weiskopf N, Ruff C, Driver J (2011) Causal evidence for frontal involvement in memory target maintenance by posterior brain areas during distracter interference of visual working memory. Proc Natl Acad Sci U S A 108(42):17510–17515

36. Ruff CC, Blankenburg F, Bjoertomt O, Bestmann S, Freeman E, Haynes J, Driver J (2006) Concurrent TMS-fMRI and psychophysics reveal frontal influences on human retinotopic visual cortex. Curr Biol 16(15):1479–1488

37. Chen AC, Oathes DJ, Chang C, Bradley T, Zhou Z-W, Williams LM, Etkin A (2013) Causal interactions between fronto-parietal

central executive and default-mode networks in humans. Proc Natl Acad Sci U S A 110(49):19944–19949

38. Yau JM, Hua J, Liao DA, Desmond JE (2013) Efficient and robust identification of cortical targets in concurrent TMS-fMRI experiments. NeuroImage 76:134–144

39. Gnadt JW, Andersen RA (1988) Memory related motor planning activity in posterior parietal cortex of macaque. Exp Brain Res 70:216–220

40. Aguirre GK, Zarahn E, D'Esposito M (1998) The variability of human, BOLD hemodynamic responses. Neuroimage 8(4):360–369

41. Handwerker DA, Ollinger JM, D'Esposito M (2004) Variation of BOLD hemodynamic responses across brain regions and subjects and their effects on statistical analyses. NeuroImage 21:1639–1651

42. Boynton GM, Engel SA, Glover GH, Heeger DJ (1996) Linear systems analysis of functional magnetic resonance imaging in human V1. J Neurosci 16:4207–4221

43. Kim SG, Richter W, Ugurbil K (1997) Limitations of temporal resolution in fMRI. Magn Reson Med 37:631–636

44. Savoy RL, Bandettini PA, O'Craven KM et al (1995) Pushing the temporal resolution of fMRI: studies of very brief stimuli, onset variability and asynchrony, and stimulucorrelated changes in noise. Proc Soc Magn Reson Med 3:450

45. Zarahn E, Aguirre GK, D'Esposito M (1997) A trial-based experimental design for functional MRI. NeuroImage 6:122–138

46. Burock MA, Buckner RL, Woldorff MG, Rosen BR, Dale AM (1998) Randomized event-related experimental designs allow for extremely rapid presentation rates using functional MRI. Neuroreport 9(16):3735–3739

47. Clark VP, Maisog JM, Haxby JV (1997) fMRI studies of visual perception and recognition using a random stimulus design. Soc Neurosci Abstr 23:301

48. Dale AM, Buckner RL (1997) Selective averaging of rapidly presented individual trials using fMRI. Hum Brain Mapp 5:1–12

49. Miezin FM, Maccotta L, Ollinger JM, Petersen SE, Buckner RL (2000) Characterizing the hemodynamic response: effects of presentation rate, sampling procedure, and the possibility of ordering brain activity based on relative timing. Neuroimage 11(6 Pt 1):735–759

50. D'Esposito M, Zarahn E, Aguirre GK (1999) Event-related functional MRI: implications for cognitive psychology. Psychol Bull 125:155–164

51. Besle J, Sanchez-Panchuelo R, Bowtell R, Francis S, Schluppeck D (2014) Event-related fMRI at 7 T reveals overlapping cortical representations for adjacent fingertips in S1 of individual subjects. Hum Brain Mapp 35:2027–2043

52. Ehses P, Bause J, Shajan G, Scheffler K. Efficient generation of T2*-weighted contrast by interslice echo-shifting for human functional and anatomacil imaging at 9.4 Tesla. Magn Reson Med (epub ahead of print)

53. Budde J, Shajan G, Zaitsev M, Scheffler K, Functional PR, MRI (2014) Human subjects with gradient-echo and spin-echo EPI at 9.4 T. Magn Reson Med 2014(71):209–218

54. Malonek D, Grinvald A (1996) Interactions between electrical activity and cortical microcirculation revealed by imaging spectroscopy: implications for functional brain mapping. Science 272:551–554

55. Kim SG, Duong TQ (2002) Mapping cortical columnar structures using fMRI. Physiol Behav 77(4–5):641–644

56. Grill-Spector K, Malach R (2001) fMR-adaptation: a tool for studying the functional properties of human cortical neurons. Acta Psychol (Amst) 107(1–3):293–321

57. Grill-Spector K, Kushnir T, Edelman S, Avidan G, Itzchak Y, Malach R (1999) Differential processing of objects under various viewing conditions in the human lateral occipital complex. Neuron 24(1):187–203

58. Posner MI, Petersen SE, Fox PT, Raichle ME (1988) Localization of cognitive operations in the human brain. Science 240:1627–1631

59. Sternberg S (1969) The discovery of processing stages: extensions of Donders' method. Acta Psychol 30:276–315

60. Petersen SE, Fox PT, Posner MI, Mintun M, Raichle ME (1988) Positron emission tomographic studies of the cortical anatomy of single word processing. Nature 331:585–589

61. Fuster J (1997) The prefrontal cortex: anatomy, physiology, and neuropsychology of the frontal lobes, 3rd edn. Raven, New York

62. Jonides J, Smith EE, Koeppe RA, Awh E, Minoshima S, Mintun MA (1993) Spatial working memory in humans as revealed by PET. Nature 363:623–625

63. Sreenivasan KK, Curtis CE, D'Esposito M (2014) Revising the role of persistent neural activity in working memory. Trends Cogn Sci 18:82–89

64. Attwell D, Iadecola C (2002) The neural basis of functional brain imaging signals. Trends Neurosci 25(12):621–625

65. Heeger DJ, Ress D (2002) What does fMRI tell us about neuronal activity? Nat Rev Neurosci 3(2):142–151

66. Friston KJ, Josephs O, Rees G, Turner R (1998) Nonlinear event-related responses in fMRI. Magn Reson Med 39(1):41–52

67. Glover GH (1999) Deconvolution of impulse response in event-related BOLD fMRI. Neuroimage 9(4):416–429

68. Miller KL, Luh WM, Liu TT et al (2001) Nonlinear temporal dynamics of the cerebral blood flow response. Hum Brain Mapp 13(1):1–12

69. Vazquez AL, Noll DC (1998) Nonlinear aspects of the BOLD response in functional MRI. NeuroImage 7(2):108–118

70. D'Esposito M, Zarahn E, Aguirre GK, Rypma B (1999) The effect of normal aging on the coupling of neural activity to the bold hemodynamic response. Neuroimage 10(1):6–14

71. Rosen BR, Buckner RL, Dale AM (1998) Event-related functional MRI: past, present, and future. Proc Natl Acad Sci U S A 95(3):773–780

72. Donaldson DI, Petersen SE, Ollinger JM, Buckner RL (2001) Dissociating state and item components of recognition memory using fMRI. Neuroimage 13(1):129–142

73. Mitchell KJ, Johnson MK, Raye CL, D'Esposito M (2000) fMRI evidence of age-related hippocampal dysfunction in feature binding in working memory. Brain Res Cogn Brain Res 10(1–2):197–206

74. Keppel G, Zedeck S (1989) Data analysis for research design. W.H. Freeman, New York

75. Worsley KJ, Friston KJ (1995) Analysis of fMRI time-series revisited – again. Neuroimage 2:173–182

76. Nichols T, Hayasaka S (2003) Controlling the familywise error rate in functional neuroimaging: a comparative review. Stat Methods Med Res 12(5):419–446

77. Eklund A, Andersson M, Josephson C, Johannesson M, Knutsson H (2012) Does parametric fMRI analysis with SPM yield valid results? An empirical study of 1484 rest datasets. Neuroimage 61(3):565–578

78. Zarahn E, Aguirre GK, D'Esposito M (1997) Empirical analyses of BOLD fMRI statistics. I Spatially unsmoothed data collected under null-hypothesis conditions. NeuroImage 5:179–197

79. Aguirre GK, Zarahn E, D'Esposito M (1997) Empirical analyses of BOLD fMRI statistics. II Spatially smoothed data collected under null-hypothesis and experimental conditions. NeuroImage 5:199–212

80. D'Esposito M, Ballard D, Zarahn E, Aguirre GK (2000) The role of prefrontal cortex in sensory memory and motor preparation: an event-related fMRI study. Neuroimage 11(5 Pt 1):400–408

81. Zarahn E, Slifstein M (2001) A reference effect approach for power analysis in fMRI. Neuroimage 14(3):768–779

82. Van Horn JD, Ellmore TM, Esposito G, Berman KF (1998) Mapping voxel-based statistical power on parametric images. Neuroimage 7(2):97–107

83. Aguirre GK, D'Esposito M (1999) Experimental design for brain fMRI. In: Moonen CTW, Bandettini PA (eds) Functional MRI. Springer, Berlin, pp 369–380

84. Logothetis NK, Pauls J, Augath M, Trinath T, Oeltermann A (2001) Neurophysiological investigation of the basis of the fMRI signal. Nature 412(6843):150–157

85. Farkas E, Luiten PG (2001) Cerebral microvascular pathology in aging and Alzheimer's disease. Prog Neurobiol 64(6):575–611

86. Fang HCH (1976) Observations on aging characteristics of cerebral blood vessels, macroscopic and microscopic features. In: Gerson S, Terry RD (eds) Neurobiology of aging. Raven, New York

87. Bentourkia M, Bol A, Ivanoiu A et al (2000) Comparison of regional cerebral blood flow and glucose metabolism in the normal brain: effect of aging. J Neurol Sci 181(1–2):19–28

88. Schultz SK, O'Leary DS, Boles Ponto LL, Watkins GL, Hichwa RD, Andreasen NC (1999) Age-related changes in regional cerebral blood flow among young to mid-life adults. Neuroreport 10(12):2493–2496

89. Yamamoto M, Meyer JS, Sakai F, Yamaguchi F (1980) Aging and cerebral vasodilator responses to hypercarbia: responses in normal aging and in persons with risk factors for stroke. Arch Neurol 37(8):489–496

90. Yamaguchi T, Kanno I, Uemura K et al (1986) Reduction in regional cerebral blood rate of oxygen during human aging. Stroke 17:1220–1228

91. Takada H, Nagata K, Hirata Y et al (1992) Age-related decline of cerebral oxygen metabolism in normal population detected with positron emission tomography. Neurol Res 14(2 Suppl):128–131

92. Claus JJ, Breteler MM, Hasan D et al (1998) Regional cerebral blood flow and cerebrovascular risk factors in the elderly population. Neurobiol Aging 19(1):57–64

93. Cunnington R, Iansek R, Bradshaw JL, Phillips JG (1995) Movement-related potentials in Parkinson's disease. Presence and predictability of temporal and spatial cues. Brain 118(Pt 4):935–950

94. Buckner RL, Snyder AZ, Sanders AL, Raichle ME, Morris JC (2000) Functional brain imaging of young, nondemented, and demented older adults. J Cogn Neurosci 12(Suppl 2):24–34

95. Huettel SA, Singerman JD, McCarthy G (2001) The effects of aging upon the hemodynamic response measured by functional MRI. Neuroimage 13(1):161–175

96. Pineiro R, Pendlebury S, Johansen-Berg H, Matthews PM (2002) Altered hemodynamic responses in patients after subcortical stroke measured by functional MRI. Stroke 33(1):103–109

97. D'Esposito M, Deouell L, Gazzaley A (2003) Alterations in the BOLD fMRI signal with ageing and disease: a challenge for neuroimaging. Nat Rev Neurosci 4:863–872

98. Handwerker DA, Gazzaley A, Inglis BA, D'Esposito M (2006) Reducing vascular variability of fMRI data across aging populations using a breath holding task. Hum Brain Mapp 28:846–859

99. Wolf RL, Detre JA (2007) Clinical neuroimaging using arterial spin-labeled perfusion magnetic resonance imaging. Neurotherapeutics 4(3):346–359

100. Brown GG, Clark C, Liu TT (2007) Measurement of cerebral perfusion with arterial spin labeling. Part 2. Applications. J Int Neuropsychol Soc 13(3):526–538

101. Aguirre GK, Detre JA, Zarahn E, Alsop DC (2002) Experimental design and the relative sensitivity of BOLD and perfusion fMRI. Neuroimage 15(3):488–500

102. Liu TT, Brown GG (2007) Measurement of cerebral perfusion with arterial spin labeling. Part 1. Methods. J Int Neuropsychol Soc 13(3):517–525

103. Fernandez-Seara MA, Wang J, Wang Z et al (2007) Imaging mesial temporal lobe activation during scene encoding: comparison of fMRI using BOLD and arterial spin labeling. Hum Brain Mapp 28(12):1391–1400

104. Feinberg DA, Beckett A, Chen L (2013) Arterial spin labeling with simultaneous multi-slice echo planar imaging. Magn Reson Med 70(6):1500–1506

105. Kanwisher N, McDermott J, Chun MM (1997) The fusiform face area: a module in human extrastriate cortex specialized for face perception. J Neurosci 17:4302–4311

106. Haxby JV, Gobbini MI, Furey ML, Ishai A, Schouten JL, Pietrini P (2001) Distributed and overlapping representations of faces and objects in ventral temporal cortex. Science 293(5539):2425–2430

107. Polyn SM, Natu VS, Cohen JD, Norman KA (2005) Category-specific cortical activity precedes retrieval during memory search. Science 310(5756):1963–1966

108. Zarahn E, Rakitin BC, Abela D, Flynn J, Stern Y (2006) Distinct spatial patterns of brain activity associated with memory storage and search. Neuroimage 33(2):794–804

109. Kay KN, Naselaris T, Prenger RJ, Gallant JG (2008) Identifying natural images from human brain activity. Nature 452:352–355

110. Haxby JV, Connolly AC, Swaroop GJ (2014) Decoding neural representational spaces using multivariate pattern analysis. Annu Rev Neurosci 37:435–456

111. Tong F, Pratte MS (2012) Decoding patterns of human brain activity. Annu Rev Psychol 63:483–509

112. Todd MT, Nystrom LE, Cohen JE (2013) Confounds in multivariate pattern analysis: theory and rule representation case study. NeuroImage 77:157–165

113. Chen AJW, Britton MS, Thompson TW, Turner GR, Vytlacil J, D'Esposito M (2012) Goal-directed attention alters the tuning of object-based representations in extrastriate cortex. Front Neurosci 6:187

114. Çukur TNishimoto S, Huth A, Gallant J (2013) Attention during natural vision warps semantic representation across the human brain. Nat Neurosci 16:763–770

115. Buchel C, Coull JT, Friston KJ (1999) The predictive value of changes in effective connectivity for human learning. Science 283(5407):1538–1541

116. McIntosh AR, Grady CL, Haxby JV, Ungerleider LG, Horwitz B (1996) Changes in limbic and prefrontal functional interactions in a working memory task for faces. Cereb Cortex 6(4):571–584

117. Gerstein GL, Perkel DH, Subramanian KN (1978) Identification of functionally related neural assemblies. Brain Res 140(1):43–62

118. Penny WD, Stephan KE, Mechelli A, Friston KJ (2004) Modelling functional integration: a comparison of structural equation and dynamic causal models. Neuroimage 23(Suppl 1):S264–S274

119. Sun FT, Miller LM, D'Esposito M (2004) Measuring interregional functional connectivity using coherence and partial coherence analyses of fMRI data. Neuroimage 21(2):647–658

120. Sun FT, Miller LM, D'Esposito M (2005) Measuring temporal dynamics of functional networks using phase spectrum of fMRI data. Neuroimage 28(1):227–237

121. Sun FT, Miller LM, Rao AA, D'Esposito M (2007) Functional connectivity of cortical networks involved in bimanual motor sequence learning. Cereb Cortex 17(5):1227–1234

122. Gazzaley A, Rissman J, D'Esposito M (2004) Functional connectivity during working

memory maintenance. Cogn Affect Behav Neurosci 4(4):580–599

123. Fuhrmann Alpert G, Sun FT, Handwerker D, D'Esposito M, Knight RT (2007) Spatio-temporal information analysis of event-related BOLD responses. Neuroimage 34(4):1545–1561

124. Rees G, Frith CD, Lavie N (1997) Modulating irrelevant motion perception by varying attentional load in an unrelated task. Science 278(5343):1616–1619

125. Treisman AM (1969) Strategies and models of selective attention. Psychol Rev 76(3):282–299

126. Lavie N, Tsal Y (1994) Perceptual load as a major determinant of the locus of selection in visual attention. Percept Psychophys 56(2):183–197

127. McCarthy RA, Warrington EK (1994) Disorders of semantic memory. Philos Trans R Soc Lond B Biol Sci 346(1315):89–96

128. Warrington EST (1984) Category specific semantic impairments. Brain 107:829–854

129. Thompson-Schill SL (2003) Neuroimaging studies of semantic memory: inferring "how" from "where". Neuropsychologia 41(3):280–292

130. Thompson-Schill SL, D'Esposito M, Aguirre GK, Farah MJ (1997) Role of left inferior prefrontal cortex in retrieval of semantic knowledge: a reevaluation. Proc Natl Acad Sci U S A 94(26):14792–14797

131. Thompson-Schill SL, Swick D, Farah MJ, D'Esposito M, Kan IP, Knight RT (1998) Verb generation in patients with focal frontal lesions: a neuropsychological test of neuroimaging findings. Proc Natl Acad Sci U S A 95(26):15855–15860

132. Pascual-Leone A, Tarazona F, Keenan J, Tormos JM, Hamilton R, Catala MD (1999) Transcranial magnetic stimulation and neuroplasticity. Neuropsychologia 37(2):207–217

133. Rushworth MF, Hadland KA, Paus T, Sipila PK (2002) Role of the human medial frontal cortex in task switching: a combined fMRI and TMS study. J Neurophysiol 87(5):2577–2592

134. Ruff CC, Bestmann S, Blankenburg F et al (2008) Distinct causal influences of parietal versus frontal areas on human visual cortex: evidence from concurrent TMS fMRI. Cereb Cortex 18(4):817–827

135. Dehaene S, Spelke E, Pinel P, Stanescu R, Tsivkin S (1999) Sources of mathematical thinking: behavioral and brain-imaging evidence. Science 284(5416):970–974

136. Gibbs SE, D'Esposito M (2005) Individual capacity differences predict working memory performance and prefrontal activity following dopamine receptor stimulation. Cogn Affect Behav Neurosci 5(2):212–221

137. Gibbs SE, D'Esposito M (2005) A functional MRI study of the effects of bromocriptine, a dopamine receptor agonist, on component processes of working memory. Psychopharmacology (Berl) 180(4):644–653

138. Gibbs SE, D'Esposito M (2006) A functional magnetic resonance imaging study of the effects of pergolide, a dopamine receptor agonist, on component processes of working memory. Neuroscience, 139:359–71

139. Cools R, Sheridan M, Jacobs E, D'Esposito M (2007) Impulsive personality predicts dopamine-dependent changes in frontostriatal activity during component processes of working memory. J Neurosci 27(20):5506–5514

140. Kastner S, Pinsk MA (2004) Visual attention as a multilevel selection process. Cogn Affect Behav Neurosci 4(4):483–500

141. Gazzaley A, Cooney JW, McEvoy K, Knight RT, D'Esposito M (2005) Top-down enhancement and suppression of the magnitude and speed of neural activity. J Cogn Neurosci 17(3):507–517

142. Arnemann KL, Chen AJW, Novakovic-Agopian T, Gratton C, Nomura EM, D'Esposito M (2015) Functional brain network modularity predicts response to cognitive training after brain injury. Neurology 84:1568–74

143. Chen AJW, Novakovic-Agopian T, Nycum TJ, Song S, Turner G, Rome S, Abrams G, D'Esposito M (2011) Training of goal-directed attention regulation enhances control over neural processing for individuals with brain injury. Brain 134(5):1541–1554

144. Poldrack RA (2000) Imaging brain plasticity: conceptual and methodological issues – a theoretical review. Neuroimage 12(1):1–13

145. Aron AR, Gluck MA, Poldrack RA (2006) Long-term test-retest reliability of functional MRI in a classification learning task. Neuroimage 29(3):1000–1006

146. Wei X, Yoo SS, Dickey CC, Zou KH, Guttmann CR, Panych LP (2004) Functional MRI of auditory verbal working memory: long-term reproducibility analysis. Neuroimage 21(3):1000–1008

147. Yoo SS, Wei X, Dickey CC, Guttmann CR, Panych LP (2005) Long-term reproducibility analysis of fMRI using hand motor task. Int J Neurosci 115(1):55–77

Chapter 12

fMRI of Language Systems

Jeffrey R. Binder

Abstract

Language refers to the uniquely human capacity for communication through productive combination of symbolic representations. Functional neuroimaging studies have in recent decades greatly expanded our knowledge of the brain systems supporting language, producing a dramatic reawakening of interest in this topic and a call to revise and extend the nineteenth century neuroanatomical model formulated by Broca, Wernicke, and others. This chapter presents some theoretical issues regarding functional imaging of language systems, a model of the functional neuroanatomy of language based on recent empirical results in several selected processing domains, and a survey of language mapping paradigms in common clinical use. A central theme is that interpretation of fMRI language studies depends on an informed analysis of the cognitive processes engaged during scanning. This analytic approach can help avoid common pitfalls in task design that limit the sensitivity and specificity of language mapping studies and should encourage the development of a standardized methodological and conceptual framework for such studies.

Key words fMRI, Language, Semantics, Phonology, Orthography

1 Language and Language Processes

The central role of language in human culture and social interaction is self-evident. In addition to providing a formal system for overt communication, the symbolic structures of language enable such uniquely human cognitive capacities as the ability to manipulate concepts, plan the future, and invent technology. Scientific investigation of the neural basis of language began in earnest with the work of Broca, Wernicke, and other nineteenth-century neurologists, leading to the classical *Wernicke-Lichtheim* model of language and aphasia that remains with us today [1–4]. Over the past two decades, however, functional imaging methods, particularly fMRI, have greatly expanded our knowledge of the brain systems supporting language, producing a dramatic reawakening of interest in this topic and a call to revise and extend the classical model [5, 6]. This chapter provides a brief survey of some of this work, together with a discussion of theoretical issues central to the design and interpretation of language mapping studies. The goal is to

Massimo Filippi (ed.), *fMRI Techniques and Protocols*, Neuromethods, vol. 119,
DOI 10.1007/978-1-4939-5611-1_12, © Springer Science+Business Media New York 2016

provide a basic theoretical foundation and practical suggestions for designing effective and interpretable clinical protocols.

What is language? One definition often cited is that language processes are those that enable communication. In biological terms, however, this definition is overly inclusive, in that many bodily functions (e.g., cardiac, pulmonary, general arousal, and sustained attention systems) provide critical support for communication but are not linguistic in nature. Communication typically requires neural systems that process auditory or visual sensory information, hold this information in a short-term store, direct attention to specific features or aspects of the information, perform comparisons and other general operations on the information, select a response based on such operations, and carry out the response. The extent to which any of these systems is specialized for use in language behavior is a matter of debate. Careful consideration of these domain-general systems is especially relevant for interpreting and designing language mapping studies, which often employ relatively complex tasks that engage motor, sensory, attention, memory, and "central executive" functions in addition to language. Should these other components be considered part of the language system because they are so critical for adequate task performance, or should they be delineated from language processes per se? In this chapter, I assume that the goal of language mapping is to identify neural systems involved specifically in language processes, i.e., to distinguish these brain networks from early sensory, motor, and general executive systems.

A more precise definition of language is that it is a system of communication based on the symbolic representation and manipulation of information. Languages are also, by definition, *generative*, in that the symbols of a language can be productively combined to make a virtually limitless number of new expressions. In formulating a general definition of this kind, however, it is critical to keep in mind that language is not a unitary process, but rather a collection of processes operating at distinct levels and on distinct types of information. Clinicians working with aphasic patients historically have focused on the dichotomy between "expressive" and "receptive" language functions, but a more useful taxonomy of component language processes is available from the field of linguistics. For spoken languages, these processes include: (1) phoneme perception, the processes serving recognition of speech sounds; (2) phonology, the processes by which speech sounds are represented and manipulated in abstract form; (3) speech articulation, the processes by which speech movements are planned and executed; (4) orthography, the processes by which written characters are represented and manipulated in abstract form; (5) semantics, the processing of word meanings, names, and other declarative knowledge about the world; and (6) syntax, the processes by which words are combined to make sentences and sentences analyzed to reveal underlying relationships between words. A basic assumption of language mapping is that activation tasks can be designed to make varying demands on these processing subsystems. For example, a task requiring careful listening to word-like nonwords (called *pseudowords*, e.g., "nurdle") would make great demands on phoneme perception (and on pre-phonetic auditory processing and attention) but very little demand on semantic or syntactic processing, given that the stimuli have no (or very little) meaning. In contrast, a task requiring semantic categorization of printed words (e.g., Is it an animal or not?) would make great demands on orthographic and semantic processing but relatively little on phonetic, phonological, or syntactic processing.

On the other hand, the processing subcomponents of language often act together. The extent to which each component can be examined in isolation remains a major methodological issue, as it is not yet clear to what extent the systems responsible for these processes become active "automatically" when presented with linguistic stimuli [7]. One familiar example of this interaction is the Stroop effect, in which orthographic and phonological processing of printed words occurs even when subjects are instructed to attend to the color of the print, and even when this processing interferes with task performance [8]. Other examples include semantic priming effects during word recognition, picture–word interference effects, lexical effects on phonetic perception, orthographic effects on letter perception, and semantic–syntactic interactions during sentence comprehension [9–17]. If linguistic stimuli such as words and pictures evoke obligatory, automatic language processing, these effects need to be considered in the design and interpretation of language activation experiments. Use of such stimuli in a "baseline" condition could result in undesirable subtraction (or partial subtraction) of language-related activation. Because investigators frequently try to match stimuli in control and language tasks very closely, such inadvertent subtraction is relatively commonplace in functional imaging studies of language processing.

A final theoretical issue is the extent to which language processes occur during "resting" states or states with minimal task requirements (e.g., visual fixation or "passive" stimulation). Language involves interactive systems for manipulating *internally stored knowledge* about words and word meanings. In examining these systems we typically use familiar stimuli or cues to engage processing, yet it seems likely that activity in these systems could occur independently of external stimulation and task demands. The idea that the conscious mind can be internally active independent of external events has a long history in psychology and neuroscience [18–22]. When asked, subjects in experimental studies frequently report experiencing seemingly unprovoked thoughts (including words and recognizable images) that are unrelated to the task at hand [21, 23–26]. The extent to which such "thinking" engages linguistic knowledge remains unclear [27, 28], but many researchers have demonstrated close parallels between behavior and language content, suggesting that at least some internal thought processes make use of verbally encoded semantic knowledge and other linguistic representations [27, 29, 30]. Many authors have argued that "rest" and similar conditions are actually active states in which subjects frequently are engaged in processing linguistic and other information [25, 26, 31–38]. Use of such states as control conditions for language imaging studies may thus obscure similar processes that occur during the language task of interest. This is a particularly difficult problem for language studies because the internal processes in question cannot be directly measured or precisely controlled.

2   Functional Neuroanatomy of Component Language Systems

Considering that neuroimaging studies on language processing now number in the thousands, the following review is inevitably incomplete and somewhat cursory. Nor is it possible to cover every topic that might be of interest. This review focuses on single word and sublexical processes, with less attention to studies of sentence comprehension and syntax. Several interesting topics, including bilingualism and sign language processing, are not touched on here; the interested reader is referred to reviews on these topics [39–48]. A visual summary of the following discussion is provided in Fig. 1.

2.1   Phoneme Perception

Traditional clinical models of aphasia often treat comprehension of speech as a single function [49], but it is important to distinguish at least two processes engaged during speech comprehension. Spoken words not only possess meanings, they are also composed of very complex and rapidly changing auditory signals. Thus, useful models of auditory word recognition include not only a semantic stage in which words are mapped onto their meanings, but also a stage prior to semantic access in which consonant and vowel sounds—known collectively as *phonemes*—are identified. The distinction between these stages becomes clearer if one considers the differences between listening to a tone (for example, a note played on a piano), a pseudoword (such as "dap"), and a word (such as "tap"). The tone has no phonemic value; it cannot be identified as any vowel or consonant. In contrast, the pseudoword "dap" contains three phonemes—/d/, /æ/, /p/—although it has no meaning. Finally, the word "tap" conveys both phonemic and semantic information. The importance of the phoneme perception stage is illustrated by the fact that "dap" and "tap" differ at a physical level only in the presence of a brief (typically 20–30 ms) noise at the beginning of "tap" but not "dap," produced by release of the tongue from the roof of the mouth slightly prior to the onset of vocal cord vibration. These and many other subtle acoustic cues must be rapidly and continuously

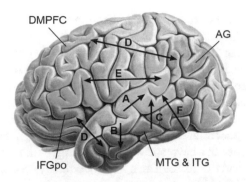

Fig. 1 A schematic model of some major language regions and networks. *Yellow* indicates a bilaterally represented phoneme (speech sound) perception system. *Blue* indicates the posterior perisylvian area (posterior superior temporal and supramarginal gyri, roughly equivalent to the traditional Wernicke area), which supports pre-articulatory phonological access. *Red* indicates the temporal and parietal components of a distributed system that stores and processes word meaning (semantic memory) information. *ORANGE* indicates several prefrontal components of the semantic processing network, including the pars orbitalis of the inferior frontal gyrus (IFGpo) and the dorsomedial prefrontal cortex (DMPFC), which are proposed to control the activation and selection of information in the posterior semantic memory store. *GREEN* indicates a more general language control system, made up of the pars triangularis and opercularis of the IFG (roughly equivalent to the traditional Broca area) and adjacent cortex in the inferior frontal sulcus, which is proposed to control the retrieval and maintenance of phonological information, a process that is critical for word retrieval, verbal working memory, and sentence production. Speech repetition requires the pathway designated *A* in the figure, linking phoneme perception with phonological access systems, as well as more anterior sensorimotor regions (not shown) that support articulatory preparation and execution. Spoken word comprehension involves pathway *B* in the figure, which maps perceived phoneme sequences to word concepts. Communicative speech production, in which the speaker retrieves words and formulates sentences to express concepts, requires control of the semantic system by the pathways marked *D*, as well as pathway *C*, which maps concept representations onto phonological representations, and pathway *E*, which controls and maintains the activation of phonological codes. Pathway *F* indicates a direct mapping from visual word forms to phonological representations, required for reading aloud

processed in real time for accurate speech comprehension to occur. The absence of this important linguistic process in traditional clinical models of language can be attributed to at least two factors. First, very little was known about the physical acoustic properties of phonemes prior to the mid-twentieth century; the scientific study of speech perception has developed only in the last 60–70 years. Second, cases of isolated phoneme perception difficulty, known as *pure word deafness*, are rare, resulting in a relative lack of familiarity with this field of study on the part of many clinicians.

Over the past several decades, scientists using functional neuroimaging methods have identified a region in the superior temporal lobes that responds more strongly to speech than to nonspeech sounds like tones and noise [50–57]. These speech-related activations are found consistently in the middle portion of the superior temporal sulcus (STS, see yellow region in Fig. 1), i.e., the sulcus separating the superior temporal gyrus (STG) from the middle temporal gyrus (MTG). This activation is often found in both the left and right STS, though usually with leftward lateralization. These activations are identical whether the stimuli are words or word-like pseudowords, thus they reflect processing of phonemes and not word meaning [54]. Though some of the activation in this region could be explained by the fact that speech sounds are more acoustically complex than the tones and noises used as nonspeech controls, more recent experiments using acoustically matched speech and nonspeech sounds (e.g., rotated speech, sinewave speech) have shown convincingly that at least some of the activation in this region is due specifically to activation of phoneme codes [58–63].

These observations are fully consistent with localization data from patients with pure word deafness, who typically have lesions restricted to the STG and STS [64–70]. Most of these cases have bilateral lesions, though rarely a large left temporal lobe lesion can produce the syndrome [71, 72]. These patients show impairments in recognizing speech phonemes and may have other deficits of higher-order auditory perception, especially when the lesions are bilateral, but they have no deficits in written comprehension, naming, or propositional speech production that would indicate any loss of word concepts. Taken together, these functional imaging and lesion data make it clear that the left STG plays a relatively specific role in language processing, i.e., that it contains general auditory systems and specialized networks for recognizing speech phonemes, regardless of whether these phonemes form words or have meaning. This conceptualization stands in stark contrast to the traditional clinical model of aphasia, which identifies the left STG as "Wernicke's area," the principal site for "language comprehension."

2.2 Grapheme Perception and Orthographic Processing

Traditional neuroanatomical models of written word recognition derive mainly from the late 19th century descriptions by Déjerine of alexia with and without agraphia [73–75]. According to these models, visual perception of letters occurs in the primary visual cortex of both hemispheres, which then transmit this information to the left angular gyrus, where "memories" of written words are activated. Lesions of the left angular gyrus destroy these visual word codes, producing both inability to read and inability to write. Alexia without agraphia (also known as pure alexia, peripheral alexia, or letter-by-letter reading) results when an occipital lobe lesion destroys both the left visual cortex and the decussating white matter pathway from right visual cortex to left angular gyrus, effectively disconnecting the angular gyrus from visual input without destruction of the visual word codes themselves [76].

In recent decades it has become clear, however, that pure alexia can result from ventral occipital–temporal lesions that damage neither the primary visual cortex nor the angular gyrus. Though initially ascribed to involvement of white matter pathways projecting to the angular gyrus [77–79], it is now clear that most of these cases are due to focal damage to the left ventral occipital-temporal cortex, particularly the mid-portion of the left fusiform gyrus [80–84]. Thus, normal reading requires the participation of a left-lateralized visual association area in or near the mid-fusiform gyrus, which receives input from earlier visual processing stages in both hemispheres.

Functional neuroimaging research has strongly confirmed this model. Numerous studies have demonstrated a focal region in the lateral left fusiform gyrus that responds more strongly to words and word-like pseudowords than to consonant letter strings or non-sense characters [85–92]. This focal region of cortex has consequently been named the "visual word form area." Activation in this area increases as a direct function of how frequently the letter combinations comprising the stimulus occur in the reader's language, and this activation is also correlated with efficiency of letter perception during tachistoscopic presentation [93]. It thus appears that during the many hours spent learning to read fluently, neurons in this region become "tuned" to detect familiar letter combinations, resulting in a high degree of perceptual expertise that allows multiletter fragments and even whole words to be processed in parallel. Destruction of these "expert" neurons prevents the patient from recognizing multiletter fragments efficiently, forcing the adoption of a much slower, letter-by-letter decoding process [94–96].

2.3 Phonological Access and Phonological Working Memory

The term *paraphasia* refers to speech production that is fluent but contains errors, such as substitution of incorrect phonemes or words, or rearrangement of the order of phonemes within a word. Paraphasia is characteristic of many forms of aphasia, particularly the Wernicke and conduction syndromes, and typically affects both spoken and written output. Paraphasia indicates an inability to retrieve or properly use a mental representation of word sounds—what nineteenth century theorists called "sound images" and what in modern parlance are referred to as *phonological representations*. Patients who cannot access (i.e., activate, compute) correct phonological representations show paraphasic errors on all speech output tasks, including conversing, naming objects, reading aloud, and repeating, as well as on a variety of other tasks that require phonological access. For example, patients with phonological impairments may be unable to determine whether two printed words rhyme. Writing normally involves a mapping from phonological (sound-based) to orthographic (grapheme-based) representations, which is why patients with impaired phonology also typically show paraphasic errors in their writing.

The brain regions most strongly implicated in phonological access (see blue region in Fig. 1) are in the left posterior perisylvian area, especially the posterior STG, posterior STS, and supramarginal gyrus (SMG). For example, patients with conduction aphasia—a relatively isolated disorder of phonological access featuring phonemic paraphasia in naming, reading, and repetition tasks—have lesions confined to this region [97–104], as do patients with phonological deficits in written production [105–107]. A recent voxel-based lesion correlation study linked damage in this posterior perisylvian region with inability to silently judge whether written word pairs rhyme [108], indicating that lesions in this region impair phonological processing prior to and independent of any overt articulation processes. A number of functional imaging results also support this model. For example, contrasts between visual stimuli that can be named and those that cannot (e.g., pictures vs. nonsense shapes, pronounceable vs. unpronounceable letter strings, letters vs. unfamiliar characters) reliably produce activation in the left posterior STS, STG, and inferior SMG [109–116] as do silent "word generation" tasks [117–119].

Temporary activation of phonological codes is also central to the concept of *verbal working memory*. The standard model of verbal working memory includes a "phonological loop" responsible for maintaining phonological sequences in short-term memory [120]. The phonological loop is further subdivided into a "phonological store" that represents the phonological information itself and an "articulatory rehearsal" mechanism that reactivates the information before it fades from the store. A number of neuroimaging studies have linked the phonological store with the left SMG [121] and with the posterior STG and STS [122–125].

These regions implicated in phonological access and temporary storage of phonological representations overlap partly with those implicated in speech perception, though the evidence suggests that the phoneme perception system is situated more anteriorly along the STG and STS [126], whereas the phonological access system is more posterior and extends more dorsally, involving dorsal STG (planum temporale) and SMG. These systems cannot be entirely overlapping, since most patients with conduction aphasia (phonemic paraphasia) do not have speech perception deficits. As noted above, the speech perception system is also bilaterally represented, which may explain why it is more resistant to left STG damage than is the phonological access system, which is more strongly left-lateralized.

2.4 Semantic Memory and Semantic Processing

The human brain has an enormous capacity to acquire knowledge from experience. The characteristic shapes, colors, textures, movements, sounds, smells, and actions associated with objects in the environment, for example, must all be learned from experience. In addition, consider the enormous variety of verb concepts (*build, celebrate, discuss, throw*, etc.), which depend on knowledge of how particular kinds of events happen, or social/emotional concepts (*anger, deceit, love, trust*, etc.), which depend on knowledge of how human beings behave and why. Much of this knowledge is represented symbolically in language and underlies our understanding of word meanings. These relationships between words and the stores of knowledge they signify are known collectively as the *semantics* of a language [127]. The term *semantic processing* refers to the cognitive act of accessing stored knowledge about the world through words. The stored knowledge itself is often called *semantic memory*.

Semantic properties of words are readily distinguished from their structural properties. For example, words can have both spoken (phonological) and written (orthographic) forms, but these surface forms are typically related to word meanings only through the arbitrary conventions of a particular language. There is nothing, for example, about the letter sequences D-O-G or C-H-I-E-N that inherently links these sequences to a particular concept. Conversely, it is trivial to construct surface forms (e.g., CHOG) that possess all of the phonological and orthographic properties of words in a particular language, but which have no meaning in that language. A simple, operational distinction can thus be made between the processes involved in analyzing the surface form (phonology, orthography) of words, and semantic processes, which concern access to knowledge that is *not directly represented in the surface form*.

Semantic processing is a defining feature of human behavior, central not only to language, but also to our capacity to access acquired knowledge in reasoning, planning, and problem solving. Impairments of semantic processing figure in a variety of brain disorders, such as Alzheimer disease, semantic dementia, fluent aphasia, schizophrenia, and autism. The neural basis of semantic processing has been studied extensively by analyzing patterns of brain damage in such patients [128–135]. This topic has also been addressed in a large number of functional neuroimaging studies on healthy volunteers (see [133, 136–140] for reviews). Of greatest interest for the present review are studies that focused specifically on semantic processing by incorporating control tasks that make comparable demands on surface form (phonological or orthographic) processing and on general executive processes such as attention, working memory, and response production (see [25, 50, 141–151] for some examples).

Binder et al. [138] performed a quantitative meta-analysis of 120 of these well-controlled studies. The results revealed a widely distributed, left-lateralized network underlying semantic memory storage and retrieval (see red and orange regions in Fig. 1). Seven major brain regions were implicated: (1) the angular gyrus; (2) the middle and inferior temporal gyri, extending into the lateral anterior temporal lobe; (3) the anterior fusiform and parahippocampal gyri; (4) the anterior aspect (pars orbitalis) of the inferior frontal gyrus (IFGpo); (5) dorsomedial prefrontal cortex, including the superior frontal gyrus and the posterior aspect of the middle frontal gyrus; (6) ventromedial prefrontal cortex; and (7) the posterior cingulate gyrus. These are all regions considered to be supramodal cortex, distant from primary sensory and motor areas, and therefore likely to be involved in processing highly abstracted (i.e., nonperceptual) information. Activation in these regions tends to be left-lateralized, though most studies show at least some activation in homologous regions of the right hemisphere. Some evidence suggests that the right hemisphere semantic system contributes to processing concrete, imageable concepts, and much less to processing abstract concepts [150, 152, 153].

These functional neuroimaging results are very consistent with pathological data from patients with semantic disorders. For example, lesion localization studies in patients with transcortical sensory aphasia, a syndrome characterized by multimodal semantic impairment with intact phonological processing, implicate widely distributed regions of the left ventral temporal lobe and angular gyrus [128, 154–156]. Semantic dementia, a degenerative disorder characterized by gradual loss of semantic knowledge, is associated with progressive neuronal loss in the anterior and ventral temporal lobes bilaterally [135, 157–161]. Other pathological conditions that affect the ventral temporal lobes, such as herpes encephalitis and Alzheimer disease, often produce focal semantic memory loss, particularly loss of knowledge about living things [132, 162–165], while inferior parietal and posterior temporal lobe damage may produce selective loss of knowledge about man-made objects, particularly tools [132, 137]. Whereas these temporal and parietal lesions damage the semantic memory store itself (red regions in Fig. 1), dorsal left prefrontal lesions seem to impair the ability to retrieve information from the semantic store. These latter lesions produce transcortical motor aphasia, a syndrome characterized by inability to initiate spontaneous speech [155, 166].

2.5 Sentence Comprehension and Syntax Processing

The work reviewed so far focused on processing of single word structure and meaning, which can be thought of as the basic building blocks of language. Natural language, however, consists almost entirely of sentences. At least two phenomena distinguish processing at the sentence level from processing of single words. First, in sentence processing, the meanings of individual words are combined to create more complex and context-specific meanings. For example, consider:

1. The tigers lost their jungle habitat.
2. The tigers lost in extra innings.

It is the combination of words that specifies in each case the meaning of "tigers" and "lost." This process of *conceptual combination* is a fundamental phenomenon in language production and comprehension. A second distinguishing feature of sentence processing is the use of syntactic information—word order, grammatical function words, and word inflections—to indicate the thematic roles played by constituent content words. In the two example sentences above, for example, "the" marks the beginning of a noun phrase, which can be followed by either a noun or a modifier phrase. The plural inflection of "tigers" then identifies this second word as a noun, which because of its position is likely to be the subject of the sentence, and so on.

One type of neuroimaging study used to examine these processes compares processing of sentences with word lists that do not form a sentence, the latter sometimes created simply by randomly rearranging the order of words in sentences ("scrambled sentences"). Common areas of activation in these contrasts (sentences vs. scrambled sentences) include the left anterior superior temporal lobe, left IFG, and left angular gyrus [167–173]. Debate has ensued over whether these activations represent syntactic or semantic processes, as sentences possess both more syntactic structure and more complex meanings than lists of isolated words. This question has focused particularly on the anterior temporal lobe, a region often activated in studies using sentence materials but rarely in studies using isolated words. Humphries et al. [172] showed that activation in the anterior STS is modulated by the presence of syntactic structure independently of the meaningfulness of constituent content words, suggesting that this region may play a role in early parsing processes (e.g., role assignment). These authors also examined the processing of combinatorial semantic structure in word lists and sentences by manipulating the degree to which words in the stimuli were thematically related. Remarkably, this contrast showed widespread regions in the ventral left temporal lobe, angular gyrus, and inferior frontal lobe that were activated when words were thematically related (and thus could be combined to form more complex and specific meanings) compared to when words were unrelated [172, 174]. This effect of combinatorial semantic structure was largely unaffected by whether the stimuli were syntactically correct sentences or word lists.

Many other fMRI and PET studies have focused on specific syntactic operations, such as repair of syntactic and morphosyntactic violations [175–180] and comprehension of object-extracted relative clauses, passive voice, and other noncanonical or derived syntactic structures [181–189]. While these studies have generally implicated regions in the left IFG and left superior temporal lobe, the precise localization of specific syntactic operations remains a source of debate. Another ongoing discussion centers on whether these activations reflect operations specific to syntax processing or instead more domain-general working memory and executive processes [187, 190–197]. The reader is referred to reviews that cover this work in detail [198–202].

2.6 Retrieval, Selection, and Maintenance

Using language depends on a variety of executive "control" processes, including the ability to voluntarily activate phonological or semantic information as needed for a given task, the ability to select the correct name or concept when a number of competing alternatives are activated, and the ability to maintain the selected item(s) in short-term memory while the task is completed. For example, if the task is to answer a question, such as, "What farm animal gives milk?", it is necessary to use the content words in the question (i.e., farm, animal, give, milk) to activate a field of associated concepts, select from among several activated alternatives (e.g., cow, goat, sheep), use the selected concept to retrieve an associated name, and maintain the concept and name in an activated state during production of the response. These control processes depend mainly on the left prefrontal cortex (green and orange regions in Fig. 1) [195, 203].

This modern view of the left prefrontal cortex contrasts with the traditional concept of "Broca's area" as a region involved only in speech production. In fact, the same retrieval, selection, and maintenance operations are required for many tasks in which no speech production occurs, such as silently naming a picture, or comprehending a sentence. Damage to the prefrontal cortex produces obvious impairments on a range of language production tasks, but usually not because speech articulation or motor sequencing is impaired. Rather, frontal lesions impair the ability to voluntarily retrieve concepts and verbal labels and to maintain these in short-term memory. The contribution of these regions increases as the need for these control processes increases, for example as sentences become more complex or ambiguous, or items to be retrieved become less familiar.

3   An Analysis of Some Language Mapping Paradigms in Common Use

The variety of possible stimuli and tasks that could be used to induce language processing is vast, and a coherent, concise discussion is difficult. Table 1 lists some of the broad categories of stimuli that have been used and some of the brain systems they engage. "Auditory Nonspeech" refers to noises or tones that are not perceived as speech. Such stimuli can be variably "complex" in their temporal or spectral features, and possess to varying degrees the acoustic properties of speech (see [54, 204–207] for some examples). "Auditory Phonemes" are speech sounds that do not comprise words in the listener's language; these may be simple consonant-vowel monosyllables or longer sequences (e.g., pseudowords). "Visual Nonletter" refers to any unfamiliar visual shape. Examples include characters from unfamiliar alphabets, nonsense shapes, and "false font." Such stimuli can be variably complex and possess to varying degrees the visual properties of familiar letters. "Visual Letterstrings" are random strings of letters that do not form familiar or easily pronounceable letter combinations (e.g., FCJVB). "Visual Pseudowords" are letterstrings that are not words but possess the orthographic and phonological characteristics of real words (e.g., SNADE).

The degree to which these stimuli engage the processes listed in Table 1 depend partly on the task that the subject is asked to perform, though the processes in Table 1 are activated "automatically" to some degree even when subjects are given no explicit task. This is less true for the processing systems listed in Table 2, which seem to be strongly task-dependent. The semantic system appears to be partly active even during "rest" or when stimuli are presented

Table 1
Effects of auditory and visual stimuli on sensory and linguistic processing systems

Stimuli	Early sensory	Phoneme perception	Visual wordform	Object recognition	Syntax
Auditory nonspeech	Aud	–	–	–	–
Auditory phonemes	Aud	+	–	–	–
Auditory words	Aud	+	–	–	–
Auditory sentences	Aud	+	–	–	+
Visual nonletters	Vis	–	–	–	–
Visual letterstrings	Vis	–	+/–	–	–
Visual pseudowords	Vis	–	+	–	–
Visual words	Vis	–	+	–	–
Visual sentences	Vis	–	+	–	+
Visual objects	Vis	–	–	+	–

Table 2
Effects of task states on some linguistic processing systems

Tasks	Semantics	Phonological Access	Speech Articulation	Working Memory	Other language
Rest or "passive"	+	–	–	–	–
Sensory discrimination	–	–	–	+/–	–
Read or repeat covert	+	+	–	+/–	–
Read or repeat overt	+	+	+	+/–	–
Phonetic decision	–	+	–	+	–
Phonological decision	–	+	–	+	–
Orthographic decision	–	+/–	–	–	–
Semantic decision	+	+/–	–	+	Semantic search
Word generation covert	+	+	–	+	Lexical search
Word generation overt	+	+	+	+	Lexical search
Naming covert	+	+	–	–	Lexical search
Naming overt	+	+	+	–	Lexical search

"passively" to the subject [25, 33–35, 37, 208]. Other tasks suppress semantic processing by requiring a focusing of attention on perceptual, orthographic, or phonological properties of stimuli [25, 26, 35, 208]. Examples include "Sensory Discrimination" tasks (e.g., intensity, size, color, frequency, and other discriminations based on physical features), "Phonetic Decision" tasks in which the subject must detect a target phoneme or phonemes, "Phonological Decision" tasks requiring a decision based on the phonological properties of a stimulus (e.g., detection of rhymes, judgment of syllable number), and "Orthographic Decision" tasks requiring a decision based on the letters in the stimulus (e.g., case matching, letter identification). Other tasks, such as reading and repeating, make no overt demands on semantic systems but may elicit automatic semantic processing. The extent to which this occurs probably depends on how meaningful the stimulus is: sentences likely elicit more semantic processing than isolated words, which in turn elicit more than pseudowords. Finally, many tasks make overt demands on retrieval and use of semantic knowledge. These include "Semantic Decision" tasks requiring a decision based on the meaning of the stimulus (e.g., "Is it living or nonliving?"), "Word Generation" tasks requiring retrieval of a word or series of words related in meaning to a cue word, and "Naming" tasks requiring retrieval of a verbal label for an object or object description.

As noted earlier, "Phonological Access" refers to the processes engaged in retrieving a phonological (sound-based) representation of a word (or pseudoword). In addition to speech output and phonological tasks, any task using printed words, including orthographic and semantic tasks, will be accompanied to some degree by obligatory phonological access [8, 15, 111]. In contrast, "Speech Articulation" processes are engaged fully only when an overt spoken response is produced [119]. "Verbal Working Memory" is required whenever a written or spoken stimulus must be held in memory. Some degree of short-term phonological memory is needed for most language tasks, and particularly in cases where the stimulus is relatively long (i.e., sentences more than single words) or has multiple components, or must be held in memory while a response is generated (e.g., word generation tasks involving multiple responses for each cue). Finally, semantic decision, word generation, and naming tasks make strong demands on frontal mechanisms involved in searching for and retrieving information associated with a stimulus [118, 195, 209, 210].

With these somewhat over-simplified stimulus and task characterizations in mind, it is possible to make some general predictions about the processing systems whose level of activation will differ when two task conditions are contrasted, and thus the likely pattern of brain activation that will be observed in a simple subtraction analysis. Some commonly encountered examples are listed below and in Table 3.

Table 3
Some task contrasts used for language mapping and the regions in which robust activations are typically observed

	Ventrolateral Prefrontal	Dorsal Prefrontal	Superior Temporal	Ventrolateral Temporal	Ventral Occipital	Angular Gyrus
1. Hearing words vs. Rest			B			
2. Hearing words vs. Nonspeech sounds			L>R			
3. Word generation vs. Rest	L>R			L>R	B	
4. Word generation vs. Reading	L					
5. Object naming vs. Rest	B			L>R	B	
6. Semantic decision vs. Sensory discrimination	L	L	L>R	L		L
7. Semantic decision vs. Phonological decision		L		L		L
8. Reading sentences vs. Letterstrings	L>R			L>R	L>R	

L=left hemisphere, R=right hemisphere, B=bilateral

Paradigm 1

Language task: Passively Listening to Words or Sentences

Control task: Rest

As shown in Table 1, auditory words activate early auditory cortices and phoneme perception areas. Since both rest and passive stimulation are accompanied by spontaneous semantic processes and make no other overt cognitive demands, no other language-related activation should appear in this contrast. The resulting activation pattern involves mainly auditory cortex in the superior temporal gyri bilaterally (Fig. 2a) [54, 167, 208, 211, 212]. The magnitude and extent of this activation increase with rate of word presentation [213, 214]. This STG activation is relatively symmetrical and is not correlated with language dominance as measured by Wada testing [215]. Although some authors have equated this STG activation with "Wernicke's area for receptive language," most of this

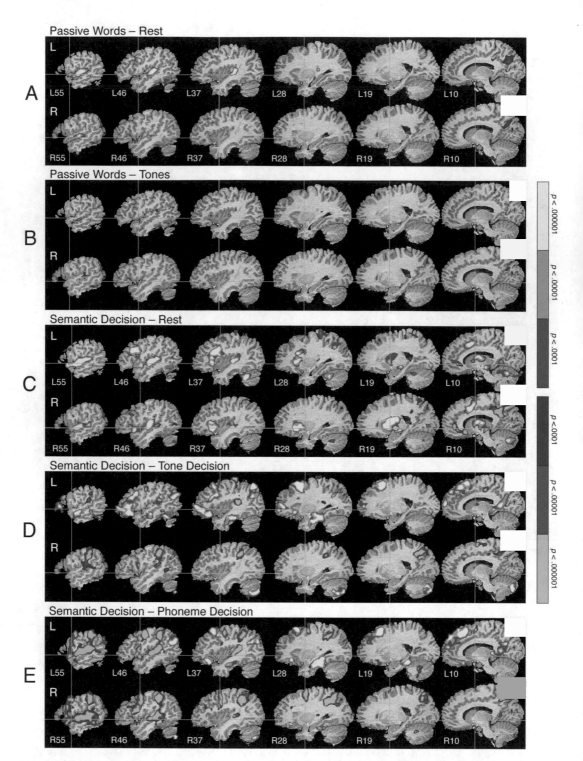

Fig. 2 Group average fMRI activation patterns in 26 neurologically normal, right-handed volunteers during five fMRI language paradigms (see [208] for details). Auditory word and tone stimuli were equivalent in each of the five paradigms. (**a**) Passive listening to words contrasted with resting. Superior temporal activation occurs bilaterally. (**b**) Passive listening to words contrasted with passive listening to tones. A small region in the left STS shows activation specifically related to speech processing. (**c**) Semantic decision on words contrasted with resting. Activation occurs in bilateral auditory (STG) and attentional/working memory (dorsolateral prefrontal, anterior cingulate, anterior insula, IPS, and subcortical) networks, with left lateralization in the IFG.

activation represents early auditory processing rather than language-specific processes per se.

Paradigm 2

Language task: Passively Listening to Words or Sentences

Control task: Passively Listening to Auditory Nonspeech

Because there are no differences in task requirements, and because semantic processing occurs in all passive conditions, the activation pattern associated with this contrast mainly reflects activation of the phoneme perception system (Table 1). As mentioned earlier, studies employing such contrasts reliably show activation in the STS, with leftward lateralization, and little or no activation elsewhere (Fig. 2b) [51, 52, 54, 56, 204, 208]. When sentences are used, this STS activation extends more anteriorly into the dorsal temporal pole region, possibly reflecting early syntactic parsing processes [167, 171, 172, 204, 216–219].

Paradigm 3

Language task: Word Generation

Control task: Rest

Because the rest state includes no control for sensory processing, early auditory or visual cortices may be activated bilaterally depending on the sensory modality of the cue stimulus (Table 1). The strength of this sensory activation depends on the rate of stimulus presentation: in some protocols, a single cue (e.g., a letter or a semantic category) is provided only at the beginning of an activation period; in others, a new cue is provided every few seconds. Unlike rest, word generation makes demands on lexical search, phonological access, and working memory systems (Table 2). Speech articulation systems will also be activated if an overt spoken response is required. These predictions are confirmed by many studies employing this contrast, which results primarily in activation of the left IFG and left > right premotor cortex, systems thought to be involved in phonological production, working memory, and lexical search [118, 211, 215, 220–225]. There may be activation of left posterior temporal regions (posterior MTG and STG) due to engagement of the phonological access system [117, 119, 226].

Fig. 2 (continued) (**d**) Semantic decision on words contrasted with a tone decision task. Activation is strongly left-lateralized in prefrontal, lateral and ventral temporal, angular, and posterior cingulate cortices. (**e**) Semantic decision on words contrasted with a phoneme decision task on pseudowords. Activation is strongly left-lateralized in dorsal prefrontal, angular, ventral temporal, and posterior cingulate cortices. Data are displayed as serial sagittal sections through the brain at 9-mm intervals. X-axis locations for each slice are given in the *top panel*. *Green lines* indicate the stereotaxic Y and Z origin planes. *Hot colors* (*red–yellow*) indicate positive activations and cold colors (*blue–cyan*) negative activations for each contrast. All maps are thresholded at a whole-brain corrected $P < 0.05$ using voxel-wise $P < 0.0001$ and cluster extent >200 mm^3. Adapted, with permission, from [208]

Paradigm 4

Language task: Word Generation

Control task: Reading or Repeating

Here we assume that the same stimulus modality (auditory or visual) is used for both tasks. The stimuli in both cases are single words, thus no difference in activation of sensory, phoneme perception, or visual word form systems is expected. Both tasks are accompanied by semantic processing (automatic semantic access in the case of the control task, effortful semantic retrieval in the case of word generation) and by phonological access processes. The word generation task makes greater demands on lexical search and on working memory; consequently greater activation is expected in left inferior frontal areas associated with these processes. These predictions match findings in many studies using this contrast, which show primarily left-lateralized activation in the IFG [118, 210, 226, 227].

Paradigm 5

Language task: Visual Object Naming

Control task: Rest

Compared to resting, visual object perception activates early visual sensory cortices and higher-level object recognition systems bilaterally (Table 1) [228–230]. There may be additional, left-lateralized activation in semantic systems of the ventrolateral posterior temporal lobe [231–235]. Unlike resting, naming requires lexical search and phonological access, and, when overt, speech articulation (Table 2). These predictions match findings in several studies using this contrast, which show extensive bilateral visual system activation and modest left lateralized inferior frontal activation [223, 234–236].

Paradigm 6

Language task: Semantic Decision

Control task: Sensory Discrimination

We again assume that the same stimulus modality is used for both tasks. If the stimuli in the sensory discrimination task are non-linguistic (e.g., tones or nonsense shapes), then the semantic decision task will produce relatively greater activation in phoneme perception or visual wordform systems, depending on the sensory modality. In addition, there will be greater activation of semantic memory and semantic search mechanisms in the semantic decision task. Note that unlike the resting and passive control tasks used in the protocols described so far, effortful sensory discrimination tasks interrupt ongoing semantic processes, providing a control state that is relatively free of conceptual or semantic processing [25, 34, 35, 37, 208]. Working memory systems may or may not be activated in this contrast, depending on whether or not the control task also has a working memory component. These predictions match findings in studies using this contrast, which show left lateralized activation of phoneme perception (middle and anterior superior temporal sulcus) or visual wordform (mid-fusiform gyrus) regions, and extensive activation of left prefrontal, lateral and ventral left temporal, and left posterior parietal systems involved in semantic memory and semantic access (Fig. 2d) [5, 50, 208, 237–240].

Paradigm 7

Language task: Semantic Decision

Control task: Phonological Decision

These tasks can also be given in either the visual or auditory modality. Stimuli in the phonological decision task can be either words or pseudowords, and these can be matched to the words used in the semantic task on all structural (physical, orthographic, phonological) variables. Thus, there should be no activation of sensory or wordform systems in this contrast. There will be greater activation of semantic memory and semantic search systems in the semantic decision task. These predictions match findings in many studies using this contrast, which show activation of left prefrontal, lateral and ventral left temporal, and left posterior parietal systems believed to be involved in semantic processing (Fig. 2e) [25, 50, 142–144, 146, 148, 208, 241, 242].

Paradigm 8

Language task: Sentence or Word Reading

Control task: Passively Viewing Letterstrings

Compared to letterstrings, sentences engage visual word-form, syntactic, and phonological access systems, and make variable demands on working memory. Both reading and passive viewing probably involve semantic processing. There should be left-lateralized activation of the fusiform gyrus (visual word-form system), posterior STG and STS (phonological access), and IFG (orthographic-phonological mapping, working memory, syntax). These predictions are consistent with several studies using this contrast [111, 243–246].

In many clinical settings, the main goal of language mapping is simply to identify as many language-related areas as possible and to assess hemispheric lateralization of language. A review of Table 3 suggests that the "Semantic Decision vs. Sensory Discrimination" paradigm may offer advantages for this purpose in terms of the sheer number of regions activated and leftward lateralization of activation. Binder et al. put this prediction to a quantitative test by comparing the extent and lateralization of activation produced by five language-related task contrasts, conducted on the same 26 participants during a single scanning session [208]. These contrasts included: (1)

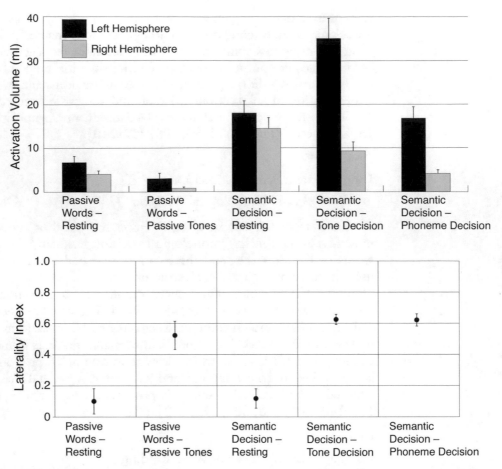

Fig. 3 Group average activation volumes (*top graph*) and laterality indexes (*bottom graph*) for five fMRI language paradigms [208]. Laterality indexes can vary from −1 (all activation in the right hemisphere) to +1 (all activation in the left hemisphere). *Error bars* represent standard error. The Semantic Decision–Tone Decision paradigm produces the greatest left hemisphere activation as well as a strongly left-lateralized pattern

passively listening to words vs. resting, (2) passively listening to words vs. passively listening to tones, (3) performing a semantic decision task with words vs. resting, (4) performing a semantic decision task with words vs. a sensory discrimination task with tones, and (5) performing a semantic decision task with words vs. a phonological task with pseudowords. As shown in Fig. 3, the Semantic Decision—Tone Decision contrast produced by far the largest activation volume in the left hemisphere, as well as an optimal combination of extensive activation and strong left-lateralization.

The example paradigms discussed here cover only a small sample of all possible language activation protocols. There are also numerous published studies employing designs that do not fit neatly into the schema provided here. Many of these represent attempts to further define or fractionate a particular language process, or to define further the functional role of a specific brain region. The reader should appreciate that the review given here is merely a coarse outline of some of the most commonly used types of stimuli and tasks. Above all, it is important to note that activations in a particular part of the language system are seldom "all or none," but vary in a graded way depending on the particular stimuli and tasks used.

4   Conclusions and Future Directions

Functional neuroimaging techniques have enhanced profoundly our understanding of how language processes are implemented in the human brain. This work has led to a number of new discoveries, such as a more precise localization of cortical networks underlying phoneme and grapheme perception, phonological access, and semantic processing. Not all of the claims made here with regard to these component language systems are uncontroversial. In particular, there are ongoing debates and a number of unresolved issues concerning localization of semantic memory and semantic retrieval systems, especially with regard to the role played by the left IFG in semantic processes [146, 209, 210, 247–250]. The notion that semantic processes are actively engaged during the resting state, though gaining traction in some quarters, is far from universal acceptance or recognition. For example, many authors continue to regard with suspicion any activation associated with semantic tasks that is not also observed in comparison to a resting baseline [147, 251, 252]. As in other areas of cognitive neuroscience, neuroimaging research on language processing has been to some extent clouded by an incomplete understanding of task demands and inadequate recognition of potential confounding factors. For example, many task contrasts are confounded by differences in task difficulty, which are well known to cause differential activation of domain-general networks involved in arousal, attention, working memory, decision, response selection, and error monitoring [151, 253–259]. Despite these well-documented effects, many researchers continue to advocate the use of covert tasks and passive conditions that provide no information about task performance, level of attention, or degree of difficulty. These deficiencies are particularly troubling in clinical studies, where the interpretation of brain activation (or lack of activation) can substantially influence clinical decision-making and patient outcome.

As the field of functional neuroimaging continues to mature, it is likely that these potential pitfalls will eventually be universally recognized and that an increasingly standardized methodological and conceptual framework for language mapping studies will emerge. FMRI practitioners, whether working in clinical or research fields, should continue to strive toward these goals.

References

1. Broca P (1861) Remarques sur le siège de la faculté du langage articulé; suivies d'une observation d'aphemie. Bulletin de la Société Anatomique de Paris 6:330–357

2. Wernicke C (1874) Der aphasische Symptomenkomplex. Cohn & Weigert, Breslau

3. Lichtheim L (1885) On aphasia. Brain 7:433–484

4. Geschwind N (1971) Aphasia. N Engl J Med 284(12):654–656

5. Binder JR, Frost JA, Hammeke TA, Cox RW, Rao SM et al (1997) Human brain language areas identified by functional MRI. J Neurosci 17(1):353–362

6. Démonet J-F, Thierry G, Cardebat D (2005) Renewal of the neurophysiology of language: functional neuroimaging. Physiol Rev 85(1):49–95

7. Binder JR, Price CJ (2001) Functional imaging of language. In: Cabeza R, Kingstone A (eds) Handbook of functional neuroimaging of cognition. MIT Press, Cambridge, MA, pp 187–251

8. Macleod CM (1991) Half a century of research on the Stroop effect: an integrative review. Psychol Bull 109:163–203

9. Reicher GM (1969) Perceptual recognition as a function of meaningfulness of stimulus material. J Exp Psychol 81:274–280

10. Warren RM, Obusek CJ (1971) Speech perception and phonemic restorations. Percept Psychophys 9:358–362

11. Ganong WF (1980) Phonetic categorization in auditory word perception. J Exp Psychol Hum Percept Perform 6:110–115

12. Marslen-Wilson WD, Tyler LK (1981) Central processes in speech understanding. Philos Trans R Soc Lond B 295:317–332

13. Carr TH, McCauley C, Sperber RD, Parmalee CM (1982) Words, pictures, and priming: on semantic activation, conscious identification, and the automaticity of information processing. J Exp Psychol Hum Percept Perform 8:757–777

14. Marcel AJ (1983) Conscious and unconscious perception: Experiments on visual masking and word recognition. Cogn Psychol 15:197–237

15. Van Orden GC (1987) A ROWS is a ROSE: spelling, sound, and reading. Mem Cogn 15(3):181–198

16. Burton MW, Baum SR, Blumstein SE (1989) Lexical effects on phonetic categorization of speech: the role of acoustic structure. J Exp Psychol Hum Percept Perform 15:567–575

17. Glaser WR (1992) Picture naming. Cognition 42:61–105

18. James W (1890) Principles of psychology. Dover, New York

19. Hebb DO (1954) The problem of consciousness and introspection. In: Adrian ED, Bremer F, Jasper HH (eds) Brain mechanisms and consciousness. A symposium. Charles C. Thomas, Springfield, IL, pp 402–421

20. Miller GA, Galanter E, Pribram K (1960) Plans and the structure of behavior. Holt, New York

21. Pope KS, Singer JL (1976) Regulation of the stream of consciousness: Toward a theory of ongoing thought. In: Schwartz GE, Shapiro D (eds) Consciousness and self-regulation. Plenum, New York, pp 101–135

22. Picton TW, Stuss DT (1994) Neurobiology of conscious experience. Curr Opin Neurobiol 4:256–265

23. Antrobus JS, Singer JL, Greenberg S (1966) Studies in the stream of consciousness: experimental enhancement and suppression of spontaneous cognitive processes. Percept Mot Skills 23:399–417

24. Teasdale JD, Proctor L, Lloyd CA, Baddeley AD (1993) Working memory and stimulus-independent thought: effects of memory load and presentation rate. Eur J Cogn Psychol 5(4):417–433

25. Binder JR, Frost JA, Hammeke TA, Bellgowan PSF, Rao SM et al (1999) Conceptual processing during the conscious resting state: a functional MRI study. J Cogn Neurosci 11(1):80–93

26. McKiernan KA, Kaufman JN, Kucera-Thompson J, Binder JR (2003) A parametric manipulation of factors affecting task-induced deactivation in functional neuroimaging. J Cogn Neurosci 15(3):394–408

27. Révész G (ed) (1954) Thinking and speaking: a symposium. North Holland Publishing, Amsterdam

28. Weiskrantz L (ed) (1988) Thought without language. Clarendon, Oxford

29. Vygotsky LS (1962) Thought and language. Wiley, New York

30. Karmiloff-Smith A (1992) Beyond modularity: a developmental perspective on cognitive science. MIT Press, Cambridge, MA

31. Andreasen NC, O'Leary DS, Cizadlo T, Arndt S, Rezai K et al (1995) Remembering the past: two facets of episodic memory explored with positron emission tomography. Am J Psychiatry 152:1576–1585

32. Shulman GL, Fiez JA, Corbetta M, Buckner RL, Meizin FM et al (1997) Common blood flow changes across visual tasks: II. Decreases in cerebral cortex. J Cogn Neurosci 9(5):648–663

33. Mazoyer B, Zago L, Mellet E, Bricogne S, Etard O et al (2001) Cortical networks for working memory and executive functions sustain the conscious resting state in man. Brain Res Bull 54(3):287–298

34. Stark CE, Squire LR (2001) When zero is not zero: the problem of ambiguous baseline conditions in fMRI. Proc Natl Acad Sci U S A 98(22):12760–12766

35. McKiernan KA, D'Angelo BR, Kaufman JN, Binder JR (2006) Interrupting the "stream of consciousness": an fMRI investigation. Neuroimage 29(4):1185–1191

36. Smallwood J, Schooler JW (2006) The restless mind. Psychol Bull 132(6):946–958

37. Mason MF, Norton MI, Van Horn JD, Wegner DM, Grafton ST et al (2007) Wandering minds: the default network and stimulus-independent thought. Science 315(5810):393–395

38. Andrews-Hanna JR (2012) The brain's default network and its adaptive role in internal mentation. Neuroscientist 18(3):251–270

39. Abutalebi J, Cappa S, Perani D (2005) What can functional neuroimaging tell us about the bilingual brain? In: Kroll J, de Groot AMB (eds) Handbook of bilingualism: psycholinguistic approaches. Oxford University Press, New York, pp 497–515

40. Hernandez A, Li P, MacWhinney B (2005) The emergence of competing modules in bilingualism. Trends Cogn Sci 9(5):220–225

41. Corina DP, Knapp H (2006) Sign language processing and the mirror neuron system. Cortex 42(4):529–539

42. MacSweeney M, Capek CM, Campbell R, Woll B (2008) The signing brain: the neurobiology of sign language. Trends Cogn Sci 12(11):432–440

43. Corina D, Singleton J (2009) Developmental social cognitive neuroscience: insights from deafness. Child Dev 80(4):952–967

44. Emmorey K, McCullough S (2009) The bimodal bilingual brain: effects of sign language experience. Brain Lang 109:124–132

45. Kotz SA (2009) A critical review of ERP and fMRI evidence on L2 syntactic processing. Brain Lang 109(2):68–74

46. van Heuven WJ, Dijkstra T (2010) Language comprehension in the bilingual brain: fMRI and ERP support for psycholinguistic models. Brain Res Rev 64(1):104–122

47. Buchweitz A, Prat C (2013) The bilingual brain: flexibility and control in the human cortex. Phys Life Rev 10(4):428–443

48. Li P, Legault J, Litcofsky KA (2014) Neuroplasticity as a function of second language learning: anatomical changes in the human brain. Cortex 58:301–324

49. Bogen JE, Bogen GM (1976) Wernicke's region – where is it? Ann NY Acad Sci 290:834–843

50. Démonet JF, Chollet F, Ramsay S, Cardebat D, Nespoulous JL et al (1992) The anatomy of phonological and semantic processing in normal subjects. Brain 115:1753–1768

51. Zatorre RJ, Evans AC, Meyer E, Gjedde A (1992) Lateralization of phonetic and pitch discrimination in speech processing. Science 256:846–849

52. Mummery CJ, Ashburner J, Scott SK, Wise RJS (1999) Functional neuroimaging of speech perception in six normal and two aphasic subjects. J Acoust Soc Am 106:449–457

53. Belin P, Zatorre RJ, Lafaille P, Ahad P, Pike B (2000) Voice-selective areas in human auditory cortex. Nature 403:309–312

54. Binder JR, Frost JA, Hammeke TA, Bellgowan PSF, Springer JA et al (2000) Human temporal lobe activation by speech and nonspeech sounds. Cereb Cortex 10:512–528

55. Blumstein SE, Myers EB, Rissman J (2005) The perception of voice onset time: an fMRI investigation of phonetic category structure. J Cogn Neurosci 17(9):1353–1366

56. Desai R, Liebenthal E, Possing ET, Waldron E, Binder JR (2005) Volumetric vs surface-based alignment for localization of auditory cortex activation. Neuroimage 26(4):1019–1029

57. Turkeltaub PE, Coslett HB (2010) Localization of sublexical speech perception components. Brain Lang 114:1–15

58. Dehaene-Lambertz G, Pallier C, Serniclaes W, Sprenger-Charolles L, Jobert A et al (2005) Neural correlates of switching from auditory to speech perception. Neuroimage 24:21–33

59. Liebenthal E, Binder JR, Spitzer SM, Possing ET, Medler DA (2005) Neural substrates of phonemic perception. Cerebral Cortex 15:1621–1631

60. Möttönen R, Calvert GA, Jaaskelainen IP, Matthews PM, Thesen A et al (2006) Perceiving identical sounds as speech or non-speech modulates activity in the left posterior superior temporal sulcus. Neuroimage 30:563–569

61. Obleser J, Zimmerman J, Van Meter J, Rauschecker JP (2007) Multiple stages of auditory speech perception reflected in event-related fMRI. Cereb Cortex 17:2251–2257

62. Desai R, Liebenthal E, Waldron E, Binder JR (2008) Left posterior temporal regions are sensitive to auditory categorization. J Cogn Neurosci 20:1174–1188

63. Liebenthal E, Desai R, Ellingson MM, Ramachandran B, Desai A et al (2010) Specialization along the left superior temporal sulcus for phonemic and non-phonemic categorization. Cerebral Cortex 20:2958–2970

64. Barrett AM (1910) A case of pure word-deafness with autopsy. J Nerv Ment Dis 37(2):73–92

65. Henschen SE (1918–1919) On the hearing sphere. Acta Otolaryngol 1:423–486

66. Wohlfart G, Lindgren A, Jernelius B (1952) Clinical picture and morbid anatomy in a case of "pure word deafness". J Nerv Ment Dis 116:818–827

67. Lhermitte F, Chain F, Escourolle R, Ducarne B, Pillon A et al (1972) Etude des troubles perceptifs auditifs dans les lésions temporales bilatérales. Revue Neurologique 24:327–351

68. Kanshepolsky J, Kelley J, Waggener JD (1973) A cortical auditory disorder: clinical, audiologic and pathologic aspects. Neurology 23:699–705

69. Buchman AS, Garron DC, Trost-Cardamone JE, Wichter MD, Schwartz D (1986) Word deafness: one hundred years later. J Neurol Neurosurg Psychiatry 49:489–499

70. Poeppel D (2001) Pure word deafness and the bilateral processing of the speech code. Cogn Sci 25:679–693

71. Liepmann H, Storch E (1902) Der mikroskopische gehirnbefund bei dem fall gorstelle. Monatsschrift fur Psychiatrie und Neurologie 11:115–120

72. Stefanatos GA, Gershkoff A, Madigan S (2005) On pure word deafness, temporal processing and the left hemisphere. J Int Neuropsychol Soc 11(4):456–470

73. Déjerine J (1891) Sur un cas de cécité verbal avec agraphie, suivi d'autopsie. C R Seances Soc Biol 3:197–201

74. Déjerine J (1892) Contribution à l'étude anatomo-pathologique et clinique des différentes variétés de cécité verbale. C R Seances Soc Biol 44(2):61–90

75. Déjerine J, Vialet N (1893) Contribution a l'étude de la localisation anatomique de la cécité verbale pure. C R Seances Soc Biol 45:790–793

76. Geschwind N (1965) Disconnection syndromes in animals and man. Brain 88:237–294, 585-644

77. Greenblatt SH (1976) Subangular alexia without agraphia or hemianopsia. Brain Lang 3:229–245

78. Vincent FM, Sadowsky CH et al (1977) Alexia without agraphia, hemianopia, or color-naming defect: a disconnection syndrome. Neurology 27:689–691

79. Henderson VW (1986) Anatomy of posterior pathways in reading: a reassessment. Brain Lang 29:119–133

80. Binder JR, Mohr JP (1992) The topography of callosal reading pathways: a case-control analysis. Brain 115:1807–1826

81. Beversdorf DQ, Ratcliffe NR, Rhodes CH, Reeves AG (1997) Pure alexia: clinical-pathological evidence for a lateralized visual language association cortex. Clin Neuropathol 16(6):328–331

82. Sakurai Y, Takeuchi S, Takada T, Horiuchi E, Nakase H et al (2000) Alexia caused by a fusiform or posterior inferior temporal lesion. J Neuro Sci 178:42–51

83. Leff AP, Crewes H, Plant GT, Scott SK, Kennard C et al (2001) The functional anatomy of single-word reading in patients with hemianopic and pure alexia. Brain 124:510–521

84. Cohen L, Martinaud O, Lemer C, Lehericy S, Samson Y et al (2003) Visual word recognition in the left and right hemispheres: anatomical and functional correlates of peripheral alexias. Cereb Cortex 13:1313–1333

85. Tarkiainen A, Helenius P, Hansen PC, Cornelissen PL, Salmelin R (1999) Dynamics of letter string perception in the human occipitotemporal cortex. Brain 122:2119–2131

86. Cohen L, Dehaene S, Naccache L, Lehéricy S, Dehaene-Lambertz G et al (2000) The visual word form area. Spatial and temporal characterization of an initial stage of reading in normal subjects and posterior split-brain patients. Brain 123:291–307

87. Dehaene S, Naccache L, Cohen L, Bihan DL, Mangin JF et al (2001) Cerebral mechanisms of word masking and unconscious repetition priming. Nat Neurosci 4(7):752–758

88. Polk TA, Farah MJ (2002) Functional MRI evidence for an abstract, not perceptual, word-form area. J Exp Psychol Gen 131:65–72

89. Cohen L, Jobert A, Le Bihanc D, Dehaene S (2004) Distinct unimodal and multimodal regions for word processing in the left temporal cortex. Neuroimage 23:1256–1270

90. Vinckier F, Dehaene S, Jobert A, Dubus J, Sigman M et al (2007) Hierarchical coding of letter strings in the ventral stream: dissecting the inner organization of the visual word-form system. Neuron 55(1):143–156

91. Glezer LS, Jiang X, Riesenhuber M (2009) Evidence for highly selective neuronal tuning of whole words in the "Visual Word Form Area". Neuron 62:199–204

92. Mano QR, Humphries CJ, Desai R, Seidenberg MS, Osmon DC et al (2013) The role of left occipitotemporal cortex in reading: reconciling stimulus, task, and lexicality effects. Cereb Cortex 23(4):988–1001

93. Binder JR, Medler DA, Westbury CF, Liebenthal E, Buchanan L (2006) Tuning of the human left fusiform gyrus to sublexical orthographic structure. Neuroimage 33:739–748

94. Patterson KE, Kay J (1982) Letter-by-letter reading: psychological descriptions of a neurological syndrome. Q J Exp Psychol 34A:411–442

95. Reuter-Lorenz PA, Brunn JL (1990) A prelexical basis for letter-by-letter reading: a case study. Cognit Neuropsychol 7:1–20

96. Behrmann M, Plaut DC, Nelson J (1998) A literature review and new data supporting an interactive activation account of letter-by-letter reading. In: Coltheart M (ed) Pure alexia (letter-by-letter reading). Pscyhology, Hove, UK, pp 7–51

97. Liepmann H, Pappenheim M (1914) Über einem Fall von sogenannter Leitungsaphasie mit anatomischer Befund. Z Gesamte Neurol Psychiatr 27:1–41

98. Benson DF, Sheremata WA, Bouchard R, Segarra JM, Price D et al (1973) Conduction aphasia. A clinicopathological study. Arch Neurol 28:339–346

99. Damasio H, Damasio AR (1980) The anatomical basis of conduction aphasia. Brain 103:337–350

100. Anderson JM, Gilmore R, Roper S, Crosson B, Bauer RM et al (1999) Conduction aphasia and the arcuate fasciculus: a reexamination of the Wernicke-Geschwind model. Brain Lang 70:1–12

101. Quigg M, Fountain NB (1999) Conduction aphasia elicited by stimulation of the left posterior superior temporal gyrus. J Neurol Neurosurg Psychiatry 66:393–396

102. Axer H, Keyserlingk AG, Berks G, Keyserlingk DF (2001) Supra- and infrasylvian conduction aphasia. Brain Lang 76:317–331

103. Fridriksson J, Kjartansson O, Morgan PS, Hjaltason H, Magnusdottir S et al (2010) Impaired speech repetition and left parietal lobe damage. J Neurosci 30:11057–11061

104. Buchsbaum BR, Baldo J, D'Esposito M, Dronkers N, Okada K et al (2011) Conduction aphasia, sensory-motor integration, and phonological short-term memory: an aggregate analysis of lesion and fMRI data. Brain Lang 119:119–128

105. Roeltgen DP, Sevush S, Heilman KM (1983) Phonological agraphia: writing by the lexical-semantic route. Neurology 33:755–765

106. Alexander MP, Friedman RB, Loverso F, Fischer RS (1992) Lesion localization of phonological agraphia. Brain Lang 43:83–95

107. Rapcsak SZ, Beeson PM, Henry ML, Leyden A, Kim E et al (2009) Phonological dyslexia and dysgraphia: cognitive mechanisms and neural substrates. Cortex 45(5):575–591

108. Pillay SB, Stengel BC, Humphries C, Book DS, Binder JR (2014) Cerebral localization of impaired phonological retrieval during rhyme judgment. Ann Neurol 76:738–746

109. Howard D, Patterson K, Wise R, Brown WD, Friston K et al (1992) The cortical localization of the lexicons. Brain 115:1769–1782

110. Price CJ, Wise RJS, Watson JDG, Patterson K, Howard D et al (1994) Brain activity during reading. The effects of exposure duration and task. Brain 117:1255–1269

111. Price CJ, Wise RSJ, Frackowiak RSJ (1996) Demonstrating the implicit processing of visually presented words and pseudowords. Cereb Cortex 6:62–70

112. Hickok G, Erhard P, Kassubek J, Helms-Tillery AK, Naeve-Velguth S et al (2000) A functional magnetic resonance imaging study of the role of left posterior superior temporal gyrus in speech production: implications for the explanation of conduction aphasia. Neurosci Lett 287:156–160

113. Booth JR, Burman DD, Meyer JR, Gitelman DR, Parrish TB et al (2002) Functional anatomy of intra- and cross-modal lexical tasks. Neuroimage 16:7–22

114. Indefrey P, Levelt WJM (2004) The spatial and temporal signatures of word production components. Cognition 92(1-2):101–144

115. Burton MW, Locasto PC, Krebs-Noble D, Gullapalli RP (2005) A systematic investigation of the functional neuroanatomy of auditory and visual phonological processing. Neuroimage 26(3):647–661

116. Callan AM, Callan DE, Masaki S (2005) When meaningless symbols become letters: neural activity change in learning new phonograms. Neuroimage 28:553–562

117. Fiez JA, Raichle ME, Balota DA, Tallal P, Petersen SE (1996) PET activation of posterior temporal regions during auditory word presentation and verb generation. Cereb Cortex 6:1–10

118. Warburton E, Wise RJS, Price CJ, Weiller C, Hadar U et al (1996) Noun and verb retrieval by normal subjects. Studies with PET. Brain 119:159–179

119. Wise RSJ, Scott SK, Blank SC, Mummery CJ, Murphy K et al (2001) Separate neural subsystems within 'Wernicke's area'. Brain 124:83–95

120. Baddeley AD (1986) Working memory. Oxford University Press, Oxford

121. Paulesu E, Frith CD, Frackowiak RSJ (1993) The neural correlates of the verbal component of working memory. Nature 362:342–345

122. Hickok G, Buchsbaum B, Humphries C, Muftuler T (2003) Auditory-motor interaction revealed by fMRI: speech, music, and working memory in area Spt. J Cognit Neurosci 15(5):673–682

123. Buchsbaum BR, Olsen RK, Koch P, Berman KF (2005) Human dorsal and ventral auditory streams subserve rehearsal-based and echoic processes during verbal working memory. Neuron 48(4):687–697

124. Buchsbaum BR, D'Esposito M (2008) The search for the phonological store: from loop to convolution. J Cogn Neurosci 20(5):762–778

125. Acheson DJ, Hamidi M, Binder JR, Postle BR (2011) A common neural substrate for language production and verbal working memory. J Cogn Neurosci 23:1358–1367

126. DeWitt I, Rauschecker JP (2012) Phoneme and word recognition in the auditory ventral stream. Proc Natl Acad Sci U S A 109:E505–514

127. Bréal M (1897) Essai de sémantique (science des significations). Librairie Hachette, Paris

128. Alexander MP, Hiltbrunner B, Fischer RS (1989) Distributed anatomy of transcortical sensory aphasia. Arch Neurol 46:885–892

129. Hart J, Gordon B (1990) Delineation of single-word semantic comprehension deficits in aphasia, with anatomic correlation. Ann Neurol 27(3):226–231

130. Chertkow H, Bub D, Deaudon C, Whitehead V (1997) On the status of object concepts in aphasia. Brain Lang 58:203–232

131. Tranel D, Damasio H, Damasio AR (1997) A neural basis for the retrieval of conceptual knowledge. Neuropsychologia 35:1319–1327

132. Gainotti G (2000) What the locus of brain lesion tells us about the nature of the cognitive defect underlying category-specific disorders: a review. Cortex 36:539–559

133. Damasio H, Tranel D, Grabowski T, Adolphs R, Damasio A (2004) Neural systems behind word and concept retrieval. Cognition 92:179–229

134. Dronkers NF, Wilkins DP, Van Valin RD, Redfern BB, Jaeger JJ (2004) Lesion analysis of the brain areas involved in language comprehension. Cognition 92:145–177

135. Patterson K, Nestor PJ, Rogers TT (2007) Where do you know what you know? The representation of semantic knowledge in the human brain. Nat Rev Neurosci 8:976–987

136. Thompson-Schill SL (2003) Neuroimaging studies of semantic memory: inferring "how" from "where". Neuropsychologia 41:280–292

137. Martin A (2007) The representation of object concepts in the brain. Annu Rev Psychol 58:25–45

138. Binder JR, Desai R, Conant LL, Graves WW (2009) Where is the semantic system? A critical review and meta-analysis of 120 functional neuroimaging studies. Cereb Cortex 19:2767–2796

139. Kiefer M, Pulvermüller F (2012) Conceptual representations in mind and brain: theoretical developments, current evidence and future directions. Cortex 48:805–825

140. Meteyard L, Rodriguez Cuadrado S, Bahrami B, Vigliocco G (2012) Coming of age: a review of embodiment and the neuroscience of semantics. Cortex 48:788–804

141. Mummery CJ, Patterson K, Hodges JR, Wise RJS (1996) Generating 'tiger' as an animal name or a word beginning with T: differences in brain activation. Proc R Soc Lond B 263:989–995

142. Price CJ, Moore CJ, Humphreys GW, Wise RJS (1997) Segregating semantic from phonological processes during reading. J Cognit Neurosci 9(6):727–733

143. Cappa SF, Perani D, Schnur T, Tettamanti M, Fazio F (1998) The effects of semantic category and knowledge type on lexical-semantic access: a PET study. Neuroimage 8(4):350–359

144. Roskies AL, Fiez JA, Balota DA, Raichle ME, Petersen SE (2001) Task-dependent modulation of regions in the left inferior frontal cortex during semantic processing. J Cognit Neurosci 13(6):829–843

145. Binder JR, McKiernan KA, Parsons M, Westbury CF, Possing ET et al (2003) Neural correlates of lexical access during visual word recognition. J Cognit Neurosci 15(3):372–393

146. Devlin JT, Matthews PM, Rushworth MFS (2003) Semantic processing in the left inferior prefrontal cortex: a combined functional magnetic resonance imaging and transcranial

magnetic stimulation study. J Cogn Neurosci 15(1):71–84

147. Rissman J, Eliassen JC, Blumstein SE (2003) An event-related fMRI investigation of implicit semantic priming. J Cogn Neurosci 15(8):1160–1175

148. Scott SK, Leff AP, Wise RJS (2003) Going beyond the information given: a neural system supporting semantic interpretation. Neuroimage 19:870–876

149. Ischebeck A, Indefrey P, Usui N, Nose I, Hellwig F et al (2004) Reading in a regular orthography: an fMRI study investigating the role of visual familiarity. J Cognit Neurosci 16(5):727–741

150. Binder JR, Westbury CF, Possing ET, McKiernan KA, Medler DA (2005) Distinct brain systems for processing concrete and abstract concepts. J Cognit Neurosci 17(6):905–917

151. Binder JR, Medler DA, Desai R, Conant LL, Liebenthal E (2005) Some neurophysiological constraints on models of word naming. Neuroimage 27:677–693

152. Sabsevitz DS, Medler DA, Seidenberg M, Binder JR (2005) Modulation of the semantic system by word imageability. Neuroimage 27:188–200

153. Vandenbulcke M, Peeters R, Fannes K, Vandenberghe R (2006) Knowledge of visual attributes in the right hemisphere. Nat Neurosci 9(7):964–970

154. Damasio H (1989) Neuroimaging contributions to the understanding of aphasia. In: Boller F, Grafman J (eds) Handbook of neuropsychology. Elsevier, Amsterdam, pp 3–46

155. Rapcsak SZ, Rubens AB (1994) Localization of lesions in transcortical aphasia. In: Kertesz A (ed) Localization and neuroimaging in neuropsychology. Academic, San Diego, pp 297–329

156. Berthier ML (1999) Transcortical aphasias. Psychology, Hove

157. Mummery CJ, Patterson K, Price CJ, Ashburner J, Frackowiak RS et al (2000) A voxel-based morphometry study of semantic dementia: relationship between temporal lobe atrophy and semantic memory. Ann Neurol 47:36–45

158. Rosen HJ, Gorno-Tempini ML, Goldman WP et al (2002) Patterns of brain atrophy in frontotemporal dementia and semantic dementia. Neurology 58:198–208

159. Davies RR, Hodges JR, Krill JJ, Patterson K, Halliday GM et al (2005) The pathological basis of semantic dementia. Brain 128:1984–1995

160. Rohrer JD, Warren JD, Modat M, Ridgway GR, Douiri A et al (2009) Patterns of cortical thinning in the language variants of frontotemporal lobar degeneration. Neurology 72:1562–1569

161. Mion M, Patterson K, Acosta-Cabronero J, Pengas G, Izquierdo-Garcia D et al (2010) What the left and right anterior fusiform gyri tell us about semantic memory. Brain 133:3256–3268

162. Warrington EK, Shallice T (1984) Category specific semantic impairments. Brain 107:829–854

163. Gonnerman LM, Andersen ES, Devlin JT, Kempler D, Seidenberg MS (1997) Double dissociation of semantic categories in Alzheimer's disease. Brain Lang 57:254–279

164. Chan AS, Salmon DP, De La Pena J (2001) Abnormal semantic network for "animals" but not "tools" in patients with Alzheimer's disease. Cortex 37:197–217

165. Fung TD, Chertkow H, Whatmough C, Murtha S, Péloquin L et al (2001) The spectrum of category effects in object and action knowledge in dementia of the Alzheimer's type. Neuropsychology 15(3):371–379

166. Alexander MP, Benson DF, Stuss DT (1989) Frontal lobes and language. Brain Lang 37:656–691

167. Mazoyer BM, Tzourio N, Frak V, Syrota A, Murayama N et al (1993) The cortical representation of speech. J Cognit Neurosci 5(4):467–479

168. Stowe LA, Paans AMJ, Wijers AA, Zwarts F, Mulder G et al (1999) Sentence comprehension and word repetition: a positron emission tomography investigation. Psychophysiology 36:786–801

169. Friederici AD, Meyer M, von Cramon DY (2000) Auditory language comprehension: an event-related fMRI study on the processing of syntactic and lexical information. Brain Lang 74:289–300

170. Vandenberghe R, Nobre AC, Price CJ (2002) The response of left temporal cortex to sentences. J Cogn Neurosci 14(4):550–560

171. Humphries C, Swinney D, Love T, Hickok G (2005) Response of anterior temporal cortex to syntactic and prosodic manipulations during sentence processing. Hum Brain Mapp 26:128–138

172. Humphries C, Binder JR, Medler DA, Liebenthal E (2006) Syntactic and semantic modulation of neural activity during auditory sentence comprehension. J Cognit Neurosci 18:665–679

173. Pallier C, Devauchelle AD, Devauchelle AD, Dehaene S (2011) Cortical representation of the constituent structure of sentences. Proc Natl Acad Sci U S A 108(6):2522–2527

174. Humphries C, Binder JR, Medler DA, Liebenthal E (2007) Time course of semantic processes during sentence comprehension: an fMRI study. Neuroimage 36(3):924–932

175. Kang AM, Constable RT, Gore JC, Avrutin S (1999) An event-related fMRI study of implicit phrase-level syntactic and semantic processing. Neuroimage 10(5):98–110

176. Embick D, Marantz A, Miyashita Y, O'Neil W, Sakai KL (2000) A syntactic specialization for Broca's area. Proc Natl Acad Sci USA 97:6150–6154

177. Meyer M, Friederici AD, von Cramon DY (2000) Neurocognition of auditory sentence comprehension: event related fMRi reveals sensitivity to syntactic violations and task demands. Cogn Brain Res 9:19–33

178. Ni W, Constable RT, Mencl WE, Pugh KR, Fullbright RK et al (2000) An event-related neuroimaging study distinguishing form and content in sentence processing. J Cognit Neurosci 12:120–133

179. Newman AJ, Pancheva R, Ozawa K, Neville HJ, Ullman MT (2001) An event-related fMRI study of syntactic and semantic violations. J Psycholinguist Res 30:339–364

180. Kuperberg GR, Holcomb PJ, Sitnikova T, Greve D, Dale AM et al (2003) Distinct patterns of neural modulation during the processing of conceptual and syntactic anomalies. J Cognit Neurosci 15(2):272–293

181. Caplan D, Alpert N, Waters GS (1998) Effects of syntactic structure and prepositional number on patterns of regional cerebral blood flow. J Cognit Neurosci 10(4):541–552

182. Fiebach CJ, Schlesewsky M, Friederici AD (2001) Syntactic working memory and the establishment of filler-gap dependencies: insights from ERPs and fMRI. J Psycholinguist Res 30:321–338

183. Ben-Shachar M, Hendler T, Kahn I, Ben-Bashat D, Grodzinsky Y (2003) The neural reality of syntactic transformations: evidence form fMRI. Psychol Sci 14:433–440

184. Friederici AD, Rüschemeyer SA, Hahne A, Fiebach CJ (2003) The role of left inferior frontal gyrus and superior temporal cortex in sentence comprehension: localizing syntactic and semantic processes. Cerebr Cortex 13:170–177

185. Ben-Shachar M, Palti D, Grodzinsky Y (2004) The neural correlates of syntactic movement: converging evidence from two fMRI experiments. Neuroimage 21:1320–1336

186. Wartenburger I, Heekeren HR, Burchert F, Heinemann S, De Bleser R et al (2004) Neural correlates of syntactic transformations. Hum Brain Mapp 22:72–81

187. Fiebach CJ, Schlesewsky M, Lohmann G (2005) Revisiting the role of Broca's area in sentence processing: Syntactic integration versus syntactic working memory. Hum Brain Mapp 24:79–91

188. Chen E, West WC, Waters G, Caplan D (2006) Determinants of BOLD signal correlates of processing object-extracted relative clauses. Cortex 42:591–604

189. Caplan D, Stanczak L, Waters G (2008) Syntactic and thematic constraint effects on blood oxygenation level dependent signal correlates of comprehension of relative clauses. J Cogn Neurosci 20(4):643–656

190. Just MA, Carpenter PA, Keller TA, Eddy WF, Thulborn KR (1996) Brain activation modulated by sentence comprehension. Science 274:114–116

191. Stowe LA, Broere CA, Paans AM, Wijers AA, Mulder G et al (1998) Localizing components of a complex task: Sentence processing and working memory. Neuroreport 9:2995–2999

192. Caplan D, Waters GS (1999) Verbal working memory and sentence comprehension. Behav Brain Sci 22(1):77–94

193. Keller TA, Carpenter PA, Just MA (2001) The neural bases of sentence comprehension: a fMRI examination of syntactic and lexical processing. Cereb Cortex 11:223–237

194. Cooke A, Zurif EB, DeVita C, Alsop D, Koenig P et al (2002) Neural basis for sentence comprehension: grammatical and short-term memory components. Hum Brain Mapp 15:80–94

195. Novick JM, Trueswell JC, Thompson-Schill SL (2005) Cognitive control and parsing: reexamining the role of Broca's area in sentence comprehension. Cognit Affect Behav Neurosci 5(3):263–281

196. Rogalsky C, Hickok G (2011) The role of Broca's area in sentence comprehension. J Cogn Neurosci 23(7):1664–1680

197. Caplan D, Waters G (2013) Memory mechanisms supporting syntactic comprehension. Psychon Bull Rev 20(2):243–268

198. Caplan D (2001) Functional neuroimaging studies of syntactic processing. J Psycholinguist Res 30(3):297–320

199. Friederici AD, Kotz SA (2003) The brain basis of syntactic processes: functional imaging and lesion studies. Neuroimage 20:S8–S17

200. Martin RC (2003) Language processing: functional organization and neuroanatomical basis. Annu Rev Psychol 54:55–89

201. Grodzinsky Y, Friederici AD (2006) Neuroimaging of syntax and syntactic processing. Curr Opin Neurobiol 16:240–246

202. Bornkessel-Schlesewsky I, Schlesewsky M (2013) Reconciling time, space and function: a new dorsal-ventral stream model of sentence comprehension. Brain Lang 125(1):60–76

203. Badre D, Poldrack RA, Pare-Blagoev EJ, Insler RZ, Wagner AD (2005) Dissociable controlled retrieval and generalized selection mechanisms in ventrolateral prefrontal cortex. Neuron 47:907–918

204. Scott SK, Blank C, Rosen S, Wise RJS (2000) Identification of a pathway for intelligible speech in the left temporal lobe. Brain 123:2400–2406

205. Davis MH, Johnsrude IS (2003) Hierarchical processing in spoken language comprehension. J Neurosci 23(8):3423–3431

206. Specht K, Reul J (2003) Functional segregation of the temporal lobes into highly differentiated subsystems for auditory perception: an auditory rapid event-related fMRI task. Neuroimage 20:1944–1954

207. Uppenkamp S, Johnsrude IS, Norris D, Marslen-Wilson W, Patterson RD (2006) Locating the initial stages of speech-sound processing in human temporal cortex. Neuroimage 31:1284–1296

208. Binder JR, Swanson SJ, Hammeke TA, Sabsevitz DS (2008) A comparison of five fMRI protocols for mapping speech comprehension systems. Epilepsia 49(12):1980–1997

209. Thompson-Schill SL, Aguirre GK, D'Esposito M, Farah MJ (1997) Role of left inferior prefrontal cortex in retrieval of semantic knowledge: a reevaluation. Proc Natl Acad Sci U S A 94:14792–14797

210. Thompson-Schill SL, D'Esposito M, Kan IP (1999) Effects of repetition and competition on activity in left prefrontal cortex during word generation. Neuron 23:513–522

211. Wise R, Chollet F, Hadar U, Friston K, Hoffner E et al (1991) Distribution of cortical neural networks involved in word comprehension and word retrieval. Brain 114:1803–1817

212. Price CJ, Wise RJS, Warburton EA, Moore CJ, Howard D et al (1996) Hearing and saying. The functional neuro-anatomy of auditory word processing. Brain 119:919–931

213. Price C, Wise R, Ramsay S, Friston K, Howard D et al (1992) Regional response differences within the human auditory cortex when listening to words. Neurosci Lett 146:179–182

214. Binder JR, Rao SM, Hammeke TA, Frost JA, Bandettini PA et al (1994) Effects of stimulus rate on signal response during functional magnetic resonance imaging of auditory cortex. Cogn Brain Res 2:31–38

215. Lehéricy S, Cohen L, Bazin B, Samson S, Giacomini E et al (2000) Functional MR evaluation of temporal and frontal language dominance compared with the Wada test. Neurology 54:1625–1633

216. Humphries C, Willard K, Buchsbaum B, Hickok G (2001) Role of anterior temporal cortex in auditory sentence comprehension: an fMRI study. Neuroreport 12:1749–1752

217. Crinion JT, Lambon-Ralph MA, Warburton EA, Howard D, Wise RJS (2003) Temporal lobe regions engaged during normal speech comprehension. Brain 126:1193–1201

218. Spitsyna G, Warren JE, Scott SK, Turkheimer FE, Wise RJS (2006) Converging language streams in the human temporal lobe. J Neurosci 26(28):7328–7336

219. Awad M, Warren JE, Scott SK, Turkheimer FE, Wise RJS (2007) A common system for the comprehension and production of narrative speech. J Neurosci 27(43):11455–11464

220. Eulitz C, Elbert T, Bartenstein P, Weiller C, Müller SP et al (1994) Comparison of magnetic and metabolic brain activity during a verb generation task. NeuroReport 6:97–100

221. Ojemann JG, Buckner RL, Akbudak E, Snyder AZ, Ollinger JM et al (1998) Functional MRI studies of word-stem completion: reliability across laboratories and comparison to blood flow imaging with PET. Hum Brain Mapp 6:203–215

222. Yetkin FZ, Swanson S, Fischer M, Akansel G, Morris G et al (1998) Functional MR of frontal lobe activation: comparison with Wada language results. Am J Neuroradiol 19:1095–1098

223. Benson RR, FitzGerald DB, LeSeuer LL, Kennedy DN, Kwong KK et al (1999) Language dominance determined by whole brain functional MRI in patients with brain lesions. Neurology 52:798–809

224. Palmer ED, Rosen HJ, Ojemann JG, Buckner RL, Kelley WM et al (2001) An event-related fMRI study of overt and covert word stem completion. Neuroimage 14:182–193

225. Liégois F, Connelly A, Salmond CH, Gadian DG, Vargha-Khadem F et al (2002) A direct test for lateralization of language activation using fMRI: comparison with invasive assess-

ments in children with epilepsy. Neuroimage 17:1861–1867

226. Raichle ME, Fiez JA, Videen TO, MacLeod AM, Pardo JV et al (1994) Practice-related changes in human brain functional anatomy during non-motor learning. Cereb Cortex 4(1):8–26

227. Petersen SE, Fox PT, Posner MI, Mintun M, Raichle ME (1988) Positron emission tomographic studies of the cortical anatomy of single-word processing. Nature 331:585–589

228. Malach R, Reppas JB, Benson RR, Kwong KK, Jiang H et al (1995) Object-related activity revealed by functional magnetic resonance imaging in human occipital cortex. Proc Natl Acad Sci USA 92:8135–8139

229. Kanwisher N, Woods R, Iacoboni M, Mazziotta J (1996) A locus in human extrastriate cortex for visual shape analysis. J Cognit Neurosci 91:133–142

230. Grill-Spector K, Kushnir T, Edelman S, Avidian-Carmel G, Itzchak Y et al (1999) Differential processing of objects under various viewing conditions in the human lateral occipital complex. Neuron 24:187–203

231. Martin A, Wiggs CL, Ungerleider LG, Haxby JV (1996) Neural correlates of category-specific knowledge. Nature 379(6566):649–652

232. Price CJ, Moore CJ, Humphreys GW, Frackowiak RSJ, Friston KJ (1996) The neural regions sustaining object recognition and naming. Proc Roy Soc Lond B 263:1501–1507

233. Zelkowicz BJ, Herbster AN, Nebes RD, Mintun MA, Becker JT (1998) An examination of regional cerebral blood flow during object naming tasks. J Int Neuropsychol Soc 4:160–166

234. Murtha S, Chertkow H, Beauregard M, Evans A (1999) The neural substrate of picture naming. J Cognit Neurosci 11(4):399–423

235. Price CJ, Devlin JT, Moore CJ, Morton C, Laird AR (2005) Meta-analyses of object naming: effect of baseline. Hum Brain Mapp 25:70–82

236. Kiasawa M, Inoue C, Kawasaki T, Tokoro T, Ishii K et al (1996) Functional neuroanatomy of object naming: a PET study. Graefes Arch Clin Exp Ophthalmol 234:110–115

237. Vandenberghe R, Price C, Wise R, Josephs O, Frackowiak RSJ (1996) Functional anatomy of a common semantic system for words and pictures. Nature 383:254–256

238. Carpentier A, Pugh KR, Westerveld M, Studholme C, Skrinjar O et al (2001) Functional MRI of language processing: dependence on input modality and temporal lobe epilepsy. Epilepsia 42:1241–1254

239. Devlin JT, Russell RP, Davis MH, Price CJ, Moss HE et al (2002) Is there an anatomical basis for category-specificity? Semantic memory studies with PET and fMRI. Neuropsychologia 40:54–75

240. Xu B, Grafman J, Gaillard WD, Spanaki M, Ishii K et al (2002) Neuroimaging reveals automatic speech coding during perception of written word meaning. Neuroimage 17(2):859–870

241. Mummery CJ, Patterson K, Hodges JR, Price CJ (1998) Functional neuroanatomy of the semantic system: divisible by what? J Cogn Neurosci 10(6):766–777

242. Miceli G, Turriziani P, Caltagirone C, Capasso R, Tomaiuolo F et al (2002) The neural correlates of grammatical gender: an fMRI investigation. J Cognit Neurosci 14:618–628

243. Bavelier D, Corina D, Jezzard P, Padmanabhan S, Clark VP et al (1997) Sentence reading: a functional MRI study at 4 tesla. J Cognit Neurosci 9(5):664–686

244. Herbster AN, Mintun MA, Nebes RD, Becker JT (1997) Regional cerebral blood flow during word and nonword reading. Hum Brain Mapp 5:84–92

245. Indefrey P, Kleinschmidt A, Merboldt KD, Krüger G, Brown C et al (1997) Equivalent responses to lexical and nonlexical visual stimuli in occipital cortex: a functional magnetic resonance imaging study. Neuroimage 5:78–81

246. Chee MW, Caplan D, Soon CS, Sriram N, Tan EWL et al (1993) Processing of visually presented sentences in Mandarin and English studied with fMRI. Neuron 23:127–137

247. Démonet JF, Wise R, Frackowiak RSJ (1993) Language functions explored in normal subjects by positron emission tomography: a critical review. Hum Brain Mapp 1(1):39–47

248. Fiez JA (1997) Phonology, semantics and the role of the left inferior prefrontal cortex. Hum Brain Mapp 5:79–83

249. Poldrack RA, Wagner AD, Prull MW, Desmond JE, Glover GH et al (1999) Functional specialization for semantic and phonological processing in the left inferior prefrontal cortex. Neuroimage 10:15–35

250. Gold BT, Buckner RL (2002) Common prefrontal regions coactivate with dissociable posterior regions during controlled semantic and phonological tasks. Neuron 35:803–812

251. Henson RNA, Price CJ, Rugg MD, Turner R, Friston KJ (2002) Detecting latency differences in event-related BOLD responses:

application to words versus nonwords and initial versus repeated face presentations. Neuroimage 15(1):83–97

252. Mechelli A, Gorno-Tempini ML, Price CJ (2003) Neuroimaging studies of word and pseudoword reading: consistencies, inconsistencies, and limitations. J Cogn Neurosci 15(2):260–271

253. Braver TS, Cohen JD, Nystrom LE, Jonides J, Smith EE et al (1997) A parametric study of prefrontal cortex involvement in human working memory. Neuroimage 5:49–62

254. Honey GD, Bullmore ET, Sharma T (2000) Prolonged reaction time to a verbal working memory task predicts increased power of posterior parietal cortical activation. Neuroimage 12(5):495–503

255. Adler CM, Sax KW, Holland SK, Schmithorst V, Rosenberg L et al (2001) Changes in neuronal activation with increasing attention demand in healthy volunteers: An fMRI study. Synapse 42:266–272

256. Braver TS, Barch DM, Gray JR, Molfese DL, Snyder A (2001) Anterior cingulate cortex and response conflict: effects of frequency, inhibition and errors. Cereb Cortex 11:825–836

257. Ullsperger M, Von Cramon DY (2001) Subprocesses of performance monitoring: a dissociation of error processing and response competition revealed by event-related fMRI and ERPs. Neuroimage 14:1387–1401

258. Binder JR, Liebenthal E, Possing ET, Medler DA, Ward BD (2004) Neural correlates of sensory and decision processes in auditory object identification. Nat Neurosci 7(3):295–301

259. Desai R, Conant LL, Waldron E, Binder JR (2006) FMRI of past tense processing: the effects of phonological complexity and task difficulty. J Cognit Neurosci 18(2):278–297

Chapter 13

Neuroimaging Approaches to the Study of Visual Attention

George R. Mangun, Yuelu Liu, Jesse J. Bengson, Sean P. Fannon,
Nicholas E. DiQuattro, and Joy J. Geng

Abstract

Selective attention is a core cognitive ability that enables organisms to effectively process and act upon relevant information while ignoring distracting events. Elucidating the neural bases of selective attention remains a key challenge for neuroscience and represents an essential aim in translational efforts to ameliorate attentional deficits in a wide variety of neurological and psychiatric disorders. Moreover, knowledge about the cognitive and neural mechanisms of attention is essential for developing and refining brain–machine interfaces, and for advancing methods for training and education. We will discuss how functional imaging methods have helped us to understand fundamental aspects of attention: How attention is controlled, focused on relevant inputs, and reoriented, and how this control results in the selection of relevant information. Work from our groups and from others will be reviewed. We will focus on fMRI methods, but where appropriate will include related discussion of electromagnetic recording methods used in conjunction with fMRI, including simultaneous EEG/fMRI methods.

Key words Attention, Control, fMRI, EEG, ERPs, DCM, Human, Vision

---

1 Introduction

Attention is a key cognitive ability that supports our momentary awareness, and affects how we analyze sensory inputs, retain information in memory, process it for meaning, and, finally, act upon it. In this review chapter we will consider the role of attention in sensory processing and perception. We begin with a definition of attention as it will be investigated in the studies to follow, starting with nineteenth century insights provided by psychologist William James [1]:

> "Everyone knows what attention is. It is the taking possession by the mind, in clear and vivid form, of one out of what seem several simultaneously possible objects or trains of thought. … It implies withdrawal from some things in order to deal effectively with others, …"

Relying on introspection, James noted key characteristics of attention that frame the theoretical landscape of attentional mechanisms. He noted the voluntary aspects of attention and its selection of relevant from irrelevant information, as well as its capacity limitations.

Massimo Filippi (ed.), *fMRI Techniques and Protocols*, Neuromethods, vol. 119,
DOI 10.1007/978-1-4939-5611-1_13, © Springer Science+Business Media New York 2016

Our review will focus on studies of visual attention, and therefore, we also consider the work of James' contemporary, Hermann von Helmholtz [2], who in his studies of visual psychophysics made observations and speculations on the mechanisms of visual attention. He wrote, in studies investigating the limits of visual perception, the following:

"... by a voluntary kind of intention, even without eye movements, and without changes of accommodation, one can concentrate attention on the sensation from a particular part of our peripheral nervous system and at the same time exclude attention from all other parts."

Helmholtz's observations are in line with today's knowledge about attention mechanisms, which include detailed understanding of how attention influences stimulus processing at early and late stages of sensory analysis.

1.1 Varieties of Attention

Attention is neither a single capability, nor is it supported by a single mechanism or brain system. Theoretically, we can consider two main forms of attention—voluntary attention and reflexive attention [3]. Voluntary attention is goal-directed and suggests a top-down influence that is under intentional control. In contrast, reflexive attention is a stimulus-driven process involving bottom-up effects, for example, as when a salient sensory signal grabs our attention. These two general categories of attention differ in their properties and perhaps their neural mechanisms. In this chapter, we will concentrate on studies of voluntary attention to illustrate how functional imaging has been used to elucidate brain attention mechanisms.

Another way of thinking about attention mechanisms is to consider the domain of information processing on which attention operates. For example, attention can operate within or between sensory modalities. That is, we can attend to visual inputs at the expense of auditory ones, or vice versa, or may attend to one aspect of visual input (e.g., stimulus location) at the expense of other stimulus attributes (e.g., color or motion). Thus, one may ask whether the mechanisms of attentional control are similar or different for attention to different sensory modalities compared with attention to different stimulus attributes [4]. One may also consider whether the mechanisms supporting stimulus selection are modality and attribute independent, being instead specialized for the particular items to be attended [5]. In order to constrain the discussion in this chapter, for the most part, we will focus on visual attention, emphasizing attentional control and selection for attention based on stimulus location (spatial attention) and elementary stimulus features (e.g., color and motion).

1.2 Early and Late Selection Models

One of the key questions in attention research has been where in information processing attention has its influence. If as Helmholtz suggested, attention could select some information coming from the peripheral nervous system, then it is essential to ask where within the ascending sensory pathways attention can alter stimulus processing to achieve selective processing. In the 1950s, Broadbent [6] described the idea of an attentional gate that could be opened for attended information and closed for ignored information. Like Helmholtz, Broadbent suggested that information selection might occur early in sensory processing. This idea has been termed *early selection*, and it is the idea that a stimulus need not be completely perceptually analyzed before it can be selected for further processing or rejected by a gating mechanism.

In contrast, so-called *late selection* models hold that all (attended and unattended) sensory inputs are processed equivalently by the perceptual system to a very high level of coding before they are selected by attention [7]. This high level is generally considered to be the level of categorical, or semantic, encoding where the elementary feature codes (e.g., orientation, contrast, color, form) are replaced by conceptual codes (e.g., that is a chair). Late selection models posit that selection takes place on these higher-level codes and thereafter representation in awareness may take place. This view, then, holds that selective attention does not influence our perceptions of stimuli by changing the low-level sensory-perceptual processing of the stimulus.

A long debate and many studies have addressed the early versus late selection controversy [8–11], and physiological approaches including functional imaging have provided key information about the stages of sensory processing that are influenced by selective attention. We will thus begin our consideration of attention with its role in stimulus selection processes and how functional imaging has been used to investigate these mechanisms. But first, in the next section, some design issues in the study of selective attention deserve consideration as they are of paramount importance for physiological studies of attention, including functional imaging.

1.3 Methodological Issues in Experimental Studies of Selective Attention

The focus of the vast majority of studies of effects of attention on perception involves selective attention, attention to one thing at the expense of another. This is to be contrasted with nonselective attention, which includes generalized behavioral arousal (e.g., the classic orienting response) that does not necessarily involve attending one input while ignoring another. These latter nonselective attention mechanisms are certainly interesting, but selective mechanisms have generated the greatest interest and we review only these studies. Therefore, it is critical to understand the design parameters that permit selective versus nonselective attention to be isolated and studied, and to make note of some of the confounding influences that might contaminate studies of selective attention when nonselective factors (e.g., arousal) are not properly controlled. We turn then to an example from the early history of the physiology of attention.

In the 1950s, the great Mexican neurophysiologist Raúl Hernández Peón and his colleagues [12] studied the neuroanatomy and neurophysiology of the ascending sensory pathways and how top-down attention might modulate sensation. Their work in the auditory system was motivated by the well-known neuroanatomical substrate for top-down modulation, the olivocochlear bundle (OCB), which involves centrifugal neural projections from higher levels of the central nervous system downward to earlier processing stages especially the peripheral nervous system out to the level of the cochlea. The OCB provided a very strong neuroanatomical mechanism by which top-down effects of mental processes like attention might gate early auditory processing, in a fashion suggested by Helmholtz and later Broadbent, among others.

Hernández Peón's group recorded the activity in neurons in the subcortical auditory pathway in cats while they were either passively listening to the sounds from a speaker or were not attending the sounds. By showing the cats two live mice safely contained in a closed bottle, the cats were induced to ignore the sounds while they attended the mice. The researchers found that the amplitude of activity recorded from electrodes implanted in the cochlear nucleus was larger when the animals attended the sounds versus ignored the sounds while attending the bottled mice. This was interpreted as evidence that selective attention influences stimulus processing as early as the subcortical sensory pathways via the influence of the top–down neural control inputs.

This work, published in the journal *Science*, would however, later be shown to include a fatal flaw. The flaw was that the sounds presented to the cats to evoke auditory activity were delivered by speakers near the cats. When the cats attended the mice, they also oriented their ears and heads toward the mice, and away from the speakers, thereby altering the amplitudes of the auditory sounds at the ears due to simple physical differences in the relation of the ears to the sounds. Changes in the amplitudes of the sounds at the ear lead, in and of themselves, to changes in the amplitudes of auditory responses in the ascending pathways, and this tells us nothing about attention, although it mimics the kind of effect one would expect from an attentional mechanism. This problem can be eliminated by controlling the amplitudes and qualities of the sensory stimuli when attended and ignored so that an ensuing difference in neural responses would be attributable to internal attentional modulations of sensory processing, not differences in the physical stimuli themselves across conditions. Similarly, the two conditions of Hernández Peón and colleagues likely also differed in nonspecific behavioral arousal that may also have influenced the neuronal recordings since when cats see mice they are undoubtedly more aroused than when they listen to Beethoven, although the opposite would hopefully hold for at least some human listeners. In either case, differences in behavioral arousal between comparison conditions could confound the effects of selective attention, and therefore, like physical stimulus differences, must be rigorously controlled [13]. Failure to do so properly could lead to changes in, for example, hemodynamic signals that would present a serious confound in functional imaging studies of attention.

2   Imaging Attentional Selection Mechanisms

Studies using electrophysiological recordings have provided evidence that voluntary selective attention can influence the processing of sensory inputs at early stages of neural analysis. Hillyard and colleagues [14] in what is considered the first properly controlled study of auditory selective attention in humans, showed that scalp-recorded auditory cortical event-related potentials (ERPs) were enhanced with selective attention. In vision, the Hillyard group [15] similarly demonstrated in humans that by 70-ms post stimulus onset, the electrical brain response to a stimulus presented to an attended spatial location was enhanced compared with an identical stimulus presented when that location was ignored; similar findings reported previously could not rule out nonspecific effects of arousal or differential preparation for relevant targets based on learning the stimulus sequence [16, 17].

The timing of the ERP method (in the order of milliseconds) provides strong evidence that early stages of information processing were influenced by visual selective attention because the first inputs to the human visual cortex arrive only at about 40 ms after stimulus onset (based on intracranial recordings in patients). However, the scalp recordings are limited in the spatial localization they can provide in humans because the ERPs are recorded outside the skull and are, therefore, distant from the neuroelectric sources generating the signals. The volume conduction of the electrical signal through the brain, skull, and scalp is, on the one hand, an advantage that permits scalp recordings in the first place, but it also means that the electrical currents spread on the scalp (and are also filtered and spatially blurred by the intervening tissues, especially the skull) and are hard to track backward to their three-dimensional intracranial site of generation [18]. Related studies in animals showed that area V4 in extrastriate cortex in the macaque monkey showed increased neuronal firing rates to attended-location stimuli as compared with unattended stimuli [19], and subsequent work has extended this to other extrastriate areas [20–22], as well as striate cortex [23, 24] and the subcortical visual relays in the thalamus [25].

In humans, however, studies using functional imaging methods have provided the most detailed information about where in the ascending pathways visual selective attention modulates stimulus processing, because methods such as positron emission tomography (PET) and functional magnetic resonance imaging (fMRI) provide precise information about the neuroanatomical loci of attention effects and attention processes in the brain.

2.1 Functional Imaging of Visual– Spatial Attention

Functional imaging methods provide neuroanatomical information about the mechanisms of human attention. That is, together with information from neurophysiological recordings, imaging approaches help to locate the neuroanatomical stages of information processing influenced by attention. For visual-spatial attention (*see* [26] for studies of nonspatial attention), we began using O_{15} PET to provide functional anatomical information to complement our earlier ERP studies. Indeed, in these early studies that began in 1991, we first combined ERP and PET methods to provide a spatial–temporal approach to studying human attention [27, 28]. In subsequent studies beginning in the mid-1990s, we incorporated fMRI methods.

The goal of these functional imaging studies was to identify where within the visual system visual–spatial selective attention first influenced sensory analysis. We presented subjects with bilateral stimulus arrays of nonsense symbols (two in each hemifield) flashed at a rapid varying rate, averaging about two stimulus arrays per second (Fig. 1). In order to control for physical stimulus differences between conditions, the subjects were required to maintain fixation on a central fixation point and their compliance was ensured using high-resolution infrared photometric monitoring

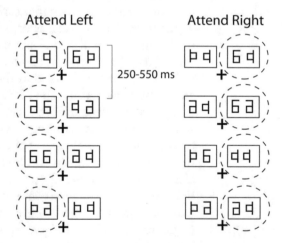

Fig. 1 Stimuli and task used in a functional imaging and event-related potential (ERP) study of spatial attention [105]. Two blocked conditions of attention are shown: Attend left condition (*left column*) and attend right condition (*right column*). Subjects viewed rapid sequences of arrays (about 2.5/s) of nonsense symbols (flashed for 100 ms) while maintaining fixation of their eyes on a central fixation spot (*plus sign*). There were always two symbols in the left and two in the right visual hemifield in locations each demarcated by an outline rectangle. The task was to detect and press a button to pairs of identical symbols at the attended location and to ignore all stimuli in the opposite hemifield. In the figure (but not in the actual experiment) the focus of covert spatial attention is indicated by a *dashed circle*

and the horizontal electro-oculograms (EOGs). In order to control for differences in nonspecific behavioral arousal, we included two main attention conditions that were equated in task difficulty and arousal. The subjects had to covertly pay attention to the left half of the array in one condition and ignore the right (attend left condition), or attend the right half of the array while ignoring the left (attend right condition). Hemodynamic responses (activations) in the brain could be compared for attend-left versus attend-right conditions. To be clear, the relevant design features were: (1) the stimulus arrays were identical, striking the same regions of the retina in the two attention conditions, (2) the two attention conditions (attend left and attend right) were also equated for nonspecific arousal because the discrimination task was identical for both conditions, and (3) the stimuli were presented randomly in a nonpredictable sequence, so that the subjects could not adjust the global level of attentiveness on a trial-by-trial basis. As a result, any changes we observed in visual cortex could not be attributed to either differences in visual stimulation or to nonspecific factors such as arousal, but instead could be attributed to spatially-specific modulations of visual processing with the direction of spatial attention.

We observed that spatial attention led to activations in extrastriate cortex in the cerebral hemisphere that was contralateral to the

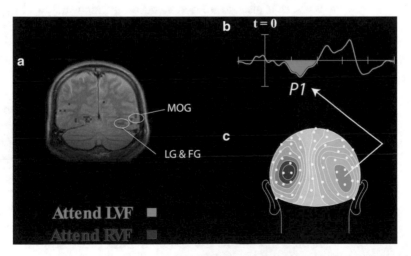

Fig. 2 fMRI and event-related potential (ERP) data from a study of visual-spatial selective attention. (**a**) Coronal structural scan of a single subject showing activations in contralateral visual cortex with spatial attention in the extrastriate cortex (left hemisphere is on the left). The activations were focused in the lingual gyrus (LG) and posterior fusiform gyrus (FG) and the middle occipital gyrus (MOG). (**b**) ERP attention effects shown as difference waves from a single lateral occipital electrode site in the right hemisphere (attend left minus attend right). The vertical scale is 2 μV per side (positive plotted downward). The onset of the array is indicated at time zero (*t*=0), and the tick marks are 100 ms. (**c**) Topographic voltage attention difference map (110-ms latency) on the scalp surface viewed from the rear (left side of head on left side of figure) (adapted from [105])

attended side of the stimulus arrays (Fig. 2a). These activations were highly statistically significant in the posterior fusiform gyrus on the ventral cortical surface and in lateral-ventral occipital regions. In these early studies, no effects of visual-spatial attention were observed in primary visual cortex (area V1) but as we shall see later, such effects have since been observed using fMRI. The finding of localized brain activations corresponding to the action of spatial attention alone is in accord with evidence from prior ERP studies [15, 29, 30]. Also in this study, in a separate session, testing the same subjects, we recorded ERPs for the same stimuli and task. The scalp-recorded ERPs showed the expected P1 attention effects (Fig. 2b, c), and by using neuroelectric dipole modeling we investigated whether intracranial neural generators at the loci of functional activations could produce activity in the model scalp that was similar to what we actually recorded in the ERPs (not shown in figure). We showed that the ERP and fMRI attention effects in the posterior fusiform gyrus (i.e., extrastriate visual cortex) were strongly related.

This combined use of ERPs and functional imaging provided evidence for short-latency (around 100 ms after stimulus onset) changes in responses to visual stimuli as a function of spatial attention that were generated early in extrastriate visual cortex. In several studies, we and others have followed up these effects [31, 32] and have observed that these modulations with spatial attention affect multiple stages of visual cortical processing, from V1 toward inferotemporal cortex in the ventral visual stream. The next section reviews related work that also combined structural, functional, and cognitive imaging to detail the structure of spatial attention effects in visual cortex.

2.2 Mapping Spatial Attention in Vision

A beautiful illustration of how spatial attention influences sensory processing in human visual cortex comes from the work of Tootell et al. [33]. They used fMRI to identify the borders of the first few visual areas in humans (retinotopic mapping) and then conducted a spatial attention study similar to what was described earlier. In the Tootell study, subjects performed a simple spatial attention task that required subjects to covertly and selectively attend stimuli located in one visual field quadrant while ignoring those in the other quadrants; different quadrants were attended in different conditions. Attentional activations were then mapped onto the flattened representations of the visual cortex, permitting the attention effects to be related directly to the multiple visual areas of human visual cortex, showing that spatial attention led to robust modulations of activity in striate cortex and multiple extrastriate visual areas (Fig. 3).

By combining different methods for recording electrical activity, imaging brain structure, defining functional anatomy (i.e., retinotopic maps), and combining this with functional imaging in carefully controlled studies of selective attention, we can learn a great deal about the effects exerted by attention on sensory processes in humans. Specifically, for spatial attention we now

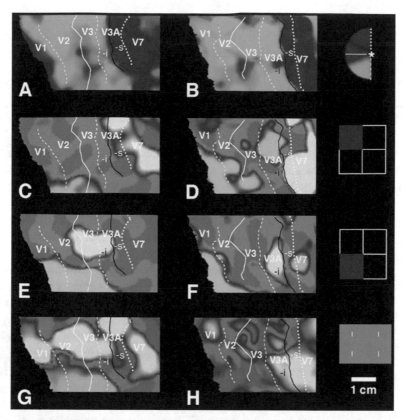

Fig. 3 Spatial attention effects in multiple visual cortical areas in humans as demonstrated by fMRI. Activations with spatial attention to left-field stimuli are shown in the flattened right visual cortices of two subjects (one in the left column and the other on the right). The *white lines (dotted and solid)* indicate the borders of the visual areas as defined by representations of the horizontal and vertical meridians; each area is labeled from V1 (striate cortex) through V7, a retinotopic area adjacent to V3A. The *solid black line* is the representation of the horizontal meridian in V3A. Panels (**a**) and (**b**) show the retinotopic mappings of the left visual field for each subject, with colored activations corresponding to the *polar angles* shown at right (which represent the left visual field). Panels (**c**) and (**d**) show the attention-related modulations (attended vs. unattended) of sensory responses to a target in the upper left quadrant (the quadrant of the stimuli is shown at right). Panels (**e**) and (**f**) show the same for stimuli in the lower left quadrant. In (**c**) through (**f**), the *yellow* to *red colors* indicate areas where activity was greater when the stimulus was attended to than when it was ignored; the *bluish colors* represent the opposite, where the activity was greater when the stimulus was ignored than when attended. The attention effects in (**c**) through (**f**) can be compared to the pure sensory responses to the target bars when passively viewed [(**g**) and (**h**)]. Note the retinotopic pattern of the attention effects in (**c**) through (**f**): The attention effects to targets in the lower left quadrant produced activity in several lower field representations, which included the appropriate half of V3A (inferior V3A labeled with an "i") in both subjects, and V3 and V2 in one subject. In contrast, attention to the upper left quadrant produced activity in the upper field representation of V3A (S) and in the adjacent upper field representation of area V7 (from [33])

understand that it exerts powerful influences over the processing of visual inputs: Attended stimuli produce greater neural responses than do unattended stimuli, even when arousal and physical stimulus differences are rigorously controlled, and this happens in multiple visual cortical areas beginning at short latencies after stimulus onset (as short as 70 ms in humans). In part, such effects may help to bias competition between attended and ignored sensory inputs at the level of neuronal receptive fields as well as higher order representations of the stimuli [34].

---

3 Neuroimaging of Attentional Control Networks

3.1 Isolating Attentional Control Mechanisms

A major goal in studies of selective attention is to understand how attention is controlled in the brain. Given that activity in sensory-specific cortex is modulated by attention, and therefore that the neuronal response properties of sensory neurons have been temporarily altered, what mechanisms lead to these changes in information processing? Most models posit that some brain systems form executive control systems that, as a function of momentary behavioral goals, are able to alter sensory-neural activity. Some models have referred to attentional control networks as the "sources" of attentional signals, and the sensory (and motor) systems that are affected as the "sites" where attention is implemented [35].

This source versus site dichotomy is useful as it helps to distinguish between the role attention plays in modulating sensory processing in the sensory systems and the neural mechanisms that produce this effect (Fig. 4). Presumably, neuronal projections from executive attentional control systems contact and influence neurons in sensory-specific cortical areas in order to alter their excitability. As a result, the response in sensory areas to a stimulus may be either enhanced if the stimulus is given high priority (i.e., is relevant to the behavioral goal) or attenuated if it is irrelevant to the current goal. More generally, though, attentional control systems could be involved in modulating thoughts and actions, as well as sensory processes.

Studies of patients with brain damage, animal recordings and lesion studies, and functional imaging converge to show that a large network of cortical and subcortical areas is activated during attentional orienting and selection [36, 37]. How can we measure the activity of the different brain regions during the execution of attentional selection tasks in order to determine which networks involve attention control (sources of attention) and which are the sites of selection? In part, the answer is to implement tasks that, at least theoretically, dissociate attentional control processes from selection in time, as do trial-by-trial attentional cuing paradigms. In such paradigms, attention is cued at time 1, which is followed by a delay period of several hundred milliseconds or seconds, and then by the target stimuli at time 2 [38]. Such designs are different from that described earlier (Fig. 1), which used blocked attention conditions and rapid streams of stimuli to study attentional mechanisms, because theoretically, the cue (e.g., an arrow) triggers the action of the attentional control network at time 1, whereas selective stimulus

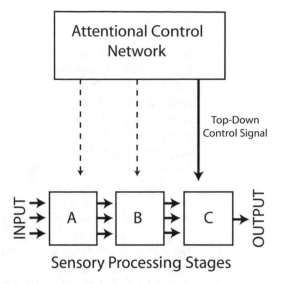

Fig. 4 Diagrammatic models of the influence of top-down attention control networks (*top*) on sensory processing (*bottom*). Sensory inputs are transduced at left and processed in multiple stages of analysis (*A*, *B*, and *C*). Top-down influences are shown as *vertical arrows* from the attentional control network. Here, the *dashed line arrows* indicated no influence on sensory processing stages *A* or *B*, and the *heavy solid line arrow* indicates an influence of attention on processing in sensory processing stage *C*, with the result being selection, which is represented by three arrows coming into *C*, but only one arrow leaving as output (*bottom right*). If this were to refer to spatial attention, then the *arrows* in the sensory processing stream could correspond to different parallel visual field location inputs in a retinotopic fashion, and the *single output arrow* at C reflecting that selective attention to one location was relatively facilitated (selected) with respect to the other locations

processing would not take place until the target is presented later at time 2. Hence, by recording brain responses to the cue and target separately, one can identify the networks for control and selection, which was first utilized in ERP studies by Russell Harter and his colleagues in the late 1980s [39]. Blocked-design fMRI attention studies make this difficult because the activations produced by attentional control and selection processes cannot be easily distinguished given that they are co-occurring in the images. Although functional brain imaging, which taps hemodynamic changes, is a rather poor method for tracking the time course of brain activity, when used in an event-related design, and appropriate analytic strategies are employed to separate overlapping hemodynamic responses, it is possible to conduct an experiment like that just described; we and others have done so in several studies of selective attention.

In our initial studies [31], we used event-related fMRI to investigate attentional control mechanisms during visual-spatial attention. An arrow, presented at the center of the display, indicated the side to which attention should be directed for that trial. Eight seconds later, a bilateral target display (flickering black and white checkerboards) appeared for 500 ms. The participants' task was to press a button if some of the checks on the cued sides only were gray rather than white. The 8-s gap between the cue and the target stimuli allowed us to extract the hemodynamic responses linked to the attention-directing cues separately from those linked to the subsequent targets. We were thus able to identify a top–down attentional control network triggered by the presentation of the cue.

The attentional control network consisted of regions in the superior frontal cortex, inferior parietal cortex, superior temporal cortex, and portions of the posterior cingulate cortex and insula (Fig. 5). These areas were not involved in the sensory analysis of the cue that was reflected by activity in the visual cortex (identified in control scans where the cue was passively viewed). Therefore, this attentional network of frontal, parietal, and temporal brain areas can be considered the sources of attention control signals in the brain [40], and has come to be known as the frontoparietal attention network or the dorsal attention network [41, 42]. The result of attentional control on the activity of visual cortex before and during target processing helps us to understand the relationship between control signals and selective sensory effects of spatial attention, as described next.

Figure 6 shows coronal sections through the visual cortex at two time points: first in the cue-to-target period, prior to the appearance of the lateral target arrays, and then second, during target processing. Two contrasts are shown for each time point: the case for activity when left cue was greater than right cue (attend left > right), and the inverse (attend right > left). Following the attention-directing cue and prior to target onset, one can see significant contralateral activation in multiple regions of visual cortex. These changes are spatially selective, being in the right visual cortex for leftward attention, and in the left cortex for rightward attention. Importantly, they occur prior to the onset of the targets, which occurred more than 8 s later, and were not related to the simple visual features of the cue, which stimulated different regions of visual cortex. These contralateral activations to the cues represent a kind of attentional priming of sensory cortex, which is thought to form the basis for later selective processing of target inputs [43, 44]. Indeed, as also shown in the bottom half of Fig. 6, the selective processing of subsequent target stimuli produced similar contralateral activations in visual cortex. Another way to describe these patterns of activity to cues and targets is to say that regions of visual cortex that code the spatial locations of the expected target stimuli showed increases in background activity levels when attention was directed to those locations by the cues, even before the targets appeared.

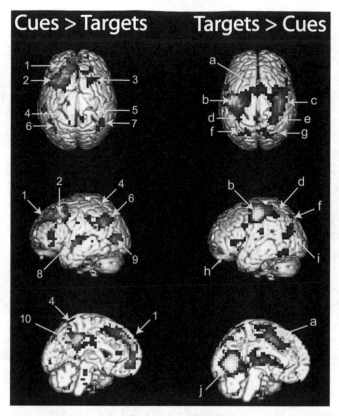

Fig. 5 Activations during attention control and target selection. Group ($N=6$ subjects) average activations time-locked to the cue (cues > targets contrast) isolating attentional control regions are shown on the *left*. Those time locked to the targets (targets > cues contrast), indicating target processing and motor responses, are on the *right*. The activated areas are shown in different colors to reflect the statistical contrasts that revealed the activations: *bluish* for brain regions that were more activated to cues than targets, and *reddish-yellow* to indicate regions that were more active to targets than cues. The *top row* shows a view of the dorsal surface of the brain, the *middle row* shows a lateral view of the left hemisphere, and the *bottom row* shows its medial surface; the activity was the same for the right hemisphere, which is not shown. Attentional control involved the frontal-parietal attention network involving superior and middle frontal gyri (labeled *1–3*), and the regions in and around the IPS (labeled *4–7*), the superior temporal cortex (labeled *8*), and the posterior cingulate cortex (labeled *10*). In contrast, during target selection, the main areas of activation were now the supplementary motor area (labeled *a*), the motor and somatosensory cortex (labeled *b–e*), posterior superior parietal lobule (labeled *f, g*), the ventral-lateral prefrontal cortex (labeled *h*) and the cuneus (labeled *i*) and the visual cortex (labeled *j*) (adapted from [31])

Similar effects have been observed in neurophysiological studies in monkeys [22]. Computationally, such effects may well be a mechanism for selective sensory processing by changing the baseline gain of sensory neurons such that when later stimulated, they produced enhanced responses [45]. This illustrates the mechanism put forward at the beginning of this section that top–down spatial attentional control may lead to selective changes in sensory processing by changing the background firing rates of neurons, thereby improving their sensitivity and/or selectivity.

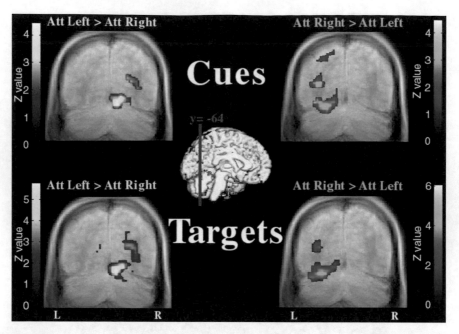

Fig. 6 Priming of visual cortex by spatial attention. The *top row* shows increased baseline activity in six subjects to the cues in visual cortex. The *bottom row* shows the same effects to targets (adapted from [31])

3.2 Specializations in Attentional Control

One key question about top–down attentional control systems is the extent to which such mechanisms are generalized for the control of attention regardless of the modality or specific stimulus attributes to which selective attention is directed [46]. For example, does preparatory attention for spatial selective attention involve the same or different control networks as would selective attention for color or motion? We have addressed this in studies using similar methods to those described in the foregoing.

In one study [32] we undertook a direct test of whether the neural mechanisms for the top-down control of both spatial and nonspatial attention were the same. In this study, we compared spatial selective attention to nonspatial color-selective attention. That is, subjects were either cued to select the target stimulus based on location or color. Our design was as follows. We randomly intermingled trials in which either the location or the color of an upcoming target was cued. The cues were letters (e.g., "L" = attend left and "B" = attend blue) located at fixation in one task or the periphery above fixation in another version of the task. The stimuli to be discriminated were rectangles (outlines) that when following spatial cues were located in the left or right hemifields or when following color cues were overlapping outline rectangles located at fixation in one task version or above fixation in the second task version. The participants were instructed to covertly direct attention to select an upcoming rectangle stimulus based on the cued feature (location or color) with the task of discriminating its orientation. That is, in the spatial cue condition, if the cue signified left, the subjects were to maintain fixation on the central fixation point, but focus covert attention to the left and to discriminate the orientation of the rectangle presented there (vertical or horizontal), and to press the appropriate response key. When the cue indicated blue, however, this meant that the subjects should focus covert attention in preparation for discriminating the blue rectangle location in the same place as the cue (i.e., either at fixation or in the periphery just above fixation).

As in our studies [31] described earlier, event-related fMRI measures were obtained to the cues and to the targets, and separately for spatial location and color attention trials. Because we jittered the stimulus-onset-asynchrony between cues and targets from 1000 to 8000 ms, it was possible to deconvolve the overlapping hemodynamic responses [47–49]. This permitted us to utilize a paradigm that was more similar to cued attention designs in the cognitive psychology literature where the time between cues and targets was not too long for subjects to maintain a strong attentional set (i.e., to sustain the covert allocation of attention to the cued color or location). In our prior work [31] and that of others, the need to separate the sluggish hemodynamic responses to adjacent stimuli typically resulted in long stimulus-onset-asynchronies in order to avoid overlap of the responses to cues and targets. By using specially designed stimulus sequences and analytic strategies it is possible to design cued attention studies that optimize the design parameters for investigating attentional mechanisms. Indeed, in cued attention designs, the speed of stimulus presentation can be even faster than the 1000–8000-ms lag between cues and target described here [50, 51].

In this study, comparing preparatory attention for locations and colors we found that large regions of the frontoparietal network were commonly activated by the spatial and nonspatial cues alike (Fig. 7). Such related patterns of activity reflect those aspects of the task that the two attentional control conditions shared, such as low-level sensory processing of the cues, decoding of linguistic information in the cue letter and matching that to the task instruction, establishing the appropriate attentional set and holding this information in working memory during the cue-to-target period, and finally, preparing to respond.

To test whether any of the areas activated in response to the cues were selective for spatial or nonspatial orienting, we directly statistically compared activity in response to location and color cues. The results of this direct comparison are shown in Fig. 8a for the activations where locations cues produce more activity than color cues. We found nonoverlapping regions of superior frontal and parietal cortex that were activated during orienting to location versus color. Orienting attention based on stimulus location activated regions of the dorsal frontal cortex [posterior middle frontal

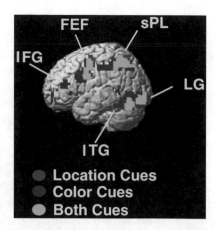

Fig. 7 Cue-related activity. Group-averaged data for brain regions significantly activated to attention-directing cues, overlaid onto a brain rendered in 3D. Areas activated in response to location cues are shown in *blue*, color cues in *red*, and those areas activated by both cues are shown in *green*. Maps are displayed using a height threshold of *p* < 0.005 (uncorrected) and an extent threshold of ten contiguous voxels (modified from [32])

gyrus and frontal eye fields (FEF)], posterior parietal cortex [intra-parietal sulcus (IPS) and precuneus], and supplementary motor cortex. The inverse statistical contrast (color > location cues) produced no significant activations in the superior frontal or parietal regions, and only showed activity in the ventral occipital cortex (OCC) (posterior fusiform, posterior middle temporal cortex, and left insula) (not shown in the figure).

The pattern of selectivity in the frontoparietal attentional control network for location cuing suggests that neural specializations exist for the control of orienting attention to locations or for coding some aspect of space tapped in spatial attention tasks [52]. But a question that must be addressed is whether these superior frontal and parietal regions are sensitive only for spatial orienting, and we have tested this by investigating specializations in attentional control for other presumably nonspatial features.

We investigated the idea that specializations in superior frontal and parietal cortex for preparatory spatial selective attention might also be involved in other aspects of visual attentional processing. In one study we compared preparatory attention for stimulus motion, a dorsal stream process, to the same nonspatial ventral stream feature, color used in Giesbrecht and colleagues [32]. In this study each trial began with an auditory word cue that instructed subjects to attend to a target of a particular stimulus feature (i.e., involving either color or motion attention). If cued to a color, then they were to prepare for and detect brief color flashes within a display of randomly moving dots presented during a subsequent test period. If cued to motion, then they were to prepare for and detect the brief coherent motion stimulus in the display of randomly moving dots [53, 54].

In the study of preparatory attention for motion and color we observed activity in the frontal-parietal attention network for both motion and color cues that was similar to prior work [31, 32, 40, 42, 43, 50, 55–58]. As in our study comparing location versus color processing (Fig. 8a) we conducted direct statistical contrasts for motion versus color cues (Fig. 8b). Again, as with the location versus color analyses, for motion versus color we found subregions of the attentional control network that were more active for motion than for color, and vice versa. Bilateral posterior superior frontal cortex and the left superior parietal cortex were selectively active to motion cues, but no superior cortical areas were selectively activated for the inverse contrast of color versus motion (color > motion). The activations for the motion versus color attentional control contrast were very similar to three regions of the attention control network we previously found to be selectively activated for preparatory spatial attention. There were also some differences, however, including failure to see activity in the right superior parietal cortex for the motion versus color contrast, suggesting that this right parietal region may also be involved in directing attention to locations.

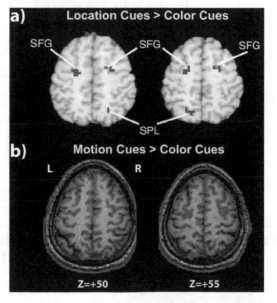

Fig. 8 Attentional control activations for location and motion cues versus color cues. (*Top*) Results of the direct comparison between location and color cue conditions overlaid on an axial slice. Greater activity to location cues than color cues is seen in the superior frontal gyrus (SFG) and the superior parietal lobule (SPL) bilaterally (modified from [32]). (*Bottom*) Results of the direct comparison between motion and color cue conditions overlaid on an axial slice. Greater activity to motion cues than color cues is seen in the superior frontal gyrus (SFG) bilaterally, but the superior parietal lobule (SPL) only on the left (modified from [54])

It is important to point out that in Fig. 8 we are comparing results across different studies in different populations of volunteers, and indeed using slightly different imaging methods and analyses (although conceptually the same). In order to know whether preparatory attention for location and motion activated identical subregions in the frontal-parietal attention network, one would have to conduct the within subjects experiment contrasting attentional cuing to location, motion, and color. Nonetheless, the similar findings across these studies are highly suggestive and require some interpretation. How can one conceptualize these fMRI results of similar specializations in location and motion attention?

One view is that the dorsal visual stream projecting from visual cortex to parietal cortex represents, in part, visual information required for generating actions [59–62], rather than merely coding location per se as originally conceived by Ungerleider and Mishkin [63]. In line with this view is the evidence that the dorsal stream attentional control areas overlap with regions implicated in the control of voluntary eye movements [52, 64–66]. This opens the possibility that the close correspondence of superior frontal and parietal activity present for location and motion selective attention more than for nonspatial (color) attention is related to the role of both of these forms of attention in preparing actions, specifically those involved in oculomotor output. In the case of attention to locations the activity in oculomotor areas may represent preparation for *unexecuted* eye movements toward the attended location, and in the case of attention to directions of motion the activity may represent analogous preparation for ocular pursuit. Such a view is supported by the study of [67] who investigated brain activations in cued covert attention, overt saccade tasks, and pointing tasks. They observed overlapping activity in the superior frontal and parietal cortex for all three tasks. As noted, future studies will be required to address the speculations we have provided here and must include studies designed to compare and contrast different forms of preparatory attention.

3.3 The Ventral Attention Network

So far we have focused our discussion of attentional control on the dorsal frontal and parietal cortex, the so-called dorsal attention network. The idea is that top-down signals from this system exert attentional control over perception based on behavioral goals and strategies. But when attention must be reoriented from an attended location to another location or object, a different network of brain areas has been shown to be activated (Fig. 9). This network has been called the *ventral attention network*, and involves the temporal-parietal junction (TPJ), together with ventral frontal regions including the insula, the portions of the inferior frontal gyrus and middle frontal gyrus of the right hemisphere [41, 68, 69]. Evidence from various sources, such as resting-state fMRI [70] indicates that the dorsal and ventral attention networks are

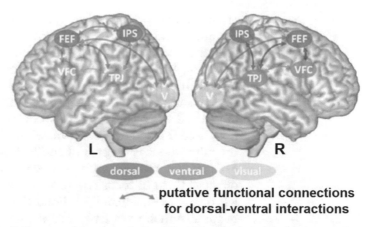

**putative functional connections
for dorsal-ventral interactions**

Fig. 9 Diagram of the nodes in the dorsal attention network (*blue*), ventral attention network (*orange*), and visual cortex (*light blue*) in the left and right hemispheres of the human brain. The dorsal system is clearly bilateral in organization, but the ventral system is believed to be more right lateralized (noted by the difference in saturation of the *orange-colored nodes*). Putative connections within each hemisphere are shown by the *arrows*. *FEF* frontal eye fields, *IPS* intraparietal sulcus, *VFC* ventral frontal cortex, *TPJ* temporoparietal junction, *V* visual cortex [76]

largely independent from one another anatomically, interacting via presumed frontal nodes (middle frontal gyrus) where the two networks show common activity in resting-state imaging.

In early models [40] activity of the ventral attention network had been likened to a circuit breaker that upon detecting a potentially task relevant item, sends a signal to the dorsal attention network to disrupt its current attention focus and reorient attention to the new event. This model would require that the ventral attention network or some portion of it respond with short latency to stimulus events that might be task relevant. However, research using magnetoencephalography (MEG) has shown that both the parietal and frontal areas of the dorsal attention network have shorter latency responses [71] than those in the ventral attention network, such as in ERP recordings of the P300 that originate in TPJ [72]. These data, therefore, raise some doubts that the ventral attention network supports a fast circuit breaking mechanism.

Despite evidence against the idea that the ventral attention network acts as a circuit breaker permitting the dorsal attention network to reorient, it is nonetheless the case that the integrity of the ventral network is important for successful reorienting of attention to relevant events. Patients with lesions in the right ventral attention network, and who have contralateral neglect, have disruption in the interhemispheric coordination of the left and right IPS as shown by resting-state fMRI [73]. Transcranial magnetic stimulation (TMS) studies has shown that disruption of the ventral attention network in healthy volunteers leads to disordered orienting and target detection [74]. Data such as these support the idea that the dorsal and ventral attention networks work together in attentional control, but this interaction may be different depending on the specific tasks demands [75] (for reviews see [76, 77]).

In order to evaluate the directionality of influences between the dorsal and ventral attention networks (or nodes in the network) we [78] used fMRI and dynamic causal modeling (DCM) to investigate the interactions between TPJ and FEF during spatial attention [79, 80]. In the task, subjects received a spatial cue at fixation that directed them to focus covert attention on a location in either the left or right visual hemifield. Following the cue a few seconds later, bilateral stimuli were flashed to the left and right visual field locations. The subjects' task was to make a button press response (yes/no) indicating whether a predesignated color target appeared at the cued (attended) location. Three different stimulus conditions were possible: a neutral condition where neither stimuli were the color target (response = no); a target condition where the color target appeared at the cued location (response = yes); and a target-colored distractor condition where the predesignated color stimulus appeared at the uncued location (response = no). The design, therefore, included a condition where a distractor stimulus that was the same color as the predesignated target color could appear at the uncued location. We reasoned that if the ventral attention network regions of the right hemisphere were involved only in reorienting spatial attention towards task-relevant stimuli, they would only respond with significant changes in activation when the target-colored distracter was present. This target-colored distractor condition is the only condition where a relevant color stimulus appeared in the unattended field; neither of the other two conditions were expected to reorient spatial attention based on feature relevance because the target either appeared in the cued location or was absent.

We found that reaction times (RT) were about 50 ms slower in the target-colored distractor condition, indicating that the relevant color stimulus presented to the uncued location resulted in distraction, as would be expected based on the literature on contingent attentional capture [81, 82]. The question then, is what brain regions are sensitive to this attentional capture? Significant activations in regions of the dorsal attention network were obtained to the target-colored distractor, including FEF, IPS, and the precuneus, and these activations were greater than for the target-present or neutral distractor conditions. This is in line with the idea that the dorsal attention network was more engaged during stimulus conditions where attentional capture by distractors had to be overcome. Regions within the dorsal network, including right FEF, were also significantly activated by the cues, suggesting involvement during shifts of spatial attention reorienting, in response to the cue and target-colored objects in the uncued location.

In contrast, the right TPJ was significantly activated in all three stimulus conditions, with activation being largest when the target-colored distractor was presented. Interestingly, right TPJ was not selectively activated in the condition where the relevant color stimulus appeared at the uncued location, providing evidence that this node of the ventral attention network is not only engaged to when attention is reoriented to task-relevant stimuli at unattended locations [69]. Importantly, the right TPJ did not respond to merely any stimulus: all responses to the 3 stimulus conditions were greater than that evoked by the spatial cues, which were not different from zero. From these patterns of activity, it appears the right FEF, but not the right TPJ was most engaged by the need to reorient spatial attention to potential or actual target stimuli; TPJ was activated by all stimulus conditions that required a response.

In order to assess the dynamics of interactions between the dorsal and ventral attention networks, we used DCM with time series data extracted from the right FEF and TPJ. The hypothesis that TPJ sends a signal to FEF to initiate reorienting of attention by the dorsal attention network predicts that our model should display an early TPJ response to stimulus driven inputs from the colored distractor, and a positive modulatory parameter on the connection from TPJ to FEF. However, if instead TPJ is involved in post-perceptual attentional or decisional processes, we would expect stimulus driven inputs to FEF, with modulation of the connection from FEF to TPJ. First, none of the models that best fit the data were consistent with the idea that right TPJ is activated rapidly by the potentially relevant (target-colored) distractor and sends a reorienting signal to the dorsal attention network. Instead, the model best supported by the DCM exercise was that right FEF receives stimulus inputs first and then modulates activation in right TPJ (Fig. 10). This supports the notion that right TPJ is involved in post-perceptual processes rather than as a stimulus-driven circuit breaker that interrupts the current focus of attention and reorients attention by signaling the dorsal attention network [68, 77].

3.4 Attentional Control Mechanisms Revealed by Simultaneous EEG-fMRI

Models of attentional control, as laid out in the foregoing, hold that activity in the sensory cortex is under top-down influences from control areas in the frontoparietal cortex during voluntary attention [35, 41, 83]. Despite recent advancements in our understanding of visual selective attention, one unresolved issue about the nature of top-down attentional control is whether attentional biasing is achieved primarily by enhancing the sensitivity of task-relevant sensory cortices or alternatively, via the inhibition of task-irrelevant areas. Further, a related question is whether areas residing outside of the sensory cortex but known to mediate other task-independent processes, such as the default mode network [84], are also "sites" that receive top–down modulation from the attentional control system as part of goal-directed behavior.

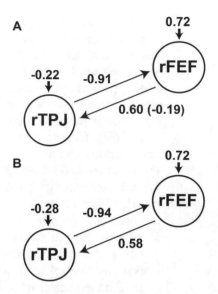

Fig. 10 Dynamic causal models of interactions between rTPJ and rFEF showing driving inputs and intrinsic connections. (**a**) Model for the condition where the stimulus included the target-colored distractor, which shows positive stimulus-driven influences on rFEF (0.72) but not rTPJ (-0.22), and positive (.60) intrinsic connectivity from FEF to TPJ: the only significant modulatory parameter (−0.19) was on the connection from rFEF to rTPJ. (**b**) Model for the condition where there was a target present (on the cued side). Here again the pattern is for a positive stimulus driven input to rFEF, and a positive intrinsic connection from rFEF to rTPJ. These models are consistent with FEF begin active first, followed by activity in TPJ

The concurrent recording of EEG and fMRI open up new avenues to address the above questions related to attentional control mechanisms. Recording EEG simultaneously with fMRI enables the examination of trial-by-trial functional relationship between well-established attention-related neural markers in EEG/ERP and BOLD activities in the attentional control system within an individual, thereby largely alleviating the issue of inter-session variability associated with multisession recordings.

In a recent study, we [85] utilized simultaneous EEG-fMRI to examine the mechanisms of visuospatial attention. Participants performed a standard visuospatial attention task during which they were cued to attend either the left or right spatial location in their peripheral vision. Following a random cue-target interval (2–8 s), the target stimulus was flashed at one of the peripheral spatial location for 100 ms. Only when the target stimulus appeared in the attended hemifield were participants required to discriminate the spatial frequency of the target stimulus (5.0 vs. 5.5 cycles per degree of visual angle) and make a two-alternate forced choice response via a button press. EEG and fMRI data from the cue-target interval, when participants allocated their attention but before the target onset, were analyzed to isolate processes specific to attentional control.

EEG studies of visuospatial attention have shown modulations of the posterior alpha rhythm (8–12 Hz) as a reliable signature of selective sensory biasing during lateralized attention allocation [86]. Following an attention-directing cue the alpha rhythm is more strongly suppressed over the occipital scalp regions contralateral to the direction of attention than over the hemisphere ipsilateral to the attended direction [86–89] (Fig. 11a). This topographically-specific modulation of alpha is thought to reflect an increase in baseline cortical excitability in the task-relevant sensory cortex and at the same time might also reflect an inhibition of task-irrelevant sensory areas in anticipation of upcoming stimulus events [90, 91]. To investigate the relationship between attention-related alpha lateralization and top–down attentional control networks we used the post-cue alpha modulation as an index of top-down sensory facilitation and correlated single-trial alpha power on each hemisphere with concurrently recorded BOLD signals to study mechanisms of attentional control.

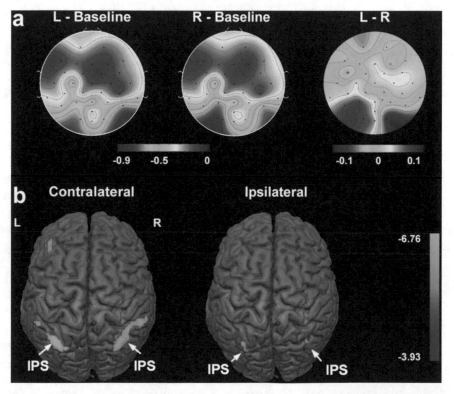

Fig. 11 Attentional modulation of EEG alpha and its coupling with attentional control structures. (**a**) Scalp topography of alpha desynchronization showed that relative to a pre-cue baseline, alpha power was more strongly suppressed over the hemisphere contralateral to the attended hemifield for both the attend-left (*top left panel*) and attend-right (*top-middle*) conditions. The asymmetry in alpha desynchronization was further demonstrated by the difference topography contrasting attend-left and attend-right (*top-right*). (**b**) BOLD activities in bilateral intraparietal sulci (IPS) showed inverse coupling with alpha power measured on the hemispheres both contralateral (*bottom-left*) and ipsilateral (*bottom-right*) to the attended hemifield (adapted from [85])

We first established a direct link between attentional modulation of alpha power (Fig. 11a) and BOLD activity in the dorsal attention network (Fig. 11b). The bilateral intraparietal sulci (IPS) were inversely correlated with single-trial alpha power measured in the hemispheres both contralateral and ipsilateral to the attended hemifield. This indicates that on a trial-by-trial basis, when the IPS was more active, alpha was reduced more over the occipital regions on both hemispheres, providing direct evidence that the sensory cortex excitability is modulated by the dorsal attention system. Further, a ROI-based analysis within the IPS revealed that the inverse coupling was stronger for alpha contralateral to the attended hemifield than for alpha ipsilateral to the attended hemifield. Specifically, the number of voxels in IPS showing inverse coupling with alpha contralateral to the attended hemifield were significantly higher than those coupled with alpha ipsilateral to the attended hemifield (lIPS-contralateral alpha: 222 voxels vs. lIPS-ipsilateral alpha: 184 voxels; rIPS-contralateral alpha: 262 voxels vs. rIPS-ipsilateral alpha: 55 voxels; $p < 0.001$, one-sided, paired t-test). In addition to the differences in cluster size, the level of inverse coupling between alpha and BOLD in lIPS was stronger (more negative) for alpha on the hemisphere contralateral than ipsilateral to the attended hemifield (coupling strength for contralateral alpha: −1.26; ipsilateral alpha: −0.96; $p < 0.05$, one-sided, paired t-test). In rIPS, we also found a similar but insignificant trend toward stronger inverse coupling for contralateral alpha compared with ipsilateral alpha (coupling strength for contralateral alpha: −1.04; ipsilateral alpha: −0.96; $p > 0.05$). Since post-cue reduction of alpha power was observed in both hemispheres relative to a pre-cue baseline in our study, the stronger inverse coupling between IPS and alpha contralateral to the attended hemifield suggests that visuospatial attention is enhancing neuronal activity within the task-relevant visual cortex, instead of inhibiting the task-irrelevant visual cortex.

Next, we examined whether other task-irrelevant networks outside of the visual system, or even outside of the sensory/motor system, were also modulated by visuospatial attention. For this purpose, we examined regions showing positive trial-by-trial coupling with occipital alpha following the cue onset. This positive BOLD-alpha coupling would indicate a decreased level of BOLD activity when alpha power also decreased during attention deployment, and hence would identify regions potentially inhibited by attentional mechanisms. We found that areas in the sensorimotor cortices including the pre- and post-central gyri, as well as nodes in the default mode network including the middle temporal gyrus (MTG), and the medial prefrontal cortex (MPFC) showed this positive BOLD-alpha coupling (Fig. 12a, b). This might suggest a "push–pull" mechanism similar to that observed in studies involving attention to multiple sensory modalities [92, 93], where increased visual alpha (decreased visual cortical excitability) was

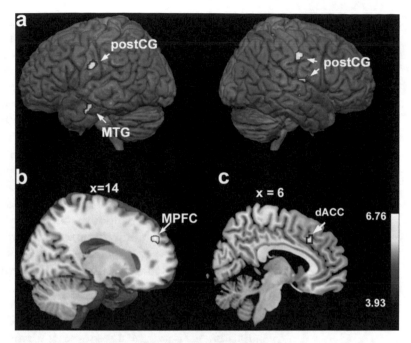

Fig. 12 Regions showing positive coupling between BOLD and alpha power measured on hemispheres contralateral (**a**) and ipsilateral (**b**) to the attended hemifield. (**c**) A sagittal slice showing a region in the dorsal anterior cingulate cortex (dACC) with BOLD being positively correlated with the alpha hemispheric lateralization. *MPFC* medial prefrontal cortex, *MTG* middle temporal gyrus, *postCG* post-central gyrus (adapted from [85])

seen when attention is directed either to the somatosensory [94, 95] or to the auditory domain [96–98]. Here we extend this "push–pull" mechanism to incorporate the default mode network and suggest that the level of task-independent self-referential processes were also inversely coupled with attentional processes on a trial-by-trial level.

Finally, as described in Sect. 1.1 of this chapter, voluntary attention is a type of goal-directed behavior under intentional control. One question related to this definition that is still not well-understood is what brain structures are responsible of representing the goal and mediating the voluntary allocation of attentional resources. More specifically, is the goal (or attentional-set) represented intrinsically within the attentional control system, or alternatively, are other executive structures also engaged in this process? Past studies have proposed the degree of alpha lateralization across two hemispheres as an index of voluntary attention allocation during spatial attention [88, 99–101], with greater alpha lateralization signifying more efficient allocation of attentional resources according to the task rules. We thus correlated the trial-by-trial alpha lateralization index with BOLD and found that activity in the dorsal anterior cingulate cortex (dACC) and adjacent regions within the MPFC and the left dorsolateral prefrontal cortex (DLPFC) are positively correlated with this alpha lateralization (Fig. 12c). This positive correlation indicates that when alpha is more strongly lateralized, BOLD activity in these regions are also higher, implicating these areas in the maintenance of the attentional-set and the modulation of goal-directed behavior [102–104]. Our results suggest that the voluntary allocation of attentional resources also involves prefrontal executive regions other than the dorsal attention system.

In summary, the simultaneous EEG-fMRI recording technique has emerged as a powerful tool that can contribute significantly to our understanding of the top-down mechanisms of attention. By mapping regions that were either positively or inversely coupled with posterior alpha over the hemispheres contralateral or ipsilateral to the attended hemifield, we revealed that attentional mechanisms act as a combination of selective enhancement of the task-relevant visual cortex and inhibition of task-irrelevant sensory and cognitive modalities. The voluntary allocation of attentional resources is further mediated by prefrontal executive cortices outside of the dorsal attention system.

4 Conclusions

In this chapter we have reviewed the use of functional imaging in the study of selective attention. We addressed this topic from four perspectives: First, we laid out the theoretical and experimental design issues that are critical for studying attention, especially selective attention. Second, we illustrated that the attention system can be conceptualized as consisting of different cognitive-neural components (i.e., sources and sites of attention), and therefore, functional imaging methods that permit these different components to be isolated and studied are necessary. Third, we described evidence that specializations in attention systems can be investigated by experimental design approaches that capitalized on comparisons between different forms of attention such as spatial versus nonspatial, and top-down voluntary attention and stimulus driven attention. In describing how specialized networks can be revealed we presented an application of the analysis of functional connectivity to probe the nodes within and between attention networks, in this case, using DCM as described in detail in a chapter of this volume by Karl Friston. Fourth, we pointed to the use of combined imaging and electrophysiological recording methods (in parallel or simultaneously) to provide temporal information and single-trial information (from electroencephalography) that is not available using functional imaging based on hemodynamic or metabolic methods, as well as to shown how electrophysiological information can be used to inform the analysis of imaging data.

We took our examples primarily from work in our laboratories but also presented the work of others where useful. We focused on visual attention, principally spatial attention, but the methodological approaches are germane to all studies of attention, across modalities, and including translational efforts in disorders of attention and other psychiatric and neurological disorders.

Methodological advances have helped to reveal the neural mechanisms of attention. It is clear that attention mechanisms help to refine the processing of sensory information beginning as early as sensory-specific cortex, a concept that was hotly debated until the 1990s. For voluntary attention, a frontal parietal attentional control system provides biasing signals to sensory cortex that result in selective sensory processing, and functional imaging has proven crucial for understanding these relationships in humans. Other networks, such as the ventral attention network and the default mode network have also been identified by functional imaging in humans. This work is leading to an increasingly rich understanding of the principles, organization, and mechanisms of human attention.

References

1. James W (1890) Principles of psychology. H Holt, New York
2. Hv H (1867) Handbuch der physiologischen optik. Voss, Leipzig
3. Posner MI, Cohen Y (1984) Components of visual orienting. In: Bouma H, Bouwhis D (eds) Attention and performance. Erlbaum, Hillsdale, NJ, pp 531–556
4. Harter MR, Aine CJ (1984) Brain mechanisms of visual selective attention. In: Parasuraman R, Davies DR (eds) Varieties of attention. Academic, Orlando, pp 293–321
5. Malcolm GL, Shomstein S (2015) Object-based attention in real-world scenes. J Exp Psychol Gen 144:257–263
6. Broadbent DE (1958) Perception and communication. Pergamon, New York
7. Deutsch JA, Deutsch D (1963) Attention – some theoretical considerations. Psychol Rev 70:80–90
8. Johnston WA, Heinz SP (1979) Depth of nontarget processing in an attention task. J Exp Psychol Hum Percept Perform 5:168–175
9. Hawkins HL, Hillyard SA, Luck SJ, Mouloua M, Downing CJ, Woodward DP (1990) Visual attention modulates signal detectability. J Exp Psychol Hum Percept Perform 16:802–811
10. Luck SJ, Hillyard SA, Mouloua M, Woldorff MG, Clark VP, Hawkins HL (1994) Effects of spatial cuing on luminance detectability: psychophysical and electrophysiological evidence for early selection. J Exp Psychol Hum Percept Perform 20:887–904
11. Palmer J, Ames CT, Lindsey DT (1993) Measuring the effect of attention on simple visual search. J Exp Psychol Hum Percept Perform 19:108–130
12. Hernández-Peón R, Scherrer R, Jouvet M (1956) Modification of electric activity in cochlear nucleus during "attention" in unanesthetized cats. Science 123:331–332
13. Naatanen R (1975) Selective attention and evoked potentials in humans–a critical review. Biol Psychol 2:237–307
14. Hillyard SA, Hink RF, Schwent VL, Picton TW (1973) Electrical signs of selective attention in the human brain. Science 182:177–180
15. Van Voorhis S, Hillyard SA (1977) Visual evoked potentials and selective attention to points in space. Percept Psychophys 22:54–62
16. Eason R, Harter M, White C (1969) Effects of attention and arousal on visually evoked cortical potentials and reaction time in man. Physiol Behav 4:283–289

17. Spong P, Haider M, Lindsley DB (1965) Selective attentiveness and cortical evoked responses to visual and auditory stimuli. Science 148:395–397

18. Nunez PL, Srinivasan R (2006) Electric fields of the brain: the neurophysics of EEG, 2nd edn. Oxford University Press, Oxford, New York

19. Moran J, Desimone R (1985) Selective attention gates visual processing in the extrastriate cortex. Science 229:782–784

20. Chelazzi L, Miller EK, Duncan J, Desimone R (1993) A neural basis for visual search in inferior temporal cortex. Nature 363:345–347

21. Chelazzi L, Miller EK, Duncan J, Desimone R (2001) Responses of neurons in macaque area V4 during memory-guided visual search. Cereb Cortex 11:761–772

22. Luck SJ, Chelazzi L, Hillyard SA, Desimone R (1997) Neural mechanisms of spatial selective attention in areas V1, V2, and V4 of macaque visual cortex. J Neurophysiol 77:24–42

23. McAdams CJ, Reid RC (2005) Attention modulates the responses of simple cells in monkey primary visual cortex. J Neurosci 25:11023–11033

24. Briggs F, Mangun GR, Usrey WM (2013) Attention enhances synaptic efficacy and the signal-to-noise ratio in neural circuits. Nature 499:476–480

25. McAlonan K, Cavanaugh J, Wurtz RH (2008) Guarding the gateway to cortex with attention in visual thalamus. Nature 456:391–394

26. Corbetta M, Miezin FM, Dobmeyer S, Shulman GL, Petersen SE (1991) Selective and divided attention during visual discriminations of shape, color, and speed: functional anatomy by positron emission tomography. J Neurosci 11:2383–2402

27. Heinze HJ, Mangun GR, Burchert W et al (1994) Combined spatial and temporal imaging of brain activity during visual selective attention in humans. Nature 372:543–546

28. Mangun GR, Hopfinger JB, Kussmaul CL, Fletcher EM, Heinze HJ (1997) Covariations in ERP and PET measures of spatial selective attention in human extrastriate visual cortex. Hum Brain Mapp 5:273–279

29. Hillyard SA, Munte TF (1984) Selective attention to color and location: an analysis with event-related brain potentials. Percept Psychophys 36:185–198

30. Mangun GR, Hillyard SA (1991) Modulations of sensory-evoked brain potentials indicate changes in perceptual processing during visual-spatial priming. J Exp Psychol Hum Percept Perform 17:1057–1074

31. Hopfinger JB, Buonocore MH, Mangun GR (2000) The neural mechanisms of top-down attentional control. Nat Neurosci 3:284–291

32. Giesbrecht B, Woldorff MG, Song AW, Mangun GR (2003) Neural mechanisms of top-down control during spatial and feature attention. Neuroimage 19:496–512

33. Tootell RBH, Hadjikhani N, Hall EK et al (1998) The retinotopy of visual spatial attention. Neuron 21:1409–1422

34. Desimone R, Duncan J (1995) Neural mechanisms of selective visual-attention. Annu Rev Neurosci 18:193–222

35. Posner MI, Petersen SE (1990) The attention system of the human brain. Annu Rev Neurosci 13:25–42

36. Gitelman DR, Nobre AC, Parrish TB et al (1999) A large-scale distributed network for covert spatial attention: further anatomical delineation based on stringent behavioural and cognitive controls. Brain 122(Pt 6):1093–1106

37. Mesulam MM (1981) A cortical network for directed attention and unilateral neglect. Ann Neurol 10:309–325

38. Posner MI, Snyder CR, Davidson BJ (1980) Attention and the detection of signals. J Exp Psychol 109:160–174

39. Harter MR, Miller SL, Price NJ, Lalonde ME, Keyes AL (1989) Neural processes involved in directing attention. J Cogn Neurosci 1:223–237

40. Corbetta M, Kincade JM, Ollinger JM, McAvoy MP, Shulman GL (2000) Voluntary orienting is dissociated from target detection in human posterior parietal cortex. Nat Neurosci 3:292–297

41. Corbetta M, Shulman GL (2002) Control of goal-directed and stimulus-driven attention in the brain. Nat Rev Neurosci 3:201–215

42. Corbetta M, Shulman GL (2011) Spatial neglect and attention networks. Annu Rev Neurosci 34(34):569–599

43. McMains SA, Fehd HM, Emmanouil TA, Kastner S (2007) Mechanisms of feature- and space-based attention: response modulation and baseline increases. J Neurophysiol 98:2110–2121

44. Kastner S, Pinsk MA, De Weerd P, Desimone R, Ungerleider LG (1999) Increased activity in human visual cortex during directed attention in the absence of visual stimulation. Neuron 22:751–761

45. Chawla D, Rees G, Friston KJ (1999) The physiological basis of attentional modulation in extrastriate visual areas. Nat Neurosci 2:671–676

46. Greenberg AS, Esterman M, Wilson D, Serences JT, Yantis S (2010) Control of spatial and feature-based attention in frontoparietal cortex. J Neurosci 30:14330–14339

47. Burock MA, Buckner RL, Woldorff MG, Rosen BR, Dale AM (1998) Randomized event-related experimental designs allow for extremely rapid presentation rates using functional MRI. Neuroreport 9:3735–3739

48. Ollinger JM, Corbetta M, Shulman GL (2001) Separating processes within a trial in event-related functional MRI – II analysis. Neuroimage 13:218–229

49. Ollinger JM, Shulman GL, Corbetta M (2001) Separating processes within a trial in event-related functional MRI - I. The method. Neuroimage 13:210–217

50. Woldorff MG, Hazlett CJ, Fichtenholtz HM, Weissman DH, Dale AM, Song AW (2004) Functional parcellation of attentional control regions of the brain. J Cogn Neurosci 16:149–165

51. Walsh BJ, Buonocore MH, Carter CS, Mangun GR (2011) Integrating conflict detection and attentional control mechanisms. J Cogn Neurosci 23:2211–2221

52. Corbetta M (1998) Frontoparietal cortical networks for directing attention and the eye to visual locations: identical, independent, or overlapping neural systems? Proc Natl Acad Sci U S A 95:831–838

53. Fannon SP, Saron CD, Mangun GR (2007) Baseline shifts do not predict attentional modulation of target processing during feature-based visual attention. Front Hum Neurosci 1:7

54. Mangun GR, Fannon SP (2007) Networks for attentional control and selection in spatial vision. In: Mast F, Jäncke L (eds) Spatial processing in navigation, imagery and perception. Springer, New York, pp 411–432

55. Kastner S, Ungerleider LG (2000) Mechanisms of visual attention in the human cortex. Annu Rev Neurosci 23:315–341

56. Kincade JM, Abrams RA, Astafiev SV, Shulman GL, Corbetta M (2005) An event-related functional magnetic resonance imaging study of voluntary and stimulus-driven orienting of attention. J Neurosci 25:4593–4604

57. Wilson KD, Woldorff MG, Mangun GR (2005) Control networks and hemispheric asymmetries in parietal cortex during attentional orienting in different spatial reference frames. Neuroimage 25:668–683

58. Geng JJ, Mangun GR (2009) Anterior intraparietal sulcus is sensitive to bottom-up attention driven by stimulus salience. J Cogn Neurosci 21:1584–1601

59. Cohen YE, Andersen RA (2002) A common reference frame for movement plans in the posterior parietal cortex. Nat Rev Neurosci 3:553–562

60. Goodale MA, Milner AD (1992) Separate visual pathways for perception and action. Trends Neurosci 15:20–25

61. Goodale MA, Westwood DA (2004) An evolving view of duplex vision: separate but interacting cortical pathways for perception and action. Curr Opin Neurobiol 14:203–211

62. Goodale MA (2014) How (and why) the visual control of action differs from visual perception. Proc Biol Sci 281:20140337

63. Ungerleider LG, Mishkin M (1982) Two cortical visual systems. In: Ingle DJ, Goodale MA, Mansfield RJW (eds) Analysis of visual behavior. MIT Press, Cambridge, pp 549–586

64. Corbetta M, Tansy AP, Stanley CM, Astafiev SV, Snyder AZ, Shulman GL (2005) A functional MRI study of preparatory signals for spatial location and objects. Neuropsychologia 43:2041–2056

65. Moore T, Fallah M (2004) Microstimulation of the frontal eye field and its effects on covert spatial attention. J Neurophysiol 91:152–162

66. Nobre AC, Sebestyen GN, Miniussi C (2000) The dynamics of shifting visuospatial attention revealed by event-related potentials. Neuropsychologia 38:964–974

67. Astafiev SV, Shulman GL, Stanley CM, Snyder AZ, Van Essen DC, Corbetta M (2003) Functional organization of human intraparietal and frontal cortex for attending, looking, and pointing. J Neurosci 23:4689–4699

68. Corbetta M, Patel G, Shulman GL (2008) The reorienting system of the human brain: from environment to theory of mind. Neuron 58:306–324

69. Shulman GL, Pope DLW, Astafiev SV, McAvoy MP, Snyder AZ, Corbetta M (2010) Right hemisphere dominance during spatial selective attention and target detection occurs outside the dorsal frontoparietal network. J Neurosci 30:3640–3651

70. Fox MD, Corbetta M, Snyder AZ, Vincent JL, Raichle ME (2006) Spontaneous neuronal activity distinguishes human dorsal and ventral attention systems. Proc Natl Acad Sci U S A 103:10046–10051

71. Sestieri C, Pizzella V, Cianflone F, Romani GL, Corbetta M (2008) Sequential activation of human oculomotor centers during planning of visually guided eye movements: a combined fMRI-MEG study. Front Hum Neurosci 1:10.3389/neuro.3309/3001.2007

72. Yamaguchi S, Knight RT (1991) Anterior and posterior association cortex contributions to the somatosensory P300. J Neurosci 11:2039–2054

73. He BJ, Snyder AZ, Vincent JL, Epstein A, Shulman GL, Corbetta M (2007) Breakdown of functional connectivity in frontoparietal networks underlies behavioral deficits in spatial neglect. Neuron 53:905–918

74. Chica AB, Bartolomeo P, Valero-Cabré A (2011) Dorsal and ventral parietal contributions to spatial orienting in the human brain. J Neurosci 31:8143–8149

75. Shulman GL, McAvoy MP, Cowan MC et al (2003) Quantitative analysis of attention and detection signals during visual search. J Neurophysiol 90:3384–3397

76. Vossel S, Geng J, Fink GR (2014) Dorsal and ventral attention systems: distinct neural circuits but collaborative roles. Neuroscientist 20:150–159

77. Geng JJ, Vossel S (2013) Re-evaluating the role of TPJ in attentional control: contextual updating? Neurosci Biobehav Rev 37:2608–2620

78. DiQuattro NE, Sawaki R, Geng JJ (2014) Effective connectivity during feature-based attentional capture: evidence against the attentional reorienting hypothesis of TPJ. Cereb Cortex 24:3131–3141

79. Friston KJ, Harrison L, Penny W (2003) Dynamic causal modelling. Neuroimage 19:1273–1302

80. Friston KJ, Li B, Daunizeau J, Stephan KE (2011) Network discovery with DCM. Neuroimage 56:1202–1221

81. Folk CL, Remington RW, Johnston JC (1992) Involuntary covert orienting is contingent on attentional control settings. J Exp Psychol Hum Percept Perform 18:1030–1044

82. Serences JT, Shomstein S, Leber AB, Golay X, Egeth HE, Yantis S (2005) Coordination of voluntary and stimulus-driven attentional control in human cortex. Psychol Sci 16:114–122

83. Petersen SE, Posner MI (2012) The attention system of the human brain: 20 years after. Annu Rev Neurosci 35:73–89

84. Buckner RL, Andrews-Hanna JR, Schacter DL (2008) The brain's default network: anatomy, function, and relevance to disease. Ann N Y Acad Sci 1124:1–38

85. Liu Y, Bengson J, Huang H, Mangun GR, Ding M (2016) Top-down modulation of neural activity in anticipatory visual attention: control mechanisms revealed by simultaneous EEG-fMRI. Cereb Cortex 26:517–529

86. Worden MS, Foxe JJ, Wang N, Simpson GV (2000) Anticipatory biasing of visuospatial attention indexed by retinotopically specific alpha-band electroencephalography increases over occipital cortex. J Neurosci 20:RC63

87. Sauseng P, Klimesch W, Stadler W et al (2005) A shift of visual spatial attention is selectively associated with human EEG alpha activity. Eur J Neurosci 22:2917–2926

88. Thut G, Nietzel A, Brandt SA, Pascual-Leone A (2006) Alpha-band electroencephalographic activity over occipital cortex indexes visuospatial attention bias and predicts visual target detection. J Neurosci 26:9494–9502

89. Rajagovindan R, Ding M (2011) From prestimulus alpha oscillation to visual-evoked response: an inverted-U function and its attentional modulation. J Cogn Neurosci 23:1379–1394

90. Romei V, Brodbeck V, Michel C, Amedi A, Pascual-Leone A, Thut G (2008) Spontaneous fluctuations in posterior alpha-band EEG activity reflect variability in excitability of human visual areas. Cereb Cortex 18:2010–2018

91. Romei V, Gross J, Thut G (2010) On the role of prestimulus alpha rhythms over occipito-parietal areas in visual input regulation: correlation or causation? J Neurosci 30:8692–8697

92. Klimesch W, Sauseng P, Hanslmayr S (2007) EEG alpha oscillations: the inhibition-timing hypothesis. Brain Res Rev 53:63–88

93. Jensen O, Mazaheri A (2010) Shaping functional architecture by oscillatory alpha activity: gating by inhibition. Front Hum Neurosci 4:186

94. Haegens S, Osipova D, Oostenveld R, Jensen O (2010) Somatosensory working memory performance in humans depends on both engagement and disengagement of regions in a distributed network. Hum Brain Mapp 31:26–35

95. Anderson KL, Ding M (2011) Attentional modulation of the somatosensory mu rhythm. Neuroscience 180:165–180

96. Foxe JJ, Simpson GV, Ahlfors SP (1998) Parieto-occipital approximately 10 Hz activity reflects anticipatory state of visual attention mechanisms. Neuroreport 9:3929–3933

97. Fu KM, Foxe JJ, Murray MM, Higgins BA, Javitt DC, Schroeder CE (2001) Attention-dependent suppression of distracter visual input can be cross-modally cued as indexed by anticipatory parieto-occipital alpha-band oscillations. Brain Res Cogn Brain Res 12:145–152

98. Bollimunta A, Chen Y, Schroeder CE, Ding M (2008) Neuronal mechanisms of cortical

alpha oscillations in awake-behaving macaques. J Neurosci 28:9976–9988

99. Handel BF, Haarmeier T, Jensen O (2011) Alpha oscillations correlate with the successful inhibition of unattended stimuli. J Cogn Neurosci 23:2494–2502

100. Haegens S, Handel BF, Jensen O (2011) Top-down controlled alpha band activity in somatosensory areas determines behavioral performance in a discrimination task. J Neurosci 31:5197–5204

101. Kelly SP, Gomez-Ramirez M, Foxe JJ (2009) The strength of anticipatory spatial biasing predicts target discrimination at attended locations: a high-density EEG study. Eur J Neurosci 30:2224–2234

102. Dosenbach NU, Visscher KM, Palmer ED et al (2006) A core system for the implementation of task sets. Neuron 50:799–812

103. Dosenbach NU, Fair DA, Cohen AL, Schlaggar BL, Petersen SE (2008) A dual-networks architecture of top-down control. Trends Cogn Sci 12:99–105

104. Sakai K (2008) Task set and prefrontal cortex. Annu Rev Neurosci 31:219–245

105. Mangun GR, Buonocore MH, Girelli M, Jha AP (1998) ERP and fMRI measures of visual spatial selective attention. Hum Brain Mapp 6:383–389

Chapter 14

fMRI of Memory

Federica Agosta, Indre V. Viskontas, Maria Luisa Gorno-Tempini, and Massimo Filippi

Abstract

Numerous fMRI studies have investigated the network of brain regions critical for memory. Whereas neuropsychological techniques can delineate the brain regions that are necessary for intact memory function, neuroimaging techniques can be used to investigate which regions are recruited during healthy memory formation, storage, and retrieval. For example, fMRI studies have shown that lateral prefrontal cortex (PFC) supports some components of working memory function. However, working memory is not localized to a single brain region but is likely a property of the functional interaction between the PFC and posterior brain regions. The medial temporal lobe (MTL) and its connections with neocortical, prefrontal, and limbic structures are implicated in episodic memory. Semantic memory is mediated by a network of neocortical structures, including lateral and anterior temporal lobes, and inferior frontal cortex, possibly to a greater extent in the left hemisphere. Memory for semantic information benefits from the MTL for only a limited time, and can be acquired, albeit slowly and with difficulty, without it. To date, most of the emphasis has been on exploring the unique aspects of these different types of memory. Some evidence, however, of functional overlap in general retrieval processes does exist.

Key words Working memory, Encoding, Retrieval, Episodic memory, Semantic memory, Prefrontal cortex, Medial temporal lobe

1 Introduction

Memory shapes our behavior by allowing us to store, retain, and retrieve past experiences, and thus enables us to imagine the consequences of our actions. These processes influence and are modified by the type of information that is to be remembered, the duration of time over which it must be retained, and the way in which the brain will use the information in the future. The neural circuits underlying these processes are dynamic, reflecting the flexibility of memory itself. fMRI enables detailed study of the neural networks that support memory function. To delineate the neural circuitry underlying memory, it is helpful to breakdown memory into simpler components. In this chapter, we focus on memory that is consciously accessible, and use the duration of retention to

Massimo Filippi (ed.), *fMRI Techniques and Protocols*, Neuromethods, vol. 119,
DOI 10.1007/978-1-4939-5611-1_14, © Springer Science+Business Media New York 2016

dictate its parcellation: beginning with working memory, which holds information in mind for seconds to minutes, and moving on to long-term memory, which can be further divided into episodic and semantic memory.

2   Working Memory

Working memory (WM) refers to the temporary storage and manipulation of information that was most recently experienced or retrieved from long-term memory but is no longer available in the external environment [1–3]. Most models of WM separate two components of WM: temporary stores of information, in the form of "buffers" or "slave systems," that are usually modality-specific, and a central "executive" or set of processes that manipulate the information [2, 4]. Items in WM are stored only as long as the information is either being rehearsed (subvocally) or manipulated in some other fashion (i.e., rotated or integrated with existing information in semantic memory). The capacity of WM is limited by attention to about seven (plus or minus two) meaningful "bits" (or chunks) of information—these bits can be manipulated and either discarded or associated and transferred into long-term memory [5]. WM is central to everyday functioning and contributes significantly to other areas of cognition. Baddeley and Hitch [1] proposed a model of memory that has influenced virtually all subsequent research in the area. Their model is composed of a three-component system: (1) the "phonological loop," comprising a limited capacity phonological store in which verbal information is stored temporarily and maintained by subvocal rehearsal (e.g., repeated subvocal articulation when trying to keep a phone number in mind), (2) the "visuospatial sketch pad," a storage buffer for nonverbal material, such as the visual representations of objects, and (3) the "central executive," which is responsible for strategic manipulation and execution of the aforementioned "slave" systems. The original model has subsequently been updated [2] to include an "episodic buffer" that provides an interface between the subsystems of WM and long-term memory.

The first evidence for a role of prefrontal cortex (PFC) in WM came from lesion and electrophysiological studies in nonhuman primates [6, 7]. The first neurons discovered – showing persistent activity during the delay period of WM—were found in the monkey PFC using single-unit neuron recording techniques [8–10]. This sustained activity is thought to provide a bridge between the stimulus cue, for instance, the location of a flash of light, and its contingent response, such as a saccade to the remembered location. Persistent activity during blank memory intervals is a very powerful observation and established a strong link implicating the PFC as a critical node supporting WM [3]. Since then, physiological studies in nonhuman primates have revealed active delay neurons in a large number of brain regions, including the dorsolateral (DL) and ventrolateral (VL) PFC, the intraparietal sulcus (IPS), posterior perceptual areas, and subcortical structures, such as the caudate and the thalamus (for a review, *see* Ref. [11]). In humans, however, the advent of modern functional brain imaging techniques, such as positron emission tomography (PET) and fMRI, enabled a more detailed study of the functional neuroanatomy of WM processes. Many of the neuroimaging experiments designed to elucidate the neural underpinnings of WM separate executive control processes from storage. For example, a number of studies have been designed such that WM maintenance is the task of interest (e.g., delayed response, delayed recognition, delayed alternation, and delayed match-to-sample tasks). In a typical delayed recognition trial, the subject is first required to remember a stimulus presented during a *cue* period and then to maintain this information for a brief *delay* interval when the stimulus is absent. Then, the subject responds to a *probe* stimulus to determine whether the information was successfully retained. Thus, the maintenance and recognition decision processes are temporally segregated and can be investigated in relative isolation. In this way, using fMRI, investigators can record neural activity during these distinct stages of WM. Brain regions exhibiting persistent activity above resting baseline during the delay period are often interpreted as being involved in WM maintenance processes [12–20]. Functional neuroimaging studies have revealed that the same collection of regions that were shown to be involved in WM in nonhuman primates displays significantly increased activity during delay tasks in humans [21–29]. For instance, in one fMRI study, subjects were scanned while they performed an oculomotor delayed matching-to-sample task that required maintenance of the spatial position of single dot of light over a delay period after which a memory-guided saccade was generated [27]. Both frontal eye fields and IPS showed activity that spanned the entire delay period. Moreover, the magnitude of the activity correlated positively with the accuracy of the memory-guided saccade that followed later [27]. Despite the relative simplicity of the delay period, however, multiple cognitive processes remain engaged concurrently, including information maintenance, suppression of distraction, motor response preparation, mental timing, expectancy, monitoring of internal and external states, and preservation of alertness. As a result, even the maintenance period of WM is likely to be mediated by a distributed network of distinct brain regions, rather than to be localized to a single brain region [20].

2.1 Organization of the WM Network

The extensive reciprocal connections from the PFC to virtually all cortical and subcortical structures place the PFC in an unique neuroanatomical position to monitor and manipulate diverse cognitive processes [20]. Marklund et al. [28] employed a mixed block and event-related design in an fMRI study of episodic, semantic, and working memories contrasted with sustained attention. This approach identified transient activity, particularly in the left DL PFC, that appears to reflect the operation of WM during retrieval from long-term memory [28]. Furthermore, it has been found that the PFC shows activity during retention interval of delay task regardless of the type of information (e.g., spatial, faces, objects, words) [22, 24, 29]. In fact, there is a critical mass of functional neuroimaging studies emphasizing the stable persistent neural activity in selective lateral PFC neurons during the delay period in humans ("the fixed-selectivity model") (for review, *see* Ref. [3]).

A controversial issue in the literature is the extent to which WM can be segregated anatomically according to the type of to-be-retained material (e.g., verbal, space, object, visual, and auditory) or the component processes (e.g., maintenance vs. manipulation of information) (for a review, *see* Refs. [3, 30–34]).

2.1.1 Organization of WM by Material Type

First, let us evaluate the evidence supporting anatomical segregation on the basis of stimulus category. Three types of material have been most commonly studied: verbal, spatial, and object information. A prevalent theory of material-type segregation in the frontal cortex suggests that there are dorsal and ventral memory streams for spatial and object information, respectively, similarly to the "where" and "what" pathways of the visual system [35]. The dorsal stream projects from the extrastriate cortex to the inferior parietal lobule (IPL) and the IPS and is involved in processing spatial information [35]. The ventral stream extends from the extrastriate cortex to the inferior surface of the frontal pole and processes object information [35]. Within the frontal cortex, WM for spatial information involves the superior DL PFC or the superior frontal sulcus, whereas object WM relies upon several mid- and inferior frontal regions (VL PFC) (for review, *see* Ref. [3]). Furthermore, there is a tendency for verbal and object WM to recruit more left-hemisphere areas, and for spatial tasks to recruit more right-hemisphere areas (for review, *see* Ref. [3]).

The processing of nonverbal spatial information is right lateralized and associated with the activation of a fronto-parietal network (e.g., *see* Ref. [36–41]). Using event-related fMRI, Courtney et al. [37] demonstrated a neuroanatomical dissociation between delay period activity during WM maintenance for either the identity (object memory) or location (spatial memory) of a set of three face stimuli. Greater activity during the delay period on face identity trials was observed in the left inferior frontal gyrus (IFG), whereas greater activity during the delay period of the location task was observed in dorsal frontal cortex (bilateral superior frontal sulcus) [37] (Fig. 1).

WM for objects, mostly visually presented faces, houses, and line drawings that are not easily verbalizable, seems to be right-lateralized and activates the temporal-occipital regions [Brodmann area (BA) 37] (e.g., *see* Refs. [21, 26, 29, 36, 39, 42–44]).

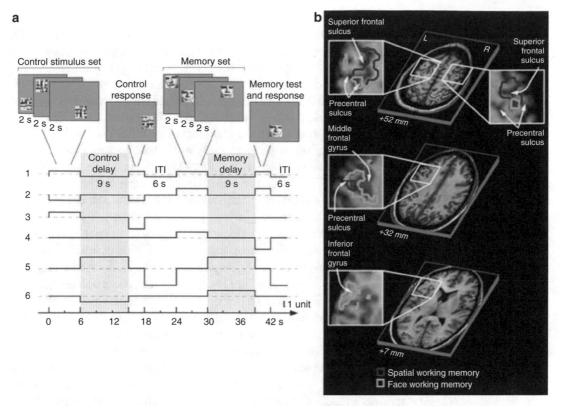

Fig. 1 An event-related fMRI study of delay period activity during working memory (WM) maintenance for either the identity (object memory) or location (spatial memory) of a set of three face stimuli. (**a**) Schematic depiction of the fMRI tasks: subjects saw a series of three faces, each presented for 2 s in a different location on the screen, followed by a 9-s memory delay. Then a single test face appeared in some location on the screen for 3 s, followed by a 6-s intertrial interval. Before each series, subjects were instructed to remember the locations or the identities of the three faces in the memory set. For the spatial task, the subject indicated with a left or right button whether the test location was the same as one of the three locations presented in the memory set, regardless of the face that marked that location. For the face memory task, the subject indicated whether the test face was the same as one of the three faces observed in the memory set, regardless of the location where the face appeared. For the sensorimotor control task, scrambled faces appeared (control stimulus set), and when the fourth scrambled picture appeared after the delay, subjects pressed both buttons (control response). Contrasts between task components are shown below the task diagram: (*1*) visual stimulation vs. no visual stimulation, (*2*) memory stimuli vs. control stimuli, (*3*) control stimulus set vs. control response, (*4*) memory stimulus set vs. test stimulus and response, (*5*) delays during anticipation of response vs. intertrial intervals, and (*6*) memory delays vs. control delays. (**b**) Areas with significant sustained recruitment in a single subject during the WM delay for faces (*blue outline*) and for spatial locations (*red outline*) overlaid onto the subject's Talairach normalized anatomical MR image. Level above the bicommissural plane is indicated for each axial section (from Courtney et al. [37])

In contrast to object, processing of verbal WM activates regions in the left hemisphere. Broca's area (BAs 44/45), premotor areas (supplementary motor area and premotor cortex) [45–48], and the cerebellum [24, 47, 49, 50] are critical for the articulatory subvocal rehearsal. Phonological maintenance has been associated with activity in parietal areas, particularly the IPL (i.e., BAs 39 and 40) [24, 45, 46, 49–53] but also in the superior parietal lobe [47, 49, 52, 53]. For instance, using an event-related design, Chein and Fiez [45] designed a delayed serial recall task requiring subjects to encode, maintain, and overtly recall sets of verbal items for which phonological similarity, articulatory length, and lexical status were manipulated and reported that Broca's area and BA 40 showed patterns of sustained activity during the delay period of a verbal WM task.

More recently, Raye et al. [48] provided evidence that left VL PFC appears associated with subvocal rehearsal of single words, whereas left DL PFC activation is associated with participants "refreshing" (simply thinking about) the visual appearance of a recently presented word [48].

Despite this growing evidence for segregation by material type, there are also several human functional imaging studies that have failed to find evidence for segregation in PFC WM activity (e.g., *see* Refs. [16, 54, 55]). For instance, using an event-related design, Postle and D'Esposito [16] evaluated the organization of WM for the identity and location of visually presented stimuli (target stimuli for all object trials were 16 abstract polygon stimuli, determined in normative testing to be difficult to associate with real-world objects). Although the task produced considerable delay-period activity in VL PFC, DL PFC, and superior frontal cortex, in no subject, PFC activity was greater for one stimulus domain than for the other [16]. Moreover, in a large meta-analysis based on 60 neuroimaging (both PET and fMRI) studies of WM [55], analyses of material type showed the expected dorsal–ventral dissociation between spatial and nonspatial storage in the posterior cortex, but not in the frontal lobe. Some support was found for left frontal dominance in verbal WM, but only for tasks with low executive demand. Executive demand increased right lateralization in the frontal cortex for spatial WM [55].

2.1.2   Organization of WM by Process Type

The other axis along which investigators have suggested that human lateral PFC involvement in WM is segregated is according to the type of operation performed upon the contents of WM, rather than the type of information being maintained. In particular, several fMRI studies have focused on the distinction between two fundamental WM processes, namely the passive *maintenance* of information in short-term memory and the active *manipulation* of this information, within the PFC (for review, *see* Ref. [3]). This model received initial support from a PET study by Owen et al. [56] in which dorsal PFC activation was found during three spatial WM tasks thought to require greater monitoring of remembered information (i.e., a mnemonic variant of modified Tower of London planning task requiring short-term retention and reproduction of problem solutions) than two other WM tasks (i.e., the modified Tower of London planning task, and a control condition that involved identical visual stimuli and motor responses) that activated only ventral PFC.

This model was also tested using fMRI (for review, *see* Ref. [3]). The VL PFC has been suggested to be primarily involved only in WM mechanisms that support simple retrieval of information for sensory-guided sequential behavior (maintenance) [57–60], whereas DL PFC (particularly BA 46) has been found to serve mechanisms of active monitoring and manipulation of information (or generalized executive processing) in WM [61–63]. For instance, in an event-related fMRI study [61], subjects were presented with two types of trials in random order in which they were required to either maintain a sequence of letters across a delay period or manipulate (alphabetize) this sequence during the delay in order to respond correctly to a probe. The authors found that dorsal PFC activity was greater in trials during which actively maintained information was manipulated, providing further support for a process-specific PFC organization. However, this functional PFC division has not been consistently replicated and seems to vary with materials and difficulty [55]. Moreover, this distinction is not compatible with evidence of continuous DL PFC activity in tasks without any manipulation [32, 64]. A large meta-analysis [55] showed that tasks requiring executive processing generally produce more dorsal frontal activations than do storage-only tasks, but not all executive processes show this pattern. For instance, superior frontal cortex (BAs 6, 8, and 9) responded most when WM must be continuously updated and when memory for temporal order has to be maintained [55]. Right ventral frontal cortex (BAs 10 and 47) responded more frequently with demand for manipulation (including dual-task requirements or mental operations) [55]. Posterior parietal cortex (BA 7) was found to be involved in all types of executive functions [55].

2.2 Medial Temporal Lobe Involvement in WM

fMRI studies of WM have also found that the fronto-parietal network is not the only region that is active during the temporary retention of task-relevant information. PFC and parietal cortex do not seem to be sufficient to perform WM for novel stimuli when parahippocampal regions are lesioned [65], although they are sufficient to maintain normal WM for familiar stimuli [66]. Surprisingly, early fMRI studies of WM did not report activity within parahippocampal regions such as perirhinal (PrC) or entorhinal cortex [67]. An fMRI study by Stern et al. [68] demonstrated differential activation for novel vs. familiar stimuli during performance of a 2-back WM task. This study showed that WM for a highly familiar set of complex visual images primarily activated prefrontal and parietal cortices, whereas the same task using novel (trial-unique) visual images strongly activated parahippocampal structures in addition to prefrontal and parietal cortices. Activation of parahippocampal structures associated with WM for novel stimuli has also been shown in an event-related fMRI study using novel face stimuli [25, 69].

2.3 Connectivity Analysis Within the WM Network

By varying experimental design (e.g., parametric memory load variation [23, 43]), attempts have been made to associate identified brain regions with different processes occurring during the delay. The first fMRI study aimed at characterizing regional interactions in WM has used a block-design fMRI paradigm with a graded *n*-back verbal WM task [70]. Changes of effective connectivity within a fronto-parietal WM network related to different levels of WM load were studied. The results revealed enhanced inferior fronto-parietal connectivity and also increased interhemispheric communication between DL PFC regions as correlates of increasing WM load [70]. A following event-related fMRI study characterized the neural network mediating the online maintenance of faces [71] (Fig. 2). The fusiform face area (FFA) was defined as a seed and was then used to generate whole-brain correlation maps. A random effects analysis revealed a network of brain regions exhibiting significant correlations with the FFA seed during the WM delay period. This maintenance network included the DL and VL PFC, the premotor cortex, the IPS, the caudate nucleus, the thalamus, the hippocampus, and occipitotemporal regions [71].

These findings support the notion that the coordinated functional interaction between nodes of a widely distributed network underlies the active maintenance of a perceptual representation and

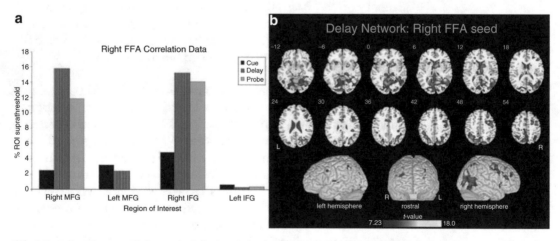

Fig. 2 Functional connectivity analysis between brain regions associated with the maintenance of a representation of a visual stimulus (face recognition task) over a short delay interval. The fusiform face area (FFA) was used as the exploratory seed. (**a**) Quantification of right FFA correlation effects in the prefrontal cortex (PFC). Bars indicate the number of significant voxels present during each stage (cue, delay, and probe) in four prefrontal regions of interest. *MFG* middle frontal gyrus, *IFG* inferior frontal gyrus, *ROI* region of interest. (**b**) Delay period right FFA seed correlation map including bilateral regions in the dorsolateral and ventrolateral PFC, premotor cortex, IPS, caudate nucleus, thalamus, hippocampus, and occipitotemporal regions. Activations are thresholded at $p < 0.05$ (corrected) and are shown overlaid on both axial slices and a three-dimensionally rendered Montreal Neurological Institute (MNI) template brain. The color scale indicates the magnitude of *t* values (from Gazzaley et al. [71])

provides convergent evidence that fronto-parietal and interhemispheric frontal connectivities are central in WM [70, 71]. More recently, it has been demonstrated that visual WM involves top-down signals from the lateral PFC to parietal cortex [72, 73] as well as communication from fronto-parietal regions to visual cortex [74]. Additionally, interactions between basal ganglia and the lateral PFC are thought to mediate the filtering of task-irrelevant information and the updating of task-relevant information in WM [75]. Future studies are warranted to understand further the functions resulting from interactions between these regions to build a complete picture of WM.

3 Long-Term Memory for Consciously Accessible Material: Episodic and Semantic

Within the declarative or consciously accessible long-term memory system, "episodic" and "semantic" memory can be distinguished. Episodic memory (EM) allows the recollection of unique personal experiences: rich, vivid reexperiencing of past events. Semantic memory (SM), in contrast, refers to generic information that is acquired across many different instances and accessed independently of the details of the context in which the information was first encountered [76]. This fractionation of declarative memory is supported by evidence that episodic and semantic memory have distinctive anatomical substrates [77–80]. Therefore, we will consider each memory type individually. Of note, autobiographical memory can be either semantic, as in one's knowledge of the names of all the schools that one attended, or episodic, as in one's memory for a particular birthday: what binds autobiographical memories to each other is self-awareness.

4 Episodic Memory

EM enables us to access and reexperience the sights, sounds, smells, and other details of a specific event [76]. Most EMs are available for several minutes or hours but over time access to their details degrades [81]. Others remain accessible with their details for a lifetime [82]. This temporal difference in storage highlights the complexity of EM: episodic remembering is composed of several component processes, including the retrieval of information from across sensory domains and the reconstruction of an event from a set of individual details.

4.1 Distinguishing Encoding and Retrieval Processes

One of the major advances in the study of EM has been the application of neuroimaging techniques to distinguish the component processes of encoding and retrieval. In neuropsychological studies, it is often difficult or even impossible to separate failures of EM due to encoding or retrieval processes. Functional imaging affords researchers the opportunity to separate the neural underpinnings of these processes and is particularly helpful in understanding the distinct roles of medial temporal lobe (MTL) subregions, which are often damaged equivocally in patients with lesions [83]. Furthermore, neuroimaging evidence has shown that prefrontal and other cortical areas are engaged during episodic remembering, a fact that had not been clear from patient data alone [84].

Neuroimaging studies have indicated that individual elements of EMs may be permanently stored within the same neocortical regions that are involved in initial processing and analyzing of the information [85, 86]. Several studies have demonstrated the reactivation of the visual cortex during retrieval of visual details [85, 87], activation in the auditory cortex during auditory memory retrieval [88], and activation in the motor cortex during the retrieval of memory for actions [89]. Insofar as episodic remembering involves reexperiencing the details of an event, it is not surprising that brain regions involved in the initial perception of these details are reactivated during their retrieval.

According to several influential memory models, each different cortical region makes a unique contribution to the storage of a given memory and all regions participate together in the creation of a complete memory representation [90–92]. The MTL, then, is saddled with the task of binding together these different regional contributions into a coherent memory trace [90]. In the MTL, the hippocampal formation receives processed sensory information from association areas in the frontal, parietal, and occipital lobes via the parahippocampal cortex [93]. Given its anatomical placement and architecture, the hippocampus has the unique ability to bind "what happened," "when it happened," and "where it happened" together [90]. The architecture of the hippocampus includes a circular pathway of neurons from the entorhinal cortex to the dentate gyrus, CA3 and CA1 neurons of the hippocampus to the subiculum, and back to the entorhinal cortex [94]. The connections within the hippocampal formation and between the MTL and neocortical regions are formed more rapidly than are the connections between disparate cortical regions [92]. Therefore, when a particular cue in the environment or the mental state of the person activates cells in the cortical regions, the MTL network that is associated with that cue is reactivated and the entire neocortical representation is strengthened. As multiple reactivations occur, the connections between the relevant neocortical regions are slowly strengthened until the memory trace no longer depends on the activity of the MTL, but may be entirely represented in the neocortex [90]. Using both fMRI and neuropsychological techniques, consistent evidence suggests that the hippocampus remains involved in the retrieval of EMs regardless of the age of the memory [95]: several authors, then, have proposed that the hippocampus acts as a pointer or index, recreating the activation in the neocortex that represents the individual elements of the memory [96].

4.1.1   Encoding

Simply perceiving and attending to information in the world is not sufficient to create a lasting long-term memory trace. Some other process or processes are engaged in order to bind elements of an episode into a coherent memory trace. There is some disagreement concerning whether all memories are initially episodic and later become semantic, or whether some memories are semantic in nature from the outset [97, 98]. One might argue that every learning episode is encoded as such, and as items from an episode get integrated into the network of semantic information, those items become disassociated from the episode itself and associated instead with the information already in semantic memory [97]. In fact, several memory models suggest that episodic and semantic information is learned via different mechanisms: elements of an episode are rapidly bound together by autoassociative processes in the dentate gyrus and CA3 field [99], while semantic information is gradually acquired over many repetitions by reorganization of Hebbian synapses in the neocortex [91, 100, 101].

Pioneering neuroimaging studies in the 1990s implicated the PFC, particularly the left lateral PFC, in semantic or associative encoding, by showing that this region, in addition to the MTL, shows greater neural activity during semantic encoding than during more superficial or perceptual encoding [102, 103]. Furthermore, novel stimuli have been shown to elicit greater neural activity in the MTL than familiar stimuli [104, 105], providing more evidence for the involvement of the MTL in encoding processes.

A direct link to episodic encoding processes, however, required the advent of event-related fMRI designs to investigate these cognitive processes [106]. Encoding trials were binned according to whether information presented on a given trial is subsequently remembered or forgotten. Subsequent memory studies have shown that activation in the left and right PFC, the parahippocampal gyrus [107, 108], and hippocampus [109] during encoding predicts successful memory retrieval (Fig. 3). Since these original studies, many more studies have replicated the finding that MTL activity correlates with episodic encoding [110–113]. Furthermore, fMRI studies have shown that hippocampal activity correlates with the subsequent retrieval of contextual details presented during encoding, while activity in the PrC correlates with successful retrieval of an individual item, but not with retrieval of episodic details [111–115]. Staresina and Davachi [116] have shown that since the PrC receives inputs mainly from cortical areas devoted to the processing of visual information, recruitment of this region also correlates with the retrieval of visual features of items, such as the color in which they were presented, but not with other

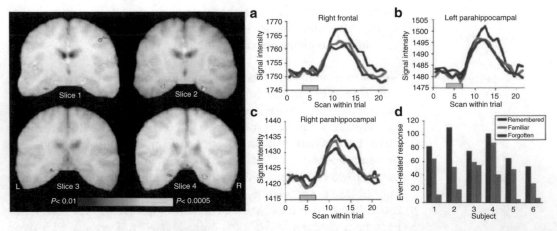

Fig. 3 Composite statistical activation maps displaying voxels with significant positive correlations between event-related activations to pictures and subsequent memory for those pictures. Areas activated are right dorsolateral prefrontal cortex (*upper right* in slice 1) and bilateral parahippocampal cortex (*lower left* in slices 1 and 3; *left* and *right* in slices 2 and 4). Examples of average signal magnitude during study from six subjects in (**a**) right frontal (slice 1), (**b**) left parahippocampal (slice 4), and (**c**) right parahippocampal (slice 2) regions for remembered (*red*), familiar (*green*), and forgotten (*blue*) pictures. *Gray block* depicts onset and offset of picture presentation. (**d**) Mean voxel response in parahippocampal areas showing significant correlation with subsequent memory in each subject for remembered, familiar, and forgotten pictures (from Brewer et al. [107])

contextual details. These data suggest that while the hippocampus binds elements from all domains, the more specialized regions of the MTL and surrounding cortices may play domain-specific roles in EM encoding. Furthermore, the authors report that the activation in the hippocampus increased stepwise with increasing numbers of successfully encoded associations.

A number of studies, however, have also failed to find reliable MTL activation related to successful episodic encoding [117–120]. While neuropsychological data demonstrate that the MTL is necessary for new episodic encoding, processes following encoding such as retrieval, consolidation, interference, and so on may have a larger impact on subsequent remembering than encoding-related MTL activity.

4.1.2 Retrieval

Work on EM retrieval points to a set of highly flexible retrieval operations that are differentially engaged depending on the particular demands of the situation. In general, the network that supports these retrieval operations includes PFC, MTL structures, and posterior sensory cortices. Identifying the contributions of these brain regions to episodic retrieval processes has demanded new methodologies capable of distinguishing subtle dissociations across similar retrieval tasks. This section explores three emerging themes: the role of PFC and posterior sensory cortices, the role of MTL subregions and the role of parietal cortex in retrieval success.

PFC and Posterior Sensory Cortices

Early neuroimaging work contrasting encoding and retrieval processes revealed a consistent asymmetry in the activations of prefrontal regions [121, 122]: specifically, greater activation of the left PFC during encoding, and relatively greater activation of the right PFC during episodic retrieval [121, 122]. This pattern, termed "hemispheric encoding-retrieval asymmetry" (HERA), provided early insights into the neural basis of EM retrieval and a framework for the design of future studies [122].

Further exploration of this relationship, however, demonstrated that the HERA model was insufficient. The current view is that HERA reflects the maintenance of a retrieval mode, or a background cognitive state in which one is mentally attuned to the retrieval process, is sensitive to incoming cues, and is capable of becoming consciously aware of successful retrieval. Lepage et al. [123] compared data across four PET studies and concluded that six PFC regions were consistently recruited by episodic recognition. These included bilateral posterior ventrolateral areas, bilateral frontopolar regions, right dorsal PFC, and midline cingulate area near the supplementary motor area. How and when these regions are recruited in EM retrieval depends on task demands. As proposed in the source-monitoring framework [124], retrieval attempts differ in the strategies used to access different cortical representations *depending on the nature and source of those representations*. For example, visuospatial, semantic, or emotional cues are often called upon to aid in episodic retrieval, and each of these cues stimulates a slightly different set of cortical regions.

Using event-related fMRI, Dobbins and Wagner [125] showed that the recollection of conceptual or perceptual details of an episode results in greater activation in the left frontopolar and posterior PFC than the detection of novelty. The authors interpret this finding as an evidence that a domain-general control network is engaged during contextual remembering. In contrast, left anterior VL PFC coactivated with a left middle temporal region associated with semantic representation, during conceptual recollection, while right VL PFC and bilateral occipito-temporal cortices were coactivated during recollection of perceptual details. Therefore, whereas left frontopolar and posterior PFC may be involved in domain-general retrieval processes, the middle temporal, right VL PFC, and occipito-temporal regions may be more domain-specific.

Interestingly, emerging data suggest that PFC is not involved in distinguishing correct from incorrect retrieval. Dobbins et al. [126] found that activation in the left MTL was greater for correct than for incorrect source attributions, or episodic retrieval, but that numerous PFC regions that showed increased activation during source memory retrieval did not distinguish between correct and incorrect trials. Moreover, activation in some of these regions was actually numerically greater for failed retrieval attempts. The authors interpret these findings as evidence that the PFC is involved in elaborative or monitoring operations, which may be enhanced when retrieval fails. Furthermore, Velanova et al. [127], using a mixed block/event-related design to explore sustained and transient control processing during EM retrieval, found that left PFC activation was associated trial-by-trial with retrieval when high control was required, whereas right PFC and several right posterior regions were associated with sustained control processes, which the authors viewed as a reflection of attentional set or retrieval mode. In line with this idea, Kahn et al. [128] also failed to find differences in prefrontal responses as a function of source retrieval outcome, and PFC regions were more active during trials for old items than for new ones. In addition, Wheeler and Buckner [129] found that activation in PFC correlates with retrieval processes and not with retrieval success: they found that left posterior and mid-VL PFC activity for items that were highly rehearsed and therefore easily identified as old was not different than for new items. By contrast, left PFC activity for items previously encountered only once was greater than for both the old and new items, suggesting that left PFC activity does not correspond to retrieval success. The authors concluded that left VL PFC activity reflects a processing control operation that is selectively engaged during demanding retrieval.

MTL Activation in Episodic Remembering

Investigating MTL activation using fMRI can be difficult not only because the region is susceptible to artifacts attributed to the ear canal, but also because the region seems to be active during a wide variety of tasks, including undirected "rest" [130]. Therefore, many studies of episodic remembering do not report greater activation in the MTL, even though this region is known to be involved. In order to observe MTL activity, a baseline task such as an odd/even digit judgment may be used to deactivate the MTL during the control trials [130].

Unlike data from the PFC, activation in the MTL has been found to correlate with episodic retrieval success [131, 132] (Fig. 4). Several neuroimaging studies investigating the role of MTL subregions in EM have relied on the remember/know procedure to distinguish episodic retrieval, or the processes of "remembering or recollecting" (R), from retrieval based on familiarity, called the "knowing or recalling" (K) [133]. This technique was based on the finding that patients with MTL damage, especially those showing selective loss of hippocampal function, show impairments in EM and a comparatively intact SM [80, 134]. In addition, patients with degeneration of extrahippocampal temporal cortex show the opposite pattern: impaired SM combined with a relatively intact EM [78]. Using the remember/know procedure, Eldridge et al. [131, 132] found that hippocampal activity is primarily associated with "remembering" rather than "knowing": the hippocampus was more active during retrieval of "R" items than

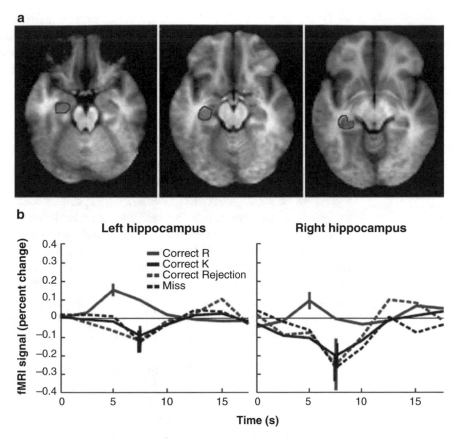

Fig. 4 Results from anatomically defined hippocampal regions of interest. (**a**) Sections from the anatomical template with the left hippocampal region of interest outlined in *red*. (**b**) Averaged event-related responses in the hippocampus from 11 subjects. *Error bars* represent one standard error (between subjects) of estimated response amplitudes. *Correct R* correct remember (R) response (when subjects could recollect the moment the item was studied), *Correct K* correct know (K) response (when the word is familiar but unaccompanied by the recollection of the specific moment the word was presented), *Correct rejection* correct response for nonrecognized items, *Miss* miss response (in which subjects did not recognize old items) (from Eldridge et al. [131])

during retrieval of "K" items. In a follow-up study, Eldridge et al. [135] found that during encoding of items that were subsequently correctly recognized, there was more activity in the dentate gyrus and CA2/3 fields of the hippocampus than during encoding of items that were subsequently forgotten. During retrieval, activity in the subiculum correlated uniquely with "R" responses. These data are particularly compelling because the dentate gyrus and CA2/3 fields are located early in the hippocampal circuit, while the subiculum is the major output region of the hippocampus.

Parietal Cortex

The medial parietal cortex exhibits opposite levels of fMRI activity during encoding and retrieval, a pattern dubbed the encoding/retrieval (E/R) flip. These opposing effects were originally reported by Daselaar et al. [136] who observed this pattern within the same participants for a variety of stimuli and memory paradigms. Since then, the E/R-flip pattern has been replicated in several other studies (for review, *see* Ref. [137]). It has been shown that this pattern occurs regardless of the type of information (words, faces, spatial scenes), stimulus modality (auditory or visual), and memory test (item or relational memory) (Fig. 5). Several hypotheses have been formulated to explain the E/R-flip pattern (for review, *see* Ref. [137]). The *internal orienting account* asserts that medial parietal cortex involvement in encoding and retrieval is dependent on the internal versus external orientation of attentional. The *self-referential processing account* explains medial parietal cortex activity in terms of orienting toward self-relevant thoughts versus the external environment. The *reallocation account* states that the activation differences in medial parietal cortex depend on task-demands and follows response times. Finally, the *bottom-up attention account* asserts that activity within medial parietal regions reflects bottom-up orienting of attention towards information retrieved from memory. Yet none of these cognitive accounts seems to provide a full explanation for the E/R-flip pattern in the PMC [137].

4.2 Temporal Processing

One somewhat less-studied component of EM is the process of assigning a temporal order to a series of events. While this part of the field is sparse in terms of neuroimaging data, there are a handful of studies that can shed some light on this process. Bilateral middle prefrontal areas near BA 9, left inferior prefrontal (near BA 44/45), left anterior prefrontal (near BA 10/46), and bilateral medial temporal areas show more activation during "high" temporal order retrieval trials (that is, when choosing between two words that were close together on a list: i.e., which came first, word # 6 or word #3?) than during "low" trials (that is, when choosing between words that were spaced far apart on a list: i.e., which came first, word #1 or word # 9?) [138]. Activation in the middle frontal gyrus (MFG) near BA 9 is especially interesting because there is convergent evidence for the involvement of this region in the human neuropsychological [139] and monkey literature [140]. There may also be a hemispheric specialization in temporal processing. Suzuki et al. [141] asked participants to study pictures during two separate sessions: one in the morning and another in the afternoon. In the scanner, participants were asked to judge whether an item was studied in the morning or in the afternoon, or which of two items in the same list was studied earlier. They found that right prefrontal activity was associated with temporal order judgments of items between lists (morning vs. afternoon) while left prefrontal activity was associated with the retrieval of temporal order information within a list.

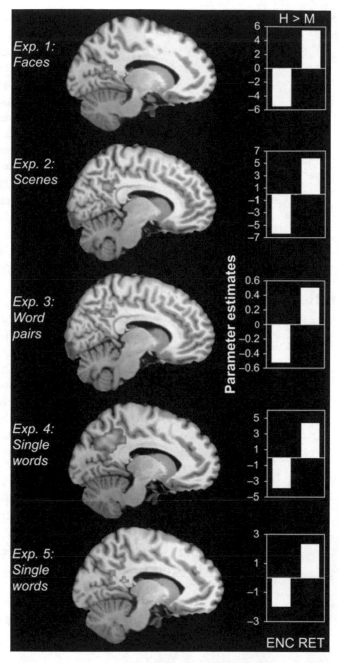

Fig. 5 Figure shows the encoding/retrieval flip in four different fMRI experiments using (*1*) faces, (*2*) spatial scenes, (*3*) word pairs, and (*4* and *5*) single words. During encoding, activity in the posterior midline region was greater for misses (*M*) than for hits (*H*), whereas during retrieval, activity was greater for hits than for misses. *Bar graphs* indicate mean cluster activity for the comparison between hits and misses during encoding and retrieval, respectively (from Daselaar et al.[136])

4.3 Summary of Episodic Memory

Prior to the advent of fMRI, declarative memory, particularly EM, was thought to be dependent almost exclusively on the MTL. Both animal and human lesion studies provided the bulk of this evidence. Neuroimaging techniques have demonstrated, however, that PFC regions are recruited time and time again to support EM retrieval. Interestingly, PFC does not seem to be sensitive to the outcome of retrieval, but rather seems to support retrieval intention and strategy. The MTL, in contrast, seems to be driven largely by "bottom-up" processes: being more data-driven and less voluntary [142]. Whereas details of an event are thought to be eventually stored in the same regions that were initially involved in their perception, the hippocampus creates and stores an index of the memory that can regenerate the original pattern of activation during EM retrieval. Therefore, in healthy adults, EM involves a large network of prefrontal, neocortical, and MTL regions acting in concert to support this complex reconstructive process.

4.4 Future Directions

The difference between EM and SM is the difference between the active reconstruction of an event to extract information specific to that occurrence and the abstraction of statistical regularities and general properties about the world over multiple experiences. The constructive nature of memory, therefore, is most easily observed in EM: the reexperiencing of past events requires the reconstruction of a narrative sequence via the reactivation of stored sensory information. A growing interest in the constructive aspects of EM has led to the postulation of the *constructive episodic simulation* hypothesis [143–145], which suggests that the EM system is built, in part, to enable the simulation or imagination of future events. Support for this view comes from several neuroimaging studies demonstrating that the network of brain regions involved in EM retrieval overlaps substantially with that supporting the imagination of future events [146–148]. Furthermore, as Hassabis and Maguire [149, 150] have eloquently described, EM retrieval can be thought of as relying heavily on scene construction as a key component process. This approach may help explain why many other cognitive tasks such as imagining a fictitious event seem to employ many of the same regions involved in episodic remembering.

5 Semantic Memory

In contrast to EM, SM corresponds to the general knowledge of objects, words, facts, and people: declarative memories that are laid down independently of the original encoding context [76]. This information is often encountered over multiple repetitions, in a variety of contexts, and thus can be retrieved without regenerating details of the original learning event. While initially limited to a memory system for "words and other verbal symbols, their meanings and referents, about relations among them, and about rules, formulas and algorithms" [76], SM currently refers to a broader knowledge set that includes facts, concepts, and beliefs [151].

Reports of patients with focal deficits in SM that were category-specific [152–155] led to the hypothesis that SM is organized by taxonomic categories [156]. In addition, a highly influential theory proposed by Allport [157] suggests that the same sensorimotor areas that are involved in the initial processing of experience are also used to represent abstractions related to the experience. Much like findings from neuroimaging studies of EM, this theory predicts that modality-specific information is represented in the cortex that processes that modality. Finally, studies of patients with progressive neural degeneration leading to deficits in SM suggest that information in SM is also organized hierarchically, such that the more specific a concept is, the more vulnerable it may be to brain damage [158]. Therefore, we will consider evidence from neuroimaging studies that address these three components of SM: category specificity, reactivation of sensorimotor areas, and hierarchical organization.

5.1 Organization of SM Network

Before the advent of functional brain imaging, our knowledge of the neural bases of SM was dependent on studies of patients with brain injury. Investigations of semantic impairment arising from brain disease suggested that the anterior temporal lobes (ATLs) are critical for semantic abilities in humans, across all stimulus modalities and for all types of conceptual knowledge [154, 155, 159–165]. As mentioned earlier, patients with semantic dementia and progressive degeneration of the anterior temporal cortex are impaired on all tasks requiring knowledge about the meanings of words, objects, and people, although possibly to different degrees for each category depending on the lateralization of atrophy [162, 164, 166, 167]. Other brain diseases that can affect the ATLs, such as Alzheimer's disease [168] and herpes simplex viral encephalitis [155], also often disrupt SM. Finally, it is worth noting that patients with damage to the left PFC often have difficulty in retrieving words in response to specific cues, even in the absence of aphasia [169].

Although ATLs' activation has been associated with a few semantic tasks (i.e., sentence comprehension, and famous face naming or identification, e.g., see Refs. [170, 171]), the vast majority of the functional imaging experiments on SM have reported posterior temporal, typically stronger in the left than in the right hemisphere, and/or frontal activations in the left VL PFC, with no mention to the ATLs (for review, see Refs. [151, 172–175]). Intersubject variability and the occurrence of fMRI susceptibility artifacts in the ATLs may be possible reasons for the lack of fMRI activations detected in ATLs [176].

Functional imaging results have also indicated that semantic knowledge is encoded in a widely distributed cortical network, with different regions specialized to represent particular types of information [177], particular categories of objects [178], or both [179], leading to the concept that no single region supports semantic abilities for all modalities and categories. However, in addition to these modality-specific regions and connections, the various different surface representations (such as shape) connect to, and communicate through, a shared, amodal "hub" in the ATLs [175]. At the hub stage, associations between different pairs of attributes (such as shape and name, shape and action, or shape and color) are all processed by a common set of neurons and synapses, regardless of the task. Damasio [180] was the first to argue for unified conceptual representations that abstract away from modality-specific attributes. He proposed the existence of "convergence zones" that associate different aspects of knowledge clearly articulating the importance of such zones for semantic processing. The convergence-zone hypothesis proposes that there is no multimodal cortical area that would build an integrated and independent semantic representation from its low-level sensory representations. Instead, the representation takes place only in the low-level cortices, with the different parts bound together by a hierarchy of convergence zones. A semantic representation can be recreated by activating its corresponding binding pattern in the convergence zone.

Finally, unlike EM, remote SM is not dependent on the involvement of the MTL [95]. The MTL is needed only temporarily, until the knowledge itself is represented permanently by the neocortical structures specialized in processing the acquired information.

5.2 Category-Specific Organization of SM

Functional neuroimaging studies in which the cortical organization for semantic knowledge has been addressed have revealed dissociations in the processing of different object categories (for review, *see* Refs. [156, 172, 173, 175, 177, 181]). The most frequently documented distinction is between "living" and "nonliving" items, body parts and numerals, and the most studied categories have been human faces, houses, animals, and tools. Various theoretical models have been proposed to explain the cognitive mechanisms underlying category specificity. (1) The sensory and functional/motor theory states that categories are defined by the type of information needed to recognize items as belonging to that category. "Living" items require object-related information appreciable through perceptual channels (shape, color, sound, etc.), whereas tools and body parts are more recognizable from information concerning action, activity, or the motor scheme to use them [154, 155, 160], (2) The "domain-specific theory" suggests that evolutionary pressure has led to specific adaptations for recognizing and responding to animals and plants, but not to objects [156]. (3) The "correlated-structure principle theory" proposes that conceptual organization reflects the statistical co-occurrence of the properties of objects rather than an explicit division into "living" and "nonliving" categories [182].

An impairment for processing items in the "living things" category has mainly been described in patients with damage to the anterior portions of the temporal lobe bilaterally [162, 163, 183, 184]. Nevertheless, the majority of functional neuroimaging studies (both PET and fMRI) failed to show consistent ATLs' activations for "living item" stimuli (see earlier). PET studies in normal subjects have shown that "living items" tend to activate predominantly posterior visual association cortices [185–188]. Only a meta-analysis of seven individual PET studies [189] found activations for "living" objects in the temporal poles bilaterally. This large multistudy dataset provided sufficient sensitivity to detect ATLs' activations despite their inconsistency across subjects and lack of significance in each study taken in isolation [189]. In contrast, patients with deficits in the "nonliving" category show damage in the left dorsolateral perisylvian regions [163, 183, 184] Consistently, PET studies found activations specific to "nonliving" stimuli in the left posterior middle and superior temporal gyri and in the left inferior frontal cortex [185, 186, 189–191]. In an event-related fMRI study in which words belonging to the categories "living" and "nonliving" were presented visually, common areas of activation during processing of both categories included the inferior occipital gyri bilaterally, the left IFG, and the left IPL [192]. During processing of "living" minus "nonliving" items, signal changes were present in the right inferior frontal, middle temporal, and fusiform gyri.

Numerous fMRI studies have shown that different object categories elicit activity in different regions of the ventral temporal cortex (for review, see Refs. [156, 172, 173, 175, 177, 181]). Although it is more likely that these differences are due to perceptual, object recognition process rather than to semantic memory function, we will briefly discuss them here. Perceiving animals showed heightened, bilateral activity in the more lateral region of the fusiform gyrus, whereas tools show heightened, bilateral activity in the medial region of the fusiform gyrus and in the posterior middle temporal gyrus (MTG) [193]. A similar pattern of activations was found for viewing faces (in the lateral fusiform) relative to viewing houses (the medial fusiform) [193]. The so-called FFA [194] responds more strongly to faces than to other object categories, but is not exclusive for faces [195, 196] (Fig. 6). House-related activity was reported in more medial regions, including the fusiform and lingual gyri [197], and parahippocampal cortex (the parahippocampal place area [191], especially for landmarks [198]). More interestingly, a meta-analysis by Joseph [195] revealed that the recognition

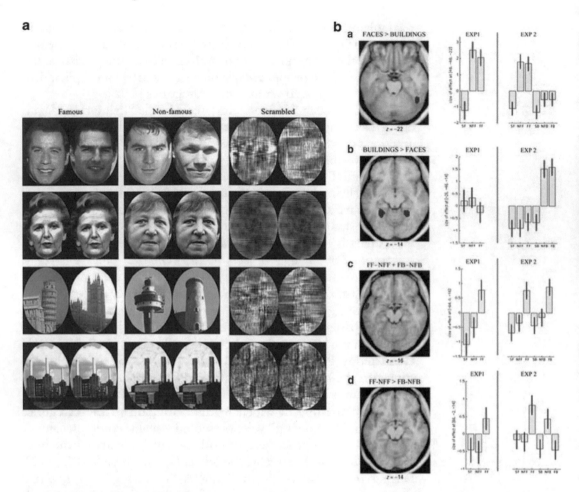

Fig. 6 A PET study investigating brain responses to famous and nonfamous faces and buildings. The results showed category-specific effects in the right fusiform and bilateral parahippocampal/lingual gyri for faces and buildings, respectively, but no effect of fame. In contrast, the left anterior middle temporal gyrus showed an effect of fame for both faces and buildings, but no effect of category. (**a**) Examples of the stimuli used in experiments 1 and 2 for the face conditions and in experiment 2 for the building conditions. (**b**) From *top* to *bottom*, this figure illustrates areas of activation and parameter estimates for regions that were more activated for (*a*) faces than for buildings, (*b*) buildings than for faces, (*c*) famous than for nonfamous faces and buildings, and (*d*) famous than for nonfamous faces only. In the left column, all activations are superimposed on axial slices of the mean of the nine subjects' normalized structural MRIs and thresholded at $p < 0.001$ (uncorrected). In the right column, the plots indicate the value of the normalized regional cerebral blood flow at the indicated voxel (*y*-axis) for each of the experimental conditions in experiments 1 and 2 (*x*-axis). *EXP* experiment, *FF* famous faces, *NFF* nonfamous faces, *FB* famous buildings, *NFB* nonfamous buildings (from Gorno-Tempini et al. [196])

task used (i.e., viewing, matching, or naming) also predicted brain activation patterns. Specifically, matching tasks recruit more inferior occipital regions than either naming or viewing tasks do, whereas naming tasks recruit more anterior ventral temporal sites than either viewing or matching tasks do, thus indicating that the cognitive demands of a particular recognition task are as predictive of cortical activation patterns as is category membership.

5.3 Modality-Specific Organization of SM

Several of the early PET studies of semantic processes focused on the question of whether words (both auditory and visual) and pictures are interpreted by a common semantic system [170, 188, 199, 200], or whether distinct systems are needed to support these two domains. These studies reached similar conclusions providing evidence for a distributed semantic system that is shared by visual/auditory and verbal modalities [170, 188, 199, 200] and that it is distributed throughout inferior temporal and frontal cortices with a few areas uniquely activated by pictures only (left posterior inferior temporal sulcus) or words (left anterior MTG and left inferior frontal sulcus) [200]. More recently, fMRI studies have provided additional support for this view by demonstrating that regions of the left posterior temporal cortex, known to be active during conceptual processing of pictures and words (fusiform gyrus and inferior and middle temporal gyri), are also active during auditory sentence comprehension [201–203]: activity was modulated by speech intelligibility [201, 202] and semantic ambiguity [203]. A fMRI study used the phenomenon of semantic ambiguity to identify regions within the fronto-temporal language network that are involved in the processes of activating, selecting, and integrating contextually appropriate word meanings [203]. Subjects heard sentences containing ambiguous words (e.g., "The shell was fired towards the tank") and well-matched low-ambiguity sentences (e.g., "Her secrets were written in her diary"). Although these sentences had similar acoustic, phonological, syntactic, and prosodic properties, the high-ambiguity sentences required additional processing by those brain regions involved in activating and selecting contextually appropriate word meanings. The ambiguity in these sentences went largely unnoticed, and yet high-ambiguity sentences resulted in increased recruitment of the left posterior inferior temporal cortex and the IFG bilaterally [203].

5.4 Representations of Object Properties

While the neural systems involved in SM may be modulated both by categories and modalities there is also evidence that the sensorimotor regions that are involved in the initial processing of particular information are recruited during SM retrieval (for review, *see* Refs. [172, 173, 175, 177, 181, 204]). In particular, semantic decisions involving object properties suggest a broad relationship between perceptual knowledge retrieval and sensory brain mechanisms, though recent work also suggests that this observation may be itself category-specific [165].

Activation of the left or bilateral ventral temporal cortex (fusiform gyrus) when retrieving *color* information, relative to other properties, has been replicated by several fMRI studies [174, 205, 206]. Beauchamp et al. [205] showed that neural activity is limited to the occipital lobes when color perception was tested by a passive viewing. When the task was made more demanding by requiring subjects to use color information to perform a color-sequencing task, several areas in the ventral cortex were identified: the most posterior, located in the posterior fusiform gyrus, corresponded to an area activated by passive viewing of colored stimuli, and more anterior and medial color-selective areas located in the collateral sulcus and fusiform gyrus [205]. These more anterior areas were also most active when visual color information was behaviorally relevant, suggesting that attention influences activity in color-selective areas [205].

Parietal cortex appears to be involved in retrieval of *size* [174]. For instance, in Oliver and Thompson-Schill [174], seven subjects made binary decisions about the shape, color, and size of named objects during fMRI. Bilateral parietal activity was significantly greater during retrieval of shape and size than during retrieval of color [174].

An area in the posterior superior temporal cortex, adjacent to the auditory-association cortex, is activated when participants are asked to judge the *sound* that an object makes [207]. *Action knowledge* involves the left lateral temporal cortex, particularly the medial and superior temporal (MT/MST) regions, anterior to an area associated with motion perception [208–210]. For instance, in two fMRI experiments Kourtzi and Kanwisher [209] found stronger activation of the MT/MST regions during viewing of static photographs with implied motion compared with viewing of photographs without it. Taken together, these data provide strong evidence that information about a particular object property is stored in the same neural system engaged when the property is perceived.

6 Conclusions

Working, episodic, and semantic memory systems engage multiple brain regions and rely upon a number of cognitive processes. WM is likely a property of the functional interaction between the PFC and posterior brain regions. The MTL and its connections with neocortical, prefrontal, and limbic structures are implicated in EM. SM is mediated through a network of neocortical structures, including the lateral and ATLs, and the inferior frontal cortex, possibly to a greater extent in the left hemisphere. Taken together, these memory systems do have overlapping neuroanatomical underpinnings, and even share some of the same component processes, such as item and information retrieval. One emergent similarity between the systems is the finding that those sensory and association areas that are engaged during perceptual and sensorimotor processing of item information are tapped once again when the information is required for memory processing. Future work by researchers using neuroimaging to study memory will require even further refinement in the definitions of memory systems and their components, as exemplified by the current work in EM, where the retrieval component is being further reduced into components such as scene construction.

References

1. Baddeley A, Hitch G (1974) Working memory. The psychology of learning and motivation. B. GH Academic, San Diego, pp 47–90

2. Baddeley A (2000) The episodic buffer: a new component of working memory? Trends Cogn Sci 4:417–423

3. Curtis CE, D'Esposito, M (2006) Working memory. In: Cabeza R, Kingstone A (eds) Handbook of functional neuroimaging of cognition. MIT Press, Cambridge, MA. 269–306

4. Miyake A, Shah P (1999) Models of working memory. Cambridge University Press, New York

5. Miller GA (1956) The magical number seven, plus or minus two: some limits on our capacity for processing information. Psych Rev 63:81–97

6. Goldman-Rakic PS (1994) Working memory dysfunction in schizophrenia. J Neuropsychiatry Clin Neurosci 6:348–357

7. Muller NG, Machado L, Knight RT (2002) Contributions of subregions of the prefrontal cortex to working memory: evidence from brain lesions in humans. J Cogn Neurosci 14:673–686

8. Fuster JM, Alexander GE (1971) Neuron activity related to short-term memory. Science 173:652–654

9. Kubota K, Niki H (1971) Prefrontal cortical unit activity and delayed alternation performance in monkeys. J Neurophysiol 34:337–347

10. Funahashi S, Bruce CJ, Goldman-Rakic PS (1989) Mnemonic coding of visual space in the monkey's dorsolateral prefrontal cortex. J Neurophysiol 61:331–349

11. Owen AM, Herrod NJ, Menon DK, Clark JC, Downey SP, Carpenter TA, Minhas PS et al (1999) Redefining the functional organization of working memory processes within human lateral prefrontal cortex. Eur J Neurosci 11:567–574

12. Jonides J, Smith EE, Koeppe RA, Awh E, Minoshima S, Mintun MA (1993) Spatial working memory in humans as revealed by PET. Nature 363:623–625

13. Petrides M, Alivisatos B, Meyer E, Evans AC (1993) Functional activation of the human frontal cortex during the performance of verbal working memory tasks. Proc Natl Acad Sci U S A 90:878–882

14. McCarthy G, Blamire AM, Puce A, Nobre AC, Bloch G, Hyder F, Goldman-Rakic P et al (1994) Functional magnetic resonance imaging of human prefrontal cortex activation during a spatial working memory task. Proc Natl Acad Sci U S A 91:8690–8694

15. Awh E, Jonides J, Smith EE, Buxton RB, Frank LR, Love T, Wong EC et al (1999) Rehearsal in spatial working memory: evidence from neuroimaging. Psych Sci 10:433–437

16. Postle BR, D'Esposito M (1999) "What"-Then-Where" in visual working memory: an event-related fMRI study. J Cogn Neurosci 11:585–597

17. Rowe JB, Passingham RE (2001) Working memory for location and time: activity in prefrontal area 46 relates to selection rather than maintenance in memory. Neuroimage 14:77–86

18. Corbetta M, Kincade JM, Shulman GL (2002) Neural systems for visual orienting and their relationships to spatial working memory. J Cogn Neurosci 14:508–523

19. Davachi L, Maril A, Wagner AD (2001) When keeping in mind supports later bringing to mind: neural markers of phonological rehearsal predict subsequent remembering. J Cogn Neurosci 13:1059–1070

20. D'Esposito M (2008). Working memory. In: Goldenberg G, Miller B (eds) Handbook of clinical neurology. Neuropsychology and Behavioral Neurology, Vol 88. Elsevier, Amsterdam, The Netherlands. 237–247

21. Courtney SM, Ungerleider LG, Keil K, Haxby JV (1997) Transient and sustained activity in a distributed neural system for human working memory. Nature 386:608–611

22. Courtney SM, Petit L, Haxby JV, Ungerleider LG (1998) The role of prefrontal cortex in working memory: examining the contents of consciousness. Philos Trans R Soc Lond B Biol Sci 353:1819–1828

23. Jha AP, McCarthy G (2000) The influence of memory load upon delay-interval activity in a working-memory task: an event-related functional MRI study. J Cogn Neurosci 12(Suppl 2):90–105

24. Gruber O (2001) Effects of domain-specific interference on brain activation associated with verbal working memory task performance. Cereb Cortex 11:1047–1055

25. Ranganath C, D'Esposito M (2001) Medial temporal lobe activity associated with active maintenance of novel information. Neuron 31:865–873

26. Postle BR, Druzgal TJ, D'Esposito M (2003) Seeking the neural substrates of visual working memory storage. Cortex 39:927–946

27. Curtis CE, Rao VY, D'Esposito M (2004) Maintenance of spatial and motor codes during oculomotor delayed response tasks. J Neurosci 24:3944–3952

28. Marklund P, Fransson P, Cabeza R, Petersson KM, Ingvar M, Nyberg L (2007) Sustained and transient neural modulations in prefrontal cortex related to declarative long-term memory, working memory, and attention. Cortex 43:22–37

29. Druzgal TJ, D'Esposito M (2003) Dissecting contributions of prefrontal cortex and fusiform face area to face working memory. J Cogn Neurosci 15:771–784

30. D'Esposito M, Aguirre GK, Zarahn E, Ballard D, Shin RK, Lease J (1998) Functional MRI studies of spatial and nonspatial working memory. Brain Res Cogn Brain Res 7:1–13

31. D'Esposito M, Postle BR, Rypma B (2000) Prefrontal cortical contributions to working memory: evidence from event-related fMRI studies. Exp Brain Res 133:3–11

32. Curtis CE, D'Esposito M (2003) Persistent activity in the prefrontal cortex during working memory. Trends Cogn Sci 7:415–423

33. Passingham D, Sakai K (2004) The prefrontal cortex and working memory: physiology and brain imaging. Curr Opin Neurobiol 14:163–168

34. Sreenivasan KK, Curtis CE, D'Esposito M (2014) Revisiting the role of persistent neural activity during working memory. Trends Cogn Sci 18:82–89

35. Ungerleider LG, Haxby JV (1994) 'What' and 'where' in the human brain. Curr Opin Neurobiol 4:157–165

36. McCarthy G, Puce A, Constable RT, Krystal JH, Gore JC, Goldman-Rakic P (1996) Activation of human prefrontal cortex during spatial and nonspatial working memory tasks measured by functional MRI. Cereb Cortex 6:600–611

37. Courtney SM, Petit L, Maisog JM, Ungerleider LG, Haxby JV (1998) An area specialized for spatial working memory in human frontal cortex. Science 279:1347–1351

38. Munk MH, Linden DE, Muckli L, Lanfermann H, Zanella FE, Singer W, Goebel R (2002) Distributed cortical systems in visual short-term memory revealed by event-related functional magnetic resonance imaging. Cereb Cortex 12:866–876

39. Sala JB, Rama P, Courtney SM (2003) Functional topography of a distributed neural system for spatial and nonspatial information maintenance in working memory. Neuropsychologia 41:341–356

40. Walter H, Wunderlich AP, Blankenhorn M, Schafer S, Tomczak R, Spitzer M, Gron G (2003) No hypofrontality, but absence of prefrontal lateralization comparing verbal and spatial working memory in schizophrenia. Schizophr Res 61:175–184

41. Leung HC, Seelig D, Gore JC (2004) The effect of memory load on cortical activity in the spatial working memory circuit. Cogn Affect Behav Neurosci 4:553–563

42. Courtney SM, Ungerleider LG, Keil K, Haxby JV (1996) Object and spatial visual working memory activate separate neural systems in human cortex. Cereb Cortex 6:39–49

43. Druzgal TJ, D'Esposito M (2001) Activity in fusiform face area modulated as a function of working memory load. Brain Res Cogn Brain Res 10:355–364

44. Rama P, Sala JB, Gillen JS, Pekar JJ, Courtney SM (2001) Dissociation of the neural systems for working memory maintenance of verbal and nonspatial visual information. Cogn Affect Behav Neurosci 1:161–171

45. Chein JM, Fiez JA (2001) Dissociation of verbal working memory system components using a delayed serial recall task. Cereb Cortex 11:1003–1014

46. Gruber O, von Cramon DY (2003) The functional neuroanatomy of human working memory revisited. Evidence from 3-T fMRI studies using classical domain-specific interference tasks. Neuroimage 19:797–809

47. Ravizza SM, Delgado MR, Chein JM, Becker JT, Fiez JA (2004) Functional dissociations within the inferior parietal cortex in verbal working memory. Neuroimage 22:562–573

48. Raye CL, Johnson MK, Mitchell KJ, Greene EJ, Johnson MR (2007) Refreshing: a minimal executive function. Cortex 43:135–145

49. Chen SH, Desmond JE (2005) Cerebrocerebellar networks during articulatory rehearsal and verbal working memory tasks. Neuroimage 24:332–338

50. Kirschen MP, Chen SH, Schraedley-Desmond P, Desmond JE (2005) Load- and practice-dependent increases in cerebro-cerebellar activation in verbal working memory: an fMRI study. Neuroimage 24:462–472

51. Smith EE, Jonides J (1999) Storage and executive processes in the frontal lobes. Science 283:1657–1661

52. Henson RN, Burgess N, Frith CD (2000) Recoding, storage, rehearsal and grouping in verbal short-term memory: an fMRI study. Neuropsychologia 38:426–440

53. Crottaz-Herbette S, Anagnoson RT, Menon V (2004) Modality effects in verbal work-

ing memory: differential prefrontal and parietal responses to auditory and visual stimuli. Neuroimage 21:340–351

54. Postle BR (2006) Working memory as an emergent property of the mind and brain. Neuroscience 139:23–38

55. Wager TD, Smith EE (2003) Neuroimaging studies of working memory: a meta-analysis. Cogn Affect Behav Neurosci 3:255–274

56. Owen AM, Doyon J, Petrides M, Evans AC (1996) Planning and spatial working memory: a positron emission tomography study in humans. Eur J Neurosci 8:353–364

57. Thompson-Schill SL, D'Esposito M, Aguirre GK, Farah MJ (1997) Role of left inferior prefrontal cortex in retrieval of semantic knowledge: a reevaluation. Proc Natl Acad Sci U S A 94:14792–14797

58. Thompson-Schill SL, D'Esposito M, Kan IP (1999) Effects of repetition and competition on activity in left prefrontal cortex during word generation. Neuron 23:513–522

59. D'Esposito M, Postle BR, Jonides J, Smith EE (1999) The neural substrate and temporal dynamics of interference effects in working memory as revealed by event-related functional MRI. Proc Natl Acad Sci U S A 96:7514–7519

60. Jonides J, Smith EE, Marshuetz C, Koeppe RA, Reuter-Lorenz PA (1998) Inhibition in verbal working memory revealed by brain activation. Proc Natl Acad Sci U S A 95:8410–8413

61. D'Esposito M, Postle BR, Ballard D, Lease J (1999) Maintenance versus manipulation of information held in working memory: an event-related fMRI study. Brain Cogn 41:66–86

62. Postle BR, Berger JS, D'Esposito M (1999) Functional neuroanatomical double dissociation of mnemonic and executive control processes contributing to working memory performance. Proc Natl Acad Sci U S A 96:12959–12964

63. Bunge SA, Klingberg T, Jacobsen RB, Gabrieli JD (2000) A resource model of the neural basis of executive working memory. Proc Natl Acad Sci U S A 97:3573–3578

64. Cohen JD, Perlstein WM, Braver TS, Nystrom LE, Noll DC, Jonides J, Smith EE (1997) Temporal dynamics of brain activation during a working memory task. Nature 386:604–608

65. Squire LR, Stark CE, Clark RE (2004) The medial temporal lobe. Annu Rev Neurosci 27:279–306

66. Corkin S (1984) Lasting consequences of bilateral medial temporal lobectomy: clinical course and experimental findings in H.M. Semin Neurol 4:249–259

67. Braver TS, Cohen JD, Nystrom LE, Jonides J, Smith EE, Noll DC (1997) A parametric study of prefrontal cortex involvement in human working memory. Neuroimage 5:49–62

68. Stern CE, Sherman SJ, Kirchhoff BA, Hasselmo ME (2001) Medial temporal and prefrontal contributions to working memory tasks with novel and familiar stimuli. Hippocampus 11:337–346

69. Ranganath C, Rainer G (2003) Neural mechanisms for detecting and remembering novel events. Nat Rev Neurosci 4:193–202

70. Honey GD, Fu CH, Kim J, Brammer MJ, Croudace TJ, Suckling J, Pich EM et al (2002) Effects of verbal working memory load on corticocortical connectivity modeled by path analysis of functional magnetic resonance imaging data. Neuroimage 17:573–582

71. Gazzaley A, Rissman J, D'Esposito M (2004) Functional connectivity during working memory maintenance. Cogn Affect Behav Neurosci 4:580–599

72. Crowe DA, Goodwin SJ, Blackman RK, Sakellaridi S, Sponheim SR, MacDonald AW 3rd, Chafee MV (2013) Prefrontal neurons transmit signals to parietal neurons that reflect executive control of cognition. Nat Neurosci 16:1484–1491

73. Edin F, Klingberg T, Johansson P, McNab F, Tegner J, Compte A (2009) Mechanism for top-down control of working memory capacity. Proc Natl Acad Sci U S A 106:6802–6807

74. Chadick JZ, Gazzaley A (2011) Differential coupling of visual cortex with default or frontal-parietal network based on goals. Nat Neurosci 14:830–832

75. Hazy TE, Frank MJ, C O'Reilly R (2007) Towards an executive without a homunculus: computational models of the prefrontal cortex/basal ganglia system. Philos Trans R Soc Lond B Biol Sci 362:1601–1613

76. Tulving E (1972) Episodic and semantic memory. Organisation of memory (Tulving E and Donaldson W (eds). Academic, New York, pp 381–403

77. Vargha-Khadem F, Gadian DG, Watkins KE, Connelly A, Van Paesschen W, Mishkin M (1997) Differential effects of early hippocampal pathology on episodic and semantic memory. Science 277:376–380

78. Hodges JR, Graham KS (2001) Episodic memory: insights from semantic dementia. Philos Trans R Soc Lond B Biol Sci 356:1423–1434

79. Levine B, Black SE, Cabeza R, Sinden M, McIntosh AR, Toth JP, Tulving E et al (1998)

Episodic memory and the self in a case of isolated retrograde amnesia. Brain 121(Pt 10):1951–1973

80. Viskontas IV, McAndrews MP, Moscovitch M (2000) Remote episodic memory deficits in patients with unilateral temporal lobe epilepsy and excisions. J Neurosci 20:5853–5857

81. Dudukovic NM, Knowlton BJ (2006) Remember-know judgments and retrieval of contextual details. Acta Psychol (Amst) 122:160–173

82. Levine B, Svoboda E, Hay JF, Winocur G, Moscovitch M (2002) Aging and autobiographical memory: dissociating episodic from semantic retrieval. Psychol Aging 17:677–689

83. Bayley PJ, Hopkins RO, Squire LR (2006) The fate of old memories after medial temporal lobe damage. J Neurosci 26:13311–13317

84. Nolde SF, Johnson MK, D'Esposito M (1998) Left prefrontal activation during episodic remembering: an event-related fMRI study. Neuroreport 9:3509–3514

85. Wheeler ME, Petersen SE, Buckner RL (2000) Memory's echo: vivid remembering reactivates sensory-specific cortex. Proc Natl Acad Sci U S A 97:11125–11129

86. Heil M, Rosler F, Hennighausen E (1996) Topographically distinct cortical activation in episodic long-term memory: the retrieval of spatial versus verbal information. Mem Cognit 24:777–795

87. O'Craven KM, Kanwisher N (2000) Mental imagery of faces and places activates corresponding stiimulus-specific brain regions. J Cogn Neurosci 12:1013–1023

88. Halpern AR, Zatorre RJ (1999) When that tune runs through your head: a PET investigation of auditory imagery for familiar melodies. Cereb Cortex 9:697–704

89. Nyberg L, Petersson KM, Nilsson LG, Sandblom J, Aberg C, Ingvar M (2001) Reactivation of motor brain areas during explicit memory for actions. Neuroimage 14:521–528

90. Marr D (1971) Simple memory: a theory for archicortex. Philos Trans R Soc Lond B Biol Sci 262:23–81

91. Hasselmo ME, McClelland JL (1999) Neural models of memory. Curr Opin Neurobiol 9:184–188

92. Aggleton JP, Brown MW (1999) Episodic memory, amnesia, and the hippocampal-anterior thalamic axis. Behav Brain Sci 22:425–444, discussion 444-489

93. Insausti R, Amaral DG, Cowan WM (1987) The entorhinal cortex of the monkey: II. Cortical afferents. J Comp Neurol 264:356–395

94. Amaral DG, Insausti R, Cowan WM (1987) The entorhinal cortex of the monkey: I. Cytoarchitectonic organization. J Comp Neurol 264:326–355

95. Moscovitch M, Rosenbaum RS, Gilboa A, Addis DR, Westmacott R, Grady C, McAndrews MP et al (2005) Functional neuroanatomy of remote episodic, semantic and spatial memory: a unified account based on multiple trace theory. J Anat 207:35–66

96. Teyler TJ, Rudy JW (2007) The hippocampal indexing theory and episodic memory: updating the index. Hippocampus 17:1158–1169

97. Squire LR, Knowlton B, Musen G (1993) The structure and organization of memory. Annu Rev Psychol 44:453–495

98. Baddeley A, Vargha-Khadem F, Mishkin M (2001) Preserved recognition in a case of developmental amnesia: implications for the acquisition of semantic memory? J Cogn Neurosci 13:357–369

99. Rolls ET (1996) A theory of hippocampal function in memory. Hippocampus 6:601–620

100. Gluck MA, Myers CE (1993) Hippocampal mediation of stimulus representation: a computational theory. Hippocampus 3:491–516

101. Norman KA, O'Reilly RC (2003) Modeling hippocampal and neocortical contributions to recognition memory: a complementary-learning-systems approach. Psychol Rev 110:611–646

102. Fletcher PC, Shallice T, Dolan RJ (1998) The functional roles of prefrontal cortex in episodic memory. I. Encoding. Brain 121(Pt 7):1239–1248

103. Shallice T, Fletcher P, Frith CD, Grasby P, Frackowiak RS, Dolan RJ (1994) Brain regions associated with acquisition and retrieval of verbal episodic memory. Nature 368:633–635

104. Dolan RJ, Fletcher PC (1997) Dissociating prefrontal and hippocampal function in episodic memory encoding. Nature 388:582–585

105. Gabrieli JD, Brewer JB, Desmond JE, Glover GH (1997) Separate neural bases of two fundamental memory processes in the human medial temporal lobe. Science 276:264–266

106. Wagner AD, Schacter DL, Rotte M, Koutstaal W, Maril A, Dale AM, Rosen BR et al (1998) Building memories: remembering and forgetting of verbal experiences as predicted by brain activity. Science 281:1188–1191

107. Brewer JB, Zhao Z, Desmond JE, Glover GH, Gabrieli JD (1998) Making memories: brain activity that predicts how well visual experience will be remembered. Science 281:1185–1187

108. Wagner AD, Poldrack RA, Eldridge LL, Desmond JE, Glover GH, Gabrieli JD (1998) Material-specific lateralization of prefrontal activation during episodic encoding and retrieval. Neuroreport 9:3711–3717

109. Fernandez G, Weyerts H, Schrader-Bolsche M, Tendolkar I, Smid HG, Tempelmann C, Hinrichs H et al (1998) Successful verbal encoding into episodic memory engages the posterior hippocampus: a parametrically analyzed functional magnetic resonance imaging study. J Neurosci 18:1841–1847

110. Davachi L, Wagner AD (2002) Hippocampal contributions to episodic encoding: insights from relational and item-based learning. J Neurophysiol 88:982–990

111. Davachi L, Mitchell JP, Wagner AD (2003) Multiple routes to memory: distinct medial temporal lobe processes build item and source memories. Proc Natl Acad Sci U S A 100:2157–2162

112. Kirwan CB, Stark CE (2004) Medial temporal lobe activation during encoding and retrieval of novel face-name pairs. Hippocampus 14:919–930

113. Ranganath C, Yonelinas AP, Cohen MX, Dy CJ, Tom SM, D'Esposito M (2004) Dissociable correlates of recollection and familiarity within the medial temporal lobes. Neuropsychologia 42:2–13

114. Kensinger EA, Schacter DL (2006) Amygdala activity is associated with the successful encoding of item, but not source, information for positive and negative stimuli. J Neurosci 26:2564–2570

115. Uncapher MR, Otten LJ, Rugg MD (2006) Episodic encoding is more than the sum of its parts: an fMRI investigation of multifeatural contextual encoding. Neuron 52:547–556

116. Staresina BP, Davachi L (2008) Selective and shared contributions of the hippocampus and perirhinal cortex to episodic item and associative encoding. J Cogn Neurosci 20:1478–1489

117. Baker JT, Sanders AL, Maccotta L, Buckner RL (2001) Neural correlates of verbal memory encoding during semantic and structural processing tasks. Neuroreport 12:1251–1256

118. Buckner RL, Wheeler ME, Sheridan MA (2001) Encoding processes during retrieval tasks. J Cogn Neurosci 13:406–415

119. Henson RN, Hornberger M, Rugg MD (2005) Further dissociating the processes involved in recognition memory: an FMRI study. J Cogn Neurosci 17:1058–1073

120. Otten LJ, Rugg MD (2001) Task-dependency of the neural correlates of episodic encoding as measured by fMRI. Cereb Cortex 11:1150–1160

121. Nyberg L, McIntosh AR, Cabeza R, Habib R, Houle S, Tulving E (1996) General and specific brain regions involved in encoding and retrieval of events: what, where, and when. Proc Natl Acad Sci U S A 93:11280–11285

122. Tulving E, Kapur S, Craik FI, Moscovitch M, Houle S (1994) Hemispheric encoding/retrieval asymmetry in episodic memory: positron emission tomography findings. Proc Natl Acad Sci U S A 91:2016–2020

123. Lepage M, Ghaffar O, Nyberg L, Tulving E (2000) Prefrontal cortex and episodic memory retrieval mode. Proc Natl Acad Sci U S A 97:506–511

124. Johnson MK, Hashtroudi S, Lindsay DS (1993) Source monitoring. Psychol Bull 114:3–28

125. Dobbins IG, Wagner AD (2005) Domain-general and domain-sensitive prefrontal mechanisms for recollecting events and detecting novelty. Cereb Cortex 15:1768–1778

126. Dobbins IG, Rice HJ, Wagner AD, Schacter DL (2003) Memory orientation and success: separable neurocognitive components underlying episodic recognition. Neuropsychologia 41:318–333

127. Velanova K, Jacoby LL, Wheeler ME, McAvoy MP, Petersen SE, Buckner RL (2003) Functional-anatomic correlates of sustained and transient processing components engaged during controlled retrieval. J Neurosci 23:8460–8470

128. Kahn I, Davachi L, Wagner AD (2004) Functional-neuroanatomic correlates of recollection: implications for models of recognition memory. J Neurosci 24:4172–4180

129. Wheeler ME, Buckner RL (2003) Functional dissociation among components of remembering: control, perceived oldness, and content. J Neurosci 23:3869–3880

130. Stark CE, Squire LR (2001) When zero is not zero: the problem of ambiguous baseline conditions in fMRI. Proc Natl Acad Sci U S A 98:12760–12766

131. Eldridge LL, Knowlton BJ, Furmanski CS, Bookheimer SY, Engel SA (2000) Remembering episodes: a selective role for the hippocampus during retrieval. Nat Neurosci 3:1149–1152

132. Wheeler ME, Buckner RL (2004) Functional-anatomic correlates of remembering and knowing. Neuroimage 21:1337–1349

133. Tulving E (1985) Memory and consciousness. Can Psychol 1:1–12

134. Tulving E, Schacter DL, McLachlan DR, Moscovitch M (1988) Priming of semantic autobiographical knowledge: a case study of retrograde amnesia. Brain Cogn 8:3–20

135. Eldridge LL, Engel SA, Zeineh MM, Bookheimer SY, Knowlton BJ (2005) A dissociation of encoding and retrieval processes in the human hippocampus. J Neurosci 25:3280–3286

136. Daselaar SM, Prince SE, Dennis NA, Hayes SM, Kim H, Cabeza R (2009) Posterior midline and ventral parietal activity is associated with retrieval success and encoding failure. Front Hum Neurosci 3:13

137. Huijbers W, Vannini P, Sperling RA, Pennartz CM, Cabeza R, Daselaar SM (2012) Explaining the encoding/retrieval flip: memory-related deactivations and activations in the posteromedial cortex. Neuropsychologia 50:3764–3774

138. Konishi S, Uchida I, Okuaki T, Machida T, Shirouzu I, Miyashita Y (2002) Neural correlates of recency judgment. J Neurosci 22:9549–9555

139. Milner B, Corsi P, Leonard G (1991) Frontal-lobe contribution to recency judgements. Neuropsychologia 29:601–618

140. Petrides M (1991) Functional specialization within the dorsolateral frontal cortex for serial order memory. Proc Biol Sci 246:299–306

141. Suzuki M, Fujii T, Tsukiura T, Okuda J, Umetsu A, Nagasaka T, Mugikura S et al (2002) Neural basis of temporal context memory: a functional MRI study. Neuroimage 17:1790–1796

142. Moscovitch M (1995) Recovered consciousness: a hypothesis concerning modularity and episodic memory. J Clin Exp Neuropsychol 17:276–290

143. Schacter DL, Addis DR (2007) The cognitive neuroscience of constructive memory: remembering the past and imagining the future. Philos Trans R Soc Lond B Biol Sci 362:773–786

144. Dudai Y, Carruthers M (2005) The Janus face of Mnemosyne. Nature 434:567

145. Suddendorf T, Corballis MC (1997) Mental time travel and the evolution of the human mind. Genet Soc Gen Psychol Monogr 123:133–167

146. Okuda J, Fujii T, Ohtake H, Tsukiura T, Tanji K, Suzuki K, Kawashima R et al (2003) Thinking of the future and past: the roles of the frontal pole and the medial temporal lobes. Neuroimage 19:1369–1380

147. Addis DR, Wong AT, Schacter DL (2007) Remembering the past and imagining the future: common and distinct neural substrates during event construction and elaboration. Neuropsychologia 45:1363–1377

148. Szpunar KK, Watson JM, McDermott KB (2007) Neural substrates of envisioning the future. Proc Natl Acad Sci U S A 104:642–647

149. Hassabis D, Maguire EA (2007) Deconstructing episodic memory with construction. Trends Cogn Sci 11:299–306

150. Hassabis D, Maguire EA (2009) The construction system of the brain. Philos Trans R Soc Lond B Biol Sci 364:1263–1271

151. Martin A (2001) Functional neuroimaging of semantic memory, Handbook of functional neuroimaging of cognition (Cabeza R and Kingstone A (eds)). The MIT Press, Cambridge, pp 153–186

152. Damasio AR, McKee J, Damasio H (1979) Determinants of performance in color anomia. Brain Lang 7:74–85

153. McKenna P, Warrington EK (1980) Testing for nominal dysphasia. J Neurol Neurosurg Psychiatry 43:781–788

154. Warrington EK, McCarthy R (1983) Category specific access dysphasia. Brain 106(Pt 4):859–878

155. Warrington EK, Shallice T (1984) Category specific semantic impairments. Brain 107(Pt 3):829–854

156. Caramazza A, Shelton JR (1998) Domain-specific knowledge systems in the brain the animate-inanimate distinction. J Cogn Neurosci 10:1–34

157. Allport DA (1985) Distributed memory, modular systems and dysphagia, Current perspectives in dysphagia (Newman SK and Epstein R (eds)). Churchill Livingstone, Edinburgh, pp 32–60

158. Hodges JR, Graham N, Patterson K (1995) Charting the progression in semantic dementia: implications for the organisation of semantic memory. Memory 3:463–495

159. Warrington EK (1975) The selective impairment of semantic memory. Q J Exp Psychol 27:635–657

160. Warrington EK, McCarthy RA (1987) Categories of knowledge. Further fractionations and an attempted integration. Brain 110(Pt 5):1273–1296

161. Hart J Jr, Gordon B (1990) Delineation of single-word semantic comprehension deficits in aphasia, with anatomical correlation. Ann Neurol 27:226–231

162. Hodges JRPK, Oxbury S, Funnell F (1992) Semantic dementia. Progressive fluent aphasia with temporal lobe atrophy. Brain 115:1783–1806

163. Gainotti G (2000) What the locus of brain lesion tells us about the nature of the cognitive defect underlying category-specific disorders: a review. Cortex 36:539–559

164. Gorno-Tempini ML, Dronkers NF, Rankin KP, Ogar JM, Phengrasamy L, Rosen HJ, Johnson JK et al (2004) Cognition and anatomy in three variants of primary progressive aphasia. Ann Neurol 55:335–346

165. Brambati SM, Myers D, Wilson A, Rankin KP, Allison SC, Rosen HJ, Miller BL et al (2006) The anatomy of category-specific object naming in neurodegenerative diseases. J Cogn Neurosci 18:1644–1653

166. Snowden JS, Thompson JC, Neary D (2004) Knowledge of famous faces and names in semantic dementia. Brain 127:860–872

167. Gorno-Tempini ML, Rankin KP, Woolley JD, Rosen HJ, Phengrasamy L, Miller BL (2004) Cognitive and behavioral profile in a case of right anterior temporal lobe neurodegeneration. Cortex 40:631–644

168. Hodges JR, Patterson K (1995) Is semantic memory consistently impaired early in the course of Alzheimer's disease? Neuroanatomical and diagnostic implications. Neuropsychologia 33:441–459

169. Baldo JV, Shimamura AP (1998) Letter and category fluency in patients with frontal lobe lesions. Neuropsychology 12:259–267

170. Gorno-Tempini ML, Price CJ, Josephs O, Vandenberghe R, Cappa SF, Kapur N, Frackowiak RS (1998) The neural systems sustaining face and proper-name processing. Brain 121(Pt 11):2103–2118

171. Damasio H, Grabowski TJ, Tranel D, Hichwa RD, Damasio AR (1996) A neural basis for lexical retrieval. Nature 380:499–505

172. Bookheimer S (2002) Functional MRI of language: new approaches to understanding the cortical organization of semantic processing. Annu Rev Neurosci 25:151–188

173. Martin A (2007) The representation of object concepts in the brain. Annu Rev Psychol 58:25–45

174. Oliver RT, Thompson-Schill SL (2003) Dorsal stream activation during retrieval of object size and shape. Cogn Affect Behav Neurosci 3:309–322

175. Patterson K, Nestor PJ, Rogers TT (2007) Where do you know what you know? The representation of semantic knowledge in the human brain. Nat Rev Neurosci 8:976–987

176. Devlin JT, Russell RP, Davis MH, Price CJ, Wilson J, Moss HE, Matthews PM et al (2000) Susceptibility-induced loss of signal: comparing PET and fMRI on a semantic task. Neuroimage 11:589–600

177. Martin A, Chao LL (2001) Semantic memory and the brain: structure and processes. Curr Opin Neurobiol 11:194–201

178. Caramazza A, Mahon BZ (2003) The organization of conceptual knowledge: the evidence from category-specific semantic deficits. Trends Cogn Sci 7:354–361

179. Thompson-Schill SL, Aguirre GK, D'Esposito M, Farah MJ (1999) A neural basis for category and modality specificity of semantic knowledge. Neuropsychologia 37:671–676

180. Damasio AR (1989) Time-locked multiregional retroactivation: a systems-level proposal for the neural substrates of recall and recognition. Cognition 33:25–62

181. Thompson-Schill SL (2003) Neuroimaging studies of semantic memory: inferring "how" from "where". Neuropsychologia 41:280–292

182. Tyler LK, Moss HE (2001) Towards a distributed account of conceptual knowledge. Trends Cogn Sci 5:244–252

183. Hillis AE, Caramazza A (1991) Category-specific naming and comprehension impairment: a double dissociation. Brain 114(Pt 5):2081–2094

184. Tranel D, Damasio H, Damasio AR (1997) A neural basis for the retrieval of conceptual knowledge. Neuropsychologia 35:1319–1327

185. Martin A, Wiggs CL, Ungerleider LG, Haxby JV (1996) Neural correlates of category-specific knowledge. Nature 379:649–652

186. Mummery CJ, Patterson K, Hodges JR, Wise RJ (1996) Generating 'tiger' as an animal name or a word beginning with T: differences in brain activation. Proc Biol Sci 263:989–995

187. Perani D, Cappa SF, Bettinardi V, Bressi S, Gorno-Tempini M, Matarrese M, Fazio F (1995) Different neural systems for the recognition of animals and man-made tools. Neuroreport 6:1637–1641

188. Perani D, Schnur T, Tettamanti M, Gorno-Tempini M, Cappa SF, Fazio F (1999) Word and picture matching: a PET study of semantic category effects. Neuropsychologia 37:293–306

189. Devlin JT, Moore CJ, Mummery CJ, Gorno-Tempini ML, Phillips JA, Noppeney U, Frackowiak RS et al (2002) Anatomic constraints on cognitive theories of category specificity. Neuroimage 15:675–685

190. Gorno-Tempini ML, Cipolotti L, Price CJ (2000) Category differences in brain activation studies: where do you come from? Proc Biol Sci 267:1253–1258

450   Federica Agosta et al.

191. Cappa SF, Perani D, Schnur T, Tettamanti M, Fazio F (1998) The effects of semantic category and knowledge type on lexical-semantic access: a PET study. Neuroimage 8:350–359

192. Leube DT, Erb M, Grodd W, Bartels M, Kircher TT (2001) Activation of right fronto-temporal cortex characterizes the 'living' category in semantic processing. Brain Res Cogn Brain Res 12:425–430

193. Chao LL, Haxby JV, Martin A (1999) Attribute-based neural substrates in temporal cortex for perceiving and knowing about objects. Nat Neurosci 2:913–919

194. Kanwisher N, McDermott J, Chun MM (1997) The fusiform face area: a module in human extrastriate cortex specialized for face perception. J Neurosci 17:4302–4311

195. Joseph JE (2001) Functional neuroimaging studies of category specificity in object recognition: a critical review and meta-analysis. Cogn Affect Behav Neurosci 1:119–136

196. Gorno-Tempini ML, Price CJ (2001) Identification of famous faces and buildings: a functional neuroimaging study of semantically unique items. Brain 124:2087–2097

197. Aguirre GK, Zarahn E, D'Esposito M (1998) An area within human ventral cortex sensitive to "building" stimuli: evidence and implications. Neuron 21:373–383

198. Epstein R, Harris A, Stanley D, Kanwisher N (1999) The parahippocampal place area: recognition, navigation, or encoding? Neuron 23:115–125

199. Petersen SE, Fox PT, Posner MI, Mintun M, Raichle ME (1988) Positron emission tomographic studies of the cortical anatomy of single-word processing. Nature 331:585–589

200. Vandenberghe R, Price C, Wise R, Josephs O, Frackowiak RSJ (1996) Functional anatomy of a common semantic system for words and pictures. Nature 383:254–256

201. Davis MH, Johnsrude IS (2003) Hierarchical processing in spoken language comprehension. J Neurosci 23:3423–3431

202. Giraud AL, Kell C, Thierfelder C, Sterzer P, Russ MO, Preibisch C, Kleinschmidt A (2004) Contributions of sensory input, auditory search and verbal comprehension to cortical activity during speech processing. Cereb Cortex 14:247–255

203. Rodd JM, Davis MH, Johnsrude IS (2005) The neural mechanisms of speech comprehension: fMRI studies of semantic ambiguity. Cereb Cortex 15:1261–1269

204. Kartsounis LD, Shallice T (1996) Modality specific semantic knowledge loss for unique items. Cortex 32:109–119

205. Beauchamp MS, Haxby JV, Jennings JE, DeYoe EA (1999) An fMRI version of the Farnsworth-Munsell 100-Hue test reveals multiple color-selective areas in human ventral occipitotemporal cortex. Cereb Cortex 9:257–263

206. Goldberg RF, Perfetti CA, Schneider W (2006) Perceptual knowledge retrieval activates sensory brain regions. J Neurosci 26:4917–4921

207. Kellenbach ML, Brett M, Patterson K (2001) Large, colorful, or noisy? Attribute- and modality-specific activations during retrieval of perceptual attribute knowledge. Cogn Affect Behav Neurosci 1:207–221

208. Puce A, Allison T, Bentin S, Gore JC, McCarthy G (1998) Temporal cortex activation in humans viewing eye and mouth movements. J Neurosci 18:2188–2199

209. Kourtzi Z, Kanwisher N (2000) Activation in human MT/MST by static images with implied motion. J Cogn Neurosci 12:48–55

210. Kable JW, Lease-Spellmeyer J, Chatterjee A (2002) Neural substrates of action event knowledge. J Cogn Neurosci 14:795–805

Chapter 15

fMRI of Emotion

Simon Robinson, Ewald Moser, and Martin Peper

Abstract

Recent brain imaging work has expanded our understanding of the mechanisms of perceptual, cognitive, and motor functions in human subjects, but research into the cerebral control of emotional and motivational function is at a much earlier stage. Important concepts and theories of emotion are briefly introduced, as are research designs and multimodal approaches to answering the central questions in the field. We provide a detailed inspection of the methodological and technical challenges in assessing the cerebral correlates of emotional activation, perception, learning, memory, and emotional regulation behavior in healthy humans. fMRI is particularly challenging in structures such as the amygdala as it is affected by susceptibility-related signal loss, image distortion, physiological and motion artifacts, and colocalized Resting State Networks (RSNs). We review how these problems can be mitigated by using optimized echo-planar imaging (EPI) parameters, alternative MR sequences, and correction schemes. High-quality data can be acquired rapidly in these problematic regions with gradient-compensated multiecho EPI or high-resolution EPI with parallel imaging and optimum gradient directions, combined with distortion correction. Although neuroimaging studies of emotion encounter many difficulties regarding the limitations of measurement precision, research design, and strategies of validating neuropsychological emotion constructs, considerable improvement in data quality and sensitivity to subtle effects can be achieved. The methods outlined offer the prospect for fMRI studies of emotion to provide more sensitive, reliable, and representative models of measurement that systematically relate the dynamics of emotional regulation behavior with topographically distinct patterns of activity in the brain. This will provide additional information as an aid to assessment, categorization, and treatment of patients with emotional and personality disorders.

Key words Emotion, fMRI, Research design, Reliability, Validity, Amygdala, Signal loss, Distortion, Resting state networks

1 Introduction

While recent brain imaging work has expanded our understanding of the mechanisms of perceptual, cognitive, and motor functions in human subjects, research into the cerebral control of emotional and motivational functions has been less intense. For several years, however, a growing body of fMRI and positron emission tomography (PET) work has been assessing the cerebral correlates of emotional activation, perception, learning and memory, and emotional regulation behavior in healthy humans [1–6].

Massimo Filippi (ed.), *fMRI Techniques and Protocols*, Neuromethods, vol. 119,
DOI 10.1007/978-1-4939-5611-1_15, © Springer Science+Business Media New York 2016

Current brain imaging work is based on the concepts and hypotheses of the multidisciplinary field of "affective neuroscience" [7–9]. The endeavors of the subdisciplines of affective neuroscience have not only complemented but also promoted each other, stimulating a rapid growth of knowledge in the functional neuroanatomy of emotions. It is increasingly recognized that these areas also share similar methodological problems.

The expanding area of emotion neuroimaging has provided new methods for validating neurocognitive models of emotion processing that are crucial for many areas of research and clinical application. Progress is being made in disentangling the cerebral correlates of interindividual differences, personality, as well as of abnormal conditions such as, for example, anxiety, depression, psychoses, and personality disorders [10–12]. It has been recognized that psychological assessment, categorization procedures, and psychotherapy treatment may profit from models that integrate functional connectivity information. The relevance and usefulness of valid neurocognitive models of emotion processing have recently been recognized by many fields of applied research such as, for example, psychotherapy research [13], criminology [14], as well as areas such as "neuroeconomics" [15] and "neuromarketing" [16].

Several human lesion studies have pointed to the deficits of neurological patients in recognizing emotions in faces, particularly often for the decoding of fearful faces especially after bilateral amygdala damage [17–20]. Other studies have reported impairments not only for fear but also for other negative emotions such as anger, disgust, and sadness [21, 22]. Recent functional imaging studies have confirmed the importance of the amygdala in emotion processing. Due to the multiple connections between the amygdala and various cortical and subcortical areas, and the fact that the amygdala receives processed input from all the sensory systems, its participation is essential during the initial phase of stimulus evaluation [23]. The appraisal function of the amygdala, combining external cues with an internal reaction, reflects the starting point for a differential emotional response and is hence the basis for emotional learning. Involvement of the amygdala during classical conditioning especially during the initial stages of learning [24, 25] as well as during processing signals of strong emotions has been documented repeatedly with fMRI. However, a problem in verifying amygdala activation with neuroimaging tools may be the rapid habituation of its responses [10, 26].

Although the need for brain imaging data is not unequivocally acknowledged by all researchers in their specialties, the increasing body of neuroimaging data has value in challenging and constraining existing theories. Followers of cognitive emotion theory must face the fact that their results need to be compatible with or at least not contradict with established neuroscience (neuroimaging) findings [27]. However, to appropriately evaluate and integrate this knowledge, it is necessary to deal with the basic methodological problems of the field.

Therefore, this chapter is organized around the two major issues of the neuroimaging of emotional function. First, it addresses underlying conceptual issues and difficulties associated with operationalizing and measuring emotion (for more detailed reviews, *see* Refs. [28–31]). The problems and limitations of brain imaging work that are associated with measurement precision, response scaling, reproducibility, as well as validity and generalizability are discussed corresponding to general principles of behavioral research [32].

Second, the complexities of neuroimaging methods are examined to supplement recent quantitative meta-analyses (for a summary of findings of the emotional neuroimaging literature, *see* Refs. [2, 4, 5]). We raise here some grounds for reflection about current measurement in neuroimaging of emotions, and to encourage the adoption of recent methodological advances of fMRI technology. In summary, it is suggested that additional interdisciplinary efforts are needed to advance measurement quality and validity, and to accomplish an integration of brain imaging technology and neuropsychological assessment theory.

2 Psychological Methods

2.1 Emotion Theories and Constructs

2.1.1 Definitions

Emotions have been defined as episodes of temporarily coupled, coordinated changes in component functions as a response of the organism to external or internal events of major significance. These component functions entail subjective feelings, physiological activation processes, cognitive processes, motivational changes, motor expression, and action tendencies [33, 34]. Emotions represent functions of fast and flexible systems that provide basic response tendencies for adaptive action [35].

Emotions can be differentiated from mood changes (extended change in subjective feeling with low intensity), interpersonal stances (affective positions during interpersonal exchange), attitudes (enduring, affectively colored beliefs, preferences, and predispositions toward objects or persons), and personality traits (stable dispositions and behavior tendencies) [29, 34].

The frequently used concept of "emotional activation" characterizes a relatively broad class of physiological or mental phenomena (e.g., strain, stress, physiological activation, arousal, etc.). It can be specified with respect to a variety of dimensions such as valence (quality of emotional experience), intensity or arousal (global organismic change), directedness (motivational and orientating functions), and selectivity (specific patterns of change) [36]. In contrast, the terms emotional reactivity or arousability, and psychophysical reactivity refer to the *dispositional* variability of the above activation processes under defined test conditions [37, 38].

Environmental objects possess a latent meaning structure of emotional information, which is represented by a hierarchy of constructs with relatively fixed intra- and interclass relations [29, 39, 40]. Accordingly, physical stimulus properties or surface cues serve as a basis for "universal" emotion categories such as happiness, surprise, fear, anger, sadness, and disgust that originate at a primary level [41]. On a secondary level, dimensions such as valence and arousal arise from the preceding levels [42]. Table 1 suggests a potential structure of emotional concepts or domains that integrates both discrete (primary) and secondary emotions [29].

Emotional activation has also been characterized as a process with a sequence of stages [29, 35]: following an initial evaluation of novelty, familiarity, and self-relevance, a stimulus object or context is fully encoded. This involves detection of physical stimulus features, recognition of object identity, and identification of higher-order emotional dimensions such as pleasantness or need significance. During the subsequent stages, cognitive appraisal processes are initiated to evaluate the significance of the event. These evaluation checks include an appraisal of whether the stimulus is relevant for personal needs or achieving certain goals. Finally, the potential to overcome or cope with the event and the compatibility of behavior with the self or social norms is evaluated [35].

2.1.2 Operationalization

The measurement of emotions crucially depends on an appropriate operationalization of the construct of interest and definition of response parameters. Such considerations have typically been elaborated in the context of psychological assessment theory [30, 38,

Table 1
Hierarchical organization of emotion concepts (modified from [29])

Emotion concepts or domains	Example constructs	Basis for higher-order grouping
Dimensional concepts	Valence (positive/negative emotions), approach/withdrawal, activity (active/passive), control, etc.	Conceptual or meaning space for subjective experience and verbal labels
Basic, fundamental, discrete, modal emotions or emotion families	Anger, fear, sadness, joy, etc.	Similarity of appraisal, motivational consequences, and response patterns; convenient label for appropriate description and communication
Specific appraisal/response configurations for recurring events/ situations	Righteous anger, jealousy, mirth, fright, etc.	Temporal coordination of different response systems for a limited period of time as produced by a specific appraisal pattern
Continuous adaptational changes	Orienting reflex, defense reflex, startle, sympathetic arousal, etc.	Automatic activations and coordination of basic biobehavioral units

43]. The latter explains how psychological and physiological measures can be empirically assessed, decomposed, and used as indicators of the psychological constructs of interest. It organizes the assumptions concerning measurement, segmentation, and aggregation of activation measures, and evaluates the distribution characteristics and reliability of the data. It also determines the range of the construct of interest by localizing it according to variables, subjects or settings/situations, or combinations of these sources of variation. Since most current operationalizations are confined to one of these aspects, the range of conclusions to be drawn from the findings is also limited.

A particular problem associated with measuring emotional reactions is a certain lack of covariation of response measures. A frequent finding is that the expected synchronization of verbal, motor, and physiological response systems during an emotional episode is the exception rather than the rule. Although emotional episodes supposedly give rise to a synchronization of central, autonomic, motor, and behavioral variables [44], most emotional response measures only show imperfect coupling [45]. This response incoherence may be attributed to a temporary decoupling or dissociation of function [46]. This has led authors to suggest a triple response measurement strategy that suggests a multimodal assessment of emotion including responses in the verbal, gross motor, and physiological (autonomic, cortical, neuromuscular) response systems [47].

Research on human emotion has illustrated how the broad concept of emotion is subdivided into several component functions that dynamically interact during an emotional episode. Diverse operationalizations have been suggested to assess these subconstructs, many of which are highly correlated and form clusters or families of similar functions. Emotional activation processes are embedded in a multicomponential system of situational and personal determinants. Factors that shape the level and pattern of the emotional activation process are the following [29, 48]: the functional context of the task (e.g., cognitive processing, motor responses, autonomic functions, etc.); the direction and extension of effects (e.g., global versus selective activation); the intensity and the degree of emotional strain (e.g., low, middle, or traumatic intensity; degree of threat; intensity of physical/mental load; stimulus intensities below or above threshold); the time characteristics (e.g., duration, structure, and variability of a stimulus; effects of stimulus repetition or pre-exposure); the informational content (e.g., the degree of information and dimensions inherent in the experimental stimuli such as emotional valence or arousal, preparedness, novelty, safety, predictability, contingency information, etc.); the implications for action (conduciveness, implications for instrumental reactions; artificial vs. realistic nature of the procedure); the coping potential (e.g., active coping vs. passive enduring, degree of controllability, helplessness, social support, specific coping strategies); and, compatibility with self or social norms (e.g., personal relevance).

These different aspects have led to a large number of operationalizations. These include procedures to elicit orienting or startle reactions, basic emotions or "stress," as well as stimulus-response paradigms and conditioning procedures. For example, one such standard procedure is to elicit orienting reactions (OR) by emotionally meaningful stimuli. The OR is a nonassociative process being modulated by excitatory (sensitization) and inhibitory (habituation) mechanisms. Pavlovian (classical) or instrumental conditioning of excitatory or inhibitory reactions has traditionally been investigated in autonomic reactions (cardiovascular, vasomotor, and electrodermal conditioning), motor responses (eye blink), and endocrine or immune system reactions [1].

Emotional experience is strongly influenced by cognitive activities which modulate attention and alertness (avoidance and escape), vigilance processes (information search and problem solving), person–situation interactions (denial, distancing, cognitive restructuring, positive reappraisal, etc.), and actions, which change the person–environment relationship [49]. Coping research has identified typical cognitive strategies to regulate arousal during an emotional episode such as rejection (venting, disengagement) and accommodation strategies (relaxation, cognitive work) [50]. Cognitive activities subsume engagement (reconceptualization, reevaluation strategies such as rationalization or reappraisal) and distraction techniques.

These behavioral and cognitive regulation processes have been studied for many decades [51]. This research has shown that the outcome of coping processes crucially depends upon the valence, ambiguity, controllability, and changeability of a stressor. Input-related regulation (denial, distraction, defense, or cognitive restructuring; [52]) or antecedent-focused regulation (selection, modification, or cognitive restructuring of situational antecedents; [53]) have been differentiated from response-focused processes (suppression of expressive behavior and physiological arousal; [53]).

While the behavioral procedures mentioned above are mostly unstandardized, a vast number of standardized psychometric instruments are available to assess the higher-order emotional processes (for a review see Ref. [31]). Questionnaires are the most frequently used method, being followed by behavior ratings by experts or significant others. However, these data assess subjective representations, that is, personal constructs and may be obscured by biased responding

2.2 Research Design and Validity

2.2.1 Research Design

The requirements for experimental research [32] are not always fulfilled by many early research designs of emotional neuroimaging work. This is typical for the pilot stage of scientific progress. In many cases, only preliminary or correlational interpretations are possible due to incomplete or missing control conditions (e.g., with respect to the "awareness" of emotional stimuli; [54]). In contrast, more recent work increasingly makes use of full factorial designs or applies parametric variations of the independent variable [55]. Moreover, new techniques of covariance analysis are available to explore the causal predictive value of structural data on emotional brain activation. The relationship of structural and functional connectivity data has been explored by means of Structural Equation Modeling [56–58] and Dynamic Causal Modeling [59]. Moreover, functional brain imaging has been successfully combined with the lesion approach to elucidate the modulating influences of interconnected brain regions [60]. Thus, by means of appropriate research plans and advanced techniques of analysis, an "effective connectivity" can be identified that elucidates the causal relations of one neural system to another [61]. For example, the functional connectivity of the prefrontal cortex (PFC) that is supposed to modulate amygdala activity [62] might thus be better evaluated in terms of causality.

To avoid operationalization errors, the quality of the emotion induction procedure needs to be scrutinized, that is, it must be evaluated whether the intended emotion has actually been elicited. For example, since a variety of emotional and nonemotional stimulus situations may trigger amygdala activations [5], it is necessary to evaluate whether the intended emotion (such as fear) has actually been elicited. Since subjective report is not always an appropriate manipulation check, additional psychophysiological criteria are needed to validate the intended emotion. Sympathetic activity as indexed by electrodermal activity (EDA) has been assessed during imaging procedures for this purpose. Nevertheless, this does not validate fear since skin conductance responses represent the endpoint of many different processes [63].

2.2.2 Construct Validation

Brain imaging work implements specific neuropsychological construct validation strategies by associating behavioral measurement of emotion with functional brain activation data for different localizations [31]. Here, functional (physiological) data are related to but still remain categorically distinct from the psychological data that emerge from a particular behavioral paradigm. During the process of construct validation, indicators of connectional or neurophysiological constructs are related to the indicators of psychological constructs. Thus, different operationalizations of a certain psychological construct (procedures or task) are expected to be correlated with activations of a certain area or cluster of areas. A different construct is expected to correlate with another but not the previous area and vice versa. This corresponds to the double dissociation approach, which inspects task by localization interactions. This process of neuropsychological concept formation typically starts at a relatively broad level and proceeds downward in the above hierarchy finally specifying within-systems localization constructs [64].

However, depending on limitations of the measurement device described below, the reliability of psychological or activation data declines at lower levels of structural constructs complicating this validation process. The diverse validation attempts typically draw upon convergent or divergent associations of constructs that are located at quite different levels of generality. However, successful construct validation very much depends upon whether brain activation and psychological measures are analyzed on the same level of generality. In cases of asymmetry, low relationships may result that provoke misinterpretations and confuse the validation process. Thus, successful construct validation in the affective neurosciences requires emotional constructs and brain activation data to be measured on the same (symmetrical) level of generality or aggregation [31].

Emotional neuroimaging is typically guided by neuropsychological construct validation strategies. Here, the constructs are operationally defined by the complementary methods of emotion psychology and of neurophysiology. Both construct types are embedded in hierarchically organized networks with lower- and higher-order levels of generality. Both types of data are associated with each other during validation. However, it is necessary to define neural and emotional constructs on the same level of generality. For example, when a relatively broad behavioral category or set of functions ("emotion regulation") is being associated with isolated cerebral substructures, the relationship is likely to be asymmetrical and disappointing low correlations might result confusing the validation process.

2.2.3 Internal Validity

FMRI is known to be a highly reactive measure because the scanner setting (gradient noise and the supine position) causes the subject to respond to the experimental situation as a stressor. Unless habituation sessions are included in the procedure, tonic stress and arousal effects may be induced that modulate responding as discussed above. For example, a decreasing rate of response of the amygdala to a conditioned stimulus during the late phase of acquisition [10, 24, 26, 65] may also be attributable to testing effects (sensitization to the setting, acquaintance with the procedure, and type of unconditioned stimulation) rather than fast amygdala habituation per se (other factors might also explain reduced amygdala perfusion measures such as potential ceiling effects, baseline dependencies, and regression to the mean). In general, familiarity with emotionally activating procedures in the scanner induces states of expectation, sensitizing or desensitizing effects that may confound follow-up measurement. In addition to these testing effects, history, that is, occurrences other than the treatment and individual experiences between a first and a second measurement are likely to endanger the assessment of emotion (e.g., when assessing psychotherapy effects).

Changes in the observational technique, the measurement device or sequence and other instrumentation effects may also obscure emotion-related treatment variance during an fMRI session or across sessions. From the discussion of MR methods it is clear that longitudinal changes of measurement precision are also to be expected from inconsistent acquisition geometry and shim, as well as system instabilities and hardware changes.

It is well known from psychophysiological research that the interpretation of repeated measurement factors is complicated by initial value dependencies [66]. When the hemodynamic response is fitted relative to the prestimulus baseline, a physiological or statistical dependency of tonic perfusion levels and the phasic reaction may prevail [67]. While the first experimental blocks may show extreme effects, subsequent measurements are likely to be closer to the mean. Moreover, it has been pointed out above that the reliability of blood oxygen level-dependent (BOLD) measurements may be compromised by distortions or signal loss. When emotional paradigms with inconsistent effects are used or when subjects with an extreme variability of emotional responsivity are investigated, experimental effects are likely to show "regression to the mean."

Subjects change as a function of time and these maturation effects may occur during the time range of the experiment (psychophysiological changes of organismic state or psychological stance, in particular during aversive paradigms). State-dependent influences or maturation effects may hamper within-subject replication or evaluations of long-term psychotherapy effects.

Subject groups with an elevated emotionality are more likely to show greater dropout rates in stressful experiments, that is, subjects of one group drop out as a consequence of their specific reactivity to the emotionally strain of the challenge paradigm. If exit from an emotionally activating study is not random, this effect of "experimental mortality" may confound comparison between groups.

Selection effects, that is, group differences from the outset of the study, are likely in functional imaging studies with very small numbers of participants. Selective recruitment of volunteers or drop out of participants may lead to decreased reactivity and lower emotionality in the remaining study group. Poor recruitment techniques (e.g., drafting subjects from the social circle of the lab partially acquainted with the procedures) or lack of random assignment to groups may further limit the validity of emotional fMRI studies.

Interactions of selection with maturation may occur when groups that differ with respect to maturation processes are compared (e.g., administering a social stress test for cortisol stimulation at different times of the day). Gender, personality traits, or psychopathology are all associated with specific individual differences of emotional regulation behavior. When these behaviors change over time as a function of personal development, follow-up measurements may be confounded by this type of effect. Thus, poor randomization or lack of control of personality-specific variance may jeopardize brain activation studies of emotional behavior. Finally, an interaction of selection with instrumentation occurs when experimental subjects and controls show pre-experimental differences with respect to the shape of their responses such as floor or ceiling effects.

In general, emotional responses show an intraindividual instability due to measurement artifacts (*see* Sect. 3), state-dependent influences, or characteristics of the subject (age, gender, experience, temperament) all impose additional effects on functional neuroimaging results [5]. A considerable degree of within- and between-subject variation in the time course of emotional responding depends on habitual, subject-specific mechanisms. First, the phasic activation pattern reflects the short-term modulation in response to the emotional stimulus. Due to the temporal within-trial variability of BOLD responses in different brain regions, averaging across subjects may obscure the detection of activation in a specific region and reduce effect sizes specifically for higher-level reactions. Second, activation also varies across the time course of the experiment. Most subjects show a constant increase in autonomic arousal depending on the degree of emotional stimulation. This is not only accompanied by a systemic response (tonic increase of sympathetic activation including blood pressure, cardiac contractility, and variability), but also by variations of tonic perfusion. These changes may show divergent trends for cortical and limbic regions imposing an unknown error on the measurement of the phasic BOLD reaction. These tonic and phasic variations appear to reflect the subject-specific mechanisms of emotional regulation behavior.

2.2.4 External Validity and Generalizability

Generalization to Other Procedures and Paradigms

The majority of current paradigms have focused on lower-level perceptual or learning processes pertaining to basic or secondary emotional categories. Since the results depend on the selected task parameters (degree of induced arousal, hedonic strength, and motivational value; degree of involvement of memory processes; reinforcement schedule; conditioning to cues or contexts; etc.), a comparison with and generalization to other operationalizations remains difficult. Systematic neuroimaging approaches to higher-level appraisal processes are still sparse. These involve evaluations of the motivational conditions and coping potential, that is, the ability to overcome obstructions or to adapt to unavoidable consequences [29]. An expanded range of constructs would involve an assessment of social communication processes, beliefs, preferences, predispositions, high-level evaluation checks, as well as modulating sociocultural influences. Higher-order appraisal processes involve the evaluation of whether stimulus events are compatible with social standards and values or with the self-concept. Another function to be explored concerns the degree to which a stimulus event may increase, decrease, or even block goal attainment or need satisfaction, and activate a reorientation of the individual's goal/need hierarchy and behavioral planning (goal/need priority setting) [29].

Whereas frontostriatal mechanisms of motor control have been increasingly investigated, recent work has made efforts toward developing an understanding of how emotion and motivation are linked to the frontal mechanisms controlling the preparation and execution of behavior [68, 69]. Behavior preparation and execution represent closely integrated components within an emotional episode. Mobilization of energy is required to prepare for a certain class of behavior. Action planning and motor preparation requires sequencing of actions and generation of movements. However, an emotion preceding behavior is only one of a number of factors, including situational pressures, strategic concerns, or instrumentality, involved in eliciting the concrete action. Additional research is needed to trace the information flow from motivational to motor systems.

Another component is the verbal or nonverbal communication of emotions such as facial expression or vocal prosody [70]. The ability to verbally conceptualize emotions and to communicate emotional experiences plays an important role in the regulation of an ongoing emotional episode. For example, explicit emotion-labeling tasks have been shown to decrease the activation level of the amygdala [71, 72].

Finally, sociocultural factors may shape attitudes (relatively enduring, affectively colored beliefs, preferences, and predispositions toward objects or persons) as well as interpersonal stances (affective stance taken toward another person in a specific interaction). The ability of the individual to form representations of beliefs, intentions, and affective states of others has a considerable importance for affective and interpersonal interaction. However, the effects of beliefs, preferences, and predispositions on lower levels of emotional responding have attracted little attention. Top-down processes may induce considerable variations of task and stimulus parameters by modulating lower-level automatic processes and by controlling the late behavior preparation stages during the emotional process. Thus, generalization to other paradigms and constructs has limitations because higher-level behavioral and cognitive strategies that are part of the individual emotion regulation system ([50]; see later) modulate the emotion process.

Generalization to Other Subjects and Populations

The study groups of many fMRI studies have been relatively small and poorly described with respect to personality dimensions. Since several studies provide evidence for trait-dependent differences in responding [73–76], it remains unclear to what extent the results may have been influenced by interindividual differences of the participating subjects. The representativeness of results is particularly poor if members of the social circle of the lab serve as participants instead of independently recruited participants. Thus, when the effects of an emotional paradigm interact with characteristics of the study groups (such as a low level of emotionality in subjects willing to participate in an activating scanning condition), this selection × treatment effect may endanger generalizations to other populations.

Generalization to Other
Times and Settings

The prediction of future emotional or psychopathological disorders on the basis of emotional behavior assessed in the scanner remains difficult [77]. Eliciting emotions in the imaging scanner is a highly artificial situation. It remains unclear to what extent these results can be generalized to other settings and, in particular, to real life settings. Small and Nusbaum [78] have criticized the unnatural MRI scanner setting and suggested an "ecological functional brain imaging approach" that includes monitoring of natural behaviors using a multimodal assessment and environmental context of presentation or behavior. Nevertheless, in contrast to the scanner, emotion in real settings is not restricted to simple reactions but includes the full range of regulatory actions. By correlating fMRI and field data, such as, for example, generated by emotion monitoring during everyday life [79], the "ecological validity," that is, the predictive value of cerebral perfusion patterns for real-life emotions could be better evaluated.

3   fMRI Methods

3.1   Methodological Challenges

3.1.1   Introduction

A host of fMRI studies have identified the amygdalae as central structures in emotion processing (*see* Sect. 1 and Zald et al. [5], for example, for a review). The amygdalae lie in the anterior medial temporal lobe (MTL), bounded ventrolaterally by the lateral ventricles and medially by the sphenoid sinuses (Fig. 1). The differing magnetic susceptibilities of these tissues cause large deviations in the static magnetic field, B_0. There is also a strong gradient in B_0 in the MTL, and differing precession frequencies lead to dephasing of

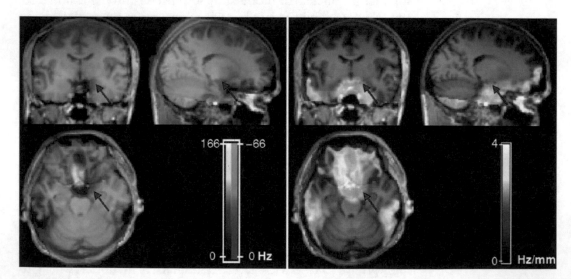

Fig. 1 The amygdalae, central brain structures in emotion processing, lie in a region of moderate deviation from the static magnetic field (*left*) and very high static magnetic field gradients (*right*). The planes intersect in the amygdala at MNI coordinate (18, −2, −18), marked by *arrows*. Single subject measurement at 4.0 T

the bulk magnetization and loss of signal in images. This problem is not restricted to the amygdala, however. Inferior frontal and orbitofrontal regions, likewise involved in emotion processing [80], are also zones of high static magnetic field gradient. In addition to signal loss, static magnetic field gradients also lead to echo times (TE) becoming shifted, so that BOLD sensitivity may be reduced, or signal may not be acquired at all (termed "Type 2" loss [81]). These problems are examined in Sect. 3.1.2.

Local variations in the static magnetic field strength confound spatial encoding of the MR signal, leading to image distortion. Particular considerations for the MTL in this regard are discussed in Sect. 3.1.3. Even at high field, deviations from B_0 immediately in the amygdala are relatively moderate (Fig. 1 left; 10 Hz measured at the arrow position, for data acquired at 4.0 T) but the field gradient is high (2 Hz/mm at the same position), leading to very large distortions in neighboring structures, which can cause signal to encroach into the amygdalae.

The ventral brain is also prone to physiological artifacts of cardiac and respiratory origin, as described in Sect. 3.1.4, which may be mitigated to some extent by simultaneous measurement of cardiac and respiratory processes and the application of postprocessing corrections. In addition to the measurement challenges of ventral brain imaging, the presence of large magnetic field gradients makes the ventral brain susceptible to stimulus-correlated motion (SCM) artifacts, as discussed in Sect. 3.1.5. These can lead to the appearance of neuronal activation (Fig. 2) arising from subtle head movements which are time locked to stimuli.

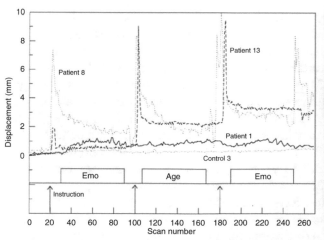

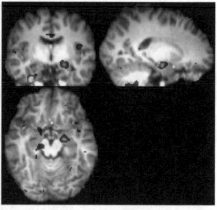

Fig. 2 Large static magnetic field gradients make the amygdala region prone to the artifactual appearance of neuronal activation when stimulus-correlated motion (SCM) is present. *Left* : Observed patterns of SCM of schizophrenic patients and controls in a 3.0-T experiment with three stimulus blocks (facial emotion and age discrimination "EMO" and "AGE"). *Right* : a baseline (no stimulus) study in which a subject executed submillimeter SCM similar to that of Patient 1. The contrast corresponds to the "EMO" periods (uncorrected $p < 0.0001$; t threshold $= 5$, Montreal Neurological Institute coordinates 22, −6, −16)

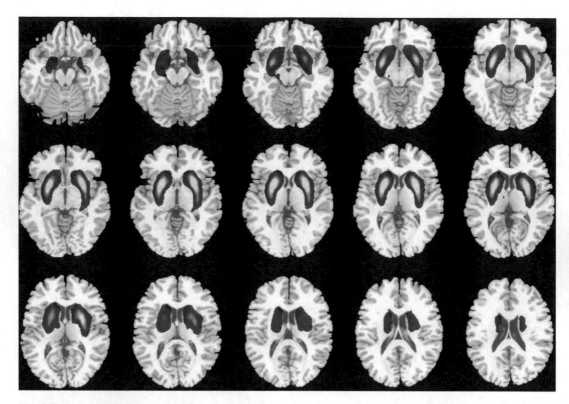

Fig. 3 Signal changes in the amygdala in emotion experiments have to be measured against a background of resting state fluctuations. A resting state network recently been reported, covering the amygdala and basal ganglia (3.0 T, group independent component analysis of 26 young healthy adults). Adapted from [106] with permission from the ISMRM

A further potential confound is the presence of RSNs which colocalize with regions under study. These show slow fluctuations in the absence of stimuli and constitute sources of unmodeled noise and intertrial variation. The existence of a RSN in the amygdalae (Fig. 3) offers a possible explanation of why small signal changes are generally recorded in these structures, despite the high neurovascular reactivity of deep gray matter nuclei. This and other RSNs which may involve the amygdala are described in Sect. 3.1.6.

In Sect. 3.1 we expand on the problems outlined here, and go on in Sect. 3.2 to detail approaches to optimizing conventional single-shot 2D gradient-recalled echo-planar imaging (EPI) to mitigate their effects, alternative sequences which are less sensitive to static magnetic field gradients and, in Sect. 3.3, methods to correct for image distortion, physiological noise, and SCM artifacts.

3.1.2   Signal Loss
and BOLD Sensitivity Loss

It is worthwhile to briefly review the problem of signal loss from an empirical perspective. A temporal resolution of 1–3 s is usually desirable in fMRI. The whole brain may be covered in this time by acquiring images with voxels of typically 3-mm size (or 27 µl).

Relatively long TEs are employed, partly also as a technical necessity—to allow time for gradient switching and echo sampling—but also to confer T_2^* weighting. As well as providing sensitivity to BOLD effects, however, this allows time for dephasing from macroscopic inhomogeneities to develop. The severe signal loss seen in EPI in the anterior MTL with typical parameters is illustrated Fig. 4 in the lower left two images.

In gradient-echo imaging, the MR signal decays with a time constant T_2^*, comprising the transverse relaxation time, T_2 (reflecting irreversible decay arising from time-varying microscopic spin-spin processes), and T_2', the reversible contribution to the transverse decay rate and the major source of BOLD contrast. T_2' itself can be separated into "mesoscopic" contributions (which operate on a scale smaller than the voxel, e.g., dephasing in the capillary bed), and "macroscopic" contributions (meaning larger than the voxel) which stem from bulk field inhomogeneities and which are dependent on the tissues present, on the quality of shim, and on the scanning parameters such as voxel size and slice orientation. Separating these effects, the MR signal S in a gradient-echo experiment decays such that at the TE it can be expressed [82] as:

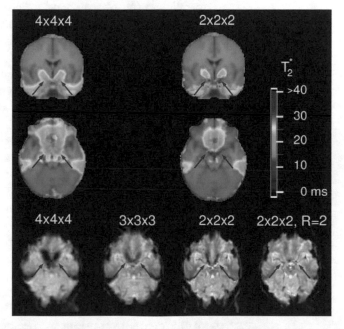

Fig. 4 Effects of voxel size and acceleration factor on T_2^* and echo-planar imaging (EPI) image quality at high field (4.0 T). *Top*: T_2^* in coronal and axial slices through the amygdala at two voxel sizes. *Bottom*: corresponding EPI in slices through the amygdala with acquisition voxel sizes of $4 \times 4 \times 4$ mm, $3 \times 3 \times 3$ mm, $2 \times 2 \times 2$ mm, and $2 \times 2 \times 2$ with GRAPPA acceleration of factor 2, all with echo time (TE) = 32 ms

$$S(\text{TE}) = S(0) \times \exp\left(\frac{-\text{TE}}{T_2}\right) \times F(\text{TE}), \qquad (1)$$

where an approximation to $F(\text{TE})$ for linear field variations over voxels, ΔBi, in the x, y, and z directions is

$$F(\text{TE}) = \text{sinc}\left(\frac{\ddagger\ B_x\text{TE}}{2}\right) \times \text{sinc}\left(\frac{\ddagger\ B_y\text{TE}}{2}\right) \times \text{sinc}\left(\frac{\ddagger\ B_z\text{TE}}{2}\right). \quad (2)$$

This illustrates that the signal decay rate may be reduced by decreasing the voxel size—to reduce the gradients across voxels, ΔBi—or by reducing the TE.

The aim of any attempt to optimize an EPI sequence is not just to maximize signal, described above, but also BOLD sensitivity (BS), which is equal to the product of image intensity and TE; for magnetically homogeneous regions is a maximum when the EPI effective TE is equal to the T_2^* of the target region [83]. In homogeneous regions, however, the presence of field gradients shifts the location of signal in k-space, mainly in the phase-encode direction (because of the low bandwidth), changing the local TE [81]. Through-plane field gradients lead to signal loss and reduce BS. If the component of the in-plane susceptibility gradient in the phase-encode direction is antiparallel to the phase-encode gradient "blip" direction, then the TE is also reduced, reducing BS further. Conversely, if it is parallel to the phase-encode "blips" then TE increases. While this increases BS, to some extent compensating for signal loss, if the shift of TE is too large the echo will fall outside the acquisition window, leading to complete signal dropout. This is commonly observed in the anterior MTL for a negative-going phase-encode scheme.

This description motivates the optimization approaches to EPI in susceptibility-affected regions which will be outlined later in this section; compensating through-plane gradients, selecting image orientation and gradient direction to minimize echo shifts, and reducing voxel sizes to reduce field gradients. These techniques will be shown to increase both signal and BS.

3.1.3 Image Distortion

Accurate spatial encoding in MRI is founded upon a homogeneous static magnetic field in the object. The location of signal is deduced from the local field strength under the application of small orthogonal, linear magnetic fields in directions usually referred to as *slice select*, *readout*, and *phase-encode*. The method is confounded if there are regional variations in the static magnetic field, which lead to signal mislocalization (distortion). Typical field offsets are illustrated in Fig. 1 (left) and lead to EPI distortions of the image shown in Fig. 4.

The extent of distortion, expressed as the number of pixels by which signal is mislocalized, is equal to the local magnetic field deviation divided by the bandwidth per pixel (the reciprocal of the time between measuring adjacent points in k-space), expressed in the same units. The bandwidth per pixel in the readout direction (rBWread/pix) is equal to the total imaging bandwidth (the signal sampling rate) divided by the image matrix size in the readout direction. In EPI, the pixel bandwidth in the phase-encode direction is smaller than this again by a factor of the image matrix size in the phase-encode direction. The fact that total bandwidth is often increased in proportion with the readout matrix dimension in order to keep rBWread/pix constant means that distortion (in distance rather than number of pixels) is approximately constant as a function of matrix size (and thereby resolution, at constant matrix size). To illustrate the size of expected distortions, in a 64×64 matrix acquisition, a typical rBWread might be 1500 Hz/pixel, giving (as 1500/64) a rBWphase of 23 Hz/pixel. A value of ΔB_0 of 50 Hz (common at high fields, *see* Fig. 1) would lead to a shift of 0.03 voxels in the readout direction, but 2 voxels in the phase-encode direction, or 7 mm for a typical field of view for brain imaging. In a higher resolution acquisition with a 128×128 matrix and the same rBWread, rBWphase would be 12 Hz/pixels and the distortion 4 voxels, but also 7 mm because of the proportionately smaller voxel size.

The relationship between EPI distortion and field strength is not simple, depending both on hardware and usage. Susceptibility-induced field changes increase linearly with static magnetic field strength while gradient amplitude (the factor which limits sampling rate) is approximately constant in the standard to high field regimes. While theoretically this leads to an approximate proportionality between distortion and field strength, in practice higher acquisition bandwidths are often used at high field to the achieve shorter effective TEs, to match reduced T_2^* times.

Image distortion frustrates attempts to coregister data from many subjects to a common probabilistic atlas [84], which can reduce significance in fMRI even in relatively homogeneous areas [85]. Established methods for correcting image distortion are compared for their performance in the amygdala in Sect. 3.3.1.

3.1.4   Physiological Artifacts

A number of physiological processes give rise to fluctuations in the MR signal which are unrelated to neuronal activation, and should therefore be corrected for or modeled in a statistical analysis. The amygdala area is particularly prone to cardiac artifacts due to the proximity of the arteries in the Circle of Willis, and to respiratory artifacts because of the susceptibility gradients.

Respiration leads to head motion, changes in the magnetic field distribution in the head due to changes of gas volume or oxygen concentration in the chest [86], and variation in the local oxyhemoglobin concentration, probably due to flow changes in draining veins [87]. Subtle changes in respiration rate and depth are thought to be the origin of spontaneous changes in arterial carbon dioxide level at about 0.03 Hz which have been shown to lead to significant low-frequency variations in BOLD signal [88]. The lag of 6 s in this process corresponds to the time taken for blood to transit from the lungs to the brain, and for cerebral blood flow volume to respond to CO_2, a cerebral vasodilator. Magnetic field changes in the head particularly affect ventral brain imaging due high field gradients. Respiration-related artifacts typically affect the image periphery, making them problematic for the amygdala, which is usually at the anterior boundary of the signal-providing region.

Cardiac pulsatility causes expansion of the arteries, bulk motion of the brain, and cerebrospinal fluid flow and leads to the influx of fully relaxed spins into an imaging slice. As a consequence, the signal may increase in many of the arteries that lie close to the amygdala, such as the middle cerebral artery and other elements of the Circle of Willis [89]. Cardiac artifacts are particularly complex with regard to emotion studies as the amygdala innervates the autonomic nervous system via the hypothalamus and brainstem, increasing heart rate, as has been shown in fMRI [90], and human depth electrode studies [91]. Recently, fluctuations in cardiac rate have been shown to explain almost as much variation in the BOLD signal as the oscillations related to each cardiac cycle, as revealed by shifted cardiac rate regressors [92].

Cardiac and respiratory cycles are connected by a number of processes [93], leading to many regions showing BOLD fluctuations of cardiac origin [92] being also observed in studies of respiratory effects [94].

Cardiac and respiratory artifacts may be corrected for by a number of approaches, some of which require additional measurements at the time of imaging. The effectiveness of these techniques in the ventral brain is outlined in Sect. 3.3.2.

3.1.5   Motion Artifacts

Motion artifacts affect all regions of the brain, but are particularly problematic in emotion studies because the nature of the task material is prone to induce SCM as a startle, attention, or repulse response. Patients with disorders with emotional components (such as schizophrenia and posttraumatic stress disorder) are less likely to remain still throughout the experiment and the interaction between motion and distortion in regions of high susceptibility gradient produces nonlinear pixel shifts that are not well corrected with rigid-body methods. Partial brain coverage protocols, such as those that may be used to allow z-shimming or high spatial and temporal resolution fMRI in the amygdala, are also more prone to partial voluming in the outermost slices and spin history effects, in which motion between the acquisition of adjacent slices leads to some spins being excited twice within one repetition time (TR) while others are not excited at all.

Head motion can be minimized using bite bars, vacuum cushions, thermoplastic masks, or plaster head casts. As well as effective immobilization, casts allow for repositioning in longitudinal studies [95]. Such devices are not appropriate for emotion studies, however, due to the added degree of discomfort and distraction they provide.

SCM was originally investigated by Hajnal et al. [96] in hybrid simulations with quite large (3 mm) introduced pixel shifts, which led to peripheral correlations. A study by Field et al. [97] found that small-amplitude motion can lead to false positive results, particularly in regions of high field gradient. Likewise, larger motions can reduce significance and lead to false negative results. Two distinct patterns of SCM are often observed in fMRI experiments. As in the example of identified motion with sample schizophrenic patients and controls (Fig. 2, left), patients may execute large motions at the first presentation of a stimulus, and many patients and controls show very small displacements which endure for entire blocks. Reproducing the submillimeter head motions observed in that experiment in a separate session (without stimuli), these have been shown to lead to highly significant correlations in the amygdala which are difficult to distinguish from genuine activation (Fig. 2, right), a problem not mitigated by standard motion correction methods [98].

3.1.6 Colocalized Resting State Networks

An additional methodological confound comes in the form of RSNs, which constitute additional sources of signal fluctuations unrelated to experimental task. In the absence of tasks or stimuli, the brain undergoes slow (0.01–0.1 Hz) fluctuations in functionally related networks of brain regions [99, 100]. These endure during task execution, and have been shown to account not only for much of the intertrial variation in the BOLD response in evoked brain response [101], but also to the intertrial variability in behavior [102]. Approximately ten such RSNs have been discovered over the past decade [99, 100, 103–105] in networks relating to sensory or cognitive function. A network with similar low-frequency characteristics has recently been identified in the amygdala and basal ganglia [106].

The network illustrated in Fig. 3 shows the results from a group of independent component analysis (ICA), performed with MELODIC [107], of resting state data acquired from 26 subjects. It is continuous, fully incorporating symmetrically the striate nuclei (pallidum, puitamen, and caudate nuclei), extending inferiorly to the amygdaloid complexes. The network is weaker than those previously reported (measured by the amount of variance it explains in the data), but is reproducible across subgroups of subjects, runs, and resting state conditions (fixation and eyes closed) and offers a tantalizing explanation as to why, despite the fact that neurovascular reactivity is high in deep gray nuclei, BOLD signal changes are weaker and less consistent in the amygdalae and basal ganglia than in the cortex.

This may not be the only RSN in which the amygdala is involved. Correlations were observed between the amygdalae, and between the amygdalae and hippocampi and anterior temporal lobes in one of the earliest resting state analyses, using functional connectivity [100]. The amygdala was also listed as an element in the "default mode" network [108], when originally reported as regions showing deactivations across a number of tasks in PET [109]. The fact that the amygdala has not been observed as part of this network in this context may relate to the technical challenges of measurement discussed in this chapter.

3.2   MR Methods, Sequences, and Protocols

3.2.1   Field Strength

While the signal to noise ratio (SNR), the magnitude of BOLD signal changes, and the specificity of the BOLD response to micro-vascular contributions all increase with field strength, so do physiological noise, field inhomogeneities, and physiological artifacts which specifically affect the anterior MTL. The advantages of high field for emotion studies are therefore restricted to particular regimes and methods in which these problems are minimized. Human emotion fMRI studies have been carried out at field strengths from 1.0 to 7.0 T. In line with the development of sequences and approaches to EPI in susceptibility-affected area which are discussed in Sect. 3.2.2–3.2.8 (high-resolution single and multishot EPI, multiecho and spiral acquisitions, gradient compensation, and parallel imaging), emotion fMRI in the high field regime (3.0–4.0 T) has become commonplace, although applied studies have generally used standard sequences and parameters despite the problems which have received attention in the MR literature [110] and a number of promising remedies (see the following sections). Ultra-high field strength studies of emotion are still sparse, however, and it is likely that they will be restricted to highly specific questions during the next 5–10 years of hardware and sequence development.

Theoretical gains in SNR at high field are limited by physiological noise, which increases both with field strength and voxel size, and causes time-series SNR (tSNR) to reach as asymptotic limit with voxel volume [111]. This limit was found to increase only modestly with field strength, being 65 at 1.5 T, 75 at 3 T, and 90 at 7 T, so that for large ($5 \times 5 \times 3$ mm) voxels, tSNR was only 11 % higher at 3 T than at 1.5 T, and only 25 % higher at 7 T than 1.5 T. The tendency toward asymptotic behavior began at relatively small volume volumes, with 80 % of the asymptotic maximum being reached at 28.6, 15.0, and 11.7 mm^3 at 1.5, 3, and 7 T, respectively. For small voxels, however, where thermal noise dominates, tSNR gains were almost linear with field strength. In the same study, the authors found that with $1.5 \times 1.5 \times 3$ mm^3 voxels, tSNR increased by 110 % at 3 T compared to 1.5 T, and by 245 % at 7 T compared to 1.5 T [111]. This study clearly shows that tSNR gains are to be made at high field in the small voxel volume regime.

These tSNR results also explain the often modest gains achieved in fMRI studies at higher field, particularly in regions affected by signal dropout. Krasnow et al. [112] compared activation in response to perceptual, cognitive, and affective tasks at 1.5 and 3 T with a relatively large voxel protocol ($3 \times 3 \times 4$ mm) and observed only moderate increases in activated volume at 3 T for the perceptual and cognitive tasks (23 and 36%, respectively), but no significant improvement in the activated amygdala volume due to increased susceptibility-related signal loss. A high-resolution, high-field approach has been exemplified in the only human study of amygdala function at 7 T to date of which we are aware, which was carried out at submillimeter resolution [113].

These studies define the regime in which field strength gains are to be made, but it is fair to ask why one should move to high-resolution measurements if the neuroscience question does not require, for instance, subnuclei of the amygdala to be resolved, but—as is more commonly the case—the study of interactions between the amygdalae and the cortex, for which whole brain coverage is essential. The use of high resolution here is not principally to distinguish activation in small structures, but to reduce both physiological noise and susceptibility artifacts. A number of works have shown the value of averaging thin slices, downsampling, and smoothing data acquired at high resolution [114–116] and using multichannel coils [115] to regain losses in SNR inherent to small voxels generally and yielding net gains in susceptibility affected areas [115, 117].

3.2.2 z-Shimming, Gradient Compensation, Tailored RF Pulses

The effect of signal dephasing arising from through-plane gradients may be reduced by creating a composite image from a number of acquisitions in which different slice-select gradients are applied [118], a process known as z-shimming. In each image the applied gradient pulse is appropriate to counteract susceptibility gradients in particular regions. The method is effective in regaining signal in the anterior MTL, but clearly reduces temporal resolution by a factor equal to the number of images acquired, usually a minimum of 3. Alternatively, a single, moderate preparation pulse may be used. This reduces through-plane dephasing in affected areas at limited cost to BS and signal in homogeneous areas, and allows slices to be orientated so that TE shifts are small, reducing signal loss due to in-plane gradients [119]. z-Shimming and other compensation schemes have been applied in a number of other sequences described in this section.

Spins may also be refocused using tailored radio frequency pulses which create uniform in-plane phase but quadratic phase variation through the slice, allowing dephasing to be "precompensated" [120]. Analogous to z-shimming, in the original implementation a number of acquisitions with different precompensations were required, suited to different regions. More recently 3D versions have been developed, and while these are promising the pulse lengths are long, and the distribution of susceptibilities must be known [121], or calculated iteratively online [122]. These are, however, important steps toward single-shot compensation of susceptibility dropout.

3.2.3 *Slice Orientation and Gradient Directions*

Divergent findings and recommendations for the optimum slice orientation for amygdala fMRI are due to the absence, until relatively recently, of an adequate description of signal loss and BS in the presence of field gradients [81, 119, 123].

In many early studies, quite nonisotropic voxels were used to achieve short TR while minimizing demands on scanner hardware, with slice thickness being substantially larger than the in-plane voxel size. Gradients across voxels were highest then, and signal loss most severe, if the direction of strongest field gradient was along the slice (through-plane) direction [124]. With many studies finding that the direction of the field vector across the amygdala was principally superior-inferior [125], this prescription precluded an axial orientation. As bilateral structures, the amygdala could be imaged in the same slice in the coronal but not the sagittal planes, leading to the coronal orientation being preferred by many [110].

The optimum imaging plane is also dependent on whether gradient compensation is used [81]. If so, through-plane gradients may be compensated for with a moderate gradient in the slice direction, although this will lead to a small decrease in BS in unaffected areas. The slice can then be orientated so that in-plane gradients are below the critical threshold for Type 2 signal loss. The value of this has been demonstrated in the orbitofrontal cortex [119] but the approach yields lower rewards in the amygdala region [126] as gradients are higher (making it more difficult to find a suitable value for compensation), and are more variable between subjects.

The simulations of Chen et al. [125] for the amygdala suggested that the maximum BS was to be achieved by orienting the slice direction perpendicular to the maximum gradient vector and the readout direction parallel to it, indicating an (oblique) coronal orientation with superior–inferior readout. The angle between the gradient vector and the superior–inferior direction was shown to vary widely between subjects (from $-7°$ to $+26°$ at 1.5 T, from $-5°$ to $+34°$ at 3 T), meaning that field gradients need to be mapped for each subject before measurement. This scheme also invokes distortions which are asymmetric about the midline (left–right). If erroneous conclusions about lateralization are to be avoided, residual distortions in the amygdala should be symmetric, requiring the phase–encode direction to be superior–inferior for coronal slices or anterior–posterior for axial slices.

As well as the direction of imaging gradients, the sign of phase-encode blips is important for signal loss and BS [123]. Encoding in

EPI can be either with a large positive phase-encode "prewinder" followed by a succession of small negative "blips," or a negative prewinder followed by positive blips. In homogeneous fields these schemes are equivalent, but we have seen that in the presence of susceptibility gradients echo positions are shifted away from the center of k-space, along the phase-encode axis. Positive and negative blip schemes have quite different properties, therefore, depending on whether the component of susceptibility gradient in the phase-encode direction is itself positive or negative [123]. The phase-encode direction (PE), slice angle, and z-shimming prepulse gradient moments (PP) that lead to maximum BS for EPI with otherwise standard EPI parameters (TE = 50/30 ms at 1.5 T/3 T, $3 \times 3 \times 2$ mm^3 voxels) have been measured throughout the brain by Weiskopf et al. at 1.5 and 3 T [126]. They define positive slice angles as being those in which, beginning from the axial plane, the anterior edge is tilted toward the feet, and a positive PE as being that in which the prewinder gradient points from the posterior to the anterior of the brain. In the amygdala they find that the highest BS is achieved with positive PE, a –45° slice tilt and a PP = +0.6 mT/m ms at 3 T, and positive PE, –45° slice tilt and PP = 0.0 mT/m ms at 1.5 T. These values led to a 14 % increase in BS at 3 T over a standard acquisition (with positive PE, a –0° slice tilt and a PP = –0.4 mT/m ms) but only 5 % at 1.5 T. This indicates that BS can be increased by selecting optimum geometry parameters and compensations gradients, although improvement is more modest than that which has been demonstrated with the more technically challenging or time-consuming strategies described in this chapter. The gradient and geometry values suggested in Weiskopf et al. [126] should be adopted for EPI with standard parameters at these field strengths. At other field strengths their analysis could be followed, or interpolated values adopted from the trends evident in that study.

3.2.4   Voxel Size

Among many solutions to the problem of signal loss in the anterior MTL, reduced voxel size was established very early as an effective means of mitigating susceptibility-related signal loss [127, 128]. Equation (2) describes how the rate of signal decay is reduced with voxel size by lowering field gradients across voxels. The effectiveness of this can be seen in the 4-T images of Fig. 4 over a range of resolutions, with T_2^* in the amygdala (measured with a multiple gradient-echo sequence with the same geometry as the EPI) increasing from 22 to 38 ms when the voxel size is reduced from 64 to 8 mm^3, with corresponding EPI signal increase apparent in the anterior MTL.

Reducing voxel size comes at the expense of temporal resolution (or brain coverage) and SNR. The relationship between image SNR and voxel volume, ΔV, is

$$\text{SNR} = V\sqrt{\frac{N_x N_y N_z}{\text{rBW}}}, \tag{3}$$

where Ni is the number of samples in direction i and rBW the receiver bandwidth [129]. The commonly held view that voxel volume is simply proportional to SNR is premised on changing the volume via the field of view [130], or that, in addition to increasing Nx and Ny by a factor f, (considering only in-plane resolution) receiver bandwidth is also increased by the same factor. If receiver bandwidth and field of view are held constant, however, then we see from Eq. (3) (because $N_x N_y N_z = k / V|_{\text{FOV}}$, where k is the total imaged volume) that SNR is proportional to the square root of the voxel volume, and SNR may be restored by downsampling high-resolution images. In this time-consuming scheme, partial k-space acquisition may be used to achieve the desired TE, SNR can be increased with multichannel coils, as has been validated for the MTL [115] and parallel imaging used to reduce an otherwise long TR.

While this analysis provides the basis for the dependence of image signal on imaging parameters, it neglects the effects of physiological noise. The most important measure of signal in this context is tSNR, which translates into the feasibility of detecting a specified signal change in fMRI [131] and has been shown to be useful in assessing the viability of amygdala fMRI in individual subjects [132]. In a study of optimum parameters for GE-EPI for 3-T amygdala EPI with a volume coil, a protocol with approximately 2-mm isotropic voxels was found to yield 60% higher tSNR than a protocol with standard parameters (with approximately 4-mm isotropic voxels) [117], despite having been measured at twice the receiver bandwidth. Additional gains with smaller voxels (thinner slices) were not large, because T_2^* had already increased to a value close to that in homogeneous regions. This is in concordance with models calculations which suggest that 2 mm represents the smallest voxel size that should be used for amygdala imaging providing the activated size is itself at least 2 mm [125].

There are many differences between the conditions and metrics of the methodological work cited and typical fMRI studies. It is encouraging, therefore, that these findings have been confirmed in the significance and extent of amygdala activation in fMRI experiments [133, 134].

In summary, small voxels should be used in high field strength studies in order to operate in a regime dominated by technical, rather than physiological noise. In inhomogeneous regions this results in reduced field gradients, reducing signal loss and echo shifts, making BS more uniform in the volume. Time-series SNR may be increased by using multichannel coils and downsampling small voxels.

3.2.5   *Echo Time*

Taking the simplest approach of matching effective echo time (TE_{eff}) to the T_2^* of the structures of interest in GE-EPI might be seen as being problematic in large voxel size acquisitions, with T_2^* s varying quite widely (e.g., between the amygdala and the fusiform face area). One solution is to use a multiecho sequence, in which the each time of each image is appropriate for regions with particular field gradients, as will be described in more detail in Sect. 3.2.8. A novel solution to matching TE_{eff} to T_2^* in the amygdala without sacrificing BS in more dorsal slices is to use an axial acquisition with slice-specific TE, demonstrated at 1.5 T with $TE_{eff} = 60$ ms in dorsal slices, $TE_{eff} = 40$ ms in ventral slices, and a transition zone with intermediate effective TE [135].

It should be remembered, though, that the maximum of BS is quite flat as a function of TE, and TE is itself not well defined in EPI. In the previous sections, we also saw that in-plane susceptibility gradients change local TE [81]. This exposes the limitation of the approach of simply reducing the TE_{eff} of the sequence. In the common, negative blip scheme, signal in the anterior MTL will in fact be shifted to a longer TE. Using a short TE_{eff} makes the sequence more prone to complete (type 2) signal loss.

This explains the experimental findings of Gorno-Tempini et al. [136] and Morawetz et al. [134]. In 2-T dual-echo EPI with large voxels, Gorno-Tempini et al. found that although signal loss was reduced at the short TE (26 ms) BOLD activation was significantly greater in the hippocampus at the longer TE (40 ms). Morawetz et al. [134] studied four EPI protocols in their efficacy at mapping amygdala activation, using variants with two different TE (27 and 36 ms) and slices thicknesses (2 and 4 mm), all with high in-plane resolution (2 mm). Activation results were poor in the 4-mm protocols, even at the shorter TE.

A more effective approach than reducing TE_{eff} is to reduce susceptibility gradients, and thereby signal dephasing and echo shifts, using the techniques described earlier; gradient compensation, selection of appropriate gradient direction and slice orientation, and the use of smaller voxels. This increases T_2^* in susceptibility-affected regions and, by reducing echo shifts, makes BS more homogeneous throughout the imaging volume. Conditions then approach those with a homogeneous static field, where BS is maximized by using $TE_{eff} = T_2^*$.

The increase in T_2^* in the amygdala with reduced voxel size is illustrated at 4 T in Fig. 4; from 22 ms in a $4 \times 4 \times 4$-mm acquisition to 38 ms in $2 \times 2 \times 2$-mm data, consistent with previous results at 3 T [117]. Likewise, increase in BS was illustrated in the Morawetz et al. study [134], in which robust amygdala activation was only detectable in the high-resolution acquisition.

3.2.6   *Parallel Imaging*

The previous sections have shown that many of the techniques which mitigate susceptibility-related signal loss in the amygdala, hypothalamus, and MTL are also time consuming, limiting either temporal resolution or brain coverage. This is undesirable where brain coverage cannot be reduced to the amygdala. Parallel imaging allows acceleration by undersampling k-space and using the sensitivity profiles of a number of receiver channel to reconstruct data without image fold-over [137, 138]. By this means it is possible to reduce TE_{eff}, which reduces susceptibility loss, and to reduce TR by the acceleration factor. Image distortions and echo shifts are likewise reduced by the acceleration factor so that even at the same effective TE as in a conventional acquisition, signal loss in the amygdala region is lower (Fig. 4, bottom right). The noise properties of images reconstructed from parallel acquisition lead to BS reductions of the order of 15–20 % in other regions, however [139].

The effectiveness of parallel imaging and suitable acceleration factors for the MTL have been studied by Schmidt et al. [140]. Statistical power in the study of MTL activation was higher in the parallel-acquisition data with an acceleration factor of 2 than in the acquisition without acceleration, but neither image quality nor statistical power improved with higher acceleration factors, as noise and reconstruction artifacts reduced tSNR prohibitively. Particular gains in BS can be made in the MTL using parallel imaging with a modest acceleration factor combined with high-resolution imaging [115]. Combining parallel imaging, high-resolution and high field has even allowed differential response of the hypothalamus to be recorded in response to funny as opposed to neutral stimuli at 3 T [141, 142], which could potentially be used to diagnose narcolepsy and cataplexy.

3.2.7 Flip Angle

The following is a consideration which is common to fMRI studies in all brain regions. The flip angle that should be used in a sequence is that which maximizes the signal with a particular experimental TR. In a spoiled gradient-echo sequence this is the Ernst angle, θ_E, given by

$$\theta_E = \arccos\left(e^{-\frac{TR}{T_1}} \right).$$

T_1 values can be taken from the literature, if available, or mapped in a single study of a representative group of subjects, mostly simply using an inversion recovery sequence and a range of inversion times. At high (3.0–4.0 T) and very high field (7.0 T or higher), dielectric effects lead to B_1 inhomogeneity, and flip angles achieved deviate from nominal values. Particularly at 7.0 T it is worthwhile to map the RF field [e.g., using the 180° signal null point using a simple spoiled gradient-echo sequence [143] to calibrate nominal flip angles].

3.2.8 Alternatives to 2D, Single-Shot, Gradient-Echo EPI

If multiple echo images are acquired following a single excitation, the range of TE_{eff} in these provides near-optimum BS for a number of regions [144, 145]. Images acquired at different TEs may be analyzed separately, or combined to maximize BOLD contrast-to-noise ratio [145]. Acquiring multiple images in a single shot also allows additional features to be built into the sequence, such as 3D gradient compensation, in which different combinations of compensation gradients are applied to each echo [146], leading to excellent signal recovery in the amygdala in the combined image [147]. Alternatively, the phase-encoding gradient polarity may be reversed to yield images with distortions in opposite directions, allowing for their correction [148].

Similar multiecho and compensation techniques have been applied to spiral acquisitions. A spiral-in trajectory has been shown to reduce signal loss compared to a conventional spiral–out scheme with the same TE, and SNR and BS could be increased with a spiral in–out scheme by combining images optimally from the two acquisitions [149]. A number of variants of this have been developed to further reduce susceptibility artifacts, including applying a z-shim gradient to the second echo [150] or subject-dependent slice-specific z-shims to both echoes [151].

A number of segmented methods are being developed to overcome the temporal constraints of multiecho and high-resolution acquisitions. In conventional segmented EPI, subsets of interleaved k-space lines are acquired after successive excitations. The higher phase-encode bandwidth leads to reduced distortions and smaller echo shifts, but the method is inherently slow and prone to motion and physiological fluctuations, as each image is built up over a number of TRs. In the MESBAC sequence, navigator echoes are acquired in both the readout and phase-encode directions between each segment. Multiple echoes are acquired with different amounts of compensation for each echo [152], and combined to give impressive signal in inferior frontal areas.

3.2.9   Summary

In the subsections of Sect. 3.2 we have looked at the influence of field strength, gradient compensation, slice orientation, voxel size, TE, and acquisition acceleration factor on susceptibility-related signal and BS reduction in the anterior MTL, as well as discussing some variants of multiecho and spiral schemes which have been tailored for this region. While the interdependent nature of EPI parameters and changing considerations at different field strength necessarily make some considerations complex, we would like to pick out two lines of approach presented here as being particularly effective, and clarify recommendations.

The first approach is high-field, high-resolution single-shot EPI with gradient compensation and acceleration. BOLD signal changes are greater at high field (3.0–4.0 T), and the tSNR advantages of high field strength are capitalized upon by measuring with small (circa 8-μl voxels), where thermal noise rather than physiological noise dominates. Measuring with small voxels reduces signal dephasing, making T_2^* more homogeneous. Shifts in local TE are also less, reducing Type 2 signal loss and increasing BOLD sensitivity. Moderate slice select gradient compensation and an oblique axial acquisition with a tilt between 20 and 45° (anterior slice edge toward the head) reduces in-plane gradients and echo shifts further. With susceptibility gradients reduced—evidenced by T_2^* values close to those in magnetically homogeneous regions— BS can be maximized by setting the $TE_{eff} = T_2^*$. The TE_{eff} can be reached using parallel imaging acceleration (e.g., factor 2), which further reduces both TE shifts and image distortion. Images acquired with these parameters have high signal in the anterior MTL, low distortion, and quite homogenous BS. Time-series SNR can be increased before statistical analysis by downsampling or smoothing images. This approach is attractive in that it may be achieved on most modern high field systems.

Not only the value of gradient compensation was discussed in Sect. 3.2.2, but also the high cost in temporal resolution, if images with a number of compensation gradients are acquired. The second approach we wish to highlight involves the application of a range of compensation gradients to each of a number of echoes acquired after a single excitation, so reducing the time penalty. Both the multiecho echo-planar [146] and multiecho spiral acquisitions [151] described in Sect. 3.2.8 have been shown to be effective in reducing susceptibility-related signal loss in the anterior MTL.

3.3 Correction Methods

3.3.1 Distortion Correction with the Field Map and Point-Spread Function Methods

The field map (FM) method was first described by Weisskoff and Davis [153] and developed by Jezzard and Balaban [154]. In Sect. 3.1.2 we saw that distortion in EPI is only significant in the phase-encode direction and that the number of pixels by which signal is mislocated is equal to the local field offset divided by the bandwidth per pixel in the phase-encode direction. In the fieldmap method, static magnetic field deviations, ΔB, are calculated from the phase difference, $\Delta \phi$, between two scans with TE separated by ΔTE (or a dual-echo scan), using the relation $\Delta B = 2\pi\gamma$ $TE\Delta\varphi$. This map is distorted (*forward-warped*) to provide a map of the voxel shifts required to reverse the distortion at each EPI location. Gaps in the corrected image are filled by interpolation.

While undemanding from the sequence perspective, considerable postprocessing is required to produce FMs that do not contain errors. Phase imaging is only capable of encoding phase values in a 2π range, with values outside this range being aliased, causing "wraps" in the image. These can be removed in the spatial domain using a number of freely available algorithms (e.g., PRELUDE [155] or ΠUN [156]), or by examining voxel-wise phase evolution in time if three or more echoes are acquired [157]. If imaging is being carried out with a multichannel radiofrequency receive coil, phase images created via the sum-of-squares reconstruction [158] will show nonphysical discontinuities from arbitrary phase offsets between the coil channels (incongruent wraps) unless these offsets are removed [159, 160]. Alternatively, images from channels may be processed separately and individual FMs, weighted by coil sensitivities, combined. In 2D spatial unwrapping, additional global, erroneous 2π phase changes are occasionally inferred between TE when the algorithm begins to unwrap from different sides of a phase wrap at the two TE. In multichannel imaging, these slice phase shifts may be identified by examining the consistency between coil channels [161], as may unreliable voxels at the image edge and in regions of high-field gradient. The FM may finally need to be smoothed to remove high frequency features and dilated to ensure that it extends to the periphery of the brain.

In the point spread function (PSF) approach [162] applied to distortion correction [163], the imaging sequence is similar to EPI, but with the initial phase prewinder gradient replaced by a phase gradient table, the values are applied in a loop. The PSF of each voxel is the Fourier transform of the acquired data, and the displacement of the voxel is the shift of the center of the PSF (e.g., if the center of this is at zero additional phase, this corresponds to no local field offset). For one major scanner manufacturer, this method has been robustly implemented with the flexibility to be used for parallel imaging with high acceleration factors [164].

The FM and PSF methods have been compared at 1.5 T [163]. The PSF was found to be generally superior, although some conclusions were based on deficiencies in FMs in regions of high field gradient which may be improved upon.

The effectiveness of the two methods in correcting larger distortions at 4.0 T is shown in Fig. 5, focusing on a section

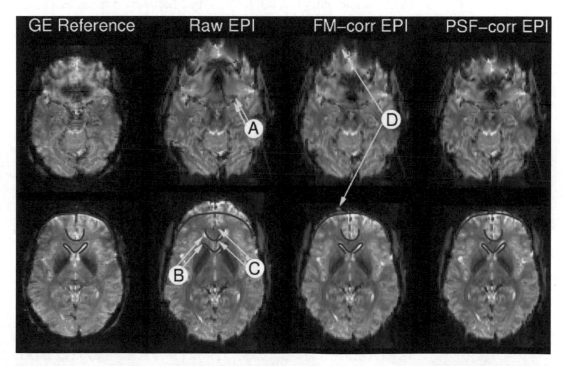

Fig. 5 Distortion correction of echo-planar imaging (EPI) at high field (4.0 T). A comparison of field-map (column 3) and point-spread function (column 4) correction of distortion in EPI (column 2) at the level of the amygdala (*top row*) compared to a more dorsal section (*bottom row*). Salient features have been copied from a gradient-echo geometric reference scan (column 1)

through the amygdala (top row), and comparing this with the situation in a more dorsal slice (bottom row). Raw and corrected EPIs are compared to a gradient-echo reference which has the same (subvoxel) distortion in the readout direction, but no distortion in the phase-encode direction. The distortion at the anterior boundary of the amygdala (A) is circa 3 mm—moderate compared to the displacement of the ventricles (9 mm at B) and the frontal gray-white matter border indicated at C (12 mm). If the multiplicity of phase information available from multichannel coils is used in the FM method [161], both FM and PSF methods perform very well in all areas, with only minor errors at the periphery of the FM-corrected images due to residual field map inaccuracies at those locations (at D, not present in the PSF-corrected images).

The choice of correction method is often a pragmatic one based on which is more robustly and conveniently implemented.

3.3.2 Correction of Physiological Artifacts

Physiological fluctuation in a sequence of gradient-echo images can be corrected using a navigator echo technique [165]. A single echo is acquired before the encoding scheme is begun and used to amend the phase changes in the image data which arise from susceptibility effects. This "global" correction approach, using the central k-space point only, can be extended to 1D [166] and 2D [167]. These methods are effective, but have as drawbacks an increase in TR.

To avoid them being aliased in EPI time series, respiratory fluctuations (circa 0.2–0.3 Hz) and cardiac fluctuations (circa 1 Hz) would need to be sampled at least at 2 Hz. That is, the TR of the sequence would need to be 500 ms or less. Typical TRs in whole-brain fMRI are 1–4 s, and the previous sections have indicated that many of the strategies that should be implemented to improve data quality in fMRI for emotion studies lead to longer repetition times. Respiratory and cardiac fluctuations will normally be aliased, then, and not generally into a particular frequency band [168]. Simple band-pass filtering is therefore not generally possible; although a range of alternative correction methods have been developed.

A class of correction methods requires additional physiological measurement to be made concurrent with the fMRI time-series, using a respiration belt to monitor breathing and an electrocardiogram or pulse oximeter to monitor heart rate. Applied in image space, the RETROICOR correction method involves plotting pixels according to their acquisition time within the respiratory cycle (classified also by respiration depth) and subtracting a fit to fluctuations over the cycle [169]. Despite the many reasons why physiological artifacts are expected to particularly affect amygdala fMRI, their correction with RETROICOR was found to bring only modest improvements in group fMRI results in an emotion processing task; up to 13% in t statistic values depending on the degree of smoothing [170]. Those improvements were mostly due to correction of cardiac effects. Recent findings that cardiac rate changes lead to signal changes of similar size to the effect due to cardiac action itself [88], which are not modeled in the RETROICOR approach, suggest that further gains are possible.

Modeling physiological fluctuations [171] by including measured signal as "Nuisance Variable Regressors (NVRs)" is a convenient alternative to fitting and removing them. A detailed examination of these and other sources of noise showed respiratory-induced noise particularly at the edge of the brain, larger veins and ventricles, and cardiac-induced noise focused on the middle cerebral artery and Circle of Willis, close to the amygdala [168], which could be well modeled.

A number of image-based methods for physiological artifact correction have been developed, which do not require physiological monitoring data. Physiological fluctuations can be modeled with NVRs based on ventricular and white matter ROI values [172]. Alternatively, the data can be decomposed using ICA (e.g., MELODIC [173] or GIFT [174]) and components relating to physiological processes identified with automated or semiautomated methods. These can be based on experimental thresholds [175], statistical testing [176], automatic thresholding [177], or supervised classifiers [178]. Once identified, these components can be removed from the data. While in their infancy, these methods are very promising, particularly for the ventral brain. Tohka et al., for instance, demonstrated marked Z-score increases in frontal ventral regions and other areas close to susceptibility artifacts.

3.3.3 Correction of Stimulus-Correlated Motion Artifacts

In patient group studies, Bullmore et al. [179] have shown the need to compare the extent to which SCM explains variance between the groups, and suggest that this be identified using an analysis of covariance (ANCOVA). Without this approach, differences between the groups arising from higher SCM in the schizophrenic group in their study would have been attributed to differential activation in response to the task.

In the example of Fig. 2 (left), realignment of the time series in the motion-only replication did not substantially reduce the amygdala SCM artifact (right), but including identified motion parameters in the model as NVRs was effective [168, 180]. Alternatively, a boxcar NVR corresponding to presentation and response periods can be included in the model [181]. This and a number of other studies [182] have shown that the temporal shift in response introduced by the hemodynamic response function (HRF) makes it possible to separate motion from activation for short presentation periods, making event-related designs less sensitive to motion than block designs.

4   Summary and Discussion

Emotional neuroimaging is a rapidly expanding area that provides an interface between neurobiological work and psychophysiological emotion research. One important view that has emerged from the area of behavioral neuroscience is that emotional processes play a central role in the adaptive modulation of perceptual encoding, learning and memory, attention, decision-making, and control of action [9]. Many of neuroimaging studies have demonstrated that amygdala activation, for example, modulates attention and memory storage in other brain regions such as the hippocampus, striatum, and neocortex. Such interactions may occur as facilitations or modulations of neurocognitive function at several levels of processing. Conversely, recent work has shown that the organism is prevented from excessive emotional activation not only by low-level habituation or negative feedback mechanisms but also as a result of protective inhibition processes. Diverse behavioral and cognitive strategies have been identified that modulate and downregulate the ongoing emotion process [6]. The modulating effects on emotional arousal during an emotional episode such as rejection (venting and disengagement) or accommodation (relaxation, distraction, reconceptualization, rationalization, or reappraisal) deserve further inspection with respect to the involved neural mechanisms.

Although important advances have been made in the area of human emotion perception, learning, and autonomic conditioning, research has typically been limited to a small number of primary and mostly negative emotions such as fear, anger, or disgust. Limiting the range of investigated categories (neglecting shame, guilt, interest, etc.), dimensions (neglecting positive emotions such as care, support, etc.) and behavioral procedures does not do justice to the complexity of the multistage emotional appraisal process described above [29]. It is equally important but more difficult to identify the correlates of complex emotions such as those resulting from beliefs, preferences, predispositions, or interpersonal exchange. Not only the social dimensions such as untrustworthiness or dishonesty [183, 184], but also positive aspects such as social fairness [185], trust, and supportiveness play a role. Moreover, an understanding of modulating sociocultural influences is essential for a comprehensive conceptualization of human emotion [29].

Current neuroimaging research on emotion can be described as an ongoing construct validation process [186], which draws upon convergent and divergent associations of local activation variables and psychological constructs. The experimental measures (operationalizations of psychological constructs) are expected to be correlated with regional brain activations. It is evaluated whether topographically distinct patterns of activation in a certain region consistently predict engagement of different processes (for an example in the area of cognitive processing, *see* Ref. [187]). Indicators of a different construct are expected to correlate with activations of different areas. This corresponds to the well-known double dissociation strategy that inspects task by localization interactions in neuropsychology [64].

This validation process typically starts at a relatively broad construct level and proceeds downward in the hierarchy of constructs to finally specify within-systems constructs. Previous studies have demonstrated a relatively high cross-laboratory repeatability of emotional brain activation patterns at a higher systems level. At lower levels, however, the reliability of psychological or activation data may decline depending on limitations of the instruments.

Nonetheless, high-field fMRI scanners permit an improved discrimination of activations, for example, within the different subnuclei of the amygdala [5]. It is evident that increased discrimination on the neural side must be accompanied by a refined technology to assess more fine-grained emotional constructs on the behavioral side.

Neuropsychological construct validation requires additional physiological data to obtain some kind of convergent information about the indicator variable. At the neurophysiological level, the perfusion mechanisms has been elucidated by combining the greater spatial resolution of fMRI with the real-time resolution of intracortical local field ERP (LFP) recordings. The neurophysiological coupling mechanisms of neural activity and the BOLD response can thus be assessed [188]. An application of both fMRI methods and electrophysiological approaches (e.g., surface and deep electrode recordings from limbic brain structures) is useful [189]. The combination of brain perfusion changes and electrophysiological correlates of oscillatory coupling will foster the understanding of the neural interaction processes within frontal and temporal networks [190].

On the level of the autonomic nervous system, multivariate coregistrations of psychophysiological response patterns including emotion modulated startle, heart rate variability, or cortisol secretion alleviate the validation of experimentally induced emotions or presence of specific emotional disorders.

Emotional neuroimaging has continuously profited from improvement in scanning techniques and the adaptation and standardization of signal processing strategies. However, this area has not only benefited from the diverse contributions of its subdisciplines but also inherited their methodological problems. An inspection of brain imaging studies of emotion showed that measurement quality may be influenced by many factors: by a rapid and differential habituation of responses to emotional stimuli in some regions; by artifacts of certain signal scaling techniques that are applied by default; by situational or state-dependent influences; and by insufficient validation of the emotion to be elicited (manipulation check). Interindividual differences of emotional regulation behavior appear to modulate event-related reactions during the time course of the experiment.

Some of the many approaches to reducing signal loss in EPI in the anterior MTL have been outlined here, as well as some of the methods for identifying and correcting artifacts arising from SCM, distortion, and physiological artifacts. Despite the gravity of the problem and the effectiveness of some of these strategies, the overwhelming majority of fMRI studies of the emotions use the same measurement protocols and analysis methods as have been applied to study cognitive function over the last decade.

Combining many of the simpler strategies described here—high field strength, small voxel volumes, partial k-space acquisition with the correction of physiological and SCM artifacts—allows reliable results to be achieved in the anterior MTL [191]. Figure 6 demonstrates such an example; the detection of subtle differences in amygdala activation between explicit and implicit emotion processing [192].

Moreover, new research designs and analysis methods such as Structural Equation Modeling or Dynamic Causal Modeling are now available to inspect the effective or causal connectivity that, for example, permits the PFC to modulate amygdala activity [62]. The influences of individual brain regions on each another can also been studied by combining functional brain imaging with the lesion approach or transcranial magnetic stimulation [193].

We have raised a number of caveats that highlight some of the limitations of emotion assessment in a scanner environment. As has been argued above, a lack of representativeness must be noted, that is, emotion includes a much broader conceptual network than

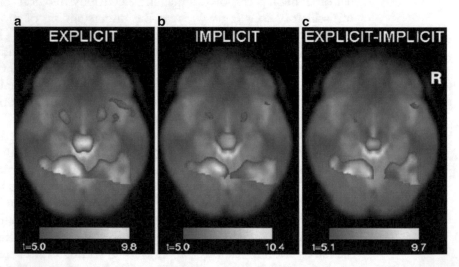

Fig. 6 High-resolution imaging detailed in this chapter allows the acquisition of low-artifact echo-planar imaging (EPI) and allows subtle processing effects to be distinguished. Group results from 29 subjects for the conditions (**a**) emotion recognition (**b**) implicit emotion processing (age discrimination) and (**c**) the difference between the two conditions (3.0 T). Results, showing activation in the amygdala and fusiform gyrus (as well as cerebellum and brainstem) are overlaid on mean EPI and thresholded at $p = 0.05$, family-wise error corrected. Reprinted from [192], with permission

currently covered by neuroimaging research. Thus, generalizations to other areas of functioning remain difficult. Representative designs are needed that pay greater attention to high-level strategies that depend on sociocultural factors and initiate, modulate, or regulate emotions. Moreover, the representativeness of results is limited due to small, selected, and poorly described study groups. Finally, since emotion elicitation in the scanner has been highly artificial, the power to predict emotions outside the neuroimaging context remains questionable. An ecological functional brain imaging approach that includes natural behaviors and environmental contexts of presentation may help to obtain a more representative view of real-life emotions.

Subject-specific mechanisms regulating the strength and temporal pattern of response to emotional stimuli and the balance of excitatory and inhibitory processes are of particular interest. The variability of the BOLD response between trials and across the time course of the experiment needs to be explained. Future research may therefore examine individual and group differences with a view to resolving inconsistencies in the literature [5, 12, 77]. Investigations into personality disorders or psychiatric diseases will provide further insight into the dispositional factors modifying the response to situational stressors. Paradigms specifically adapted to the investigated disorder may help to identify prefrontal dysfunction and associated failure to tonically inhibit amygdala output or to recognize safety signals eventually inducing sympathetic overactivity [194]. It may be that—as is the case in motor tasks—a large proportion of the intertrial variation not only in the behavioral response [102], but also in the BOLD signal [101] is explained by fluctuations in underlying RSNs.

Eliciting emotions in the environment of an imaging scanner remains a highly artificial process. This raises the question as to the predictive value of current neuroimaging data for explaining the emotional modulations in real-life contexts. This is particularly important for applied areas such as psychotherapy and coping research. Thus, in addition to identifying the neurobiological basis of emotional regulation behavior, the generalizability or predictive validity of imaging data for real-life emotions should be systematically evaluated.

5 Conclusions

Neuroimaging has replicated and extended earlier findings of neuropsychological studies in brain damaged subjects. It has significantly contributed to unraveling the organization of neural systems subserving the different components of emotional stimulus-response mediation along the neuraxis in healthy human subjects. Improved operational definitions and paradigms have contributed to differentiating subcomponents of emotional functions such as, for example, perceptual decoding, anticipation, associative learning, awareness, and response mediation. However, despite obvious advances, a comprehensive model integrating the diverse emotional behaviors on the basis of involved cerebral mechanisms is still unavailable. Moreover, the interpretation of findings is complicated by technical and methodological difficulties.

Research advances not only depend upon the technical refinement of imaging methodology but also on the improvement of behavioral procedures and measurement models. Neuropsychological construct validation procedures imply that an increase of localization precision of the imaging technology would also require an enhanced precision on the side of behavioral operationalizations. However, this seems not to be case as many studies still use unsophisticated stimulus materials or global instructions involving multiple or undefined subfunctions. As much as relatively global operationalizations are applied, however, the obtained neuropsychological correlations (for example, regarding activations of the PFC) will remain incomprehensible.

We have suggested here the framework of a lense-type assessment model, wherein activations in well-characterized neural structures may be used as predictors of particular emotional processes. According to this, a hierarchy of latent constructs constitutes the behavioral level, an idea, which is largely accepted in psychology. On the level of brain activity, patterns or families of topographically distinct activity can be identified in a similar way and used as a predictor of behavioral function. Following the assumptions of a methodological parallelism, neuropsychological construct validation procedures make uses of this framework of activity–behavior associations on different levels of the hierarchy. It can be extrapolated from multivariate personality theory, that the prediction of behavior will only be successful if activation measures and psychological data are analyzed on a similar level of generality or aggregation.

In view of the complexities of emotional regulation behavior in human subjects, it is equally important to advance assessment theory, psychological conceptualization, and behavioral methodology [29]. Future work should therefore more closely inspect issues related to model construction, symmetry of neural and behavioral variables, and their aggregation levels. Multidisciplinary approaches that combine improvement in brain activation measurement with enhanced psychological data theory may thus foster construct validity, reliability, and predictive power of emotional neuroimaging.

Knowledge pertaining to the localization of brain activations and its functional connectivity is also an important input to inform and constrain cognitive theories of emotion psychology. Thus, insights from the brain will thus help to explain the incoherences of psychophysiological, behavioral, and subjective indicators of emotion that are so frequently observed in psychophysiological studies. Activation data may also help to establish models that possess a better "breakdown compatibility," that is, power to predict behavioral change as a consequence of brain damage.

The introduction of structural/connectional and functional data has considerably bolstered scientific construct validation processes in the affective neurosciences and emotion psychology. Topographically distinct activity patterns are increasingly identified that possess a certain incremental validity, that is, an increasing power to predict the individual dynamics of emotional regulation behavior. Establishing a representative and valid model of emotional functioning is a necessary precondition for many areas of application such as the categorization of patients with emotional disorders and the assessment of psychotherapy.

Greater attention to methodological issues may help to bring more rigors to experimentation in the field of emotional neuroimaging, promote interdisciplinary research, and alleviate cross-laboratory replication. A wealth of approaches have been presented to countering BS loss in the amygdala, many of which are available as standard on commercial scanners or simply require the adoption of suitable imaging parameters [117, 125, 134]. Also, in the absence of a measurement theory that describes validated procedures or instruments for assessing emotional constructs, single findings cannot be trusted. Although absence of validation is acceptable for early stages of the research cycle, current emotional neuroimaging work has only just begun to approach the confirmatory stage. To establish confidence in the suggested models, additional efforts are required to empirically validate assessment strategies and instrumentation.

Acknowledgments

The author's own work reported in this chapter was supported by the Austria FWF grant P16669-B02, grant 11437 from the Austrian National Bank, the government of the Provincia Autonoma di Trento, Italy, the private foundation Fondazione Cassa di Risparmio di Trento e Rovereto, the University of Trento, Italy, and by grant Pe 499/3–2 from the Deutsche Forschungsgemeinschaft to M.P. J. Jovicich is thanked for helpful comments.

References

1. Büchel C, Dolan RJ (2000) Classical fear conditioning in functional neuroimaging. Curr Opin Neurobiol 10:219–223

2. Phan KL, Wager T, Taylor SF, Liberzon I (2002) Functional neuroanatomy of emotion: a meta-analysis of emotion activation studies in PET and fMRI. NeuroImage 16:331–348

3. Dolan RJ, Vuilleumier P (2003) Amygdala automaticity in emotional processing. Ann N Y Acad Sci 985:348–355

4. Wager TD, Phan KL, Liberzon I, Taylor SF (2003) Valence, gender, and lateralization of functional brain anatomy in emotion: a meta-analysis of findings from neuroimaging. NeuroImage 19:513–531

5. Zald DH (2003) The human amygdala and the emotional evaluation of sensory stimuli. Brain Res Brain Res Rev 41:88–123

6. Ochsner KN, Gross JJ (2005) The cognitive control of emotion. Trends Cogn Sci 9:242–249

7. Panksepp J (1998) Affective neuroscience: the foundations of human and animal emotions. Oxford University Press, Oxford

8. Davidson RJ, Jackson DC, Kalin NH (2000) Emotion, plasticity, context, and regulation: perspectives from affective neuroscience. Psychol Bull 126:890–909

9. Dolan RJ (2002) Emotion, cognition, and behavior. Science 298:1191–1194

10. Breiter HC, Rauch SL (1996) Functional MRI and the study of OCD: from symptom provocation to cognitive-behavioral probes of cortico-striatal systems and the amygdala. NeuroImage 4:127–138

11. Johnson PA, Hurley RA, Benkelfat C, Herpertz SC, Taber KH (2003) Understanding emotion regulation in borderline personality disorder: contributions of neuroimaging. J Neuropsychiatry Clin Neurosci 15:397–402

12. Hamann S, Canli T (2004) Individual differences in emotion processing. Curr Opin Neurobiol 14:233–238

13. Paquette V, Levesque J, Mensour B et al (2003) Change the mind and you change the brain: effects of cognitive-behavioral therapy on the neural correlates of spider phobia. Neuroimage 18:401–409

14. Popma A, Raine A (2006) Will future forensic assessment be neurobiologic? Child Adolesc Psychiatr Clin N Am 15:429–444

15. Bräutigam S (2005) Neuroeconomics – from neural systems to economic behaviour. Brain Res Bull 67:355–360

16. Walter H, Abler B, Ciaramidaro A, Erk S (2005) Motivating forces of human actions: neuroimaging reward and social interaction. Brain Res Bull 67:368–381

17. Adolphs R, Tranel D, Damasio H, Damasio A (1994) Impaired recognition of emotion in facial expressions following bilateral damage to the human amygdala. Nature 372:669–672

18. Broks P, Young AW, Maratos EJ et al (1998) Face processing impairments after encephalitis: amygdala damage and recognition of fear. Neuropsychologia 36:59–70

19. Sprengelmeyer R, Young AW, Schröder U et al (1999) Knowing no fear. Proc R Soc Lond B Biol Sci 266:2451–2456

20. Adolphs R, Gosselin F, Buchanan T, Tranel D, Schyns P, Damasio A (2005) A mechanism for impaired fear recognition after amygdala damage. Nature 433:68–72

21. Adolphs R, Tranel D, Hamann S et al (1999) Recognition of facial emotion in nine individuals with bilateral amygdala damage. Neuropsychologia 37:1111–1117

22. Schmolck H, Squire LR (2001) Impaired perception of facial emotions following bilateral damage to the anterior temporal lobe. Neuropsychology 15:30–38

23. LeDoux JE (1995) Emotion: clues from the brain. Annu Rev Psychol 46:209–235

24. LaBar KS, Gatenby JC, Gore JC, LeDoux JE, Phelps EA (1998) Human amygdala activation during conditioned fear acquisition and extinction: a mixed-trial fMRI study. Neuron 20:937–945

25. Morris JS, Büchel C, Dolan RJ (2002) Parallel neural responses in amygdala subregions and sensory cortex during implicit fear. Neurobiol 12:169–177

26. Büchel C, Morris J, Dolan RJ, Friston KJ (1998) Brain systems mediating aversive conditioning: an event-related fMRI study. Neuron 20:947–957

27. Krech D (1950) Dynamic systems as open neurological systems. Psychol Rev 57:345–361

28. Cacioppo JT, Tassinary LG, Berntson GG (eds) (2000) Handbook of psychophysiology. Cambridge University Press, Cambridge

29. Scherer KR, Peper M (2001) Psychological theories of emotion and neuropsychological research. In: Gainotti G (ed) Handbook of neuropsychology. Vol. 5: Emotional behavior and its disorders, 2nd edn. Elsevier, Amsterdam, pp 17–48

30. Coan JA, Allen JJB (eds) (2007) Handbook of emotion elicitation and assessment. Oxford University Press, New York, NY

31. Peper M, Vauth R (2008) Socio-emotional processing competences: assessment and clinical application. In: Vandekerckhove M, von Scheve C, Ismer S, Jung S, Kronast S (eds) Regulating emotions (ch. 9). Wiley, Hoboken, NJ

32. Kerlinger FN, Lee HB (2000) Foundations of behavioral research (4th edition). Harcourt College Publishers, Fort Worth, TX

33. Scherer KR (1993) Neuroscience projections to current debates in emotion psychology. Cogn Emot 7:1–41

34. Scherer KR (2000) Psychological models of emotion. In: Borod J (ed) The neuropsychology of emotion. Oxford University Press, Oxford, pp 137–162

35. Scherer KR (1999) Appraisal theories. In: Dalgleish T, Power M (eds) Handbook of cognition and emotion. Wiley, Chichester, pp 637–663

36. Peper M, Fahrenberg J (2008) Psychophysiologie. In: Sturm W, Herrmann M, Münte TF (eds) Lehrbuch der Klinischen Neuropsychologie. Spektrum Akademischer Verlag, Heidelberg

37. Stemmler G, Fahrenberg J (1989) Psychophysiological assessment: conceptual, psychometric, and statistical issues. In: Turpin G (ed) Handbook of clinical psychophysiology. Wiley, Chichester, pp 71–104

38. Stemmler G (1992) Differential psychophysiology: persons in situations. Springer, Heidelberg

39. Johnsen BH, Thayer JF, Hugdahl K (1995) Affective judgment of the Ekman faces: a dimensional approach. J Psychophysiol 9:193–202

40. Peper M, Irle E (1997) The decoding of emotional concepts in patients with focal cerebral lesions. Brain Cogn 34:360–387

41. Ekman P (1994) Strong evidence for universals in facial expressions. Psychol Bull 115:268–287

42. Russell JA (1994) Is there universal recognition of emotion from facial expressions? Psychol Bull 115:102–141

43. Stemmler G (1998) Emotionen. In: F Rösler (Ed) Enzyklopädie der Psychologie: Themenbereich C Theorie und Forschung, Serie 1 Biologische Psychologie, Band 5 Ergebnisse und Anwendungen der Psychophysiologie. Göttingen: Hogrefe, pp 95–163

44. Scherer KR (1984) On the nature and function of emotion: a component process approach. In: Scherer KR, Ekman P (eds) Approaches to emotion. Erlbaum, Hillsdale, NJ, pp 203–317

45. Lacey JI (1967) Somatic response patterning and stress: some revisions of activation theory. In: Appley MH, Trumbull R (eds) Psychological stress: issues in research. Appleton-Century Crofts, New York, NY, pp 14–42

46. Peper M (2000) Awareness of emotions: a neuropsychological perspective. Adv Consciousness Stud 16:245–270

47. Lang P, Rice DG, Sternbach RA (1972) The psychophysiology of emotion. In: Greenfield NS, Sternbach RA (eds) Handbook of psychophysiology. Holt, New York, NY, pp 623–643

48. Fahrenberg J (1983) Psychophysiologische Methodik. In: Groffman K-J, Michel L (eds) Enzyklopädie der Psychologie: Themenbereich B Methodologie und Methoden, Serie II Psychologische Diagnostik, Band 4 Verhaltensdiagnostik. Hogrefe, Göttingen, pp 1–192

49. Lazarus RS (1991) Emotion and adaptation. Oxford University Press, New York, NY

50. Parkinson B, Totterdell P (1999) Classifying affect-regulation strategies. Cogn Emot 13:277–303

51. Gross JJ, Levenson RW (1991) Emotional suppression: physiology, self-report, and expressive behavior. J Pers Soc Psychol 64:970–986

52. Lazarus RS, Folkman S (1984) Stress, appraisal, and coping. Springer, New York, NY

53. Gross JJ (1998) The emerging field of emotion regulation: an integrative review. Rev Gen Psychol 2:271–299

54. Whalen PJ, Rauch SL, Etcoff NL, McInerney SC, Lee MB, Jenike MA (1998) Masked presentations of emotional facial expressions modulate amygdala activity without explicit knowledge. J Neurosci 18:411–418

55. Pessoa L, Japee S, Sturman D, Ungerleider LG (2006) Target visibility and visual awareness modulate amygdala responses to fearful faces. Cereb Cortex 16:366–375

56. McIntosh AR, Gonzalez-Lima F (1994) Network interactions among limbic cortices, basal forebrain, and cerebellum differentiate a tone conditioned as a Pavlovian excitor or inhibitor: fluorodeoxyglucose mapping and covariance structural modeling. J Neurophysiol 72:1717–1733

57. Büchel C, Friston KJ (1997) Modulation of connectivity in visual pathways by attention: cortical interactions evaluated with structural equation modelling. Magn Reson Imaging 15:763–770

58. Büchel C, Friston KJ (2000) Assessing interactions among neuronal systems using functional neuroimaging. Neural Netw 13:871–882

59. Friston KJ, Harrison L, Penny W (2003) Dynamic causal modelling. NeuroImage 19:1273–1302

60. Vuilleumier P, Richardson MP, Armony JL, Driver J, Dolan RJ (2004) Distant influences of amygdala lesion on visual cortical activation during emotional face processing. Nat Neurosci 7:1271–1278

61. Friston KJ (1994) Functional and effective connectivity in neuroimaging: a synthesis. Hum Brain Mapp 2:56–78

62. Ochsner KN, Ray RD, Cooper JC et al (2004) For better or for worse: neural systems supporting the cognitive down- and up-regulation of negative emotion. Neuroimage 23:483–499

63. Fahrenberg J, Peper M (2000) Psychophysiologie. In: Sturm W, Herrmann M, Wallesch CW (eds) Lehrbuch der neuropsychologie, chapter 1. 10. Swets and Zeitlinger, Amsterdam, pp 154–68

64. Passingham RE, Stephan KE, Kötter R (2002) The anatomical basis of functional localization in the cortex. Nat Rev Neurosci 3:606–616

65. Büchel C, Dolan RJ, Armony JL, Friston KJ (1999) Amygdala-hippocampal involvement in human aversive trace conditioning revealed through event-related functional magnetic resonance imaging. J Neurosci 19:10869–10876

66. Foerster F (1995) On the problems of initial-value-dependencies and measurement of change. J Psychophysiol 9:324–341

67. Peper M, Herpers M, Spreer J, Hennig J, Zentner J (2006) Functional neuroimaging studies of emotional learning and autonomic reactions. J Physiol Paris 99:342–354

68. Cardinal RN, Parkinson JA, Hall J, Everitt BJ (2002) Emotion and motivation: the role of the amygdala, ventral striatum, and prefrontal cortex. Neurosci Biobehav Rev 26:321–352

69. O'Doherty J, Dayan P, Schultz J, Deichmann R, Friston K, Dolan RJ (2004) Dissociable roles of ventral and dorsal striatum in instrumental conditioning. Science 304:452–454

70. Grandjean D, Sander D, Pourtois G et al (2005) The voices of wrath: brain responses to angry prosody in meaningless speech. Nat Neurosci 8:145–146

71. Critchley H, Daly E, Phillips M et al (2000) Explicit and implicit neural mechanisms for processing of social information from facial expressions: a functional Magnetic Resonance Imaging study. Hum Brain Mapp 9:93–105

72. Hariri S, Bookheimer SY, Mazziotta JC (2000) Modulating emotional responses: effects of a neocortical network on the limbic system. Neuroreport 11:43–48

73. Furmark T, Fischer H, Wik G, Larsson M, Fredrikson M (1997) The amygdala and individual differences in human fear conditioning. Neuroreport 8:3957–3960

74. Canli T, Zhao Z, Desmond JE, Kang E, Gross J, Gabrieli JDE (2001) An fMRI study of personality influences on brain reactivity to emotional stimuli. Behav Neurosci 115:33–42

75. Canli T, Sivers H, Whitfield SL, Gotlib IH, Gabrieli JDE (2002) Amygdala response to happy faces as a function of extraversion. Science 296:2191

76. Schienle A, Schafer A, Stark R, Walter B, Vaitl D (2005) Relationship between disgust sensitivity, trait anxiety and brain activity during disgust induction. Neuropsychobiology 51:86–92

77. Canli T, Amin Z (2002) Neuroimaging of emotion and personality: scientific evidence and ethical considerations. Brain Cogn 50:414–431

78. Small SL, Nusbaum HC (2004) On the neurobiological investigation of language understanding in context. Brain Lang 89:300–311

79. Fahrenberg J, Myrtek M (eds) (2001) Progress in ambulatory assessment. Hogrefe and Huber, Seattle, WA

80. Adolphs R (2002) Neural systems for recognizing emotion. Curr Opin Neurobiol 12:169–177

81. Deichmann R, Josephs O, Hutton C, Corfield DR, Turner R (2002) Compensation of susceptibility-induced BOLD sensitivity losses in echo-planar fMRI imaging. NeuroImage 15:120–135

82. Yablonskiy DA (1998) Quantitation of intrinsic magnetic susceptibility-related effects in a tissue matrix. Phantom study. Magn Reson Med 39:417–428

83. Lipschutz B, Friston KJ, Ashburner J, Turner R, Price CJ (2001) Assessing study-specific regional variations in fMRI signal. Neuroimage 13:392–398

84. Toga AW, Thompson PM (2001) Maps of the brain. Anat Rec 265:37–53

85. Hutton C, Bork A, Josephs O, Deichmann R, Ashburner J, Turner R (2002) Image distortion correction in fMRI: a quantitative evaluation. NeuroImage 16:217–240

86. Raj D, Paley DP, Anderson AW, Kennan RP, Gore JC (2000) A model for susceptibility artefacts from respiration in functional echo-planar magnetic resonance imaging. Phys Med Biol 45:3809–3820

87. Windischberger C, Langenberger H, Sycha T et al (2002) On the origin of respiratory artifacts in BOLD-EPI of the human brain. Magn Reson Imaging 20:575–582

88. Wise RG, Ide K, Poulin MJ, Tracey I (2004) Resting fluctuations in arterial carbon dioxide induce significant low frequency variations in BOLD signal. Neuroimage 21:1652–1664

89. Dagli MS, Ingeholm JE, Haxby JV (1999) Localization of cardiac-induced signal change in fMRI. Neuroimage 9:407–415

90. Critchley HD, Rotshtein P, Nagai Y, O'Doherty J, Mathias CJ, Dolan RJ (2005) Activity in the human brain predicting differential heart rate responses to emotional facial expressions. Neuroimage 24:751–762

91. Frysinger RC, Harper RM (1989) Cardiac and respiratory correlations with unit discharge in human amygdala and hippocampus. Electroencephalogr Clin Neurophysiol 72:463–470

92. Shmueli K, van Gelderen P, de Zwart JA et al (2007) Low-frequency fluctuations in the cardiac rate as a source of variance in the resting-state fMRI BOLD signal. Neuroimage 38:306–320

93. Cohen MA, Taylor JA (2002) Short-term cardiovascular oscillations in man: measuring and modelling the physiologies. J Physiol 542:669–683

94. Birn RM, Diamond JB, Smith MA, Bandettini PA (2006) Separating respiratory-variation-related fluctuations from neuronal-activity-related fluctuations in fMRI. Neuroimage 31:1536–1548

95. Edward V, Windischberger C, Cunnington R et al (2000) Quantification of fMRI artifact reduction by a novel plaster cast head holder. Hum Brain Mapp 11:207–213

96. Hajnal J, Myers R, Oatridge A, Schwieso J, Young I, Bydder G (1994) Artifacts due to stimulus correlated motion in functional imaging of the brain. Magn Reson Med 31:283–291

97. Field A, Yen Y, Burdette J, Elster A (2000) False cerebral activation on BOLD functional MR images: study of low-amplitude motion weakly correlated to stimulus. AJNR Am J Neuroradiol 21:1388–1396

98. Robinson S, Moser E (2004) Positive results in amygdala fMRI: Emotion or head motion? NeuroImage 22:S47, WE 294

99. Biswal B, Yetkin FZ, Haughton VM, Hyde JS (1995) Functional connectivity in the motor cortex of resting human brain using echoplanar MRI. Magn Reson Med 34:537–541

100. Lowe MJ, Mock BJ, Sorenson JA (1998) Functional connectivity in single and multislice echoplanar imaging using resting-state fluctuations. Neuroimage 7:119–132

101. Fox MD, Snyder AZ, Zacks JM, Raichle ME (2006) Coherent spontaneous activity accounts for trial-to-trial variability in human evoked brain responses. Nat Neurosci 9:23–25

102. Fox MD, Snyder AZ, Vincent JL, Raichle ME (2007) Intrinsic fluctuations within cortical systems account for intertrial variability in human behavior. Neuron 56:171–184

103. Beckmann CF, De Luca M, Devlin JT, Smith SM (2005) Investigations into resting-state connectivity using independent component analysis. Philos Trans R Soc Lond B Biol Sci 360:1001–1013

104. Damoiseaux JS, Rombouts SA, Barkhof F et al (2006) Consistent resting-state networks across healthy subjects. Proc Natl Acad Sci U S A 103:13848–13853

105. De Luca M, Beckmann CF, De Stefano N, Matthews PM, Smith SM (2006) fMRI resting state networks define distinct modes of long-distance interactions in the human brain. Neuroimage 29:1359–1367

106. Robinson S, Soldati N, Basso G et al (2008) A resting state network in the basal ganglia. Proc Intl Soc Magn Res Med 16:746

107. Beckmann CF, Smith SM (2005) Tensorial extensions of independent component analysis for multisubject FMRI analysis. Neuroimage 25:294–311

108. Raichle M, MacLeod A, Snyder A, Powers W, Gusnard D, Shulman G (2001) A default mode of brain function. Proc Natl Acad Sci U S A 98:676–682

109. Shulman G, Fiez J, Corbetta M et al (1997) Common blood flow changes across visual tasks: II. Decreases in cerebral cortex. J Cogn Neurosci 9:648–663

110. Merboldt KD, Fransson P, Bruhn H, Frahm J (2001) Functional MRI of the human amygdala? Neuroimage 14:253–257

111. Triantafyllou C, Hoge RD, Krueger G et al (2005) Comparison of physiological noise at 1.5 T, 3 T and 7 T and optimization of fMRI acquisition parameters. Neuroimage 26:243–250

112. Krasnow B, Tamm L, Greicius MD et al (2003) Comparison of fMRI activation at 3 and 1.5 T during perceptual, cognitive, and affective processing. Neuroimage 18:813–826

113. Dickerson BC, Wright CI, Miller S et al (2006) Ultrahigh-field differentiation of medial temporal lobe function: sub-millimeter fMRI of amygdala and hippocampal activation at 7 Tesla. NeuroImage 31:S154

114. Merboldt KD, Finsterbusch J, Frahm J (2000) Reducing inhomogeneity artifacts in functional MRI of human brain activation-thin sections vs gradient compensation. J Magn Reson 145:184–191

115. Bellgowan PS, Bandettini PA, van Gelderen P, Martin A, Bodurka J (2006) Improved BOLD detection in the medial temporal region using parallel imaging and voxel volume reduction. Neuroimage 29:1244–1251

116. Triantafyllou C, Hoge RD, Wald LL (2006) Effect of spatial smoothing on physiological noise in high-resolution fMRI. Neuroimage 32:551–557

117. Robinson S, Windischberger C, Rauscher A, Moser E (2004) Optimized 3 T EPI of the amygdalae. NeuroImage 22:203–210

118. Frahm J, Merboldt KD, Hänicke W (1988) Direct FLASH MR imaging of magnetic field inhomogeneities by gradient compensation. Magn Reson Med 6:474–480

119. Deichmann R, Gottfried JA, Hutton C, Turner R (2003) Optimized EPI for fMRI studies of the orbitofrontal cortex. Neuroimage 19:430–441

120. Cho Z, Ro Y (1992) Reduction of susceptibility artifact in gradient-echo imaging. Magn Reson Med 23:193–200

121. Stenger VA, Boada FE, Noll DC (2000) Three-dimensional tailored RF pulses for the reduction of susceptibility artifacts in T(*)(2)-weighted functional MRI. Magn Reson Med 44:525–531

122. Yip CY, Fessler JA, Noll DC (2006) Advanced three-dimensional tailored RF pulse for signal recovery in T2*-weighted functional magnetic resonance imaging. Magn Reson Med 56:1050–1059

123. De Panfilis C, Schwarzbauer C (2005) Positive or negative blips? The effect of phase encoding scheme on susceptibility-induced signal losses in EPI. Neuroimage 25:112–121

124. Ojemann JG, Akbudak E, Snyder AZ, McKinstry RC, Raichle ME, Conturo TE (1997) Anatomic localization and quantitative analysis of gradient refocused echo-planar fMRI susceptibility artifacts. Neuroimage 6:156–167

125. Chen N, Dickey CC, Guttman CRG, Panych LP (2003) Selection of voxel size and slice orientation for fMRI in the presence of susceptibility field gradients: application to imaging of the amygdala. NeuroImage 19:817–825

126. Weiskopf N, Hutton C, Josephs O, Deichmann R (2006) Optimal EPI parameters for reduction of susceptibility-induced BOLD sensitivity losses: a whole-brain analysis at 3 T and 1.5 T. Neuroimage 33:493–504

127. Young IR, Cox IJ, Bryant DJ, Bydder GM (1988) The benefits of increasing spatial resolution as a means of reducing artifacts due to field inhomogeneities. Magn Reson Imaging 6:585–590

128. Hyde S, Biswal B, Jesmanowicz A (2001) High-resolution fMRI using multislice partial k-space GR-EPI with cubic voxels. Magn Reson Med 46:114–125

129. Haacke E, Brown R, Thompson M, Venkatesan R (1999) Magnetic resonance imaging: physical principles and sequence design. Wiley-Liss, New York, NY

130. Scouten A, Papademetris X, Constable RT (2006) Spatial resolution, signal-to-noise ratio, and smoothing in multi-subject functional MRI studies. Neuroimage 30:787–793

131. Parrish T, Gitelman D, LaBar K, Mesulam M (2000) Impact of signal-to-noise on functional MRI. Magn Reson Med 44:925–932

132. LaBar K, Gitelman D, Mesulam M, Parrish T (2001) Impact of signal-to-noise on functional MRI of the human amygdala. Neuroreport 12:3461–3464

133. Robinson S, Hoheisel B, Windischberger C, Habel U, Lanzenberger R, Moser E (2005) FMRI of the emotions, towards an improved understanding of amygdala function. Curr Med Imaging Rev 1:115–129

134. Morawetz C, Holz P, Lange C et al (2008) Improved functional mapping of the human amygdala using a standard functional magnetic resonance imaging sequence with simple modifications. Magn Reson Imaging 26:45–53

135. Stocker T, Kellermann T, Schneider F et al (2006) Dependence of amygdala activation on echo time: results from olfactory fMRI experiments. Neuroimage 30:151–159

136. Gorno-Tempini M, Hutton C, Josephs O, Deichmann R, Price C, Turner R (2002) Echo time dependence of BOLD contrast and susceptibility artifacts. NeuroImage 15:136–142

137. Sodickson DK, Manning WJ (1997) Simultaneous acquisition of spatial harmonics (SMASH): fast imaging with radiofrequency coil arrays. Magn Reson Med 38:591–603

138. Pruessmann K, Weiger M, Scheidegger M, Boesiger P (1999) SENSE: sensitivity encoding for fast MRI. Magn Reson Med 42:952–962

139. Lutcke H, Merboldt KD, Frahm J (2006) The cost of parallel imaging in functional MRI of the human brain. Magn Reson Imaging 24:1–5

140. Schmidt CF, Degonda N, Luechinger R, Henke K, Boesiger P (2005) Sensitivity-encoded (SENSE) echo planar fMRI at 3 T in the medial temporal lobe. NeuroImage 25:625–641

141. Fürsatz M, Windischberger C, Karlsson KÆ, Moser E (2008) Successful fMRI of the hypothalamus at 3T. Proc Intl Soc Magn Reson Med 16:2501

142. Fürsatz M, Windischberger C, Karlsson KÆ, Mayr W, Moser E. Valence-dependent modulation of hypothalamic activity. Neuroimage (in press)

143. Dowell NG, Tofts PS (2007) Fast, accurate, and precise mapping of the RF field in vivo using the 180 degrees signal null. Magn Reson Med 58:622–630

144. Speck O, Hennig J (1998) Functional imaging by I0- and T2*-parameter mapping using multi-image EPI. Magn Reson Med 40:243–248

145. Posse S, Wiese S, Gembris D et al (1999) Enhancement of BOLD-contrast sensitivity by single-shot multi-echo functional MR imaging. Magn Reson Med 42:87–97

146. Posse S, Shen Z, Kiselev V, Kemna LJ (2003) Single-shot T(2)* mapping with 3D compensation of local susceptibility gradients in multiple regions. Neuroimage 18:390–400

147. Posse S, Holten D, Gao K, Rick J, Speck O (2006) Evaluation of interleaved XYZ-shimming with multi-echo EPI in prefrontal cortex and amygdala at 4 Tesla. NeuroImage 31:S154

148. Weiskopf N, Klose U, Birbaumer N, Mathiak K (2005) Single-shot compensation of image distortions and BOLD contrast optimization using multi-echo EPI for real-time fMRI. Neuroimage 24:1068–1079

149. Glover GH, Law CS (2001) Spiral-in/out BOLD fMRI for increased SNR and reduced susceptibility artifacts. Magn Reson Med 46:515–522

150. Guo H, Song AW (2003) Single-shot spiral image acquisition with embedded z-shimming for susceptibility signal recovery. J Magn Reson Imaging 18:389–395

151. Truong TK, Song AW (2008) Single-shot dual-z-shimmed sensitivity-encoded spiral-in/out imaging for functional MRI with reduced susceptibility artifacts. Magn Reson Med 59:221–227

152. Li Z, Wu G, Zhao X, Luo F, Li SJ (2002) Multiecho segmented EPI with z-shimmed background gradient compensation (MESBAC) pulse sequence for fMRI. Magn Reson Med 48:312–321

153. Weisskoff RM, Davis TL (1992) Correcting gross distortion on echo planar images. Paper presented at the SMRM, Berlin

154. Jezzard P, Balaban RS (1995) Correction for geometric distortion in echo planar images from B0 field variations. Magn Reson Med 34:65–73

155. Jenkinson M (2003) Fast, automated, N-dimensional phase-unwrapping algorithm. Magn Reson Med 49:193–197

156. Witoszynskyj S, Rauscher A, Reichenbach JR, Barth M (2007) ΠUN (Πhase UNwrapping) validation of a 2D region-growing phase unwrapping program. Proc Intl Soc Magn Reson Med 15:3436

157. Windischberger C, Robinson S, Rauscher A, Barth M, Moser E (2004) Robust field map generation using a triple-echo acquisition. J Magn Reson Imaging 20:730

158. Roemer PB, Edelstein WA, Hayes CE, Souza SP, Mueller OM (1990) The NMR phased array. Magn Reson Med 16:192–225

159. Bernstein MA, Grgic M, Brosnan TJ, Pelc NJ (1994) Reconstructions of phase contrast, phased array multicoil data. Magn Reson Med 32:330–334

160. Hammond KE, Lupo JM, Xu D et al (2008) Development of a robust method for generating 7.0 T multichannel phase images of the brain with application to normal volunteers and patients with neurological diseases. Neuroimage 39:1682–1692

161. Robinson S, Jovicich J (2008) EPI distortion corrections at 4 T: Multi-channel field mapping and a comparison with the point-spread function method. Proc Intl Soc Magn Reson Med 16:3031

162. Robson MD, Gore JC, Constable RT (1997) Measurement of the point spread function in MRI using constant time imaging. Magn Reson Med 38:733–740

163. Zeng H, Constable RT (2002) Image distortion correction in EPI: Comparison of field mapping with point spread function mapping. Magn Reson Med 48:137–146

164. Zaitsev M, Hennig J, Speck O (2004) Point spread function mapping with parallel imaging techniques and high acceleration factors: fast, robust, and flexible method for echo-planar imaging distortion correction. Magn Reson Med 52:1156–1166

165. Hu X, Kim SG (1994) Reduction of signal fluctuation in functional MRI using navigator echoes. Magn Reson Med 31:495–503

166. Bruder H, Fischer H, Reinfelder HE, Schmitt F (1992) Image reconstruction for echo planar imaging with nonequidistant k-space sampling. Magn Reson Med 23:311–323

167. Barry RL, Klassen LM, Williams JM, Menon RS (2008) Hybrid two-dimensional navigator correction: a new technique to suppress respiratory-induced physiological noise in multi-shot echo-planar functional MRI. Neuroimage 39:1142–1150

168. Lund TE, Madsen KH, Sidaros K, Luo WL, Nichols TE (2006) Non-white noise in fMRI:

does modelling have an impact? Neuroimage 29:54–66

169. Glover GH, Li TQ, Ress D (2000) Image-based method for retrospective correction of physiological motion effects in fMRI: RETROICOR. Magn Reson Med 44:162–167

170. Windischberger C, Friedreich S, Hoheisel B, Moser E (2004) The importance of correcting for physiological artifacts for functional MRI in deep brain structures. NeuroImage 22:S28

171. Josephs O, Howseman A, Friston K, Turner R (1997) Physiological noise modelling for multi-slice EPI fMRI using SPM. Proc Intl Soc Magn Reson Med 5:1682

172. Weissenbacher A, Windischberger C, Lanzenberger R, Moser E (2008) Efficient correction for artificial signal fluctuations in resting-state fMRI-data. Proc Intl Soc Magn Reson Med 16:2467

173. Beckmann CF, Smith SM (2004) Probabilistic independent component analysis for functional magnetic resonance imaging. IEEE Trans Med Imaging 23:137–152

174. Calhoun V, Adali T, Stevens M, Kiehl K, Pekar J (2005) Semi-blind ICA of fMRI: a method for utilizing hypothesis-derived time courses in a spatial ICA analysis. Neuroimage 25:527–538

175. Thomas CG, Harshman RA, Menon RS (2002) Noise reduction in BOLD-based fMRI using component analysis. Neuroimage 17:1521–1537

176. Kochiyama T, Morita T, Okada T, Yonekura Y, Matsumura M, Sadato N (2005) Removing the effects of task-related motion using independent-component analysis. Neuroimage 25:802–814

177. Perlbarg V, Bellec P, Anton JL, Pelegrini-Issac M, Doyon J, Benali H (2007) CORSICA: correction of structured noise in fMRI by automatic identification of ICA components. Magn Reson Imaging 25:35–46

178. Tohka J, Foerde K, Aron AR, Tom SM, Toga AW, Poldrack RA (2008) Automatic independent component labeling for artifact removal in fMRI. Neuroimage 39:1227–1245

179. Bullmore ET, Brammer MJ, Rabe-Hesketh S et al (1999) Methods for diagnosis and treatment of stimulus-correlated motion in generic brain activation studies using fMRI. Hum Brain Mapp 7:38–48

180. Morgan VL, Dawant BM, Li Y, Pickens DR (2007) Comparison of fMRI statistical software packages and strategies for analysis of images containing random and stimulus-correlated motion. Comput Med Imaging Graph 31:436–446

181. Preibisch C, Raab P, Neumann K et al (2003) Event-related fMRI for the suppression of speech-associated artifacts in stuttering. Neuroimage 19:1076–1084

182. Birn RM, Bandettini PA, Cox RW, Shaker R (1999) Event-related fMRI of tasks involving brief motion. Hum Brain Mapp 7:106–114

183. Phelps EA, O'Connor KJ, Cunningham WA et al (2000) Performance on indirect measures of race evaluation predicts amygdala activation. J Cogn Neurosci 12:729–738

184. Winston JS, Strange BA, O'Doherty J, Dolan RJ (2002) Automatic and intentional brain responses during evaluation of trustworthiness of faces. Nat Neurosci 5:277–283

185. Singer T, Kiebel SJ, Winston JS, Dolan RJ, Frith CD (2004) Brain responses to the acquired moral status of faces. Neuron 41:653–662

186. Campbell DT, Stanley JC (1966) Experimental and quasi-experimental designs for research. Houghton Mifflin, Boston

187. Poldrack RA, Wagner AD (2004) What can neuroimaging tell us about the mind? Insights from prefrontal cortex. Curr Dir Psychol Sci 13:177–181

188. Logothetis NK, Pauls J, Augath M, Trinath T, Oeltermann A (2001) Neurophysiological investigation of the basis of the fMRI signal. Nature 412:150–157

189. Janz C, Heinrich SP, Kornmayer J, Bach M, Hennig J (2001) Coupling of neural activity and BOLD fMRI response: new insights by combination of fMRI and VEP experiments in transition from single events to continuous stimulation. Magn Reson Med 46:482–486

190. Baas D, Aleman A, Kahn RS (2004) Lateralization of amygdala activation: a systematic review of functional neuroimaging studies. Brain Res Brain Res Rev 4:96–103

191. Robinson S, Pripfl J, Bauer H, Moser M (2005) Empirical evidence for the minimum voxel size required for reliable 3 T fMRI of the amygdala. NeuroImage 26:S795

192. Habel U, Windischberger C, Derntl B et al (2007) Amygdala activation and facial expressions: explicit emotion discrimination versus implicit emotion processing. Neuropsychologia 45:2369–2377

193. Sack AT, Linden DE (2003) Combining transcranial magnetic stimulation and functional imaging in cognitive brain research: possibilities and limitations. Brain Res Brain Res Rev 43:41–56

194. Thayer JF, Brosschot JF (2005) Psychosomatics and psychopathology: looking up and down from the brain. Psychoneuroendocrinology 30:1050–1058

Chapter 16

fMRI of Pain

Emma G. Duerden, Roberta Messina, Maria A. Rocca,
Massimo Filippi, and Gary H. Duncan

Abstract

Pain was first considered to be a hard-wired system in which noxious input was passively transmitted along sensory channels to the brain. However, today it is generally accepted that the experience of pain is not simply driven by noxious stimulus characteristics, but that the brain is the structure where the subjective perception of pain emerges and is critically linked with other cognitive processes.

The field of pain research has progressed immensely due to the advancement of brain imaging techniques. The initial goal of this research was to expand our understanding of the cerebral mechanisms underlying the perception of pain; more recently the research objectives have shifted toward chronic pain—understanding its origins, developing methods for its diagnosis, and exploring potential avenues for its treatment. While several different neuroimaging approaches have certain advantages for the study of pain, fMRI has ultimately become the most widely utilized imaging technique over the past decade because of its noninvasive nature, high-temporal and spatial resolution, and general availability; thus, the following chapter will focus on fMRI and the special aspects of this technique that are particular to pain research.

Key words Pain, Functional neuroimaging, Brain, Perception

1 Introduction

The history of pain imaging is relatively short, although it has advanced immensely within the last decade due to improvements in imaging techniques, statistical analysis, and specialized equipment for the delivery of painful stimuli. Initially, brain-imaging studies sought simply to examine the brain areas that are involved in pain processing, to make comparisons with the long established neurophysiological studies reported in this field. Many of these initial imaging studies were prompted by electrophysiological data from patients undergoing brain surgery in the early part of the twentieth century [1], which had questioned the role of the cortex in nociceptive processing. It was initially believed that the thalamus was primarily responsible for nociceptive processing as suggested by deficits in pain perception observed in patients with thalamic lesions [2].

Massimo Filippi (ed.), *fMRI Techniques and Protocols*, Neuromethods, vol. 119,
DOI 10.1007/978-1-4939-5611-1_16, © Springer Science+Business Media New York 2016

In the early 1990s, activation in the human brain evoked by experimental pain stimuli was studied using positron emission tomography (PET) [3, 4] and single photon emission tomography (SPECT) [5]. Then in 1995 the first fMRI studies examining the cortical representation of pain [6] were conducted largely to confirm the findings of previous PET studies and to examine whether the cortical nociceptive signal could be detected using fMRI. In more recent years, the field of pain imaging has expanded immensely, allowing researchers to answer complex questions concerning pain processing, such as how cortical regions are connected and modified during the perception of pain and, most importantly, how the cortex responds during the modulation of pain. These experimental studies were conducted in healthy humans in order to answer broad questions regarding pain processing, with the eventual goal of applying this knowledge to a better understanding and alleviation of pain and suffering associated with chronic pain syndromes. The use of fMRI and other imaging techniques have revealed a number of cortical and subcortical changes that may occur as a result of prolonged exposure to pain—or possibly as causal factors in chronic pain conditions [7–9]. Indeed, with the advent of high-speed image acquisition and computational processing, not only has the technology of fMRI revealed areas of cortical plasticity associated with chronic pain, but it is also now possible to use fMRI in real-time to furnish feedback to subjects (and patients) to teach them how to modulate their cortical activation in response to chronic pain [10, 11].

This chapter reviews and discusses the various advances in our knowledge of cerebral pain processing that have been achieved using fMRI, the response properties of cortical nociceptive neurons in relation to both imaging techniques and stimuli used to evoke pain, the applications of this research to treat clinical pain in patients, and the future of pain research using fMRI.

2   Use of fMRI to Study Nociceptive Processing

Compared to other brain mapping techniques currently used to study pain experimentally in humans—such as PET, electroencephalography (EEG), magnetoencephalography (MEG), or optical imaging—fMRI is the tool of choice, given its high spatial resolution, noninvasiveness, and reasonable temporal resolution, which allow the study of rapid dynamic processes involved in pain processing. However, a number of methodological issues concerning the use of the BOLD signal in research involving cortical, and more recently, spinal mechanisms of pain perception has to be considered.

2.1   Nociceptive BOLD Signal

For cortical nociceptive processing related to cutaneous heat stimuli, the hemodynamic response function (HRF) peaks slightly later and lasts longer in comparison to innocuous stimuli. Chen et al. [12] performed a direct comparison of the temporal properties of the HRF in response to noxious thermal heat pain and innocuous brushing stimuli in SI and SII. While both stimuli were of the same duration, the time course for innocuous stimuli peaked ~10 s after the onset of the stimulus and dissipated quickly after its removal. However, noxious thermal heat stimuli produced a time course peaking at ~15 s after the onset of the stimulus and the response was sustained for several seconds. Similar results have been reported in response to painful electrical stimuli [13]; identical trains of noxious and innocuous stimuli produced differential time courses, with the HRF for painful stimulation lasting twice as long as that produced by nonpainful stimuli.

Time course information on the BOLD response to noxious stimuli is crucial for interpreting data analyzed using the standard canonical HRFs available in the majority of fMRI analysis software, which approximate this time period at ~6 s. Ideally to establish a more representative model of painful stimuli, a canonical HRF should be created based on data from independent studies employing similar noxious stimulation. The BOLD signal can then be regressed against this canonical HRF to reveal activation more specific to the nociceptive signal.

A related issue in analyzing data recorded during experimental pain studies is the critical importance of considering the rise time of thermal stimuli when establishing time periods in the event design matrix. As the temperature of the thermode gradually increases, warm and pain fibers will become increasingly activated. In order to maximize sensitivity for detection of the pain-related BOLD signal, it is important to enter into the design matrix solely the period of time during which the thermode has exceeded the subjects' pain threshold—not the initial rise-time of the stimulus period, which would be associated with the innocuous warm sensations perceived before the actual onset of pain.

2.2 BOLD fMRI of Spinal Nociceptive Signals

A newly developing field in pain fMRI is spinal cord imaging, which is crucial for a better understanding of central nervous system (CNS) pain processing. The spinal cord and brainstem receive input from the periphery before relaying this information on to the cortex. These subcortical regions are involved in the modulation of nociceptive input and the potentially abnormal processing of that input that may lead to chronic pain syndromes. Therefore, knowledge concerning the peripheral mechanisms of nociceptive processing is crucial to understanding a number of pathological pain conditions resulting from nerve injury or inflammation. These factors contribute to the generation and maintenance of two key components of chronic pain, namely hyperalgesia and allodynia. Hyperalgesia is the phenomenon where an exaggerated response occurs after exposure to a noxious stimulus. Allodynia is an exaggerated response toward nonpainful mechanical stimuli. Both occur when nociceptive fibers become sensitized, after exposure to a noxious stimulus, causing the release of "painful" substances in the periphery. Peripheral sensitization can occur due to inflammation of peripheral tissues as a result of a burn or cut. Because of this barrage of input, peripheral nociceptors can become hyperexcitable. This peripheral sensitization can also occur due to ectopic firing of peripheral nerves resulting from an amputation or injury. Central sensitization can occur in the dorsal horns of the spinal cord, when peripheral nerves that were once insensitive to nociceptive input switch their firing patterns and begin to transmit nociceptive information, causing the area of affected skin to become painful to the slightest touch. Much research in this area is focused at the periphery, although these processes have been shown to have supra-spinal effects resulting in aberrant cortical activity and the reorganization of body maps in somatosensory cortices.

To fulfill this need to study spinal mechanisms of nociception, experimental models directed toward spinal fMRI have begun in humans [14, 15].

Applications of spinal fMRI to the study of chronic pain could have vast clinical applications. Use of a noninvasive functional imaging modality could shed light on the spinal mechanisms involved in the generation of neuropathic pain, such as dysesthetic pain in patients with spinal cord injury or syringomyelia. In addition to understanding the effects of chronic pain on neuroplasticity of the spinal cord, spinal fMRI could provide insight into the potential mechanisms of medications and their efficacy at treating chronic pain.

3   Methods for fMRI Pain Experiments

3.1   Pain Assessment

A key issue in functional imaging of the cortical nociceptive signal is to ensure that the stimuli delivered to the subjects are perceived as noxious. Pain thresholds are commonly determined during a separate session prior to the scan. This procedure also serves to familiarize participants with the stimuli and reduce anxiety, thereby minimizing anxiety-related fluctuations in cardiovascular activity [16]. Stimuli utilized for the scanning session are frequently tailored to each individual's pain threshold; conversely, all subjects can be administered the same level of noxious stimulation, which has been determined to evoke the perception of pain in all subjects. A corollary to the appropriate choice of noxious stimuli is the confirmation that predetermined levels of stimulation are actually perceived as painful, within the scanning environment. A number of contextual factors can alter the perception of stimuli that were originally considered painful during a pre-scanning test, including the temperature of the scanning suite, the position of the body in the scanner, and distractions of noise, possible feelings of claustrophobia, and other conditions specific to the scanning paradigm.

It is also important to note that the perception of pain can change during the course of a scanning session, due either to habituation, sensitization, or the potential changes in attention during a long scanning experiment [17, 18].

To address this issue, pain assessment ratings can be obtained *during* the fMRI scanning session through subjective reports from participants using a variety of methods. Subjects can rate their perception after each stimulus, continuously during the stimuli, or at the end of the scanning run by giving an average rating of all the stimuli. Subjects' scores are recorded typically using numerical or visual analog scales (VAS) [19]. In fMRI experiments, ratings can be obtained during or immediately after the presentation of each stimulus. Conversely, due to methodological issues, pain ratings within the context of PET studies can be taken only at the end of a scanning session several minutes after stimulus presentation. Increased time between stimulus presentation and assessment can cause inaccuracies in subject responses [20]. This is a special consideration in studies examining mechanisms of analgesic relief since retrospective ratings can be inflated with increased time after stimulus presentation [21, 22].

In addition to ensuring that the noxious stimuli are actually painful, pain assessment ratings (and other behavioral measures) can be used as regressors in the fMRI design matrix to aid in identifying cortical regions involved in various aspects of pain processing. Behavioral data can be incorporated into the fMRI design matrix as a weighting factor applied to the canonical HRF. Alternatively, continuous pain ratings (recorded during the stimulus presentations) can be modeled in the design matrix (e.g., *see* ref. [23]). The resulting contrasts produce activation sites that are more closely based on the degree to which a region's activity correlates with the perceived intensity of the stimuli rather than with the physical intensity of the stimulus—in other words a "percept-related" activation as opposed to a "stimulus-related" activation [24].

This experimental approach may have important implications for studying the dissociation that sometimes occurs between the intensity of peripheral stimulation and the perception of pain. For example, presentation of noxious mechanical stimuli over longer durations (~2 min) has been shown to disrupt the relationship between the firing frequency of nociceptive afferents and the perceived intensity of pain evoked by the stimuli [25, 26]. This paradoxical relationship may be explained by the process of temporal summation—a disproportionate increase in the firing rate of dorsal horn neurons over time, whereby their response threshold to sensory input is substantially lowered. Additionally, repeated exposure to short-duration heat pain stimuli can cause habituation to both the perceived intensity and unpleasantness of the stimuli [27]. Therefore, subjective pain ratings can play a key role in the interpretation of nociceptive processing in the cortex, as opposed to utilizing simply the duration or intensity of the noxious stimuli that may not aptly reflect the resulting activations.

Several studies have explored the possible cerebral mechanisms underlying habituation or sensitization to painful stimuli, showing activation of cerebral regions involved in anti-nociception. However, these gave conflicting results concerning any specific association between cerebral activity and ratings of pain intensity [17, 28–30].

On the whole, however, these ambiguities in the correspondence between stimulus delivery, evoked nociceptive signal, and subjective reports of pain intensity, underscore the importance of accessing the level of perceived pain during scanning sessions, rather than assuming a fixed relationship between stimulation and percept.

A number of advantages and potential disadvantages are associated with obtaining continuous pain ratings of stimuli during fMRI experiments. Clearly, participants' perceptual evaluations will rely less on memory and will tend to be more accurate, compared with evaluations made after the scanning run. In turn, the resulting brain activation will be less reflective of mnemonic or error detection processes. Additionally, continuous ratings can be used to deduce the time lag between the application of the stimulus and the onset of pain perceived by the subject, and to provide further details about the time course of pain perception and the underlying neural activity.

While continuous ratings provide real-time information about a subject's perception of the stimuli, a clear disadvantage to their use is that the motor activity and motor-related activation can produce a confound that complicates interpretation of sensory-related activity. However, this can be accounted for by including the movements as covariates in the fMRI design matrix.

4   Neuroanatomy of Pain Processing

Before describing how fMRI measures the cortical and spinal nociceptive signal, it is important to understand how this signal is transferred to the cortex. In the periphery, a painful stimulus applied to the body is transmitted to the CNS through nociceptors [31]. Myelinated A-delta fibers transmit sharp pricking pain [32], while unmyelinated C-fibers transmit slow burning pain, often referred to as second pain [33]. The cell bodies of A-delta and C-fibers are located in the dorsal root ganglia, receiving afferent input from the periphery and then sending the information into the spinal cord to terminate in the dorsal horn [34, 35]. Axons from the second-order dorsal horn neurons rise through several ascending pathways that transmit nociceptive information to the thalamus, reticular formation, and cortex [8, 36]. Pain and temperature information applied to the face is relayed through cranial nerves to the spinal nucleus V terminating in the thalamus via the trigeminothalamic tract, which is then relayed to the cortex.

A number of spinal and cortical neurons respond to noxious stimuli including nociceptive-specific (NS) and wide dynamic range (WDR) projection neurons, the latter of which respond to both noxious and innocuous stimuli. Additionally, the dorsal horns and cortical somatosensory regions contain neurons responsive solely to innocuous stimuli called low threshold mechanical (LTM) neurons and thermoreceptive neurons responsive to temperatures in the warm and cold range. However, recent studies demonstrated that nociceptive, tactile, auditory and visual stimuli can elicit spatially indistinguishable cortical responses, thus indicating that the bulk of the brain responses to nociceptive stimuli reflects multimodal neural activity (i.e., activity that can be triggered by any kind of stimulus independently of sensory modality) [37]. This range of responses is an important consideration when interpreting results from fMRI studies of pain in terms of exactly what the activation pattern is reflecting.

Typically, pain-evoked brain activation is achieved by applying contact thermodes to the skin. This technique involves an increase in temperature at the rate of 1–10 °C s^{-1}. Depending on the baseline temperature it can take several seconds to reach perceived pain threshold. In addition to activating NS neurons with noxious heat, contact thermodes may activate both LTM and WDR neurons through innocuous mechanical and thermal stimulation of the skin as the stimulation temperature rises towards pain threshold. Therefore, to examine pain-specific cortical activations, it is necessary to compare pain-related activations to those associated with the presentation of innocuous warm stimuli.

In addition to conductive heating of the skin using contact thermodes, nociceptive afferents can be activated using thermal radiation administered through infrared laser stimulators [38, 39]. Lasers can deliver heat stimuli without the need for a contact probe, thus selectively stimulating C-fibers and A-delta fibers without contaminant activation of A-beta fibers that transmit touch information. Additionally, laser stimuli can activate nociceptive nerve endings at rapid rates for short durations (1 ms) [40, 41] and are therefore well suited for rapid event-related fMRI studies. However, an important consideration associated with the use of laser stimuli is the difficulty of measuring and controlling skin temperature, which is the primary factor triggering the cascade of neural responses that culminate in the processing of heat-related nociceptive information in the brain and likewise the assessment of pain by the subjects [42]. Laser and contact heat stimuli have been shown to produce similar patterns of BOLD activation in anterior cingulate cortex (ACC), insula, primary motor cortex, prefrontal cortex (PFC), parahippocampal gyrus, thalamus, basal ganglia, periaqueductal gray (PAG), and cerebellum. However, stronger activation in response to contact heat stimuli was noted in secondary somatosensory cortex (SII), posterior insula, posterior ACC, and regions in parietal and frontal cortices [43]. Thus, these two modes of delivering noxious heat stimulation cannot be considered identical in terms of the evoked pain-related BOLD activations, and the advantages and disadvantages of each should be weighed in relation to the research questions and appropriate stimulation paradigms.

4.1 Supraspinal Processing of Nociceptive Stimuli

During the past 3–5 years, neuroimaging studies have extensively investigated the neural basis of pain perception, thus showing that nociceptive stimuli commonly elicit activity within a very wide array of subcortical and cortical brain structures [44, 45]. Regions most frequently activated by painful stimuli include primary somatosensory cortex (SI), SII, ACC, the insula, the PFC, and the thalamus.

Regions responsible for pain processing are categorized along two functional lines—the first being the sensory-discriminative (lateral pain system) component involved in the perception of temporal, intensity, and localization aspects of pain processing, and the second, the affective-motivational (medial) component associated with the emotional aspects of pain [46, 47]. Dissociations between the two systems are made through subjective reports on pain scales. After exposure to noxious stimuli, subjects are asked to quantify separately how intense and how unpleasant is the perceived pain. Regions implicated in the lateral pain system include SI, SII, posterior insula, and lateral thalamus, while the medial pain system consists of the medial thalamic nuclei, the ACC, and the PFC. Much of what is known regarding the two components in pain processing was initially explored through single-unit recordings in nonhuman primates and lesion studies in humans. However, the more recent ability to study these functional components noninvasively in humans using fMRI and other brain mapping techniques has allowed pain researchers to advance rapidly in their understanding of the role of these cortical regions in pain processing and how they interact.

4.1.1 Primary Somatosensory Cortex

SI is located in the postcentral gyrus, is composed of four areas (areas 3a, 3b, 1, and 2) [48], and is involved in the processing of both tactile and noxious stimuli [49]. It was long debated whether SI was necessary to perceive pain. Early studies of patients with brain lesions suggested that deficits in nociceptive processing were rather common following lesions to the thalamus, but were very rare when damage was restricted to the area believed to incorporate SI [2]. Likewise, later studies, using electrical stimulation of the human cortex during awake brain surgery, reported that direct stimulation of SI rarely evoked any perception of pain in patients [1].

The advent of imaging technology allowed a more global exploration of the role of SI and other cortical regions involved in pain processing, and these studies could be conducted in healthy volunteers, rather than in patients with brain injuries that might alter normal function. The first of these studies involved PET and demonstrated that noxious stimuli applied to the hands were associated with robust activation in SI [3]. Several other early studies failed to detect SI activation [4, 5], and subsequent reports, using either PET or fMRI, have resulted in contradictory findings (for SI activation, see for example: [50–53]; for absence of SI activation, *see* refs. [54, 55]).

In particular, several studies showed that the anterior portion of SI (the area 3a) receives input originating predominantly from unmyelinated nociceptors, distinguishing it from posterior SI (areas 3b and 1), long recognized as receiving input predominantly from myelinated afferents, including nociceptors [56].

The inconsistency of SI activation reported across imaging studies could be due to several factors. Wide variations in the location of the central sulcus across subjects may lead to a wash out in signal across averaged group data. In addition, a reduction in SI activity below statistically significant levels could be caused in some paradigms by inhibitory effects induced by noxious stimuli on tactile inputs [4, 57]. In a review discussing the issue of pain-related activation of SI, Bushnell et al. concluded that the BOLD signal in SI largely depends on task design that is likely to influence the attentional state of the subject [58]. Results from subsequent studies have likewise indicated that pain-related BOLD activation of SI is increased when subjects attend to pain and decreased when they are distracted [59].

On the contrary, attention may also show a deleterious effect on SI activation as noted by Oshiro et al. [60] in their fMRI study examining the neural correlates involved in processing spatial localization of pain. The authors failed to find activation in SI in response to painful stimulation of the calf. However, the authors noted that this lack of activation may have been a result of the response properties of the cortical nociceptive neurons.

Nociceptive input to SI is somatotopically organized [61–63], and the small receptive fields of SI [64] suggest that this region is well suited to make fine spatial discriminations of noxious stimuli applied to the body. Oshiro et al. [60] required subjects to focus on stimulation applied to their calves, and this increased attention on the leg area may have caused a reduction in the receptive field sizes of nociceptive neurons, which would enhance spatial acuity needed to perform the task—but cause deterioration in resulting brain activation. In another study using a discrimination task, Albanese et al. [65] explored short-term memory for the spatial location and intensity of painful thermal stimuli applied to the palms. In contrast to the study by Oshiro et al. [60], Albanese et al. [65] reported robust pain-related activation in SI/posterior parietal cortex, which was sustained during the memory period of the trial, suggesting that this region has a role in the encoding and retention of noxious stimuli. Differences between the two studies may be due to the larger somatotopic organization of the hand representation of SI. Additionally, subjects in the Albanese study were required to detect the end of each stimulus, a strategy that may have heightened attention toward the stimuli and contributed to a temporal summation of the BOLD signal in SI.

Somatosensory brain regions have also been found to play a role during pain anticipation. A recent study [66] revealed greater changes in activation of SI in nocebo responders compared to controls, supporting the notion that anticipatory activation of a prefrontal-limbic network is involved in nocebo hyperalgesia.

4.1.2 Secondary Somatosensory Cortex (SII)

SII is also considered to be an important region for processing the sensory-discriminative component of pain. SII is located in the parietal operculum in the dorsal bank of the lateral sulcus. Like SI, this region receives projections from the ventroposterior lateral nucleus (VPL) of the thalamus, but its major nociceptive input comes directly from the ventroposterior inferior (VPI) nucleus [52]. Studies of patients with lesions that include SII have demonstrated deficits in the perception of pain intensity [67, 68]; however, lesions comprised additional cortical regions that may work in concert with SII to process this piece of information. In addition to these clinical findings, converging evidence from a number of studies supports the notion that SII possesses a functional capacity to discriminate between different intensities of noxious stimuli presented to the contralateral side of the body. Evidence from PET provides a role for this region in intensity processing in that subjects' ratings of pain intensity in response to thermal heat pain have been shown to be highly correlated with activation of SII [69]. Additionally, an fMRI study by Maihofner et al. [70] found increased activation in SII in response to painful mechanical stimuli compared to thermal heat pain. In turn, ratings of subjective intensity were correlated with the intensity of mechanical pain. However, dissociative processing was noted in this region as ratings of unpleasantness were not found to correlate with SII activation.

Contrary to these findings, evidence from fMRI suggests this region may be involved in some emotional aspects of pain processing. For example, Gracely et al. [71] found that fibromyalgia patients who scored higher on a pain catastrophizing questionnaire showed increased activation in both the ACC and SII in response to noxious stimuli. Catastrophizing (and in turn anxiety about painful stimuli) is inherently linked with pain perception, where the individual's emotional state augments neural processing of these stimuli. In line with these findings are data that show increased activity in SII during the anticipation of painful stimuli, indicative of an enhanced emotional response [66, 72].

4.1.3 Insular Cortex

The insula is extensively connected to other brain regions such as the prefrontal cortex, cingulum, amygdala, SI, SII, and also thalamic nuclei (VPI, the centromedian-parafasicular, the medial dorsal [MD], and the ventral medial posterior [Vmpo] nuclei). It may therefore act as a relay integrating afferent nociceptive information with working memory, affect, and attention, and may selectively gate nociceptive information at the cortical level to modulate varying levels of appreciation of the nociceptive stimulus [73].

Functional neuroimaging studies suggested that the anterior, mid, and posterior division of the insula subserve different functions in the perception of pain [74].

Early clinical reports and quantitative studies [75–77], have indicated that patients with lesions encompassing the insula do not exhibit normal withdrawal or emotional responses to noxious stimuli, indicating an altered or deficient perception of pain affect. Accordingly, fMRI activity in the anterior insula in response to noxious stimuli is correlated with subjective ratings of pain unpleasantness [78, 79].

The insula, in particular its posterior region, has also been found to process sensory-discriminative features of nociceptive information, making it a likely area of convergence of the two pain systems. Evidence for the role of the insula in sensory-discriminative processing comes from direct electrical stimulation to the region during awake brain surgery, demonstrating evoked painful sensations in the body [80], and also from fMRI studies that showed a significant correlation between posterior insula activity and painful stimulus intensity [79, 81]. Furthermore, several other lines of evidence indicate that this region may be involved in the localization of painful stimuli, as it contains a somatotopic map of the body. The dorsal posterior insula receives pain and temperature information from a somatotopically organized region of the thalamus—the VMpo [82], which in turn receives projections from thermoreceptive and nociceptive neurons residing in lamina I of the spinal cord [79, 83] (Fig. 1).

Neuroimaging studies of pain perception frequently report insular activation, making it difficult to dissociate it from activation seen in adjacent regions of SII [84]. Resolving the precise somatotopic organization of the insula using fMRI has become feasible with the availability of high-field strength magnets. Several fMRI studies at 3 T have revealed a nociceptive somatotopic organization in the dorsal posterior insula in response to both cutaneous and muscle pain [85, 86].

It is further notable that the posterior insula has been suggested to be involved also in directing pain-related motor responses [73].

4.1.4 Anterior Cingulate Cortex (ACC)

The ACC plays a prominent role in pain processing. This region receives thalamo-cortical input from nociceptive neurons in the thalamus and contains nociceptive-specific neurons responsive to noxious stimuli [87]. Additionally, the ACC is implicated in mediating antinociceptive responses as it contains high numbers of opiate receptors [88, 89].

Historically, the ACC was considered key to affective processing, as it was classified along with the retrosplenial cortex, hippocampus, amygdala, and several basal forebrain structures as part of

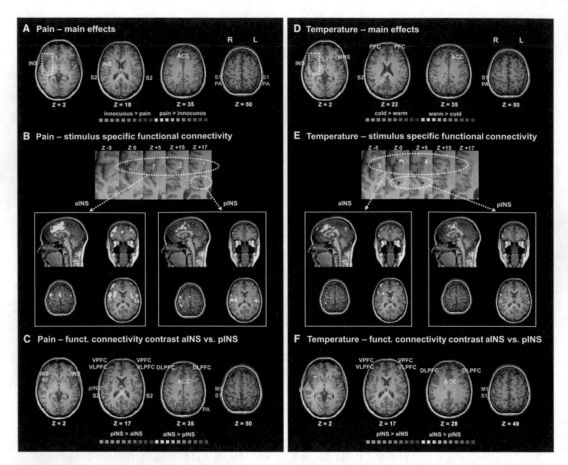

Fig. 1 Pain (**a**) and temperature (**d**) processing brain regions: brain areas with significantly increased activation during noxious than innocuous stimulation and during warm stimulation are coded *red*, while regions coded *blue* show significantly increased activation during innocuous than noxious stimulation and during cold stimulation. Insular clusters (seed clusters) with pain- (**b**) and temperature- (**e**) specific activity divided in aINS (*green* and *yellow*) and pINS (*red* and *blue*), that were used for the insular functional connectivity analysis: both aINS and pINS were functionally connected to a large brain network, which predominantly includes areas involved in nociception and thermoception (SI, SII, cingulate gyrus, PFC, and parietal association cortices). Comparison of pain- (**c**) and temperature- (**f**) specific functional connectivity of the two insular areas (areas with significantly stronger functional connectivity to aINS than to pINS are coded *red–yellow*, while areas with significantly stronger functional connectivity to pINS than to aINS are coded *blue–green*): the aINS was more strongly connected to PFC and to ACC than was pINS; pINS meanwhile was more strongly connected to SI and to the primary motor cortex. From [79] with permission

the limbic lobe, which was considered central in mediating emotion [90]. Likewise, the ACC was targeted for surgical lesions to alleviate the suffering of chronic pain [91]; patients reported that they still experienced the pain they felt prior to surgery, but its emotional unpleasantness was dampened [92].

The ACC is subdivided cytoarchitectonically into several Brodmann areas (BA), namely 24 and 32 [93], with two further subdivisions: BA33 located in the perigenual region, and BA25 located in the subcallosal region. The ACC is functionally divided, rather independent of the cytoarchitectonic borders, into a caudal cognitive division involved in attention (BA24 and BA32) and a rostral affective division, which is more involved in emotional processes (BA24, 25, 33) [94, 95]. Dissociation between the cognitive division and pain-related processing region was elegantly demonstrated using fMRI by Davis et al. [96], who compared BOLD activation evoked by noxious stimuli to that seen during a demanding cognitive task. Activation associated with the noxious stimuli was found to be inferior and caudal to that produced by the cognitive task.

The first direct evidence for the role of ACC in processing affective components of pain came from a PET study, in which subjects under hypnosis were instructed to modulate the perceived unpleasantness of a painful stimulus while maintaining perceived pain intensity [97]. Results showed that activation of the ACC was highly correlated with the subjects' ratings of pain unpleasantness, while activation of the SI was unaltered by emotional processes. More recently, a resting state (RS) functional connectivity (FC) study [98] demonstrated that the dorsal ACC is connected with regions which comprise the affective/motivational network (anterior insula and medial thalamus), the cognitive/evaluative network (dorsolateral prefrontal cortex, inferior parietal lobule), and the motor network (pre-supplementary motor area (SMA), and SMA).

Nevertheless, these imaging and lesion data should not be interpreted too rigidly, since the ACC has been shown to have some sensory-discriminative characteristics, such as a crude nociceptive somatotopic organization [99]. Furthermore, reductions in both pain intensity and unpleasantness have been described following a neurosurgical capsulotomy—interruption of fiber tracts to the ACC [3]. The rostral ACC has also been supposed to play a relevant role in coding the variability of pain perception [100].

4.1.5 Prefrontal Cortex (PFC)

Regions of the PFC have been implicated in both pain processing and pain modulation. PFC activation seen in brain imaging studies of pain is believed to reflect attention toward the stimuli [69, 101], but it has also been shown to be directly involved in modulating responses to painful stimuli. Recently, Lobanov et al. [102] demonstrated that attention to both spatial and intensity feature of the noxious stimulus was associated with activation of fronto-parietal areas, including the PFC.

Functional imaging studies have shown that activity in the PFC reduces the pain magnitude or hyperalgesia, suggesting that the PFC can regulate the amount of pain an individual perceives. Activity in the PFC is also associated with episodes of emotional detachment, when the "suffering" element of pain is absent. Negative emotional responses can heighten the experience of pain, and the ventrolateral PFC seems to regulate these responses via interactions with the nucleus accumbens and the amygdala [103].

Using fMRI, Wager et al. have demonstrated increased PFC activity during the anticipation of pain, which was interpreted as a preemptive anticipatory response triggering a descending modulation of the pending nociceptive signals via activation of midbrain structures [104].

PFC activity is consistently seen in studies employing experimental models of chronic pain. Most commonly, sensitization was associated with a signal increase in the DLPFC. The functional significance of this activation is, however, still under debate: a positive correlation with the unpleasantness of pain indicates that DLPFC activation reflects altered cognitive-affective processing in the pathological pain state. Increased DLPFC activations might also reflect the recruitment of endogenous mechanisms of pain control [77].

4.1.6 Amygdala

The amygdala, buried beneath the uncus and located at the tail of the caudate nucleus, is a key limbic structure involved in the processing of emotional stimuli. The amygdala is suited for such processing as it is the sole subcortical structure to receive projections from every sensory area.

Functional neuroimaging studies utilizing various types of aversive stimuli including pain, habitually report amygdala activation [105]. Studies using fMRI have demonstrated that amygdala activation is associated with extremely unpleasant noxious stimuli, suggesting an involvement of this region in processing the affective component of pain [106, 107]. Other evidence from fMRI has implicated the amygdala in processing uncertainty associated with painful stimuli [108].

4.1.7 Brainstem

In addition to cortical regions, a host of midbrain structures are also involved in processing pain affect including the PAG, superior colliculus, red nucleus, nucleus cuneiformis, Edinger-Westphal nucleus, nucleus of Darkschewitsch, pretectal nuclei, interstitial nucleus, and intercolliculus nucleus [109].

Several of these structures are involved in pain modulation—the best characterized being the PAG. The PAG surrounds the cerebral aqueduct in the midbrain. Inhibitory encephalin-containing neurons in the PAG disinhibit local interneurons and in turn excite neurons in the rostral ventral medulla (RVM) and/or the locus coeruleus (LC). The aminergic efferents from the RVM and LC then project to the spinal cord and dampen pain transmission in dorsal horn neurons through several different mechanisms [103, 110–112]. Activity within the RVM can provide important antinociceptive effects, which can be beneficial during stressful circumstances. However, the RVM can also enhance nociception following inflammation and injury. Clearly, this pronociceptive effect is protective during recovery from injury and promotes tissue repair, but the failure of such an effect to resolve after the tissue has healed may result in chronic pain [103].

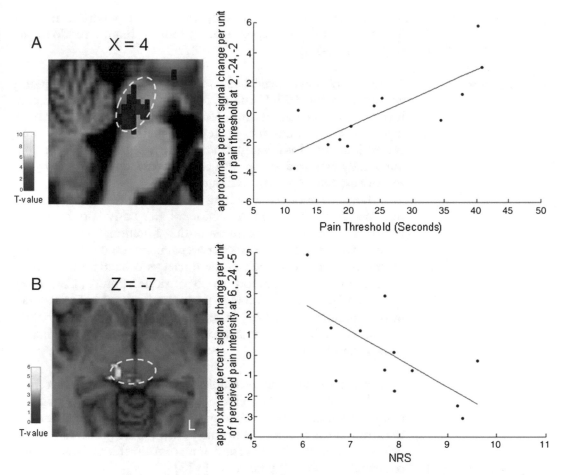

Fig. 2 (a) Correlation analysis between the BOLD response in the PAG and pain threshold, recorded in seconds, during the cold pressor test: the pain threshold directly correlated with BOLD activation in the PAG (cluster level corrected threshold $p < 0.05$, Pearson's $r = 0.63$). **(b)** Correlation analysis between the BOLD response in the PAG and pain intensity ratings, as assessed by the numerical rating scale, during the cold pressor test: the pain rating inversely correlated with BOLD activation in the PAG (cluster level corrected threshold $p < 0.05$, Pearson's $r = 0.45$). From [113] with permission

Activity within the PAG and the mesencephalic pontine reticular formation (MPRF) has been strongly associated with the development of hyperalgesia, and evidence suggests that the MPRF is specifically involved in the maintenance of central sensitization [103].

However, recently, La Cesa et al. showed that the greater the PAG activation the higher the pain threshold and the weaker the pain intensity perceived, thus highlighting the key role of the PAG in inhibiting the pain afferent pathway function [113] (Fig. 2).

The sensitivity and in-plane resolution of 1.5 and 3.0 T MRI scanners are limited in their ability to resolve fine spatial localization of many brainstem structures. In addition, brainstem functional imaging is also limited by image distortion and is susceptible to local magnetic field inhomogeneities and pulsation artifacts. Therefore, optimized approaches to study brainstem fMRI are needed [114–116].

4.1.8 Motor Cortices

A number of other cortical and subcortical regions are commonly activated during fMRI studies of pain including many regions involved in motor processing. Motor regions include the primary motor cortex, premotor cortex, supplementary motor area, cerebellum, and basal ganglia. Frequently, these regions are concomitantly activated along with those involved with affective and sensory aspects of pain processing [117].

The perception of a painful stimulus involves an orienting response and subsequent retraction of the body part being targeted. Activation of motor areas during functional neuroimaging studies is believed to reflect motor preparatory responses. However, several of these areas, such as the nuclei associated with the basal ganglia, are directly responsive to noxious stimuli [118, 119]. Using fMRI, a reliable somatotopic organization has been shown in the putamen [120, 121] in response to noxious stimuli, which indicates that this region may be involved in sensory-discriminative processing of pain.

4.1.9 Thalamus

Higher-resolution imaging studies coupled to surgical investigations have confirmed the relevance of thalamic nuclei in nociceptive processing. As a critical relay site, it is not surprising that the thalamus is implicated in chronic pain [8]. In particular, recent evidence indicates that neuroinflammation in the thalamus might contribute to chronic pain states [122].

4.2 Spinal Cord Processing of Nociceptive Stimuli

To date, only a few reports have assessed the feasibility of studying nociception using fMRI of the spinal cord [123]. One study by Brooks et al. [14] examined the spinal nociceptive signal at 1.5 T in response to noxious heat pain stimuli. Using a tailored, high-resolution scanning protocol and postprocessing techniques for controlling physiological noise, they demonstrated reliable pain-related activation in the ipsilateral dorsal horn.

A recent connectivity analysis revealed functional coupling between the spinal cord dorsal horn and typical ascending thalamo-cortical pain pathways. More importantly, the spinal cord was also functionally connected with brain regions involved in descending pain modulation, such as the PAG. A positive correlation between the individual strength of connectivity within this descending pain modulatory pathway and the behavioral pain ratings was found, thus pointing to the functional relevance of this system during the processing of physiological nociceptor pain [124]. In a similar study, BOLD fMRI response in the spinal cord was correlated with individual pain ratings, further supporting the contribution of spinal cord activity to the perception of pain [125].

5   fMRI and the Study of Higher Cognitive Pain Processing

5.1   Pain Modulation   What is clear from several studies is that nociceptive information processing, and consequent pain perception, is subject to significant pro- and anti-nociceptive modulations that can be influenced dramatically by cognitive, emotional, and contextual factors [126].

Pain modulation can occur through both endogenous mechanisms and as a result of exogenously administered agents. One final common pathway for analgesic mechanisms is believed to be through the release of endogeneous opioids [127] acting on sites in the brainstem and midbrain that block the nociceptive signal through their descending pathways; the final effects of this descending modulation are exerted either on the spinal cord and/or at the site of peripheral nerves that transmit the nociceptive stimuli. Additionally, recent research has implicated endocannibinoids in pain modulation, which may act on similar descending pathways [128].

fMRI is a useful tool for examining cerebral mechanisms of pain modulation, whereby subjects experience either analgesia or hyperalgesia—a decrease or increase in perceived pain, respectively. First, the anatomical resolution of fMRI is sufficient to localize some of the small brain regions involved in pain modulation, such as the RVM or PAG [112, 129], and the temporal resolution allows an assessment of the time course of activations within those regions. fMRI is also well suited to study procedures that evoke changes in pain perception since it accommodates the use of parametric data, whereby experimental parameters such as pain ratings (intensity, expectation, unpleasantness) can be correlated with brain activations and thus used to characterize cortical structures according to their response profile to various experimental parameters. As a corollary of increased temporal resolution, a major advantage of using fMRI to study pain modulation is the possibility of utilizing event-related designs whereby the time course of brain activations over different phases of the modulation period can be studied—the anticipation of the noxious stimulus, the onset of pain perception, changes in pain perception over time, and post-stimulus ratings. Anticipation of the painful stimulus is a crucial phase of the pain modulation process, since at this time point neural mechanisms act on descending modulatory systems to diminish or enhance the response to the stimulus [130].

fMRI has been widely applied to study modulatory processes triggered either through endogenous mechanisms utilizing cognitive strategies, such as attention [77, 131], hypnosis [132–134], placebo and nocebo effects [104, 135], or through exogenous agents, such as pharmacological [126, 136–138] and non pharmacological interventions [139].

Several lines of evidence strengthen the notion that pain modulation occurs via an integrated "frontal to brainstem to spinal cord" system. Disruption of the descending pain modulatory system may represent a point of vulnerability for the development and maintenance of chronic pain [126].

5.2   Pain Empathy

Inherent to processing the emotional component of pain is the ability to understand the emotional reactions of other people who are experiencing pain—i.e., pain empathy [140]. This rapidly growing field of empathy research is directed toward studying the mental representation of pain—both that which is perceived to be experienced by others, as well as that which is perceived as one's own. Several different types of experimental stimuli implicating other people in pain have been used in these fMRI paradigms, including photographic images [141–145], or short animations [146] of body parts in potentially tissue-damaging situations, viewing the faces of actors evoking facial expressions of pain [147], or subjects actually receiving painful stimuli [148], or those of chronic pain patients [149], or being cued that a loved one in the room was receiving painful stimuli [150].

A common finding from these studies is that the processing of pain in others recruits brain regions involved in affective processing—namely the ACC and insula. In a recent meta-analysis, Lamm et al. [151] compiled brain activation coordinates from 32 studies that had investigated empathy for pain using fMRI. Authors identified a core network, consisting of bilateral anterior insular cortex and medial/anterior cingulate cortex, that was associated with empathy for pain. Activation in these areas overlapped with activation during directly experienced pain, thus linking their involvement to representing global feeling states and the guidance of adaptive behavior for both self- and other-related experiences. Moreover, the analysis demonstrates that—depending on the type of experimental paradigm—this core network was coactivated with distinct brain regions: viewing pictures of body parts in painful situations recruited areas underpinning action understanding (inferior parietal/ventral premotor cortices) to a stronger extent; eliciting empathy by means of abstract visual information about the other's affective state more strongly engaged areas associated with inferring and representing mental states of self and others (precuneus, ventral medial prefrontal cortex, superior temporal cortex, and temporo-parietal junction).

Several transcranial magnetic stimulation studies reported modulation of sensory-discriminative regions associated with pain empathy, thus suggesting a somatotopic specificity in the perceived pain of others. Although early functional imaging studies suggested that somatosensory areas contribute very little to the neural response when seeing others' pain or empathizing with it, Morrison et al. have recently demonstrated that just viewing others' painful actions biases participants to report tactile stimulation even when none occurred [145]. Such discrepancies might have resulted from differences in experimental paradigms, since it was shown that only the picture-based paradigms activated somatosensory areas during empathy for pain [151].

fMRI has provided considerable insight into the neural mechanisms of processing pain in others, and suggests a number of interesting clinical implications. Since pain is a sensory and emotional phenomenon that is primarily experienced by the patient—as opposed to an easily measured sign of illness, such as fever or weight loss, for example—health-care professionals who are confronted with patients in pain must be able to infer their discomfort accurately and treat them accordingly. Further understanding of the neural mechanisms underlying how we interpret pain in others is an initial step toward how these neural circuits can change—depending on the clinical context or after years of repeated exposure to those in pain.

6   Combining fMRI with Morphometry

For the study of nociceptive processing and pain perception, MRI-based morphometric analyses can be used to examine neuroanatomical changes that are correlated with particular chronic pain states or to examine differences in the anatomy of specific brain regions that may underlie the variability in pain perception that is observed within a population in healthy volunteers. A number of recent studies have reported changes in cortical and subcortical brain regions in individuals with chronic pain [152, 153].

To date, a few studies have combined morphometric and functional neuroimaging analysis. The combination of functional and structural brain measures has revealed that patients with fibromyalgia had overlapping decreases of cortical thickness, brain volumes, and regional functional activity in the rACC [154]. These findings provide a neuroanatomical basis for reduced cortical activity, strengthening the importance of relating anatomical structure to physiological function.

However the future of pain fMRI lies in the development of complimentary brain imaging analysis techniques to improve our understanding of pain processing.

7   fMRI as a Therapy for Chronic Pain

Recent improvements in the speed of analysis of fMRI data have led to the possibility that "real-time" fMRI (rt-fMRI) can be developed as a potential "therapy" for chronic pain patients. In principle, if patients can be given feedback regarding the level of activity in specific areas of the brain that are associated with the perception of pain or its unpleasantness, then learning to (self)-regulate this activity can allow them to control their own chronic pain—in much the same way as neurosurgeons attempt to control a patient's pain by stimulating a specific area of the brain or by placing a lesion in a targeted area. Self-regulation training with EEG has provided the basis for much of the neurofeedback research; however, due to several methodological limitations, EEG offers relatively poor spatial specificity within the brain [155, 156]. By contrast, fMRI offers superior spatial resolution, especially for deeper brain regions, and is more suitable for targeting activity in a small, localized brain region [157, 158].

Neurofeedback, using real-time analysis of fMRI data, was initially developed by Cox et al. [159], and several groups have used this technology to study learned control over brain activity during a number of tasks [160–163]. Recently, the use of rt-fMRI has been applied to several clinical conditions whose etiology or symptoms might be linked to abnormal activity in known areas of the brain, including pain syndromes [11, 164]. In one study testing the feasibility of rt-fMRI as a neuroimaging therapy for chronic pain patients [10], normal subjects receiving experimental noxious stimuli were trained to control activity in a targeted region within the ACC. Results demonstrated that these subjects were able to use the feedback provided by rt-fMRI to either increase or decrease, on command, ACC activity, and that the level of this activity correlated with their estimates of pain evoked by the experimental stimuli. Likewise, a small cohort of chronic pain patients, following a similar rt-fMRI training paradigm, reported a significant reduction in their level of chronic pain in comparison to that of a control patient group, which received feedback training based only on autonomic measures. Furthermore, the patients in the rt-fMRI group demonstrated a direct correlation between their ability to control ACC activation and their degree of pain reduction.

In the future, rt-fMRI could also be applied to modify cortical hyperactivity that has been described for a number of other pain syndromes.

8   Future of Pain Imaging

The last decade has seen a considerable improvement in the sensitivity of fMRI in both the spatial and temporal localization of regions of activation. Moreover, the shift to higher field strengths of 4.0 and 7.0 T scanners has been shown to significantly enhance the SNR, compared to that observed with the 1.5–3.0 T scanners, which have been used in most pain studies. Pain imaging is poised to benefit from these advances more than other disciplines, because—unlike visual or motor tasks, for example, which produce changes in CBF on the order of ~40 % [165], BOLD response to nociceptive stimuli produce signal changes only in the range of ~5 % [54]. Improved spatial localization of fMRI pain protocols would provide better information regarding the specificity of somatosensory regions involved in noxious processing and their somatotopic organization; likewise, improved spatial localization and SNR will aid greatly to investigations of small brainstem structures that have been implicated in modulating pain processing at both spinal and supra-spinal levels.

Another burgeoning field in pain imaging is that of arterial spin labeling (ASL) perfusion MRI, which was first described more than a decade ago [166]. ASL directly measures CBF by magnetically labeling water molecules in inflowing arteries. Recent application of ASL to study experimental pain in healthy subjects [167, 168] has shown that this technique offers several advantages. ASL gives a precise localization of neuronal structures and has demonstrated great inter-individual reliability of activation. Additionally, compared with BOLD fMRI, ASL is well suited for pain imaging studies, since it is less susceptible to signal loss and image distortions [169] due to magnetic field inhomogeneities at the air–tissue interface around frontal, medial, and inferior temporal lobes [170, 171]. Although several methods are available to reduce these susceptibility artifacts in the BOLD signal, ASL is nevertheless an attractive alternative for pain studies that target the limbic system where signal loss from susceptibility artifacts is troublesome for such regions as the orbitofrontal cortex and amygdala. ASL also has the additional advantage of permitting longer acquisition times and is thus well suited for studying neuronal processing that may take longer to develop, such as pain modulation through hypnotic induction; fMRI, on the contrary, is limited in terms of its length of acquisition due to drifts in the baseline. However, ASL is limited in that it cannot detect changes occurring faster than 30 s and is therefore not suited for event-related designs. Additionally, the technique is limited by its temporal resolution and slice coverage in which whole brain imaging is not possible using current methods. These issues should be resolved with advances made in fast echo planar imaging sequences.

9   Conclusions

The experience of pain is complex: both sensory and cognitive components depend on a network of neural processing spread throughout many cortical and subcortical regions of the CNS. The advent of noninvasive imaging techniques has allowed us to gain a deep understanding of this multifaceted phenomenon in humans— the experimental preparation that is most relevant to our ultimate goal of understanding, managing, and alleviating pain in patients. Pain is a characteristic common to many diseases and injuries, a consequence of many medical and dental procedures, and chronic pain is essentially a syndrome in its own right—an insufferable sensation that many times has no obvious stimulus. fMRI in human subjects is helping us to understand the cerebral mechanisms of pain processing and the modulation of pain by both endogenous and exogenous factors. The results of these studies are making substantial contributions to the development of efficacious interventions for treating and alleviating pain.

References

1. Penfield W, Boldrey E (1937) Somatic motor and sensory representation in the cerebral cortex of man as studied by electrical stimulation. Brain 60(4):389–443

2. Head H, Holmes G (1911) Sensory disturbances from cerebral lesions. Brain 34(2–3):102–254

3. Talbot J et al (1991) Multiple representations of pain in human cerebral cortex. Science 251(4999):1355–1358

4. Jones AKP et al (1991) Cortical and subcortical localization of response to pain in man using positron emission tomography. Proc R Soc B Biol Sci 244(1309):39–44

5. Apkarian AV et al (1992) Persistent pain inhibits contralateral somatosensory cortical activity in humans. Neurosci Lett 140(2):141–147

6. Davis KD et al (1995) fMRI of human somatosensory and cingulate cortex during painful electrical nerve stimulation. Neuroreport 7(1):321–325

7. Flor H (2000) The functional organization of the brain in chronic pain. Prog Brain Res 129:313–322

8. Tracey I, Mantyh PW (2007) The cerebral signature for pain perception and its modulation. Neuron 55(3):377–391

9. Wager TD et al (2013) An fMRI-based neurologic signature of physical pain. N Engl J Med 368(15):1388–1397

10. deCharms RC et al (2005) Control over brain activation and pain learned by using real-time functional MRI. Proc Natl Acad Sci 102(51):18626–18631

11. Rance M et al (2014) Real time fMRI feedback of the anterior cingulate and posterior insular cortex in the processing of pain. Hum Brain Mapp 35(12):5784–5798

12. Chen JI et al (2002) Differentiating noxious- and innocuous-related activation of human somatosensory cortices using temporal analysis of fMRI. J Neurophysiol 88(1):464–474

13. Iramina K et al (1999) Effects of stimulus intensity on fMRI and MEG in somatosensory cortex using electrical stimulation. IEEE Trans Magn 35(5):4106–4108

14. Brooks J, Tracey I (2005) From nociception to pain perception: imaging the spinal and supraspinal pathways. J Anat 207(1):19–33

15. Mackey S et al (2006) FMRI evidence of noxious thermal stimuli encoding in the human spinal cord. J Pain 7(4):S25

16. Rollnik JD, Schmitz N, Kugler J (1999) Anxiety moderates cardiovascular responses to painful stimuli during sphygmomanometry. Int J Psychophysiol 33(3):253–257

17. Mobascher A et al (2010) Brain activation patterns underlying fast habituation to painful laser stimuli. Int J Psychophysiol 75(1):16–24

18. Becerra LR et al (1999) Human brain activation under controlled thermal stimulation and habituation to noxious heat: an fMRI study. Magn Reson Med 41(5):1044–1057

19. Price DD et al (1994) A comparison of pain measurement characteristics of mechanical visual analogue and simple numerical rating scales. Pain 56(2):217–226

20. Rainville P et al (2004) Rapid deterioration of pain sensory-discriminative information in short-term memory. Pain 110(3):605–615

21. Charron J, Rainville P, Marchand S (2006) Direct comparison of placebo effects on clinical and experimental pain. Clin J Pain 22(2):204–211

22. Price DD et al (1999) An analysis of factors that contribute to the magnitude of placebo analgesia in an experimental paradigm. Pain 83(2):147–156

23. Apkarian AV et al (1999) Differentiating cortical areas related to pain perception from stimulus identification: temporal analysis of fMRI activity. J Neurophysiol 81(6):2956–2963

24. Porro CA et al (2004) Percept-related activity in the human somatosensory system: functional magnetic resonance imaging studies. Magn Reson Imaging 22(10):1539–1548

25. Andrew D, Greenspan JD (1999) Peripheral coding of tonic mechanical cutaneous pain: comparison of nociceptor activity in rat and human psychophysics. J Neurophysiol 82(5):2641–2648

26. Adriaensen H et al (1984) Nociceptor discharges and sensations due to prolonged noxious mechanical stimulation--a paradox. Hum Neurobiol 3(1):53–58

27. Gallez A et al (2005) Attenuation of sensory and affective responses to heat pain: evidence for contralateral mechanisms. J Neurophysiol 94(5):3509–3515

28. Bingel U et al (2007) Habituation to painful stimulation involves the antinociceptive system. Pain 131(1):21–30

29. Valeriani M et al (2003) Reduced habituation to experimental pain in migraine patients: a CO2 laser evoked potential study. Pain 105(1):57–64

30. Nickel FT et al (2013) Brain correlates of short-term habituation to repetitive electrical noxious stimulation. Eur J Pain 18(1):56–66

31. Willis WD Jr (1985) The pain system. The neural basis of nociceptive transmission in the mammalian nervous system. Pain Headache 8:1–346

32. Adriaensen H et al (1983) Response properties of thin myelinated (A-delta) fibers in human skin nerves. J Neurophysiol 49(1):111–122

33. Ochoa J, Torebjörk E (1989) Sensations evoked by intraneural microstimulation of C nociceptor fibres in human skin nerves. J Physiol 415(1):583–599

34. Cervero F, Iggo A (1980) The substantia gelatinosa of the spinal cord: a critical review. Brain 103(4):717–772

35. Wilson P, Kitchener PD (1996) Plasticity of cutaneous primary afferent projections to the spinal dorsal horn. Prog Neurobiol 48(2):105–129

36. Craig AD et al (1994) A thalamic nucleus specific for pain and temperature sensation. Nature 372(6508):770–773

37. Legrain V et al (2011) The pain matrix reloaded: a salience detection system for the body. Prog Neurobiol 93(1):111–124

38. Bromm B, Treede RD (1984) Nerve fibre discharges, cerebral potentials and sensations induced by CO2 laser stimulation. Hum Neurobiol 3(1):33–40

39. Carmon A, Dotan Y, Sarne Y (1978) Correlation of subjective pain experience with cerebral evoked responses to noxious thermal stimulations. Exp Brain Res 33(3–4):445–453

40. Iannetti GD et al (2004) Aδ nociceptor response to laser stimuli: selective effect of stimulus duration on skin temperature, brain potentials and pain perception. Clin Neurophysiol 115(11):2629–2637

41. Spiegel J, Hansen C, Treede RD (2000) Clinical evaluation criteria for the assessment of impaired pain sensitivity by thulium-laser evoked potentials. Clin Neurophysiol 111(4):725–735

42. Leandri M et al (2006) Measurement of skin temperature after infrared laser stimulation. Neurophysiol Clin 36(4):207–218

43. Helmchen C et al (2008) Common neural systems for contact heat and laser pain stimulation reveal higher-level pain processing. Hum Brain Mapp 29(9):1080–1091

44. Iannetti GD, Mouraux A (2010) From the neuromatrix to the pain matrix (and back). Exp Brain Res 205(1):1–12

45. Apkarian AV et al (2005) Human brain mechanisms of pain perception and regulation in health and disease. Eur J Pain 9(4):463–484

46. Melzack R, Casey KL (1968) Sensory, motivational and central control determinants of pain: a new conceptual model. In: Kenshalo DR (ed) The skin senses. Charles C. Thomas Publishers, Springfield, IL, pp 423–443

47. Tracey I (2008) Imaging pain. Br J Anaesth 101(1):32–39

48. Kaas J et al (1979) Multiple representations of the body within the primary somatosensory cortex of primates. Science 204(4392):521–523

49. Kenshalo DR Jr, Isensee O (1983) Responses of primate SI cortical neurons to noxious stimuli. J Neurophysiol 50(6):1479–1496

50. Casey KL et al (1996) Comparison of human cerebral activation pattern during cutaneous warmth, heat pain, and deep cold pain. J Neurophysiol 76(1):571–581

51. Gelnar PA et al (1998) Fingertip representation in the human somatosensory cortex: an fMRI study. Neuroimage 7(4):261–283

52. Liang M, Mouraux A, Iannetti GD (2011) Parallel processing of nociceptive and non-nociceptive somatosensory information in the human primary and secondary somatosensory cortices: evidence from dynamic causal modeling of functional magnetic resonance imaging data. J Neurosci 31(24):8976–8985

53. Cheng JC et al (2015) Individual differences in temporal summation of pain reflect pronociceptive and antinociceptive brain structure and function. J Neurosci 35(26):9689–9700

54. Derbyshire GSW, Jones PAK (1998) Cerebral responses to a continual tonic pain stimulus measured using positron emission tomography. Pain 76(1):127–135

55. Disbrow E et al (1998) Somatosensory cortex: a comparison of the response to noxious thermal, mechanical, and electrical stimuli using functional magnetic resonance imaging. Hum Brain Mapp 6(3):150–159

56. Vierck CJ et al (2013) Role of primary somatosensory cortex in the coding of pain. Pain 154(3):334–344

57. Tommerdahl M et al (1996) Anterior parietal cortical response to tactile and skin-heating stimuli applied to the same skin site. J Neurophysiol 75(6):2662–2670

58. Bushnell MC et al (1999) Pain perception: is there a role for primary somatosensory cortex? Proc Natl Acad Sci 96(14):7705–7709

59. Seminowicz DA, Mikulis DJ, Davis KD (2004) Cognitive modulation of pain-related brain responses depends on behavioral strategy. Pain 112(1):48–58

60. Oshiro Y et al (2007) Brain mechanisms supporting spatial discrimination of pain. J Neurosci 27(13):3388–3394

61. Andersson JLR et al (1997) Somatotopic organization along the central sulcus, for

pain localization in humans, as revealed by positron emission tomography. Exp Brain Res 117(2):192–199

62. DaSilva AF et al (2002) Somatotopic activation in the human trigeminal pain pathway. J Neurosci 22(18):8183–8192

63. Ogino Y, Nemoto H, Goto F (2005) Somatotopy in human primary somatosensory cortex in pain system. Anesthesiology 103(4):821–827

64. Kaas JH (1983) What, if anything, is SI? Organization of first somatosensory area of cortex. Physiol Rev 63(1):206–231

65. Albanese MC et al (2007) Memory traces of pain in human cortex. J Neurosci 27(17):4612–4620

66. Schmid J et al (2015) Neural underpinnings of nocebo hyperalgesia in visceral pain: a fMRI study in healthy volunteers. Neuroimage 120:114–122

67. Greenspan JD, Lee RR, Lenz FA (1999) Pain sensitivity alterations as a function of lesion location in the parasylvian cortex. Pain 81(3):273–282

68. Ploner M, Freund HJ, Schnitzler A (1999) Pain affect without pain sensation in a patient with a postcentral lesion. Pain 81(1):211–214

69. Coghill RC et al (1999) Pain intensity processing within the human brain: a bilateral, distributed mechanism. J Neurophysiol 82(4):1934–1943

70. Maihöfner C, Herzner B, Otto Handwerker H (2006) Secondary somatosensory cortex is important for the sensory-discriminative dimension of pain: a functional MRI study. Eur J Neurosci 23(5):1377–1383

71. Gracely RH et al. (2004) Pain catastrophizing and neural responses to pain among persons with fibromyalgia. Brain 127(4):835–843

72. Sawamoto N et al (2000) Expectation of pain enhances responses to nonpainful somatosensory stimulation in the anterior cingulate cortex and parietal operculum/posterior insula: an event-related functional magnetic resonance imaging study. J Neurosci 20(19):7438–7445

73. Oertel BG et al (2012) Separating brain processing of pain from that of stimulus intensity. Hum Brain Mapp 33(4):883–894

74. Wiech K et al (2014) Differential structural and resting state connectivity between insular subdivisions and other pain-related brain regions. Pain 155(10):2047–2055

75. Schilder P, Stengel E (1932) Asymbolia for pain. Arch Neurol Psychiatry 25(3):598–600

76. Berthier M, Starkstein S, Leiguarda R (1988) Asymbolia for pain: a sensory-limbic disconnection syndrome. Ann Neurol 24(1):41–49

77. Wiech K, Ploner M, Tracey I (2008) Neurocognitive aspects of pain perception. Trends Cogn Sci 12(8):306–313

78. Maihöfner C, Handwerker HO (2005) Differential coding of hyperalgesia in the human brain: a functional MRI study. Neuroimage 28(4):996–1006

79. Peltz E et al (2011) Functional connectivity of the human insular cortex during noxious and innocuous thermal stimulation. Neuroimage 54(2):1324–1335

80. Mazzola L, Isnard J, Mauguiere F (2006) Somatosensory and pain responses to stimulation of the second somatosensory area (SII) in humans. A comparison with SI and insular responses. Cereb Cortex 16(7):960–968

81. Segerdahl AR et al (2015) The dorsal posterior insula subserves a fundamental role in human pain. Nat Neurosci 18(4):499–500

82. Craig AD (2002) New and old thoughts on the mechanisms of spinal cord injury pain. In: Yezierski RP, Burchiel KJ (eds) Spinal cord injury pain: assessment, mechanisms, management. IASP Press, Seattle 237–264

83. Blomqvist A (2000) Cytoarchitectonic and immunohistochemical characterization of a specific pain and temperature relay, the posterior portion of the ventral medial nucleus, in the human thalamus. Brain 123(3):601–619

84. Peyron R et al (2002) Role of operculoinsular cortices in human pain processing: converging evidence from PET, fMRI, dipole modeling, and intracerebral recordings of evoked potentials. Neuroimage 17(3):1336–1346

85. Brooks JCW et al (2005) Somatotopic organisation of the human insula to painful heat studied with high resolution functional imaging. Neuroimage 27(1):201–209

86. Henderson LA, Rubin TK, Macefield VG (2011) Within-limb somatotopic representation of acute muscle pain in the human contralateral dorsal posterior insula. Hum Brain Mapp 32(10):1592–1601

87. Hutchison WD et al (1999) Pain-related neurons in the human cingulate cortex. Nat Neurosci 2(5):403–405

88. Jones AKP et al (1991) In vivo distribution of opioid receptors in man in relation to the cortical projections of the medial and lateral pain systems measured with positron emission tomography. Neurosci Lett 126(1):25–28

89. Baumgärtner U et al (2007) High opiate receptor binding potential in the human lateral pain system: a (FEDPN)PET study. Clin Neurophysiol 118(4):e12

90. Pessoa L (2008) On the relationship between emotion and cognition. Nat Rev Neurosci 9(2):148–158

91. Pillay PK, Hassenbusch SJ (1992) Bilateral MRI-guided stereotactic cingulotomy for intractable pain. Stereotact Funct Neurosurg 59(1–4):33–38

92. Gybels JM, Sweet WH (1989) Neurosurgical treatment of persistent pain. Physiological and pathological mechanisms of human pain. Pain Headache 11:1–402

93. Vogt BA et al (1995) Human cingulate cortex: surface features, flat maps, and cytoarchitecture. J Comp Neurol 359(3):490–506

94. Devinsky O, Morrell MJ, Vogt BA (1995) Contributions of anterior cingulate cortex to behaviour. Brain 118(1):279–306

95. Vogt BA (2005) Pain and emotion interactions in subregions of the cingulate gyrus. Nat Rev Neurosci 6(7):533–544

96. Davis KD et al (1997) Functional MRI of pain- and attention-related activations in the human cingulate cortex. J Neurophysiol 77(6):3370–3380

97. Rainville P et al (1997) Pain affect encoded in human anterior cingulate but not somatosensory cortex. Science 277(5328):968–971

98. Wilcox CE et al (2015) The subjective experience of pain: an FMRI study of percept-related models and functional connectivity. Pain Med 16(11):2121–33

99. Arienzo D et al (2006) Somatotopy of anterior cingulate cortex (ACC) and supplementary motor area (SMA) for electric stimulation of the median and tibial nerves: an fMRI study. Neuroimage 33(2):700–705

100. Kroger IL, Menz MM, May A (2015) Dissociating the neural mechanisms of pain consistency and pain intensity in the trigemino-nociceptive system. Cephalalgia [published online before print October 22, 2015, doi:10.1177/0333102415612765]

101. Casey KL (1999) Forebrain mechanisms of nociception and pain: analysis through imaging. Proc Natl Acad Sci 96(14):7668–7674

102. Lobanov OV et al (2013) Frontoparietal mechanisms supporting attention to location and intensity of painful stimuli. Pain 154(9):1758–1768

103. Tracey I (2011) Can neuroimaging studies identify pain endophenotypes in humans? Nat Rev Neurol 7(3):173–181

104. Wager TD et al (2004) Placebo-induced changes in FMRI in the anticipation and experience of pain. Science 303(5661):1162–1167

105. Zald DH (2003) The human amygdala and the emotional evaluation of sensory stimuli. Brain Res Rev 41(1):88–123

106. Schneider F et al (2001) Subjective ratings of pain correlate with subcortical-limbic blood flow: an fMRI study. Neuropsychobiology 43(3):175–185

107. Berna C et al (2010) Induction of depressed mood disrupts emotion regulation neurocircuitry and enhances pain unpleasantness. Biol Psychiatry 67(11):1083–1090

108. Bornhovd K et al (2002) Painful stimuli evoke different stimulus–response functions in the amygdala, prefrontal, insula and somatosensory cortex: a single-trial fMRI study. Brain 125(6):1326–1336

109. Schulte LH, Sprenger C, May A (2015) Physiological brainstem mechanisms of trigeminal nociception: an fMRI study at 3T. Neuroimage 124(Pt A):518–525

110. Mason P (2005) Deconstructing endogenous pain modulations. J Neurophysiol 94(3):1659–1663

111. Fields HL, Heinricher MM (1985) Anatomy and physiology of a nociceptive modulatory system. Philos Trans R Soc Lond B Biol Sci 308(1136):361–374

112. Fields HL (2000) Pain modulation: expectation, opioid analgesia and virtual pain. Prog Brain Res 122:245–253

113. La Cesa S et al (2014) fMRI pain activation in the periaqueductal gray in healthy volunteers during the cold pressor test. Magn Reson Imaging 32(3):236–240

114. Dunckley P et al (2005) A comparison of visceral and somatic pain processing in the human brainstem using functional magnetic resonance imaging. J Neurosci 25(32):7333–7341

115. Tracey I, Iannetti GD (2006) Brainstem functional imaging in humans. Suppl Clin Neurophysiol 58:52–67

116. Guimaraes AR et al (1998) Imaging subcortical auditory activity in humans. Hum Brain Mapp 6(1):33–41

117. Farina S et al (2003) Pain-related modulation of the human motor cortex. Neurol Res 25(2):130–142

118. Chudler EH, Dong WK (1995) The role of the basal ganglia in nociception and pain. Pain 60(1):3–38

119. Borsook D et al (2010) A key role of the basal ganglia in pain and analgesia--insights gained through human functional imaging. Mol Pain 6:27

120. Tomycz ND, Friedlander RM (2011) The experience of pain and the putamen: a new link found with functional MRI and diffusion tensor imaging. Neurosurgery 69(4):N12–N13

121. Bingel U et al (2004) Somatotopic representation of nociceptive information in the

putamen: an event-related fMRI study. Cereb Cortex 14(12):1340–1345

122. Loggia ML et al (2015) Evidence for brain glial activation in chronic pain patients. Brain 138(Pt 3):604–615

123. Cahill CM, Stroman PW (2011) Mapping of neural activity produced by thermal pain in the healthy human spinal cord and brain stem: a functional magnetic resonance imaging study. Magn Reson Imaging 29(3):342–352

124. Sprenger C, Finsterbusch J, Buchel C (2015) Spinal cord-midbrain functional connectivity is related to perceived pain intensity: a combined spino-cortical FMRI study. J Neurosci 35(10):4248–4257

125. Khan HS, Stroman PW (2015) Inter-individual differences in pain processing investigated by functional magnetic resonance imaging of the brainstem and spinal cord. Neuroscience 307:231–241

126. Bingel U, Tracey I (2008) Imaging CNS modulation of pain in humans. Physiology (Bethesda) 23:371–380

127. Levine JD et al (1978) The narcotic antagonist naloxone enhances clinical pain. Nature 272(5656):826–827

128. Hohmann AG, Suplita RL (2006) Endocannabinoid mechanisms of pain modulation. AAPS J 8(4):E693–E708

129. Tracey I et al (2002) Imaging attentional modulation of pain in the periaqueductal gray in humans. J Neurosci 22(7):2748–2752

130. Porro CA (2003) Functional imaging and pain: behavior, perception, and modulation. Neuroscientist 9(5):354–369

131. Bantick SJ et al (2002) Imaging how attention modulates pain in humans using functional MRI. Brain 125(2):310–319

132. Roder CH et al (2007) Pain response in depersonalization: a functional imaging study using hypnosis in healthy subjects. Psychother Psychosom 76(2):115–121

133. Schulz-Stübner S et al (2004) Clinical hypnosis modulates functional magnetic resonance imaging signal intensities and pain perception in a thermal stimulation paradigm. Reg Anesth Pain Med 29(6):549–556

134. Nakata H, Sakamoto K, Kakigi R (2014) Meditation reduces pain-related neural activity in the anterior cingulate cortex, insula, secondary somatosensory cortex, and thalamus. Front Psychol 5:1489

135. Tracey I (2010) Getting the pain you expect: mechanisms of placebo, nocebo and reappraisal effects in humans. Nat Med 16(11):1277–1283

136. Maihöfner C et al (2007) Brain imaging of analgesic and antihyperalgesic effects of cyclooxygenase inhibition in an experimental human pain model: a functional MRI study. Eur J Neurosci 26(5):1344–1356

137. Wise RG et al (2007) The anxiolytic effects of midazolam during anticipation to pain revealed using fMRI. Magn Reson Imaging 25(6):801–810

138. Sanders D et al (2015) Pharmacologic modulation of hand pain in osteoarthritis: a double-blind placebo-controlled functional magnetic resonance imaging study using naproxen. Arthritis Rheumatol 67(3):741–751

139. Li K et al (2015) The effects of acupuncture treatment on the right frontoparietal network in migraine without aura patients. J Headache Pain 16:518

140. Thompson E (2001) Empathy and consciousness. J Conscious Stud 8(5–7):1–32

141. Jackson PL et al (2006) Empathy examined through the neural mechanisms involved in imagining how I feel versus how you feel pain. Neuropsychologia 44(5):752–761

142. Jackson PL, Meltzoff AN, Decety J (2005) How do we perceive the pain of others? A window into the neural processes involved in empathy. Neuroimage 24(3):771–779

143. Lamm C et al (2007) What are you feeling? Using functional magnetic resonance imaging to assess the modulation of sensory and affective responses during empathy for pain. PLoS One 2(12):e1292

144. Moriguchi Y et al (2006) Empathy and judging other's pain: an fMRI study of alexithymia. Cereb Cortex 17(9):2223–2234

145. Morrison I et al (2013) "Feeling" others' painful actions: the sensorimotor integration of pain and action information. Hum Brain Mapp 34(8):1982–1998

146. Morrison I, Peelen MV, Downing PE (2007) The sight of others' pain modulates motor processing in human cingulate cortex. Cereb Cortex 17(9):2214–2222

147. Simon D et al (2006) Brain responses to dynamic facial expressions of pain. Pain 126(1):309–318

148. Botvinick M et al (2005) Viewing facial expressions of pain engages cortical areas involved in the direct experience of pain. Neuroimage 25(1):312–319

149. Saarela MV et al (2007) The compassionate brain: humans detect intensity of pain from another's face. Cereb Cortex 17(1):230–237

150. Singer T et al (2004) Empathy for pain involves the affective but not sensory components of pain. Science 303(5661):1157–1162

151. Lamm C, Decety J, Singer T (2011) Meta-analytic evidence for common and distinct neural networks associated with directly experienced pain and empathy for pain. Neuroimage 54(3):2492–2502

152. Apkarian AV et al (2004) Chronic back pain is associated with decreased prefrontal and thalamic gray matter density. J Neurosci 24(46):10410–10415

153. Schmidt-Wilcke T et al (2006) Affective components and intensity of pain correlate with structural differences in gray matter in chronic back pain patients. Pain 125(1):89–97

154. Jensen KB et al (2013) Overlapping structural and functional brain changes in patients with long-term exposure to fibromyalgia pain. Arthritis Rheum 65(12):3293–3303

155. Vernon DJ (2005) Can neurofeedback training enhance performance? An evaluation of the evidence with implications for future research. Appl Psychophysiol Biofeedback 30(4):347–364

156. Tao JX et al (2005) Intracranial EEG substrates of scalp EEG interictal spikes. Epilepsia 46(5):669–676

157. Lantz G et al (2001) Localization of distributed sources and comparison with functional MRI. Epileptic Disord, Special Issue:45–58.

158. Stern JM (2006) Simultaneous electroencephalography and functional magnetic resonance imaging applied to epilepsy. Epilepsy Behav 8(4):683–692

159. Cox RW, Jesmanowicz A, Hyde JS (1995) Real-time functional magnetic resonance imaging. Magn Reson Med 33(2):230–236

160. Yoo S-S, Jolesz FA (2002) Functional MRI for neurofeedback: feasibility study on a hand motor task. Neuroreport 13(11):1377–1381

161. Weiskopf N et al (2003) Physiological self-regulation of regional brain activity using real-time functional magnetic resonance imaging (fMRI): methodology and exemplary data. Neuroimage 19(3):577–586

162. Posse S et al (2003) Real-time fMRI of temporolimbic regions detects amygdala activation during single-trial self-induced sadness. Neuroimage 18(3):760–768

163. Weiskopf N (2012) Real-time fMRI and its application to neurofeedback. Neuroimage 62(2):682–692

164. Guan M et al (2015) Self-regulation of brain activity in patients with postherpetic neuralgia: a double-blind randomized study using real-time FMRI neurofeedback. PLoS One 10(4):e0123675

165. Ramsey NF et al (1996) Functional mapping of human sensorimotor cortex with 3D BOLD fMRI correlates highly with H215O PET rCBF. J Cereb Blood Flow Metab 16(5):755–759

166. Detre JA et al (1992) Perfusion imaging. Magn Reson Med 23(1):37–45

167. Owen DG et al (2008) Quantification of pain-induced changes in cerebral blood flow by perfusion MRI. Pain 136(1):85–96

168. Maleki N et al (2013) Pain response measured with arterial spin labeling. NMR Biomed 26(6):664–673

169. Wang J et al (2004) Reduced susceptibility effects in perfusion fMRI with single-shot spin-echo EPI acquisitions at 1.5 tesla. Magn Reson Imaging 22(1):1–7

170. Devlin JT et al (2000) Susceptibility-induced loss of signal: comparing PET and fMRI on a semantic task. Neuroimage 11(6):589–600

171. Ojemann JG et al (1997) Anatomic localization and quantitative analysis of gradient refocused echo-planar fMRI susceptibility artifacts. Neuroimage 6(3):156–167

Chapter 17

fMRI of the Sensorimotor System

Massimo Filippi, Roberta Messina, and Maria A. Rocca

Abstract

The extensive application of fMRI to the assessment of the human sensorimotor system has disclosed a complexity that is largely beyond our original understanding. From the available data, it is accepted that this system consists of a large, and somewhat yet unknown, number of cortical and subcortical areas, with a precise location and a specialized function. In particular, a large number of regions in the frontal and parietal lobes contribute to different aspects of motor act performance. It is also evident that the properties and potentialities of this network still need to be fully elucidated by further research. Defining how the human sensorimotor system works is of outmost importance for understanding its dysfunction in case of diseases and also to develop potential therapeutic strategies capable to enhance its functional plasticity and reserve.

Key words Sensorimotor system, Mirror-neuron system, fMRI, Motor training

1 Introduction

During the past 15 years, fMRI has become a valuable tool to study normal brain function, due to the development of revolutionary methods for data acquisition and postprocessing, as well as for paradigm design. Due to its noninvasiveness and relatively high spatial and temporal resolution, fMRI has rapidly substituted other techniques, such as positron emission tomography (PET), in the assessment of brain function. In addition, the combination of fMRI with neurophysiological techniques, such as transcranial magnetic stimulation (TMS), is providing important pieces of information for the understanding of brain function in healthy individuals, which, on turn, is critical for the interpretation of functional changes in diseased people.

This chapter summarizes the major contributions of fMRI for the in vivo assessment of the sensorimotor network in healthy human subjects, with a specific focus on studies of performance of a simple motor task with the dominant upper limb.

Massimo Filippi (ed.), *fMRI Techniques and Protocols*, Neuromethods, vol. 119,
DOI 10.1007/978-1-4939-5611-1_17, © Springer Science+Business Media New York 2016

2 Sensorimotor Paradigms

Activity of the sensorimotor system has been investigated by using several experimental paradigms. The majority of the studies analyzed the performance of active tasks consisting of movement of the hand, using tasks that require flexion-extension of the hand and/or fingers, tapping the hand or fingers, closing–opening the hand, hand manipulation, and squeezing. A few studies investigated the movement of the foot, leg, arm, shoulder, and tongue, with the main goal of defining the somatotopic location and hemispheric lateralization of these body parts and assessing the complex interplay between multiple sensorimotor areas [1–3]. Other studies analyzed the fMRI correlates of interlimb coordination [4, 5].

The brain activations associated to the performance of passive tasks have also been evaluated [2, 6–9]. This strategy has mainly been prompted by the need of obtaining meaningful comparisons between controls and patients with neurological diseases that might impair the "ability" to perform active tasks correctly [6–8]. The use of passive tasks is also supported by the fact that there are reciprocal projections between the motor and the related sensory cortices; hence, patterns of brain activations that reflect local field potentials from presynaptic activity primarily [10], even with entirely passive movements, may identify those brain regions involved in active voluntary movements. This hypothesis has indeed been confirmed by fMRI studies of healthy controls which have demonstrated that activations associated to active and passive limb movements are similar in localization and size [6, 7, 11]. However, recent studies revealed a greater activation of brain areas responsible for motor planning and visuomotor coordination during active-movements execution and a selective recruitment of regions involved in motor response inhibition during passive movements [9, 12]. Finally, it is now established that motor network recruitment can be elicited also by the imagination of movements [13–15].

One of the major caveats in the set up of fMRI experiments of the sensorimotor system is an adequate monitoring of subjects' performance during task execution, which might require to be corrected during the statistical analysis. Several variables have been shown to influence the observed patterns of movement-associated cortical activations in healthy subjects during motor task execution, including:

1. Movement rate, which has been positively correlated with the recruitment of the contralateral primary sensorimotor cortex (SMC) [16], supplementary motor area (SMA) [17] and ipsilateral cerebellar cortex [11, 18].

2. Force, as suggested by the load-dependent effect observed in the primary SMC [16, 19] and cerebellum [20], and the different pattern of brain activity associated with the production of static or dynamic force pulses [21].

3. Movement complexity, which has been shown to modulate activity of the primary SMC [22, 23], SMA, and premotor cortex [11, 24], as well as several regions of the parietal lobes [16].

Several strategies can be adopted to minimize these possible confounding factors, including an accurate monitoring of task performance during fMRI acquisition either visually or using more sophisticated techniques, such as position and force sensors.

Other variables that need to be considered when dealing with motor task investigations include:

1. Hemispheric dominance. Approximately 90% of the population has a left-hemispheric dominance for processing motor acts [25]. In line with this, fMRI studies have demonstrated that motor-related activations are usually lateralized to the left hemisphere in right-handers and bilateralized or lateralized to the right hemisphere in left-handers [26–30].

2. Gender. Women have larger activations of cortical motor areas during motor tasks, while men exhibited significantly stronger activation in the striatal regions [31].

3. Age. There appears to be greater motor task-related brain activity in a wider network of brain regions in older compared to younger subjects [32, 33]. A study of healthy individuals has demonstrated an age-related increased functional connectivity of motor cortices between the two hemispheres [34]. These results are consistent with a more general reduction of functional lateralization of the motor cortex recruitment with aging, which has been interpreted as a compensatory response to increased functional demands (Fig. 1) [35, 36]. However, other studies suggested that the increased cerebral recruitment might reflect an inefficient response to an age-related higher difficulty of task [33]. An age-specific pattern of activations of cerebral areas during motor imagery has also been demonstrated [37].

3   Components of the Human Sensorimotor Network

The control of motor acts is a complex process that involves several motor, sensory, and association areas, including the primary SMC, secondary sensorimotor cortex (SII), SMA, cingulum, basal ganglia, cerebellum, and several regions located in the frontal and parietal lobes. Thus, the sensorimotor network is a relatively complex system, with a hierarchic organization. Anatomically, this view is supported by the presence of large, somatotopically organized, primary cortices with converging projections to smaller association areas. The extensive application of functional techniques to the assessment of this system's function in healthy subjects is contributing to increase our knowledge of its behavior and connections.

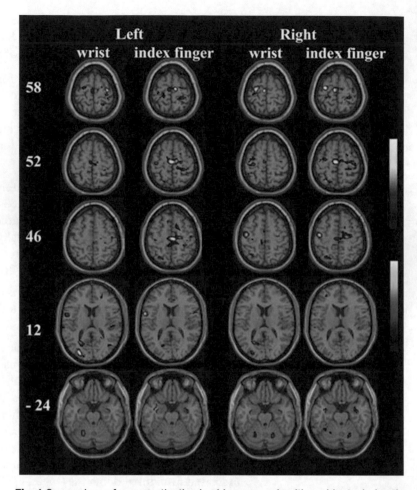

Fig. 1 Comparison of mean activation in old vs. young healthy subjects during the performance of wrist extension/flexion and index finger abduction/adduction with the left and right upper limb, respectively. Areas more significantly activated in old subjects are coded in *red spectrum*, while areas more significantly activated in young subjects are coded in *blue spectrum*. Activations have been overlaid on a standard T1 brain image in neurological view. For each motor task, the contralateral primary sensorimotor cortex and the premotor cortex had significantly greater activation in the young group and caudal supplementary motor area had significantly greater activation in the old group. Ipsilateral sensorimotor cortex was more significantly activated in the old group for index finger motor tasks of both hands (From ref. [35], with permission)

In addition, this has allowed to define the role that the different components of the network have during the performance of a motor act.

3.1 The Primary Sensorimotor Cortex

Anatomically, the primary SMC is the cortex lying within the anterior and posterior banks of the central sulcus [1]. In line with clinical and electrical stimulation studies, fMRI studies confirmed the somatotopic organization of the primary SMC of the left hemisphere, with distinct subregions controlling movements of the foot, arm, and face [38, 39]. As already mentioned, there is a large body of evidence supporting the prominent role of the primary SMC of the dominant hemisphere in the performance of simple motor acts. In healthy subjects, the role of the ipsilateral primary SMC in the control of movements is still controversial, since several studies have reported conflicting results with respect to the occurrence of ipsilateral primary SMC activation [16, 26, 40, 41]. In particular, while there is a general agreement on consistent activation of the primary SMC of the ipsilateral hemisphere with increasing motor task complexity [16, 24], only a few studies reported its activation during simple task performance [26, 38, 40]. A mechanism that has been advocated to shed lights on primary SMC behavior is transcallosal inhibition. In healthy individuals, a transcallosal inhibitory pathway between the primary motor cortices of the two hemispheres has been demonstrated by neurophysiological studies [42, 43] and has been postulated to be responsible for the control of homologous hand muscles during unilateral movements [42, 43]. These data are supported by fMRI studies that have shown a decreased activation of the ipsilateral primary SMC during sequential finger movements [41, 44–46]. The reason for the ipsilateral inhibition during unilateral hand movements remains speculative. However, this decreased excitability could improve the capacity to perform fine movements of the fingers, for which a high level of dexterity is needed. Usually, such movements are carried out unilaterally. A suppression of excitability of the ipsilateral SMC would then minimize the risk of contralateral interference, and improve the cortical focus on unilateral activation [47]. The inhibitory interhemispheric interactions decrease with age [48] and this loss of inhibition appears to be ameliorated by physical activity [49].

3.2 The Supplementary Motor Area

Another important component of the motor network is the SMA, which is the cortex lying above the cingulate sulcus and anteriorly within the paracentral lobule [1]. The SMA contributes to the preparation, coordination, temporal course, and execution of movements [50–52]. Studies of healthy individuals suggest that the movement-related activity of the primary SMC might be mediated by the extensive input it receives from the SMA [53], which might act as facilitator or suppressor according to the task conditions [54], and that the SMA recruitment might increase by increasing task difficulty and complexity [16, 21, 24]. The extent of SMA activation has been inversely related to the amount of training an individual has gained with that specific task [50, 52, 55]. Inter- and intrahemispheric connections between the primary SMC, the premotor cortex, and the SMA are likely to be mediated primarily by the SMA [56]. In addition, strong bilateral connections exist between bilateral SMA and the basal ganglia, thus playing an important role in motor planning and behavior [57]. In agreement with this notion, lesions of the SMA typically result in alterations of bimanually coordinated movements [58–60]. Efferents from the SMA project directly to the brainstem and the cervical cord; as a consequence, an increased SMA activation might represent recruitment of motor pathways that can function in parallel with the contralateral corticospinal tract [61]. A recent study showed that spinal cord injury (SCI) can lead to network functional changes, including increased neural activity within the SMA, that might reflect a compensatory mechanism [62].

Functionally, the SMA can be divided into the pre-SMA (located more rostrally) and the SMA-proper (located more caudally), and event-related fMRI studies have shown that the pre-SMA is activated preferentially during movement preparation [51, 63] and contributes to motor response inhibition [64]. In addition, it has been shown that pre-SMA recruitment precedes primary SMC activation by several seconds [65].

3.3 The Frontal Cortex

The frontal cortex contains many areas contributing to the motor network [66, 67]. In addition to the primary SMC, these areas include the ventral premotor areas (including the inferior frontal gyrus [IFG]), the dorsal premotor cortex (sometimes divided into a caudal and a rostral part) (PMd), and a set of motor areas on the medial wall of the hemispheres, such as the SMA and the cingulate motor area (CMA). The premotor areas in the frontal lobe influence motor output through connections with the primary SMC and direct projections to the spinal cord [68]. All previous premotor areas contain corticospinal neurons that give a substantial contribution to corticospinal projections, which have a high degree of topographic organization [39, 69].

The role of the left inferior frontal lobe (ventral premotor cortex/Broca's area) in motor sequence control is well documented by several studies [70–74]. Activation of Broca's area has been reported in various functional imaging studies based on finger movements [71, 72], movement imagination, and motor learning [70]. This area is supposed to receive rich sensory information originating from the parietal lobe (including the SII) and to use it for action [75]. In addition, modulation of this area's activity by task complexity has been clearly documented [73]. Studies in humans have shown that this region is important for encoding hand/object interactions [3, 75] and mediating motor response inhibition via connectivity with the preSMA [74].

The PMd has an important role in motor preparation, selection, and initiation of voluntary actions [76–78]. Imaging and TMS experiments suggest that the PMd cortex of the left hemisphere is dominant in right-handed people [79]. This area is reciprocally connected with the ipsilateral and contralateral primary SMC, as well as with the parietal cortex and the contralateral PMd

[79–81]. Using a labeling retrograde strategy, Marconi et al. [82] showed transcallosal homotopic and heterotopic connections between different portions of the PMd of the two hemispheres and between the two PMd and the SMA. A correlation between preservation of motor performance after disruption of the left PMd activity by means of TMS and increased activation of the right PMd cortex, the SMA, and the cingulate cortex has been demonstrated in healthy individuals. This pattern was not seen after TMS inhibition of the left SMC, while TMS of the reorganized right PMd disrupted motor performance. These findings suggest that adaptive changes of PMd function might contribute to maintain motor behavior despite the presence of structural damage [83]. A recent experiment confirmed the same results, revealing a stronger TMS-related increase in activity in medial and premotor areas in association with external cues [81].

The anatomical variability of the cingulate sulcus in humans hampers functional analysis of this region. The caudal CMA is considered to be primarily involved in movement execution [1, 66, 84], while the rostral portion of the CMA has been shown to have a role in action selection [85], initiation, motivation, and goal-directed behaviors [86]. A recent study demonstrated that the CMA has a key role in realizing intentional motor control (Fig. 2) [87]. Activation of this region has also been found to be related to the presentation of new motor tasks and perhaps its recruitment reflects relative task difficulty [24, 88]. This cortical area is involved in attentional tasks and subserves several executive functions [89]. In addition, the CMA has been suggested to play an important role in conflict monitoring [90, 91]. The role of the CMA in the execution of spatially complex coordination tasks has been underlined by a study of Wenderoth et al. [92], where an increased CMA activation was detected during the performance of a bimanual task.

3.4 The Parietal Cortex

The parietal cortex is formed by a multiplicity of independent areas, each of which deals with specific aspects of sensory information [93]. Physiological and imaging techniques have been extensively applied to define the location and functional specialization of parietal cortex regions in humans. Although this effort resulted in the identification of several areas related to the sensorimotor network, understanding their precise function and relationship still require further experiments.

Among the regions of the parietal cortex, the SII is considered to function as a high-order processing area for somatosensory perception, and its activation seems also to be related to attention, manual dexterity, and coordination [94, 95]. SII activity has been associated with processing of the temporal features of somatic sensations, sensorimotor integration [96, 97], tactile recognition, and tactile learning and memory [98]. In addition, neurons from SII project directly to the spinal cord [99], indicating that this region

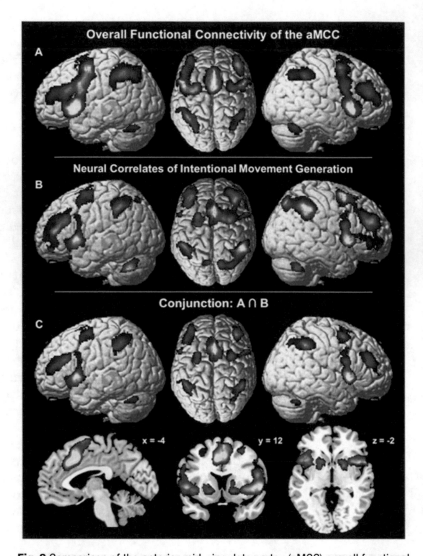

Fig. 2 Comparison of the anterior mid-cingulate cortex (aMCC) overall functional connectivity (FC) with neural correlates of intentional movement generation. Intentional movement initiation yielded increased neural activity in virtually the same brain regions involved in the FC network of the aMCC. This observation was quantitatively verified by the conjunction of the FC analysis of the aMCC (**a**) and the neural network of intentional movement initiation (**b**), revealing a "core network" that comprised the aMCC, supplementary motor area (SMA), pre-SMA, dorsolateral prefrontal cortex, dorsolateral premotor cortex, area 44, anterior insula, inferior parietal lobule, intraparietal sulcus, cerebellum, anterior putamen, and the right caudate nucleus (**c**) (From ref. [87], with permission)

might provide alternative pathways for motor control in case of primary SMC injury. SII is known to have extensive connections with the prefrontal cortex, the parietal lobe, and the insula. Similarly to the primary SMC, SII has a somatotopic representation of different body parts, with the upper limb areas located more anteriorly and more inferiorly than the lower limb areas [39, 100].

Numerous areas along the intraparietal sulcus (IPS) have also been associated with processing of sensorimotor tasks. The anterior part of the IPS contains neurons that discharge in response to 3D object presentation and during grasping movements [101, 102], and it is connected to the IFG for control of action in object manipulation [3, 71]. This area has a central role for visuomotor integration (crossmodal information process) [93]. In addition, increased activity of the IPS has also been described in normal subjects during complex finger movement sequences [22]. Activity of the IPS has been associated to object matching and grasping, with a selective involvement in processing intrinsic object attributes, such as size and shape, for the execution of an efficient grasp [93, 102]. The precuneus has been related to the execution of spatially complex coordination tasks [92], which require shifting attention between different locations in space. Finally, the superior parietal gyrus (SPG), that has a well-demonstrated hand/finger representation [71], is thought to be involved in the elaboration of somatosensory modalities and computations underlying a transformation from spatial target to movement vector [103].

3.5 The Basal Ganglia, Insula, and Thalamus

The basal ganglia have extensive connections to the motor and somatosensory cortices and are involved in motor programming, execution, and control [104, 105]. In particular, basal ganglia activity has been associated with motor program selection and suppression at early stages of motor planning, as well as with control of movement simulation [57, 106]. In addition, they are implicated in the formation of motor skills and are part of subsystems whose activity has been associated with timing of motor acts [107–110].

The thalamus [104, 111] and the insula [112] have extensive connections with the motor and somatosensory cortices and are involved in motor execution [104, 112]. Interestingly, the thalamus is an important relay station of the complex re-entrant circuitry that links the motor and the prefrontal cortices to the basal ganglia, which is part of the feedback loops of the limbic system able to modulate the cortical motor output [57, 113].

The insular cortex has been shown to play a role in crossmodal transfer of information [114]. In addition, the insular cortex, which has connections with numerous cortical and subcortical motor regions, is involved in the synchronization of movement kinematic [115] and skeletomotor body orientation [116].

3.6 The Cerebellum

The cerebellum integrates sensory information and motor programs to coordinate fine movements. Functionally, the cerebellum is organized in modules arranged in the medio-lateral direction, being the medial part responsible for control of posture and the lateral regions for coordination and movements. Anatomically, the cerebellum is divided along the rostro-caudal axis in the anterior lobe, which contains a somatotopic representation of movement of the ipsilateral side and contributes to motor control [117], and the posterior lobe, which is thought to be related to motor imagery [75] and motor learning [118, 119]. The posterior lobe of the cerebellum has projections from and to regions of the parietal cortex, involved in the processing of sensory information [120–122], which is then used to correct movements. However, recent studies revealed functionally distinct areas within and across cerebellar lobules, thus demonstrating a functional parcellation that is independent of anatomical lobular divisions [122].

Several imaging studies have reported a cerebellar recruitment associated to timing of rhythmic movements [123]. Some studies also described increased cerebellar activation corresponding to increase in movement frequency and velocity [50, 110, 124, 125]. The cerebellum has also been involved in the "automatization" (improvement of motor performance) of learned skills, establishment of movement strategies, and consolidation of such a motor knowledge [110, 126, 127]. Further evidence supporting the role of the cerebellum in motor learning is based on data from patients with focal cerebellar lesions, who have shown impairment in learning new motor skills [126, 128, 129], imaging studies that highlighted its contribution to motor recovery after SCI [62] and other studies of motor learning in healthy individuals, who showed prominent cerebellar recruitment [130, 131].

4 Cortical Reorganization During Motor Training and Motor Skill Learning

Psychophysiological studies have demonstrated that the acquisition of motor skills follows two distinct stages. The first is a fast learning stage during which considerable improvement in performance can be observed within a single training session; the second is a later, slow learning stage, during which further gains can be observed across several sessions of practice [23]. Karni et al. [23] used a simple finger-opposition task, during which healthy individuals were trained over the course of several weeks and were scanned at weekly interval using fMRI. Repetition of the task after 3 weeks of practice showed that there was a significant larger activation of the contralateral primary SMC as compared with the activation obtained with a control, untrained finger-opposition sequence. These results support the notion that motor practice induces recruitment of additional M1 units into a local network specifically representing the motor trained sequence. These findings are in agreement with the demonstration that, in healthy individuals, the recruitment of the primary SMC can be modified by previous activities, such as playing musical instruments [124, 132] or sports [133–135]. In the previous experiment, changes in primary SMC recruitment were also observed in the early scan session, reflecting a sort of initial habituation-like effect, in which the second sequence performed in a set evoked a smaller response than the first sequence [23]. Different type of motor training might induce different cortical reorganization. For instance, music-related motor training usually requires auditory feedback. The coupling between auditory input and motor output increases during motor training, thus leading to a cortical reorganization not only in motor areas but also in auditory areas [55]. On the other hand, sport-related motor expertise might induce cortical reorganization in motor, visual, and sensory-motor areas.

Although several studies demonstrated that learning of motor tasks is associated with an increased activation of the contralateral primary SMC, the pattern of functional reorganization is still unclear. The evaluation of activation patterns associated with repetition of simple movements gave conflicting results, since some studies reported reductions, and others increases of task-related activations [23, 136–138]. These discrepancies among studies might be related to variability in number and length of sessions, length of training, as well as monitoring of motor performance. Recent evidence suggests that the repetition of a simple sequence within a brief time window typically results in a reduced recruitment of the primary SMC, due to habituation [23, 136, 139]. Activity decreases may also indicate a more efficient organization of task-related brain networks through intensive training [55, 140].

In addition, a change in the degree of activation of the parietal lobe from healthy volunteers has also been described after motor training [139].

Dynamic activation changes during acquisition of motor skills have also been seen in different subcortical areas [135, 141]. In details, the dorsal parts of the putamen and the more rostral striatal areas have been shown to be active only during the early learning stage. On the contrary, activations of the posteroventral regions of the putamen and globus pallidus increase with practice (Fig. 3) [141].

Recent research has indicated that motor expertise influences not only brain activity during motor execution, but also during motor observation [142], motor imagery and planning [143] and motor prediction [144].

5 The Mirror-Neuron System

The mirror-neuron system (MNS) is an observation-execution matching system. Several neurophysiological [145, 146] and neuroimaging [75, 147–149] studies have demonstrated that MNS neurons discharge not only when an individual performs a specific goal-directed action, but also when an individual observes actions made by other individuals, implying an involvement of this system in imitation and motor learning [55, 140, 150]. The main role of the MNS is postulated to be the understanding of actions [151, 152]. This system is also thought to be involved in motor imagery [153] and empathy [154, 155]. In humans, neurons of this system have been described in

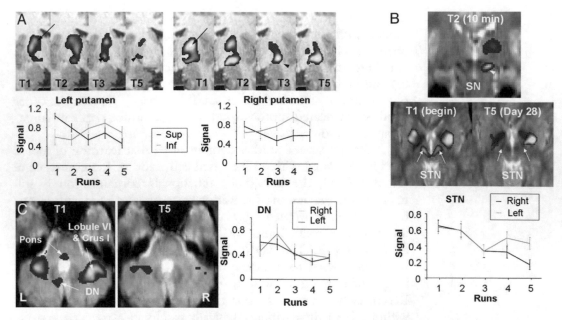

Fig. 3 Activation patterns in the basal ganglia and cerebellum during acquisition of motor skills. (**a** *Upper*) Activation maps obtained in the putamen superimposed on a coronal T1-weighted image. There was a progressive activation decrease in the dorsal part of the putamen (*arrows*) and an increase in a more ventrolateral area (*arrowheads*) bilaterally, which persisted after 4 weeks of training. (**a** *Lower*) Percentage signal increase ± SEM averaged across all subjects for each run of the trained sequence confirmed the activation decrease in the dorsal putamen and increase in the ventral putamen. (**b** *Top*) Activation maps obtained in the substantia nigra (SN) and subthalamic nucleus (STN) superimposed on EPI images. During session 1, STN activation was observed during the first run of T-sequence (T1). After 4 weeks of training, these areas were no more activated during the T-sequence. There was no significant signal change in the SN across runs. (**b** *Bottom*) Signal-to-time curves ± SEM in the STN averaged across all subjects and epochs confirm the activation decrease. (**c** *Left*) Activation maps obtained in the cerebellum during the T-sequence (T1 on day 1 and T5 on day 28). Activation in the lateral cerebellar hemispheres, the left dentate nucleus (DN), and the pons decreased with training. (**c** *Right*) Percentage signal increase ± SEM averaged across all subjects for each run of the trained sequence in the left and right DN. In the right DN, activation increased transiently during T2 (10 min of practice) and returned to pretraining values (From ref. [141], with permission)

the IFG, the adjacent premotor cortex [156], and the rostral part of the inferior parietal lobule [157]. The MNS is connected with the superior temporal sulcus (STS) that provides a higher-order visual description of the observed action [152, 158]. Mirror neurons are likely to be multimodal, as they respond to both the visual observation of an action as well as the sound associated with specific actions [159]. A recent fMRI study showed that regions supporting visuomotor integration and MNS abilities are multimodal convergent zones of the visual and motor streams (Fig. 4). In addition, the MNS seems to be a privilege position for incorporating and integrating basic sensory-motor information into higher-order cognitive centers [160].

In humans, mirror neurons are part of a system serving the imitation of actions and speech generation. Therefore, the MNS

Stepwise Convergence of
Visual and **Motor** Cortices

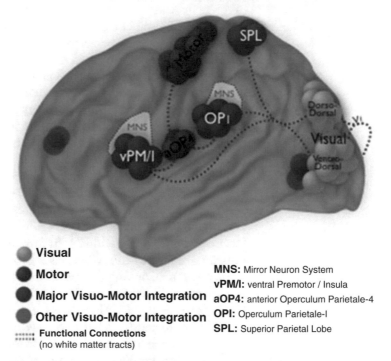

Visual

Motor

Major Visuo-Motor Integration

Other Visuo-Motor Integration

Functional Connections
(no white matter tracts)

MNS: Mirror Neuron System
vPM/I: ventral Premotor / Insula
aOP4: anterior Operculum Parietale-4
OPI: Operculum Parietale-I
SPL: Superior Parietal Lobe

Fig. 4 Diagram showing the convergence of the visual (*green nodes*) and motor (*red nodes*) systems into the multimodal integration network (*blue nodes*) or mirror neuron system. Visual cortex streams converge into a common destiny in the brain network. The main motor functional stream connects motor areas with the same network as the visual system does. Therefore, motor and visual functional-related streams meet in a multimodal integration network. Other regions such as premotor, dorsolateral prefrontal, anterior lateral occipital, or even in situ primary motor areas may be relevant for visuomotor integration as well, although they engage fewer convergent functional pathways than the multimodal integration core (*light blue nodes*) (From ref. [160] with permission)

might constitute a bridge between action and language processing and might represent the neuronal substrate from which human language evolved [161, 162]. Activity of this system is elicited by both the execution and observation of object-related transitive and intransitive actions [148, 158, 163]. These observations suggest that the MNS is a network that has been preserved during evolution and has developed "new" functions. Therefore, the MNS seems to be an extremely plastic system, with the capacity to adapt to new cognitive, social, or behavioral requirements to which an individual is exposed [162, 164]. This hypothesis assumes that these "evolutionary" changes have occurred over an extremely long time window (phylogenetic plasticity). It is plausible that, as shown for other brain networks, including the motor one, disease-related changes of such a plastic system might occur in case of CNS injury (adaptive plasticity). This hypothesis has indeed been supported by the results of a recent study in patients with multiple sclerosis, which demonstrated that these patients tend to activate regions that are part of the MNS during the performance of a simple motor task [165]. Defining the role of the MNS after brain injury may be central to a better understanding of the clinical manifestations of various neurological conditions and, as a consequence, to develop new rehabilitative strategies. Indeed, mirror therapy has been administered to treat various neurological conditions leading to improvement in motor function [166]. A recent study suggested that mirror therapy represents an appropriate method to recruit the contralesional motor areas to promote functional recovery through interhemispheric transfer of information (Fig. 5) [167].

The majority of the MNS studies has been focused on the attempt to better define the exact role of this system and its precise location in healthy individuals. In this perspective, it has been demonstrated that: (1) the MNS has a bilateral representation [168]; (2) mirror neurons in the premotor cortex have a somatotopic organization, as shown for the classical motor cortex homunculus [163]; and, finally, (3) there are gender differences in MNS function [169]. Recently, it has been suggested that there might be a relation between activity of the MNS and handedness [165].

Functional studies have suggested a role of MNS dysfunction, in combination with limbic system impairment, in patients with autism [170, 171], suggesting that this neuronal system may play a role in autistic social impairment. These studies described a reduced activity in the IFG and premotor cortex during action/face imitation and observation in adults [170] and children [171] with autism spectrum disorders. However, recently, some studies showed that the capacity of children with autism to understand the goal of observed motor acts is preserved, thus bringing the mirror hypothesis of autism into question [152].

6   Conclusions

Functional neuroimaging has dramatically changed our understanding of the human sensorimotor system by showing that it is constituted by a large number of cortical and subcortical areas, with a precise location and a specialized function. It is also evident that this system functions in cooperation with other brain networks in order to integrate all the information coming from the environment and to finalize the performance of motor acts. Defining this system's behavior is of the outmost importance for the understanding of its dysfunction in case of disease and to develop potentially successful therapeutic strategies capable to enhance its plasticity.

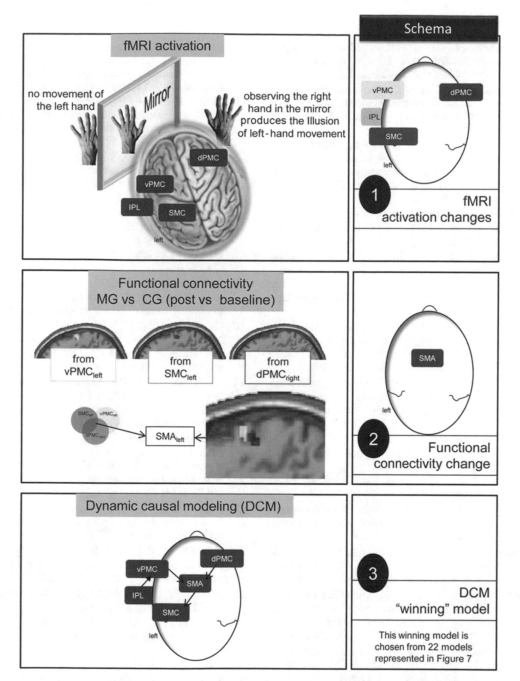

Fig. 5 Brain activity patterns between healthy participants who were trained on simple motor tasks with their right hand with (mirror training group—MG) and without a mirror (control training group—CG). The fMRI analysis revealed activation changes within the right dorsal premotor cortex (dPMC), left inferior parietal lobule (vPMC), left ventral premotor cortex (IPL) and left primary sensorimotor cortex (SMC) in the MG in comparison with the CG (*schema 1*). The functional connectivity (FC) analysis revealed an increased FC between previous regions and the left supplementary motor area (SMA) in the MG over the CG (*schema 2*). The dynamic causal modeling, which estimates and makes inferences about the coupling among brain areas, revealed that the right dPMC and left vPMC interacted with the left SMA, which in turn had access to the left SMC (and the left IPL interacted with the left vPMC) (*schema 3*) (From ref. [167], with permission)

References

1. Fink GR et al (1997) Multiple nonprimary motor areas in the human cortex. J Neurophysiol 77(4):2164–2174

2. Ciccarelli O et al (2005) Identifying brain regions for integrative sensorimotor processing with ankle movements. Exp Brain Res 166(1):31–42

3. Nowak DA, Glasauer S, Hermsdorfer J (2013) Force control in object manipulation—a model for the study of sensorimotor control strategies. Neurosci Biobehav Rev 37(8):1578–1586

4. Debaere F et al (2001) Brain areas involved in interlimb coordination: a distributed network. Neuroimage 14(5):947–958

5. Rocca MA et al (2007) Influence of body segment position during in-phase and antiphase hand and foot movements: a kinematic and functional MRI study. Hum Brain Mapp 28(3):218–227

6. Reddy H et al (2001) Altered cortical activation with finger movement after peripheral denervation: comparison of active and passive tasks. Exp Brain Res 138(4):484–491

7. Reddy H et al (2002) Functional brain reorganization for hand movement in patients with multiple sclerosis: defining distinct effects of injury and disability. Brain 125(Pt 12):2646–2657

8. Petsas N et al (2013) Evidence of impaired brain activity balance after passive sensorimotor stimulation in multiple sclerosis. PLoS One 8(6):e65315

9. Jaeger L et al (2014) Brain activation associated with active and passive lower limb stepping. Front Hum Neurosci 8:828

10. Logothetis N K et al (2001) Neurophysiological investigation of the basis of the fMRI signal. Nature 412(6843):150–157

11. Mehta JP et al (2012) The effect of movement rate and complexity on functional magnetic resonance signal change during pedaling. Motor Control 16(2):158–175

12. Francis S et al (2009) fMRI analysis of active, passive and electrically stimulated ankle dorsiflexion. Neuroimage 44(2):469–479

13. Decety J et al (1994) Mapping motor representations with positron emission tomography. Nature 371(6498):600–602

14. Porro CA et al (1996) Primary motor and sensory cortex activation during motor performance and motor imagery: a functional magnetic resonance imaging study. J Neurosci 16(23):7688–7698

15. van der Meulen M et al (2014) The influence of individual motor imagery ability on cerebral recruitment during gait imagery. Hum Brain Mapp 35(2):455–470

16. Wexler BE et al (1997) An fMRI study of the human cortical motor system response to increasing functional demands. Magn Reson Imaging 15(4):385–396

17. Deiber MP et al (1999) Mesial motor areas in self-initiated versus externally triggered movements examined with fMRI: effect of movement type and rate. J Neurophysiol 81(6):3065–3077

18. VanMeter JW et al (1995) Parametric analysis of functional neuroimages: application to a variable-rate motor task. Neuroimage 2(4):273–283

19. Dettmers C et al (1995) Relation between cerebral activity and force in the motor areas of the human brain. J Neurophysiol 74:802–815

20. Keisker B et al (2009) Differential force scaling of fine-graded power grip force in the sensorimotor network. Hum Brain Mapp 30(8):2453–2465

21. Neely KA et al (2013) Segregated and overlapping neural circuits exist for the production of static and dynamic precision grip force. Hum Brain Mapp 34(3):698–712

22. Schlaug G, Knorr U, Seitz R (1994) Intersubject variability of cerebral activations in acquiring a motor skill: a study with positron emission tomography. Exp Brain Res 98(3):523–534

23. Karni A et al (1995) Functional MRI evidence for adult motor cortex plasticity during motor skill learning. Nature 377(6545):155–158

24. Rao SM et al (1993) Functional magnetic resonance imaging of complex human movements. Neurology 43(11):2311–2318

25. Annett M (1973) Handedness in families. Ann Hum Genet 37(1):93–105

26. Kim SG et al (1993) Functional magnetic resonance imaging of motor cortex: hemispheric asymmetry and handedness. Science 261(5121):615–617

27. Singh LN et al (1998) Comparison of ipsilateral activation between right and left handers: a functional MR imaging study. Neuroreport 9(8):1861–1866

28. Solodkin A et al (2001) Lateralization of motor circuits and handedness during finger movements. Eur J Neurol 8(5):425–434

29. Verstynen T et al (2005) Ipsilateral motor cortex activity during unimanual hand

movements relates to task complexity. J Neurophysiol 93(3):1209–1222

30. Pool EM et al (2015) Functional resting-state connectivity of the human motor network: differences between right- and left-handers. Neuroimage 109:298–306

31. Lissek S et al (2007) Sex differences in cortical and subcortical recruitment during simple and complex motor control: an fMRI study. Neuroimage 37(3):912–926

32. Ward NS (2006) Compensatory mechanisms in the aging motor system. Ageing Res Rev 5(3):239–254

33. Loibl M et al (2011) Non-effective increase of fMRI-activation for motor performance in elder individuals. Behav Brain Res 223(2):280–286

34. Taniwaki T et al (2007) Age-related alterations of the functional interactions within the basal ganglia and cerebellar motor loops in vivo. Neuroimage 36(4):1263–1276

35. Hutchinson S et al (2002) Age-related differences in movement representation. Neuroimage 17(4):1720–1728

36. Mattay VS et al (2002) Neurophysiological correlates of age-related changes in human motor function. Neurology 58(4):630–635

37. Wang L et al (2014) Age-specific activation of cerebral areas in motor imagery—a fMRI study. Neuroradiology 56(4):339–348

38. Alkadhi H et al (2002) Reproducibility of primary motor cortex somatotopy under controlled conditions. AJNR Am J Neuroradiol 23(9):1524–1532

39. Cunningham DA et al (2013) Functional somatotopy revealed across multiple cortical regions using a model of complex motor task. Brain Res 1531:25–36

40. Chiou SY et al (2013) Co-activation of primary motor cortex ipsilateral to muscles contracting in a unilateral motor task. Clin Neurophysiol 124(7):1353–1363

41. McGregor KM et al (2015) Reliability of negative BOLD in ipsilateral sensorimotor areas during unimanual task activity. Brain Imaging Behav 9(2):245–254

42. Netz J, Ziemann U, Homberg V (1995) Hemispheric asymmetry of transcallosal inhibition in man. Exp Brain Res 104(3):527–533

43. Liepert J et al (2001) Inhibition of ipsilateral motor cortex during phasic generation of low force. Clin Neurophysiol 112(1):114–121

44. Allison JD et al (2000) Functional MRI cerebral activation and deactivation during finger movement. Neurology 54(1):135–142

45. Nirkko AC et al (2001) Different ipsilateral representations for distal and proximal movements in the sensorimotor cortex: activation and deactivation patterns. Neuroimage 13(5):825–835

46. Stefanovic B, Warnking JM, Pike GB (2004) Hemodynamic and metabolic responses to neuronal inhibition. Neuroimage 22(2):771–778

47. Geffen GM, Jones DL, Geffen LB (1994) Interhemispheric control of manual motor activity. Behav Brain Res 64(1–2):131–140

48. Davidson T, Tremblay F (2013) Age and hemispheric differences in transcallosal inhibition between motor cortices: an ispsilateral silent period study. BMC Neurosci 14:62

49. McGregor KM et al (2011) Physical activity and neural correlates of aging: a combined TMS/fMRI study. Behav Brain Res 222(1):158–168

50. Sadato N et al (1997) Role of the supplementary motor area and the right premotor cortex in the coordination of bimanual finger movements. J Neurosci 17(24):9667–9674

51. Lee KM, Chang KH, Roh JK (1999) Subregions within the supplementary motor area activated at different stages of movement preparation and execution. Neuroimage 9(1):117–123

52. Ohara S et al (2000) Movement-related change of electrocorticographic activity in human supplementary motor area proper. Brain 123(Pt 6):1203–1215

53. Arai N et al (2012) Effective connectivity between human supplementary motor area and primary motor cortex: a paired-coil TMS study. Exp Brain Res 220(1):79–87

54. Gao Q et al (2014) Differential contribution of bilateral supplementary motor area to the effective connectivity networks induced by task conditions using dynamic causal modeling. Brain Connect 4(4):256–264

55. Yang J (2015) The influence of motor expertise on the brain activity of motor task performance: a meta-analysis of functional magnetic resonance imaging studies. Cogn Affect Behav Neurosci 15(2):381–394

56. Rouiller EM et al (1994) Transcallosal connections of the distal forelimb representations of the primary and supplementary motor cortical areas in macaque monkeys. Exp Brain Res 102(2):227–243

57. Marchand WR et al (2013) Functional architecture of the cortico-basal ganglia circuitry during motor task execution: correlations of strength of functional connectivity with neuropsychological task performance

among female subjects. Hum Brain Mapp 34(5):1194–1207

58. Brinkman C (1981) Lesions in supplementary motor area interfere with a monkey's performance of a bimanual coordination task. Neurosci Lett 27(3):267–270

59. Brinkman C (1984) Supplementary motor area of the monkey's cerebral cortex: short- and long-term deficits after unilateral ablation and the effects of subsequent callosal section. J Neurosci 4(4):918–929

60. Akkal D, Dum RP, Strick PL (2007) Supplementary motor area and presupplementary motor area: targets of basal ganglia and cerebellar output. J Neurosci 27(40):10659–10673

61. Martino AM, Strick PL (1987) Corticospinal projections originate from the arcuate premotor area. Brain Res 404(1–2):307–312

62. Hou JM et al (2014) Alterations of resting-state regional and network-level neural function after acute spinal cord injury. Neuroscience 277:446–454

63. Humberstone M et al (1997) Functional magnetic resonance imaging of single motor events reveals human presupplementary motor area. Ann Neurol 42(4):632–637

64. Zhang S, Ide JS, Li CS (2012) Resting-state functional connectivity of the medial superior frontal cortex. Cereb Cortex 22(1):99–111

65. Weilke F et al (2001) Time-resolved fMRI of activation patterns in M1 and SMA during complex voluntary movement. J Neurophysiol 85(5):1858–1863

66. Picard N, Strick PL (1996) Motor areas of the medial wall: a review of their location and functional activation. Cereb Cortex 6(3):342–353

67. Rizzolatti G, Luppino G (2001) The cortical motor system. Neuron 31(6):889–901

68. Dum RP, Strick PL (1991) The origin of corticospinal projections from the premotor areas in the frontal lobe. J Neurosci 11(3):667–689

69. Dum RP, Strick PL (2002) Motor areas in the frontal lobe of the primate. Physiol Behav 77(4–5):677–682

70. Stephan KM et al (1995) Functional anatomy of the mental representation of upper extremity movements in healthy subjects. J Neurophysiol 73:373–386

71. Binkofski F et al (1999) A fronto-parietal circuit for object manipulation in man: evidence from an fMRI-study. Eur J Neurosci 11(9):3276–3286

72. Harrington DL et al (2000) Specialized neural systems underlying representations of sequential movements. J Cogn Neurosci 12:56–77

73. Haslinger B et al (2002) The role of lateral premotor-cerebellar-parietal circuits in motor sequence control: a parametric fMRI study. Brain Res Cogn Brain Res 13(2):159–168

74. Duann JR et al (2009) Functional connectivity delineates distinct roles of the inferior frontal cortex and presupplementary motor area in stop signal inhibition. J Neurosci 29(32):10171–10179

75. Grafton ST et al (1996) Localization of grasp representations in humans by positron emission tomography. 2. Observation compared with imagination. Exp Brain Res 112(1):103–111

76. Scott SH, Sergio LE, Kalaska JF (1997) Reaching movements with similar hand paths but different arm orientations. II. Activity of individual cells in dorsal premotor cortex and parietal area 5. J Neurophysiol 78(5):2413–2426

77. Grafton ST, Fagg AH, Arbib MA (1998) Dorsal premotor cortex and conditional movement selection: a PET functional mapping study. J Neurophysiol 79(2):1092–1097

78. Bestmann S et al (2008) Dorsal premotor cortex exerts state-dependent causal influences on activity in contralateral primary motor and dorsal premotor cortex. Cereb Cortex 18(6):1281–1291

79. Schluter ND et al (2001) Cerebral dominance for action in the human brain: the selection of actions. Neuropsychologia 39(2):105–113

80. Schluter ND et al (1998) Temporary interference in human lateral premotor cortex suggests dominance for the selection of movements. A study using transcranial magnetic stimulation. Brain 121(Pt 5):785–799

81. Moisa M et al (2012) Uncovering a context-specific connectional fingerprint of human dorsal premotor cortex. J Neurosci 32(21):7244–7252

82. Marconi B et al (2003) Callosal connections of dorso-lateral premotor cortex. Eur J Neurosci 18(4):775–788

83. O'Shea J et al (2007) Functionally specific reorganization in human premotor cortex. Neuron 54(3):479–490

84. Cunnington R et al (2006) The selection of intended actions and the observation of others' actions: a time-resolved fMRI study. Neuroimage 29(4):1294–1302

85. Deiber MP et al (1991) Cortical areas and the selection of movement: a study with positron emission tomography. Exp Brain Res 84(2):393–402

86. Devinsky O, Morrell MJ, Vogt BA (1995) Contributions of anterior cingulate cortex to behaviour. Brain 118:279–306

87. Hoffstaedter F et al (2014) The role of anterior midcingulate cortex in cognitive motor control: evidence from functional connectivity analyses. Hum Brain Mapp 35(6):2741–2753

88. Paus T et al (1993) Role of the human anterior cingulate cortex in the control of oculomotor, manual, and speech responses: a positron emission tomography study. J Neurophysiol 70(2):453–469

89. Vogt BA, Finch DM, Olson CR (1992) Functional heterogeneity in cingulate cortex: the anterior executive and posterior evaluative regions. Cereb Cortex 2(6):435–443

90. Botvinick M et al (1999) Conflict monitoring versus selection-for-action in anterior cingulate cortex. Nature 402(6758):179–181

91. Carter CS et al (1998) Anterior cingulate cortex, error detection, and the online monitoring of performance. Science 280(5364):747–749

92. Wenderoth N et al (2005) The role of anterior cingulate cortex and precuneus in the coordination of motor behaviour. Eur J Neurosci 22(1):235–246

93. Rizzolatti G, Fogassi L, Gallese V (1997) Parietal cortex: from sight to action. Curr Opin Neurobiol 7(4):562–567

94. Karhu J, Tesche CD (1999) Simultaneous early processing of sensory input in human primary (SI) and secondary (SII) somatosensory cortices. J Neurophysiol 81(5):2017–2025

95. Hamalainen H, Hiltunen J, Titievskaja I (2000) fMRI activations of SI and SII cortices during tactile stimulation depend on attention. Neuroreport 11(8):1673–1676

96. Huttunen J et al (1996) Significance of the second somatosensory cortex in sensorimotor integration: enhancement of sensory responses during finger movements. Neuroreport 7(5):1009–1012

97. Shergill SS et al (2013) Modulation of somatosensory processing by action. Neuroimage 70:356–362

98. Mima T et al (1998) Attention modulates both primary and second somatosensory cortical activities in humans: a magnetoencephalographic study. J Neurophysiol 80(4):2215–2221

99. Dobkin BH (2003) Functional MRI: a potential physiologic indicator for stroke rehabilitation interventions. Stroke 34(5):e23–e28

100. Del Gratta C et al (2002) Topographic organization of the human primary and secondary somatosensory cortices: comparison of fMRI and MEG findings. Neuroimage 17(3):1373–1383

101. Culham JC, Kanwisher NG (2001) Neuroimaging of cognitive functions in human parietal cortex. Curr Opin Neurobiol 11(2):157–163

102. Monaco S et al (2015) Neural correlates of object size and object location during grasping actions. Eur J Neurosci 41(4):454–465

103. Barany DA et al (2014) Feature interactions enable decoding of sensorimotor transformations for goal-directed movement. J Neurosci 34(20):6860–6873

104. Parent A, Hazrati LN (1995) Functional anatomy of the basal ganglia. I. The cortico-basal ganglia-thalamo-cortical loop. Brain Res Brain Res Rev 20(1):91–9127

105. Choi EY, Yeo BT, Buckner RL (2012) The organization of the human striatum estimated by intrinsic functional connectivity. J Neurophysiol 108(8):2242–2263

106. Kessler K et al (2006) Investigating the human mirror neuron system by means of cortical synchronization during the imitation of biological movements. Neuroimage 33(1):227–238

107. Harrington DL, Haaland KY, Knight RT (1998) Cortical networks underlying mechanisms of time perception. J Neurosci 18(3):1085–1095

108. Ivry RB, Keele SW, Diener HC (1988) Dissociation of the lateral and medial cerebellum in movement timing and movement execution. Exp Brain Res 73(1):167–180

109. Jantzen KJ, Steinberg FL, Kelso JAS (2004) Brain networks underlying human timing behavior are influenced by prior context. Proc Natl Acad Sci U S A 101(17):6815–6820

110. Walz AD et al (2014) Changes in cortical, cerebellar and basal ganglia representation after comprehensive long term unilateral hand motor training. Behav Brain Res 278C:393–403

111. Brooks DJ (1995) The role of the basal ganglia in motor control: contributions from PET. J Neurol Sci 128(1):1–13

112. Mesulam MM (1998) From sensation to cognition. Brain 121(Pt 6):1013–1052

113. Chaudhuri A, Behan PO (2000) Fatigue and basal ganglia. J Neurol Sci 179(S 1–2):34–42

114. Hadjikhani N, Roland PE (1998) Cross-modal transfer of information between the tactile and the visual representations in the human brain: a positron emission tomographic study. J Neurosci 18(3):1072–1084

115. Mosier K, Bereznaya I (2001) Parallel cortical networks for volitional control of swallowing in humans. Exp Brain Res 140(3):280–289

116. Taylor KS, Seminowicz DA, Davis KD (2009) Two systems of resting state connectiv-

ity between the insula and cingulate cortex. Hum Brain Mapp 30(9):2731–2745

117. Nitschke MF et al (1996) Somatotopic motor representation in the human anterior cerebellum. A high-resolution functional MRI study. Brain 119(Pt 3):1023–1029

118. Sakai K et al (1998) Separate cerebellar areas for motor control. Neuroreport 9(10):2359–2363

119. Kim JJ, Thompson RF (1997) Cerebellar circuits and synaptic mechanisms involved in classical eyeblink conditioning. Trends Neurosci 20(4):177–181

120. Ehrsson HH, Kuhtz-Buschbeck JP, Forssberg H (2002) Brain regions controlling nonsynergistic versus synergistic movement of the digits: a functional magnetic resonance imaging study. J Neurosci 22(12):5074–5080

121. Allen GI, Tsukahara N (1974) Cerebrocerebellar communication systems. Physiol Rev 54(4):957–951006

122. Kipping JA et al (2013) Overlapping and parallel cerebello-cerebral networks contributing to sensorimotor control: an intrinsic functional connectivity study. Neuroimage 83:837–848

123. Ramnani N, Passingham RE (2001) Changes in the human brain during rhythm learning. J Cogn Neurosci 13(7):952–966

124. Jancke L, Shah NJ, Peters M (2000) Cortical activations in primary and secondary motor areas for complex bimanual movements in professional pianists. Brain Res Cogn Brain Res 10(1–2):177–183

125. Wenzel U et al (2014) Functional and structural correlates of motor speed in the cerebellar anterior lobe. PLoS One 9(5):e96871

126. Doyon J et al (1998) Role of the striatum, cerebellum and frontal lobes in the automatization of a repeated visuomotor sequence of movements. Neuropsychologia 36(7):625–641

127. Jueptner M, Weiller C (1998) A review of differences between basal ganglia and cerebellar control of movements as revealed by functional imaging studies. Brain 121(Pt 8):1437–1449

128. Sanes JN, Dimitrov B, Hallett M (1990) Motor learning in patients with cerebellar dysfunction. Brain 113(Pt 1):103–120

129. Bracha V et al (2000) The human cerebellum and associative learning: dissociation between the acquisition, retention and extinction of conditioned eyeblinks. Brain Res 860(1–2):87–94

130. Jenkins IH, Frackowiak RS (1993) Functional studies of the human cerebellum with positron emission tomography. Rev Neurol (Paris) 149:647–653

131. Jenkins IH et al (1994) Motor sequence learning: a study with positron emission tomography. J Neurosci 14(6):3775–3790

132. Krings T et al (2000) Cortical activation patterns during complex motor tasks in piano players and control subjects. A functional magnetic resonance imaging study. Neurosci Lett 278(3):189–193

133. Pearce AJ et al (2000) Functional reorganisation of the corticomotor projection to the hand in skilled racquet players. Exp Brain Res 130(2):238–243

134. Milton J et al (2007) The mind of expert motor performance is cool and focused. Neuroimage 35(2):804–813

135. Bishop DT et al (2013) Neural bases for anticipation skill in soccer: an FMRI study. J Sport Exerc Psychol 35(1):98–109

136. Dirnberger G et al (2004) Habituation in a simple repetitive motor task: a study with movement-related cortical potentials. Clin Neurophysiol 115(2):378–384

137. Loubinoux I et al (2001) Within-session and between-session reproducibility of cerebral sensorimotor activation: a test–retest effect evidenced with functional magnetic resonance imaging. J Cereb Blood Flow Metab 21(5):592–607

138. Tracy JI et al (2001) A comparison of 'Early' and 'Late' stage brain activation during brief practice of a simple motor task. Brain Res Cogn Brain Res 10(3):303–316

139. Morgen K et al (2004) Kinematic specificity of cortical reorganization associated with motor training. Neuroimage 21(3):1182–1187

140. Kruger B et al (2014) Parietal and premotor cortices: activation reflects imitation accuracy during observation, delayed imitation and concurrent imitation. Neuroimage 100:39–50

141. Lehericy S et al (2005) Distinct basal ganglia territories are engaged in early and advanced motor sequence learning. Proc Natl Acad Sci U S A 102(35):12566–12571

142. Kim YT et al (2011) Neural correlates related to action observation in expert archers. Behav Brain Res 223(2):342–347

143. Baeck JS et al (2012) Brain activation patterns of motor imagery reflect plastic changes associated with intensive shooting training. Behav Brain Res 234(1):26–32

144. Balser N et al (2014) Prediction of human actions: expertise and task-related effects on neural activation of the action observation network. Hum Brain Mapp 35(8):4016–4034

145. Fadiga L et al (1995) Motor facilitation during action observation: a magnetic stimulation study. J Neurophysiol 73(6):2608–2611

146. Hari R et al (1998) Activation of human primary motor cortex during action observation: a neuromagnetic study. Proc Natl Acad Sci U S A 95(25):15061–15065

147. Grezes J et al (2003) Activations related to "mirror" and "canonical" neurones in the human brain: an fMRI study. Neuroimage 18(4):928–937

148. Rizzolatti G et al (1996) Localization of grasp representations in humans by PET: 1. Observation versus execution. Exp Brain Res 111(2):246–252

149. Mengotti P, Corradi-Dell'acqua C, Rumiati RI (2012) Imitation components in the human brain: an fMRI study. Neuroimage 59(2):1622–1630

150. Buccino G et al (2004) Neural circuits underlying imitation learning of hand actions: an event-related fMRI study. Neuron 42(2):323–334

151. Rizzolatti G, Craighero L (2004) The mirror-neuron system. Annu Rev Neurosci 27:169–192

152. Rizzolatti G, Sinigaglia C (2010) The functional role of the parieto-frontal mirror circuit: interpretations and misinterpretations. Nat Rev Neurosci 11(4):264–274

153. Johnson SH et al (2002) Selective activation of a parietofrontal circuit during implicitly imagined prehension. Neuroimage 17(4):1693–1704

154. Leslie KR, Johnson-Frey SH, Grafton ST (2004) Functional imaging of face and hand imitation: towards a motor theory of empathy. Neuroimage 21(2):601–607

155. Filippi M et al (2010) The brain functional networks associated to human and animal suffering differ among omnivores, vegetarians and vegans. PLoS One 5(5):e10847

156. Cerri G et al (2015) The mirror neuron system and the strange case of Broca's area. Hum Brain Mapp 36(3):1010–1027

157. Rizzolatti G, Fogassi L, Gallese V (2001) Neurophysiological mechanisms underlying the understanding and imitation of action. Nat Rev Neurosci 2(9):661–670

158. Iacoboni M (2005) Neural mechanisms of imitation. Curr Opin Neurobiol 15(6):632–637

159. Kohler E et al (2002) Hearing sounds, understanding actions: action representation in mirror neurons. Science 297:846–848

160. Sepulcre J (2014) Integration of visual and motor functional streams in the human brain. Neurosci Lett 567:68–73

161. Rizzolatti G, Arbib MA (1998) Language within our grasp. Trends Neurosci 21(5):188–194

162. Oztop E, Kawato M, Arbib MA (2013) Mirror neurons: functions, mechanisms and models. Neurosci Lett 540:43–55

163. Buccino G et al (2001) Action observation activates premotor and parietal areas in a somatotopic manner: an fMRI study. Eur J Neurosci 13(2):400–404

164. Filippi M et al (2013) The "vegetarian brain": chatting with monkeys and pigs? Brain Struct Funct 218(5):1211–1227

165. Rocca MA et al (2008) The mirror-neuron system and handedness: a "right" world? Hum Brain Mapp 29(11):1243–1254

166. Buccino G (2014) Action observation treatment: a novel tool in neurorehabilitation. Philos Trans R Soc Lond B Biol Sci 369(1644):20130185

167. Hamzei F et al (2012) Functional plasticity induced by mirror training: the mirror as the element connecting both hands to one hemisphere. Neurorehabil Neural Repair 26(5):484–496

168. Aziz-Zadeh L et al (2006) Lateralization of the human mirror neuron system. J Neurosci 26(11):2964–2970

169. Cheng Y-W et al (2006) Gender differences in the human mirror system: a magnetoencephalography study. Neuroreport 17(11):1115–1119

170. Theoret H et al (2005) Impaired motor facilitation during action observation in individuals with autism spectrum disorder. Curr Biol 15(3):R84–R85

171. Dapretto M et al (2006) Understanding emotions in others: mirror neuron dysfunction in children with autism spectrum disorders. Nat Neurosci 9(1):28–30

Chapter 18

Functional Imaging of the Human Visual System

Guy A. Orban and Stefania Ferri

Abstract

The human visual system consists of a large, yet unknown number of cortical areas. We summarize the efforts which have led to the identification of 19 retinotopic areas in human occipital cortex, using the macaque visual cortex as a guide. In this process retinotopic mapping has proven far superior to the study of functional properties. Macaques and humans share early areas (V1, V2, and V3), a motion-sensitive middle temporal (MT/V5) cluster as well as six other areas. The remaining human occipital areas either result from reorganization of a group of monkey areas or seem to be specifically human. Several regions sensitive to motion and even higher-order motion have been described in parietal cortex, the retinotopic organization of which is still under debate. On the other hand, both dorsal and ventral regions are sensitive to shape, which is most pronounced in the lateral occipital complex (LOC) extending into the fusiform gyrus. The anterior part of this complex is flanked by specialized regions devoted to processing faces and bodies and represents "visual objects" rather than image properties. Its exact organization requires further investigation.

Key words Vision, Retinotopy, Cortical area, Visual field, Motion, 2D and 3D Shape, Actions

1 Introduction

The human visual system is located in the occipital lobe and extends rostrally into the parietal and temporal lobes. It is estimated to encompass 30 % of human cortex [1]. Functional imaging gives us direct access to the function of this important part of human cortex. One way to study this system is to consider a number of perceptual or visual cognitive functions and to localize their neural correlates. An alternative is to consider the visual system as an anatomically organized collection of cortical areas and subcortical centers that process retinal information and transform it into messages appropriate for processing in the nonvisual cerebral regions to which the visual system projects. A critical aim in visual neuroscience is to define the different cortical areas that make up the human visual system. In other species, such as the nonhuman primates, cortical areas are defined by the combination of four criteria: (1) cyto- and myeloarchitectonics, (2) anatomical connections with other (known) areas, (3) topographic organization, i.e., retinotopic organization, and

Massimo Filippi (ed.), *fMRI Techniques and Protocols*, Neuromethods, vol. 119,
DOI 10.1007/978-1-4939-5611-1_18, © Springer Science+Business Media New York 2016

(4) functional properties. It is important to note that while not all criteria may apply to each area, it is essential to obtain as much converging information as possible. In the nonhuman primate, 30 or more visual cortical areas have been identified using these criteria, although it is fair to state that even in these species there is discussion about the exact definition of areas, especially those at the higher levels in the system [1]. The definition of the visual cortical areas is only a first step in understanding the visual system; next is the investigation of the nature of the processing performed by these areas and the flow of information through the areas as a function of the behavioral context and task demands.

Recent advances in brain imaging have provided powerful tools for the definition and mapping of cortical areas. Functional magnetic resonance imaging (fMRI) provides insights into the functional characteristics of cortical areas by means of specific contrasts of brain activity that isolate a functional property. For example, in the monkey in which a number of visual areas have been identified using anatomical and neurophysiological measurements, fMRI has shown that a small number of functional characteristics, defined by a few subtractions, allow the definition of six motion-sensitive regions in the monkey superior temporal sulcus (STS) [2]. Functional MRI can also provide evidence for retinotopic organization. It actually is more powerful than single cell studies in this respect, as it is less biased in its sampling and the measure required is simply responsiveness. It has been suggested that the topology of an area, that is, its localization with respect to neighboring areas, might be a valuable addition for the identification of areas [3]. Imaging has not yet provided clear means to obtain histological structure, although at high field (7.0 T) the stria of Gennari becomes visible, and myelin density can be measured indirectly at 3 T [4]. The situation is slightly better for anatomical connections, as diffusion tensor imaging (DTI) [5, 6] is increasingly seen as a potential measure of connectivity between areas, although the methodological issues remain formidable [7]. In the present chapter we will provide an overview of how these two fMRI strategies, functional specialization and retinotopic organization, have been used for defining cortical areas.

Despite all its strengths functional imaging has severe limitations due to its limited temporal and spatial resolution. With the present 3 T systems a few millimeters can be resolved. While this is ample to define cortical regions it is a long way from the resolution of the single neuron. In fact, fMRI signals are only indirect, hemodynamic reflections of average activity of thousands of neurons. Hence, fMRI is very sensitive at detecting average activity levels, but it has great difficulty in measuring neuronal selectivity. It has been proposed that repetition suppression can be used to measure neuronal tuning, but the case for it might be overstated as in single neurons the tuning of the adaptation is narrower than the response tuning [8–10]. Recent developments using multivoxel pattern analysis (MVPA) [11] provide sensitive tools for studying neural representations beyond the resolution of conventional fMRI approaches. Yet the estimation provided by this analysis depends heavily on the clustering of neurons with similar properties, like those in cortical columns, and the discrimination provided falls quite short of what single neurons can achieve. For example, single V1 neurons can signal orientation differences of 5°–10° with an 84 % chance of success [12]. MVPA of human V1 has so far yielded values of 35° [13]. Therefore, much can be gained by combining functional imaging in humans with knowledge derived from invasive studies, such as single cell recordings in nonhuman primates. The combination has become possible with the advent of fMRI in the awake monkey [14]. Indeed this allows parallel imaging experiments leading to the definition of cortical regions and their characteristics in the two species, paving the way for establishing homologies. Once a homology is established, one can test whether the neuronal properties in that area apply to the human homolog. Indeed, comparison of the single cell recordings and fMRI in the monkey using similar stimuli allows one to derive an fMRI signature of a neuronal property. One can then verify that the human homolog also exhibits this fMRI signature [15, 16]. Hence, the definition of cortical areas in both species is a critical step for knowledge transfer from animal models to the human visual system.

2   Methodological Issues

2.1   Stimulus Definition

Definition of the visual stimulus is important as it determines to a large degree the brain activation pattern and thus the experimental findings reported. It is important to note that a precise stimulus description is crucial for repeating an experiment and replicating the results. For example, very different stimuli are used for defining motion-responsive areas. A motion localizer used to localize human middle temporal (hMT/V5) region often consists of random dots, but may also consist of gratings, either rectangular or circular. Random dots may have different densities, sizes, luminance, etc., or the whole pattern may be of a different size. Random dots may translate in one or several directions, but may also rotate or move radially. All these paradigms, using very different stimuli, are referred to as motion localizers, but because of their differences they may result in activation of different cortical regions, reducing the value of the localization.

2.2   Tasks

One of the main challenges in brain imaging is investigating the link between neural activity and human behavior. Recent studies using parametric stimulus manipulation employ detection or discrimination tasks [17–19] rather than passive viewing of the stimuli. These paradigms allow correlation between behavioral data (psychometric functions) and fMRI activations. This approach is important for discerning the functional role of different cortical areas and evaluating their contribution to behavior. Further, attentionally demanding tasks (e.g., detection of changes in the fixation target, 1-back matching task) are used during scanning to ensure that observers pay attention across all stimulus conditions and that activation differences across conditions are not simply due to differences in the general arousal of the participants or the task difficulty across conditions. For example, when mapping the lateral occipital complex (LOC), participants view intact and scrambled images of objects. It is possible that higher activations for intact images of objects are due to the fact that these images attract the participants' attention more than scrambled images. To control for this potential confound observers are instructed to perform a task on different properties of the fixation target or the stimulus (e.g., dimming of the fixation point or part of the shape) [20] that entail similar attention across all stimulus conditions. Another task that has been adopted for controlling attentional confounds is the 1-back matching task (detect a repeat of an intact or scrambled image) [21]. This task is more demanding for scrambled than intact images, thus excluding the possibility that higher activations for intact images of objects are due to attentional differences.

2.3 Control of Eye Movements

Control of fixation is mandatory in motion response studies, retinotopic mapping experiments, and in spatial attention studies. Although in the past it was acceptable to show that the subjects fixated well based on off-line measurements, standards have evolved. In addition, precise eye movement records, provided by infrared corneal reflection methods, allow one to remove the effect of residual eye movements that occur despite fixation. In general, in all visual experiments, control of fixation will ensure that the part of visual field stimulated is known and will remove eye movements as a source of unwanted and uncontrolled activations.

2.4 fMRI Designs and Paradigms

The conventional fMRI approach for identifying cortical areas involved in different processes and cognitive tasks entails a subtraction of activations between different stimulus types that are presented in blocked or event-related designs.

One of the limitations of these fMRI paradigms is that they average across neural populations that may respond homogeneously across stimulus properties or may be differentially tuned to different stimulus attributes. Thus, in most cases, it is impossible to infer the properties of the underlying imaged neural populations. fMRI adaptation (or repetition suppression) paradigms [22–27] have recently been employed to study the properties of neuronal populations beyond the limited spatial resolution of fMRI. These paradigms capitalize on the reduction of neural responses for stimuli that have been presented for prolonged time or repeatedly [28, 29]. A change in a specific stimulus dimension that elicits increased responses (i.e., rebound of activity) identifies neural populations that are tuned to the modified stimulus attributes. fMRI adaptation paradigms have been used in both monkey and human fMRI studies as a sensitive tool that allows us to investigate: (a) the sensitivity of the neural populations to stimulus properties, and (b) the invariance of their responses within the imaged voxels. Adaptation across a change between two stimuli suggests a common neural representation invariant to that change, while recovery from adaptation suggests neural representations sensitive to specific stimulus properties. For example, recent imaging studies tested whether fMRI measurements can reveal neural populations in early visual areas sensitive to elementary visual features, e.g., orientation, color, and direction of motion [30–34]. Consider the case of motion direction: after prolonged exposure to the adapting motion direction, observers were tested with the same stimulus in the same or in an orthogonal motion direction. Decreased fMRI responses were observed in MT when the test stimuli were at the same motion direction as the adapting stimulus. However, recovery from this adaptation effect was observed for stimuli presented at an orthogonal direction. These studies suggest that the neural populations in human MT are sensitive to direction of motion [31, 34]. Using the same procedure in the monkey, Nelissen et al. [2] indeed observed adaptation in MT/V5 but also in other motion-sensitive regions, such as the medial superior temporal (MST) region. Similarly, recent studies have shown stronger adaptation in hMT/V5$_+$ for coherently than transparently moving plaid stimuli. These findings provide evidence that fMRI adaptation responses are linked to the activity of pattern-motion rather than component-motion cells in MT/MST [32]. Thus, these studies suggest that the fMRI signal can reveal neural selectivity consistent with the selectivity established by neurophysiological methods. However, recent studies comparing fMRI adaptation and neurophysiology in monkeys call for cautious interpretation of the relationship between fMRI adaptation effects and neural selectivity or invariance at higher levels in the system [10]. In particular, fMRI adaptation in a given cortical area may be the result of adaptation at earlier or later stages of processing that is propagated along the visual areas. Hence in higher-order areas receiving from multiple inputs fMRI adaptation might reflect adaptation of one of the inputs, while recordings show that local neuronal responses driven by the other inputs are not adapted.

Interestingly, novel MVPA methods [11, 35, 36] provide an alternative approach for investigating neural selectivity based on fMRI signals. Unlike conventional univariate analysis, MVPA takes advantage of the information across multiple voxels in a cortical area and allows us to characterize neural representations of features that are encoded at a higher spatial resolution in the brain than the typical resolution of fMRI. These classification analyses have been used successfully for the decoding of elementary visual features (e.g., orientation [13, 37], motion direction [38], and object categories [39–42]). The weakness of the MVPA approach is its dependence on the clustering of neurons with similar properties. This is also the case for a third technique which is has been proposed to infer neuronal selectivity from fMRI measurements: measuring the tuning of individual voxels [43]. Just as MVPA, tuning of voxels is prone to false negative results, as the grouping of neurons for higher-order selectivity is frequently unknown. In contrast adaptation fMRI is prone to false positives as inputs may adapt and not the local neuronal activity. For all these methods greater caution is required at higher level in the cortex.

2.5 Whole Brain Versus Region of Interest Analyses

The statistical evaluation of activation differences between stimulus and tasks is typically conducted by comparing responses for each voxel using the general linear model. Analysis of activation patterns across the whole brain (whole brain analysis) reveals clusters of activations in different anatomical regions that show significant differences in their functional processing. This approach has allowed researchers to identify and localize cortical regions with different functions and evaluate their involvement in various cognitive tasks. In contrast, region of interest (ROI) analysis focuses on specific cortical areas identified anatomically or functionally following standard mapping procedures (e.g., retinotopic mapping). The advantage of this approach is that it allows us to zoom in on specific cortical regions and investigate their neural computations using parametric stimulus manipulations. Such manipulations result in fine stimulus variations and differences in behavioral performance. Identifying fMRI activations that reflect these fine differences in neural processing may require the high signal-to-noise ratio that is possible when scanning and analyzing smaller regions of cortex. However, ROI analyses are limited in two respects: (a) the ROI may be outside the volume scanned or analyzed, (b) the voxels of interest (i.e., voxels that show differential activations across conditions) may cover a smaller cortical volume than the ROI; as a result, the differential activations may be averaged out within the ROI. Taken together, whole brain and ROI analyses can be used as complementary tools for studying the functional roles of cortical regions. Whole brain analyses search the entire brain for regions involved in the analysis of a given stimulus or a cognitive task, while ROI methods are more appropriate for finer investigation of the neural processing in these cortical regions [44, 45].

3 Retinotopic Organization

3.1 Early Visual Areas (V1, V2, V3)

Initially, positron emission tomography (PET) studies have concentrated on the retinotopy of V1 [46], which is a large area of known localization in the calcarine sulcus. With the advent of fMRI, mapping was extended to areas neighboring V1 [47] (but see also [48]). An additional step was the introduction of angular and eccentricity periodic sweeping stimuli that generate eccentricity and polar angle maps based on phase encoding of stimulus position [49]. This allowed the mapping of all three early areas (V1, V2, V3, Fig. 1) [51–53], in which polar angle and eccentricity vary along orthogonal directions on the cortical surface. The eccentricity varies from the central representation at the posterior tip of the calcarine fissure to that of large eccentricities rostrally along the calcarine. Polar angle varies in dorsoventral direction with the lower field being represented above the calcarine and the upper field below (Fig. 1). The three early visual areas are also shown on the flatmaps of Fig. 2 which cover a smaller eccentricity range (0.25°–7.75°) compared to that in Fig. 1 (0°–12°). Figures 1 and 2 show retinotopic maps of individual subjects and these maps exhibit quite some variability. To derive a more general representation one generates maximum probability maps (MPM) which plot in each voxel the area with the highest probability for a given set of subjects. These maps depend heavily on the quality of the inter-subject alignment [20, 56, 57] and this is dramatically improved by using the novel multimodal surface matching technique [57]. The resulting MPM of the left hemisphere is shown on the inflated brain in Fig. 3a, b, and on the flatmap in Fig. 3c. These maps are freely available in Caret [56], and can be used to identify activations without the need to spend valuable time mapping the retinotopic areas in the subjects

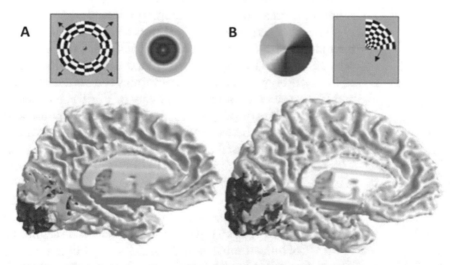

Fig. 1 The sweeping stimulus retinotopy paradigm. Two stimuli are used to measure the retinotopic maps in the cortex. Expanding ring stimuli map eccentricity, and rotating wedge stimuli map polar angle. The phase of the best-fitting sinusoid for each voxel indicates the position in the visual field that produces maximal activation for that voxel. Thus, these pseudocolor phase maps are used to visualize the retinotopic maps. Data area is shown for the left hemisphere (medial view) of one subject. Because of the heavy folding of human cortex, these retinotopic maps are best seen on flattened hemispheres (from Dougherty et al. [50])

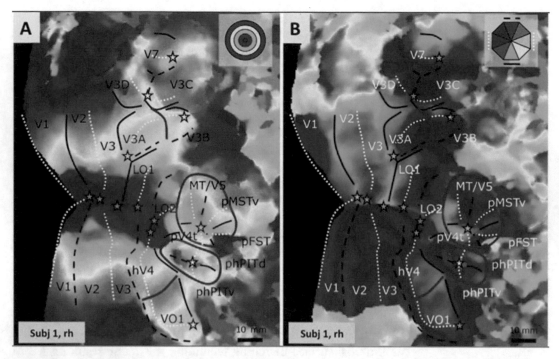

Fig. 2 The 18 retinotopic areas defined in the polar angle (**a**) and eccentricity (**b**) maps by Georgieva et al. [54], and Kolster et al. [55]; right hemisphere of subject 1. *Stars*: central visual field, *purple*: eccentricity ridge, *white dotted lines*: horizontal meridian, *black full* and *dashed lines*: lower and upper vertical meridian (from Abdollahi et al. [56])

under study. Of course, when investigating the actual properties of retinotopic areas, the direct mapping remains a superior strategy.

The three early visual cortical areas all have a large, complete representation of the contralateral hemifield, with the upper quadrant projecting ventrally and lower quadrant dorsally. The representation of the vertical meridian (VM) constitutes the boundary between V1 and V2 as well as the anterior boundary of V3. The representations of the horizontal meridian (HM) split the V1 representation and constitute the boundary between V2 and V3. The central representations of the three areas are fused in the central confluence (Figs. 1, 2, and 3). This retinotopic organization is very similar in humans and macaques (Fig. 4). This is not surprising as the presence of three early visual areas is a feature of primates [59, 60]. In all three areas the central representation is magnified compared to that of the periphery [51]. Duncan and Boynton [61] observed a correlation between magnification factor in V1 of human subjects and Vernier acuity but not grating acuity. The surface of V1 has been estimated from histological specimens to range between 2000 and 4500 mm^2, while the central 12° occupy 2200 mm^2 according to one imaging study [50]. Comparison between histological and fMRI estimates is difficult because of the difficulty of estimating the shrinkage in the histological specimens and the portion of V1 occupied by the central representation [62]. In

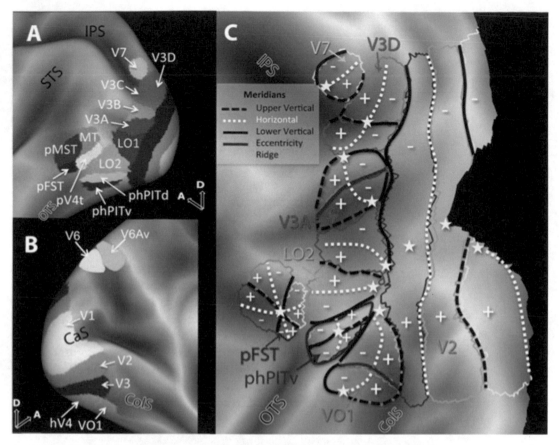

Fig. 3 (**a** and **b**) MPMs of the retinotopic areas (MSM-retino registration) in the left hemisphere on the inflated cortical surface, lateral (**a**) and medial (**b**) views; (**c**) schematic representation of retinotopic organization of the 18 areas shown on the retinotopic MPM: upper (+) and lower (−) fields and central vision (*stars*); same color code as in (**a**). The location of V6 and V6A in the parieto-occipital sulcus is indicated in (**b**); Note that the MPM of areas V2 and V3 does not include the large eccentricities, otherwise V3 would abut V6

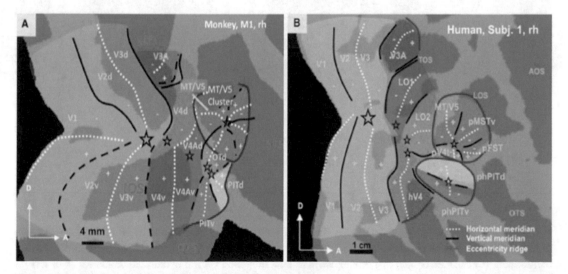

Fig. 4 Comparison of retinotopic layout of monkey (**a**) and human (**b**) visual cortex. Adapted with permission from Vanduffel et al. [58]. *Stars* indicate central representations

both types of studies large variation between individuals (a factor of 2) were observed. A similar range of variation has been observed in the macaque, in which the average surface of V1 is roughly half the size of its human counterpart [63]. The surface of human V2 is estimated to be 80% of that of V1—that of V3 60%. Hence, cortical magnification is somewhat lower in V2 and V3 than in V1 [50, 51], but magnification factors decrease with eccentricity at similar rates in V1, V2, and V3 [50]. In fact the relative size of V1, V2, and V3 depend on the eccentricity range explored, e.g. in the Abdollahi et al. study [56] V2 is actually slightly larger than V1.

The retinotopic maximum probability maps allow also comparison with other parcellations of the same region, in particular those based on morphological features. Figure 5a comparers the retinotopic regions with the average myelin density maps based on the T1/T2 ratio [4]. It shows that the three early visual cortical areas are heavily myelinated. The three early areas correspond relatively closely to the cytoarchitectonic areas hOc1, hOc2, and the combination of hOc3d and hOc3v, respectively (Fig. 5b). On the other hand the retinotopic parcellation has little in common with earlier attempts to parcel occipital cortex using DTI [65]. The comparisons in Fig. 5 also indicate that the central 7.75° of the visual field are represented in roughly half of the V1–V3 surface.

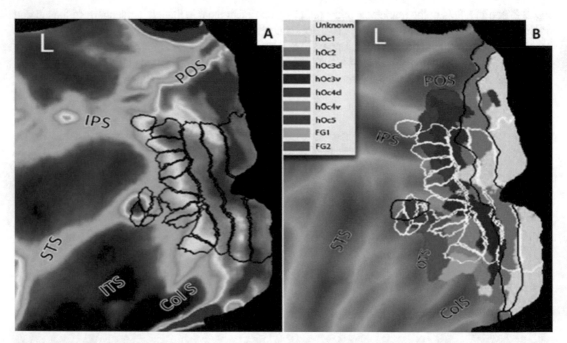

Fig. 5 (**a**) Outlines (*black lines*) of retinotopic MPMs superimposed on myelin density(*blue* to *red color*) maps of left hemisphere (L) of 196 subjects. Color code: myelin content in percentiles of the normalized T1w/T2w distribution. (**b**) outlines (*white*) of retinotopic MPM superimposed on the cytoarchitectonic MPM of left (L) hemisphere. *Black lines*: 50% contours of PAs of hOc1, hOc2, hOc5 from Fischl et al. [64]. *Inset* color code of cytoarchitectonic areas (from Abdollahi et al. [56])

3.2 Two Middle-Level Areas: Human MT/V5 and V3A

V5 or the Middle Temporal (MT) area in humans was initially localized in the ascending branch of the inferior temporal sulcus (ITS) [66, 67]. This identification was supported by the fMRI study of Tootell et al. [68], showing that this region of human cortex has properties, such as luminance and color contrast sensitivity, similar to those of macaque MT/V5. Subsequently this region has been referred to as human MT/V5+ [52] to indicate that probably it corresponds not just to MT/V5 of the macaque but also to several of its satellites. It has proven difficult to demonstrate a retinotopic organization in this region. Huk et al. [69] have suggested that the MT/V5 complex in humans contains a posterior retinotopic part, considered the homolog of MT/V5, and an anterior part driven by ipsilateral stimuli [70], considered the homolog of MST. One of the drawbacks of this parcellation was the absence of an homolog of the fundus of the superior temporal (FST) area. Also the retinotopic organization of what was believed to be MT in humans [69, 71] seems opposite to that of macaque MT in which the lower visual field projects in the dorsal part of MT [72, 73]. The breakthrough occurred when refining the sweeping technique proved that MT and its satellites could be mapped in the macaque (Fig. 4) [74]. Applying the same strategy to humans yielded a MT cluster organized exactly as in the monkey and including four retinotopic areas, considered homologues of MT, MSTv, FST, and V4t (Figs. 2 and 3). The critical point was to identify the central visual field representation in the eccentricity maps, as it corresponds to the center of the cluster from which the four areas radiate. This center is distinct from the central confluence (Fig. 2) and separated from it by a representation of the periphery, the so-called peripheral edge (purple in Fig. 2), which was initially noted by Tootell and coworkers [75]. It is noteworthy that in both species the cluster does not included MSTd involved in optic flow processing [76]. There is at present little consensus on the criteria to define the human counterpart of this MST component [69, 77].

In humans, V3A has a similar retinotopic organization as in macaque: it is defined by a hemifield representation in which the representations of the two quadrants, separated by the HM, are neighbors and occupies the banks of the transverse sulcus [78]. The posterior quadrant is the lower quadrant, separated from that of V3d by a lower VM. In contrast to macaque V3A, hV3A is motion sensitive [14, 78, 79]. In the initial mapping study [68] the central representation of V3A was considered to be fused with that of V1–V2–V3. Subsequent studies [80–82] have shown that the central representation is separated from and located more dorsal than that of the V1–3 confluence, as it generally is in monkeys (5/8 hemispheres in [73]). It has also been noted in humans that this foveal projection, which V3A shares with V3B (see below), can vary considerably in clarity, being well defined in about half (13/30) hemispheres [83]. In fact the retinotopic organization of dorsal occipital cortex is more complex than initially assumed based on the monkey model, and this part of cortex includes one or more retinotopic areas in addition to V3A (see below). It is noteworthy that in all primates the visual cortex includes an MT area, but that the presence of an area V3A in new world monkeys in unclear [60, 84].

3.3 The Fate of V4 in Human Visual Cortex

In their 1995 study, Sereno et al. [51] reported an upper quadrant representation anterior to V3v, that they labeled V4v as it occupied the same position as ventral V4 in macaque. Many studies have replicated that finding of a lower quadrant in front of V3v, but it has proven difficult to identify a corresponding dorsal V4 quadrant in front of dorsal V3 [75]. One possible explanation was that standard mapping technique locating meridians did not apply. Indeed, in the macaque the horizontal meridian, which represents the anterior border of ventral V4, forms the boundary of dorsal V4 only over a short distance, as it curves to join the HM splitting MT/V5 into two halves [73, 85]. Hence, we [3] and others [75] have suggested that the region between V3/V3A and hMT/V5$_+$, which we refer to as LOS [20], is the homolog of macaque dorsal V4. Indeed it is located in a position similar to that of dorsal V4 and has functional properties relatively similar to those of macaque dorsal V4, for example, is sensitive to 3D shape from motion (Fig. 7), to 2D shape [20], and kinetic boundaries [87, 88].

Yet, subsequent mapping studies concentrating on the central 6° of the visual field have suggested that the two halves of macaque V4 have become separated in humans and are each integrated into a separate representation of the contralateral hemifield. Brewer et al. [89] have shown that a lower quadrant was located in front of the upper quadrant initially labeled V4v, with the eccentricity running at right angle to the polar variations. They proposed that this hemifield, located in front of V3v (Figs. 2 and 3) should be considered human V4. They went on to describe two additional maps located in front of hV4: ventral occipital (VO)1 and VO2, each supposedly containing a hemifield representation. Interestingly, the two face areas, the fusiform and occipital face areas are located just lateral to hV4 and VO2, respectively. In the same vein, Larsson and Heeger [83] have described a complete hemifield representation in front of V3d, which they refer to as lateral occipital (LO)1. The posterior half of this region is a lower quadrant that was initially described by Smith et al. [90] as V3B. Thus, the posterior parts of hV4 and LO1 apparently seem more responsive, explaining why they were discovered first. Just as is the case ventrally, a second hemifield representation has been described in front of LO1: LO2, of which the anterior border is close to hMT/V5$^+$. The LO1-2/hV4 scheme led to the suggestion that beyond V1–3 the monkey occipital cortex was not an adequate model for human cortex [81], prompting some [91] to attempt to rescue the monkey model by suggesting that human V4 was similar to that of the monkey.

Our mapping results also favor the LO1–2/hV4 scheme [54, 92]. The final resolution of this problem came with the recognition that the retinotopic organization of occipito-temporal cortex in the monkey is more complex than initially appreciated. A recent study showed that cytoarchitectonic area TEO, located just in front of V4, and which initially was thought to contain a single retinotopic map [93] in fact corresponds to four retinotopic areas: V4A, OTd and PITv and PITd (Fig. 4 [94, 95]). This resolved the problem in the sense that human cortex also includes a cluster similar to the PIT cluster [55] and considered homologous to the monkey PITs, and that the region between V3 and the PIT cluster contains six quadrant representations in both species. In the monkey four of these quadrants are part of a split organization and only two combine into a hemifield, while in humans all quadrants form hemifield representations.

3.4 Dorsal Occipital and Intraparietal Areas

Human V3A has been suggested to share its central representation with an area referred to as V3B, located in front of V3A and dorsally from the LO1/LO2 pair [83, 96]. V3B occupied in this scheme a position initially referred to as V7 [97]. V7 is now instead described as an area rostro-dorsal to a complex of dorsal occipital areas, the V3A complex, which includes four hemifields organized pairwise (Figs. 3 and 4). The lower pair, V3A/V3B shares its peripheral representation (P-cluster) like hV4 and VO1, while the upper pair V3C/V3D shares a central representation (C-cluster). Area V7 instead is a parietal area corresponding to IPS0 of Swisher et al. [98] and seems to correspond to the ventral intraparietal sulcus (VIPS) motion-sensitive region [79, 82, 99], located in the most ventral part of the occipital part of human intraparietal sulcus (IPS) [100]. In fact V7 is part of another C-cluster sharing its center with V7A [101], corresponding to IPS1 and likely the homologue of the pair CIP1–2 described in the monkey [102, 103].

In the human parieto-occipital sulcus (POS) Pitzalis et al. [104] have described human V6, which borders the dorsal parts of V2 and V3, representing large eccentricities in the lower visual field (Fig. 3b), and seems to be homologous in both species. It represents the contralateral hemifield, but with an emphasis on the periphery of the visual field rather than the center. Pitzalis et al. [105] described lower-field only representation in the opposite bank of the POS (Fig. 3b), which they labeled human V6A. This area shows strong pointing responses, unlike V6, and likely belongs to parietal cortex.

Finally, several attempts have been made to parcel visual regions in human IPS. Using standard retinotopic mapping, Swisher et al. [98] described four retinotopic maps, labeled IPS1–4, separated by VM representations. Konen and Kastner [106] added IPS5 and SPL1, relying again only on polar angle maps. Responses to standard retinotopic stimuli are weak in this region, and within anterior parts of IPS moving stimuli are more appropriate to map retinotopic organization than are black and white flickering checkerboards (Fig. 1). Others have used attentional stimuli [107], delayed saccade stimuli [108–110] or stereoscopic stimuli [101] to map retinotopic organization. Progress will come not just from using more appropriate stimuli but also recognizing that eccentricity has to be mapped in addition to polar angle in order to identify correctly retinotopic clusters, which seem to be the dominant organization. Our preliminary results suggest that human IPS includes two additional C-clusters including together four to eight areas. Further work is need to understand the retinotopic organization of this part of human parietal cortex and its relationship to the monkey organization in four areas (LIPv, LIPd, VIP, and AIP)

3.5 Conclusions

Human occipital cortex is now almost completely mapped and includes 19 areas: early areas V1–V3, middle areas LO1–2, hV4, ventral areas VO1–2, dorsal areas V3A–D and hV6, plus the occipito-temporal MT and PIT clusters. The competing scheme using only polar angle maps to define areas [111] only lists 12 occipital retinotopic areas.

Most (13/19) areas are similar to those in the monkey (Fig. 4), if we admit the proposal of Orban et al. [112] that TFO1–2 located ventrally to V4/PITv in the monkey are the homologues of VO1–2. The main inter-species differences are the reorganization of V4/V4A/OTd into LO1–2/hV4, perhaps related to the separation of the PITs from the central confluence [55], and the emergence of areas V3B–D. These latter areas seem to have no counterpart in the monkey in which V3A neighbors CIP1–2, and may relate to the expansion of IPL in humans giving rise to the occipital part of IPS. It is noteworthy that clear homologies are present both at early and high-order level in the occipital cortex, refuting the idea that the human visual system divergence more and more from its monkey counterpart as one ascends into the hierarchy. Also homologous areas may differ in functional properties, e.g. V3A is motion sensitive in humans and not in monkeys.

In human occipital cortex all areas beyond V1–3 have a hemifield organization, while in macaque hemifield representations seemed for a long time the exception and split representations, with separate dorsal and ventral quadrants, the rule. Indeed most initially known areas (V1–4) had split organization with MT/V5 and V3A being the exceptions. With most areas mapped, only 5/16 areas have a split representation in the monkey (Fig. 4), still a larger proportion than in humans (3/19). What is the benefit of the hemifield arrangement? As noticed earlier the dorsal region between V3/V3A and hMT/V5+, in macaque as well as in human, has some particular functional characteristics, such as 3D shape from motion sensitivity. The advantage of the human arrangement is that this sensitivity applies to the whole visual field, while in macaque it applies only to the lower field. This might be an evolutionary advantage explaining the changes in this region, which has expanded considerably in humans. More generally the hemifield organization shortens the distance between neurons with RFs in upper and lower field allowing a better integration across the visual field. This apparently outweighs the need for shorter distances across neighboring areas which favors split representations.

Finally it is worth mentioning that most if not all of occipital cortex is retinotopically organized in both species, and that this organization, again in both species, is maintained in the visual parts of parietal cortex but not temporal cortex [103], with the exception of parahippocampal cortex [113].

4   Motion-Sensitive Regions

4.1   Low-Level Motion Regions

The two most prominent motion-sensitive regions in human visual cortex are human MT/V5+ and V3A (see earlier). They display the highest z scores in a contrast between moving and static random dots. Their activation remains significant at low stimulus contrasts typical of the magnocellular stream [78]. In the occipital cortex motion responses have also been noted in lingual gyrus, probably corresponding to ventral V2, V3, and in parts of LOS [20, 79, 114, 115]. This activation pattern depends heavily on the size of the stimuli. With large stimuli, lower-order motion additionally recruits hV6 [116].

In the early studies it was noted that some parietal regions were also responsive to motion in a contrast between moving and static random dots. Sunaert et al. [79] described four motion-sensitive regions in the IPS. The ventral IPS (VIPS) region is located at the bottom of the IPS near hV3A. This region, we believe corresponds to V7 (see above). The parieto-occipital IPS (POIPS) region is located dorsally with respect to VIPS, at the junction of the parieto-occipital sulcus and IPS, in the vicinity of hV6. Not surprisingly, it represents mainly the peripheral visual field [82] (Fig. 6). The dorsal IPS medial and anterior (DIPSM and DIPSA) regions are located in the horizontal part of IPS, and both represent mainly the central visual field [82] (Fig. 6). They are considered the homolog of anterior part of lateral intraparietal (LIP) region (DIPSM) and posterior part of anterior intraparietal (AIP) region (DIPSA), and indeed DIPSA is located just behind the region referred to as human homolog of AIP based on activation by grasping actions [117]. All these regions are also activated by 3D shape from motion [100], which just as motion itself has a much more extensive representation in human IPS than in macaque IPS (Fig. 7) [86, 118]. We have speculated that this might in part be due to the more extensive tool use in humans than in monkeys, and using a tool indeed activates DIPSM and DIPSA [119]. These different parietal regions may be engaged in different visuomotor

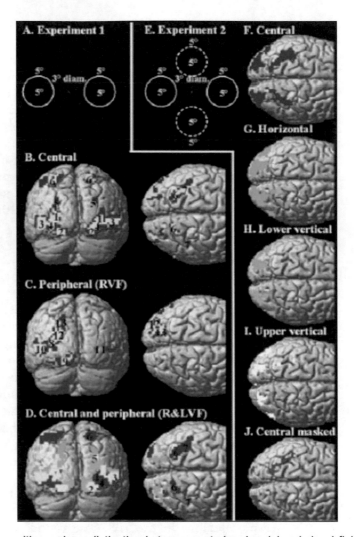

Fig. 6 Human motion-sensitive regions: distinction between central and peripheral visual field. (a) Stimulus configuration in experiment 1: the randomly textured pattern (RTP) was positioned either centrally or 5° into left and right visual field (*red dot* indicates fixation point). (**b** and **c**) Statistical parametric maps (SPMs) showing voxels significant (*yellow*: $p < 0.0001$ uncorrected for multiple comparisons, corresponding to a false discovery rate of less than 5 % false positives; *red*: $p < 0.001$ uncorrected) in the group random-effects analysis (experiment 1, $n = 16$) for the subtraction moving minus stationary conditions for the centrally (**b**) and peripherally (right visual field) (**c**) positioned stimulus, rendered on the posterior and superior views of the standard human brain. Further statistical testing revealed that the interaction between type of stimulus (motion, stationary) and location (center, periphery) was significant (random effects analysis) in DIPSA ($Z = 3.12$, $p < 0.001$ uncorrected and $Z = 3.58$, $p < 0.001$ uncorrected for right and left, respectively), DIPSM ($Z = 3.58$, $p < 0.001$ uncorrected and $Z = 4.35$, $p < 0.0001$ uncorrected for right and left, respectively) and weakly in POIPS ($Z = 2.69$, $p < 0.01$ uncorrected and $Z = 2.24$, $p < 0.01$ uncorrected for right and left, respectively). (**d**) Overlap of voxels ($p < 0.001$ uncorrected; *yellow*) in the group random-effects analysis for the subtraction moving minus stationary conditions for the centrally (*red*) and peripherally (right and left visual field; *green*) positioned stimulus (experiment 1), rendered on the posterior and superior views of the standard human brain. (**e**) Stimulus configuration in experiment 2: RTP was positioned centrally or at 5° eccentricity on upper or lower vertical or horizontal meridian. (**f–i**) SPMs showing voxels significant ($p < 0.05$ corrected) in experiment 2 ($n = 3$) for the subtraction moving minus stationary conditions for the stimuli positioned in the central visual field (**f**, *red*), peripherally left and right on the horizontal meridian (**g**, *green*), and on the lower (**h**, *blue*) and upper vertical meridian (**i**, *white*), rendered on the superior view of the standard human brain (posterior part). (**j**) SPM showing voxels that are active only in the central condition (obtained by exclusive masking of the subtraction in (**f**) with those in (**g–i**)).

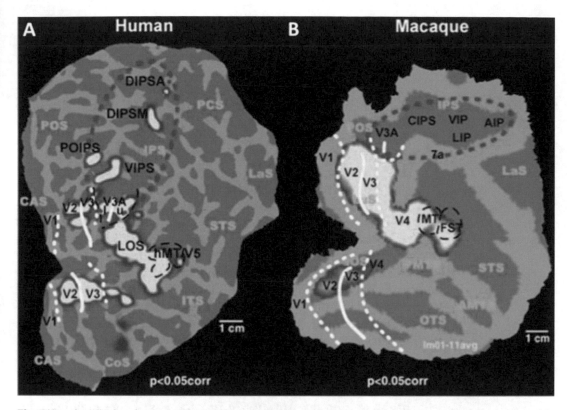

Fig. 7 Visual cortical regions sensitive to 3D shape from motion in human and macaque. Statistical parametric maps (SPMs) for the subtraction viewing of 3D rotating lines minus viewing of 2D translating lines ($p < 0.05$, corrected) of a single human (**a**) and monkey (M4) (**b**) subject projected on the posterior part of the flattened right hemisphere. *White stippled and solid lines*: vertical and horizontal meridian projections (from separate retinotopic mapping experiments); *black stippled lines*: motion-responsive regions from separate motion localizing tests; *purple stippled lines*: region of interspecies difference encompassing V3 and intraparietal sulcus. *PCS* post-central sulcus, *IPS* intraparietal sulcus, *LaS* lateral sulcus, *POS* parieto-occipital sulcus, *CAS* calcarine sulcus, *STS* superior temporal sulcus, *ITS* inferior temporal sulcus, *CoS* collateral sulcus, *IOS* inferior occipital sulcus, *OTS* occipito-temporal sulcus, *PMTS* posterior middle temporal sulcus, *AMTS* anterior middle temporal sulcus (modified from Vanduffel et al. [86])

control circuits, for example, in the control of heading [120], or tracking [121]. Furthermore, it has been shown that flicker is rejected gradually from hMT/V5+ to the more anterior IPS regions [72, 99, 122].

Further regions sensitive to motion, but not to 3D shape from motion, are V6 and premotor regions corresponding to the frontal eye field (FEF) [79, 100, 116], as well as a region in the posterior insula, caudal to the somatosensory opercular complex

---

Fig. 6 (continued) The opposite procedure, subtractions (**g–i**) masked by that in (**f**), yielded no active voxels. *R* right, *L* left, *VF* visual field. *White and yellow numbers* in (**a**) and (**e**) indicate eccentricity and diameter (diameter), respectively. Numbers in (**b–d**) correspond to the activation sites listed: *1* and *8*: hV3A; *2* and *9*: lingual gyrus; *3, 10*, and *11*: hMT/V5+; *4*: LOS; *5* and *12*: VIPS; *13*: POIPS; *6*: DIPSM; and *7*: DIPSA [82]

[123], which we refer to as posterior insular cortex (PIC) region [79, 118] and which might be the homolog of a visual region located next to the posterior insular vestibular cortex (PIVC) in macaques [14, 124, 125].

4.2 The Kinetic Occipital (KO) Region

Using kinetic gratings, that is, stimuli in which random dots move in opposite directions in alternate stripes, and comparing them to luminance gratings or uniform motion, our group [87, 126, 127] discovered a region located between V3/V3A and hMT/V5+ that appeared selective for kinetic boundaries and that we referred to as the kinetic occipital (KO) region. Recent work by Zeki et al. [128] has proposed that KO responds to boundaries defined by other cues (e.g., colors). These findings do not dispute the responsiveness of KO to kinetic gratings as several groups have observed these responses [83, 129]. Although they have been presented differently, these findings are in fact consistent with our PET [127] and fMRI studies [87] showing responses in KO for both kinetic and luminance gratings, suggesting that KO responds to contours of different nature, not just kinetic contours. However, it is important to emphasize that in contrast with responses in hMT/V5+ and other motion-sensitive regions, KO is selective for kinetic contours as opposed to uniform motion. Thus, we meant selectivity in the motion domain, not in the domain of cues defining contours, when we stated [87] that KO is selective for kinetic boundaries. In the Van Oostende et al. study [87] we observed overlap of the KO region with response to the LO localizer. Indeed, Larsson and Heeger [83] in their study identifying LO1/2 showed that the maximal response to kinetic gratings compared to transparent motion, the contrast most sharply defining KO [87], was strongest in LO1 and V3A/B. The coordinates of LO1 [83] are very similar to those of KO (± 31, -91, 0, and -32, -92, 0 [87]), supporting the identifying LO1 as the core region of KO. Thus KO is another functionally defined region that is incorporated into retinotopic regions, as those become known, the human motion area [66], or hMT/V5+, being the primary example, and EBA [130] another one [92].

4.3 High-Level Motion Area

All these motion-sensitive regions are low-level motion regions in the sense that they are driven by motion of light over the retina. Claeys et al. [99] provided evidence for an attention-based motion-sensitive region in the inferior parietal lobule (IPL). This region has activated equiluminant color gratings in which one of the colors is more salient than the other, a paradigm tapping third-order motion [131, 132]. In addition this region has a bilateral representation of the visual field, while all other motion-sensitive areas have mainly a contralateral representation.

5   Shape-Sensitive Regions

There is accumulating evidence that neuronal processes supporting object recognition are coarsely localized in the ventral visual stream [133] that contains a hierarchy of cortical processing stages (V1 → V2 → V4 → IT). The highest stages of this stream (i.e., anterior inferior temporal cortex, AIT or anterior TE in the monkey, and the rostral part of LOC in the human [20, 21, 134, 135]) are thought to be involved in shape processing and support object recognition (Fig. 8). But how are these neuronal representations

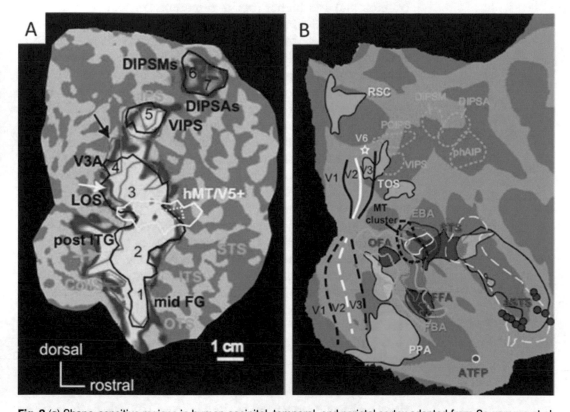

Fig. 8 (**a**) Shape-sensitive regions in human occipital, temporal, and parietal cortex adapted from Sawamura et al. [136]: *yellow*: voxels significant ($p < 0.05$; corrected) in the subtraction 32-objects minus identical condition; *black and blue lines*: borders of shape-sensitive regions [i.e., voxels significant in the subtractions intact vs. scrambled images] obtained by Sawamura et al. [136] and Denys et al. [20] respectively. *Numbers*: local maxima listed in *black*. (**b**) Face, place, and body patches in human occipital and occipito-temporal cortex: activations are projected onto flattened right hemisphere of fsaverage atlas. Faces (*red*)- and body (*dark blue*) selective regions: approximate probabilistic data of Engell and McCarthy [137]; Dynamic facial expressions (*ocre*): real data from Zhu et al. [138]; ATFP: approximate data from (Rajimehr et al. [139]; aSTS: approximate data from (Pitcher et al. [140], showing Talairach coordinates of individual activations (local maxima). Scene patches (*light blue*) from Nasr et al. [141]. Biological motion sensitivity (*green shape* main effect, *white* kinematics main effect, *white interaction*): approximate data from Jastorff and Orban [142]. Regions sensitive for primate vocalizations (*black outlines*) and intelligible speech (*dashed yellow outlines*): real data from Joly et al. [143]. Retinotopy: approximate probabilistic data from PALS-B12 atlas; VIPS, POIPS, DIPSM, DIPSA, and phAIP (*green dashed outlines*): are approximate locations from Jastorff et al. [144],. *White star* is central representation in V6

that support object recognition constructed in the brain? In the monkey, the visual system has been suggested to recruit a hierarchical network of areas across the ventral visual pathway [133, 145] with selectivity for features of increasing complexity from early to later stages of processing [146]. Recent neuroimaging studies suggest a similar organization in the human brain. That is, local image features (e.g., position, orientation) are shown to be processed at the first stage of cortical processing (V1) [11, 37] while complex shapes and even abstract object categories (faces, bodies, places) are represented toward the end of the pathway in the LOC [147–150]. Combined monkey and human fMRI studies showed that the perception of global shapes involves both early (retinotopic) and higher (occipito-temporal) visual areas that may integrate local elements to global shapes at different spatial scales [151, 152]. However, unlike neurons in early visual areas that integrate local information about global shapes within the neighborhood of their receptive fields, neural populations in the LOC represent the perceived global form of objects. In particular, recent imaging studies [153] have shown fMRI adaptation in LOC when the perceived shape of visual stimuli was identical but the image contours differed (because occluding bars occurred in front of the shape in one stimulus and behind the shape in the other). In contrast, recovery from adaptation was observed when the contours were identical but the perceived shapes were different (because of a figure-ground reversal).

The idea of a single, general ventral stream processing objects, has been contradicted by the recent findings of multiple specialized regions processing faces, bodies, and scenes (Fig. 8b [58])This has led to the view that in addition to a general purpose object processing system housed in LOC, the human ventral pathway includes also category specific processing regions [154]. This compromise is not very satisfactory as it implied a dissociation between semantic and visual definition of categories and the fact that general purpose mechanisms for categorization have been located in prefrontal and parietal cortex [155] and not in inferotemporal cortex of the monkey [156]. Therefore, we have recently proposed that the ventral visual pathway is organized in three stages [103]: first a retinotopic stage which included the phPIT cluster, processing visual features of the image; second the anterior part of LOC, corresponding to monkey TE, processing real world entities (RWE), a general term covering objects, faces, and bodies, and third the temporal pole, processing known, complete RWEs. Furthermore the second stage operates in parallel with more dorsal regions processing actions and more ventral regions processing scenes. This middle stage is subdivided into a more dorsal substream processing shape and a more ventral substream processing material properties (color, texture etc.).

The parallel streams and substreams for general shape, faces, bodies, and material properties start in the rostral part of the retinotopic cortex, as shown by the overlap between the caudal face, body and color patches, and retinotopic cortex. For example OFA overlaps with retinotopic cortex but also the posterior two thirds of EBA [92]. At more anterior levels, i.e. the second and third stage, retinotopy is absent, as stated above. Central-periphery organization has been reported at this level [157, 158] but this is the simple consequence of the fact that faces require detail available in central vision while scenes require at least moderately large eccentricities. Finally along these streams and substreams the visual information is gradually abstracted away from the image properties. This is best documented for the LOC, and its monkey counterpart TE [23, 136, 146, 148, 159], i.e. the general shape substream, but likely applies to all (sub)streams [160]. In particular, representations in the anterior subregion of the LOC in the fusiform gyrus (pFs) were shown to be largely invariant to size and position, but not invariant to the direction of illumination and rotation around the vertical axis. In contrast, representations in the posterior subregion of the LOC in the lateral occipital (LO) cortex did not show size or position invariance [23, 24].

6   Depth Processing and 3D Shape Perception

Neurophysiological studies have revealed selectivity for binocular disparity at multiple levels of the visual hierarchy in the monkey brain from early visual areas, to object- and motion-selective areas and the parietal cortex (for reviews: [161–164]). Imaging studies have identified multiple human brain areas in the visual, temporal, and parietal cortex that show stronger activations for stimuli defined by binocular or monocular depth cues than for 2D versions of these stimuli. In particular, areas V3A [165–168] and V3B/LO1/KO [87, 128, 129, 169] have been implicated in the analysis of disparity-defined surfaces and boundaries. Furthermore, studies have employed parametric manipulations to investigate the neural correlates of surface depth (i.e., near vs. far) judgments [167, 170] and 3D shape perception [19]. Finally, several recent studies suggest that areas involved in disparity processing, primarily in the temporal and parietal cortex, are also engaged in the processing of monocular cues to depth (e.g., texture, motion, shading) [20, 86, 100, 171–178] and the combination of binocular and monocular cues for depth perception [179].

Depth relates to the distance from the fixation point and needs to be combined with eye position information to yield distance from the observer. The derivatives of depth provide information about surface orientation and object shape. Gradient selective neurons extracting these derivatives from monocular or binocular image(s) shave been amply documented in parietal and temporal cortex of the monkey [15]. A systematic set of fMRI studies [15] have documented the parietal and temporal regions involved in this extraction in humans. While 3D shape from texture, motion, and disparity is extracted both in dorsal and ventral pathways, 3D shape from shading is predominantly processed in ventral regions close to the phPIT cluster. Systematic combination of single cell recordings, monkey and human fMRI with identical stimuli have allowed to infer the presence of gradient selective neurons in some of the human regions such as pFST or DIPSA [15, 58].

7   Processing of Observed Actions

The visual processing of actions performed by others has been largely neglected in studies of the visual system [180]. Recent studies [144, 181] have shown that this information is processed in regions homologous to the upper and lower bank of middle and rostral STS of the monkey [112, 182]: posterior MTG/pSTG and posterior OTS/posterior fusiform cortex respectively. These areas are also involved in processing biological motion [112, 142, 183].

8   Conclusions

The human visual system likely includes about 40–45 cortical areas. About two/thirds of these have been identified so far, using retinotopic mapping, which proved more efficient than functional properties or morphological features. Further progress can be expected from mapping retinotopic organization with functionally more specific stimuli than black and white checkerboards and from mapping higher-order visual attributes, such as 3D shape or actions, combined with detection of gradients in maps relying on morphological features and/or connections [184].

References

1. Van Essen DC (2004) Organization of visual areas in macaque and human cerebral cortex. In: Chalupa LM, Werner JS (eds) The visual neurosciences, vol 1. MIT Press, Cambridge, MA, pp 507–521

2. Nelissen K, Vanduffel W, Orban GA (2006) Charting the lower superior temporal region, a new motion-sensitive region in monkey superior temporal sulcus. J Neurosci 26:5929–5947

3. Orban GA, Van Essen D, Vanduffel W (2004) Comparative mapping of higher visual areas in monkeys and humans. Trends Cogn Sci 8:315–324

4. Glasser MF, Van Essen DC (2011) Mapping human cortical areas in vivo based on myelin content as revealed by t1- and t2-weighted MRI. J Neurosci 31:11597–11616

5. Dougherty RF et al (2005) Occipital-callosal pathways in children validation and atlas development. Ann N Y Acad Sci 1064:98–112

6. Schmahmann JD et al (2007) Association fibre pathways of the brain: parallel observations from diffusion spectrum imaging and autoradiography. Brain 130:630–653

7. Van Essen DC, Jbabdi S, Sotiropoulos SN, Chen C, Dikranian K, Coalson T, Harwell J, Behrens TE, Glasser MT (2013) Mapping connections in humans and non-human primates: aspirations and challenges for diffusion imaging. In: Johansen-Berg H, Behrens

TE (eds) Diffusion MRI from quantitative measurement to in vivo neuroanatomy. Academic, Amsterdam

8. Krekelberg B, Boynton GM, van Wezel RJA (2006) Adaptation: from single cells to BOLD signals. Trends Neurosci 29:250–256

9. Kovács G, Kaiser D, Kaliukhovich DA, Vidnyánszky Z, Vogels R (2013) Repetition probability does not affect fMRI repetition suppression for objects. J Neurosci 33:9805–9812

10. Sawamura H, Orban GA, Vogels R (2006) Selectivity of neuronal adaptation does not match response selectivity: a single-cell study of the fMRI adaptation paradigm. Neuron 49:307–318

11. Haynes J-D, Rees G (2006) Decoding mental states from brain activity in humans. Nat Rev Neurosci 7:523–534

12. Vogels R, Orban GA (1990) How well do response changes of striate neurons signal differences in orientation: a study in the discriminating monkey. J Neurosci 10:3543–3558

13. Haynes J-D, Rees G (2005) Predicting the orientation of invisible stimuli from activity in human primary visual cortex. Nat Neurosci 8:686–691

14. Vanduffel W et al (2001) Visual motion processing investigated using contrast agent-enhanced fMRI in awake behaving monkeys. Neuron 32:565–577

15. Orban GA (2011) The extraction of 3D shape in the visual system of human and nonhuman primates. Annu Rev Neurosci 34:361–388

16. Orban GA (2002) Functional MRI in the awake monkey: the missing link. J Cogn Neurosci 14:965–969

17. Boynton GM, Demb JB, Glover GH, Heeger DJ (1999) Neuronal basis of contrast discrimination. Vision Res 39:257–269

18. Zenger-Landolt B, Heeger DJ (2003) Response suppression in v1 agrees with psychophysics of surround masking. J Neurosci 23:6884–6893

19. Chandrasekaran C, Canon V, Dahmen JC, Kourtzi Z, Welchman AE (2007) Neural correlates of disparity-defined shape discrimination in the human brain. J Neurophysiol 97:1553–1565

20. Denys K et al (2004) The processing of visual shape in the cerebral cortex of human and nonhuman primates: a functional magnetic resonance imaging study. J Neurosci 24:2551–2565

21. Kourtzi Z, Kanwisher N (2000) Cortical regions involved in perceiving object shape. J Neurosci 20:3310–3318

22. Buckner RL et al (1998) Functional-anatomic correlates of object priming in humans revealed by rapid presentation event-related fMRI. Neuron 20:285–295

23. Grill-Spector K et al (1999) Differential processing of objects under various viewing conditions in the human lateral occipital complex. Neuron 24:187–203

24. Grill-Spector K, Malach R (2001) fMR-adaptation: a tool for studying the functional properties of human cortical neurons. Acta Psychol (Amst) 107:293–321

25. Grill-Spector K, Henson R, Martin A (2006) Repetition and the brain: neural models of stimulus-specific effects. Trends Cogn Sci 10:14–23

26. Koutstaal W et al (2001) Perceptual specificity in visual object priming: functional magnetic resonance imaging evidence for a laterality difference in fusiform cortex. Neuropsychologia 39:184–199

27. Vuilleumier P, Henson RN, Driver J, Dolan RJ (2002) Multiple levels of visual object constancy revealed by event-related fMRI of repetition priming. Nat Neurosci 5:491–499

28. Lisberger SG, Movshon JA (1999) Visual motion analysis for pursuit eye movements in area MT of macaque monkeys. J Neurosci 19:2224–2246

29. Müller JR, Metha AB, Krauskopf J, Lennie P (1999) Rapid adaptation in visual cortex to the structure of images. Science 285:1405–1408

30. Tootell RB et al (1995) Visual motion aftereffect in human cortical area MT revealed by functional magnetic resonance imaging. Nature 375:139–141

31. Huk AC, Ress D, Heeger DJ (2001) Neuronal basis of the motion aftereffect reconsidered. Neuron 32:161–172

32. Huk AC, Heeger DJ (2002) Pattern-motion responses in human visual cortex. Nat Neurosci 5:72–75

33. Engel SA, Furmanski CS (2001) Selective adaptation to color contrast in human primary visual cortex. J Neurosci 21:3949–3954

34. Tolias AS, Smirnakis SM, Augath MA, Trinath T, Logothetis NK (2001) Motion processing in the macaque: revisited with functional magnetic resonance imaging. J Neurosci 21:8594–8601

35. Norman KA, Polyn SM, Detre GJ, Haxby JV (2006) Beyond mind-reading: multi-voxel pattern analysis of fMRI data. Trends Cogn Sci 10:424–430

36. Cox DD, Savoy RL (2003) Functional magnetic resonance imaging (fMRI) "brain reading": detecting and classifying distributed

patterns of fMRI activity in human visual cortex. Neuroimage 19:261–270

37. Kamitani Y, Tong F (2005) Decoding the visual and subjective contents of the human brain. Nat Neurosci 8:679–685

38. Kamitani Y, Tong F (2006) Decoding seen and attended motion directions from activity in the human visual cortex. Curr Biol 16:1096–1102

39. Williams MA, Dang S, Kanwisher NG (2007) Only some spatial patterns of fMRI response are read out in task performance. Nat Neurosci 10:685–686

40. O'Toole AJ, Jiang F, Abdi H, Haxby JV (2005) Partially distributed representations of objects and faces in ventral temporal cortex. J Cogn Neurosci 17:580–590

41. Hanson SJ, Matsuka T, Haxby JV (2004) Combinatorial codes in ventral temporal lobe for object recognition: Haxby (2001) revisited: is there a "face" area? Neuroimage 23:156–166

42. Haxby JV et al (2001) Distributed and overlapping representations of faces and objects in ventral temporal cortex. Science 293:2425–2430

43. Serences JT, Saproo S, Scolari M, Ho T, Muftuler LT (2009) Estimating the influence of attention on population codes in human visual cortex using voxel-based tuning functions. Neuroimage 44:223–231

44. Friston KJ, Rotshtein P, Geng JJ, Sterzer P, Henson RN (2006) A critique of functional localisers. Neuroimage 30:1077–1087

45. Saxe R, Brett M, Kanwisher N (2006) Divide and conquer: a defense of functional localizers. Neuroimage 30:1088–1096

46. Fox PT et al (1986) Mapping human visual cortex with positron emission tomography. Nature 323:806–809

47. Schneider W, Noll DC, Cohen JD (1993) Functional topographic mapping of the cortical ribbon in human vision with conventional MRI scanners. Nature 365:150–153

48. Shipp S, Watson JD, Frackowiak RS, Zeki S (1995) Retinotopic maps in human prestriate visual cortex: the demarcation of areas V2 and V3. Neuroimage 2:125–132

49. Engel SA et al (1994) fMRI of human visual cortex. Nature 369:525

50. Dougherty RF et al (2003) Visual field representations and locations of visual areas V1/2/3 in human visual cortex. J Vis 3:586–598

51. Sereno MI et al (1995) Borders of multiple visual areas in humans revealed by functional magnetic resonance imaging. Science 268:889–893

52. DeYoe EA et al (1996) Mapping striate and extrastriate visual areas in human cerebral cortex. Proc Natl Acad Sci U S A 93:2382–2386

53. Engel SA, Glover GH, Wandell BA (1997) Retinotopic organization in human visual cortex and the spatial precision of functional MRI. Cereb Cortex 7:181–192

54. Georgieva S, Peeters R, Kolster H, Todd JT, Orban GA (2009) The processing of three-dimensional shape from disparity in the human brain. J Neurosci 29:727–742

55. Kolster H, Peeters R, Orban GA (2010) The retinotopic organization of the human middle temporal area MT/V5 and its cortical neighbors. J Neurosci 30:9801–9820

56. Abdollahi RO et al (2014) Correspondences between retinotopic areas and myelin maps in human visual cortex. Neuroimage 99:509–524

57. Robinson EC et al (2013) Multimodal surface matching: fast and generalisable cortical registration using discrete optimisation. In: Lect Notes Comput Sci (including Subser. Lect Notes Artif Intell Lect Notes Bioinformatics) 7917 LNCS. pp 475–486

58. Vanduffel W, Zhu Q, Orban GA (2014) Monkey cortex through fMRI glasses. Neuron 83:533–550

59. Lyon DC, Kaas JH (2002) Evidence for a modified V3 with dorsal and ventral halves in macaque monkeys. Neuron 33:453–461

60. Rosa MGP, Tweedale R (2005) Brain maps, great and small: lessons from comparative studies of primate visual cortical organization. Philos Trans R Soc Lond B Biol Sci 360:665–691

61. Duncan RO, Boynton GM (2003) Cortical magnification within human primary visual cortex correlates with acuity thresholds. Neuron 38:659–671

62. Adams DL, Sincich LC, Horton JC (2007) Complete pattern of ocular dominance columns in human primary visual cortex. J Neurosci 27:10391–10403

63. Van Essen DC, Newsome WT, Maunsell JHR (1984) The visual field representation in striate cortex of the macaque monkey: asymmetries, anisotropies, and individual variability. Vision Res 24:429–448

64. Fischl B et al (2008) Cortical folding patterns and predicting cytoarchitecture. Cereb Cortex 18:1973–1980

65. Hagmann P et al (2008) Mapping the structural core of human cerebral cortex. PLoS Biol 6:1479–1493

66. Zeki S et al (1991) A direct demonstration of functional specialization in human visual cortex. J Neurosci 11:641–649

67. Watson JD et al (1993) Area V5 of the human brain: evidence from a combined study using positron emission tomography and magnetic resonance imaging. Cereb Cortex 3:79–94

68. Tootell RB et al (1995) Functional analysis of human MT and related visual cortical areas using magnetic resonance imaging. J Neurosci 15:3215–3230

69. Huk AC, Dougherty RF, Heeger DJ (2002) Retinotopy and functional subdivision of human areas MT and MST. J Neurosci 22:7195–7205

70. Dukelow SP et al (2001) Distinguishing subregions of the human MT+ complex using visual fields and pursuit eye movements. J Neurophysiol 86:1991–2000

71. Smith AT, Wall MB, Williams AL, Singh KD (2006) Sensitivity to optic flow in human cortical areas MT and MST. Eur J Neurosci 23:561–569

72. Van Essen DC, Maunsell JH, Bixby JL (1981) The middle temporal visual area in the macaque: myeloarchitecture, connections, functional properties and topographic organization. J Comp Neurol 199:293–326

73. Fize D et al (2003) The retinotopic organization of primate dorsal V4 and surrounding areas: a functional magnetic resonance imaging study in awake monkeys. J Neurosci 23:7395–7406

74. Kolster H et al (2009) Visual field map clusters in macaque extrastriate visual cortex. J Neurosci 29:7031–7039

75. Tootell RB, Hadjikhani N (2001) Where is "dorsal V4" in human visual cortex? Retinotopic, topographic and functional evidence. Cereb Cortex 11:298–311

76. Tanaka K et al (1986) Analysis of local and wide-field movements in the superior temporal visual areas of the macaque monkey. J Neurosci 6:134–144

77. Morrone MC et al (2000) A cortical area that responds specifically to optic flow, revealed by fMRI. Nat Neurosci 3:1322–1328

78. Tootell R et al (1997) Functional analysis of V3A and related areas in human visual cortex. J Neurosci 17:7060–7078

79. Sunaert S, Van Hecke P, Marchal G, Orban GA (1999) Motion-responsive regions of the human brain. Exp Brain Res 127:355–370

80. Press WA, Brewer AA, Dougherty RF, Wade AR, Wandell BA (2001) Visual areas and spatial summation in human visual cortex. Vision Res 41:1321–1332

81. Wandell BA, Dumoulin SO, Brewer AA (2007) Visual field maps in human cortex. Neuron 56:366–383

82. Orban GA et al (2006) Mapping the parietal cortex of human and non-human primates. Neuropsychologia 44:2647–2667

83. Larsson J, Heeger DJ (2006) Two retinotopic visual areas in human lateral occipital cortex. J Neurosci 26:13128–13142

84. Lyon DC, Kaas JH (2002) Evidence from V1 connections for both dorsal and ventral subdivisions of V3 in three species of new world monkeys. J Comp Neurol 449:281–297

85. Gattass R, Sousa AP, Gross CG (1988) Visuotopic organization and extent of V3 and V4 of the macaque. J Neurosci 8:1831–1845

86. Vanduffel W et al (2002) Extracting 3D from motion: differences in human and monkey intraparietal cortex. Science 298:413–415

87. Van Oostende S, Sunaert S, Van Hecke P, Marchal G, Orban GA (1997) The kinetic occipital (KO) region in man: an fMRI study. Cereb Cortex 7:690–701

88. Nelissen K, Vanduffel W, Sunaert S, Janssen P, Tootell RB, Orban GA (2000) Processing of kinetic boundaries investigated using fMRI and double-label deoxyglucose technique in awake monkeys. Soc Neurosci Abstr 26:1584

89. Brewer AA, Liu J, Wade AR, Wandell BA (2005) Visual field maps and stimulus selectivity in human ventral occipital cortex. Nat Neurosci 8:1102–1109

90. Smith AT, Greenlee MW, Singh KD, Kraemer FM, Hennig J (1998) The processing of first- and second-order motion in human visual cortex assessed by functional magnetic resonance imaging (fMRI). J Neurosci 18:3816–3830

91. Hansen KA, Kay KN, Gallant JL (2007) Topographic organization in and near human visual area V4. J Neurosci 27:11896–11911

92. Ferri S, Kolster H, Jastorff J, Orban GA (2013) The overlap of the EBA and the MT/V5 cluster. Neuroimage 66:412–425

93. Boussaoud D, Desimone R, Ungerleider LG (1991) Visual topography of area TEO in the macaque. J Comp Neurol 306:554–575

94. Janssens T, Zhu Q, Popivanov ID, Vanduffel W (2014) Probabilistic and single-subject retinotopic maps reveal the topographic organization of face patches in the macaque cortex. J Neurosci 34:10156–10167

95. Kolster H, Janssens T, Orban GA, Vanduffel W (2014) The retinotopic organization of macaque occipitotemporal cortex anterior to V4 and caudoventral to the middle temporal (MT) cluster. J Neurosci 34:10168–10191

96. Wandell BA, Brewer AA, Dougherty RF (2005) Visual field map clusters in human cortex. Philos Trans R Soc Lond B Biol Sci 360:693–707

97. Tootell RBH, Tsao D, Vanduffel W (2003) Neuroimaging weighs in: humans meet macaques in "primate" visual cortex. J Neurosci 23:3981–3989

98. Swisher JD, Halko MA, Merabet LB, McMains SA, Somers DC (2007) Visual topography of human intraparietal sulcus. J Neurosci 27:5326–5337

99. Claeys KG, Lindsey DT, De Schutter E, Orban GA (2003) A higher order motion region in human inferior parietal lobule: evidence from fMRI. Neuron 40:631–642

100. Orban GA, Sunaert S, Todd JT, Van Hecke P, Marchal G (1999) Human cortical regions involved in extracting depth from motion. Neuron 24:929–940

101. Kolster H, Peeters R, Orban GA (2011) Ten retinotopically organized areas in human paritetal cortex. Soc Neurosci Abstr 851.10

102. Arcaro MJ, Pinsk MA, Li X, Kastner S (2011) Visuotopic organization of macaque posterior parietal cortex: a functional magnetic resonance imaging study. J Neurosci 31:2064–2078

103. Orban GA, Zhu Q, Vanduffel W (2014) The transition in the ventral stream from feature to real-world entity representations. Front Psychol 5:695

104. Pitzalis S et al (2006) Wide-field retinotopy defines human cortical visual area v6. J Neurosci 26:7962–7973

105. Pitzalis S et al (2013) The human homologue of macaque area V6A. Neuroimage 82:517–530

106. Konen CS, Kastner S (2008) Two hierarchically organized neural systems for object information in human visual cortex. Nat Neurosci 11:224–231

107. Silver MA, Ress D, Heeger DJ (2005) Topographic maps of visual spatial attention in human parietal cortex. J Neurophysiol 94:1358–1371

108. Sereno MI, Pitzalis S, Martinez A (2001) Mapping of contralateral space in retinotopic coordinates by a parietal cortical area in humans. Science 294:1350–1354

109. Schluppeck D, Curtis CE, Glimcher PW, Heeger DJ (2006) Sustained activity in topographic areas of human posterior parietal cortex during memory-guided saccades. J Neurosci 26:5098–5108

110. Schluppeck D, Glimcher P, Heeger DJ (2005) Topographic organization for delayed saccades in human posterior parietal cortex. J Neurophysiol 94:1372–1384

111. Wang L, Mruczek REB, Arcaro MJ, Kastner S (2015) Probabilistic maps of visual topography in human cortex. Cereb Cortex 25(10):3911–3931

112. Orban GA, Jastorff J (2014) Functional mapping of motion regions in human and nonhuman primates. In: Chalupa LM, Werner JS (eds) The new visual neuroscience, vol 1. MIT Press, Cambridge, MA, pp 777–791

113. Arcaro MJ, McMains SA, Singer BD, Kastner S (2009) Retinotopic organization of human ventral visual cortex. J Neurosci 29:10638–10652

114. Sunaert S, Van Hecke P, Marchal G, Orban GA (2000) Attention to speed of motion, speed discrimination, and task difficulty: an fMRI study. Neuroimage 11:612–623

115. Rees G, Friston K, Koch C (2000) A direct quantitative relationship between the functional properties of human and macaque V5. Nat Neurosci 3:716–723

116. Pitzalis S et al (2010) Human V6: the medial motion area. Cereb Cortex 20:411–424

117. Binkofski F et al (1998) Human anterior intraparietal area subserves prehension: a combined lesion and functional MRI activation study. Neurology 50:1253–1259

118. Orban GA et al (2003) Similarities and differences in motion processing between the human and macaque brain: evidence from fMRI. Neuropsychologia 41:1757–1768

119. Stout D, Chaminade T (2007) The evolutionary neuroscience of tool making. Neuropsychologia 45:1091–1100

120. Peuskens H et al (2001) Human brain regions involved in heading estimation. J Neurosci 21:2451–2461

121. Gori M et al (2012) Long integration time for accelerating and decelerating visual, tactile and visuo-tactile stimuli. Multisens Res 26:53–68

122. Braddick OJ, O'Brien JMD, Wattam-Bell J, Atkinson J, Turner R (2000) Form and motion coherence activate independent, but not dorsal/ventral segregated, networks in the human brain. Curr Biol 10:731–734

123. Eickhoff SB, Grefkes C, Zilles K, Fink GR (2007) The somatotopic organization of cytoarchitectonic areas on the human parietal operculum. Cereb Cortex 17:1800–1811

124. Grüsser OJ, Pause M, Schreiter U (1990) Vestibular neurones in the parieto-insular cortex of monkeys (Macaca fascicularis): visual and neck receptor responses. J Physiol 430:559–583

125. Grüsser O-J, Guldin WO, Mirring S, Salah-Eldin A (1994) Comparative physiological and anatomical studies of the primate vestibular cortex. In: Albowitz B, Albus K, Kuhnt

U, Nothdurft H-C, Wahle P (eds) Structural and functional organization of the neocortex. Proceedings of a Symposium in the Memory of Otto D. Creutzfeldt, May 1993, Exp Brain Res Series 24. pp 358–371

126. Orban GA et al (1995) A motion area in human visual cortex. Proc Natl Acad Sci U S A 92:993–997

127. Dupont P et al (1997) The kinetic occipital region in human visual cortex. Cereb Cortex 7:283–292

128. Zeki S, Perry RJ, Bartels A (2003) The processing of kinetic contours in the brain. Cereb Cortex 13:189–202

129. Tyler CW, Likova LT, Kontsevich LL, Wade AR (2006) The specificity of cortical region KO to depth structure. Neuroimage 30:228–238

130. Downing PE, Jiang Y, Shuman M, Kanwisher N (2001) A cortical area selective for visual processing of the human body. Science 293:2470–2473

131. Lu ZL, Sperling G (1995) Attention-generated apparent motion. Nature 377:237–239

132. Lu ZL, Lesmes LA, Sperling G (1999) The mechanism of isoluminant chromatic motion perception. Proc Natl Acad Sci U S A 96:8289–8294

133. Ungerleider LG, Mishkin M (1982) Two cortical visual systems. Anal Vis Behav 549:549–586

134. Malach R et al (1995) Object-related activity revealed by functional magnetic resonance imaging in human occipital cortex. Proc Natl Acad Sci U S A 92:8135–8139

135. Kanwisher N, Chun MM, McDermott J, Ledden PJ (1996) Functional imaging of human visual recognition. Cogn Brain Res 5:55–67

136. Sawamura H, Georgieva S, Vogels R, Vanduffel W, Orban GA (2005) Using functional magnetic resonance imaging to assess adaptation and size invariance of shape processing by humans and monkeys. J Neurosci 25:4294–4306

137. Engell AD, McCarthy G (2013) Probabilistic atlases for face and biological motion perception: an analysis of their reliability and overlap. Neuroimage 74:140–151

138. Zhu Q et al (2012) Dissimilar processing of emotional facial expressions in human and monkey temporal cortex. Neuroimage 66C:402–411

139. Rajimehr R, Young JC, Tootell RBH (2009) An anterior temporal face patch in human cortex, predicted by macaque maps. Proc Natl Acad Sci U S A 106:1995–2000

140. Pitcher D, Dilks DD, Saxe RR, Triantafyllou C, Kanwisher N (2011) Differential selectiv-ity for dynamic versus static information in face-selective cortical regions. Neuroimage 56:2356–2363

141. Nasr S et al (2011) Scene-selective cortical regions in human and nonhuman primates. J Neurosci 31:13771–13785

142. Jastorff J, Orban GA (2009) Human functional magnetic resonance imaging reveals separation and integration of shape and motion cues in biological motion processing. J Neurosci 29:7315–7329

143. Joly O et al (2012) Processing of vocalizations in humans and monkeys: a comparative fMRI study. Neuroimage 62:1376–1389

144. Jastorff J, Begliomini C, Fabbri-Destro M, Rizzolatti G, Orban GA (2010) Coding observed motor acts: different organizational principles in the parietal and premotor cortex of humans. J Neurophysiol 104:128–140

145. Felleman DJ, Van Essen DC (1991) Distributed hierarchical processing in the primate cerebral cortex. Cereb Cortex 1:1–47

146. Tanaka K, Saito H, Fukada Y, Moriya M (1991) Coding visual images of objects in the inferotemporal cortex of the macaque monkey. J Neurophysiol 66:170–189

147. Grill-Spector K, Malach R (2004) The human visual cortex. Annu Rev Neurosci 27:649–677

148. Quiroga RQ, Reddy L, Kreiman G, Koch C, Fried I (2005) Invariant visual representation by single neurons in the human brain. Nature 435:1102–1107

149. Reddy L, Kanwisher N (2006) Coding of visual objects in the ventral stream. Curr Opin Neurobiol 16:408–414

150. Privman E et al (2007) Enhanced category tuning revealed by intracranial electroencephalograms in high-order human visual areas. J Neurosci 27:6234–6242

151. Altmann CF, Bülthoff HH, Kourtzi Z (2003) Perceptual organization of local elements into global shapes in the human visual cortex. Curr Biol 13:342–349

152. Kourtzi Z, Tolias AS, Altmann CF, Augath M, Logothetis NK (2003) Integration of local features into global shapes: monkey and human fMRI studies. Neuron 37:333–346

153. Kourtzi Z, Kanwisher N (2001) Human lateral occipital complex representation of perceived object shape by the human lateral occipital complex. Science 293:1506–1509

154. Downing PE, Chan AW-Y, Peelen MV, Dodds CM, Kanwisher N (2006) Domain specificity in visual cortex. Cereb Cortex 16:1453–1461

155. Freedman DJ, Riesenhuber M, Poggio T, Miller EK (2003) A comparison of primate

prefrontal and inferior temporal cortices during visual categorization. J Neurosci 23:5235–5246

156. Vogels R (1999) Categorization of complex visual images by rhesus monkeys. Part 2: Single-cell study. Eur J Neurosci 11:1239–1255

157. Levy I, Hasson U, Avidan G, Hendler T, Malach R (2001) Center-periphery organization of human object areas. Nat Neurosci 4:533–539

158. Hasson U, Levy I, Behrmann M, Hendler T, Malach R (2002) Eccentricity bias as an organizing principle for human high-order object areas. Neuron 34:479–490

159. Rolls ET (2000) Functions of the primate temporal lobe cortical visual areas in invariant visual object and face recognition. Neuron 27:205–218

160. Freiwald WA, Tsao DY (2010) Functional compartmentalization and viewpoint generalization within the macaque face-processing system. Science 330:845–851

161. Cumming BG, DeAngelis GC (2001) The physiology of stereopsis. Annu Rev Neurosci 24:203–238

162. Parker AJ (2007) Binocular depth perception and the cerebral cortex. Nat Rev Neurosci 8:379–391

163. Neri P, Bridge H, Heeger DJ (2004) Stereoscopic processing of absolute and relative disparity in human visual cortex. J Neurophysiol 92:1880–1891

164. Orban GA, Janssen P, Vogels R (2006) Extracting 3D structure from disparity. Trends Neurosci 29:466–473

165. Gulyas B, Roland PE (1994) Processing and analysis of form, colour and binocular disparity in the human brain: functional anatomy by positron emission tomography. Eur J Neurosci 6:1811–1828

166. Mendola JD, Dale AM, Fischl B, Liu AK, Tootell RB (1999) The representation of illusory and real contours in human cortical visual areas revealed by functional magnetic resonance imaging. J Neurosci 19:8560–8572

167. Backus BT, Fleet DJ, Parker AJ, Heeger DJ (2001) Human cortical activity correlates with stereoscopic depth perception. J Neurophysiol 86:2054–2068

168. Tsao DY et al (2003) Stereopsis activates V3A and caudal intraparietal areas in macaques and humans. Neuron 39:555–568

169. Brouwer GJ, van Ee R, Schwarzbach J (2005) Activation in visual cortex correlates with the awareness of stereoscopic depth. J Neurosci 25:10403–10413

170. Gilaie-Dotan S, Ullman S, Kushnir T, Malach R (2002) Shape-selective stereo processing in human object-related visual areas. Hum Brain Mapp 15:67–79

171. Orban GA (2007) Three-dimensional shape: cortical mechanisms of shape extraction. In: Masland RH, Albright T (eds) Handbook of the senses, vol 5, Vision. Elsevier, Amsterdam

172. Durand JB et al (2007) Anterior regions of monkey parietal cortex process visual 3D shape. Neuron 55:493–505

173. James TW et al (2002) Haptic study of three-dimensional objects activates extrastriate visual areas. Neuropsychologia 40:1706–1714

174. Kourtzi Z, Erb M, Grodd W, Bülthoff HH (2003) Representation of the perceived 3-D object shape in the human lateral occipital complex. Cereb Cortex 13:911–920

175. Murray SO, Olshausen BA, Woods DL (2003) Processing shape, motion and three-dimensional shape-from-motion in the human cortex. Cereb Cortex 13:508–516

176. Sereno ME, Trinath T, Augath M, Logothetis NK (2002) Three-dimensional shape representation in monkey cortex. Neuron 33:635–652

177. Shikata E et al (2001) Surface orientation discrimination activates caudal and anterior intraparietal sulcus in humans: an event-related fMRI study. J Neurophysiol 85:1309–1314

178. Taira M, Nose I, Inoue K, Tsutsui K (2001) Cortical areas related to attention to 3D surface structures based on shading: an fMRI study. Neuroimage 14:959–966

179. Welchman AE, Deubelius A, Conrad V, Bülthoff HH, Kourtzi Z (2005) 3D shape perception from combined depth cues in human visual cortex. Nat Neurosci 8:820–827

180. Perrett DI et al (1985) Visual analysis of body movements by neurones in the temporal cortex of the macaque monkey: a preliminary report. Behav Brain Res 16:153–170

181. Abdollahi RO, Jastorff J, Orban GA (2013) Common and segregated processing of observed actions in human SPL. Cereb Cortex 23:2734–2753

182. Jastorff J, Popivanov ID, Vogels R, Vanduffel W, Orban GA (2012) Integration of shape and motion cues in biological motion processing in the monkey STS. Neuroimage 60:911–921

183. Grossman E et al (2000) Brain areas involved in perception of biological motion. J Cogn Neurosci 12:711–720

184. Glasser MF, Robinson EC, Coalson TS, Smith SM, Jenkinson M, Hacker CS, Laumann TO, Van Essen DC (2014) Partial correlation functional connectivity gradients for cortical parcellation: methods and multi-modal comparisons SFN abstract WCC 147B

Chapter 19

fMRI of the Central Auditory System

Deborah Ann Hall and Aspasia Eleni Paltoglou

Abstract

Over the years, blood oxygen level-dependent (BOLD) fMRI has made important contributions to the understanding of central auditory processing in humans. Although there are significant technical challenges to overcome in the case of auditory fMRI, the unique methodological advantage of fMRI as an indicator of population neural activity lies in its spatial precision. It can be used to examine the neural basis of auditory representation at a number of spatial scales, from the micro-anatomical scale of population assemblies to the macro-anatomical scale of cortico-cortical circuits. The spatial resolution of fMRI is maximized in the case of mapping individual brain activity, and here it has been possible to demonstrate known organizational features of the auditory system that have hitherto been possible only using invasive electrophysiological recording methods. Frequency coding in the primary auditory cortex is one such example that we shall discuss in this chapter. Of course, noninvasive procedures for neuroscience are the ultimate aim and as the field moves towards this goal by recording in awake, behaving animals so human neuroimaging techniques will be increasingly relied upon to provide an interpretive link between animal neurophysiology at the multi-unit level and the operation of larger neuronal assemblies, as well as the mechanisms of auditory perception itself. For example, the neural effects of intentional behavior on stimulus-driven coding have been explored both in animals, using electrophysiological techniques, and in humans, using fMRI. While the feature-specific effects of selective attention are well established in the visual cortex, the effect of auditory attention in the auditory cortex has generally been examined at a very coarse spatial scale. Ongoing research in our laboratory has started to address this question and here we present preliminary evidence for frequency-specific effects of attentional enhancement in the human auditory cortex. We end with a brief discussion of several future directions for auditory fMRI research.

Key words Technical challenges, Frequency coding, Selective attention, Perceptual representation, Task specificity

1 Challenges of Auditory fMRI

The construction of a brain image using MR imaging depends upon the magnetic properties of hydrogen ions that, when placed in a static magnetic field, can absorb pulses of radiowave energy of a specific frequency. The time taken for the ion alignments to return to equilibrium after the radiofrequency (RF) pulse differs according to the surrounding tissue, thus providing the image contrast, for example between gray matter, white matter, cerebrospinal fluid, and bone. The use of MR techniques for detecting

Massimo Filippi (ed.), *fMRI Techniques and Protocols*, Neuromethods, vol. 119,
DOI 10.1007/978-1-4939-5611-1_19, © Springer Science+Business Media New York 2016

functional brain activation relies on two factors: first that local neural activity is a metabolically demanding process that is closely associated with a local increase in the supply of oxygenated blood to those active parts of the brain, and second that the different paramagnetic properties of oxygenated and deoxygenated blood produce measurable effects on the MR signal. The functional signal detected during fMRI is known as the blood oxygen level-dependent (BOLD) response. Essentially, the functional image represents the spatial distribution of blood oxygenation levels in the brain, and the small fluctuations in these levels over time are correlated with the stimulus input or cognitive task.

MR scanners operate using three different types of electromagnetic fields: a very high static field generated by a superconducting magnet, time-varying gradient magnetic fields, and pulsed RF fields. The latter two fields are much weaker than the first, but all pose a number of unique and considerable technical challenges for conducting auditory fMRI research within this hostile environment. In the first place, the static and time-varying magnetic fields preclude the use of many types of electronic sound presentation equipment, as well as preventing the safe scanning of patients who are wearing listening devices such as hearing aids or implants. Additionally, the high levels of scanner noise generated by the flexing of the gradient coils in the static magnetic field can potentially cause hearing difficulties. The scanner noise masks the perception of the acoustic stimuli presented to the subject in the scanner making it difficult to calibrate audible hearing levels and adding to the difficulty of the listening task. And finally, the scanner noise not only activates parts of the auditory brain, but also interacts with the patterns of activity evoked by experimental stimuli. Auditory fMRI poses a number of other challenges, not related to the hostile environment of the MR scanner, but related instead to the nature of the neural coding in the auditory cortex. The response of auditory cortical neurons to a particular class of sound is determined not only by the acoustic features of that sound, but also by its presentation context. For example, neurons respond strongly to the onset of sound events and thereafter tend to show rapid adaptation to that sound in terms of a reduction in their firing rate. Thus, the result of any particular auditory fMRI experiment will depend not only on the physical attributes of a stimulus, but also on the way in which the stimuli are presented. In this first section, we shall take each one of these issues in turn, introducing the problems in more detail as well as proposing some solutions.

1.1 Use of Electronic Equipment for Sound Presentation in the MR Scanner

1.1.1 Problems

The ideal requirement is a sound presentation system that produces a range of sound levels [up to 100-dB sound pressure level (SPL)], with low distortion, a flat frequency response, and a smooth and predictable phase response. The first commercially available solution utilized loudspeakers, placed away from the high static magnetic field, from which the sound was delivered through plastic tubes inserted into the ear canal (Fig. 1a) through a protective ear

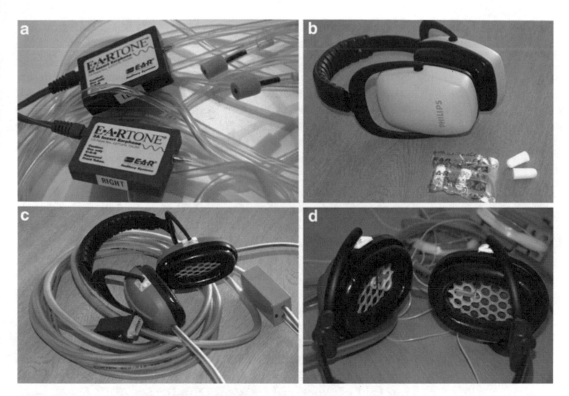

Fig. 1 MR compatible headsets for sound delivery and noise reduction: (**a**) tube phones system with foam ear inserts, (**b**) circum-aural ear defenders, plus foam ear plugs for passive noise reduction, (**c**) MRC IHR sound presentation headset combining commercially available electrostatic transducers in an industry standard ear defender, and (**d**) modified MRC IHR headset for sound presentation and for active noise cancellation (ANC), including an optical error microphone positioned underneath the ear defender

defender (Fig. 1b). One general disadvantage of the tube phone system is that the tubing distorts both the phase and amplitude of the acoustic signal, for example, by imposing a severe ripple on the spectra and reducing sound level, especially at higher frequencies. Another limitation is the leak of the scanner noise through the pipe walls to the pipe inner and hence the ear. Despite alternative systems now being readily available, tube phone systems are still commercially manufactured (e.g. Avotec Inc. Stuart, Florida, USA, www.avotec.org/). The Avotec system has been specifically designed for fMRI use and boasts an equalizer to provide a reasonably flat audio output (±5 dB) across its nominal bandwidth (150 Hz–4.5 kHz) and a procedure for acoustic calibration that feeds a known electrical input signal to the audio system input and makes a direct acoustic output measurement at the headset.

Alternative electronic systems often used for psychoacoustical research deliver high-quality signals, but these systems are generally unsuitable for use in the MR environment because most headphones use an electromagnet to push and pull on a diaphragm to vibrate the air and generate sound. Of course, this electromagnet is rendered inoperable by the magnetic fields in the MR scanner. Headphone components constructed from ferromagnetic material also disrupt the magnetic fields locally and induce signal loss or spatial distortion in areas close to the ears. In addition, the electronic components can be damaged by the static magnetic field, while electromagnetic interference generated by the equipment is detected by the MR receiver head coil. Electronic sound delivery systems for use in auditory fMRI research have been designed specifically to overcome these difficulties.

1.1.2 Solutions

Despite the restriction on the materials that can be used in a scanner, a number of different MR-compatible active headphone driving units have been produced. An ingenious system has been developed and marketed by one auditory neuroimaging research group (MR confon GmbH, Magdeburg, Germany, www.mr-confon.de). This system incorporates a unique, electrodynamic driver that uses the scanner's static magnetic field in place of the permanent magnets that are found in conventional headphones and loudspeakers. It produces a wide frequency range (less than 200 Hz–35 kHz) with a flat frequency response (±6 dB). Another company manufactures and supplies high-quality products for MRI, with a special focus on the fast-growing field of functional imaging (NordicNeuroLab AS, Bergen, Norway, www.nordicneurolab.com/). Their audio system uses electrostatic transducers to ensure high performance. Electrostatic headphones generate sound using a conductive diaphragm placed next to a fixed conducting panel. A high voltage polarizes the fixed panel and the audio signal passing through the diaphragm rapidly switches between a positive and a negative signal, attracting or repelling it to the fixed panel and thus vibrating the air. Their technical specification claims a flat frequency response from 8 Hz to 35 kHz. The signal is transferred from the audio source to the headphones in the RF screened scanner room using either filters through a filter panel or fiber-optic cable through the waveguide.

Here at the MRC Institute of Hearing Research, we became engaged in auditory fMRI research well before such commercial systems were widely available and so, for our own purposes, we developed an MR-compatible headset (Fig. 1c) based on commercially available electrostatic headphones, modified to remove or replace their ferromagnetic components, and combined with standard industrial ear defenders to provide good acoustic isolation [1]. Our custom-built system delivers a flat frequency response (±10 dB) across the frequency range 50 Hz–10 kHz and has an output level capability up to 120-dB SPL. Again, the digital audio source, electronics, and power supply that drive the system are housed outside the RF screened scanner room to avoid electromagnetic interference with MR scanning, and all electrical signals passing into the screened scanner room are RF filtered.

1.2 Risk to Patients Who Are Wearing Listening Devices in the MR Scanner

1.2.1 Problems

No ferromagnetic components can be placed in the scanner bore as they would experience a strong attraction by the static magnetic field and potentially cause damage not only to the scanner and the listening device, but also to the patient. Induced currents in the electronics, caused directly by the time-varying gradient magnetic fields or the RF pulses, are an additional hazard to the electronic devices themselves, while some materials can also absorb the RF energy causing local tissue heating and even burns if in contact with soft tissue. For these reasons, there are restrictions on scanning people who have electronic listening devices. These include hearing aids, cochlear implants, and brainstem implants. Hearing aids amplify sound for people who have moderate to profound hearing loss. The aid is battery-operated and worn in or around the ear. Hearing aids are available in different shapes, sizes, and types, but they all work in a similar way. They all have a built-in microphone that picks up sound from the environment. These sounds are processed electronically and made louder, either by analogue circuits or digitally, and the resulting signals are passed to a receiver in the hearing aid where they are converted back into audible sounds. In contrast, cochlear and brainstem implants are both small, complex electronic devices that can help to provide a sense of sound to people who are profoundly deaf or severely hard-of-hearing. Cochlear implants bypass damaged portions of the inner ear (the cochlea) and directly stimulate the auditory nerve, while auditory brainstem implants bypass the vestibulocochlear nerve in cases when it is damaged by tumors or surgery and directly stimulate the lower part of the auditory brain (the cochlear nucleus). In general, both types of implant consist of an external portion that sits behind the ear and a second portion that is surgically placed under the skin. They contain a microphone, a sound processor (which converts sounds picked up by the microphone into an electrical code), a transmitter and receiver/stimulator (which receive signals from the processor and convert them into electric impulses), and finally an electrode array (which is a set of electrodes that collect the impulses from the stimulator and stimulate groups of auditory neurons). Coded information from the sound processor is delivered across the skin via electromagnetic induction to the implanted receiver/stimulator, which is surgically placed on a bone behind the ear.

1.2.2 Solutions

Official approval for the manufacture of implant devices requires rigorous testing for susceptibility to electromagnetic fields, radiated electromagnetic fields, and electrical safety testing (including susceptibility to electrical discharge). However, such tests are conducted under normal conditions, not in the magnetic fields of an MR scanner. Some implant designs have been proven to be MR compatible [2–5], but they are not routinely supplied in clinical practice. Standard listening devices do not meet MR compatibility criteria and, for the patient, risks include movement of the device and localized heating of brain tissue, whereas, for the device, the electronic components may be damaged. Magnetic Resonance Safety Testing Services (MRSTS) is a highly experienced testing company that conducts comprehensive evaluations of implants, devices, objects, and materials in the MR environment (MRSTS, Los Angeles, CA, www.magneticresonancesafetytesting.com/). Testing includes approved assessment of magnetic field interactions, heating, induced electrical currents, and artifacts. A database of the devices and results of implant testing is accessible to the interested reader (www.mrisafety.com/). However, auditory devices have generally been tested only at low magnetic fields (up to 1.5 T) because most clinical MR systems operate at this field strength. Since research systems typically operate at 3.0 T (for improved BOLD signal-to-noise ratio, BOLD SNR) it may be necessary for individual research teams to ensure the safety of their patients. For example, here at the MRC Institute of Hearing Research, we have recently assessed the risks of movement and localized tissue heating for two middle ear piston devices [6]. For the safety reasons discussed in this subsection, listeners who normally wear hearing aids could be scanned without their aid but, to compensate, have been presented with sounds amplified to an audible level. Given that implanted devices cannot be removed without surgical intervention, clinical imaging research of implantees has generally used other brain imaging methods, namely positron emission tomography [7].

1.3   Intense MR Scanner Noise and Its Effects on Hearing

1.3.1   Problems

The scanning sequence used to measure the BOLD fMRI signal requires rapid on and off switching of electrical currents through the three gradient coils of wire in order to create time-varying magnetic fields that are required for selecting and encoding the three-dimensional image volume (in the x, y, and z planes). This rapid switching in the static magnetic field induces bending and buckling of the gradient coils during MRI. As a result, the gradient coils act like a moving coil loudspeaker to produce a compression wave in the air, which is heard as acoustic noise during the image acquisition. Scanner noise increases nonlinearly with static magnetic field strength, such that ramping from 0.5 to 2 T could account for a rise in sound level of as much as 11-dB SPL [8]. A brain scan is composed of a set of two-dimensional "slices" through the brain. Gradient switching is required for each slice acquisition and so an intense scanner "ping" occurs each time a brain slice is collected. Each ping lasts about 50 ms and so during fMRI, each scan is audible as a rapid sequence of such "pings" (see inset in Fig. 2 for an example of the amplitude envelope of the scanner noise).

The dominant components of the noise spectrum are composed of a peak of sound energy at the gradient switching frequency plus its higher harmonics. Most of the energy lies below 3 kHz. Secondary acoustic noise can be produced if the vibration of the

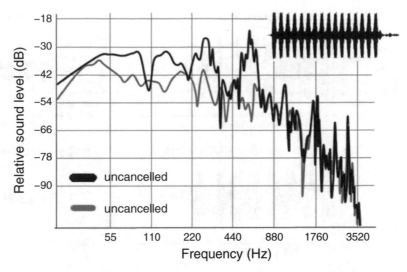

Fig. 2 Typical frequency spectrum of the scanner noise generated during blood oxygen level-dependent (BOLD) fMRI. This example was measured in the bore of a Philips Intera 3.0 Tesla scanner. The *black line* (uncancelled) indicates the acoustic energy of the noise recorded under normal scanning conditions. The *gray line* (canceled) indicates the residual acoustic energy at the ear when the active noise cancellation (ANC) system is operative. The *inset* (*upper right*) shows an example of the amplitude envelope of the scanner noise for a brain scan consisting of 16 slices corresponding to a sequence of 16 intense "pings"

coils and the core on which they are wound conducts through the core supports to the rest of the scanner structure. These secondary noise characteristics depend more on the mechanical resonances of the coil assemblies than on the type of imaging sequence and they tend to be the dominant contributor to the bandwidth and the spectral envelope of the noise. In this example of the frequency spectrum captured from a BOLD fMRI scanning sequence that was run on a Philips 3 Tesla Intera (Fig. 2), the spectrum has a peak component at 600 Hz with several other prominent pseudo-harmonics at 300, 1080, and 1720 Hz. The sound level measured in the bore of the scanner is typically 99-dB SPL [98 dB(A) using an A-weighting], measured using the maximum "fast" root-mean-square (RMS) time constant (125 ms). Clearly, exposure to such an intense sound levels without protection is likely to cause a temporary threshold shift in hearing and tinnitus, and it could be permanently damaging over a prolonged dosage [9].

1.3.2   Solutions

The simplest way to treat the intense noise is to use ear protection in the form of ear defenders and/or ear plugs (shown in Fig. 1b). Foam ear plugs can compromise the acoustic quality of the experimental sounds delivered to the subject and so ear defenders are preferable.

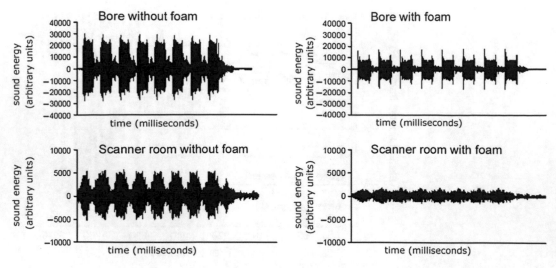

Fig. 3 Acoustic waveforms of the scanner noise measured with and without a lining of acoustic damping foam in the bore of the scanner. Our data demonstrate that the foam reduces the sound pressure level (SPL) at the position of the subject's head and in scanner room by a significant margin (about 8 dB). The segment of scanner noise that is illustrated here has a duration of approximately 1 s

Typically, transducers are fitted into sound attenuating earmuffs to reduce the ambient noise level at the subject's ears. Attenuation of the external sound by up to 40 dB can be achieved in this manner, although the level of reduction drops off at the high-frequency end of the spectrum. Commercial sound delivery systems all incorporate passive noise attenuation of this sort. An additional method of noise reduction is to line the bore of the scanner with a sound-energy absorbing material ([10]; see also www.ihr.mrc.ac.uk/research/technical/soundsystem/). The results of a set of measurements directly comparing the sound intensity of the scanner noise with and without the foam lining are shown in Fig. 3. However, this strategy does not provide a feasible solution because neither the design of the scanner bore nor the automated patient table are suited to the permanent installation of a foam lining and some types of acoustic foam can present risks of noxious fumes if they catch fire.

Some manufacturers have attempted to minimize scanner sound levels by modifying the design of the scanner hardware. For example, MR scanners manufactured by Toshiba (Toshiba America Medical systems, Inc., www.medical.toshiba.com/) incorporate Pianissimo technology—employing a solid foundation for gradient support, integrating sound dampening material in the gradient coils and enclosing them in a vacuum to reduce acoustic noise, even at full gradient power. This technology claims to reduce scanner noise by up to 90 % [11]. Subjects are reported to hear sounds at the volume of gentle drumming instead of the jackhammer noise level of other MR systems.

Another solution is to run modified pulse sequences that reduce acoustic noise by slowing down the gradient switching. This approach is based on the premise that the spectrum of the acoustic noise is determined by the product of the frequency spectrum of the gradient waveforms and the frequency response function of the gradient system [12]. The frequency response function is generally substantially reduced at low frequencies (i.e. below 200 Hz) and so the sound level can be reduced by using gradient pulse sequences whose spectra are band limited to this low-frequency range using pulse shapes with smooth onset and offset ramps [13]. A low-noise fast low-angle shot (FLASH) sequence can be modified to have a long gradient ramp time (6000 μs) and it generates a peak sound level of 48-dB SPL measured at the position of the ear. This type of sequence has been used for mapping central auditory function [14]. However, the low noise is achieved at the expense of slower gradient switching, extending the acquisition time. Low-noise sequences are not suitable for rapid BOLD imaging in which the fundamental frequency of the gradient waveform is greater than 200 Hz.

1.4 The Effect of Scanner Noise on Stimulus Audibility

1.4.1 Problems

Not only is the intense scanner noise a risk for hearing, but it also masks the perception of the acoustic stimuli presented to the subject. The exact specification of the acoustic signal-to-scanner-noise ratio (acoustic SNR) in fMRI studies using auditory stimuli is a potentially complicated matter. Nevertheless, we have sought to establish the relative difference between the stimulus level and the scanner noise level at the ear, by measuring these signals using a reference microphone placed inside the cup of the ear defender while participants perform a signal detection in noise task. Detection thresholds for a narrow band noise centered at the peak frequency of the scanner noise (600 Hz) are elevated when the target coincides with the scanner noise. We have demonstrated an average 11-dB shift in the 71 % detection threshold for the 600-Hz target when we modulate the perceived level of the scanner noise using active noise cancelation (ANC) methods (see later).

This evidence suggests that even with hearing protection, whenever the scanner noise coincides with the presented sound stimulus it produces changes in task performance and probably also increases the attentional demands of the listening task. The frequency range of the scanner acoustic noise is crucial for speech intelligibility, and speech experiments can be particularly compromised by a noisy environment ([15]; for review, see [16]). A recent study has quantified the effect of acoustic SNR using four listening tasks: pitch discrimination of complex tones, same/different judgments of minimal-pair nonsense syllables, lexical decision, and judgement of sentence plausibility [17]. Across these tasks, performance was assessed in silence (acoustic SNR=infinity) and in a background of MR scanner noise at the three acoustic SNR levels

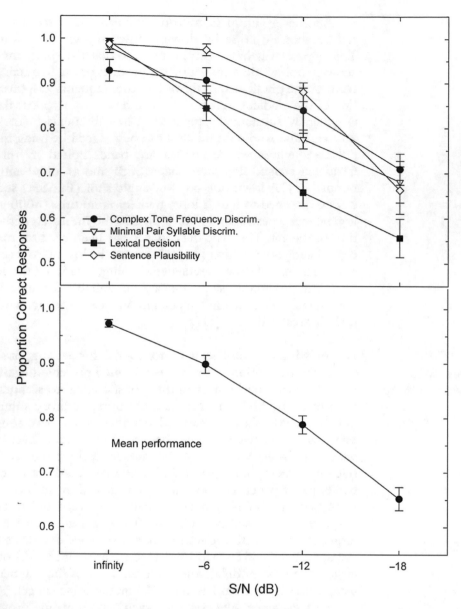

Fig. 4 Mean performance in a simulated scanning environment across four acoustic signal-to-noise ratios [17]. The *top panel* plots the proportion of correct responses on the individual tasks, while the *bottom panel* shows the overall mean performance (*SNR* signal-to-noise ratio, *dB* decibels)

(−6, −12, and −18 dB). Performance of normally hearing listeners significantly decreased as a function of the noise (Fig. 4). Even at −6 dB acoustic SNR, participants made many more errors than in quiet listening conditions ($p < 0.01$). Thus, across a range of auditory tasks that vary in linguistic complexity, listeners are highly susceptible to the disruptive impact of the intense noise associated with fMRI scanning.

1.4.2   Solutions

The aggregate noise dosage can be reduced by acquiring either a single or at least very few brain slices, but at the expense of only a partial view of brain activity [18]. For whole brain fMRI, other strategies are required.

One novel method that has been developed and evaluated at our Institute combines optical microphone technology with an active noise controller for significant attenuation of ambient noise received at the ears [19]. The canceller is based upon a variation of the single channel feed-forward filtered-x adaptive controller and uses a digital signal processor to achieve the noise reduction in real time. The canceler minimizes the noise pressure level at a specific control point in space that is defined by the position of the error microphone, positioned underneath the circum-aural ear defender of the headset (*see* Fig. 1d). In 2001, we published a psychophysical assessment of the system using a prototype system built in the laboratory that utilized a loudspeaker as the noise generator [19]. This system produced 10–20 dB of subjective noise reduction between 250 Hz and 1 kHz and smaller amounts at higher frequencies. More recently, we have obtained psychophysical threshold data in a Philips 3 Tesla scanner confirming that the same level of cancellation is achieved in the real scanner environment (Fig. 5; [20]). Again, the subjective impression of the scanner noise is the volume of gentle drumming when the sound system is operating in its canceled mode. Thus, it is possible to achieve a high level of noise attenuation by combining both passive and active methods.

A much more common strategy for reducing the masking influence of the concomitant scanner noise combines a passive method of ear protection with an experimental protocol that carefully controls

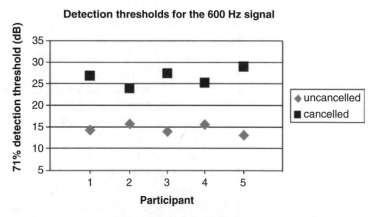

Fig. 5 Performance on a signal detection in noise task measured in a real scanning environment [Philips 3.0 Tesla MR scanner during blood oxygen level-dependent (BOLD) fMRI]. The data show that when the noise canceller was operative, the sound level of the signal could be 8–16 dB softer (depending upon the listener) in order to achieve the same detection performance

the timing between stimulus presentation and image acquisition so that sound stimuli can be delivered during brief periods of quiet in between successive brain scans [21]. Specific details of several pulse sequence protocols that reduce the masking effects of scanner noise are discussed in more detail in the next subsection.

1.5   The Effect of Scanner Noise on Sound-Related Activation in the Brain

1.5.1   Problems

To increase the BOLD SNR, it is necessary to acquire a large number of scans in each condition in an fMRI experiment. Typically, an experimenter would collect many hundreds of brain scans in a single session, with the time in between each scan chosen to be as short as the scanner hardware and software will permit. Remember that, for fMRI, an intense "ping" is generated for each slice of the scan and so of course this means that the participant can easily be subjected to several thousand repeated "pings" of noise during the experiment. Not only does this scanner noise acoustically mask the presented sound stimuli, but the elevated baseline of sound-evoked activation due to the ambient scanner noise also makes the experimentally induced auditory activation more difficult to detect statistically. Much of the work examining the influence of acoustic scanner noise has been directed toward its capacity to interfere with the study of audition or speech perception by producing activation of various brain regions, especially the auditory cortex [22–25]. Several studies highlight the reduced activation signal (i.e. the difference between stimulation and baseline conditions) in the auditory cortex when the amount of prior scanner noise is increased, demonstrating that the scanner noise effectively masks the detection of auditory activation [22, 26, 27]. In another example, taken from one of the early fMRI experiments conducted at the MRC Institute of Hearing Research, we used a specially tailored scanning protocol to measure the amplitude and the time course of the BOLD response to a high-quality recording of a single burst of scanner noise presented to participants over headphones [24]. Our results revealed a reliable transient increase in the BOLD signal across a large part of the auditory cortex. As in many other brain regions, the evoked response to this single brief stimulus event was smoothed and delayed in time. It rose to a peak by 4–5 s after stimulus onset and decayed by 5–8 s after stimulus offset [24]. Its amplitude reached about 1.5 % of the overall signal change, which is considerable considering that stimulus-related activation usually accounts for a BOLD signal change of approximately 2–5 %. Figure 6 illustrates the canonical BOLD response to a noise onset.

In many fMRI experimental paradigms, regions of stimulus-evoked activation are detected by comparing the BOLD scans acquired during one sound condition with the BOLD scans acquired during another condition, which could be either a condition in which a different type of sound was presented or no sound (known as a baseline "silent" condition) was presented. Activation is defined as those parts of the brain that demonstrate a statistically significant

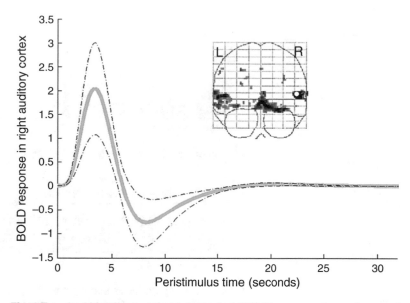

Fig. 6 Transient blood oxygen level-dependent (BOLD) response to a noise onset. The graph shows the fitted response and the 90% confidence interval. This example illustrates all the characteristic features of the transient BOLD response—a peak at 4-s post-stimulus onset followed by an undershoot and then return to baseline at 16 s

difference between the two conditions. For example, let us consider the simplest case in which one condition contains a sound and the other does not. Since the scanner noise is present throughout, the sound condition effectively contains both stimulus and scanner noise, while the baseline condition also contains the scanner noise (i.e. it is not silent). Given the spectrotemporal characteristics of the scanner noise, it generates widespread sound-related activity across the auditory cortex. Thus, the subtraction analysis for detecting activation is sensitive only to whatever is the small additional contribution of the sound stimulus to auditory neural activity.

1.5.2 Solutions

A number of different scanning protocols have been used to minimize the effect of the scanner acoustic noise on the measured patterns of auditory cortical activation. In this section, we will describe two of these, but before we do, we need to consider some important details about the time course of the BOLD response to the scanner noise and introduce some new terms.

During an fMRI experiment, the BOLD response to the scanner noise spans two different temporal scales. First, the "ping" generated by the acquisition of one slice early in the scan may induce a BOLD response in a slice, which is acquired later in the same scan if that later scan is positioned over the auditory cortex. We shall call this inter-slice interference. Inter-slice interference is maximally reduced when all slices in the scan are acquired in rapid

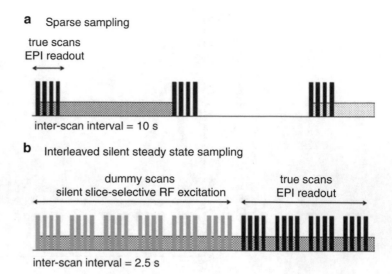

Fig. 7 Two scanning protocols that have been used to minimize the effect of the scanner acoustic noise on the measured patterns of auditory cortical activation. See text for further explanation (*s* seconds, *EPI* echo-planar imaging, *RF* radiofrequency)

succession and the total duration of the scan is not more than 2 s [26]. A common term for the scanning protocol that uses a minimum inter-slice interval is a clustered-acquisition sequence. Edmister et al. [28] found that the clustered-acquisition sequence provides an advantageous auditory BOLD SNR compared with a conventional scanning protocol. The second form of interference is called inter-scan interference. This occurs when the scanner noise evokes an auditory BOLD response that extends across time to subsequent scans, predominantly when the interval between scans is as short as the MR system will permit. Reducing the inter-scan interference can easily be achieved by extending the period between scans (the inter-scan interval). By separately manipulating the timing between slices and between scans, we can reduce the inter-slice and inter-scan interference independently of one another. When the clustered-acquisition sequence is combined with a long (e.g. 10 s) inter-scan interval, the activation associated with the experimental sound can be separated from the activation associated with the scanner sound (Fig. 7a). Furthermore, because the scanner sound is temporally offset, it does not produce acoustical masking and does not distract the listener. This scanning protocol is commonly known as sparse sampling [21]. Sparse sampling is often the scanning protocol of choice for identifying auditory cortical evoked responses in the absence of scanner noise (*see* e.g. [29–33]). However, it requires a scanning session that is longer than that of conventional "continuous" protocols in order to acquire the same amount of imaging data, and participants can be intolerant of long sessions. It also relies upon certain assumptions about the time to peak of the BOLD response after stimulus onset and a sustained plateau of evoked activity for the duration of the stimulus.

A second type of scanning protocol acquires a rapid set of scans following each silent period in order to avoid some of the afore-mentioned difficulties—"interleaved silent steady state" sampling [34]. The increased number of scans permits a greater proportion of scanning time to be used for data acquisition and at least partial mapping of the time course of the BOLD response (Fig. 7b). However, some pulse programming is required to avoid T1-related signal decay during the data acquisition, hence ensuring that signal contrast is constant across successive scans. The software modification maintains the longitudinal magnetization in a steady state throughout the scanning session by applying a train of slice-selective excitation pulses (quiet dummy scans) during each silent period.

1.6 The Effect of Stimulus Context: Neural Adaptation to Sounds

1.6.1 Problems

The acoustic environment is typically composed of one or more sound sources that change over time. Over the years, both psycho-physical and electrophysiological studies have amply demonstrated that stimulus context strongly influences the perception and neural coding of individual sounds, especially in the context of stream seg-regation and grouping [35–37]. A simple example of the influence of stimulus context is forward masking, which occurs when the presence of one sound increases the detection threshold for the subsequent sound. The perceptual effects of forward masking are strongest when the spectral content of the first sound is similar to the second sound, when there is no delay between the two sounds, and when the masker duration is long [38]. Forward inhibition typically lasts from 70 to 200 ms. This type of suppression has not only been demonstrated in anesthetized preparations, but also in awake primates. In the latter case, suppression was seen to extend up to 1 s in time [39]. As well as tone–tone interactions, neural fir-ing rate is sensitive to stimulus duration. Neurons respond strongly to the onset of a sound and their response decays thereafter. Many illustrative examples can be found in the literature, especially in cases where longer duration sounds are presented (e.g. 750–1500 ms in the case of Bartlett and Wang [38], see their Fig. 4).

By transporting these well-established paradigms into a neuro-imaging experiment, researchers are beginning to address the con-text dependency of neural coding in humans. One way in which the effect of sound context on the auditory BOLD fMRI signal has been examined is in terms of different repetition rates [19, 40]. This is conceptually analogous to the presentation rate manipula-tions of the forward masking studies described earlier, but goes beyond the simple case of two-tone interactions. In the fMRI stud-ies, stimuli were long trains of noise bursts presented at different rates. The slowest rate was 2 Hz and the fastest rate was 35 Hz, with intermediate rates being 10 and 20 Hz. Noise bursts at each repetition rate were presented in prolonged blocks of 30 s, each followed by a 30-s "silent" period. During sound presentation, scans were acquired at a short inter-scan interval (approximately 2 s) so that the experimenters could reconstruct the 30-s time course of the BOLD response to each of the different repetition rates, hence determining the multi-second time pattern of neural activity. The scans were positioned so that a number of different auditory sites in the ascending auditory system could be measured: (1) the inferior colliculus in the midbrain, (2) the medial geniculate nucleus in the thalamus, and (3) Heschl's gyrus and the superior temporal gyrus in the cortex. The plots of the BOLD time course demonstrated a systematic change in its shape from midbrain up to cortex. In the inferior colliculus, the amplitude of the BOLD response increased as a function of repetition rate while its shape was sustained throughout the 30-s stimulus period. In the medial geniculate body, increasing rate also produced an increase in BOLD amplitude with a moderate peak in the BOLD shape just after stimulus onset. Repetition rate exerted its largest effect in the auditory cortex. The most striking change was in the shape of the BOLD response. The low repetition rate (2 Hz) elicited a sustained response, whereas the high rate (35 Hz) elicited a phasic response with prominent peaks just after stimulus onset and offset. The follow-up study [40] confirmed that it was the temporal envelope characteristics of the acoustic stimulus, not its sound level or bandwidth, that strongly influenced the shape of the BOLD response. The authors offer a perceptual interpretation of the neural response to different repetition rates. The shift in the shape of the cortical BOLD response from sustained to phasic corresponds to a shift from a stimulus in which component noise bursts are perceptually distinct to one in which successive noise bursts fuse to become individually indistinguishable. The onset and offset responses of the phasic response coincide with the onset and offset of a distinct, meaningful event. The logical conclusion to this argument is that the succession of individual perceptual events in the low repetition rate conditions defines the sustained BOLD response observed at the 2-Hz rate. It is clear from these results that while the amplitude of the BOLD response to sound can inform us about the tuning properties of the underlying neural population (e.g. sensitivity to repetition rate), other properties of the BOLD response, such as its shape, provide different information about neural coding (e.g. segmentation of the auditory environment into perceptual events).

It is crucial that these contextual influences on the BOLD signal are accounted for in the design and/or interpretation of auditory fMRI experiments. To illustrate this case in point, I use a set of our own experimental data [41]. In this experiment, one of the sound conditions was a diotic noise (identical signal at the two ears) presented continuously for 32 s at a constant sound level

(~86-dB SPL) and at a fixed location in the azimuthal plane. Scans were acquired every 4 s throughout the stimulus period. When the scans acquired during this sound condition were combined together and contrasted against the scans acquired during the "silent" baseline condition, no overall significant activation was obtained ($p > 0.001$). We interpret this lack of activation as evidence that the auditory response had rapidly habituated to a static signal. This conclusion is confirmed by plotting out the time course of the response at one location within the auditory cortex. The initial transient rise in the BOLD response at the onset of the sound begins to decay at about 4 s and this reduction continues across the stimulus epoch. The end of the epoch is characterized by a further rise in the BOLD response, elicited by the other types of sound stimuli that were presented in the experiment (Fig. 8a).

1.6.2   Solutions

It is common for auditory fMRI experiments to use a blocked design in which a sound condition is presented over a prolonged time period that extends over many seconds, even tens of seconds. Indeed as we described in Sect. 1.5, the blocked design is at the core of the sparse sampling protocol, and so the risk of neural adaptation is a legitimate one. The BOLD signal detection problem caused by neural adaptation is often circumvented by presenting the stimulus of interest as a train of stimulus bursts at a repetition rate that elicits the sustained cortical response (e.g. 2 Hz). Many of the auditory fMRI experiments that have been conducted over the years in our research group have taken this form [30, 31, 42–44]. Alternatively, if the stimulus contains dynamic spectrotemporal changes, then it is not always necessary to pulse the stimulus on and off. To illustrate this case in point, I return to a set of our own experimental data [41]. In this experiment, one of the sound conditions was a broadband noise convolved with a generic head-related transfer function to give the perceptual impression of a sound source that was continuously rotating around the azimuthal plane of the listener. Although the sound was presented continuously for 32 s, the filter functions of the pinnae imposed a changing frequency spectrum and the head shadow effect imposed low-rate amplitude modulations in the sound envelope presented to each ear. When the scans acquired during this sound condition were combined together and contrasted against the scans acquired during the "fixed sound source" condition, widespread activation was obtained ($p < 0.001$) across the posterior auditory cortex (planum temporale): an area traditionally linked with spatial acoustic analysis. The time course of activation demonstrates a sustained BOLD response across the entire duration of the epoch (Fig. 8b). The sustained response contrasts with the transient response observed for the fixed sound source condition (Fig. 8a).

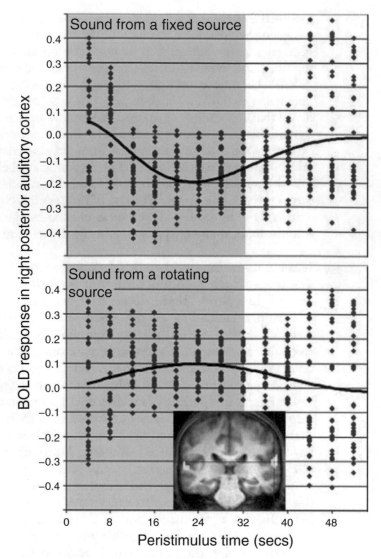

Fig. 8 Adjusted blood oxygen level-dependent (BOLD) response (measured in arbitrary units) across the 32-s stimulus epoch shaded in *gray* (**a**) for a sound from a fixed source and (**b**) for a sound from a rotating source. Adjusted values are combined for all six participants and the trend line is indicated using a polynomial sixth order function. The response for both stimulus types is plotted using the same voxel location in the planum temporale region of the right auditory cortex (coordinates x 63, y −30, z 15 mm). The position of this voxel is shown in the *inserted panel*. The activation illustrated in this *insert* represents the subtraction of the fixed sound location from the rotating sound conditions ($p < 0.001$)

2 Examples of Auditory Feature Processing

2.1 The Representation of Frequency in the Auditory Cortex

Within the inner ear, an incoming sound is separated into its individual frequency components by the way in which the energy at different frequencies travels along the cochlear partition [45]. High-frequency tones maximally stimulate those nerve fibers near the base of the cochlea while low-frequency tones are best coded towards the apex. This cochleotopic representation persists throughout the auditory pathway where it is referred to as a tonotopic map. Within the mammalian auditory cortex, electrophysiological recordings have revealed many tonotopic maps [46, 47]. Within each map, neurons tuned to the same sound frequency are colocalized in a strip across the cortical surface, with an orderly progression of frequency tuning across adjacent strips. Frequency tuning is sharper in the primary auditory fields than it is in the surrounding nonprimary fields, and so the most complete representations of the audible frequency range are found in the primary fields. Primates have at least two tonotopic maps in primary auditory cortex, adjacent to one another and with mirror-reversed frequency axes. It is possible to demonstrate tonotopy by fMRI as well as by electrophysiology, even though frequency selectivity deteriorates at the moderate to high sound intensities required for fMRI sound presentation. As a recent example, mirror-symmetric frequency gradients have been confirmed across primary auditory fields using high-resolution fMRI at 4.7 T in anesthetized macaques and at 7.0 T in awake behaving macaques [48]. This section describes results from several fMRI experiments that have sought to demonstrate tonotopy in the human auditory cortex.

fMRI is an ideal tool for exploring the spatial distribution of the frequency-dependent responses across the human auditory cortex because it provides good spatial resolution and the analysis requires few a priori modeling assumptions (*see* [49] for a review). In addition, it is possible to detect statistically significant activation using individual fMRI analysis. This is important when determining fine-grained spatial organization because averaging data across different listeners would inevitably blur the subtle distinctions. A number of recent studies have sought to determine the organization of human tonotopy [29, 33, 50–52]. To avoid the problem of neural adaptation discussed in Sect. 1.6, experimenters chose stimuli that would elicit robust auditory cortical activation. For example, Talavage et al. [51, 52] presented amplitude-modulated signals, while Schönwiesner et al. [50] presented sine tones that were frequency modulated across a narrow bandwidth. Langers et al. [33] used a signal detection task in which the tone targets at each frequency were briefly presented (0.5 s). In agreement with the primate literature, evidence for the presence of tonotopic organization is at its most apparent within the primary auditory cortex while frequency preferences in the surrounding nonprimary areas are more erratic [33]. Thus, we shall consider in more detail the precise arrangement of tonotopy in the primary region.

In their first study, Talavage et al. [51] contrasted pairs of low (<66 Hz) and high (>2490 Hz) frequency stimuli of moderate intensity and sufficient spectral separation to produce spatially resolvable differences in activation (low > high and high > low) across the auditory cortical surface. These activation foci were considered to define the endpoints of a frequency gradient. In total,

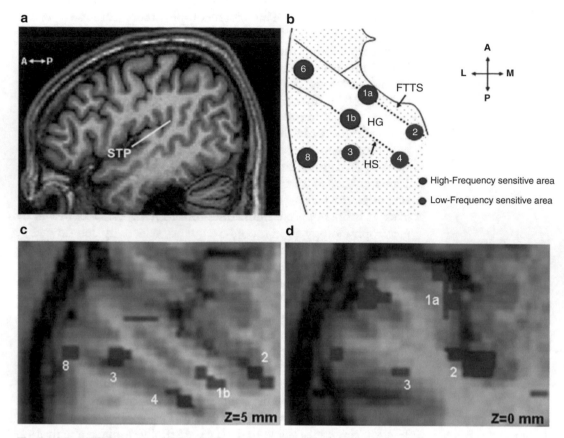

Fig. 9 (a) Sagittal view of the brain with the *oblique white line* denoting the approximate location and orientation of the schematic view shown in panel (**b**) along the supratemporal plane. (**b**) Schematic representation of the most consistently found high (*red*) and low (*blue*) frequency-sensitive areas across the human auditory cortex reported by Talavage et al. [50, 51]. The primary area is shown in *white* and the nonprimary areas are shown by *dotted shading*. Panels (**c**) and (**d**) illustrate the high- (*red*) and low- (*blue*) frequency sensitive areas across the left auditory cortex of one participant (unpublished data). Two planes in the superior-inferior dimension are shown ($z = 5$ mm and $z = 0$ mm above the CA-CP line). *A* anterior, *P* posterior, *M* medial, *L* lateral, *HG* Heschl's gyrus, *HS* Heschl's sulcus, *FTTS* first transverse temporal sulcus, *STP* supratemporal plane

Talavage et al. identified eight frequency-sensitive sites across Heschl's gyrus (HG, the primary auditory cortex) and the surrounding superior temporal plane (STP, the nonprimary auditory cortex). Each site was reliably identified across listeners and the sites were defined by a numerical label [1–8].

Foci 1–4 occurred around the medial two-thirds of HG and are good candidates for representing frequency coding within the primary auditory cortex (Fig. 9). Finding several endpoints does not provide direct confirmation of tonotopy because tonotopy necessitates a linear gradient of frequency sensitivity. Nevertheless, Talavage et al. argued that the foci 1–3 were at least consistent with predictions from primate electrophysiology.

The arrangement of the three foci encompassed the primary auditory cortex, suggested a common low-frequency border, and had a mirror-image reversed pattern. This interpretation was criticized by Schönwiesner et al. [50] who stated that it was wrong to associate these foci with specific tonotopic fields because pairs of low- and high-frequency foci could not clearly be attributed to specific frequency axes nor to anatomically defined fields. Indeed, in their own study, Schönwiesner et al. [50] did not observe the predicted gradual decrease in frequency-response amplitude at locations away from the best-frequency focus, but instead found a rather complex distribution of response profiles. Their explanation for this finding was that the regions of frequency sensitivity reflected not tonotopy, but distinct cortical areas that each preferred different acoustic features associated with a limited bandwidth signal.

Increasing the BOLD SNR might be necessary for characterizing some of the more subtle changes in the response away from best frequency and more recent evidence using more sophisticated scanning techniques does support the tonotopy viewpoint. Frequency sensitivity in the primary auditory cortex was studied using a 7-T ultra-high field MR scanner to improve the BOLD SNR and to provide reasonably fine-grained (1 mm^3) spatial resolution [29]. Formisano et al. [29] sought to map the progression of activation as a smooth function of tone frequency across HG. Frequency sensitivity was mapped by computing the locations of the best response to single frequency tones presented at a range of frequencies ($0.3, 0.5, 0.8, 1, 2$, and 3 kHz). Flattened cortical maps of best frequency revealed two mirror-symmetric gradients (high-to-low and low-to-high) traveling along HG from an anterolateral point to the posteromedial extremity. In general, the amplitude of the BOLD response decreased as the stimulating tone frequency moved away from the best frequency tuning characteristics of the voxel. A receiver coil placed close to the scalp over the position of the auditory cortex is another way to achieve a good BOLD SNR and this was the method used by Talavage et al. [52]. Talavage et al. measured best-frequency responses to an acoustic signal that was slowly modulated in frequency across the range 0.1–8 kHz. Again, the results confirmed the presence of two mirror-symmetric maps that crossed HG (extending from the anterior first transverse temporal sulcus to the posterior Heschl's sulcus) and shared a low-frequency border.

Although more evidence will be required before a clear consensus is established, the studies presented in this section have made influential contributions to the understanding of frequency representation in the human auditory cortex and its correspondence to primate models of auditory coding.

2.2 The Influence of Selective Attention on Frequency Representations in Human Auditory Cortex

We live in a complex sound environment in which many different overlapping auditory sources contribute to the incoming acoustical signal. Our brains have a limited processing capacity and so one of the most important functions of neural coding is to separate out these competing sources of information. One way to achieve this is by filtering out the uninformative signals (the "ground") and attending to the signal of interest (the "figure"). Competition between incoming signals can be resolved by a bottom-up, stimulus-driven process (such as a highly salient stimulus that evokes an involuntary orienting response), or it can be resolved by a top-down, goal-directed process (such as selective attention). Selective attention provides a modulatory influence that enables a listener to focus on the figure and to filter out or attenuate the ground [53].

Visual scientists have shown that attention can be directed to the features of the figure (feature-based attention, for a review *see* [54]) or to the entire figure (object-based attention, for a review *see* [55]). Given that so little is known about the mechanisms by which auditory objects are coded [56], we shall focus on those studies of auditory feature-based attention. A sound can be defined according to many different feature dimensions including frequency spectrum, temporal envelope, periodicity, spatial location, sound level, and duration. The experimenter can instruct listeners to attend to any feature dimension in order to investigate the effect of selective attention on the neural coding of that feature. Different listening conditions have been used for comparison with the "attend" condition. The least controlled of these is a passive listening condition in which participants are not given any explicit task instructions [30, 57, 58]. Even if there are cases where a task is required, but the cognitive demand of that task is low, participants are able to divide their attention across both relevant and irrelevant stimulus dimensions (*see* [59] for a review on attentional load). Again, this leads to an uncontrolled experimental situation. For greater control, some studies have employed a visual distractor task to compete for attentional resources and pull selective attention away from the auditory modality [60, 61]. However, there is some evidence that the mere presence of a visual stimulus exerts a significant influence on auditory cortical responses [62, 63] and hence modulation related to selective attention might interact with that related to the presence of visual stimuli in a rather complex manner. This can make comparison between the results from bimodal studies [60, 61] and unimodal auditory studies [32, 64] somewhat problematic.

One paradigm that has been commonly used to examine feature-based attention manipulates two different feature dimensions independently within the same experimental session and listeners are required to make a discrimination judgement to one feature or the other. Studies have compared attention to spatial features such as location, motion, and ear of presentation with attention to nonspatial features such as pitch and phonemes [60,

64]. Results typically demonstrate a response enhancement in nonprimary auditory regions. For example, Degerman et al. [60] found auditory enhancement in left posterior nonprimary regions, but only for attending to location relative to pitch and not the other way round. Ahveninen et al. [64] used a novel paradigm in which they measured the effect of attention on neural adaptation. Their fMRI results showed smaller adaptation effects in the right posterior nonprimary auditory cortex when attending to location (relative to phonemes), but again not the other way round. Both studies reported enhancement for attending to location in additional nonauditory regions, notably the prefrontal and right parietal areas. This asymmetry in the effects observed across spatial and nonspatial attended domains is worthy of further exploration since spatial analysis is well known to engage the right posterior auditory and right parietal cortex [65].

Another experimental design that has been used to examine feature-based attention presents concurrent visual and auditory stimuli and participants are required to make a discrimination judgement to stimuli in one modality or the other. One example of this design used novel melodies and geometric shapes, and participants were required to respond to either long note targets or vertical line targets [57, 58]. When "attending to the shapes" was subtracted from "attending to the melodies" the results revealed relative enhancement bilaterally in the lower boundary of the superior temporal gyrus. This finding supports the view that there is sensory enhancement when attending to the auditory modality. In addition, it was shown that when "attending to the shapes," the auditory response was suppressed relative to a bimodal passive condition. This is tentative evidence for neural suppression when ignoring the auditory modality. A novel feature of the experiment by Degerman et al. [66] was that in one selective attention condition, participants had to respond to a target defined by a particular combination of cross-modal features (e.g. high pitch and red circle). The conventional general linear analysis did not show any significant difference in the magnitude of the auditory response in the cross-modal condition compared with a condition in which participants simply attended to the high- and low-pitch targets in the audiovisual stimulus. However, a region of interest analysis (defining a region in the posterolateral superior temporal gyrus) did suggest some enhancement for the audiovisual attention condition compared with the auditory attention condition. Thus, it is possible that nonprimary auditory regions are involved in attention-dependent binding of synchronous auditory and visual events into coherent audio–visual objects.

In audition, it has long been established behaviourally that when participants expect a tone at a specific frequency, their ability to detect a tone in a noise masker is significantly better when the tone is at the expected frequency than when it is at an unexpected frequency (the probe-signal paradigm [67]). The benefit of selective attention for signal detection thresholds can be plotted as a function of frequency. The ability to detect tones at frequencies close to the expected frequency is also enhanced, and this benefit drops off smoothly with the distance away from the expected frequency [67, 68]. The width of this attention-based listening band is comparable to the width of the critical band related to the frequency-tuning curve, which can be measured psychophysically using notched noise maskers [68]. This equivalence suggests that selective attention might be operating at the level of the sensory representation of tone frequency.

Evidence from electrophysiological recordings demonstrates frequency-specific attentional modulation at the level of the primary auditory cortex, consistent with a neural correlate of the psychophysical phenomena found in the probe-signal paradigm. In a series of experiments, awake behaving ferrets were trained to perform a number of spectral tasks including tone detection and frequency discrimination [69]. In the tone detection task, ferrets were trained to identify the presence of a tone against a background of broadband rippled noise. The spectro-temporal receptive fields measured during the noise for frequency-tuned neurons showed strong facilitation around the target frequency that persisted for 30–40 ms. In the two-tone discrimination task, ferrets performed an oddball task in which they responded to an infrequent target frequency. Again, the spectro-temporal receptive fields showed an enhanced and persistent response for the target frequency, plus a *decreased* response for the reference frequency. These opposite effects serve to magnify the contrast between the two center frequencies, and thus facilitate the selection of the target. The results of these two tasks confirm that the acoustic filter properties of auditory cortical neurons can dynamically adapt to the attentional focus of the task.

Recently, we have addressed the question of attentional enhancement for selective attention to frequency using a high-resolution scanning protocol ($1.5 \text{ mm}^2 \times 2.5 \text{ mm}$) (unpublished data). To control for the demands on selective attention, we presented two concurrent streams (low- and high-frequency tones). Participants were requested to attend to one frequency stream or the other and these attend conditions were presented in an interleaved manner throughout the experiment. Behavioral testing confirmed that performance significantly deteriorated when these sounds were presented in a divided attention task. To be able to identify high- and low-frequency sensitive areas around the primary auditory cortex we designed two types of stimuli using different rhythms for each of the two streams. For example, one stimulus contained a "fast" high-frequency rhythm and a "slow" low-frequency rhythm so that the stimulus contained a majority of high-frequency tones. The other stimulus was the converse. Areas of high-frequency sensitivity were identified by subtracting the low-frequency majority stimulus from the high-frequency majority stimulus, and vice-versa (Fig. 9c, d). For each of the three

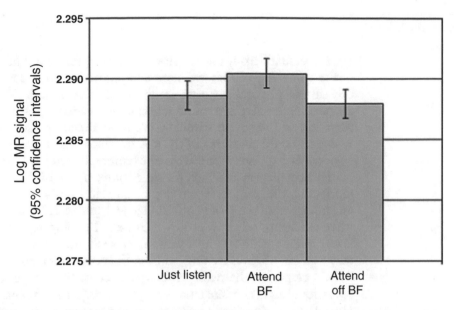

Fig. 10 Response to the three listening conditions: just listen, attend to the best frequency (BF) tones, and attend OFF BF. The data shown are for those stimuli in which BF tones formed the majority (80 %) of the tones in the sound sequence, combining responses across areas 1–3. The error bars denote the 95 % confidence intervals

participants, we selected those frequency-specific areas that best corresponded to areas 1–4 (defined by Talavage et al. [51, 52]; *see* Sect. 2.1). Within these areas, we extracted the BOLD signal time course for every voxel and performed a log transform to standardize the data. We collapsed the data across low- and high-frequency sensitive areas [1–4] according to their "best frequency" (BF). The best frequency of an area corresponds to the frequency that evokes the largest BOLD response. A univariate ANOVA showed response enhancement when participants were attending to the BF of that area, compared with attending to the other frequency ($p < 0.01$). In addition, response enhancement was also found when attending to the BF of that area, compared with passive listening ($p < 0.05$) (Fig. 10). Note that for these results area 4 was excluded from the analysis, because it showed different pattern of attentional modulation. The response profile of area 4 might differ from that of areas 1–3 in other ways because it is not consistently present in all listeners [51]. Our finding of frequency-specific attentional enhancement in primary auditory regions contrasts with that of Petkov et al. [61], who reported attention-related modulation to be independent of stimulus frequency and to engage mainly the nonprimary auditory areas. However, our result is more in keeping with the predictions made by the neurophysiological data reported by Fritz et al. [69].

3   Future Directions

It is increasingly likely that auditory cortical regions compute aspects of the sound signal that are more complicated in their nature than the simple physical acoustic attributes of the sound. Thus, the encoded features of the sound reflect an increasingly abstract representation of the sound stimulus. We have already presented some evidence for this in terms of the way in which the auditory cortical response is exquisitely sensitive to the temporal context of the sound, particularly the way in which the time course of the BOLD response represents the temporal envelope characteristics of the sound, including sound onsets and offsets [18, 40], (*see* Sect. 1.4). However, there are many other ways in which neural coding reflects higher level processing. In this final section, we shall introduce two important aspects of the listening context that determine the auditory BOLD response: the perceptual experience of the listener and the operational aspects of the task. A number of fMRI studies have demonstrated ways in which activity within the human auditory cortex is modulated by auditory sensations, including loudness, pitch, and spatial width. Other studies have revealed that task relatedness is also a significant determining factor for the pattern of activation. These findings highlight how future auditory fMRI studies could usefully investigate these contributory factors in order to provide a more complete picture of the neural basis of the listening process.

3.1   Cortical Activation Reflects Perceptually Relevant Coding

One approach used in auditory fMRI to investigate perceptually relevant coding imposes systematic changes to the listener's perception of a sound signal by parametrically manipulating certain acoustic parameters and subsequently correlating the perceptual change with the variation in the pattern of activation. For example, by increasing sound intensity (measured in SPL), one also increases its perceived loudness (measured in phons). Loudness is a perceptual phenomenon that is a function of the auditory excitation pattern induced by the sound, integrated across frequency. Sound intensity and loudness are measures of different phenomena. For example, if the bandwidth of a broadband signal is increased while its intensity is held constant, then loudness nevertheless increases because the signal spans a greater number of frequency channels. In an early fMRI study, Hall et al. [31] presented single-frequency tones and harmonic-complex tones that were matched either in intensity or in loudness. The results showed that the complex tones produced greater activation than did the single-frequency tones, irrespective of the matching scheme. This result indicates that bandwidth had a greater effect on the pattern of auditory activation than sound level. Nevertheless, when the data were collapsed across stimulus class, the amount of activation was significantly correlated with the loudness scale, not with the intensity scale.

In people with elevated hearing thresholds, the perception of sound level is distorted. They typically experience the same dynamic range of loudness as normally hearing listeners despite having a compressed range of sensitivity to sound level. The BOLD response to sound level is reflected in a disproportionate increase in loudness with intensity. A recent study has characterized the BOLD response to frequency-modulated tones presented at a broad range of intensities (0–70 dB above the normal hearing threshold) [33]. Both normally hearing and hearing impaired groups showed the same steepness in the linear increase in auditory activation as a function of loudness, but not of intensity (Fig. 11). The results from this study clearly demonstrate that the BOLD response can be interpreted as a correlate of the subjective strength of the stimulus percept.

Pitch can be defined as the sensation whose variation is associated with musical melodies. Together with loudness, timbre, and spatial location, pitch is one of the primary auditory sensations. The salience of a pitch is determined by several physical properties of the pitch signal, one being the numbered harmonic components comprising a harmonic-complex tone. The cochlea separates out the frequency components of sounds to a limited extent, so that the first eight harmonics of a harmonic-complex tone excite distinct places in the cochlea and are said to be "resolved," whereas the higher harmonics are not separated and are said to be "unresolved." Pitch discrimination thresholds for unresolved harmonics are substantially higher than those for resolved harmonics, consistent with the former type of stimulus evoking a less salient pitch

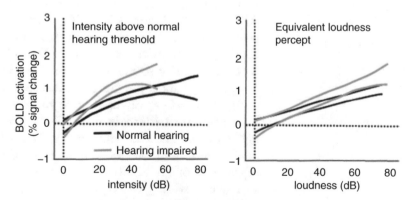

Fig. 11 Growth in the blood oxygen level-dependent (BOLD) response (measured in % signal change) for high (4–8 kHz) frequency-modulated tone presented across a range of sound levels for ten normally hearing participants (*black lines*) and ten participants with a high-frequency hearing loss (*gray lines*). The *left hand panel* shows the growth as a function of sound intensity, while the *left hand panel* plots the same data as a function of the equivalent loudness percept. The *upper and lower lines* denote the 95 % confidence interval of the quadratic polynomial fit to the data. This graph summarizes data presented by Langers et al [33]

[70]. A pairwise comparison between the activation patterns for resolved (strong pitch) and unresolved (weak pitch) harmonic-complex tones has identified differential activation in a small, spatially localized region of nonprimary auditory cortex, overlapping the anterolateral end of Heschl's gyrus [71]. The authors claim that this finding reflects the cortical representation for pitch salience. Another way to determine the salience of a pitch is by the degree of fine temporal regularity in the stimulus (i.e. the monaural repeating pattern within frequency channels). This is true even for signals in which there are no distinct frequency peaks in the cochlear excitation pattern from which to calculate the pitch. A range of pitch saliencies can be created by parametrically varying the degree of temporal regularity in an iterated-ripple noise stimulus (using 0, 1, 2, 4, 8, and 16 add-and-delay iterations during stimulus generation) [72]. Again the anterolateral end of Heschl's gyrus appeared highly responsive to the change in pitch salience, in a linear manner.

Spatial location is another important auditory sensation that is determined by the fine temporal structure in the signal, this time it being the binaural temporal characteristics across the two ears. The interaural correlation (IAC) of a sound represents the similarity between the signals at the left and right ears. Changes in the IAC of a wideband signal result in changes in sound's perceived "width" when presented through headphones. A noise with an IAC of 1.0 is typically perceived as sound with a compact source located at the center of the head. As the IAC is reduced the source broadens. For an IAC of 0.0, it eventually splits into two separate sources, one at each ear [73]. Again the parametric approach has been employed to measure activation across a range of IAC values (1.00, 0.93, 0.80, 0.60, 0.33, and 0.00) [74]. The authors found a significant positive relationship between BOLD activity and IAC, which was confined to the anterolateral end of Heschl's gyrus, the region that is also responsive to pitch salience. The slope of the function was not precisely linear but the BOLD response was more sensitive to changes in IAC at values near to unity than at values near zero. This response pattern is qualitatively compatible with previous behavioral measures of sensitivity to IAC [75].

There is some evidence to support the claim that the neural representations of auditory sensations (including loudness, pitch, and spatial width) evolve as one ascends the auditory pathway. Budd et al. [74] examined sensitivity to values of IAC associated with spatial width within the inferior colliculus, the medial geniculate nucleus, as well as across different auditory cortical regions, but the effects were significant only within the nonprimary auditory cortex. Griffiths et al. [72, 76] also examined sensitivity to the increases in temporal regularity associated with pitch salience within the cochlear nucleus, inferior colliculus, medial geniculate nucleus, as well as across different auditory cortical regions. Some degree of sensitivity to pitch salience was found at all sites, but the preference appeared greater in the higher centers than in the cochlear nucleus [76]. Thus, the evidence supports the notion of an increasing responsiveness to percept attributes of sound throughout the ascending auditory system, culminating in the nonprimary auditory cortex. These findings are consistent with the hierarchical processing of sound attributes.

Encoding the perceptual properties of a sound is integral to identifying the object properties of that sound source. The nonprimary auditory cortex probably plays a key role in this process because it has widespread cortical projections to frontal and parietal brain regions and is therefore ideally suited to access distinct higher level cortical mechanisms for sound identification and localization. Recent trends in auditory neuroscience are increasingly concerned with auditory coding beyond the conventional limits of the auditory cortex (the superior temporal gyrus in humans), particularly with respect to the hierarchical organization of sensory coding via dorsal and ventral auditory processing routes. At the top of this hierarchy stands the brain's representation of an auditory "object." The concept of an auditory object still remains controversial [56]. Although it is clear that the brain needs to code information about the invariant properties of a sound source, research in this field is considerably underdeveloped. Future directions are likely to begin to address critical issues such as the definition of an auditory object, whether the concept is informative for auditory perception, and optimal paradigms for studying object coding.

3.2 Cortical Activation Also Reflects Behaviourally Relevant Coding

Listeners interact with complex auditory environments that, at any one time point, contain multiple auditory objects located at dynamically varying spatial locations. One of the primary challenges for the auditory system is to analyze this external environment in order to inform goal-directed behavior. In Sect. 2.2 we introduced some of the neurophysiological evidence for the importance of the attentional focus of the task in determining the pattern of auditory cortical activity [69]. Here, we consider the contribution of human auditory fMRI research to this question. In particular, we present the interesting findings of one group who have started to address how the auditory cortex responds to the context and the procedural and cognitive demands of the listening task (*see* [77] for a review).

In that review, Scheich and colleagues report a series of research studies in which they suggest that the function of different auditory cortical areas is not determined so much by stimulus features (such as timbre, pitch, motion, etc.), but rather by the task that is performed. For example, one study reported the results of two fMRI experiments in which the same frequency-modulated stimuli were presented under different task conditions [78]. Top-down influences strongly affected the strength of the auditory response.

When a pitch-direction categorization task was compared with passive listening, a greater response was found in right posterior nonprimary auditory areas (planum temporale). Moreover, hemispheric differences were also found when comparing the response to two different tasks. The right nonprimary areas responded more strongly when the task required a judgement about pitch direction (rising or falling), whereas the left nonprimary areas responded more strongly when the task required a judgement about the sound duration. It is not the case that the right posterior nonprimary areas were *only* engaged by sound categorization because this region was more responsive to the critical sound feature (frequency modulation) during passive listening than were surrounding auditory areas (*see* also ref. [32]). These results broadly indicate an interaction between the stimulus and the task, which influences the pattern of auditory cortical activity. The precise characteristics of this interaction are worthy of future studies.

References

1. Palmer AR, Bullock DC, Chambers JD (1998) A high-output, high-quality sound system for use in auditory fMRI. Neuroimage 7:S357

2. Chou CK, McDougall JA, Chan KW (1995) Absence of radiofrequency heating from auditory implants during magnetic-resonance imaging. Bioelectromagnetics 16(5):307–316

3. Heller JW, Brackmann DE, Tucci DL, Nyenhuis JA, Chou CK (1996) Evaluation of MRI compatibility of the modified nucleus multichannel auditory brainstem and cochlear implants. Am J Otol 17(5):724–729

4. Shellock FG, Morisoli S, Kanal E (1993) MR procedures and biomedical implants, materials, and devices—1993 update. Radiology 189(2):587–599

5. Weber BP, Neuburger J, Battmer RD, Lenarz T (1997) Magnetless cochlear implant: relevance of adult experience for children. Am J Otol 18(6):S50–S51

6. Wild DC, Head K, Hall DA (2006) Safe magnetic resonance scanning of patients with metallic middle ear implants. Clin Otolaryngol 31(6):508–510

7. Giraud AL, Truy E, Frackowiak R (2001) Imaging plasticity in cochlear implant patients. Audiol Neurootol 6(6):381–393

8. Moelker A, Wielopolski PA, Pattynama PM (2003) Relationship between magnetic field strength and magnetic-resonance-related acoustic noise levels. MAGMA 16:52–55

9. Foster JR, Hall DA, Summerfield AQ, Palmer AR, Bowtell RW (2000) Sound-level measurements and calculations of safe noise dosage during fMRI at 3T. J Magn Reson Imaging 12:157–163

10. Ravicz ME, Melcher JR (2001) Isolating the auditory system from acoustic noise during functional magnetic resonance imaging: examination of noise conduction through the ear canal, head, and body. J Acoust Soc Am 109(1):216–231

11. Price DL, De Wilde JP, Papadaki AM, Curran JS, Kitney RI (2001) Investigation of acoustic noise on 15 MRI scanners from 0.2 T to 3 T. J Magn Reson Imaging 13(2):288–293

12. Hedeen RA, Edelstein WA (1997) Characterization and prediction of gradient acoustic noise in MR imagers. Magn Reson Med 37(1):7–10

13. Hennel F, Girard F, Loenneker T (1999) "Silent" MRI with soft gradient pulses. Magn Reson Med 42:6–10

14. Brechmann A, Baumgart F, Scheich H (2002) Sound-level-dependent representation of frequency modulations in human auditory cortex: a low-noise fMRI study. J Neurophysiol 87:423–433

15. Sumby WH, Pollack I (1954) Visual contribution to speech intelligibility in noise. J Acoust Soc Am 26(2):212–215

16. Assmann P, Summerfield Q (2004) Perception of speech under adverse conditions. In: Greenberg S, Ainsworth WA, Popper AN, Fay RR (eds) Speech processing in the auditory system. Springer, New York, pp 231–308

17. Healy EW, Moser DC, Morrow-Odom KL, Hall DA, Fridriksson J (2007) Speech perception in MRI scanner noise by persons with aphasia. J Speech Lang Hear Res 50:323–334

18. Harms MP, Melcher JR (2002) Sound repetition rate in the human auditory pathway: representations in the waveshape and amplitude of fMRI activation. J Neurophysiol 88:1433–1450

19. Chambers JD, Akeroyd MA, Summerfield AQ, Palmer AR (2001) Active control of the volume acquisition noise in functional magnetic resonance imaging: method and psychoacoustical evaluation. J Acoust Soc Am 110(6):3041–3054

20. Hall DA, Chambers J, Foster J, Akeroyd MA, Coxon R, Palmer AR (2009) Acoustic, psychophysical, and neuroimaging measurements of the effectiveness of active cancellation during auditory functional magnetic resonance imaging. J Acoust Soc Am 125(1):347–359

21. Hall DA, Haggard MP, Akeroyd MA, Palmer AR, Summerfield AQ, Elliott MR, Gurney EM, Bowtell RW (1999) 'Sparse' temporal sampling in auditory fMRI. Hum Brain Mapp 7:213–223

22. Bandettini PA, Jesmanowicz A, Van Kylen J, Birn RM, Hyde JS (1998) Functional MRI of brain activation induced by scanner acoustic noise. Magn Reson Med 39:410–416

23. Bilecen D, Scheffler K, Schmid N, Tschopp K, Seelig J (1998) Tonotopic organization of the human auditory cortex as detected by BOLD-FMRI. Hear Res 126:19–27

24. Hall DA, Summerfield AQ, Gonçalves MS, Foster JR, Palmer AR, Bowtell RW (2000) Time-course of the auditory BOLD response to scanner noise. Magn Reson Med 43:601–606

25. Shah NJ, Jäncke L, Grosse-Ruyken M-L, Müller-Gärtner HW (1999) Influence of acoustic masking noise in fMRI of the auditory cortex during phonetic discrimination. J Magn Reson Imaging 9(1):19–25

26. Talavage TM, Edmister WB, Ledden PJ, Weisskoff RM (1999) Quantitative assessment of auditory cortex responses induced by imager acoustic noise. Hum Brain Mapp 7(2):79–88

27. Elliott MR, Bowtell RW, Morris PG (1999) The effect of scanner sound in visual, motor, and auditory functional MRI. Magn Reson Med 41(6):1230–1235

28. Edmister WB, Talavage TM, Ledden PJ, Weisskoff RM (1999) Improved auditory cortex imaging using clustered volume acquisitions. Hum Brain Mapp 7:89–97

29. Formisano E, Kim DS, Di Salle F, van de Moortele PF, Ugurbil K, Goebel R (2003) Mirror-symmetric tonotopic maps in human primary auditory cortex. Neuron 40(4):859–869

30. Hall DA, Haggard MP, Akeroyd MA, Summerfield AQ, Palmer AR, Elliott MR, Bowtell RW (2000) Modulation and task effects in auditory processing measured using fMRI. Hum Brain Mapp 10(3):107–119

31. Hall DA, Haggard MP, Summerfield AQ, Akeroyd MA, Palmer AR, Bowtell RW (2001) Functional magnetic resonance imaging measurements of sound-level encoding in the absence of background scanner noise. J Acoust Soc Am 109(4):1559–1570

32. Hart HC, Palmer AR, Hall DA (2004) Different areas of human non-primary auditory cortex are activated by sounds with spatial and nonspatial properties. Hum Brain Mapp 21:178–190

33. Langers DRM, Backes WH, Van Dijk P (2007) Representation of lateralization and tonotopy in primary versus secondary human auditory cortex. Neuroimage 34:264–273

34. Schwarzbauer C, Davis MH, Rodd JM, Johnsrude I (2006) Interleaved silent steady state (ISSS) imaging: a new sparse imaging method applied to auditory fMRI. Neuroimage 29(3):774–782

35. Bregman AS (1990) Auditory scene analysis: the perceptual organisation of sound. MIT, Cambridge, MA

36. Fishman YI, Arezzo JC, Steinschneider M (2004) Auditory stream segregation in monkey auditory cortex: effects of frequency separation, presentation rate, and tone duration. J Acoust Soc Am 116(3):1656–1670

37. Fishman YI, Reser DH, Arezzo JC, Steinschneider M (2001) Neural correlates of auditory stream segregation in primary auditory cortex of the awake monkey. Hear Res 151:167–187

38. Brosch M, Schreiner CE (1997) Time course of forward masking tuning curves in cat primary auditory cortex. J Neurophysiol 77:923–943

39. Bartlett EL, Wang X (2005) Long-lasting modulation by stimulus context in primate auditory cortex. J Neurophysiol 94:83–104

40. Harms MP, Guinan JJ, Sigalovsky IS, Melcher JR (2005) Short-term sound temporal envelope characteristics determine multisecond time patterns of activity in human auditory cortex as shown by fMRI. J Neurophysiol 93:210–222

41. Palmer AR, Hall DA, Sumner C, Barrett DJK, Jones S, Nakamoto K, Moore DR (2007) Some investigations into non-passive listening. Hear Res 229:148–157

42. Hall DA, Edmondson-Jones M, Fridriksson J (2006) Periodicity and frequency coding in human auditory cortex. Eur J Neurosci 24:3601–3610

43. Hall DA, Johnsrude IS, Haggard MP, Palmer AR, Akeroyd MA, Summerfield AQ (2002) Spectral and temporal processing in human auditory cortex. Cereb Cortex 12:140–149

44. Hart HC, Hall DA, Palmer AR (2003) The sound-level-dependent growth in the extent of fMRI activation in Heschl's gyrus is different for low- and high-frequency tones. Hear Res 179(1–2):104–112

45. Von Békésy G (1947) The variations of phase along the basilar membrane with sinusoidal vibrations. J Acoust Soc Am 19:452–460

46. Kosaki H, Hashikawa T, He J, Jones EG (1997) Tonotopic organization of auditory cortical fields delineated by parvalbumin immunoreactivity in macaque monkeys. J Comp Neurol 386:304–316

47. Merzenich MM, Brugge JF (1973) Representation of the cochlear partition on the superior temporal plane of the macaque monkey. Brain Res 50:275–296

48. Petkov CL, Kayser C, Augath M, Logothetis NK (2006) Functional imaging reveals numerous fields in the monkey auditory cortex. PLoS Biol 4(7):213–226

49. Hall DA, Hart HC, Johnsrude IS (2003) Relationships between human auditory cortical structure and function. Audiol Neurootol 8(1):1–18

50. Schönwiesner M, Von Cramon DY, Rubsamen R (2002) Is it tonotopy after all? Neuroimage 17:1144–1161

51. Talavage TM, Ledden PJ, Benson RR, Rosen BR, Melcher JR (2000) Frequency-dependent responses exhibited by multiple regions in human auditory cortex. Hear Res 150:225–244

52. Talavage TM, Sereno MI, Melcher JR, Ledden PJ, Rosen BR, Dale AM (2004) Tonotopic organization in human auditory cortex revealed by progressions of frequency sensitivity. J Neurophysiol 91:1282–1296

53. Kastner S, Ungerleider LG (2000) Mechanisms of visual attention in the human cortex. Annu Rev Neurosci 23(1):315–341

54. Maunsell JHR, Treue S (2006) Feature-based attention in visual cortex. Trends Neurosci 29(6):317–322

55. Scholl BJ (2001) Objects and attention: the state of the art. Cognition 80(1–2):1–46

56. Griffiths TD, Warren JD (2004) What is an auditory object? Nat Rev Neurosci 5(11):887–892

57. Johnson JA, Zatorre RJ (2005) Attention to simultaneous unrelated auditory and visual events: behavioral and neural correlates. Cereb Cortex 15(10):1609–1620

58. Johnson JA, Zatorre RJ (2006) Neural substrates for dividing and focusing attention between simultaneous auditory and visual events. Neuroimage 31(4):1673–1681

59. Lavie N (2005) Distracted and confused? Selective attention under load. Trends Cogn Sci 9(2):75–82

60. Degerman A, Rinne T, Salmi J, Salonen O, Alho K (2006) Selective attention to sound location or pitch studied with fMRI. Brain Res 1077(1):123–134

61. Petkov CI, Kang X, Alho K, Bertrand O, Yund EW, Woods DL (2004) Attentional modulation of human auditory cortex. Nat Neurosci 7(6):658–663

62. Kayser C, Petkov CI, Augath M, Logothetis NK (2007) Functional imaging reveals visual modulation of specific fields in auditory cortex. J Neurosci 27(8):1824–1835

63. Lehmann C, Herdener M, Esposito F, Hubl D, di Salle F, Scheffler K, Bach DR, Federspiel A, Kretz R, Dierks T, Seifritz E (2006) Differential patterns of multisensory interactions in core and belt areas of human auditory cortex. Neuroimage 31(1):294–300

64. Ahveninen J, Jaaskelainen IP, Raij T, Bonmassar G, Devore S, Hamalainen M, Levanen S, Lin F-H, Sams M, Shinn-Cunningham BG, Witzel T, Belliveau JW (2006) Task-modulated "what" and "where" pathways in human auditory cortex. Proc Natl Acad Sci U S A 103(39):14608–14613

65. Lewald J, Meister IG, Weidemann J, Topper R (2004) Involvement of the superior temporal cortex and the occipital cortex in spatial hearing: evidence from repetitive transcranial magnetic stimulation. J Cogn Neurosci 16(5):828–838

66. Degerman A, Rinne T, Pekkola J, Autti T, Jaaskelainen IP, Sams M, Alho K (2007) Human brain activity associated with audiovisual perception and attention. Neuroimage 34(4):1683–1691

67. Greenberg GZ, Larkin WD (1968) Frequency-response characteristic of auditory observers detecting signals of a single frequency in noise: the probe-signal method. J Acoust Soc Am 44(6):1513–1523

68. Schlauch RS, Hafter ER (1991) Listening bandwidths and frequency uncertainty in pure-tone signal detection. J Acoust Soc Am 90(3):1332–1339

69. Fritz JB, Elhilali M, David SV, Shamma SA (2007) Does attention play a role in dynamic receptive field adaptation to changing acoustic salience in A1? Hear Res 229:186–203

70. Shackleton TM, Carlyon RP (1994) The role of resolved and unresolved harmonics in pitch perception and frequency modulation discrimination. J Acoust Soc Am 95:3529–3540

71. Penagos H, Melcher JR, Oxenham AJ (2004) A neural representation of pitch salience in nonprimary human auditory cortex revealed with functional magnetic resonance imaging. J Neurosci 24(30):6810–6815

72. Griffiths TD, Büchel C, Frackowiak RSJ, Patterson RD (1998) Analysis of temporal structure in sound by the human brain. Nat Neurosci 1:422–427

73. Blauert J, Lindemann W (1986) Spatial mapping of intracranial auditory events for various degrees of interaural coherence. J Acoust Soc Am 79(3):806–813

74. Budd TW, Hall DA, Goncalves MS, Akeroyd MA, Foster JR, Palmer AR, Head K, Summerfield AQ (2003) Binaural

specialisation in human auditory cortex: an fMRI investigation of interaural correlation sensitivity. Neuroimage 20(3):1783–1794

75. Culling JF, Colburn HS, Spurchise M (2001) Interaural correlation sensitivity. J Acoust Soc Am 110(2):1020–1029

76. Griffiths TD, Uppenkamp S, Johnsrude I, Josephs O, Patterson RD (2001) Encoding of the temporal regularity of sound in the human brainstem. Nat Neurosci 4:633–637

77. Scheich H, Brechmann A, Brosch M, Budinger E, Ohl FW (2007) The cognitive auditory cortex: task-specificity of stimulus representations. Hear Res 229:213–224

78. Brechmann A, Scheich H (2005) Hemispheric shifts of sound representation in auditory cortex with conceptual listening. Cereb Cortex 15(5):578–587

Part III

fMRI Clinical Application

Chapter 20

Application of fMRI to Multiple Sclerosis and Other White Matter Disorders

Massimo Filippi and Maria A. Rocca

Abstract

The variable effectiveness of reparative and recovery mechanisms following tissue damage is among the factors that might contribute to explain, at least partially, the paucity of the correlation between clinical and magnetic resonance imaging (MRI) findings in patients with white matter disorders. Among the mechanisms of recovery, brain plasticity is likely to be one of the most important with several possible different substrates (including increased axonal expression of sodium channels, synaptic changes, increased recruitment of parallel existing pathways or "latent" connections, and reorganization of distant sites). The application of fMRI has shown that plastic cortical changes do occur after white matter injury of different etiology, that such changes are related to the extent of white matter damage, and that they can contribute in limiting the clinical consequences of brain damage. Conversely, the failure or exhaustion of the adaptive properties of the cerebral cortex might be among the factors responsible for the accumulation of "fixed" neurological deficits in patients with white matter disorders.

Key words Multiple sclerosis, Functional magnetic resonance imaging, White matter, Adaptation, Maladaptation, Myelitis, Vasculitides

1 Introduction

Over the past decade, modern structural magnetic resonance imaging (MRI) techniques have been extensively used to study patients with white matter disorders with the ultimate goal of increasing the understanding of the mechanisms responsible for the accumulation of irreversible disability [1–3]. Although the application of these techniques has provided important insight into the pathobiology of many of these disorders, the magnitude of the correlation between MRI and clinical findings remains suboptimal [1–3]. This might be explained, at least partially, by the variable effectiveness of reparative and recovery mechanisms following tissue damage. Cortical reorganization has been suggested as a potential contributor to the recovery or to the maintenance of function in the presence of irreversible white matter damage [4, 5]. Brain plasticity is a

Massimo Filippi (ed.), *fMRI Techniques and Protocols*, Neuromethods, vol. 119,
DOI 10.1007/978-1-4939-5611-1_20, © Springer Science+Business Media New York 2016

well-known feature of the human brain, which is likely to have several different substrates (including increased axonal expression of sodium channels, synaptic changes, increased recruitment of parallel existing pathways or "latent" connections, and reorganization of distant sites) [6]. The application of functional MRI (fMRI) has shown that plastic cortical changes do occur after central nervous system (CNS) white matter injury of different etiology, that such changes are related to the extent of WM damage, and that they can contribute in limiting the clinical consequences of widespread disease-related tissue damage [4, 5]. Conversely, the failure or the exhaustion of the adaptive properties of the cerebral cortex might be among the factors responsible for the accumulation of "fixed" neurological deficits in patients affected by white matter disorders (WMD) [4, 5].

This chapter summarizes the major contributions of fMRI for the in vivo monitoring of several white matter diseases. Since fMRI has been mostly applied to improve our understanding of the pathophysiology of multiple sclerosis (MS), a special focus is devoted to this condition and allied WMD.

2   fMRI in MS

2.1   General Considerations

The main problem in the interpretation of fMRI studies in diseased people is that the observed changes might be biased by differences in task performance between patients and controls. Clearly, this is a major issue in MS, which typically causes impairment of various functional systems. Therefore, despite providing several important pieces of information, the value of the earliest fMRI studies of patients with MS [7–13] has to be weight against this background. For this reason, more recent fMRI studies in MS have been based on larger and more selected patients' groups than the seminal studies. These studies have investigated the brain patterns of cortical activations during the performance of a number of motor, visual, and cognitive tasks in patients with all the major clinical phenotypes of the disease. Another appealing strategy which has been introduced for the study of functional network rewiring in clinically impaired patients is based on the assessment of functional abnormalities at rest in the main brain functional networks (resting state networks). One of the most solid conclusions that can be drawn from fMRI studies of MS is that cortical reorganization does occur in patients affected by this condition. The correlation between various measures of structural MS damage and the extent of cortical activations also suggests an adaptive role of such cortical changes in contributing to clinical recovery and maintaining a normal level of functioning in patients with MS, despite the presence of irreversible axonal/neuronal loss.

2.2 Visual System

The method usually applied to investigate the visual system consists of the application of a 8 Hz photic stimulation to one or both eyes [8, 12–18]. A study of the visual system [12] in patients who had recovered from a single episode of acute unilateral optic neuritis demonstrated that these patients, relative to healthy volunteers, had an extensive activation of the visual network, including the claustrum, lateral temporal and posterior parietal cortices, and thalamus, in addition to the primary visual cortex, when the clinically affected eye was studied. When the unaffected eye was stimulated, only activations of the visual cortex and the right insula/claustrum were observed. A strong correlation was found in these patients between the volume of the extra-occipital activation and the latency of the visual evoked potential (VEP) P100, suggesting that the functional reorganization of the cortex might represent an adaptive response to a persistently abnormal visual input. The results of this preliminary study have been confirmed and extended by subsequent studies [14, 15, 17]. Using fMRI and VEP to monitor the functional recovery after an acute unilateral optic neuritis, Russ et al. [15] found a strong relationship between fMRI and VEP latencies, suggesting that fMRI might contribute to the assessment of the temporal evolution of the visual deficits during recovery. Levin et al. [17] showed reduced activation of the primary visual cortex and increased activation of the lateral occipital complex (LOC) in eight subjects who recovered clinically from an episode of optic neuritis, but who still had prolonged VEP latencies in comparison with healthy controls.

Structural MRI, electrophysiology, and fMRI have been combined in another study to investigate why ON patients exhibit a wide variation in severity of acute visual loss. Optic nerve lesion length and VEP amplitude were associated with visual loss. Bilateral activation in the extra-striate occipital cortex correlated directly with vision, after adjusting for optic nerve lesion length, VEP amplitude, and demographic characteristics [19] (Fig. 1).

These data suggest that acute visual loss is associated with the extent of inflammation and conduction block in the optic nerve, but not with pathology in the optic radiations or occipital cortex. The association of better vision with greater fMRI responses, after accounting for factors which reduce afferent input, suggested a role for adaptive neuroplasticity within the association cortex of the dorsal stream of higher visual processing [19]. In a 1-year follow-up study, Toosy et al. [14], using a novel technique that modeled the fMRI response and optic nerve structure together with clinical function, demonstrated a potential adaptive role of cortical reorganization within the extra-striate visual areas. An increased optic nerve gadolinium-enhanced lesion length at baseline was associated with a reduced functional activation within the visual cortex and poorer vision. At 3 months, more severe optic nerve damage was associated with an increased fMRI response in

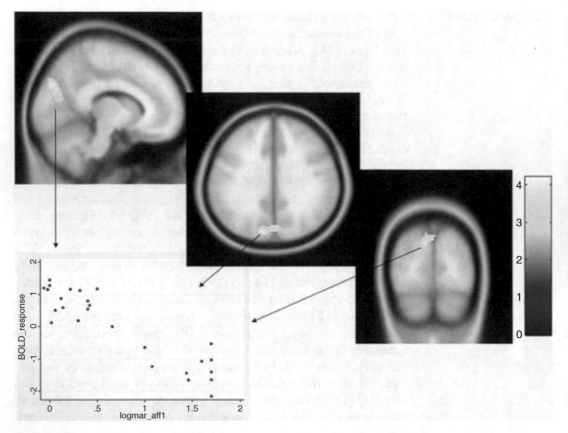

Fig. 1 Statistical parametric maps showing group correlations between functional magnetic resonance imaging (fMRI) response and visual acuity, after correcting for age, gender, side affected, gadolinium-enhanced lesion length, fast spin-echo lesion length, and visual evoked potential (VEP) amplitude. A correlation is seen in the region of the cuneus bilaterally, where better visual acuity is associated with a greater fMRI response. The graph plots logMAR visual acuity against the mean corrected fMRI response (approximate percentage blood oxygenated level dependent signal change), at the peak voxel. The statistical parametric maps are thresholded at cluster level $p < 0.05$ (corrected), and the scale bar indicates the voxel level t-scores (from ref. [19])

the bilateral temporal cortices, whereas at 1 year, the right temporal cortex correlation reversed. These results illustrate how the same regions may play different roles at different times during recovery, reflecting the complexity of brain plasticity and the MS process. This notion has been supported by a region-of-interest longitudinal study [18] that demonstrated dynamic changes in the fMRI response following visual stimulation not only in V1, V2, and the LOC, but also in the lateral geniculate nucleus (LGN) in patients with isolated acute optic neuritis. In this study, abnormal LGN response was found not only following stimulation of the affected eye, but also after stimulation of the unaffected one, indicating that the visual pathways undergo early functional changes following tissue injury.

Another 1-year longitudinal study demonstrated that early functional abnormalities in the LOC contribute to predict visual outcome after a clinically isolated ON [20]. Specifically, greater baseline fMRI responses in the LOCs were associated with better visual outcome at 12 months. This was evident on stimulation of either eye and was independent of measures of demyelination and neuroaxonal loss. A negative fMRI response in the LOCs at baseline was associated with a relatively worse visual outcome. These findings suggest that early neuroplasticity in higher visual areas is likely to be an important determinant of recovery from ON, independent of tissue damage in the anterior or posterior visual pathways.

In patients with established MS and a relapsing-remitting (RR) course with a unilateral optic neuritis, a reduced recruitment of the visual cortex after stimulation of the affected and the unaffected eyes was found when compared with healthy subjects. On average, patients with optimal clinical recovery showed increased visual cortex activation than those with poor or no recovery, although the extent of the activation remained reduced compared with controls [8]. Another study [13] of nine patients with previous optic neuritis confirmed the previous results [8] and showed that these patients not only have a reduced activation of the primary visual cortex, but also a reduced fMRI percentage signal change in this region, again suggesting an abnormality of the synaptic input.

2.3 Motor System

The investigation of the motor system in patients with MS has mainly focused on the analysis of the performance of simple motor tasks with the dominant right upper limbs [9–11, 21–40]. Such tasks were either self-paced or paced by a metronome. A few studies assessed the performance of simple motor tasks with the dominant right lower limbs [23, 27, 33], while even fewer studies have investigated the performance of more complex tasks, including phasic movements of dominant hand and foot [27, 33], object manipulation [41], and visuomotor integration tasks [42].

An altered brain pattern of movement-associated cortical activations, characterized by an increased recruitment of the contralateral primary sensorimotor cortex (SMC) during the performance of simple tasks [23, 27] and by the recruitment of additional "classical" and "higher-order" sensorimotor areas during the performance of more complex tasks [27], has been demonstrated in patients with clinically isolated syndrome (CIS) suggestive of MS. The clinical and conventional MRI follow-up of these patients has shown that, at disease onset, CIS patients with a subsequent evolution to clinically definite MS tend to recruit a more widespread sensorimotor network than those without short-term disease evolution [36]. These findings suggest that in CIS patients the extent of early cortical reorganization might be a factor associated with a different clinical evolution. This would support the notion that, whereas increased recruitment of a widespread sensorimotor network contributes to limiting the impact of structural damage during the course of MS, its early activation might be counterproductive, as it might result in an early exhaustion of the adaptive properties of the brain. This notion is also supported by studies of stroke patients, where a persistent overrecruitment of a widespread cortical network has been related to an unfavorable clinical outcome [43].

An increased recruitment of several sensorimotor areas, mainly located in the cerebral hemisphere ipsilateral to the limb that performed the task, has also been demonstrated in patients with early RRMS and a previous episode of hemiparesis [29]. In patients with similar characteristics, but who presented with an episode of optic neuritis, this increased recruitment involved sensorimotor areas that were mainly located in the contralateral cerebral hemisphere [30].

In patients with established MS and a RR course, functional cortical changes, mainly characterized by an increased recruitment of "classical" motor areas, including the primary SMC, the supplementary motor area (SMA), and the secondary sensorimotor cortex (SII), have been shown during the performance of simple motor [9–11, 22] and visuomotor integration tasks [42]. Movement-associated cortical changes, characterized by the activation of highly specialized cortical areas, have also been described in patients with secondary progressive (SP) MS [24] during the performance of a simple motor task and in patients with primary progressive (PP) MS during the performance of active [21, 33] and passive [44] motor experiments. Interestingly enough, contrary to what happens in SPMS, the movement-associated pattern of activations seen in benign (B) MS resembles that of healthy people, and its abnormalities are restricted to the sensorimotor network [45]. These results suggest that the long-term preservation of brain functional adaptive mechanisms might be among the factors contributing to the favorable clinical course of BMS.

The concept that movement-associated cortical reorganization varies across patients at different stages of the disease has been shown by a fMRI study of patients with different disease phenotypes [28] (including 16 patients with a CIS suggestive of MS, 14 with RRMS and no disability, 15 with RRMS and mild clinical disability, and 12 with SPMS) acquired during the performance of a simple motor task with their unimpaired dominant hand. CIS patients had an increased activation of the contralateral primary SMC when compared with those with RRMS and no disability, whereas patients with RRMS and no disability had an increased activation of the SMA when compared with those with CIS (Fig. 2). Patients with RRMS and no disability had an increased activation of the primary SMC, bilaterally, and ipsilateral SMA when compared with patients with RRMS and mild clinical disability. Conversely, patients with RRMS and mild clinical disability had an increased activation of the contralateral SII, inferior frontal gyrus (IFG), and ipsilateral precuneus. Patients with

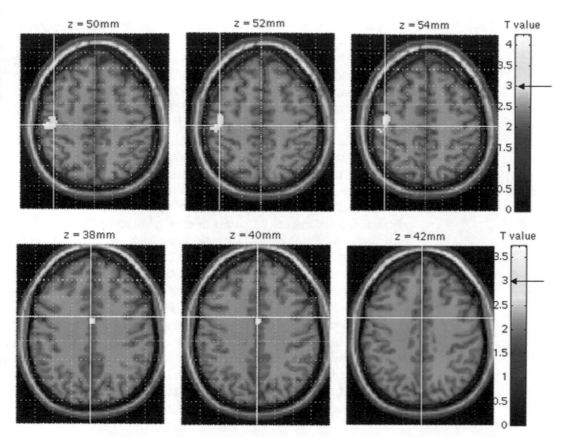

Fig. 2 Comparisons of patients at presentation with clinically isolated syndrome (CIS) suggestive of MS and patients with relapsing-remitting (RR) MS and no disability during a simple, right-hand, motor task. Patients with CIS showed an increased activation of the contralateral primary sensorimotor cortex when compared with patients with RRMS and no disability (*top row*). Patients with RRMS and no disability had a more significant activation of the supplementary motor area, bilaterally, when compared with patients with a CIS (*bottom row*). Images are color-coded for activation and *arrows* show *t* cut-off values. Activations were superimposed on a high-resolution T1-weighted scan obtained from one healthy individual and normalized into a standard statistical parametric mapping space (neurological convention) (from ref. [28])

RRMS and mild clinical disability had an increased activation of the contralateral thalamus and ipsilateral SII when compared with those with SPMS. The opposite contrast showed that patients with SPMS had an increased activation of the IFG, bilaterally, middle frontal gyrus (MFG), bilaterally, contralateral precuneus, and ipsilateral cingulate motor area (CMA) and inferior parietal lobule. This study suggests that early in the disease course more areas typically devoted to motor tasks are recruited, then a bilateral activation of these regions is seen, and late in the disease course, areas that healthy people recruit to perform novel or complex tasks are activated [28], perhaps in an attempt to limit the functional consequences of accumulating tissue damage.

To provide a more comprehensive view of abnormalities of cerebral response to task stimulation in MS patients at different stages of the disease, activation and deactivation patterns during passive hand movements have also been investigated [46]. In line with previous findings, analysis of activations showed a progressive extension to the ipsilateral brain hemisphere according to the group and the clinical form (HC < RRMS < SPMS). Compare to controls, MS patients had reduced deactivation of the ipsilateral cortical sensorimotor areas. Deactivation of posterior cortical areas belonging to the default mode network (DMN) was increased in RRMS, but not in SPMS, with respect to HC.

As described in another chapter of this book, fMRI has been also applied for the investigation of cervical cord neuronal activity during a proprioceptive and a tactile stimulation of the right upper limb from patients with the main MS clinical phenotypes [47–51]. During the application of both stimulations, healthy controls and MS patients had significant activations of the cervical cord between C5 and C8. On average, MS patients had 20% higher cord fMRI signal changes during either stimulations, suggesting an abnormal cord function in these patients. Compared to controls, MS patients tend to show a more distributed pattern of cord recruitment with additional activations of regions located in the anterior and contralateral portions of the cord. Such a behavior was more pronounced in patients with SPMS vs. those with RRMS and PPMS [49, 50].

2.4 Cognition

Several fMRI studies have suggested that functional cortical changes might have an adaptive role also in limiting MS-related cognitive impairment [52–68]. Therefore, brain plasticity might, in part, explain the weak relationship found in MS between neuropsychological deficits and conventional MRI measures of disease burden [69].

Several cognitive domains have been investigated in MS patients with fMRI. Working memory has been the most extensively studied by means of the Paced Auditory Serial Addition Test (PASAT) or the Paced Visual Serial Addition Task (PVSAT) [52–55, 60, 64, 70] (which also involve sustained attention, information processing speed, and simple calculation), the n-back task [59, 61–63, 65, 71], or a task adapted from the Sternberg paradigm [57]. Additional cognitive domains including attention [58], episodic memory [72], planning [68], and emotional processing [73] have also been interrogated.

In patients at presentation with CIS suggestive of MS, an altered pattern of cortical activations has been described during the performance of the PASAT [55, 56, 70], confirming the presence of cortical reorganization at the earliest clinical stage of the disease. Staffen et al. [52] found that, during the performance of the PVSAT, MS patients with intact task performance had an increased activation of several regions located in the frontal and parietal lobes, bilaterally, compared with healthy volunteers, suggesting the

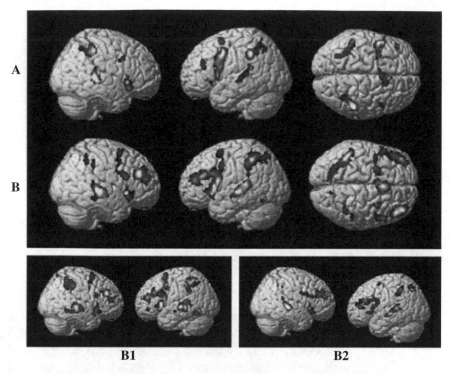

Fig. 3 Brain patterns of cortical activations on a rendered brain during the execution of the Paced Auditory Serial Addition Task (PASAT) in (**a**) 22 healthy controls and in (**b**) 22 patients with MS. (**b1**, **b2**) Rendered images for patients with MS subgrouped according to their performance at the PASAT during fMRI showing significant activated foci in (**b1**) for 12 patients whose performance was similar to that of healthy controls and in (**b2**) the activations found in the 10 patients who exhibited lower scores (from ref. [60])

presence of functional compensatory mechanisms. An increased recruitment of several cortical areas during the performance of a simple cognitive task has also been shown in patients with RRMS and mild clinical disability (Fig. 3) [60].

An increased activation of regions exclusively located in the right cerebral hemisphere (in particular in the frontal and temporal lobes) has also been found in MS patients when testing rehearsal within working memory [57]. The degree of right hemisphere recruitment was strongly related to patient neuropsychological performance [57]. In patients with RRMS and no cognitive deficits, using fMRI during an *n*-back test, a reduced activation of the "core" areas of the working memory circuitry (including prefrontal and parietal regions) and an increased activation of other regions within and beyond the typical working memory circuitry (including areas in the frontal, parietal, temporal, and occipital lobes) have been found [63]. This shift of activation was most prominent with increased working memory demands. These findings suggest that, as shown for motor and visual tasks, dynamic changes of brain activation patterns can occur in RRMS patients during cognitive tasks. Other studies [61, 62, 66], which also investigated working memory performance in MS

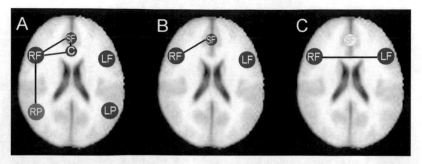

Fig. 4 Analysis of functional connectivity during the performance of the *n*-back task with different levels of difficulty in healthy individuals and patients with MS. The most significant correlations between activation in regions involved in processing increasing task demand are indicated in (**a**). In (**b**) are those connections more significant for controls ($p<0.05$). The image in (**c**) shows connections that were more significant in patients than controls ($p<0.05$). *C* cingulate, *SF* superior medial frontal, *RF* right dorsolateral prefrontal, *LF* left dorsolateral prefrontal, *RP* right parietal, *LP* left parietal (from ref. [65])

patients, demonstrated: (1) an increased recruitment of regions related to sensorimotor functions and anterior attentional/executive components of the working memory system in patients compared with healthy controls, and (2) a reduced recruitment of several regions in the right cerebellar hemisphere in patients compared with healthy individuals [66], thus suggesting that the cerebellum might play a role in the working memory impairment of MS.

In a fMRI study [65], working memory was investigated with an *n*-back task and functional connectivity analysis in a group of 21 RRMS patients and 16 age- and sex-matched healthy controls. With similar task performances, activations were found in similar regions for both groups. However, patients had relatively reduced activations of the superior frontal and anterior cingulate gyri. Patients also showed a variable, but generally substantially smaller increase of activation than healthy controls with greater task complexity, depending on the specific brain regions assessed. These findings suggest that, despite similar brain regions were recruited in both groups, patients have a reduced functional reserve for cognition relevant to memory. The functional connectivity analysis revealed increased correlations between right dorsolateral prefrontal and superior frontal/anterior cingulate activations in controls, and increased correlations between activations in the right and left prefrontal cortices in patients (Fig. 4), suggesting that altered interhemispheric interactions between dorsal and lateral prefrontal regions may yet be an additional adaptive mechanism distinct from recruitment of novel processing regions [65].

Using a 3 T scanner, more significant activations of several areas of the cognitive network involved in the performance of the Stroop test have also been demonstrated in a group of 15 cognitively preserved patients with benign MS (BMS) when compared with 19 healthy controls (Fig. 5) [67]. BMS patients also showed an

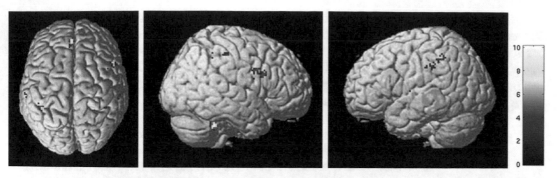

Fig. 5 Areas showing increased activations in patients with BMS in comparison with healthy controls during the analysis of the Stroop facilitation condition (random effect interaction analysis, ANOVA, $p < 0.05$ corrected for multiple comparisons). BMS patients had increased activations of several areas located in the frontal and parietal lobes, bilaterally, including the anterior cingulate cortex, the superior frontal sulcus, the inferior frontal gyrus, the precuneus, the secondary sensorimotor cortex, the bilateral visual cortex, and the cerebellum, bilaterally. Note that the color-encoded activations have been superimposed on a rendered brain and normalized into standard SPM space (neurological convention) (*see* ref. [67])

increased connectivity of several cortical areas of the sensorimotor network, including the left IFG, the anterior cingulated cortex and the left SII, with the right IFG and the right cerebellum, as well as a decreased connectivity between some areas (including the left SII, the prefrontal cortex, and the right cerebellum), and the anterior cingulate cortex. These results suggest an altered interhemispheric balance in favor of the right hemisphere in BMS patients in comparison with healthy controls, when performing cognitive tasks.

fMRI during the performance of the Stroop task has also been used to determine whether modification of the connections between cerebellar and prefrontal areas might vary among MS phenotypes and might be associated with cognitive failure [74]. Activation and effective connectivity analyses showed that, compared with the other groups, RRMS patients had abnormal recruitment of regions of the left frontoparietal lobes, whereas compared with RRMS, SPMS patients had abnormal recruitment of the posterior cingulum/precuneus. BMS patients had increased activation of the right prefrontal cortex, and increased interaction between these regions and the right cerebellum. In healthy controls, reaction times (RTs) inversely correlated with activity of right cerebellum and several frontoparietal regions. In MS, RTs inversely correlated with bilateral cerebellar activity and directly correlated with right precuneus activity. In MS, disease duration inversely correlated with right cerebellar activity and directly correlated with left inferior frontal gyrus and right precuneus activity. Higher T2 lesion volume and lower brain volumes were related to activity in these areas. These results suggest that MS patients who have various clinical phenotypes experience different abnormalities in activation and effective connectivity between the right cerebellum and frontoparietal areas, which contribute to inefficient cortical reorganization, with increasing cognitive load.

2.5 Correlations Between the Extent of Functional Cortical Reorganization and the Extent of Brain Damage in MS

In addition to an abnormal pattern of functional activations, the majority of the previous studies described a variable relationship between the extent of fMRI activation and several measures of tissue damage [7, 8, 10, 11, 21–25, 30, 33, 42, 53, 56, 58, 60].

An increased recruitment of several brain areas with increasing T2 lesion load has been shown in patients with RR [7, 8, 54, 60, 72] and PPMS. The severity of intrinsic T2-visible lesion damage, measured using T1-weighted images [30], magnetization transfer (MT), and diffusion tensor (DT) MRI [22], has been found to modulate the activity of some cortical areas in these patients. The severity of normal appearing brain tissue (NABT) injury, measured using proton MR spectroscopy [10, 11, 25], MT MRI [21, 22, 53, 56], and DT MRI [21, 23, 25], is another important factor associated to an increased recruitment of motor- and cognitive-related brain regions, as shown by studies of patients at presentation with CIS suggestive of MS [23, 53, 56], patients with RRMS and variable degrees of clinical disability [10, 11, 22, 25], patients with PPMS [21], and with SPMS [10, 11]. Finally, subtle GM damage, which goes undetected when using conventional MRI, may also influence functional cortical recruitment, as demonstrated, for the motor system, in patients with RRMS [42], SPMS [24], and patients with clinically definite MS and nonspecific (less than three focal white matter lesions) conventional MRI findings [25]. In cognitively intact MS patients, the increased activation of a left prefrontal region during the counting Stroop task has been correlated with the normalized brain parenchymal volume [58].

2.6 Functional Cortical Reorganization and Regional Damage in MS

Structural damage of white matter pathways that connect functional relevant areas for a given task has been shown to modify the observed brain patterns of cortical activations in patients with MS. Damage to the corticospinal tract [30, 31] (Fig. 6) as well as damage to the corpus callosum (CC) [75, 76] has been related to a more bilateral movement-associated brain pattern of cortical activations.

The role of the CC in interhemispheric connectivity and in eliciting functional cortical changes has been underpinned by a study by Lowe et al. [32], who showed, by measuring low-frequency BOLD fluctuations, a reduced functional connectivity between the right and the left hemisphere primary motor cortices in MS patients.

The recent development of diffusion-based tractography methods that allow to define with precision the pathways connecting different CNS structures and their application to patients with MS resulted in an improvement of the correlation between structural and functional abnormalities. Several studies combined measures of abnormal functional connectivity with DT MR measures of damage within selected white matter fiber bundles in patients with RRMS [40], BMS [67] and PPMS [77]. In patients with RRMS and no clinical disability [40], measures of abnormal connectivity inside the motor network were correlated with

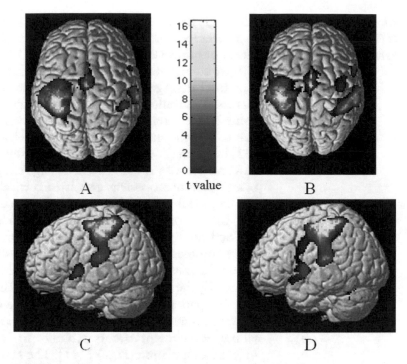

Fig. 6 Brain patterns of cortical activations on a rendered brain from MS patients without (**a** and **c**) and with (**b** and **d**) lesions in the left corticospinal tracts, during the performance of a simple motor task with their clinically unimpaired and fully normal functioning, dominant right hands. In patients with corticospinal tract lesions, a more bilateral pattern of activations is visible. Note that the activations are color-coded according to their *t* values. Images are in neurological convention (*see* ref. [31])

structural MRI metrics of tissue damage of the corticospinal and the dentatorubrothalamic tracts, while no correlation was found with measures of damage within "not-motor" white matter fiber bundles. These findings suggest an adaptive role of functional connectivity changes in limiting the clinical consequences of structural damage to selected white matter pathways in RRMS patients [40]. In patients with BMS [67], measures of abnormal connectivities inside the cognitive network were moderately correlated with structural MRI metrics of tissue damage within intra- and inter-hemispheric cognitive-related white matter fiber bundles, while no correlations were found with the remaining fiber bundles studied, suggesting that functional cortical changes in patients with BMS might represent an adaptive response driven by damage to specific white matter structures [67].

In patients with PPMS [21], a relationship has been demonstrated between the severity of spinal cord pathology, measured using MT MRI, and the extent of movement-associated cortical activations.

2.7 Adaptive Role of Functional Cortical Reorganization in MS

Although the actual role of cortical reorganization on the clinical manifestations of MS remains to be established, there are several pieces of evidence which suggest that cortical adaptive changes are likely to contribute in limiting the clinical consequences of MS-related structural damage. In nondisabled patients with RRMS [22], an increased activation of several motor regions, mainly located in the contralateral cerebral hemisphere, has been seen during the performance of a simple motor task. The correlations found in this study [22] between the extent of fMRI activations and several MT and DT MRI metrics of structural brain damage suggested that an increased recruitment of movement-associated cortical network contributes to limiting the functional impact of MS-related damage.

The notion that an increased recruitment of areas that are usually activated by healthy individuals when performing different/more complex motor tasks might be one of the mechanisms playing a role in MS recovery/maintenance of function has been highlighted by the results of two experiments [39, 41]. The first showed that MS patients, during the performance of a simple motor task, activate some regions that are part of a fronto-parietal circuit, whose recruitment occurs typically in healthy subjects during object manipulation (Fig. 7) [41]. The second, which assessed the fMRI patterns of activation during the performance of a simple motor task and of a task aimed at investigating the mirror-neuron system, demonstrated activations of regions that are part of the mirror-neuron system in patients with MS during the performance of the simple motor task [39].

The compensatory role of cortical reorganization has also been demonstrated by studies investigating the cognitive domains, which showed increased recruitment of several cortico-subcortical areas in cognitively preserved MS patients [52–58]. In patients complaining of fatigue, when compared with matched nonfatigued MS patients [78], a reduced activation of a complex movement-associated cortical/subcortical network, including the cerebellum, the thalamus, and regions in the frontal lobes, has been shown. The correlation found in these patients between the reduction of thalamic activity and the clinical severity of fatigue indicates that a "pseudoreduction" of brain functional recruitment might be associated with the appearance of MS symptomatology. Additional work has shown that the pattern of movement-associated cortical activations in MS is determined by both the extent of brain injury and disability and that these changes are distinct [26, 33].

2.8 Maladaptive Role of Functional Cortical Reorganization

The results of several studies suggest that an increased cortical recruitment might not always be beneficial for patients with MS. As already mentioned, disease progression and accrual of disability has been observed in patients with SPMS, despite the widespread activations of regions in the frontal and parietal lobes during the performance of simple motor tasks [24, 28]. fMRI studies of the

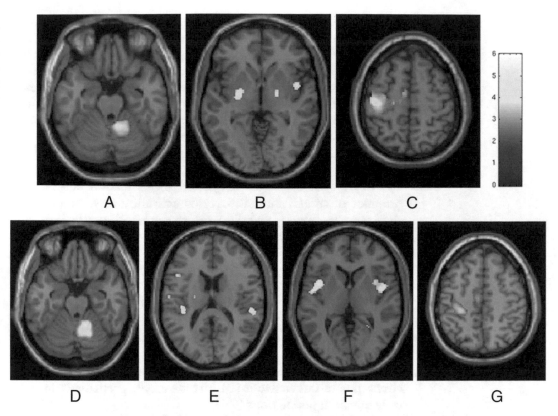

Fig. 7 Comparison of simple vs. complex task with the dominant right hands in healthy subjects (*top row;* **a–c**) and patients with MS (*bottom row;* **d–g**) (paired *t* test for each group, corrected *p* value <0.05). The ipsilateral anterior lobe of the cerebellum (**a** and **d**), bilateral insula/basal-ganglia (**b**, **e**, **f**) and contralateral primary sensorimotor cortex and supplementary motor area (**c** and **g**) were identified in both groups. Compared with healthy subjects, MS patients also had a significant activation of the contralateral inferior frontal gyrus and bilateral secondary sensorimotor cortex (**e**). Note that the activations are color-coded according to their *t* values. Images are in neurological convention (*see* ref. [41])

motor system [21, 33, 44] of patients with PPMS suggested a lack of "classical" adaptive mechanisms as a potential additional factor contributing to the accumulation of disability. In these patients during the performance of different motor tasks with the nonimpaired dominant limbs, a recruitment of a widespread movement-associated cortical network usually considered to function in motor, sensory, and multimodal integration processing (i.e., the frontal and temporal lobes, and the insula) was detected [21, 33, 44]. The absence of a concomitant recruitment of the "classical" motor areas, including the primary SMC, the SMA, the infraparietal sulcus, and the SII, was interpreted as a failure of part of the adaptive capacity of the cerebral cortex in this severely disabling phenotype of the disease [21, 33]. The notion that multimodal integration areas might have a critical role in PPMS patients has been strengthened by another study which showed increased activations of these regions in PPMS patients, in comparison with healthy controls, during passive movements [44]. Similar findings have led to similar conclusions in patients with cognitive decline [59, 64], in whom a "poor" pattern of cortical activations in the expected areas [59] and the activation of regions that are not normally devoted to the performance of the investigated task [64] were detected during the performance of the administered tasks.

The comparison of the movement-associated brain patterns of cortical activations between RRMS patients complaining of reversible fatigue after weekly interferon (IFN) beta-1a administration and those without fatigue suggested an association between the presence of fatigue and an increased activation of several areas of the motor network, including the thalamus, the cingulum, and several regions located in the frontal lobes, including the SMA and the primary SMC bilaterally [37] (Fig. 8). These results suggest that the overrecruitment of brain networks in MS might, at least to some degree, have a detrimental effect.

The assessment and interpretation of fMRI results (cerebral activity) during cognitive tasks can be difficult when task performance differs across patients (e.g., poorer performance in cognitively impaired patients). As such, the analysis of resting state (RS) brain functional connectivity has been proposed as a valid alternative to task-active fMRI investigations, particularly in clinically impaired populations [79].

The analysis of brain activity at rest has shown an increased synchronization of the majority of the resting-state networks in CIS patients [80]. In patients with cognitive impairment a reduced functional connectivity of anterior regions of the brain, mostly located in the frontal lobes [81, 82] is related not only to the severity of cognitive impairment, but also to structural disruption of connecting WM tracts [81]. In another study [83] better cognitive performance in MS patients was associated with an increased functional connectivity among several regions of the attention network, thus supporting the adaptive role of RS functional connectivity modifications. However, findings from other studies [84–86] seem to contradict the adaptive/compensation hypothesis since a correlation between increased functional connectivity and worse cognitive performance was found.

A recent study assessing intra- and inter-network functional connectivity at rest in the brain from RRMS patients has demonstrated a distributed pattern of abnormal functional connectivity, which was correlated to the extent of T2 lesions and the severity of disability [87]. If such functional abnormalities confer a systematic vulnerability to disease progression or, conversely, protect against the onset of clinical deficits needs to be investigated. The "cognitive reserve hypothesis" has also been considered to explain the incomplete relationship between brain disease and cognitive status in MS; it has been demonstrated that MS patients with a greater

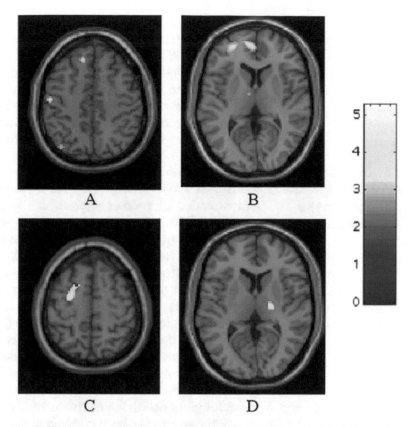

Fig. 8 Relative cortical activations of MS patients with reversible fatigue after interferon (IFN) beta-1a injection during the performance of a simple motor task with their clinically unimpaired and fully normal functioning, dominant right hands. At entry (**a** and **b**) (when they did not complain of fatigue), compared with MS patients without reversible fatigue, these patients showed an increased recruitment of the contralateral primary SMC (**a**), the thalamus (**b**), the superior frontal sulcus (**a** and **b**), and the cingulate motor area (**b**). At day 1 (**c** and **d**) (after IFN beta-1a administration, when fatigue was present), compared with MS patients without fatigue, these patients showed increased recruitment of the ipsilateral thalamus (**d**), and contralateral middle frontal gyrus (**c**). Note that the color-coded activations have been superimposed on a high-resolution T1-weighted scan obtained from a single, healthy subject and normalized into standard statistical parametric mapping space (neurological convention) (from ref. [37])

intellectual enrichment require less deactivation of the brain at rest and less recruitment of prefrontal cortices to perform tasks in a way equal to that of patients with lesser enrichment [88].

Despite the clear advantages of RS fMRI, it is important to note that there might be some mechanisms related to cognitive network function which are unlikely to be captured properly by the use of a non task-active/resting paradigm. In particular, it has been suggested that, compared to healthy controls, MS patients might have a limited ability to increase the activation/deactivation within cognitive-related networks with increasing task difficulty. Such an inability to optimize cognitive network recruitment, which reflects an impaired cognitive functional reserve (that is the ability to match brain activity to increasing cognitive demand) [65, 89–91], is likely to be a maladaptive mechanism contributing to the clinical manifestations of the disease, since it is more pronounced in patients with SPMS [89] as well as in those with cognitive impairment [92].

2.9 Use of fMRI to Assess Longitudinal Changes of Cortical Reorganization

Dynamic functional changes have been described in an MS patient following an acute relapse [10]. These results have been confirmed and extended by another study which assessed the early cortical changes following acute motor relapses secondary to pseudotumoral lesions in 12 MS patients and the evolution over time of cortical reorganization in a subgroup of these patients [38]. Short-term cortical changes were mainly characterized by the recruitment of pathways in the unaffected hemisphere. A recovery of function of the primary SMC of the affected hemisphere was found in patients with clinical improvement, while in patients without clinical recovery, there was a persistent recruitment of the primary SMC of the unaffected hemisphere, suggesting that the restoration of function of motor areas of the affected hemisphere might be a critical factor for a favorable recovery (Fig. 9).

A longitudinal (time interval of 15–26 months) fMRI study of the motor system has been conducted in a group of patients with early RRMS [93]. Patients exhibited greater bilateral activations than controls in both fMRI studies. Although no significant differences between the two fMRI scans were observed in controls, a reduction of the functional activity of the ipsilateral SMC and the contralateral cerebellum was seen in patients at follow-up. Moreover, activation changes in ipsilateral motor areas correlated inversely with age, extent, and progression of T1 lesion load, and occurrence of a new relapse, suggesting that younger patients with less structural brain damage and a favorable clinical course demonstrate brain plasticity that follows a more lateralized pattern of brain activations [93].

Longitudinal modifications of cognitive networks' recruitment and their impact on patients' cognitive status have been marginally explored. A 1-year longitudinal study in patients with early MS found an association between increased levels of activation in the right dorsolateral prefrontal cortex during a cognitive task and improved working memory and processing speed performance [94]. A 20 month longitudinal study [95] showed that worsening of SDMT performance in RRMS patients is correlated with increased activity of the left inferior parietal lobule over time, probably reflecting a maladaptive mechanism.

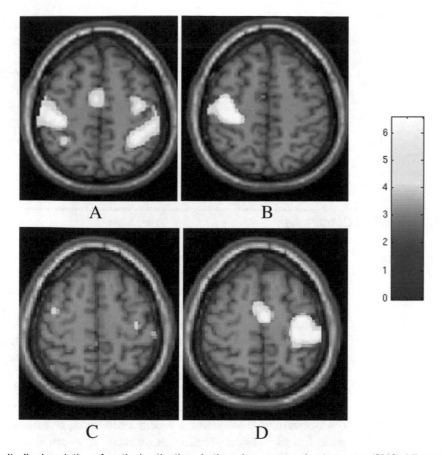

Fig. 9 Longitudinal evolution of cortical activations in the primary sensorimotor cortex (SMC), bilaterally, during task performance with impaired hand compared with unimpaired hand in one patient with good clinical recovery during follow-up (**a** and **b**) and in one patient with poor/absent clinical recovery (**c** and **d**). Scans obtained during left hand motor task have been flipped to keep the left hemisphere contralateral to movement. At baseline, both patients showed an increased activation of the primary SMC of the unaffected (ipsilateral) hemisphere (**a** and **c**). During follow-up, the patient with good clinical recovery showed an increased recruitment of the primary SMC of the affected hemisphere (**b**), while the patient with poor/absent clinical recovery continued to show an increased recruitment of the primary SMC of the unaffected hemisphere (**d**). Note that the activations are color-coded according to their *t* values (*see* ref. [38])

2.10   fMRI to Monitor Treatment

The potential of fMRI in a multicentre setting has been explored by a few studies of the motor [96, 97] and cognitive [92] networks.

Only a few fMRI studies have been performed to monitor the effect of treatments in MS [58, 98, 99]. Different patterns of brain response to lower dose acute and higher dose chronic administration of rivastigmine have been demonstrated in MS patients using cognitive tasks [58, 99]. Rivastigmine was shown to enhance the prefrontal function and alter the functional connectivity associated with cognition. In MS patients, increased activation in the ipsilateral primary SMC and SMA has been observed after a single dose of 3,4-diaminopyridine (a potassium channel blocker), suggesting that this treatment may modulate brain motor activity in patients with MS, probably by enhancing excitatory synaptic transmission [98].

Active and RS fMRI have also been applied to assess modifications of the patterns of activations and functional connectivity following motor [100] and cognitive rehabilitation in a few single-center studies [101–103]. A recent study demonstrated that changes in RS functional connectivity of cognitive-related networks contributes to the persistence of the effects of cognitive rehabilitation after 6 months in RRMS patients [104].

3   fMRI in MS-Allied Conditions

Among the known MS-allied conditions, only patients with neuromyelitis optica (NMO) have been assessed using fMRI. Rocca et al. [35] investigated the performance of a simple motor task with the dominant and nondominant upper limbs in ten patients with NMO and found that, compared with matched controls, NMO patients had an increased recruitment of several regions of the sensorimotor network (primary SMC, postcentral gyrus, MFG, rolandic operculum, SII, precuneus, and cerebellum) and of several other regions mainly in the temporal and occipital lobes, such as MT/V5, the fusiform gyrus, the cuneus, and the parahippocampal gyrus during the performance of both tasks. For both tasks, strong correlations were found between relative activations of cortical sensorimotor areas and the severity of cervical cord damage, suggesting that the observed functional cortical changes might have an adaptive role in limiting the clinical outcome of NMO structural pathology.

Using RS fMRI, abnormal functional connectivity in the majority of RS networks has also been detected in NMO patients and has been correlated to the severity of clinical disability [105].

4   fMRI in Other WMD and Conditions Associated with "Significant" White Matter Damage

4.1   Isolated Spinal Cord Injury

Studies of patients with spinal cord injury of different etiology (i.e., traumatic and/or demyelinating) with no or only partial clinical recovery have shown movement-associated cortical changes, consisting of an abnormal location of the activated areas and in a more widespread recruitment of motor areas, mainly located in the hemisphere contralateral to the limb used to perform the task [106–108].

In patients with isolated myelitis of probable demyelinating origin and normal function in the investigated limbs, an abnormal pattern of movement-associated cortical activation has been described [34, 109, 110] and has been related to the degree of daily hand use [109], the severity of cervical cord damage [34, 110], and the level of spinal cord involvement [110].

4.2 CNS Vasculitides   In patients with neuropsychiatric systemic lupus erythematosus (NPSLE) without overt motor impairment, movement-associated functional cortical changes, characterized by more significant activations of the contralateral primary SMC, putamen, dentate nucleus, several regions located in the frontal and parietal lobes, MT/V5, and the middle occipital gyrus, bilaterally, have been observed when compared with matched healthy controls [111] (Fig. 10). The correlations found in these patients between relative activations of sensorimotor areas and the extent and severity of brain damage suggest that also in these patients functional cortical changes might contribute to the maintenance of their normal functional capacities [111] (Fig. 10).

4.3 Migraine   Several fMRI studies have contributed to the classification of migraine as a neurovascular or even a brain disorder. The pioneering study by Hadjikhani and coworkers [112] has shown that during induced and spontaneous visual aura a focal increase of BOLD signal (possibly reflecting vasodilation) developed within occipital extrastriate cortex (area V3A) (Fig. 11).

This BOLD change progressed contiguously and slowly over the cortex, congruent with the retinotopy of the visual percept and, then, following the same retinotopic progression, it diminished (possibly reflecting vasoconstriction). This spreading signal

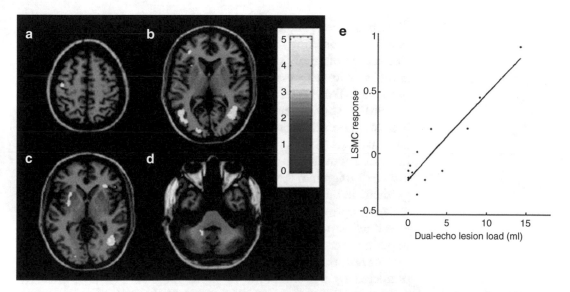

Fig. 10 Relative cortical activations in patients with neuropsychiatric systemic lupus erythematosus during a simple motor task with the right hand in comparison with healthy volunteers (color-coded *t* values). (**a**) Contralateral primary sensorimotor cortex. (**b**) Contralateral putamen, contralateral middle frontal gyrus (MFG), bilateral MT/V5 complex, contralateral middle occipital gyrus (MOG). (**c**) Contralateral putamen, ipsilateral inferior frontal gyrus, bilateral MT/V5 complex, ipsilateral MOG. (**d**) Contralateral dentate nucleus. The relative activation of the contralateral primary sensorimotor cortex was significantly correlated with brain dual-echo lesion load (**e**) ($p < 0.001$, $r = 0.79$) (from ref. [111])

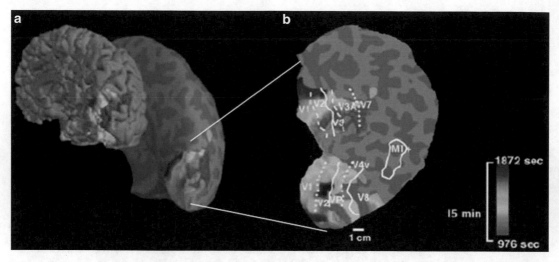

Fig. 11 Source localization and time of onset of the blood oxygenation level-dependent (BOLD) signal changes during induced and spontaneous visual aura attack in patients with migraine: (**a**) data are represented on inflated cortical surface shown from a posterior-medial view; (**b**) a fully flattened view of the cortical surface of the involved region. Cortical regions showing the first BOLD perturbations are coded in *red* (according to the color-coded scale representing variation in time) and locations showing the BOLD perturbations at progressively later times are coded by *green* and *blue* (according to the color-coded scale representing variation in time). The aura-related changes appeared first in extrastriate cortex (area V3A) and then progressed contiguously and slowly over the cortex following the same retinotopic progression of visual disturbance. From [112] with permission

disturbance had striking similarity with cortical spreading depression (CSD) phenomenon, thus supporting the theory that CSD was the electrophysiological correlate of visual aura. Notably, a spreading neural activity suppression has also been described in migraine without aura (MWoA) patients during triggered migraine attacks [113].

FMRI studies have also confirmed dysfunctional activity of brainstem nuclei involved in pain modulation during both the ictal and interictal phases [114]. Stankewitz et al. [115] confirmed an increased activation of the dorsal pons in migraine patients during induced migraine attacks. This study also revealed a selective gradient-like activity in the spinal trigeminal nucleus: after trigeminal nociceptive stimulation, interictal migraine patients exhibited lower activations of this nucleus; however, shortly before the migraine attack, patients had an increased activation at this level. Of interest, the time interval to the next headache attack could be predicted by the amplitude of signal intensities in the spinal trigeminal nuclei, suggesting that this oscillating behavior may represent a key phenomenon in migraine pathogenesis.

Studies which have explored cerebral activity with the pain network in migraine using experimental pain stimulation have shown abnormalities of a widespread subcortical and cortical brain network involved in pain processing in these subjects. However, one of the main challenges in the interpretation of these results is to differentiate findings consistent with a general pain response from those that might be specific to migraine [116]. Using a contact thermode as a noxious stimulation paradigm, migraine patients were found to exhibit a greater activation in the anterior cingulate cortex at 51 °C and less activation in the bilateral somatosensory cortex at 53 °C [117], thus supporting the presence of an increased antinociceptive activity in these patients, which could represent a compensatory functional reorganization aimed at modulating pain perception to the intensity of healthy controls. Other fMRI studies have demonstrated alterations in pain modulatory/inhibitory circuits, which may be related to the lack of habituation after repetitive painful stimulation and increased cortical excitability to painful stimuli, that may lead to the development of allodynia [118]. The thalamus is now considered to play a pivolat role in the manifestation of allodynia. Burstein et al. [119] showed that brush and heat stimulation at the skin of the dorsum of the hand produced larger BOLD responses in the posterior thalamus of patients undergoing a migraine attack with extracephalic allodynia than the corresponding responses registered when the same patients were free of migraine and allodynia.

An increased activation of cortical regions mediating the affective dimension of pain has also been demonstrated in migraineurs. During spontaneous and induced migraine, these patients had increased BOLD signal intensities in limbic structures (e.g., the amygdala and insula) and exhibited a stronger recruitment of affective cortical areas when exposed to emotional inputs [118]. Based on these data, a model of migraine as a dysfunction of a "neurolimbic" pain network has been proposed [120].

Abnormalities of function of pain processing regions have also been investigated in patients with chronic migraine, particularly those with medication-overuse headache (MOH). Before the withdrawal of the offending medications, these patients had reduced pain related activity in areas of the lateral pain pathway, including the primary and secondary sensorimotor area. Such abnormalities regressed after treatment withdrawal [121]. In addition, patients with MOH presented dysfunctional activity of the meso-cortico-limbic dopamine circuit, including the ventro-medial prefrontal cortex and the substantia nigra/ventral tegmental area complex, during the execution of a decision-making under risk paradigm. The ventro-medial prefrontal cortex dysfunctions were reversible and attributable to the headache, whereas the substantia nigra/ventral tegmental area complex dysfunctions were persistent despite treatment withdrawal, suggesting that MOH may share some neurophysiological features with addiction [118].

It is well established that migraine patients show also hyperresponsiveness of the primary visual cortex and a lack of habituation to visual stimuli [114]. These phenomena are more pronounced in patients with migraine with aura (MWA) [122]. Hyper-responsiveness of the visual cortex in migraine extends beyond primary visual areas, even in the interictal period. Antal et al. [123] demonstrated significantly stronger activation of the extrastriate, motion-responsive MT area, representing the medial-superior temporal area, in migraine patients vs. healthy controls in response to coherent/incoherent moving dot stimuli. This cortical hyperexcitability may represent the biological basis for the clinical observation of heightened vulnerability to motion sickness that migraine sufferers often report [114].

Numerous studies provided a conceptual framework for understanding vestibular migraine as a variant of MWA produced by the convergence of vestibular information within migraine circuits. Several fMRI studies showed that vestibular stimulation activate cerebral regions that are generally involved in migraine and pain perception, such as the posterior and anterior insula, orbitofrontal cortex, and the posterior and anterior cingulate gyri, thus suggesting that central constituents of the migraine circuit might include components of central vestibular pathways [124]. However, so far, fMRI has not been applied yet to investigate functional cortical abnormalities in patients with vestibular migraine.

RS fMRI studies have shown that functional connectivity is generally increased in pain-processing networks in migraineurs, whereas it is decreased in pain modulatory circuits [125]. In particular, migraineurs with a history of allodynia exhibit significantly reduced RS functional connectivity between periacqueductal gray matter (PAG), prefrontal regions, and anterior cingulate cortex compared with migraineurs without allodynia [126]. This RS functional connectivity abnormalities have been related to the frequency of migraine attacks and disease duration [125]. Significant abnormalities of RS functional connectivity occur also in affective networks [125], the DMN [127] and the executive network [116].

5 Conclusions

Taken all together, fMRI studies of patients with various white matter disorders demonstrate the potential of this technique to provide important insights into the mechanisms of cortical reorganization following white matter injury. As a consequence, fMRI holds promise to improve our understanding of the factors associated with the accumulation of irreversible disability in MS and other white matter conditions. Although the role of cortical reorganization in limiting the functional impact of white matter structural damage is still not proved definitively, the available data support the concept that cortical adaptive responses may have an important role in compensating for irreversible tissue damage, such as axonal loss. Thereby, it can be concluded that the presently available fMRI data suggest that the rate of accumulation of disability in MS and other white matter disorders might be a function not only of tissue loss, but also of the progressive failure of the adaptive capacity of the brain with increasing tissue damage.

References

1. Filippi M, Rocca MA (2004) Magnetization transfer magnetic resonance imaging in the assessment of neurological diseases. J Neuroimaging 14(4):303–313

2. Filippi M, Rocca MA, Comi G (2003) The use of quantitative magnetic-resonance-based techniques to monitor the evolution of multiple sclerosis. Lancet Neurol 2(6):337–346

3. Hesselink JR (2006) Differential diagnostic approach to MR imaging of white matter diseases. Top Magn Reson Imaging 17:243–263

4. Rocca MA, Filippi M (2006) Functional MRI to study brain plasticity in clinical neurology. Neurol Sci 27(Suppl 1):S24–S26

5. Rocca MA, Filippi M (2007) Functional MRI in multiple sclerosis. J Neuroimaging 17(Suppl 1):36S–41S

6. Waxman SG (1998) Demyelinating diseases—new pathological insights, new therapeutic targets. N Engl J Med 338(5):323–325

7. Clanet M, Berry I, Boulanouar K (1997) Functional imaging in multiple sclerosis. Int MS J 4:26–32

8. Rombouts SA et al (1998) Visual activation patterns in patients with optic neuritis: an fMRI pilot study. Neurology 50(6):1896–1899

9. Lee M et al (2000) The motor cortex shows adaptive functional changes to brain injury from multiple sclerosis. Ann Neurol 47(5):606–613

10. Reddy H et al (2000) Relating axonal injury to functional recovery in MS. Neurology 54(1):236–239

11. Reddy H et al (2000) Evidence for adaptive functional changes in the cerebral cortex with axonal injury from multiple sclerosis. Brain 123(Pt 11):2314–2320

12. Werring DJ et al (2000) Recovery from optic neuritis is associated with a change in the distribution of cerebral response to visual stimulation: a functional magnetic resonance imaging study. J Neurol Neurosurg Psychiatry 68(4):441–449

13. Langkilde AR et al (2002) Functional MRI of the visual cortex and visual testing in patients with previous optic neuritis. Eur J Neurol 9(3):277–286

14. Toosy AT et al (2002) Functional magnetic resonance imaging of the cortical response to photic stimulation in humans following optic neuritis recovery. Neurosci Lett 330(3):255–259

15. Russ MO et al (2002) Functional magnetic resonance imaging in acute unilateral optic neuritis. J Neuroimaging 12(4):339–350

16. Toosy AT et al (2005) Adaptive cortical plasticity in higher visual areas after acute optic neuritis. Ann Neurol 57(5):622–633

17. Levin N et al (2006) Normal and abnormal fMRI activation patterns in the visual cortex after recovery from optic neuritis. Neuroimage 33(4):1161–1168

18. Korsholm K et al (2007) Recovery from optic neuritis: an ROI-based analysis of LGN and visual cortical areas. Brain 130(Pt 5):1244–1253

19. Jenkins T et al (2010) Dissecting structure-function interactions in acute optic neuritis to investigate neuroplasticity. Hum Brain Mapp 31(2):276–286

20. Jenkins TM et al (2010) Neuroplasticity predicts outcome of optic neuritis independent of tissue damage. Ann Neurol 67(1):99–113

21. Filippi M et al (2002) Correlations between structural CNS damage and functional MRI changes in primary progressive MS. Neuroimage 15(3):537–546

22. Rocca MA et al (2002) Adaptive functional changes in the cerebral cortex of patients with nondisabling multiple sclerosis correlate with the extent of brain structural damage. Ann Neurol 51(3):330–339

23. Rocca MA et al (2003) Evidence for axonal pathology and adaptive cortical reorganization in patients at presentation with clinically isolated syndromes suggestive of multiple sclerosis. Neuroimage 18(4):847–855

24. Rocca MA et al (2003) A functional magnetic resonance imaging study of patients with secondary progressive multiple sclerosis. Neuroimage 19(4):1770–1777

25. Rocca MA et al (2003) Functional cortical changes in patients with multiple sclerosis and nonspecific findings on conventional magnetic resonance imaging scans of the brain. Neuroimage 19(3):826–836

26. Reddy H et al (2002) Functional brain reorganization for hand movement in patients with multiple sclerosis: defining distinct effects of injury and disability. Brain 125(Pt 12):2646–2657

27. Filippi M et al (2004) Simple and complex movement-associated functional MRI changes in patients at presentation with clinically isolated syndromes suggestive of multiple sclerosis. Hum Brain Mapp 21(2):108–117

28. Rocca MA et al (2005) Cortical adaptation in patients with MS: a cross-sectional functional MRI study of disease phenotypes. Lancet Neurol 4(10):618–626

29. Pantano P et al (2002) Cortical motor reorganization after a single clinical attack of multiple sclerosis. Brain 125(Pt 7):1607–1615

30. Pantano P et al (2002) Contribution of corticospinal tract damage to cortical motor reorganization after a single clinical attack of multiple sclerosis. Neuroimage 17(4):1837–1843

31. Rocca MA et al (2004) Pyramidal tract lesions and movement-associated cortical recruitment in patients with MS. Neuroimage 23(1):141–147

32. Lowe MJ et al (2002) Multiple sclerosis: low-frequency temporal blood oxygen level-dependent fluctuations indicate reduced functional connectivity initial results. Radiology 224(1):184–192

33. Rocca MA et al (2002) Evidence for widespread movement-associated functional MRI changes in patients with PPMS. Neurology 58(6):866–872

34. Rocca MA et al (2003) Cord damage elicits brain functional reorganization after a single episode of myelitis. Neurology 61(8):1078–1085

35. Rocca MA et al (2004) A functional MRI study of movement-associated cortical changes in patients with Devic's neuromyelitis optica. Neuroimage 21(3):1061–1068

36. Rocca MA et al (2005) A widespread pattern of cortical activations in patients at presentation with clinically isolated symptoms is associated with evolution to definite multiple sclerosis. AJNR Am J Neuroradiol 26(5):1136–1139

37. Rocca MA et al (2007) fMRI changes in relapsing-remitting multiple sclerosis patients complaining of fatigue after IFNbeta-1a injection. Hum Brain Mapp 28(5):373–382

38. Mezzapesa DM et al (2008) Functional cortical changes of the sensorimotor network are associated with clinical recovery in multiple sclerosis. Hum Brain Mapp 29(5):562–573

39. Rocca MA et al (2008) The "mirror-neuron system" in MS: a 3 tesla fMRI study. Neurology 70(4):255–262

40. Rocca MA et al (2007) Altered functional and structural connectivities in patients with MS: a 3-T study. Neurology 69(23):2136–2145

41. Filippi M et al (2004) A functional MRI study of cortical activations associated with object manipulation in patients with MS. Neuroimage 21(3):1147–1154

42. Cerasa A et al (2006) Adaptive cortical changes and the functional correlates of visuomotor integration in relapsing-remitting multiple sclerosis. Brain Res Bull 69(6):597–605

43. Calautti C, Baron J-C (2003) Functional neuroimaging studies of motor recovery after stroke in adults: a review. Stroke 34(6):1553–1566

44. Ciccarelli O et al (2006) Functional response to active and passive ankle movements with clinical correlations in patients with primary progressive multiple sclerosis. J Neurol 253(7):882–891

45. Rocca MA et al (2010) Preserved brain adaptive properties in patients with benign multiple sclerosis. Neurology 74(2):142–149

46. Petsas N et al (2013) Evidence of impaired brain activity balance after passive sensorimotor stimulation in multiple sclerosis. PLoS One 8(6):e65315

47. Agosta F et al (2008) Tactile-associated recruitment of the cervical cord is altered in patients with multiple sclerosis. Neuroimage 39(4):1542–1548

48. Agosta F et al (2008) Evidence for enhanced functional activity of cervical cord in relapsing multiple sclerosis. Magn Reson Med 59(5):1035–1042

49. Valsasina P et al (2010) Cervical cord functional MRI changes in relapse-onset MS patients. J Neurol Neurosurg Psychiatry 81(4):405–408

50. Valsasina P et al (2012) Cervical cord fMRI abnormalities differ between the progressive forms of multiple sclerosis. Hum Brain Mapp 33(9):2072–2080

51. Rocca MA et al (2012) Abnormal cervical cord function contributes to fatigue in multiple sclerosis. Mult Scler 18(11):1552–1559

52. Staffen W et al (2002) Cognitive function and fMRI in patients with multiple sclerosis: evidence for compensatory cortical activation during an attention task. Brain 125(Pt 6):1275–1282

53. Au Duong MV et al (2005) Altered functional connectivity related to white matter changes inside the working memory network at the very early stage of MS. J Cereb Blood Flow Metab 25(10):1245–1253

54. Au Duong MV et al (2005) Modulation of effective connectivity inside the working memory network in patients at the earliest stage of multiple sclerosis. Neuroimage 24(2):533–538

55. Audoin B et al (2003) Compensatory cortical activation observed by fMRI during a cognitive task at the earliest stage of MS. Hum Brain Mapp 20(2):51–58

56. Audoin B et al (2005) Magnetic resonance study of the influence of tissue damage and cortical reorganization on PASAT performance at the earliest stage of multiple sclerosis. Hum Brain Mapp 24(3):216–228

57. Hillary FG et al (2003) An investigation of working memory rehearsal in multiple sclerosis using fMRI. J Clin Exp Neuropsychol 25(7):965–978

58. Parry AM et al (2003) Potentially adaptive functional changes in cognitive processing for patients with multiple sclerosis and their acute modulation by rivastigmine. Brain 126(Pt 12):2750–2760

59. Penner IK et al (2003) Analysis of impairment related functional architecture in MS patients during performance of different attention tasks. J Neurol 250(4):461–472

60. Mainero C et al (2004) fMRI evidence of brain reorganization during attention and memory tasks in multiple sclerosis. Neuroimage 21(3):858–867

61. Sweet LH et al (2004) Functional magnetic resonance imaging of working memory among multiple sclerosis patients. J Neuroimaging 14(2):150–157

62. Sweet LH et al (2006) Functional magnetic resonance imaging response to increased verbal working memory demands among patients with multiple sclerosis. Hum Brain Mapp 27(1):28–36

63. Wishart HA et al (2004) Brain activation patterns associated with working memory in relapsing-remitting MS. Neurology 62(2):234–238

64. Chiaravalloti N et al (2005) Cerebral activation patterns during working memory performance in multiple sclerosis using FMRI. J Clin Exp Neuropsychol 27(1):33–54

65. Cader S et al (2006) Reduced brain functional reserve and altered functional connectivity in patients with multiple sclerosis. Brain 129(Pt 2):527–537

66. Li Y et al (2004) Differential cerebellar activation on functional magnetic resonance imaging during working memory performance in persons with multiple sclerosis. Arch Phys Med Rehabil 85(4):635–639

67. Rocca MA et al (2009) Structural and functional MRI correlates of Stroop control in benign MS. Hum Brain Mapp 30(1):276–290

68. Lazeron RHC et al (2004) An fMRI study of planning-related brain activity in patients with moderately advanced multiple sclerosis. Mult Scler 10(5):549–555

69. Comi G et al (2001) Clinical and MRI assessment of brain damage in MS. Neurol Sci 22(Suppl 2):123–127

70. Forn C et al (2012) Functional magnetic resonance imaging correlates of cognitive performance in patients with a clinically isolated syndrome suggestive of multiple sclerosis at presentation: an activation and connectivity study. Mult Scler 18(2):153–163

71. Cerasa A et al (2010) The effects of BDNF Val66Met polymorphism on brain function in controls and patients with multiple sclerosis: an imaging genetic study. Behav Brain Res 207(2):377–386

72. Bobholz JA et al (2006) fMRI study of episodic memory in relapsing-remitting MS: correlation with T2 lesion volume. Neurology 67(9):1640–1645

73. Jehna M et al (2011) Cognitively preserved MS patients demonstrate functional differences in processing neutral and emotional faces. Brain Imaging Behav 5(4):241–251

74. Rocca MA et al (2012) Differential cerebellar functional interactions during an interference task across multiple sclerosis phenotypes. Radiology 265(3):864–873

75. Lenzi D et al (2007) Effect of corpus callosum damage on ipsilateral motor activation in patients with multiple sclerosis: a functional and anatomical study. Hum Brain Mapp 28(7):636–644

76. Manson SC et al (2006) Loss of interhemispheric inhibition in patients with multiple sclerosis is related to corpus callosum atrophy. Exp Brain Res 174(4):728–733

77. Ceccarelli A et al (2010) Structural and functional magnetic resonance imaging correlates of motor network dysfunction in primary progressive multiple sclerosis. Eur J Neurosci 31(7):1273–1280

78. Filippi M et al (2002) Functional magnetic resonance imaging correlates of fatigue in multiple sclerosis. Neuroimage 15(3):559–567

79. Raichle ME, Snyder AZ (2007) A default mode of brain function: a brief history of an evolving idea. Neuroimage 37(4):1083–1090, discussion 1097–1099

80. Roosendaal SD et al (2010) Resting state networks change in clinically isolated syndrome. Brain 133(Pt 6):1612–1621

81. Rocca MA et al (2010) Default-mode network dysfunction and cognitive impairment in progressive MS. Neurology 74(16):1252–1259

82. Bonavita S et al (2011) Distributed changes in default-mode resting-state connectivity in multiple sclerosis. Mult Scler 17(4):411–422

83. Loitfelder M et al (2012) Abnormalities of resting state functional connectivity are related to sustained attention deficits in MS. PLoS One 7(8):e42862

84. Hawellek DJ et al (2011) Increased functional connectivity indicates the severity of cognitive impairment in multiple sclerosis. Proc Natl Acad Sci U S A 108(47):19066–19071

85. Faivre A et al (2012) Assessing brain connectivity at rest is clinically relevant in early multiple sclerosis. Mult Scler 18(9):1251–1258

86. Schoonheim MM et al (2015) Thalamus structure and function determine severity of cognitive impairment in multiple sclerosis. Neurology 84(8):776–783

87. Rocca M et al (2012) Large-scale neuronal network dysfunction in relapsing-remitting multiple sclerosis. Neurology 79(14):1449–1457

88. Sumowski JF et al (2010) Intellectual enrichment is linked to cerebral efficiency in multiple sclerosis: functional magnetic resonance imaging evidence for cognitive reserve. Brain 133(Pt 2):362–374

89. Loitfelder M et al (2011) Reorganization in cognitive networks with progression of multiple sclerosis: insights from fMRI. Neurology 76(6):526–533

90. Tortorella C et al (2013) Load-dependent dysfunction of the putamen during attentional processing in patients with clinically isolated syndrome suggestive of multiple sclerosis. Mult Scler 19(9):1153–1160

91. Amann M et al (2011) Altered functional adaptation to attention and working memory tasks with increasing complexity in relapsing-remitting multiple sclerosis patients. Hum Brain Mapp 32(10):1704–1719

92. Rocca MA et al (2014) Functional correlates of cognitive dysfunction in multiple sclerosis: a multicenter fMRI Study. Hum Brain Mapp 35(12):5799–5814

93. Pantano P et al (2005) A longitudinal fMRI study on motor activity in patients with multiple sclerosis. Brain 128(Pt 9):2146–2153

94. Audoin B et al (2008) Efficiency of cognitive control recruitment in the very early stage of multiple sclerosis: a one-year fMRI follow-up study. Mult Scler 14(6):786–792

95. Loitfelder M et al (2014) Brain activity changes in cognitive networks in relapsing-remitting multiple sclerosis—insights from a longitudinal FMRI study. PLoS One 9(4):e93715

96. Wegner C et al (2008) Relating functional changes during hand movement to clinical parameters in patients with multiple sclerosis in a multi-centre fMRI study. Eur J Neurol 15(2):113–122

97. Rocca MA et al (2009) Abnormal connectivity of the sensorimotor network in patients with MS: a multicenter fMRI study. Hum Brain Mapp 30(8):2412–2425

98. Mainero C et al (2004) Enhanced brain motor activity in patients with MS after a single dose of 3,4-diaminopyridine. Neurology 62(11):2044–2050

99. Cader S, Palace J, Matthews PM (2009) Cholinergic agonism alters cognitive processing and enhances brain functional connectivity in patients with multiple sclerosis. J Psychopharmacol 23(6):686–696

100. Tomassini V et al (2012) Relating brain damage to brain plasticity in patients with multiple sclerosis. Neurorehabil Neural Repair 26(6):581–593

101. Filippi M et al (2012) Effects of cognitive rehabilitation on structural and functional MRI measures in multiple sclerosis: an explorative study. Radiology 262(3):932–940

102. Sastre-Garriga J et al (2011) A functional magnetic resonance proof of concept pilot trial of cognitive rehabilitation in multiple sclerosis. Mult Scler 17(4):457–467

103. Cerasa A et al (2013) Computer-assisted cognitive rehabilitation of attention deficits for multiple sclerosis: a randomized trial with fMRI correlates. Neurorehabil Neural Repair 27(4):284–295

104. Parisi L et al (2014) Changes of brain resting state functional connectivity predict the persistence of cognitive rehabilitation effects in patients with multiple sclerosis. Mult Scler 20(6):686–694

105. Liu Y et al (2011) Abnormal baseline brain activity in patients with neuromyelitis optica: a resting-state fMRI study. Eur J Radiol 80(2):407–411

106. Mikulis DJ et al (2002) Adaptation in the motor cortex following cervical spinal cord injury. Neurology 58(5):794–801

107. Curt A et al (2002) Changes of non-affected upper limb cortical representation in paraplegic patients as assessed by fMRI. Brain 125(Pt 11):2567–2578

108. Sabbah P et al (2002) Sensorimotor cortical activity in patients with complete spinal cord injury: a functional magnetic resonance imaging study. J Neurotrauma 19(1):53–60

109. Cramer SC et al (2001) Changes in motor cortex activation after recovery from spinal cord inflammation. Mult Scler 7(6):364–370

110. Rocca MA et al (2006) The level of spinal cord involvement influences the pattern of

movement-associated cortical recruitment in patients with isolated myelitis. Neuroimage 30(3):879–884

111. Rocca MA et al (2006) An fMRI study of the motor system in patients with neuro-psychiatric systemic lupus erythematosus. Neuroimage 30(2):478–484

112. Hadjikhani N et al (2001) Mechanisms of migraine aura revealed by functional MRI in human visual cortex. Proc Natl Acad Sci U S A 98(8):4687–4692

113. Cao Y et al (1999) Functional MRI-BOLD of visually triggered headache in patients with migraine. Arch Neurol 56(5):548–554

114. Lakhan SE, Avramut M, Tepper SJ (2013) Structural and functional neuroimaging in migraine: insights from 3 decades of research. Headache 53(1):46–66

115. Stankewitz A et al (2011) Trigeminal nociceptive transmission in migraineurs predicts migraine attacks. J Neurosci 31(6):1937–1943

116. Tedeschi G et al (2013) The role of BOLD-fMRI in elucidating migraine pathophysiology. Neurol Sci 34(Suppl 1):S47–S50

117. Russo A et al (2012) Pain processing in patients with migraine: an event-related fMRI study during trigeminal nociceptive stimulation. J Neurol 259(9):1903–1912

118. Sprenger T, Borsook D (2012) Migraine changes the brain: neuroimaging makes its mark. Curr Opin Neurol 25(3):252–262

119. Burstein R et al (2010) Thalamic sensitization transforms localized pain into widespread allodynia. Ann Neurol 68(1):81–91

120. Maizels M, Aurora S, Heinricher M (2012) Beyond neurovascular: migraine as a dysfunctional neurolimbic pain network. Headache 52(10):1553–1565

121. Chiapparini L et al (2010) Neuroimaging in chronic migraine. Neurol Sci 31(Suppl 1):S19–S22

122. Bhaskar S et al (2013) Recent progress in migraine pathophysiology: role of cortical spreading depression and magnetic resonance imaging. Eur J Neurosci 38(11):3540–3551

123. Antal A et al (2011) Differential activation of the middle-temporal complex to visual stimulation in migraineurs. Cephalalgia 31(3):338–345

124. Furman JM, Marcus DA, Balaban CD (2013) Vestibular migraine: clinical aspects and pathophysiology. Lancet Neurol 12(7):706–715

125. Sprenger T, Magon S (2013) Can functional magnetic resonance imaging at rest shed light on the pathophysiology of migraine? Headache 53(5):723–725

126. Mainero C, Boshyan J, Hadjikhani N (2011) Altered functional magnetic resonance imaging resting-state connectivity in periaqueductal gray networks in migraine. Ann Neurol 70(5):838–845

127. Tessitore A et al (2013) Disrupted default mode network connectivity in migraine without aura. J Headache Pain 14(1):89

Chapter 21

fMRI in Cerebrovascular Disorders

Nick S. Ward

Abstract

Stroke is a major cause of long-term disability worldwide. One of the key factors underpinning recovery of function is reorganization of surviving neural networks. Noninvasive techniques such as fMRI allow this reorganization to be studied in humans. However, the design of experiments involving patients with impairment requires careful consideration and is often constrained. Difficulty with some tasks can lead to a number of performance confounds, and so tasks and task parameters that avoid or minimize this should be selected. Furthermore, when studying patients with cerebrovascular disease, it is important to consider the possibility that the blood oxygen level-dependent signal may be altered and affect interpretation of results. Despite these potential problems, careful experimental design can provide real insights into system-level reorganization after stroke and how it is related to functional recovery. Currently, results suggest that functionally relevant reorganization does occur in cerebral networks in human stroke patients. For example, it is apparent that initial attempts to move a paretic limb following stroke are associated with widespread activity within the distributed motor system in both cerebral hemispheres. This reliance on nonprimary motor output pathways is unlikely to support full recovery, but improved efficiency of the surviving networks is associated with behavioral gains. This reorganization can only occur in structurally and functionally intact brain regions. Understanding the dynamic process of system-level reorganization will allow greater understanding of the mechanisms of recovery and potentially improve our ability to deliver effective restorative therapy.

Key words fMRI, Stroke, Blood oxygen level-dependent, Motor cortex, Premotor cortex, Plasticity, Rehabilitation

1 Introduction

Studying patients who have suffered from stroke with functional brain imaging is difficult for a variety of reasons. The motivation behind such studies is a desire to understand and subsequently improve the process of functional recovery. Stroke and other forms of neurological damage account for nearly half of all severely disabled adults [1–3]. Longitudinal studies of recovery suggest that only 50% of stroke survivors with significant initial upper limb paresis recover useful function of the limb [4]. Furthermore, those with poor recovery of arm function have dramatically impaired quality of life and sense of well-being [5, 6]. It is clear that effective treatment of motor impairment after stroke is critically important to many people.

Massimo Filippi (ed.), *fMRI Techniques and Protocols*, Neuromethods, vol. 119,
DOI 10.1007/978-1-4939-5611-1_21, © Springer Science+Business Media New York 2016

The mainstay of treatment is neurorehabilitation. The overall approach is effective and the benefit of strategies aimed at helping patients *adapt* to impairment well proven [7]. Treatments aimed at *reducing* impairment, remain poorly developed. There is an over-riding assumption that one way to tackle impairment in those patients with focal brain damage is to attempt to promote functionally relevant reorganization within surviving neural networks [8]. Over the last decade, advances in our understanding of how the normal brain is organized at the molecular, cellular, and systems level have improved enormously. Advances in our understanding of the mechanisms of impairment after brain injury, including stroke, are way behind. Translating findings from proof-of-principle studies into real treatments remains problematic and requires urgent attention [9]. Thus, the clinical neurosciences have the potential to make a unique contribution toward developing rehabilitation strategies designed to reduce impairment.

The tools available for studying the working human brain are different to those used in animal models. In human subjects, experiments are performed at the level of neural systems rather than single cells or molecules. This chapter will concentrate on the way that fMRI can be used to contribute. The first half will consider the specific difficulties involved in performing fMRI experiments in stroke patients, in particular how fMRI signals might be affected in cerebrovascular disease and how studying patients with impairment should influence experimental design. In the second half, examples of studies using fMRI in stroke patients will be discussed to illustrate advances in our understanding of post-stroke functional brain reorganization. Many studies have been performed in the somatosensory [10–12] and language systems [13–15], but studies of the motor system are particularly numerous and will be used to illustrate how fMRI may be used.

2   Blood Oxygen Level-Dependent Signal in Cerebrovascular Disease

fMRI relies on the blood oxygen level-dependent (BOLD) signal. In brief, the BOLD signal relies on the close coupling between blood flow and metabolism. During an increase in neuronal activation there is an increase in local cerebral blood flow, but only a small proportion of the greater amount of oxygen delivered locally to the tissue is used. This results in a net increase in the tissue concentration of oxyhemoglobin and a net reduction in paramagnetic deoxyhemoglobin in the local capillary bed. The magnetic properties of hemoglobin depend on its level of oxygenation so that this change results in an increase in local tissue-derived signal intensity on T2*-weighted MR images [16].

The mechanism of neurovascular signaling to the blood vessels controlling cerebral blood flow is still unclear, although it may involve metabolic [17, 18] or neurochemical [19, 20] mechanisms. In addition, the generation of BOLD signal is still reliant on venous blood volume, blood flow, blood oxygenation, and oxygen consumption. Thus, it is possible that any disease state that changes these parameters will potentially modify the BOLD signal. It is therefore legitimate to be concerned whether the BOLD signal is reliable in patients who have suffered stroke and in subjects with evidence of both large and small vessel atherosclerosis. The potential problem arises because in general, the BOLD signal is assumed to have the same shape in all subjects and in all brain regions. In one case, the canonical hemodynamic response function (HRF) has been derived by principle component analysis of empirical data with a peak magnitude occurring 6 s after the neuronal activity [21].

There is evidence to suggest that the shape of the hemodynamic response might be altered after stroke. Newton et al. [22] demonstrated a greater time to peak BOLD response in ipsilateral compared with contralateral primary motor cortex (M1) in controls. In three chronic stroke patients, the time to peak BOLD response was increased in ipsilesional (contralateral) M1 compared with controls. Interestingly, in these patients the time to peak BOLD response in contralesional M1 was equivalent or less than that for ipsilesional M1, representing a finding opposite to that seen in healthy controls. Pineiro et al. [23] have also described a slower time to peak BOLD response in sensorimotor cortex bilaterally in 12 chronic stroke patients with lacunar infarcts. Thus, modeling the BOLD response with a canonical HRF might be less efficient in stroke patients. It is worth considering what the effect of this would be in the context of a standard functional imaging analysis using the general linear model approach. If the canonical HRF was a poor fit for the actual response, then the residual error of the analytical model would be greater (than if the fit was good), thus *lowering* t- and Z-scores and depressing sensitivity to detection of differences. In fact, in general, most studies of stroke patients have found increased activation in a number of brain regions over and above healthy controls, so it might be the case that these overactivations have been *underestimated*. However, modeling differences in measured HRF from the canonical is likely to be beneficial. The use of temporal and dispersion derivatives of the canonical HRF to specifically capture differences in the timing or duration of the peak response, for example, is likely to increase sensitivity.

In addition to changes in the shape of the HRF, there is evidence that in patients with impaired cerebrovascular reserve or advanced narrowing of the cerebral arteries, the BOLD fMRI signal may be reduced, or even become negative [24–26]. Röther et al. [27] describe a single patient, which illustrates the point. The patient was found to have bilateral occluded internal carotid arteries and an occluded vertebral artery. The cerebrovascular reactivity, as determined by reduced change in T2* signal during hypercapnia, was severely impaired in the left hemisphere. The finding of importance was that during a motor task with the right hand, the BOLD response in the left motor cortex was negative for the duration of the task. This suggests that the initial dip in BOLD signal due to a relative decrease in oxyhemoglobin was not followed by the normal vascular response (which would have resulted in an increase in oxygenated and decrease in deoxygenated hemoglobin). This subject had previously suffered from a transient ischemic attack involving the right arm. It is likely that these symptoms were related to hemodynamic insufficiency, and it is interesting to speculate that the presence of a prolonged negative dip in BOLD signal represents a marker for those at risk from such symptoms. Others have made the point that this impaired cerebrovascular reactivity might be due to either large or small vessel disease [26]. Further investigation will reveal whether this idea has genuine potential as a clinical tool.

Several studies have now suggested that impairment of normal vasodilatation in response to hypercapnia is associated with diminished magnitude of BOLD signal [25, 26, 28, 29]. Thus, in patients with severely impaired cerebrovascular reactivity neuronal activation may not translate into a BOLD response in the conventional sense, and standard models using the canonical HRF may not be sufficient.

However, the scale of the problem is not yet clear. For example, the cerebrovascular reactivity in the right hemisphere of the patient studied by Röther et al. [27] was moderately impaired and the BOLD response during a motor task with the left hand was entirely normal. Patients with hemodynamic symptoms are rare, and are likely to be excluded from fMRI studies. In addition, patients with severe stenosis of ipsilesional internal carotid arteries are usually also excluded, although it is not clear that this is necessary. It may also be the case that small vessel disease may also make a significant contribution to impaired cerebrovascular reactivity.

Thus, although there is evidence that impaired cerebrovascular reactivity can diminish the BOLD response, there is no evidence that the BOLD signal is erroneously detected in these patients, i.e., this is largely a problem of false negative results. In general, the literature concerning differences between stroke patients and healthy controls is dominated by the finding of overactivity in patients compared with controls, and once again, it is possible that the issue of cerebrovascular disease has led to an underestimation of changes in cortical organization after stroke. It is clear that when examining for differences between a group of patients and a group of healthy controls, the nonneural factors that can influence the BOLD response will contribute significantly to whether a difference is found or not. However, several studies have begun to use a correlation approach. That is to say, to attempt to explain variability in the task-related BOLD signal with some other parameters, such as a measure of recovery [30] or corticospinal tract integrity [31, 32]. It is unlikely that changes in cerebrovascular responsiveness will correlate with recovery or an anatomical measure of corticospinal tract integrity, and so it is unlikely to account for any significant (and biologically plausible) results. As we have already discussed, however, the ability to detect a real finding is likely to be diminished by alterations in nonneural contributors to the BOLD signal. A multimodal approach using different imaging techniques (BOLD, perfusion, hypercapnic challenge) and concurrent neurophysiological methods (electroencephalography [EEG], magnetoencephalography [MEG], transcranial magnetic stimulation [TMS]) may be useful when addressing the influence of multiple physiological variables. These issues will require further empirical study.

The discussion regarding differences in hemodynamic coupling is also of relevance when considering the effects of age, given that stroke is commoner with advancing age, and that often age-matched controls are used in studies of stroke patients. D'Esposito et al. [33] examined the effects of age on the BOLD signal generated during a button press task in response to a visual cue, using a sparse event-related design. Using a standard fixed effects analysis, task-related activation was detected in M1 above the chosen threshold in only 75% of the older subjects but 100% of the younger subjects. Furthermore, for those in whom activation was detected, four times the number of suprathreshold voxels was present in the younger compared with the older subjects. Thus, on the face of it, M1 appears to be less active during a button press task in older subjects. However, the most important finding was in relation to the signal-to-noise ratio (SNR), which was reduced in elderly subjects. Results from single subject or group fixed effects analyses of functional imaging data are generally presented as t statistics for each voxel (volume element) of the brain. The result is therefore dependent on both the magnitude of the signal change and the residual variance after this has been accounted for. Thus, an increased SNR will lead to a lower t statistic, and therefore fewer suprathreshold voxels. In fact, D'Esposito et al. [33] found no difference in the magnitude of task-related signal change in M1, supporting the notion that the diminished number of suprathreshold voxels was largely attributable to the decreased SNR in the older subjects.

The problem of reduced SNR can be effectively dealt with by employing the statistical technique of random effects analysis as opposed to fixed effects analysis. Random effects analysis of functional imaging data treats each subject as a random variable. The experimental variance is dominated by between subjects variability (as opposed to within subject variability in the case of fixed effects models). The data for each subject comprise the voxel-wise parameter estimate for the task under consideration, which reflects the magnitude of the signal change in each voxel. Appropriate statistics can be performed on these data, which are less likely to be influenced by differences in SNR [33]. Using a random effects analysis, and employing both temporal and dispersion derivatives of the HRF, Ward et al. [34] demonstrated no change in the shape of the hemodynamic response during a hand grip task with advancing age, in keeping with the findings described earlier.

3   Issues in Experimental Design

The results from any functional imaging study are only as reliable as the care with which the experiment is constructed and executed, but studies involving patients who have had a stroke raise some specific issues. The selection of patients, choice of experimental paradigm, within scanner monitoring of performance, and the approaches to data analysis all require careful consideration.

3.1   Subject Selection

In general, stroke patients are a heterogeneous group differing in several important ways, not least the site and size of infarct, patency of the vascular system, age, comorbidities, and concurrent medication. The criteria for patient selection will to an extent depend on the experimental question. It is unlikely that averaging the results from a wide variety of patient types will prove useful because of this variability, but there are two other ways of approaching the experimental design. First, it may be desirable to use a group of patients highly selected on the basis of lesion location, for example. Results from this type of controlled study are powerful, but do not generalize outside the subgroup selected. Alternatively, it might be more useful to study a group of patients who vary in a specific factor of interest (e.g., outcome), to explore the relationship between this factor and task-related brain activity. Results from studies using this approach can be generalized more easily.

3.2   Performance Confounds

The choice of experimental task is critical and is dependent on the experimental question. For example, a study of the relationship between brain activation and outcome after stroke will by necessity involve patients with different performance abilities. Similarly, a longitudinal study will require that patients are studied at different stages of recovery. For an active motor or language task this can result in the problem of performance confounds, because the ability to perform the task is not the same across patients or sessions. A change in experimental task performance can have significant effects upon the pattern of brain activation. In other words, comparison across patients or time points is made difficult if the patients are performing the task differently. Thus, each patient must perform the *same task* during the fMRI experiment, so that a meaningful comparison can be made across subjects or scanning sessions. Maintaining a consistent task is therefore of great importance, but in stroke recovery studies equality of task may be interpreted in a number of ways. In particular, a task may be consistent across patients with different abilities in terms of *absolute* or *relative* parameters.

This can be illustrated by considering a simple motor experiment. The absolute task parameters can be fixed (by setting the same target force and rate for each subject or session), but performing a task may be experienced as more or less effortful depending on the level of recovery. Consequently, any differences in results between subjects or sessions could be attributed to differences in "effort" exerted. Alternatively, the relative task parameters (i.e., the level of task difficulty) can be fixed across subjects/sessions. In this scenario, patients will perform the task at different absolute forces and rates, and so differences in results across subjects/sessions could be attributed to differences in the absolute task parameters. When using an "active task," these factors must always be considered and results interpreted with these confounds in mind. Increased effort is a potentially useful strategy for overcoming motor, language, or cognitive impairment in a real world setting. As described earlier, some patients may use less effortful as their performance improves. Is the focus of interest the reorganization that might be the substrate for recovery, or is it the strategy that each patient uses to perform a task to a certain level, given the constraints of their impairment? Both may be of interest, but the choice of experimental design has an impact on which process is being studied. The problem of performance confounds is avoided with passive tasks (e.g., passive limb movements and passive listening), but these are complementary approaches to active tasks, not substitutes for them.

3.3 Task Frequency

The rate of task performance is also something that has implications for both data analysis and interpretation of results. Consider once again a simple motor task such as finger tapping or hand squeezing. The rate of performance of a repetitive task will influence how effortful the task is, in the same way as the target force. Most experiments are conducted in a "block design"; that is to say a period of activity (usually for 16–30 s) followed by a period of rest. If subjects are asked to perform at the same rate (even if the target force is scaled according to each subject's own performance abilities), for example, finger tapping at 1 Hz for 20 s, then differences across subjects/sessions could be due to differences in perceived effort, just as with equal absolute target forces. Some investigators have varied the rate at which subjects are asked to perform a task to try to control for effort exerted. However, comparing blocks with different numbers of "events" within them is problematic because the BOLD signal summates depending on how many events there are. The BOLD response needs approximately 10 s and longer to return to baseline, but "events" are usually more frequent than this (e.g., 1 Hz finger tapping). It is usually assumed that there is a summation of the overlapping BOLD responses, which is largely (but not entirely) linear [35, 36]. The basis function (boxcar design) in the general linear model will have the same "height," and so more frequent events will result in a larger parameter estimate, for the same amount of event-related activity. In fact, it is the quantity of events, not their frequency, which will modulate BOLD signal in motor cortices [37]. Thus, if a subject performs the task at 1 Hz and then at 0.5 Hz in both cases for 20 s, and each period is modeled with the same boxcar basis function, then the parameter estimate (or magnitude of activity) will appear to be roughly twice as much during the more frequent task. This reflects the quantity of "events," but not a change in the way the brain is organized. Thus, if patients perform a task more slowly than healthy controls to control for effort, then the brain activity associated with that movement will be underestimated in comparison to those subjects who perform the task at a faster rate. One way around this problem is to use an event-related design in which the intertrial interval is long enough for the task to be performed repetitively without increasing the sense of effort. This design may be less efficient in terms of fMRI design, but avoids the confounds described earlier [31, 32].

3.4 Task Complexity

Investigators are often tempted to use more complex tasks when studying patients with impairment in the hope that this will maximize differences between patients and control subjects. This is sometimes done in the hope of exploring a more *ecologically valid* task, i.e., one which is relevant to function in the real world. However, it is never possible to study the neural correlates of a task that a subject cannot themselves perform. By introducing more complexity into the task, patients with significant impairment are more likely to adopt new operational strategies toward these experimental tasks in an attempt to adapt to their impairment. These differences in strategy could therefore account for differences between subject groups. Although of clinical interest, differences in strategy across a group represent a potential experimental confound if they are unexpected and not measured. One approach is therefore to use a simple task that minimizes difference in strategic approach to the task so that valid comparisons can then be made across subjects/sessions.

3.5 Task Monitoring

Once a paradigm has been selected it is important that task performance is monitored during the experiment. Intersubject variability may be greater after stroke and new sources of variability can arise, such as mirror or associated movements. To take account of this, some investigators record behavior during a prescan rehearsal, whilst others incorporate the increasingly available instrumentation that is compatible with the MRI setting. Prescan rehearsal provides some idea of whether a task can be performed correctly, or whether mirror movements are present, for example. However, in-scanner recordings allow this information to be incorporated into image analysis as a covariate, and thus improving statistical power by accounting for correlated variance in the measured scan signal.

The experimental approach is therefore dictated by the experimental question. Not all investigators will have the same question, but the issues discussed earlier need to be considered in all cases. For most questions, this approach is entirely appropriate and standardization of experimental paradigms, patient selection, and method of analysis across experiments is not required. In the case of experimental questions that require a multicenter approach that is technically feasible, standardization of such factors would be required.

4 Reorganization in the Motor System After Stroke

4.1 Residual Functional Architecture After Stroke: The Story So Far?

Early studies of motor system organization after stroke compared brain activation during movement in well-recovered patients and normal controls. Early group studies of stroke patients with subcortical lesions described greater activation within a number of motor-related cortical regions compared with controls during a finger tapping task [38–42]. It was suggested that nonprimary (or secondary) cortical motor regions were thus responsible for recovery of motor function in these patients. Strick [43] had proposed this as a potential mechanism of restoration of function, some years before based on an understanding of the organization of the cortical motor system in primates. Normal distal motor function is facilitated largely through the corticospinal pathway, from the cortical motor system to the spinal cord motor neurons. The majority of corticospinal fibers originate in the M1, but there are contributions from other cortical regions [44]. In primates, the M1, arcuate (or lateral) premotor cortex (PM), and supplementary motor area (SMA) are each part of parallel, independent motor networks with (1) separate projections to spinal cord motoneuron and (2) interactions at the level of the cortex [43]. There is some similarity between the corticospinal projections from the hand regions of M1, PM, and SMA. Thus, it seemed feasible that a number of motor networks acting in parallel could generate an output to the spinal cord necessary for movement, and that damage in one of these networks could be at least partially compensated for by activity in another [45, 46]. Subsequently, many studies have demonstrated that the performance of a simple motor task with the affected limb is associated with greater bilateral brain activation in a number of cortical motor-related areas compared with healthy volunteers, including dorsal PM (PMd) and ventral PM (PMv), SMA, and cingulate motor areas (CMA) [23, 38–42, 47–52].

A critical question is whether these differences are related to recovery. As discussed previously in this chapter, this question requires that the group of patients examined have a wide variety of outcomes, or else longitudinal studies should be performed. In the first such cross-sectional study, a group of chronic stroke patients with infarcts sparing M1 were scanned during a hand grip with visual feedback task using fMRI [30]. The target forces used were always a proportion of each subject's own maximum grip force, so

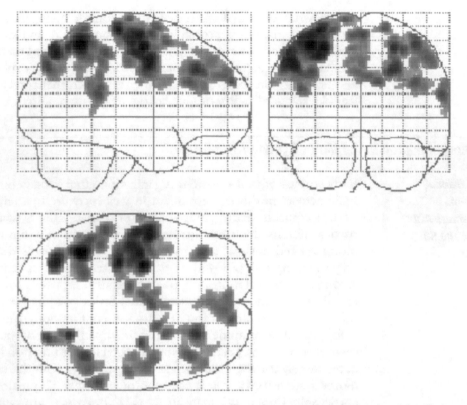

Fig. 1 Brain regions in which there is a negative correlation between corticospinal system integrity (as assessed with transcranial magnetic stimulation) and task-related signal change during hand grip with the affected hand. Increasing task-related activity is seen in a number of secondary motor areas including premotor regions and supplementary motor area as damage to the corticospinal system increases. The affected hand was on the left side. Results are displayed on a "glass brain" shown from the right side (*top left image*), from behind (*top right image*), and from above (*bottom left image*). Voxels are significant at $P < 0.001$ (uncorrected), and clusters are significant at $P < 0.05$ (corrected) (Reproduced from ref. [31], Oxford University Press.)

that any differences were unlikely to be due to differences in effort. The more affected patients had greater task-related activity in secondary motor regions in both affected and unaffected hemispheres, whereas patients with the best motor scores had activation patterns that were indistinguishable from healthy age-matched volunteers. A similar result was observed in a group of patients studied at approximately 10 days post-stroke [53]. It was hypothesized that, secondary motor areas are recruited in response to damage to corticospinal output. A subsequent study demonstrated a strong positive correlation between secondary motor area recruitment in both hemispheres and corticospinal system damage (Fig. 1) [31]. A more "normal" corticospinal system was associated with greater task-related activity in contralesional M1 (hand area), suggesting a progressive shift away from primary to secondary motor areas with increasing disruption to corticospinal system. This has also been described as an increase in either bilateral or contralesional activity, but the exact pattern is likely to depend on the anatomy of the damage. Furthermore, several secondary motor areas have bilateral projections to motor output systems [54, 55]. Thus, after stroke, the brain will use what is available (i.e., what is intact and connected so that motor output can be influenced) in an attempt to generate motor output to spinal cord motoneurons.

These results do not immediately support the idea that secondary motor areas are the substrate for motor recovery. Labeling corticospinal neurons with retrograde tracers has revealed multiple nonprimary corticospinal output zones in both the lateral and the medial areas of the frontal lobe (SMA, CMA, PMd, and PMv) [56–58]. These output zones contain large numbers of corticospinal neurons that project to the intermediate zone and ventral horn of the spinal cord suggesting their potential for direct control of spinal motoneurons in a way paralleling corticospinal output from M1 [45]. In primates, however, projections from secondary motor areas to spinal cord motor neurons are less numerous and less efficient at exciting spinal cord motoneurons than those from M1 [59, 60]. Moreover, unlike M1, facilitation of distal muscles from SMA, PMd, and PMv is not significantly stronger than facilitation of proximal muscles. These brain regions probably exert their influence via indirect descending motor pathways such as the reticulospinal or rubrospinal pathways [61]. These pathways often have bihemispheric origins, hence bihemispheric task-related activity is more common after stroke. Furthermore, they are also more likely to supply the upper limb flexors, hence flexor synergistic patterns of movement in patients who are reliant on these indirect rather than direct pathways. Alternatively, or possibly in addition, cortico-cortical interactions, presumably with surviving M1 output, may also play an important role in supporting recovered motor function.

What is the evidence for the idea that the secondary motor areas of both hemispheres are contributing to recovered function? There are two ways to investigate the functional relevance of secondary motor region recruitment. One is to measure how task-related activity co-varies with modulation of task parameters. In healthy humans, for example, increasing force production is associated with linear increases in BOLD signal in contralateral M1 and medial motor regions, implying that they have a functional role in force production [62–64]. A recent study examined specifically for regional changes in the control of force modulation after stroke [32]. In patients with greater corticospinal system damage, force-related signal changes were seen mainly in contralesional dorsolateral PM, bilateral ventrolateral premotor cortices, and contralesional cerebellum, but not ipsilesional M1 (Fig. 2).

Thus, not only do premotor cortices become increasingly active as corticospinal system integrity diminishes [31], but they can take on a new "M1-like" role during modulation of force output, which implies a new and functionally relevant role in motor control.

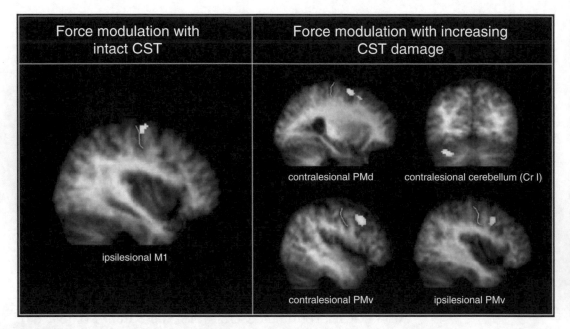

Fig. 2 Brain regions in which the blood oxygen level-dependent (BOLD) signal varies linearly with force exerted during hand grip change as a function of corticospinal tract (CST) integrity (as assessed with transcranial magnetic stimulation). The affected hand was on the left side. In the *left panel*, increasing force leads to greater modulation of BOLD signal in ipsilesional M1 in patients with less damage to CST. In the *right panel*, increasing force leads to greater modulation of BOLD signal in contralesional dorsal premotor cortex, contralesional cerebellum, contralesional, and ipsilesional ventral premotor cortex. This demonstrates that brain regions involved in force modulation shift away from primary motor cortex to premotor regions with increasing CST damage. Results are overlaid onto the average T1-weighted structural scan obtained from all stroke patients in the study (Adapted from ref. [32], Blackwell Publishing.)

Second, experiments in which premotor activity is transiently disrupted with TMS can lead to worsening of recovered motor behaviors in patients with no effect on the performance of control subjects [51, 65, 66], again implying new and functionally relevant roles. Furthermore, TMS to contralesional PMd is more disruptive in patients with greater impairment [51], whereas TMS to ipsilesional PMd is more disruptive in less impaired patients [65] in keeping with a general shift toward functionally relevant activity in the contralesional hemisphere of patients with greater damage to motor output pathways.

These results are important because they tell us that the response to focal injury does not involve simple substitution of one cortical region for another. It is clear that nodes within remaining motor networks can take on new functional roles. The contribution of contralesional M1 to recovery is surprisingly unresolved. An early view was that the contralesional M1 might be viewed much like an extra secondary motor area, contributing what descending signals it could. An alternative view suggested that contralesional M1 impairs motor recovery through excessive inhibitory drive to ipsilesional M1 [8]. The more mundane truth is likely to be that the role played by contralesional M1, secondary motor areas and indeed ipsilesional M1 is likely to depend on a number of factors, most importantly the profile of anatomical damage and the actual task being performed.

4.2 Connectivity Based Analysis of Post-stroke Motor Networks

Subsequent studies have turned to examining how connectivity between nodes in the motor network is affected by stroke [67]. Techniques such as Dynamic Causal Modelling have examined task-related networks, whereas other approaches have looked at "large-scale" whole brain networks (using for example, graph-theory). In general, both approaches have followed the same theme in looking for differences between patients and healthy controls as well as testing for correlations between brain imaging measures, particularly interhemispheric connectivity, and clinical scores. Most findings at rest point to lower connectivity between motor cortices in patients with more impairment [68] and greater corticospinal tract damage [69]. During movement of the affected hand the influence of contralesional to ipsilesional M1 is more inhibitory, but once again, only in more impaired patients [70]. Even the interhemispheric influence of ipsilesional dorsolateral premotor cortex is different depending on the level of motor impairment (itself dependent on residual structural anatomy) [71].

In summary, in the chronic stroke brain, there is a reconfiguration of the cerebral motor system. Task-related brain activation varies across chronic stroke patients in a way that appears to be predictable. It is important to stress that this reorganization is often not successful in returning motor function to normal. It is less effective than that in the intact brain but will nevertheless attempt to generate some form of motor signal to spinal cord motoneurons in the most efficient way. The exact configuration of this new motor system will be determined most obviously by the extent of the anatomic damage. This includes the extent to which the damage affects cortical motor regions, white matter pathways, and even which hemisphere is affected [72]. The more of the normal functional architecture that survives, the greater will be the potential for full recovery. In patients with damage to primary sensorimotor cortex, for example, tests of fractionated finger movement correlated more strongly with the proportion of surviving "normal" sensorimotor cortex (as defined by functional activation maps in normal controls) than with total infarct volume [73].

This anatomic explanation accounts for why some patients do better than others, but it does not account for the recovery of function that occurs over weeks and months in individual patients. How does the reorganized state evolve? Longitudinal fMRI studies have shed further light on the process [47, 50, 74–76], although only a handful have studied patients on more than two occasions. One study scanned subcortical stroke patients on average eight times over 6 months after stroke [76] and demonstrated an early overactivation in primary and many nonprimary motor regions. Thereafter, functional recovery was associated with a focusing of task-related brain activation patterns toward a "normal" lateralized pattern. In general, longitudinal studies have demonstrated a focusing of activity toward the lesioned hemisphere motor regions that is associated with improvement in motor function [47, 74], with some patients showing persistent recruitment [50].

5 Conclusions

In summary, the brain activation pattern of an individual patient represents the state of reorganization within that system at the time of study. This pattern is highly influenced by a number of methodological factors as previously discussed. However, in appropriately controlled experiments, these activation patterns tell us something about how that brain is functionally organized. Functional improvement with treatment is likely to be associated with changes within this network. The potential for functionally relevant change to occur will depend on a number of other factors, not least the biologic age of the subject and the premorbid state of their brain, but also current drug treatments. Furthermore, levels of neurotransmitters and growth factors that are able to influence the ability of the brain to respond to afferent input (i.e., how plastic it is) might be determined by their genetic status [77]. All of these factors will influence the potential for activity driven change within the intact motor networks, the putative mechanism of therapy driven improvements in motor performance.

References

1. Hoffman C, Rice D, Sung HY (1996) Persons with chronic conditions. Their prevalence and costs. JAMA 276(18):1473–1479

2. Office of Population Censuses and Surveys (1988) OPCS surveys of disability in Great Britain. I. The prevalence of disability among adults. HMSO, London

3. Wade DT, Hewer RL (1987) Epidemiology of some neurological diseases with special reference to work load on the NHS. Int Rehabil Med 8(3):129–137

4. Wade DT (1989) Measuring arm impairment and disability after stroke. Int Disabil Stud 11(2):89–92

5. Nichols-Larsen DS, Clark PC, Zeringue A, Greenspan A, Blanton S (2005) Factors influencing stroke survivors' quality of life during subacute recovery. Stroke 36(7):1480–1484

6. Wyller TB, Sveen U, Sodring KM, Pettersen AM, Bautz-Holter E (1997) Subjective well-being one year after stroke. Clin Rehabil 11(2):139–145

7. Stroke Unit Trialists' Collaboration (2000) Organised inpatient (stroke unit) care for stroke (Cochrane Review). The Cochrane Library, Issue 2. Oxford: Update Software

8. Ward NS, Cohen LG (2004) Mechanisms underlying recovery of motor function after stroke. Arch Neurol 61(12):1844–1848

9. The Academy of Medical Sciences (2004) Restoring neurological function: putting the neurosciences to work in neurorehabilitation. Academy of Medical Sciences, London

10. Loubinoux I, Carel C, Pariente J, Dechaumont S, Albucher JF, Marque P et al (2003) Correlation between cerebral reorganization and motor recovery after subcortical infarcts. Neuroimage 20(4):2166–2180

11. Tombari D, Loubinoux I, Pariente J, Gerdelat A, Albucher JF, Tardy J et al (2004) A longitudinal

fMRI study: in recovering and then in clinically stable sub-cortical stroke patients. Neuroimage 23(3):827–839

12. Ward NS, Brown MM, Thompson AJ, Frackowiak RS (2006) Longitudinal changes in cerebral response to proprioceptive input in individual patients after stroke: an FMRI study. Neurorehabil Neural Repair 20(3):398–405

13. Lee A, Kannan V, Hillis AE (2006) The contribution of neuroimaging to the study of language and aphasia. Neuropsychol Rev 16(4):171–183

14. Price CJ, Crinion J (2005) The latest on functional imaging studies of aphasic stroke. Curr Opin Neurol 18(4):429–434

15. Wise RJ (2003) Language systems in normal and aphasic human subjects: functional imaging studies and inferences from animal studies. Br Med Bull 65:95–119

16. Buxton RB (2002) An introduction to functional magnetic resonance imaging: principles and techniques. Cambridge University Press, Cambridge

17. Magistretti PJ, Pellerin L (1999) Cellular mechanisms of brain energy metabolism and their relevance to functional brain imaging. Philos Trans R Soc Lond B Biol Sci 354(1387):1155–1163

18. Magistretti PJ, Pellerin L, Rothman DL, Shulman RG (1999) Energy on demand. Science 283(5401):496–497

19. Iadecola C (2004) Neurovascular regulation in the normal brain and in Alzheimer's disease. Nat Rev Neurosci 5(5):347–360

20. Attwell D, Iadecola C (2002) The neural basis of functional brain imaging signals. Trends Neurosci 25(12):621–625

21. Friston KJ, Josephs O, Rees G, Turner R (1998) Nonlinear event-related responses in fMRI. Magn Reson Med 39(1):41–52

22. Newton J, Sunderland A, Butterworth SE, Peters AM, Peck KK, Gowland PA (2002) A pilot study of event-related functional magnetic resonance imaging of monitored wrist movements in patients with partial recovery. Stroke 33(12):2881–2887

23. Pineiro R, Pendlebury S, Johansen-Berg H, Matthews PM (2001) Functional MRI detects posterior shifts in primary sensorimotor cortex activation after stroke: evidence of local adaptive reorganization? Stroke 32(5):1134–1139

24. Carusone LM, Srinivasan J, Gitelman DR, Mesulam MM, Parrish TB (2002) Hemodynamic response changes in cerebrovascular disease: implications for functional MR imaging. AJNR Am J Neuroradiol 23(7):1222–1228

25. Hamzei F, Knab R, Weiller C, Rother J (2003) The influence of extra- and intracranial artery disease on the BOLD signal in fMRI. Neuroimage 20(2):1393–1399

26. Rossini PM, Altamura C, Ferretti A, Vernieri F, Zappasodi F, Caulo M et al (2004) Does cerebrovascular disease affect the coupling between neuronal activity and local haemodynamics? Brain 127(Pt 1):99–110

27. Rother J, Knab R, Hamzei F, Fiehler J, Reichenbach JR, Buchel C et al (2002) Negative dip in BOLD fMRI is caused by blood flow-oxygen consumption uncoupling in humans. Neuroimage 15(1):98–102

28. Krainik A, Hund-Georgiadis M, Zysset S, von Cramon DY (2005) Regional impairment of cerebrovascular reactivity and BOLD signal in adults after stroke. Stroke 36(6):1146–1152

29. Murata Y, Sakatani K, Hoshino T, Fujiwara N, Kano T, Nakamura S et al (2006) Effects of cerebral ischemia on evoked cerebral blood oxygenation responses and BOLD contrast functional MRI in stroke patients. Stroke 37(10):2514–2520

30. Ward NS, Brown MM, Thompson AJ, Frackowiak RS (2003) Neural correlates of outcome after stroke: a cross-sectional fMRI study. Brain 126(Pt 6):1430–1448

31. Ward NS, Newton JM, Swayne OB, Lee L, Thompson AJ, Greenwood RJ et al (2006) Motor system activation after subcortical stroke depends on corticospinal system integrity. Brain 129(Pt 3):809–819

32. Ward NS, Newton JM, Swayne OB, Lee L, Frackowiak RS, Thompson AJ et al (2007) The relationship between brain activity and peak grip force is modulated by corticospinal system integrity after subcortical stroke. Eur J Neurosci 25(6):1865–1873

33. D'Esposito M, Zarahn E, Aguirre GK, Rypma B (1999) The effect of normal aging on the coupling of neural activity to the bold hemodynamic response. Neuroimage 10(1):6–14

34. Ward NS, Swayne OB, Newton JM (2008) Age-dependent changes in the neural correlates of force modulation: an fMRI study. Neurobiol Aging 29(9):1434–1446

35. Pollmann S, Dove A, Yves von Cramon D, Wiggins CJ (2000) Event-related fMRI: comparison of conditions with varying BOLD overlap. Hum Brain Mapp 9(1):26–37

36. Wager TD, Vazquez A, Hernandez L, Noll DC (2005) Accounting for nonlinear BOLD effects in fMRI: parameter estimates and a model for prediction in rapid event-related studies. Neuroimage 25(1):206–218

37. Kim JA, Eliassen JC, Sanes JN (2005) Movement quantity and frequency coding in human motor areas. J Neurophysiol 94(4):2504–2511

38. Cao Y, D'Olhaberriague L, Vikingstad EM, Levine SR, Welch KM (1998) Pilot study of functional MRI to assess cerebral activation of motor function after poststroke hemiparesis. Stroke 29(1):112–122

39. Chollet F, DiPiero V, Wise RJ, Brooks DJ, Dolan RJ, Frackowiak RS (1991) The functional anatomy of motor recovery after stroke in humans: a study with positron emission tomography. Ann Neurol 29(1):63–71

40. Cramer SC, Nelles G, Benson RR, Kaplan JD, Parker RA, Kwong KK et al (1997) A functional MRI study of subjects recovered from hemiparetic stroke. Stroke 28(12):2518–2527

41. Weiller C, Chollet F, Friston KJ, Wise RJ, Frackowiak RS (1992) Functional reorganization of the brain in recovery from striatocapsular infarction in man. Ann Neurol 31(5):463–472

42. Weiller C, Ramsay SC, Wise RJ, Friston KJ, Frackowiak RS (1993) Individual patterns of functional reorganization in the human cerebral cortex after capsular infarction. Ann Neurol 33(2):181–189

43. Strick PL (1988) Anatomical organization of multiple motor areas in the frontal lobe: implications for recovery of function. Adv Neurol 47:293–312

44. Porter R, Lemon RN (1993) Corticospinal function and voluntary movement. Oxford University Press, Oxford, UK

45. Dum RP, Strick PL (1996) Spinal cord terminations of the medial wall motor areas in macaque monkeys. J Neurosci 16(20):6513–6525

46. Rouiller EM, Moret V, Tanne J, Boussaoud D (1996) Evidence for direct connections between the hand region of the supplementary motor area and cervical motoneurons in the macaque monkey. Eur J Neurosci 8(5):1055–1059

47. Calautti C, Leroy F, Guincestre JY, Baron JC (2001) Dynamics of motor network overactivation after striatocapsular stroke: a longitudinal PET study using a fixed-performance paradigm. Stroke 32(11):2534–2542

48. Calautti C, Leroy F, Guincestre JY, Baron JC (2003) Displacement of primary sensorimotor cortex activation after subcortical stroke: a longitudinal PET study with clinical correlation. Neuroimage 19(4):1650–1654

49. Cramer SC, Shah R, Juranek J, Crafton KR, Le V (2006) Activity in the peri-infarct rim in relation to recovery from stroke. Stroke 37(1):111–115

50. Feydy A, Carlier R, Roby-Brami A, Bussel B, Cazalis F, Pierot L et al (2002) Longitudinal study of motor recovery after stroke: recruitment and focusing of brain activation. Stroke 33(6):1610–1617

51. Johansen-Berg H, Rushworth MF, Bogdanovic MD, Kischka U, Wimalaratna S, Matthews PM (2002) The role of ipsilateral premotor cortex in hand movement after stroke. Proc Natl Acad Sci U S A 99(22):14518–14523

52. Seitz RJ, Hoflich P, Binkofski F, Tellmann L, Herzog H, Freund HJ (1998) Role of the premotor cortex in recovery from middle cerebral artery infarction. Arch Neurol 55(8):1081–1088

53. Ward NS, Brown MM, Thompson AJ, Frackowiak RS (2004) The influence of time after stroke on brain activations during a motor task. Ann Neurol 55(6):829–834

54. Dancause N, Barbay S, Frost SB, Plautz EJ, Stowe AM, Friel KM et al (2006) Ipsilateral connections of the ventral premotor cortex in a new world primate. J Comp Neurol 495(4):374–390

55. Dancause N, Barbay S, Frost SB, Mahnken JD, Nudo RJ (2007) Interhemispheric connections of the ventral premotor cortex in a new world primate. J Comp Neurol 505(6):701–715

56. Dum RP, Strick PL (1991) The origin of corticospinal projections from the premotor areas in the frontal lobe. J Neurosci 11(3):667–689

57. He SQ, Dum RP, Strick PL (1993) Topographic organization of corticospinal projections from the frontal lobe: motor areas on the lateral surface of the hemisphere. J Neurosci 13(3):952–980

58. He SQ, Dum RP, Strick PL (1995) Topographic organization of corticospinal projections from the frontal lobe: motor areas on the medial surface of the hemisphere. J Neurosci 15(5 Pt 1):3284–3306

59. Boudrias MH, Belhaj-Saif A, Park MC, Cheney PD (2006) Contrasting properties of motor output from the supplementary motor area and primary motor cortex in rhesus macaques. Cereb Cortex 16(5):632–638

60. Maier MA, Armand J, Kirkwood PA, Yang HW, Davis JN, Lemon RN (2002) Differences in the corticospinal projection from primary motor cortex and supplementary motor area to macaque upper limb motoneurons: an anatomical and electrophysiological study. Cereb Cortex 12(3):281–296

61. Baker SN, Zaaimi B, Fisher KM, Edgley SA, Soteropoulos DS (2015) Pathways mediating functional recovery. Prog Brain Res 218:389–412

62. Dettmers C, Fink GR, Lemon RN, Stephan KM, Passingham RE, Silbersweig D et al (1995) Relation between cerebral activity and

force in the motor areas of the human brain. J Neurophysiol 74(2):802–815

63. Thickbroom GW, Phillips BA, Morris I, Byrnes ML, Sacco P, Mastaglia FL (1999) Differences in functional magnetic resonance imaging of sensorimotor cortex during static and dynamic finger flexion. Exp Brain Res 126(3):431–438

64. Ward NS, Frackowiak RS (2003) Age-related changes in the neural correlates of motor performance. Brain 126(Pt 4):873–888

65. Fridman EA, Hanakawa T, Chung M, Hummel F, Leiguarda RC, Cohen LG (2004) Reorganization of the human ipsilesional premotor cortex after stroke. Brain 127(Pt 4):747–758

66. Lotze M, Markert J, Sauseng P, Hoppe J, Plewnia C, Gerloff C (2006) The role of multiple contralesional motor areas for complex hand movements after internal capsular lesion. J Neurosci 26(22):6096–6102

67. Grefkes C, Fink GR (2014) Connectivity-based approaches in stroke and recovery of function. Lancet Neurol 13(2):206–216

68. Carter AR, Patel KR, Astafiev SV, Snyder AZ, Rengachary J, Strube MJ et al (2012) Upstream dysfunction of somatomotor functional connectivity after corticospinal damage in stroke. Neurorehabil Neural Repair 26(1):7–19

69. Carter AR, Astafiev SV, Lang CE, Connor LT, Rengachary J, Strube MJ et al (2010) Resting interhemispheric functional magnetic resonance imaging connectivity predicts performance after stroke. Ann Neurol 67(3):365–375

70. Grefkes C, Nowak DA, Eickhoff SB, Dafotakis M, Küst J, Karbe H et al (2008) Cortical connectivity after subcortical stroke assessed with functional magnetic resonance imaging. Ann Neurol 63(2):236–246

71. Bestmann S, Swayne O, Blankenburg F, Ruff CC, Teo J, Weiskopf N et al (2010) The role of contralesional dorsal premotor cortex after stroke as studied with concurrent TMS-fMRI. J Neurosci 30(36):11926–11937

72. Zemke AC, Heagerty PJ, Lee C, Cramer SC (2003) Motor cortex organization after stroke is related to side of stroke and level of recovery. Stroke 34(5):e23–e28

73. Crafton KR, Mark AN, Cramer SC (2003) Improved understanding of cortical injury by incorporating measures of functional anatomy. Brain 126(Pt 7):1650–1659

74. Marshall RS, Perera GM, Lazar RM, Krakauer JW, Constantine RC, DeLaPaz RL (2000) Evolution of cortical activation during recovery from corticospinal tract infarction. Stroke 31(3):656–661

75. Small SL, Hlustik P, Noll DC, Genovese C, Solodkin A (2002) Cerebellar hemispheric activation ipsilateral to the paretic hand correlates with functional recovery after stroke. Brain 125(Pt 7):1544–1557

76. Ward NS, Brown MM, Thompson AJ, Frackowiak RS (2003) Neural correlates of motor recovery after stroke: a longitudinal fMRI study. Brain 126(Pt 11):2476–2496

77. Kleim JA, Chan S, Pringle E, Schallert K, Procaccio V, Jimenez R et al (2006) BDNF val-66met polymorphism is associated with modified experience-dependent plasticity in human motor cortex. Nat Neurosci 9(6):735–737

Chapter 22

fMRI in Psychiatric Disorders

Erin L. Habecker, Melissa A. Daniels, Elisa Canu, Maria A. Rocca, Massimo Filippi, and Perry F. Renshaw

Abstract

Functional neuroimaging has become an important tool for clinical research, with the potentiality to provide information on psychiatric disease pathology and treatment response. We review functional magnetic resonance imaging (fMRI) research findings for five psychiatric disorders: schizophrenia, major depressive disorder, bipolar disorder, obsessive-compulsive disorder, and posttraumatic stress disorder. Brain functional abnormalities and possible underlying mechanisms for disease symptoms are discussed, with a focus on future clinical implications for fMRI in psychiatric disease.

Key words fMRI, Blood oxygen level dependent, Psychiatric disorders, Schizophrenia, Major depressive disorder, Bipolar disorder, Obsessive-compulsive disorder, Posttraumatic stress disorder

1 Introduction

1.1 Overview of fMRI   Functional magnetic resonance imaging (fMRI) is a unique, noninvasive method of measuring neural activation through changes in oxidation and regional blood flow. An important clinical research tool that has been used more and more frequently in recent years, fMRI is able to indirectly detect brain activity in the working brain, allowing for the assessment of psychiatric disease physiology and treatment effects. fMRI does not involve exposure to radioactive tracers, thus allowing patients and subjects to undergo multiple scans over a short period of time, if necessary. Most fMRI studies involve the measurement of signal arising from hydrogen nuclei [1, 2]. Common types of fMRI used in psychiatric neuroimaging include blood oxygen level dependent (BOLD) and arterial spin labeling (ASL).

Instead of incorporating a radioactive tracer as in positron emission tomography (PET) or single photon emission computed tomography (SPECT), fMRI makes use of the unique properties of hemoglobin (BOLD and BOLD contrast methods) or the water molecules of flowing blood (ASL) to produce images of neural activation. Most fMRI studies today are BOLD studies that make

Massimo Filippi (ed.), *fMRI Techniques and Protocols*, Neuromethods, vol. 119,
DOI 10.1007/978-1-4939-5611-1_22, © Springer Science+Business Media New York 2016

use of T_2* mechanisms. ASL fMRI depends on T_1 effects, which will be further described below. Hemoglobin is present in the body in two forms: the oxygenated form, oxyhemoglobin; and the deoxygenated form, deoxyhemoglobin. T_2* weighed images of each form of hemoglobin are distinctive because the two have different magnetic properties. Neuronal activity results in greater cerebral blood flow (CBF) to the specific brain areas involved in the processing of a particular task, leading to an increase in T_2* signal and a more intense MR signal on the images created. Tasks and stimuli are used during fMRI to elicit a predicted brain response—they are intended to alter neural activity in brain regions thought to be impacted by the disorder in question.

1.2 BOLD

The advantages of BOLD fMRI—the most common psychiatric fMRI modality—as compared with other functional imaging techniques, such as PET, include the greater sensitivity of the fMRI signal to event-related changes in neuronal blood flow and the increased spatial resolution of fMRI images [3]. Temporal resolution, which was historically poor in previous imaging methods, has been improved drastically through the use of high-speed MR scanners with the ability to perform echo planar imaging, acquiring single image planes in 50–100 ms [4]. However, one distinct disadvantage of this mode of functional imaging that must be taken into consideration during experimental construction is the inability of BOLD signal to differentiate between changes in CBF that are correlated with neuronal activity and changes that are independent of it. Such changes include activity-related signal changes in draining veins away from the brain activity [5], incidental neural activations that are unrelated to the task at hand [6], and changes in CBF caused by changes in respiration. Even small respiration changes can alter blood arterial carbon dioxide tension (PCO_2), which has a large effect on CBF [7, 8]. Subjects with an anxiety disorder, or state anxiety induced by the MRI environment, are particularly susceptible to variations in respiration, and this must be taken into account during experiment planning using BOLD. The effects of respiration changes on PCO_2 may be managed by acquiring continuous measurements of PCO_2, or end-tidal CO_2, during the experimental protocol [3]. When this function is available, investigators have the option of either acquiring data only during steady-state CO_2 levels [9], or attempting to adjust for the effect of PCO_2 on global CBF [10, 11] and integrate this modification into fMRI indices.

1.3 ASL

ASL differs from BOLD in that it depends on T_1 mechanisms and the magnetic labeling of water molecules to generate images. Water molecules in flowing blood are tagged through the saturation or inversion of the longitudinal component of the MR signal [12]; these molecules then diffuse from capillaries into brain tissue where they alter the magnetization of the local tissue [1]. As blood flow into the imaging slice increases, there is a more significant difference between the magnetized condition and the control condition, during which the magnetization of arterial blood is fully relaxed [1]. Control and tagged images are then taken, and the difference between them is proportional to the CBF. ASL can be used to measure global CBF changes dynamically; its use of water molecules as an endogenous blood flow tracer means that the images generated by ASL are not susceptible to neurovascular changes that are not related to neuronal activation. ASL has another advantage over BOLD in that effects of frequency drifts tend to be minimized in ASL, making this method more suitable for longer duration scans [12]. However, BOLD acquisitions tend to have greater temporal resolution, greater maximum number of slices, and appear to be more sensitive to parametric manipulations of task demands [1, 13, 14]. BOLD maps also usually have larger activation areas than ASL maps [15, 16], which could either be due to the decreased sensitivity or improved signal localization inherent in ASL [13]. There are three classes of ASL methods: pulsed ASL, continuous ASL, and velocity selective ASL [1]. A discussion of the relative methodologies and merits of the three techniques is beyond the scope of this chapter.

2 fMRI in Psychiatry

2.1 Clinical Disorders

A wide range of neuropsychiatric disorders have now been investigated using fMRI techniques and protocols. This review will explore the paradigms employed, imaging results, and future research opportunities in five mental disorders: schizophrenia, major depression, bipolar disorder (BD), obsessive-compulsive disorder (OCD), and posttraumatic stress disorder (PTSD). These particular disorders were selected due to the fact that they represent a subset of psychotic, mood, and anxiety disorders; have a significant prevalence in the general population (0.4–14% depending on age and gender of the sample); are popular candidates for fMRI research; and have each been the subject of research on diagnosis, disease progression, and treatment using imaging. Table 1 summarizes the nature, range, and prevalence of the selected disorders in the general population.

Modern imaging techniques have been crucial to the delineation of the brain structures and functions that are negatively impacted in psychiatric disorders such as the ones reviewed here. Traditionally, such disorders have been characterized primarily via clinical psychiatric evaluation of abnormal symptoms, and treatments consist of a trial-and-error strategy combined with patient self-selection of treatment options or option combinations [17]. The use of fMRI to evaluate the underlying cognitive disturbances

Table 1
Summary of psychiatric disorders discussed in this chapter

Disorder	Type	Characteristic symptoms	Subtypes	Prevalence	Prognosis	Typical treatment
Schizophrenia	Psychotic disorder	Delusions, hallucinations (often auditory), disorganized speech, grossly disorganized or catatonic behavior, negative symptoms (affective flattening, alogia, aviolotion)	Subtypes are defined by the predominant symptom at the time of evaluation (paranoid, disorganized, catatonic, undifferentiated, residual). However, symptoms are not stable during the course of the disorder and often change from one subtype to another and/or present overlapping within subtypes	0.5–1.5% among adults	Complete remission uncommon. Some individuals display exacerbations and remissions of symptoms while others remain chronically ill, either on a stable course or a progressively worsening one	Antipsychotic medications, including clozapine, risperidone, olanzapine, quetiapine, ziprasidone, and aripiprazole
Major depressive disorder	Mood disorder	One or more major depressive episodes accompanied by changes in appetite, weight, or sleep; decreased energy; feelings of worthlessness or guilt; difficulty thinking, concentrating, or making decisions; or recurrent thoughts of death or suicide	Single episode, recurrent	5–9% among adult women, 2–3% among adult men	Major depressive episodes may end completely (2/3 of cases) or partially or not at all (1/3 of cases). Up to 15% of affected individuals die by suicide	SSRIs, tricyclics, and MAOIs
Bipolar disorder	Mood disorder	One or more manic or mixed episodes (Bipolar I) or one or more hypomanic episodes (Bipolar II) accompanied by one or more major depressive episodes. Cyclothymia characterized by periods of hypomanic and depressive symptoms	Bipolar I, bipolar II, cyclothymia, bipolar disorder not otherwise specified	0.4–5% among adults	Recurrent disorder: interval between episodes tends to decrease with age. 20–30% of affected individuals do not return to full functionality between episodes. Up to 15% of affected individuals die by suicide	Mood stabilizers, including lithium, valproate, carbamazepine, lamotrigine, gabapentin, and topiramate

| Obsessive-compulsive disorder | Anxiety disorder | Recurrent obsessions (persistent ideas, thoughts, impulses, or images) or compulsions (repetitive behaviors) that are time-consuming or cause marked distress or significant impairment. Affected adults realize that obsessions or compulsions are excessive or unreasonable | Checkers, washers, touchers, counters, and arrangers; Insight needs to be verified. | 1.5–2.5% among adults | Majority of individuals have chronic disease course with waxing and waning symptoms. 15% show progressive deterioration of occupational and social functioning. 5% have episodic course with minimal symptoms between episodes | SSRIs, tricyclics (clomipramine only), MAOIs, and benzodiazepines |
| Posttraumatic stress disorder | Anxiety disorder | Development of characteristic symptoms following an extreme traumatic stressor. Symptoms include persistent reexperiencing of the trauma, avoidance of stimuli associated with the trauma, numbing of general responsiveness, and increased arousal | Acute, chronic, delayed onset | 1–14% among adults; up to 58% for at risk populations (combat victims, victims of criminal violence, etc.) | Symptomatic onset usually occurs within 3 months after trauma. Complete recovery occurs within 3 months in 50% of cases; symptoms persist longer than 12 months in many cases | SSRIs, including sertraline and paroxetine |

MAOIs monoamine oxidase inhibitors, *NMDA* N-methyl d-aspartate, *SSRIs* selective serotonin reuptake inhibitors present across a heterogeneous psychiatric disorder, or even a range of psychiatric disorders, is important in that it allows for the investigation of core dysfunctions that might highlight more effective treatment options. Studying differences in neural response between psychiatric patients and normal subjects with respect to affected brain regions, and incorporating a variety of emotional and cognitive challenges to examine localized activation, investigators have the opportunity to evaluate subtle differences in the ways that the brains of patients process different types of information and perform tasks. In addition, functional imaging can be used to evaluate the efficacy of a psychiatric medication, especially in conjunction with the usual clinical symptom assessments. New findings are also highlighting the ways in which fMRI could be utilized to aid in the diagnosis of psychiatric disease [17–19].

3   Psychotic Disorders

3.1   Schizophrenia

Schizophrenia is a lifelong illness associated with a high rate of morbidity and disability due to the severity and neurologically disruptive nature of its symptoms. It is often thought of as the most serious psychiatric disorder, and the afflicted population (about 1 % of the general population) is impaired in one or more major areas of functioning: interpersonal relations, work or education, or self-care (American Psychiatric Association, Diagnostic and Statistical Manual of Mental Disorders, 5th edition, 2014—DSMV 2014). The disorder, which typically manifests sometime in an individual's mid-20s, is diagnosed through a number of characteristic symptoms falling into positive (an excess or distortion of normal function) or negative (attenuation or loss of normal function) categories. Positive symptoms include delusions, hallucinations, disorganized speech, and disorganized or catatonic behavior, while the negative symptoms encompass affective flattening and abulia (DSM-V, 2014). Because schizophrenia is a heterogeneous disorder with a wide range of associated impairments, the range of tasks employed during fMRI studies has been similarly broad. Table 2 summarizes the most common paradigms, which include verbal fluency [31], affective pictures [22, 27, 28], working memory (WM) [20, 23, 26], and inhibitory control [21, 33]. These studies have reported attenuation and deactivation of fMRI signal, as compared to healthy control groups, in prefrontal and temporal lobe structures including the amygdala, hippocampus, and parahippocampal gyrus. In addition, increased activation of the basal ganglia and striatum has been observed during WM and inhibitory control tasks [20, 21, 26, 33].

fMRI assessment of cognitive verbal and memory tasks in schizophrenic patients allows for an analysis of cognitive deficits and enables investigation of the abnormal language functionality

Table 2

Summary of fMRI research in schizophrenia

Authors	Subjects	fMRI paradigm	Results
Schizophrenia			
Manoach et al. [20]	9 schizophrenic subjects (8 under stable dose of antipsychotic medication, 1 unmedicated) and 9 healthy controls	Working memory: Sternberg Item Recognition Paradigm adapted to include monetary reward for correct responses	Schizophrenic patients exhibited deficiencies in working memory, despite similar activity in the DLPFC compared to healthy controls. Compared with controls, during the task, schizophrenic patients showed activation in the basal ganglia and thalamus
Rubia et al. [21]	6 male patients with schizophrenia (all under atypical antipsychotic medication) and 7 matched healthy controls	Inhibitory control: "stop" and "go/no-go" tasks	Schizophrenic patients exhibited reduced left prefrontal activation compared to healthy controls
Hempel et al. [22]	10 partially remitted schizophrenic patients (all under atypical antipsychotic medication) and 10 healthy controls	Facial affect discrimination and labeling	Compared with controls, schizophrenic patients showed a significantly decreased activation in the ACC during the discrimination task, and decreased activity in the amygdala-hippocampal complex bilaterally during labeling
Hofer et al. [23]	10 male outpatients with schizophrenia (all under atypical antipsychotic medication) and 10 male healthy controls	Episodic encoding/ recognition of words	Compared with controls, patients with schizophrenia demonstrated similar cognitive performance in word recognition, but decreased activation in the bilateral DLPFC and lateral temporal cortices during both tasks
Kubicki et al. [24]	9 male chronic schizophrenic patients (unspecified medication) and 9 control subjects	Semantic encoding	Schizophrenic patients had decreased activation of the left inferior prefrontal cortex and increased activation of the left superior temporal gyrus compared to healthy controls
Habel et al. [25]	13 male patients with schizophrenia (8 under typical and 4 under atypical antipsychotic medication, and 1 unmedicated), 13 of their non affected brothers (asymptomatic, unmedicated), and 26 unrelated matched healthy controls	Positive and negative mood evocation	Schizophrenic patients and brothers of schizophrenic patients showed reduced amygdala activity compared to healthy controls
Ragland et al. [26]	14 patients with schizophrenia (2 under typical and 9 under atypical antipsychotic medication, and 3 under both typical/atypical medication) and 15 healthy controls	Word encoding/recognition	Compared with controls, schizophrenic patients demonstrated reduced activation of the prefrontal cortex and increased activation of the parahippocampal gyri during encoding

(continued)

Table 2
(continued)

Authors	Subjects	fMRI paradigm	Results
Takahashi et al. [27]	15 schizophrenic patients (11 under atypical antipsychotic medication; 4 unmedicated) and 15 healthy volunteers	Affective pictures	Compared with controls, patients with schizophrenia had decreased activity in the right amygdala and medial prefrontal cortex
Williams et al. [28]	27 schizophrenic patients and 22 healthy controls	Facial expressions of fear	Compared with controls, schizophrenic patients had enhanced arousal responses coupled with reduction in amygdala and medial prefrontal activity
Honey et al. [29]	12 healthy volunteers were administered with 100 ng/ml plasma ketamine or placebo	Episodic memory task	Subjects on ketamine showed left frontal activation during semantic processing at encoding and in these subjects successful encoding was supplemented by additional nonverbal processing
Morey et al. [30]	52 subjects: 10 ultra-high risk for schizophrenia, 15 with early schizophrenia, 11 with chronic schizophrenia and 16 healthy controls	Visual oddball task	Compared with controls, early and chronic schizophrenic patients showed lower activation associated with the target stimuli in the ACC, inferior frontal gyrus, and the medial frontal gyrus. Although the ultra-high-risk group did not reach the significance when compared with controls, they showed a trend toward the early group
Yurgelun-Todd et al. [31]	12 schizophrenic patients under 8 weeks of d-cycloserine treatment or placebo	Word fluency task	Compared with patients receiving placebo, patients receiving d-cycloserine showed a significant increased activation, associated with a reduction in negative symptoms, in the temporal lobe
Juckel et al. [32]	20 schizophrenic patients (10 under typical and 10 under atypical antipsychotic medication) and 10 age-matched male healthy controls	Incentive monetary delay task	Schizophrenic patients showed reduced ventral striatal activation during the presentation of reward-indicating cues as compared to healthy controls. Decreased activation of the left ventral striatum was inversely correlated with the severity of symptoms
Vink et al. [33]	21 schizophrenic patients (all under stable atypical antipsychotic medication), 15 unaffected siblings, and 36 matched healthy controls	Inhibitory control: stop cues	Compared with controls, schizophrenic patients and unaffected siblings did not activate the striatum when responding to motor cues

ACC anterior cingulate cortex, *DLPFC* dorsolateral prefrontal cortex seen in some individuals with the disorder. WM performance, as commonly measured by encoding and recognition tasks, continuous performance paradigms, and the Sternberg test, to name a few, appears to be impacted by schizophrenia. Difficulties in encoding and free recall are common, indicating the possible involvement of the prefrontal cortex and hippocampus, both of which have been shown to participate in the neural mechanisms of working, episodic, and semantic memory [34–36]. Ragland et al. [26], using a word encoding and recognition task, observed dorsolateral prefrontal cortex (DLPFC) dysfunction manifested by bilateral defects during encoding, left hemisphere hypoactivity during recognition, and right side signal attenuation during successful retrieval, as compared to a healthy control group. This finding has been replicated using a visual oddball continuous task [30]. In addition, marked increase in parahippocampal activation in schizophrenic patients as compared to healthy controls has been observed during task performance, suggesting a core deficit in the reciprocal connections between the hippocampus and the neocortex [26]. Other WM fMRI studies of schizophrenia have reported attenuated activations in the anterior cingulate cortex (ACC) and cerebellum [29], and significantly increased activity in the basal ganglia and thalamus [20]. It has been hypothesized that frontostriatal circuitry defects could account for these deviations, and the anatomical as well as functional normality of this circuitry in schizophrenic patients has been targeted for further study [20].

Investigations of emotional processing, long a part of psychiatric research into mood disorders, have begun in recent years to be used in the study of schizophrenia. fMRI protocols have used primarily affective facial expressions, as well as emotional pictures and words, to evoke cortical responses in the temporal and frontal lobes. Attenuation of amygdala response, as well as that of the amygdala-hippocampal complex, has been noted by several investigators in response to emotional faces, discrimination of facial affect, and the evocation of negative mood [22, 25, 27, 28]. Some of these studies have also shown attenuation in the ACC and medial prefrontal cortex (mPFC) as compared to healthy controls [22, 27]. This reduced activation is often accompanied by increased activation in another area, such as the middle frontal gyrus (MFG) in one study [22]. This suggests that the observed increased activations are secondary mechanisms evoked to compensate for the dysfunction in related areas.

Other fMRI studies of schizophrenia have focused on the neural underpinnings of the episodic memory, language, and learning deficits common in the disorder. Hofer et al. [23] found decreased activations in the bilateral DLPFC and lateral temporal cortices in schizophrenic patients, despite the fact that recognition performance in the schizophrenic patients was intact. A language processing task highlighted underactivity in the temporal lobe which improved after treatment with d-cycloserine, in conjunction with negative symptom improvement [31]. Reduced activation of the left inferior prefrontal cortex was observed, accompanied by exaggerated activation in the left superior temporal gyrus, during a semantic encoding learning task [24]. These findings are particularly relevant in a clinical sense, as encoding, learning, and language skills are important for normal social interaction and functioning. These studies increase our level of understanding of the neuropsychological dysfunctions that underlie some of the most disruptive symptoms of schizophrenia, and point to a need for further study to determine optimal treatment options.

3.1.1 fMRI Features of Subjects at Risk to Develop Schizophrenia

In the last decades, several studies have attempted to define the prodromal traits predictive of future conversion to schizophrenia in at-risk subjects, especially after demonstration that early intervention improves disease prognosis [37]. The eventual "risk" to develop schizophrenia is defined in terms of genetic aspects, such as a positive family history for schizophrenia (e.g., individuals who have an affected first-degree family member-FHR), or in terms of clinical (prodromal) features. Among individuals with a clinical high risk (CHR) to develop schizophrenia, subjects 'at risk mental state' (ARMS) [38] are those having clinically defined sub-psychotic symptoms and subjects at ultra-high risk (UHR) are defined by the presence of prevalent positive clinical symptoms [38, 39].

Although altered activation of frontal and prefrontal cortices has been amply reported in people with increased risk of psychosis, at present it is still not clear if this neurofunctional alteration increases in line with the level of psychosis risk. On this purpose, a study observed a relationship between the level of working memory task-related deactivation in the mPFC and precuneus and the level of psychosis risk, with deactivation weakest in the UHR group, at an intermediate level in the FHR group, and greatest in healthy controls [40]. In another study, while controls showed a negative association between age and frontal functional activation during verbal working memory, clinical high risk youth who converted to psychosis showed the opposite [41], likely reflecting an emerging hyperactivity in frontal regions for compensative purposes [41]. During an executive task, a significant reduction in the topological centrality of the ACC in ARMS subjects which later converted into a psychotic disorder suggested this as a potential biomarker for the transition to psychosis [42].

In subjects at risk, functional alterations at the frontal regions are likely to subtend not only their cognitive but also their emotional features. In FHR subjects, it has been reported a reduced coupling between amygdala and prefrontal cortex during facial expression processing [43]. Adolescent FHR subjects have shown reduced ACC activation during emotional processing [44] and specific hyperactivation in the right superior frontal gyrus and right precentral gyrus during fearful face presentations [45]. Patients with schizophrenia and their nonpsychotic relatives showed limbic system hypoactivation similarities during facial affect attribution tasks [46, 47] and similar hypoactivation in the amygdala while a sad mood was elicited [25].

In terms of metacognitive abilities, while processing a decision, subjects at risk did not demonstrate "jumping to conclusion" biases, typical of patients with schizophrenia. However, ARMS subjects showed a significant hypoactivation in the right ventral striatum, similar to that of schizophrenic patients, during the decision making [48].

Finally, since poor social functioning is a hallmark of schizophrenia and may precede the onset of illness, a number of studies focused on the investigation of the theory of mind (ToM) in subjects at high risk of psychosis [49]. While performing a ToM task, UHR subjects experiencing psychotic symptoms in the past had lower activation of the right inferior parietal lobule and parts of the prefrontal cortex [50]. Moreover, those subjects who at the day of scanning had psychotic symptoms displayed activations more similar to patients with manifest schizophrenia [50]. In contrast, subjects at high risk who had never experienced psychotic symptoms showed significantly greater activation in the MFG compared to high risk subjects who did experience psychotic symptoms in the past.

4 Mood Disorders

4.1 Major Depression

Major depression is diagnosed in patients who experience one or more major depressive episodes without any associated mania. A major depressive episode occurs when a patient presents with persistent feelings of deep despair and loss of pleasure or interest in nearly all activities for at least 2 weeks, accompanied by at least five of the following symptoms: two are core features, i.e., depressed mood and markedly diminished or loss of interests and pleasure; the other three could be among sleep disturbances, disruption of appetite, feelings of hopelessness or worthlessness, difficulty concentrating, or suicidal thoughts (DSM V, 2014). Individuals with major depressive disorder may present a range of heterogeneous symptoms within this framework, implying that the disease impacts more than one brain region or neurotransmitter system. Studying mood and cognition-induced brain activations in affected individuals represents a powerful way to unlock the functional discrepancies between the depressed and normal nervous system. fMRI studies have revealed a wide-range network of limbic and paralimbic neural regions and circuitry whose interactions appear to be disrupted in major depressive disorder [51–53].

The limbic-cortical model of depression advanced by Mayberg et al. hypothesizes major depression as a dysfunction among discrete, but functionally integrated pathways in the dorsal, ventral, and rostral compartments of the brain [17]. Respective dysfunctions in components of each of these compartments, which include the dorsolateral and dorsomedial prefrontal cortex, dorsal ACC, and posterior cingulate (dorsal compartment); subgenual anterior cingulate, ventral prefrontal cortex, insula, hippocampus, and amygdala (ventral compartment); and rostral ACC (rostral compartment), can all be associated with the collection symptoms seen in major depression (Fig. 1) [53]. In addition, it has been theorized that the failure of the healthy elements in the system to maintain homeostasis of emotionality during times of stress when a part of the system is compromised is a contributor to major depressive episodes [17]. Depressed subjects demonstrate abnormalities in regional cerebral blood flow (rCBF) and regional cerebral glucose metabolism (regional cerebral metabolic rate for glucose, rCMRglc) in the dorsal and ventral compartments. Decreases rCBF and rCMRglc that have been observed in the DLPFC [52, 54], dorsomedial and dorsal anterolateral prefrontal cortex, as well as the dorsal ACC [54, 55] in depressed subjects during PET and SPECT studies have highlighted these areas for exploration with fMRI BOLD paradigms—as do the observed increases in rCBF and rCMRglc that have been found in

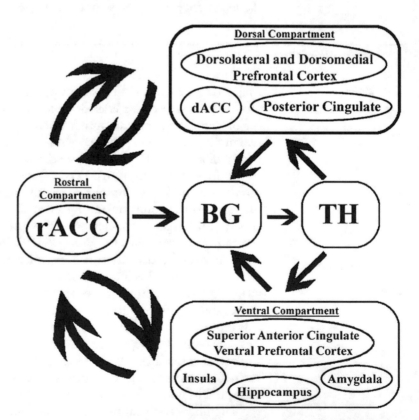

Fig. 1 A limbic-cortical model of depression adapted and modified from Mayberg et al. [53]. It involves three compartments: a dorsal, a ventral, and a rostral compartment. *BG* basal ganglia, *TH* thalamus, *dACC* dorsal anterior cingulate cortex, *rACC* rostral anterior cingulate cortex

components of the ventral compartment, including the ventrolateral, ventromedial, orbitofrontal cortex (OFC), the subgenal prefrontal cortex, the amygdala, and the insular cortex [56, 57].

The amygdala has been extensively investigated for its involvement in the processing of emotional stimuli and abnormal response in individuals with major depressive disorder. Emotional paradigms used in these fMRI studies include the exhibition of emotional film clips, facial photographs, or the presentation of audio cues during the scan to induce feelings of sadness, happiness, or fear [4]. During negative emotional tasks (negative words and sad faces), it has been shown that the amygdala in the depressed brain displays abnormally sustained activations as compared to the amygdala in the normal brain [19, 58, 59]. Other studies utilizing emotional facial expression stimuli with depressed subjects have also shown increased activation in the hippocampus, left parahippocampal gyrus, and other regions of the left brain (Table 3) [19, 60], as well as increased dynamic range bilaterally in the cerebellum and ACC extending to the rostral prefrontal cortex in response to sad faces, as compared to healthy controls [19].

Cognitive disturbances associated with depression, such as concentration difficulties, explicit memory impairment, and impairment in executive functioning, have been examined with fMRI paradigms that incorporate various WM and executive control tasks. In comparison to healthy control subjects, depressed individuals demonstrate slower reaction times and decreased accuracy with regards to executive control challenges in conjunction with decreased DLPFC activity [59]. Interestingly, depressed subjects were also shown, in another study, to have increased activation in the DLPFC in response to cognitive load in a WM task [64].

Ideally, identifying brain regions and circuitry impacted by depressive disorder will lead to a greater understanding of the effects of drug treatment and yield important information regarding the feasibility and success of various treatment options. In order to delineate the neurocircuitry involved in processing emotional cues and gather information about the pharmacodynamics of various antidepressants, studies have been undertaken that examine the effects of antidepressant treatment in a variety of scenarios. PharmacoMRI (pMRI) studies of a single dose of an antidepressant with healthy control subjects allow any focal changes in brain activity induced by the drug to be observed during BOLD scanning. Pre- and posttreatment studies of antidepressants incorporate structural MRI and fMRI to combine activation paradigms with antidepressant treatment and identify brain functional correlates of antidepressant treatment and symptomatic response [19, 78]. Two studies investigating, respectively, the effects of citalopram and mirtazapine on the healthy nervous system in a pMRI format found that each antidepressant enhanced activations right OFC during a Go/No-go task [61, 63]. In addition, one study incorporating the emotional faces paradigm found that the

Table 3

Summary of fMRI research in mood disorders

Authors	Subjects	fMRI paradigm	Results
Major depressive disorder			
Siegle et al. [58]	7 depressed and 10 never-depressed subjects	Alternating 15 s emotional and nonemotional processing trials	Depressed subjects displayed sustained amygdala responses to negative words that lasted throughout the subsequent nonemotional processing trial (25 s later). Never-depressed subjects displayed amygdala responses to all emotional stimuli, decaying within 10 s
Fu et al. [19]	19 medication free, acutely depressed subjects and 19 matched healthy volunteers	Facial expressions of sadness: low, medium, and high intensity. Subjects asked to identify sex of face	Compared with controls, depressed subjects had decreased activation in regions of the left brain: hippocampus and parahippocampal gyrus, amygdala, insula, caudate nucleus, thalamus, dorsal cingulate gyrus, inferior parietal cortex. They also showed negative relationship between increased differential response to variable affective intensity and activation in the rostral prefrontal cortex
Surguladze et al. [60]	16 individuals with major depressive disorder and 14 healthy controls	Facial expressions: happy and sad	Healthy individuals displayed increased activation in bilateral fusiform gyri and right putamen in response to increasing happiness. Depressed subjects demonstrated activation in left putamen, left parahippocampal gyrus/amygdala, and right fusiform gyrus in response to increasing sadness
Del-Ben et al. [61]	12 healthy male subjects, under 7.5 mg IV citalopram or placebo, single blind crossover design	Go/No-go, Loss/No-loss, covert (averse) face emotion recognition	Citalopram enhanced activations in the right BA47 during Go/No-go task but attenuated BA47 response to aversive faces. Citalopram attenuated the right amygdala response to aversive faces and the BA11 response during Loss/No-loss task
Wagner et al. [62]	16 patients with unipolar depression (free of psychotropic medication for a week at the time of the fMRI); 16 matched healthy controls	Adapted version of Stroop task	No differences in reaction time and accuracy between groups. Compared with controls, patients showed increased activation in the rostral ACC and left DLPFC
Siegle et al. [59]	27 unmedicated, unipolar depressive subjects (free of antidepressant medication for 2 weeks before testing) and 25 never-depressed healthy controls	Executive control task: digit sorting; emotional information processing: personal relevance rating of words	Compared with controls, depressed subjects displayed sustained amygdala reactivity on emotional tasks and decreased DLPFC activity on the digit-sorting task
Vollm et al. [63]	45 healthy male subjects under Mirtazapine or placebo, double blind, placebo controlled	Go/No-go, Reward/No-reward, Loss/No-loss	The Mirtazapine treatment was associated with enhanced activation of the right orbitofrontal cortex during Go/No-go and Reward/No-reward and of the bilateral parietal cortex during Reward/No-reward

Walter et al. [64]	12 partially remitted, medicated inpatients with major depressive disorder; 17 healthy control	Working memory: delayed match to sample	Compared with controls, depressed patients were slower and less accurate in task. They showed increased activation in the left DLPFC during highest cognitive load and in the VMPFC during control condition
Bipolar disorder			
Yurgelun-Todd et al. [65]	14 BD subjects (12 under mood stabilizers; 13 under atypical and 1 under typical antipsychotic medication) and 10 healthy controls	Faces: happy and fearful affect recognition paradigm	Compared with controls, BD subjects had reduced DLPFC activation and increased amygdala response to fearful facial affect
Blumberg et al. [66]	36 BD subjects: 11 elevated mood state, 10 depressed mood state, 15 euthymic mood state (13 unmedicated; 11 under lithium; 11 under anticonvulsants; 13 under antidepressants; 3 under atypical and 2 typical antipsychotic medication); 20 matched healthy controls	Color-word Stroop task	Compared with controls, elevated mood group showed small signal increase in right frontal cortex; and depressed mood group showed large signal increase in the left frontal cortex. Regardless of the mood state, BD subjects showed increased activation in rostral region of left VPFC
Adler et al. [67]	12 euthymic BD patients (8 medicated with mood stabilizers and/or antipsychotic) and 10 healthy controls	Working Memory: Two-back task, zero-back control/attention task	BD patients performed more poorly than healthy controls. Compared with controls, BD patients had increased activation in fronto-polar prefrontal cortex, temporal cortex, basal ganglia, thalamus, and posterior parietal cortex
Chang et al. [68]	12 young (12–18 years old) male BD subjects off medication for 24 h; 10 age-matched healthy controls	Working Memory: two-back task	Compared with controls, BD subjects showed increased activations in bilateral ACC, left putamen, left thalamus, left DLPFC, and right inferior frontal gyrus
Gruber et al. [69]	14 BD patients on stable pharmacotherapy regimen (11 under mood stabilizers; 7 under antipsychotic medication; 4 under antidepressants; 3 unmedicated) 10 healthy controls	Stroop test	Compared with controls, BD patients had reduced activations in the right subdivision of the ACC and increased activation in DLPFC
Lawrence et al. [70]	12 euthymic BD patients (5 under SSRIs medication; 5 under atypical antipsychotics; 9 under mood stabilizers) 9 major depressive disorder patients (all under antidepressants), and 11 healthy controls	Faces: fear, happiness, and sadness	Compared with controls, BD patients had increased activations in the ventral striatal, thalamic, hippocampal, and ventral prefrontal cortical areas in response to intense fear, mild happiness, and mild sadness

(continued)

Table 3
(continued)

Authors	Subjects	fMRI paradigm	Results
Malhi et al. [71]	10 hypomanic-state female subjects with BD (on mood-stabilizing psychotropic medications only, with dosages unaltered over the preceding 2 weeks), 10 matched healthy controls	Negative, positive, or neutral captioned pictures	Compared with controls, BD patients had increased activations in the caudate and thalamus in response to negative-captioned pictures
Malhi et al. [72]	10 depressed-state female subjects with BD (all under mood stabilizers with dosage unaltered during the preceding 2 weeks), 10 matched healthy controls	Negative, positive or neutral captioned pictures	Compared with controls, BD patients had increased activation in the amygdala, thalamus, hypothalamus, and medial globus pallidus in response to positive-captioned pictures
Monks et al. [73]	12 male BD subjects on lithium carbonate monotherapy and 12 healthy controls	Working Memory: Two-back task and Sternberg test	During the Two-back task, compared with controls, BD group had attenuated activity in bilateral frontal, temporal, and parietal regions; increased activity in left precentral, right medial frontal and left supramarginal gyri
Blumberg et al. [66]	17 BD patients (5 unmedicated; 8 under anticonvulsants; 3 under antidepressant; 4 under mood stabilizers; 1 under atypical antipsychotic) and 17 healthy controls	Faces: fear, happiness, sadness, and neutral	Compared with controls, patients showed increased amygdala activation, with greatest activations in unmedicated patients. Rostral ACC activations were attenuated in unmedicated patients as compared to healthy controls and medicated patients
Frangou et al. [74]	7 euthymic BD subjects on monotherapy and seven healthy controls	Working Memory: N-back task and Iowa Gambling task	Compared with controls, BD patients exhibited decreased activation in VPFC bilaterally and in the left DLPFC
Strakowski et al. [75]	16 euthymic BD patients and 16 healthy controls	Stroop test	Compared with controls, BD patients showed lower performance at the task; greater activation in medial occipital cortex and reduced activation in temporal cortical regions, middle frontal gyrus, putamen, and midline cerebellum
Pavuluri et al. [76]	10 euthymic, unmedicated BD subjects and 10 matched healthy controls	Faces: angry, happy, and neutral	Compared with controls, BD patients showed reduced activation in the right rostral VLPFC and increased activation in right pregenual ACC, amygdala, and paralimbic cortex in response to angry and happy faces
Lagopoulos et al. [77]	10 euthymic BD patients (7 under mood stabilizers) and 10 healthy controls	Working memory	Compared with controls, BD patients had slower reaction times and they exhibited attenuated or absent activation of frontal brain regions, including the DLPFC, superior frontal gyri, ACC, and intraparietal sulcus and increased activation in the inferior frontal gyrus during task conditions

ACC anterior cingulate cortex, *BD* bipolar disorder, *DLPFC* dorsolateral prefrontal cortex, *VPFC* ventral prefrontal cortex, *VLPFC* administration of a serotonergic drug attenuated right amygdala response to aversive faces [61].

Pre- and posttreatment studies of depressed subjects have posed the question of whether clinical response to antidepressant drugs can be predicted by indicators present at the baseline scan. One study found that more positive activation in the ACC at the baseline scan (in response to facial affect processing) was associated with faster rates of symptom improvement as measured by Hamilton Depression Rating Scale [79]. This finding needs to be replicated, but clear differences between pre- and posttreatment have been shown in the fMRI results: significantly attenuated activations after treatment were seen in limbic-subcortical systems, including the amygdala, that had shown enhanced activations in depressed subjects at baseline in response to an emotional face processing task [19]. In addition, as the treatment decreased this capacity for over-activation in the limbic-subcortical region, à corresponding increase was seen in prefrontal cortex activation in response to the highest levels of affective load [19]. This could be explained by the fact that treatment exposure induced changes in mood state have a proposed association with reciprocal changes in limbic-subcortical systems and frontoparietal circuitry: as limbic-subcortical regions activations to sadness are selectively lowered by drug treatment, greater dynamic range is available for high levels of affective load and increased activation in the prefrontal cortex is seen [17]. These results have great implications for the future of fMRI studies in major depressive disorder: in theory, it should be possible in the years to come to use quantitative measurements of brain function to determine optimal treatment and predict treatment response patterns for a person presenting with a major depressive episode, thereby increasing positive outcomes and chances of eventual recovery [17].

4.1.1 fMRI Features of Subjects at Risk to Develop Major Depression

As in schizophrenia, subjects at risk to develop major depression have been the focus of a number of fMRI studies in the field with the main aim to emphasize the adverse impact of a positive family history (FH+) and/or early psychosocial stressors on cerebral vulnerability and risk for depression. In this perspective, in a sample of 120 adolescent girls at the age of 11 and 12 years, a study observed that low parental warmth was associated with increased response to potential rewards in the mPFC, striatum, and amygdala and with increased depressive symptoms at the age of 16 years [80]. While performing a similar task, compared with healthy controls, right-sided ventral striatum activation was reduced in both currently depressed and high-risk girls who were daughters of mothers affected by major depression [81]. This ventral striatal activity correlated significantly with maternal depression scores suggesting this as a vulnerable factor for major depression [81]. Another study demonstrated an overactivity of the bilateral insula (also associated with subject personality) in response to increasing executive and language task difficulty in FH+ subjects who developed major depression 2 years later [82]. This pattern differentiated them from healthy controls and from other individuals at high risk who did not become unwell [82]. During an encoding task, FH+ subjects showed an overrecruitment (likely reflecting a compensatory mechanism for the task performance maintaining) of the dorsal ACC, insula, and putamen, which are all regions involved in processing the salience of stimuli [83]. While performing an emotional face processing, never-depressed monozygotic twins with a co-twin history of depression showed increased neural response in dorsal ACC, dorsomedial PFC, and occipito-parietal regions; a stronger negative coupling between the hyperactive regions and amygdala; increased attention vigilance for fearful faces and slowness at recognizing facial expressions compared to low-risk twins (monozygotic twins without a co-twin history of depression) [84].

4.2 Bipolar Disorder

BD, a prevalent neuropsychiatric illness manifesting as depressed and manic episodes in affected individuals, is among the leading worldwide causes of disability (DSM V, 2014). Bipolar depressed patients exhibit symptomatology that overlaps with that of unipolar depressed patients: feelings of despair, lack of motivation and goal-setting behavior, social isolation, lethargy, and sleep disturbances. Bipolar patients in the manic state, meanwhile, experience elevated mood, heightened energy levels, altered thought processes, and sometimes irritability, while bipolar euthymic individuals show neither depressive or manic symptoms (DSM V, 2014).

FMRI studies coupled with analysis of bipolar clinical manifestations have led to the hypothesis that, much like depression, the mechanism of the disorder involves abnormalities in the limbic, frontal, and subcortical cortical areas, perhaps caused by a dysfunction in the neural networks present in these regions [75, 85]. Anterior limbic networks, incorporating the amygdala, ACC, DLPFC, and midline cerebellum, have been implicated in this disorder, as these areas contribute to behavioral functions seen to be abnormal in individuals with BD [75]. Functional imaging studies have been performed on bipolar patients in the euthymic state, the manic state, and the depressed state to compare brain activations across groups, and found comparatively increased activations in prefrontal and dorsal ACC in all states that were not consistent bilaterally [86]. There was evidence of a small increase in signal on the right side of the ventral prefrontal cortex for patients in the manic state, while patients in the depressed state showed much greater signal increases as compared to healthy controls and manic patients in the left ventral PFC [86]. This indicates that differential signal changes across brain hemispheres may be connected with the type of mood episode experienced by the bipolar individual. Traditionally, two types of tasks have been used to generate neural activations thought to be altered in BD: fronto-executive function cognitive studies and emotional processing studies [75, 85, 86], as bipolar patients are known to have emotional regulatory impairments and impaired cognitive control.

Cognitive tasks employed in fMRI study of BD include tests of WM, interference tasks, encoding tasks, and other performance related paradigms [85]. Lagopoulos et al. [77], noting that deficits in WM seem to be of particular significance in bipolar individuals, examined cortical activations in euthymic bipolar patients and healthy controls during a parametric WM task with three load conditions. As compared to the healthy subjects, bipolar patients exhibited attenuation of activation across several frontal brain regions. The DLPFC, which was activated across all WM conditions in healthy controls, failed to activate under the same conditions in bipolar patients, although these did not have significantly poorer task performance. As the group noted that bipolar subjects recruited the inferior frontal gyrus during all WM components, while healthy subjects did not, it was theorized that this could represent a compensatory mechanism for normal DLPFC performance [77]. This group also noted a failure of BD patients to activate the parahippocampal gyrus during the delay condition. It should be noted that these cognitive defects were observed in euthymic bipolar patients, suggesting that BD-associated cognitive defects are not restricted to state conditions of mania or depression. Similarly, Monks et al. [73] found that euthymic bipolar patients on lithium therapy showed reduced activations bilaterally in frontal, parietal, and temporal regions during two WM tasks, coupled with increased activations as compared to the control group in the left precentral, right medial frontal, and left supramarginal gyri. This and similar data [74] suggest that fronto-executive region function is compromised during WM tasks in bipolar patients, which could lead to the recruitment of other neural areas in task performance. Two other studies, also using WM, found exaggerated task-induced activations in the left DLPFC, ACC, and thalamus of bipolar patients [67, 68]. Unification of these data will require further fMRI study with increased, heterogeneous sample sizes, but all the studies consistently point to an underlying dysfunction in the region of the prefrontal-subcortical circuitry [68].

Stroop tasks have been utilized in several fMRI studies of BD to examine variations in local activations induced by cognitive interference. As mentioned above, Blumberg et al. [86] found bilateral differences in the cortical activations of bipolar patients that were state-dependent and significantly different from those of healthy controls. In addition, Gruber et al. showed that stable bipolar patients had reduced signal intensity in the right anterior amygdaloid area (AAA) subdivision of the ACC, accompanied by an increase in DLPFC activation during the interference condition in what was hypothesized to be a compensatory manner [69]. In a study of medicated bipolar subjects, unmedicated bipolar subjects, and healthy controls, both groups of patients were seen to exhibit relatively increased activations as compared to the control group in the medial occipital cortex, as well as reduced activations in the temporal cortical regions, MFG, putamen, and midline cerebellum [87]. These studies illustrate the various differences in neural structures involved in interference processing in the bipolar vs. healthy brain, as well as differences in magnitude of MR signal intensity in identical areas.

Emotional processing studies of BD involve the evocation of transient mood reactions by presenting subjects in the scanner with cues, such as charged facial expressions or auditory stimuli. Four fMRI paradigms involving the presentation of some combination of happy, fearful, sad, and neutral faces yielded varied results. Lawrence et al. [70] noted increased subcortical and ventrolateral PFC activation to all categories of emotional expression, as compared to the healthy group and a group with major depressive disorder. Yurgelun-Todd et al. [65] found increased amygdala activation in BD patients in response to fearful facial affect, accompanied by a reduction in DLPFC signal. Individuals with BD also demonstrated an impaired ability to identify fearful faces as compared to their ability to identify faces carrying positive emotion. An increased amygdala response was also found by Blumberg et al. [66], this time in response to happy faces, as well as decreased rostral ACC activation in unmedicated BD patients–medicated BD patients, meanwhile, exhibited an attenuation of emotional response differences across the two groups, demonstrating that mood-stabilizing medications have the ability to ameliorate BD-induced functional abnormalities. Increased activity in the right amygdala, right pregenual ACC, and paralimbic cortex was noted by Pavuluri et al. [76] in pediatric BD in response to faces displaying both a positive and a negative emotional state. Face stimuli presentation to bipolar patients, then, appears to elicit a range of abnormal frontotemporal responses, with consistent over-activation of the amygdala seen across several studies.

State-dependent differences in brain activation in response to emotional cues have been investigated using charged pictures as well as facial affect paradigms. Malhi et al. [71] showed positive, negative, and reference captioned pictures to hypomanic and depressed female patients, finding that the hypomanic patients restricted response to the negative-captioned pictures to subcortical regions while healthy controls displayed a more widespread pattern of cortical activation. In depressed-state patients, positive-captioned pictures significantly increased similar subcortical region reactions, including activations in the thalamus and amygdala [72]. The depressed patients also showed relatively increased right-side brain activity as compared to the healthy control group [72]. These results suggest that subcortical limbic systems are involved to a much greater extent in emotional processing in the bipolar hypomanic and the bipolar depressed individual, as well as further illustrating the need to study euthymic, manic, and depressed bipolar subjects as separate groups.

BD involves dysfunction in several key limbic and cortical networks, evidence for which is summarized above. One other additional feature reflected by BD fMRI research is the consistent findings of abnormal PFC activations across state and trait boundaries. Similar findings in major depressive disorder and schizophrenia could indicate that the dysfunction caused by BD may share certain underlying characteristics with other psychiatric disorders [85].

4.2.1 fMRI Features of Subjects at Risk to Develop Bipolar Disorder

Given that the strongest risk factor for developing mania is a positive family history (FH+) for BD [88], several studies have attempted to characterize the fMRI features in non symptomatic subjects at risk using cognitive and emotional processing tasks. In a motor inhibition task, compared to healthy controls and BD patients, asymptomatic youths with a first-degree BD relative exhibited increased activation of the putamen during unsuccessful inhibition [89]. Compared with their low-risk peers, children without disorders born from parents with BD showed aberrant prefrontal neural responses to reward, aberrant connectivities among reward-related regions, and neural correlates in mesolimbic regions to novelty seeking and impulsive traits [90]. During a WM task, compared to controls, both BD patients and non symptomatic FH+ subjects exhibited failure to suppress emotional arousal and functional activity of the anterior insular and frontopolar cortices [91]. While processing facial expressions, relative to HC, both BD patients and FH+ subjects rated anger faces as less hostile and they showed decreased modulation in the amygdala and inferior frontal gyrus during anger face presentation [92]. Youth at increased genetic risk for BD demonstrated reduced brain signal of the left inferior frontal gyrus when inhibiting responses to fearful face stimuli, compared with subjects from control families [93].

5 Anxiety Disorders

5.1 Obsessive-Compulsive Disorder

OCD is a complex and clinically heterogeneous disorder characterized by obsessions (intrusive, unwanted, and repetitive thoughts), compulsions (repetitive behaviors), or both. These dysfunctional "solutions" represent the patient intention to diminish the levels of anxiety in specific daily situations, however they cyclically lead to always higher levels of distress (DSM V, 2014). The intensity of symptoms is generally varied throughout a patient's lifetime, but complete and spontaneous remission is rare [94]. Some studies have pointed to an association between different dimensions of the disorder and different treatment responses: in particular, the hoarding impulse has been associated with poorer behavioral and pharmacologic treatment response [95]. These findings illustrate the heterogeneity of the disease; however, fMRI and other neuro-imaging studies have tended to group together patients with a range of symptoms, allowing for the delineation of underlying neural mechanisms present across all categories.

Resting-state studies of functional neuroanatomy using PET and SPECT have uncovered elements of the neurobiology of OCD that should be taken into consideration when undertaking and analyzing results of fMRI research. Specifically, examinations of CBF in OCD patients during a resting state have revealed that these patients exhibit increased metabolism in the OFC and head of the caudate nucleus compared to healthy controls [94]. In addition, one study involving OCD patients with comorbid major depression showed reduced CBF in the hippocampus, caudate, and thalamus as compared to both the group of "pure" OCD patients and the control group [96]. These resting state variations should be taken into account when conducting BOLD research on this disorder, as the sensitivity of fMRI to changes in brain metabolism that are independent of task-evoked activation can be a significant confounding factor [3]. Accordingly, fMRI techniques have been used to investigate a number of states in patients with OCD (Table 4): studies comparing local activations in the brains of OCD patients and healthy controls during cognitive tasks; pre- and post-treatment studies; and symptom-provocation studies during which transient OCD-related anxiety symptoms are incited through pictures or contact with "contaminated" objects [108, 109]. This research has generally supported the involvement of frontal-striatal-thalamic-cortical circuitry in OCD symptomatology, with OCD patients demonstrating functional deviations from healthy controls in the affected brain regions [98, 99, 108].

Cognitive challenge studies examine abnormal activations in the brains of OCD patients as compared to healthy controls during a variety of learning and inhibition control tasks. The proposal of the frontal cortex and striatum as possible sites of dysfunction in the disease suggests the use of tasks that have previously been found to require processing by the frontal and subcortical systems during performance by healthy subjects [110]. Using a Tower of London task, van den Heuvel et al. [97] found decreased frontal-striatal responsiveness in OCD patients as compared to the control group, noting that this was accompanied by increased involvement of the ACC, the ventrolateral PFC, and the parahippocampal cortex in a possibly "compensatory" mechanism. Roth et al., using a response inhibition "Go/no-go" task, demonstrated that the OCD group had reduced activations in the right thalamus during response inhibition [99], a finding consistent with a significant body of literature reporting structural [111, 112] and functional [96, 113] thalamus abnormalities in OCD patients. This study also reported a reduced activation of the right OFC and dorsal cingulate gyrus in patients

Table 4
Summary of fMRI research in anxiety disorders, including posttraumatic stress disorder

Authors	Subjects	fMRI paradigm	Results
Obsessive-compulsive disorder			
van den Heuvel et al. [97]	22 unmedicated OCD patients and 22 healthy controls	Tower of London	Compared with healthy controls, OCD patients showed decreased frontal-striatal responsiveness, mainly in dorsolateral prefrontal cortex and caudate nucleus; and increased involvement of anterior cingulate, ventrolateral prefrontal, and parahippocampal cortices
Remijnse et al. [98]	20 unmedicated OCD patients and 27 healthy controls	Reversal learning task	Compared with healthy controls, patients with OCD showed reduced number of correct responses but similar responses to receipt of punishment and demonstrated normal affective switching. During the task patients showed reduced activations in the right medial and lateral OFC and in right caudate nucleus. Patients recruited the left posterior OFC, bilateral insular cortex, bilateral dorsolateral, and bilateral anterior prefrontal cortex to a lesser extent than control subjects
Roth et al. [99]	12 adults with OCD (6 under SSRIs, and 6 had not taken any psychotropic medication for at least 6 weeks before scanning) and 14 healthy control subjects	Response inhibition: Go/No-go	During response inhibition, healthy controls demonstrated right-hemisphere activation while the patient group showed a more diffuse and bilateral pattern of activation. The OCD group had less activation than control group during response inhibition in several right-hemisphere regions. Severe OCD symptoms were positively correlated with thalamic and posterior cortical activations and inversely correlated with right OFC and anterior cingulate gyri activations
Posttraumatic stress disorder			
Rauch et al. [100]	8 Vietnam combat veterans with PTSD and 8 Vietnam combat veterans free from PTSD	Masked-fearful vs. masked-happy faces	Subjects with PTSD had an increased amygdala response to the masked-fearful faces as compared to control group and to the PTSD group's responses to the masked-happy faces
Lanius et al. [101]	9 traumatized subjects with PTSD, 9 traumatized subjects without PTSD	Script-driven symptom provocation	Compared with controls, PTSD subjects showed reduced activation of the thalamus, anterior cingulate gyrus, and medial frontal gyrus
Shin et al. [102]	8 Vietnam veterans with PTSD, 8 Vietnam veterans without PTSD	Emotional counting Stroop task	Compared with controls, PTSD group exhibited diminished response in rostral anterior cingulate cortex

(continued)

Table 4
(continued)

Authors	Subjects	fMRI paradigm	Results
Hendler et al. [103]	21 male veterans, 10 with PTSD (1 unmedicated; 7 under antidepressants; 1 under antipsychotics; 1 under mood stabilizers; 6 under benzodiazepams;, and 11 (all unmedicated) without	Parametric factorial design with combat slides and noncombat slides	Compared with veterans without PTSD, PTSD group showed increased activation in amygdala in response to all images and increased activation in visual cortex of PTSD group in response to combat content
Lanius et al. [104]	10 traumatized subjects with PTSD, 10 traumatized subjects without PTSD	Script-driven symptom provocation	Compared with controls, PTSD subjects had less activation of the thalamus and the anterior cingulate gyrus
Driessen et al. [105]	12 traumatized female patients with BPD, 6 with PTSD and 6 without PTSD	Autobiographical recall of traumatic vs. negative but nontraumatic events	Subgroup without PTSD: predominant bilateral activation of OFC and Broca's area. Subjects with PTSD: predominant activation of right anterior temporal lobes, mesiotemporal areas, amygdala, posterior cingulate gyrus, occipital areas, and cerebellum
Protopopescu et al. [106]	11 patients with assault-related PTSD, 21 healthy controls	Trauma and nontrauma-related emotional words	Compared with controls, PTSD patients had increased initial amygdala response to trauma-related emotional words and did not become habituated to negative stimuli
Shin et al. [107]	13 traumatized subjects with PTSD, 13 traumatized subjects without PTSD	Emotional faces	Compared with traumatized subjects without PTSD, the PTSD group exhibited increased amygdala response and diminished medial prefrontal cortex response to fearful facial expressions and decreased ability to habituate right amygdala response to fearful faces

BPD borderline personality disorder, *OCD* obsessive-compulsive disorder, *OFC* orbitofrontal cortex, *PTSD* posttraumatic stress disorder with the most severe symptoms during response inhibition, while another group also observed reduced activations in the right OFC and in the right caudate nucleus during a reversal learning task [98]. These findings support the involvement of the OFC, thalamus, and cortical circuitry in the abnormal patterns of response inhibition that characterize OCD, as well as indicating that some of the behavioral impairments associated with the disorder may be attributed to dysfunction in this region [98].

Treatment studies in OCD incorporating fMRI to track local activation changes across a course of medication have traditionally combined symptom-provocation tasks with longitudinal study designs [108]. The caudate nucleus has shown decreased glucose metabolism following treatment with serotonin reuptake inhibitors (SRIs) such as clomipramine and fluoxetine, suggesting that the right anteriolateral OFC plays a role in the mediation of OCD symptoms and response of OCD patients to pharmacotherapy [114]. During symptom-provocation experiments, increased activations have been observed in the OFC, cingulate cortex, striatum, thalamus, lateral PFC, amygdala, caudate, and insula among unmedicated OCD patients [98, 114]. These results add support to the theory that dysfunctions of the OFC and frontal-subcortical circuitry are responsible for a great extent of OCD symptomatology. One theory with the potential to unify a great deal of functional imaging data to date involves the orbitofrontal-subcortical circuitry, which has classically been described as having a "direct" and an "indirect" pathway (Fig. 2) [115]. It has been hypothesized that OCD symptoms could be caused by a captured signal in the direct pathway creating a positive feedback loop and increasing activity in the OFC, ventromedial caudate, and medial dorsal thalamus—leading to an excessive fixation on issues of hygiene, order, danger, violence, and sex coupled with an inability to distract oneself from these thoughts or change behavior patterns [114]. Interventions that alter and functional imaging experiments that study the direct–indirect pathway balance within the orbitofrontal-subcortical circuits would be particularly beneficial to the future of OCD research, as they could directly test these theories and help to advance the understanding of OFC functionality in the brains of OCD patients [114].

5.1.1 fMRI Features of Subjects at Risk to Develop Obsessive-Compulsive Disorder

Genetic epidemiological studies have revealed that OCD has a significant familial aggregation [116]. The aggregate risk in first-degree relative of probands with OCD has been estimated at approximately 8–23% [116]. As relatives of patients are at a significantly higher risk of developing OCD symptoms than the general population, young relatives at risk represent a valuable group to examine potential neurobiological precursors of the disorder, however up to date few fMRI studies have been carried on in this population. One of these studies observed that a WM increased task

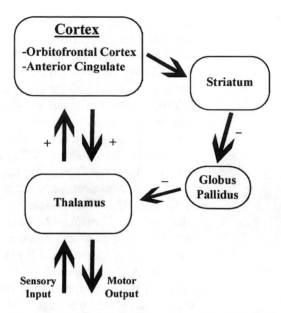

Fig. 2 The cortico-striatal model of obsessive-compulsive disorder (adapted from ref. [115]). The striatum projects via direct and indirect pathways through the globus pallidus to the thalamus, which, in turn, projects to the neocortex. Excitatory connections are labeled "+"; inhibitory connections are labeled "−"

load was associated with increased task-related brain activity in OCD and in their unaffected siblings compared with controls, however this increase was smaller while reaching the highest task load [117]. These findings indicate that compensatory frontoparietal brain activity in OCD patients and their unaffected relatives preserves task performance at low task loads but is insufficient to maintain performance at high task loads [117]. Authors concluded that the frontoparietal dysfunction may constitute a neurocognitive endophenotype for OCD [117]. In another study, patients with OCD and their siblings performed a stop-signal task during the MRI [118]. In this task, reaction times provide a behavioral measure of response inhibition [118]. Compared with controls, patients with OCD and their siblings showed greater activity in the presupplementary motor area during (successful) inhibition [118]. The presupplementary motor area hyperactivity was negatively correlated with stop-signal reaction time, likely representing an OCD neurocognitive endophenotype which may contribute to their inhibition deficit [118].

5.2 Posttraumatic Stress Disorder

PTSD is an anxiety disorder caused by the onset of an extreme stressor such as combat, childhood physical/sexual abuse, motor vehicle accidents, rape, and natural disasters. Symptoms vary across subtype of disorder and can include one or several of the following: sleep disturbance with dreams characterized by negative emotions, intrusive memories, flashbacks, avoidance of traumatic stimuli, numbing of emotions, and social dysfunction (DSM V, 2014).

There are four basic neural mechanisms that appear to function abnormally in the brains of patients with PTSD: the fear response, fear extinction, behavioral sensitization, and memory [119]. Fear response in PTSD patients appears greatly exaggerated: in the normal brain, the pairing of potentially dangerous stimuli with fear is an important survival mechanism, decreasing response time and initiating fight or flight mechanisms when such a stimuli is presented. In the brain of a PTSD patient, it seems that there is an overgeneralization of danger cues such that nonthreatening stimuli is seen as dangerous and can then be linked to past traumatic memories, which come hand-in-hand with intrusive memories and flashbacks. As a result of such nonthreatening yet triggering stimuli, patients with PTSD often exhibit avoidance of such stimuli or numbing of emotional reactions [120]. In functional neuroimaging studies of PTSD, symptom-provocation paradigms measure brain activity when subjects are exposed to visual or auditory stimulation reminiscent of their past experienced trauma to determine the mechanism of the abnormal response [2]. The main structure involved in the response to fearful stimuli is the central nucleus of the amygdala, with additional involvement by the sensory cortex, thalamus, and the mPFC [120]. PTSD patients also often exhibit a failure of fear extinction: normally, the brain is able to process stimuli from a dangerous situation from which there were no adverse outcomes such that the representation of these stimuli elicits a smaller fear response than the initial situation. However, in patients with PTSD, repeated encounters with fearful stimuli can continue to result in a consistent, heightened fear response, regardless of the actual danger of the situation [2]. The main neuroanatomical structures involved in the extinction of fear response overlap with those involved in fear conditioning, and their functionality may be studied concurrently [121].

PTSD patients often have increased sensitivity to stress, which leads to an increase in responses such as arousal and vigilance in response to stressful stimuli [119] known as behavioral sensitization. The neuroanatomy of the stress response is not centralized but involves a wide range of structures and mechanisms—many of which overlap with those involved in fear conditioning and fear response [122]. Many of those afflicted with PTSD also exhibit memory deficiencies thought to be connected to hippocampal function and reduction in hippocampal volume [123]. It is unclear whether a stressful event leading to PTSD causes volume reductions in the hippocampus through exposure to elevated glucocorticoids accompanied by a reduction of brain-derived neurotrophic factor [124, 125], or if persons with reduced hippocampal volume from birth are simply more prone to developing PTSD [126]. However, hippocampal volume and functionality is an important area of PTSD study.

FMRI research into PTSD has focused on the examination of the brain structures outlined above using symptom-provocation paradigms to examine fear response and stress sensitivity, as well as cognitive task paradigms to investigate memory deficiencies and other cognitive problems potentially associated with the disorder. It has been theorized that neurocircuitry links the amygdala to the mPFC, the hippocampus, and the thalamus through excitatory and inhibitory connections that are dysfunctional in PTSD [115] (Fig. 3). Consistent with this, fMRI studies have found alterations in BOLD signal in OFC, ACC, anterior temporal cortex, and amygdala [100, 127], as well as in the hippocampus, parahippo-campus, and thalamus [2, 128, 129]. Symptom-provocation stud-ies find exaggerated amygdala responses and decreased activation within medial frontal areas [101, 130, 131]. Rauch et al. [100] used an fMRI paradigm incorporating a happy vs. fearful faces task in healthy combat veterans and combat veterans suffering from PTSD. The study found increased activation of the amygdala in the PTSD group in response to fearful faces, which could be positively correlated with PTSD symptom severity. Similarly, Hendler et al. [103] presented combat and noncombat related slides to PTSD and non-PTSD Israeli soldiers, and found that activity in the amyg-dala was significantly increased, although their results demonstrated

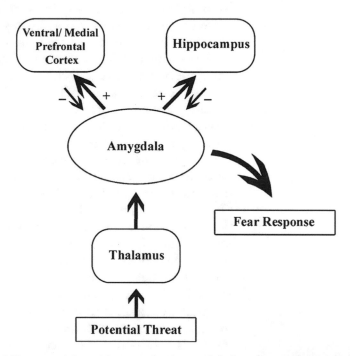

Fig. 3 The amygdalocentric neurocircuitry model of posttraumatic stress disor-der (adapted from ref. [115]). Excitatory connections are labeled "+"; inhibitory connections are labeled "−"

this increased activity regardless of whether the slide presented was traumatic (combat-related) or nontraumatic (noncombat-related). One study presenting autobiographical cues to borderline personality disorder patients with and without PTSD found increased activation in the amygdala in the group with PTSD only [105], and another showed increased amygdala activation in response to traumatic stimuli correlating with PTSD symptom severity [106].

The inappropriate fear responses associated with PTSD may be linked to dysfunction of the mPFC, as well as the amygdala. The mPFC is associated with fear response and fear extinction [132–134]. A number of studies have reported diminished activations in the mPFC as compared to groups of healthy control subjects [101, 102, 104, 107, 135]. These abnormal activations have been reported during such symptom-provocation tasks as traumatic narratives, combat-related stimuli, emotional word tasks, and emotional Stroop tasks. For example, using an autobiographical script, two studies by Lanius et al. [101, 104] found reduced activity as compared to healthy controls in both the mPFC and thalamus of PTSD subjects, suggesting an alteration of normal neurocircuitry in those regions. In a fMRI paradigm incorporating a happy vs. fearful faces task, Shin et al. [107] noted that the observed decrease in mPFC activity in the PTSD group as compared to controls was accompanied by increased amygdala activations. The theory has been advanced that PTSD symptoms are caused by an overactive amygdala in PTSD patients accompanied by a failure of the mPFC, including the ACC, and the hippocampus to inhibit this abnormal activation [136–139]. The fact that the functionality and structure of the hippocampus is also often altered in this disorder contributes evidence to this hypothesis. In one study combining fMRI and PET with verbal declarative memory tasks, the hippocampus of PTSD patients failed to activate entirely during the same tasks that caused activations in two control groups [125]. Other PET studies have noted hypoactivations of the hippocampus during memory tasks [140, 141], but one other fMRI study found increased hippocampal activation during a Stroop task [78], and another PET study noted increased resting state blood flow in the hippocampus and parahippocampal gyrus that was positively correlated with symptom severity [142]. Overall, reduced volume and increased resting blood flow in the hippocampus have been seen in many PTSD patients during both PET and fMRI studies. More study is needed to bring a greater degree of consistency to the imaging data with regards to this structure, incorporating subjects that are consistent for age, gender, and subtype of PTSD (primarily manifesting as flashback symptoms or primarily involving dissociation) [104].

The majority of functional PTSD fMRI studies to date provide evidence for hyperactivation of the amygdala coupled with a relative decrease in mPFC activity. It is hypothesized that the hippocampus, parahippocampus, and thalamus also represent participatory areas and that dysfunction in regional neurocircuitry leads to the symptoms of the disorder [115]. More research with larger subject groups and standardized study guidelines is needed to unify the fMRI data and uncover more consistent trends in PTSD imaging research.

5.2.1   fMRI Features of Subjects at Risk to Develop Posttraumatic Stress Disorder

Only one study so far investigated the fMRI features related to the risk to develop PTSD [143]. Authors assessed the fMRI differences between combat-exposed veterans with PTSD and their identical combat-unexposed co-twins vs. combat-exposed veterans without PTSD and their identical combat-unexposed co-twins. During a Multi-Source Interference Task, combat-exposed veterans with PTSD and their unexposed co-twins had significantly greater activation in the dorsal ACC and tended to have larger response time difference scores, as compared to combat-exposed veterans without PTSD and their co-twins [143]. This cerebral activation in the unexposed twins was positively correlated with their combat exposed co-twins' PTSD symptom severity [143] leading authors to conclude that hyperresponsivity in the dorsal ACC appears to be a familial risk factor for the development of PTSD following psychological trauma [143].

6   Summary and Future Directions

Table 5 summarizes the overall neural activation differences between affected psychiatric patients and healthy subjects during a sampling of common fMRI paradigms designed to test the normality of the patients' emotional, memory, inhibitory, learning, language, and executive functionality. As expected, decreased activity in cortical regions is common in the diseased brain, but neural structure overactivity is just as prevalent in certain disorders. Many researchers have theorized about the possibility of secondary effects, whereby the functional deficits that underlie the symptomatology of certain disorders are compensated for by involvement of accessory structures or overactivation in another area [17, 22]. Clearly, the effects of these disorders cannot be summarized by a single regional deficit, and current fMRI research is moving toward investigation of neural circuitry and abnormal systemic interactions. It is hoped that the documentation of these disorders' underlying dysfunctionality will yield common threads that connect the ranges of heterogeneous symptoms and indicate new clinical strategies. However, the use of fMRI as a diagnostic imaging method in psychiatric practice has thus far been limited. Although some studies are beginning to hone in on early indicators of disease, as well as the potential treatment response of the newly diagnosed, clinical method development is hampered by cost, complexity of typical research paradigms and protocols, and the known and unknown effects of drug treatments of CBF [4].

Table 5

Summary of fMRI paradigms and protocols in psychiatric disorders

Function	Stimulation paradigms	Observed activation differences in affected group	Disorder
Emotional processing	Evocation of happy and sad mood (autobiographical scripts, positive, and negative words, etc.), affective pictures, facial expressions	↓ Anterior cingulate	Schizophrenia
		↓ Amygdala–hippocampal complex (Hempel et al. [22])	
		↓ Amygdala (Habel et al. [24], Williams et al. [28])	
		↓ Right amygdala	
		↓ mPFC (Takahashi et al. [27])	
		↑ Left amygdala	Major depressive disorder
		↑ Ventral striatum	
		↑ Left hippocampus, insula, caudate nucleus, thalamus, dorsal cingulate gyrus, inferior parietal cortex (Fu et al. [19])	
		↑ Amygdala (Siegle et al. [58, 59])	
		↓ DLPFC (Yurgulen-Todd et al. [65])	Bipolar disorder
		↑ Amygdala (Yurgulen-Todd et al. [65], Malhi et al. [72], Blumberg et al. [66], Pavuluri et al. [76])	
		↑ Thalamus (Lawrence et al. [70], Malhi et al. [71, 72])	
		↑ Ventral striatum, ventral PFC	
		↑ Hippocampus (Lawrence et al. [70])	
		↑ Caudate nucleus (Malhi et al. [71])	
		↑ Hypothalamus, medial globus pallidus (Malhi et al. [72])	
		↓ Rostral anterior cingulate (Blumberg et al. [66])	
		↑ Right rostral ventrolateral PFC (Pavuluri et al. [76])	
		↑ Right pregenual anterior cingulate, paralimbic cortex (Pavuluri et al. [76])	
		↑ Amygdala (Rauch et al. [100], Hendler et al. [103], Driessen et al. [105], Protopopescu et al. [106], Shin et al. [107])	Posttraumatic stress disorder
		↓ Thalamus, anterior cingulate gyrus (Lanius et al. [104])	
		↑ Right anterior temporal lobes, mesiotemporal areas, posterior cingulate gyrus, occipital areas, cerebellum (Driessen et al. [105])	
		↓ Medial prefrontal cortex (Shin et al. [107])	

(continued)

Table 5
(continued)

Function	Stimulation paradigms	Observed activation differences in affected group	Disorder
Working memory/ attention	Word encoding/ recognition, delayed match to sample, Sternberg test, N-back, continuous performance	↑ Basal ganglia	Schizophrenia
		↑ Thalamus (Manoach et al. [20])	
		↑ Parahippocampus (Ragland et al. [26])	
		↓ Prefrontal cortex (Ragland et al. [26], Morey et al. [30])	
		↓ Anterior cingulate cortex	
		↓ Cerebellum (Honey et al. [29])	
		↑ Left DLPFC	Major depressive disorder
		↑ Ventromedial PFC (Walter et al. [64])	
		↓ Left DLPFC (Rubia et al. [21], Chang et al. [68], Frangou et al. [74])	Bipolar disorder
		↑ Fronto-polar PFC, temporal cortex, posterior parietal cortex	
		↑ Thalamus, basal ganglia (Adler et al. [67])	
		↑ Bilateral ACC, left thalamus, left putamen, right inferior frontal gyrus (Chang et al. [68])	
		↑ Frontal, temporal, parietal regions (Monks et al. [73])	
		↑ Left precentral, right medial frontal, left supramarginal gyri (Monks et al. [73])	
		↓ Bilateral VPFC (Frangou et al. [74])	
		↓ DLPFC (Lagopoulos et al. [77])	
		↑ Inferior frontal gyrus (Lagopoulos et al. [77])	
Inhibitory control	Go/No-go, stop cues, Stroop task	↑ Rostral anterior cingulate gyrus	Major depressive disorder
		↑ Left DLPFC (Wagner et al. [62])	
		↑ Rostral left VPFC	
		↑ Left frontal cortex (depressed mood group)	Bipolar disorder
		↑ Right frontal cortex (elevated mood group) (Blumberg et al. [86])	
		↓ ACC (Gruber et al. [69])	
		↑ DLPFC (Gruber et al. [69])	
		↓ Temporal cortical regions, medial frontal gyrus, putamen, midline cerebellum (Strakowski et al. [87])	
		↑ Medial occipital cortex (Strakowski et al. [87])	
		↓ Right inferior and medial frontal gyri	
		↓ Right orbitofrontal cortex, dorsal anterior cingulate gyrus (Roth et al. [99])	Obsessive-compulsive disorder
		↓ Rostral anterior cingulate cortex (Shin et al. [102])	Posttraumatic stress disorder

Executive function	Digit sorting, Tower of London	↓ DLPFC (Siegle et al. [59]) ↓ DLPFC; ↓ Caudate nucleus (van den Heuvel et al. [97]); ↑ Anterior cingulate, ventrolateral PFC, parahippocampal cortex (van den Heuvel et al. [97])	Major depressive disorder Obsessive-compulsive disorder
Episodic memory	Episodic encoding/recognition	↓ Bilateral DLPFC; ↓ Lateral temporal cortex (Hofer et al. [23])	Schizophrenia
Language processing	Word fluency, letter fluency	↓ Temporal lobe (Yurgelun-Todd et al. [85])	Schizophrenia
Reward and punishment	Loss/No loss, reward/no reward, incentive monetary tasks	↓ Ventral striatum (Juckel et al. [32])	Schizophrenia
Learning	Picture encoding, semantic encoding	↓ Left inferior PFC; ↑ Left superior temporal gyrus (Kubicki et al. [24])	Schizophrenia
Affective switching	Reversal learning task	↓ Right medial and right lateral OFC, right caudate nucleus; ↓ Left posterior OFC, insular cortex, DLPFC, anterior PFC (Remijnse et al. [98])	Obsessive-compulsive disorder

ACC anterior cingulate cortex, *DLPFC* dorsolateral prefrontal cortex, *VPFC* ventral prefrontal cortex

An important future direction for fMRI research into these disorders is the development of simple paradigms that can be relied upon to produce distinct and consistent patterns of activation in healthy and affected brains. In addition, the integration of current trends in clinical and molecular genetics, cellular biology, and social health factors such as population vulnerability factors and life events into the study of these disorders has the potential to delineate disease causality, risk factors, and course to an increasingly accurate extent.

References

1. Brown GG, Perthen JE, Liu TT, Buxton RB (2007) A primer on functional magnetic resonance imaging. Neuropsychol Rev 17(2): 107–125

2. Francati V, Vermetten E, Bremner JD (2007) Functional neuroimaging studies in posttraumatic stress disorder: review of current methods and findings. Depress Anxiety 24(3):202–218

3. Giardino ND, Friedman SD, Dager SR (2007) Anxiety, respiration, and cerebral blood flow: implications for functional brain imaging. Compr Psychiatry 48(2):103–112

4. Yurgelun-Todd DA, Renshaw PF, Femia LA (2006) Applications of fMRI to psychiatry. In: Faro SH, Mohamed FB (eds) Functional MRI: basic principles and clinical applications. Springer, New York, pp 183–220

5. Lai S, Hopkins AL, Haacke EM, Li D, Wasserman BA, Buckley P, Friedman L, Meltzer H, Hedera P, Friedland R (1993) Identification of vascular structures as a major source of signal contrast in high resolution 2D and 3D functional activation imaging of the motor cortex at 1.5T: preliminary results. Magn Reson Med 30(3):387–392

6. Saad ZS, Ropella KM, DeYoe EA, Bandettini PA (2003) The spatial extent of the BOLD response. Neuroimage 19(1):132–144

7. Ide K, Eliasziw M, Poulin MJ (2003) Relationship between middle cerebral artery blood velocity and end-tidal PCO2 in the hypocapnic-hypercapnic range in humans. J Appl Physiol (1985) 95(1):129–137

8. Poulin MJ, Liang PJ, Robbins PA (1996) Dynamics of the cerebral blood flow response to step changes in end-tidal PCO2 and PO2 in humans. J Appl Physiol (1985) 81(3):1084–1095

9. Rostrup E, Knudsen GM, Law I, Holm S, Larsson HBW, Paulson OB (2005) The relationship between cerebral blood flow and volume in humans. Neuroimage 24(1):1–11

10. Grubb RL, Raichle ME, Eichling JO, Ter-Pogossian MM (1974) The effects of changes in PaCO2 on cerebral blood volume, blood flow, and vascular mean transit time. Stroke 5(5):630–639

11. Reiman EM, Raichle ME, Robins E, Butler FK, Herscovitch P, Fox P, Perlmutter J (1986) The application of positron emission tomography to the study of panic disorder. Am J Psychiatry 143(4):469–477

12. Aguirre GK, Detre JA, Wang J (2005) Perfusion fMRI for functional neuroimaging. Int Rev Neurobiol 66:213–236

13. Liu TT, Brown GG (2007) Measurement of cerebral perfusion with arterial spin labeling: Part 1. Methods. J Int Neuropsychol Soc 13(3):517–525. doi:10.1017/S1355617707070646

14. Rao SM, Salmeron BJ, Durgerian S, Janowiak JA, Fischer M, Risinger RC, Conant LL, Stein EA (2000) Effects of methylphenidate on functional MRI blood-oxygen-level-dependent contrast. Am J Psychiatry 157(10):1697–1699

15. Mildner T, Zysset S, Trampel R, Driesel W, Moller HE (2005) Towards quantification of blood-flow changes during cognitive task activation using perfusion-based fMRI. Neuroimage 27(4):919–926

16. Tjandra T, Brooks JCW, Figueiredo P, Wise R, Matthews PM, Tracey I (2005) Quantitative assessment of the reproducibility of functional activation measured with BOLD and MR perfusion imaging: implications for clinical trial design. Neuroimage 27(2):393–401

17. Mayberg HS (2003) Modulating dysfunctional limbic-cortical circuits in depression: towards development of brain-based algorithms for diagnosis and optimised treatment. Br Med Bull 65:193–207

18. Deckersbach T, Dougherty DD, Rauch SL (2006) Functional imaging of mood and anxiety disorders. J Neuroimaging 16(1):1–10

19. Fu CHY, Williams SCR, Cleare AJ, Brammer MJ, Walsh ND, Kim J, Andrew CM, Pich EM, Williams PM, Reed LJ, Mitterschiffthaler MT,

Suckling J, Bullmore ET (2004) Attenuation of the neural response to sad faces in major depression by antidepressant treatment: a prospective, event-related functional magnetic resonance imaging study. Arch Gen Psychiatry 61(9):877–889

20. Manoach DS, Gollub RL, Benson ES, Searl MM, Goff DC, Halpern E, Saper CB, Rauch SL (2000) Schizophrenic subjects show aberrant fMRI activation of dorsolateral prefrontal cortex and basal ganglia during working memory performance. Biol Psychiatry 48(2):99–9109

21. Rubia K, Russell T, Bullmore ET, Soni W, Brammer MJ, Simmons A, Taylor E, Andrew C, Giampietro V, Sharma T (2001) An fMRI study of reduced left prefrontal activation in schizophrenia during normal inhibitory function. Schizophr Res 52(1–2):47–55

22. Hempel A, Hempel E, Schonknecht P, Stippich C, Schroder J (2003) Impairment in basal limbic function in schizophrenia during affect recognition. Psychiatry Res 122(2):115–124

23. Hofer A, Weiss EM, Golaszewski SM, Siedentopf CM, Brinkhoff C, Kremser C, Felber S, Fleischhacker WW (2003) An FMRI study of episodic encoding and recognition of words in patients with schizophrenia in remission. Am J Psychiatry 160(5):911–918

24. Kubicki M, McCarley RW, Nestor PG, Huh T, Kikinis R, Shenton ME, Wible CG (2003) An fMRI study of semantic processing in men with schizophrenia. Neuroimage 20(4):1923–1933

25. Habel U, Klein M, Shah NJ, Toni I, Zilles K, Falkai P, Schneider F (2004) Genetic load on amygdala hypofunction during sadness in nonaffected brothers of schizophrenia patients. Am J Psychiatry 161(10):1806–1813

26. Ragland JD, Gur RC, Valdez J, Turetsky BI, Elliott M, Kohler C, Siegel S, Kanes S, Gur RE (2004) Event-related fMRI of frontotemporal activity during word encoding and recognition in schizophrenia. Am J Psychiatry 161(6):1004–1015

27. Takahashi H, Koeda M, Oda K, Matsuda T, Matsushima E, Matsuura M, Asai K, Okubo Y (2004) An fMRI study of differential neural response to affective pictures in schizophrenia. Neuroimage 22(3):1247–1254

28. Williams LM, Das P, Harris AWF, Liddell BB, Brammer MJ, Olivieri G, Skerrett D, Phillips ML, David AS, Peduto A, Gordon E (2004) Dysregulation of arousal and amygdala-prefrontal systems in paranoid schizophrenia. Am J Psychiatry 161(3):480–489

29. Honey GD, Honey RAE, O'Loughlin C, Sharar SR, Kumaran D, Suckling J, Menon DK, Sleator C, Bullmore ET, Fletcher PC (2005) Ketamine disrupts frontal and hippocampal contribution to encoding and retrieval of episodic memory: an fMRI study. Cereb Cortex 15(6):749–759

30. Morey RA, Inan S, Mitchell TV, Perkins DO, Lieberman JA, Belger A (2005) Imaging frontostriatal function in ultra-high-risk, early, and chronic schizophrenia during executive processing. Arch Gen Psychiatry 62(3):254–262

31. Yurgelun-Todd DA, Coyle JT, Gruber SA, Renshaw PF, Silveri MM, Amico E, Cohen B, Goff DC (2005) Functional magnetic resonance imaging studies of schizophrenic patients during word production: effects of D-cycloserine. Psychiatry Res 138(1):23–31

32. Juckel G, Schlagenhauf F, Koslowski M, Filonov D, Wustenberg T, Villringer A, Knutson B, Kienast T, Gallinat J, Wrase J, Heinz A (2006) Dysfunction of ventral striatal reward prediction in schizophrenic patients treated with typical, not atypical, neuroleptics. Psychopharmacology (Berl) 187(2):222–228

33. Vink M, Ramsey NF, Raemaekers M, Kahn RS (2006) Striatal dysfunction in schizophrenia and unaffected relatives. Biol Psychiatry 60(1):32–39

34. Braver TS, Barch DM, Kelley WM, Buckner RL, Cohen NJ, Miezin FM, Snyder AZ, Ollinger JM, Akbudak E, Conturo TE, Petersen SE (2001) Direct comparison of prefrontal cortex regions engaged by working and long-term memory tasks. Neuroimage 14(1 Pt 1):48–59

35. Cohen NJ, Ryan J, Hunt C, Romine L, Wszalek T, Nash C (1999) Hippocampal system and declarative (relational) memory: summarizing the data from functional neuroimaging studies. Hippocampus 9(1):83–98

36. Nyberg L, Marklund P, Persson J, Cabeza R, Forkstam C, Petersson KM, Ingvar M (2003) Common prefrontal activations during working memory, episodic memory, and semantic memory. Neuropsychologia 41(3):371–377

37. Klosterkotter J, Schultze-Lutter F, Bechdolf A, Ruhrmann S (2011) Prediction and prevention of schizophrenia: what has been achieved and where to go next? World Psychiatry 10(3):165–174

38. Yung AR, Phillips LJ, Yuen HP, McGorry PD (2004) Risk factors for psychosis in an ultra high-risk group: psychopathology and clinical features. Schizophr Res 67(2–3):131–142. doi:10.1016/S0920-9964(03)00192-0

39. Yung AR, Phillips LJ, Yuen HP, Francey SM, McFarlane CA, Hallgren M, McGorry PD (2003) Psychosis prediction: 12-month follow up of a high-risk ("prodromal") group. Schizophr Res 60(1):21–32, S0920996402001676 [pii]

40. Falkenberg I, Chaddock C, Murray RM, McDonald C, Modinos G, Bramon E, Walshe M, Broome M, McGuire P, Allen P (2015) Failure to deactivate medial prefrontal cortex in people at high risk for psychosis. Eur Psychiatry 30(5):633–640. doi:10.1016/j.eurpsy.2015.03.003

41. Karlsgodt KH, van Erp TG, Bearden CE, Cannon TD (2014) Altered relationships between age and functional brain activation in adolescents at clinical high risk for psychosis. Psychiatry Res 221(1):21–29. doi:10.1016/j.pscychresns.2013.08.004

42. Lord LD, Allen P, Expert P, Howes O, Broome M, Lambiotte R, Fusar-Poli P, Valli I, McGuire P, Turkheimer FE (2012) Functional brain networks before the onset of psychosis: a prospective fMRI study with graph theoretical analysis. NeuroImage Clin 1(1):91–98. doi:10.1016/j.nicl.2012.09.008

43. Pulkkinen J, Nikkinen J, Kiviniemi V, Maki P, Miettunen J, Koivukangas J, Mukkala S, Nordstrom T, Barnett JH, Jones PB, Moilanen I, Murray GK, Veijola J (2015) Functional mapping of dynamic happy and fearful facial expressions in young adults with familial risk for psychosis—Oulu brain and mind study. Schizophr Res 164(1–3):242–249. doi:10.1016/j.schres.2015.01.039

44. Hart SJ, Bizzell J, McMahon MA, Gu H, Perkins DO, Belger A (2013) Altered fronto-limbic activity in children and adolescents with familial high risk for schizophrenia. Psychiatry Res 212(1):19–27. doi:10.1016/j.pscychresns.2012.12.003

45. Li HJ, Chan RC, Gong QY, Liu Y, Liu SM, Shum D, Ma ZL (2012) Facial emotion processing in patients with schizophrenia and their non-psychotic siblings: a functional magnetic resonance imaging study. Schizophr Res 134(2–3):143–150. doi:10.1016/j.schres.2011.10.019

46. Barbour T, Pruitt P, Diwadkar VA (2012) fMRI responses to emotional faces in children and adolescents at genetic risk for psychiatric illness share some of the features of depression. J Affect Disord 136(3):276–285. doi:10.1016/j.jad.2011.11.036

47. Habel U, Chechko N, Pauly K, Koch K, Backes V, Seiferth N, Shah NJ, Stocker T, Schneider F, Kellermann T (2010) Neural correlates of emotion recognition in schizophrenia. Schizophr Res 122(1–3):113–123. doi:10.1016/j.schres.2010.06.009

48. Rausch F, Mier D, Eifler S, Esslinger C, Schilling C, Schirmbeck F, Englisch S, Meyer-Lindenberg A, Kirsch P, Zink M (2014) Reduced activation in ventral striatum and ventral tegmental area during probabilistic decision-making in schizophrenia. Schizophr Res 156(2–3):143–149. doi:10.1016/j.schres.2014.04.020

49. Chung YS, Kang DH, Shin NY, Yoo SY, Kwon JS (2008) Deficit of theory of mind in individuals at ultra-high-risk for schizophrenia. Schizophr Res 99(1–3):111–118. doi:10.1016/j.schres.2007.11.012

50. Marjoram D, Job DE, Whalley HC, Gountouna VE, McIntosh AM, Simonotto E, Cunningham-Owens D, Johnstone EC, Lawrie S (2006) A visual joke fMRI investigation into Theory of Mind and enhanced risk of schizophrenia. Neuroimage 31(4):1850–1858. doi:10.1016/j.neuroimage.2006.02.011

51. Drevets WC (2000) Neuroimaging studies of mood disorders. Biol Psychiatry 48(8):813–829

52. Mayberg HS (1997) Limbic-cortical dysregulation: a proposed model of depression. J Neuropsychiatry Clin Neurosci 9(3):471–481

53. Mayberg HS, Liotti M, Brannan SK, McGinnis S, Mahurin RK, Jerabek PA, Silva JA, Tekell JL, Martin CC, Lancaster JL, Fox PT (1999) Reciprocal limbic-cortical function and negative mood: converging PET findings in depression and normal sadness. Am J Psychiatry 156(5):675–682

54. Baxter LR, Schwartz JM, Phelps ME, Mazziotta JC, Guze BH, Selin CE, Gerner RH, Sumida RM (1989) Reduction of prefrontal cortex glucose metabolism common to three types of depression. Arch Gen Psychiatry 46(3):243–250

55. Bench CJ, Friston KJ, Brown RG, Scott LC, Frackowiak RS, Dolan RJ (1992) The anatomy of melancholia—focal abnormalities of cerebral blood flow in major depression. Psychol Med 22(3):607–615

56. Drevets WC, Price JL, Simpson JR, Todd RD, Reich T, Vannier M, Raichle ME (1997) Subgenual prefrontal cortex abnormalities in mood disorders. Nature 386(6627):824–827

57. Liotti M, Mayberg HS, Brannan SK, McGinnis S, Jerabek P, Fox PT (2000) Differential limbic–cortical correlates of sadness and anxiety in healthy subjects: implications for affective disorders. Biol Psychiatry 48(1):30–42

58. Siegle GJ, Steinhauer SR, Thase ME, Stenger VA, Carter CS (2002) Can't shake that feeling: event-related fMRI assessment of sustained amygdala activity in response to emotional information in depressed individuals. Biol Psychiatry 51(9):693–707

59. Siegle GJ, Thompson W, Carter CS, Steinhauer SR, Thase ME (2007) Increased amygdala and decreased dorsolateral prefrontal BOLD responses in unipolar depression: related and independent features. Biol Psychiatry 61(2):198–209

60. Surguladze SA, Young AW, Senior C, Brebion G, Travis MJ, Phillips ML (2004) Recognition accuracy and response bias to happy and sad facial expressions in patients with major depression. Neuropsychology 18(2):212–218

61. Del-Ben CM, Deakin JF, McKie S, Delvai NA, Williams SR, Elliott R, Dolan M, Anderson IM (2005) The effect of citalopram pretreatment on neuronal responses to neuropsychological tasks in normal volunteers: an FMRI study. Neuropsychopharmacology 30(9):1724–1734. doi:10.1038/sj.npp.1300728

62. Wagner G, Sinsel E, Sobanski T, Kohler S, Marinou V, Mentzel H-J, Sauer H, Schlosser RGM (2006) Cortical inefficiency in patients with unipolar depression: an event-related FMRI study with the Stroop task. Biol Psychiatry 59(10):958–965

63. Vollm B, Richardson P, McKie S, Elliott R, Deakin JFW, Anderson IM (2006) Serotonergic modulation of neuronal responses to behavioural inhibition and reinforcing stimuli: an fMRI study in healthy volunteers. Eur J Neurosci 23(2):552–560

64. Walter H, Wolf RC, Spitzer M, Vasic N (2007) Increased left prefrontal activation in patients with unipolar depression: an event-related, parametric, performance-controlled fMRI study. J Affect Disord 101(1–3):175–185

65. Yurgelun-Todd DA, Gruber SA, Kanayama G, Killgore WD, Baird AA, Young AD (2000) fMRI during affect discrimination in bipolar affective disorder. Bipolar Disord 2(3 Pt 2): 237–248

66. Blumberg HP, Donegan NH, Sanislow CA, Collins S, Lacadie C, Skudlarski P, Gueorguieva R, Fulbright RK, McGlashan TH, Gore JC, Krystal JH (2005) Preliminary evidence for medication effects on functional abnormalities in the amygdala and anterior cingulate in bipolar disorder. Psychopharmacology (Berl) 183(3):308–313

67. Adler CM, Holland SK, Schmithorst V, Tuchfarber MJ, Strakowski SM (2004) Changes in neuronal activation in patients with bipolar disorder during performance of a working memory task. Bipolar Disord 6(6):540–549

68. Chang K, Adleman NE, Dienes K, Simeonova DI, Menon V, Reiss A (2004) Anomalous prefrontal-subcortical activation in familial pediatric bipolar disorder: a functional magnetic resonance imaging investigation. Arch Gen Psychiatry 61(8):781–792

69. Gruber SA, Rogowska J, Yurgelun-Todd DA (2004) Decreased activation of the anterior cingulate in bipolar patients: an fMRI study. J Affect Disord 82(2):191–201

70. Lawrence NS, Williams AM, Surguladze S, Giampietro V, Brammer MJ, Andrew C, Frangou S, Ecker C, Phillips ML (2004) Subcortical and ventral prefrontal cortical neural responses to facial expressions distinguish patients with bipolar disorder and major depression. Biol Psychiatry 55(6):578–587

71. Malhi GS, Lagopoulos J, Sachdev P, Mitchell PB, Ivanovski B, Parker GB (2004) Cognitive generation of affect in hypomania: an fMRI study. Bipolar Disord 6(4):271–285. doi:10.1111/j.1399-5618.2004.00123.x

72. Malhi GS, Lagopoulos J, Ward PB, Kumari V, Mitchell PB, Parker GB, Ivanovski B, Sachdev P (2004) Cognitive generation of affect in bipolar depression: an fMRI study. Eur J Neurosci 19(3):741–754

73. Monks PJ, Thompson JM, Bullmore ET, Suckling J, Brammer MJ, Williams SCR, Simmons A, Giles N, Lloyd AJ, Harrison CL, Seal M, Murray RM, Ferrier IN, Young AH, Curtis VA (2004) A functional MRI study of working memory task in euthymic bipolar disorder: evidence for task-specific dysfunction. Bipolar Disord 6(6):550–564

74. Frangou S, Raymont V, Bettany D (2002) The Maudsley bipolar disorder project. A survey of psychotropic prescribing patterns in bipolar I disorder. Bipolar Disord 4(6): 378–385

75. Strakowski SM, Delbello MP, Adler CM (2005) The functional neuroanatomy of bipolar disorder: a review of neuroimaging findings. Mol Psychiatry 10(1):105–116

76. Pavuluri MN, O'Connor MM, Harral E, Sweeney JA (2007) Affective neural circuitry during facial emotion processing in pediatric bipolar disorder. Biol Psychiatry 62(2):158–167

77. Lagopoulos J, Ivanovski B, Malhi GS (2007) An event-related functional MRI study of working memory in euthymic bipolar disorder. J Psychiatry Neurosci 32(3):174–184

78. Sheline YI, Barch DM, Donnelly JM, Ollinger JM, Snyder AZ, Mintun MA (2001) Increased amygdala response to masked emotional faces in depressed subjects resolves with antidepressant treatment: an fMRI study. Biol Psychiatry 50(9):651–658

79. Chen C-H, Ridler K, Suckling J, Williams S, Fu CHY, Merlo-Pich E, Bullmore E (2007) Brain imaging correlates of depressive symptom severity and predictors of symptom improvement after antidepressant treatment. Biol Psychiatry 62(5):407–414

80. Casement MD, Guyer AE, Hipwell AE, McAloon RL, Hoffmann AM, Keenan KE, Forbes EE (2014) Girls' challenging social experiences in early adolescence predict neural response to rewards and depressive symptoms. Dev Cogn Neurosci 8:18–27. doi:10.1016/j.dcn.2013.12.003

81. Sharp C, Kim S, Herman L, Pane H, Reuter T, Strathearn L (2014) Major depression in mothers predicts reduced ventral striatum activation in adolescent female offspring with and without depression. J Abnorm Psychol 123(2):298–309. doi:10.1037/a0036191

82. Whalley HC, Sussmann JE, Romaniuk L, Stewart T, Papmeyer M, Sprooten E, Hackett S, Hall J, Lawrie SM, McIntosh AM (2013) Prediction of depression in individuals at high familial risk of mood disorders using functional magnetic resonance imaging. PLoS One 8(3):e57357. doi:10.1371/journal.pone.0057357

83. Mannie ZN, Filippini N, Williams C, Near J, Mackay CE, Cowen PJ (2014) Structural and functional imaging of the hippocampus in young people at familial risk of depression. Psychol Med 44(14):2939–2948. doi:10.1017/S0033291714000580

84. Miskowiak KW, Glerup L, Vestbo C, Harmer CJ, Reinecke A, Macoveanu J, Siebner HR, Kessing LV, Vinberg M (2015) Different neural and cognitive response to emotional faces in healthy monozygotic twins at risk of depression. Psychol Med 45(7):1447–1458. doi:10.1017/S0033291714002542

85. Yurgelun-Todd DA, Ross AJ (2006) Functional magnetic resonance imaging studies in bipolar disorder. CNS Spectr 11(4):287–297

86. Blumberg HP, Leung HC, Skudlarski P, Lacadie CM, Fredericks CA, Harris BC, Charney DS, Gore JC, Krystal JH, Peterson BS (2003) A functional magnetic resonance imaging study of bipolar disorder: state- and trait-related dysfunction in ventral prefrontal cortices. Arch Gen Psychiatry 60(6):601–609. doi:10.1001/archpsyc.60.6.601

87. Strakowski SM, Adler CM, Holland SK, Mills NP, DelBello MP, Eliassen JC (2005) Abnormal FMRI brain activation in euthymic bipolar disorder patients during a counting Stroop interference task. Am J Psychiatry 162(9):1697–1705

88. Goodwin FK, Jamison KR (1990) Manic-depressive illness. Oxford University Press, New York

89. Deveney CM, Connolly ME, Jenkins SE, Kim P, Fromm SJ, Brotman MA, Pine DS, Leibenluft E (2012) Striatal dysfunction during failed motor inhibition in children at risk for bipolar disorder. Prog Neuropsychopharmacol Biol Psychiatry 38(2):127–133. doi:10.1016/j.pnpbp.2012.02.014

90. Singh MK, Kelley RG, Howe ME, Reiss AL, Gotlib IH, Chang KD (2014) Reward processing in healthy offspring of parents with bipolar disorder. JAMA Psychiatry 71(10):1148–1156. doi:10.1001/jamapsychiatry.2014.1031

91. Thermenos HW, Goldstein JM, Milanovic SM, Whitfield-Gabrieli S, Makris N, Laviolette P, Koch JK, Faraone SV, Tsuang MT, Buka SL, Seidman LJ (2010) An fMRI study of working memory in persons with bipolar disorder or at genetic risk for bipolar disorder. Am J Med Genet B Neuropsychiatr Genet 153B(1):120–131. doi:10.1002/ajmg.b.30964

92. Brotman MA, Deveney CM, Thomas LA, Hinton KE, Yi JY, Pine DS, Leibenluft E (2014) Parametric modulation of neural activity during face emotion processing in unaffected youth at familial risk for bipolar disorder. Bipolar Disord 16(7):756–763. doi:10.1111/bdi.12193

93. Roberts G, Green MJ, Breakspear M, McCormack C, Frankland A, Wright A, Levy F, Lenroot R, Chan HN, Mitchell PB (2013) Reduced inferior frontal gyrus activation during response inhibition to emotional stimuli in youth at high risk of bipolar disorder. Biol Psychiatry 74(1):55–61. doi:10.1016/j.biopsych.2012.11.004

94. Whiteside SP, Port JD, Abramowitz JS (2004) A meta-analysis of functional neuroimaging in obsessive-compulsive disorder. Psychiatry Res 132(1):69–79

95. Mataix-Cols D, Rosario-Campos MC, Leckman JF (2005) A multidimensional model of obsessive-compulsive disorder. Am J Psychiatry 162(2):228–238. doi:10.1176/appi.ajp.162.2.228

96. Saxena S, Brody AL, Ho ML, Alborzian S, Ho MK, Maidment KM, Huang SC, Wu HM, Au SC, Baxter LR Jr (2001) Cerebral metabolism in major depression and obsessive-compulsive disorder occurring separately and concurrently. Biol Psychiatry 50(3):159–170

97. van den Heuvel OA, Veltman DJ, Groenewegen HJ, Cath DC, van Balkom AJ, van Hartskamp J, Barkhof F, van Dyck R (2005) Frontal-striatal dysfunction during planning in obsessive-compulsive disorder. Arch Gen Psychiatry 62(3):301–309. doi:10.1001/archpsyc.62.3.301

98. Remijnse PL, Nielen MMA, van Balkom AJLM, Cath DC, van Oppen P, Uylings HBM,

Veltman DJ (2006) Reduced orbitofrontal-striatal activity on a reversal learning task in obsessive-compulsive disorder. Arch Gen Psychiatry 63(11):1225–1236

99. Roth RM, Saykin AJ, Flashman LA, Pixley HS, West JD, Mamourian AC (2007) Event-related functional magnetic resonance imaging of response inhibition in obsessive-compulsive disorder. Biol Psychiatry 62(8):901–909

100. Rauch SL, Whalen PJ, Shin LM, McInerney SC, Macklin ML, Lasko NB, Orr SP, Pitman RK (2000) Exaggerated amygdala response to masked facial stimuli in posttraumatic stress disorder: a functional MRI study. Biol Psychiatry 47(9):769–776

101. Lanius RA, Williamson PC, Densmore M, Boksman K, Gupta MA, Neufeld RW, Gati JS, Menon RS (2001) Neural correlates of traumatic memories in posttraumatic stress disorder: a functional MRI investigation. Am J Psychiatry 158(11):1920–1922

102. Shin LM, Whalen PJ, Pitman RK, Bush G, Macklin ML, Lasko NB, Orr SP, McInerney SC, Rauch SL (2001) An fMRI study of anterior cingulate function in posttraumatic stress disorder. Biol Psychiatry 50(12):932–942

103. Hendler T, Rotshtein P, Yeshurun Y, Weizmann T, Kahn I, Ben-Bashat D, Malach R, Bleich A (2003) Sensing the invisible: differential sensitivity of visual cortex and amygdala to traumatic context. Neuroimage 19(3):587–600

104. Lanius RA, Williamson PC, Hopper J, Densmore M, Boksman K, Gupta MA, Neufeld RWJ, Gati JS, Menon RS (2003) Recall of emotional states in posttraumatic stress disorder: an fMRI investigation. Biol Psychiatry 53(3):204–210

105. Driessen M, Beblo T, Mertens M, Piefke M, Rullkoetter N, Silva-Saavedra A, Reddemann L, Rau H, Markowitsch HJ, Wulff H, Lange W, Woermann FG (2004) Posttraumatic stress disorder and fMRI activation patterns of traumatic memory in patients with borderline personality disorder. Biol Psychiatry 55(6):603–611

106. Protopopescu X, Pan H, Tuescher O, Cloitre M, Goldstein M, Engelien W, Epstein J, Yang Y, Gorman J, LeDoux J, Silbersweig D, Stern E (2005) Differential time courses and specificity of amygdala activity in posttraumatic stress disorder subjects and normal control subjects. Biol Psychiatry 57(5):464–473

107. Shin LM, Wright CI, Cannistraro PA, Wedig MM, McMullin K, Martis B, Macklin ML, Lasko NB, Cavanagh SR, Krangel TS, Orr SP, Pitman RK, Whalen PJ, Rauch SL (2005) A functional magnetic resonance imaging study of amygdala and medial prefrontal cortex responses to overtly presented fearful faces in posttraumatic stress disorder. Arch Gen Psychiatry 62(3):273–281

108. Mitterschiffthaler MT, Ettinger U, Mehta MA, Mataix-Cols D, Williams SCR (2006) Applications of functional magnetic resonance imaging in psychiatry. J Magn Reson Imaging 23(6):851–861

109. Saxena S, Brody AL, Schwartz JM, Baxter LR (1998) Neuroimaging and frontal-subcortical circuitry in obsessive-compulsive disorder. Br J Psychiatry Suppl 35:26–37

110. Purcell R, Maruff P, Kyrios M, Pantelis C (1998) Cognitive deficits in obsessive-compulsive disorder on tests of frontal-striatal function. Biol Psychiatry 43(5):348–357

111. Gilbert AR, Moore GJ, Keshavan MS, Paulson LA, Narula V, Mac Master FP, Stewart CM, Rosenberg DR (2000) Decrease in thalamic volumes of pediatric patients with obsessive-compulsive disorder who are taking paroxetine. Arch Gen Psychiatry 57(5):449–456

112. Kim JJ, Lee MC, Kim J, Kim IY, Kim SI, Han MH, Chang KH, Kwon JS (2001) Grey matter abnormalities in obsessive-compulsive disorder: statistical parametric mapping of segmented magnetic resonance images. Br J Psychiatry 179:330–334

113. Lacerda ALT, Dalgalarrondo P, Caetano D, Camargo EE, Etchebehere ECSC, Soares JC (2003) Elevated thalamic and prefrontal regional cerebral blood flow in obsessive-compulsive disorder: a SPECT study. Psychiatry Res 123(2):125–134

114. Saxena S, Rauch SL (2000) Functional neuroimaging and the neuroanatomy of obsessive-compulsive disorder. Psychiatr Clin North Am 23(3):563–586

115. Rauch SL, Savage CR, Alpert NM, Fischman AJ, Jenike MA (1997) The functional neuroanatomy of anxiety: a study of three disorders using positron emission tomography and symptom provocation. Biol Psychiatry 42(6):446–452

116. Hettema JM, Neale MC, Kendler KS (2001) A review and meta-analysis of the genetic epidemiology of anxiety disorders. Am J Psychiatry 158(10):1568–1578

117. de Vries FE, de Wit SJ, Cath DC, van der Werf YD, van der Borden V, van Rossum TB, van Balkom AJ, van der Wee NJ, Veltman DJ, van den Heuvel OA (2014) Compensatory frontoparietal activity during working memory: an endophenotype of obsessive-compulsive disorder. Biol Psychiatry 76(11):878–887. doi:10.1016/j.biopsych.2013.11.021

118. de Wit SJ, de Vries FE, van der Werf YD, Cath DC, Heslenfeld DJ, Veltman EM, van Balkom AJ, Veltman DJ, van den Heuvel OA (2012) Presupplementary motor area hyperactivity during response inhibition: a candidate endophenotype of obsessive-compulsive disorder. Am J Psychiatry 169(10):1100–1108. doi:10.1176/appi.ajp.2012.12010073

119. Charney DS, Deutch AY, Krystal JH, Southwick SM, Davis M (1993) Psychobiologic mechanisms of posttraumatic stress disorder. Arch Gen Psychiatry 50(4):295–305

120. Charney DS (2004) Psychobiological mechanisms of resilience and vulnerability: implications for successful adaptation to extreme stress. Am J Psychiatry 161(2):195–216

121. Quirk GJ, Gehlert DR (2003) Inhibition of the amygdala: key to pathological states? Ann N Y Acad Sci 985:263–272

122. Stein MB, Simmons AN, Feinstein JS, Paulus MP (2007) Increased amygdala and insula activation during emotion processing in anxiety-prone subjects. Am J Psychiatry 164(2):318–327

123. Geuze E, Vermetten E, Bremner JD (2005) MR-based in vivo hippocampal volumetrics: 2. Findings in neuropsychiatric disorders. Mol Psychiatry 10(2):160–184

124. Bremner JD (2002) Neuroimaging of childhood trauma. Semin Clin Neuropsychiatry 7(2):104–112

125. Bremner JD, Vythilingam M, Vermetten E, Southwick SM, McGlashan T, Nazeer A, Khan S, Vaccarino LV, Soufer R, Garg PK, Ng CK, Staib LH, Duncan JS, Charney DS (2003) MRI and PET study of deficits in hippocampal structure and function in women with childhood sexual abuse and posttraumatic stress disorder. Am J Psychiatry 160(5):924–932

126. Gilbertson MW, Shenton ME, Ciszewski A, Kasai K, Lasko NB, Orr SP, Pitman RK (2002) Smaller hippocampal volume predicts pathologic vulnerability to psychological trauma. Nat Neurosci 5(11):1242–1247

127. Whalen PJ, Rauch SL, Etcoff NL, McInerney SC, Lee MB, Jenike MA (1998) Masked presentations of emotional facial expressions modulate amygdala activity without explicit knowledge. J Neurosci 18(1):411–418

128. Lanius RA, Williamson PC, Bluhm RL, Densmore M, Boksman K, Neufeld RWJ, Gati JS, Menon RS (2005) Functional connectivity of dissociative responses in posttraumatic stress disorder: a functional magnetic resonance imaging investigation. Biol Psychiatry 57(8):873–884

129. Vermetten E, Vythilingam M, Southwick SM, Charney DS, Bremner JD (2003) Long-term treatment with paroxetine increases verbal declarative memory and hippocampal volume in posttraumatic stress disorder. Biol Psychiatry 54(7):693–702

130. Pitman RK, Shin LM, Rauch SL (2001) Investigating the pathogenesis of posttraumatic stress disorder with neuroimaging. J Clin Psychiatry 62(Suppl 17):47–54

131. Villarreal G, King CY (2001) Brain imaging in posttraumatic stress disorder. Semin Clin Neuropsychiatry 6(2):131–145

132. Morgan MA, LeDoux JE (1995) Differential contribution of dorsal and ventral medial prefrontal cortex to the acquisition and extinction of conditioned fear in rats. Behav Neurosci 109(4):681–688

133. Quirk GJ, Russo GK, Barron JL, Lebron K (2000) The role of ventromedial prefrontal cortex in the recovery of extinguished fear. J Neurosci 20(16):6225–6231

134. Santini E, Ge H, Ren K, Pena de Ortiz S, Quirk GJ (2004) Consolidation of fear extinction requires protein synthesis in the medial prefrontal cortex. J Neurosci 24(25):5704–5710

135. Williams LM, Kemp AH, Felmingham K, Barton M, Olivieri G, Peduto A, Gordon E, Bryant RA (2006) Trauma modulates amygdala and medial prefrontal responses to consciously attended fear. Neuroimage 29(2):347–357

136. Golier JA, Yehuda R, Lupien SJ, Harvey PD, Grossman R, Elkin A (2002) Memory performance in Holocaust survivors with posttraumatic stress disorder. Am J Psychiatry 159(10):1682–1688

137. Rauch SL, Shin LM, Phelps EA (2006) Neurocircuitry models of posttraumatic stress disorder and extinction: human neuroimaging research—past, present, and future. Biol Psychiatry 60(4):376–382

138. Shin LM, Shin PS, Heckers S, Krangel TS, Macklin ML, Orr SP, Lasko N, Segal E, Makris N, Richert K, Levering J, Schacter DL, Alpert NM, Fischman AJ, Pitman RK, Rauch SL (2004) Hippocampal function in posttraumatic stress disorder. Hippocampus 14(3):292–300

139. Vermetten E, Bremner JD (2002) Circuits and systems in stress. II. Applications to neurobiology and treatment in posttraumatic stress disorder. Depress Anxiety 16(1):14–38

140. Bremner JD, Narayan M, Staib LH, Southwick SM, McGlashan T, Charney DS (1999) Neural correlates of memories of childhood sexual abuse in women with and

without posttraumatic stress disorder. Am J Psychiatry 156(11):1787–1795

141. Shin LM, McNally RJ, Kosslyn SM, Thompson WL, Rauch SL, Alpert NM, Metzger LJ, Lasko NB, Orr SP, Pitman RK (1999) Regional cerebral blood flow during script-driven imagery in childhood sexual abuse-related PTSD: a PET investigation. Am J Psychiatry 156(4):575–584

142. Semple WE, Goyer PF, McCormick R, Donovan B, Muzic RF, Rugle L, McCutcheon K, Lewis C, Liebling D, Kowaliw S, Vapenik K, Semple MA, Flener CR, Schulz SC (2000) Higher brain blood flow at amygdala and lower frontal cortex blood flow in PTSD patients with comorbid cocaine and alcohol abuse compared with normals. Psychiatry 63(1):65–74

143. Shin LM, Bush G, Milad MR, Lasko NB, Brohawn KH, Hughes KC, Macklin ML, Gold AL, Karpf RD, Orr SP, Rauch SL, Pitman RK (2011) Exaggerated activation of dorsal anterior cingulate cortex during cognitive interference: a monozygotic twin study of posttraumatic stress disorder. Am J Psychiatry 168(9):979–985. doi:10.1176/appi.ajp.2011.09121812

Chapter 23

fMRI in Neurodegenerative Diseases: From Scientific Insights to Clinical Applications

Bradford C. Dickerson, Federica Agosta, and Massimo Filippi

Abstract

fMRI is a technology with great promise as a tool to probe abnormalities of brain activity in neurodegenerative diseases. The detection of functional brain abnormalities may be useful, in the appropriate clinical context, for early diagnosis, differential diagnosis, or prognostication. Prediction of response to treatment or therapeutic monitoring may also be possible with fMRI. In addition, fMRI has the potential to provide a variety of scientific insights that may have clinical relevance, including compensatory hyperactivation of brain circuits or genetic modulation of functional brain activity.

Key words Alzheimer's disease, Amyotrophic lateral sclerosis, Functional MRI, Huntington disease, Magnetic resonance imaging, Neurodegenerative diseases, Parkinson's disease

1 Introduction

Neurodegenerative diseases are a major medical and social burden in many societies, particularly with the growth of older population segments. Neurodegenerative diseases include many dementias, movement disorders, cerebellar, and motor neuron diseases. In many cases, these diseases involve the pathologic accumulation of abnormal protein forms. As the biology of these diseases is elucidated, hope is beginning to emerge for specific treatments targeted at modification of fundamental pathophysiologic processes [1, 2]. For this hope to be realized, methods for early detection of specific disease processes need to be identified. Furthermore, reliable methods for monitoring the progression of the diseases will likely be critical in demonstrating the effects of putative disease-modifying therapies. Although molecular biomarkers measured via positron emission tomography (PET) or cerebrospinal fluid (CSF) have emerged in the past 5–10 years as a major class of biomarker, magnetic resonance imaging (MRI) continues to offer great potential in assessment and monitoring of neurodegenerative diseases [3]. Given the growing body of evidence that alterations in synaptic function are present very early in the course of

Massimo Filippi (ed.), *fMRI Techniques and Protocols*, Neuromethods, vol. 119,
DOI 10.1007/978-1-4939-5611-1_23, © Springer Science+Business Media New York 2016

neurodegenerative disease processes, possibly long before the development of clinical symptoms and even significant neuropathology [4, 5], functional MRI (fMRI) may be particularly useful for detecting alterations in brain function that may be present very early in the trajectory of neurodegenerative diseases. In addition, fMRI may be a critical biomarker for the detection of physiological alterations over a short period of time, thus serving as a measure of target engagement in clinical trials. The uses of fMRI in neurodegenerative dementias will be reviewed, with a focus on Alzheimer's disease (AD), to illustrate many points of relevance to other neurodegenerative diseases.

Before specifically discussing fMRI, though, it is worth considering the current concepts of clinicopathologic constructs of neurodegenerative diseases, since the interpretation of imaging data in patients depends critically on a detailed understanding of the clinical characteristics of the patient population(s) being studied.

2 Constructs of Neurodegenerative Disease: Clinical, Prodromal, and Presymptomatic Phases

The field of neurodegenerative dementias has increasingly shifted as molecular biomarkers have become available to considering neuropathology separately from clinical syndrome. The major neuropathologies of neurodegenerative dementias include Alzheimer's disease (AD), frontotemporal lobar degeneration (including FTLD-TDP43 and FTLD-tau, which many experts consider to encompass not only classical Pick's disease but also progressive supranuclear palsy [PSP] and corticobasal degeneration [CBD]), the Lewy body diseases, and Huntington's disease. The clinical syndromes include AD dementia, the primary progressive aphasias (PPA), behavioral variant frontotemporal dementia (bvFTD), the PSP syndromes, corticobasal syndrome, dementia with Lewy bodies (DLB) and Parkinson's disease dementia (PDD), amyotrophic lateral sclerosis with cognitive/behavioral impairment (ALS-FTD), and Huntington's disease [6, 7]. Clinicopathologic relationships within the family of neurodegenerative dementias is immensely complex, with growing data supporting many types of overlap between conditions traditionally thought of as distinct.

Many neurodegenerative diseases contributing to dementia are thought to arise from pathophysiologic processes that take place over a decade or more prior to the development of symptoms. For example, the clinical diagnosis of AD has traditionally been made after a patient has developed impairment in multiple cognitive domains that is substantial enough to interfere with routine social and/or occupational function (dementia). Previously, it was only after this point that FDA-approved medications were currently indicated—that is, in patients with clinically probable AD dementia. By this time, substantial neuronal loss and neuropathologic change have damaged many brain regions. Furthermore, clinical trials of amyloid-modifying drugs have failed in patients with AD dementia, indicating that to date it is not possible to slow progression. Thus, it would be ideal to initiate treatment with neuroprotective medications at a time when—or even before—AD is mildly symptomatic [3, 8]. We are moving rapidly toward this goal with the 2011 revision of the diagnostic criteria for AD to include prodromal and preclinical phases [9–11]. This scenario is true for many neurodegenerative diseases, and is even more compelling in diseases in which known genetic abnormalities can be identified that predict future disease, as in FTD [12], Huntington's disease [13], and autosomal dominant forms of AD [14].

To approach the goal of early intervention in neurodegenerative diseases, we must improve our capability to identify individuals in the earliest symptomatic phases of the diseases prior to significant functional impairment. For example, individuals are categorized as having mild cognitive impairment (MCI) when symptoms suggestive of AD are present but mild enough that traditional diagnostic criteria (which require functional impairment consistent with dementia) are not fulfilled. This gradual transitional state often lasts for a number of years, offering opportunities for intervention when the disease is in its prodromal stages. Diagnostic criteria for MCI have been developed [15] and operationalized [16] and subsequently revised including biomarkers to enable the identification of patients with MCI highly likely due to AD [10], or prodromal AD, which has been the target population in a number of clinical trials to date. If the pathophysiologic process of AD can be slowed at this stage of the disease, then it may be possible to preserve cognitive function and delay the ultimate development of dementia for a period of time, which is clearly clinically meaningful.

Finally, the presymptomatic phase of neurodegenerative diseases is the phase when pathologic alterations are developing but symptoms are not yet apparent. In the case of AD, this phase has been studied through the identification of cohorts with particular risk factors, such as genetic determinants (e.g., amyloid precursor protein (APP) or presenilin mutations, Down syndrome) or susceptibility factors (e.g., apolipoprotein E (*APOE*) ε4) or through biomarkers of the underlying disease process (e.g., amyloid imaging or spinal fluid). The development of research diagnostic criteria for preclinical AD has been transformative [9], and in the last few years therapeutic clinical trials of medications in preclinical AD have begun using all of these strategies [17].

3   Strengths and Weaknesses of fMRI as a Tool to Probe Brain Activity in Neurodegenerative Diseases

Since functional neuroimaging tools assess inherently dynamic processes that may change over short time intervals in relation to a host of factors, these measures have unique characteristics that may offer both strengths and weaknesses as potential biomarkers of neurologic disease. Functional neuroimaging measures may be affected by transient brain and body states at the time of imaging, such as arousal, attention, sleep deprivation, sensory processing of irrelevant stimuli, or the effects of substances with pharmacologic central nervous system activity. Imaging measures of brain function may also be more sensitive than structural measures to constitutional or chronic differences between individuals, such as genetics, intelligence or educational level, learning, mood, or medication use. While these may be effects of interest in certain experimental settings, they need to be controlled when the focus is on disease-related changes and differences between subject groups or within individuals over time.

Among functional neuroimaging techniques, fMRI has many potential advantages in studying patients with neurodegenerative disorders, as it is a noninvasive imaging technique that does not require the injection of a contrast agent. It can be repeated many times over the course of a longitudinal study and thus lends itself well as a measure in clinical drug trials. It has relatively high spatial and temporal resolution, and the use of event-related designs enables the hemodynamic correlates of specific behavioral events, such as successful memory formation [18] or momentary ratings of emotional intensity [19], to be measured.

A caveat essential to the interpretation of task-related functional neuroimaging data is that healthy individuals of any age demonstrate differences in brain activation depending on how well they are able to perform the particular task. For example, when cognitively intact individuals learn new information during fMRI scanning, the strength of this signal is related to subsequent ability to remember the information [20–24]. AD patients typically perform less well on the memory tasks, which complicates the interpretation of these data [25]. Conversely, the recruitment of additional brain regions during task performance by patients with neurodegenerative or other neurologic disease may indicate the presence of processes attempting to compensate for damaged networks [26, 27]. While the task performance factor is important to consider when designing or interpreting functional neuroimaging studies of neurodegenerative diseases, it also indicates that these imaging biomarkers may be particularly sensitive to changes in cognitive or sensorimotor function, which not only provides face validity for these measures but also supports their potential use in short-term, early proof-of-concept drug trials.

There are additional challenges to performing fMRI studies in patients with neurodegenerative diseases. The technique is particularly sensitive to even small amounts of head motion. Finally, although a growing number of test-retest reliability studies of task-related [28–39] and resting state [31, 40–44] fMRI have been published in the last 5–10 years, most of these are in healthy young adults. A small but growing number of longitudinal fMRI studies have been performed in patient populations; some of these have been focused on investigating reliability of signal [45–50] and others have been aimed at identifying changes in brain activation or connectivity associated with longitudinal clinical change [51, 52].

4   Clinical Applications of fMRI in Neurodegenerative Diseases

Functional MRI has been applied in a number of ways in studies of patients with neurodegenerative diseases. fMRI has been used to identify abnormal patterns of brain activity during the performance of a variety of tasks in patients with neurodegenerative diseases; these abnormal patterns may reveal new insights into the disruption of brain circuits by such diseases. They may also be useful in differential diagnosis. Similarly, studies of resting state fMRI have begun to demonstrate how this relatively newer method can provide insights into disrupted brain circuitry in neurodegenerative diseases. Hyperactivation or hyperconnectivity has been identified in many of these studies, which is a fascinating area of ongoing research [53]. fMRI has been used to assess the modulatory effects of genetic factors on brain activation in patients with or at risk for neurodegenerative disorders, and has been used to monitor the effects of therapeutic interventions and is beginning to be used to try to help predict the course of the diseases.

4.1   Patterns of Abnormal Regional Brain Activation During Task Performance

fMRI has been used to investigate abnormalities in patterns of regional brain activation during a variety of cognitive tasks in patients diagnosed with mild AD compared with control subjects. It is important to keep in mind that the particular abnormalities found in an fMRI study of an AD or other patient group are heavily dependent on the type of behavioral task used in the study—if the task does not engage a particular circuit, functional abnormalities will not likely be observed. Also, the nature of functional abnormalities may depend on whether the activated brain regions are directly affected by the disease, are indirectly affected via connectivity, or are not pathologically affected. Tools are now available to directly investigate the overlap of disease-related alterations in brain structure and task-related functional activity (Fig. 1). Yet it should also be kept in mind that even brain regions not usually thought to be affected by a particular neurodegenerative disease (e.g., sensorimotor areas in AD) have been shown to exhibit abnormal function [54, 55].

In addition to memory, which is discussed next, aspects of language and attention have been studied. Altered patterns of frontal and temporal activation have been observed in AD patients performing language tasks [56–58]. Similarly, although temporo-parietal activation was diminished in AD during performance of semantic memory task, increased activation in temporal and frontal

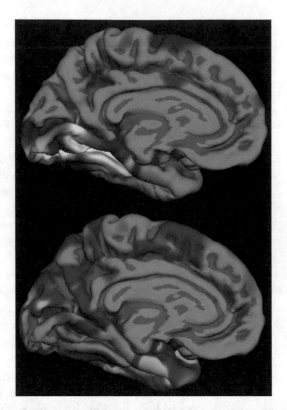

Fig. 1 The localization, magnitude, and extent of abnormalities observed in fMRI studies of patients with neurologic diseases depend on both localization and severity of pathology and on functional networks engaged by the particular fMRI task, as well as participant performance on the task. In this illustration, regions of cortical thinning in Alzheimer's disease from structural MRI (*bottom; brighter blue colors* indicate greater degree of thinning) are compared with cortical areas activated, as measured with fMRI, in normals during an event-related study of successful learning of new information that was able to later be freely recalled (*top, yellow colors* indicate greater blood oxygen level-dependent (BOLD) signal in the contrast of recalled items vs. fixation) (reproduced with permission from [76])

regions was also observed, suggesting possible compensatory processes [59]. Increased activation of the semantic memory circuits was observed in MCI patients when task performance remains at levels comparable to healthy controls [60]. This increased activation was particularly evident in the posterior cingulate cortex, posterolateral parietal cortex, and frontal cortex [60]. During performance of a visual attention task, AD patients showed alterations in parietal activation; increased prefrontal activation was also observed compared with controls, again suggesting possible compensatory mechanisms [61]. fMRI has been used to explore the neural basis of interesting phenomena in AD including abnormalities in discriminating foreground from background sounds (the "cocktail party effect" in auditory scene analysis) [62].

With respect to memory, a number of pioneering fMRI studies in patients with clinically diagnosed AD, using a variety of visually presented stimuli, have identified decreased activation in hippocampal and parahippocampal regions than in control subjects during episodic encoding tasks [63–67]. Neocortical abnormalities in AD have also been demonstrated using fMRI. A meta-analysis of both fMRI and fluorodeoxyglucose positron emission tomography (FDG-PET) memory activation studies of AD identified several cortical regions as showing greater encoding-related activation in controls than in AD patients, including the ventrolateral prefrontal cortex, precuneus, cingulum and lingual cortex [68]. In addition to AD-related differences in task-related blood oxygen level-dependent (BOLD) signal amplitude or spatial extent, the temporal dynamics of activation appear to be altered in patients with AD [69]. Increased activation in prefrontal and other regions has also been found in AD patients performing memory tasks [67].

While memory-task related fMRI data regarding medial temporal lobe (MTL) activation in individuals with MCI are less consistent than data from patients diagnosed with AD, with reports of both decreased and increased activation [26, 64, 66, 70–75], they indicate that differences are present in comparison to older controls. Some of the variability in fMRI data in MTL activation appears to relate to degree of impairment along the spectrum of MCI, which suggests that fMRI may be sensitive to relatively subtle clinical differences in disease severity [76, 77]. Similarly to studies of AD patients, investigations of posteromedial cortical deactivation during encoding in MCI patients has demonstrated reduced deactivation [78]. Other tasks have been used to demonstrate functional brain activation abnormalities in patients with MCI, including abnormalities of lateral temporal activation during word reading [79].

Since the advent of molecular markers of brain amyloid, including both amyloid imaging and CSF amyloid assays, research has intensified investigating task-related fMRI in cognitively normal (CN) older adults with brain amyloid, who are often viewed as having preclinical AD. One of the most widely replicated findings in CN amyloid-positive individuals is impaired posterior cingulate/precuneus deactivation during memory encoding [80–83], as well as impaired activation of this region during retrieval [81], and also in an attentional control task [84]. The MTL hyperactivates during encoding in association with increasing amyloid load in CN older adults [82, 85]. Prefrontal regions may also hyperactivate during successful encoding [82, 85], although not all studies demonstrate this finding [83]; in one study, the degree of hyperactivation was associated with memory performance [85], supporting the interpretation of possible compensatory hyperactivation. As tau PET imaging is developed [86], it will be exciting to test hypotheses regarding the effects of local hyperphosphorylated tau deposition on regional activation and connectivity.

Very little work has been done with task-related fMRI in patients with bvFTD, probably in large part due to the difficulties these patients often have in cooperating with task instructions and in the setting of MRI. During the viewing of faces conveying emotional expressions, bvFTD patients exhibit abnormally reduced activation in ventrolateral prefrontal cortex, insula, and a variety of other brain regions, but elevated activation in posterior parietal cortex [87]. fMRI has also been used to investigate the neural basis for changes in understanding of emotion conveyed through music [88] and in reasoning for personal vs. impersonal moral dilemmas [89] in bvFTD.

In PPA, an early fMRI study using simple phonological and semantic tasks demonstrated relatively normal activation within canonical regions of the language network but with increased recruitment in areas not typically recruited by controls; these increases correlated with greater language impairment [90]. Sonty et al. subsequently used dynamic causal modeling to show that effective connectivity between canonical language regions was impaired in PPA [91].

In a study of syntactic comprehension in nonfluent variant PPA patients, the caudal inferior frontal cortex did not show relatively greater activation for complex sentences than for simple sentences, as it did in controls [92].

Vandenbulcke et al. [93] used an associative-semantic fMRI paradigm to demonstrate that patients with the semantic variant of PPA show an abnormal rightward lateralization of anterior temporal activation, which is reminiscent of a similar language laterality shift that has been reported in patients with stroke aphasia. This paradigm was also used in nonfluent variant PPA patients to identify a similar effect [94]. The phenomenon of surface dyslexia was investigated in semantic variant PPA; in patients but not controls the inferior parietal cortex was recruited for irregular words, but mid-fusiform and superior temporal cortex was under-recruited in PPA patients [95].

Patients with the semantic variant of PPA have also been studied using a paradigm comparing meaningful to meaningless sounds. Compared with controls, patients showed abnormal activation of dorsolateral temporal cortical areas for both meaningless sounds as well as for meaningful sounds (animal sounds versus tool sounds), suggesting that aberrant processing of sounds in semantic PPA extends to pre-semantic perceptual processing [96]; prior work by the same authors demonstrated abnormal auditory processing in patients with the nonfluent variant of PPA as well [97].

fMRI has been used to study abnormal patterns of brain activation in a variety of tasks in patients with Parkinson's disease (PD), most commonly in tasks engaging the motor system [98, 99]. fMRI studies investigating brain activations in PD patients during self-initiated movements demonstrated evidence of reduced neural responses in the pre-supplementary motor areas (SMA), along with hyperactivation in both the lateral premotor cortex and parietal cortex

[100–102]. Some authors interpret these findings as a shift from medial to lateral premotor loops, possibly indicating impairment in generating internal cues for movement and greater dependence on external cues [103–105]. In addition, cognitive tasks have been employed, such as a working memory paradigm that demonstrated hypoactivation in fronto-striatal regions in PD patients with cognitive impairment compared with those patients who were not cognitively impaired [106]. These findings were confirmed by a large fMRI study showing a frontostriatal hypoactivation during a verbal two-back working memory task in PD patients [107]. In the PD group, patients with MCI (PD-MCI) had an additional hypoactivation of the bilateral anterior cingulate cortex and the right caudate nucleus compared with non-MCI PD patients [107] (Fig. 2). Interestingly, PD-MCI patients also showed a lower single-photon emission computed tomography dopamine presynaptic uptake in the right caudate than patients without MCI, which correlated with striatal fMRI signal [107]. The relationship between cognitive impairment and frontostriatal dysfunction was also suggested by a study showing that patients with PD-MCI had a reduced activity of the premotor cortex and cognitive corticostriatal loop, which includes the caudate nucleus and prefrontal cortex, while planning a set shift during fMRI, whereas non-MCI patients experienced activation patterns similar to those of healthy participants [108]. Impulsive behavior is an important problem in some patients with PD, and fMRI paradigms have been developed to understand the neural basis of these symptoms as well; behavioral abnormalities and their underlying neural substrates are exacerbated with dopaminergic therapy [109, 110].

Very little work using fMRI has been performed in patients with diffuse Lewy body disease (DLB), with one recent study demonstrating a complex set of differences between DLB and AD patients in visual cortical activation during face, color, and motion perceptual tasks, many of which were explainable by differences in task performance [111].

In Huntington's disease, a growing body of work has employed attentional or executive tasks to probe fronto-parietal or fronto-striatal systems using fMRI. In an early study, reduced activation was found in multiple cortical regions in HD patients performing a serial reaction time task, compared to controls [112]. Both increased and decreased recruitment has been observed on working memory tasks in HD patients [113, 114]. On a visual attention/response inhibition task, HD patients showed reduced prefrontal-anterior cingulate interhemispheric functional connectivity, and reduced connectivity predicted slower reaction times and increased numbers of errors on the task [115]. On a set-shifting task, HD patients showed greater prefrontal activation than controls; a lesser degree of activation was associated with more prominent performance impairment on the task and neuropsychiatric symptoms [116]. Reduced error-related activation was also found in an anti-saccade task [117].

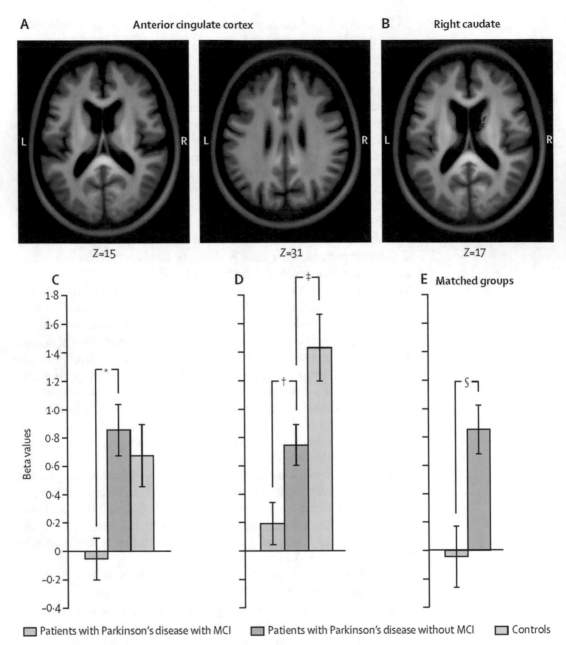

Fig. 2 Pattern of fMRI signal during a verbal two-back working memory task in PD patients and cognitive impairment. Significant under-recruitment occurred in (**a**) the bilateral anterior cingulate cortex and (**b**) the right caudate in PD patients with MCI ($n=30$) compared with those without MCI ($n=26$; *red*). Mean beta values in (**c**) the anterior cingulate cortex cluster and (**d**) right caudate, contrasting two-back with baseline conditions for PD patients with (*blue*) and without (*red*) MCI and control individuals (*green*). (**e**) Mean beta values in the right caudate, contrasting two-back with baseline for motor-matched (Unified Parkinson's disease Rating Scale III) groups for PD patients with MCI and without MCI. The group sizes were held constant ($n=18$ in each group) and matched by scanner model. Error bars are 1 standard error. Figure reprinted from [107] with permission

In ALS, fMRI has been used to demonstrate abnormalities in motor cortical activation. Furthermore, it has been used to investigate whether there are different patterns of cortical activation during a simple motor task in ALS patients with a primarily upper motor neuron (UMN) pattern of clinical deficits vs. those with a lower motor neuron (LMN) pattern [118]. The UMN patient subgroup showed relatively greater activation of the anterior cingulate and caudate than the LMN subgroup, suggesting that fMRI may provide insights into the differentiation of disease subtypes. Since many patients with ALS develop cognitive and/or affective symptoms (and in some cases full-blown frontotemporal dementia), fMRI paradigms including verbal fluency [119], inhibitory control [120], and an emotional word viewing task [121] have been used to demonstrate functional brain activation abnormalities in relevant circuits. Further work is necessary to better understand whether fMRI task activation abnormalities may predict the emergence of symptoms impacting function in daily life.

4.2 Abnormalities in Functional Connectivity in Patients with Neurodegenerative Diseases: Resting State fMRI

Hypoperfusion/metabolism is typically seen with nuclear medical imaging techniques (such as FDG-PET or SPECT) in temporoparietal/posterior cingulate cortical regions in AD patients during the "resting" state. The medial parietal/posterior cingulate cortex, along with medial frontal and lateral parietal regions, compose a "default mode" network (DMN) that is generally more active when individuals are not engaged in specific tasks [122], and which is thought to play a role in memory, self-appraisal, self-referential planning, and a variety of other mental states [123]. Multiple studies in AD patients have demonstrated alterations in the deactivation and functional connectivity of these regions [124–127]. Substantial overlap is present between the DMN and the localization of amyloid PET tracer binding [128].

These early observations have been markedly enriched in the past decade with the explosion of research using resting state fMRI. More than 20 years ago, Biswal originally observed [129] that low-frequency fluctuations in BOLD signal amplitude could be measured while subjects lie "at rest" in the scanner without engaging in a specific task, and that correlations between the oscillations in a "seed" region of interest could be used to identify correlated signal in other areas that are likely part of the same large-scale circuit, and that share the same topography as regions activated during a finger-tapping task. Similar findings can be obtained using data-driven analysis methods such as independent component analysis. Although maps of networks obtained using resting state fMRI methods are thought to represent connections beyond strict monosynaptic connections, the topography of networks identified using these methods has been validated against traditional tract tracing methods in nonhuman primates as reflecting the major gray matter sites of origin and termination of large-scale networks [130]. One advantage of using resting state fMRI methods in patients with cognitive impairment or dementia is that the challenges of engaging patients in specific tasks can be avoided.

A growing body of work has developed confirming Biswal's original observation across multiple networks, showing the correspondence between the topography of distributed neural networks identified at rest and those engaged by task performance in traditional task-related fMRI [131]. Major networks have been identified using these methods including somatosensory, visual, auditory, language, attention/executive function, affective, memory, DMN, and other networks [131, 132].

In resting state fMRI studies of AD dementia, one of the most reproduced findings from is the consistent reduction of connectivity within the DMN [133–138]. As has been pointed out more than a decade ago, this should not be surprising given that amyloid deposition, hypometabolism, and atrophy convergently affect many key nodes of the DMN in patients with AD [128]. The magnitude of reduced connectivity within the DMN correlates with the severity of symptoms [133, 135, 136, 139]. The sensitivity of resting state fMRI measures in differentiating AD patients from healthy elderly controls ranges from 72 to 85% and specificity from 77 to 80% [125, 140, 141]. Longitudinal data have begun to suggest that some subnetworks within the DMN may develop hyperconnectivity early in the disease course while others show hypoconnectivity, but all decline over time [142]. A few studies have begun to compare connectivity at rest to connectivity during tasks; typically, controls show greater connectivity within the DMN at rest than they do while performing a task (presumably because some nodes of the DMN may contribute to the task). One recent study showed that AD patients fail to modulate connectivity within the DMN during a simple visual attention task [143]. Another study showed that AD patients had greater memory encoding task-related prefrontal activation and greater prefrontal resting state connectivity than controls [144].

Since AD dementia is a syndrome of multiple domains of cognitive impairment, it stands to reason that other networks would be disrupted as well. However, evidence to date is not consistent. There is some evidence demonstrating reduced connectivity in multiple networks in addition to the DMN [136], but not all studies demonstrate such widespread effects. This is likely in part related to the heterogeneity of AD itself and of the patient samples in these studies, particularly with regard to the types and severity of symptoms. In addition, some findings point to increased connectivity in the salience network [133] and frontoparietal network [144], although conflicting data have been reported showing reduced rather than increased connectivity within these networks [136, 145]. The connectivity within the executive control network, in contrast, appears more consistently increased in AD [144–146], which is surprising given the common occurrence of executive dysfunction in these patients, although in one of these studies the increased connectivity was correlated with neuropsychological performance [146].

Although AD dementia is typically a multidomain amnesic dementia, major subtypes are well recognized. Two of these have received preliminary investigation with regard to the overlap of the topography of atrophy with the topography of healthy functional connectivity network. In posterior cortical atrophy (PCA), the atrophy pattern mapped most closely onto higher visual networks, while in logopenic variant PPA (lvPPA), the atrophy pattern mapped most closely onto the language network [147]. It is also well-recognized that early-onset AD differs in many respects from late-onset AD; whereas in late-onset AD the DMN is most prominently affected as described above, in early-onset AD the salience and executive control networks seem most robustly affected [148, 149]. This fits with the atypical clinical phenotypes commonly seen in early-onset AD [150].

Decreased DMN connectivity was also found to be predictive of clinical conversion to AD in patients with MCI [139, 151]. With the rise of biomarkers of amyloid, an increasing number of studies have investigated the functional connectivity abnormalities associated with preclinical AD (asymptomatic cerebral amyloidosis). There is consistent evidence for disrupted DMN connectivity in preclinical AD [152–156]. See below for a discussion of *APOE* and other genetic effects.

In patients with bvFTD, the resting state fMRI connectivity of the salience network is consistently abnormal [133, 145]; in one study, this was seen in both bvFTD and semantic dementia [157]. The degree of abnormality of functional connectivity in the salience network correlates with symptom severity [133, 157]. Interestingly, some investigations have reported increased DMN connectivity in bvFTD [133], although other studies have reported reduced DMN connectivity [145, 158]. In another study of semantic variant PPA, graph theoretical analyses of resting state fMRI data demonstrate loss of integrity in multimodal ventral temporal cortex as well as in the modality-specific caudal ventral visual stream [159]. Surprisingly, there have been no other studies of functional connectivity in PPA.

In patients with the typical "Richardson" PSP clinical syndrome, resting state fMRI has been used to identify a dorsal midbrain tegmentum-thalamocortical network in healthy individuals that shows reduced connectivity in patients with PSP, with the magnitude of reduced connectivity correlating with symptom severity [160]. The topography of this network bears remarkable similarity to the distributed set of regions known to accrue tau pathology in PSP [161]. In a separate study, PSP patients showed reduced thalamo-cortical connectivity which correlated with cognitive and motor symptoms [162]. To date, there have been no resting state fMRI studies of CBD.

In ALS, most resting state fMRI studies have focused on the sensorimotor network, with findings of either reduced or mixed changes in connectivity relative to controls [163]. Very little work in this area has included participants with cognitive or behavioral impairment and there is not yet a study using resting state fMRI in patients with ALS-FTD. In one study, a subset of the ALS patients exhibited cognitive or behavioral impairment [164]. As a group, the ALS patients showed both reductions and increases in connectivity within the DMN and frontoparietal network with no differences in executive or salience networks. Within the group of ALS patients, there was an inverse correlation between performance on the Wisconsin Card Sort Test and connectivity within the DMN as well as within the frontoparietal network, suggesting that at least some of the increases in connectivity relate to impairments in cognitive performance.

Early resting state fMRI studies of DLB produced substantially differing results. One study showed that connectivity was reduced between the precuneus seed and medial prefrontal cortex and hippocampus, while it was increased between the precuneus seed and regions of the dorsal attention network and putamen [165]. Another study found increased connectivity between the posterior cingulate cortex seed and anterior cingulate, culmen, cerebellum, and putamen with no areas of decreased connectivity [166]. In another analysis focused on subcortical connectivity, seed regions in bilateral caudate, putamen, and thalamus showed greater connectivity with other structures than controls [167], with no reduced connectivity compared to controls. Using independent component analysis, another analysis demonstrated reduced functional connectivity within the DMN, salience, and executive networks with increased basal ganglia connectivity compared to controls [168].

Several resting state fMRI studies have been conducted of PD with mild cognitive impairment (PD-MCI) or PD dementia (PDD). In one study which employed seed-based analysis, the DMN showed no differences but the caudate seed showed reduced connectivity in PDD with dorsolateral prefrontal cortex and putamen [169]. Another group used the posterior cingulate cortex/ precuneus as a seed for a DMN analysis and reported decreased connectivity in the right inferior frontal gyrus in PDD as compared to PD and controls [170]. In another study, PDD patients demonstrated reduced connectivity within the DMN compared to PD patients without cognitive impairment, and widespread reductions compared with controls [171]. In another study of PD with or without MCI, the DMN was found using independent component analysis to be reduced in both groups compared to controls, while the connectivity of bilateral prefrontal cortex within the frontoparietal network was reduced and the strength of connectivity correlated with visuospatial cognitive test performance [172]. Finally, another study examined local resting signal fluctuations in PD-MCI or PDD and found reduced spontaneous brain activity in regions important for motor control (e.g., caudate, supplementary motor area, precentral gyrus, thalamus), attention and executive functions (e.g., lateral prefrontal cortex), and episodic memory (e.g., precuneus, angular gyrus, hippocampus) [173].

Another study employed both independent component and seed-based analyses to study PD-MCI. Independent component analysis results revealed reduced connectivity between the dorsal attention network and frontoinsular regions, associated with worse performance in attention/executive functions. The DMN displayed increased connectivity with medial and lateral occipito-parietal regions, associated with worse visuospatial performance. The seed-based analyses mainly revealed reduced within both the DMN and dorsal attention network, along with disruptions of dorsal attention-frontoparietal and DMN-dorsal attention network interactions [174].

A series of recent resting state fMRI studies have investigated group differences in PD and DLB compared to each other or to patients with PD but no cognitive impairment. In a study investigating both PDD and DLB, PDD was associated with local connectivity reductions in frontal regions while DLB was associated with local connectivity reductions in posterior cortical regions [175]. Another study compared PDD and DLB and found reductions in both frontoparietal networks and supplementary motor network in both diseases [176].

Other recent studies have focused on resting state fMRI correlates of specific symptoms. Fluctuations in cognition in DLB were associated with reduced connectivity in the left frontoparietal network [177]. Apathy in PD was found to be associated with reduced connectivity in limbic striatal and frontal circuits, but surprisingly, the severity of apathy was inversely correlated with connectivity in these circuits [178]. Visual hallucinations in PD were associated with increased occipito-striatal [179] and DMN connectivity [180].

In HD, several resting state fMRI studies have been conducted with mixed results in analysis of sensorimotor and basal ganglia–thalamocortical networks [181–184]. With regard to cognitive networks, HD patients showed widespread reduction in synchrony in the dorsal attention network, which was associated with poorer cognitive performance [182]. Widespread DMN changes, not correlating with the atrophy of the involved nodes, also appear to be present in symptomatic HD patients [181], and correlate with cognitive disturbances [183].

In part from functional connectivity MRI studies (as well as from diffusion tensor imaging structural connectivity MRI studies), the hypothesis that neurodegeneration progresses along connectional pathways has received greater support. Seeley et al. [185] provided a compelling set of results relating the spatial topographic patterns of atrophy in five neurodegenerative syndromes (AD, bvFTD, semantic variant PPA, nonfluent variant PPA, and cortico-basal syndrome) to the topography of large-scale networks in the healthy brain as measured by resting state fMRI and gray matter structural covariance. This seminal observation has spurred a growing body of research and links with studies of the potential cell-to-cell transmission of misfolded proteins [186]. Technical advances in graph theoretical and related mathematical modeling approaches to functional and diffusion MRI data are leading to new ideas about how the human brain connectome may provide important context for neurodegeneration [187].

4.3 Compensatory Hyperactivation: A Universal Adaptation Response to Brain Injury?

Aside from neocortical hyperactivation in AD, consistent data suggest that there is a phase of increased MTL activation in MCI (Fig. 3) [188]. This increase, which also may be present in cognitively intact carriers of the *APOE-ε4* allele (for review, *see* [189]), may represent, at least in part, an attempted compensatory response to AD neuropathology [190–192], given that some MCI individuals with smaller hippocampal volume perform similarly on memory tasks to MCI individuals with larger hippocampal volume but have relatively greater MTL activation [26, 74]. Additional studies employing event-related fMRI paradigms [18, 24, 75] will be very helpful in determining whether increased MTL activation in MCI patients is specifically associated with successful memory, as opposed to a general effect that is present regardless of success (possibly indicating increased effort). Whether or not hyperactivation is associated with better memory performance and thus could be viewed as behaviorally compensatory, it appears to be associated with more prominent neurodegeneration [193] and poorer prognosis [194]. It is possible that MTL hyperactivation reflects cholinergic or other neurotransmitter upregulation in MCI patients [195]. Alternatively, increased regional brain activation may be a marker of the pathophysiologic process of AD itself, such as aberrant sprouting of cholinergic fibers [196] or inefficiency in synaptic

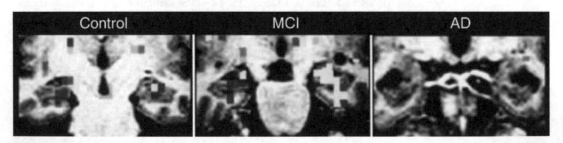

Fig. 3 A phase of compensatory hyperactivation appears to occur in the medial temporal lobe (MTL) in mild cognitive impairment (MCI), prior to the clinical onset of Alzheimer's disease (AD) dementia. Representative single subjects from each group, showing normal memory-related MTL activation measured with fMRI in normal older controls, hyperactivation, and very mild atrophy in MCI, and hypoactivation and more prominent atrophy in mild AD (reproduced with permission from [76])

transmission [197] or aberrant synchronous firing along the lines of epileptiform discharges [198]. In fact, the treatment of MTL hyperactivation using the anticonvulsant levitiracetam in MCI is associated with improved memory performance [199]. It is important, however, to acknowledge that multiple nonneural factors may confound the interpretation of changes in the hemodynamic response measured by BOLD fMRI, such as age- and disease-related changes in neurovascular coupling [54, 55], AD-specific alterations in vascular physiology [200], and resting hypoperfusion and metabolism in MCI and AD [201], which may result in an amplified BOLD fMRI signal during activation [202, 203]. Further research to determine the specificity of hyperactivation with respect to particular brain regions and behavioral conditions will be valuable to better characterize this phenomenon [76, 77].

"Compensation" is typically defined as greater regional brain activity (hyperactivation) in an MCI/AD group in the setting of task performance accuracy that is similar to that of a matched control group. Regional hyperactivation may involve greater magnitude of activity in brain regions typically active during performance of the task (when performed by controls), or the recruitment of additional brain regions not normally engaged by controls. However, it is also clear that greater task difficulty may provoke similar alterations in regional brain activity in healthy individuals [204–207]. It is challenging to know to what degree MCI/AD groups find memory tasks to be "more difficult" than they would in the absence of disease. This has led some investigators to attempt to match task difficulty between MCI/AD patients and controls [192]. It is also possible that different cognitive strategies during memory task performance (e.g., semantic elaborative encoding strategies vs. visualization strategies) may contribute to differences in the recruitment of particular brain regions [208] and that this may vary between patient and control groups. Further work in this area, including longitudinal studies in MCI/AD patients [52], ideally including detailed behavioral measures of reaction time as well as accuracy and possibly self-report of task difficulty, will be important to better clarify the situations in which activity increases can be reasonably interpreted as compensatory for brain disease.

In PD, a potential compensatory response was demonstrated in an fMRI study showing that maintenance of movement is accompanied by hyperactivation in lateral premotor areas [209]. In the same subjects, dopaminergic therapy normalized these activation patterns in the setting of constant motor performance (see later for additional discussion of pharmacologic fMRI). Monchi et al. [210] showed that, during a set-shifting task, PD patients have reduced ventrolateral prefrontal activity relative to controls, but greater dorsolateral prefrontal activity, suggesting not only that frontostriatal circuits are dysfunctional in PD but that there also may be attempted compensatory activity. An elegant study of motor imagery in PD patients with strongly lateralized symptoms demonstrated that when patients judged the laterality of hand images in different orientations, occipito-parietal cortex hyperactivation was most prominent for imagery employing the affected hand compared with the unaffected hand [211].

In HD, hyperactivation was observed in multiple regions during a visual attention/interference task [212]. Notably, greater premotor activation correlated with a greater degree of clinical impairment, supporting the conjecture that response that is at least attempting to compensate for the disease. The authors suggest that the HD patients may have required increased effort to inhibit inappropriate motor responses. As in most studies identifying possible compensatory hyperactivation, there are a number of other interpretations of how hyperactivation may relate to severity of illness.

In ALS, sensorimotor cortical hyperactivation is present in comparison to both healthy normal controls and to controls with peripheral motor weakness, indicating that the hyperactivation is not purely a reflection of weakness [213]. Schoenfeld et al. [214] used a button-press sequencing task to investigate whether task difficulty level was primarily responsible for greater activation within motor circuits. Although during the simple task ALS patients showed motor hyperactivation compared with controls, when the task was manipulated such that controls had to respond more rapidly and thus made more errors (equivalent to those of ALS patients in the simpler task), motor activation was similar between the two groups. Functional recruitment of cerebral regions involved in motor learning has also been noted in ALS patients, including basal ganglia, cerebellum, and brainstem [215].

Despite the caveats mentioned earlier with regard to many of the studies of hyperactivation in neurodegenerative diseases, accumulating evidence suggests that task-related regional brain hyperactivation may be a universal neural response to insult, as it occurs in sleep deprivation [216], aging [217], and a variety of neuropsychiatric disorders and conditions, including AD/MCI, PD, ALS, cerebrovascular disease [218, 219], multiple sclerosis [220, 221], traumatic brain injury [222], human immunodeficiency virus (HIV) [223], alcoholism [224], and schizophrenia [225]. In many of these studies, task-related regional brain hyperactivation was associated with the relative preservation of performance on the task, suggesting that hyperactivation may be serving, at least in part, a compensatory role for neurologic insult. The evidence discussed earlier also indicates that increased MTL activation can be seen in MCI in the setting of minimal MTL atrophy [70], which provides in vivo support for laboratory data suggesting that physiologic alterations may precede significant structural abnormalities very early in the course of a neurodegenerative disease such as AD [4, 226] and may represent inefficient neural circuit function [197]. Thus, fMRI may provide a means to detect changes in

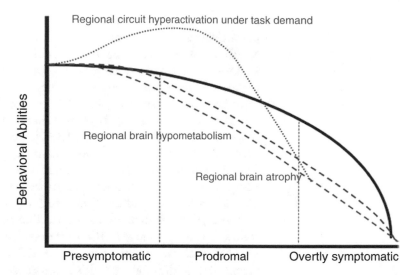

Fig. 4 Illustration of degenerative-compensatory model, proposed as a universal response to brain insult. In the case of neurodegenerative diseases, the model proposes that there is a phase of task-related hyperactivation of regional brain circuits subserving task performance, followed by the gradual development of regional hypometabolism and atrophy as the disease progresses from presymptomatic to prodromal to overtly symptomatic phases

human brain circuit function that underlie the earliest symptoms of neurodegenerative diseases and may be useful in identifying groups of subjects at high risk for future decline prior to a clinical diagnosis of these diseases (Fig. 4).

It is possible, however, that hyperactivation reflects inefficient function of neural circuits in the face of injury, and that such a response may be deleterious in the long run. Thus, it will be critical to elucidate the relationships between behavioral performance, neural circuit function, and clinical course of disease, with the ultimate goal of determining how best to use these fMRI measures as biomarkers of putative therapeutic response in clinical trials. Numerous resting state fMRI studies have also found increased connectivity [227], and efforts are also ongoing to better interpret these findings in the context of clinical or longitudinal data.

4.4 The Modulatory Effects of Genetic Risk Factors for Neurologic Disease on Brain Activation

In the last decade, there has been an explosion in literature on imaging and genetics, primarily in psychiatric disorders [228] and the basic science of genetic modulators of brain function [229, 230]. This is an area that is ripe for study in neurologic disease, with a number of studies having been done in populations at elevated genetic risk for AD, FTD, ALS, and HD.

The *APOE* ε4 allele is a major genetic susceptibility factor associated with increased risk for AD. Several fMRI studies have investigated regional brain activation during task performance in cognitively intact subjects stratified by their *APOE* allele status. Smith et al. reported decreased activation in inferior temporal regions on a visual naming and a letter fluency fMRI paradigm (there was no hippocampal or other medial temporal activation reported with these tasks) in *APOE* ε4 carriers [231]. In a subsequent report, this group reported increased parietal activation in women with an *APOE* ε4 allele [232]. Bookheimer et al. reported increased activation in left hippocampal, parietal, and prefrontal regions among *APOE* ε4 carriers, compared with noncarriers, using a word-pair associative memory paradigm [233]. In addition, an increased number of activated regions in the left hemisphere at baseline was associated with a decline in memory at the 2-year follow-up among the *APOE* ε4 carriers. The authors hypothesized that this increase in activation in the *APOE* ε4 carriers might represent the additional cognitive effort or neuronal recruitment required to adequately perform the task. Similarly increased activation in multiple brain regions was recently reported in cognitively intact older *APOE* ε4 carriers compared with ε3 carriers, although the effect was lateralized to the right MTL region (left hippocampal activation was greater in ε3 carriers) [234]. Among a group of 29 controls, MCI subjects, and AD patients, Dickerson et al. reported that 13 *APOE* ε4 carriers demonstrated greater entorhinal activation than noncarriers, in the absence of genotype-related differences in the volumes of these regions [70]. Other studies suggest that decreased medial temporal activation may also be seen in *APOE* ε4 carriers [235].

With regard to functional connectivity, a number of studies have identified reduced DMN connectivity in cognitively normal middle-aged or older adults who carry the *APOE* ε4 allele [236–239], suggesting the possibilities that either early preclinical pathological changes associated with AD may be present or that this genetic risk factor itself may alter brain network function (Fig. 5). In these studies, the amyloid status of participants was unknown. In another investigation, amyloid-negative *APOE* ε4 carriers were compared to noncarriers and found to have reduced DMN connectivity [240]. This latter finding raises the question of whether *APOE* itself influences the efficiency of this neural network independently of its promotion of amyloid pathology. In support of the hypothesis that *APOE* modulates network efficiency, two studies of healthy young adults have found increased DMN connectivity in *APOE ε4* carriers [241, 242]. These findings are challenging to interpret at present, but raise the possibility that *APOE ε4* may have both developmental effects on the connectivity of the DMN as well as its well-known effect of increasing the likelihood of amyloid pathology as a function of age.

A number of studies of functional activation and connectivity in autosomal dominant forms of AD have been conducted in the past

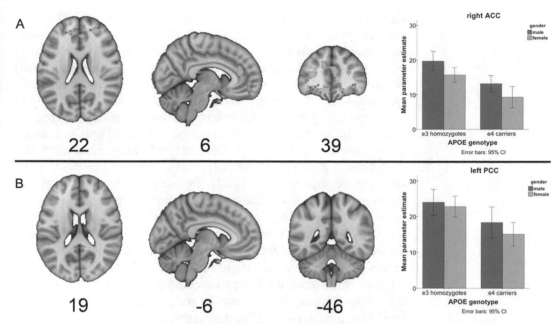

Fig. 5 *APOE* ε4 status reduces DMN connectivity in healthy older adults. Functional connectivity is decreased in ε4 carriers compared to ε3 homozygotes in both the (**a**) anterior, and (**b**) posterior DMN. The bar graphs depict the distribution of the effects across subgroups. They show the mean parameter estimates of a selected region within the anterior DMN (the right anterior cingulate gyrus) and the left posterior cingulate cluster within the posterior DMN, for male and female ε3 homozygotes and male and female ε4 carriers. The difference across both genotype and gender is significant for the anterior cingulate gyrus. For the posterior cingulate gyrus only the difference across genotype is significant. The statistical maps are overlaid on the Montreal Neurological Institute (MNI) 152 brain; MNI coordinates (in mm) of the slices are displayed. Figure modified from [239] with permission

decade. During memory encoding, presymptomatic carriers of *presenilin* (*PSEN*) mutations show hippocampal hyperactivation [243, 244] and reduced precuneus deactivation [244, 245]. DMN functional connectivity is reduced in presymptomatic carriers of autosomal dominant mutations as they approach estimated age of onset in their family [134, 246]. Yet children who carry these mutations show increased DMN functional connectivity [245], again suggesting a possible inverse U-shaped curve of hyperconnectivity prior to loss of connectivity. These findings are supported by similar previous results in a small study of two *PSEN* mutation carriers compared to three noncarriers [247]. In summary, these findings suggest that the increased activation seen in the *APOE* ε4 allele carriers may not be due solely to compensatory mechanisms but may indicate, in part, an independent physiological mechanism.

With regard to FTD, a few studies of resting state fMRI have been performed in presymptomatic carriers of autosomal dominant mutations associated with these FTD and/or ALS. In eight presymptomatic *MAPT* carriers, reduced DMN connectivity was observed with no change in salience network connectivity [158]. In a mixed group of *MAPT* and *GRN* carriers, presymptomatic mutation carriers showed an age-related reduction in connectivity between the anterior insula and the anterior cingulate cortex [248] which was not seen in controls, suggesting that as carriers approach symptom onset the integrity of this network declines. In a group of *GRN* carriers, presymptomatic carriers showed increased connectivity in the salience network on one study [249] and in another study showed reduced connectivity in the frontoparietal network/dorsal attention network with increased connectivity within the executive control network [250]. In this same study, the authors investigated the modulatory effects of a polymorphism in *TMEM106B* which has been shown to influence age of onset in *GRN* carriers; they found that carriers of the increased risk allele showed reduced connectivity in the frontoparietal network and ventral salience network compared to noncarriers [250]. One resting state fMRI study has been conducted of symptomatic carriers of the *C9orf72* repeat expansion, a major genetic factor contributing to FTD and ALS. FTD patients with this mutation show reductions in connectivity in the salience and sensorimotor networks; salience network connectivity reduction correlated with atrophy in the pulvinar thalamic nucleus, which is known to exhibit pathology in this form of the disease [251].

A few task-related fMRI studies have been performed in individuals with presymptomatic HD. An attentional interference task was used to identify reduced anterior cingulate activation in presymptomatic *huntingtin* mutation carriers compared to controls; carriers also had subtle performance abnormalities on the task [252]. Two studies have focused on mood-related abnormalities. In the first, negative feedback on performance was given to participants, and abnormally reduced amygdala activation and aberrant amygdala-orbitofrontal coupling was observed in presymptomatic HD individuals [253]. In the other study, a mood induction task was used to identify increased activation in presymptomatic HD individuals relative to controls of pulvinar, cingulate cortex, and somatosensory association cortex; pulvinar activation correlated inversely with putaminal gray matter volume and directly with clinical ratings of irritability [254].

Reduced dorsolateral prefrontal cortical activation was observed in presymptomatic HD individuals on a working memory task [255]; this reduction was stable over a 2-year period [256]. Working-memory task-related functional connectivity was reduced in presymptomatic HD [257]. Presymptomatic HD individuals showed increased recruitment of prefrontal regions but reduced task-related connectivity between these regions longitudinally over an 18-month period [258].

In a resting state functional connectivity study, presymptomatic HD individuals showed decreased correlated activity in the sensorimotor and dorsal attention networks; decreased level of synchrony in the sensorimotor network was associated with poorer motor performance [182]. In presymptomatic HD individuals, DMN connectivity was reduced [259]. In another study of presymptomatic HD, the authors found no longitudinal change in resting-state connectivity over 3-year period [260].

4.5   fMRI as a Predictive Biomarker

Very little work has been done on the use of fMRI as a predictive biomarker for prognosis. Miller et al. pursued such a study of a group of 25 senior citizens spanning the spectrum of MCI, none of whom were demented at the time of baseline assessment, but who exhibited varying degrees of mild symptoms of cognitive impairment clinically [as measured using the CDR sum-of-boxes (CDR-SB)] [194]. At baseline, subjects performed a visual scene-encoding task during fMRI scanning and were clinically followed longitudinally after scanning. Over about 6 years of follow-up after scanning, subjects demonstrated a wide range of cognitive decline, with some showing no change and others progressing to dementia (change in CDR-SB ranged from 0 to 6). The degree of cognitive decline was predicted by hippocampal activation at the time of baseline scanning, with greater hippocampal activation predicting

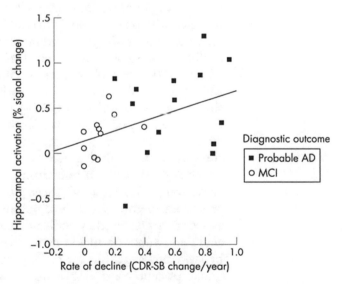

Fig. 6 fMRI as a predictive quantitative imaging biomarker. In a group of mild cognitive impairment patients, hippocampal activation at baseline predicts the degree of cognitive decline over 6 years after scanning. Scatterplot shows, on the *Y*-axis, parameter estimates (representing percent BOLD signal change) of differential hippocampal activation in novel versus repeated contrast. The *X*-axis shows estimated rate of change in CDR SB score per year after fMRI scan in participants who remained classified as having MCI and in those who were diagnosed with probable AD during the follow up interval. Figure reprinted from [194] with permission

greater decline (Fig. 6). This finding was present even after controlling for baseline degree of impairment (CDR-SB), age, education, and hippocampal volume, as well as gender and *APOE* status. Similarly, a longitudinal fMRI study showed that healthy subjects with more rapid cognitive decline over a 2-year period had both the highest hippocampal activation at baseline and the greatest loss of hippocampal activation over follow up, where the rate of activation loss correlated with the rate of cognitive decline [52]. These data suggest that fMRI may provide a physiologic imaging biomarker useful for identifying the subgroup of MCI individuals at highest risk of cognitive decline for potential inclusion in disease-modifying clinical trials. Thus, hyperactivation might represent an early response to AD pathology, which may predict forthcoming hippocampal failure and memory decline.

4.6 Uses of fMRI in Understanding and Monitoring Neurotherapeutics

FMRI may be particularly valuable in evaluating acute and subacute effects of therapeutic interventions—whether pharmacologic, non-pharmacologic (e.g., transcranial magnetic stimulation), or behavioral/rehabilitative—on neural activity [261]. This may be useful for showing that the intervention modulates targeted circuits and may help elucidate mechanisms of action. Additionally, it may help identify or even predict treatment responders or nonresponders.

Alterations in memory-related activation related to the administration of pharmacologic agents known to impair memory can be detected with pharmacologic fMRI [262–264]. The effects of cognitive enhancing drugs on brain activation during cognitive task performance have shown that fMRI can detect changes after administration of cholinesterase inhibitors in patients with AD and MCI [265, 266]. In one of the earliest studies, after receiving a single dose of galanthamine, AD patients demonstrated increased fusiform activity during a face encoding task and increased prefrontal activity during a working memory task [265]. Another study showed that acute dosing increased hippocampal activation during a memory task while chronic dosing was associated with decreased activation [267]. Although these pilot studies did not include placebo-control groups to reduce potential confounding factors, such as learning effects, they indicate that fMRI is sensitive to both acute and subacute medication effects, some of which relate to behavioral change. Similar findings have been observed in subsequent placebo-controlled acute dosing studies [268]. Another placebo-controlled acute dose physostigmine study demonstrated normalization of both visual stimulus-specific activation in occipital and attention-dependent activation in frontoparietal regions in patients with AD [269].

Pharmacological fMRI has also been used to evaluate the effects of chronic treatment over 2–6 months in patients with AD dementia. For the most part, these studies have demonstrated treatment-related increased activation in brain regions engaged by the variety of tasks used [270–272]. Whereas some studies have found decreased activation in areas engaged by the task at baseline [267, 273, 274]. However, some studies have demonstrated a more complex effect, with relatively greater prefrontal activation increases being associated with a more advanced stage of cognitive impairment [268], supporting the perspective that task-related activation in the MCI-AD dementia spectrum whether on or off drug changes with the course of the disease, presenting challenges for attempts at group comparisons. Nevertheless, at least two studies have demonstrated that increased task-related activation in association with cholinesterase inhibitor therapy is associated with improved cognitive function in patients with AD dementia [275, 276].

Pharmacological fMRI can be used to extend our understanding of the effects of standard AD therapies on brain function. During the auditory encoding of sentences, healthy individuals show a suppression of activity in auditory cortex; AD patients do not, but treatment with donepezil normalizes this activity and is associated with better memory [277]. A subsequent analysis of these data also demonstrated attenuated activity within the executive function network during verbal recall, which was also partially normalized after donepezil treatment [278].

Fewer pharmacological fMRI investigations have been done of MCI. In one of the first, MCI patients who received 6 weeks of donepezil showed increased prefrontal activation after this course of medication, which related to improvements in performance of a working memory task [266]. Similarly, after MCI patients received galanthamine for approximately 1 week, performance on a working memory task was improved in conjunction with increased activation in precuneus and middle frontal regions. In addition, increases in hippocampal, prefrontal, cingulate, and occipital regions were seen during an episodic encoding task, although performance did not improve on this memory task [279]. In another study, increased hippocampal activation in MCI patients after 7 days of galathamine treatment was associated with neuropsychological improvement [280]. In a more recent study, investigators compared task-related activity during a memory encoding paradigm in an MCI group before and after 3 months of treatment with donepezil. After treatment, the medial temporal lobe hypoactivation and medial parietal hypo-deactivation seen at baseline were both normalized in association with improved cognitive performance [281], similar to findings observed in other chronic cholinergic therapy pharmacological fMRI studies [282, 283]. In the latter study, after 3 months of donepezil treatment, functional connectivity during episodic encoding increased between the fusiform cortex and hippocampus [283].

Several recent studies have begun to employ resting-state fMRI as a probe to assess the modulatory effects of pharmacologic therapies on AD and MCI, in part because of the ease of its implementation. Functional connectivity of the hippocampus at rest with other brain regions was shown to be modulated by 6 weeks of donepezil therapy in patients with AD, with connectivity changes in parahippocampal and frontal regions being correlated with cognitive improvement [284]. A follow-up analysis of the same dataset along with arterial spin labeled perfusion MRI showed increases in regional cerebral blood flow after therapy, with functional connectivity changes in the medial prefrontal cortex being correlated with cingulate perfusion and cognition [285]. Furthermore, increased functional connectivity in the medial prefrontal areas demonstrated an association with Alzheimer's disease Assessment Scale-Cognitive subscale (ADAS-cog) score changes (Fig. 7). In a similar study of mild AD dementia patients before and after 8 weeks of treatment with donepezil, the investigators observed increases in prefrontal connectivity [286]. Lorenzi et al. [287] studied AD patients before and after 6 months of treatment with memantine relative to placebo, showing a greater DMN connectivity in the precuneus in the group treated with memantine than the placebo group. A recent study incorporated both task-related visual scene encoding and resting-state functional MRI in eight patients with AD dementia receiving 3 months of donepezil therapy. Resting-state fMRI showed changes mainly in the parahippocampal cortex while task-related activation fMRI showed stable middle temporal gyrus activation in contrast with the control group who showed declining middle temporal gyral activation [288]. Another study found that *APOE* genotype modulated resting-state functional connectivity measures in patients with AD dementia being treated with cholinesterase inhibitors. They found that *APOE* e4 carriers showed greater responses to donepezil in functional connectivity of multiple cognitive networks compared to *APOE* e4 noncarriers with AD dementia [289]. Taken together, these studies suggest that functional connectivity can be a feasible and valuable biomarker for tracking treatment-related changes in AD.

In addition to studies of pharmacological interventions, resting state fMRI has been used to study the effects of acupuncture, meditation, and cognitive rehabilitative training in the MCI/AD spectrum. In one study of patients with AD, baseline resting state scans showed reduced connectivity (compared to controls) between the hippocampus and frontal and temporal cortex which was increased following 3 min of acupuncture [290]. Ongoing work is attempting to determine whether acupuncture stimulation at particular acupoints is associated with stronger effects than stimulation at other acupoints [291].

In AD dementia patients who participated in meditation training, there was increased functional connectivity between the posterior cingulate cortex and bilateral medial prefrontal cortex and left hippocampus compared to controls [292]. An fMRI encoding and recognition paradigm was used before and after 8 weeks of cognitive rehabilitation training in a group of seven patients with AD dementia, compared with a control group of eight AD patients who received either relaxation therapy or no intervention [293].

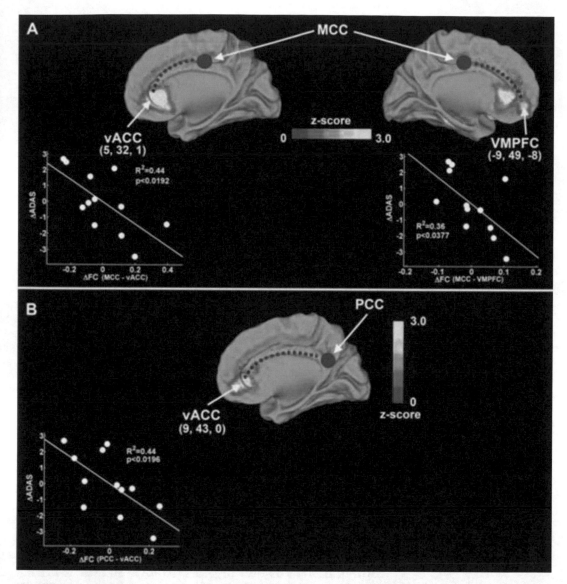

Fig. 7 Behavioral significance of functional connectivity changes in the middle and posterior cingulate networks in AD patients after 12-week donepezil treatment. In the middle congulate connectivity network (**a**), functional connectivity changes in the ventral anterior cingulate cortex and the ventral prefrontal cortex were significantly correlated with the ADAS-cog score change. Functional connectivity change in the ventral anterior cingulate cortex was correlated with the ADAS-cog score change also in the posterior cingulate connectivity network (**b**). The *blue solid circles* and *dotted lines* represent the seed regions and the functional connections, respectively. Figure modified from [285] with permission

On the recognition task, the group who received cognitive rehabilitation showed increased activation in bilateral prefrontal cortex and insula, while the control group showed decreases in these areas. Another study used memory strategy training in patients with MCI, and found that following training, hippocampal activation during a memory task was increased compared to the control group [294].

PD is an excellent clinical scenario in which to apply pharmacological fMRI, given the typical rapid responsiveness of symptoms to dopaminergic therapy. In drug-naïve patients with mild PD, SMA and contralateral motor cortex hypoactivation during a simple finger movement task normalized with l-dopa therapy [295]. Motor performance was constant across conditions, suggesting that any change could be ascribed to pharmacological modulation within basal ganglia–thalamocortical loops. A l-dopa-induced spatial remapping of the cortico-striatal resting state fMRI connectivity has been detected in chronically treated PD patients [296, 297]. Recent studies have suggested that dopaminergic-related changes in resting state functional connectivity occur also in drug-naïve PD cases [298–300].

In addition to motor behavior, the modulatory effects of dopaminergic therapy on cognition and emotion have also been studied in PD. In one well-controlled study, the modulatory effects of dopamine replacement were studied using both sensorimotor and working memory tasks [301]. The cortical motor regions activated during the motor task showed greater activation during the dopamine-replete state, but the cortical regions subserving working memory displayed greater activation during the hypodopaminergic state. Interestingly, the greater cortical activation during the working memory task in the hypodopaminergic state correlated with errors in task performance, while the increased activation in the cortical motor regions during the dopamine-replete state was correlated with improvement in motor function. These results are consistent with evidence that the hypodopaminergic state is associated with decreased efficiency of prefrontal cortical information processing and that dopaminergic therapy improves the physiological efficiency of this region, and also indicate that hyperactivation does not necessarily reflect better behavioral performance.

In a study of emotional face processing in PD, amygdala activation was reduced compared with controls during a hypodopaminergic state, and partly normalized with dopaminergic therapy [302].

Moving beyond its uses in pharmacologic studies, fMRI has been used to study the effects of deep brain stimulation (DBS). DBS is a well-accepted therapeutic modality for PD and is finding a growing number of applications in other neurologic and psychiatric disorders. Initial studies focused on safety and the measurement of BOLD signal changes during on versus off stimulator activity [303, 304]. More recently, fMRI is elucidating mechanisms through which DBS modulates neural circuits [305, 306].

5 Beyond Exclusion: The Use of Imaging Measures as Disease Biomarkers

At present, the potential efficacy of disease-modifying therapies for AD and other neurodegenerative diseases is evaluated primarily using clinical measures of cognition, movement, and other behaviors.

In clinical trials, outcome measures are typically performance-based instruments, such as the ADAS-Cog, or structured surveys of clinician/caregiver impression of change. Although the efficacy of disease-modifying treatments for AD and other neurodegenerative diseases must ultimately be demonstrated using clinically meaningful outcome measures such as the slowing of decline in progression of symptoms or functional impairment, such trials will likely require hundreds of patients studied for a minimum of 1–2 years. Thus, surrogate markers of efficacy with less variability than clinical assessments are desperately needed to reduce the number of subjects. These markers may also prove particularly valuable in the early phases of drug development to detect a preliminary "signal of efficacy" over a shorter time period.

Since the pathophysiologic process underlying cognitive decline in AD and other neurodegenerative diseases involves the progressive degeneration of particular brain regions, repeatable in vivo neuroimaging measures of brain anatomy, chemistry, physiology, and pathology hold promise as an important class of potential biomarkers [3]. A growing body of data indicates that the natural history of gradually progressive cognitive decline in AD can be reliably related to changes in such imaging measures. Furthermore, regionally specific changes in brain anatomy, chemistry, and physiology can be detected by imaging prior to the point at which the disease is symptomatic enough to make a typical clinical diagnosis. Finally, evidence is accumulating that alterations in synaptic function are present very early in the disease process, possibly long before the development of clinical symptoms and significant cell loss, which may relate closely to symptomatic progression in manifest disease [4, 5, 307, 308]. Thus, potential disease-modifying therapies may act by impeding the accumulation of neuropathology, slowing the loss of neurons, altering neurochemistry, or preserving synaptic function; neuroimaging modalities exist to measure each of these putative therapeutic goals, and fMRI could potentially play a valuable role in the development of new scientific insights into functional brain abnormalities early in the course of these diseases and in the development of therapeutic agents.

Although measures of brain structure (MRI) and brain metabolism (FDG-PET) are well-accepted as imaging biomarkers of neurodegeneration in AD and other neurodegenerative diseases and are used clinically, fMRI received approval for current procedural technology (CPT) codes by the American Medical Association, one application of which is toward the assessment of complex cognitive and sensorimotor function in patients with neuropsychiatric disorders [309]. It will be extremely valuable for studies to continue to collect fMRI data in the context of structural MRI, FDG-PET, and other multimodal imaging data to begin to understand the relationships between these data types and the ultimate clinical or research utility of fMRI in neurodegenerative diseases.

6   Conclusions

Functional MRI is a particularly attractive method for use by clinical investigators to study task-related brain activation or large-scale functional network connectivity in patients with neurodegenerative diseases. There has been a dramatic growth of promising fMRI studies in various neurodegenerative disorders that highlight the potential uses of fMRI in both basic and clinical spheres of investigation. Functional MRI may provide novel insights into the neural correlates of cognitive, affective, and sensorimotor abilities, and how they are altered by neurologic disease and by medications. The technique may help elucidate fundamental aspects of brain-behavior relationships, such as the genetic influences on task-related brain physiology. Functional MRI measures hold promise for multiple clinical applications, including the early detection and differential diagnosis, predicting future change in clinical status, and as a marker of alterations in brain physiology related to neurotherapeutic agents. The greatest potential of fMRI may lie in the study of very early and preclinical stages of progressive neurologic diseases, at the point of subtle neuronal dysfunction prior to overt anatomic pathology, or as an early readout of target engagement in intervention studies. There is a need for further validation and reliability studies and continued technical advances to fully realize the potential of fMRI.

References

1. Perry D et al (2015) Building a roadmap for developing combination therapies for Alzheimer's disease. Expert Rev Neurother 15(3):327–333

2. Selkoe DJ (2013) The therapeutics of Alzheimer's disease: where we stand and where we are heading. Ann Neurol 74(3):328–336

3. DeKosky ST, Marek K (2003) Looking backward to move forward: early detection of neurodegenerative disorders. Science 302(5646):830–834

4. Selkoe DJ (2002) Alzheimer's disease is a synaptic failure. Science 298(5594):789–791

5. Coleman P, Federoff H, Kurlan R (2004) A focus on the synapse for neuroprotection in Alzheimer disease and other dementias. Neurology 63(7):1155–1162

6. Dickerson BC, Atri A (2014) Dementia: comprehensive principles and practice. Oxford University Press, New York

7. Dickerson BC (2015) Hodges' frontotemporal dementia, 2nd edn. Cambridge University Press, Cambridge, UK

8. Sperling RA, Jack CR Jr, Aisen PS (2011) Testing the right target and right drug at the right stage. Sci Transl Med 3(111):111cm33

9. Sperling RA et al (2011) Toward defining the preclinical stages of Alzheimer's disease: recommendations from the National Institute on Aging and the Alzheimer's Association workgroup. Alzheimers Dement 7(3):280–292

10. Albert MS et al (2011) The diagnosis of mild cognitive impairment due to Alzheimer's disease: recommendations from the National Institute on Aging-Alzheimer's Association workgroups on diagnostic guidelines for Alzheimer's disease. Alzheimers Dement 7(3):270–279

11. McKhann GM et al (2011) The diagnosis of dementia due to Alzheimer's disease: recommendations from the National Institute on Aging and the Alzheimer's Association workgroup. Alzheimers Dement 7(3):263–269

12. Boxer AL et al (2013) The advantages of frontotemporal degeneration drug development (part 2 of frontotemporal degeneration: the next therapeutic frontier). Alzheimers Dement 9(2):189–198

13. Paulsen JS et al (2014) Clinical and biomarker changes in premanifest Huntington disease show trial feasibility: a decade of the PREDICT-HD study. Front Aging Neurosci 6:78

14. Mills SM et al (2013) Preclinical trials in autosomal dominant AD: implementation of the DIAN-TU trial. Rev Neurol (Paris) 169(10):737–743

15. Petersen RC et al (1999) Mild cognitive impairment: clinical characterization and outcome. Arch Neurol 56(3):303–308

16. Grundman M et al (2004) Mild cognitive impairment can be distinguished from Alzheimer disease and normal aging for clinical trials. Arch Neurol 61(1):59–66

17. Sperling R, Mormino E, Johnson K (2014) The evolution of preclinical Alzheimer's disease: implications for prevention trials. Neuron 84(3):608–622

18. Dickerson BC et al (2007) Prefrontal-hippocampal-fusiform activity during encoding predicts intraindividual differences in free recall ability: an event-related functional-anatomic MRI study. Hippocampus 17(11): 1060–1070

19. Touroutoglou A et al (2014) Amygdala task-evoked activity and task-free connectivity independently contribute to feelings of arousal. Hum Brain Mapp 35(10):5316–5327

20. Brewer JB et al (1998) Making memories: brain activity that predicts how well visual experience will be remembered. Science 281(5380):1185–1187

21. Kirchhoff BA et al (2000) Prefrontal-temporal circuitry for episodic encoding and subsequent memory. J Neurosci 20(16):6173–6180

22. Wagner AD et al (1998) Building memories: remembering and forgetting of verbal experiences as predicted by brain activity. Science 281(5380):1188–1191

23. Daselaar SM et al (2003) Neuroanatomical correlates of episodic encoding and retrieval in young and elderly subjects. Brain 126(Pt 1):43–56

24. Sperling R et al (2003) Putting names to faces: successful encoding of associative memories activates the anterior hippocampal formation. Neuroimage 20(2):1400–1410

25. Price CJ, Friston KJ (1999) Scanning patients with tasks they can perform. Hum Brain Mapp 8(2–3):102–108

26. Dickerson BC et al (2004) Medial temporal lobe function and structure in mild cognitive impairment. Ann Neurol 56(1):27–35

27. Grady CL et al (2003) Evidence from functional neuroimaging of a compensatory prefrontal network in Alzheimer's disease. J Neurosci 23(3):986–993

28. Bennett CM, Miller MB (2013) fMRI reliability: influences of task and experimental design. Cogn Affect Behav Neurosci 13(4):690–702

29. Brandt DJ et al (2013) Test-retest reliability of fMRI brain activity during memory encoding. Front Psychiatry 4:163

30. Brown GG et al (2011) Multisite reliability of cognitive BOLD data. Neuroimage 54(3):2163–2175

31. Cao H et al (2014) Test-retest reliability of fMRI-based graph theoretical properties during working memory, emotion processing, and resting state. Neuroimage 84:888–900

32. Chase HW et al (2015) Accounting for dynamic fluctuations across time when examining fMRI test-retest reliability: analysis of a reward paradigm in the EMBARC study. PLoS One 10(5):e0126326

33. de Bertoldi F et al (2015) Improving the reliability of single-subject fMRI by weighting intra-run variability. Neuroimage 114:287–293

34. Frassle S et al (2015) Test-retest reliability of dynamic causal modeling for fMRI. Neuroimage 117:56–66

35. Gee DG et al (2015) Reliability of an fMRI paradigm for emotional processing in a multisite longitudinal study. Hum Brain Mapp 36(7):2558–2579

36. Golestani AM et al (2015) Mapping the end-tidal CO2 response function in the resting-state BOLD fMRI signal: spatial specificity, test-retest reliability and effect of fMRI sampling rate. Neuroimage 104:266–277

37. Lukasova K et al (2014) Test-retest reliability of fMRI activation generated by different saccade tasks. J Magn Reson Imaging 40(1):37–46

38. McGonigle DJ (2012) Test-retest reliability in fMRI: or how I learned to stop worrying and love the variability. Neuroimage 62(2):1116–1120

39. Plichta MM et al (2012) Test-retest reliability of evoked BOLD signals from a cognitive-emotive fMRI test battery. Neuroimage 60(3):1746–1758

40. Aurich NK et al (2015) Evaluating the reliability of different preprocessing steps to estimate graph theoretical measures in resting state fMRI data. Front Neurosci 9:48

41. Birn RM et al (2013) The effect of scan length on the reliability of resting-state fMRI connectivity estimates. Neuroimage 83:550–558

42. Braun U et al (2012) Test-retest reliability of resting-state connectivity network characteristics using fMRI and graph theoretical measures. Neuroimage 59(2):1404–1412

730   Bradford C. Dickerson et al.

43. Patriat R et al (2013) The effect of resting condition on resting-state fMRI reliability and consistency: a comparison between resting with eyes open, closed, and fixated. Neuroimage 78:463–473

44. Wisner KM et al (2013) Neurometrics of intrinsic connectivity networks at rest using fMRI: retest reliability and cross-validation using a meta-level method. Neuroimage 76:236–251

45. Atri A et al (2011) Test-retest reliability of memory task functional magnetic resonance imaging in Alzheimer disease clinical trials. Arch Neurol 68(5):599–606

46. Turner JA et al (2012) Reliability of the amplitude of low-frequency fluctuations in resting state fMRI in chronic schizophrenia. Psychiatry Res 201(3):253–255

47. Eaton KP et al (2008) Reliability of fMRI for studies of language in post-stroke aphasia subjects. Neuroimage 41(2):311–322

48. Kurland J et al (2004) Test-retest reliability of fMRI during nonverbal semantic decisions in moderate-severe nonfluent aphasia patients. Behav Neurol 15(3–4):87–97

49. Clement F, Belleville S (2009) Test-retest reliability of fMRI verbal episodic memory paradigms in healthy older adults and in persons with mild cognitive impairment. Hum Brain Mapp 30(12):4033–4047

50. Zanto TP, Pa J, Gazzaley A (2014) Reliability measures of functional magnetic resonance imaging in a longitudinal evaluation of mild cognitive impairment. Neuroimage 84:443–452

51. Poudel GR et al (2015) Functional changes during working memory in Huntington's disease: 30-month longitudinal data from the IMAGE-HD study. Brain Struct Funct 220(1):501–512

52. O'Brien JL et al (2010) Longitudinal fMRI in elderly reveals loss of hippocampal activation with clinical decline. Neurology 74(24):1969–1976

53. Scheller E et al (2014) Attempted and successful compensation in preclinical and early manifest neurodegeneration—a review of task FMRI studies. Front Psychiatry 5:132

54. Buckner RL et al (2000) Functional brain imaging of young, nondemented, and demented older adults. J Cogn Neurosci 12(Suppl 2):24–34

55. D'Esposito M, Deouell LY, Gazzaley A (2003) Alterations in the BOLD fMRI signal with ageing and disease: a challenge for neuroimaging. Nat Rev Neurosci 4(11):863–872

56. Grossman M et al (2003) Neural basis for verb processing in Alzheimer's disease: an fMRI study. Neuropsychology 17(4):658–674

57. Johnson SC et al (2000) The relationship between fMRI activation and cerebral atrophy: comparison of normal aging and Alzheimer disease. Neuroimage 11(3):179–187

58. Saykin AJ et al (1999) Neuroanatomic substrates of semantic memory impairment in Alzheimer's disease: patterns of functional MRI activation. J Int Neuropsychol Soc 5(5):377–392

59. Grossman M et al (2003) Neural basis for semantic memory difficulty in Alzheimer's disease: an fMRI study. Brain 126(Pt 2):292–311

60. Woodard JL et al (2009) Semantic memory activation in amnestic mild cognitive impairment. Brain 132(Pt 8):2068–2078

61. Thulborn KR, Martin C, Voyvodic JT (2000) Functional MR imaging using a visually guided saccade paradigm for comparing activation patterns in patients with probable Alzheimer's disease and in cognitively able elderly volunteers. AJNR Am J Neuroradiol 21(3):524–531

62. Golden HL et al (2015) Functional neuroanatomy of auditory scene analysis in Alzheimer's disease. Neuroimage Clin 7:699–708

63. Kato T, Knopman D, Liu H (2001) Dissociation of regional activation in mild AD during visual encoding: a functional MRI study. Neurology 57(5):812–816

64. Machulda MM et al (2003) Comparison of memory fMRI response among normal, MCI, and Alzheimer's patients. Neurology 61(4):500–506

65. Rombouts SA et al (2000) Functional MR imaging in Alzheimer's disease during memory encoding. AJNR Am J Neuroradiol 21(10):1869–1875

66. Small SA et al (1999) Differential regional dysfunction of the hippocampal formation among elderly with memory decline and Alzheimer's disease. Ann Neurol 45(4):466–472

67. Sperling RA et al (2003) fMRI studies of associative encoding in young and elderly controls and mild Alzheimer's disease. J Neurol Neurosurg Psychiatry 74(1):44–50

68. Schwindt GC, Black SE (2009) Functional imaging studies of episodic memory in Alzheimer's disease: a quantitative meta-analysis. Neuroimage 45(1):181–190

69. Rombouts SARB et al (2005) Delayed rather than decreased BOLD response as a marker for early Alzheimer's disease. Neuroimage 26(4):1078–1085

70. Dickerson BC et al (2005) Increased hippocampal activation in mild cognitive impairment compared to normal aging and AD. Neurology 65(3):404–411

71. Johnson SC et al (2006) Activation of brain regions vulnerable to Alzheimer's disease: the effect of mild cognitive impairment. Neurobiol Aging 27(11):1604–1612

72. Johnson SC et al (2004) Hippocampal adaptation to face repetition in healthy elderly and mild cognitive impairment. Neuropsychologia 42(7):980–989

73. Petrella JR et al (2006) Mild cognitive impairment: evaluation with 4-T functional MR imaging. Radiology 240(1):177–186

74. Hamalainen A et al (2007) Increased fMRI responses during encoding in mild cognitive impairment. Neurobiol Aging 28(12):1889–1903

75. Kircher T et al (2007) Hippocampal activation in MCI patients is necessary for successful memory encoding. J Neurol Neurosurg Psychiatry 78(8):812–818

76. Dickerson BC, Sperling RA (2008) Functional abnormalities of the medial temporal lobe memory system in mild cognitive impairment and Alzheimer's disease: insights from functional MRI studies. Neuropsychologia 46(6):1624–1635

77. Dickerson BC, Sperling RA (2009) Large-scale functional brain network abnormalities in Alzheimer's disease: insights from functional neuroimaging. Behav Neurol 21(1):63–75

78. Petrella JR et al (2007) Cortical deactivation in mild cognitive impairment: high-field-strength functional MR imaging. Radiology 245(1):224–235

79. Vandenbulcke M et al (2007) Word reading and posterior temporal dysfunction in amnestic mild cognitive impairment. Cereb Cortex 17(3):542–551

80. Vannini P et al (2012) Age and amyloid-related alterations in default network habituation to stimulus repetition. Neurobiol Aging 33(7):1237–1252

81. Vannini P et al (2013) The ups and downs of the posteromedial cortex: age- and amyloid-related functional alterations of the encoding/retrieval flip in cognitively normal older adults. Cereb Cortex 23(6):1317–1328

82. Sperling RA et al (2009) Amyloid deposition is associated with impaired default network function in older persons without dementia. Neuron 63(2):178–188

83. Kennedy KM et al (2012) Effects of beta-amyloid accumulation on neural function during encoding across the adult lifespan. Neuroimage 62(1):1–8

84. Hedden T et al (2012) Failure to modulate attentional control in advanced aging linked to white matter pathology. Cereb Cortex 22(5):1038–1051

85. Mormino EC et al (2012) Abeta Deposition in aging is associated with increases in brain activation during successful memory encoding. Cereb Cortex 22(8):1813–1823

86. Villemagne VL et al (2015) Tau imaging: early progress and future directions. Lancet Neurol 14(1):114–124

87. Virani K et al (2013) Functional neural correlates of emotional expression processing deficits in behavioural variant frontotemporal dementia. J Psychiatry Neurosci 38(3):174–182

88. Agustus JL et al (2015) Functional MRI of music emotion processing in frontotemporal dementia. Ann N Y Acad Sci 1337:232–240

89. Chiong W et al (2013) The salience network causally influences default mode network activity during moral reasoning. Brain 136(Pt 6):1929–1941

90. Sonty SP et al (2003) Primary progressive aphasia: PPA and the language network. Ann Neurol 53(1):35–49

91. Sonty SP et al (2007) Altered effective connectivity within the language network in primary progressive aphasia. J Neurosci 27(6):1334–1345

92. Wilson SM et al (2010) Neural correlates of syntactic processing in the nonfluent variant of primary progressive aphasia. J Neurosci 30(50):16845–16854

93. Vandenbulcke M et al (2005) Anterior temporal laterality in primary progressive aphasia shifts to the right. Ann Neurol 58(3):362–370

94. Nelissen N et al (2011) Right hemisphere recruitment during language processing in frontotemporal lobar degeneration and Alzheimer's disease. J Mol Neurosci 45(3):637–647

95. Wilson SM et al (2009) The neural basis of surface dyslexia in semantic dementia. Brain 132(Pt 1):71–86

96. Goll JC et al (2012) Nonverbal sound processing in semantic dementia: a functional MRI study. Neuroimage 61(1):170–180

97. Goll JC et al (2010) Non-verbal sound processing in the primary progressive aphasias. Brain 133(Pt 1):272–285

98. Elsinger CL et al (2003) Neural basis for impaired time reproduction in Parkinson's disease: an fMRI study. J Int Neuropsychol Soc 9(7):1088–1098

99. Rowe JB, Siebner HR (2012) The motor system and its disorders. Neuroimage 61(2):464–477

100. Sabatini U et al (2000) Cortical motor reorganization in akinetic patients with Parkinson's disease: a functional MRI study. Brain 123(Pt 2):394–403

101. Wu T, Hallett M (2005) A functional MRI study of automatic movements in patients with Parkinson's disease. Brain 128(Pt 10):2250–2259

102. Wu T et al (2010) Neural correlates of bimanual anti-phase and in-phase movements in Parkinson's disease. Brain 133(Pt 8):2394–2409

103. Rowe JB et al (2010) Dynamic causal modelling of effective connectivity from fMRI: are results reproducible and sensitive to Parkinson's disease and its treatment? Neuroimage 52(3):1015–1026

104. Ceballos-Baumann AO (2003) Functional imaging in Parkinson's disease: activation studies with PET, fMRI and SPECT. J Neurol 250(Suppl 1):I15–I23

105. Grafton ST (2004) Contributions of functional imaging to understanding parkinsonian symptoms. Curr Opin Neurobiol 14(6):715–719

106. Lewis SJ et al (2003) Cognitive impairments in early Parkinson's disease are accompanied by reductions in activity in frontostriatal neural circuitry. J Neurosci 23(15):6351–6356

107. Ekman U et al (2012) Functional brain activity and presynaptic dopamine uptake in patients with Parkinson's disease and mild cognitive impairment: a cross-sectional study. Lancet Neurol 11(8):679–687

108. Nagano-Saito A et al (2014) Effect of mild cognitive impairment on the patterns of neural activity in early Parkinson's disease. Neurobiol Aging 35(1):223–231

109. Voon V et al (2011) Dopamine agonists and risk: impulse control disorders in Parkinson's disease. Brain 134(Pt 5):1438–1446

110. Voon V et al (2011) Impulse control disorders in Parkinson disease: a multicenter case–control study. Ann Neurol 69(6):986–996

111. Sauer J et al (2006) Differences between Alzheimer's disease and dementia with Lewy bodies: an fMRI study of task-related brain activity. Brain 129(Pt 7):1780–1788

112. Kim JS et al (2004) Functional MRI study of a serial reaction time task in Huntington's disease. Psychiatry Res 131(1):23–30

113. Wolf RC et al (2009) Cortical dysfunction in patients with Huntington's disease during working memory performance. Hum Brain Mapp 30(1):327–339

114. Georgiou-Karistianis N et al (2014) Functional magnetic resonance imaging of working memory in Huntington's disease: cross-sectional data from the IMAGE-HD study. Hum Brain Mapp 35(5):1847–1864

115. Thiruvady DR et al (2007) Functional connectivity of the prefrontal cortex in Huntington's disease. J Neurol Neurosurg Psychiatry 78(2):127–133

116. Gray MA et al (2013) Prefrontal activity in Huntington's disease reflects cognitive and neuropsychiatric disturbances: the IMAGE-HD study. Exp Neurol 239:218–228

117. Rupp J et al (2011) Abnormal error-related antisaccade activation in premanifest and early manifest Huntington disease. Neuropsychology 25(3):306–318

118. Tessitore A et al (2006) Subcortical motor plasticity in patients with sporadic ALS: an fMRI study. Brain Res Bull 69(5):489–494

119. Abrahams S et al (2004) Word retrieval in amyotrophic lateral sclerosis: a functional magnetic resonance imaging study. Brain 127(Pt 7):1507–1517

120. Witiuk K et al (2014) Cognitive deterioration and functional compensation in ALS measured with fMRI using an inhibitory task. J Neurosci 34(43):14260–14271

121. Palmieri A et al (2010) Right hemisphere dysfunction and emotional processing in ALS: an fMRI study. J Neurol 257(12):1970–1978

122. Raichle ME et al (2001) A default mode of brain function. Proc Natl Acad Sci U S A 98(2):676–682

123. Buckner RL, Andrews-Hanna JR, Schacter DL (2008) The brain's default network: anatomy, function, and relevance to disease. Ann N Y Acad Sci 1124:1–38

124. Lustig C et al (2003) Functional deactivations: change with age and dementia of the Alzheimer type. Proc Natl Acad Sci U S A 100(24):14504–14509

125. Greicius MD et al (2004) Default-mode network activity distinguishes Alzheimer's disease from healthy aging: evidence from functional MRI. Proc Natl Acad Sci U S A 101(13):4637–4642

126. Rombouts SARB et al (2005) Altered resting state networks in mild cognitive impairment and mild Alzheimer's disease: an fMRI study. Hum Brain Mapp 26(4):231–239

127. Celone KA et al (2006) Alterations in memory networks in mild cognitive impairment and Alzheimer's disease: an independent component analysis. J Neurosci 26(40):10222–10231

128. Buckner RL et al (2005) Molecular, structural, and functional characterization of Alzheimer's disease: evidence for a relationship between default activity, amyloid, and memory. J Neurosci 25(34):7709–7717

129. Biswal B et al (1995) Functional connectivity in the motor cortex of resting human brain

using echo-planar MRI. Magn Reson Med 34(4):537–541

130. Vincent JL et al (2007) Intrinsic functional architecture in the anaesthetized monkey brain. Nature 447(7140):83–86

131. Smith SM et al (2009) Correspondence of the brain's functional architecture during activation and rest. Proc Natl Acad Sci U S A 106(31):13040–13045

132. Yeo BT et al (2011) The organization of the human cerebral cortex estimated by intrinsic functional connectivity. J Neurophysiol 106(3):1125–1165

133. Zhou J et al (2010) Divergent network connectivity changes in behavioural variant frontotemporal dementia and Alzheimer's disease. Brain 133(Pt 5):1352–1367

134. Thomas JB et al (2014) Functional connectivity in autosomal dominant and late-onset Alzheimer disease. JAMA Neurol 71(9):1111–1122

135. Binnewijzend MA et al (2012) Resting-state fMRI changes in Alzheimer's disease and mild cognitive impairment. Neurobiol Aging 33(9):2018–2028

136. Brier MR et al (2012) Loss of intranetwork and internetwork resting state functional connections with Alzheimer's disease progression. J Neurosci 32(26):8890–8899

137. Zhang HY et al (2010) Resting brain connectivity: changes during the progress of Alzheimer disease. Radiology 256(2):598–606

138. Sheline YI et al (2009) The default mode network and self-referential processes in depression. Proc Natl Acad Sci U S A 106(6):1942–1947

139. Petrella JR et al (2011) Default mode network connectivity in stable vs progressive mild cognitive impairment. Neurology 76(6):511–517

140. Li SJ et al (2002) Alzheimer disease: evaluation of a functional MR imaging index as a marker. Radiology 225(1):253–259

141. Supekar K et al (2010) Development of functional and structural connectivity within the default mode network in young children. Neuroimage 52(1):290–301

142. Damoiseaux JS et al (2012) Functional connectivity tracks clinical deterioration in Alzheimer's disease. Neurobiol Aging 33(4):828.e19–30

143. Schwindt GC et al (2013) Modulation of the default-mode network between rest and task in Alzheimer's Disease. Cereb Cortex 23(7):1685–1694

144. Zamboni G et al (2013) Resting functional connectivity reveals residual functional activity in Alzheimer's disease. Biol Psychiatry 74(5):375–383

145. Filippi M et al (2013) Functional network connectivity in the behavioral variant of frontotemporal dementia. Cortex 49(9):2389–2401

146. Agosta F et al (2012) Resting state fMRI in Alzheimer's disease: beyond the default mode network. Neurobiol Aging 33(8):1564–1578

147. Lehmann M et al (2013) Intrinsic connectivity networks in healthy subjects explain clinical variability in Alzheimer's disease. Proc Natl Acad Sci U S A 110(28):11606–11611

148. Gour N et al (2014) Functional connectivity changes differ in early and late-onset Alzheimer's disease. Hum Brain Mapp 35(7):2978–2994

149. Lehmann M et al (2015) Loss of functional connectivity is greater outside the default mode network in nonfamilial early-onset Alzheimer's disease variants. Neurobiol Aging 36(10):2678–2686

150. Barnes J et al (2015) Alzheimer's disease first symptoms are age dependent: evidence from the NACC dataset. Alzheimers Dement 11(11):1349–1357

151. Petrella JR et al (2007) Prognostic value of posteromedial cortex deactivation in mild cognitive impairment. PLoS One 2(10):e1104

152. Hedden T et al (2009) Disruption of functional connectivity in clinically normal older adults harboring amyloid burden. J Neurosci 29(40):12686–12694

153. Drzezga A et al (2011) Neuronal dysfunction and disconnection of cortical hubs in non-demented subjects with elevated amyloid burden. Brain 134(Pt 6):1635–1646

154. Brier MR et al (2014) Functional connectivity and graph theory in preclinical Alzheimer's disease. Neurobiol Aging 35(4):757–768

155. Wang L et al (2013) Cerebrospinal fluid Abeta42, phosphorylated Tau181, and resting-state functional connectivity. JAMA Neurol 70(10):1242–1248

156. Sheline YI et al (2010) Amyloid plaques disrupt resting state default mode network connectivity in cognitively normal elderly. Biol Psychiatry 67(6):584–587

157. Farb NA et al (2013) Abnormal network connectivity in frontotemporal dementia: evidence for prefrontal isolation. Cortex 49(7):1856–1873

158. Whitwell JL et al (2011) Altered functional connectivity in asymptomatic MAPT subjects: a comparison to bvFTD. Neurology 77(9):866–874

159. Agosta F et al (2014) Disrupted brain connectome in semantic variant of primary progressive aphasia. Neurobiol Aging 35(11):2646–2655

160. Gardner RC et al (2013) Intrinsic connectivity network disruption in progressive supranuclear palsy. Ann Neurol 73(5):603–616

161. Williams DR et al (2007) Pathological tau burden and distribution distinguishes progressive supranuclear palsy-parkinsonism from Richardson's syndrome. Brain 130(Pt 6):1566–1576

162. Whitwell JL et al (2011) Disrupted thalamo-cortical connectivity in PSP: a resting-state fMRI, DTI, and VBM study. Parkinsonism Relat Disord 17(8):599–605

163. Filippi M et al (2015) Progress towards a neuroimaging biomarker for amyotrophic lateral sclerosis. Lancet Neurol 14(8):786–788

164. Agosta F et al (2013) Divergent brain network connectivity in amyotrophic lateral sclerosis. Neurobiol Aging 34(2):419–427

165. Galvin JE et al (2011) Resting bold fMRI differentiates dementia with Lewy bodies vs Alzheimer disease. Neurology 76(21):1797–1803

166. Kenny ER et al (2012) Functional connectivity in cortical regions in dementia with Lewy bodies and Alzheimer's disease. Brain 135(Pt 2):569–581

167. Kenny ER et al (2013) Subcortical connectivity in dementia with Lewy bodies and Alzheimer's disease. Br J Psychiatry 203(3):209–214

168. Lowther ER et al (2014) Lewy body compared with Alzheimer dementia is associated with decreased functional connectivity in resting state networks. Psychiatry Res 223(3):192–201

169. Seibert TM et al (2012) Interregional correlations in Parkinson disease and Parkinson-related dementia with resting functional MR imaging. Radiology 263(1):226–234

170. Rektorova I et al (2012) Default mode network and extrastriate visual resting state network in patients with Parkinson's disease dementia. Neurodegener Dis 10(1–4):232–237

171. Chen B et al (2015) Changes in anatomical and functional connectivity of Parkinson's disease patients according to cognitive status. Eur J Radiol 84(7):1318–1324

172. Amboni M et al (2015) Resting-state functional connectivity associated with mild cognitive impairment in Parkinson's disease. J Neurol 262(2):425–434

173. Possin KL et al (2013) Rivastigmine is associated with restoration of left frontal brain activity in Parkinson's disease. Mov Disord 28(10):1384–1390

174. Baggio HC et al (2015) Cognitive impairment and resting-state network connectivity in Parkinson's disease. Hum Brain Mapp 36(1):199–212

175. Borroni B et al (2015) Structural and functional imaging study in dementia with Lewy bodies and Parkinson's disease dementia. Parkinsonism Relat Disord 21(9):1049–1055

176. Peraza LR et al (2015) Resting state in Parkinson's disease dementia and dementia with Lewy bodies: commonalities and differences. Int J Geriatr Psychiatry 30(11):1135–1146

177. Peraza LR et al (2014) fMRI resting state networks and their association with cognitive fluctuations in dementia with Lewy bodies. Neuroimage Clin 4:558–565

178. Baggio HC et al (2015) Resting-state fronto-striatal functional connectivity in Parkinson's disease-related apathy. Mov Disord 30(5):671–679

179. Yao N et al (2015) Resting activity in visual and corticostriatal pathways in Parkinson's disease with hallucinations. Parkinsonism Relat Disord 21(2):131–137

180. Yao N et al (2014) The default mode network is disrupted in Parkinson's disease with visual hallucinations. Hum Brain Mapp 35(11):5658–5666

181. Dumas EM et al (2013) Reduced functional brain connectivity prior to and after disease onset in Huntington's disease. Neuroimage Clin 2:377–384

182. Poudel GR et al (2014) Abnormal synchrony of resting state networks in premanifest and symptomatic Huntington disease: the IMAGE-HD study. J Psychiatry Neurosci 39(2):87–96

183. Werner CJ et al (2014) Altered resting-state connectivity in Huntington's disease. Hum Brain Mapp 35(6):2582–2593

184. Quarantelli M et al (2013) Default-mode network changes in Huntington's disease: an integrated MRI study of functional connectivity and morphometry. PLoS One 8(8):e72159

185. Seeley WW et al (2009) Neurodegenerative diseases target large-scale human brain networks. Neuron 62(1):42–52

186. Sanders DW et al (2014) Distinct tau prion strains propagate in cells and mice and define different tauopathies. Neuron 82(6):1271–1288

187. Filippi M et al (2013) Assessment of system dysfunction in the brain through MRI-based connectomics. Lancet Neurol 12(12):1189–1199

188. Sperling R (2011) Potential of functional MRI as a biomarker in early Alzheimer's disease. Neurobiol Aging 32(Suppl 1):S37–S43

189. Wierenga CE, Bondi MW (2007) Use of functional magnetic resonance imaging in

the early identification of Alzheimer's disease. Neuropsychol Rev 17(2):127–143

190. Backman L et al (1999) Brain regions associated with episodic retrieval in normal aging and Alzheimer's disease. Neurology 52(9):1861–1870

191. Becker JT et al (1996) Compensatory reallocation of brain resources supporting verbal episodic memory in Alzheimer's disease. Neurology 46(3):692–700

192. Stern Y et al (2000) Different brain networks mediate task performance in normal aging and AD: defining compensation. Neurology 55(9):1291–1297

193. Putcha D et al (2011) Hippocampal hyperactivation associated with cortical thinning in Alzheimer's disease signature regions in non-demented elderly adults. J Neurosci 31(48):17680–17688

194. Miller SL et al (2008) Hippocampal activation in adults with mild cognitive impairment predicts subsequent cognitive decline. J Neurol Neurosurg Psychiatry 79(6):630–635

195. DeKosky ST et al (2002) Upregulation of choline acetyltransferase activity in hippocampus and frontal cortex of elderly subjects with mild cognitive impairment. Ann Neurol 51(2):145–155

196. Hashimoto M, Masliah E (2003) Cycles of aberrant synaptic sprouting and neurodegeneration in Alzheimer's and dementia with Lewy bodies. Neurochem Res 28(11):1743–1756

197. Stern EA et al (2004) Cortical synaptic integration in vivo is disrupted by amyloid-beta plaques. J Neurosci 24(19):4535–4540

198. Palop JJ, Chin J, Mucke L (2006) A network dysfunction perspective on neurodegenerative diseases. Nature 443(7113):768–773

199. Bakker A et al (2012) Reduction of hippocampal hyperactivity improves cognition in amnestic mild cognitive impairment. Neuron 74(3):467–474

200. Mueggler T et al (2002) Compromised hemodynamic response in amyloid precursor protein transgenic mice. J Neurosci 22(16):7218–7224

201. El Fakhri G et al (2003) MRI-guided SPECT perfusion measures and volumetric MRI in prodromal Alzheimer disease. Arch Neurol 60(8):1066–1072

202. Cohen ER, Ugurbil K, Kim SG (2002) Effect of basal conditions on the magnitude and dynamics of the blood oxygenation level-dependent fMRI response. J Cereb Blood Flow Metab 22(9):1042–1053

203. Davis TL et al (1998) Calibrated functional MRI: mapping the dynamics of oxidative metabolism. Proc Natl Acad Sci U S A 95(4):1834–1839

204. Gur RC et al (1988) Effects of task difficulty on regional cerebral blood flow: relationships with anxiety and performance. Psychophysiology 25(4):392–399

205. Grasby PM et al (1994) A graded task approach to the functional mapping of brain areas implicated in auditory-verbal memory. Brain 117(Pt 6):1271–1282

206. Grady CL (1996) Age-related changes in cortical blood flow activation during perception and memory. Ann N Y Acad Sci 777:14–21

207. Rypma B, D'Esposito M (1999) The roles of prefrontal brain regions in components of working memory: effects of memory load and individual differences. Proc Natl Acad Sci U S A 96(11):6558–6563

208. Kirchhoff BA, Buckner RL (2006) Functional-anatomic correlates of individual differences in memory. Neuron 51(2):263–274

209. Haslinger B et al (2001) Event-related functional magnetic resonance imaging in Parkinson's disease before and after levodopa. Brain 124(Pt 3):558–570

210. Monchi O et al (2004) Neural bases of set-shifting deficits in Parkinson's disease. J Neurosci 24(3):702–710

211. Helmich RC et al (2007) Cerebral compensation during motor imagery in Parkinson's disease. Neuropsychologia 45(10):2201–2215

212. Georgiou-Karistianis N et al (2007) Increased cortical recruitment in Huntington's disease using a Simon task. Neuropsychologia 45(8):1791–1800

213. Stanton BR et al (2007) Altered cortical activation during a motor task in ALS: evidence for involvement of central pathways. J Neurol 254(9):1260–1267

214. Schoenfeld MA et al (2005) Functional motor compensation in amyotrophic lateral sclerosis. J Neurol 252(8):944–952

215. Konrad C et al (2006) Subcortical reorganization in amyotrophic lateral sclerosis. Exp Brain Res 172(3):361–369

216. Drummond SP et al (2000) Altered brain response to verbal learning following sleep deprivation. Nature 403(6770):655–657

217. Cabeza R et al (2002) Aging gracefully: compensatory brain activity in high-performing older adults. Neuroimage 17(3):1394–1402

218. Carey JR et al (2002) Analysis of fMRI and finger tracking training in subjects with chronic stroke. Brain 125(Pt 4):773–788

219. Johansen-Berg H et al (2002) Correlation between motor improvements and altered fMRI activity after rehabilitative therapy. Brain 125(Pt 12):2731–2742

220. Morgen K et al (2004) Training-dependent plasticity in patients with multiple sclerosis. Brain 127(Pt 11):2506–2517

221. Reddy H et al (2000) Evidence for adaptive functional changes in the cerebral cortex with axonal injury from multiple sclerosis. Brain 123(Pt 11):2314–2320

222. McAllister TW et al (1999) Brain activation during working memory 1 month after mild traumatic brain injury: a functional MRI study. Neurology 53(6):1300–1308

223. Ernst T et al (2002) Abnormal brain activation on functional MRI in cognitively asymptomatic HIV patients. Neurology 59(9):1343–1349

224. Desmond JE et al (2003) Increased frontocerebellar activation in alcoholics during verbal working memory: an fMRI study. Neuroimage 19(4):1510–1520

225. Callicott JH et al (2003) Complexity of prefrontal cortical dysfunction in schizophrenia: more than up or down. Am J Psychiatry 160(12):2209–2215

226. Walsh DM, Selkoe DJ (2004) Deciphering the molecular basis of memory failure in Alzheimer's disease. Neuron 44(1):181–193

227. Pievani M et al (2014) Brain connectivity in neurodegenerative diseases—from phenotype to proteinopathy. Nat Rev Neurol 10(11):620–633

228. Winterer G et al (2005) Neuroimaging and human genetics. Int Rev Neurobiol 67:325–383

229. Hariri AR, Weinberger DR (2003) Functional neuroimaging of genetic variation in serotonergic neurotransmission. Genes Brain Behav 2(6):341–349

230. Nikolova YS, Hariri AR (2015) Can we observe epigenetic effects on human brain function? Trends Cogn Sci 19(7):366–373

231. Smith CD et al (1999) Altered brain activation in cognitively intact individuals at high risk for Alzheimer's disease. Neurology 53(7):1391–1396

232. Smith CD et al (2002) Women at risk for AD show increased parietal activation during a fluency task. Neurology 58(8):1197–1202

233. Bookheimer SY et al (2000) Patterns of brain activation in people at risk for Alzheimer's disease. N Engl J Med 343(7):450–456

234. Bondi MW et al (2005) fMRI evidence of compensatory mechanisms in older adults at genetic risk for Alzheimer disease. Neurology 64(3):501–508

235. Johnson SC et al (2006) The influence of Alzheimer disease family history and apolipoprotein E epsilon4 on mesial temporal lobe activation. J Neurosci 26(22):6069–6076

236. Patel KT et al (2013) Default mode network activity and white matter integrity in healthy middle-aged ApoE4 carriers. Brain Imaging Behav 7(1):60–67

237. Machulda MM et al (2011) Effect of APOE epsilon4 status on intrinsic network connectivity in cognitively normal elderly subjects. Arch Neurol 68(9):1131–1136

238. Heise V et al (2014) Apolipoprotein E genotype, gender and age modulate connectivity of the hippocampus in healthy adults. Neuroimage 98:23–30

239. Damoiseaux JS et al (2012) Gender modulates the APOE epsilon4 effect in healthy older adults: convergent evidence from functional brain connectivity and spinal fluid tau levels. J Neurosci 32(24):8254–8262

240. Sheline YI et al (2010) APOE4 allele disrupts resting state fMRI connectivity in the absence of amyloid plaques or decreased CSF Abeta42. J Neurosci 30(50):17035–17040

241. Filippini N et al (2009) Distinct patterns of brain activity in young carriers of the APOE-{varepsilon}4 allele. Proc Natl Acad Sci U S A 106(17):7209–7214

242. Dennis NA et al (2010) Temporal lobe functional activity and connectivity in young adult APOE varepsilon4 carriers. Alzheimers Dement 6(4):303–311

243. Quiroz YT et al (2010) Hippocampal hyperactivation in presymptomatic familial Alzheimer's disease. Ann Neurol 68(6):865–875

244. Reiman EM et al (2012) Brain imaging and fluid biomarker analysis in young adults at genetic risk for autosomal dominant Alzheimer's disease in the presenilin 1 E280A kindred: a case-control study. Lancet Neurol 11(12):1048–1056

245. Quiroz YT et al (2015) Brain imaging and blood biomarker abnormalities in children with autosomal dominant Alzheimer disease: a cross-sectional study. JAMA Neurol 72(8):912–919

246. Chhatwal JP et al (2013) Impaired default network functional connectivity in autosomal dominant Alzheimer disease. Neurology 81(8):736–744

247. Mondadori CR et al (2006) Enhanced brain activity may precede the diagnosis of Alzheimer's disease by 30 years. Brain 129(Pt 11):2908–2922

248. Dopper EG et al (2013) Structural and functional brain connectivity in presymptomatic

familial frontotemporal dementia. Neurology 80(9):814–823

249. Borroni B et al (2012) Granulin mutation drives brain damage and reorganization from preclinical to symptomatic FTLD. Neurobiol Aging 33(10):2506–2520

250. Premi E et al (2014) Effect of TMEM106B polymorphism on functional network connectivity in asymptomatic GRN mutation carriers. JAMA Neurol 71(2):216–221

251. Lee SE et al (2014) Altered network connectivity in frontotemporal dementia with C9orf72 hexanucleotide repeat expansion. Brain 137(Pt 11):3047–3060

252. Reading SA et al (2004) Functional brain changes in presymptomatic Huntington's disease. Ann Neurol 55(6):879–883

253. Kloppel S et al (2010) Irritability in pre-clinical Huntington's disease. Neuropsychologia 48(2):549–557

254. Van den Stock J et al (2015) Functional brain changes underlying irritability in premanifest Huntington's disease. Hum Brain Mapp 36(7):2681–2690

255. Wolf RC et al (2007) Dorsolateral prefrontal cortex dysfunction in presymptomatic Huntington's disease: evidence from event-related fMRI. Brain 130(Pt 11):2845–2857

256. Wolf RC et al (2011) Longitudinal functional magnetic resonance imaging of cognition in preclinical Huntington's disease. Exp Neurol 231(2):214–222

257. Wolf RC et al (2008) Altered frontostriatal coupling in pre-manifest Huntington's disease: effects of increasing cognitive load. Eur J Neurol 15(11):1180–1190

258. Georgiou-Karistianis N et al (2013) Functional and connectivity changes during working memory in Huntington's disease: 18 month longitudinal data from the IMAGE-HD study. Brain Cogn 83(1):80–91

259. Wolf RC et al (2012) Default-mode network changes in preclinical Huntington's disease. Exp Neurol 237(1):191–198

260. Odish OF et al (2015) Longitudinal resting state fMRI analysis in healthy controls and premanifest Huntington's disease gene carriers: a three-year follow-up study. Hum Brain Mapp 36(1):110–119

261. Hampel H et al (2014) Perspective on future role of biological markers in clinical therapy trials of Alzheimer's disease: a long-range point of view beyond 2020. Biochem Pharmacol 88(4):426–449

262. Sperling R et al (2002) Functional MRI detection of pharmacologically induced memory impairment. Proc Natl Acad Sci U S A 99(1):455–460

263. Thiel CM, Henson RN, Dolan RJ (2002) Scopolamine but not lorazepam modulates face repetition priming: a psychopharmacological fMRI study. Neuropsychopharmacology 27(2):282–292

264. Leslie RA, James MF (2000) Pharmacological magnetic resonance imaging: a new application for functional MRI. Trends Pharmacol Sci 21(8):314–318

265. Rombouts SA et al (2002) Alterations in brain activation during cholinergic enhancement with rivastigmine in Alzheimer's disease. J Neurol Neurosurg Psychiatry 73(6):665–671

266. Saykin AJ et al (2004) Cholinergic enhancement of frontal lobe activity in mild cognitive impairment. Brain 127(Pt 7):1574–1583

267. Goekoop R et al (2006) Cholinergic challenge in Alzheimer patients and mild cognitive impairment differentially affects hippocampal activation—a pharmacological fMRI study. Brain 129(Pt 1):141–157

268. Miettinen PS et al (2011) Effect of cholinergic stimulation in early Alzheimer's disease—functional imaging during a recognition memory task. Curr Alzheimer Res 8(7):753–764

269. Bentley P, Driver J, Dolan RJ (2008) Cholinesterase inhibition modulates visual and attentional brain responses in Alzheimer's disease and health. Brain 131(Pt 2):409–424

270. Shanks MF et al (2007) Regional brain activity after prolonged cholinergic enhancement in early Alzheimer's disease. Magn Reson Imaging 25(6):848–859

271. Kircher TT et al (2005) Cortical activation during cholinesterase-inhibitor treatment in Alzheimer disease: preliminary findings from a pharmaco-fMRI study. Am J Geriatr Psychiatry 13(11):1006–1013

272. Thiyagesh SN et al (2010) Treatment effects of therapeutic cholinesterase inhibitors on visuospatial processing in Alzheimer's disease: a longitudinal functional MRI study. Dement Geriatr Cogn Disord 29(2):176–188

273. McGeown WJ, Shanks MF, Venneri A (2008) Prolonged cholinergic enrichment influences regional cortical activation in early Alzheimer's disease. Neuropsychiatr Dis Treat 4(2):465–476

274. Bokde AL et al (2009) Decreased activation along the dorsal visual pathway after a 3-month treatment with galantamine in mild Alzheimer disease: a functional magnetic resonance imaging study. J Clin Psychopharmacol 29(2):147–156

275. McLaren DG et al (2012) Tracking cognitive change over 24 weeks with longitudi-

nal functional magnetic resonance imaging in Alzheimer's disease. Neurodegener Dis 9(4):176–186

276. Venneri A, McGeown WJ, Shanks MF (2009) Responders to ChEI treatment of Alzheimer's disease show restitution of normal regional cortical activation. Curr Alzheimer Res 6(2):97–111

277. Dhanjal NS et al (2013) Auditory cortical function during verbal episodic memory encoding in Alzheimer's disease. Ann Neurol 73(2):294–302

278. Dhanjal NS, Wise RJ (2014) Frontoparietal cognitive control of verbal memory recall in Alzheimer's disease. Ann Neurol 76(2):241–251

279. Goekoop R et al (2004) Challenging the cholinergic system in mild cognitive impairment: a pharmacological fMRI study. Neuroimage 23(4):1450–1459

280. Gron G et al (2006) Inhibition of hippocampal function in mild cognitive impairment: targeting the cholinergic hypothesis. Neurobiol Aging 27(1):78–87

281. Risacher SL et al (2013) Cholinergic enhancement of brain activation in mild cognitive impairment during episodic memory encoding. Front Psychiatry 4:105

282. Petrella JR et al (2009) Effects of donepezil on cortical activation in mild cognitive impairment: a pilot double-blind placebo-controlled trial using functional MR imaging. AJNR Am J Neuroradiol 30(2):411–416

283. Pa J et al (2013) Cholinergic enhancement of functional networks in older adults with mild cognitive impairment. Ann Neurol 73(6):762–773

284. Goveas JS et al (2011) Recovery of hippocampal network connectivity correlates with cognitive improvement in mild Alzheimer's disease patients treated with donepezil assessed by resting-state fMRI. J Magn Reson Imaging 34(4):764–773

285. Li W et al (2012) Changes in regional cerebral blood flow and functional connectivity in the cholinergic pathway associated with cognitive performance in subjects with mild Alzheimer's disease after 12-week donepezil treatment. Neuroimage 60(2):1083–1091

286. Zaidel L et al (2012) Donepezil effects on hippocampal and prefrontal functional connectivity in Alzheimer's disease: preliminary report. J Alzheimers Dis 31(Suppl 3):S221–S226

287. Lorenzi M et al (2012) Effect of memantine on resting state default mode network activity in Alzheimer's disease. Drugs Aging 28(3):205–217

288. Sole-Padulles C et al (2013) Donepezil treatment stabilizes functional connectivity during resting state and brain activity during memory encoding in Alzheimer's disease. J Clin Psychopharmacol 33(2):199–205

289. Wang L et al (2014) The effect of APOE epsilon4 allele on cholinesterase inhibitors in patients with Alzheimer disease: evaluation of the feasibility of resting state functional connectivity magnetic resonance imaging. Alzheimer Dis Assoc Disord 28(2):122–127

290. Wang Z et al (2014) Acupuncture modulates resting state hippocampal functional connectivity in Alzheimer disease. PLoS One 9(3):e91160

291. Jia B et al (2015) The effects of acupuncture at real or sham acupoints on the intrinsic brain activity in mild cognitive impairment patients. Evid Based Complement Alternat Med 2015:529675

292. Wells RE et al (2013) Meditation's impact on default mode network and hippocampus in mild cognitive impairment: a pilot study. Neurosci Lett 556:15–19

293. van Paasschen J et al (2013) Cognitive rehabilitation changes memory-related brain activity in people with Alzheimer disease. Neurorehabil Neural Repair 27(5):448–459

294. Hampstead BM et al (2012) Mnemonic strategy training partially restores hippocampal activity in patients with mild cognitive impairment. Hippocampus 22(8):1652–1658

295. Buhmann C et al (2003) Pharmacologically modulated fMRI—cortical responsiveness to levodopa in drug-naive hemiparkinsonian patients. Brain 126(Pt 2):451–461

296. Kwak Y et al (2010) Altered resting state cortico-striatal connectivity in mild to moderate stage Parkinson's disease. Front Syst Neurosci 4:143

297. Wu T et al (2009) Regional homogeneity changes in patients with Parkinson's disease. Hum Brain Mapp 30(5):1502–1510

298. Agosta F et al (2014) Cortico-striatal-thalamic network functional connectivity in hemiparkinsonism. Neurobiol Aging 35(11):2592–2602

299. Choe IH et al (2013) Decreased and increased cerebral regional homogeneity in early Parkinson's disease. Brain Res 1527:230–237

300. Esposito F et al (2013) Rhythm-specific modulation of the sensorimotor network in drug-naive patients with Parkinson's disease by levodopa. Brain 136(Pt 3):710–725

301. Mattay VS et al (2002) Dopaminergic modulation of cortical function in patients with

Parkinson's disease. Ann Neurol 51(2): 156–164

302. Tessitore A et al (2002) Dopamine modulates the response of the human amygdala: a study in Parkinson's disease. J Neurosci 22(20):9099–9103

303. Arantes PR et al (2006) Performing functional magnetic resonance imaging in patients with Parkinson's disease treated with deep brain stimulation. Mov Disord 21(8):1154–1162

304. Phillips MD et al (2006) Parkinson disease: pattern of functional MR imaging activation during deep brain stimulation of subthalamic nucleus—initial experience. Radiology 239(1): 209–216

305. Kahan J et al (2012) Therapeutic subthalamic nucleus deep brain stimulation reverses cortico-thalamic coupling during voluntary movements in Parkinson's disease. PLoS One 7(12):e50270

306. Kahan J et al (2014) Resting state functional MRI in Parkinson's disease: the impact of deep brain stimulation on 'effective' connectivity. Brain 137(Pt 4):1130–1144

307. Scheff SW et al (2006) Hippocampal synaptic loss in early Alzheimer's disease and mild cognitive impairment. Neurobiol Aging 27(10):1372–1384

308. Jack CR Jr et al (2013) Tracking pathophysiological processes in Alzheimer's disease: an updated hypothetical model of dynamic biomarkers. Lancet Neurol 12(2):207–216

309. Bobholz JA et al (2007) Clinical use of functional magnetic resonance imaging: reflections on the new CPT codes. Neuropsychol Rev 17(2):189–191

Chapter 24

fMRI in Epilepsy

Rachel C. Thornton, Louis André van Graan, Robert H. Powell, and Louis Lemieux

Abstract

This chapter provides an overview of the application of functional MRI applied to the field of Epilepsy and is divided into two sections, covering cognitive mapping and imaging of paroxysmal activity, respectively. In addition to a review of the most scientifically and clinically relevant findings, technical and methodological background information is provided to help the reader better understand the data acquisition process. We show how both approaches may play a role in the presurgical evaluation of patients with drug-resistant focal epilepsy and provide opportunities for new insights into the neuropathological processes that underlie both focal and generalized epilepsy.

Key words Epilepsy, Focal epilepsy, Generalized epilepsy, Interictal, Ictal, Imaging, Functional magnetic resonance imaging, fMRI, Electroencephalography, EEG, Multi-modal imaging, EEG-correlated fMRI, Cognitive mapping, Functional mapping, Brain activity mapping, Language lateralization, Memory mapping, Presurgical evaluation

1 Cognitive fMRI in Epilepsy

The commonest surgical procedure for patients with drug resistant temporal lobe epilepsy (TLE) is anterior temporal lobe resection (ATLR). Complications of this operation include a decline in language and memory abilities, and an important part of the presurgical assessment lies in the careful selection of patients to minimize these adverse cognitive sequelae. This has traditionally been the role of neuropsychology and the intracarotid amytal test (IAT). Since the advent of functional MRI (fMRI), however, there has been much interest in its possible role in the presurgical assessment of those with epilepsy, principally in the identification of eloquent cortex to be spared during surgery.

There is a substantial body of literature reporting on cognitive function in epilepsy—employing a spectrum of indices and parameters, including task performance, lesion type and location, functional and effective connectivity measures and activation study

Massimo Filippi (ed.), *fMRI Techniques and Protocols*, Neuromethods, vol. 119,
DOI 10.1007/978-1-4939-5611-1_24, © Springer Science+Business Media New York 2016

outcomes that investigate and report cognitive impairment and idiosyncrasies in epilepsy, including working memory [1–3], long-term memory [4] and language organization [5, 6].

Neuropsychology has played a prominent role throughout the modern era of epilepsy surgery, mainly because of the importance of the temporal lobes in memory function. The principal role of baseline neuropsychological assessments is in establishing a baseline quantification of relevant cognitive function whilst providing an indication of the potential impact of surgery. It does so in the context of a conventional understanding of lateralization and localization of cerebral disturbance, predicting the impact of surgery on memory, providing data on lateralization and localization of cerebral disturbance, and providing evidence for cerebral reorganization.

The IAT plays a role in the presurgical assessment of TLE in some centers. Its uses are in assessing the capacity of the contralateral temporal lobe to maintain useful memory functions thus guarding against a severe postoperative amnesic syndrome, and provides a means of lateralizing language function. The procedure involves the injection of sodium amytal into one carotid artery, inactivating the corresponding hemisphere for around 10 min, and thus crudely mimicking the effects of surgery on the medial temporal lobe (MTL) structures. During this time, the patient's language and memory abilities are tested. Although still commonly used, the IAT has considerable disadvantages, not least the fact that it is an expensive, invasive procedure with potentially serious complications. Doubts also exist about its reliability and validity in predicting postoperative amnesia. In contrast to the traditional neuropsychological assessment, which relies on standardized tests of cognitive abilities and yields results that are easily validated, IAT procedures vary significantly between institutions with respect to the testing protocol used, choice of behavioral stimuli, dosage, and administration of amytal, all of which can lead to variations in the results [7]. The IAT is also poor at predicting verbal memory decline as deactivation of the language dominant hemisphere may impede verbal process that are constituent to verbal memory, thereby causing increased errors on verbal memory testing [8].

fMRI has the potential for replacing the IAT and for providing additional information to that provided by baseline neuropsychological assessment in the lateralization and localization of language and memory function. Practically, fMRI is cheaper than the IAT, noninvasive, and repeatable. There are, however, important potential caveats when considering the role of fMRI. First, areas activated by a particular fMRI paradigm are not necessarily crucial for the performance of that task. Second, it does not necessarily follow that all areas involved in a task will be activated by a particular fMRI paradigm. Third, the extent of activation seen in a task, in terms of both the area activated and the magnitude of the peak, may bear no relation to the competence with which that task is performed. The development of insight and greater appreciation of engagement of multiple haemodynamic networks in cognitive tasks together with technical improvement in acquisition and processing is likely to elaborate the value of fMRI paradigms in mapping cognitive functions.

Whilst being neither very sensitive nor specific with regards the neurobiological substrates of these cognitive functions—psychometric data provides some measurement and quantification. When understood that the psychometric testing itself provides a behavioral sample of cognitive processes rather than a true ecological measure it compounds the interpretation of the correlation between fMRI data and the test scores, as sensitive or specific measure of changes in brain function. Caution will also be needed in the interpretation of results bearing in mind that fMRI techniques, while useful for the localization of cognitive function, may not reliably indicate the capacity of unilateral temporal lobe structures.

In the following section, we review how fMRI is used to lateralize language function. We then discuss the current state of research efforts to localize brain regions involved in language and memory, and study the effects of epilepsy upon these. Furthermore, the efforts to assess the reliability of fMRI in the prediction of postoperative language and memory deficits following ATLR are discussed.

1.1 Language fMRI

1.1.1 Paradigm Design and Analysis

The aims of preoperative language fMRI are primarily to lateralize and localize language functions and to use this information to predict and avert postoperative complications. A number of task paradigms to engage anterior/expressive as well as posterior/comprehension language areas [9–14] have been used to identify language representation. Complementary indices provided by verbal fluency, verb generation and semantic decision tasks, are variously and commonly employed in the clinical context [15]. The most widely used tasks in language fMRI experiments are verbal fluency tasks. These are generally strongly lateralizing and reliably identify "expressive language functions" in the dominant inferior frontal gyrus (IFG) (Brodmann Areas [BA], 44, 45). Specifically, verbal fluency tasks show more prominent activity in left frontal regions, corresponding to Broca's area, than in the medial temporal lobe in healthy controls and TLE patients [16, 17]. Although these tasks are usually covert (i.e. performed silently without performance monitoring), they have been reliably replicated in numerous studies in both normal and patient populations. Their within-subject reproducibility has been demonstrated, with frontal activations shown to be more reliable than temporoparietal ones [18]. In addition they can be applied to patients with a wide range of cognitive abilities, with language lateralization results appearing to be relatively unaffected by patients' performance levels [19].

Although tasks of verbal fluency do show language-related activations, they are not pure language tasks, containing substantial components of executive processing and of working and verbal memory. These activations are typically seen in the middle frontal gyrus (MFG) (BA 46, 49) (Fig. 1).

Paradigms that specifically recruit the areas that are removed during ATLR should be established [17, 20]: Activation paradigms for naming functions that could provide greater specificity in the context of ATLR [21] include object naming paradigms involving visual [22, 23] and auditory stimuli [24, 25]. The multidimensional structure of language representation is illustrated by discrete contribution of different task paradigms. For example, verbal fluency identifies areas that are not activated with verb generation. In turn verb generation causes more discrete activation than verbal fluency [15]. Fluency tasks are also less reliable in identifying "receptive" language areas located in the dominant temporal lobe. These processing areas are best assessed by tasks that probe language comprehension such as reading sentences or stories, which tend to activate superior temporal cortex extending to supramarginal gyrus (BA 20, 21, 39) [26, 27], but are less strongly lateralizing than verbal fluency tasks (Fig. 1). Using a panel of fMRI tasks (verbal fluency, reading comprehension, and auditory comprehension) was shown to be helpful in reducing inter-rater variability and helped in the evaluation of language laterality in patients with focal epilepsy [28].

Verbal fluency   Reading comprehension

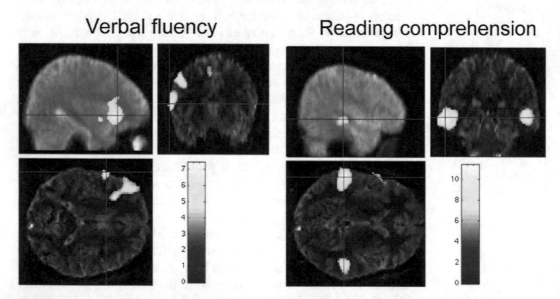

Fig. 1 Verbal fluency and reading comprehension: Typical fMRI findings in a patient performing tasks of verbal fluency (*left*) and reading comprehension (*right*) showing activation in the dominant frontal lobe and bilateral superior temporal lobes, respectively. Areas of activation are overlaid on a distortion matched high resolution echo-planar image

These functional imaging experiments have used block design paradigms to detect regions of the brain showing greater activation during task blocks when compared with rest blocks. The advantage of block designs over event-related designs is that they are efficient in detecting differences between two conditions; however, they offer less flexibility in the experimental design required for studying complex cognitive functions.

The degree of lateralization is often quantified using a lateralization (LI) or asymmetry index, $(AI) = (L-R)/(L+R)$, where L and R represent the strength of activation for the left (L) and right (R) sides, respectively, based on the number of activated voxels for the whole hemisphere or using regions of interest (ROIs) targeted to known as language areas [29]. A positive value represents left language lateralization and a negative, right-sided dominance, although AI values between -0.2 and $+0.2$ are often classified as bilateral. This can be determined by counting the number of voxels exceeding a specified threshold of significance. This type of AI calculation has some problems, in particular, the fact that the AI can differ according to the significance threshold chosen for the activation map from which it is calculated [30]. One suggested solution to this is to calculate AIs from all voxels that correlated positively with a task, but with each weighted by their own statistical significance [31]. Using this method, AIs were less variable than those calculated from suprathreshold voxels only. Alternative approaches to estimating the degree of asymmetry include measuring the mean signal intensity change induced by the task within a brain volume of interest [30], and performing a statistical comparison of the magnitude of task-induced activation in homotopic regions of the two hemispheres [32]. These methods measure the magnitude of the mean signal change and have the advantage of not being threshold dependent. Language networks in focal epilepsy have, in fact, been localized using data driven approaches: These algorithms yielded significant observations not seen with conventional fMRI, specifically in relation to the effects of epilepsy on language representation [33–35]. Other studies have suggested that visual rating appears to work as well as calculating AIs [36] by using pattern classifying algorithms that have demonstrated concordance with existing LI and visual rating classification methods [37, 38].

The simplest forms of study designs (including those described earlier) employ cognitive subtraction designs. These involve selecting a task that activates the cognitive process of interest and a baseline task that controls for all but the process of interest. One problem of this type of design is that it depends on an assumption known as pure insertion, which supposes that a new cognitive component can be inserted without affecting those processes that are also engaged by the baseline task [39]. Another problem with cognitive subtraction is in finding baseline tasks that activate all but the process of interest. These problems can be overcome by using more complex experimental designs, such as factorial designs and cognitive conjunctions.

Factorial designs use two or more variables (e.g. sentence vs. word presentation and auditory vs. visual presentation) and allow the effect that one variable has on the other to be measured explicitly. The analysis of this type of design involves calculating the main effects of each variable and the interaction between them [39]. Cognitive conjunctions are an extension of cognitive subtraction paradigms. Cognitive subtraction looks for activation differences between a single pair of tasks, while cognitive conjunction looks at two or more task pairs, which share a common processing difference [40]. The advantages of this approach are that it allows greater freedom in selecting the baseline task as it is not necessary to control for all but the component of interest, and that it does not depend on the assumption of pure insertion.

1.1.2 Language Lateralization in Epilepsy

Focal epilepsy may be associated with disrupted lateralization and localization of language regions; therefore, one would expect a higher probability of abnormal language lateralization. Nevertheless, significant differences have been reported between centers in the relative proportions of right and left hemisphere dominant patients using the IAT, some of which may be due to the different criteria used for assessing dominance. The percentage of left hemisphere dominant right-handed patients has ranged from 63 to 96% [41] while for left-handers a similar variation has been reported between 38 and 70% [42, 43]. Results of fMRI studies have also shown greater atypical language dominance in patients. In a comparison between 100 right-handed healthy subjects and 50 right-handed epilepsy patients, 94% of the normal subjects were considered as left hemisphere dominant and 6% had bilateral representation. The epilepsy group showed greater variability of language dominance, with 78% showing left hemisphere dominance, 16% symmetric activation, and 6% showing right hemisphere dominance. Atypical language dominance was associated with an earlier age of brain injury and with weaker right hand dominance [44].

The localization of the epileptogenic lesion and epileptic activity [45] has also been shown to influence language organization. In a retrospective study of patients with hippocampal sclerosis (HS) who had undergone presurgical evaluation, atypical speech dominance occurred in 24% of those with left-sided HS, whereas all those with right-sided HS had left-sided speech dominance. In addition, atypical speech representation was associated with higher spiking frequency and in those with sensory auras suggesting ictal involvement of the lateral temporal structures. No association was demonstrated between either age at epilepsy onset or age at initial precipitating injury and atypical speech representation [46].

Comparing the degree of reorganization of frontal and temporal lobe language functions has shown a significantly more left lateralized pattern of language activation in controls and right TLE patients than left TLE patients [47]. In patients with atypical language representation, the degree of reorganization towards the right hemisphere was greater in the temporal lobes than in the frontal lobes. In a study of 50 patients with focal epilepsy, greater atypical language dominance was seen in those with left hemisphere seizure focus [48]. Left TLE patients who did not have atypical language also had lower asymmetry indices in both frontal and temporal ROIs, mainly because of greater activation in homologous right hemisphere regions. Atypical language representation in Wernicke's area is more frequently observed in TLE patients, whereas FLE, conversely, appears to have a greater effect on the organization of anterior language areas [49].

The degree of language lateralization has also been related to the nature of the epileptogenic lesion with early acquired lesions, such as HS, considered to be associated with greater incidence of atypical language lateralization compared with developmental lesions originating in utero, such as malformations of cortical development (MCDs). Atypical language organization has been observed in a number of other clinical presentations: including stroke, medial temporal sclerosis, focal cortical lesions as well as patients with a normal structural MRI [50]. A higher degree of atypical language dominance, in both frontal and temporal language areas, has been demonstrated in patients with left HS compared with patients with left frontal and lateral temporal lesions [51], suggesting that the hippocampus itself may play an important role in the establishment of language dominance. Another study, however, demonstrated no difference in the frequency of atypical language lateralization between left TLE patients with HS and those with developmental tumors [52]. Interestingly, HS has been associated with altered functional organization of cortical networks involved in lexical and semantic processing [53] in TLE patients.

It is interesting to speculate on how TLE affects language lateralization, and it is possible that strong connectivity between inferior frontal and temporal areas make frontal lobe functions particularly sensitive to temporal pathology. Language lateralization is not associated with type and location of lesion (acquired or developmental), symptoms and gender [54] or age of seizure onset [49]. However, an association between language lateralization and handedness, location and nature of pathologic substrate and duration of epilepsy has been identified [55]. Verbal memory scores on psychometric tests have been associated with lateralization of language implicating connectivity between inferior frontal cortex and hippocampus [56, 57]. The increased incidence of atypical language dominance in epilepsy illustrates the importance of establishing language dominance prior to performing surgical resection and as a consequence much of the work on fMRI in epilepsy has been directed towards trying to replace the IAT as a means of doing this.

Studies comparing fMRI and the IAT are summarized in Table 1. Just as IAT protocols differ between centers, a number of fMRI paradigms to determine language dominance have been employed but agreement of approximately 80–90% is seen between the two techniques [11, 58]. The remaining cases generally exhibit partial disparity where one method shows bilateral language representation and the other lateralized language dominance and outright disagreement between fMRI and IAT is rare. In an interesting study that assessed the relative accuracy of Wada and fMRI in discordant cases fMRI provided a more accurate prediction of naming at postsurgical outcome in seven patients, Wada was more accurate in two patients. The two methods provided comparable accuracy in one patient [58].

One study suggested that fMRI may be less reliable in left-sided neocortical epilepsy (25% disparity) in comparison with left-sided medial TLE (3% disparity) [36]. Another showed that concordance between fMRI-based laterality and IAT was much lower in left TLE patients than in patients with right TLE [59]. One interesting case of false lateralization of language function in a post-ictal patient with left HS also illustrates the need for caution in the interpretation of results in individual patients. No activation was seen in the left temporal lobe during multiple language tasks after a cluster of left temporal lobe seizures but in a repeat fMRI experiment 2 weeks later, activation was seen predominantly over the left temporal region [60]. However, bearing in mind the previously mentioned limitations of the IAT, it is even debatable whether fMRI and IAT are directly comparable as they probe different aspects of language. fMRI language localization can replace Wada test in the majority of patients. However, the Wada test is still a valuable adjunct and can be employed when a patient cannot undergo fMRI. It can also be used for validation of fMRI results or for the assessment of selective language areas near structural abnormalities [61].

Comparisons have also been performed between fMRI activation maps and regions showing disruption of function during intraoperative electrocortical stimulation (ECS). In order for fMRI to be used instead of ECS, it must demonstrate a high predictive power for the presence as well as the absence of critical language function in regions of the brain. As with IAT, these studies show strong, but incomplete agreement with fMRI, with high sensitivity but lower specificity [69–71]. Although false-positive activation (fMRI activation but no ECS disruption) is relatively common, this is not surprising given that fMRI activates whole networks of regions, not all of which are essential for the task in question. False-negative findings (regions showing disruption by ECS but no fMRI activation) are more critical when planning a surgical resection, and these were identified in 2 patients out of 21 reported in two series. Activation and disruption was typically within 5 mm in frontal regions and 10 mm in temporal areas.

Table 1
Concordance between fMRI language lateralization and the IAT

Authors	Sample size[a]	fMRI language lateralization tasks	Concordance
Desmond et al. (1995) [62]	7	Semantic decision task	100%
Binder et al. (1996) [63]	22	Semantic decision task	$r=0.96$
Hertz-Pannier et al. (1997) [64]	6[b]	Verbal fluency paradigm	100%
Yetkin et al. (1998) [65]	13	Word generation task	$r=0.93$
Benson et al. (1999) [66]	12	Verb generation task	100%
Lehericy et al. (2000) [27]	10	Semantic fluency Sentence repetition Story listening	Semantic fluency > story listening > sentence repetition Greater concordance between IAT results and activation asymmetry in frontal than temporal lobes
Carpentier et al. (2001) [67]	10	Identification of syntactic/semantic errors in target sentences	80%
Gaillard et al. (2002) [29]	21	Reading paradigm	85%
Woermann et al. (2003) [36]	100	Word generation	91%
Sabbah et al. (2003) [68]	20[c]	Word generation Semantic decision	95%
Benke et al. (2006) [59] Arora et al. (2009) [11] Janecek et al. (2013) [58]	68 40 229	Semantic decision Reading sentence comprehension/ auditory sentence comprehension and a verbal fluency task. Semantic decision/tone decision	89%—right TLE 72%—left TLE 91.3% 86%

IAT intracarotid amytal test, *TLE* temporal lobe epilepsy

[a]Some of these studies report fMRI data on larger samples. However, only the patients with fMRI and IAT data are included here

[b]Age range 8–18

[c]Patients with suspected atypical language lateralization were selected

A combination of four different language tasks has shown more reliable and robust lateralization in normal subjects by targeting brain regions common to different tasks, thereby focusing on areas critical to language function. Regions of activation detected in this way corresponded well with ECS findings in the temporoparietal region [72]. Sensitivity was 100% in all but one patient. This high negative predictive value suggested that areas where no significant fMRI activity was present could be safely resected without using ECS. fMRI activity, however, was not always absent at noncritical language areas limiting its positive predictive value for the presence of critical language. Although this suggests that fMRI is not yet ready to replace ECS, it could be used to speed up intracranial mapping procedures and to guide the extent of the craniotomy.

1.1.4   Language Localization and Prediction of Postoperative Language Deficits

Language deficits have been reported following language-dominant ATLR, with naming the most commonly affected function [73, 74]. It has also been suggested that the risk for post-operative decline in naming abilities increases with age of seizure onset and the extent of lateral temporal neocortex resected [75]. Preoperative cortical stimulation via subdural grid electrodes has been used to localize language function, suggesting that early onset of dominant temporal lobe seizure foci leads to a more widespread or atypical distribution of language areas, particularly naming and reading areas [76]. A subsequent study also reported that markers of early left hemisphere damage (such as early seizure onset, poor verbal IQ, left handedness, and right hemisphere memory dominance) increase the chances of essential language areas being located in more anterior temporal regions. Again these areas were identified using naming and reading tasks [77].

These findings suggest that naming and reading abilities are the language skills most at risk following dominant temporal lobe surgery. In patients with right TLE, preserved naming function is associated with activation of the left hippocampus by the verbal fluency task. Patients with left TLE, who show preservation of naming, conversely appear to involve the left frontal lobe, in an apparent compensatory response to epileptic activity in the left hippocampus [13]. Although the IAT may provide a useful index of language laterality, it does not provide detailed information on the localization of these specific language skills. As these may also vary in location between individuals, the role IAT can play in the prediction of postoperative deficits in individual patients is therefore limited. Designing fMRI paradigms that specifically probe naming and reading skills would provide a useful clinical tool for mapping relevant language skills that could be used in the prediction of postoperative deficits. Specifically, auditory and visual naming paradigms may yield greater predictive specificity with regard to naming difficulties after ATLR [21].

One study has used preoperative functional neuroimaging to predict language deficits following left ATLR: Temporal lobe fMRI asymmetry was found to be predictive of deficits seen on a postoperative naming test with a greater degree of language lateralization toward the left hemisphere related to poorer naming outcome and language lateralization towards the right hemisphere associated with less or no decline. The correlation between temporal lobe fMRI AI and naming deficits was stronger than that seen in the frontal lobes and also stronger than that between IAT and naming deficits [78]. In this regard, fMRI activation of the left middle frontal gyrus (MFG) with a verbal fluency task was seen to predict significant postsurgical naming decline in patients with left TLE, showing good sensitivity but rather poor specificity [17].

Interestingly, many patients do not suffer any language deficits following ATLR, suggesting that multiple sets of neural systems may exist that are capable of performing the same cognitive function, and that some of these may be engaged following focal brain injuries. In a study of patients who had undergone left ATLR but did not have deficits in sentence comprehension, decreased activation was demonstrated in undamaged areas of the normal left hemisphere system but increased activation was seen in several right frontal and temporal regions not usually engaged by normal subjects [79]. This suggests that there is more than one neural system capable of sustaining sentence comprehension. This study was, however, unable to tell whether this functional reorganization to the right IFG occurred pre- or postoperatively. A separate study looked at the role of the right IFG by comparing its functional activation on a verbal fluency task in controls with left TLE patients [80]. The patients were shown to activate a more posterior right IFG region compared with controls, although left IFG activation did not differ significantly between the two groups. Further, verbal fluency-related activation in the right IFG was not anatomically homologous to left IFG activation in either patients or controls. This suggests that reorganization takes place preoperatively in patients with chronic left TLE, and that the prediction of language outcome following left ATLR may depend not only on the extent of preoperative right hemisphere activation, but also its location.

1.1.5 Combination of MRI and MR-Tractography

Complex behaviors such as language and memory rely upon networks of neurons, which integrate the functions of spatially remote brain regions. The combination of fMRI to identify cortical regions involved in language function and MR-tractography to visualize white matter pathways connecting these regions offers an opportunity to study the relationship between structure and function in the language system (Fig. 2). Studies have revealed structural asymmetries in controls, with greater left-sided frontotemporal connections in the dominant hemisphere [81]. Patients with left TLE had reduced left-sided and greater right-sided connections than both

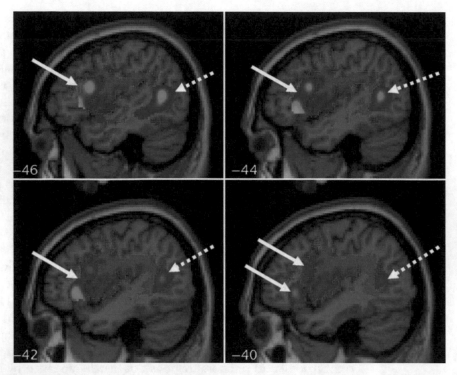

Fig. 2 Combined MR tractography and functional mapping. Frontal lobe connections overlaid on a structural template along with group fMRI effects for word generation (*solid arrows*) and reading comprehension (*dashed arrows*), showing how the tracts connect together the frontal and temporal lobe functionally active regions

controls and right TLE patients, reflecting the altered functional lateralization seen in left TLE patients, and significant correlations were demonstrated between structure and function in controls and patients, with subjects with more highly lateralized language function having a more lateralized pattern of connections [82]. The combination of fMRI with information on the structural connections of these normally and abnormally functioning areas offers the opportunity to improve understanding of the relationship between brain structure and function and may improve the planning of surgical resections to maximize the chance of seizure remission and to minimize the risks of cognitive impairment.

1.2 Memory fMRI

A range of memory functions are commonly affected in epilepsy including modality specific processes in working and long term memory. fMRI can reliably localize and assess the impact of surgery on memory networks [83, 84]. In addition LIs have been established to evaluate the effects of epilepsy and surgical intervention on memory [85]. MTL structures are associated with memory functioning, and surgical resection is known to cause reduced memory function in some cases. The study of patients following temporal lobe surgery has provided considerable evidence supporting the critical role that the hippocampi play in memory functioning. Bilateral injury to these areas leads to a characteristic amnesic syndrome [86], while unilateral lesions lead to material-specific deficits, and a decline in verbal memory following surgery to the language-dominant hemisphere has been consistently reported and studied [87], along with deficits in topographical memory following nondominant ATLR [88]. Although rare, some patients have a severe anterograde amnesic syndrome following a unilateral ATLR. Most of these, however, have subsequently been found to have evidence of contralateral hippocampal pathology, either on postoperative electroencephalography (EEG) [89], post-mortem pathological findings [90], or post-operative volumetric MRI [91].

Immediate recall in the verbal and visual modalities demonstrate significantly less activation in patients with HS as compared to healthy subjects [92]. A recent study shows that patients employ the contralateral hippocampus and the ipsilateral parahippocampal gyrus, reflecting mechanisms of functional adaptation [93]. Two different models of hippocampal function have previously been proposed to explain memory deficits following unilateral ATLR: hippocampal reserve and functional adequacy [94]. According to the hippocampal reserve theory, postoperative memory decline depends on the capacity or reserve of the contralateral hippocampus to support memory following surgery, while the functional adequacy model suggests that it is the capacity of the hippocampus that is to be resected that determines whether changes in memory function will be observed. Evidence from baseline neuropsychology [95], the IAT [96], histological studies of hippocampal cell density [97], and MRI volumetry [98] has suggested that of the two, it is the functional adequacy of the ipsilateral MTL, rather than the functional reserve of the contralateral MTL that is most closely related to the typical material-specific memory deficits seen following ATLR. However, compensatory reorganization in the context of HS have been shown to be elaborate involving temporal and extra temporal structures [92, 93, 99, 100].

The assessment of ability to sustain memory is critical for planning ATLR as memory decline is not an inevitable consequence of temporal lobe surgery. Accurate prediction of likelihood and severity of postoperative memory decline is necessary to make an informed decision regarding surgical treatment. Much work has been focused on the identification of prognostic indicators for risk of memory loss after ATLR. Language lateralization, verbal as well as visual memory fMRI activation patterns, age at onset of epilepsy and memory performance all serve as predictors of verbal memory decline in left ATLR. However, these factors appear to be less sensitive in prediction of postsurgical visual memory impairment in right ATLR [101]. Recent results [102] confirm age at onset of epilepsy, shorter duration of epilepsy and lower seizure frequency as critical factors that influenced verbal memory encoding in patients with TLE. Conversely this study showed that longer duration and higher seizure frequency were associated with greater inefficient, extra-temporal reorganization. The severity of HS on MRI is an important determinant, being inversely correlated with a decline in verbal memory following left ATLR, with less severe HS increasing the risk of memory decline [98, 103, 104]. Specifically, the extent of verbal memory decline after left ATLR is correlated with greater BOLD activation of the diseased left hippocampus and its connectivity to ipsilateral posterior cingulate [105]. Preoperative memory performance has been related to degree of postoperative memory impairment, with better performance increasing the risk of memory decline [95, 106, 107]. These risk factors reflect the functional integrity of the resected temporal lobe and suggest that patients with residual memory function in the pathological hippocampus are at greater risk of memory impairment postoperatively. Recently, fMRI has also been shown to be a potential predictor of postoperative material-specific memory decline following ATLR. Comparison of pre- and postoperative fMRI activation and correlation with better verbal memory outcome after left ATLR indicate preoperative reorganization of verbal memory function to the ipsilateral posterior medial temporal lobe [85]. Other results indicate that visual and verbal memory function following ATLR is correlated to activity of the contralateral medial temporal lobe and its connectivity to the posterior cingulate cortex ipsilateral to the damaged hippocampus [105, 108].

1.2.1 The Difficulty in Seeing Anterior Hippocampal Activation

Impairment in memory encoding following ATLRs suggests that anterior MTL regions are critical for successful memory encoding, and in the patient HM, who was rendered amnesic following bilateral temporal lobe resections, more posterior MTL structures remained intact [109]. Intracranial electrophysiological recordings during verbal encoding tasks have also shown greater responses in anterior hippocampal and parahippocampal regions for words remembered than those forgotten [110]. However, functional imaging studies have proved contradictory, with many showing encoding-related activations in posterior hippocampal and parahippocampal regions, which would be left intact following ATLRs.

One possible explanation for this apparent conflict is that anterior temporal regions are subject to signal loss during fMRI sequences. It has been demonstrated that signal loss due to susceptibility artifact is most prominent in the inferior frontal and inferolateral temporal regions [111], and as the hippocampus rises from anterior to posterior, one would expect greater susceptibility-induced signal loss in the anterior (inferior) relative to posterior (superior) hippocampus. This may have been one reason for the relative lack of anterior hippocampal activation in early fMRI studies of memory [112]. One study has directly examined the effects of susceptibility artifact on hippocampal activation by demonstrating its differential effect on the anterior vs. the posterior hippocampus. The averaged resting voxel intensity in an anterior hippocampal ROI was significantly less than in a posterior hippocampal ROI and intensity decreases were substantial enough to leave many voxels below the threshold at which BOLD effects could be detected [112]. Moreover, it has been shown that the sensitivity to BOLD changes is proportional to signal intensity at rest so that voxels with a lower baseline signal (such as those in anterior hippocampal regions) would be more difficult to activate than those with higher baseline signals [113].

An alternative explanation for the lack of anterior hippocampal activation seen in many early memory fMRI experiments is that the paradigms used were not optimal for detecting subsequent memory effects. The use of fMRI in studying memory function is more challenging than for language. This is partly due to the different components involved in memory processing, such as encoding and retrieval, and the fact that the nature of the material being encoded or retrieved influences which brain areas are activated. A further difficulty is how to separate brain activity related specifically to memory from that related to other cognitive processes. In consequence, more complex paradigms are required when studying memory than for examining language function.

Standard fMRI experiments initially used block design paradigms looking for regions of the brain showing greater activation during task blocks compared with rest blocks. A problem when designing memory fMRI experiments was how to separate brain activity specifically due to memory from that due to other cognitive processes being used in the task. Early fMRI studies of memory encoding employed block experimental designs to contrast tasks promoting differing memory performance, using the "depth of encoding" principle [114]. This states that if you manipulate material in a "deep" way (e.g. make a semantic decision about a word), then it is more likely to be recalled successfully than material manipulated in a "shallow" way (e.g. make a decision of whether the first letter of a word is alphabetically before the last letter). These studies tended to show consistent activation in left prefrontal cortical regions along with less reliable MTL activation [115–118]. Similar assumptions underlie the use of "novelty" paradigms in probing memory encoding. During these experiments, alternating blocks of novel and repeated stimuli are presented, with the hypothesis being that more memory encoding takes place while viewing a block of novel stimuli than when viewing the same repeated stimulus [119].

The advantage of block designs is that they are generally the most efficient in detecting differences between two conditions. The main problem in their interpretation, however, lies in the inference that the effects shown by these contrasts reflect differences in memory encoding, rather than any other differences between the two conditions (e.g. response to novelty and semantic processing) that are independent from differences in memory encoding. Attempts were made to overcome this problem using parametric block designs but were soon superseded by the advent of event-related studies.

Event-related fMRI is defined as the detection of transient hemodynamic responses to brief stimuli or tasks. This technique, derived from those used by electrophysiologists to study event-related potentials, enables trial-based rather than block-based experiments to be carried out. Trial-based designs have a number of methodological advantages, in particular that trials can be categorized post-hoc according to a subject's performance on a subsequent test to obtain fMRI data at the individual item level. Therefore, when studying memory encoding, activations for individual items presented can be contrasted according to whether they are remembered or forgotten in a subsequent memory test. This type of analysis allows the identification of brain regions showing greater activation during the encoding of items that are subsequently remembered compared with items subsequently forgotten (subsequent memory effects), which are then taken as candidate neural correlates of memory encoding [120]. Although event-related designs are less powerful than block designs at detecting differences between two brain states and may be more vulnerable to alterations in the hemodynamic response function (e.g. due to pathology), they have the advantage of permitting specifically the detection of subsequent memory effects due to successful encoding.

One study looking at encoding of words, pictures, and faces in healthy controls employed an experimental design, which allowed data analysis either as a block design, or as an event-related design of successful encoding [121]. The results demonstrated a functional dissociation between anterior and posterior hippocampus. The main effects of memory encoding, demonstrated specifically using an event-related analysis, were seen in the anterior hippocampus (Fig. 3), with the main effects of viewing stimuli, demonstrated using a block analysis, being located in more posterior regions.

1.2.2 The Effect of TLE on Memory Processes

Deficits in verbal memory following left ATLR and topographical memory following right ATLR suggest a material-specific lateralization of function in MTL structures. Functional imaging studies have been used to look for lateralization of cerebral activation patterns during episodic memory processes. Many have shown material-specific lateralization in prefrontal regions but this has been more difficult to demonstrate in the MTL [115, 119, 121, 122]. Working memory can be affected in TLE patients with HS, with indication of altered connectivity between regions [123]. Specifically, a disruption of the regional balance between task-positive and task-negative functional networks is associated with working memory dysfunction in TLE [124]. Reduced right superior parietal lobe activity is associated with suppression of activity in the healthy hippocampus in the context of an increasing WM load with

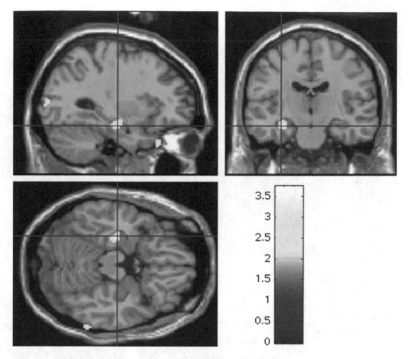

Fig. 3 Left hippocampal activation in a single subject performing a word encoding task

maintained levels of performance [3]. In FLE, patients particularly recruit the frontal lobe contralateral to the seizure focus in a compensatory pattern that sees patients employ a wider range of networks than healthy controls [84].

A number of studies have used fMRI to look at the lateralization of memory in patients with TLE compared with that seen in normal subjects, and also compared the findings with the results of the IAT. These are summarized in Table 2. These employed block design studies, demonstrating predominantly posterior MTL activation, and therefore cannot claim that subsequent memory effects have been specifically examined.

Studies performed in patients with TLE showing patient groups studied, experimental design employed, and principal findings *HS* hippocampal sclerosis, *IAT* intracarotid amytal test, *MTL* medial temporal lobe, *TLE* temporal lobe epilepsy.

More recently event-related studies have demonstrated a material-specific lateralization of memory encoding within anterior MTL regions that would be resected during standard ATLR (Fig. 4) [121]. In addition, a reorganization of function has been demonstrated in patients with unilateral TLE due to HS, with reduced ipsilateral activation, and increased contralateral activation in patients compared with controls [128, 129] (Fig. 5). Comparing groups of patients with controls demonstrated a functional reorganization away from the pathological hemisphere; however, it is not clear whether

Table 2
fMRI memory studies in TLE

Authors	Sample size	fMRI tasks	Findings
Detre et al. (1998) [122]	Controls $n=8$ Patients $n=10$	Block design Complex visual scenes vs. abstract pictures	Symmetric MTL activation in normal subjects. Lateralization of memory concordant with IAT in 9/10 subjects
Bellgowan et al. (1998) [125]	Patients $n=28$, 14 left TLE, 14 right TLE	Block design Semantic decision vs. auditory perception task	Greater activation in the left MTL in right TLE compared to left TLE group
Dupont et al. (2000) (79)	Controls $n=10$ Patients $n=7$, left HS	Block design Verbal encoding and retrieval vs. fixation on the letter A	Left occipitotemporoparietal network activated in controls. Reduced MTL activation and increased activation in left dorsolateral frontal cortex in patients
Jokeit et al. (2001) [126]	Controls $n=17$ Patients $n=30$	Block design Roland's Hometown Walking vs. baseline	No asymmetry of MTL activation in controls, greater activation in the MTL contralateral to seizure focus in 90% of patients
Golby et al. (2002) [127]	Patients $n=9$	Block design comparing novel vs. repeated stimuli Four encoding stimuli used—patterns, faces, scenes and words	Group level—greater activation in the MTL contralateral to seizure focus for all encoding stimuli Single subjects—lateralization of memory concordant with IAT in 8/9 subjects

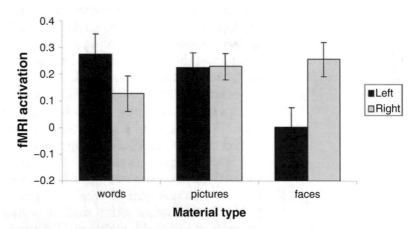

Fig. 4 Material-specific lateralization of memory encoding in the anterior hippocampus. fMRI activation within left and right hippocampal ROIs in healthy controls demonstrating left lateralized activation for word encoding, right-lateralized activation for face encoding and bilateral activation for picture encoding

this represents an effective way of maintaining memory function in individual patients. By correlating fMRI activation and performance on standard neuropsychological memory tests, it has been shown that MTL activation ipsilateral to the pathology is correlated with better performance while contralateral, compensatory activation correlates with poorer performance [129]. The conclusion that memory function in unilateral TLE is better when sustained by the activation within the damaged hippocampus is consistent with the observation that preoperative memory performance is a predictor of postoperative memory decline, with better performance predicting worse decline [106, 107], and adds further support to the functional adequacy model of hippocampal function.

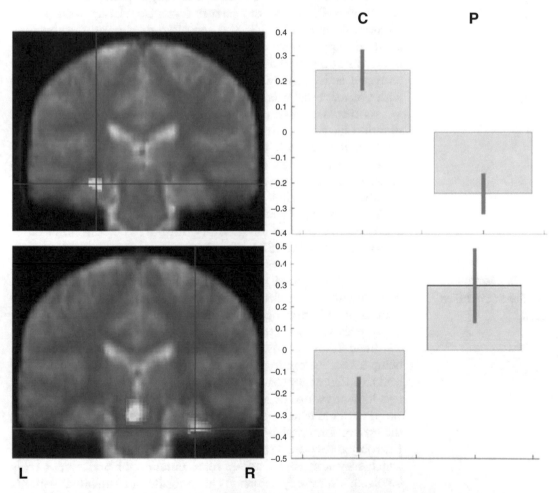

Fig. 5 fMRI memory encoding experiment: Left TLE patients vs. healthy controls. Regions showing significant differences in activation between left temporal lobe epilepsy (TLE) patients and controls are highlighted. Contrast estimates are shown on the right of the images. Controls (C) are on the *left* and patients (P) on the *right*. A reorganization of function is seen in the left TLE patients with reduced activation in the left hippocampus, and greater activation in the right hippocampus, compared with healthy controls

1.2.3 The Prediction of Postoperative Memory Changes

Prediction of postoperative memory decline is necessary to make an informed decision regarding surgical treatment. To date a small number of studies have used fMRI to predict the effect of left or right ATLR on verbal and nonverbal memory. In patients with left HS, greater verbal memory encoding activity in the left hippocampus compared with the right hippocampus predicted the extent of verbal memory decline following left ATLR [130]. In a further analysis of the same patients, it was demonstrated that greater activation within the left hippocampus predicted a greater postoperative decline in verbal memory [131]. These findings have since been replicated and extended to patients undergoing right ATLR [132]. Other groups have demonstrated correlations between MTL activation asymmetry ratios and postsurgical memory outcome in patients with both left and right TLE, with increased activation ipsilateral to the seizure focus correlating with greater memory decline [133, 134]. A recent study showed that bilateral posterior hippocampal activation correlated with less verbal memory decline postoperatively whereas left frontal and anterior medial temporal activations in left TLE patients correlated significantly with greater verbal memory decline [102].

As discussed earlier, two different models of hippocampal function have been proposed to explain memory deficits following unilateral ATLR: hippocampal reserve and functional adequacy [94]. Studies using asymmetry indices are unable to address this important issue; however, the findings of some of the above studies that greater preoperative activation within the ipsilateral, to-be-resected hippocampus, correlated with greater postoperative decline in memory support the functional adequacy theory [131, 132].

1.3 Challenges of Clinical Cognitive fMRI

When designing paradigms for patients with neurological deficits, it is important to use tasks that they are able to perform. A differential pattern of activation between patients and normal subjects is only interpretable if patients are performing the task adequately [135].

In addition, one must be aware of differences in the questions being asked by cognitive neuroscientists and clinicians, which can lead to different approaches to data analysis. Generally, neuroscientists look at groups of matched controls performing the same task and determine which brain regions are commonly activated across the group. The emphasis is on avoiding false-positive results (Type I errors) and conservative statistical thresholds need to be used, which may lead to an under representation of brain areas truly involved. Conversely, clinicians are considering individual patients where the priority is to identify all brain regions involved in a task, i.e., avoiding false negatives (Type II errors). As a result, less stringent statistical thresholds are required and indeed thresholds used may need to vary on an individual basis.

1.4 Cognitive fMRI in Epilepsy: Summary

fMRI is a noninvasive and widely available tool, which has had a dramatic impact on cognitive neuroscience. Much of the progress made will benefit clinical neuroimaging, although some problems exist in the application of fMRI to patients with neurological deficits. fMRI allows the noninvasive assessment of language function to be performed and offers a valid alternative to the IAT for establishing language dominance. By tailoring paradigms towards the localization of the specific language skills most at risk following temporal and frontal resections, it will be possible to map relevant language functions in the epilepsy surgery population. This in turn will allow better assessment of the risks posed by surgery in each individual patient.

Considerable effort is also being made in the development of memory paradigms that can lateralize MTL functions and provide meaningful data at the single subject level. This information, in combination with structural MRI to evaluate hippocampal pathology and baseline neuropsychology, will enable preoperative prediction of likely material specific memory impairments seen following unilateral ATLR to be made with greater accuracy. In consequence, it will be possible to modify surgical approaches in those patients most at risk and to improve preoperative patient counseling.

Clinically it is what happens to individual patients that is important and the next step in the validation of these techniques will involve similar studies with larger numbers of patients. These should include more heterogeneous samples, including both left and right TLE undergoing ATLR. As well as showing group level correlations either at the voxel-level or within a predefined ROI, it will be important to establish methods for using this data to predict language and memory changes in individual cases. Investigating how the brain sustains memory postoperatively also requires further investigation. Longitudinal fMRI studies with pre- and postoperative imaging, including correlations with neuropsychological measures of language and memory, will be required to look at functional reorganization following surgery, and it is anticipated that these will offer valuable insights into brain plasticity.

2 fMRI of Paroxysmal Activity

Despite major developments in the field of neuroimaging over the last two decades, the localization of the brain regions involved in seizure onset is problematic in a significant proportion of patients with focal epilepsy, thereby precluding surgical treatment. Furthermore, our understanding of the neurobiological mechanisms underlying epileptogenic networks in focal and generalized epilepsies is incomplete.

Scalp EEG and magnetoencephalography (MEG) are comparable in their ability to detect and measure synchronized neuronal activity taking place mainly over relatively superficial parts of the brain with exquisite temporal resolution. Although both remain extremely active areas of investigation, the interpretation of EEG and MEG data and in particular their utility in localizing brain generators is severely limited as a consequence of the principle of superposition and its corollary, the non-unicity of the inverse solution [136, 137]. This is in contrast to tomographic functional imaging modalities such as positron emission tomography (PET) or fMRI, which do not suffer from the problem of non-unicity and sampling bias is relatively minor. Although much superior to PET, the temporal resolution of fMRI, which is essentially governed by the local hemodynamic response, remains inferior to that of EEG by roughly three orders of magnitude. Nonetheless, fMRI allows hemodynamic changes linked to brief (~ms) neuronal events to be detected and localized with a fair degree of reliability. Although our understanding of the blood oxygen level-dependent (BOLD) fMRI signal is constantly improving, in part due to combined EEG and fMRI experimental data, as a general rule it remains an indirect and relative measure of neuronal activity. Although combined MEG-MRI seems a distant prospect, combined EEG-fMRI experiments were performed only a few years following the advent of fMRI [138].

Often presented as combining the advantages of its constituent parts, a concept that motivated the technique's pioneers, inevitably combined EEG-fMRI also suffers from some of their individual limitations. Whatever the technique's pros and cons, it will soon become clear to the reader that EEG-fMRI is unique in allowing the hemodynamic correlates of brief, unpredictable bursts of neuronal activity observed on scalp EEG, such as interictal spikes, to be investigated.

Prior to the possibility of EEG-fMRI experiments, studies of paroxysmal activity using fMRI were limited to ictal events and often relied on the correlation of the image time-series with observed clinical manifestations but sometimes did not [139–142].

The first studies of paroxysmal brain activity using fMRI were predominantly in patients with focal epilepsy, clinically motivated by the possibility of noninvasively localizing seizure focus. This continues to be an important source of motivation for this rapidly moving field, but much current research focuses its attention on the understanding of the networks underlying the generation of seizures in both focal and generalized epilepsies.

Although an exciting development with potential clinical value, the technique currently remains within the realm of advanced, exploratory imaging modalities that require resources not available in most epilepsy clinics. We therefore begin this review by discussing some of the technique's key technological and methodological aspects. We will then present an overview of the state of EEG-fMRI applied to the investigation of focal and generalized epilepsies.

The analysis and interpretation of fMRI data acquired from patients lying in the resting state with simultaneous EEG recording differs fundamentally from that of paradigm-driven fMRI in at least two ways: a lack of a prior experimental control and uncertainty about the nature of the relationship between EEG event and putative hemodynamic effects. This important topic will be the subject of a discussion.

2.1   EEG-Correlated fMRI in Epilepsy: Technical Issues

The recording of EEG inside the MR scanner still presents safety, image data quality and EEG data quality challenges. Historically, the issue of EEG data quality has been the determining factor in the technique's evolution, from interleaved to simultaneous EEG-fMRI. This reflects in part the fact that a gradual degradation in EEG quality mainly linked to cardiac activity can be readily observed in most subjects as they are moved inside the MR scanner (without scanning), posing an immediate challenge ahead of any other considerations such as safety (albeit this should also be at the forefront of the considerations of investigators introducing any new equipment in the scanner room) or the effect of scanning on EEG quality and the possible impact of the EEG recording equipment on image quality. In the following, we provide an overview of the state of EEG-fMRI technology, which remains an active area of research in particular in the area of EEG quality, although mostly for the purpose of evoked response recordings. The focus will be on the implications for studies in epilepsy and in particular at field strengths commonly used in neurological studies (≤ 3.0 T); the reader interested in the implementation of combined EEG and fMRI recordings at higher field strengths (e.g. 7 T) is directed towards two recent specialized reports that address data quality and safety [143, 144].

2.1.1   Physical Principles of EEG-MR System Interactions

The electro-magnetic processes that take place during MR image acquisition and that are susceptible to interactions with the EEG system are: strong static magnetic field (~ 1.0–3.0 T), switching magnetic gradient fields (~ 100 T/m/s), and radio frequency (RF) pulses (~ 10 µT and 100 MHz). In addition, although MR scanners are designed to optimize the magnetic component of the RF pulses, an electrical component is unavoidable. This may lead to linear antenna effects with possible safety implications [145]. EEG recording, on the other hand, requires electrodes and leads to be placed within the imaging field of view and electronic components, depending on the exact equipment and setup, in proximity to the scanner coils and antenna(s).

Four main mechanisms are at the origin of EEG-MR instrumentation interactions:

1. Magnetic induction: any change in magnetic flux (essentially the component of the magnetic field that is perpendicular to a surface) over time through a conducting medium (loop, surface, volume) gives rise to an electromotive force in the mate-

rial and hence an induced current. This phenomenon is governed by Faraday's law of induction. Changes in magnetic flux, and the associated induced currents, can be caused by movement (change of position, orientation, or shape) of the conducting medium in a magnetic field or change in the magnetic field to which the conducting medium is exposed;

2. Magnetic susceptibility differences: static interactions due to the magnetic properties of the components of the EEG recording system;

3. RF radiation emanating from active components of the EEG recording system;

4. Magnetic force on ferro- or paramagnetic components with associated risks of foreign instruments or their elements becoming projectiles in the scanner room; in the following we will assume that all usual design and manipulation precautions have been taken to avoid these effects.

Magnetic induction can result in EEG quality degradation in the form of pulse-related and image acquisition (gradient-switching and RF)-related artifacts. Magnetic susceptibility differences and RF radiation linked to the EEG system can give rise to image artifacts.

2.1.2 Safety

Health hazards not normally encountered when MR or EEG are performed separately can arise due to induced currents flowing through loops or the heating of EEG components in proximity or contact with the subject. For a specific 1.5 T scanner, and based on a worst case scenario, this study recommended that one 10 kΩ current-limiting resistor be inserted serially at each electrode lead and the possibility of large (EEG lead-electrode-head-electrode-lead-amplifier circuit) loops being formed reduced to a minimum by lead twisting. In experiments using a different custom-made EEG system, no significant heating was observed [146]. An important general consideration when placing wires in contact with the body is the type of RF transmit coil used and length of wire exposed to the electrical component of the RF field [147]. A number of MR-compatible EEG system or electrode cap vendors have incorporated current-limiting resistors in their product design. To the authors' knowledge, no adverse incident linked specifically to EEG-fMRI data acquisition has been formally reported to date.

2.1.3 Image Quality

Image quality remains an important issue throughout the field of MRI and the subject of investigation, particularly for echo-planar imaging (EPI), which is particularly prone to distortion and local signal dropout [148]. Artifacts caused by electrodes and leads were observed in early EEG-fMRI experiments [138]. Therefore, one must consider carefully the choice of materials and components placed within the field-of-view [146, 149–152]. It has been shown that the presence of high-density (256 channels) EEG caps can significantly impair structural MR imaging [153].

2.1.4 EEG Quality

In the literature it is common to categorize the artifacts observed on EEG recorded inside the MR scanner into two types: heart beat-related (whether scanning is taking place or not) and MR image acquisition-related. These can be considered distinct in terms of their generating mechanism, and deserve to be addressed separately in terms of remedies to minimize them, as reflected in the structure of this section. However, in practice they are linked by a third phenomenon, widely recognized as a nuisance in fMRI, namely subject motion. This is in part because the heart beat-related artifact is thought to mostly originate from the body motion linked to the heart beat, but also because subject (and EEG electrode and lead) motion can have an important impact on the ability to correct both types of artifact, depending on the approach taken. Therefore body motion is nefarious for EEG recording quality (and almost without saying, fMR image quality) and should be minimized. This will be the subject of Sect. 2.1.4.3.

EEG Quality: Pulse Artifact, Reduction, and Correction Methods

The first attempts at recording EEG inside MR scanners revealed the presence of pulse-related artifacts delayed in relation to the QRS complexes on ECG [138]. This effect has been shown to be common across subjects and has a slight frontal emphasis [154]. The pulse artifacts can have amplitude of the order 50 μV (at 1.5 T) and resemble epileptic spikes. Because of natural heart beat variability, it is considered a more challenging problem than that of image acquisition artifacts. EEG artifacts linked to subject movement are also amplified in the scanner's strong static magnetic field.

The precise mechanism through which the circulatory system exposed to a strong magnetic field gives rise to these artifacts remains uncertain, but it is thought to represent a combination of the motion of the electrodes and leads (induction) and the Hall effect (voltage induced by flow of conducting blood in proximity of electrodes) [155]. Electrode motion can result from local arterial pulsation, brain and head motion or whole-body motion (ballistocardiogram, or BCG, in the latter case) [156, 157].

Methods to reduce artifacts at the source include: careful laying out and immobilization of the leads, twisting of the leads, bipolar electrode chain arrangement [158], head vacuum cushion [159] and the introduction of a reference electrode layer insulated from the EEG-measuring electrodes to capture and subtract the artifact from the EEG prior to amplification [160]. Such measures do not eliminate the problem completely resulting in degraded EEG quality, impeding the identification of epileptiform discharges. The first pulse artifact reduction algorithm published, and to this day still the gold standard against which most methods are compared, is based on subtraction of a running average estimate of the artifact based on automatic QRS detection, and is commonly referred to as the average artifact subtraction (AAS) method [154]. Using this method, the residual artifact is of the order of a few microvolts. The reliance of the algorithm on ECG is a common, though not universal, feature among subsequently developed techniques (some of which use the signal from the standard scanner pulse oxymeter). The method has been and continues to be used successfully in our lab allowing the satisfactory identification of ictal and interictal epileptiform discharges (IED) in real time (at 1.5 T) [161] and for the purpose of source analysis [162], and has been implemented in widely used commercial MR-compatible EEG recording systems (*see* Fig. 6).

The artifact amplitude is theoretically directly proportional to the scanner static field strength (B_0). This phenomenon, and an increasing interest in recording evoked potentials in the MR scanner, has motivated an important research effort towards improving existing pulse-related artifact reduction methods and the development of new ones. Variants of the AAS method have been proposed, ranging from different ways of estimating the artifact waveform, for example to account for a greater degree of inter-beat variability [158, 166–168], more general motion effects [151, 169, 170], to improving QRS detection [171] and removing the need for ECG recording [172].

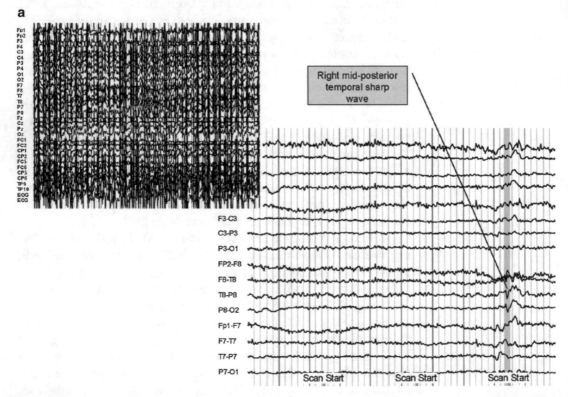

Fig. 6 IED-related BOLD pattern in patient with drug-resistant focal epilepsy. The patient had refractory focal epilepsy, lateralized to the right with a normal structural MRI. Frequent mid and posterior temporal sharp waves were recorded on EEG.(**a**) Representative segment of 32-channel EEG showing a sharp wave, maximum at the right mid-posterior temporal region, recorded during two 20-min fMRI sessions. *Top left*: EEG prior to artifact correction [154, 163].

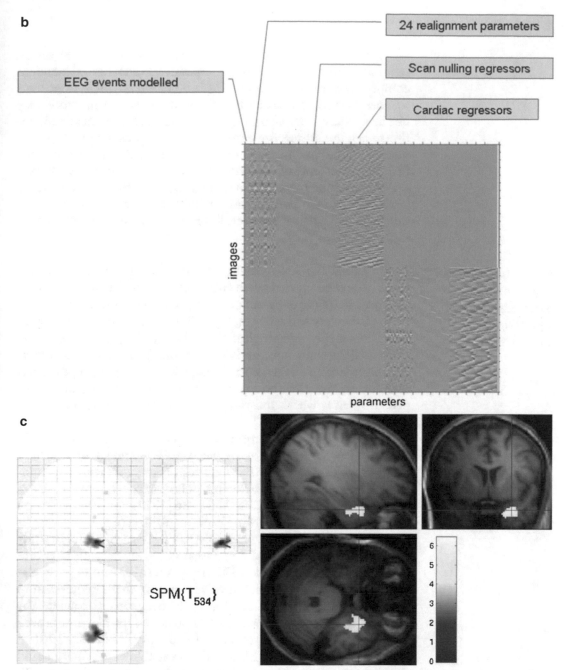

Fig. 6 (continued) (**b**) Design matrix: BOLD signal changes related to 40 sharp waves were modeled by convolution of the EEG event onsets with a canonical HRF and its time-derivative. Signal changes linked to head motion and heartbeat were modeled as nuisance effects [164, 165]. (**c**) *Top*: SPM showing significant sharp wave-related BOLD response in glass brain display (*p* < 0.05 corrected for multiple comparisons). The *red arrow* marks the global maximum, located in the BA 28 (superior temporal gyrus). *Bottom*: BOLD response overlaid onto the patient's normalized T1-weighted volumetric scan. Intracranial recording confirmed a right posterior temporal lobe onset. No significant sharp wave-related deactivation was revealed. The activation clusters were labeled using the Talairach Daemon, http://ric.uthscsa.edu/project/talairachdaemon.html

Spatial EEG filtering methods have been proposed based on temporal principal components analysis (PCA) or independent components analysis (ICA) [152, 159, 173–180]. It is important to note that some PCA and ICA-based correction methods may not be applicable in real time, making it difficult to visualize epileptiform activity during the experiments with possible practical and safety implications. In studies in which residual noise was quantified, improvements of the order of 0.1–1 μV compared with various implementations of the AAS method have been demonstrated [172, 179, 181]. Two important pulse artifact correction methods, AAS and optimal basis set, have been evaluated and compared independently [182].

EEG Quality: Image Acquisition Artifact, Special EEG Recording Equipment, and Correction Methods

In the absence of any special measures, the EEG recorded inside the MR scanner becomes un-interpretable during image acquisition because of the presence of repetitive artifact waveforms caused by the time-varying fields employed in the scanning process superimposed on the physiological signal [163, 183]. In addition, on some MR instruments, the helium cooling pump can introduce significant amount of noise in the signals measured using the EEG equipment, and a method to remedy this problem has been proposed in cases where the pump cannot be switched off for the duration of the scan [184].

One way of circumventing this problem is to leave time gaps in the fMRI acquisition (e.g. between EPI volumes) of sufficient duration to capture the EEG features of interest (assuming sufficient data quality, e.g., following pulse artifact removal); this is *interleaved EEG-fMRI* [158, 185, 186]. This approach relies on artifact not persisting following each acquisition (e.g. due to amplifier saturation). Interleaved EEG-fMRI can be most useful to study predictable events (evoked responses) or slowly varying phenomena, such as brain rhythms. EEG-triggered fMRI and in particular spike-triggered fMRI, which involves limiting fMRI acquisition to single or multi-volume blocks, each triggered following the identification of an EEG event of interest is a form of interleaved EEG-fMRI with obvious relevance to epilepsy [183, 187–190].

Although interleaved EEG-fMRI is capable of providing useful data in many circumstances, it imposes a limit on experimental efficiency due to EEG quality degradation during scanning.

We now review the technical developments that have made it possible to record EEG of sufficient quality throughout fMRI acquisition (so-called *continuous EEG-fMRI*), by the image acquisition artifact to be corrected. For all practical purposes, and assuming that the time gap between volumes is the same as between slices, the artifact's spectral signature ranges from 1/TR (TR: slice acquisition repetition time) to around 1 kHz (corresponding to the readout gradient). In fact it extends into the mega-hertz (RF) range, well beyond the recording capability of any EEG equipment. It can appear artificially benign when captured using standard EEG equipment [138]. Only using equipment with sufficient bandwidth, sampling rate and dynamic range, can one capture the artifact with adequate accuracy [163, 191].

Experiments have shown that the gradient switching-related effects generally dominate over RF in terms of amplitude and extent in time, although the balance between the two mechanisms will vary depending on the specific MR sequence used [163, 171, 191, 192]. For standard EPI sequences, the pattern of gradients is repeated exactly across slices. Compared with the problem of cardiac-related artifact reduction, this determinism greatly facilitates the task of image acquisition artifact removal; however, the induced waveforms will be subjected to variations in time due to changes in the electrode/lead configuration caused by subject motion.

Before discussing artifact reduction postprocessing techniques, let us review some of the hardware modifications and other measures that can facilitate the recording of good quality EEG during fMRI. First, some of the tricks described previously to reduce the pulse-related artifact at the source, and in particular those to limit the area of loops formed by EEG leads and head motion, can also help to lessen the image acquisition artifact problem. Second, low-pass filtering at the front end of the EEG system may be used to reduce the artifact significantly, although not sufficiently to result in adequate EEG quality [163]. Third, a scheme has been devised to reduce the amplitude of the artifact at the source by modification of the MR sequence and careful synchronization with EEG sampling [191]. It has also been noted that the subject's head position can have an important impact on the magnitude of the gradient-switching artifact and therefore this can be reduced at the source through simple manipulation [193].

In the studies by Allen et al. [163] and Anami et al. [191], custom-built EEG recording systems with high sampling rates (1–20 KHz) and large dynamic range (~20 mV), based on the notion that the artifact must be captured accurately to be understood, were measured and eliminated. Specially designed "MR-compatible" EEG recording hardware has now become the norm in the field, with a number of commercial products now available on the market (*see also* [158, 192, 194] for a description of other modified or purpose-built apparatus).

Postprocessing methods to reduce image acquisition artifacts can be categorized as filtering, template subtraction methods or PCA/ICA. Filtering based on the identification and subsequent suppression of frequencies linked to the image acquisition process can lead to an improvement in EEG quality and be used for simultaneous EEG-fMRI [168, 194, 195]. However, it is severely limited by the spectral overlap between the artifact and physiological signals, ringing effects and has been shown to be inferior to template subtraction [159].

The most commonly used image acquisition artifact reduction method is based on average template artifact subtraction (sometimes referred to as AAS) method [163] (*see* Fig. 6). It relies on the lack of correlation between physiological signals and the artifacts, enabling the latter to be estimated by averaging the EEG over a number of epochs, corresponding to individual scan repetitions, for example. The success of the artifact (template) estimation and its subsequent subtraction from the ongoing EEG depend critically on the sampling rate, the number of averaging epochs, and the precision of their timing. In Allen's original implementation, this is addressed by the use of the scanner's scan trigger pulse to mark each scan acquisition and interpolation. Following subtraction, residual artifacts are reduced using Adaptive Noise Cancellation. The method can be used in real time, allowing continuous EEG-fMRI studies in patients with epilepsy [196–199]. Possibly the most important practical development has been the demonstration that synchronized MR acquisition and EEG digitization lead to significantly improved EEG quality, and in particular over a wider frequency range, when combined with an AAS-like method [225]. The method has been found to perform well for spiral EPI [226]. As noted previously, changes in the artifact waveform due to subject motion will lead to suboptimal template estimation. To address this, refinements of Allen's method which incorporate PCA of the residual artifact have been proposed [171, 227]. The shape of the image acquisition artifact may be captured in a separate experiment for subsequent subtraction [192]. In the study by Wan et al. [228], a method designed to bypass the requirement for a slice acquisition signal from the scanner is proposed. As is commonly the case for artifact reduction methods based on ICA, identification of the components containing artifact is mainly done visually [179]. A difficulty encountered when comparing the various methods available for artifact reduction is the range of methodologies used. This has been addressed to some degree in a rigorous comparative study [229].

EEG Quality: The Impact of Subject Motion

Subject (head) motion, and almost inevitably motion of the EEG recording circuit (formed by the head, electrodes and leads) which results from it, will cause fluctuations in the recorded signal additive to the effects discussed in the previous sections. In most situations, this motion is not synchronized with either the scanning process or heartbeat; due to the associated change in geometry of the EEG circuit in relation to the scanner, it therefore can introduce random fluctuations in the magnitude of both types of artifact. In terms of the AAS algorithm, this phenomenon imposes a limit on the quality of the artifact correction: the greater and more random the motion, the less efficient AAS will be. The previously mentioned measures to reduce the pulse artifacts based on subject immobilization and minimization of EEG loops are obviously worth reiterating at this point [158, 159]. More recently, a method to measure head motion and use the information to reduce pulse and scanning-related artifacts has been proposed [230, 231].

**2.2 Application
of fMRI to the Study
of Paroxysmal Activity**

fMRI can be used to investigate the hemodynamic correlates of paroxysmal activity, and in particular to reveal regional changes in the BOLD signal thereby potentially providing new localizing information. *See* Table 3 for a list of notable published studies.

The conventional approach to the analysis of fMRI data is predicated on the correlation of the fMRI time series with experimentally-determined stimuli within the framework of the general linear model (GLM). This methodology allows voxel-by-voxel testing of the degree of fit of predicted and observed BOLD time courses, and subsequent inferences. The application of fMRI for the assessment of spontaneous paroxysmal activity in epilepsy offers a number of additional challenges, namely: the lack of experimental paradigm (subjects scanned in the resting state), the identification of paroxysmal activity in relation to the fMRI time series, and the representation and translation of this activity into a GLM. The latter point is crucial and as we will see, has been an area of continuing investigation in the field.

In addition to the difficulties associated with the observation of subjects within the confined space of an MR scanner, and perhaps more importantly the use of behaviourally derived (seizure manifestations) time markers when possible, although crucial, does not provide as complete a picture of the event as one would wish given the importance of putative concomitant EEG abnormalities.

The advent of EEG-correlated fMRI has been a major advance in this respect, providing an established, albeit imperfect, marker of paroxysmal activity and more generally brain state. Importantly, it allows the study of interictal activity, particularly IED, which only manifest on EEG. Although this type of data allows the experimentalist to study hemodynamic changes linked to specific EEG events, it presents a number of challenges linked to the subjective nature of EEG interpretation. The investigator is also soon confronted with a "chicken and egg" type problem of not knowing precisely what the time course of the changes is, and in particular whether the hemodynamic "response" function associated with paroxysmal discharges deviates from normality, a necessary element of the modeling, or its spatial substrate, the latter being precisely the motivation for undertaking the experiment given its clinical implication.

*2.2.1 Ictal fMRI in Focal
Epilepsy*

As mentioned previously, a number of case studies of ictal events captured using fMRI alone were published prior to the advent of EEG-fMRI [139–142]. Despite the fact that these contain interesting observations, particularly with regard to the signal change around the time of seizure onset, the availability of simultaneously recorded EEG would have contributed important information. The development of the ability to record good quality EEG inside the MR scanner has offered the possibility of improved models of ictal fMRI signal changes by the inclusion of precisely timed

Table 3 EEG-fMRI studies of paroxysmal activity: early milestones and important series

Study	Number of subjects/no. in which IED recorded	Results	Conclusion	Comment
Focal epilepsy Warach et al. 1996 [183]	1/1 frequent IED	Bilateral activation where EEG suggested left temporal localization and anterior cingulate activation in relation to generalized epileptiform activity	"We cannot make conclusions about the source of the discharge from the present data"	
Seeck et al. 1998 [200]	1/1 frequent IED	Multiple areas of signal enhancement on fMRI. Confirmed on 3D-EEG source localization with evidence of a focal onset. Focus later confirmed on subdural recordings	The combination of EEG-triggered fMRI and 3D EEG source analysis, represents a promising additional tool for presurgical epilepsy evaluation allowing precise noninvasive identification of the epileptic foci	
Symms et al. 1999 [201]	1/1 frequent IED	Reproducible and concordant activation across four sessions		
Patel et al. 1999 [202]	20/10 frequent IED	9/10 overall reported as showing "activation corresponding to the EEG focus"		
Krakow et al. 1999 [187]	10 frequent IED	Reproducible activations (same lobe and overlapping) obtained in 6/10 patients in close spatial relation to EEG focus		
Krakow et al. 1999 [203]	1 frequent IED	Focal activation within a large malformation of cortical development in response to focal epileptiform discharges		
Lazeyras et al. 2000 [204]	11 frequent IED	Activation confirmed clinical diagnosis in 7/11. In 5/6 intracranial EEG confirmed result		
Lazeyras et al. 2000 [205]	1/1 frequent IED	Area of signal enhancement concordant with hyperintensity seen on ictal FLAIR images in a patient with nonlesional partial epilepsy		
Krakow et al. 2001 [189]	24/14 frequent IED	12/24 patients showed activations concordant with EEG focus, 7/12 of which also had concordant structural lesions. 2/24 were discordant and 10/24 showed no significant activation		
Lemieux et al. 2001 [197]	1/1 frequent, stereotyped IED	In a case with stereotyped frequent IED, BOLD signal change concordant with the seizure onset zone was recorded	Localization of BOLD activation was consistent with previous findings and EEG source modeling	First description of application of continuous EEG-fMRI
Jager et al. 2002 [206]	10/5 frequent IED, focal epilepsy	Focal activation in 5/5 patients, concordant with EEG amplitude mapping. Mean signal increase was 15 ± 9%. Spike amplitude correlated with volume of activation		

Benar et al. 2002 [207]	4/4 frequent IED	The average HRF presented a wider positive lobe in three patients and a longer undershoot in two.	There was no clear correlation between the amplitudes of individual BOLD responses and EEG spikes
Al-Asmi et al. 2003 [190]	48/31 frequent IED	BOLD activation in 39% of studies. Concordant with seizure focus in almost all. four patients had concordant intracranial recording (by lobe)	Combining EEG and fMRI in focal epilepsy yields regions of activation that are presumably the source of spiking activity and these are high
Benar et al. 2006 [208]	5/5 presurgical candidates having sEEG	When an intracranial electrode is in the vicinity of an EEG or fMRI peak, it usually includes one active contact	Largest series of intracranial EEG correlated fMRI activation
Aghakhani et al. 2006 [209]	64/40 focal epilepsy with frequent unilateral or bilateral IED	A positive thalamic response was seen in 12.5% of studies with unilateral and 55% with bilateral spikes. Cortical acitvation was more concordant with focus than deactivation	The thalamus is involved in partial epilepsy during interictal discharges. This involvement and also cortical deactivation are more commonly seen with bilateral spikes than focal discharges
Salek-Hadaddi et al. 2006 [199]	63/34 focal epilepsy frequent IED	Significant hemodynamic correlates were detectable in over 68% of patients and were highly, but not entirely, concordant with site of presumed seizure onset	These findings provide important new information on the optimal use and interpretation of EEG-fMRI in focal epilepsy
Zijlmans et al. 2007 [210] Thornton et al. 2011 [211] Pittau et al. 2012 [212] Chaudhary et al. 2012 [213] Coan et al. 2015 [214]	29/15 focal epilepsy, declined for surgery; 23/12 focal epilepsy, focal cortical dysplasia, intracranial EEG and surgery; 43/33 focal epilepsy; 20/15 seizures in focal epilepsy; 30 patients who underwent surgery for TLE; in 14 cases with no IED during fMRI, a topographic method was used [215]	8/15 subjects: IED correlated BOLD response at site of focus. Multifocal in 4, unifocal in 4. Concordant with IC data in 2 11/12 showed significant IED-correlated BOLD. BOLD matched icEEG SOZ and outcome was >50% seizure reduction in 5; BOLD was widespread in 6 and outcome poor in 5/6 21/33 IED-related BOLD contributed to the delineation of the focus compared to scalp EEG; icEEG validation was positive in 12/14 patients Widespread preictal BOLD changes followed by more focused early ictal and spread Good surgical outcome in 13/16 patients with concordant BOLD changes and in 3/14 patients with discordant BOLD	First evaluation of impact on presurgical evaluation EEG-fMRI provides additional information about the epileptic source in the presurgical work-up of complex cases Widely distributed discordant regions of IED-related hemodynamic change appear to be associated with a widespread SOZ and poor postsurgical outcome EEG-fMRI may contribute to the localization of the interictal epileptic generator in patients with focal epilepsy Preictal and ictal haemodynamic changes in refractory focal seizures can noninvasively localize seizure onset at sublobar/gyral level when ictal scalp electroencephalography is not helpful Interictal EEG-fMRI retrospectively confirmed the epileptogenic zone of TLE patients EEG-fMRI (with video) of seizures can be done with acceptable risk, and provides important new localizing information

(continued)

Table 3 (continued)

Study	Number of subjects/no. in which IED recorded	Results	Conclusion	Comment
Tousseyn et al. 2015 [216]	28/27 refractory focal epilepsy ictal SPECT and interictal EEG-fMRI	High congruence between maps derived from the two techniques; some discrepancies	Hemodynamic changes related to seizures and spikes varied spatially within a common network. Overlap observed nearby and distant from discharge origin	Largest comparison of interictal EEG-fMRI and ictal SPECT
Generalized epilepsy Salek-Haddadi et al. 2003 [217]	1/1 prolonged GSW epochs (ictal)	Thalamic activation and widespread cortical deactivation	Supports thalamo-cortical model of GSW	
Baudewig et al. 2001 [218]	1/1 frequent GSW	Unilateral insular activation shown in relation to generalized epileptiform discharges	Strategy resulted in robust BOLD MRI responses to epileptic activity that resemble those commonly observed for functional challenges	
Hamandi et al. 2006 [196]	46/30 interictal GSW in IGE and SGE	Thalamic activation and cortical deactivation observed at group level. Deactivation in the default brain areas. Cortical pattern mixed at individual level	Observed cortical deactivation may represent correlate of clinical absence seizure.	
Aghakhani et al. 2004 [219]	15/14 interictal GSW	Bilateral thalamic activation in 80 % of BOLD response. Cortical deactivation in 93 %	Cortical deactivation mediated by hyperpolarization of the thalamus	
Hamandi et al. 2007	4/4 interictal GSW	Qualitatively reproducible BOLD and blood perfusion patterns; Cortical deactivation corresponds to decrease in blood flow	Consistent with preserved neurovascular coupling in GSW and decreased cortical activity	
Yang et al. 2013 [220]	10/10 drug-naïve CAE patients; interictal GSW	GSW discharge-related alterations in the default mode network (DMN), cognitive control network (CCN), and affective networks	Interictal GSWDs can cause dysfunction in specific networks important for psychosocial function	First study specifically focused on functional connectivity alterations during GSWD in drug-naïve patients

Study	Population	Findings	Comments	
Vaudano et al., 2014 [221]	15 with eyelid myoclonus with absences (EMA) + 14 with IGE	Elevated eye closure-related BOLD signal in the visual cortex, posterior thalamus, and eye motor control areas in EMA group compared to IGE	Supports concept of EMA as a distinct epileptic condition	
Children De Tiege et al. 2007 [222]	6/6 focal epilepsy (lesional and nonlesional)	Concordant activation with presumed focus in four cases. IC recording corroborative in 1	EEG-fMRI is a promising tool to noninvasively localize epileptogenic regions in children with pharmacoresistant focal epilepsy	First series specifically addressing use of EEG-fMRI in children
Jacobs et al. 2007 [223]	9/9, mixed focal epilepsy, sedated	All had BOLD activation or deactivation concordant with seizure focus. Deactivation appears more common in adults	EEG-fMRI at 3 T could be useful to localize focus in children with focal epilepsies. Nature and origin of the negative BOLD requires investigation	
Jacobs et al. 2007 [224]	13/13 symptomatic (lesional) epilepsy), sedated	Activation corresponding with the lesion was seen in 20% and deactivation in 52% of the studies	"Good results could be obtained from the EEG-fMRI recordings, performed in sedated children"	

EEG-derived information in the model, and also by providing an indication of head motion due to artifacts on the EEG which may assist interpretation. Clearly the problem of motion is particularly pertinent to the acquisition of seizure data, as it can severely degrade image data quality and consequently the ability to detect and map regional hemodynamic changes accurately in relation to those manifestations [232–234].

Given the above limitations and concerns regarding patient safety, EEG-fMRI data of seizures is often acquired incidentally [198, 205, 235–237]; however, a systematic study of patients recruited specifically in the expectation of capturing seizures has provided new information on the transition from the interictal state, and ictal onset and evolution (spread) [213]. A noteworthy observation was the common involvement of multiple regions beyond the conventionally defined EZ, during the preictal period. The electrophysiological substrate of ictal BOLD increases and decreases were specifically investigated in a case who underwent SEEG [238], showing distinct signatures.

2.2.2 Interictal fMRI: EEG-fMRI

Although the study of ictal events may be useful in pursuing the aim of noninvasively identifying the seizure onset zone and thereby contributing most to presurgical evaluation, its acquisition remains challenging. Attention has, therefore, focused on interictal activity and the insight gained from it into the function of epileptic networks. Spike-triggered fMRI acquisition was used in the first applications of EEG-fMRI in epilepsy, which tended to focus on the study of subjects pre-selected for the large number of IED observed in prior routine surface EEG or as part of presurgical assessment. The acquisition and analysis of spike-triggered fMRI data generally rests on the assumption that interictal spikes are associated with a BOLD signal change pattern similar to the so-called canonical HRF. In these series, regions of positive BOLD signal change associated with IED were observed in approximately 50 % of the cases overall, and occasional negative changes [183, 187, 189, 190, 200–202, 204, 239, 240]. Using bursts of BOLD EPI scans, Krakow et al. made an initial attempt at estimating the shape of the IED-related HRF [189]. We note that in the work by Seeck and colleagues, Clonazepam was used to suppress interictal discharges, thereby creating a control scan state [200, 204]. Interleaved EEG-fMRI, whereby fMRI data are acquired in blocks with inter-block gaps of a sufficient duration to allow interpretation of part of the EEG was employed in a patient with IGE for the mapping of BOLD changes related to spike-wave complexes [218].

Following the implementation of image acquisition artifact removal [163], the technique of continuous, simultaneous EEG-fMRI acquisition was demonstrated along with estimation of the shape of the IED-related HRF in a subject with Raussmussen's encephalitis and stereotyped high amplitude sharp waves on the

EEG [197]. *See* Fig. 6 for example of EEG-fMRI in patient with focal epilepsy. When applied to relatively large case series of subjects with predominantly focal epilepsy selected on frequent interictal discharges on routine EEG [190, 199], a yield (proportion of significant BOLD activations or deactivations) of 60–70% was observed where IED were recorded. This result has been reproduced in further studies despite variation in the analysis strategies and patient groups. Sensitivity can be increased using spike template matching based on EEG recorded outside the MRI scanner [241]. Positive BOLD changes tended to correspond with the site of the presumed seizure onset zone (based on electroclinical localization), but occasionally appeared at sites distant from this region, the significance of which remains unclear. Negative BOLD changes were more often observed remote from the presumed focus, with a striking pattern of retrosplenial deactivation in a significant proportion of cases [199]. It was shown that activations were more likely when there was good electroclinical localization, frequent stereotyped spikes, less head motion, and less background EEG abnormality. Furthermore, the findings suggest that significant activation is more likely for runs of spikes than for isolated discharges, when the event duration is taken into account in the modeling [199, 242].

In TLE, IED-related activation of the MTL ipsi-lateral to the presumed focus was found to be common, and is reminiscent of the thalamic pattern observed in generalized spike and wave discharges (GSW) [243].

The possible explanations for lack of activation include: combination of insufficient number of events and limited fMRI sensitivity, incorrect model due to suboptimal EEG event identification and classification, choice of HRF and limited extent to which scalp EEG reflects ongoing activity (throughout the brain). Automated and semi-automated spike detection methods have been proposed to attempt to reduce the level of subjectivity of EEG event identification [162]. Nonetheless, although the ability to record EEG during fMRI is necessary to study the hemodynamic correlates of EEG events observed on the scalp, total reliance on scalp EEG can also be seen as a limitation when attempting to interpret the BOLD signal throughout the brain and in particular the part which may not be linked to activity reflected on the scalp. In the absence of clear epileptiform discharges, the presence of slow activity may provide useful localizing information [244, 245]. However, if data from clinical video-telemetry EEG are available showing IED, the corresponding topographic map can be correlated with that of the intra-MRI EEG as a function of time to attempt to reveal BOLD changes corresponding to the epileptic topographic pattern [246]. Data-driven or region-based fMRI analysis techniques may offer another way forward in this respect, although interpretation of the observed patterns in the former is significantly impaired by the lack of an a priori model and numerous possible confounds [247, 248].

2.2.3   The HRF in Focal Epilepsy

The normal hemodynamic response associated with neuronal activity arising from brief external stimuli in humans has a characteristic shape with a peak at around 5–6 s following the event, the so-called canonical HRF, with a significant degree of inter-subject variability [249]. The shape of the HRF is a key element of fMRI signal modeling with an important impact on sensitivity, and it is therefore important to attempt to characterize it. In epilepsy, particularly in relation to deactivations it has been proposed that neurovascular coupling is abnormal with possible consequences on the shape of the HRF, and this was suggested to be one possible explanation for the significant widespread BOLD deactivations observed in both focal and generalized epilepsy [250, 251].

This has lead to increased interest in estimating the shape of the HRF in epilepsy. An efficient way of estimating the shape of the hemodynamic changes linked with epileptiform discharges is by using a set of functions (sometimes referred to as "basis set") that can, by linear superposition, fit signal changes with almost any time course. Various such schemes have been used in studies of epilepsy including sets of gamma response functions, Fourier basis sets and a linear combination of canonical HRF, its time derivative and a dispersion derivative [197, 199, 207, 252, 253]. Although a degree of variability has been observed, the HRF linked to IED was found to be principally canonical in shape [254]. In some cases deviation from the norm at locations distant to the IED may reflect artifacts, while in others, deviant activation in proximity to the presumed focus, early responses may reflect brain activity that systematically precedes the event captured on the scalp or propagation [252, 255]. In a recent study, the development of penicillin-induced epileptic activity was correlated with increase in BOLD signal prior to the onset of IED [256]. These studies coupled with those combining EEG-fMRI with multiple source analysis [257] suggest that in some cases the maximum BOLD response may be detected just prior to seizure or even IED onset. It may be, however, that this reflects BOLD changes linked to interictal abnormalities, which are not included in the model by virtue of the fact they are not seen on the scalp EEG.

2.2.4   Clinical Relevance of fMRI of Paroxysmal Activity in Focal Epilepsy

Having shown that EEG-fMRI is capable of providing a unique form of localizing information, the issue of its clinical value arises. The evaluation of new noninvasive imaging modalities in the presurgical assessment of patients with drug-resistant epilepsy is a complex issue in part due to the lack of an established methodological consensus (baseline) across centers. Added value and clinical relevance are a function of the new test's sensitivity and specificity. In the case of EEG-fMRI, neither has been properly assessed to date. Thus far the field has focused on proof of principle demonstrations, usually in patients selected based on high rates of EEG abnormalities, with a success rate of roughly 50–60 % [183,

187, 189, 199, 202, 204, 240]. Therefore, one may anticipate a lower success rate in the most clinically challenging cases for which the need for noninvasive assessment is most pressing.

When available, the new localizing information must be evaluated relative to a surgically confirmed irritative and epileptogenic zone [258] using the current gold standard of invasive recording and outcome data. In practice, a gradual approach to validation is often taken, whereby the face validity of new localizing information is tested against other existing techniques in cases with well characterized syndromes, such as mesial TLE.

In our laboratory, we have assessed the value of the fMRI findings by comparing the localization of the BOLD cluster containing the most significant activation to the seizure onset zone defined electro-clinically, when possible, and found a very good degree of concordance at the lobar level [187, 189, 199]. Additional regions of activation were observed in roughly 50% of the cases with significant activation. The finding of localized BOLD activation in cases with poor electroclinical localization suggests a means of obtaining target areas for intracranial EEG [199]. In TLE, the yield has been characterized as relatively high [259] and the degree of concordance of BOLD activations with the presumed focus generally good in one study [199], but more varied in another [259].

Comparison of EEG-fMRI with source localization suggests that IED-related BOLD activation are often in proximity to those detected by conventional EEG source localization, but some studies have suggested a distance of up to 50 mm. The possible sources of discrepancy between BOLD and electrical (or magnetic) source localization include: differences in the nature of the observed phenomena and neurovascular coupling, vascular architecture and scanner field-related effects on sensitivity, instrumental and physiological noise, source reconstruction limitations, fMRI sensitivity limitations [208, 260–262].

Presurgically, intracerebral EEG data are widely recognized as the localization gold standard. Comparison of scalp EEG-fMRI against invasive EEG has been performed in small groups and using widely varying fMRI analysis techniques and comparison criteria [190, 200, 250], noting a degree of concordance between the epileptogenic zone and the area of maximal BOLD activation in some, but not all subjects [204, 208]. The potential role of EEG-fMRI in presurgical evaluation was assessed in a series of patients with focal epilepsy in whom surgery was not offered following conventional electroclinical evaluation [210]. The impact of the EEG-fMRI findings was assessed in eight cases with unclear foci or suspected mutifocality (based on the center's usual battery of tests) in whom significant IED-related BOLD activation were observed: IED-related changes suggested a more restricted seizure onset zone in four, and multifocality in a further patient. The EEG-fMRI finding was concordant with intracranial recordings in two. The authors suggest that this supports a possible role for EEG-fMRI in presurgical evaluation where a potential, but possibly widespread focus is identified. In a series of 23 patients with FCD who underwent invasive EEG, EEG-fMRI mapping of IED was successful in 50%; the pattern of IED-related BOLD changes tended to reflect the invasive EEG findings and in particular widespread involvement was associated with poor surgical outcome or widespread SOZ, suggesting a significant clinical role [211]. In another substantial retrospective case series, Pittau et al., found a high degree of concordance of the interictal EEG-fMRI localization with the confirmed or presumed IZ, and showed the test to provide added value ("contributory" to the definition of the epileptic focus) in a large proportion of cases [212]. A comparison of the results of IED mapping and ictal SPECT in a series of 28 patients has revealed a common high degree of spatial overlap often beyond the presumed EZ, possibly reflecting common propagation pathways, with some notable mismatches which the authors explain by the different nature of the two techniques and inter-session effects [216].

2.2.5 Relationship with Pathology in Focal Epilepsy

Focal epilepsy can be divided by pathological subtype. Given the knowledge that the irritative and epileptogenic zones may extend beyond the area of pathological abnormality and that animal models suggest abnormal subpopulations of neurons within dysplastic areas, there have been studies of EEG-fMRI aimed at evaluating the hemodynamic response in abnormal tissue revealed on MRI [203, 211, 244, 263–266]. In a series of 14 cases with heterotopia notable variability of the BOLD response across abnormal tissue was observed, but the area of BOLD signal increase was often concordant with the area of pathology, with a more mixed pattern of deactivations [266]. In MCDs and particularly Taylor type focal cortical dysplasia, BOLD signal increase was observed within the lesion while peri-close and distant from the lesional activity displayed a negative BOLD response in four out of six cases [237]. Other studies in cases with MCD have supported these findings. IED-related BOLD signal changes have been observed in patients with cavernomas [267] close and distant from the lesion. The frequently observed negative BOLD responses, particularly in MCD have been attributed to loss of neuronal inhibition (in the presence of normal neurovascular coupling) in the regions surrounding the abnormality or abnormalities in neurovascular coupling itself. The significance of these deactivations will be discussed in more detail later.

2.2.6 fMRI in the Investigation of Epilepsy Syndromes

EEG-fMRI has been used to attempt to localize sources in specific epilepsy syndromes, in addition to adding evidence to the understanding of the differences in subtypes of focal epilepsy. TLE is of particular interest in this context as surface EEG may not detect the deep sources involved in TLE and surgery for TLE, where it is correctly localized, is associated with excellent outcome [268].

A large series of subjects with temporal and extra-temporal lobe epilepsy showed that temporal lobe spikes are seen at the presumed seizure focus, but also unsurprisingly on the opposite homologous cortex, but did not give further localizing information [259].

A further more recent study compared IED-related BOLD responses between subjects with temporal and extra-temporal lobe epilepsy and found that those brain areas involved in the so-called "default mode network" [269] were commonly deactivated during temporal lobe IED whereas other areas were involved in BOLD activation and deactivation in extra-temporal lobe epilepsy [243].

Other subtypes of epilepsy studied include benign rolandic epilepsy of childhood [262, 270] and other focal epilepsies in children, which have demonstrated concordant IED-correlated BOLD activation in approximately 60% of cases [222, 223]. A study of lesional epilepsy in children not only revealed positive BOLD response concordant with the seizure onset zone in a significant number of subjects, but also a higher number of deactivations than those observed in adult studies [223]. However, sedation was used in this study, the effect of which has not been investigated in detail to date. EEG-fMRI is a particularly attractive option for the study of childhood epilepsies as it is noninvasive, but experiments are long and require a high degree of subject cooperation.

In a recent study comparing the BOLD patterns (and cortical morphology) associated with eye closure and (spontaneous and triggered) GSW in patients with eyelid myoclonia with absences (a form of reflex epilepsy) has provided evidence of specific alterations in the visual (and other, related) system, thereby supporting the notion of a distinct epileptic syndrome [221].

2.2.7   EEG-fMRI in Generalized Epilepsy

The generalized epilepsies (IGE) (that is the syndromes of Absences, Myoclonic epilepsies, and primary generalized tonic clonic seizures coupled with generalized spike and wave discharges on the EEG [271] are not currently amenable to surgical treatment. Therefore, the primary focus of EEG-fMRI investigations of generalized epilepsy has been the exploration of hypotheses developed in vitro and animal models [196, 217, 219, 261, 272–276].

Specific BOLD activation and deactivation of resting state networks are found to be correlated with interictal GSW activity. In particular, the so-called Default Mode Network (DMN), normally found to be particularly active at rest, is commonly deactivated and thalamic-cortical network activated [273, 277, 278]. Moeller and colleagues [279] found changes in functional connectivity during GSW and GSW-free period.

It is commonly assumed that decreased DMN activity is a consequence of the epileptic discharges, perhaps reflecting alterations in awareness. This may be the case. However, a study using connectivity analysis has suggested that the deactivation of the DMN drives thalamic-cortical network changes and can lead to the production of generalized spikes waves (GSW) suggesting a large-scale interaction between different brain networks. In particular, it has been proposed that the precuneus has a crucial role in facilitating the production of GSW and in alter the state of consciousness [280]. Other evidence based on the analysis of the temporal sequence of BOLD response suggesting an early frontal activation in support of the cortical focus theory [281]. Moreover, other results have shown that the thalamus is the pacemaker of the abnormal status in brain networks in this patient population [282]. As demonstrated, there is no clear evidence regarding the direction of the interaction between resting state networks and GSW.

Simultaneous EEG-Arterial Spin Labelling (ASL) MRI in patients with spike and wave discharges revealed degrees of correlation between BOLD and CBF consistent with normal neurovascular coupling [277]. This study also revealed a remarkable degree of within subject, inter-session (and inter-scanner) reproducibility, albeit in a small group.

2.2.8 The Significance of BOLD Deactivation in Epilepsy

BOLD signal deactivation has been observed in the monkey visual system and found to have a linear relationship with CBF in the same way as positive BOLD signal change in normal physiological conditions [283]. The pattern of cortical deactivation commonly observed in relation to GSW using EEG-fMRI in humans has been discussed earlier. Gotman et al. proposed that the reason for observing the widespread cortical deactivation may be hypersynchronization of the thalamus, supporting Avoli's proposed model of the thalamus driving the cortex during GSW [273]. The results of Hamandi et al. using EEG-ASL are consistent with GSW-related deactivation reflecting decreased cortical activity [277].

Focal IED-related deactivations are less common than activations and seem particularly linked to the presence of activation, possibly reflecting a smaller hemodynamic effect of individual events [199, 251]. In cases with MCD IED-related negative BOLD signal change adjacent to the seizure onset zone or area of dysplastic cortex, has led to several possible explanations including "vascular steal" from more metabolically active regions, abnormal neuronal coupling, or perhaps more likely, loss of inhibitory neuronal activity in these regions supporting the link between BOLD deactivation and underlying decrease in neuronal activity [251, 264].

Deactivation of the default mode areas is a typical feature of IED-related BOLD changes in TLE in contrast to cases with extra-TLE [243]. The observed pattern may reflect IED-related effects in areas involved in cognition, with a possible link to transient cognitive impairment [284].

2.2.9 fMRI and the Neurobiology of Epileptic Networks

Although limited by the sluggish BOLD response, we have seen that EEG-fMRI is able to reveal multiple regions more or less simultaneously activated or deactivated in relation to EEG events

[196, 199, 243]. The involvement of the thalamus in GSW has been discussed earlier, but it has also been observed that subcortical BOLD activation can be seen in focal epilepsy [209, 259, 264]. A spike-triggered EEG-fMRI study of subjects with MCD identified BOLD activation concordant with seizure onset in all patients studied, and subcortical activations in various structures were observed in 50% of subjects [264], and in Agakhani et al.'s study of focal epilepsies, thalamic activations were observed correlated with interictal IED, particularly where these were bilateral and synchronous [209].

2.2.10 Limitations, Challenges, and Future Work

The study of spontaneous, pathological brain activity using fMRI presents the experimentalist numerous challenges. The advent of simultaneously recorded EEG greatly facilitates this endeavor. EEG-fMRI was originally held to be an excellent tool incorporating the spatial coverage and resolution of imaging techniques and the temporal resolution of EEG. However, a number of challenges remain.

Technically, postprocessing methods are now available to offer EEG quality sufficient to allow a degree of abnormality identification reliability comparable to that of routine clinical EEG. Although EEG quality is sufficient to detect most IED with a good degree of certainty (at 1.5 T) [161], pulse-related artifacts can sometimes interfere with EEG interpretation particularly at 3 T. Improved artifact correction and the use of techniques based on EEG source reconstruction may yield more localized and reproducible results [162, 257]. In addition, very little work has been done on this aspect of EEG interpretation and in particular the differences between the clinical and experimental approaches, with the former using a summary of the abnormalities observed over the whole EEG rather than categorization and quantification of all IED required for fMRI. Concerning the effects of motion, which is particularly problematic in patients with epilepsy (and other neurological conditions), on MR image and EEG quality, the advent of prospective motion correction offers a possible way forward [285].

Reliance on scalp EEG for fMRI modeling is both a strength, as it allows to answer the question "what, if any, are the BOLD correlates of EEG pattern X?", and weakness because it is burdened with certain limiting aspects of scalp EEG, such as sensitivity bias subjectivity in interpretation, and the unpredictable nature of the phenomena of interest. This presents the investigator with significant challenges in terms of unpredictable experimental efficiency (and consequently yield) and EEG interpretation for the purpose of GLM building. The former may require a more aggressive approach possibly using drug management as a tool for modulating EEG activity [200]. Although EEG event classification remains problematic, possible solutions may give the opportunity of using fMRI to inform EEG interpretation [162].

The "yield" of EEG-fMRI studies is to some extent a function of the previous points. A particular problem is that subjects are scanned for a limited period of time, and interictal discharges are not always observed during this period. A significant BOLD response is also only present in 50% of cases giving an overall "yield" of the order of only 25% in many studies.

Attempts at interpreting resting state fMRI patterns in patients with epilepsy without reference to EEG (or clinical manifestations) using data-driven analysis techniques have had mixed success [248], but one can envisage the possibility of being able to reliably identify sets of patterns that are potentially related to epileptic activity, with 100% yield [286].

Despite providing whole-brain coverage of putative BOLD changes, the technique is effectively limited to the study of the events seen on the scalp EEG. We know from intracranial EEG that a large amount of pathological activity is not seen on the scalp, despite originating from a fixed location [287]. Therefore, the possible mismatch between scalp-derived model and ongoing activity raises the question of the nature of the baseline, with implications for sensitivity. This may be reflected in event onset time offset reflected in "early responses" [252, 253]. Furthermore, this means that EEG-fMRI cannot be used to exclude regions from epileptogenic zone.

An additional degree of modeling uncertainty arises with regards to the IED-related hemodynamic response function linked with the potential effects of pathology and the nature of scalp EEG [250]. However, the latest evidence points to a preserved neurovascular coupling and shape of the HRF in line with the physiological response in healthy brains [252, 277].

The sluggishness of the BOLD response in relation to EEG means that activation patterns effectively represent a time averaged picture of a sequence of neuronal events, such as propagation and loops. The causal relationship between the activity in these regions on the one hand and the EEG on the other cannot be untangled based on the correlation-based machinery that is the GLM. Nonetheless, these sets of regions may be thought of as a static or averaged picture of evolving networks, which may be subjected to further investigation. This may be particularly relevant to spontaneous brain activity compared with experimentally controlled experiments because of the greater uncertainty in the origin and sequence of neuronal events. The neurophysiology underlying the BOLD response is not fully understood. Much remains to be elucidated on the spatial-temporal relationship between IED and time-locked BOLD signal changes, for example the significance of responses distant from the presumed focus for which extensive comparison of the fMRI with the gold standards of intracranial EEG and postsurgical outcome will be required. The possibility of using MRI to directly detect local changes in magnetic field is an exciting prospect [288].

The interpretation of BOLD activation patterns consisting of multiple clusters in relation to focal IED remains an area of active investigation. Recently, an attempt has been made to increase the technique's ability to temporally resolve multiple regions of BOLD activation by combining EEG-fMRI with multiple source analysis [262].

Finally, EEG-fMRI as implemented and analyzed currently may be considered purely as a hemodynamic imaging technique, with EEG simply acting as a time marker for scan classification. However, we envisage a more symmetrical approach whereby the two forms of data are used to infer neuronal activity, which must be the ultimate aim of the entire functional imaging enterprise [289].

2.2.11 fMRI of Paroxysmal Activity: Conclusion

The possibility of recording EEG of good quality during fMRI has created a new instrument, which to date has been mainly used in an exploratory fashion.

EEG-fMRI of ictal and interictal activity in focal epilepsy has demonstrated the capability to provide new localization information in a large proportion of cases. In focal epilepsy, varied patterns have been identified often suggestive of "functional lesions" homologous to abnormalities seen on structural imaging, but additionally the involvement of possibly less disease-specific regions. The clinical value of this information remains uncertain and is the subject of ongoing investigations. At the very least, EEG-fMRI may provide complementary data useful for planning of invasive recordings where conventional localization of the seizure focus is unsuccessful. However, there are signs that it may be able to offer more.

In generalized epilepsy, EEG-fMRI has revealed activation and deactivation patterns that are mainly suggestive of less specific effects linked to generalized spike-wave, rather than reflecting syndrome, the spatial distribution of the EEG generators or more specifically a putative focus responsible for initiating GSW due in part to the averaging effect of BOLD.

2.2.12 Data Acquisition and Modeling Challenges in Clinical fMRI

Some of the brain regions of most interest in epilepsy are subject to susceptibility artifact, leading to geometric distortions and signal loss during fMRI acquisition. Ideally in the absence of an applied gradient, the magnetic field would be homogenous throughout the bore of an MRI scanner. Unfortunately, the different magnetic properties of bone, tissue, and air introduce inhomogeneities in the field when a head is introduced into the bore. Brain regions closest to borders between sinuses and brain or bone and brain, for example the inferior frontal and MTLs, are most affected, and therefore especially likely to suffer geometric distortions or signal loss [290]. This can result in reduced sensitivity and anatomical uncertainties when interpreting the images. Most epilepsy studies have been performed on 1.5 T clinical MRI scanners. Scanning at higher field strength improves signal-to-noise ratio but increases distortions and dropout [291].

Geometric distortions of the EPI data make it difficult to directly overlay fMRI activations on coregistered high-resolution scans. They can be unwarped using techniques that map the local field in the head [292], though it has been shown that approaches of this kind can introduce extra noise into the corrected EPI data [293]. Alternative acquisition sequences that do not experience geometric distortions are available [294] though these rarely have the temporal resolution or high signal to noise ratio (SNR) per unit time of EPI.

The second artifact in EPI data is more serious, as signal loss leads to sensitivity loss, which is unrecoverable by image processing techniques. Some of these artifacts can be corrected by shimming, a process whereby the static magnetic field is made more homogenous over the region of interest [290]. Some will remain, however, leading to distortion and dropout in echo-planar images. Other approaches to removing dropout often involve acquiring extra images, leading to a loss of temporal resolution [295], but more recent work has shown that dropouts and distortions can be reduced without incurring time penalties if regions of reduced spatial extent are imaged [296]. The use of high-performance gradients and thin slice acquisitions ameliorate these problems and improve fMRI quality. fMRI is also extremely sensitive to motion, although generally epilepsy patients are familiar with MRI scanners and may move less than control subjects [29].

Motion remains problematic for all applications of fMRI [234], but the problem may be particularly harmful in patient studies in view of two factors: first, patients may be more prone to movement than selected and highly motivated healthy subjects; second, each patient dataset may have greater value than that from inter-changeable healthy subjects. Therefore, measures have been proposed to extract as much information as possible from what are effectively suboptimal fMRI studies [164, 165, 297].

Acknowledgments

Thanks to Suejen Perani for help with the second edition and to Philip Allen for his comments on parts of the manuscript and to Dr. Anna Vaudano and Dr. Serge Vulliemoz for supplying some of the illustrations. Some of the work reported in this chapter was funded through a grant from the Medical Research Council (MRC grant number G0301067) and by the Wellcome Trust. We are grateful to the Big Lottery Fund, Wolfson Trust, and National Society for Epilepsy for supporting the NSE MRI scanner. This work was carried out under the auspices of the UCL/UCLH Biomedical Comprehensive Research Centre.

References

1. Wagner DD, Sziklas V, Garver KE, Jones-Gotman M (2009) Material-specific lateralization of working memory in the medial temporal lobe. Neuropsychologia 47:112–122

2. Campo P, Garrido MI, Moran RJ, Garcia-Morales I, Poch C, Toledano R, Gil-Nagel A et al (2013) Network reconfiguration and working memory impairment in mesial temporal lobe epilepsy. NeuroImage 72:48–54

3. Stretton J, Winston G, Sidhu M, Centeno M, Vollmar C, Bonelli S, Symms M et al (2012) Neural correlates of working memory in Temporal Lobe Epilepsy--an fMRI study. NeuroImage 60:1696–1703

4. Hoppe C, Elger CE, Helmstaedter C (2007) Long-term memory impairment in patients with focal epilepsy. Epilepsia 48(Suppl 9):26–29

5. Waites AB, Briellmann RS, Saling MM, Abbott DF, Jackson GD (2006) Functional connectivity networks are disrupted in left temporal lobe epilepsy. Ann Neurol 59:335–343

6. Vlooswijk MC, Jansen JF, Majoie HJ, Hofman PA, de Krom MC, Aldenkamp AP, Backes WH (2010) Functional connectivity and language impairment in cryptogenic localization-related epilepsy. Neurology 75:395–402

7. Baxendale S (2002) The role of functional MRI in the presurgical investigation of temporal lobe epilepsy patients: a clinical perspective and review. J Clin Exp Neuropsychol 24:664–676

8. Kirsch HE, Walker JA, Winstanley FS, Hendrickson R, Wong ST, Barbaro NM, Laxer KD et al (2005) Limitations of Wada memory asymmetry as a predictor of outcomes after temporal lobectomy. Neurology 65:676–680

9. Abbott DF, Waites AB, Lillywhite LM, Jackson GD (2010) fMRI assessment of language lateralization: an objective approach. NeuroImage 50:1446–1455

10. Appel S, Duke ES, Martinez AR, Khan OI, Dustin IM, Reeves-Tyer P, Berl MB et al (2012) Cerebral blood flow and fMRI BOLD auditory language activation in temporal lobe epilepsy. Epilepsia 53:631–638

11. Arora J, Pugh K, Westerveld M, Spencer S, Spencer DD, Todd Constable R (2009) Language lateralization in epilepsy patients: fMRI validated with the Wada procedure. Epilepsia 50:2225–2241

12. Binder JR, Gross WL, Allendorfer JB, Bonilha L, Chapin J, Edwards JC, Grabowski TJ et al (2011) Mapping anterior temporal lobe language areas with fMRI: a multicenter normative study. NeuroImage 54:1465–1475

13. Bonelli SB, Powell R, Thompson PJ, Yogarajah M, Focke NK, Stretton J, Vollmar C et al (2011) Hippocampal activation correlates with visual confrontation naming: fMRI findings in controls and patients with temporal lobe epilepsy. Epilepsy Res 95:246–254

14. Gartus A, Foki T, Geissler A, Beisteiner R (2009) Improvement of clinical language localization with an overt semantic and syntactic language functional MR imaging paradigm. AJNR Am J Neuroradiol 30:1977–1985

15. Sanjuan A, Bustamante JC, Forn C, Ventura-Campos N, Barros-Loscertales A, Martinez JC, Villanueva V et al (2010) Comparison of two fMRI tasks for the evaluation of the expressive language function. Neuroradiology 52:407–415

16. Friedman L, Kenny JT, Wise AL, Wu D, Stuve TA, Miller DA, Jesberger JA et al (1998) Brain activation during silent word generation evaluated with functional MRI. Brain Lang 64:231–256

17. Bonelli SB, Thompson PJ, Yogarajah M, Vollmar C, Powell RH, Symms MR, McEvoy AW et al (2012) Imaging language networks before and after anterior temporal lobe resection: results of a longitudinal fMRI study. Epilepsia 53:639–650

18. Fernandez G, Specht K, Weis S, Tendolkar I, Reuber M, Fell J, Klaver P et al (2003) Intrasubject reproducibility of presurgical language lateralization and mapping using fMRI. Neurology 60:969–975

19. Weber B, Wellmer J, Schur S, Dinkelacker V, Ruhlmann J, Mormann F, Axmacher N et al (2006) Presurgical language fMRI in patients with drug-resistant epilepsy: effects of task performance. Epilepsia 47:880–886

20. Duncan J (2009) The current status of neuroimaging for epilepsy. Curr Opin Neurol 22:179–184

21. Rosazza C, Ghielmetti F, Minati L, Vitali P, Giovagnoli AR, Deleo F, Didato G et al (2013) Preoperative language lateralization in temporal lobe epilepsy (TLE) predicts peri-ictal, pre- and post-operative language performance: an fMRI study. NeuroImage Clin 3:73–83

22. Hamberger MJ, McClelland S 3rd, McKhann GM 2nd, Williams AC, Goodman RR (2007) Distribution of auditory and visual naming sites in nonlesional temporal lobe epilepsy patients and patients with space-occupying temporal lobe lesions. Epilepsia 48:531–538

23. Hermann BP, Wyler AR (1988) Effects of anterior temporal lobectomy on language function: a controlled study. Ann Neurol 23:585–588

24. Hamberger MJ, Seidel WT (2009) Localization of cortical dysfunction based on auditory and visual naming performance. J Int Neuropsychol Soc 15:529–535

25. Specht K, Osnes B, Hugdahl K (2009) Detection of differential speech-specific processes in the temporal lobe using fMRI and a dynamic "sound morphing" technique. Hum Brain Mapp 30:3436–3444

26. Schlosser MJ, Aoyagi N, Fulbright RK, Gore JC, McCarthy G (1998) Functional MRI studies of auditory comprehension. Hum Brain Mapp 6:1–13

27. Lehericy S, Cohen L, Bazin B, Samson S, Giacomini E, Rougetet R, Hertz-Pannier L et al (2000) Functional MR evaluation of temporal and frontal language dominance compared with the Wada test. Neurology 54:1625–1633

28. Gaillard WD, Balsamo L, Xu B, McKinney C, Papero PH, Weinstein S, Conry J et al (2004) fMRI language task panel improves determination of language dominance. Neurology 63:1403–1408

29. Gaillard WD, Balsamo L, Xu B, Grandin CB, Braniecki SH, Papero PH, Weinstein S et al (2002) Language dominance in partial epilepsy patients identified with an fMRI reading task. Neurology 59:256–265

30. Adcock JE, Wise RG, Oxbury JM, Oxbury SM, Matthews PM (2003) Quantitative fMRI assessment of the differences in lateralization of language-related brain activation in patients with temporal lobe epilepsy. NeuroImage 18:423–438

31. Branco DM, Suarez RO, Whalen S, O'Shea JP, Nelson AP, da Costa JC, Golby AJ (2006) Functional MRI of memory in the hippocampus: Laterality indices may be more meaningful if calculated from whole voxel distributions. NeuroImage 32:592–602

32. Liegeois F, Connelly A, Cross JH, Boyd SG, Gadian DG, Vargha-Khadem F, Baldeweg T (2004) Language reorganization in children with early-onset lesions of the left hemisphere: an fMRI study. Brain 127:1229–1236

33. Karunanayaka P, Kim KK, Holland SK, Szaflarski JP (2011) The effects of left or right hemispheric epilepsy on language networks investigated with semantic decision fMRI task and independent component analysis. Epilepsy BehavB 20:623–632

34. Mbwana J, Berl MM, Ritzl EK, Rosenberger L, Mayo J, Weinstein S, Conry JA et al (2009) Limitations to plasticity of language network reorganization in localization related epilepsy. Brain 132:347–356

35. You X, Adjouadi M, Wang J, Guillen MR, Bernal B, Sullivan J, Donner E et al (2013) A decisional space for fMRI pattern separation using the principal component analysis--a comparative study of language networks in pediatric epilepsy. Hum Brain Mapp 34:2330–2342

36. Woermann FG, Jokeit H, Luerding R, Freitag H, Schulz R, Guertler S, Okujava M et al (2003) Language lateralization by Wada test and fMRI in 100 patients with epilepsy. Neurology 61:699–701

37. Wang J, You X, Wu W, Guillen MR, Cabrerizo M, Sullivan J, Donner E et al (2014) Classification of fMRI patterns--a study of the language network segregation in pediatric localization related epilepsy. Hum Brain Mapp 35:1446–1460

38. You X, Adjouadi M, Guillen MR, Ayala M, Barreto A, Rishe N, Sullivan J et al (2011) Sub-patterns of language network reorganization in pediatric localization related epilepsy: a multisite study. Hum Brain Mapp 32:784–799

39. Friston KJ, Price CJ, Fletcher P, Moore C, Frackowiak RS, Dolan RJ (1996) The trouble with cognitive subtraction. NeuroImage 4:97–104

40. Price CJ, Friston KJ (1997) Cognitive conjunction: a new approach to brain activation experiments. NeuroImage 5:261–270

41. Risse GL, Gates JR, Fangman MC (1997) A reconsideration of bilateral language representation based on the intracarotid amobarbital procedure. Brain Cogn 33:118–132

42. Serafetinides EA, Hoare RD, Driver M (1965) Intracarotid sodium amylobarbitone and cerebral dominance for speech and consciousness. Brain 88:107–130

43. Rasmussen T, Milner B (1977) The role of early left-brain injury in determining lateralization of cerebral speech functions. Ann N Y Acad Sci 299:355–369

44. Springer JA, Binder JR, Hammeke TA, Swanson SJ, Frost JA, Bellgowan PS, Brewer CC et al (1999) Language dominance in neurologically normal and epilepsy subjects: a functional MRI study. Brain 122(Pt 11):2033–2046

45. Janszky J, Mertens M, Janszky I, Ebner A, Woermann FG (2006) Left-sided interictal epileptic activity induces shift of language lateralization in temporal lobe epilepsy: an fMRI study. Epilepsia 47:921–927

46. Janszky J, Jokeit H, Heinemann D, Schulz R, Woermann FG, Ebner A (2003) Epileptic activity influences the speech organization in medial temporal lobe epilepsy. Brain 126:2043–2051

47. Thivard L, Hombrouck J, du Montcel ST, Delmaire C, Cohen L, Samson S, Dupont S et al (2005) Productive and perceptive language reorganization in temporal lobe epilepsy. NeuroImage 24:841–851

48. Berl MM, Balsamo LM, Xu B, Moore EN, Weinstein SL, Conry JA, Pearl PL et al (2005) Seizure focus affects regional language networks assessed by fMRI. Neurology 65:1604–1611

49. Duke ES, Tesfaye M, Berl MM, Walker JE, Ritzl EK, Fasano RE, Conry JA et al (2012) The effect of seizure focus on regional language processing areas. Epilepsia 53:1044–1050

50. Wilke M, Pieper T, Lindner K, Dushe T, Staudt M, Grodd W, Holthausen H et al (2011) Clinical functional MRI of the language domain in children with epilepsy. Hum Brain Mapp 32:1882–1893

51. Weber B, Wellmer J, Reuber M, Mormann F, Weis S, Urbach H, Ruhlmann J et al (2006) Left hippocampal pathology is associated with atypical language lateralization in patients with focal epilepsy. Brain 129:346–351

52. Briellmann RS, Labate A, Harvey AS, Saling MM, Sveller C, Lillywhite L, Abbott DF et al (2006) Is language lateralization in temporal lobe epilepsy patients related to the nature of the epileptogenic lesion? Epilepsia 47:916–920

53. Jensen EJ, Hargreaves IS, Pexman PM, Bass A, Goodyear BG, Federico P (2011) Abnormalities of lexical and semantic processing in left temporal lobe epilepsy: an fMRI study. Epilepsia 52:2013–2021

54. Fakhri M, Oghabian MA, Vedaei F, Zandieh A, Masoom N, Sharifi G, Ghodsi M et al (2013) Atypical language lateralization: an fMRI study in patients with cerebral lesions. Funct Neurol 28:55–61

55. Wellmer J, Weber B, Urbach H, Reul J, Fernandez G, Elger CE (2009) Cerebral lesions can impair fMRI-based language lateralization. Epilepsia 50:2213–2224

56. Sanjuan A, Bustamante JC, Garcia-Porcar M, Rodriguez-Pujadas A, Forn C, Martinez JC, Campos A et al (2013) Bilateral inferior frontal language-related activation correlates with verbal recall in patients with left temporal lobe epilepsy and typical language distribution. Epilepsy Res 104:118–124

57. Everts R, Harvey AS, Lillywhite L, Wrennall J, Abbott DF, Gonzalez L, Kean M et al (2010) Language lateralization correlates with verbal memory performance in children with focal epilepsy. Epilepsia 51:627–638

58. Janecek JK, Swanson SJ, Sabsevitz DS, Hammeke TA, Raghavan M, E Rozman M M, Binder JR (2013) Language lateralization by fMRI and Wada testing in 229 patients with epilepsy: rates and predictors of discordance. Epilepsia 54:314–322

59. Benke T, Koylu B, Visani P, Karner E, Brenneis C, Bartha L, Trinka E et al (2006) Language lateralization in temporal lobe epilepsy: a comparison between fMRI and the Wada Test. Epilepsia 47:1308–1319

60. Jayakar P, Bernal B, Santiago Medina L, Altman N (2002) False lateralization of language cortex on functional MRI after a cluster of focal seizures. Neurology 58:490–492

61. Wagner K, Hader C, Metternich B, Buschmann F, Schwarzwald R, Schulze-Bonhage A (2012) Who needs a Wada test? Present clinical indications for amobarbital procedures. J Neurol Neurosurg Psychiatry 83:503–509

62. Desmond JE, Sum JM, Wagner AD, Demb JB, Shear PK, Glover GH, Gabrieli JD et al (1995) Functional MRI measurement of language lateralization in Wada-tested patients. Brain 118(Pt 6):1411–1419

63. Binder JR, Swanson SJ, Hammeke TA, Morris GL, Mueller WM, Fischer M, Benbadis S et al (1996) Determination of language dominance using functional MRI: a comparison with the Wada test. Neurology 46:978–984

64. Hertz-Pannier L, Gaillard WD, Mott SH, Cuenod CA, Bookheimer SY, Weinstein S, Conry J et al (1997) Noninvasive assessment of language dominance in children and adolescents with functional MRI: a preliminary study. Neurology 48:1003–1012

65. Yetkin FZ, Swanson S, Fischer M, Akansel G, Morris G, Mueller W, Haughton V (1998) Functional MR of frontal lobe activation: comparison with Wada language results. AJNR Am J Neuroradiol 19:1095–1098

66. Benson RR, FitzGerald DB, LeSueur LL, Kennedy DN, Kwong KK, Buchbinder BR, Davis TL et al (1999) Language dominance determined by whole brain functional MRI in patients with brain lesions. Neurology 52:798–809

67. Carpentier A, Pugh KR, Westerveld M, Studholme C, Skrinjar O, Thompson JL, Spencer DD et al (2001) Functional MRI of language processing: dependence on input modality and temporal lobe epilepsy. Epilepsia 42:1241–1254

68. Sabbah P, Chassoux F, Leveque C, Landre E, Baudoin-Chial S, Devaux B, Mann M et al (2003) Functional MR imaging in assessment of language dominance in epileptic patients. NeuroImage 18:460–467

69. FitzGerald DB, Cosgrove GR, Ronner S, Jiang H, Buchbinder BR, Belliveau JW, Rosen BR et al (1997) Location of language in the cortex: a comparison between functional MR imaging and electrocortical stimulation. AJNR Am J Neuroradiol 18:1529–1539

70. Schlosser MJ, Luby M, Spencer DD, Awad IA, McCarthy G (1999) Comparative localization of auditory comprehension by using functional magnetic resonance imaging and cortical stimulation. J Neurosurg 91:626–635

71. Pouratian N, Bookheimer SY, Rex DE, Martin NA, Toga AW (2002) Utility of preoperative functional magnetic resonance imaging for identifying language cortices in patients with vascular malformations. J Neurosurg 97:21–32

72. Rutten GJ, Ramsey NF, van Rijen PC, Noordmans HJ, van Veelen CW (2002) Development of a functional magnetic resonance imaging protocol for intraoperative localization of critical temporoparietal language areas. Ann Neurol 51:350–360

73. Davies KG, Bell BD, Bush AJ, Hermann BP, Dohan FC Jr, Jaap AS (1998) Naming decline after left anterior temporal lobectomy correlates with pathological status of resected hippocampus. Epilepsia 39:407–419

74. Saykin AJ, Stafiniak P, Robinson LJ, Flannery KA, Gur RC, O'Connor MJ, Sperling MR (1995) Language before and after temporal lobectomy: specificity of acute changes and relation to early risk factors. Epilepsia 36:1071–1077

75. Hermann BP, Perrine K, Chelune GJ, Barr W, Loring DW, Strauss E, Trenerry MR et al (1999) Visual confrontation naming following left anterior temporal lobectomy: a comparison of surgical approaches. Neuropsychology 13:3–9

76. Devinsky O, Perrine K, Llinas R, Luciano DJ, Dogali M (1993) Anterior temporal language areas in patients with early onset of temporal lobe epilepsy. Ann Neurol 34:727–732

77. Schwartz TH, Devinsky O, Doyle W, Perrine K (1998) Preoperative predictors of anterior temporal language areas. J Neurosurg 89:962–970

78. Sabsevitz DS, Swanson SJ, Hammeke TA, Spanaki MV, Possing ET, Morris GL 3rd, Mueller WM et al (2003) Use of preoperative functional neuroimaging to predict language deficits from epilepsy surgery. Neurology 60:1788–1792

79. Noppeney U, Price CJ, Duncan JS, Koepp MJ (2005) Reading skills after left anterior temporal lobe resection: an fMRI study. Brain 128:1377–1385

80. Voets NL, Adcock JE, Flitney DE, Behrens TE, Hart Y, Stacey R, Carpenter K et al (2006) Distinct right frontal lobe activation in language processing following left hemisphere injury. Brain 129:754–766

81. Powell HW, Parker GJ, Alexander DC, Symms MR, Boulby PA, Wheeler-Kingshott CA, Barker GJ et al (2006) Hemispheric asymmetries in language-related pathways: a combined functional MRI and tractography study. NeuroImage 32:388–399

82. Powell HW, Parker GJ, Alexander DC, Symms MR, Boulby PA, Wheeler-Kingshott CA, Barker GJ et al (2007) Abnormalities of language networks in temporal lobe epilepsy. NeuroImage 36:209–221

83. Stretton J, Thompson PJ (2012) Frontal lobe function in temporal lobe epilepsy. Epilepsy Res 98:1–13

84. Centeno M, Thompson PJ, Koepp MJ, Helmstaedter C, Duncan JS (2010) Memory in frontal lobe epilepsy. Epilepsy Res 91:123–132

85. Bonelli SB, Thompson PJ, Yogarajah M, Powell RH, Samson RS, McEvoy AW, Symms MR et al (2013) Memory reorganization following anterior temporal lobe resection: a longitudinal functional MRI study. Brain 136:1889–1900

86. Scoville WB, Milner B (2000) Loss of recent memory after bilateral hippocampal lesions. 1957. J Neuropsychiatry Clin Neurosci 12:103–113

87. Ivnik RJ, Sharbrough FW, Laws ER Jr (1987) Effects of anterior temporal lobectomy on cognitive function. J Clin Psychol 43:128–137

88. Spiers HJ, Burgess N, Maguire EA, Baxendale SA, Hartley T, Thompson PJ, O'Keefe J (2001) Unilateral temporal lobectomy patients show lateralized topographical and episodic memory deficits in a virtual town. Brain 124:2476–2489

89. Penfield W, Milner B (1958) Memory deficit produced by bilateral lesions in the hippocampal zone. AMA Arch Neurol Psychiatry 79:475–497

90. Warrington EK, Duchen LW (1992) A reappraisal of a case of persistent global amnesia following right temporal lobectomy: a clinico-pathological study. Neuropsychologia 30:437–450

91. Loring DW, Hermann BP, Meador KJ, Lee GP, Gallagher BB, King DW, Murro AM et al (1994) Amnesia after unilateral temporal lobectomy: a case report. Epilepsia 35:757–763

92. Alessio A, Pereira FR, Sercheli MS, Rondina JM, Ozelo HB, Bilevicius E, Pedro T et al (2013) Brain plasticity for verbal and visual memories in patients with mesial temporal lobe epilepsy and hippocampal sclerosis: an fMRI study. Hum Brain Mapp 34:186–199

93. Banks SJ, Sziklas V, Sodums DJ, Jones-Gotman M (2012) fMRI of verbal and nonverbal mem-

ory processes in healthy and epileptogenic medial temporal lobes. Epilepsy Behav 25:42–49

94. Chelune GJ (1995) Hippocampal adequacy versus functional reserve: predicting memory functions following temporal lobectomy. Arch Clin Neuropsychol 10:413–432

95. Chelune GJ, Naugle RI, Luders H, Awad IA (1991) Prediction of cognitive change as a function of preoperative ability status among temporal lobectomy patients seen at 6-month follow-up. Neurology 41:399–404

96. Kneebone AC, Chelune GJ, Dinner DS, Naugle RI, Awad IA (1995) Intracarotid amobarbital procedure as a predictor of material-specific memory change after anterior temporal lobectomy. Epilepsia 36:857–865

97. Sass KJ, Spencer DD, Kim JH, Westerveld M, Novelly RA, Lencz T (1990) Verbal memory impairment correlates with hippocampal pyramidal cell density. Neurology 40:1694–1697

98. Trenerry MR, Jack CR Jr, Ivnik RJ, Sharbrough FW, Cascino GD, Hirschorn KA, Marsh WR et al (1993) MRI hippocampal volumes and memory function before and after temporal lobectomy. Neurology 43:1800–1805

99. Guedj E, Bettus G, Barbeau EJ, Liegeois-Chauvel C, Confort-Gouny S, Bartolomei F, Chauvel P et al (2011) Hyperactivation of parahippocampal region and fusiform gyrus associated with successful encoding in medial temporal lobe epilepsy. Epilepsia 52:1100–1109

100. Sidhu MK, Stretton J, Winston GP, Bonelli S, Centeno M, Vollmar C, Symms M et al (2013) A functional magnetic resonance imaging study mapping the episodic memory encoding network in temporal lobe epilepsy. Brain 136:1868–1888

101. Bonelli SB, Powell RH, Yogarajah M, Samson RS, Symms MR, Thompson PJ, Koepp MJ et al (2010) Imaging memory in temporal lobe epilepsy: predicting the effects of temporal lobe resection. Brain 133:1186–1199

102. Sidhu MK, Stretton J, Winston GP, Symms M, Thompson PJ, Koepp MJ, Duncan JS (2015) Memory fMRI predicts verbal memory decline after anterior temporal lobe resection. Neurology 84:1512–1519

103. Hermann BP, Wyler AR, Somes G, Berry AD 3rd, Dohan FC Jr (1992) Pathological status of the mesial temporal lobe predicts memory outcome from left anterior temporal lobectomy. Neurosurgery 31:652–656, discussion 656-657

104. Sass KJ, Westerveld M, Buchanan CP, Spencer SS, Kim JH, Spencer DD (1994) Degree of hippocampal neuron loss determines severity of verbal memory decrease after left anteromesiotemporal lobectomy. Epilepsia 35:1179–1186

105. McCormick C, Quraan M, Cohn M, Valiante TA, McAndrews MP (2013) Default mode network connectivity indicates episodic memory capacity in mesial temporal lobe epilepsy. Epilepsia 54:809–818

106. Jokeit H, Ebner A, Holthausen H, Markowitsch HJ, Moch A, Pannek H, Schulz R et al (1997) Individual prediction of change in delayed recall of prose passages after left-sided anterior temporal lobectomy. Neurology 49:481–487

107. Helmstaedter C, Elger CE (1996) Cognitive consequences of two-thirds anterior temporal lobectomy on verbal memory in 144 patients: a three-month follow-up study. Epilepsia 37:171–180

108. Cheung MC, Chan AS, Lam JM, Chan YL (2009) Pre- and postoperative fMRI and clinical memory performance in temporal lobe epilepsy. J Neurol Neurosurg Psychiatry 80:1099–1106

109. Corkin S, Amaral DG, Gonzalez RG, Johnson KA, Hyman BT (1997) H. M'.s medial temporal lobe lesion: findings from magnetic resonance imaging. J Neurosci 17:3964–3979

110. Fernandez G, Effern A, Grunwald T, Pezer N, Lehnertz K, Dumpelmann M, Van Roost D et al (1999) Real-time tracking of memory formation in the human rhinal cortex and hippocampus. Science 285:1582–1585

111. Ojemann JG, Akbudak E, Snyder AZ, McKinstry RC, Raichle ME, Conturo TE (1997) Anatomic localization and quantitative analysis of gradient refocused echo-planar fMRI susceptibility artifacts. NeuroImage 6:156–167

112. Greicius MD, Krasnow B, Boyett-Anderson JM, Eliez S, Schatzberg AF, Reiss AL, Menon V (2003) Regional analysis of hippocampal activation during memory encoding and retrieval: fMRI study. Hippocampus 13:164–174

113. Lipschutz B, Friston KJ, Ashburner J, Turner R, Price CJ (2001) Assessing study-specific regional variations in fMRI signal. NeuroImage 13:392–398

114. Craik FIM, Lockhart RS (1972) Levels of processing: A framework for memory research. J Verbal Learn Verbal Behav 11:671–684

115. Kelley WM, Miezin FM, McDermott KB, Buckner RL, Raichle ME, Cohen NJ, Ollinger JM et al (1998) Hemispheric specialization in human dorsal frontal cortex and medial temporal lobe for verbal and nonverbal memory encoding. Neuron 20:927–936

116. Demb JB, Desmond JE, Wagner AD, Vaidya CJ, Glover GH, Gabrieli JD (1995) Semantic encoding and retrieval in the left inferior prefrontal cortex: a functional MRI study of task difficulty and process specificity. J Neurosci 15:5870–5878

117. Wagner AD, Schacter DL, Rotte M, Koutstaal W, Maril A, Dale AM, Rosen BR et al (1998) Building memories: remembering and forgetting of verbal experiences as predicted by brain activity. Science 281:1188–1191

118. Buckner RL, Kelley WM, Petersen SE (1999) Frontal cortex contributes to human memory formation. Nat Neurosci 2:311–314

119. Golby AJ, Poldrack RA, Brewer JB, Spencer D, Desmond JE, Aron AP, Gabrieli JD (2001) Material-specific lateralization in the medial temporal lobe and prefrontal cortex during memory encoding. Brain 124:1841–1854

120. Wagner AD, Koutstaal W, Schacter DL (1999) When encoding yields remembering: insights from event-related neuroimaging. Phil Trans Roy Soc Lond B Biol Sci 354:1307–1324

121. Powell HW, Koepp MJ, Symms MR, Boulby PA, Salek-Haddadi A, Thompson PJ, Duncan JS et al (2005) Material-specific lateralization of memory encoding in the medial temporal lobe: blocked versus event-related design. NeuroImage 27:231–239

122. Detre JA, Maccotta L, King D, Alsop DC, Glosser G, D'Esposito M, Zarahn E et al (1998) Functional MRI lateralization of memory in temporal lobe epilepsy. Neurology 50:926–932

123. Doucet G, Osipowicz K, Sharan A, Sperling MR, Tracy JI (2013) Hippocampal functional connectivity patterns during spatial working memory differ in right versus left temporal lobe epilepsy. Brain Connect 3:398–406

124. Stretton J, Winston GP, Sidhu M, Bonelli S, Centeno M, Vollmar C, Cleary RA et al (2013) Disrupted segregation of working memory networks in temporal lobe epilepsy. NeuroImage Clin 2:273–281

125. Bellgowan PS, Binder JR, Swanson SJ, Hammeke TA, Springer JA, Frost JA, Mueller WM et al (1998) Side of seizure focus predicts left medial temporal lobe activation during verbal encoding. Neurology 51:479–484

126. Jokeit H, Okujava M, Woermann FG (2001) Memory fMRI lateralizes temporal lobe epilepsy. Neurology 57:1786–1793

127. Golby AJ, Poldrack RA, Illes J, Chen D, Desmond JE, Gabrieli JD (2002) Memory lateralization in medial temporal lobe epilepsy assessed by functional MRI. Epilepsia 43:855–863

128. Richardson MP, Strange BA, Duncan JS, Dolan RJ (2003) Preserved verbal memory function in left medial temporal pathology involves reorganisation of function to right medial temporal lobe. NeuroImage 20(Suppl 1):S112–S119

129. Powell HW, Richardson MP, Symms MR, Boulby PA, Thompson PJ, Duncan JS, Koepp MJ (2007) Reorganization of verbal and nonverbal memory in temporal lobe epilepsy due to unilateral hippocampal sclerosis. Epilepsia 48:1512–1525

130. Richardson MP, Strange BA, Thompson PJ, Baxendale SA, Duncan JS, Dolan RJ (2004) Pre-operative verbal memory fMRI predicts post-operative memory decline after left temporal lobe resection. Brain 127:2419–2426

131. Richardson MP, Strange BA, Duncan JS, Dolan RJ (2006) Memory fMRI in left hippocampal sclerosis: optimizing the approach to predicting postsurgical memory. Neurology 66:699–705

132. Powell HW, Richardson MP, Symms MR, Boulby PA, Thompson PJ, Duncan JS, Koepp MJ (2008) Preoperative fMRI predicts memory decline following anterior temporal lobe resection. J Neurol Neurosurg Psychiatry 79:686–693

133. Rabin ML, Narayan VM, Kimberg DY, Casasanto DJ, Glosser G, Tracy JI, French JA et al (2004) Functional MRI predicts post-surgical memory following temporal lobectomy. Brain 127:2286–2298

134. Janszky J, Jokeit H, Kontopoulou K, Mertens M, Ebner A, Pohlmann-Eden B, Woermann FG (2005) Functional MRI predicts memory performance after right mesiotemporal epilepsy surgery. Epilepsia 46:244–250

135. Price CJ, Friston KJ (1999) Scanning patients with tasks they can perform. Hum Brain Mapp 8:102–108

136. von Helmholtz HLF (2004) Some laws concerning the distribution of electric currents in volume conductors with applications to experiments on animal electricity. Proc IEEE 92:868–870

137. Geselowitz DB (2004) Introduction to "some laws concerning the distribution of electric currents in volume conductors with applications to experiments on animal electricity". Proc IEEE 92:864–867

138. Ives JR, Warach S, Schmitt F, Edelman RR, Schomer DL (1993) Monitoring the patient's EEG during echo planar MRI. Electroencephalogr Clin Neurophysiol 87:417–420

139. Detre JA, Alsop DC, Aguirre GK, Sperling MR (1996) Coupling of cortical and thalamic ictal activity in human partial epilepsy: demonstration by functional magnetic resonance imaging. Epilepsia 37:657–661

140. Krings T, Topper R, Reinges MH, Foltys H, Spetzger U, Chiappa KH, Gilsbach JM et al (2000) Hemodynamic changes in simple partial epilepsy: a functional MRI study. Neurology 54:524–527

141. Connelly A (1995) Ictal imaging using functional magnetic resonance. Magn Reson Imaging 13:1233–1237

142. Jackson GD, Connelly A, Cross JH, Gordon I, Gadian DG (1994) Functional magnetic resonance imaging of focal seizures. Neurology 44:850–856

143. Jorge J, Grouiller F, Ipek O, Stoermer R, Michel CM, Figueiredo P, van der Zwaag W et al (2015) Simultaneous EEG-fMRI at ultra-high field: artifact prevention and safety assessment. NeuroImage 105:132–144

144. Arrubla J, Neuner I, Dammers J, Breuer L, Warbrick T, Hahn D, Poole MS et al (2014) Methods for pulse artefact reduction: experiences with EEG data recorded at 9.4 T static magnetic field. J Neurosci Methods 232:110–117

145. Lemieux L, Allen PJ, Franconi F, Symms MR, Fish DR (1997) Recording of EEG during fMRI experiments: patient safety. Magn Reson Med 38:943–952

146. Mirsattari SM, Lee DH, Jones D, Bihari F, Ives JR (2004) MRI compatible EEG electrode system for routine use in the epilepsy monitoring unit and intensive care unit. Clin Neurophysiol 115:2175–2180

147. Konings MK, Bartels LW, Smits HF, Bakker CJ (2000) Heating around intravascular guidewires by resonating RF waves. J Magn Reson Imaging 12:79–85

148. Fischer H, Ladebeck R (1998) Echo-planar imaging image artifacts. In: Schmitt F, Stehling MK, Turner R (eds) Echo-planar imaging: theory, technique, and application. Springer, Berlin, pp 179–200

149. Krakow K, Allen PJ, Symms MR, Lemieux L, Josephs O, Fish DR (2000) EEG recording during fMRI experiments: image quality. Hum Brain Mapp 10:10–15

150. Bonmassar G, Anami K, Ives J, Belliveau JW (1999) Visual evoked potential (VEP) measured by simultaneous 64-channel EEG and 3T fMRI. Neuroreport 10:1893–1897

151. Bonmassar G, Purdon PL, Jaaskelainen IP, Chiappa K, Solo V, Brown EN, Belliveau JW (2002) Motion and ballistocardiogram artifact removal for interleaved recording of EEG and EPs during MRI. NeuroImage 16:1127–1141

152. Scarff CJ, Reynolds A, Goodyear BG, Ponton CW, Dort JC, Eggermont JJ (2004) Simultaneous 3-T fMRI and high-density recording of human auditory evoked potentials. NeuroImage 23:1129–1142

153. Klein C, Hanggi J, Luechinger R, Jancke L (2015) MRI with and without a high-density EEG cap--what makes the difference? NeuroImage 106:189–197

154. Allen PJ, Polizzi G, Krakow K, Fish DR, Lemieux L (1998) Identification of EEG events in the MR scanner: the problem of pulse artifact and a method for its subtraction. NeuroImage 8:229–239

155. Wendt RE 3rd, Rokey R, Vick GW 3rd, Johnston DL (1988) Electrocardiographic gating and monitoring in NMR imaging. Magn Reson Imaging 6:89–95

156. Poncelet BP, Wedeen VJ, Weisskoff RM, Cohen MS (1992) Brain parenchyma motion: measurement with cine echo-planar MR imaging. Radiology 185:645–651

157. Tenforde TS, Gaffey CT, Moyer BR, Budinger TF (1983) Cardiovascular alterations in Macaca monkeys exposed to stationary magnetic fields: experimental observations and theoretical analysis. Bioelectromagnetics 4:1–9

158. Goldman RI, Stern JM, Engel J Jr, Cohen MS (2000) Acquiring simultaneous EEG and functional MRI. Clin Neurophysiol 111:1974–1980

159. Benar C, Aghakhani Y, Wang Y, Izenberg A, Al-Asmi A, Dubeau F, Gotman J (2003) Quality of EEG in simultaneous EEG-fMRI for epilepsy. Clin Neurophysiol 114:569–580

160. Chowdhury ME, Mullinger KJ, Glover P, Bowtell R (2014) Reference layer artefact subtraction (RLAS): a novel method of minimizing EEG artefacts during simultaneous fMRI. NeuroImage 84:307–319

161. Salek-Haddadi A, Lemieux L, Merschhemke M, Diehl B, Allen PJ, Fish DR (2003) EEG quality during simultaneous functional MRI of interictal epileptiform discharges. Magn Reson Imaging 21:1159–1166

162. Liston AD, De Munck JC, Hamandi K, Laufs H, Ossenblok P, Duncan JS, Lemieux L (2006) Analysis of EEG-fMRI data in focal epilepsy based on automated spike classification and Signal Space Projection. NeuroImage 31:1015–1024

163. Allen PJ, Josephs O, Turner R (2000) A method for removing imaging artifact from continuous EEG recorded during functional MRI. NeuroImage 12:230–239

164. Lemieux L, Salek-Haddadi A, Lund TE, Laufs H, Carmichael D (2007) Modelling large motion events in fMRI studies of patients with epilepsy. Magn Reson Imaging 25:894–901

165. Liston AD, Lund TE, Salek-Haddadi A, Hamandi K, Friston KJ, Lemieux L (2006) Modelling cardiac signal as a confound in EEG-fMRI and its application in focal epilepsy studies. NeuroImage 30:827–834

166. Ellingson ML, Liebenthal E, Spanaki MV, Prieto TE, Binder JR, Ropella KM (2004) Ballistocardiogram artifact reduction in the simultaneous acquisition of auditory ERPS and fMRI. NeuroImage 22:1534–1542

167. Kruggel F, Wiggins CJ, Herrmann CS, von Cramon DY (2000) Recording of the event-

related potentials during functional MRI at 3.0 Tesla field strength. Magn Reson Med 44:277–282

168. Sijbersa J, Van Audekerke J, Verhoye M, Van der Linden A, Van Dyck D (2000) Reduction of ECG and gradient related artifacts in simultaneously recorded human EEG/MRI data. Magn Reson Imaging 18:881–886

169. Kim KH, Yoon HW, Park HW (2004) Improved ballistocardiac artifact removal from the electroencephalogram recorded in fMRI. J Neurosci Methods 135:193–203

170. Wan X, Iwata K, Riera J, Ozaki T, Kitamura M, Kawashima R (2006) Artifact reduction for EEG/fMRI recording: nonlinear reduction of ballistocardiogram artifacts. Clin Neurophysiol 117:668–680

171. Niazy RK, Beckmann CF, Iannetti GD, Brady JM, Smith SM (2005) Removal of FMRI environment artifacts from EEG data using optimal basis sets. NeuroImage 28:720–737

172. In MH, Lee SY, Park TS, Kim TS, Cho MH, Ahn YB (2006) Ballistocardiogram artifact removal from EEG signals using adaptive filtering of EOG signals. Physiol Meas 27:1227–1240

173. Eichele T, Specht K, Moosmann M, Jongsma ML, Quiroga RQ, Nordby H, Hugdahl K (2005) Assessing the spatiotemporal evolution of neuronal activation with single-trial event-related potentials and functional MRI. Proc Natl Acad Sci U S A 102:17798–17803

174. Otzenberger H, Gounot D, Foucher JR (2005) P300 recordings during event-related fMRI: a feasibility study. Brain Res Cogn Brain Res 23:306–315

175. Otzenberger H, Gounot D, Foucher JR (2007) Optimisation of a post-processing method to remove the pulse artifact from EEG data recorded during fMRI: an application to P300 recordings during e-fMRI. Neurosci Res 57:230–239

176. Srivastava G, Crottaz-Herbette S, Lau KM, Glover GH, Menon V (2005) ICA-based procedures for removing ballistocardiogram artifacts from EEG data acquired in the MRI scanner. NeuroImage 24:50–60

177. Nakamura W, Anami K, Mori T, Saitoh O, Cichocki A, Amari S (2006) Removal of ballistocardiogram artifacts from simultaneously recorded EEG and fMRI data using independent component analysis. IEEE Trans Biomed Eng 53:1294–1308

178. Briselli E, Garreffa G, Bianchi L, Bianciardi M, Macaluso E, Abbafati M, Grazia Marciani M et al (2006) An independent component analysis-based approach on ballistocardio-

gram artifact removing. Magn Reson Imaging 24:393–400

179. Mantini D, Perrucci MG, Cugini S, Ferretti A, Romani GL, Del Gratta C (2007) Complete artifact removal for EEG recorded during continuous fMRI using independent component analysis. NeuroImage 34:598–607

180. Maggioni E, Arrubla J, Warbrick T, Dammers J, Bianchi AM, Reni G, Tosetti M et al (2014) Removal of pulse artefact from EEG data recorded in MR environment at 3T. Setting of ICA parameters for marking artefactual components: application to resting-state data. PLoS One 9, e112147

181. Debener S, Strobel A, Sorger B, Peters J, Kranczioch C, Engel AK, Goebel R (2007) Improved quality of auditory event-related potentials recorded simultaneously with 3-T fMRI: removal of the ballistocardiogram artefact. NeuroImage 34:587–597

182. Harrison AH, Noseworthy MD, Reilly JP, Connolly JF (2014) Ballistocardiogram correction in simultaneous EEG/ fMRI recordings: a comparison of average artifact subtraction and optimal basis set methods using two popular software tools. Crit Rev Biomed Eng 42:95–107

183. Warach S, Ives JR, Schlaug G, Patel MR, Darby DG, Thangaraj V, Edelman RR et al (1996) EEG-triggered echo-planar functional MRI in epilepsy. Neurology 47:89–93

184. Rothlubbers S, Relvas V, Leal A, Murta T, Lemieux L, Figueiredo P (2015) Characterisation and reduction of the EEG artefact caused by the helium cooling pump in the MR environment: validation in epilepsy patient data. Brain Topogr 28:208–220

185. Bonmassar G, Schwartz DP, Liu AK, Kwong KK, Dale AM, Belliveau JW (2001) Spatiotemporal brain imaging of visual-evoked activity using interleaved EEG and fMRI recordings. NeuroImage 13:1035–1043

186. Huang-Hellinger FR, Breiter HC, McCormack G, Cohen MS, Kwong KK, Sutton JP, Savoy RL et al (1995) Simultaneous functional magnetic resonance imaging and electrophysiological recording. Hum Brain Mapp 3:13–23

187. Krakow K, Woermann FG, Symms MR, Allen PJ, Lemieux L, Barker GJ, Duncan JS et al (1999) EEG-triggered functional MRI of interictal epileptiform activity in patients with partial seizures. Brain 122(Pt 9):1679–1688

188. Krakow K, Allen PJ, Lemieux L, Symms MR, Fish DR (2000) Methodology: EEG-correlated fMRI. Adv Neurol 83:187–201

189. Krakow K, Lemieux L, Messina D, Scott CA, Symms MR, Duncan JS, Fish DR (2001) Spatio-temporal imaging of focal interictal epileptiform activity using EEG-triggered functional MRI. Epileptic Disord 3:67–74

190. Al-Asmi A, Benar CG, Gross DW, Khani YA, Andermann F, Pike B, Dubeau F et al (2003) fMRI activation in continuous and spike-triggered EEG-fMRI studies of epileptic spikes. Epilepsia 44:1328–1339

191. Anami K, Mori T, Tanaka F, Kawagoe Y, Okamoto J, Yarita M, Ohnishi T et al (2003) Stepping stone sampling for retrieving artifact-free electroencephalogram during functional magnetic resonance imaging. NeuroImage 19:281–295

192. Garreffa G, Carni M, Gualniera G, Ricci GB, Bozzao L, De Carli D, Morasso P et al (2003) Real-time MR artifacts filtering during continuous EEG/fMRI acquisition. Magn Reson Imaging 21:1175–1189

193. Mullinger KJ, Yan WX, Bowtell R (2011) Reducing the gradient artefact in simultaneous EEG-fMRI by adjusting the subject's axial position. NeuroImage 54:1942–1950

194. Hoffmann A, Jager L, Werhahn KJ, Jaschke M, Noachtar S, Reiser M (2000) Electroencephalography during functional echo-planar imaging: detection of epileptic spikes using post-processing methods. Magn Reson Med 44:791–798

195. Sijbers J, Michiels I, Verhoye M, Van Audekerke J, Van der Linden A, Van Dyck D (1999) Restoration of MR-induced artifacts in simultaneously recorded MR/EEG data. Magn Reson Imaging 17:1383–1391

196. Hamandi K, Salek-Haddadi A, Laufs H, Liston A, Friston K, Fish DR, Duncan JS et al (2006) EEG-fMRI of idiopathic and secondarily generalized epilepsies. NeuroImage 31:1700–1710

197. Lemieux L, Salek-Haddadi A, Josephs O, Allen P, Toms N, Scott C, Krakow K et al (2001) Event-related fMRI with simultaneous and continuous EEG: description of the method and initial case report. NeuroImage 14:780–787

198. Salek-Haddadi A, Merschhemke M, Lemieux L, Fish DR (2002) Simultaneous EEG-correlated Ictal fMRI. NeuroImage 16:32–40

199. Salek-Haddadi A, Diehl B, Hamandi K, Merschhemke M, Liston A, Friston K, Duncan JS et al (2006) Hemodynamic correlates of epileptiform discharges: an EEG-fMRI study of 63 patients with focal epilepsy. Brain Res 1088:148–166

200. Seeck M, Lazeyras F, Michel CM, Blanke O, Gericke CA, Ives J, Delavelle J et al (1998) Non-invasive epileptic focus localization using EEG-triggered functional MRI and electromagnetic tomography. Electroencephalogr Clin Neurophysiol 106:508–512

201. Symms MR, Allen PJ, Woermann FG, Polizzi G, Krakow K, Barker GJ, Fish DR et al (1999) Reproducible localization of interictal epileptiform discharges using EEG-triggered fMRI. Phys Med Biol 44:N161–N168

202. Patel MR, Blum A, Pearlman JD, Yousuf N, Ives JR, Saeteng S, Schomer DL et al (1999) Echo-planar functional MR imaging of epilepsy with concurrent EEG monitoring. AJNR Am J Neuroradiol 20:1916–1919

203. Krakow K, Wieshmann UC, Woermann FG, Symms MR, McLean MA, Lemieux L, Allen PJ et al (1999) Multimodal MR imaging: functional, diffusion tensor, and chemical shift imaging in a patient with localization-related epilepsy. Epilepsia 40:1459–1462

204. Lazeyras F, Blanke O, Perrig S, Zimine I, Golay X, Delavelle J, Michel CM et al (2000) EEG-triggered functional MRI in patients with pharmacoresistant epilepsy. J Magn Reson Imaging 12:177–185

205. Lazeyras F, Blanke O, Zimine I, Delavelle J, Perrig SH, Seeck M (2000) MRI, (1) H-MRS, and functional MRI during and after prolonged nonconvulsive seizure activity. Neurology 55:1677–1682

206. Jager L, Werhahn KJ, Hoffmann A, Berthold S, Scholz V, Weber J, Noachtar S et al (2002) Focal epileptiform activity in the brain: detection with spike-related functional MR imaging-preliminary results. Radiology 223:860–869

207. Benar CG, Gross DW, Wang Y, Petre V, Pike B, Dubeau F, Gotman J (2002) The BOLD response to interictal epileptiform discharges. NeuroImage 17:1182–1192

208. Benar CG, Grova C, Kobayashi E, Bagshaw AP, Aghakhani Y, Dubeau F, Gotman J (2006) EEG-fMRI of epileptic spikes: concordance with EEG source localization and intracranial EEG. NeuroImage 30:1161–1170

209. Aghakhani Y, Kobayashi E, Bagshaw AP, Hawco C, Benar CG, Dubeau F, Gotman J (2006) Cortical and thalamic fMRI responses in partial epilepsy with focal and bilateral synchronous spikes. Clin Neurophysiol 117:177–191

210. Zijlmans M, Huiskamp G, Hersevoort M, Seppenwoolde JH, van Huffelen AC, Leijten FS (2007) EEG-fMRI in the preoperative work-up for epilepsy surgery. Brain 130:2343–2353

211. Thornton R, Vulliemoz S, Rodionov R, Carmichael DW, Chaudhary UJ, Diehl B, Laufs H et al (2011) Epileptic networks

in focal cortical dysplasia revealed using electroencephalography-functional magnetic resonance imaging. Ann Neurol 70:822–837

212. Pittau F, Dubeau F, Gotman J (2012) Contribution of EEG/fMRI to the definition of the epileptic focus. Neurology 78:1479–1487

213. Chaudhary UJ, Carmichael DW, Rodionov R, Thornton RC, Bartlett P, Vulliemoz S, Micallef C et al (2012) Mapping preictal and ictal haemodynamic networks using video-electroencephalography and functional imaging. Brain 135:3645–3663

214. Coan AC, Chaudhary UJ, Grouiller F, Campos BM, Perani S, De Ciantis A, Vulliemoz S, et al. (2015) EEG-fMRI in the pre-surgical evaluation of temporal lobe epilepsy. J Neurol Neurosurg Psychiatry (in press)

215. Grouiller F, Thornton RC, Groening K, Spinelli L, Duncan JS, Schaller K, Siniatchkin M et al (2011) With or without spikes: localization of focal epileptic activity by simultaneous electroencephalography and functional magnetic resonance imaging. Brain 134:2867–2886

216. Tousseyn S, Dupont P, Goffin K, Sunaert S, Van Paesschen W (2015) Correspondence between large-scale ictal and interictal epileptic networks revealed by single photon emission computed tomography (SPECT) and electroencephalography (EEG)-functional magnetic resonance imaging (fMRI). Epilepsia 56:382–392

217. Salek-Haddadi A, Lemieux L, Merschhemke M, Friston KJ, Duncan JS, Fish DR (2003) Functional magnetic resonance imaging of human absence seizures. Ann Neurol 53:663–667

218. Baudewig J, Bittermann HJ, Paulus W, Frahm J (2001) Simultaneous EEG and functional MRI of epileptic activity: a case report. Clin Neurophysiol 112:1196–1200

219. Aghakhani Y, Bagshaw AP, Benar CG, Hawco C, Andermann F, Dubeau F, Gotman J (2004) fMRI activation during spike and wave discharges in idiopathic generalized epilepsy. Brain 127:1127–1144

220. Yang T, Luo C, Li Q, Guo Z, Liu L, Gong Q, Yao D et al (2013) Altered resting-state connectivity during interictal generalized spike-wave discharges in drug-naive childhood absence epilepsy. Hum Brain Mapp 34:1761–1767

221. Vaudano AE, Ruggieri A, Tondelli M, Avanzini P, Benuzzi F, Gessaroli G, Cantalupo G et al (2014) The visual system in eyelid myoclonia with absences. Ann Neurol 76:412–427

222. De Tiege X, Laufs H, Boyd SG, Harkness W, Allen PJ, Clark CA, Connelly A et al (2007) EEG-fMRI in children with pharmacoresistant focal epilepsy. Epilepsia 48:385–389

223. Jacobs J, Jacobs J, Boor R, Jansen O, Wolff S, Siniatchkin M, Stephani U (2007) Localization of epileptic foci in children with focal epilepsies using 3-Tesla simultaneous EEG-fMRI recordings. Clin Neurophysiol 118:e50–e51

224. Jacobs J, Kobayashi E, Boor R, Muhle H, Stephan W, Hawco C, Dubeau F et al (2007) Hemodynamic responses to interictal epileptiform discharges in children with symptomatic epilepsy. Epilepsia 48:2068–2078

225. Mandelkow H, Halder P, Boesiger P, Brandeis D (2006) Synchronization facilitates removal of MRI artefacts from concurrent EEG recordings and increases usable bandwidth. NeuroImage 32:1120–1126

226. Solana AB, Hernandez-Tamames JA, Manzanedo E, Garcia-Alvarez R, Zelaya FO, del Pozo F (2014) Gradient induced artifacts in simultaneous EEG-fMRI: Effect of synchronization on spiral and EPI k-space trajectories. Magn Reson Imaging 32:684–692

227. Negishi M, Abildgaard M, Nixon T, Constable RT (2004) Removal of time-varying gradient artifacts from EEG data acquired during continuous fMRI. Clin Neurophysiol 115:2181–2192

228. Wan X, Iwata K, Riera J, Kitamura M, Kawashima R (2006) Artifact reduction for simultaneous EEG/fMRI recording: adaptive FIR reduction of imaging artifacts. Clin Neurophysiol 117:681–692

229. Ritter P, Becker R, Graefe C, Villringer A (2007) Evaluating gradient artifact correction of EEG data acquired simultaneously with fMRI. Magn Reson Imaging 25:923–932

230. Masterton RA, Abbott DF, Fleming SW, Jackson GD (2007) Measurement and reduction of motion and ballistocardiogram artefacts from simultaneous EEG and fMRI recordings. NeuroImage 37:202–211

231. Abbott DF, Masterton RA, Archer JS, Fleming SW, Warren AE, Jackson GD (2014) Constructing carbon fiber motion-detection loops for simultaneous EEG-fMRI. Front Neurol 5:260

232. Friston KJ, Williams S, Howard R, Frackowiak RS, Turner R (1996) Movement-related effects in fMRI time-series. Magn Reson Med 35:346–355

233. Hajnal JV, Myers R, Oatridge A, Schwieso JE, Young IR, Bydder GM (1994) Artifacts due to stimulus correlated motion in functional imaging of the brain. Magn Reson Med 31:283–291

234. Lund TE, Norgaard MD, Rostrup E, Rowe JB, Paulson OB (2005) Motion or activity:

their role in intra- and inter-subject variation in fMRI. NeuroImage 26:960–964

235. Kobayashi E, Hawco CS, Grova C, Dubeau F, Gotman J (2006) Widespread and intense BOLD changes during brief focal electrographic seizures. Neurology 66:1049–1055

236. Baumgartner C, Serles W, Leutmezer F, Pataraia E, Aull S, Czech T, Pietrzyk U et al (1998) Preictal SPECT in temporal lobe epilepsy: regional cerebral blood flow is increased prior to electroencephalography-seizure onset. J Nucl Med 39:978–982

237. Federico P, Abbott DF, Briellmann RS, Harvey AS, Jackson GD (2005) Functional MRI of the pre-ictal state. Brain 128:1811–1817

238. Meletti S, Vaudano AE, Tassi L, Caruana F, Avanzini P (2015) Intracranial time-frequency correlates of seizure-related negative BOLD response in the sensory-motor network. Clin Neurophysiol 126:847–849

239. Archer JS, Briellman RS, Abbott DF, Syngeniotis A, Wellard RM, Jackson GD (2003) Benign epilepsy with centro-temporal spikes: spike triggered fMRI shows somatosensory cortex activity. Epilepsia 44:200–204

240. Archer JS, Briellmann RS, Syngeniotis A, Abbott DF, Jackson GD (2003) Spike-triggered fMRI in reading epilepsy: involvement of left frontal cortex working memory area. Neurology 60:415–421

241. Tousseyn S, Dupont P, Robben D, Goffin K, Sunaert S, Van Paesschen W (2014) A reliable and time-saving semiautomatic spike-template-based analysis of interictal EEG-fMRI. Epilepsia 55:2048–2058

242. Bagshaw AP, Hawco C, Benar CG, Kobayashi E, Aghakhani Y, Dubeau F, Pike GB et al (2005) Analysis of the EEG-fMRI response to prolonged bursts of interictal epileptiform activity. NeuroImage 24:1099–1112

243. Laufs H, Hamandi K, Salek-Haddadi A, Kleinschmidt AK, Duncan JS, Lemieux L (2007) Temporal lobe interictal epileptic discharges affect cerebral activity in "default mode" brain regions. Hum Brain Mapp 28:1023–1032

244. Diehl B, Salek-haddadi A, Fish DR, Lemieux L (2003) Mapping of spikes, slow waves, and motor tasks in a patient with malformation of cortical development using simultaneous EEG and fMRI. Magn Reson Imaging 21:1167–1173

245. Laufs H, Hamandi K, Walker MC, Scott C, Smith S, Duncan JS, Lemieux L (2006) EEG-fMRI mapping of asymmetrical delta activity in a patient with refractory epilepsy is concor-

dant with the epileptogenic region determined by intracranial EEG. Magn Reson Imaging 24:367–371

246. Grouiller F, Thornton R, Groening K, Spinelli L, Duncan JS, Schaller K, Siniatchkin M et al (2011) Localization of focal epileptic activity with EEG-FMRI informed by EEG voltage maps. Epilepsia 52:169–169

247. Morgan VL, Price RR, Arain A, Modur P, Abou-Khalil B (2004) Resting functional MRI with temporal clustering analysis for localization of epileptic activity without EEG. NeuroImage 21:473–481

248. Hamandi K, Salek Haddadi A, Liston A, Laufs H, Fish DR, Lemieux L (2005) fMRI temporal clustering analysis in patients with frequent interictal epileptiform discharges: comparison with EEG-driven analysis. NeuroImage 26:309–316

249. Aguirre GK, Zarahn E, D'Esposito M (1998) The variability of human, BOLD hemodynamic responses. NeuroImage 8:360–369

250. Salek-Haddadi A, Friston KJ, Lemieux L, Fish DR (2003) Studying spontaneous EEG activity with fMRI. Brain Res Brain Res Rev 43:110–133

251. Kobayashi E, Bagshaw AP, Grova C, Dubeau F, Gotman J (2006) Negative BOLD responses to epileptic spikes. Hum Brain Mapp 27:488–497

252. Lemieux L, Laufs H, Carmichael D, Paul JS, Walker MC, Duncan JS (2008) Noncanonical spike-related BOLD responses in focal epilepsy. Hum Brain Mapp 29:329–345

253. Lu Y, Bagshaw AP, Grova C, Kobayashi E, Dubeau F, Gotman J (2006) Using voxel-specific hemodynamic response function in EEG-fMRI data analysis. NeuroImage 32:238–247

254. Watanabe S, An D, Safi-Harb M, Dubeau F, Gotman J (2014) Hemodynamic response function (HRF) in epilepsy patients with hippocampal sclerosis and focal cortical dysplasia. Brain Topogr 27:613–619

255. Hawco CS, Bagshaw AP, Lu Y, Dubeau F, Gotman J (2007) BOLD changes occur prior to epileptic spikes seen on scalp EEG. NeuroImage 35:1450–1458

256. Makiranta M, Ruohonen J, Suominen K, Niinimaki J, Sonkajarvi E, Kiviniemi V, Seppanen T et al (2005) BOLD signal increase preceeds EEG spike activity--a dynamic penicillin induced focal epilepsy in deep anesthesia. NeuroImage 27:715–724

257. Siniatchkin M, Moeller F, Jacobs J, Stephani U, Boor R, Wolff S, Jansen O et al (2007) Spatial filters and automated spike detection

based on brain topographies improve sensitivity of EEG-fMRI studies in focal epilepsy. NeuroImage 37:834–843

258. Rosenow F, Luders H (2001) Presurgical evaluation of epilepsy. Brain 124:1683–1700

259. Kobayashi E, Bagshaw AP, Benar CG, Aghakhani Y, Andermann F, Dubeau F, Gotman J (2006) Temporal and extratemporal BOLD responses to temporal lobe interictal spikes. Epilepsia 47:343–354

260. Lemieux L, Krakow K, Fish DR (2001) Comparison of spike-triggered functional MRI BOLD activation and EEG dipole model localization. NeuroImage 14:1097–1104

261. Bagshaw AP, Kobayashi E, Dubeau F, Pike GB, Gotman J (2006) Correspondence between EEG-fMRI and EEG dipole localisation of interictal discharges in focal epilepsy. NeuroImage 30:417–425

262. Boor R, Jacobs J, Hinzmann A, Bauermann T, Scherg M, Boor S, Vucurevic G et al (2007) Combined spike-related functional MRI and multiple source analysis in the non-invasive spike localization of benign rolandic epilepsy. Clin Neurophysiol 118:901–909

263. Salek-Haddadi A, Lemieux L, Fish DR (2002) Role of functional magnetic resonance imaging in the evaluation of patients with malformations caused by cortical development. Neurosurg Clin N Am 13:63–69, viii

264. Federico P, Archer JS, Abbott DF, Jackson GD (2005) Cortical/subcortical BOLD changes associated with epileptic discharges: an EEG-fMRI study at 3 T. Neurology 64:1125–1130

265. Kobayashi E, Bagshaw AP, Jansen A, Andermann F, Andermann E, Gotman J, Dubeau F (2005) Intrinsic epileptogenicity in polymicrogyric cortex suggested by EEG-fMRI BOLD responses. Neurology 64:1263–1266

266. Kobayashi E, Bagshaw AP, Grova C, Gotman J, Dubeau F (2006) Grey matter heterotopia: what EEG-fMRI can tell us about epileptogenicity of neuronal migration disorders. Brain 129:366–374

267. Kobayashi E, Bagshaw AP, Gotman J, Dubeau F (2007) Metabolic correlates of epileptic spikes in cerebral cavernous angiomas. Epilepsy Res 73:98–103

268. Wiebe S, Blume WT, Girvin JP, Eliasziw M (2001) Effectiveness and G efficiency of surgery for temporal lobe epilepsy study. A randomized, controlled trial of surgery for temporal-lobe epilepsy. N Engl J Med 345:311–318

269. Raichle ME, MacLeod AM, Snyder AZ, Powers WJ, Gusnard DA, Shulman GL (2001) A default mode of brain function. Proc Natl Acad Sci U S A 98:676–682

270. Lengler U, Kafadar I, Neubauer BA, Krakow K (2007) fMRI correlates of interictal epileptic activity in patients with idiopathic benign focal epilepsy of childhood. A simultaneous EEG-functional MRI study. Epilepsy Res 75:29–38

271. Engel J Jr, International League Against E (2001) A proposed diagnostic scheme for people with epileptic seizures and with epilepsy: report of the ILAE Task Force on Classification and Terminology. Epilepsia 42:796–803

272. Archer JS, Abbott DF, Waites AB, Jackson GD (2003) fMRI "deactivation" of the posterior cingulate during generalized spike and wave. NeuroImage 20:1915–1922

273. Gotman J, Grova C, Bagshaw A, Kobayashi E, Aghakhani Y, Dubeau F (2005) Generalized epileptic discharges show thalamocortical activation and suspension of the default state of the brain. Proc Natl Acad Sci U S A 102:15236–15240

274. Laufs H, Lengler U, Hamandi K, Kleinschmidt A, Krakow K (2006) Linking generalized spike-and-wave discharges and resting state brain activity by using EEG/fMRI in a patient with absence seizures. Epilepsia 47:444–448

275. Meeren HK, Pijn JP, Van Luijtelaar EL, Coenen AM, Lopes da Silva FH (2002) Cortical focus drives widespread corticothalamic networks during spontaneous absence seizures in rats. J Neurosci 22:1480–1495

276. Steriade M, Dossi RC, Nunez A (1991) Network modulation of a slow intrinsic oscillation of cat thalamocortical neurons implicated in sleep delta waves: cortically induced synchronization and brainstem cholinergic suppression. J Neurosci 11:3200–3217

277. Hamandi K, Laufs H, Noth U, Carmichael DW, Duncan JS, Lemieux L (2008) BOLD and perfusion changes during epileptic generalised spike wave activity. NeuroImage 39:608–618

278. Moeller F, Siebner HR, Wolff S, Muhle H, Granert O, Jansen O, Stephani U et al (2008) Simultaneous EEG-fMRI in drug-naive children with newly diagnosed absence epilepsy. Epilepsia 49:1510–1519

279. Moeller F, Maneshi M, Pittau F, Gholipour T, Bellec P, Dubeau F, Grova C et al (2011) Functional connectivity in patients with idiopathic generalized epilepsy. Epilepsia 52:515–522

280. Vaudano AE, Laufs H, Kiebel SJ, Carmichael DW, Hamandi K, Guye M, Thornton R et al (2009) Causal hierarchy within the thalamo-cortical network in spike and wave discharges. PLoS One 4, e6475

281. Moeller F, LeVan P, Muhle H, Stephani U, Dubeau F, Siniatchkin M, Gotman J (2010) Absence seizures: individual patterns revealed by EEG-fMRI. Epilepsia 51:2000–2010

282. Moeller F, Muthuraman M, Stephani U, Deuschl G, Raethjen J, Siniatchkin M (2013) Representation and propagation of epileptic activity in absences and generalized photoparoxysmal responses. Hum Brain Mapp 34(8): 1896–1909

283. Shmuel A, Augath M, Oeltermann A, Logothetis NK (2006) Negative functional MRI response correlates with decreases in neuronal activity in monkey visual area V1. Nat Neurosci 9:569–577

284. Binnie CD (2003) Cognitive impairment during epileptiform discharges: is it ever justifiable to treat the EEG? Lancet Neurology 2:725–730

285. Todd N, Josephs O, Callaghan MF, Lutti A, Weiskopf N (2015) Prospective motion correction of 3D echo-planar imaging data for functional MRI using optical tracking. NeuroImage 113:1–12

286. Rodionov R, De Martino F, Laufs H, Carmichael DW, Formisano E, Walker M, Duncan JS et al (2007) Independent component analysis of interictal fMRI in focal epilepsy: comparison with general linear model-based EEG-correlated fMRI. NeuroImage 38:488–500

287. Merlet I, Gotman J (2001) Dipole modeling of scalp electroencephalogram epileptic discharges: correlation with intracerebral fields. Clin Neurophysiol 112:414–430

288. Liston AD, Salek-Haddadi A, Kiebel SJ, Hamandi K, Turner R, Lemieux L (2004) The MR detection of neuronal depolarization during 3-Hz spike-and-wave complexes in generalized epilepsy. Magn Reson Imaging 22:1441–1444

289. Daunizeau J, Grova C, Marrelec G, Mattout J, Jbabdi S, Pelegrini-Issac M, Lina JM et al (2007) Symmetrical event-related EEG/fMRI information fusion in a variational Bayesian framework. NeuroImage 36:69–87

290. Jezzard P, Clare S (1999) Sources of distortion in functional MRI data. Hum Brain Mapp 8:80–85

291. Bagshaw AP, Torab L, Kobayashi E, Hawco C, Dubeau F, Pike GB, Gotman J (2006) EEG-fMRI using z-shimming in patients with temporal lobe epilepsy. J Magn Reson Imaging 24:1025–1032

292. Jezzard P, Balaban RS (1995) Correction for geometric distortion in echo planar images from B0 field variations. Magn Reson Med 34:65–73

293. Hutton C, Bork A, Josephs O, Deichmann R, Ashburner J, Turner R (2002) Image distortion correction in fMRI: A quantitative evaluation. NeuroImage 16:217–240

294. Niendorf T (1999) On the application of susceptibility-weighted ultra-fast low-angle RARE experiments in functional MR imaging. Magn Reson Med 41:1189–1198

295. Deichmann R, Josephs O, Hutton C, Corfield DR, Turner R (2002) Compensation of susceptibility-induced BOLD sensitivity losses in echo-planar fMRI imaging. NeuroImage 15:120–135

296. Deichmann R, Gottfried JA, Hutton C, Turner R (2003) Optimized EPI for fMRI studies of the orbitofrontal cortex. NeuroImage 19:430–441

297. Glover GH, Li TQ, Ress D (2000) Image-based method for retrospective correction of physiological motion effects in fMRI: RETROICOR. Magn Reson Med 44:162–167

Chapter 25

fMRI in Neurosurgery

Oliver Ganslandt, Christopher Nimsky, Michael Buchfelder, and Peter Grummich

Abstract

Functional magnetic resonance imaging has evolved from a basic research application to a useful clinical tool that also has found its place in modern neurosurgery. The localization of functional important brain areas as language and sensorimotor cortex has been the focus of numerous investigations and can now be implemented in neurosurgical planning. Since the neurosurgeon must have detailed knowledge about the individual anatomy and related neurological function to resect a brain tumor with the highest safety, the need for individualized maps of brain function is essential. Advanced fMRI techniques and modern imaging methods contribute significantly to brain mapping as do already established concepts of electrophysiological monitoring and the Wada test. The implementation of functional maps into neuronavigation systems enables the surgeon to superimpose anatomy and function to the surgical site. This chapter describes our experience with the use of fMRI in neurosurgery.

Key words fMRI, Neurosurgery, Functional neuronavigation, Magnetoencephalography, Language, Somatosensory cortex

1 Introduction

The concept of using information about functionally important brain areas (also known as "eloquent cortex") to safely guide neurosurgical procedures has been established in the middle of the twentieth century by use of electrical stimulation in awake craniotomies. Based on the seminal work of Penfield [1], modern neurosurgeons used the technique of electrical stimulation to meticulously map the cerebral cortex of their patients, for instance, to delineate the borders of resection in epilepsy and tumor surgery. Today, electrical stimulation is still regarded as the "gold standard" for neurosurgical functional brain localization [2, 3]. However, these invasive direct cortical stimulation methods are not available for preoperative decision making and surgical planning. They are also time consuming and demand special resources. Therefore, in the past decade new efforts have been made to overcome the

Massimo Filippi (ed.), *fMRI Techniques and Protocols*, Neuromethods, vol. 119,
DOI 10.1007/978-1-4939-5611-1_25, © Springer Science+Business Media New York 2016

limitations of using electrical stimulation in awake craniotomy by using new techniques of brain imaging and the implementation of these data into the neurosurgical workspace. In recent years, two noninvasive techniques have been found especially suitable for presurgical localization of the eloquent cortex: magnetoencephalography (MEG) and fMRI. Studies using these techniques successfully localized functional activity [4, 5].

One of the most interesting applications was the merge of functional brain imaging with frame-based and frameless stereotaxy, also known as functional neuronavigation [6, 7]. There is strong evidence that the use of functional neuronavigation for lesions adjacent to eloquent brain areas may favor clinical outcome [8–10], but large controlled studies to support this assumption are still needed.

If surgery near eloquent brain areas is planned, a detailed knowledge about the topographic relation of a lesion to the adjacent functional brain area is crucial to avoid postoperative neurological deficits. In neurosurgery, the primary sensorimotor cortex and the cortical areas subserving language comprehension and production are considered to be the main structures of risk. In temporal lobe surgery memory function is also an important function to preserve. These structures usually cannot be depicted from conventional structural imaging techniques. Other reasons that warrant a detailed evaluation are the individual representation of these eloquent areas and the phenomenon of cortical reorganization of these areas from their original positions [11, 12]. Furthermore, normal sulcal anatomy is not often discernible because of a space occupying lesion. These situations require methods for localizing functional areas prior to surgery for decision making, planning, and avoiding crippling postoperative results.

fMRI has become indispensable in neurosurgery to easily gain knowledge about the topographic relation of a given lesion to the functional brain area at risk and thus to plan the surgical approach. Furthermore, fMRI-derived information about the extent of cortical involvement in function can be used in conjunction with image-guided surgery during resection of lesions adjacent to eloquent brain areas under general anesthesia for navigation. The almost ubiquitous availability of modern MR scanners favors the use of fMRI over other modalities as MEG or positron emission tomography (PET) that demand resources not commonly available. In addition, its noninvasiveness gives the opportunity to repeat the examinations and conduct follow-up studies on reorganization of cortical function. Advances in MRI technology, such as the introduction of higher field strengths, will undoubtedly improve signal acquisition and processing [13]. Over the last years, a substantial number of publications have described the usefulness of clinical fMRI for neurosurgical applications [14, 15]. The use of fMRI for the presurgical localization of the sensorimotor cortex is now widely appreciated and has been investigated by several groups

[14, 16, 17], which also performed comparisons with direct motor stimulation. Language fMRI has been found to be an alternative to the invasive Wada test [18–20] for language lateralization. Furthermore, fMRI has been used to predict memory localization [21]. Concerning the reliability of fMRI-localization of speech areas in the frontotemporal cortex, as compared with direct electrical stimulation, the neurosurgical community is still reluctant to rely on fMRI language alone, since inconsistent agreement has been found between activation sites by fMRI naming and verb-generation tasks and cortical stimulation [5]. As language fMRI and intraoperative electrophysiological monitoring use different physiological mechanisms the results of language fMRI are not per se to be considered wrong. The ongoing clinical use of language fMRI in our department has shown that this modality can be used with the same results as awake craniotomy.

2 Methods

In our department, all neurosurgical fMRI measurements are acquired on a 1.5 T MR scanner by echo-planar imaging (Magnetom Sonata, Siemens, Erlangen, Germany).

Measurements for localization of motor and sensory activity are performed with 16 slices of 3 mm thickness, a TR = 1580, and a TE = 60. Stimulation is done in a block paradigm with 120 stimulus presentations in six blocks. Twenty measurements during rest alternated with 20 measurements during stimulation. During the motor activation blocks, the patient is asked to perform a motor task: in particular, we are interested in localizing the cortical representation of the toes, foot, leg, fingers, hand, arm, tongue, lips, and eye lid.

Our selection of the motor tasks for each patient depends on the tumor location. Attention is paid that the patient does not move the opposite limb. With this, the possibility to detect reorganization of functional areas to the contralateral hemisphere is ensured. Each patient is also instructed not to touch anything during movement. For this reason, the motor task is usually not a finger-tapping task. Only in cases where we are interested in localizing the supplementary motor area we conduct a finger-tapping task.

For the localization of the sensory cortex, we use a tactile stimulation of different limbs whose cortical representations we wish to localize.

Measurements for language are done with 25 slices of 3 mm thickness, a TR = 2470, and a TE = 60. Stimulation is done in a block paradigm with 180 measurements in six blocks. We perform 30 measurements in an activation condition during which the patient is instructed to perform a language task; we alternate these with 30 measurements in a resting condition. We also developed a battery of several paradigms for mapping of memory function.

Two tests for factual knowledge consist of a recall of capitals of given countries. Factual knowledge can also be tested by recall of celebrities. Memorizing and recall of four digits is another test of hippocampal memory function (Fig. 1). In other tests the patients are asked to connect telephone numbers to persons. By means of a mirror that is attached to the head coil, the patient is able to observe words, numbers, or pictures projected onto a screen.

We developed several stimulation paradigms for localizing the Broca's and the Wernicke's areas. Usually, each patient is asked to undergo four different paradigms during fMRI measurements. The paradigms are selected according to tumor location and are adapted to the abilities of each patient. The length of the interstimulus interval is also adjusted according to the patient's abilities, varying between 900 and 3000 ms. The duration of the stimulus presentation lies between 600 and 1700 ms (300 ms less than the interstimulus interval). We ask the patients to perform the tasks as quickly as possible immediately after stimulus presentation and to perform the task silently to avoid artifacts from mouth movement.

Paradigms are chosen on the basis of: (a) tumor location and (b) patient cognitive abilities:

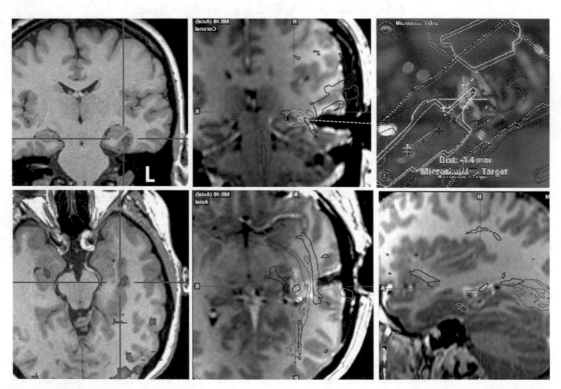

Fig. 1 fMRI-guided epilepsy surgery for ganglioglioma in the left hippocampus. We localized activity for factual knowledge in the parahippocampal gyrus (*crosshair*). The segmented area is depicted in *green* in the *right image* showing the view through the navigation microscope. Also shown in *purple* are the fiber tracts of the visual pathway and occipitofrontal fasciculus (*middle*)

(a) In case of tumor location in the inferior parietal area close to the intraparietal sulcus, we use an arithmetic task so that besides Wernicke's area (activated by reading, adding numbers, and formulating the result) the cortex for calculation in the intraparietal sulcus is also activated and so can be spared during surgery.

In case of tumor location close to Broca's area, we select language tasks that are also expressive and demand grammatical abilities, because these may increase activity in Broca's area. This happens during the verb-generation task, but also the verb conjugation task gives suitable Broca's area activations.

(b) For patients with reduced abilities, simple tasks are selected to obtain reliable results. Especially in patients who suffer from word finding disorders, we avoid the picture-naming task. For patients with better cognitive performance, we select one or more complex tasks such as verb-generation task, because these are reported to show a more clear lateralization, whereas in patients who have difficulties in this complex task the activation is usually worse than with a simple paradigm.

For motion correction, we apply an image-based prospective acquisition correction by applying interpolation in the k-space [22]. We produce activation maps by analyzing the correlation between signal intensity and a square wave reference function for each pixel according to the paradigm. Pixels exceeding a significance threshold (typical correlations above a threshold of 0.3 with $p < 0.000045$) are displayed, if at least six contiguous voxels constitute a cluster, to eliminate isolated voxels. We align the functional slices to magnetization prepared rapid acquisition gradient echo (MPRAGE) images (160 slices of 1 mm slice thickness).

3   Results

3.1   Localizations

Since 2002, we have investigated preoperative fMRI with motor or sensory stimulation in 515 cases. Of these patients, 205 underwent tumor resection and 75 had stereotactic brain biopsy. In five additional patients, invasive electrodes were implanted by fMRI guidance for chronic recording of epileptic discharges.

For language and memory testing, we examined 623 cases and used additional information from MEG studies. Of them, 465 underwent tumor resection and 53 had stereotactic brain biopsy. The remaining patients either obtained radiation therapy, endovascular treatment, or were just enrolled in a "wait-and-see" protocol.

It was possible to localize the primary motor and sensory cortex as well as the supplementary motor area (SMA) in all examined cases (Figs. 2 and 3). Only in one case, the motor activity of the toe was not detectable by fMRI because of tumor infiltration. However, in this patient it was possible to obtain motor activation from nearby muscle representations of the motor homunculus.

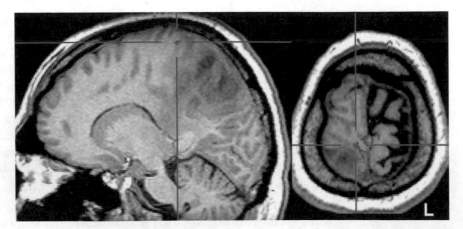

Fig. 2 fMRI activations during movement of left foot in a patient with an oligoastrocytoma (WHO III) in the right parietal lobe. In front of the activation of the motor cortex (posterior wall of precentral gyrus), activation of the supplementary motor area (SMA) is also evident

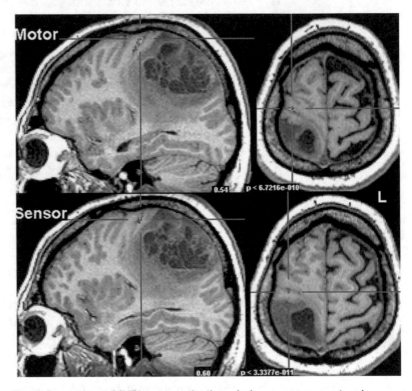

Fig. 3 Comparison of fMRI motor activations during arm movement and sensory stimulation of the arm (oligoastrocytoma WHO III, same patient as Fig. 2)

We were able to define the motor homunculus along the central sulcus in the posterior wall of the precentral gyrus with fMRI by motor activation of the respective muscle groups (Fig. 4). We localized toe, foot, leg, arm, hand, finger, thumb, tongue, lip, and eye movements. Additionally, sometimes we found activity in the

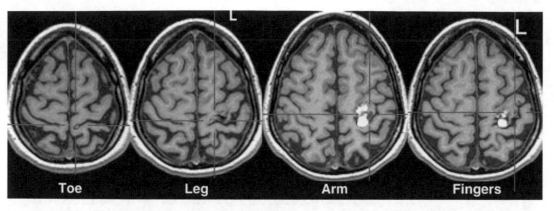

Fig. 4 fMRI activations during movement of toe, leg, arm, and fingers (note that the lesion, a cavernoma in the left motor cortex, is located between the cortical representation of the arm and that of the leg in the precentral gyrus)

ipsilateral homotopic cortex. Especially when the motor task was more complex there was also activity at the frontal wall of the precentral gyrus. Additional activity can be detected in the SMA in the interhemispheric sulcus.

Activity at the posterior wall of the postcentral gyrus (Brodman area 2) can be found and sometimes in the gyrus posteriorly, which is likely to represent proprioceptive activation due to the positions of the limbs.

Sensory activity was seen in the anterior wall (Brodman area 3) of the postcentral gyrus. Sometimes there was also blood oxygen level-dependent (BOLD) activation in the homotopic cortex of the ipsilateral hemisphere that may indicate the presence of mechanisms of cortical plasticity.

With verbal stimulation tasks, we were able to localize language, calculation, and memory activity. In the frontal lobe, we found activity in the following cortical areas: (1) at the bottom of the opercular part of the inferior frontal gyrus anteriorly to the precentral gyrus (Brodmann area 44, classical Broca's area); (2) in the adjacent part of the frontal cranial edge of the insular cortex; and (3) close to the upper end of the inferior frontal gyrus, which sometimes extends into the medial frontal gyrus. Here, there are usually three cortical areas that extend from the pars triangularis to pars opercularis (from anterior to posterior).

In the temporo-parietal region, we found language activity in the superior temporal gyrus at its bottom in the temporal sulcus and at its lateral side. Activity was also found at the top of the superior temporal gyrus in the planum temporale. Additional activity was found in the supramarginal gyrus in the frontal part of the intraparietal sulcus.

In 53% of our patients with high-grade glioma, language activity was not clearly detectable by fMRI alone, because of changes in vascular function (see below).

3.2 Laterality

Although it is generally accepted that the majority of people has left hemispheric language dominance, the true number of atypical (right) dominance is unknown. Studies using the Wada test showed an incidence of left hemispheric dominance in right handers in a range of 63–96% and a right hemispheric dominance for left handers and ambidextrous patients in 48–75% [23]. Furthermore, it is thought that there are varying degrees of language dominance in the population.

In certain circumstances, activity can be located in both hemispheres or reorganization to the other hemisphere could have been occurred. FMRI is a useful method to clarify this. If activity is only found in one hemisphere or the activity on one side is much stronger than the activity on the other side, then it is clear that the active area has to be spared during surgery.

It is important to know that certain stimulation tasks and modalities show more lateralized activations than others. In case of complex motor tasks, the ipsilateral hemisphere may also show activation.

For the localization of language activity, we found that visual stimulation shows a more accentuated lateralization than acoustic stimulation [24]. Stimulation with words, especially in a complex task, shows a stronger lateralization than a picture-naming task, a finding that was also described by Herholz et al. [25, 26]. In rare cases, it can occur that not all language areas are located on the same side.

3.3 Surgery

We perform fMRI-guided surgery by coregistering the activation maps onto a 3D MRI data set that can be used with a navigation system. Targets and areas at risk determined by fMRI are segmented and made visible for the surgeon through a navigation microscope (Figs. 5, 6, and 7). Thus functional data are visualized

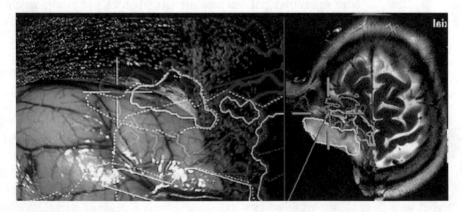

Fig. 5 Microscopic view with neuronavigation markers showing the sensory activation of the arm area in *light blue* and the pyramidal tract in *purple* (oligoastrocytoma WHO III, same patient as Fig. 2)

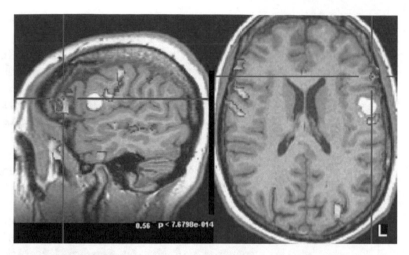

Fig. 6 fMRI activation of Broca's and Wernicke's areas and primary motor cortex after a reading paradigm. The functional mapping was requested to plan surgery of a cavernoma, which was located between Broca's area and the motor cortex

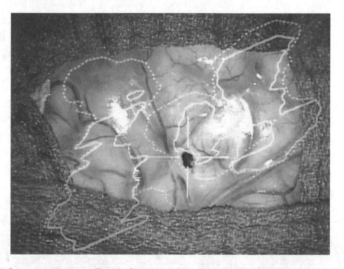

Fig. 7 Same patient as Fig. 5. Segmentation lines indicating Broca's area (*left*) and motor cortex of tongue (*right*). The figure shows the beginning of the corti-cotomy on a trajectory that spared the eloquent cortices (*cross*). Postoperatively the patient was neurologically intact

in the operation field throughout the whole surgery. Neuronavigation support is provided by the VectorVision Sky Navigation System (BrainLab AG, Heimstetten, Germany). A fiber optic connection ensures MR-compatible integration into the radiofrequency-shielded room of our intraoperative MR suite. A ceiling-mounted camera is used to monitor the positions of the operating microscope (Pentero, Zeiss, Oberkochen, Germany), which is placed outside the 5 G line, and other instruments.

A 1.0 mm isotropic 3D MPRAGE dataset (TE: 4.38 ms, TR: 2020 ms, slice thickness: 1.0 mm, FOV: 250×250 mm, measurement time: 8 min 39 s) is acquired prior to surgery with the head already fixed in the MR-compatible headholder as navigational reference dataset. For registration, five adhesive skin fiducial markers are placed in a scattered pattern on the head surface prior to imaging and registered with a pointer after their position is defined in the 3D dataset (Fig. 6). Functional data from MEG and fMRI, which were acquired preoperatively, are integrated into the 3D dataset. Furthermore, data from diffusion tensor imaging (DTI) depicting the course of major white matter tracts are integrated as well as in selected cases metabolic maps from proton magnetic resonance spectroscopy (^{1}H-MRS) are coregistered to the navigational dataset. In addition, further standard anatomical datasets, such as T2-weighted images, are coregistered. Repeated landmark checks are performed to ensure overall accuracy. In case intraoperative imaging depicts some remaining tumor, which should be further removed, intraoperative image data are used for updating the navigation system (Fig. 8). After a rigid registration of pre- and intraoperative images (ImageFusion software, BrainLAB, Heimstetten, Germany), all data are transferred to the navigation system and then the initial patient registration file is restored, so that no repeated patient registration procedure is needed.

In our series with a surgical resection close to the motor cortex, only 15 out of 205 patients (7.3%) had postoperative neurologic dysfunction. Three of them recovered within 2 days. In the other cases, the condition improved over several months. One patient, who had a hemiplegia prior to surgery, was able to move the affected side after surgery.

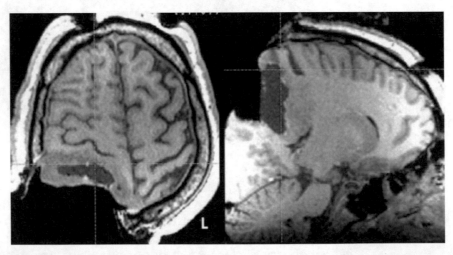

Fig. 8 Intraoperative MRI showing the outcome of the fMRI-guided tumor resection. Note that the tumor was removed sparing the sensory cortex (oligoastrocytoma WHO III *right*, same patient as Fig. 2)

No permanent postoperative deterioration of speech was observed in our patients with surgery close to the language areas (N=465 patients). However, in 53 patients, in whom surgery was conducted very close to the language areas, a transitory deterioration was observed (11% of all patients with surgery neighboring language areas). Thirty-two patients were not able to name part of the shown objects; this impairment lasted from 1 day to few weeks, but they all resolved completely.

No patient had suffered from global aphasia after surgery. The result of having no permanent speech disorder in our patients indicates that our language mapping is reliable. The presence of patients with transitory disturbances suggests that resection was conducted close to the boundaries of functional areas. This is in accordance to other series that evaluated the outcome of glioma surgery in eloquent areas with direct cortical stimulation. In a recent publication by Duffau et al., the rate of severe neurological deficits was 6.5% [27].

The safety margin that should be kept to preserve the functional areas depends from several factors: (1) the kind of functional center and the situation of reorganization; (2) the situation of blood supply; and (3) the status of the connectivity fibers. At present, no recommendations can be given for the exact distance to avoid the risk of neurologic deficits. Neurosurgeons who use fMRI-guided neuronavigation have to keep in mind that the fMRI-activation does not represent the actual extent of the functional brain areas, but rather a "center of gravity" of the functional units that are measured. Also one has to take into account that descending pathways (e.g., the pyramidal tract) have also to be spared. A recent study that investigated the accuracy between the actual location of the pyramidal tract and subcortical electric stimulation with stereotactic navigation found a mean difference in distance of 8.7 ± 3.1 mm (standard deviation) [28]. Nevertheless, there are functional areas that can be compensated for, if destroyed. These are the SMA and the area in the fusiform gyrus for word recognition.

For the language areas, Haglund et al. [29] described in an electrical stimulation study that above a resection distance of 10 mm from the eloquent areas they observed no permanent language deficits. When surgery was 7–10 mm close to the language area, they found 43% patients who suffered from permanent language deficits (severe or mild aphasia). The 9% of patients had no language deficit at all and the remaining 48% experienced transitory language deficits, which resolved within 4 weeks [29]. Two of our patients showed an amelioration of language function after surgery. One patient, who was not able to talk before surgery, was able to talk afterwards. Another patient, who had severe naming problems, showed an improvement after surgery.

3.4   Problems with the BOLD Effect

Sometimes the BOLD activations are not clearly visible in spite of the fact that the function is there, as confirmed by MEG measurements. In our experience, such a discrepancy between MEG and fMRI occurred only in the case of large tumors. Previous reports indicated similar effects of vascular conditions on the BOLD effect [30–32]. A reduction of the BOLD effect in the vicinity of a glioma but not in the vicinity of nonglial tumors was described by Schreiber et al. [33]. These findings are in agreement with our results. In our series, we found that in 53 % of the patients with high-grade gliomas the fMRI maps did not give clear indications of language areas in their vicinity.

Because of the impact of gliomas on the BOLD effect, the dominant hemisphere sometimes is more easily found by MEG measurements. This is seen for a patient with an astrocytoma (WHO Grade II) in Fig. 9. Here MEG localizations of Wernicke's and Broca's area were only on the right side. This was in accordance with the Wada test that showed right hemispheric language dominance in this left-handed patient. In this patient, fMRI localizations of Wernicke's activity were similar on both sides, in MEG

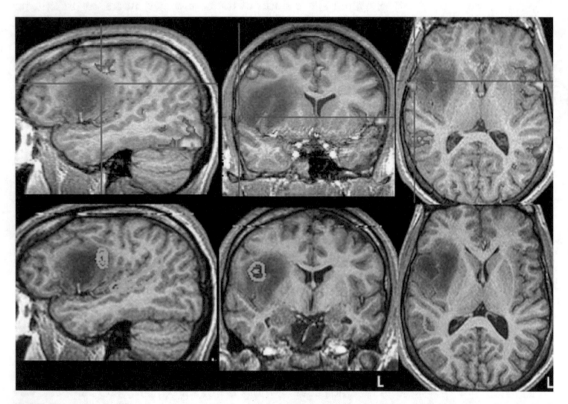

Fig. 9 Wernicke and Broca activity during reading of fragmentary sentences with mistakes. Comparison between fMRI (*orange*) and MEG beamformer localizations at 500 ms (*light blue*). With fMRI a bilateral activation in the operculum frontale and in the superior temporal sulcus can be seen. With MEG, activity is only seen in the right hemisphere. In the *first* and *second image* in the *lower row*, activity of the insula can be seen with MEG only. Left-handed patient with astrocytoma WHO II

they were only found in the right hemisphere. The activity detected by MEG in the right insula was not found by fMRI.

Other reasons that might lead to suboptimal fMRI results are continuous brain activation during rest or a very short activation of brain areas. This may be the reason why memory activity in the hippocampus is difficult to find by fMRI.

4   Conclusions

The use of preoperative fMRI brain mapping provides important information for: (1) estimating the risk of a surgical procedure; (2) planning the surgical approach; (3) indicating hemispheric dominance; and (4) revealing whether reorganization of brain function took place and at what degree. The integration of the functional markers into the navigation system is a good tool to continuously track the locations of the functional areas during surgery and enables a resection close to the eloquent areas to be performed. Thus, fMRI-guided functional navigation increases the amount of radical surgery and decreases morbidity. When using fMRI in neurosurgery, it is important to know that, in certain circumstances, the BOLD effect can be suppressed, which may lead to wrong conclusions. Beside integration of fMRI data the additional use of fiber tracking of the descending pathways as well as other paraclinical investigations (PET, proton magnetic resonance spectroscopy, MEG, etc.) should lead to a comprehensive understanding of the options and limitations of glioma surgery adjacent to important functional brain areas.

References

1. Penfield W, Rasmussen T (1950) The cerebral cortex of man. A clinical study of localization of function. Macmillan, New York

2. Berger MS, Rostomily RC (1997) Low grade gliomas: functional mapping resection strategies, extent of resection, and outcome. J Neurooncol 34:85–101

3. Duffau H, Capelle L, Denvil D, Sichez N, Gatignol P et al (2003) Usefulness of intraoperative electrical subcortical mapping during surgery for low-grade gliomas located within eloquent brain regions: functional results in a consecutive series of 103 patients. J Neurosurg 98:764–778

4. Kober H, Moller M, Nimsky C, Vieth J, Fahlbusch R et al (2001) New approach to localize speech relevant brain areas and hemispheric dominance using spatially filtered magnetoencephalography. Hum Brain Mapp 14:236–250

5. Roux FE, Boulanouar K, Lotterie JA, Mejdoubi M, LeSage JP et al (2003) Language functional magnetic resonance imaging in preoperative assessment of language areas: correlation with direct cortical stimulation. Neurosurgery 52:1335–1345, discussion 1345–1337

6. Nimsky C, Ganslandt O, Kober H, Moller M, Ulmer S et al (1999) Integration of functional magnetic resonance imaging supported by magnetoencephalography in functional neuronavigation. Neurosurgery 44:1249–1255, discussion 1255–1246

7. Rutten GJ, Ramsey N, Noordmans HJ, Willems P, van Rijen P et al (2003) Toward functional neuronavigation: implementation of functional magnetic resonance imaging data in a surgical guidance system for intraoperative identification of motor and language cortices. Technical note and illustrative case. Neurosurg Focus 15:E6

8. Rossler K, Sommer B, Grummich P, Hamer HM, Pauli E et al (2015) Risk reduction in dominant temporal lobe epilepsy surgery combining fMRI/DTI maps, neuronavigation and intraoperative 1.5-Tesla MRI. Stereotact Funct Neurosurg 93:168–177

9. Zhang J, Chen X, Zhao Y, Wang F, Li F et al (2015) Impact of intraoperative magnetic resonance imaging and functional neuronavigation on surgical outcome in patients with gliomas involving language areas. Neurosurg Rev 38:319–330, discussion 330

10. Sun GC, Chen XL, Yu XG, Zhang M, Liu G et al (2015) Functional neuronavigation-guided transparieto-occipital cortical resection of meningiomas in trigone of lateral ventricle. World Neurosurg 84(3):756–765

11. Duffau H, Denvil D, Capelle L (2002) Long term reshaping of language, sensory, and motor maps after glioma resection: a new parameter to integrate in the surgical strategy. J Neurol Neurosurg Psychiatry 72:511–516

12. Grummich P, Nimsky C, Fahlbusch R, Ganslandt O (2005) Observation of unaveraged giant MEG activity from language areas during speech tasks in patients harboring brain lesions very close to essential language areas: expression of brain plasticity in language processing networks? Neurosci Lett 380:143–148

13. Tieleman A, Vandemaele P, Seurinck R, Deblaere K, Achten E (2007) Comparison between functional magnetic resonance imaging at 1.5 and 3 Tesla: effect of increased field strength on 4 paradigms used during presurgical work-up. Invest Radiol 42:130–138

14. Matthews PM, Jezzard P (2004) Functional magnetic resonance imaging. J Neurol Neurosurg Psychiatry 75:6–12

15. Tharin S, Golby A (2007) Functional brain mapping and its applications to neurosurgery. Neurosurgery 60:185–201, discussion 201–202

16. Majos A, Tybor K, Stefanczyk L, Goraj B (2005) Cortical mapping by functional magnetic resonance imaging in patients with brain tumors. Eur Radiol 15:1148–1158

17. Roux FE, Boulanouar K, Ibarrola D, Tremoulet M, Chollet F et al (2000) Functional MRI and intraoperative brain mapping to evaluate brain plasticity in patients with brain tumours and hemiparesis. J Neurol Neurosurg Psychiatry 69:453–463

18. Desmond JE, Sum JM, Wagner AD, Demb JB, Shear PK et al (1995) Functional MRI measurement of language lateralization in Wada-tested patients. Brain 118(Pt 6):1411–1419

19. Lehericy S, Cohen L, Bazin B, Samson S, Giacomini E et al (2000) Functional MR evaluation of temporal and frontal language dominance compared with the Wada test. Neurology 54:1625–1633

20. Stippich C, Rapps N, Dreyhaupt J, Durst A, Kress B et al (2007) Localizing and lateralizing language in patients with brain tumors: feasibility of routine preoperative functional MR imaging in 81 consecutive patients. Radiology 243:828–836

21. Branco DM, Suarez RO, Whalen S, O'Shea JP, Nelson AP et al (2006) Functional MRI of memory in the hippocampus: laterality indices may be more meaningful if calculated from whole voxel distributions. Neuroimage 32:592–602

22. Thesen S, Heid O, Mueller E, Schad R (2000) Prospective acquisition correction for head motion with image-base tracking for real-time fMRI. Magn Reson Med 44:457–465

23. Springer JA, Binder JR, Hammeke TA, Swanson SJ, Frost JA et al (1999) Language dominance in neurologically normal and epilepsy subjects: a functional MRI study. Brain 122(Pt 11):2033–2046

24. Grummich P, Nimsky C, Pauli E, Buchfelder M, Ganslandt O (2006) Combining fMRI and MEG increases the reliability of presurgical language localization: a clinical study on the difference between and congruence of both modalities. Neuroimage 32:1793–1803

25. Herholz K, Reulen HJ, von Stockhausen HM, Thiel A, Ilmberger J et al (1997) Preoperative activation and intraoperative stimulation of language-related areas in patients with glioma. Neurosurgery 41:1253–1260, discussion 1260–1262

26. Lazar RM, Marshall RS, Pile-Spellman J, Duong HC, Mohr JP et al (2000) Interhemispheric transfer of language in patients with left frontal cerebral arteriovenous malformation. Neuropsychologia 38:1325–1332

27. Duffau H, Lopes M, Arthuis F, Bitar A, Sichez JP et al (2005) Contribution of intraoperative electrical stimulations in surgery of low grade gliomas: a comparative study between two series without (1985–96) and with (1996–2003) functional mapping in the same institution. J Neurol Neurosurg Psychiatry 76:845–851

28. Berman JI, Berger MS, Chung SW, Nagarajan SS, Henry RG (2007) Accuracy of diffusion tensor magnetic resonance imaging tractography assessed using intraoperative subcortical stimulation mapping and magnetic source imaging. J Neurosurg 107:488–494

29. Haglund MM, Berger MS, Shamseldin M, Lettich E, Ojemann GA (1994) Cortical localization of temporal lobe language sites in patients with gliomas. Neurosurgery 34:567–576, discussion 576

30. Hamzei F, Knab R, Weiller C, Roether J (2002) Intra- und extrakranielle Gefäßstenosen beeinflussen BOLD Antwort. Aktuelle Neurologie 29:231

31. Holodny AI, Schulder M, Liu WC, Maldjian JA, Kalnin AJ (1999) Decreased BOLD functional MR activation of the motor and sensory cortices adjacent to a glioblastoma multiforme: implications for image-guided neurosurgery. AJNR Am J Neuroradiol 20:609–612

32. Holodny AI, Schulder M, Liu WC, Wolko J, Maldjian JA et al (2000) The effect of brain tumors on BOLD functional MR imaging activation in the adjacent motor cortex: implications for image-guided neurosurgery. AJNR Am J Neuroradiol 21:1415–1422

33. Schreiber A, Hubbe U, Ziyeh S, Hennig J (2000) The influence of gliomas and nonglial space-occupying lesions on blood-oxygen-level-dependent contrast enhancement. AJNR Am J Neuroradiol 21:1055–1063

Chapter 26

Pharmacological Applications of fMRI

Paul M. Matthews

Abstract

Increasing societal expectations for new drugs, lack of confidence in short-term endpoints related to long-term outcomes for chronic neurological and psychiatric diseases and rising costs of development in an increasing cost-constrained market all have created a sense of crisis in CNS drug development. New approaches are needed. For some time, the potential of clinical functional imaging for more confident progression from preclinical to clinical development stages has been recognized. Pharmacological functional MRI (fMRI), which refers specifically to applications of fMRI to questions in drug development, provides one set of these tools. With related structural MRI measures, relatively high resolution data concerning target, disease-relevant pathophysiology and effects of therapeutic interventions can be related to brain functional anatomy. In this chapter, current and potential applications of pharmacological fMRI for target validation, patient stratification and characterization of therapeutic molecule pharmacokinetics and pharmacodynamics are reviewed. Challenges to better realizing the promise of pharmacological fMRI will be discussed. The review concludes that there is a strong rationale for greater use of pharmacological fMRI particularly for early phase studies, but also outlines the need for preclinical and early clinical development to be more seamlessly integrated, for greater harmonization of clinical imaging methodologies and for sharing of data to facilitate these goals.

Key words Pharmacological fMRI, Target validation, Patient stratification, Pharmacokinetics, Pharmacodynamics

1 Introduction

Both the pharmaceutical industry and regulators are searching for better models and for new drug development, particularly for CNS drugs [1]. Public confidence in the industry has declined in the face of what is viewed as a lack of commitment to addressing major diseases with innovative drugs, while new drug costs continue to escalate. Industry sees the risks of drug development to be high particularly for chronic CNS diseases, for which there is a notable lack of consensus regarding underlying causes and mechanisms in the scientific community. CNS drug development appears uncertain, slow, and expensive.

Pharmacological fMRI provides a relatively direct measure of CNS functions. Noninvasive imaging methods also allow the same

Massimo Filippi (ed.), *fMRI Techniques and Protocols*, Neuromethods, vol. 119,
DOI 10.1007/978-1-4939-5611-1_26, © Springer Science+Business Media New York 2016

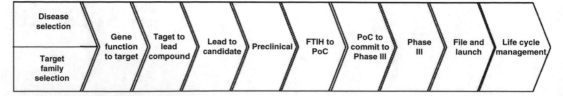

Fig. 1 The "critical path" for drug development. Pharmacological MRI has the potential to enhance the efficiency of early clinical development with better translation of biological concepts from preclinical to clinical studies, providing a new pharmacodynamic measure and enhancing potential in proof-of-mechanism studies (*FTIH* first time in human study, *PoC* proof of concept study)

endpoint measures to be used in preclinical as in clinical development. This facilitates interpretation of clinical imaging outcomes in terms of molecular and cellular changes found with invasive methods preclinically [2]. These and related considerations have embedded imaging in drug development already. Almost 30% of new molecular entities approved for neuropsychiatric indications by the Food and Drug Administration between 1995 and 2004 were developed with contributions from imaging [3]. A 2013 review identified 70 CNS drug trials registered on the registry website clinicaltrials.gov, that incorporated imaging endpoints [4]. In selected areas, such as multiple sclerosis drug development or recent trials of molecules for Alzheimer's disease, clinical imaging measures are used routinely for patient selection, for trials, or for response and safety monitoring. While most of these applications have relied on serial structural MRI, they have demonstrated the feasibility of implementing large scale, regulatory compliant clinical trials with imaging endpoints. They make the case for future use of pharmacological fMRI plausible.

Another factor that contributes to the plausibility of greater use of pharmacological fMRI in clinical drug development is the increasing premium being placed on integration of preclinical studies and early-phase development in an "experimental medicine" (sitting fluidly on the Phase I/IIa boundary) stage as part of confidence building and risk mitigation. Experimental medicine uses human experimentation to address mechanistic questions in ways that traditionally were reserved for preclinical studies. It is part of a biologically driven therapeutics development strategy involving hypothesis-led research that often is performed widely across levels of biological complexity (e.g., cells to the whole organism). A fundamental premise is that animals can be used to model biology, but cannot be expected to model human disease, which must be studied in the human. With this thinking, the classically unidirectional "critical path" from drug development (Fig. 1) is enabled by tools (e.g., from omics and imaging) to become more powerfully bidirectional (e.g., from preclinical to clinical data and "back again").

2   Principles of Functional MRI (fMRI)

FMRI is based on indirect measures of neuronal response mediated through associated changes in blood flow. Increased neuronal activity is associated with a local hemodynamic response involving both increased blood flow and blood volume. This neurovascular coupling is related to the increased local energy consumption associated with neuronal activity, which generally is believed to reflect predominantly presynaptic activity [5–7]. The hemodynamic response has a magnitude and time course that depends on contributions from both inhibitory and excitatory inputs to the local field potential [8]. It therefore can be considered as a measure of local information input. While this may be correlated with local multi-unit activity (neuronal spiking activity) under some conditions, such a relationship is not necessarily generalizable.

The neurovascular response is regulated by neuronal–glial interactions mediated by multiple signaling mechanisms. Pharmacological fMRI applications therefore need to take into account any potential impact of experimental molecules (or the disease of interest) on these coupling mechanisms. For example, the cerebrovascular effects of multiple neurotransmitter systems that may be the target for therapeutic molecules (e.g., glutamate, dopamine, norepinephrine, serotonin, acetylcholine, and prostaglandins) are well described [9]. Disorders of cerebrovascular regulation also are recognized in a number of disease states including not only primary cerebrovascular diseases such as stroke, but also, e.g., Alzheimer's disease [10–12].

The most commonly used fMRI methods rely on blood oxygen level-dependent (BOLD) imaging contrast [13, 14]. This contrast arises because the concentrations of deoxyhemoglobin, which is paramagnetic and thus locally modulates an applied static magnetic field, vary with local blood flow and oxygen consumption. In the MRI magnet, where a highly *homogeneous* (i.e., spatially invariant) magnetic field is generated, the paramagnetic deoxyhemoglobin generates small magnetic field *inhomogeneities* around blood vessels. Their magnitude increases with the amount of paramagnetic deoxyhemoglobin. These inhomogeneities reduce the MRI signal acquired with a gradient echo MRI acquisition sequence (echo planar imaging or EPI). Transient decreases in BOLD contrast associated with brain activity reflect neuronal activation because blood flow increases with greater neuronal activity to an extent that is larger than is needed simply for increased oxygen delivery with greater tissue demands. This reduces the local ratio of deoxy- to oxyhemoglobin in the blood enough to be associated with an increase in the local EPI MRI signal. While this effect is small (0.5–5 % typically at 3 T), it can be measured reliably with signal averaging.

Alternative approaches to brain functional imaging rely on measures of brain blood flow. The advent of fMRI was heralded by changes in blood flow measured by tracking a bolus of intravenously injected exogenous contrast material [15]. Arterial spin labeling MRI (ASL) has been developed more recently as an alternative, noninvasive pharmacological fMRI approach that is based on measuring brain activity associated changes in blood flow by means of noninvasive magnetic "tagging" (with a radiofrequency pulse) of blood flowing into the brain. Methods have become increasingly standardized in recent years, are widely available on commercial clinical imaging systems and can have considerable precision [16].

Both approaches to pharmacological fMRI can be applied in two general ways. In "task based" pharmacological fMRI, constrained shifts in cognitive state are induced to explore the way in which physiological differences between the states are modulated by an associated intervention. A typical experiment would involve acquisition of a series of images over the course of a controlled, periodic variation in cognitive state (e.g., performing a working memory task relative to resting) with and without the putative modulatory intervention of interest. Regions of significant change in the difference in BOLD signal between the two cognitive states then are defined by statistical analysis of the time series data.

An alternative design relies on the modulation of brain spontaneous activity in the absence of specific stimuli, i.e., in the "resting state". This approach is based on the observation that correlated, local and long-distance temporally varying signals are found with fMRI just as was previously found in the EEG [17]. This oscillatory activity appears fundamental to brain functional organization. Far field activity in the gamma range (~30–80 Hz) may be particularly relevant for the BOLD signal responses found in resting-state fMRI [7, 18]. There are multiple ways of defining the long-distance oscillatory coupling in fMRI [19], as yet without great standardization. For both task- and resting-state fMRI applications, assessment of responses to interventions involves statistical contrasts of time-courses before and after the intervention [20].

Both BOLD and ASL-based fMRI signals are low and can be confounded by other contributions to the temporally varying brain signal from subject movement (even on the order of mm), cardiorespiratory variations, image acquisition artifacts, and even difference in imaging system performance over time [21]. Some artifacts (e.g., movement) are easier to recognize and can be "edited out" *post hoc* [22]. Controlling for potential systematic variation in the parameters (e.g., increased respiratory rate in anxious subjects with a brain disease relative to healthy control subjects) as best as is possible is particularly important [21]. The potential for these factors to have an impact on outcomes emphasizes the importance of replications of results across laboratories and study populations, although this has rarely been achieved to date.

3   Target Validation

The traditional progression of drug development through target validation in preclinical models that express phenotypes plausibly related to the human disease is hugely challenged by most of the major diseases of the brain. Concepts for preclinical analogues of neuropsychiatric disorders with complex behavioral phenotypes (e.g., schizophrenia) and the validity of models for other major diseases including the chronic diseases of late life and those involving slow, progressive neurodegeneration are limited by differences in biological context and environment. New strategies for drug development are needed.

Preclinical models still provide powerful tools for detailed study of specific biological mechanisms believed to contribute to disease. With these models, pharmacological fMRI endpoints can be related to the underlying molecular changes in ways that both validate interpretation of the imaging endpoints and establish a framework in which they can be used to infer the dynamics of molecular pathogenic events. For example, the acute effects of NMDA receptor antagonism with ketamine were mapped in the rodent, demonstrating a pattern of cortico-limbic-thalamic activation and establishing a relationship between specific cognitive systems and the pharmacology [23]. Similar functional effects also were seen with other antagonists against the same target [24, 25], further confirming the specificity of the systems modulated. A framework for interpretation of these results was able to be provided by convergent studies using 2-deoxyglucose autoradiography [26] as an index of presynaptic activity, along with single unit electrophysiological recording and immediate early gene expression [27]. Analogous pharmacological fMRI experiments conducted in human studies provided mapped homologous systems in humans and to relate the pharmacology to the associated thought disorder and disturbance of consciousness in turn [28]. While indirect and insufficient alone, these clinical studies together provided important information supporting target validation of NMDA receptors for psychotic disorders; the "bi-directional" translational approach also supported the potential relevance of this preclinical pharmacology for understanding a form of human psychosis.

An exciting, emerging extension of this approach applies structural MRI and pharmacological fMRI measures together as *endophenotypes* in testing for heritable quantitative traits [29]. Consider, for example, a complex genetic disease such as schizophrenia, which shows a heritable phenotype with variable expression. Both structural and functional differences can be defined relative to the healthy brain. The concept of the endophenotype is that their *forme fruste* are heritable and can be identified in people even without clinical expression of the disease or trait. To the extent that this is true, the imaging endpoints themselves can be used as outcome measures in searches for genetic or other factors that may contribute to the disease. An endophenotype-based target validation approach also may bias detection towards causative rather than simply (possibly incidental or non-specific) associated features. Candidate genes *DISC1*, *GRM3*, and *COMT*, which are associated with altered hippocampal structure and function [30], glutamatergic fronto-hippocampal function [31], and prefrontal dopamine responsiveness [32], respectively, all have been related to imaging endophenotypes for schizophrenia in this way.

The concept of fMRI endophenotypes strengthens the rationale nosological reclassification of disease in terms of shared neurobiological system dysfunction. Applications of fMRI approaches that define neurobiological bases for general cognitive processes (such as, in the context of psychiatric disease, motivation, or reward) facilitate more holistic views of targets that may be relevant to more than one disease. For example, fMRI approaches have contributed to the current appreciation for neural mechanisms common to addictive behaviors across a wide range of substances abuse states. Studies of cue-elicited craving have defined similar activities of the mesolimbic reward circuit in addictions to nicotine [33], alcohol [34], gambling [35], amphetamine [36], cocaine [37] and opiates [38].

Combination of pharmacological fMRI with positron emission tomography (PET) receptor mapping can be used to relate systems-level dysfunction directly with the molecular targets of drug therapies in ways that enhance target validation for new pharmacological treatments faster and more cheaply than conventional clinical designs allow (see, e.g., [39]). In another example, a combined PET D3 receptor availability and resting-state pharmacological fMRI study provides a paradigmatic example of the way in which modulation of both target and system contributes to better defining fundamental mechanistic relationships between different symptoms [40]. First, D3 receptor availability was assessed in the ventral tegmentum/substantia nigra in healthy subjects using PET with the D3/D2 selective radioligand, $[^{11}C](+)$-4-propyl-9-hydroxynaphthoxazine ($[^{11}C]PHNO$). Differences in receptor expression and basal dopamine release determine binding of the $[^{11}C]PHNO$, which varied across subjects. A resting-state pharmacological fMRI study was conducted simultaneously. Parametric variation of the resting-state pharmacological fMRI functional connectivities with D3 receptor availability measured by PET showed that low midbrain D3 receptor availability (reflecting dopamine release) was associated with increased connectivity between orbitofrontal cortex (OFC) and brain networks implicated in cognitive control and salience processing. The results together further validated dopamine D3 receptor signaling as an important modulator of systems underpinning human goal-directed behavior, while highlighting differentially modulated interactions between OFC and networks implicated in cognitive control and reward.

With confidence in the relationship between a pattern of brain functional network activation and behaviors of interest, the former can be used as a clinically relevant biomarker for target validation. One of the first demonstrations of this was with the modulation of hippocampal activation with a working memory fMRI task based on allelic differences in a *BDNF* gene polymorphism [41]. This provided early evidence in humans supporting target validation of the TrkB receptor agonism for the treatment of cognitive symptoms associated with synaptic plasticity [42]. A different example illustrating how such studies can be used for decision making in drug development was provided by an imaging experimental medicine study linking to a PET receptor occupancy of a highly specific μ-opioid antagonist, GSK1521498, to pharmacological fMRI modulation of brain activation associated with palatable taste stimuli [43]. This allowed a first demonstration that salience and reward systems relevant to food intake were modulated by the target, suggesting the potential of antagonists as appetite suppressants, an inference supported by a later, larger Phase IIa study with a direct behavioral endpoint [44].

4   Patient Stratification

A critical issue in early drug development is to establish an appropriate level of confidence in the potential of a new molecule to become a therapy. One way in which this can be done is by better controlling for the substantial variations in therapeutic responses between individuals in early-phase studies. As well demonstrated in oncology [45], stratification of patients based on specific disease characteristics can enable more powerful trial designs [46]. Consider, hypothetically, the difference in outcome of trials for a population in which a new molecule has a 50% treatment effect in 20% of patients (giving a 10% *net* treatment effect, i.e., unlikely to be detected) relative to that in a stratified population enriched so that 70% are responders (a net 35% treatment effect). By predicting potential responders, imaging also can suggest ways of best selecting optimal indications for new molecules. To the extent that the enrichment is successful and any new pharmacological activity being evaluated is detectable, clinical trials may demonstrate molecule effects with fewer subjects exposed. This can be of special value in early Phase II trials when safety data is limited and the focus is on internal decision making.

An early application of imaging based stratification is expected to be for enrichment of populations for clinical trials in diseases such as Alzheimer's disease for which there is considerable phenotypic overlap between different disorders manifesting in the same population (e.g., dementia and late-life depression). The posterior cingulate and hippocampus show high functional connectivity in resting-state fMRI [47] and form the core of a so-called "default mode" network [48]. Decreases in default mode resting-state fMRI connectivity distinguish Alzheimer's patients from healthy subjects and can distinguish patients with mild cognitive impairment who undergo cognitive decline and conversion to Alzheimer's disease from those who remain stable over a medium term follow-up period [49, 50]. Distinct patterns of resting-state fMRI may distinguish patients with Parkinson's disease, for whom reduced resting state functional connectivity from the basal ganglia was reported [51]. Together, these findings suggest that resting-state fMRI (conducted in conjunction with other structural imaging measures), could be used to enrich trials for early disease modification of Alzheimer's disease.

Establishing fMRI measures for stratification of patients [52] also ultimately could aid in establishing prognosis and in patient management. Where alternative treatment approaches are available that have potentially significant individual variation in response across a population, selection of the optimal treatment for an individual patient could be assisted by fMRI (*personalized medicine*). For example, with depression, treatment responses are highly variable, e.g., only about 70% of patients respond well to a given antidepressant [53]. Higher BOLD signal in the amygdala with a simple task fMRI may be predictive of subsequent treatment response [54]. Multivariate fMRI responses that change with treatment in depression also have been proposed as candidate pharmacological fMRI markers, e.g., signal change in the ventromedial prefrontal and anterior cingulate cortices [55] or modulation of cortico-limbic functional connectivity [56].

In similar ways, there is a potential for integrated structural MRI and verbal task fMRI to distinguish people with prodromal schizophrenia from phenotypic mimics [57]. Network based analyses provide evidence for a continuous spectrum of psychosis from healthy variants to disabling expressions of schizophrenia [58]. Brain functional measures distinguishing abnormal network functions ultimately may provide more meaningful approaches to the classification of neuropsychiatric diseases for improved prognosis and for targeting of treatment [59–62], although establishing the robustness of classifiers in terms of longer term clinical outcome will demand standardization of methods and long-term, prospective studies.

Arguably fundamental changes in the understanding of chronic pain as a disease with individual differences in susceptibility have developed in recent years in part as a consequence of fMRI studies [63, 64]. Activity in the posterior insula with nociception provides a link between the subjectively "painful" experiences of pain empathy [65], hypnotically induced pain [66], and recalled pain experiences [67]. Inspired by studies showing a dopaminergic response with anticipation of benefit in Parkinson's disease, nigro-striatal pathways (as well as those associated with endogenous opioid release) have been implicated in the placebo response in pain and depression [68]. Individual variation in pain vulnerability thus is associated with alterations in wide range brain networks concerned with reward, motivation/learning, and descending modulatory control [69]. Greater functional connectivity between the PFC and nucleus accumbens explains pain persistence, suggesting that the frontal-striatal connectivity mediates the transition from acute to chronic pain; cortical-striatal connectivity explains longer term outcomes of patients with sub-acute back pain [70].

Nonetheless, despite this promise, validation and development of these concepts as clinical tools or for confident use as an enrichment strategy or as a secondary outcome measure in later-phase clinical trials appears stalled by lack of standardization of evaluations and methods for quality control and analysis [4]. A focus on longer term, well powered clinical studies is needed to validate relationships between fMRI measures and disease pathology or long-term clinical outcomes. Confident demonstrations are needed to establish that fMRI or pharmacological fMRI reliably distinguish clinically meaningfully changes.

5   Pharmacodynamics

As the previous section highlighted, applications of pharmacological fMRI to the direct assessment of drug action are expanding. *Pharmacodynamic* data (e.g., testing whether a drug at the chosen dose has an effect on brain function) can be obtained from analysis of brain imaging changes induced by the administration of a drug. The similar intrinsic brain architecture across species can support translational proof of mechanism studies with comparisons of endpoints from preclinical and imaging-supported Phase I studies using similar methodologies [71]. Additional information can come from correlation of brain activity with behavioral effects of drug administration [72] (Fig. 2) or with characterization of the way in brain activity associated with a probe-task is modulated by a drug [73–75]. This information can inform clinical dose-ranging studies. As noted earlier, correlations between fMRI measures of brain functional system response and drug receptor or receptor occupancy measurements by PET are possible [39, 43, 76]. The last, more recent study [43], demonstrated additionally how integration of time-receptor occupancy data from PET with fMRI measures can differentiate the distinct pharmacologies of different antagonists.

In some situations, by providing a measure of *endophenotype* responses, pharmacological fMRI can define effects of treatment in populations too small for behavioral effects to be discerned or where usual clinical measures are simply insensitive to drug effects [77–79]. In the simplest application, modulation of brain activation in functional anatomically plausible regions after dosing with

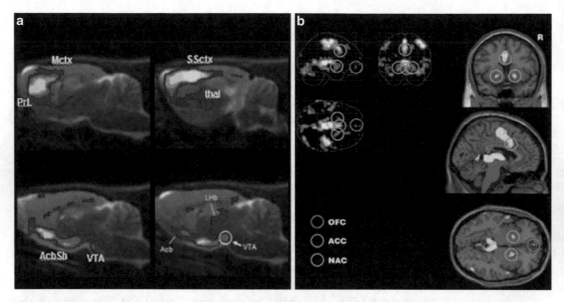

Fig. 2 Pharmacological fMRI can be performed in both animals and humans to assess correspondences in tests of mechanisms. (**a**) Pharmacological fMRI results with metamphetamine challenge of a rodent, identifying major regions in the monamine network as sites of direct or indirect action (*Mctx* motor cortex, *PrL* prelimbic medial prefrontal cortex, *thal* thalamus, *SSctx* somatosensory cortex, *AcbSh* shell of the nucleus accumbens, *VTA* ventral tegmental area) (Images courtesy of Dr. A. Bifone, GSK, Verona). (**b**) A similar pharmacological fMRI experiment with acute amphetamine infusion in human subjects performed using "mind racing" as a behavioral index of drug effects identified comparable elements of the core response network (*OFC* orbitofrontal cortex, *ACC* anterior cingulate cortex, *NAC* nucleus accumbens)

candidate molecule simply to provide supportive evidence for relevant direct CNS activity. A retrospective case study of NK-1 receptor antagonists for chronic pain proposed that early decisions based on fMRI measures could have anticipated the later failure of clinical trials [80]. However, a potential risk of such entirely pharmacological fMRI-derived pharmacodynamics markers is that they may not be specific for (or predictive of) clinically relevant changes.

One way of minimizing this risk is to frame the measures in terms of important disease symptoms based on the relationship between fMRI measures and individual symptoms. Mechanistic plausibility is suggested by the extent to which changes in the associated networks have been independently related to clinically meaningful symptoms. An illustration of this is provided by the way fMRI has been used to dissect the *subjective experience* of pain into anatomically distinct activities of different functional systems (including arousal and the somatosensory and limbic systems), the precise pattern for which may vary for an individual depending on context, mood, and cognitive state [64, 81].

As highlighted in the introduction to this review, imaging has the potential to bridge directly between preclinical and clinical studies [2]. While many behaviors cannot be translated across species, functional-anatomical correlations allow direct drug responses elicited in the brain for translation of underlying neurobiology. For example, pharmacological fMRI experiments in which unstimulated brain responses to acute compound challenges can be used to define brain regions in which activity is modulated by the same compound in animals (Fig. 2). Preclinically, these observations can be linked to results from more invasive studies, e.g., direct measurements of neurotransmitter release that distinguish direct and indirect effects of the compound [82]. Similar observations of drug modulation of brain activity can be made in human volunteers, providing a way of confirming mechanism (Fig. 2) [72]. State-dependent modulation of these regions can further contribute to this [83]. By relating plasma concentrations to brain responses, similar approaches could be used to define dose, for example. fMRI can address the need for evaluation of receptor agonists, partial agonists, and inverse agonists, as well as antagonists. Even when a receptor targeted radioligand is available, PET methods generally will not be informative with the former classes of agents [84].

However, caution is needed in the confidence with which fMRI endpoints are interpreted. There are two distinct validation issues that must be addressed. First is the "proof of biology" based on demonstration that the biological change being measured is related to the relevant target engagement. Second is the "proof of concept" that the biological change has relevance for clinical outcome [9]. Relationships seen with the natural history of the disease should not be assumed to hold after therapeutic modulations [85]. Testing for any changes in this relationship with pharmacological modulation is important to ensure that the biomarker remains plausibly related to a clinically meaningful outcome.

In general, validation of a candidate biomarker's surrogacy involves the demonstration that it possesses the properties required for its use as a substitute for a true endpoint. A surrogate can be used at the individual subject level when there is a perfect association between the surrogate and the final endpoint after adjustment for treatment. This criterion essentially requires the surrogate variable to 'capture' any relationship between the treatment and the true endpoint, a notion that can be operationalized by requiring the true endpoint rate at any follow-up time to be independent of treatment, given the preceding history of the surrogate variable [86].

6   Current Limitations and Some Future Extensions of Pharmacological fMRI

Although there is real promise for pharmacological fMRI, there are major general challenges to meaningful, quantitative interpretations of measures that need to be considered in planning applications. A first challenge is to distinguish disease or pharmacodynamic effects on hemodynamic coupling from those on neuronal activity and metabolism [11]. Some limitations to interpretation of the BOLD response can be addressed with use of complementary forms of MRI contrast. For example, direct measures of brain blood flow can be made using noninvasive "arterial spin labeling" MRI methods and the BOLD signal can be calibrated as a measure of local oxygen extraction for quantitative MRI [87]. However, even without this uncertainty, the relationship of blood flow changes to modulation of presynaptic activity can change with physiological (and, potentially, pharmacological) context. Even the relative direction of relative activation in disease states may be difficult to interpret precisely. For example, reduced activation may reflect brain functional impairment [88] or improved efficiency [89] in different contexts. Experimental designs need to recognize this uncertainty and incorporate elements that allow meaningfully specific interpretations, e.g., by studying dose–response relations, parametric activity relationships and behavioral correlates [90]. A more direct approach is to link pharmacological fMRI with electrophysiological measures [91].

General validation of methods to enable their wider use will depend on standardization across sites, reliability and repeatability, and the development of validated quantification methods, ideally largely automated to minimize needs for harmonization of user training. Practical considerations also need to be address to enable integrated use of the most accurate and efficient combination of markers and optimization of costs for the clinical trial environment [92]. Greater openness and sharing of data would be an important enabler of this. These steps, while still not yet part of routine practice in the academic laboratories in which advanced clinical imaging is most often performed, need not stifle innovation, which can progress in parallel, but is essential of translation of this promising method as a major tool for drug development is to be achieved.

Acknowledgements

The author gratefully acknowledges support from the Edmond J. Safra Foundation and from Lily Safra, the Medical Research Council, the MS Society of Great Britain, the Progressive MS Alliance and the Imperial College Biomedical Research Council. He is an NIHR Senior Investigator. He has received additional research funding from Biogen and GlaxoSmithKline, has received consultancy funding to his University from Biogen, Novartis, Adephi Communications and IXICO and honoraria or educational funding support from Biogen and Novartis.

References

1. Trusheim MR, Berndt ER, Douglas FL (2007) Stratified medicine: strategic and economic implications of combining drugs and clinical biomarkers. Nat Rev Drug Discov 6(4):287–293

2. Matthews PM et al (2013) Technologies: preclinical imaging for drug development. Drug Discov Today Technol 10(3):e343–e350

3. Uppoor RS et al (2008) The use of imaging in the early development of neuropharmacological drugs: a survey of approved NDAs. Clin Pharmacol Ther 84(1):69–74

4. Borsook D, Becerra L, Fava M (2013) Use of functional imaging across clinical phases in CNS drug development. Transl Psychiatry 3:e282

5. Mathiesen C et al (1998) Modification of activity-dependent increases of cerebral blood flow by excitatory synaptic activity and spikes in rat cerebellar cortex. J Physiol 512(Pt 2):555–566

6. Logothetis NK (2003) The underpinnings of the BOLD functional magnetic resonance imaging signal. J Neurosci 23(10):3963–3971

7. Mukamel R et al (2005) Coupling between neuronal firing, field potentials, and FMRI in human auditory cortex. Science 309(5736):951–954

8. Caesar K, Thomsen K, Lauritzen M (2003) Dissociation of spikes, synaptic activity, and activity-dependent increments in rat cerebellar blood flow by tonic synaptic inhibition. Proc Natl Acad Sci U S A 100(26):16000–16005

9. Minzenberg MJ (2012) Pharmacological MRI approaches to understanding mechanisms of drug action. Curr Top Behav Neurosci 11:365–388

10. Girouard H, Iadecola C (2006) Neurovascular coupling in the normal brain and in hypertension, stroke, and Alzheimer disease. J Appl Physiol (1985) 100(1):328–335

11. Suri S et al (2015) Reduced cerebrovascular reactivity in young adults carrying the APOE epsilon4 allele. Alzheimers Dement 11(6):648–657.e1

12. Glodzik L et al (2013) Cerebrovascular reactivity to carbon dioxide in Alzheimer's disease. J Alzheimers Dis 35(3):427–440

13. Kwong KK et al (1992) Dynamic magnetic resonance imaging of human brain activity during primary sensory stimulation. Proc Natl Acad Sci U S A 89(12):5675–5679

14. Ogawa S et al (1990) Oxygenation-sensitive contrast in magnetic resonance image of rodent brain at high magnetic fields. Magn Reson Med 14(1):68–78

15. Belliveau JW et al (1991) Functional mapping of the human visual cortex by magnetic resonance imaging. Science 254(5032):716–719

16. Mezue M et al (2014) Optimization and reliability of multiple postlabeling delay pseudo-continuous arterial spin labeling during rest and stimulus-induced functional task activation. J Cereb Blood Flow Metab 34(12):1919–1927

17. Brookes MJ et al (2011) Investigating the electrophysiological basis of resting state networks using magnetoencephalography. Proc Natl Acad Sci U S A 108(40):16783–16788

18. Goense JB, Logothetis NK (2008) Neurophysiology of the BOLD fMRI signal in awake monkeys. Curr Biol 18(9):631–640

19. Smith SM (2012) The future of FMRI connectivity. Neuroimage 62(2):1257–1266

20. FSL—FslWiki (2015) http://fsl.fmrib.ox.ac.uk/fsl/fslwiki/%5D

21. Iannetti GD, Wise RG (2007) BOLD functional MRI in disease and pharmacological studies: room for improvement? Magn Reson Imaging 25(6):978–988

22. Beckmann CF, Smith SM (2005) Tensorial extensions of independent component analysis for multisubject FMRI analysis. Neuroimage 25(1):294–311

23. Hodkinson DJ et al (2012) Differential effects of anaesthesia on the phMRI response to acute ketamine challenge. Br J Med Med Res 2(3):373–385

24. Littlewood CL et al (2006) Using the BOLD MR signal to differentiate the stereoisomers of ketamine in the rat. Neuroimage 32(4):1733–1746

25. Roberts TJ, Williams SC, Modo M (2008) A pharmacological MRI assessment of dizocilpine (MK-801) in the 3-nitroproprionic acid-lesioned rat. Neurosci Lett 444(1):42–47

26. Miyamoto S et al (2000) Effects of ketamine, MK-801, and amphetamine on regional brain 2-deoxyglucose uptake in freely moving mice. Neuropsychopharmacology 22(4):400–412

27. Homayoun H, Jackson ME, Moghaddam B (2005) Activation of metabotropic glutamate 2/3 receptors reverses the effects of NMDA receptor hypofunction on prefrontal cortex unit activity in awake rats. J Neurophysiol 93(4):1989–2001

28. Deakin JF et al (2008) Glutamate and the neural basis of the subjective effects of ketamine: a pharmaco-magnetic resonance imaging study. Arch Gen Psychiatry 65(2):154–164

29. Gottesman II, Gould TD (2003) The endophenotype concept in psychiatry: etymology and strategic intentions. Am J Psychiatry 160(4):636–645

30. Callicott JH et al (2005) Variation in DISC1 affects hippocampal structure and function and increases risk for schizophrenia. Proc Natl Acad Sci U S A 102(24):8627–8632

31. Egan MF et al (2004) Variation in GRM3 affects cognition, prefrontal glutamate, and risk for schizophrenia. Proc Natl Acad Sci U S A 101(34):12604–12609

32. Egan MF et al (2001) Effect of COMT Val108/158 Met genotype on frontal lobe function and risk for schizophrenia. Proc Natl Acad Sci U S A 98(12):6917–6922

33. David SP et al (2005) Ventral striatum/nucleus accumbens activation to smoking-related pictorial cues in smokers and nonsmokers: a functional magnetic resonance imaging study. Biol Psychiatry 58(6):488–494

34. Myrick H et al (2004) Differential brain activity in alcoholics and social drinkers to alcohol cues: relationship to craving. Neuropsychopharmacology 29(2):393–402

35. Reuter J et al (2005) Pathological gambling is linked to reduced activation of the mesolimbic reward system. Nat Neurosci 8(2):147–148

36. Paulus MP, Tapert SF, Schuckit MA (2005) Neural activation patterns of methamphetamine-dependent subjects during decision making predict relapse. Arch Gen Psychiatry 62(7):761–768

37. Kaufman JN et al (2003) Cingulate hypoactivity in cocaine users during a GO-NOGO task as revealed by event-related functional magnetic resonance imaging. J Neurosci 23(21):7839–7843

38. Forman SD et al (2004) Opiate addicts lack error-dependent activation of rostral anterior cingulate. Biol Psychiatry 55(5):531–537

39. Heinz A et al (2004) Correlation between dopamine D(2) receptors in the ventral striatum and central processing of alcohol cues and craving. Am J Psychiatry 161(10):1783–1789

40. Cole DM et al (2012) Orbitofrontal connectivity with resting-state networks is associated with midbrain dopamine D3 receptor availability. Cereb Cortex 22(12):2784–2793

41. Egan MF et al (2003) The BDNF val66met polymorphism affects activity-dependent secretion of BDNF and human memory and hippocampal function. Cell 112(2):257–269

42. Lu B, Nagappan G, Lu Y (2014) BDNF and synaptic plasticity, cognitive function, and dysfunction. Handb Exp Pharmacol 220:223–250

43. Rabiner EA et al (2011) Pharmacological differentiation of opioid receptor antagonists by molecular and functional imaging of target occupancy and food reward-related brain activation in humans. Mol Psychiatry 16(8):826–835, 785

44. Ziauddeen H et al (2013) Effects of the mu-opioid receptor antagonist GSK1521498 on hedonic and consummatory eating behaviour: a proof of mechanism study in binge-eating obese subjects. Mol Psychiatry 18(12):1287–1293

45. Engel RH, Kaklamani VG (2007) HER2-positive breast cancer: current and future treatment strategies. Drugs 67(9):1329–1341

46. Matthews PM et al (2014) The emerging agenda of stratified medicine in neurology. Nat Rev Neurol 10(1):15–26

47. Greicius MD et al (2004) Default-mode network activity distinguishes Alzheimer's disease from healthy aging: evidence from functional MRI. Proc Natl Acad Sci U S A 101(13):4637–4642

48. Raichle ME, Snyder AZ (2007) A default mode of brain function: a brief history of an evolving idea. Neuroimage 37(4):1083–1090, discussion 1097–1099

49. Petrella JR et al (2011) Default mode network connectivity in stable vs progressive mild cognitive impairment. Neurology 76(6):511–517

50. Sheline YI, Raichle ME (2013) Resting state functional connectivity in preclinical Alzheimer's disease. Biol Psychiatry 74(5):340–347

51. Szewczyk-Krolikowski K et al (2014) Functional connectivity in the basal ganglia network differentiates PD patients from controls. Neurology 83(3):208–214

52. Honey GD et al (2003) The functional neuroanatomy of schizophrenic subsyndromes. Psychol Med 33(6):1007–1018

53. Baghai TC, Moller HJ, Rupprecht R (2006) Recent progress in pharmacological and non-pharmacological treatment options of major depression. Curr Pharm Des 12(4):503–515

54. Canli T et al (2005) Amygdala reactivity to emotional faces predicts improvement in major depression. Neuroreport 16(12):1267–1270

55. Killgore WD, Yurgelun-Todd DA (2006) Ventromedial prefrontal activity correlates with depressed mood in adolescent children. Neuroreport 17(2):167–171

56. Anand A et al (2005) Antidepressant effect on connectivity of the mood-regulating circuit: an FMRI study. Neuropsychopharmacology 30(7):1334–1344

57. Allen P et al (2012) Transition to psychosis associated with prefrontal and subcortical dysfunction in ultra high-risk individuals. Schizophr Bull 38(6):1268–1276

58. Schmidt A et al (2014) Approaching a network connectivity-driven classification of the psychosis continuum: a selective review and suggestions for future research. Front Hum Neurosci 8:1047

59. del Campo N, Muller U, Sahakian BJ (2012) Neural and behavioral endophenotypes in ADHD. Curr Top Behav Neurosci 11:65–91

60. Hasler G, Northoff G (2011) Discovering imaging endophenotypes for major depression. Mol Psychiatry 16(6):604–619

61. Savitz JB, Drevets WC (2009) Imaging phenotypes of major depressive disorder: genetic correlates. Neuroscience 164(1):300–330

62. Keener MT, Phillips ML (2007) Neuroimaging in bipolar disorder: a critical review of current findings. Curr Psychiatry Rep 9(6):512–520

63. Lee MC et al (2013) Amygdala activity contributes to the dissociative effect of cannabis on pain perception. Pain 154(1):124–134

64. Lee MC, Tracey I (2013) Imaging pain: a potent means for investigating pain mechanisms in patients. Br J Anaesth 111(1):64–72

65. Mazzola V et al (2010) Affective response to a loved one's pain: insula activity as a function of individual differences. PLoS One 5(12):e15268

66. Derbyshire SW, Whalley MG, Oakley DA (2009) Fibromyalgia pain and its modulation by hypnotic and non-hypnotic suggestion: an fMRI analysis. Eur J Pain 13(5):542–550

67. Fairhurst M et al (2012) An fMRI study exploring the overlap and differences between neural representations of physical and recalled pain. PLoS One 7(10):e48711

68. Murray D, Stoessl AJ (2013) Mechanisms and therapeutic implications of the placebo effect in neurological and psychiatric conditions. Pharmacol Ther 140(3):306–318

69. Denk F, McMahon SB, Tracey I (2014) Pain vulnerability: a neurobiological perspective. Nat Neurosci 17(2):192–200

70. Baliki MN et al (2012) Corticostriatal functional connectivity predicts transition to chronic back pain. Nat Neurosci 15(8):1117–1119

71. Smucny J, Wylie KP, Tregellas JR (2014) Functional magnetic resonance imaging of intrinsic brain networks for translational drug discovery. Trends Pharmacol Sci 35(8):397–403

72. Vollm BA et al (2004) Methamphetamine activates reward circuitry in drug naive human subjects. Neuropsychopharmacology 29(9):1715–1722

73. Gerdelat-Mas A et al (2005) Chronic administration of selective serotonin reuptake inhibitor (SSRI) paroxetine modulates human motor cortex excitability in healthy subjects. Neuroimage 27(2):314–322

74. Pariente J et al (2001) Fluoxetine modulates motor performance and cerebral activation of patients recovering from stroke. Ann Neurol 50(6):718–729

75. Goekoop R et al (2004) Challenging the cholinergic system in mild cognitive impairment: a pharmacological fMRI study. Neuroimage 23(4):1450–1459

76. Farahani K et al (1999) Contemporaneous positron emission tomography and MR imaging at 1.5 T. J Magn Reson Imaging 9(3):497–500

77. Wilkinson D, Halligan P (2004) The relevance of behavioural measures for functional-imaging studies of cognition. Nat Rev Neurosci 5(1):67–73

78. Parry AM et al (2003) Potentially adaptive functional changes in cognitive processing for patients with multiple sclerosis and their acute modulation by rivastigmine. Brain 126(Pt 12):2750–2760

79. Matthews PM, Johansen-Berg H, Reddy H (2004) Non-invasive mapping of brain functions and brain recovery: applying lessons from cognitive neuroscience to neurorehabilitation. Restor Neurol Neurosci 22(3–5):245–260

80. Borsook D et al (2012) Decision-making using fMRI in clinical drug development: revisiting NK-1 receptor antagonists for pain. Drug Discov Today 17(17–18):964–973

81. Leknes S et al (2013) The importance of context: when relative relief renders pain pleasant. Pain 154(3):402–410

82. Schwarz AJ et al (2007) In vivo mapping of functional connectivity in neurotransmitter systems using pharmacological MRI. Neuroimage 34(4):1627–1636

83. Batterham RL et al (2007) PYY modulation of cortical and hypothalamic brain areas predicts feeding behaviour in humans. Nature 450(7166):106–109

84. Borsook D, Becerra L, Hargreaves R (2006) A role for fMRI in optimizing CNS drug development. Nat Rev Drug Discov 5(5):411–424

85. Cummings JL (2010) Integrating ADNI results into Alzheimer's disease drug development programs. Neurobiol Aging 31(8):1481–1492

86. Prentice RL (1989) Surrogate endpoints in clinical trials: definition and operational criteria. Stat Med 8(4):431–440

87. Hoge RD et al (1999) Linear coupling between cerebral blood flow and oxygen consumption in activated human cortex. Proc Natl Acad Sci U S A 96(16):9403–9408

88. Rombouts SA et al (2003) Loss of frontal fMRI activation in early frontotemporal dementia compared to early AD. Neurology 60(12):1904–1908

89. Floyer-Lea A, Matthews PM (2004) Changing brain networks for visuomotor control with increased movement automaticity. J Neurophysiol 92(4):2405–2412

90. Cader S et al (2006) Reduced brain functional reserve and altered functional connectivity in patients with multiple sclerosis. Brain 129(Pt 2):527–537

91. Lachaux JP et al (2007) Relationship between task-related gamma oscillations and BOLD signal: new insights from combined fMRI and intracranial EEG. Hum Brain Mapp 28(12):1368–1375

92. Merlo Pich E et al (2014) Imaging as a biomarker in drug discovery for Alzheimer's disease: is MRI a suitable technology? Alzheimers Res Ther 6(4):51

Chapter 27

Application of fMRI to Monitor Motor Rehabilitation

Steven C. Cramer and Jessica M. Cassidy

Abstract

Motor deficits contribute to disability in a number of neurological conditions. A wide range of emerging restorative therapies have the potential to reduce this by favorably modifying function. In many medical contexts, a study of target organ function improves efficacy of a therapeutic intervention. However, the optimal methods to prescribe a restorative therapy in the setting of central nervous system (CNS) disease are not clear. Brain mapping studies have the potential to provide useful insights in this regard. Examples of restorative therapies are provided, and human trials are summarized whereby brain mapping data have proven useful in promoting motor improvements in subjects with a neurological condition. A number of forms of brain mapping metrics are under study, including those emphasizing network connectivity obtained using resting-state fMRI. In some cases, brain mapping findings that correlate with better outcome with spontaneous behavioral recovery correspond to findings that predict better treatment response in the context of a clinical trial. Similarities across CNS conditions, such as stroke and multiple sclerosis, are discussed. Further studies are needed to understand which methods have the greatest value to monitor, predict, triage, and dose restorative therapies in trials that aim to reduce motor, and other neurological, deficits.

Key words Functional neuroimaging, Brain mapping, Stroke, Motor system, Recovery, Repair, Plasticity, Treatment

---

1 Motor Deficits and Restorative Therapies

Motor deficits are a major contributor to disability in the setting of a number of neurological diseases marked by focal central nervous system (CNS) injury, such as stroke, multiple sclerosis (MS), spinal cord injury (SCI), and traumatic brain injury [1]. In general, motor deficits show some degree of spontaneous improvement in the weeks following the insult. Spontaneous recovery is generally incomplete, however.

A number of therapies are in development to promote recovery in patients with motor deficits after a CNS insult. Some target the acute phase of injury, when the brain is galvanized and produces growth-related substances at levels reminiscent of development. Other therapies target patients in the chronic phase. Regardless, the goal of such therapies is not to salvage injured tissue, rather to promote repair and restore function.

Massimo Filippi (ed.), *fMRI Techniques and Protocols*, Neuromethods, vol. 119,
DOI 10.1007/978-1-4939-5611-1_27, © Springer Science+Business Media New York 2016

Many forms of restorative therapy are under study. Examples include small molecules [2–5], immune approaches such as via neutralization of the axon growth inhibitor Nogo-A with monoclonal antibodies [6], growth factors [7–15], cell-based methods [16–19], electromagnetic stimulation [20–24], neuroprosthetics [25, 26], and methods based on various forms of therapy and practice [27–37]. Issues of inter-individual variability in response to many of these therapies make it difficult to assess treatment feasibility and/or efficacy. Examination of these restorative therapies requires judicious analysis of the intended target, the brain.

A key thesis of this chapter is that optimal prescription of such restorative therapies will be achieved by probing the state of the brain. Clinical trials often enroll patients based on demographic or behavioral measures. However, these are only an approximation of the type of brain state information that is important to promoting repair and recovery.

There are examples in other medical specialties a measure of target organ function is obtained in addition to behavior and demographic data in order to maximize therapeutic gains. For example, hypothyroidism is optimally treated not by serial behavioral exam, rather by serial measures of pituitary–thyroid axis via serum TSH. Treatment of myeloproliferative and related hematological syndromes is ultimately dosed not by behavioral or demographic measures, but at least in part on the basis of serial measure of the cell population of interest. Cardiac arrhythmias and coronary artery disease are often assessed by evaluating cardiac function, e.g., in the setting of electrophysiological studies, exercise, or a sympathomimetic challenge. These practices suggest the general principle that some form of study of the therapy's target organ might be useful for optimizing therapy. In the setting of focal CNS injury, a technique such as fMRI might therefore be useful for a restorative therapy to assist with study entry criteria, to define optimal therapy dose or duration, or to serve as a biological marker of treatment effect. This issue is considered below with respect to three conditions characterized by an acute focal neurological insult.

2   Stroke

The motor system is among the most frequently affected domains by stroke [38, 39]. Duncan et al. [40, 41] found that the most dramatic improvements occurred in the first 30 days post-stroke, though significant improvement continued to occur up to 90 days after stroke in patients with more severe deficits. Nakayama et al. [42] measured arm disability and found that maximum arm function was achieved by 80% of patients within 3 weeks, and by 95% of patients within 9 weeks. Wade et al. [43] also found that significant improvement was mainly seen across the first 3 months after stroke. Despite these improvements, residual motor deficits remain in approximately half of patients in the chronic phase of stroke [38, 39].

A number of studies [44–46] have examined the brain events underlying the spontaneous recovery of motor behavior that does arise after stroke. In sum, stroke-related injury is associated with reduced activation, function, and neurophysiological responsiveness in injured (or for deep strokes, the overlying/corresponding) primary cortex. The best spontaneous return of behavior is associated with resolution of these reductions, i.e., return of activity in primary cortex, sometimes with particular shifts in the site of activation. Several compensatory responses may also contribute to spontaneous behavioral recovery, including increased activation in secondary areas that are normally connected to the injured zones in a distributed network, as well as a shift in interhemispheric laterality towards the contralesional hemisphere. The larger the injury or greater the deficits, the more these compensatory events are seen. These compensatory responses are tricks of the desperate, but in patients with injury-related deficits they are better present than absent [47–51]. These events that underlie spontaneous recovery are important because in some cases they are the same measures used to guide optimization of therapy-derived recovery.

2.1 Use of Functional Neuroimaging to Guide a Restorative Intervention in Patients with Stroke

One study used functional neuroimaging in a clinical trial of a restorative intervention to extract data from an fMRI scan in order to guide details of decision-making during therapy [24, 52]. An fMRI scan was used to identify the centroid of ipsilesional primary motor cortex activation when patients with stroke moved the affected hand. This information then guided neurosurgical placement of an investigational epidural cortical stimulation device over ipsilesional motor cortex. Using this approach, patients receiving stimulation plus rehabilitation therapy showed significantly greater arm motor gains than patients receiving rehabilitation therapy alone. A similar approach was used in studies based on transcranial magnetic stimulation (TMS) to identify the optimal physiological representation site for hand motor function. These studies found repetitive TMS to be useful for improving motor function after stroke [20, 21].

2.2 Use of Functional Neuroimaging to Predict Response to a Restorative Intervention in Patients with Stroke

An additional application of functional neuroimaging in the setting of restorative therapy is to predict behavioral response to treatment. Several studies have examined this issue [53–60]. For example, Koski et al. [56], using TMS, and Dong et al. [54], using fMRI, have found that changes in brain function early into therapy predict behavioral gains measured at the end of therapy. Note that in both cases, the findings that predicted better treatment response (improved motor evoked response in affected hand with TMS of ipsilesional hemisphere, and increased laterality of fMRI activation, i.e., towards the ipsilesional hemisphere, with movement of the affected hand, respectively) correspond to the findings correlating better outcome with spontaneous behavioral recovery. This latter pair of studies also hints at the potential use of human brain mapping measures to identify the dose of a restorative therapy in individual patients. For example, could a TMS or fMRI measure of brain function inform a clinician of the likelihood that the brain is receptive to further change that supports behavioral gains? In this regard, note that a probe of brain plasticity, such as might be used to predict treatment response to a restorative intervention, can be developed even in the setting of severe deficits, such as complete plegia [61].

Another study found that fMRI had independent value for predicting treatment response in a restorative therapy trial in patients with chronic stroke [57]. This study used a multivariate model to examine the specific ability of a baseline fMRI to predict trial-related behavioral gains, and compared this fMRI predictive ability directly to a number of other baseline measures. Patients in this study each underwent baseline clinical and functional MRI assessments, received 6 weeks of rehabilitation therapy with or without investigational motor cortex stimulation, then had repeat assessments. Across all patients, univariate analyses found that several baseline measures had predictive value for trial-related gains. However, multiple linear regression modeling found that only two variables remained significant predictors: degree of motor cortex activation on fMRI (lower motor cortex activation predicted larger gains) and arm motor function (greater arm function predicted larger gains). This study emphasized that an assessment of brain function can be a unique source of information for clinical decision-making in the setting of restorative therapy after stroke. Interestingly, clinical gains during study participation were paralleled by boosts in motor cortex activity, the latter detected via serial fMRI scanning, suggesting that lower baseline cortical activity in some patients likely represents under-use of an available cortical resource.

Burke Quinlan and colleagues [60] also utilized a multivariate model that encompassed various demographic, behavioral, and neuroimaging measures to determine which metrics best predicted behavioral gains following a three-week upper-extremity robotic therapy program in individuals with chronic stroke. Bivariate screening revealed significant correlations between improvement in upper-extremity status with baseline MRI and diffusion tensor imaging measures of brain injury (i.e. infarct volume, cortical injury, percentage injury to corticospinal tract), task-evoked fMRI measures of cortical function (i.e. ipsilesional primary motor cortex area (M1) activation), and resting fMRI measures of ipsilesional/contralesional M1 connectivity. Subsequent multivariate linear regression modeling revealed that the percentage injury to corticospinal tract and ipsilesional/contralesional M1 connectivity accounted for 44% of the variance in treatment gains. Brain-based measures, therefore, depicted better predictive quality than the more conventional behavioral and demographic measures.

A similar study focusing on lower-extremity motor gains after a course of physical therapy showed that baseline lower-extremity status and ipsilesional foot primary sensorimotor cortex activation volume contributed to 63% of the variance in gait velocity change [59]. Combined, these studies [59, 60] provide supporting evidence for the use of fMRI measures as potential biomarkers for rehabilitation gains. There are a number of important variables that differ across patients, study designs, fMRI acquisition and analysis methods, and more. As such, further studies are needed to understand the extent to which the above findings generalize across other stroke studies.

2.3   Use of Functional Neuroimaging to Investigate the Biological Mechanisms of Restorative Intervention Effects in Patients with Stroke

Carey et al. [62] found that a population of subjects with chronic stroke, when performing a finger tracking task with the stroke-affected hand, had activation within contralesional brain regions, i.e., regions that were primarily ipsilateral to movement. After training at this task, the normal pattern of laterality of brain activation was restored, with activation shifting to ipsilesional brain regions, i.e., contralateral to movement, and thereby more closely resembling findings in healthy control subjects. In this landmark study, functional neuroimaging provided insights into the mechanistic effects of treatment. Since then, other studies have shown varying modulatory effects of cortical activation following Botox [63], constraint-induced movement therapy (CIMT) [64], visuo-motor tracking task practice [65], implicit motor learning [66], real-time fMRI feedback training [67], and noninvasive brain stimulation in individuals with chronic stroke [68].

Two meta-analyses [69, 70] extend these results by examining studies that have employed functional neuroimaging as a biological marker of treatment effects targeting the motor system after stroke. A review of 24 studies utilizing sensorimotor tasks in 255 patients found higher activation in the contralesional M1 (relative to healthy controls) that decreased over time but was unrelated to motor outcome. Reorganization consistent with increased ipsilesional M1 and medial premotor cortex activation was associated with positive recovery; whereas, increased activation of the cerebellar vermis was associated with negative recovery. These conclusions highlight both beneficial and detrimental examples of neuroplastic reorganization following stroke as demonstrated by shifts in premotor and cerebellar vermis activation, respectively. An earlier meta-analysis that reviewed 13 studies of 121 patients permitted drawing a number of conclusions [70]. Motor deficits have been most often studied, in part because of their substantial contribution to overall disability after stroke, and in part because of their relatively high prevalence. Most published studies have focused on patients with good to excellent outcome at baseline since they were more able to perform the motor tasks required to probe brain function. Consequently, less is known about the functional anatomy of therapy-induced recovery processes in the large population of patients with more severe deficits after stroke despite the great need for further study of restorative interventions in this population. Very few studies have used functional imaging to examine treatment effects during the first few months after stroke, when spontaneous behavioral recovery is at its greatest. The effects that many key variables such as lesion site, recovery level, gender, and age have on the performance of functional neuroimaging in this context requires further study. Studies could be improved by incorporating measures of injury and/or physiology.

Baseline differences in the stroke population under study can have a significant impact on the informative value of functional neuroimaging measures in the setting of a clinical trial. One successful CIMT study was associated with *decreased* inter-hemispheric laterality in a study of weaker patients [71], while a second study found *increased* laterality in a study of stronger patients [72] with chronic stroke. Additionally, Könönen et al. [73] observed increased sensorimotor area activation after a 2-week CIMT program amongst individuals with poorer baseline hand function. Investigators found no change in hemispheric laterality for premotor and sensorimotor regions of interest.

This divergence in findings emphasizes how differences in a single variable, such as baseline motor status, might influence the utility of brain mapping in the setting of a clinical trial, and highlights the need for further studies in this regard.

3   Multiple Sclerosis

Motor deficits are common in MS. For example, across a broad population of subjects with MS, the median time to reach irreversible limited walking ability for more than 500 m without aid or rest is 8 years, to walk with unilateral support no more than 100 m without rest is 20 years, and to walk no more than 10 m without rest while leaning against a wall or support is 30 years [74]. Upper extremity motor deficits, such as those related to ataxia and paresis, are also a common source of disability.

Brain plasticity is an important determinant in MS in at least two contexts. First, steady destruction of myelin and of axons over years results in disability. During this period, reorganization of brain function can reduce the impact of such injury on behavioral status. Second, approximately 85 % of patients with MS have a relapsing, remitting course [75], in which a relapse peaks over 1–2 months and then improves over a similar time period. The resolution of these MS flares has been attributed to a number of brain events, such as neurological reserve and resolution of inflammatory insult [75], and a number of studies suggest that brain plasticity is also important [76]. Note too that there are numerous asymptomatic brain lesions for each symptomatic one in most patients with MS, a fact that might further support the importance of brain plasticity in maintenance of behavioral status in this condition.

Brain plasticity thus is likely important to motor status in MS, by minimizing the debilitating effects of MS injury accrual over time, and by promoting recovery from silent or symptomatic MS flares. A number of studies have provided insights into the brain events important in this regard, with substantial overlap as compared to findings in patients with stroke. This information gains importance in the current discussion because events important to maximizing behavioral status in the natural course of the disease are likely to be many of the same measures whose measurement can guide optimization of therapy-derived recovery.

Studies of brain plasticity in MS have found that, early in the course of the disease, brain activation is larger and more widespread as compared to healthy controls. Later in the disease, laterality of activation is reduced (i.e., activation is more bilateral) [77–79], akin to stroke patients who have larger infarcts or greater deficits [80, 81]. Bilateral sensorimotor cortical regions are activated to a greater extent in the setting of MS-related white matter injury [82, 83]. This increased degree of bilateral organization persists to the greatest extent in subjects with persistent deficits after an acute MS relapse, and returns to a normal, lateralized (i.e., contralateral-predominant) form of organization in subjects with the least degree of persistent disability [84, 85]. The pattern of brain activation during performance of a simple motor task in subjects recovered from stroke has been considered similar to the pattern seen in healthy subjects during performance of a complex task [86]; a similar analogy has been made in subjects with MS [87].

Interestingly, when a subset of subjects with mild MS, depicting no outward clinical deficits or disability despite long-standing diagnosis, performed a combination of sensory, cognitive, and motor tasks during an fMRI session, investigators found increased activation of cognitive-related regions relative to controls [88]. Further, when comparing mild vs. more severe relapsing-remitting MS phenotypes during an fMRI hand motor task, those subjects with mild MS showed increased activation throughout sensorimotor regions that correlated with increasing lesion volume and decreasing cortical volume [89]. These findings reaffirm the notion that preservation of physical function in MS relies on the deployment of compensatory brain plasticity mechanisms involving enhanced recruitment of cognitive and sensorimotor areas.

3.1 Use of Functional Neuroimaging to Investigate the Biological Mechanism of Restorative Intervention Effects in Patients with MS

The extent to which these spontaneous changes in brain function after MS can be used to monitor therapeutic interventions has been assessed in several small studies. One study tested the effects of increased cholinergic tone on the pattern of fMRI activation during performance of a cognitive task, the Stroop test. At baseline, patients with MS and moderate disability had similar behavioral performance as compared to controls, but on fMRI showed increased left medial prefrontal, and decreased right frontal, activation. Treatment with the cholinesterase inhibitor rivastigmine normalized both of these fMRI abnormalities in patients, but had little effect on a small cohort of healthy control subjects [90].

In another study, administration of 3,4-diaminopyridine to patients with MS was associated with increased activation in sensorimotor cortex and SMA ipsilateral to movement. This pattern is the reverse of the laterality pattern seen in normals but might correspond to effects of increased injury [80, 81] or task complexity [91, 92]. TMS measures were also affected by treatment, showing a drug-induced reduction in intracortical inhibition and increase in intracortical facilitation [93]. The relationship that these changes had with behavioral effects of drug administration was not reported.

A few studies have examined experience-dependent plasticity in MS following motor practice [94, 95]. In one study, subjects with MS practiced a visuomotor tracking task daily for approximately two weeks [94]. Investigators examined both short- and long-term (minutes and days, respectively) practice effects. They discovered an association between tracking improvement and attenuated blood-oxygen level dependent (BOLD) signal bilatera in sensorimotor, premotor, cingulate, temporal, and parahippocampal cortices for short-term practice improvement and trends of smaller BOLD signal in superior parietal lobule and occipital cortex for long-term improvement. Importantly, tracking improvement was independent of MRI-derived brain pathology metrics; meaning, that improvement was not dictated by the extent of structural brain damage. Subjects with MS were also studied with fMRI before and after 30 min of thumb flexion training [95]. Subjects with MS did not show a training-induced reduction in contralateral primary sensorimotor and parietal association cortices that healthy controls illustrated across the training period. Apart from implications regarding brain function in the setting of MS, these findings demonstrate a limitation of this paradigm for probing short-term, experience-dependent plasticity in this population [95]. Consideration of task complexity is also important, and may further explain these contrasting findings [94, 95] of short-term experience-dependent plasticity or lack thereof in MS.

4 Spinal Cord Injury

Though SCI can be associated with a range of injury patterns, motor deficits are generally a prominent feature. At the time of discharge from initial SCI, the most frequent neurologic category is incomplete tetraplegia (34.1%), followed by complete paraplegia (23.0%), complete tetraplegia (18.3%), and incomplete paraplegia (18.5%). Less than 1% of persons experience complete neurologic recovery by hospital discharge. By 10 years after SCI, 68% of persons with paraplegia, and 76% of those with tetraplegia, are unemployed [96].

Subjects with SCI generally show modest spontaneous sensory and motor improvement in the first 3–6 months following injury [97, 98], although significant improvement beyond the first year post-SCI is uncommon [99]. Motor deficits are thus common and persistent after SCI, and these impact a number of health, quality of life, and other issues in subjects with SCI [100–102].

4.1 Use of Functional Neuroimaging to Investigate the Biological Mechanism of Restorative Intervention Effects in Patients with SCI

There has been limited study of the CNS mechanisms underlying spontaneous motor improvement during the months following SCI. Jurkiewicz et al. [103] examined the acute-to-chronic time-course of post-SCI sensorimotor reorganization in four individuals with tetraplegia over a 12-month period. Shortly after injury, subjects with SCI demonstrated a similar volume of contralateral M1 activation as healthy controls when attempting ankle dorsiflexion movements. However, with increasing time post injury and persisting paralysis, contralateral M1 activation decreased along with prefrontal, premotor, supplementary, primary somatosensory, and posterior parietal cortices and cingulate motor area activation. These results depict a progressive shift in cortical reorganization further influenced by lower-extremity disuse. A related study in individuals with chronic SCI found cortical thinning in the leg area of the M1 and primary sensory cortex compared to healthy controls [104]. Moreover, subjects with SCI demonstrated increased activation of the left M1 leg area during right handgrip task compared to controls that was associated with smaller cervical cord area and impaired upper-extremity function. Lundell et al. [105] also found associations between spinal cord atrophy, motor function, and ipsilateral M1, somatosensory, and premotor cortical activation during ankle dorsiflexion movements in individuals with chronic SCI. Additional investigation is needed to further substantiate the relationship between neuroplastic reorganization, severity of SCI, and ensuing motor function.

Studies to date have more been focused on the nature of brain motor systems function in the chronic state, with some divergence of results to date. Some studies have found a broad decrease in activation [106–108], particularly in primary sensorimotor cortex, whereas others have found supranormal activation [109]. The basis for these discrepancies remains unclear but could be due to differences in age or injury pattern of the population studied, years post-SCI at the time of study, amount of motor function at the time of study, or the nature of the task used to probe motor system function, some uncovering deficient processing and others emphasizing supranormal efforts to compensate [107, 110]. A commonly described feature is a change in somatotopic organization within primary sensorimotor cortex contralateral to sensory or motor events, with representation of supralesional body regions expanding at the expense of infralesional body regions [111–115]. Spontaneous changes in laterality, so prominent in studies of stroke or MS, as above, are generally not prominent after SCI [116], perhaps due in part to the fact that injury typically affects the CNS bilaterally or perhaps due in part to the fact that SCI spares brain commissural fibers whose integrity helps maintain normal hemispheric balance. As such, laterality is unlikely to be a useful variable to examine in brain mapping studies of treatment effects in the setting of SCI.

At least two studies have evaluated changes in brain function in relation to therapy after SCI. Winchester et al. [117] studied body weight supported treadmill training in four patients with motor incomplete SCI. These authors compared fMRI during attempted unilateral foot and toe movement before vs. after training. This therapy was associated with increased activation within several bilateral areas, including primary sensorimotor cortex and cerebellum, though to a variable extent. The authors observed that, although all participants demonstrated a change in the BOLD signal following training, only those patients who demonstrated a substantial increase in activation of the cerebellum demonstrated an improvement in their ability to walk over ground, suggesting that this measure in this brain region, at least when examined using this task during fMRI, might be useful as a biological marker of successful treatment effect.

Another form of intervention that has been evaluated after SCI is motor imagery. Motor imagery normally activates many of the same brain regions as motor execution, and has been associated with improvements in motor performance [118, 119]. The effects of 1-week of motor imagery training to tongue and to foot were evaluated in ten subjects with chronic, complete tetra-/paraplegia plus ten healthy controls [61]. The behavioral outcome measure was speed of performance of a complex sequence. Motor imagery training was associated with a significant improvement in this behavior in non-paralyzed muscles (tongue for both groups, right foot for healthy subjects). In both the healthy controls and the subjects with SCI, serial fMRI scanning (before vs. after training) during attempted right foot movement was associated with increased fMRI activation in left putamen, an area associated with motor learning, despite foot movements being present in controls and absent in subjects with SCI. Behavioral training can thus result in measurable brain plasticity that is not accompanied by outward behavioral gains, a finding that might be important for designing biological markers in trials targeting severely disabled patient populations. Note that this fMRI change was absent in a second healthy control group serially imaged without training. The main conclusion from this study is that motor imagery training improves brain function whether or not sensorimotor function is present in the trained limb. An additional conclusion is that motor imagery, by virtue of its favorable effects on brain motor system organization, might have value as an adjunct motor restorative therapy. Another key point from this study is that brain plasticity related to plegic limbs can be studied in subjects with chronic SCI.

One hypothesis suggested by this study's findings is that the results of a short-term brain plasticity probe, such as this motor imagery training intervention, will predict response to a longer-term treatment [17, 18], for example, patients who show the greatest extent of brain plasticity with such a 1-week motor imagery intervention are those might be those who are most likely to respond to a more intensive intervention such as stem cell injections. Thus, at least in chronic SCI, some measure of the capacity for the brain to adapt in the short-term might predict likelihood of response to a more intense intervention.

5   Conclusions

Motor deficits are a major source of disability, across a number of conditions. A number of restorative therapies are under study to improve motor function in this regard. Optimal prescription of such therapies might benefit from an assessment of the function of the target organ, in addition to assessment of behavior or demographics. This is an approach that has often proven fruitful in general medical practice, and given the added complexities related to the CNS, is likely to be particularly important in for application of CNS restorative therapies.

Towards this goal, establishment of standardized protocols, such as for measuring motor cortex plasticity [120], to extract measures of brain function might help maximize the extent to which functional neuroimaging can be effectively applied. Dynamic protocols that incite a CNS response, such as over 30 min [121] or days [54, 56] of activity, might have particular value as compared to a single cross-sectional behavioral probe. Also, studies that provide a greater understanding of the underlying neurobiologic principles related to spontaneous recovery will also aid application of restorative therapies given that brain changes important to spontaneous recovery likely overlap substantially with changes whose measurement can effectively guide trials to maximize treatment-induced gains.

Further studies are needed to better characterize the measures that have the potential for monitoring, predicting, and dosing in the setting of a restorative trial of patients with motor deficits. Also, a minority of studies has examined language, neglect, and other domains injured in CNS disease, and further studies in these areas are also needed. Some similarities exist across diseases, such as those discussed between stroke and MS above, and further investigation of such points of similarity might prove fruitful in a broader sense to advancing restorative therapeutics.

The crux of this review centered on task-related fMRI. Examination of network connectivity using resting-state fMRI may also serve as a valuable biologic measure of disease status and/or treatment effectiveness in stroke [122–126], MS [127–129], and

SCI [130, 131]. Finally, this review focused on fMRI as a means of probing the state of the CNS. Other investigative methods might also prove useful, including functional, anatomical, and other forms of probe. Examples include positron emission tomography, diffusion tensor imaging, proton MR spectroscopy, TMS, electroencephalography, and measures of anatomy or perfusion. These can measure white matter integrity [58], injury in relation to normal anatomy [132, 133], metabolic state [134, 135], and more that might prove equally useful in models that aim to inform therapeutic approaches to restoring motor function in the setting of neurological disease.

References

1. Dobkin B (2003) The clinical science of neurologic rehabilitation. Oxford University Press, New York

2. Chen J, Cui X, Zacharek A, Jiang H, Roberts C, Zhang C et al (2007) Niaspan increases angiogenesis and improves functional recovery after stroke. Ann Neurol 62(1):49–58

3. Li L, Jiang Q, Zhang L, Ding G, Gang Zhang Z, Li Q et al (2007) Angiogenesis and improved cerebral blood flow in the ischemic boundary area detected by MRI after administration of sildenafil to rats with embolic stroke. Brain Res 1132(1):185–192

4. Chen P, Goldberg D, Kolb B, Lanser M, Benowitz L (2002) Inosine induces axonal rewiring and improves behavioral outcome after stroke. Proc Natl Acad Sci U S A 99(13):9031–9036

5. Freret T, Valable S, Chazalviel L, Saulnier R, Mackenzie ET, Petit E et al (2006) Delayed administration of deferoxamine reduces brain damage and promotes functional recovery after transient focal cerebral ischemia in the rat. Eur J Neurosci 23(7):1757–1765

6. Papadopoulos CM, Tsai SY, Cheatwood JL, Bollnow MR, Kolb BE, Schwab ME et al (2006) Dendritic plasticity in the adult rat following middle cerebral artery occlusion and Nogo-a neutralization. Cereb Cortex 16(4):529–536

7. Kawamata T, Dietrich W, Schallert T, Gotts J, Cocke R, Benowitz L et al (1997) Intracisternal basic fibroblast growth factor (bFGF) enhances functional recovery and upregulates the expression of a molecular marker of neuronal sprouting following focal cerebral infarction. Proc Natl Acad Sci U S A 94:8179–8184

8. Kawamata T, Ren J, Chan T, Charette M, Finklestein S (1998) Intracisternal osteogenic protein-1 enhances functional recovery following focal stroke. Neuroreport 9(7):1441–1445

9. Schabitz WR, Berger C, Kollmar R, Seitz M, Tanay E, Kiessling M et al (2004) Effect of brain-derived neurotrophic factor treatment and forced arm use on functional motor recovery after small cortical ischemia. Stroke 35(4):992–997

10. Wang L, Zhang Z, Wang Y, Zhang R, Chopp M (2004) Treatment of stroke with erythropoietin enhances neurogenesis and angiogenesis and improves neurological function in rats. Stroke 35(7):1732–1737

11. Tsai PT, Ohab JJ, Kertesz N, Groszer M, Matter C, Gao J et al (2006) A critical role of erythropoietin receptor in neurogenesis and post-stroke recovery. J Neurosci 26(4):1269–1274

12. Schneider UC, Schilling L, Schroeck H, Nebe CT, Vajkoczy P, Woitzik J (2007) Granulocyte-macrophage colony-stimulating factor-induced vessel growth restores cerebral blood supply after bilateral carotid artery occlusion. Stroke 38(4):1320–1328

13. Kolb B, Morshead C, Gonzalez C, Kim M, Gregg C, Shingo T et al (2007) Growth factor-stimulated generation of new cortical tissue and functional recovery after stroke damage to the motor cortex of rats. J Cereb Blood Flow Metab 27(5):983–997

14. Zhao LR, Berra HH, Duan WM, Singhal S, Mehta J, Apkarian AV et al (2007) Beneficial effects of hematopoietic growth factor therapy in chronic ischemic stroke in rats. Stroke 38(10):2804–2811

15. Ehrenreich H, Hasselblatt M, Dembowski C, Cepek L, Lewczuk P, Stiefel M et al (2002) Erythropoietin therapy for acute stroke is both safe and beneficial. Mol Med 8(8):495–505

16. Savitz SI, Dinsmore JH, Wechsler LR, Rosenbaum DM, Caplan LR (2004) Cell therapy for stroke. NeuroRx 1(4):406–414

17. Keirstead HS, Nistor G, Bernal G, Totoiu M, Cloutier F, Sharp K et al (2005) Human embryonic stem cell-derived oligodendrocyte progenitor cell transplants remyelinate and restore locomotion after spinal cord injury. J Neurosci 25(19):4694–4705

18. Cummings BJ, Uchida N, Tamaki SJ, Salazar DL, Hooshmand M, Summers R et al (2005) Human neural stem cells differentiate and promote locomotor recovery in spinal cord-injured mice. Proc Natl Acad Sci U S A 102(39):14069–14074

19. Shen LH, Li Y, Chen J, Zacharek A, Gao Q, Kapke A et al (2006) Therapeutic benefit of bone marrow stromal cells administered 1 month after stroke. J Cereb Blood Flow Metab 27:6–13

20. Khedr EM, Ahmed MA, Fathy N, Rothwell JC (2005) Therapeutic trial of repetitive transcranial magnetic stimulation after acute ischemic stroke. Neurology 65(3):466–468

21. Kim YH, You SH, Ko MH, Park JW, Lee KH, Jang SH et al (2006) Repetitive transcranial magnetic stimulation-induced corticomotor excitability and associated motor skill acquisition in chronic stroke. Stroke 37(6):1471–1476

22. Malcolm MP, Triggs WJ, Light KE, Gonzalez Rothi LJ, Wu S, Reid K et al (2007) Repetitive transcranial magnetic stimulation as an adjunct to constraint-induced therapy: an exploratory randomized controlled trial. Am J Phys Med Rehabil 86(9):707–715

23. Hummel F, Celnik P, Giraux P, Floel A, Wu WH, Gerloff C et al (2005) Effects of non-invasive cortical stimulation on skilled motor function in chronic stroke. Brain 128(Pt 3):490–499

24. Brown JA, Lutsep HL, Weinand M, Cramer SC (2006) Motor cortex stimulation for the enhancement of recovery from stroke: a prospective, multicenter safety study. Neurosurgery 58(3):464–473

25. Ring H, Rosenthal N (2005) Controlled study of neuroprosthetic functional electrical stimulation in sub-acute post-stroke rehabilitation. J Rehabil Med 37(1):32–36

26. Sheffler LR, Chae J (2007) Neuromuscular electrical stimulation in neurorehabilitation. Muscle Nerve 35(5):562–590

27. Kwakkel G, Kollen BJ, Krebs HI (2007) Effects of robot-assisted therapy on upper limb recovery after stroke: a systematic review. Neurorehabil Neural Repair 22:111–121

28. Volpe BT, Ferraro M, Lynch D, Christos P, Krol J, Trudell C et al (2005) Robotics and other devices in the treatment of patients recovering from stroke. Curr Neurol Neurosci Rep 5(6):465–470

29. Reinkensmeyer D, Emken J, Cramer S (2004) Robotics, motor learning, and neurologic recovery. Annu Rev Biomed Eng 6:497–525

30. Deutsch JE, Lewis JA, Burdea G (2007) Technical and patient performance using a virtual reality-integrated telerehabilitation system: preliminary finding. IEEE Trans Neural Syst Rehabil Eng 15(1):30–35

31. Duncan P, Studenski S, Richards L, Gollub S, Lai S, Reker D et al (2003) Randomized clinical trial of therapeutic exercise in subacute stroke. Stroke 34(9):2173–2180

32. Woldag H, Hummelsheim H (2002) Evidence-based physiotherapeutic concepts for improving arm and hand function in stroke patients: a review. J Neurol 249(5):518–528

33. French B, Thomas L, Leathley M, Sutton C, McAdam J, Forster A et al (2007) Repetitive task training for improving functional ability after stroke. Cochrane Database Syst Rev (4):CD006073

34. Kwakkel G, Wagenaar R, Twisk J, Lankhorst G, Koetsier J (1999) Intensity of leg and arm training after primary middle-cerebral-artery stroke: a randomised trial. Lancet 354(9174):191–196

35. Van Peppen RP, Kwakkel G, Wood-Dauphinee S, Hendriks HJ, Van der Wees PJ, Dekker J (2004) The impact of physical therapy on functional outcomes after stroke: what's the evidence? Clin Rehabil 18(8):833–862

36. Luft A, McCombe-Waller S, Whitall J, Forrester L, Macko R, Sorkin J et al (2004) Repetitive bilateral arm training and motor cortex activation in chronic stroke: a randomized controlled trial. JAMA 292(15):1853–1861

37. Wolf SL, Winstein CJ, Miller JP, Taub E, Uswatte G, Morris D et al (2006) Effect of constraint-induced movement therapy on upper extremity function 3 to 9 months after stroke: the EXCITE randomized clinical trial. JAMA 296(17):2095–2104

38. Rathore S, Hinn A, Cooper L, Tyroler H, Rosamond W (2002) Characterization of incident stroke signs and symptoms: findings from the atherosclerosis risk in communities study. Stroke 33(11):2718–2721

39. Gresham G, Duncan P, Stason W, Adams H, Adelman A, Alexander D et al (1995) Post-stroke rehabilitation. U.S. Department of

Health and Human Services. Public Health Service, Agency for Health Care Policy and Research, Rockville, MD

40. Duncan P, Goldstein L, Horner R, Landsman P, Samsa G, Matchar D (1994) Similar motor recovery of upper and lower extremities after stroke. Stroke 25(6):1181–1188

41. Duncan P, Goldstein L, Matchar D, Divine G, Feussner J (1992) Measurement of motor recovery after stroke. Stroke 23:1084–1089

42. Nakayama H, Jorgensen H, Raaschou H, Olsen T (1994) Recovery of upper extremity function in stroke patients: the Copenhagen Stroke Study. Arch Phys Med Rehabil 75(4):394–398

43. Wade D, Langton-Hewer R, Wood V, Skilbeck C, Ismail H (1983) The hemiplegic arm after stroke: measurement and recovery. J Neurol Neurosurg Psychiatry 46(6):521–524

44. Yozbatiran N, Cramer SC (2006) Imaging motor recovery after stroke. NeuroRx 3(4):482–488

45. Ward NS, Cohen LG (2004) Mechanisms underlying recovery of motor function after stroke. Arch Neurol 61(12):1844–1848

46. Baron J, Cohen L, Cramer S, Dobkin B, Johansen-Berg H, Loubinoux I et al (2004) Neuroimaging in stroke recovery: a position paper from the First International workshop on neuroimaging and stroke recovery. Cerebrovasc Dis (Basel, Switzerland) 18(3):260–267

47. Lotze M, Markert J, Sauseng P, Hoppe J, Plewnia C, Gerloff C (2006) The role of multiple contralesional motor areas for complex hand movements after internal capsular lesion. J Neurosci 26(22):6096–6102

48. Winhuisen L, Thiel A, Schumacher B, Kessler J, Rudolf J, Haupt WF et al (2005) Role of the contralateral inferior frontal gyrus in recovery of language function in poststroke aphasia: a combined repetitive transcranial magnetic stimulation and positron emission tomography study. Stroke 36(8):1759–1763

49. Johansen-Berg H, Rushworth M, Bogdanovic M, Kischka U, Wimalaratna S, Matthews P (2002) The role of ipsilateral premotor cortex in hand movement after stroke. Proc Natl Acad Sci U S A 99(22):14518–14523

50. Werhahn K, Conforto A, Kadom N, Hallett M, Cohen L (2003) Contribution of the ipsilateral motor cortex to recovery after chronic stroke. Ann Neurol 54(4):464–472

51. Fridman E, Hanakawa T, Chung M, Hummel F, Leiguarda R, Cohen L (2004) Reorganization of the human ipsilesional pre-

motor cortex after stroke. Brain 127(Pt 4):747–758

52. Cramer S, Benson R, Himes D, Burra V, Janowsky J, Weinand M et al (2005) Use of functional MRI to guide decisions in a clinical stroke trial. Stroke 36(5):e50–e52

53. Platz T, Kim I, Engel U, Kieselbach A, Mauritz K (2002) Brain activation pattern as assessed with multi-modal EEG analysis predict motor recovery among stroke patients with mild arm paresis who receive the Arm Ability Training. Restor Neurol Neurosci 20(1–2):21–35

54. Dong Y, Dobkin BH, Cen SY, Wu AD, Winstein CJ (2006) Motor cortex activation during treatment may predict therapeutic gains in paretic hand function after stroke. Stroke 37(6):1552–1555

55. Fritz SL, Light KE, Patterson TS, Behrman AL, Davis SB (2005) Active finger extension predicts outcomes after constraint-induced movement therapy for individuals with hemiparesis after stroke. Stroke 36(6):1172–1177

56. Koski L, Mernar T, Dobkin B (2004) Immediate and long-term changes in corticomotor output in response to rehabilitation: correlation with functional improvements in chronic stroke. Neurorehabil Neural Repair 18(4):230–249

57. Cramer S, Parrish T, Levy R, Stebbins G, Ruland S, Lowry D et al (2007) An assessment of brain function predicts functional gains in a clinical stroke trial. Stroke 38:520 (abstract)

58. Stinear CM, Barber PA, Smale PR, Coxon JP, Fleming MK, Byblow WD (2007) Functional potential in chronic stroke patients depends on corticospinal tract integrity. Brain 130(Pt 1):170–180

59. Burke E, Dobkin BH, Noser EA, Enney LA, Cramer SC (2014) Predictors and biomarkers of treatment gains in a clinical stroke trial targeting the lower extremity. Stroke 45(8):2379–2384

60. Burke Quinlan E, Dodakian L, See J, McKenzie A, Le V, Wojnowicz M et al (2015) Neural function, injury, and stroke subtype predict treatment gains after stroke. Ann Neurol 77(1):132–145

61. Cramer SC, Orr EL, Cohen MJ, Lacourse MG (2007) Effects of motor imagery training after chronic, complete spinal cord injury. Exp Brain Res 177(2):233–242

62. Carey J, Kimberley T, Lewis S, Auerbach E, Dorsey L, Rundquist P et al (2002) Analysis of fMRI and finger tracking training in subjects with chronic stroke. Brain 125(Pt 4):773–788

63. Veverka T, Hlustik P, Hok P, Otruba P, Tudos Z, Zapletalova J et al (2014) Cortical activity modulation by botulinum toxin type A in patients with post-stroke arm spasticity: real and imagined hand movement. J Neurol Sci 346(1–2):276–283

64. Laible M, Grieshammer S, Seidel G, Rijntjes M, Weiller C, Hamzei F (2012) Association of activity changes in the primary sensory cortex with successful motor rehabilitation of the hand following stroke. Neurorehabil Neural Repair 26(7):881–888

65. Bosnell RA, Kincses T, Stagg CJ, Tomassini V, Kischka U, Jbabdi S et al (2011) Motor practice promotes increased activity in brain regions structurally disconnected after subcortical stroke. Neurorehabil Neural Repair 25(7):607–616

66. Meehan SK, Randhawa B, Wessel B, Boyd LA (2011) Implicit sequence-specific motor learning after subcortical stroke is associated with increased prefrontal brain activations: an fMRI study. Hum Brain Mapp 32(2):290–303

67. Sitaram R, Veit R, Stevens B, Caria A, Gerloff C, Birbaumer N et al (2012) Acquired control of ventral premotor cortex activity by feedback training: an exploratory real-time FMRI and TMS study. Neurorehabil Neural Repair 26(3):256–265

68. Stagg CJ, Bachtiar V, O'Shea J, Allman C, Bosnell RA, Kischka U et al (2012) Cortical activation changes underlying stimulation-induced behavioural gains in chronic stroke. Brain 135(Pt 1):276–284

69. Favre I, Zeffiro TA, Detante O, Krainik A, Hommel M, Jaillard A (2014) Upper limb recovery after stroke is associated with ipsilesional primary motor cortical activity: a meta-analysis. Stroke 45(4):1077–1083

70. Hodics T, Cohen LG, Cramer SC (2006) Functional imaging of intervention effects in stroke motor rehabilitation. Arch Phys Med Rehabil 87(12 Suppl):36–42

71. Schaechter J, Kraft E, Hilliard T, Dijkhuizen R, Benner T, Finklestein S et al (2002) Motor recovery and cortical reorganization after constraint-induced movement therapy in stroke patients: a preliminary study. Neurorehabil Neural Repair 16(4):326–338

72. Johansen-Berg H, Dawes H, Guy C, Smith S, Wade D, Matthews P (2002) Correlation between motor improvements and altered fMRI activity after rehabilitative therapy. Brain 125(Pt 12):2731–2742

73. Kononen M, Tarkka IM, Niskanen E, Pihlajamaki M, Mervaala E, Pitkanen K et al (2012) Functional MRI and motor behavioral changes obtained with constraint-induced movement therapy in chronic stroke. Eur J Neurol 19(4):578–586

74. Vukusic S, Confavreux C (2007) Natural history of multiple sclerosis: risk factors and prognostic indicators. Curr Opin Neurol 20(3):269–274

75. Vollmer T (2007) The natural history of relapses in multiple sclerosis. J Neurol Sci 256(Suppl 1):S5–S13

76. Rocca MA, Filippi M (2007) Functional MRI in multiple sclerosis. J Neuroimaging 17(Suppl 1):36S–41S

77. Rocca MA, Colombo B, Falini A, Ghezzi A, Martinelli V, Scotti G et al (2005) Cortical adaptation in patients with MS: a cross-sectional functional MRI study of disease phenotypes. Lancet Neurol 4(10):618–626

78. Wang J, Hier DB (2007) Motor reorganization in multiple sclerosis. Neurol Res 29(1):3–8

79. Pantano P, Mainero C, Caramia F (2006) Functional brain reorganization in multiple sclerosis: evidence from fMRI studies. J Neuroimaging 16(2):104–114

80. Ward N, Brown M, Thompson A, Frackowiak R (2003) Neural correlates of outcome after stroke: a cross-sectional fMRI study. Brain 126(Pt 6):1430–1448

81. Cramer SC, Crafton KR (2006) Somatotopy and movement representation sites following cortical stroke. Exp Brain Res 168(1–2):25–32

82. Lenzi D, Conte A, Mainero C, Frasca V, Fubelli F, Totaro P et al (2007) Effect of corpus callosum damage on ipsilateral motor activation in patients with multiple sclerosis: a functional and anatomical study. Hum Brain Mapp 28(7):636–644

83. Rocca MA, Gallo A, Colombo B, Falini A, Scotti G, Comi G et al (2004) Pyramidal tract lesions and movement-associated cortical recruitment in patients with MS. Neuroimage 23(1):141–147

84. Reddy H, Narayanan S, Matthews P, Hoge R, Pike G, Duquette P et al (2000) Relating axonal injury to functional recovery in MS. Neurology 54(1):236–239

85. Mezzapesa DM, Rocca MA, Rodegher M, Comi G, Filippi M (2007) Functional cortical changes of the sensorimotor network are associated with clinical recovery in multiple sclerosis. Hum Brain Mapp 29(5):562–573

86. Cramer S, Nelles G, Benson R, Kaplan J, Parker R, Kwong K et al (1997) A functional MRI study of subjects recovered from hemiparetic stroke. Stroke 28(12):2518–2527

87. Filippi M, Rocca MA, Mezzapesa DM, Falini A, Colombo B, Scotti G et al (2004) A functional

MRI study of cortical activations associated with object manipulation in patients with MS. Neuroimage 21(3):1147–1154

88. Colorado RA, Shukla K, Zhou Y, Wolinsky JS, Narayana PA (2012) Multi-task functional MRI in multiple sclerosis patients without clinical disability. Neuroimage 59(1):573–581

89. Giorgio A, Portaccio E, Stromillo ML, Marino S, Zipoli V, Battaglini M et al (2010) Cortical functional reorganization and its relationship with brain structural damage in patients with benign multiple sclerosis. Mult Scler 16(11):1326–1334

90. Parry AM, Scott RB, Palace J, Smith S, Matthews PM (2003) Potentially adaptive functional changes in cognitive processing for patients with multiple sclerosis and their acute modulation by rivastigmine. Brain 126(Pt 12):2750–2760

91. Sadato N, Campbell G, Ibanez V, Deiber M, Hallett M (1996) Complexity affects regional cerebral blood flow change during sequential finger movements. J Neurosci 16(8):2691–2700

92. Verstynen T, Diedrichsen J, Albert N, Aparicio P, Ivry RB (2005) Ipsilateral motor cortex activity during unimanual hand movements relates to task complexity. J Neurophysiol 93(3):1209–1222

93. Mainero C, Inghilleri M, Pantano P, Conte A, Lenzi D, Frasca V et al (2004) Enhanced brain motor activity in patients with MS after a single dose of 3,4-diaminopyridine. Neurology 62(11):2044–2050

94. Tomassini V, Johansen-Berg H, Jbabdi S, Wise RG, Pozzilli C, Palace J et al (2012) Relating brain damage to brain plasticity in patients with multiple sclerosis. Neurorehabil Neural Repair 26(6):581–593

95. Morgen K, Kadom N, Sawaki L, Tessitore A, Ohayon J, McFarland H et al (2004) Training-dependent plasticity in patients with multiple sclerosis. Brain 127(Pt 11):2506–2517

96. Facts and Figures at a Glance—June 2006 (2007) www.spinalcord.uab.edu

97. Geisler F, Dorsey F, Coleman W (1991) Recovery of motor function after spinal-cord injury—a randomized, placebo-controlled trial with GM-1 ganglioside. N Engl J Med 324(26):1829–1838

98. Ditunno J, Stover S, Freed M, Ahn J (1992) Motor recovery of the upper extremities in traumatic quadriplegia: a multicenter study. Arch Phys Med Rehabil 73(5):431–436

99. Kirshblum S, Millis S, McKinley W, Tulsky D (2004) Late neurologic recovery after traumatic spinal cord injury. Arch Phys Med Rehabil 85(11):1811–1817

100. Jayaraman A, Gregory CM, Bowden M, Stevens JE, Shah P, Behrman AL et al (2006) Lower extremity skeletal muscle function in persons with incomplete spinal cord injury. Spinal Cord 44(11):680–687

101. DeVivo MJ, Richards JS (1992) Community reintegration and quality of life following spinal cord injury. Paraplegia 30(2):108–112

102. Frankel HL, Coll JR, Charlifue SW, Whiteneck GG, Gardner BP, Jamous MA et al (1998) Long-term survival in spinal cord injury: a fifty year investigation. Spinal Cord 36(4):266–274

103. Jurkiewicz MT, Mikulis DJ, Fehlings MG, Verrier MC (2010) Sensorimotor cortical activation in patients with cervical spinal cord injury with persisting paralysis. Neurorehabil Neural Repair 24(2):136–140

104. Freund P, Weiskopf N, Ward NS, Hutton C, Gall A, Ciccarelli O et al (2011) Disability, atrophy and cortical reorganization following spinal cord injury. Brain 134(Pt 6):1610–1622

105. Lundell H, Christensen MS, Barthelemy D, Willerslev-Olsen M, Biering-Sorensen F, Nielsen JB (2011) Cerebral activation is correlated to regional atrophy of the spinal cord and functional motor disability in spinal cord injured individuals. Neuroimage 54(2):1254–1261

106. Cramer SC, Lastra L, Lacourse MG, Cohen MJ (2005) Brain motor system function after chronic, complete spinal cord injury. Brain 128(Pt 12):2941–2950

107. Jurkiewicz MT, Mikulis DJ, McIlroy WE, Fehlings MG, Verrier MC (2007) Sensorimotor cortical plasticity during recovery following spinal cord injury: a longitudinal fMRI study. Neurorehabil Neural Repair 21(6):527–538

108. Sabbah P, de Schonen S, Leveque C, Gay S, Pfefer F, Nioche C et al (2002) Sensorimotor cortical activity in patients with complete spinal cord injury: a functional magnetic resonance imaging study. J Neurotrauma 19(1):53–60

109. Alkadhi H, Brugger P, Boendermaker S, Crelier G, Curt A, Hepp-Reymond M et al (2005) What disconnection tells about motor imagery: evidence from paraplegic patients. Cereb Cortex 15(2):131–140

110. Humphrey D, Mao H, Schaeffer E (eds) (2000) Voluntary activation of ineffective cerebral motor areas in short- and long-term paraplegics. Society for Neuroscience (abstract)

111. Topka H, Cohen L, Cole R, Hallett M (1991) Reorganization of corticospinal pathways following spinal cord injury. Neurology 41(8):1276–1283

112. Bruehlmeier M, Dietz V, Leenders K, Roelcke U, Missimer J, Curt A (1998) How does the

human brain deal with a spinal cord injury? Eur J Neurosci 10(12):3918–3922

113. Mikulis D, Jurkiewicz M, McIlroy W, Staines W, Rickards L, Kalsi-Ryan S et al (2002) Adaptation in the motor cortex following cervical spinal cord injury. Neurology 58(5):794–801

114. Turner J, Lee J, Martinez O, Medlin A, Schandler S, Cohen M (2001) Somatotopy of the motor cortex after long-term spinal cord injury or amputation. IEEE Trans Neural Syst Rehabil Eng 9(2):154–160

115. Corbetta M, Burton H, Sinclair R, Conturo T, Akbudak E, McDonald J (2002) Functional reorganization and stability of somatosensory-motor cortical topography in a tetraplegic subject with late recovery. Proc Natl Acad Sci U S A 99(26):17066–17071

116. Sabre L, Tomberg T, Korv J, Kepler J, Kepler K, Linnamagi U et al (2013) Brain activation in the acute phase of traumatic spinal cord injury. Spinal Cord 51(8):623–629

117. Winchester P, McColl R, Querry R, Foreman N, Mosby J, Tansey K et al (2005) Changes in supraspinal activation patterns following robotic locomotor therapy in motor-incomplete spinal cord injury. Neurorehabil Neural Repair 19(4):313–324

118. Lacourse MG, Turner JA, Randolph-Orr E, Schandler SL, Cohen MJ (2004) Cerebral and cerebellar sensorimotor plasticity following motor imagery-based mental practice of a sequential movement. J Rehabil Res Dev 41(4):505–524

119. Sharma N, Pomeroy VM, Baron JC (2006) Motor imagery: a backdoor to the motor system after stroke? Stroke 37(7):1941–1952

120. Kleim J, Kleim E, Cramer S (2007) Systematic assessment of training-induced changes in corticospinal output to hand using frameless stereotaxic transcranial magnetic stimulation. Nat Protoc 2:1675–1684

121. Kleim JA, Chan S, Pringle E, Schallert K, Procaccio V, Jimenez R et al (2006) BDNF val-66met polymorphism is associated with modified experience-dependent plasticity in human motor cortex. Nat Neurosci 9(6):735–737

122. Wadden KP, Woodward TS, Metzak PD, Lavigne KM, Lakhani B, Auriat AM et al (2015) Compensatory motor network connectivity is associated with motor sequence learning after subcortical stroke. Behav Brain Res 286:136–145

123. Golestani AM, Tymchuk S, Demchuk A, Goodyear BG (2013) Longitudinal evaluation of resting-state FMRI after acute stroke with hemiparesis. Neurorehabil Neural Repair 27(2):153–163

124. Park CH, Chang WH, Ohn SH, Kim ST, Bang OY, Pascual-Leone A et al (2011) Longitudinal changes of resting-state functional connectivity during motor recovery after stroke. Stroke 42(5):1357–1362

125. Carter AR, Patel KR, Astafiev SV, Snyder AZ, Rengachary J, Strube MJ et al (2012) Upstream dysfunction of somatomotor functional connectivity after corticospinal damage in stroke. Neurorehabil Neural Repair 26(1):7–19

126. Varkuti B, Guan C, Pan Y, Phua KS, Ang KK, Kuah CW et al (2013) Resting state changes in functional connectivity correlate with movement recovery for BCI and robot-assisted upper-extremity training after stroke. Neurorehabil Neural Repair 27(1):53–62

127. Petsas N, Tomassini V, Filippini N, Sbardella E, Tona F, Piattella MC et al (2015) Impaired functional connectivity unmasked by simple repetitive motor task in early relapsing-remitting multiple sclerosis. Neurorehabil Neural Repair 29(6):557–565

128. Basile B, Castelli M, Monteleone F, Nocentini U, Caltagirone C, Centonze D et al (2013) Functional connectivity changes within specific networks parallel the clinical evolution of multiple sclerosis. Mult Scler 20(8):1050–1057

129. Filippi M, Agosta F, Spinelli EG, Rocca MA (2013) Imaging resting state brain function in multiple sclerosis. J Neurol 260(7):1709–1713

130. Chisholm AE, Peters S, Borich MR, Boyd LA, Lam T (2015) Short-term cortical plasticity associated with feedback-error learning after locomotor training in a patient with incomplete spinal cord injury. Phys Ther 95(2):257–266

131. Hou JM, Sun TS, Xiang ZM, Zhang JZ, Zhang ZC, Zhao M et al (2014) Alterations of resting-state regional and network-level neural function after acute spinal cord injury. Neuroscience 277:446–454

132. Newton JM, Ward NS, Parker GJ, Deichmann R, Alexander DC, Friston KJ et al (2006) Non-invasive mapping of corticofugal fibres from multiple motor areas—relevance to stroke recovery. Brain 129(Pt 7):1844–1858

133. Crafton K, Mark A, Cramer S (2003) Improved understanding of cortical injury by incorporating measures of functional anatomy. Brain 126(Pt 7):1650–1659

134. Heiss W, Emunds H, Herholz K (1993) Cerebral glucose metabolism as a predictor of rehabilitation after ischemic stroke. Stroke 24(12):1784–1788

135. Cappa S, Perani D, Grassi F, Bressi S, Alberoni M, Franceschi M et al (1997) A PET follow-up study of recovery after stroke in acute aphasics. Brain Lang 56(1):55–67

Part IV

Future fMRI Development

Chapter 28

Multimodal Fusion of Structural and Functional Brain Imaging Data

Jing Sui and Vince D. Calhoun

Abstract

Recent years have witnessed a rapid growth of interest in moving functional magnetic resonance imaging (fMRI) beyond simple scan-length averages and into approaches that can integrate structural MRI measures and capture rich multimodal interactions. It is becoming increasingly clear that multimodal fusion is able to provide more information for individual subjects by exploiting covariation between modalities, rather an analysis of each modality alone. Multimodal fusion is a more complicated endeavor that must be approached carefully and efficient methods should be developed to draw generalized and valid conclusions out of high dimensional data with a limited number of subjects, such as patients with brain disorders. Numerous research efforts have been reported in the field based on various statistical models, including independent component analysis (ICA), canonical correlation analysis (CCA), and partial least squares (PLS). In this chapter, we survey a number of methods previously shown in multimodal fusion reports, performed with or without prior information, and with their possible strengths and limitations addressed. To examine the function–structure associations of the brain in a more comprehensive and integrated manner, we also reviewed a number of multimodal studies that combined fMRI and structural (sMRI and/or diffusion tensor MRI) measures, which could reveal important brain alterations that may not be fully detected by employing separate analysis of individual modalities, and also enable us to identify potential brain illness biomarkers.

Key words Multimodal fusion methods, Data driven, Functional magnetic resonance imaging, Structural MRI, Diffusion MRI, Independent component analysis, Canonical correlation analysis

1 Introduction

There is increasing evidence that instead of focusing on the relationship between physiological or behavioral features using a single imaging modality, multimodal brain imaging studies can help provide a better understanding of inter-subject variability from how brain structure shapes brain function, to what degree brain function feeds back to change its structure, and what functional or structural aspects of physiology ultimately drive cognition and behavior. Collecting multiple modalities of magnetic resonance imaging (MRI) data from the same individual has become a common practice, in order to search for task- or disease-related changes.

Massimo Filippi (ed.), *fMRI Techniques and Protocols*, Neuromethods, vol. 119,
DOI 10.1007/978-1-4939-5611-1_28, © Springer Science+Business Media New York 2016

Each imaging technique provides a different view in examining the brain activity. For example, functional MRI (fMRI) measures the hemodynamic response related to neural activity in the brain dynamically; structural MRI (sMRI) provides information about the tissue type of the brain [gray matter (GM), white matter (WM), cerebrospinal fluid (CSF)]. Diffusion tensor (DT) MRI can additionally provide information on structural connectivity among brain networks. A key motivation for jointly analyzing multimodal data is to take advantage of the cross-information in the existing data, thereby potentially revealing important variations that may only partially be detected by a single modality. The availability of several modal measurements allows joint analysis via the application of a number of statistical approaches, including (but not being limited to) correlational analyses [1], data integration [2, 3] or data fusion [4, 5] based on higher-order statistics and/or modern machine learning algorithms. These methods enable indirect or direct associations to be inferred on putative structure–function relationships [6], however, it is not necessary for these modalities to have been measured simultaneously or have later been processed in a concurrent fusion model.

Approaches for combining or fusing data in brain imaging can be conceptualized as having a place on an analytic spectrum with meta-analysis (highly distilled data) to examine convergent evidence at one end and large-scale computational modeling (highly detailed theoretical modeling) at the other end [7]. In between are methods that attempt to perform a direct data fusion [8]. We next review several multivariate multimodal fusion methods including their statistical assumption, possible strengths and limitations, and multimodal neuroimaging applications, especially combining fMRI with structure measures. Behavioral relevance of the assessed physiological features will be mentioned whenever possible.

2   Summary of Approaches Applied in Joint Analysis of Multimodal MRI Data

Before introduction of the statistical models, we would like to clarify the use of "feature" as input to most of the mentioned models. We note that this definition of "feature" is somewhat different than what is used in traditional machine learning algorithms [9]. Basically, a "feature" is a distilled dataset representing the interesting part of each modality [8] and it contributes as an input vector for each modality and each subject. Usually the brain imaging data is high dimensional, in order to reduce the redundancy and facilitate the identification of relationships between modalities, the raw data can be preprocessed to generate a second-level output, that is "feature," which can be a contrast map calculated from task-related fMRI by the general linear model (GLM), a component image such as the "default mode" resulting from a first-level ICA, a fractional anisotropy (FA) from DT MRI measures, or segmented gray matter (GM) from sMRI data. The main reason to use features is to provide a simpler space in which to link the data. The trade-off is that some information may be lost, e.g., GM does not directly measure volume or cortical thickness and FA does not provide directional information; however, there is considerable evidence that the use of features is quite useful and valid [8, 10]. Figure 1 provides a direct view of the current popular multimodal data analysis approaches related to fMRI.

"Data integration" is an alternative, but dissimilar approach to "data fusion." One characteristic of data integration is that it is asymmetric, e.g., using the results from one modality to constrain models of the other—as DT MRI being constrained by fMRI or sMRI data in [11, 12], or fMRI-informed EEG [13, 14]. While

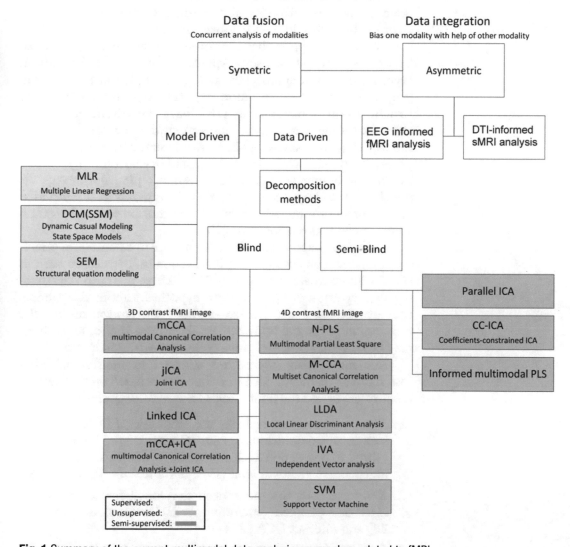

Fig. 1 Summary of the current multimodal data analysis approaches related to fMRI

these are powerful techniques, a limitation is that they may impose potentially unrealistic assumptions upon the constrained data, which are fundamentally of a different nature than the known modality. Another integration example is to analyze each data type separately and overlay them—thereby not allowing for any interaction between the data types. For example, a data integration approach would not detect a change in fMRI activation maps that is related to a change in brain structure remotely.

By contrast, multimodal fusion refers to the use of a common forward model of neuronal activity that explains different sorts of data symmetrically [15], which provides more views for individual subjects and covariation between modalities. This is a more complicated endeavor, especially when studying complex mental illnesses which impact many brain circuits. Also, in the real world, the conclusions usually need to be drawn out of high dimensional brain imaging data from a limited number of subjects. Hence efficient methods should be developed carefully.

Multivariate approaches for combing brain imaging data can be divided into two classes: hypotheses-driven and data-driven. Hypotheses-driven approaches such as multiple linear regression [16, 17], dynamic causal modeling (DCM), and confirmatory structural equation modeling [18], have the advantages of: (1) allow for testing specific hypotheses about the networks implied in the experimental paradigm; (2) allow for simultaneous assessment of several connectivity links, which would have been compromised by the one-by-one assessment of covariance [19]. However, it is possible to miss important connectivity links that were not included in a set of a priori hypotheses and it does not provide information about inter-voxel relationships [20, 21].

Data-driven approaches include, but are not limited to, principal component analysis (PCA), ICA, canonical correlation analysis (CCA), and partial least squares (PLS). These methods are attractive as they do not require a priori hypotheses about the connection of interest. Hence, these methods are convenient for the exploration of the full body of data. However, the some methods may be more demanding from a computational standpoint.

The multivariate approaches adopted in multimodal MRI fusion can be further divided into four classes based on the requirement of priori and the dimension of the MRI data used:

1. *Blind* methods that typically use second-level fMRI data (3D contrast image), including joint ICA (jICA) [22], multimodal canonical correlation analysis (mCCA) [23], linked ICA [24], independent vector analysis(IVA) and mCCA+jICA [25, 26].

2. *Blind* methods that have been developed for use with raw fMRI data (4D data), including partial least squares (PLS) [27, 28] and multiset CCA [29].

3. *Semi-blind* methods that use second-level fMRI data (3D contrast), e.g., parallel ICA [30], coefficient-constrained ICA

(CC-ICA) [31, 32], PCA with reference (PCA-R) [33, 34] and informed multimodal PLS [27].

4. Other multimodal fusion applications using 4D fMRI and EEG data, include multiple linear regression, structural equation modeling (SEM) [18] or DCM [35].

Here we will emphasize on the data-driven multivariate fusion methods due to their flexibilities and advantages. The optimization strategies of seven above mentioned multivariate models are displayed in Fig. 2. In the following sections, we will introduce five blind fusion models and their applications in combining functional and structural MRI in detail.

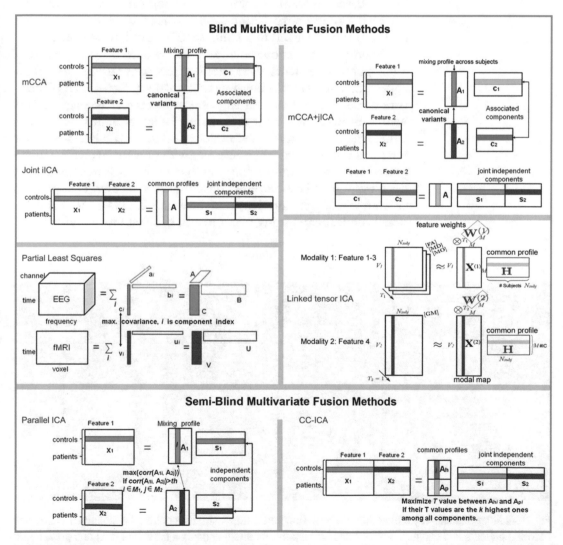

Fig. 2 A summary of seven blind and semi-blind data-driven methods for multimodal fusion. Figure modified and reprinted with permission from Sui et al. [4]

3   Review of Multivariate Models Applied in Multimodal Fusion with fMRI

3.1   Joint ICA

Joint ICA (jICA) is a second-level fMRI analysis method that assumes two or more features (modalities) share the same mixing matrix and maximizes the independence among joint components [36]. This is a straightforward yet effective method by performing ICA on the horizontally concatenated features (along voxels). It is suitable for examining the common connection among modalities and requires acceptance of the likelihood of changes in one data type (e.g., GM) being related to another one, such as functional activation. Joint ICA is feasible to many paired combinations of features, such as fMRI, sMRI, and DT MRI, or three-way data fusion [37–41]. In order to control for intensity differences in MR images based on scanner, template, and population variations, usually each feature matrix (dimension: number of subject by number of voxel) is normalized to a study specific template [42, 43].

Figure 3 [36], shows analyzed data collected from groups of schizophrenia patients and healthy controls using the jICA approach. The main finding was that group differences in bilateral parietal and frontal as well as posterior temporal regions in GM matter distinguished groups. A finding of less patient GM and less hemodynamic activity for target detection in these bilateral anterior temporal lobe regions was consistent with previous work. An unexpected corollary to this finding was that, in the regions showing the largest group differences, GM concentrations were larger in patients vs. controls, suggesting that more GM may be related to less functional connectivity during performance of an auditory oddball task.

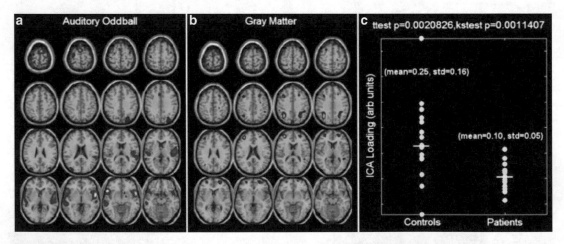

Fig. 3 Auditory oddball/gray matter jICA analysis. Only one component demonstrated a significant difference between patients and controls. The joint source map for the auditory oddball fMRI data (*left*) and gray matter (*middle*) data is presented along with the loading parameters for patients and controls (*far right*). Reproduced with permission from Calhoun et al. [36]

3.2 Multimodal CCA

Multimodal CCA allows a different mixing matrix for each modality and is used to find the transformed coordinate system that maximizes these inter-subject covariations across the two data sets [44]. This method decomposes each dataset into a set of components and their corresponding mixing profile, which is called canonical variants (CVs). The CVs have varying levels of activations for different subjects and are linked if they modulate similarly across subjects. After decomposition, the CVs correlate each other only on the same indices and their corresponding correlation values are called canonical correlation coefficients. Compared to jICA that constrains two features to have the same mixing matrix, mCCA is flexible in that it allows common as well as distinct level of connection between two features, as shown in Fig. 4, but the associated source maps may not be spatially sparse, especially when the canonical correlation coefficients are not sufficiently distinct [4].

Multimodal CCA is invariant to differences in the range of the data types and can be used to jointly analyze very diverse data

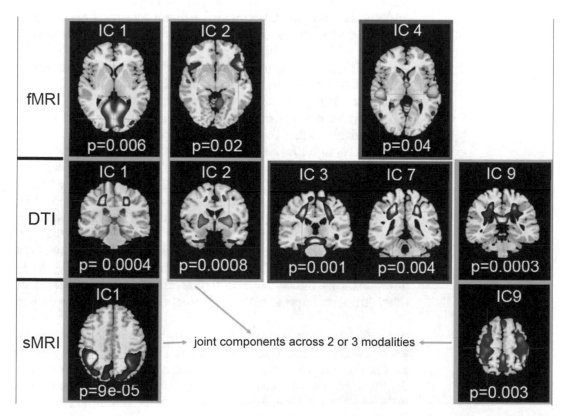

Fig. 4 MCCA + jICA enables people to capture components of interest that are either common or distinct across modalities. For example, when examining group differences across three modalities, joint ICs are significantly group-discriminative in more than two modalities (*green framed*), while modality-specific discriminative ICs (*pink framed*), i.e., fMRI_IC4, DTI_IC3, and DTI_IC7 only show significant group difference in a single modality. Reproduced with permission from Sui et al. [5]

types. It can also be extended to multiset CCA to incorporate more than two modalities [45]. Note that multimodal CCA works on the second-level fMRI feature-contrast maps, while multiset-CCA can work with 4D raw fMRI data (not shown in Fig. 1), e.g., to maximize the trial-to-trial covariation between the concurrent fMRI and EEG data [29].

3.3 Partial Least Squares

Partial Least Squares (PLS) as part of a family of multivariate data analyses, is based on the definition of a linear relationship between a dependent variable and predictor variables, and hence the goal is to determine which aspect of a set of observations (e.g., imaging data) are related directly to another set of data (e.g., experimental design, behavior) [46]. PLS was first applied to multimodal fusion by Martinen-Montes et al. [28], where the multiway PLS (N-PLS) was proposed to find correlations between fMRI time courses (dependent variables) and the spectral components of the EEG data (independent variables) from a single subject. Chen et al. [27] further proposed a multimodal PLS (MMPLS) to simultaneously characterize the linkage between patterns of PET and GM. The multimodal PLS can be performed either informed to the variable of interest such as age or agnostic to this additional information (with or without *priori*). Investigators may want to pre-specify which of these two methods to use in the data analysis. The agnostic MMPLS was used to identify the linkage between PET (dependent data block) and sMRI (independent data block), which is mainly oriented for extraction of covariance patterns and related latent variables rather than for classifications.

Even though PLS has some similarity to CCA in that they both maximize between-set correlations, PLS is based on the definition of a dependency and works well especially when the dependence among the constituents of the datasets is explicitly assessed [47]; while CCA does not assign independent/dependent labels to either of the modalities and treats both equally [48]. Hence PLS is particularly suited to the analysis of relationships between measures of brain activity and of behavior or experimental design [49], while interpretational difficulties in PLS often arise when the effects identified do not correspond to the a priori expectations of the researcher.

3.4 mCCA + jICA

According to many previous findings in brain connectivity studies which combined function and structure [19, 50], it is plausible to assume the components decomposed from each modality have some degree of correlation between their mixing profiles among subjects. mCCA + jICA is a blind data-driven model that is optimized for this situation [26, 51] and also to have excellent performance for achieving both flexible modal association and source separation. It takes advantage of two complementary approaches: mCCA and jICA, thus allowing both strong and weak inter-modality connection as well as the joint independent components. mCCA makes the jICA job more reliable by providing a closer initial match via correlation; while jICA further decomposes the remaining mixtures in the associated maps and relax the requirement of sufficiently distinction imposed on the canonical correlation. The code of jICA, mCCA+jICA can be accessed via the Fusion ICA Toolbox (FIT, http://mialab.mrn.org/software/fit/). Note that the mCCA+jICA approach does not increase the computational load appreciably and is not limited to two-way fusion, but can potentially be extended to three-way or N-way fusion of multiple data types by replacing the multimodal CCA with multiset CCA [45]. It enables robust identification of correspondence among N diverse data types and enables one to investigate the important question of whether certain disease risk factors are shared or are distinct across multiple modalities. In accordance with this notion, this approach has already been used to fuse fMRI, sMRI, and DT MRI to study schizophrenia [25, 51], as shown in Fig. 4.

3.5 Linked ICA

Linked ICA is a fully probabilistic approach based on a modular Bayesian framework, which is designed for simultaneously modeling and discovering common characteristics across multiple modalities [24]. The combined modalities can potentially have different units, noise level, spatial smoothness, and intensity distributions. Each modality group is modeled using a Bayesian tensor ICA model [52]. Note that the Bayesian ICA differs from standard methods like FastICA [53] and Infomax [54] in that it incorporates dimensionality reduction into the ICA method itself by the use of automatic relevance determination priors on components [55] and works on the full-dimensionality data directly. Linked ICA is good at detecting and isolating single-modality noise; however, it is much more computationally demanding than the above four methods, and the spatial maps of the decomposed components tend to be scattered. In a real application, linked ICA was applied to high functioning autism by combining DT MRI, voxel-based morphometry (VBM), and resting-state fMRI connectivity to assess differences in brain structure and function [56].

4 Other Multimodal Applications Based on Functional and Structural MRI Measures

All above multivariate, data-driven approaches focus on examining the inter-modality covariance, which provides a natural way to find multimodality associations. Whereas, there are many other multimodal studies in the context of cross-modal connectivity and classifications incorporating both functional and structural MRI measures.

Historically, the combination of structural and functional brain imaging has been used for the purpose of analyzing functional activity in a priori defined (based on atlas) or data-driven selected (e.g., based on ICA components) brain regions. A central assumption of systems neuroscience is that the structure of the brain can predict and/or is related to functional connectivity. The findings of Segall et al. [57] support this hypothesis, which generally show that each single structural component derived from ICA usually corresponds to several resting-state functional components. Functional information has been shown to be able to improve the correspondence of functional boundaries across subjects beyond the standard structural normalization [58]. Studies on psychopathological phenomena could also discover spatial overlaps of structural and functional alterations in schizophrenia or at risk mental state using cognitive tasks and GM volume. Salgado-Pineda et al. [59] found three regions including the thalamus, the anterior cingulate, and the inferior parietal that showed both structural and functional impairments associated with attentional processing in schizophrenia. A follow-up study of the same group [60] also found both functional alterations (facial emotion task) and GM volume reductions in the DMN in schizophrenia.

On the other hand, many of the conducted fMRI-DT MRI studies addressed disruptions of brain connectivity seen with mental illnesses such as depression, schizophrenia, Alzheimer's disease, and bipolar disorder. For example [61], evaluated interactions between measurements of anatomical and functional connectivity collected in the same subjects to study global schizophrenia-related alterations in brain connectivity. Schlosser et al. observed a direct correlation in schizophrenia between frontal FA reduction and fMRI activation in regions of the prefrontal and occipital cortices [62]. This finding highlights a potential relationship between anatomical changes in a frontal-temporal anatomical circuit and functional alterations in the prefrontal cortex. Another study [63] also illustrated that DT MRI- and fMRI-derived topologies are similar, and that the fMRI-DT MRI combination can provide additional information in order to choose reasonable seed regions for identifying functionally relevant networks and to validate reconstructed WM fibers. A recent study of Koch et al. [64] showed that WM fiber integrity in terms of increased radial diffusivity of the left superior temporal gyrus is associated with reduced neuronal activation in lateral frontal and cingulate cortices, suggesting that intact WM connectivity plays an important role for the pattern and intensity of functional activations with neuronal networks engaged in decision making.

As to the classification, there are additional studies demonstrating that the potential of the fusion of structural and functional data may improve the brain disease classification. For example, mild cognitive impairment (MCI), often an early stage of Alzheimer's disease, is difficult to diagnose due to its rather mild and nearly insignificant symptoms of cognitive impairment. When trying to identify MCI patients from healthy controls, Kim et al. [65] showed that the integration of sMRI and fMRI can provide complementary information to improve the diagnosis of MCI relative to results from one modality alone (error rate: 6% using both versus 15% using fMRI only and 35% using sMRI only) [66]. Similarly, Wee et al. [67] integrated information from DT MRI and resting fMRI by employing multiple-kernel support vector machines (SVMs), yielding statistically significant improvement (>7.4%) in classification accuracy of predicting MCI from healthy controls by using multimodal data (96.3%) compared to using each modality independently. Furthermore, with the ensemble feature selection strategy and advance support vector machine, Westlet et al. [68] combined resting-state fMRI, EEG, and sMRI data to classify schizophrenia from healthy controls and achieved the best performance with 91% accuracy compared to each single modalities, confirming the effectiveness and advantages of multimodal fusion. The above findings suggest the multimodal classification facilitated by advanced modeling techniques could provide more accurate and early detection of brain abnormalities, which may not have been revealed through separate uni-modal analyses as typically performed in the majority of neuroimaging experiments.

5 Conclusions

It is abundantly clear that there is a diverse and growing collection of scientific tools available for noninvasively studying human brain functioning and relating it to cognitive and behavioral measures. Using these technologies, substantial progress has been made in characterizing structural/functional brain abnormalities and their interactions. In addition, a major goal in integrating/fusing approaches is to capitalize on the relative strengths of each modality, providing results synergistically. All of the multimodal studies that are reviewed in this paper are summarized and separated into different modality categories in Table 1. In general, most studies we reviewed demonstrate congruent effects across measurement modalities and combining modalities does provide more differentiating power among multiple diseases.

Although recent multimodal imaging results are promising [19, 69], much work remains to be done. As the field of multimodal imaging is relatively new, most of the studies represent novel findings; however, replication is needed to draw general conclusions about structure–function relationships. Secondly, multimodal fusion proves to be fruitful for a more informative understanding of brain activity and disorders, but fusing as many modalities/features as possible in the training sample does not guarantee best discrimination or classification between groups, as reported in [8, 70]; thus it would be helpful to compare a combination of uni-modal and multimodal results, as done in [71]. This work can be pursued in future by using larger data sets and various modalities. Furthermore, a main challenge in multimodal data fusion comes from

Table 1

Summary of studies combining measures of functional and structural MRI

Modality	Focus	Papers	Subject type	Methods
fMRI-sMRI	Connectivity	[6, 57, 59, 60, 65, 72–80]	HC-MDD, HC-SZ, HC-MCI-AD HC-TBI	Correlational analysis Multiple regression
	Covariance	[36, 44, 67, 81–83]	HC-SZ, HC-MDD, HC-AD	jICA, mCCA
fMRI-DTI	Connectivity	[61–64, 84–92]	Healthy children, HC-MDD, HC-SZ, HC-BP, HC-TBI HC-AD-MCI	Correlational analysis, SEM Multiple Regression
	Covariance	[26, 38, 40]	HC-SZ, HC-SZ-BP	jICA, mCCA+jICA
Three-way fusion	connectivity	[93–97]	HC-ADHD Children-Adults HC-MDD HC-psychotic	Correlational analysis Multiple regression
	Covariance	[8, 24, 25, 48, 51, 70, 98–101]	HC-SZ HC-AD HC-MCI-AD	jICA, mCCA, mCCA+jICA Linked ICA SVM
Other fusion applications	fMRI-EEG DTI-sMRI GM-WM	[16, 27–29, 42, 48, 102–110]	HC-SZ, HC-AD HC-MCI, HC-BP	Correlational analysis Multiple regression jICA, mCCA PCA, PLS, parallel ICA

dissimilarity of the data types being fused and result interpretation. However, emerging in 2009, N-way multimodal fusion may become one of the leading directions in future neuroimaging research given the predominance of multimodal data acquisition [5]. Finally, introducing multimodal analyses in longitudinal brain studies has not been done frequently yet, which could be another new direction, as there are many possibilities for modeling the baseline and change over multiple time points.

In summary, we are just beginning to unlock the potential of multimodal imaging, which provides unprecedented opportunities to further deepen our understanding of the brain disorders [51] based on various brain imaging measures. The most promising avenues for the future may rest on developing better models that can complement and exploit the richness of our data [15]. These models may well already exist in other disciplines (such as machine learning, machine vision, computational neuroscience, and behavioral economics) and may enable the broader neurosciences to access neuroimaging so that key questions can be addressed in a theoretically grounded fashion.

References

1. Skudlarski P, Jagannathan K, Calhoun VD et al (2008) Measuring brain connectivity: diffusion tensor imaging validates resting state temporal correlations. Neuroimage 43(3):554–561

2. Savopol F, Armenakis C (2002) Mergine of heterogeneous data for emergency mapping: data integration or data fusion? Proc. ISPRS

3. Ardnt C, Loffeld O (1996) Information gained by data fusion, Proc. SPIE, vol. 2784

4. Sui J, Adali T, Yu Q et al (2012) A review of multivariate methods for multimodal fusion of brain imaging data. J Neurosci Methods 204(1):68–81

5. Sui J, Huster R, Yu Q et al (2014) Function-structure associations of the brain: evidence from multimodal connectivity and covariance studies. Neuroimage 102P1:11–23

6. Schultz CC, Fusar-Poli P, Wagner G et al (2012) Multimodal functional and structural imaging investigations in psychosis research. Eur Arch Psychiatry Clin Neurosci 262(Suppl 2):S97–S106

7. Horwitz B, Poeppel D (2002) How can EEG/MEG and fMRI/PET data be combined? Hum Brain Mapp 17(1):1–3

8. Calhoun VD, Adali T (2009) Feature-based fusion of medical imaging data. IEEE Trans Inf Technol Biomed 13(5):711–720

9. Blum AL, Langley P (1997) Selection of relevant features and examples in machine learning. Artif Intell 97(1–2):245–271

10. Smith SM, Fox PT, Miller KL et al (2009) Correspondence of the brain's functional architecture during activation and rest. Proc Natl Acad Sci U S A 106(31):13040–13045

11. Goldberg-Zimring D, Mewes AU, Maddah M et al (2005) Diffusion tensor magnetic resonance imaging in multiple sclerosis. J Neuroimaging 15(4 Suppl):68S–81S

12. Ramnani N, Lee L, Mechelli A et al (2002) Exploring brain connectivity: a new frontier in systems neuroscience. Functional Brain Connectivity, 4–6 April 2002, Dusseldorf, Germany. Trends Neurosci 25:496–497

13. Henson RN, Flandin G, Friston KJ et al (2010) A parametric empirical Bayesian framework for fMRI-constrained MEG/EEG source reconstruction. Hum Brain Mapp 31(10):1512–1531

14. Lemieux L (2004) Electroencephalography-correlated functional MR imaging studies of epileptic activity. Neuroimaging Clin N Am 14(3):487–506

15. Friston KJ (2009) Modalities, modes, and models in functional neuroimaging. Science 326(5951):399–403

16. De Martino F, Valente G, de Borst AW et al (2010) Multimodal imaging: an evaluation of univariate and multivariate methods for simultaneous EEG/fMRI. Magn Reson Imaging 28(8):1104–1112

17. Eichele T, Specht K, Moosmann M et al (2005) Assessing the spatiotemporal evolution of neuronal activation with single-trial event-related potentials and functional MRI. Proc Natl Acad Sci U S A 102(49):17798–17803

18. Astolfi L, Cincotti F, Mattia D et al (2004) Estimation of the effective and functional human cortical connectivity with structural equation modeling and directed transfer function applied to high-resolution EEG. Magn Reson Imaging 22(10):1457–1470

19. Rykhlevskaia E, Gratton G, Fabiani M (2008) Combining structural and functional neuroimaging data for studying brain connectivity: a review. Psychophysiology 45(2):173–187

20. Oakes TR, Fox AS, Johnstone T et al (2007) Integrating VBM into the General Linear Model with voxelwise anatomical covariates. Neuroimage 34(2):500–508

21. Schlosser R, Gesierich T, Kaufmann B et al (2003) Altered effective connectivity during working memory performance in schizophrenia: a study with fMRI and structural equation modeling. Neuroimage 19(3):751–763

22. Calhoun VD, Adali T, Kiehl KA et al (2006) A method for multitask fMRI data fusion applied to schizophrenia. Hum Brain Mapp 27(7):598–610

23. Correa N, Adali T, Calhoun VD (2007) Performance of blind source separation algorithms for fMRI analysis using a group ICA method. Magn Reson Imaging 25(5):684–694

24. Groves AR, Beckmann CF, Smith SM et al (2011) Linked independent component analysis for multimodal data fusion. Neuroimage 54(3):2198–2217

25. Sui J, He H, Pearlson GD et al (2012) Three-way (N-way) fusion of brain imaging data based on mCCA+jICA and its application to discriminating schizophrenia. Neuroimage 2(66):119–132

26. Sui J, Pearlson G, Caprihan A et al (2011) Discriminating schizophrenia and bipolar disorder by fusing fMRI and DTI in a multimodal CCA+ joint ICA model. Neuroimage 57(3):839–855

27. Chen K, Reiman EM, Huan Z et al (2009) Linking functional and structural brain images with multivariate network analyses: a novel application of the partial least square method. Neuroimage 47(2):602–610

28. Martinez-Montes E, Valdes-Sosa PA, Miwakeichi F et al (2004) Concurrent EEG/fMRI analysis by multiway Partial Least Squares. Neuroimage 22(3):1023–1034

29. Correa NM, Eichele T, Adali T et al (2010) Multi-set canonical correlation analysis for the fusion of concurrent single trial ERP and functional MRI. Neuroimage 50(4):1438–1445

30. Liu J, Pearlson G, Windemuth A et al (2009) Combining fMRI and SNP data to investigate connections between brain function and genetics using parallel ICA. Hum Brain Mapp 30(1):241–255

31. Sui J, Adali T, Pearlson GD et al (2009) A method for accurate group difference detection by constraining the mixing coefficients in an ICA framework. Hum Brain Mapp 30(9):2953–2970

32. Sui J, Adali T, Pearlson GD et al (2009) An ICA-based method for the identification of optimal FMRI features and components using combined group-discriminative techniques. Neuroimage 46(1):73–86

33. Caprihan A, Pearlson GD, Calhoun VD (2008) Application of principal component analysis to distinguish patients with schizophrenia from healthy controls based on fractional anisotropy measurements. Neuroimage 42(2):675–682

34. Liu J, Xu L, Calhoun VD (2008). Extracting principle components for discriminant analysis of FMRI images. ICASSP 449–452

35. Ulloa AE, Chen J, Vergara VM et al (2014) Association between copy number variation losses and alcohol dependence across African American and European American ethnic groups. Alcohol Clin Exp Res 38(5):1266–1274

36. Calhoun VD, Adali T, Giuliani NR et al (2006) Method for multimodal analysis of independent source differences in schizophrenia: combining gray matter structural and auditory oddball functional data. Hum Brain Mapp 27(1):47–62

37. Xu L, Pearlson G, Calhoun VD (2009) Joint source based morphometry identifies linked gray and white matter group differences. Neuroimage 44(3):777–789

38. Franco AR, Ling J, Caprihan A et al (2008) Multimodal and multi-tissue measures of connectivity revealed by joint independent component analysis. IEEE J Sel Top Signal Process 2(6):986–997

39. Calhoun VD, Adali T, Liu J (2006) A feature-based approach to combine functional MRI, structural MRI and EEG brain imaging data. Conf Proc IEEE Eng Med Biol Soc 1:3672–3675

40. Teipel SJ, Bokde AL, Meindl T et al (2010) White matter microstructure underlying default mode network connectivity in the human brain. Neuroimage 49(3):2021–2032

41. Eichele T, Calhoun VD, Debener S (2009) Mining EEG-fMRI using independent component analysis. Int J Psychophysiol 73(1):53–61

42. Xu L, Groth KM, Pearlson G et al (2009) Source-based morphometry: the use of independent component analysis to identify gray matter differences with application to schizophrenia. Hum Brain Mapp 30(3):711–724

43. Shenton ME, Dickey CC, Frumin M et al (2001) A review of MRI findings in schizophrenia. Schizophr Res 49(1–2):1–52

44. Correa NM, Li YO, Adali T et al (2008) Canonical correlation analysis for feature-based fusion of biomedical imaging modalities and its application to detection of associative networks in schizophrenia. IEEE J Sel Top Signal Process 2(6):998–1007

45. Li Y-O, Adali T, Wang W et al (2009) Joint blind source separation by multiset canonical correlation analysis. IEEE Trans Signal Process 57(10):3918–3929

46. Lin FH, McIntosh AR, Agnew JA et al (2003) Multivariate analysis of neuronal interactions in the generalized partial least squares framework: simulations and empirical studies. Neuroimage 20(2):625–642

47. Krishnan A, Williams LJ, McIntosh AR et al (2010) Partial Least Squares (PLS) methods for neuroimaging: a tutorial and review. Neuroimage 56(2):455–475

48. Correa NM, Adali T, Li YO et al (2010) Canonical correlation analysis for data fusion and group inferences: examining applications of medical imaging data. IEEE Signal Process Mag 27(4):39–50

49. McIntosh AR, Lobaugh NJ (2004) Partial least squares analysis of neuroimaging data: applications and advances. Neuroimage 23(Suppl 1):S250–S263

50. Camara E, Rodriguez-Fornells A, Munte TF (2010) Microstructural brain differences predict functional hemodynamic responses in a reward processing task. J Neurosci 30(34):11398–11402

51. Sui J, He H, Yu Q et al (2013) Combination of resting state fMRI, DTI, and sMRI data to discriminate schizophrenia by N-way MCCA + jICA. Front Hum Neurosci 7:235

52. Beckmann CF, Smith SM (2005) Tensorial extensions of independent component analysis for multisubject FMRI analysis. Neuroimage 25(1):294–311

53. Hyvarinen A, Oja E (2000) Independent component analysis: algorithms and applications. Neural Netw 13(4–5):411–430

54. Bell AJ, Sejnowski TJ (1995) An information-maximization approach to blind separation and blind deconvolution. Neural Comput 7(6):1129–1159

55. Bishop CM (1999) Variational principal components. Artif Neural Netw 7:509, Conference Publication No. 470

56. Liu J, Chen J, Ehrlich S et al (2014) Methylation patterns in whole blood correlate with symptoms in schizophrenia patients. Schizophr Bull 40(4):769–776

57. Segall JM, Allen EA, Jung RE et al (2012) Correspondence between structure and function in the human brain at rest. Front Neuroinform 6:10

58. Khullar S, Michael AM, Cahill ND et al (2011) ICA-fNORM: spatial normalization of fMRI data using intrinsic group-ICA networks. Front Syst Neurosci 5:93

59. Salgado-Pineda P, Junque C, Vendrell P et al (2004) Decreased cerebral activation during CPT performance: structural and functional deficits in schizophrenic patients. Neuroimage 21(3):840–847

60. Salgado-Pineda P, Fakra E, Delaveau P et al (2011) Correlated structural and functional brain abnormalities in the default mode network in schizophrenia patients. Schizophr Res 125(2–3):101–109

61. Skudlarski P, Jagannathan K, Anderson K et al (2010) Brain connectivity is not only lower but different in schizophrenia: a combined anatomical and functional approach. Biol Psychiatry 68(1):61–69

62. Schlosser RG, Nenadic I, Wagner G et al (2007) White matter abnormalities and brain activation in schizophrenia: a combined DTI and fMRI study. Schizophr Res 89(1–3):1–11

63. Staempfli P, Reischauer C, Jaermann T et al (2008) Combining fMRI and DTI: a framework for exploring the limits of fMRI-guided DTI fiber tracking and for verifying DTI-based fiber tractography results. Neuroimage 39(1):119–126

64. Koch K, Wagner G, Schachtzabel C et al (2011) Neural activation and radial diffusivity in schizophrenia: combined fMRI and diffusion tensor imaging study. Br J Psychiatry 198(3):223–229

65. Kim J, Lee JH (2012) Integration of structural and functional magnetic resonance imaging improves mild cognitive impairment detection. Magn Reson Imaging 31:718–732

66. Fan Y, Resnick SM, Wu X et al (2008) Structural and functional biomarkers of prodromal Alzheimer's disease: a high-dimensional pattern classification study. Neuroimage 41(2):277–285

67. Wee CY, Yap PT, Zhang D et al (2011) Identification of MCI individuals using structural and functional connectivity networks. Neuroimage 59(3):2045–2056

68. Westlye LT, Walhovd KB, Bjornerud A et al (2009) Error-related negativity is mediated by fractional anisotropy in the posterior cingulate gyrus—a study combining diffusion tensor imaging and electrophysiology in healthy adults. Cereb Cortex 19(2):293–304

69. Sui J, Yu Q, He H et al (2012) A selective review of multimodal fusion methods in schizophrenia. Front Hum Neurosci 6:27

70. Zhang H, Liu L, Wu H, Fan Y (2012) Feature selection and SVM classification of multiple modality images for predicting MCI, in OHBM, Beijing, China.

71. Kim DI, Sui J, Rachakonda S et al (2010) Identification of imaging biomarkers in schizophrenia: a coefficient-constrained independent component analysis of the mind multi-site schizophrenia study. Neuroinformatics 8(4):213–229

72. Tian L, Meng C, Yan H et al (2011) Convergent evidence from multimodal imaging reveals amygdala abnormalities in schizophrenic patients and their first-degree relatives. PLoS One 6(12):e28794

73. Casey BJ, Tottenham N, Liston C et al (2005) Imaging the developing brain: what have we learned about cognitive development? Trends Cogn Sci 9(3):104–110

74. Smieskova R, Allen P, Simon A et al (2012) Different duration of at-risk mental state associated with neurofunctional abnormalities. A multimodal imaging study. Hum Brain Mapp 33(10):2281–2294

75. Rasser PE, Johnston P, Lagopoulos J et al (2005) Functional MRI BOLD response to Tower of London performance of first-episode schizophrenia patients using cortical pattern matching. Neuroimage 26(3):941–951

76. Fusar-Poli P, Broome MR, Woolley JB et al (2011) Altered brain function directly related to structural abnormalities in people at ultra high risk of psychosis: longitudinal VBM-fMRI study. J Psychiatr Res 45(2):190–198

77. Michael AM, Baum SA, White T et al (2010) Does function follow form?: methods to fuse structural and functional brain images show decreased linkage in schizophrenia. Neuroimage 49(3):2626–2637

78. Michael AM, King MD, Ehrlich S et al (2011) A data-driven investigation of gray matter-function correlations in schizophrenia during a working memory task. Front Hum Neurosci 5:71

79. Rektorova I, Mikl M, Barrett J et al (2012) Functional neuroanatomy of vocalization in patients with Parkinson's disease. J Neurol Sci 313(1–2):7–12

80. Harms MP, Wang L, Csernansky JG et al (2012) Structure-function relationship of working memory activity with hippocampal and prefrontal cortex volumes. Brain Struct Funct 218(1):173–186

81. Choi K, Yang Z, Hu X et al (2008) A combined functional-structural connectivity analysis of major depression using joint independent components analysis. Psychiatric MRI/MRS 3555

82. Camchong J, MacDonald AW III, Bell C et al (2011) Altered functional and anatomical connectivity in schizophrenia. Schizophr Bull 37(3):640–650

83. Kim DJ, Park B, Park HJ (2012) Functional connectivity-based identification of subdivisions of the basal ganglia and thalamus using multilevel independent component analysis of resting state fMRI. Hum Brain Mapp 34:1371–1385

84. Olesen PJ, Nagy Z, Westerberg H et al (2003) Combined analysis of DTI and fMRI data reveals a joint maturation of white and grey matter in a fronto-parietal network. Brain Res Cogn Brain Res 18(1):48–57

85. Matthews SC, Strigo IA, Simmons AN et al (2011) A multimodal imaging study in U.S. veterans of Operations Iraqi and Enduring Freedom with and without major depression after blast-related concussion. Neuroimage 54(Suppl 1):S69–S75

86. Zhou Y, Shu N, Liu Y et al (2008) Altered resting-state functional connectivity and anatomical connectivity of hippocampus in schizophrenia. Schizophr Res 100(1–3):120–132

87. Yan H, Tian L, Yan J et al (2012) Functional and anatomical connectivity abnormalities in cognitive division of anterior cingulate cortex in schizophrenia. PLoS One 7(9):e45659

88. Soldner J, Meindl T, Koch W et al (2011) Structural and functional neuronal connectivity in Alzheimer's disease: a combined DTI and fMRI study. Nervenarzt 83(7):878–887

89. Wang F, Kalmar JH, He Y et al (2009) Functional and structural connectivity between the perigenual anterior cingulate and amygdala in bipolar disorder. Biol Psychiatry 66(5):516–521

90. Schonberg T, Pianka P, Hendler T et al (2006) Characterization of displaced white matter by brain tumors using combined DTI and fMRI. Neuroimage 30(4):1100–1111

91. Voss HU, Schiff ND (2009) MRI of neuronal network structure, function, and plasticity. Prog Brain Res 175:483–496

92. Palacios EM, Sala-Llonch R, Junque C et al (2012) White matter integrity related to functional working memory networks in traumatic brain injury. Neurology 78(12):852–860

93. Jacobson S, Kelleher I, Harley M et al (2009) Structural and functional brain correlates of subclinical psychotic symptoms in 11–13 year old schoolchildren. Neuroimage 49(2):1875–1885

94. Supekar K, Uddin LQ, Prater K et al (2010) Development of functional and structural connectivity within the default mode network in young children. Neuroimage 52(1):290–301

95. Pomarol-Clotet E, Canales-Rodriguez EJ, Salvador R et al (2010) Medial prefrontal cortex pathology in schizophrenia as revealed by convergent findings from multimodal imaging. Mol Psychiatry 15(8):823–830

96. Sexton CE, Allan CL, Le Masurier M et al (2012) Magnetic resonance imaging in late-life depression: multimodal examination of network disruption. Arch Gen Psychiatry 69(7):680–689

97. Qiu MG, Ye Z, Li QY et al (2011) Changes of brain structure and function in ADHD children. Brain Topogr 24(3–4):243–252

98. Sui J, He H, Liu J et al (2012) Three-way FMRI-DTI-methylation data fusion based on mCCA+jICA and its application to schizophrenia. Conf Proc IEEE Eng Med Biol Soc 2012:2692–2695

99. Zhang D, Wang Y, Zhou L et al (2011) Multimodal classification of Alzheimer's disease and mild cognitive impairment. Neuroimage 55(3):856–867

100. Groves AR, Smith SM, Fjell AM et al (2012) Benefits of multi-modal fusion analysis on a large-scale dataset: life-span patterns of inter-subject variability in cortical morphometry and white matter microstructure. Neuroimage 63(1):365–380

101. Sui J, Castro E, Hao H et al (2014) Combination of FMRI-SMRI-EEG data improves discrimina-

tion of schizophrenia patients by ensemble feature selection. The 36th Annual International Conference of the IEEE engineering in medicine and biology society (EMBC'14) Chicago, Illinois, USA, no. August 26–30

102. Eichele T, Calhoun VD, Moosmann M et al (2008) Unmixing concurrent EEG-fMRI with parallel independent component analysis. Int J Psychophysiol 67(3):222–234

103. Haller S, Xekardaki A, Delaloye C et al (2011) Combined analysis of grey matter voxel-based morphometry and white matter tract-based spatial statistics in late-life bipolar disorder. J Psychiatry Neurosci 36(6):391–401

104. Chen Z, Cui L, Li M et al (2011) Voxel based morphometric and diffusion tensor imaging analysis in male bipolar patients with first-episode mania. Prog Neuropsychopharmacol Biol Psychiatry 36(2):231–238

105. Meda SA, Jagannathan K, Gelernter J et al (2010) A pilot multivariate parallel ICA study to investigate differential linkage between neural networks and genetic profiles in schizophrenia. Neuroimage 53(3):1007–1015

106. Jagannathan K, Calhoun VD, Gelernter J et al (2010) Genetic associations of brain structural networks in schizophrenia: a preliminary study. Biol Psychiatry 68(7):657–666

107. Jamadar S, Powers NR, Meda SA et al (2010) Genetic influences of cortical gray matter in language-related regions in healthy controls and schizophrenia. Schizophr Res 129:141–148

108. Hao X, Xu D, Bansal R et al (2011) Multimodal magnetic resonance imaging: the coordinated use of multiple, mutually informative probes to understand brain structure and function. Hum Brain Mapp 34(2):253–271

109. Fusar-Poli P, McGuire P, Borgwardt S (2011) Mapping prodromal psychosis: a critical review of neuroimaging studies. Eur Psychiatry 27(3):181–191

110. Meda SA, Gill A, Stevens MC et al (2012) Differences in resting-state functional magnetic resonance imaging functional network connectivity between schizophrenia and psychotic bipolar probands and their unaffected first-degree relatives. Biol Psychiatry 71(10):881–889

Chapter 29

Functional MRI of the Spinal Cord

Patrick Stroman and Massimo Filippi

Abstract

Evidence to date shows that fMRI of the spinal cord (spinal fMRI) can reliably demonstrate regions involved with sensation of tactile, thermal, and painful stimuli, and with motor tasks. Spinal fMRI acquisition methods based on BOLD contrast have been recently optimized. Results have demonstrated the ability of spinal fMRI to provide objective assessments of sensory and motor function, and discriminate responses when modulated by cognitive/emotional factors. Studies have been also carried out with patients with cord trauma, and in people with multiple sclerosis (MS). The availability of essentially automated analysis, large extent coverage of the spinal cord, and spatial normalization to permit comparisons with reference results and labeling of active regions are being implemented with the aim to translate the method into a practical clinical assessment tool.

The research completed so far indicates that spinal fMRI will be able to demonstrate where the neuronal activity is altered at any level (cervical, thoracic, lumbar, or sacral), whether or not information is reaching the cord from the periphery, and whether or not there is descending modulation of the response. It may also be able to provide an objective measure of pain, and to demonstrate the extent and mechanism of changes over time after an injury.

Key words Spinal fMRI, Blood oxygen level dependent, Multiple sclerosis, Cord trauma, Pain

1 Introduction

Evidence to date shows that fMRI of the spinal cord (spinal fMRI) can reliably demonstrate regions involved with sensation of tactile, thermal, and painful stimuli, and with motor tasks. There is also reliable evidence of the descending modulation of activity in the spinal cord. While spinal fMRI has not yet been applied or verified in a clinical setting, its value is expected to be in its ability to provide objective assessments of sensory and motor function, and discriminate responses when modulated by cognitive/emotional factors, or even detect responses when a patient cannot feel the stimulus or is even conscious. Studies have been carried out with patients with cord trauma, and in people with multiple sclerosis (MS) to investigate the clinical utility of the results. Robust methods for analysis, and for displaying the results in an effective manner

Massimo Filippi (ed.), *fMRI Techniques and Protocols*, Neuromethods, vol. 119,
DOI 10.1007/978-1-4939-5611-1_29, © Springer Science+Business Media New York 2016

to facilitate their interpretation, are also necessary. At present, the usefulness and reliability of spinal fMRI as a research tool has been demonstrated, analysis and display methods have been developed, and further improvements are rapidly developing. The research completed so far indicates that spinal fMRI will be able to demonstrate where the neuronal activity is altered at any level (cervical, thoracic, lumbar, or sacral), whether or not information is reaching the cord from the periphery, and whether or not there is descending modulation of the response. It may also be able to provide an objective measure of pain, and to demonstrate the extent and mechanism of changes over time after an injury.

In the following paragraphs, we discuss the current evidence for the most effective spinal fMRI method and the points that are still under debate, the applications that have been carried out to date and the degree of reliability and sensitivity these studies demonstrate, and the proposed future developments and applications.

2 Background of Spinal fMRI

As with any fMRI method, spinal fMRI requires alternated periods of stimulation and a reference condition, while a time-series of images is acquired over several minutes. Neuronal activity is detected only in gray matter regions, and is revealed by the local MR signal intensity having a component of signal change that corresponds with the stimulation paradigm. Unlike conventional brain fMRI, unique challenges are encountered in the heterogeneous tissues of the spine and spinal cord, because differences in magnetic susceptibilities between tissues produce spatial variations in the magnetic field. The spinal cord itself lies within the spinal canal surrounded by cerebrospinal fluid, and averages 45 cm in length, with cross-sectional dimensions of roughly 15 mm × 8 mm in the largest regions of the cervical and lumbar enlargements (Fig. 1). The entire cord is therefore in close proximity to the heart and lungs, and has been observed to move with each heart-beat, presumably as a result of the pulsatile CSF flow around it [1, 2]. An inherent challenge of fMRI is that it is necessary to monitor the signal intensity changes in a specific tissue volume, and this tissue may move between adjacent voxels over the course of the fMRI time-series, obscuring the measurements that are obtained. Moreover, other sources of signal intensity change such as random noise and motion artifacts can interfere with the detection of signal changes related to neuronal activity. While these challenges may seem daunting, most have been overcome by adapting methods for analysis and motion correction that have been developed for brain fMRI.

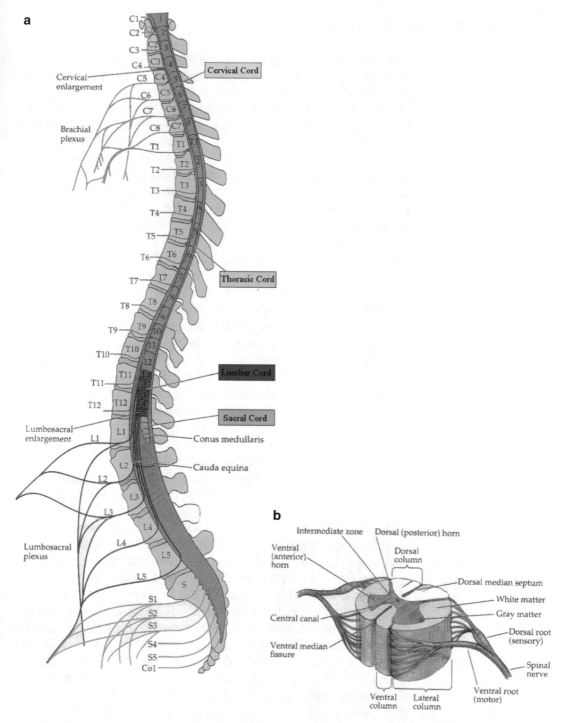

Fig. 1 Anatomy of the spinal cord (reproduced with permission from Blumenfeld H., Neuroanatomy Through Clinical Cases. Sunderland, MA: Sinauer Associates; 2002, p 22)

3   Spinal fMRI Acquisition Methods

The first published example of fMRI in the spinal cord was by Yoshizawa et al. in 1996 [3]. This work and the earliest attempts by other groups applied the established brain fMRI methods of the time to the spinal cord [4, 5]. The consistent features of the studies by Yoshizawa et al. [3], Stroman et al. [6], Madi et al. [7], and Backes et al. [8] were that they were carried out with healthy volunteers and employed a hand motor task with imaging of the cervical spinal cord. All used gradient-echo methods with echo times (TE) of 40–50 ms at 1.5 T, and 31 ms at 3 T, as is typical with brain fMRI, and all compared data obtained with transverse and sagittal slices. The areas of activity in the spinal cord that were demonstrated by these studies corresponded to areas of neuronal activity that were expected with the stimuli applied. The conclusions reported from each of these studies included the point that spinal fMRI is a feasible method for assessing neuronal activity in the cord. However, the results obtained also demonstrated variability in the areas of activity, and that it is difficult to obtain high-quality fMRI data in the spinal cord with gradient-echo methods and sensitivity to the BOLD effect.

The debate over the choice of imaging method was started with a study designed to verify that the BOLD effect occurred in the spinal cord [9], by comparing data obtained with gradient-echo and spin-echo methods at the same echo times. This was followed by a study to characterize the signal changes detected with spin-echo methods [10]. The BOLD theory shows that with the comparison in the first of these studies the gradient-echo method should produce signal intensity changes between rest and stimulation conditions that are three to four times higher than the spin-echo method [11]. The results, however, showed that the two methods produced signal intensity changes of approximately equal magnitudes, the image quality obtained with the spin-echo method was superior, and that the spin-echo method may therefore be superior for spinal fMRI. However, the apparent departure from the expected BOLD responses was unexplained.

A number of studies followed which investigated the underlying contrast mechanisms, and it was proposed that changes in tissue water content related to neuronal activity could augment the BOLD contrast at short echo time (TE) values [12–16]. However, it has been shown that the optimal contrast-to-noise ratio with spin-echo fMRI in the spinal cord is obtained when the TE value is set for optimal spin-echo BOLD contrast, at approximately 75 ms [17]. As a result, whether spin-echo or gradient-echo imaging methods are used for spinal cord fMRI, the contrast provided is BOLD, and the two methods have roughly equal sensitivity when optimized, as detailed below. The debate over the choice of imaging method continues however, between gradient-echo methods which provide greater speed, and spin-echo methods which provide better image quality. The real advantage of spin-echo methods was in its lower sensitivity to the nonuniform magnetic field environment in the spinal cord, and the ability to acquire images quickly without resorting the echo-planar imaging (EPI) spatial encoding methods.

3.1 Gradient-Echo Methods

As mentioned above, the earliest results with BOLD methods [3, 6–9, 18] showed promise that spinal fMRI is feasible. Yoshizawa et al. demonstrated areas of activity with a hand motor task (average signal changes 4.8 %) corresponded with consistent areas of the spinal cord gray matter; they also showed that the rostral-caudal distribution of activity corresponded well with the neuroanatomy [3]. This was the first spinal fMRI study, and it set the standard for the studies which followed by Stroman et al. [6, 9], Madi et al. [7], and Backes et al. [8], which had a number of similarities. As in the study by Yoshizawa et al., each of these employed gradient-echo imaging (fast gradient-echo or echo-planar encoding), relatively thick (5–10 mm) transverse slices, with the echo-time (TE) set for BOLD sensitivity. All of these studies investigated activity with motor tasks, two investigated activity with sensory stimuli as well [6, 9], and three of them [6–8] compared results obtained with sagittal (4–8 mm) and transverse slices.

The results of these studies showed a number of consistent features (Fig. 2). Signal intensity changes with hand motor tasks were consistently in the range of 4.3–4.8 % by Yoshizawa et al. [3] and Stroman et al. [6], and 0.5–7.5 % with graded force tasks by Madi et al. [7], and 8–12 % by Backes et al. [8], although there were

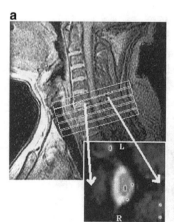

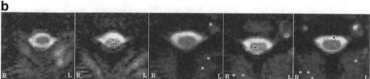

Fig. 2 Example of spinal fMRI results obtained with gradient-recalled echo EPI in transverse slices of the cervical spinal cord (reproduced from Stroman and Ryner. Magn. Reson. Imaging 19: 27–32, 2001) [9]. (**a**) The left side of the body is at the *top* of this image and the *red marks* indicate the locations which underwent intensity changes in response to sensory stimulation of the right hand. The location of the slice is indicated in relation to the sagittal view of the cervical spine shown in the larger image. (**b**) Five transverse slices corresponding to the slices indicated in the sagittal view

differences in field strength (1.5 T vs. 3 T) and image resolution which complicate any direct comparison. Each of these studies showed a rostral-caudal dependence of the areas with the task being performed, but only about half observed apparent laterality (left vs. right) of the active regions [3, 6]. Backes et al. [8] pointed out that there may be problems with the sensitivity to draining veins and the small anatomy, and the differences in spatial localization may simply be attributed to the extent of the draining vein field for the area being stimulated (ventral vs. dorsal). Another significant feature introduced by Backes et al., was the use of cardiac gating. In their results the cardiac gating can be expected to have significantly reduced the effects of cerebrospinal fluid (CSF) flow and spinal cord motion, and their results also showed the sensitivity of the BOLD methods to the draining veins leading from the gray matter to cord surface, as shown in more recent studies as well [19].

Another consistent feature of many of these studies was that they used echo-planar imaging (EPI), and made efforts to reduce the data readout time in order to reduce image distortion, including high bandwidth and sampling on the gradient ramps [7] or multishot acquisitions [8]. Maieron et al. [20] employed methods of parallel imaging (SENSE encoding) with EPI to reduce the distortion effects for spinal fMRI. This method showed improvements over previous methods but spatial distortions and variations in sensitivity in the rostral-caudal direction along the spinal cord were clearly visible. Even with extensive efforts to reduce distortions in images obtained with EPI encoding the image data suffer from severe distortions and areas of signal dropout, and they cannot be used in the presence of implanted devices to stabilize the spine after an injury.

More recent examples of applications of spinal fMRI with gradient-echo methods include one by Summers et al. who investigated details of BOLD responses to innocuous and noxious heat applied to the hand [21]. Their results demonstrated activity spread across a small S/I range in ipsilateral dorsal and contralateral ventral spinal cord, and BOLD time-course responses were shown with good correspondence with the timing of the thermal stimulation. This study was carried out an 3 T and employed a segmented gradient-echo EPI sequence, with an echo time of 23 ms, temporal resolution of 4 s, with images acquired in four segments (segment TR = 1 s). Fifteen, 4-mm-thick, axial slices were imaged with a field of view of 186×140 mm, acquisition matrix of 170×108 and a reconstructed voxel size of $0.98 \times 0.98 \times 4$ mm. The approach of dividing the acquisition into multiple segments allowed for a larger sampling matrix, reduced the distortion, and provided greater spatial precision in the results. Several other studies have also used gradient-echo EPI methods, with details below, and have focused on pain processing and changes with cognitive state such as attention, placebo, and nocebo [22–26]. These studies were focused primarily on sensory or pain processing, and effects such as placebo, nocebo, and attention modulation. Four of these examples used almost identical methods at 3 T, and used gradient-echo EPI acquisitions with 10–12 axial slices which were 5 mm thick, with slice-specific z-shimming, with in-plane resolution of 1 mm×1 mm with a 128×128 matrix, and parallel imaging (GRAPPA) with an acceleration factor of 2 [23–26]. The repetition times ranged from 1.17 to 1.5 s, and the TE was 40–42 ms. One recent study was slightly different and employed spiral encoding instead of EPI [22]. In this study a 128×128 matrix was used, with 1.25 mm×1.25 mm in-plane resolution, with 4 mm thick slices, with a 1.25 s TR and TE of 25 ms. In each of these examples, BOLD responses were detected in the spinal cord, ipsilateral to the stimulus, and were highly localized. In many cases the activity was reported at a single location in the spinal cord.

3.2 Spin-Echo Methods

One of the earliest spinal fMRI studies [10] was a comparison of the signal intensity time-course properties obtained with T_2-weighted and T_2^*-weighted acquisitions. The intent was to investigate whether the BOLD effect occurred in the spinal cord as in the brain. The T_2-weighted data had signal changes that were as large, or larger, than the T_2^*-weighted data at approximately the same echo time, and so the observations were not entirely consistent with the BOLD model. More importantly, the results indicated that spinal fMRI is feasible with both motor and sensory stimulation at 1.5 T, and can be achieved with good image quality with spin-echo imaging methods. A series of studies followed, to investigate the biophysical nature of the underlying contrast mechanism, and consistently showed significant BOLD effects, as well as a contribution from a proton-density change which was greater at shorter echo times [12–16, 27–30]. However, a key observation across these studies was that the spatial encoding method—EPI or fast spin-echo—is a critical factor in the choice of methods. A detailed analysis of the methods, including characterizations of the noise, physiological motion, and sensitivity to neuronal activity, demonstrated several key findings [17]. These included that it is important to avoid EPI methods for spinal fMRI, and that optimal sensitivity is obtained with spin-echo fMRI with an echo time (TE) of 75 ms (at 3 T), which corresponds with the T_2 of the spinal cord tissue. This finding agrees with the established BOLD theory. This increase in TE, compared to methods used in earlier studies, represented a small (20%) but significant increase in sensitivity. These findings also confirmed that the previous studies with an echo time of 38 ms were likely dominated by BOLD contrast, with only a small contribution from the proton-density change [31–37]. This point is important because it means that earlier studies done with a shorter echo time, and the more recent studies that are optimized for BOLD contrast, are dominated by the same contrast mechanism and the results are comparable.

The applications of spinal fMRI with spin-echo fMRI methods have included studies of sensory responses with thermal and vibration stimuli, pain processing, and effects of traumatic spinal cord injury and diseases such as MS. Lawrence et al. demonstrated the ability to localize activity in the spinal cord with vibration stimulation of different sensory dermatomes [38]. The spatial precision was further demonstrated with a later study of somatotopic mapping of thermal sensory responses in the cervical spinal cord and brainstem, with localization of C5 and C8 dermatomes and distinct right/left responses, as well as corresponding responses in brainstem regions [39]. A study of response characteristics compared constant thermal stimuli (i.e. block design) and a stimulus that gradually increased in intensity in a staircase pattern [40]. The static and dynamic thermal stimulation paradigms were designed to stimulate different peripheral receptors, and the comparison of the fMRI responses revealed significant differences in the spinal cord and brainstem in terms of the extent of the activity and the time-courses of the BOLD responses. These studies serve to demonstrate the spatial precision and the sensitivity of current spinal fMRI methods.

Several pain studies have also been carried out to date using spinal fMRI with spin-echo methods. One study demonstrated different responses in the spinal cord and brainstem with innocuous and noxious thermal stimuli [31]. BOLD responses were observed to be correlated with individual pain ratings in both the ipsilateral dorsal spinal cord, corresponding to the segment that was stimulated, and in the midbrain near the periaqueductal gray matter (PAG). Innocuous and noxious touch and brush stimuli were also compared in a separate study, and results demonstrated differences in fMRI responses in the spinal cord and brainstem [32]. Thermal pain responses in the spinal cord and brainstem have also been shown to be modulated by manipulating the attention focus of research participants [41], and by sensitizing the skin with capsaicin [35, 36]. Recently, the analgesic effects of listening to preferred music were investigated, and demonstrated moderate pain relief and corresponding changes in the cervical spinal cord and brainstem [42]. These studies consistently demonstrate a distribution of active regions in the cervical spinal cord with ipsilateral dorsal activity with a small rostral-caudal spread, within the stimulated segment of the cord and occasionally within adjacent segments, and contralateral ventral activity. Depending on the stimulus and study conditions, contralateral dorsal activity has also been observed. In addition, corresponding brainstem activity is consistently detected in studies that span the cervical spinal cord and brainstem. The active regions that have been detected, and their variation with study conditions, consistently demonstrate the role of descending modulation of sensory and pain responses in the spinal cord from brainstem regions. These represent important findings for our understanding of human pain processing, and the methods that have been developed show promise as tools for characterizing individual pain states.

A considerable component of the potential value of spinal cord fMRI methods depends on what they can reveal about pathological processes in the spinal cord, such as the effects of injury or disease or aberrant pain conditions. Several spinal fMRI studies have been carried out by the same group to investigate the effects of MS on spinal cord function [43–49]. These studies used tactile stimuli with spinal fMRI to probe sensory changes in MS, and demonstrated differences in responses in the cervical spinal cord depending on the MS subtype. Specifically, they observed that cord recruitment was increased in progressive MS patients compared to healthy control participants and in SPMS compared to PPMS patients. This finding also builds on prior studies that showed greater activity in RRMS and SPMS patients compared to controls, but no difference between patient groups. The effects of traumatic spinal cord injury on thermal sensory/pain processing have also been studied with spinal fMRI, even in patients with implanted devices to stabilize the spine. Cadotte et al. observed increased fMRI responses to thermal stimulation in the cervical spinal cord after injury, providing evidence of plasticity [50]. This work was built upon prior studies that demonstrated spinal cord responses to sensory stimulation or attempted movement tasks, below the site of injury [51, 52]. Again in these earlier studies, the fMRI responses were noted to be larger after spinal cord injury in many participants, compared to a group of healthy control participants.

The spin-echo methods that are widely used for spinal fMRI have also been quite similar across a number of recent studies, with an example shown in Fig. 3 [32, 35–37, 39–42, 45, 48, 50, 52].

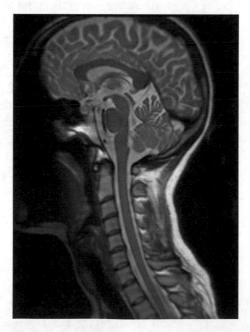

Fig. 3 Example of T_2-weighted fMRI data acquired with a half-Fourier single-shot fast spin-echo (HASTE) sequence at 3 T, with 1.5 mm × 1.5 mm in-plane resolution, 2 mm thick slices, a 192 × 144 acquisition matrix, and echo time of 76 ms

These studies were focused primarily on sensory or pain processing, and effects of attention focus, mood, or stimulation modality on activity detected in the spinal cord, or the effects of MS or spinal cord injury on sensory processing. Most of these examples used very similar methods at 3 T, and used single-shot fast spin-echo acquisitions, with partial k-space sampling (such as the HASTE sequence) with nine to ten contiguous sagittal slices which were 2 mm thick, with in-plane resolution of 1.5 mm × 1.5 mm with a 192 × 144 matrix. Parallel imaging has not been used in any of these studies to date. The repetition times ranged from 6.75 to 9 s. In earlier studies the TE was typically 32–40 ms. These studies employed BOLD contrast with a contribution from changes in proton-density, due to changes in tissue water content [15]. More recent studies have been carried out with longer TE values of 76 ms to optimize the BOLD contrast, as described above [40, 42].

4   Comparison of the Currently Most-Used Spin-Echo and Gradient-Echo fMRI Methods

The spinal fMRI methods that have been used in studies to date appear to have converged on two widely-used methods, one based on spin-echo to avoid EPI spatial encoding, and the other based on gradient-echo EPI. In terms of the magnitude of the BOLD response that is expected, when the two methods are both optimized, they can be expected to be approximately equal, as detailed in Table 1.

In Table 1 the terms $\Delta(1/T_2^*)$ and $\Delta(1/T_2)$ refer to the changes in transverse relaxation rates between two conditions, such as a "baseline" state and "active" state. It is well established that, under the same experimental conditions, $\Delta(1/T_2^*)$ is three to four times larger than $\Delta(1/T_2)$ and hence T_2^*-weighted imaging is most frequently used for fMRI [11, 28]. However, another important factor is that the echo time, TE, for optimal BOLD contrast is equal to the transverse relaxation time (i.e., T_2^* or T_2) which is roughly three times larger for spin-echo than gradient-echo. The net effect is that the magnitude of the BOLD response detected with spin-echo methods is nearly equal to that with gradient-echo methods, when both methods are set for optimal BOLD sensitivity.

The signal-to-noise ratio and image quality are also important considerations for fMRI methods. The SNR can be compared between imaging methods with the expression [53]:

$$\text{SNR} \propto \frac{\text{total imaging volume}}{\sqrt{N_x N_y \text{BW}}} e^{-\text{TE}/\left(T_2^* \text{ or } T_2\right)} \frac{1}{\sqrt{\text{acceleration factor}}}$$

This equation applies if T_1-weighting can be ignored and a roughly 90° flip angle is used for both methods. The "total imaging volume" is the image field-of-view multiplied by the slice thickness, the values

Table 1
Comparison of the expected BOLD response magnitude detected with gradient-echo and spin-echo methods

Gradient-echo	Spin-echo
$\dfrac{\Delta S}{S} \cong -TE\,\Delta\!\left(\dfrac{1}{T_2^*}\right)$	$\dfrac{\Delta S}{S} \cong -TE\,\Delta\!\left(\dfrac{1}{T_2}\right)$
$\dfrac{\Delta S}{S} \cong -0.025\,s \times -1.22\,s^{-1}$	$\dfrac{\Delta S}{S} \cong -0.075\,s \times -0.37\,s^{-1}$
$\dfrac{\Delta S}{S} \cong 0.031$	$\dfrac{\Delta S}{S} \cong 0.028$

Estimates are for data acquired at 3 T, in the spinal cord, with TE values set for optimal BOLD contrast. The values for the expected changes in relaxation rates, $\Delta(1/T_2^*)$ and $\Delta(1/T_2)$, are taken from [28]

of Nx and Ny are the image acquisition matrix dimensions, and the "acceleration factor" refers to the parallel imaging acceleration factor. With the acquisition parameters used in the most recent methods, the spin-echo method is expected to have more than three times higher SNR than the gradient-echo method (Table 2). However, another important factor is the acquisition speed, because the number of volumes that are acquired to detect the BOLD responses influences the sensitivity. Rearranging the expression described by Murphy et al. [54] to estimate the number of volumes needed, we can estimate the effect size (i.e., the % BOLD change) that can be detected for a given number of volumes (N):

$$\text{eff} = \frac{\text{erfc}^{-1}(p)}{\text{SNR}}\sqrt{\frac{2}{NR(1-R)}}$$

where "eff" is the effect size, "p" is the statistical threshold used, the function "erfc^{-1}" is the inverse complementary error function, and "R" is the proportion of time spent in the stimulation condition, assuming a block design with only two conditions. For the purposes of this comparison we can set $R=0.5$, and $p=10^{-6}$, which corresponds to erfc$^{-1}(p)=3.46$. An estimate of the corresponding t-value is given by $\sqrt{2}\,\text{erfc}^{-1}(p)$, which is equal to 5.0. Using these numbers we can compare the relative sensitivities of the methods, in terms of the % BOLD signal change that can be detected with a fixed acquisition duration (Table 2).

These estimates show that the faster sampling of the gradient-echo EPI method offsets its lower SNR and improves its sensitivity, but it does not reach the sensitivity of the spin-echo HASTE method. With either of these methods the duration of the fMRI acquisitions can be increased to provide greater sensitivity, within practical limits.

Table 2
Estimates of SNR for commonly used gradient-echo and spin-echo spinal fMRI acquisitions

	Typical brain fMRI	Gradient-echo spinal fMRI	Spin-echo spinal fMRI
Imaging parameters	3.3 mm × 3.3 mm 3.3 mm thick slice 200 kHz bandwidth 64 × 64 matrix TE = T_2^* Acceleration factor = 1 (no parallel imaging assumed)	1 mm × 1 mm 5 mm thick slice 200 kHz bandwidth 128 × 128 matrix TE: 43 ms (~1.7 T_2^*) acceleration factor = 2	1.5 mm × 1.5 mm 2 mm thick slice 151 kHz bandwidth 192 × 144 matrix TE: 75 ms (T_2) acceleration factor = 1
Estimated SNR	150	15	56
Acquisition time/ volume	3 s	1.1 s	6.75 s
Estimated effect size $p = 10^{-6}$, 12 min acquisition	0.42 %	2.6 %	1.7 %

SNR values are estimated compared to a typical brain fMRI method which is assumed to have an SNR of approximately 150 at 3 T

Ultimately, the number of volumes that is acquired influences the sensitivity for detecting BOLD responses, not the sampling rate [53].

The final factor to be considered when comparing the spin-echo and gradient-echo methods is the image quality that is provided, as shown in Fig. 4. Spinal fMRI acquisitions with gradient-echo EPI methods employ axial slices in all of the examples cited above, because axial slices appear to provide better image quality. However sagittal views are extracted from the volume spanned by axial slices, it can be seen that the images are still severely spatially distorted, and the distortion depends on the rostral-caudal position. Slice-specific shimming has been shown to improve the image quality over that shown in Fig. 4, but it does not eliminate the distortions [55]. The net trade-off of using EPI methods is that they introduce more problems than they solve, and provide higher temporal resolution at the expense of loss of spatial fidelity, less anatomical coverage, increased physiological noise, lower SNR, and lower BOLD sensitivity, compared to single-shot fast spin-echo methods.

5   Recent Developments

Recent studies have further developed spinal fMRI methods with regard to improving study designs, our understanding of the noise characteristics, and analysis methods. One of the greatest technical

Axial gradient-echo EPI
for T$_2$*-weighted BOLD

Axial gradient-echo EPI
reformatted into a sagittal view

Sagittal HASTE
for T$_2$-weighted BOLD

Rostral
Medulla

-21 mm

-36 mm

-51 mm

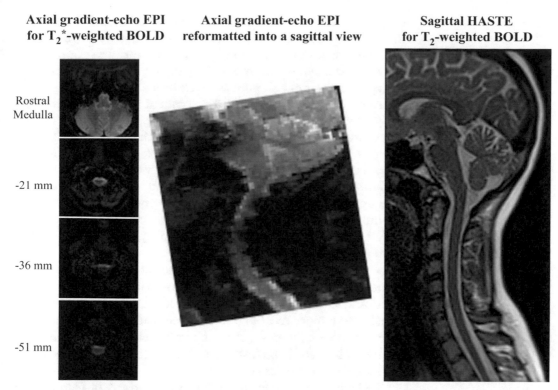

Fig. 4 Comparison of image quality obtained with gradient-echo EPI and spin-echo HASTE sequences for spinal fMRI. Gradient-echo EPI images were acquired in contiguous axial slices (*left panel*) and were reformatted into sagittal views (*middle frame*) for comparison with spin-echo HASTE images acquired in sagittal planes (*right frame*). Selected axial slices are shown for comparison, and the slice positions are indicated relative to the caudal medulla (*top slice*)

challenges encountered with spinal fMRI is the physiological noise arising from multiple sources. Cardiac-related motion has been identified as arising from CSF flow, related movement of the spinal cord within the spinal canal, and artifacts related to heart movement itself, and the motion has been shown to be a complex interaction of several sources [1, 56, 57]. A recent study has identified an additional important component arising from artifactual movement of the spinal cord in gradient-echo EPI images, in relation to respiration [58]. Changes in the static magnetic field, B_0, due to changes in lung volume have been shown to cause a shift of the apparent position of the spine/spinal cord as large as 10 mm in MR images, when acquired with EPI methods. The shift also varies with rostral/caudal position along the spine. Understanding this source of artifactual movement can be expected to improve the sensitivity and reliability of spinal fMRI methods. The noise characteristics in spin-echo HASTE spinal fMRI data has also been investigated recently, and have shown that in spite of the multiple sources of cardiac- and respiratory-related movement, the noise is essentially random with negligible auto-correlation [17]. That is, the physiological noise is uncorrelated between successive volumes due to the relatively slow sampling rate. The authors proposed that the most effective way to reduce the impact of physiological noise is therefore to simply acquire as much data as possible.

Another recent important development is the ability to accurately coregister 3D spinal cord data. The Medical Image Registration Toolbox (MIRT) includes a nonrigid 3D registration tool that has been shown to be effective at reducing noise in spinal fMRI data acquired in sagittal planes [17, 59, 60]. In addition, an automated spatial normalization method has been developed recently, and with it a normalized 3D anatomical template of the cervical spinal cord and brainstem. A previous user-guided normalization method has been described [17, 61], and this was used to create a reference template using data from 356 healthy participants. An iterative process was used to normalize the individual image data to this template using the automated method, and each iteration resulted in more precisely coregistered images from individuals and a more reliable reference template. The resulting template consists of $1 \text{ mm} \times 1 \text{ mm} \times 1 \text{ mm}$ voxels spanning the entire cervical spinal cord, brainstem, medial portion of the thalamus, and includes the corpus collosum.

The automated normalization process consists of first interpolating the data to 1 mm cubic voxels and matching the orientation to the reference template. Predefined sections of the template are then matched to sections of the image data using the location at the maximum cross-correlation, with template sections rotated over a small range of angles. The first section identified includes the corpus callosum and thalamus, because of their distinct features. In subsequent sections the position and angle are weighted towards predicted values based on prior segments resulting in a stable mapping process. The distance along the cord (moving caudally) from the pontomedullary junction is matched in the template and image data to ensure that the cord anatomy is not altered in length due to a lack of rostral/caudal features in the cord tissue [62]. The final step of the normalization process was to fine-tune the mapping to the normalized template using the MIRT toolbox [60]. The resulting template is shown in Fig. 5. The automated normalization process can be applied to data from individuals acquired in sagittal planes with spin-echo methods, to enable mapping of data to the normalized space for group analyses, group contrasts, second-level analyses, etc.

The normalized 3D template has also been used to define an anatomical region mask for the cervical spinal cord and brainstem. The region locations and extents were compiled from numerous anatomical atlases and published papers [63–65]. This mask has been used to extract fMRI data from specific anatomical regions for regions-of-interest analyses, and for effective connectivity analyses using Structural Equation Modeling (SEM) [40, 42]. A SEM

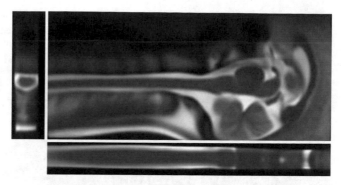

Fig. 5 A 3D-normalized reference template for the cervical spinal cord, brainstem, and medial regions of the thalamus and corpus collosum. Image views through the center of the volume are shown in axial (*upper left*), sagittal (*upper right*), and coronal (*lower right*) slices

method has been developed for spinal cord fMRI data, written in MatLab® (The Mathworks Inc.), based on prior descriptions [66, 67]. The method is based on identifying coordinated BOLD responses between regions, accounting for the fact that input to one region may arise simultaneously from multiple other regions [68, 69]. The BOLD signal time-courses in each region are expressed as a linear combination of the BOLD signal time-courses in other regions, and the weighting factors for the linear combination (i.e. the "connectivity strengths") are determined for a complete network [70]. This analysis requires a predefined anatomical model of all plausible connections between regions, which is provided by the normalized temperature and region mask described above. Possible connectivity relationships between the regions were identified based on the extensive description of the regions/networks involved in pain processing provided by Millan [71]. An example of SEM results from Bosma and Stroman [40] is shown in Fig. 6.

Recent advances in the applications of spinal fMRI serve to further demonstrate the reliability and sensitivity of the results, and their potential value for future clinical applications. The first detailed resting-state study was carried out by Barry et al. [72] and used a 3D multishot gradient-echo sequence at 7 T. They demonstrated functional connectivity between right- and left-side gray matter in the spinal cord, in the resting state. This is an important finding for spinal cord fMRI in general, because of the variations in the baseline state that may occur, even when a stimulus is not applied. Related studies have also been carried out using spin-echo methods, by using either thought directed at a particular region of the body [73] or images displayed to the participant [74, 75]. In each of these studies, signal variations were detected in the spinal cord in response to the cognitive/emotional stimuli, demonstrating the influence of descending input to the spinal cord. These findings may be related to the recent demonstrations of effects such as placebo [23]. Detailed studies of

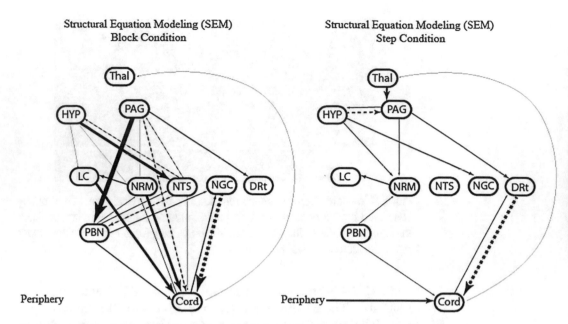

Fig. 6 SEM results obtained from healthy participants with a block-design stimulus at 45 °C, and a step-wise increase in temperature to 45 °C. *Arrows* indicate the direction of the influence, while the *line thickness* indicates the strength of the SEM connectivity. *Solid lines* represent positive path coefficients and *dotted lines* represent negative coefficients. Abbreviations are as follows: *Cord* right dorsal region of the C8 spinal cord segment, *PBN* parabrachial nucleus, *LC* locus coeruleus, *NRM* nucleus raphe magnus, *NTS* nucleus tractus solitarius, *NGC* nucleus gigantocellularis, *DRt* dorsal reticular nucleus, *PAG* periaqueductal gray matter, *HYP* hypothalamus, *Thal* thalamus. This figure is reproduced from Bosma and Stroman, J Magn Reson Imaging 2014 (DOI: 10.1002/jmri.24656) [41]

pain processing have also been carried out, showing effects of pain modulation by changes in attention focus and by listening to music [23, 24, 42]. Individual differences in pain processing have also been investigated, and demonstrate correlations between BOLD responses and individual pain ratings in the PAG, PBN, and spinal cord dorsal horn [76]. This study provides evidence that spinal fMRI methods are adequately sensitive to provide characterizations of the pain state in individuals. Moreover, a study of the effects of spinal cord injury on thermal sensory processing has demonstrated plasticity, with significant differences detected in individuals compared to a group of healthy control participants [50]. This body of recent work shows that spinal fMRI methods and applications are rapidly developing and expanding. Their future clinical potential is also further demonstrated.

6   Data Acquisition Details and Analysis Methods

The sensitivity of fMRI studies depends on how well the paradigm design matches the neural function of interest, and excludes confounding effects, possibly more than it depends on the acquisition parameters discussed earlier. The spinal cord receives tonic input from brainstem regions, even when no task is performed or stimulus is applied [71]. Therefore, as has been observed already with spinal fMRI, the "baseline" state, in the absence of a stimulus, cannot be considered an "inactive" state [31, 76]. Other important considerations for spinal fMRI study designs are the properties of sensory receptors, and the expected input they provide to the spinal cord, communication via interneurons within and between spinal cord segments, and the descending modulation of responses from brainstem regions [40]. Spinal cord responses cannot be assumed to be constant across repeated applications of stimuli, or tasks, because of the emotional/cognitive component of descending modulation [23, 24, 40–42, 74–77]. An additional consequence of these properties of the spinal cord and brainstem is that upon application of a stimulus or task, both inhibitory and facilitory input signaling to the regions can increase, or decrease. The net change in metabolic demand can also therefore be an increase, or a decrease, and negative BOLD responses are possible, and have been observed consistently across studies [17, 31–33, 35, 37, 39–42]. These observations across studies show that taking the peripheral and supraspinal input signaling into consideration in the study design can improve the sensitivity of the results and the accuracy of their interpretation. Moreover, these observations demonstrate the sensitivity of the results that can currently be achieved.

The key steps in the analysis of spinal fMRI data are essentially the same as with brain fMRI data. The preprocessing steps typically include converting the data from DICOM to an easier-to-use format such as NIfTI, then applying motion-correction by realigning the data to match one volume in the time series, slice-timing correction, spatial normalization, and spatial smoothing. The preprocessed data can then be analyzed with a number of methods such as a General Linear Model (GLM) to detect predicted BOLD responses, region-of-interest (ROI) analyses to extract time-course responses from specific regions, and connectivity analyses. A number of studies have used brain fMRI analysis software such as SPM (Statistical Parametric Mapping, Wellcome Trust Centre for Neuroimaging, London, UK) [23, 24, 26] or AFNI [72, 78] and have adapted its use for spinal fMRI. Specialized software has also been developed in MatLab for analysis of spinal fMRI data acquired with spin-echo methods [17, 40, 42]. A detailed investigation of the noise characteristics and effects of preprocessing steps has shown that efforts to model physiological noise for inclusion as regressors in GLM analysis provide only a small improvement in the ability to detect BOLD responses [17, 56, 57]. This limited effect appears to be the result of the complexity of the physiological noise. A data-driven method of extracting global sources of variance across the entire spinal cord has been shown to have a significant effect on improving the sensitivity of the results [17]. In the same study it was shown that accurate motion-correction (i.e. realignment of the data) provides the greatest improvement in measurement sensitivity of all of the preprocessing steps. Temporal filtering, on the other hand, alters the t-value distribution which artificially inflates the significance of the results, leading to false-positive results [17, 79]. Ultimately, this study showed that the best analysis approach (to date) involves only effective motion-correction, data-driven estimates of noise regressors in the GLM. Moreover, the most effective way to deal with either random or physiological noise is to collect as many volumes as possible to describe the fMRI time-series. In effect, these results to date have confirmed principles of fMRI study design and analysis that have long been known for brain fMRI, but are often over-looked in the effort to develop more "sophisticated" methods.

7   Conclusions and Future Directions

The evidence to date shows that spinal fMRI has developed rapidly over the past several years, and highly sensitive results have been obtained. Details of responses to sensory and motor paradigms, pain processing, and descending modulation related to cognitive and emotional factors have all been demonstrated. Detailed anatomical mapping of responses has also been demonstrated. There is no question that spinal fMRI is feasible and effective for research applications. It has also been shown to provide valuable information about multiple-sclerosis and traumatic spinal cord injury. However, it still has many limitations and will require more development before clinical applications can be considered. While the technical challenges and limitations of spatial and temporal resolution have been identified, and efforts to overcome these challenges are proceeding, one key limitation for the progress of spinal fMRI is the lack of "critical mass" of researchers working on it. Divisions over the best methods to use have further limited the pace of development. Consensus over the acquisition methods would contribute to the development of common software methods for analysis. This would reduce the burden of time and effort for new groups to apply spinal fMRI to new research questions, and would allow new developments to be shared more easily between groups. Fortunately, efforts are underway to facilitate sharing of ideas such as "Spinal Cord Hack 2014" organized by Dr. Paul Summers as a satellite meeting of the International Society for Magnetic Resonance in Medicine (ISMRM) 2014 annual meeting, and the 2015 version being organized by Dr Julien Cohen-Adad. In addition, the International Spinal Research Trust and the Wings for Life Foundation have organized international imaging workshops [80, 81], and are promoting a multisite diagnostic trial using spinal cord imaging.

References

1. Figley CR, Stroman PW (2007) Investigation of human cervical and upper thoracic spinal cord motion: implications for imaging spinal cord structure and function. Magn Reson Med 58(1):185–189

2. Figley CR, Yau D, Stroman PW (2008) Attenuation of lower-thoracic, lumbar, and sacral spinal cord motion: implications for imaging human spinal cord structure and function. AJNR Am J Neuroradiol 29(8):1450–1454

3. Yoshizawa T, Nose T, Moore GJ, Sillerud LO (1996) Functional magnetic resonance imaging of motor activation in the human cervical spinal cord. Neuroimage 4(3 Pt 1):174–182

4. Menon RS, Ogawa S, Kim SG, Ellermann JM, Merkle H, Tank DW, Ugurbil K (1992) Functional brain mapping using magnetic resonance imaging. Signal changes accompanying visual stimulation. Invest Radiol 27(Suppl 2):S47–S53

5. Ogawa S, Tank DW, Menon R, Ellermann JM, Kim SG, Merkle H, Ugurbil K (1992) Intrinsic signal changes accompanying sensory stimulation: functional brain mapping with magnetic resonance imaging. Proc Natl Acad Sci U S A 89(13):5951–5955

6. Stroman PW, Nance PW, Ryner LN (1999) BOLD MRI of the human cervical spinal cord at 3 tesla. Magn Reson Med 42(3):571–576

7. Madi S, Flanders AE, Vinitski S, Herbison GJ, Nissanov J (2001) Functional MR imaging of the human cervical spinal cord. AJNR Am J Neuroradiol 22(9):1768–1774

8. Backes WH, Mess WH, Wilmink JT (2001) Functional MR imaging of the cervical spinal cord by use of median nerve stimulation and fist clenching. AJNR Am J Neuroradiol 22(10):1854–1859

9. Stroman PW, Ryner LN (2001) Functional MRI of motor and sensory activation in the human spinal cord. Magn Reson Imaging 19(1):27–32

10. Stroman PW, Krause V, Malisza KL, Frankenstein UN, Tomanek B (2001) Characterization of contrast changes in functional MRI of the human spinal cord at 1.5 T. Magn Reson Imaging 19(6):833–838

11. Bandettini PA, Wong EC, Jesmanowicz A, Hinks RS, Hyde JS (1994) Spin-echo and gradient-echo EPI of human brain activation using BOLD contrast: a comparative study at 1.5 T. NMR Biomed 7(1–2):12–20

12. Figley CR, Leitch JK, Stroman PW (2010) In contrast to BOLD: signal enhancement by extravascular water protons as an alternative mechanism of endogenous fMRI signal change. Magn Reson Imaging 28(8):1234–1243

13. Figley CR, Stroman PW (2012) Measurement and characterization of the human spinal cord SEEP response using event-related spinal fMRI. Magn Reson Imaging 30(4):471–484

14. Stroman PW, Krause V, Malisza KL, Frankenstein UN, Tomanek B (2002) Extravascular proton-density changes as a non-BOLD component of contrast in fMRI of the human spinal cord. Magn Reson Med 48(1):122–127

15. Stroman PW, Lee AS, Pitchers KK, Andrew RD (2008) Magnetic resonance imaging of neuronal and glial swelling as an indicator of function in cerebral tissue slices. Magn Reson Med 59(4):700–706

16. Stroman PW, Malisza KL, Onu M (2003) Functional magnetic resonance imaging at 0.2 Tesla. Neuroimage 20(2):1210–1214

17. Bosma RL, Stroman PW (2014) Assessment of data acquisition parameters, and analysis techniques for noise reduction in spinal cord fMRI data. Magn Reson Imaging 32(5):473–481

18. Komisaruk BR, Mosier KM, Liu WC, Criminale C, Zaborszky L, Whipple B, Kalnin A (2002) Functional localization of brainstem and cervical spinal cord nuclei in humans with fMRI. AJNR Am J Neuroradiol 23(4):609–617

19. Cohen-Adad J, Gauthier CJ, Brooks JC, Slessarev M, Han J, Fisher JA, Rossignol S, Hoge RD (2010) BOLD signal responses to controlled hypercapnia in human spinal cord. Neuroimage 50(3):1074–1084

20. Maieron M, Iannetti GD, Bodurka J, Tracey I, Bandettini PA, Porro CA (2007) Functional responses in the human spinal cord during willed motor actions: evidence for side- and rate-dependent activity. J Neurosci 27(15):4182–4190

21. Summers PE, Ferraro D, Duzzi D, Lui F, Iannetti GD, Porro CA (2010) A quantitative comparison of BOLD fMRI responses to noxious and innocuous stimuli in the human spinal cord. Neuroimage 50(4):1408–1415

22. Nash P, Wiley K, Brown J, Shinaman R, Ludlow D, Sawyer AM, Glover G, Mackey S (2013) Functional magnetic resonance imaging identifies somatotopic organization of nociception in the human spinal cord. Pain 154(6):776–781

23. Eippert F, Finsterbusch J, Bingel U, Buchel C (2009) Direct evidence for spinal cord involvement in placebo analgesia. Science 326(5951):404

24. Sprenger C, Eippert F, Finsterbusch J, Bingel U, Rose M, Buchel C (2012) Attention modulates spinal cord responses to pain. Curr Biol 22(11):1019–1022

25. Geuter S, Buchel C (2013) Facilitation of pain in the human spinal cord by nocebo treatment. J Neurosci 33(34):13784–13790

26. van de Sand MF, Sprenger C, Buchel C (2015) BOLD responses to itch in the human spinal cord. Neuroimage 108:138–143

27. Figley CR, Stroman PW (2011) The role(s) of astrocytes and astrocyte activity in neurometabolism, neurovascular coupling, and the production of functional neuroimaging signals. Eur J Neurosci 33(4):577–588

28. Stroman PW, Krause V, Frankenstein UN, Malisza KL, Tomanek B (2001) Spin-echo versus gradient-echo fMRI with short echo times. Magn Reson Imaging 19(6):827–831

29. Stroman PW, Tomanek B, Krause V, Frankenstein UN, Malisza KL (2003) Functional magnetic resonance imaging of the human brain based on signal enhancement by extravascular protons (SEEP fMRI). Magn Reson Med 49(3):433–439

30. Stroman PW, Kornelsen J, Lawrence J, Malisza KL (2005) Functional magnetic resonance imaging based on SEEP contrast: response function and anatomical specificity. Magn Reson Imaging 23(8):843–850

31. Cahill CM, Stroman PW (2011) Mapping of neural activity produced by thermal pain in the healthy human spinal cord and brain stem: a functional magnetic resonance imaging study. Magn Reson Imaging 29(3):342–352

32. Ghazni NF, Cahill CM, Stroman PW (2010) Tactile sensory and pain networks in the human spinal cord and brain stem mapped by means of functional MR imaging. AJNR Am J Neuroradiol 31(4):661–667

33. Kozyrev N, Figley CR, Alexander MS, Richards JS, Bosma RL, Stroman PW (2012) Neural correlates of sexual arousal in the spinal cords of able-bodied men: a spinal fMRI investigation. J Sex Marital Ther 38(5):418–435

34. Lawrence JM, Kornelsen J, Stroman PW (2011) Noninvasive observation of cervical spinal cord activity in children by functional MRI during cold thermal stimulation. Magn Reson Imaging 29(6):813–818

35. Rempe T, Wolff S, Riedel C, Baron R, Stroman PW, Jansen O, Gierthmuhlen J (2014) Spinal fMRI reveals decreased descending inhibition during secondary mechanical hyperalgesia. PLoS One 9(11):e112325

36. Rempe T, Wolff S, Riedel C, Baron R, Stroman PW, Jansen O, Gierthmuhlen J (2014) Spinal and supraspinal processing of thermal stimuli: an fMRI study. J Magn Reson Imaging 41(4):1046–1055

37. Stroman PW (2009) Spinal fMRI investigation of human spinal cord function over a range of

innocuous thermal sensory stimuli and study-related emotional influences. Magn Reson Imaging 27(10):1333–1346

38. Lawrence JM, Stroman PW, Kollias SS (2008) Functional magnetic resonance imaging of the human spinal cord during vibration stimulation of different dermatomes. Neuroradiology 50(3):273–280

39. Stroman PW, Bosma RL, Tsyben A (2012) Somatotopic arrangement of thermal sensory regions in the healthy human spinal cord determined by means of spinal cord functional MRI. Magn Reson Med 68(3):923–931

40. Bosma RL, Stroman PW (2014) Spinal cord response to stepwise and block presentation of thermal stimuli: a functional MRI study. J Magn Reson Imaging 41(5):1318–1325

41. Stroman PW, Coe BC, Munoz DP (2011) Influence of attention focus on neural activity in the human spinal cord during thermal sensory stimulation. Magn Reson Imaging 29(1):9–18

42. Dobek CE, Beynon ME, Bosma RL, Stroman PW (2014) Music modulation of pain perception and pain-related activity in the brain, brainstem, and spinal cord: an fMRI study. J Pain 15(10):1057–1068

43. Agosta F, Valsasina P, Absinta M, Sala S, Caputo D, Filippi M (2009) Primary progressive multiple sclerosis: tactile-associated functional MR activity in the cervical spinal cord. Radiology 253(1):209–215

44. Agosta F, Valsasina P, Caputo D, Rocca MA, Filippi M (2009) Tactile-associated fMRI recruitment of the cervical cord in healthy subjects. Hum Brain Mapp 30(1):340–345

45. Agosta F, Valsasina P, Caputo D, Stroman PW, Filippi M (2008) Tactile-associated recruitment of the cervical cord is altered in patients with multiple sclerosis. Neuroimage 39(4):1542–1548

46. Agosta F, Valsasina P, Rocca MA, Caputo D, Sala S, Judica E, Stroman PW, Filippi M (2008) Evidence for enhanced functional activity of cervical cord in relapsing multiple sclerosis. Magn Reson Med 59(5):1035–1042

47. Valsasina P, Agosta F, Absinta M, Sala S, Caputo D, Filippi M (2010) Cervical cord functional MRI changes in relapse-onset MS patients. J Neurol Neurosurg Psychiatry 81(4):405–408

48. Valsasina P, Agosta F, Caputo D, Stroman PW, Filippi M (2008) Spinal fMRI during proprioceptive and tactile tasks in healthy subjects: activity detected using cross-correlation, general linear model and independent component analysis. Neuroradiology 50(10):895–902

49. Valsasina P, Rocca MA, Absinta M, Agosta F, Caputo D, Comi G, Filippi M (2012) Cervical cord FMRI abnormalities differ between the progressive forms of multiple sclerosis. Hum Brain Mapp 33(9):2072–2080

50. Cadotte DW, Bosma R, Mikulis D, Nugaeva N, Smith K, Pokrupa R, Islam O, Stroman PW, Fehlings MG (2012) Plasticity of the injured human spinal cord: insights revealed by spinal cord functional MRI. PLoS One 7(9):e45560

51. Stroman PW, Kornelsen J, Bergman A, Krause V, Ethans K, Malisza KL, Tomanek B (2004) Noninvasive assessment of the injured human spinal cord by means of functional magnetic resonance imaging. Spinal Cord 42(2):59–66

52. Kornelsen J, Stroman PW (2007) Detection of the neuronal activity occurring caudal to the site of spinal cord injury that is elicited during lower limb movement tasks. Spinal Cord 45(7):485–490

53. Stroman PW (2011) Essentials of functional MRI. Taylor & Francis Group, LLC, Boca Raton, FL

54. Murphy K, Bodurka J, Bandettini PA (2007) How long to scan? The relationship between fMRI temporal signal to noise ratio and necessary scan duration. Neuroimage 34(2):565–574

55. Finsterbusch J, Eippert F, Buchel C (2012) Single, slice-specific z-shim gradient pulses improve T2*-weighted imaging of the spinal cord. Neuroimage 59(3):2307–2315

56. Figley CR, Stroman PW (2009) Development and validation of retrospective spinal cord motion time-course estimates (RESPITE) for spin-echo spinal fMRI: improved sensitivity and specificity by means of a motion-compensating general linear model analysis. Neuroimage 44(2):421–427

57. Brooks JC, Beckmann CF, Miller KL, Wise RG, Porro CA, Tracey I, Jenkinson M (2008) Physiological noise modelling for spinal functional magnetic resonance imaging studies. Neuroimage 39(2):680–692

58. Verma T, Cohen-Adad J (2014) Effect of respiration on the B0 field in the human spinal cord at 3T. Magn Reson Med 72(6):1629–1636

59. Myronenko A, Song XB (2009) Image registration by minimization of residual complexity. Proc Cvpr IEEE. pp 49–56

60. Myronenko A, Song XB (2010) Intensity-based image registration by minimizing residual complexity. IEEE Trans Med Imag 29(11):1882–1891

61. Stroman PW, Figley CR, Cahill CM (2008) Spatial normalization, bulk motion correction and coregistration for functional magnetic resonance imaging of the human cervical spinal cord and brainstem. Magn Reson Imaging 26(6):809–814

62. Lang J, Bartram CT (1982) Fila radicularia of the ventral and dorsal radices of the human spinal cord. Gegenbaurs Morphol Jahrb 128(4):417–462

63. Gray H (1995) Gray's anatomy: the anatomical basis of medicine and surgery. In: Williams PL, Bannister LH, Berry MM, Collins P, Dyson M, Dussek JE, Ferguson MWJ (eds) Gray's anatomy: the anatomical basis of medicine and surgery. Churchill-Livingstone, London, pp 975–1011

64. Talairach J, Tournoux P (1988) Co-planar sterotaxic atlas of the human brain. Thieme Medical Publishers Inc, New York

65. Naidich TP, Duvernoy HM, Delman BN, Sorensen AG, Kollias SS, Haacke EM (2009) Internal architecture of the brain stem with key axial sections. Duvernoy's atlas of the human brain stem and cerebellum. Springer, New York, pp 79–82

66. McArdle JJ, McDonald RP (1984) Some algebraic properties of the Reticular Action Model for moment structures. Br J Math Stat Psychol 37(Pt 2):234–251

67. Craggs JG, Staud R, Robinson ME, Perlstein WM, Price DD (2012) Effective connectivity among brain regions associated with slow temporal summation of C-fiber-evoked pain in fibromyalgia patients and healthy controls. J Pain 13(4):390–400

68. Buchel C, Friston K (2001) Interactions among neuronal systems assessed with functional neuroimaging. Rev Neurol 157(8–9 Pt 1):807–815

69. Buchel C, Friston KJ (1997) Modulation of connectivity in visual pathways by attention: cortical interactions evaluated with structural equation modelling and fMRI. Cereb Cortex 7(8):768–778

70. Bollen KA (1989) Structural equations with latent variables. Wiley, New York

71. Millan MJ (2002) Descending control of pain. Prog Neurobiol 66(6):355–474

72. Barry RL, Smith SA, Dula AN, Gore JC (2014) Resting state functional connectivity in the human spinal cord. eLife 3:e02812

73. Kashkouli Nejad K, Sugiura M, Thyreau B, Nozawa T, Kotozaki Y, Furusawa Y, Nishino K, Nukiwa T, Kawashima R (2014) Spinal fMRI of interoceptive attention/awareness in experts and novices. Neural Plast 2014:679509

74. Smith SD, Kornelsen J (2011) Emotion-dependent responses in spinal cord neurons: a spinal fMRI study. Neuroimage 58(1):269–274

75. Kornelsen J, Smith SD, McIver TA (2014) A neural correlate of visceral emotional responses: evidence from fMRI of the thoracic spinal cord. Soc Cogn Affect Neurosci 10(4):584–588

76. Khan HS, Bosma RL, Beynon M, Dobek C, McIver T, Stroman PW (2013) Pain processing

networks in the brain and spinal cord mapped using Functional Magnetic Resonance Imaging. Program number II6 66.01. 2013 Meeing Planner San Diego, CA; Society for Neuroscience

77. McIver TA, Kornelsen J, Smith SD (2013) Limb-specific emotional modulation of cervical spinal cord neurons. Cogn Affect Behav Neurosci 13(3):464–472

78. Cox RW (1996) AFNI: software for analysis and visualization of functional magnetic resonance neuroimages. Comput Biomed Res 29(3):162–173

79. Friston KJ, Josephs O, Zarahn E, Holmes AP, Rouquette S, Poline J (2000) To smooth or not to smooth? Bias and efficiency in fMRI time-series analysis. Neuroimage 12(2):196–208

80. Stroman PW, Wheeler-Kingshott C, Bacon M, Schwab JM, Bosma R, Brooks J, Cadotte D, Carlstedt T, Ciccarelli O, Cohen-Adad J, Curt A, Evangelou N, Fehlings MG, Filippi M, Kelley BJ, Kollias S, Mackay A, Porro CA, Smith S, Strittmatter SM, Summers P, Tracey I (2014) The current state-of-the-art of spinal cord imaging: methods. Neuroimage 84:1070–1081

81. Wheeler-Kingshott CA, Stroman PW, Schwab JM, Bacon M, Bosma R, Brooks J, Cadotte DW, Carlstedt T, Ciccarelli O, Cohen-Adad J, Curt A, Evangelou N, Fehlings MG, Filippi M, Kelley BJ, Kollias S, Mackay A, Porro CA, Smith S, Strittmatter SM, Summers P, Thompson AJ, Tracey I (2014) The current state-of-the-art of spinal cord imaging: applications. Neuroimage 84:1082–1093

Chapter 30

Clinical Applications of the Functional Connectome

Massimo Filippi and Maria A. Rocca

Abstract

Network-based analysis of brain functional connections has provided a novel instrument to study the human brain in healthy and diseased individuals. Graph theory provides a powerful tool to describe quantitatively the topological organization of brain connectivity. Using such a framework, the brain can be depicted as a set of nodes connected by edges. Distinct modifications of brain network topology have been identified during development and normal aging, whereas disrupted functional connectivity has been associated to several neurological and psychiatric conditions, including multiple sclerosis, dementia, amyotrophic lateral sclerosis, and schizophrenia. Such an assessment has contributed to explain part of the clinical manifestations usually observed in these patients, including disability and cognitive impairment. Future network-based research might reveal different stages of the different diseases, subtypes for cognitive impairments, and connectivity profiles associated with different clinical outcomes.

Key words Brain networks, Structural connectivity, Functional connectivity, Graph theory, Multiple sclerosis, Dementias, Psychiatric conditions

1 Introduction

Brain function depends on both local information processing combined with effective global communication and integration of information within a network of neural interactions. Brain function is not only based on the properties of single regions, but rather emerges from interactions of the network as a whole. Brain areas are interconnected by large-scale bundles of axonal projections, forming a macroscopic network of white matter pathways that enable functional communication between distinct, anatomically separated brain regions. Recent advances in magnetic resonance imaging (MRI) allow the reconstruction of both the structural and functional connections of this large-scale neural system, thus enabling efficient mapping of connectivity across the entire brain [1]. Network-based analysis of brain structural and functional connections has provided a novel instrument to study the human brain in healthy and diseased individuals [2]. Using the theoretical framework of networks and graphs, the brain can be

Massimo Filippi (ed.), *fMRI Techniques and Protocols*, Neuromethods vol. 119,
DOI 10.1007/978-1-4939-5611-1_30, © Springer Science+Business Media New York 2016

represented as a set of nodes (i.e., brain regions) joined by pairs by lines (i.e., structural or functional connectivity). Graph analysis has revealed important features of brain organization, such as an efficient "small-world" architecture (which combines a high level of segregation with a high level of global efficiency) and distributed, highly connected network regions, called "hubs." In a small-world network, a high clustering coefficient indicates that nodes tend to form dense regional cliques, implying high efficiency in local information transfer/processing. Path length and global efficiency are measures of network integration, which is the ability to combine specialized information rapidly from distributed brain regions. Distinct modifications of brain network topology have been identified during development and normal aging, whereas disrupted functional network properties have been associated with several neurological and psychiatric conditions, including neurodegenerative diseases, multiple sclerosis (MS), and schizophrenia.

The methodological aspects related to graph analysis (the key mathematical framework for much of this research) have been described in details in another chapter of this book (Chapter 10). This chapter provides a summary of modifications of brain network topology associated with normal development and aging, and of how they are perturbed in course of brain diseases.

2   Normal Development

Using graph-theoretical methods, the maturation of the control network and default-mode network (DMN) have been explored by analyzing resting state (RS) fMRI data of healthy subjects from 7 to 31 years old [3, 4]. In children, the control network was more integrated and comprised a single system, contrasting with the adult configuration, which was characterized by a dual-network structure including cingulo-opercular and fronto-parietal networks. These data indicated a less between-network segregation and greater within-network connectivity in children. The DMN, which contains a set of regions usually deactivated during goal-oriented tasks, was only sparsely connected in children while a cohesive integrated network emerged in adults. Developmental changes in DMN regions were characterized by increases in correlation strengths and occurred in an anterior-posterior orientation. Conversely, the developmental pattern of the control network involved both increases and decreases of correlation strengths. Maturation of whole-brain functional networks from 8 to 25 years old has also been investigated [5]. Such a study showed that although the modular structure was well built since 8 years old, the modules changed dramatically from anatomical proximity to a more "distributed" architecture, grouping regions mainly by their functional roles. Measures of small-world structure (i.e., clustering coefficient and path length) were preserved through development, indicating that functional networks in children were efficient for both global and local information transfer as those in adults. Another RS fMRI study [6] replicated the previous results and also described a significantly decreased subcortical-cortical and increased cortico-cortical connectivity in the developing brain (Fig. 1). Both these studies observed the phenomenon of increased long-range connections and decreased short-range connections, which provides crucial support for segregation and integration processes at a system level over brain development. Another study showed that the hub locations and the core hub–hub network structure was kept consistent from 10 to 20 years of age while the main changes happened to the connectivity linking hubs and nonhub regions [7].

RS functional connectivity (FC) patterns extracted by support vector machine-based multivariate pattern analysis can make accurate predictions about individuals' brain maturity across development [8]. Gender effects on whole-brain functional networks have also been explored in healthy children from 6 to 18 years old [9]. Compared to girls, boys had higher global efficiency and shorter path length, with regional differences mainly located in the DMN, language, and visual areas. This finding is consistent with the notion that cognitive and emotional development differs between girls and boys, particularly in visuospatial, language, and emotion processing areas of the brain. A study in 12-year-old monozygotic and dizygotic twins showed that the global network efficiency was under genetic control [10]. The functional brain network in infancy was also studied. Two-week-old pediatric subjects only have primitive and incomplete DMN, whereas, after a marked increase of connectivity, the DMN in 2 years of age became similar to that observed in adults. Whole-brain functional networks in the infant brain already exhibited functional hubs mainly located in primary sensory and motor areas, which was distinct from the hub distribution of the adults in the heteromodal association cortex [11]. Another RS fMRI study showed that the functional brain networks have the small-world topology immediately after birth, followed by a remarkable improvement in the network efficiency and resilience until 2 years of age [12].

3   Normal Aging

The analysis of RS fMRI data has demonstrated a reduced cost-efficiency in older people with aging with detrimental effects mainly located to frontal and temporal cortical and subcortical regions [13]. Another study [14] found the RS FC of both DMN and dorsal attention network decreased with aging, with long-range connections showing higher vulnerability to aging effects than short-range connections. Other investigations confirmed a decrease

a

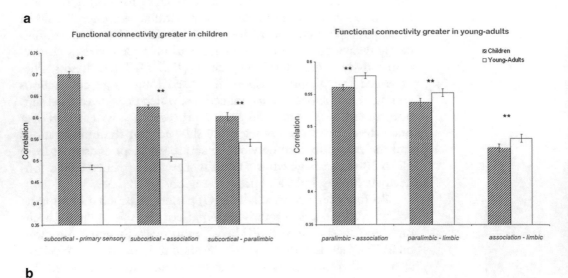

b

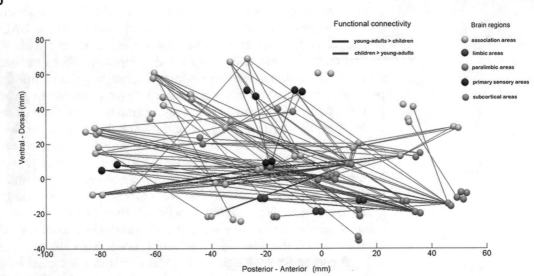

Fig. 1 Developmental changes in interregional functional connectivity. (**a**) Children had significantly greater subcortical-primary sensory, subcortical-association, subcortical-paralimbic, and lower paralimbic-association, paralimbic-limbic, association-limbic connectivity than young-adults ($p < 0.01$, indicated by **). (**b**) Graphical representation of developmental changes in functional connectivity along the posterior-anterior and ventral-dorsal axes, highlighting higher subcortical connectivity (subcortical nodes are shown in green) and lower paralimbic connectivity (paralimbic nodes are shown in gold) in children, compared to young-adults. Brain regions are plotted using the y and z coordinates of their centroids (in mm) in the MNI space. 430 pairs of anatomical regions showed significantly higher correlations in children and 321 pairs showed significantly higher correlations in young-adults ($p < 0.005$, FDR corrected). For illustration purposes, the plot shows differential connectivity that were most significant, 105 pairs higher in children (indicated in *red*) and 53 higher in young-adults (indicated in *blue*), ($p < 0.0001$, FDR corrected). From [6] with permission

in DMN connectivity (e.g., precuneus and posterior cingulate regions) with aging. Using supporting vector machine analysis [15], modifications of connectivity in the sensorimotor and cingulo-opercular networks were identified as distinguishing characteristics of age-related reorganization (Fig. 2).

4 Multiple Sclerosis

Consistent with the known multifocal distribution of structural damage to the central nervous system, MS patients experience a distributed pattern of RS FC abnormalities, which are related to the extent of structural damage and the severity of clinical disability and cognitive impairment [16]. Only a few studies have applied graph analysis methods in these patients.

Several authors have used graph-theoretical analysis of RS fMRI data to improve the understanding of the mechanisms responsible for the presence of cognitive deficits and clinical disability in patients with MS [17–19]. A study from 246 MS patients and 55 matched healthy controls [18] found that global network properties (including network degree, global efficiency, and path length) were abnormal in MS patients compared to controls and contributed to distinguish cognitively impaired MS patients from controls, but not the main MS clinical phenotypes. Compared to controls, MS patients also showed a loss of hubs in the superior frontal gyrus, precuneus, and anterior cingulum in the left hemisphere; a different lateralization of basal ganglia hubs (mostly located in the left hemisphere in controls, and in the right hemisphere in MS patients); and the formation of hubs, not seen in controls, in the left temporal pole and cerebellar lobule IV-V. Such a modification of regional network properties was found to contribute to cognitive impairment (Fig. 3) and phenotypic variability of MS.

A RS fMRI study of 16 early MS patients detected increased network modularity (i.e., diminished functional integration between separate functional modules) in patients compared to healthy controls [19]. Such modularity abnormalities in patients correlated with worse performance at a dual task. Another study explored the effects of gender on the correlation between network functional abnormalities and cognitive function [17]. Compared to male controls, male MS patients had reduced network efficiency, but normal clustering coefficient. No abnormalities were found in female patients. Decreased network efficiency in male patients was correlated with reduced visuospatial memory.

Using a pattern recognition technique, MS patients could be discriminated from healthy controls based on RS FC with a sensitivity of 82% and specificity of 86%. The most discriminative connectivity changes were found in subcortical and temporal regions, and contralateral connections were more discriminative than ipsilateral ones [20].

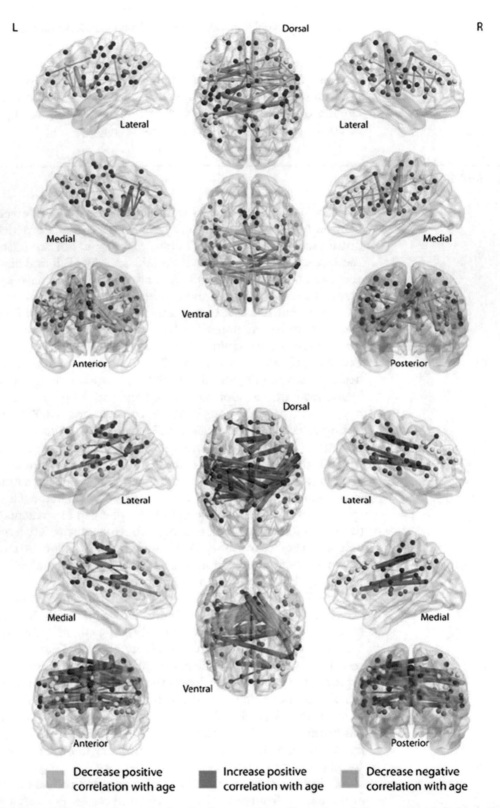

Fig. 2 Aging effect on resting-state functional connectivity. Illustration of the consensus features that decreased positive correlation with age (*top*) and the consensus features that increased positive correlation with age and decreased negative correlation with age (*bottom*). Connections are scaled by their respective feature weight, with thicker connections representing greater feature weight. From [15] with permission

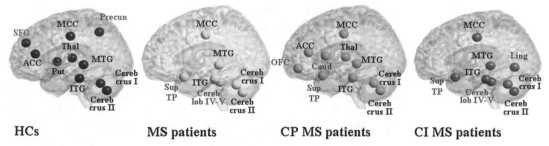

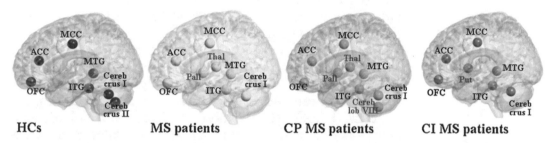

Fig. 3 Functional hub distribution in healthy controls and multiple sclerosis patients. Brain hubs (**a** left hemisphere, **b** right hemisphere) of the functional networks of healthy controls (HCs) and patients with MS as a whole and according to the presence/absence of cognitive impairment. Hubs were identified as brain regions having either integrated nodal degree or betweenness centrality one standard deviation greater than the network average. Hubs present in HC only are reported in *red*, hubs present in MS patients only are reported in *blue*, and hubs present in all groups are reported in *black*. *CP* cognitively preserved, *CI* cognitively impaired, *ACC* anterior cingulate cortex, *Caud* caudate nucleus, *Cereb* cerebellum, *ITG* inferior temporal gyrus, *Ling* lingual gyrus, *MCC* middle cingulate cortex, *MTG* middle temporal gyrus, *OFC* orbitofrontal cortex, *Pall* pallidus, *Precun* precuneus, *Put* putamen, *SFG* superior frontal gyrus, *Sup TP* superior temporal pole, *Thal* thalamus. From [18] with permission

5   Neurodegenerative Diseases

5.1   Alzheimer's Disease and Other Dementias

Many studies used graph theoretical analysis in patients with the most prevalent type of dementia, Alzheimer's disease (AD). A correlation between the site of amyloid-β deposition in AD patients and the location of major hubs as defined by graph theoretical analysis of RS FC in healthy adults has been demonstrated [21]. These regions include the posterior cingulate cortex/precuneus, the inferior parietal lobule, and the medial frontal cortex, implying that the hubs are preferentially affected in the progression of AD. There is also convergent evidence from methodologically disparate MRI studies that AD is associated with perturbations of brain small-world network organization [22–24].

Although studies showed considerable variability in reported group differences of most graph properties, the average characteristic path length has been most consistently reported to be increased in AD, as a result of loss of connectivity, while the clustering coefficient is likely to be less affected by AD pathology [25].

An fMRI graph analysis study in mild AD [24] suggested that loss of small-world network properties might provide a clinically useful diagnostic marker, since clustering was reduced at a global level and at a local level (in both hippocampi), and global clustering was able to discriminate AD patients from healthy elderly subjects with relatively high sensitivity (72%) and specificity (78%). Another fMRI study [23] showed that the characteristic path length of AD brains is closer to the theoretical values of random networks compared with controls. Decreased RS FC in the parietal and occipital regions, and increased connectivity in frontal cortices and corpus striatum were also found [23] (Fig. 4). Decreased global efficiency and increased local efficiency were also found in moderate AD cases, in whom the altered brain regions were mainly located in the DMN, temporal lobe, and subcortical regions [26].

Graph theoretical analysis was recently applied to RS fMRI data from patients with the behavioral variant of frontotemporal dementia (bvFTD) [27]. Global and local functional networks were altered in bvFTD patients, indicated by reduced mean network degree, clustering coefficient, and global efficiency and increased clustering coefficient relative to normal subjects. Altered brain regions were located in structures that are closely associated with neuropathological changes in bvFTD, such as the frontotemporal lobes and subcortical regions.

5.2 Amyotrophic Lateral Sclerosis

Consistent motor and extra-motor brain pathology supports the notion of amyotrophic lateral sclerosis (ALS) as a system failure. Thus, a simplistic (motor-based) approach to ALS is no longer tenable. Overall functional organization of the motor network was unchanged in patients with ALS compared to healthy controls; however, patients with a stronger and more interconnected motor network had a more progressive disease course [28].

6   Psychiatric Conditions

Evidence is accumulating that neural network changes are underlying structural and functional brain changes in psychiatric diseases, and may provide a more sensitive measure to detect brain abnormalities than measurements of properties of separate brain areas alone [29, 30]. The potential of network approaches to psychiatry can be illustrated by findings in schizophrenia, a severe psychiatric disease which has for long been hypothesized to reflect a disconnection syndrome. Connectomic studies have shown a reduced whole-brain functional connectivity in patients with schizophrenia [31, 32]. Functional networks in schizophrenia patients were

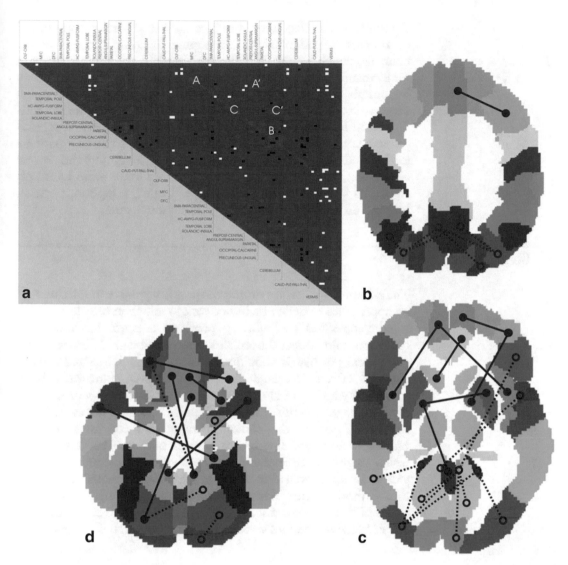

Fig. 4 Resting-state synchronization in healthy controls and Alzheimer's disease patients. (**a**) Matrix of significant differences of synchronization between Alzheimer's disease (AD) and controls (2-tail t-test, $p < 0.05$ uncorrected). The white and black dots represent brain areas pairs with increased and decreased synchronization in AD respectively. (**b-d**) A subset of connectional differences corresponding to the matrix (**a**) are plotted at three superior-to-inferior levels through the anatomical automatic labeled brain template. Lines depict synchronization between pairs of regions: solid lines = enhanced synchronization; dashed lines = reduced synchronization. Note the pattern of generalized posterior (parietal and occipital) synchronization reductions and increased frontal synchronization. From [23] with permission

characterized by reduced clustering and modularity and increased global efficiency compared with healthy controls, suggesting increased global integration and decreased local segregation [31, 33, 34]. By assessing dynamic graph properties of time-varying functional brain connectivity in RS fMRI data, a recent study demonstrated a decreased variance in the dynamic topological parameters in patients with schizophrenia compared to healthy controls, which might contribute to explain the abnormal brain performance in this mental illness [35]. The disconnectivity hypothesis in schizophrenia is also supported by studies which showed an abnormal hub organization. A less hub-dominated configuration has been observed in functional and structural schizophrenia connectomes [31, 36] Specifically, schizophrenia patients exhibit reduced regional centrality in hubs including the frontal association, parietal, limbic, and paralimbic brain areas based on structural studies, and including the frontal, temporal, parietal, limbic, and occipital areas based on functional studies.

7   Conclusions

The extensive application during the past few years of graph theoretical approaches to define brain network topology in healthy and diseased people has undoubtedly provided a novel instrument to characterize functional abnormalities associated with different neurological and psychiatric conditions and to test hypothesized disconnectivity effects in these diseases. Disrupted functional brain connectivity is present in the major neurological and psychiatric conditions discussed in this chapter and their assessment has contributed to explain part of the clinical manifestations of these patients. However, there are also inconsistencies between existing studies, which might be attributable to the clinical heterogeneity of the patient groups as well as to differences in imaging modality and analytic methods. Future network-based research might reveal different stages of the different diseases, subtypes for cognitive impairments, and connectivity profiles associated with different clinical outcomes.

References

1. Bullmore E, Sporns O (2009) Complex brain networks: graph theoretical analysis of structural and functional systems. Nat Rev Neurosci 10:186–198

2. Filippi M, van den Heuvel MP, Fornito A et al (2013) Assessing brain system dysfunction through MRI-based connectomics. Lancet Neurol 12:1189–1199

3. Fair DA, Dosenbach NU, Church JA et al (2007) Development of distinct control networks through segregation and integration. Proc Natl Acad Sci U S A 104:13507–13512

4. Fair DA, Cohen AL, Dosenbach NU et al (2008) The maturing architecture of the brain's default network. Proc Natl Acad Sci U S A 105:4028–4032

5. Fair DA, Cohen AL, Power JD et al (2009) Functional brain networks develop from a "local to distributed" organization. PLoS Comput Biol 5:e1000381

6. Supekar K, Musen M, Menon V (2009) Development of large-scale functional brain networks in children. PLoS Biol 7:e1000157

7. Hwang K, Hallquist MN, Luna B (2013) The development of hub architecture in the human functional brain network. Cereb Cortex 23:2380–2393

8. Dosenbach NU, Nardos B, Cohen AL et al (2010) Prediction of individual brain maturity using fMRI. Science 329:1358–1361

9. Wu K, Taki Y, Sato K et al (2013) Topological organization of functional brain networks in healthy children: differences in relation to age, sex, and intelligence. PLoS One 8:e55347

10. van den Heuvel MP, van Soelen IL, Stam CJ, Kahn RS, Boomsma DI, Hulshoff Pol HE

(2013) Genetic control of functional brain network efficiency in children. Eur Neuropsychopharmacol 23:19–23

11. Fransson P, Aden U, Blennow M, Lagercrantz H (2011) The functional architecture of the infant brain as revealed by resting-state fMRI. Cereb Cortex 21:145–154

12. Gao W, Gilmore JH, Giovanello KS et al (2011) Temporal and spatial evolution of brain network topology during the first two years of life. PLoS One 6:e25278

13. Achard S, Bullmore E (2007) Efficiency and cost of economical brain functional networks. PLoS Comput Biol 3:e17

14. Tomasi D, Volkow ND (2012) Aging and functional brain networks. Mol Psychiatry 17(471):549–458

15. Meier TB, Desphande AS, Vergun S et al (2012) Support vector machine classification and characterization of age-related reorganization of functional brain networks. Neuroimage 60:601–613

16. Rocca M, Valsasina P, Martinelli V et al (2012) Large-scle neuronal network dysfunction in relapsing-remitting multiple sclerosis. Neurology 79:1449–1457

17. Schoonheim MM, Hulst HE, Landi D et al (2012) Gender-related differences in functional connectivity in multiple sclerosis. Mult Scler 18:164–173

18. Rocca MA, Valsasina P, Meani A, Falini A, Comi G, Filippi M (2016) Impaired functional integration in multiple sclerosis: a graph theory study. Brain Struct Funct 221:115–131

19. Gamboa OL, Tagliazucchi E, von Wegner F et al (2014) Working memory performance of early MS patients correlates inversely with modularity increases in resting state functional connectivity networks. Neuroimage 94:385–395

20. Richiardi J, Gschwind M, Simioni S et al (2012) Classifying minimally disabled multiple sclerosis patients from resting state functional connectivity. Neuroimage 62:2021–2033

21. Buckner RL, Sepulcre J, Talukdar T et al (2009) Cortical hubs revealed by intrinsic functional connectivity: mapping, assessment of stability, and relation to Alzheimer's disease. J Neurosci 29:1860–1873

22. He Y, Chen Z, Evans A (2008) Structural insights into aberrant topological patterns of large-scale cortical networks in Alzheimer's disease. J Neurosci 28:4756–4766

23. Sanz-Arigita EJ, Schoonheim MM, Damoiseaux JS et al (2010) Loss of 'small-world' networks in Alzheimer's disease: graph analysis of FMRI resting-state functional connectivity. PLoS One 5:e13788

24. Supekar K, Menon V, Rubin D, Musen M, Greicius MD (2008) Network analysis of intrinsic functional brain connectivity in Alzheimer's disease. PLoS Comput Biol 4:e1000100

25. Tijms BM, Wink AM, de Haan W et al (2013) Alzheimer's disease: connecting findings from graph theoretical studies of brain networks. Neurobiol Aging 34:2023–2036

26. Zhao X, Liu Y, Wang X et al (2012) Disrupted small-world brain networks in moderate Alzheimer's disease: a resting-state FMRI study. PLoS One 7:e33540

27. Agosta F, Sala S, Valsasina P et al (2013) Brain network connectivity assessed using graph theory in frontotemporal dementia. Neurology 9:134–143

28. Verstraete E, van den Heuvel MP, Veldink JH et al (2010) Motor network degeneration in amyotrophic lateral sclerosis: a structural and functional connectivity study. PLoS One 5:e13664

29. Bullmore E, Sporns O (2012) The economy of brain network organization. Nat Rev Neurosci 13:336–349

30. van den Heuvel MP, Mandl RC, Stam CJ, Kahn RS, Hulshoff Pol HE (2010) Aberrant frontal and temporal complex network structure in schizophrenia: a graph theoretical analysis. J Neurosci 30:15915–15926

31. Lynall ME, Bassett DS, Kerwin R et al (2010) Functional connectivity and brain networks in schizophrenia. J Neurosci 30:9477–9487

32. Bassett DS, Nelson BG, Mueller BA, Camchong J, Lim KO (2012) Altered resting state complexity in schizophrenia. Neuroimage 59:2196–2207

33. Lo CY, Su TW, Huang CC et al (2015) Randomization and resilience of brain functional networks as systems-level endophenotypes of schizophrenia. Proc Natl Acad Sci U S A 112:9123–9128

34. Alexander-Bloch AF, Gogtay N, Meunier D et al (2010) Disrupted modularity and local connectivity of brain functional networks in childhood-onset schizophrenia. Front Syst Neurosci 4:147

35. Yu Q, Erhardt EB, Sui J et al (2015) Assessing dynamic brain graphs of time-varying connectivity in fMRI data: application to healthy controls and patients with schizophrenia. Neuroimage 107:345–355

36. Rubinov M, Bullmore E (2013) Schizophrenia and abnormal brain network hubs. Dialogues Clin Neurosci 15:339–349

INDEX

Massimo Filippi (ed.), *fMRI Techniques and Protocols*, Neuromethods, vol. 119,
DOI 10.1007/978-1-4939-5611-1, © Springer Science+Business Media New York 2016

Printed in the United States
By Bookmasters